PROCEEDINGS OF THE THIRD INTERNATIONAL BIODEGRADATION SYMPOSIUM

PROCEEDINGS OF THE

3rd INTERNATIONAL BIODEGRADATION SYMPOSIUM

KINGSTON, RHODE ISLAND, U.S.A.

17th – 23rd AUGUST, 1975

Under the auspices of The Biodeterioration Society

Sponsored by Office of Naval Research, National Bureau of Standards, U.S. Coast Guard, Naval Ship Engineering Center, Naval Facilities Engineering Command, Maritime Administration, U.S. Army Research Office

Supported by Rhode Island Agricultural Experiment Station, Biodeterioration Information Centre, Society for Industrial Microbiology

This project was supported in part by the Office of Naval Research, Microbiology Program, Naval Biology Project, under contract No. N00014-68-A-0215-0014, NR 306-098

EXECUTIVE COMMITTEE

PROCEEDINGS OF THE THIRD INTERNATIONAL BIODEGRADATION SYMPOSIUM

Editors

J. MILES SHARPLEY

Sharpley Laboratories Inc., Fredericksburg, Virginia, and Virginia Commonwealth University, U.S.A.

and

ARTHUR M. KAPLAN

U.S. Army Research and Development Command, Natick, Massachusetts, U.S.A.

APPLIED SCIENCE PUBLISHERS LTD
LONDON

APPLIED SCIENCE PUBLISHERS LTD
RIPPLE ROAD, BARKING, ESSEX, ENGLAND

This work relates to Department of the Navy Contract N00014-68-A-0215-0014 issued by the Office of Naval Research under Contract Authority NR 306-098. However, the content does not necessarily reflect the position or the policy of the Department of the Navy or the Government, and no official endorsement should be inferred.

ISBN: 0 85334 679 8

WITH 452 ILLUSTRATIONS AND 245 TABLES

Printed in Great Britain by Galliard (Printers) Ltd Great Yarmouth

FOREWORD

This Symposium, the third in a series, was designed to bring together scientists from the international community interested in the various aspects of Biodegradation and Biodeterioration. The Symposium was organized under the auspices of the Standing Committee for International Symposia of the Biodeterioration Society and was held at the University of Rhode Island, Kingston, Rhode Island, U.S.A. between 17–23 August, 1975. The majority of the papers presented at this meeting have been brought together as the Proceedings of the 3rd International Biodegradation Symposium.

Biodegradation involves fundamental as well as applied investigations on the effects of organisms on materials of interest to humans and the relationships of organisms and chemical substances in our environment. These interests centre primarily on the action of microorganisms, but also include action by insects, rodents, birds and other animals. A session was included on the use of immobilized enzymes for degradation which added a new dimension to degradation studies.

The programme was developed using fourteen symposia chosen so as to cover major topics in Biodegradation and Biodeterioration. A recognized leader for a specific topic was asked to organize a symposium by inviting speakers to discuss selected aspects of the topic. Papers contributed by scientists from the international community were organized into an additional eleven sessions based on the general direction and subject matter of the papers. Due to the high cost of publication, the Executive Committee could not include in the Proceedings papers offered but not actually presented at the Symposium.

The Chairman of the Executive Committee wishes to thank the members of the Committee for their dedication to the objectives of this Symposium, their sound advice and continuing efforts in the three and a half years of planning. Special appreciation must be given to the Society for Industrial Microbiology who chose to hold their annual meeting at the University of Rhode Island concurrent with the 3rd International Biodegradation Symposium. The Executive Committee wish to express their thanks to Dr G. A. Donovan, Dean of the College of Resource Development, University of Rhode Island, who opened the Symposium.

R. W. TRAXLER, *Chairman of the Executive Committee*

Contents

Session XVII—Symposium—Biodegradation of Fuels

CHAIRMAN: H. W. ROSSMOORE (WAYNE STATE UNIVERSITY, U.S.A.)

Session XIX—Symposium—Biodegradation of Lubricants

CHAIRMAN: E. C. HILL (UNIVERSITY COLLEGE, CARDIFF, U.K.)

DETERIORATION BY INSECTS, RODENTS, BIRDS AND ANIMALS

Session II—Symposium—Deterioration by Insects

CHAIRMAN: J. J. PRATT (U.S. ARMY NATICK RESEARCH AND DEVELOPMENT COMMAND, NATICK, U.S.A.)

Session VIII—Symposium—Deterioration by Rodents, Birds and Other Animals

CHAIRMAN: J. J. PRATT (U.S. ARMY NATICK RESEARCH AND DEVELOPMENT COMMAND, NATICK, U.S.A.)

MATERIALS DETERIORATION AND MECHANISMS OF DETERIORATION

Session III—Symposium—Materials Deterioration

CHAIRMAN: C. C. YEAGER (VENTRON CORPORATION, U.S.A.)

Session X—Contributed—Biodegradation of Materials

CHAIRMAN: A. KEMPTON (UNIVERSITY OF WATERLOO, CANADA)

Session XXIV—Contributed—Materials Degradation

CHAIRMAN: ARTHUR O. LLOYD (CATOMANCE LTD, U.K.)

Session IX—Symposium—Mechanisms of Materials Degradation in the Marine Environment

CHAIRMAN: D. R. HOUGHTON (H.M. NAVAL BASE, PORTSMOUTH, U.K.)

DETERIORATION OF AGRICULTURAL PRODUCTS AND POSTHARVEST DETERIORATION

Session IV—Contributed—Biodegradation of Agricultural Products

CHAIRMAN: H. O. W. EGGINS (BIODETERIORATION INFORMATION CENTRE, U.K.)

Session XIV—Symposium—Postharvest Deterioration and Mycotoxins

CHAIRMAN: C. W. HESSELTINE (NORTHERN REGIONAL RESEARCH LABORATORY OF U.S.D.A., U.S.A.)

Organised by R. J. Lukens

INVITED PAPERS:

CONTRIBUTED PAPERS:

GENERAL BIODETERIORATION AND DETERIORATION OF ORGANIC WASTES

Session XI—Contributed—General Papers

CHAIRMAN: R. E. KLAUSMEIER (NAVAL AMMUNITION DEPOT, U.S.A.)

Session XII—Contributed—General Biodegradation

CHAIRMAN: D. ALLSOPP (BIODETERIORATION INFORMATION CENTRE, U.K.)

Session XVI—Contributed—Biodegradation of Organic Waste

CHAIRMAN: BROOKS D. CHURCH (UNIVERSITY OF DENVER, U.S.A.)

BIODEGRADATION OF POLYMERS AND SYNTHETIC ORGANIC COMPOUNDS

Session XVIII—Symposium—Biodegradation of Polymers

CHAIRMAN: D. M. WILES (NATIONAL RESEARCH COUNCIL, CANADA)

Session XXI—Contributed—Biodegradation of Synthetic Organic Compounds

CHAIRMAN: J. J. COONEY (UNIVERSITY OF DAYTON, U.S.A.)

RECALCITRANT MOLECULES

Session VII—Symposium—Recalcitrant Molecules

CHAIRMAN: M. ALEXANDER (CORNELL UNIVERSITY, U.S.A.)

MICROBIOLOGICAL TRANSFORMATION OF METALS

Session V—Contributed—General Papers

CHAIRMAN: F. E. BRINCKMAN (NATIONAL BUREAU OF STANDARDS, U.S.A.)

Session XXII—Symposium—Microbial Transformation of Metals

CHAIRMAN: W. P. IVERSON (NATIONAL BUREAU OF STANDARDS, U.S.A.)

BIODEGRADATION OF WOOD

Session XXIII—Contributed—Biodegradation of Wood

CHAIRMAN: GÜNTHER BECKER (BUNDESANSTALT FÜR MATERIAL-PRÜFUNG, GERMAN FEDERAL REPUBLIC)

IMMOBILIZED ENZYMES FOR DEGRADATION

Session XV—Symposium—Immobilized Enzymes for Degradation

CHAIRMEN: D. L. MARSHALL and J. M. LITCHFIELD
(DOW CHEMICAL COMPANY AND BATTELLE MEMORIAL INSTITUTE, U.S.A.)

FUNGICIDE TOXICITY AND METABOLISM

Session XX—Symposium—Fungicide Toxicity and Metabolism

CHAIRMAN: H. P. BURCHFIELD (GULF SOUTH RESEARCH INSTITUTE, U.S.A.)
Organised by R. J. Lukens

INVITED PAPERS:

CONTRIBUTED PAPERS:

BIOANALYTICAL TECHNIQUES

Session VI—Contributed—Bioanalytical Techniques

CHAIRMAN: OSKAR WÄLCHLI (EIDGENÖSSISCHE MATERIALPRÜFUNGSANSTALT, SWITZERLAND)

IN THE COURSE OF NATURAL EVENTS

ROBERT L. STARKEY

Department of Biochemistry and Microbiology, Cook College, Rutgers University, New Brunswick, New Jersey, U.S.A.

As you are all aware, my being here is fortuitous and I am sure that your surprise is no greater than was mine when I was approached for this presentation.

It is a privilege to meet with you and discuss some subjects of common interest dealing with the microbial breakdown of materials, both organic and inorganic. The subject was the prime interest of my research for many years but the terms biodeterioration and biodegradation were unknown or unfamiliar at that time. Instead, the term 'microorganisms as transforming agents' was more familiar.

The fact that this is an international conference recognizes that the problems of biodeterioration and biodegradation have universal significance. As I view them, the goals are to control microbial destruction and to provide a desirable environment, one free from toxic materials in food, air, and water, free from obnoxious odours and unsightly disposal dumps, free from industrial and domestic waste, and free from conditions inimical to health, aesthetics, and plant development. The objectives are essential, lofty, opportune, and demanding. Without the control of destructive effects of microorganisms, there would be prohibitive losses of food, fabrics, wood, metal, masonry structures, and other diverse industrial raw materials and consumer products. Without the breakdown of wastes, the environment would be polluted beyond use.

MAN AND NATURE

Nature has been defined as the physical and biological world about us consisting of the landscape, animals, and plants. By definition, it presumes to exclude man, his products, and his alteration of the physical and biological features of the world. But man is an animal, and, in my opinion, this definition is unsatisfactory in that it has generated an attitude that man is an outsider who came from some place beyond the earth, that he is unnatural, and that his influences on nature are abnormal and generally devastating. There is an associated idea that all of nature that is outside of man and his influences is good. This concept is unrealistic in that it fails to recognize that man arose in the process of evolution like all other animal life and is a product of so-called 'nature'. It seems certain that at some period when man was primitive, he would have qualified as being a part of 'nature'. Why is it that now, by definition, man has become divorced from nature? It seems to me irrational to consider that man walking, eating, breathing, or otherwise occupied is an unnatural being according to the definition of nature. I prefer to consider man as part of it, and a significant part. Under this concept, whatever man does or has done is natural irrespective of the consequences. Some of the things he does are indeed undesirable in man's own judgement but they are not unnatural. Because some act has a 'natural' cause should not imply it is good or that because it was done by man it is bad. A forest fire started by lightning could be just as destructive as one due to man's negligence. Accelerating or decreasing the rates of decomposition of organic compounds or metals may be desirable or undesirable in man's judgement of his own needs whether they are due to 'natural' causes or his own actions. A devastating earthquake or flood is accepted as fate, but if there are ill consequences of man's attempt to control pests and food supplies or to develop energy resources and to dispose of wastes, he is maligned and castigated. The lack of complete success may have been due to his fallibility. For good or bad, man's actions are as natural as are those of the physical forces of gravity, the effects of water and wind, or the biological effects of the fish in the sea and the microorganisms in the soil. Therefore, from this viewpoint, whatever man does to affect biodeterioration and biodegradation is natural and it happens 'in the course of natural events'. This is the concept which I have adopted for my discourse.

Man continually faces a relentless struggle to protect food, fibres, his domestic animals, and himself from destructive agents. Most of the problems of pollution and disposal of wastes arise from the fantastic increases in population, the concentration of people in large cities, and the expansion of industrial establishments. Only slowly does there seem to be recognition that there are far too many people on the earth for human welfare. Unfortunately, there is little evidence that this has had much effect, or that real progress will be made to control populations short of calamities of war, starvation, or the depletion of the world's sources of energy. The conclusion seems obvious that our problems will continue and are more likely to increase than to diminish.

SCOPE AND VIEWS

Since the scope of the symposium embraces the broad subjects of preservation of materials as well as the disposal of wastes, even a listing of the problems would consume much time. Certainly, I could not hope to cover the problems even briefly and you could not endure it. I will confine my remarks to some wide-ranging generalizations and reference to matters with which I have been directly concerned.

Initially, I should like to consider briefly the meaning of the terms biodeterioration and biodegradation. The first term is self-explanatory in that it implies biological processes that are destructive or yield undesirable products or both. The second term is less specific although in context its meaning is clear. However, broadly inclusive, the term refers to the breakdown of substances or compounds into smaller moieties. Thus interpreted, all living things would be involved in degradation, for this occurs in all respiratory processes. It is particularly apparent in the dissimilation of organic compounds which serve as sources of energy and yield completely oxidized end-products as CO_2, H_2O, sulphate, nitrate, and, in many cases, incompletely oxidized organic products. The term biodegradation, as considered here, has a more restricted meaning, namely, the breakdown of undesirable materials or compounds to harmless or tolerable products. Some of these undesirable materials have been discharged into rivers and streams where they cause serious pollution problems. Such materials include sewage, coal mine drainage water, waste from paper pulp mills, breweries, slaughter houses, sugar and oil refineries, and plants processing milk, vegetables, and other foods. Much has been accomplished but doubtless there will be need for eternal vigilance to control pollution because these bodies of water are attractive vehicles for waste disposal.

There is an additional important aspect of the process of biodegradation. Present-day objectives in waste disposal seek not only promotion of rapid decomposition but also the associated production of useful substances such as alcohols, organic acids, combustible gases, and harvestable cell material of the degrading microorganisms. Unfortunately, this is not always attainable, but it is a challenging goal requiring ingenuity in devising processes that are both practical and economical.

Most biodegradation occurs as an unnoticed process in and on soil in the grasslands and forests and in the water of rivers, lakes, and oceans where vast quantities of plant and animal wastes and the bodies of dead animals and plants both macroscopic and microscopic are disintegrated. Fortunately, over the ages, parts of these residues have resisted decomposition, and they now constitute our indispensable fossil fuels of coal and oil. Their consumption may be likened to the expenditure of capital assets that cannot be restored.

There is no simple separation of compounds into desirable and undesirable categories. Like weeds, which are defined as plants out of place, petroleum in an oil slick is something undesirable to be eliminated. Similarly, much effort has been expended to preserve cellulose in fabrics and paper, but cellulose is one of the major components of forest waste and rubbish.

Many factors affect the rate and course of microbial decomposition. Whereas most organic materials and oxidizable minerals are rapidly transformed under favourable conditions, they persist almost indefinitely in other environments. Consider in this regard the fossil fuels, elemental sulphur, mineral sulphides, the animal carcasses engulfed in glaciers, and human artifacts of metal, wood, and bone which became buried in muds or in dry desert soils. Among the well-known factors affecting microbial attack are the following: the inherent degree of decomposability of the substance itself, coating of the material with resistant compounds such as lignins, tannins, and resins, ambient temperature, moisture, pH, E_h, and available sources of required macro- and micro-elements.

INDUSTRIAL, DOMESTIC, AND FARM WASTES

There are always wastes in nature if one defines them according to man's needs and interests as useless or undesirable materials. Incidentally, it may be well to remember that there is no such thing as waste except as man conceives it. Industrial waste may be residues or by-products of the principal manufactured products or discarded machinery and equipment. Household waste may consist of glass and plastic containers, paper, metal, and diverse items. Rocks and weeds are waste in cultivated land and even water may be considered a waste when it floods cropped fields and cities.

Wastes are of particular concern to man because they must be disposed of.

This group of industrial microbiologists is well aware of concern for waste disposal—the problems involved, methods available, and costs. It is recognized that large quantities of unaltered waste can no longer be released into nearby drainage areas. They must be freed of readily oxidized compounds, and toxic materials must be removed. As has been mentioned previously, there are alternatives for their disposal. It is possible that they can be converted into useful products or utilized directly. In the latter case, pulp of citrus fruit and sugar beets and residues from vegetables in centres where they are prepared for market might be used as animal feed. Both animal feed and fertilizer are prepared from slaughterhouse waste. With knowledge of the costs and other factors involved in all of the possible processes of disposal, the choice can be readily ascertained. Even household waste may have salvage potentials. Here again, the cost of collection, separation, and reclamation compared to the value of the potential new product are factors that influence the choice of the means of disposal.

Burial and incineration have long been used for disposal of garbage and combustible materials. Burial becomes impractical near centres of large populations where suitable land areas are no longer available. Incineration may cause unacceptable air pollution. We in New Jersey hear that the smoke from burning rubbish dumps on the New Jersey side of the Hudson River are offensive to the residents of nearby New York City. There is much available knowledge about composting, and many different systems have been used. These range from the small-scaled hand turned compost of garbage, plant waste, and soil to processes involving great quantities of waste. Here there is need for highly mechanized equipment designed for handling municipal waste and manures from packing houses and feed lots. In the Netherlands, city waste was brought to an open area by gondola freight cars and dumped into large windrows where it was kept moist and occasionally mixed. The product was acceptable for use as a soil conditioner and low-grade fertilizer. The costs of collection and composting have been so great that such an operation has had only restricted use in the United States. However, with limitation of other methods of disposal, these expenses may become acceptable.

Until recent years, animal manures of cattle, hogs, and poultry have been used on farms where they were inoffensive to residents of populated areas. Effective disposal of such waste is becoming more of a problem as concentrations of animals such as poultry and hogs are located near encroaching urban areas. New composting methods have been developed, and the permissible limits of the amounts of manures that can be broken down after being buried in soil are being explored. One problem generated by burial of large quantities of animal manure in soil is that considerable amounts of nitrate are produced. This is leached from the soil and enters water systems where it promotes excessive algal growth.

Under present social and economic conditions, there is need for new and more desirable methods for disposal and utilization of decomposable plant materials and animal manures. Among the potentialities is the conversion of the energy of the organic compounds into combustible gases, heat or other usable forms of energy, animal feed, or other useful products. Such potentialities present a challenge to one's resourcefulness. Consider, for example, the success of the unlikely fabrication from sawdust and other wood waste into chipboard suitable for construction of furniture and flooring.

As a frugal person, I am disturbed by evidence of what appears to be needless waste. One common example is the discharge of large amounts of steam from industrial plants. I will mention another item that impressed me particularly. During the process for the production of an antibiotic, the massive amounts of cell material of the antibiotic-producing culture were filtered by suction from the culture medium on a large rotating drum. It was peeled from the drum as a continuous blanket some 1–2 in thick and about 6 ft wide. Since there was no obvious use for the material, it was chopped up and discharged into the sewerage system. This cell material contained 5% or more nitrogen on the dry basis which represented 30% or more protein. In it were considerable quantities of hemicelluloses, vitamins, and other compounds. It seemed to me that some of its constituents should have sufficient value to justify their recovery. Possibly the material as a whole could be processed to make animal feed or fertilizer. It is recognized that recovery and conversion of wastes to usable products must be profitable, but I wonder if future requirements will not justify recovery processes that have been considered impractical in the past and at the present time.

FOODS AND FIBRES

The accomplishments of industry in the manufacture and fabrication of useful products is the marvel of the century. They have been made possible by man's creative imagination, his development of vast quantities of natural resources, and a favourable economic climate. But this development has exacerbated the problems which are being considered in this symposium. At great cost and expenditure of thought, time, and labour, man has developed and used various materials and processes to prevent the ravages of micro-organisms, insects, vermin, and environmental conditions. He has made drugs, preservatives, and protective coatings. He has used sterilization and desiccation and has modified reaction (pH), excluded oxygen and applied electric current for this purpose. During World War II, losses through spoilage and degradation were particularly great by what became known as tropical deterioration of stored foods, fabrics, metals, and electrical and optical equipment. This was due principally to high humidity and temperature in the tropical area, which favoured biodeterioration. Recognition of the factors involved resulted in the development of effective control measures, as can be attested by Dr Kaplan, programme chairman of this symposium.

There are great limitations to solving microbial problems at the desk or by what has been referred to as 'fireside arbeit'. It is here that experiments are planned but results of these experiments may be unexpected. The following is an example of surprising results that qualify for inclusion under the designation of serendipity. This involved an industrial process for the production of a fabric called bemberg. It was woven from a modified cellulose fibre to produce cloth with a silky sheen. To produce the thread, cellulose was dissolved in Schweitzer's reagent, ammoniacal copper hydrate. The resulting solution was passed through fine openings in spinnerettes into a solution of sulphuric acid of pH 0·5 whereby the cellulose was precipitated. The resulting filaments were spun into thread used to produce the fabric. Copper sulphate accumulated in the acid solution, called the spinning liquor, and its concentration was controlled at about 5%. Slime developed in the spinning liquor and this fouled the thread. Although it would seem that this liquor of pH 0·5 containing 5%

copper sulphate would be an unlikely place for microbial development, it was found that the slimy fouling material consisted of filaments of fungi that grew on compounds originating from the cellulose. The fungi proved to be remarkably tolerant to high concentrations of hydrogen ions and copper sulphate, the latter commonly considered to be toxic to fungi. It was even more remarkable that these fungi were able to grow in a dextrose mineral salts medium at pH 0 and saturated with copper sulphate. The saturated medium contained from 20 to 25% of dissolved copper sulphate. Additional studies with one of these fungi indicated that even though it grew at very high concentrations of copper sulphate in very acid solutions, it was sensitive to the salt at reactions close to neutrality. In fact, at nearly neutral reactions, it was even more sensitive to copper sulphate than some other common fungi. The results provide one more bit of evidence that it is difficult to formulate correct broad generalizations about the limits of factors affecting the growth of microorganisms.

DEGRADATION OF CRUDE OIL

Although crude oil is one of the major sources of energy, and we are aware of the need to conserve it, oil becomes a nuisance and can be destructive when it gains access to bodies of water and soil. It is principally a temporary nuisance in soil because it becomes degraded there if conditions are kept aerobic and adequate amounts of required nitrogen compounds and other mineral nutrients are available. Understandably, there is grave concern about the discharge of petroleum in the oceans, because oil slicks are not confined, the oil is recoverable with difficulty if at all, it is destructive to fish, fowl, and plankton, and its decomposition is slow. One of the reasons for its slow breakdown in the ocean is the lack of nutrient elements required by the degrading microorganisms. But the addition of water-soluble substances have little benefit because they are dissipated. Bartha provided what seems to be a novel and practical aid in solving this problem. He demonstrated that oil-soluble compounds containing nitrogen and other microbial nutrient elements promoted decomposition because they were retained by the oil itself and were available at the required site of decomposition.

RECALCITRANT MOLECULES

Not being satisfied with the nearly unlimited array of organic compounds in plants and animals, man has synthesized a great diversity of new plastics, drugs, pigments, adhesives, pesticides, and poisons. Many of these have become almost indispensable items of household and industry. Some of them have the limitation that they are exceedingly resistant to decomposition. Therefore, the innocuous substances persist to litter the environment and remove their constituent elements from recirculation in nature. Incineration seems to be the most suitable way to eliminate these wastes, but this fails because of the inadequacy of present methods of waste collection. In the case of toxic pesticides, residues may persist and exert harmful effects on animals or plants other than those which they were devised to control.

Conditions in the natural environment attest to the fact that *all natural compounds* undergo decomposition, although some persist longer than others or they persist under certain conditions. If this were not so, the undecomposable substances would now occur in soil in great quantities after accumulating for millions of years. Animal remains as well as the proteins, carbohydrates, and fats of plants are particularly susceptible to decomposition. The large benzene ring compounds of lignins and tannins are more resistant. They form complexes with proteins which are thus less readily broken down. This is evident in the greater resistance to decomposition of tanned leather compared to that of raw hides. Environmental conditions affect decomposition as witnessed by the fact that plant materials persist as peat in bogs where oxygen is excluded and partially decomposed leaves and other tree residues accumulate in the so-called mull layer on the surface of acid forest soils.

The microbial populations of soils are remarkable in that they harbour great numbers of versatile bacteria, actinomycetes, and filamentous fungi that are able to transform practically all organic compounds and also many mineral substances. In addition, the microorganisms undergo genetic changes as well as adaptations which result in ability to transform new compounds or to induce differences in rates of decomposition or variations in the course of the conversions. Consequently, soils have been used as a source of microorganisms for industrial processes, and it is the expectation that all kinds of organic compounds in industrial wastes will disintegrate in soil. Thus it was with some surprise that the cross-linked alkyl benzene sulphonate detergents tended to persist in soil and water. Although this resulted in their use being discontinued, I believe there is evidence now that they are less resistant to decomposition than was originally supposed.

The course of events in the use of DDT is more unusual. Its control of insect hosts of serious human diseases has saved millions of lives but its remarkable ability to destroy all insects resulted in its use in excess. This caused it to be banned. The principal disadvantage was its persistence in nature which led to its entering into the food chain and destroying birds and fish. There was little evidence that it was degraded appreciably in soil and water and it persisted even to being recovered in animal life of the Antarctic. Now it appears that it is not a completely indestructible molecule, for Alexander has recovered bacteria from soil that transform it through cometabolism. Until there is evidence that toxic compounds, detergents, and other polluting agents are degradable, their release into soil, water, and air may be questioned.

However, this does not apply invariably and decomposability of organic compounds is not always an asset or a useful property. Complete resistance to decomposition is not only desirable but even an attribute of compounds to be used for certain purposes. Such materials are needed as impervious, indestructible coatings for marine cables, pipes, cables, and other equipment buried in soil and sediments of rivers and lakes. Compounds with this property are also needed to impregnate poles and other wood that is to be covered with soil or submerged in water.

TRANSFORMATIONS OF MASONRY, METALS, AND SULPHUR

Even though major interest is focused on deterioration and degradation of organic compounds, inorganic compounds of mercury, manganese, iron, and sulphur are transformed by microorganisms directly or indirectly. Even masonry, in particular that constructed of limestone, is disintegrated by attack of nitric and sulphuric acids produced from ammonia and sulphide absorbed by the masonry from polluted air. Mercury that pollutes lakes may be methylated in sediments thus becoming soluble and posing a hazard if the water is consumed. In addition, fish in these waters may become poisoned or poisonous. It is also probable that microorganisms have been concerned with the production of manganese nodules located on marine sediments, but the reactions involved are obscure.

The effects of microorganisms on both iron and sulphur are more diverse. Their compounds and elemental sulphur support growth of autotrophic bacteria. Their microbial reactions involve oxidations and reductions, and iron compounds are affected by transformations of sulphur. These interrelations are well illustrated by the changes resulting in the undesirable qualities of coal mine drainage waters. With the mining of coal, water seeps through the overlying strata and enters shafts and galleries. During passage through the strata, sulphuric acid is produced by oxidation of metal sulphides when they are present. The oxidation is partly chemical and partly due to the activity of *Thiobacillus thiooxidans* and other sulphur bacteria. The acid dissolves iron from the rocks and drains into the open mine areas as ferrous sulphate. The strongly acid mine water containing large quantities of iron is altered further after leaving the mine; the ferrous sulphate is oxidized by the iron and sulphur bacterium *Thiobacillus ferrooxidans* and the iron precipitates as ferric hydrate. Pollution of rivers by the acid water loaded with insoluble ferric hydrate is obviously undesirable and to my knowledge there is no practical method of control. Therefore, the condition has seriously affected mining of coal in areas where iron sulphides occur in the strata overlying the coal seams.

Occasionally, high concentrations of sulphuric acid develop in certain land areas. This is generally the result of oxidation of elemental sulphur and metal sulphides, principally iron sulphides. This destructive condition develops in drained swamps, in material excavated from tidal marshes, is a major factor in the strip mining of coal, and occurs near sulphur mines where there is spillage of sulphur. It has practical value and is even promoted in waste dumps from processed ores of copper and other metals. The sulphates of copper and other dissolved metals are recovered from water drained from the dumps.

Under anaerobic conditions, sulphate-reducing bacteria convert sulphate to sulphide, much of which is retained in muds and sediments as iron sulphide. This is responsible for the large amounts of iron sulphide in the mud of tidal marshes and bogs. It is presumed that sulphate-reducing bacteria were important agents of the production of the natural mineral deposits of metal sulphides. The water of oil deposits which accompanies oil pumped from the wells is nearly saturated with hydrogen sulphide, and its disposal is a major problem in the recovery of crude oil. There is general agreement that this sulphide was produced by sulphate-reducing bacteria and it may be a by-product of the natural process of oil formation. Also, the natural deposits of elemental sulphur, which are associated frequently with oil formations, originated by oxidation of the hydrogen sulphide produced by the sulphate-reducing bacteria.

An additional effect of these bacteria is their destructive action as agents of anaerobic corrosion of metallic iron and steel. The corrosion occurs in oil and gas pipe lines, oil well pipes, pumps and other industrial equipment, hulls of ships, steel piling, and other harbour installations. In laboratory experiments, Hadley found that the rate of anaerobic corrosion increased rapidly with time. However, in studies with which I was concerned, steel specimens in seawater showed progressively decreasing rates of anaerobic corrosion.

In the absence of bacterial action, there should be negligible corrosion under anaerobic conditions. According to von Wolzogen Kühr, their role in anaerobic corrosion is depolarization in an electrochemical reaction through the removal of the cathodic hydrogen. In support of this theory, it has been found that many strains of the sulphate-reducing bacteria oxidize hydrogen with sulphate as the hydrogen acceptor and strains that oxidize hydrogen are more actively corrosive than the nonhydrogen users. These bacteria are abundant in the products of anaerobic corrosion, which consist principally of ferrous sulphide and ferrous hydrate, and the relative amounts of the two products are approximately those calculated to be present according to the proposed chemical reaction. Furthermore, corrosion may be catalysed by one of the bacterial products.

The mechanism of the corrosion process is still in doubt and it may be that it is not the major cause of the rapid rate of corrosion under anaerobic conditions. I have encountered several conditions which support this conclusion. Particularly disturbing is the fact that there was negligible corrosion under a condition where severe corrosion was expected. This was in sulphide-containing mud where steel panels were completely buried under nearly stagnant polluted water. Anaerobic conditions favouring development of the sulphate-reducing bacteria were evident from the fact that hydrogen sulphide was present in the gas released when the mud was disturbed. Nevertheless, the steel panels had not corroded appreciably after being buried for many months. In fact, calcium carbonate had been deposited on the panels and this suggested that they had served as cathodes, whereas for anaerobic corrosion, they should have been anodic. In another test, long steel rods and strips were located in seawater off the eastern coast. The lower ends were buried in sulphide-containing muds and the upper portions extended through the overlying water and above its surface. After a few months, the buried portions of the specimens failed to corrode appreciably, contrary to what would be expected since conditions were presumably favourable for anaerobic corrosion. Nearly all microbial corrosion was confined to the portions immersed in the seawater, and these were covered with corrosion products and fouling organisms. The products close to the metal surface were black from iron sulphide and the metal was severely pitted.

In other tests steel panels completely buried in sulphide-containing muds corroded much more slowly than panels partly buried and extending into the water above. The results of these tests and those with the above-mentioned panels, long rods, and strips suggested that rapid corrosion promoted by sulphate-reducing bacteria was caused by something more than depolarization through removal of cathodic hydrogen. It may be that corrosion, according to the mechanism proposed by von Wolzogen Kühr, is slow and that the rapid rate of iron loss ascribed to anaerobic corrosion is caused by some reaction superimposed on the microbial one. It is my opinion that this reaction is brought about by oxygen concentration cells. In this system the anodes would be the metal surfaces in the strongly reducing areas of sulphide production and the cathodes would be the metal surfaces bathed with oxygen-containing water. In pipe lines, the cathodes would be the portions of pipe extending into aerobic soils; in panels partly buried in muds, they would be the areas extending into oxygen-containing water; and, in the long steel specimens, they would be the areas of the specimens submerged in seawater where corrosion products were thin or where the metal had become exposed to oxygen-containing water by sloughing off of corrosion products. Whatever the mechanism of the anaerobic process, it can be controlled by the conventional methods of insulating the metal by coatings, attaching wasting electrodes, or applying direct electric current to make the metal cathodic.

CONCLUDING REMARKS

In closing, I wish to remark briefly about future prospects for the areas of our concern. We recognize there are political factors that affect your search for basic information on biodeterioration and biodegradation and practical means to control them. Future success in preserving useful materials and disposing of wastes effectively will depend in the long run upon the wisdom of leaders in industry and the central government. They provide the funds which support the research needed to obtain the information upon which new and improved practices will be based. Also, wise judgement of these leaders and those in the local governments is required to make the right decisions as to whether or not control measures will be taken and, if so, what choices will be made. To a large degree, their support and their decisions will determine the success of your efforts to make this a better place to live in. You are the ones who have the information they need, and you should make certain that this information reaches them.

It has been a pleasure to meet with you and consider certain microbial processes that affect organic and inorganic compounds 'in the course of natural events'. They are significant and justify your best thought and enthusiastic effort. May you all continue to enjoy the stimulation of your search for understanding. I assure you that I shall follow with interest your discussions of new and important information on many aspects of biodeterioration and biodegradation during the week that lies ahead.

MARINE POLLUTION AND INTERNATIONAL LAW

BOSTWICK H. KETCHUM

Woods Hole Oceanographic Institution,
Woods Hole, Massachusetts, U.S.A.

Summary

The marine pollutants of greatest concern are those which are toxic to marine organisms, are produced and reach the sea in large quantities relative to their toxicity, and persist for extended periods of time in the marine environment. International conventions provide for the prevention of pollution by oil and other hazardous substances, by artificial radioactivity, and by dumping. Organic materials decompose in seawater at rates which vary by orders of magnitude. Chlorinated hydrocarbons, for example, decompose very slowly; oil is decomposed somewhat more rapidly but tar balls may persist; domestic sewage wastes decompose more rapidly and represent a hazard only when released in excessive amounts or in confined areas. Heavy metals persist indefinitely, but may be removed by sedimentation. Radioisotopes decay at rates which vary by many orders of magnitude, but the influence of the biological system on the distribution and fate of radioisotopes is not well known. The various conventions recognize these differences, either explicitly or by implication, and provide different regulations for control.

INTRODUCTION

Marine pollution has been of increasing concern, both nationally and internationally, in recent years. For centuries people, when they thought about the problem at all, considered the ocean so vast and the available volume of water so great that it was generally thought that man's activities could have no significant or even measurable effect on the waters of the open sea. For more than a century it has been recognized that rivers, estuaries, and even the coastal waters were being degraded by man's introduction of polluting waste materials, but the sea was considered safe from these activities. Thus, even as we gradually recognized that dilution is not the solution to pollution for more limited bodies of water, the oceans were considered available as the ultimate sink for the waste products of man's population and developing technology. After all, they do contain 98·5% of all of the free water, including the ice in polar icecaps and glaciers, available on the surface of the globe.

This false sense of security was shattered dramatically with the atmospheric testing of nuclear weapons. Shortly after the early tests, radioactivity was traced throughout the Pacific Ocean and before too long it was identified in waters of all parts of the globe many, many miles from shore. These tests inadvertently provided the first global, massive geophysical experiment and the artificial radioactivity produced is still being used in evaluating oceanic water transport and mixing processes.

More recently, many other materials have been demonstrated to have a global distribution in the marine environment. These include petroleum hydrocarbons, chlorinated hydrocarbons, such as DDT and its derivatives and PCB's, and a number of heavy metals. In part, the widespread distribution of these polluting substances has been discovered because of our greatly improved analytical capabilities; in part, it has also been due to the fact that we are now concerned about these problems; and, in part, because of the rapidly accelerating growth of our technology which introduces ever-increasing amounts of waste materials into the environment.

The definition of pollution employed by the United Nations (UN 1973, p. 73) is: 'The introduction by man, directly or indirectly, of substances or energy into the marine environment (including estuaries) resulting in such deleterious effects as harm to living resources, hazards to human health, hindrance to marine activities including fishing, impairment of quality or use of seawater, and reduction of amenities.' The deleterious effects may develop gradually and insidiously and are sometimes difficult to establish in a direct cause-and-effect relationship. Because of the increasing rates of pollution and the slow reaction time of legal and especially of international mechanisms, it is not practical to wait until the deleterious effect is demonstrable in the ocean. It is necessary to recognize problems when they are still a threat and not wait for the impact to be demonstrable. In this connection, Hood (1971) has defined contamination as the addition of any foreign substance to the ocean in contrast to pollution which is the addition of a contaminant that is found to be harmful to an ocean.

For the preservation of the marine ecosystem, the pollutants of greatest concern are those which are toxic to marine organisms, or which accumulate within organisms to a higher concentration than in the environment, which reach the ocean in large amounts relative to their toxicity, and which persist for long periods of time in the marine environment. For this conference on biodegradation, the question of the persistence of pollutants in the marine environment is of primary concern, and indeed the possible impact of the pollutant on organisms in the marine environment cannot be adequately assessed without this basic knowledge. A substance which is decomposed to its elements or to intermediate products of innocuous character is obviously of far less concern than a substance which persists for years or decades so that its concentration in the environment can gradually build up to intolerable levels. Unfortunately, the basic data are lacking for most of the organic compounds of concern.

BACKGROUND

International concern with marine pollution has a brief and recent history. The Intergovernmental Maritime Consultative Organization (IMCO), an independent agency of the United Nations, convened a conference on pollution of the sea by oil in London in 1954. This resulted in the development of the 'International Convention for the Prevention of Pollution of the Sea by Oil' which was opened for signature in London in May 1954 and entered into force in July 1958. The original text has been amended in 1962, 1969, and 1971. In 1973 an 'International Convention for the Prevention of Pollution from Ships' extended the Convention to include pollution by substances other than oil. When the 1973 Convention enters into force, it will supersede the 1954 Convention as amended (Barros and Johnston, 1974; IMCO, 1974).

Further emphasis was given to international concern about pollution at the UN Conference on the Human Environment held in Stockholm in June 1972. In Recommendation 92 of the Stockholm Conference (UN, 1973, p. 23), the following statement of objectives which was agreed on at the Second Session of the Intergovernmental Working Group on Marine Pollution was quoted: 'The marine environment and all the living organisms which it supports are of vital importance to humanity, and all people have an interest in assuring that this environment is so managed that its quality and resources are not impaired. This applies especially to coastal nations, which have a particular interest in the management of coastal area resources. The capacity of the sea to assimilate wastes and to render them harmless and its ability to regenerate natural resources are not unlimited. Proper management is required and measures to prevent and control marine pollution must be regarded as an essential element in this management of the oceans and seas and their natural resources.'

The Stockholm Conference approved nine recommendations directly concerned with marine pollution which covered topics such as pollution by ocean dumping and by discharges from ships to the need for research and monitoring in order to understand the basic problems involved.

The 'London Convention on the Prevention of Marine Pollution by Dumping of Wastes and Other Matter' was produced at an intergovernmental conference held in London in October and November 1972. The Preamble to this Convention notes that 'Marine pollution originates in many sources such as dumping and discharges through the atmosphere, rivers, estuaries, outfalls, and pipelines and that it is important that states use the best practicable means to prevent such pollution and develop products and processes which will reduce the amount of harmful wastes to be disposed of'. It recognizes that the control of pollution of the sea by dumping is, thus, only part of the solution and states that action on this matter should not preclude discussion of measures to control other sources of marine pollution as soon as possible.

The results of the United Nations Third Conference on the Law of the Sea could influence greatly the international approach to marine pollution. The Informal Single Negotiating Text, produced in the spring of 1975, considers in Part III the 'Protection and Preservation of the Marine Environment'. If this text is finally adopted in substantially its present form, the Law of the Sea will provide for a comprehensive approach to the problem. The tentative text provides that the measures taken shall deal with all sources whatsoever of pollution of the marine environment and shall include, *inter alia*, the release of toxic, harmful, and noxious substances, especially those which are persistent, from land-based sources, from or through the atmosphere, and by dumping. Pollution from vessels, pollution from installations and devices used in the exploration or exploitation of the natural resources of the seabed and subsoil, and pollution from all other installations and devices operating in the marine environment are also specifically identified as being subject to the articles when finally adopted by the Conference on the Law of the Sea.

The draft convention also provides that states shall take all necessary measures to ensure that marine pollution does not spread outside their national jurisdiction and that activities under their jurisdiction or control are so conducted that they do not cause damage by pollution to other states and their environment nor cause pollution beyond the areas where states exercise sovereign rights. Other aspects of the text are concerned with cooperation, technical assistance, monitoring, environmental assessment, standards, and enforcement (UN, 1975).

This brief historical review would be incomplete without the mention of two Committees of the United Nations and its various agencies, namely GESAMP and GIPME. The Joint Group of Experts on the Scientific Aspects of Marine Pollution (GESAMP) was established by the United Nations and six independent UN agencies (IMCO, FAO, UNESCO, WMO, WHO, and IAEA). One of the continuing activities of GESAMP is the review of chemicals of potential environmental importance and hazard. This compiled information forms the basis for the specific recommendations in the Conventions to be discussed below.

The International Coordinating Group for the Global Investigations of Pollution in the Marine Environment (GIPME) was established by the International Oceanographic Commission of UNESCO and has as its objective the preparation of a plan for the assessment of the state of pollution of the world ocean. The preliminary plan presented to the IOC in New York in July 1974 (GIPME, 1974) states that 'The ultimate objective of a comprehensive investigation of marine pollution is to provide a sound scientific basis for the assessment and regulation of the pollution problem, including sensibly planned and implemented monitoring programmes'. It is further recognized that the first priority of GIPME should be the baseline studies to give the framework to facilitate collation and interpretation of data for future health of the ocean statements. GIPME itself would not conduct baseline studies or monitoring which would be done largely on a national or regional basis, but GIPME would be a coordinating group for the planning and coordination of such studies.

THE IMCO CONVENTIONS

As mentioned previously, the IMCO Convention of 1954 and succeeding amendments were concerned solely with the control of pollution of the sea by oil. The 1973 Convention extends this to other noxious substances transported in bulk, to harmful substances carried in packaged forms, and to sewage and garbage pollution from ships. The Preamble states that it is an objective to 'achieve the complete elimination of intentional pollution of the marine environment by oil and other harmful substances and the minimization of accidental discharges of substances'. As pointed out by Maurice Strong, Executive Director of the United Nations Environment Program, in his address at the opening session of the Conference, the provisions of the Convention will not provide for complete elimination, even though they would provide for substantial reduction in the course of time. Application of different parts of the regulations, which are scheduled to take effect at different times, would reduce oil pollution of the sea by tanker transport to between 8 and 16% of the 1972 rate of release from ships. These estimates do not take account of the increased transport which might be expected in the future (Srong, 1973).

A recent estimate of the amount of petroleum hydrocarbons reaching the marine environment from various sources is summarized in Table I (NAS, 1975). This table indicates that, of the total quantity of petroleum hydrocarbons reaching the oceans, about 30% is contributed by transportation and this is the part of the problem that the IMCO Conventions are designed to control.

Table I
PETROLEUM HYDROCARBONS INTRODUCED INTO THE OCEANS

Source	*Input mta*	%
Natural seeps	0·6	9·8
Offshore production	0·08	1·3
Transportation		
Tanker operations	1·33	21·8
Other ship operations	0·5	8·2
Accidental spills	0·3	4·9
Coastal refineries	0·2	3·3
Municipal and industrial wastes	0·6	9·8
Urban runoff	0·3	4·9
River runoff	1·6	26·2
Atmospheric input	0·6	9·8
Total	6·11	100

Since the danger of oil pollution has been recognized for over 20 years, one would expect that the biological effects and the fate of oil added to the marine environment would be well known. This is, unfortunately, far from the case. It is, of course, true that petroleum is a complex mixture of many different hydrocarbons; different crude oils have different compositions; various refined oils also vary in composition. It is not surprising that few broad generalizations can be drawn concerning such a complex mixture, especially when one also considers that different environmental conditions have existed and different ecosystems have been involved in the accidental releases which have been studied.

When petroleum is released to the marine environment, a variety of transformations take place involving physical, chemical, and biological processes. Oil forms a slick which will gradually spread and reduce in thickness with time. Some of the lighter oils evaporate into the atmosphere; others become dissolved in the water at low, but significant, concentrations. In rough seas, oil-in-water emulsions will be formed and material so dispersed is more readily available to the biota. Photochemical oxidation will occur and, for lighter oils, the combination of this process and evaporation may remove an oil slick naturally in a matter of a few days.

The end result of all these processes, commonly called weathering, results in the formation of tarballs from the heavier components of the oil. These tarballs have been found to be abundant in the open sea in a number of studies (Horn *et al.*, 1970; Heyerdahl, 1971; Morris and Butler, 1973; Butler *et al.*, 1973). They also wash up on beaches as is well known by anyone who has walked the beach adjacent to any major oil transportation route with onshore winds.

Once the fuel oil becomes incorporated in the sediments, it persists for extended periods of time; especially when the sediments are anoxic. NAS (1975) summarizes the results of numerous investigations which show that the concentration of hydrocarbons in sediments ranges from 1 ppm or less in the Arctic or in the deep sea sediments to several thousand ppm in polluted harbours or on polluted beaches.

The biological effects of oil are also highly dependent upon the character of the molecule being investigated. Many of the long-chain paraffinic hydrocarbons are a common constituent of marine organisms and are naturally produced. They are also the most readily metabolized and decomposed by metabolic processes. Most aromatic hydrocarbons found in petroleum are not known to be synthesized by marine organisms and these compounds are more slowly decomposed and more toxic to marine organisms. Stegeman and Teal (1973) and Anderson (1973) demonstrated that oysters may discharge 90% of the *n*-paraffins within a few days though they retain aromatic hydrocarbons longer. Anderson (1973) showed that the clam and oyster take up aromatic hydrocarbons to a greater extent than the saturated forms and also showed that oysters exposed to crude and fuel oil were free of detectable levels of aromatic and saturated hydrocarbons after maintenance for 24 to 52 days in clean seawater. It is not clear whether this is biological decomposition or excretion.

In conclusion, it appears apparent that some constituents of petroleum are highly toxic to marine organisms, particularly the aromatic compounds and that these are also the more soluble and more slowly decomposed parts of the oil mixture. The effect of an oil spill on the marine ecosystem will vary depending upon the type of oil which is released, the character of the environment and the oceanographic conditions at the time of release, the extent to which the oil is confined by the coastal configuration, the season of the year when the spill occurs, and the type of biota which is present in the area (NAS, 1975). Because of the possible deleterious effect of oil spills or intentional discharges of oil into the marine environment, it seems obvious that every effort should be made to minimize this source of pollution of the oceans.

The 1973 IMCO Convention also lists in Annex II over 150 noxious liquid substances carried in bulk in chemical tankers. These substances are categorized into four groups ranging from a major to a minor hazard to marine resources or human health, or which can cause harm to amenities or other legitimate uses of the sea. Limits for the permissible residual concentration are indicated for the 13 compounds judged to constitute a major hazard (Category A) if released to the marine environment. Basic information on the ecological effects and the fate of these compounds in the marine environment is even more meagre than it is for petroleum.

THE LONDON DUMPING CONVENTION

The London Convention on the Prevention of Marine Pollution by Dumping of Wastes and Other Matter prohibits the deliberate discharge of some materials into the ocean (Annex I) and requires permits for other materials considered to be less hazardous (Annex II). The prohibited substances include organohalogen compounds, mercury and cadmium and their compounds, persistent plastics and other persistent synthetic materials, various types of oil, high-level radioactive wastes, and materials produced for biological and chemical warfare. A specific statement is made that the prohibition of such releases 'does not apply to substances which are rapidly rendered harmless by physical, chemical, or biological processes in the sea provided they do not (1) make edible marine organisms unpalatable or (2) endanger human life or that of domestic animals'. Thus, the importance of biodegradation is specifically recognized in this Convention.

It is appropriate for each of the above substances to be included in a prohibited list for dumping at sea. The distribution of the heavy metals, mercury and cadmium, and of the radioisotopes may be modified by biological processes but they are not biodegradable. The persistent plastics are, by definition, not biodegradable to any great extent and if much is known about the biodegradation of materials produced for biological and chemical warfare, it is probably in the classified literature and not generally available. Among this list of prohibited substances, then, the biodegradation of organohalogen compounds is pertinent to this meeting. Unfortunately, very little is known about the rate of decomposition of these substances in the marine environment.

The distribution of chlorinated hydrocarbons, particularly DDT and its breakdown products, and PCB's has been extensively studied in recent years. Harvey (1974) states that observations indicate that PCB is 10 to 100 times as concentrated in marine organisms as is DDT. This was contrary to the expected results for several reasons. In the first place, ten times as much DDT had been produced during the 1960s as PCB. The DDT had been sprayed directly into the environment, while PCB's have to leak out from mostly enclosed systems or be accidentally introduced. These results indicate that DDT is not as persistent as was once believed.

The distribution of chlorinated hydrocarbons in the sea does not show any clear-cut change in distribution with distance from shore even though it is believed that the atmospheric transport is the main route of supply. Harvey (1974) shows that the PCB concentration in the atmosphere does decrease markedly with distance from the coast and he attributes the low concentrations of both PCB and DDT in the coastal waters as being the result of adsorption on particulate material, which then sediments out of the water column. Laboratory tests indicate that ocean particles can absorb a million times more PCB's than an equal quantity of water can dissolve. How long the chlorinated hydrocarbons can persist in the sediments and the potential hazard to bottom-dwelling organisms is unknown. This process has resulted in lower concentrations in the productive coastal waters and hence a smaller addition to the open sea where the density of particles is much less. Controls on DDT spraying and PCB uses have been in effect since the early 1970s. There has been a dramatic decrease in the concentrations of both PCB and DDT since 1972. Harvey (1974) quotes a fortyfold decrease in PCB concentrations in North Atlantic seawater and similar decreases in the DDT content of coastal marine shellfish. Thus, it appears that the oceans can recover rapidly from excessive pollution when the source of the pollution is eliminated.

The London Dumping Convention provides for prior special permits for the dumping of a variety of other substances which are listed in Annex II of the Convention and prior general permits are required for all other substances not listed in the previous two annexes. An application for a dumping permit must be accompanied by information (Annex III) including characterization and composition of the matter to be dumped, the characteristics of the dumping site, and the methods of deposit and general consideration of possible effects on amenities, on marine life, on other uses of the sea, and an evaluation of the practical availability of alternative land-based methods of treatment, disposal, or elimination, or of treatment to render the matter less harmful for dumping at sea. A working group of GESAMP defined the scientific criteria in more detail in a report which was presented to the seventh meeting held in February 1975 (GESAMP, 1975).

CONCLUSIONS

I hope that it is apparent that international concern about the pollution of the sea has been aroused and that some steps have been taken to attempt to control the further deterioration of the ocean as a result of pollution. However, pollution is largely land-based in origin and most of the sources are thus within the sovereign boundaries of individual states, each of which must exercise its own controls. International conventions can urge that this be done, but cannot infringe upon the sovereign rights of individual states within their own boundaries.

In addition to being able to control only a part of ocean pollution, international deliberations and decisions are ponderously slow. Even after the wording of a convention is agreed upon at an intergovernmental conference, a period of several years generally elapses before the convention is ratified by enough states for it to enter into effect. It is for this reason that it is important for the scientific community to anticipate those substances or human activities which are potentially hazardous long before the problem becomes acute in the environment. Even after it is scientifically demonstrable that a given substance or process may be hazardous, there will be a considerable lapse of time before the problem becomes apparent enough to spur the political scene to action.

While scientific information is greatly needed in order to develop sensible and useful regulations about pollution, it should also be recognized that political considerations may outweigh the best scientific evidence. For example, the developing nations commonly feel that pollution is a problem of the highly industrialized, developed nations who have reached their present state of development, in part at least, by disregarding the effects of pollution of the environment. The developing nations thus argue that they should not be forced to adopt expensive pollution control technology which would, in effect, delay or retard their rate of development. Even in the developed countries, marginal industries use a similar argument, namely, that the introduction of adequate pollution controls will eliminate their profits. It is easy to say, but less easy to prove, that cleaning up a damaged environment is more costly than keeping it clean in the first place. Also, forecasts of doom in decades or centuries do not greatly concern a man who is afraid that he will starve in the next month or year.

To one who has been concerned about pollution for many years, it is encouraging that today more people are concerned with the problems than was true 15 or 20 years ago. The fact that the progress is slow is discouraging, but the fact that some progress is being made, both on the national and international scale, does offer hope for the future.

REFERENCES

Anderson, J. W. (1973). Uptake and depuration of specific hydrocarbons from fuel oil by the bivalves *Rangia cuneata* and *Crassostrea virginica*. In *Background papers, workshop on petroleum in the marine environment*, pp. 690–708. National Academy of Sciences, Washington, D.C.

Barros, J., and Johnston, D. M. (1974). *The International Law of Pollution*. The Free Press, Macmillan Publ. Co. Inc., New York. 476 pp.

Butler, J. N., Morris, B. F., and Sass, J. (1973). *Pelagic tar from Bermuda and the Sargasso Sea*. Spec. Publ. No. 10. Bermuda Biological Station for Research, St. George's West. AD 769 873.

GESAMP (1975). *Scientific criteria for the selection of sites for dumping of wastes into the sea*. Joint Group of Experts on the Scientific Aspects of Marine Pollution, VII Session, February, 1975.

GIPME (1974). International Co-ordinating Group for the Global Investigation of Pollution in the Marine Environment. Second Session, United Nations, New York, 15–19 July 1974. IOC/GIPME II/3 Annex II.

Harvey, G. R. (1974). DDT and PCB in the Atlantic. *Oceanus*, **18** (1), 18–23.

Heyerdahl, T. (1971). Atlantic Ocean pollution and biota observed by the 'Ra' expedition. *Biol. Conserv.*, **3**, 164–167.

Hood, D. W. (Ed.) (1971). *Impingement of Man on the Oceans*. Wiley-Interscience of John Wiley & Sons, Inc., New York. 738 pp.

Horn, M. H., Teal, J. M., and Backus, R. H. (1970). Petroleum lumps on the surface of the sea. *Science*, **168**, 245–246.

IMCO (1954). International Convention for Prevention of Pollution of the Sea by Oil (1954)—as amended (1962, 1969, 1971). In *The International Law of Pollution*. The Free Press, Macmillan Publ. Co. Inc., New York.

IMCO (1974). International Conference on Marine Pollution, 1973. International Convention for the Prevention of Pollution from Ships, 1973. London.

London Convention (1972). London Convention on the Prevention of Marine Pollution by Dumping of Wastes and Other Matter. In *The International Law of Pollution*. The Free Press, Macmillan Publ. Co. Inc., New York.

Morris, B. F., and Butler, J. N. (1973). Petroleum residues in the Sargasso Sea and on Bermuda beaches. In *Proceedings, Joint Conference on Prevention and Control of Oil Spills*, pp. 521–530. American Petroleum Institute, Washington, D.C.

NAS (1975). *Petroleum in the Marine Environment*. National Academy of Sciences, Washington, D.C. 107 pp.

Stegeman, J. J., and Teal, J. M. (1973). Accumulation, release, and retention of petroleum hydrocarbons by the oyster *Crassostrea virginica*. *Mar. Biol.*, **22**, 37–44.

Strong, M. (1973). Statement at the Opening Session of the International Conference on Marine Pollution at the Church House, London, on 8 October 1973. (Unpublished manuscript) MP/CONF/INF. 16.

U.N. (1973). *Report of the United Nations Conference on the Human Environment*. Stockholm, 5–16 June, 1972. United Nations, New York. 77 pp.

U.N. (1975). *Third Conference on the Law of the Sea; Informal Single Negotiating Text*. Reproduced by Department of State, Washington, D.C. May 9, 1975.

THE TWO FACES OF BIODETERIORATION

CARL LAMANNA

Food and Drug Administration, Washington, D.C., U.S.A.

Biodeterioration from the point of view of technological requirements can be defined as decreases in the economic value of materials caused by biological agents. From the point of view of the physical scientist, biodeterioration can be defined as the transition of a material from a higher to a lower energy level, or chemical transition from more complex to less complex states caused by biological processes. The biologist does not deny the validity of these definitions of biodeterioration, but he does bring a special point of view to a definition of biodeterioration. He relates it to the cyclic changes of materials that characterize the biological world. Thus his definition of biodeterioration is that it is any process that returns a material for biological recycling which incidentally involves an attack on materials making up commodities. A commodity, of course, is an object of economic value.

These differences in definition are not a play on words nor a meaningless parochial competition for notice by specialists in different areas of science. I would argue for the greater merit of the biologists' definition for three reasons. First, it places biodeterioration in proper intellectual perspective as a common essential part of the total picture of the natural world. Second, by so doing it suggests areas of scientific investigation which otherwise are neglected. Third, it relates fundamental biological phenomena to significant economic problems.

Let us admit that how a scientist attacks a problem is constrained by his view of how the problem fits into the scene that attracts his attention. The greater the sweep of the scene, the greater the likelihood that individual parts and interrelationships of a multifaceted object will be recognized. In practical terms, this means a greater likelihood for recognition of the varied and even obscure potentials for scientific knowledge to contribute to applications. By increasing the size of your target, you increase the probability of desired hits.

The biological view places the biodeterioration phenomenon squarely as a natural extension of normal recycling of material. This immediately suggests that if the aim is to produce materials immune to biodegradation, all that is needed is the synthesis of compounds composed of elements not at all or only poorly or slowly subject to natural biological recycling. It happens that the elements which comprise quantitatively the dominant compounds of living matter are few in number: carbon, nitrogen, oxygen, hydrogen, phosphorus, and sulphur. Consequently, these elements are the major ones involved in biological cycling of materials. This fact arises from necessity since biological cycling is an adaptive phenomenon that ensures the continuity of life by return of elements in highly organized compounds to lower energy states to provide work and substance for resynthesis to higher energy and complex organizational states. Biological cycling is overwhelmingly a phenomenon of carbon or organic chemistry. Other biologically important elements such as iron, calcium, magnesium, manganese, and zinc generally function dynamically in their ionic state though they may be associated with organic compounds. They are not parts of the backbone structure of organic compounds for which elaborate specific enzyme systems have had to evolve in order to break covalent bonds to return them for recycling. An attack on the organic moiety of compounds with which these metals are associated is generally sufficient for their recycling.

To repeat, biological cycling is overwhelmingly a phenomenon of organic chemistry and associated enzymology. Since it so happens that the versatility of carbon as an element for synthesis into compounds is unparalleled, it is not surprising that man's inventiveness in developing new compounds is in the majority of cases confined to the realm of organic chemistry. Thus if one were to generalize in seeking a guidepost for outflanking biodeterioration, one would look away from carbon compounds to other compounds whose elements are not prominent in biological recycling. An obvious candidate for such an enterprise is silicon because silicon is the closest relative to carbon in the periodic table of elements. Unfortunately, silicon is not known to be as versatile as carbon in its ability to form compounds. Yet where biodeterioration is a serious economic problem, good cause exists for considering silicon compounds to replace carbon compounds. There is reason for those interested in biodeterioration prevention to become the sponsors of an investment in long-term, large-scale, imaginative basic research on the chemistry of silicon. The aim would be to provide knowledge to support an inventory of silicon compounds to substitute for biosensitive organic compounds as materials.

With these background statements in mind, it is appropriate to define two specific aspects of biodeterioration. One face is enzymology, the chemistry of biological breakdown of specific materials. The other face is biochemical evolution, the development of the enzymatic systems in elaborate cyclings of materials which ensure the return of components of organic compounds to the biological economy. Each face presents its own set of opportunities for understanding and control of biodeterioration.

Enzymological attack on materials can be labelled the phenotypic, extrinsic, or external face of biodeterioration. The properties of a material relating to enzyme action determine susceptibility to biodeterioration. The evolutionary aspect of biodeterioration is genetic and can be labelled as the genotypic, inherent, developmental, adaptive, or intrinsic face. The genetic makeup of organisms and their evolutionary potential determine whether or not and how a given material placed in a particular habitat will be subject to biodeterioration.

THE EXTRINSIC FACE OF BIODETERIORATION

To control biodeterioration, enzyme activity can be balked. This can be done by preventing synthesis of enzymes, inhibiting enzyme activity, or by averting contact between enzymes and their substrates.

The obvious requirement for biodeterioration is that enzymes make actual physical contact with a material susceptible to attack. On this basis, the simplest thought for preventing biodeterioration is to place an impenetrable barrier between enzyme and susceptible material. In theory, there are a number of ways by which this might be accomplished. An electrical charge could be induced or built up on the surface of a material to repel enzymes.

This takes advantage of the fact that enzymes carry a net electrical charge except at the isoelectric point. Another way would be to coat the susceptible material with a nonbiodegradable substance. Such a barrier can be a thin layer, and in the interests of economy as slight as monomolecular. A monomolecular layer should suffice since enzymes are not generally believed to be able to work at a distance from a substrate. If a barrier substance were also resistant to nonbiological influences which degrade materials such as temperature extremes, radiant energy, salts, acids, and alkalis, there would be assurance against breakdown of the barrier from physical environment influences as well as biological influences. Longevity might be limited to fixed periods by using barriers of limited life which when exceeded would result in biodeterioration of the formerly protected material. There are situations where pollution abatement requirements call for limits in time in the exercise of means for protection against biodeterioration.

Ceramics, including glass, are candidates for such coating. The technological advances surrounding development of means for spinning flexible inorganic fibres such as carbon fibres could be pursued. It is desirable to sponsor research aimed at learning how to deposit monomolecular layers of elemental carbon or other inert substances on such materials as wood and on fibres as textiles are spun.

Contact with enzymes can be prevented by making surfaces nonwettable, a special case of the barrier idea. Enzymes must act in the presence of water. Thus the hydrophobic surface resists enzymatic attack. Biodeterioration can be controlled if organic molecules are oriented so that only their hydrophobic portions are exposed at surfaces. It helps if these hydrophobic groups are so placed as to prevent enzyme molecules from slipping between them to contact the interior of a substrate molecule. An alternative is to coat a biodegradable material with hydrophobic compounds such as the silicones.

The state of crystallinity of a material affects biodeterioration probably in two ways. The pristine crystalline state could favour a regularity of molecular organization least subject to enzymatic attack as in the case of cellulose. Crystallinity is influential in determining surface characteristics such as surface electrical charge and wettability. Also crystals possess the most perfect smooth surfaces. Such surfaces would be least marred by fractures, cracks, fissures, pits, or scars in which water could condense or otherwise collect. Once water accumulates on a surface, it will gradually concentrate dust and volatile substances from the air. Such pools of solutions and detritus will inevitably include nutrient substances supporting the growth of microorganisms. The products of microbial metabolism accompanying growth such as acids, alkalis, aldehydes, and organic solvents could act to attack otherwise nonbiodegradable surfaces and materials. Here we have illustrated the principle that biodeterioration need not be a direct attack on a material by enzymes but rather can also be an attack by products of enzymatic activity associated with normal growth and catabolism. The ultimate source of these metabolic products foreign to the material attacked is often a contaminant from the environment adhering to the material or dissolved in pools or layers of water resting on the material in which microbes can grow.

Alternatives to controlling biodeterioration by frustrating contact between enzymes and materials are to prevent the appearance of enzymes or to poison enzymatic activity. Traditionally, much activity has focused on these objectives. There are many examples of compounds used to impregnate biodegradable material in order to prevent microbial growth or to inhibit enzymatic extracellular activity associated with growth.

There are several ways to prevent microbial growth. The usual approach is to poison metabolic reactions. An alternative not well represented in past research on biodeterioration prevention is to tie up an essential nutrient before it gets to the microbe. Such an approach is particularly feasible with essential minerals. An increasing number of chelating agents is available which can serve to bind metals known to be essential for growth. Also technology is now well advanced for the commercial production of controlled slow release preparations of drugs and other substances. The potential for incorporating into materials slow releasers of chelating agents for essential metals deserves serious consideration. Chelation of iron and magnesium is a prime candidate for such studies. Iron is known to be essential for most organisms, and chelation of iron has been discovered as a natural means the animal body employs to inhibit *in vivo* growth of pathogens. Chelating agents may also be useful in decreasing the stability of enzymes toward denaturation and adverse environmental factors such as temperature.

The idea of built-in obsolescence of a factor preventing biodeterioration, such as a chelating agent, or exhaustion of the supply of a slowly released fixed-life biocide has a special attraction because it provides a means for biological disposal of an item after its useful economic life has been exceeded. As an antipollution measure, there is a need to consider initiation of biodegradation of specific consumer goods which have served their purpose but which must resist biodeterioration during manufacture, storage, transit, and use. Packaging materials are notable candidates for this category.

Any means for decreasing sources of microbial nutrients aid in preventing biodeterioration. Any factor which adds to the kinds and quantities of nutrients in the environment of a material increases the chances for enhanced production of enzymes in quantities and variety. This is true whether the nutrients serve as sources of assimilation or energy. Thus 'cleanliness is next to godliness' has meaning for preventing biodeterioration as well as for theologians. Reduction in the variety and quantity of extraneous matter in an environment results in lower population densities of microbes, reduces the accumulation of metabolic end products capable of reacting with materials, and decreases the probability of appearance of an inducer for an enzyme attacking an exotic material.

In the case of animal pests attacking materials, a course of action has been to seek repellents. Compounds offensive to the tactile, taste, and olfactory senses can be employed. The approach has been ignored in the case of bacterial and fungal pests since these organisms are not known to possess the five senses of animals. But the senses of olfaction, taste, touch, sight, and sound are only specialized evolvants of the fundamental property of irritability possessed by all living organisms. It would be useful to think of both exclusion from materials of attractants and inclusion of repellents in the sense of negative chemotaxis. There is a growing body of scientific literature on the chemotaxis of bacteria and fungi. The time has come for workers on biodeterioration control to take this work into account and to make their own contributions to the field. Imaginative research can lead to the discovery of techniques and substances that introduce or enhance negative chemotactic properties in materials. The study of chemotaxis also has the promise of specificity. It is possible to visualize situations where chemotaxis is used against specific miscreants. There are situations where it may be desirable to exclude only certain species of microorganisms from the mixed population surrounding an object.

It might be argued that repellents would not succeed with microbes since the repellents do not act on extracellular enzymes synthesized at a distance from an object but capable of diffusing to the site of the susceptible material. Yet this thought should not enjoin consideration of the value of repellents. The fact is that defeating or decreasing growth on the surface of a material may prevent induction of a specific extracellular enzyme or prevent buildup of high concentrations of extracellular enzyme because of dilution by diffusion. Also certain microbial species causing biodeterioration can be expected to be limited to the ecological niche provided by the surface of a solid. Such species would include those with special adaptative structures for surface growth such as holdfasts. Negative chemotaxis directed against organisms possessing holdfast structures should be useful in controlling deterioration.

An alternative to measures directed against the synthesis and functioning of enzymes is to modify the chemical structure of substrates. Biodeterioration can be defeated by building into molecules substituents and linkages not easily attacked by enzymes if such chemical measures can be taken without reducing the economic value of a material. Success with this approach has already been attained in the use of methylation and polymerization. It is worth noting that such a use for polymerization apes cross-linkaging in naturally resistant substances such as kerogens. Such cross-linkaging is thought to explain the persistence of certain compounds in ancient geological formations, the so-called paleobiochemicals or archeochemicals. Parenthetically, it may be mentioned that the success of cross-linking in polymers and substituent addition in preventing biodeterioration can be special instances of the value of adding to the hydrophobic character of a molecule, or of adding barriers to the penetration of enzymes into sites of sensitive chemical bonds. Both the composition and the configuration of a molecule can be manipulated to affect sensitivity to enzyme actions.

THE INTRINSIC FACE OF BIODETERIORATION

The capability of a cell to produce an enzyme is determined by its genome. The genome of a species arises by evolution. Evolution has resulted in genotypes which provide for recycling of materials in the natural environment of organisms. This assures the continuity of life by preventing the exhaustion of supplies of nutrients needed for maintenance of viability and reproduction. Recycling of materials and continuous input of solar energy by photosynthesis to make up for increases in entropy are the bases for the continuity of life on earth. One caveat must be stated. Continuity is assured only if man does not throw into the equation irreversible disequilibrium by reason of stupidity, short-sighted greed, or catastrophic violence.

The genome of a species is not fixed and so evolution is possible. The genome has a plasticity which permits adaptation to changes in environment. But it should not go unrecorded that there are limits to biological adaptation. The plasticity that exists rests on only a few known genetic phenomena. These are gene mutation, recombination of genes, and gene amplification. Polyploidy is an example of amplification, the multiplication of the numbers

support of large and diverse biological populations, and should slow down the rate of appearance of mutants capable of attacking alien materials.

Plasmid transfer is recognized as a factor in bacterial evolution. Thus, any means for preventing plasmid transfer would be working against adaptation. Such means could be directed toward destruction of extracellular DNA or prevention of uptake of DNA by a recipient. Inhibition of the synthesis or functioning of pili would be one way to inhibit uptake of DNA.

A more remote possibility for working against adaptation would be to add decoys to the environment which would favour the selection of organisms not attacking the material to be protected. Such a decoy could act, for example, when an essential nutrient required by the organism biodegrading a material was exhausted by growth of a more rapidly multiplying organism competing for the same essential nutrient but not able to attack the material.

What has been said is predicated on the belief that students of biodeterioration also can be students of genetics and evolution. A portion of their thoughts and actions should be directed to learning how adaptation occurs with the hope that such studies would ultimately suggest practical means for thwarting adaptation.

A ready tool is available for development to pursue such studies in the laboratory, namely, continuous culture. An increasingly refined technology and theory of continuous culture are available. The phenomenon of periodic selection in such cultures can be instructive for the student of biodeterioration. Continuous cultures can be compared to learn how a particular set of pure or mixed cultures with variously naturally occurring nutrients affect population densities and periodic selection in the presence and absence of a material of interest to the student of biodegradation. The inoculum can be taken from the environment within which the material will be exposed in normal use. The source of replenishable nutrient for the continuous culture can be an infusion, extract, or perfusate of soil if biodeterioration is being studied in soil or the soil-atmosphere interface.

It is not necessary for continuous culture studies to be done only with soluble substrates even if this is traditional. The solid material often of interest to students of biodegradation can be blendorized and dispersed as small particles or colloid in the menstruum of the continuous culture. It is true that the application of scientific imagination will be required to modify the usual continuous culture technique confined to soluble substrates in order to meet the special needs of study of biodegradation of particulate materials. Marrying the theory and techniques of soil perfusion of the soil bacteriologist to those of continuous culture may be feasible.

THE ROLE OF THE ENVIRONMENTAL PROTECTION AGENCY IN ENVIRONMENTAL AFFAIRS

JAMES L. AGEE

Assistant Administrator for Water and Hazardous Materials,
Environmental Protection Agency, Washington, D.C., U.S.A.

It is indeed an honour for me to be invited to address the attendees to the Third International Biodegradation Symposium. The struggle by man to reach a viable compatibility with his environment has reached a significant and accelerating degree of concern in recent times. It is fitting that an interchange of scientific thought on the complex subject of biodegradation should share a part of such concern. Words such as 'ecology', 'eutrophication', and 'biodegradation' were relatively unknown by the general public a short time ago. Today, not only is there an awareness of such terms but also there is a general understanding of the concepts underlying usage of such words and the importance of such concepts to a vigorous and dynamic society.

The decade of the 1970s may well be known as the decade of the environment. The early days of this decade produced a Federal governmental organization to deal with environmental problems, as well as laws appropriate to laying the ground work for a comprehensive and penetrating attack on water pollution. History has yet to record the approach that society will take in the full implementation of many of the goals and concepts espoused by these legislative acts.

Environmental pollution can be achieved through (1) strong governmental organizations at all levels to propose, pass, and enforce laws, standards, and regulations; (2) a citizen concern and consciousness in the form of an environmental ethic with an inherent desire to improve the environment and leave it a better place in which to live, although doing so may mean changed life-styles; and (3) the identification, investigation, research, demonstration, and implementation of appropriate treatment or management controls for pollution sources.

Laws to control water pollution were introduced in the early 1900s and by the 1930s and 1940s, relatively good pollution laws began to appear, particularly as state statutes. Federal law, likewise, has become more stringent and comprehensive through time.

On 1 January 1970, the National Environmental Policy Act (P.L. 91-190) became law. This Act established a national policy to 'maintain conditions under which man and nature can exist in productive harmony, and fulfil the social, economic, and other requirements of present and future generations of Americans'.

The Act directed all agencies of the Federal government to take into account the environmental consequences of their proposed actions in decision-making. Specifically, it required a detailed statement of the environmental impact of each proposed action significantly affecting the quality of the human environment. Such a statement, available for other governmental agency and public scrutiny, must address the environmental impact of the proposed action, any adverse environmental effects which cannot be avoided should the proposal be implemented, alternatives to the proposed action, the relationship between local short-term uses of man's environment and the maintenance and enhancement of long-term productivity, and any irreversible and irretrievable commitments of resources involved should the proposed action be implemented.

Under this law the President appointed a Council on Environmental Quality. This Council assists and advises the President and prepares an annual Environmental Quality Report. The Council is required to gather information concerning conditions and trends in the quality of the environment, both current and prospective, and report this information to the President. The Council develops and recommends to the President national policy to foster and promote the improvement of environmental quality.

Late in 1970, Reorganization Plan Number 3 established the Environmental Protection Agency. The Environmental Protection Agency was composed of a number of individual environmental control groups and it brought together in one organization, for the first time at the Federal level, the capability for a multi-media attack on environmental problems spoken with one voice from an agency representing the management interests of water, air, solid waste, pesticides, and other hazardous substances.

The environment of man is a complex system of interacting and interrelated biological, chemical, physical, social, and economic forces. Its management in the interest of mankind and for the preservation and enhancement of society is the challenge of the century. The EPA shares with the states, local governments, and the citizenry of the United States that great challenge. The task is huge and in certain respects borders on the incomprehensible. Nevertheless, there is but one acceptable result—society cannot afford failure.

The Federal Water Pollution Control Act Amendments of 1972 (P.L. 92-500) were enacted into law on 18 October 1972. The stated objective of this Act is to restore and maintain the chemical, physical, and biological integrity of the nation's waters. The prescribed goals included water quality clean enough for swimming and for the protection and propagation of fish, shellfish, and wildlife by 1 July 1983, and an additional future national goal that the discharge of pollutants into the navigable waters be eliminated by 1985.

The Act established a nationwide system of permits designed to control the discharge of pollutants from point sources into the nation's waterways. About 40 000 of the nation's 300 000 industrial water users and municipal sewage treatment plants in 10 000 to 13 000 communities are subject to regulation under this system. Discharges from various other point sources such as feedlots, irrigation conduits, mining activities, and oil rigs also are addressed. Each establishment discharging effluent into the water must apply for a permit. In applying, data about the discharge must be provided. Each permit issued will specify conditions limiting the amount of pollution, the volume of wastewater, and the amount of solid waste permitted for each outfall. In addition, the dischargers must meet a schedule for compliance with permit conditions, water quality standards within the receiving waters, any appropriate controls on toxic pollutants, and new source performance standards, if applicable. Eventually, the states will operate the permit programme and EPA will maintain review authority with the option to object to the issuance of particular permits under certain conditions.

P.L. 92-500 authorized the EPA to award grants in the amount of 75% of the allowable project costs for the construction of publicly owned and operated wastewater treatment works. In general, projects eligible for Federal assistance include those to construct new treatment plants, to expand or improve existing plants, to construct interceptor and outfall sewer lines, or to provide pumping, power, and other equipment necessary to operate a sewage treatment system. Under certain conditions, sewage collection systems and projects to control pollution from combined sewers also may receive Federal assistance. Additional grants may be made to states for the conduct of state water pollution control programmes, to colleges and universities for training purposes, and to institutions and persons for the conduct of research, demonstration, or development of new or innovative techniques to treat or control water pollution when those techniques are deemed generally applicable to a large segment of the particular industry or to a large geographic segment of the United States.

The Act mandates the promulgation of standards and regulations and the development of criteria to prevent water pollution for a great many separate entities. Point source waste discharges, excluding municipal sewage treatment plants, must adopt the best practicable control technology currently available by 1 July 1977, and the best available technology economically achievable by 1 July 1983. Sewage treatment plants must adopt secondary treatment by 1 July 1977. The identification of best practicable control technology currently available and best available technology economically feasible, as well as pretreatment standards, is developed within effluent limitations that apply to existing industrial waste sources and to standards of performance for new sources. Industries have been separated into categories based upon the raw materials used, the manufacturing processes, or other similar characteristics. Waste characteristics for each category have been identified. The control and treatment technologies for each category are delineated, taking into the account the age and the size of the industrial plant. The economic impact of the industry is given consideration in the development of effluent limits. The development of effluent limits has entailed the awarding of contracts for technical studies and economic impact analyses on particular industrial categories. Draft reports resulting from such studies and analyses are reviewed by EPA technical staff, industrial associations, professional organizations, other governmental agencies, public interest groups, and a statutory advisory committee. Following such reviews and an assessment of comments, a proposed guideline is published in the *Federal Register* and a development document is prepared that represents the technical basis for the guideline. The proposed guideline is then subjected to public scrutiny with an appropriate period for comment. The comments received from industry and the public are assessed. Informal discussions with industry may occur to clarify such comments and a final effluent limitation is promulgated.

Water quality standards which, in the past, have existed for interstate streams only are now being reviewed and revised and are applicable to all waters of the states. As a basis of consideration for such standards, the Act provides that the Agency shall develop water quality criteria to meet the goals of the Act. About 65 water quality constituents currently are being addressed in this action. The drafting and initial review of water quality criteria are now completed and the criteria will be reviewed by the government agencies and the states prior to their release for use by the public. The Act provides that water quality standards shall be reviewed every three years and that the states shall establish load allocations from sources to ensure that the specified water quality is achieved.

Section 208 of the Act provides for the development and implementation of areawide waste treatment management plans. Areawide planning generally is confined to those areas where there are complex urban and industrial or nonpoint source water quality problems that cannot be solved by secondary treatment or equivalent industrial technology. In most cases, the governor of a state designates areas for areawide planning. The programme provides for the development of a special management structure to carry out the planning and the Federal government provided a 100% grant during FY-1975 to help local governments with their planning costs. In less complex water quality management areas which may entail nonpoint source management problems, but may not entail complex urban or industrial point source waste discharges, the state will act as a planning agency in such nondesignated areas. There is now no provision for special Federal funding in nondesignated areas.

The subject of nonpoint source wastes was highlighted in the 1972 Act. In response to Congressional mandates, the Agency has developed reports identifying and evaluating the nature and extent of agricultural and silvicultural activities, mining activities, construction activities, disposal of pollutants in wells or in subsurface excavation, saltwater intrusion, and changes in movement, flow, or circulation of water caused by construction or flow diversion facilities. Although there is no specific authorization for the implementation or enforcement of nonpoint source regulatory measures, the states must provide for management of such sources in developing their section 208 and other water quality planning activities.

The states are mandated to identify and classify all publicly owned lakes according to their state of eutrophy and to identify the cause of particular lake problems. EPA, in concert with the states, may financially assist in methods and procedures to restore the quality of such lakes.

Section 404 of the Act addresses the development of guidelines for the disposal of dredged or fill materials into navigable waters. In cooperation with the United States Corps of Engineers, the Agency published proposed guidelines on 6 May 1975. Such efforts are expected to provide for protection and enhancement of the nation's 80 million acres of wetlands in addition to assuring that water quality standards will not be violated within navigable waters through discharge of dredged or fill materials.

Specific sections of the Act address toxic and hazardous substances. The Agency was mandated by Congress to publish a list of toxic materials and soon thereafter to develop effluent limitations that would manage their discharge to navigable waters from point sources. Such a list containing nine toxic substances has been published and the Agency is engaged in implementing that portion that refers to the promulgation of effluent limitations. Likewise, the Agency has proposed a list of substances that are considered to be hazardous if discharged in significant amounts accidentally or intentionally as a spill or a dump. The Act provides that rates of penalty for such spills or dumps will be developed and published by the Agency to encourage prevention of such environmental insults.

The Marine Protection, Research, and Sanctuaries Act of 1972 (P.L.

92-532) was enacted on 23 October 1972. This Act prohibits the dumping of any radiological, chemical, or biological warfare agents or any high-level radioactive wastes into the oceans and establishes an EPA permit system for the control of dumping of other materials in designated dump sites. Permits are based on developed criteria and take into account the effect of such dumping on fisheries resources, plankton, fish, shellfish, wildlife, shore life, and beaches. This Act provides that the Corps of Engineers shall permit the dumping of dredged materials pursuant to the criteria developed by this Agency for the protection of the oceanic resources.

On 16 December 1974, P.L. 93-523, the Safe Drinking Water Act, became law. Like other recent water legislation, this Act is a milestone in Federal legislation for water quality. The Act provides for the issuance of primary drinking water regulations for the protection of public health, as well as secondary drinking water regulations for the protection of public welfare. National interim primary drinking water regulations already have been published in the *Federal Register*. A significant portion of the Act is devoted to a mandate for regulations to provide protection of underground aquifers if threatened by the injection of fluids. The Act contains a number of stringent congressionally mandated deadlines for certain actions by the Environmental Protection Agency. We are striving to meet each of these deadlines for the promulgation of regulations and for other activities as each becomes due.

Man stresses the water environment in a number of ways. Chemical constituents that are discharged in effluents on a more or less continuing basis, that are spilled periodically through accident or otherwise, or that may be introduced into water by man to fulfil a particular purpose become the causative agents in a substantial number of separable water quality effects. Stresses associated with such chemical constituents and abnormal concentrations may be manifest in actual or potential development of cancers within water users. There are other health effects associated with consumers of the water or with consumers of aquatic life associated with the water. The most famous of these are those associated with the bioaccumulation of mercury, diseases associated with the ingestion of harmful quantities of cadmium, or death in fur-bearing or bird predators because of the consumption of fish containing PCB's or certain chlorinated hydrocarbons. Chemicals introduced in effluents may be acutely toxic to important species within the aquatic food web. In lower concentrations they may exhibit chronic toxicity or impairment to reproduction or physiological activities when the organisms are exposed to such concentrations over an extended period of time. Within the past couple of years only we have begun to examine test species in constant-flow bioassays through one or more generations of a test organism to determine chronic effects and to identify levels that can be considered to have no substantial effect within the aquatic environment. These data are extremely valuable in formulating quality criteria.

Nontoxic chemicals such as nitrogen and phosphorus or other nutrients may give rise to prolific biological growths that in themselves become nuisances to man or interfere with water uses. Nuisance algal blooms, prolific growths of vascular aquatic plants, or abundant slime accumulations are common sights in highly enriched waters and are associated with the term, 'eutrophication.'

Physical stresses may be caused by solids that float, become suspended, or settle to blanket the bottom or form sludge beds. Suspended solids give rise to turbidity that in turn reduces light transmission and curtails natural photosynthetic action by aquatic plants. Sediment erosion is by volume our largest pollutant. Over one billion tons of sediment reach the nation's rivers annually, 10% of which is from land erosion, undergoing highway construction or land development. Heat is another pollutant that can create irreparable changes within the aquatic environment whether such heat results from the direct discharge of cooling waters such as in Biscayne Bay, Florida, or as a result of destruction of bank cover along waterways.

Hydrographic modifications have resulted in many irreversible changes within waterways. Such modifications include channelization, the construction of dams, dredging, and developments such as draining wetlands and providing residential access to a major waterway through a system of channels. The operation of reservoirs may contribute to a chemical stress such as that of nitrogen gas supersaturation that has been responsible for fish mortalities within the Snake and Columbia rivers in the Northwest.

Biological stresses on water quality may occur in the form of pathogenic organisms that create a health hazard to man or other users of the water when the decomposition of organic materials by bacteria causes life-giving oxygen to be removed in copious quantities, and in large numbers of introduced species that have resulted in aquatic liabilities instead of aquatic assets. Notably among the latter are the German carp, Eurasian water millfoil, water hyacinth, walking catfish, and others. The amount of money that is expended annually to control biological nuisances, introduced or developed as a result of water quality degradation, is enormous.

It is to the management and control of a myriad of water pollution problems that the various pieces of legislation discussed earlier are addressed. However, when such controls are accomplished, many waterways still will need to recover from the abuses placed on the system in the past. The biodegradation of a pollutant is an important contributing process toward recovery, but the products of the recovery process may remain within the water system. The recovery of the waterway requires time—time for a flushing or cleansing of the system of the pollutants that it contained. It requires time also for the introduction through natural or other means of the aquatic life components that represent a 'balanced' aquatic life community within the system. Streams and rivers recover relatively rapidly once pollutants are removed because most are flushed by spring floods. Lakes recover much less rapidly and in some lake systems the physical flushing time, or biodegradation time, may require many years.

It seems to me there is a growing awareness currently that man is an integral part of his environment. Perhaps there is a developing awareness that although he may not be the predominant species within the system, often he is the ecologically dominant species. Man has the power to modify, to make better, to maintain as a supportable entity for those life forms that we now recognize, or ultimately, to destroy many of the uses within our aquatic resources. This growing awareness of man as an integral part of a living ecosystem can lead only to the better of available options for environmental management in the best interests of society.

THE MICROBIAL UTILIZATION OF BUTADIENE

R. J. WATKINSON and H. J. SOMERVILLE

Shell Research Limited, Sittingbourne Research Centre, Sittingbourne, Kent, U.K.

Summary

A Nocardia species was isolated with the ability to utilize butadiene as sole carbon and energy source. Isocitrate lyase concentrations, lipid fatty acid composition patterns, growth yield and respiratory activities indicate that in *Nocardia* sp. 249, butadiene is metabolized via acetate during washed cell experiments. Simultaneous adaptation studies indicate that butadiene monoepoxide is an intermediate of butadiene catabolism, and the stoichiometry of oxygen consumption by washed cell suspensions suggest a stereospecific production of the epoxide. These results and cell free activities are consistent with a catabolic sequence involving the monoepoxide, a β,γ-unsaturated α-ketoacid, acrylate, lactate, and pyruvate.

INTRODUCTION

Microorganisms attack a wide range of hydrocarbons (Klug and Markovetz, 1971), including both saturated and unsaturated compounds. The mechanisms of attack have been studied to a less extent with the unsaturated hydrocarbons. Results with *Candida lipolytica* indicate incorporation of a single atom of oxygen by oxidation of the terminal double bond, with a 1,2 epoxide as a probable precursor of the 1,2 diol isolated from the system (Ishikura and Foster, 1961). Formation of a 1,2 diol could also result from dioxetane formation in which case both atoms of oxygen would be incorporated. Such a mechanism has been implicated in the oxidation of benzene by a *Pseudomonas* sp. (Gibson *et al.*, 1968).

Although butadiene does not, to our knowledge, occur naturally it is produced on a large scale and is used widely in chemical manufacture. It has chemical characteristics which increase the possible routes of biological attack beyond those expected for an unsaturated hydrocarbon, as any attack will involve an unsaturated centre, whether the double bonds act as separate entities or, more likely, as a conjugated system. The first likely attack by a microorganism would be to introduce oxygen and a number of possible routes can be visualized.

The present work was undertaken to demonstrate the microbial breakdown of butadiene and to attempt to identify the degradative pathway.

MATERIALS AND METHODS

Isolation, growth, and maintenance of organisms

The growth medium contained Na_2HPO_4: 4·2 g; KH_2PO_4: 2·8 g;

$MgSO_4 \cdot 7H_2O$: 0·2 g; NH_4NO_3: 2·0 g; $FeCl_3$: 17 mg; $CaCl_2$: 0·6 mg; $ZnSO_4$: 0·2 mg; $CuSO_4 \cdot 7H_2O$: 0·2 mg; $MnSO_4$: 0·2 mg; $CoCl_2 \cdot H_2O$: 0·2 mg; H_3BO_3: 0·1 mg; Na_2MnO_4: 0·3 mg in 1000 ml distilled water.

Growth yields were determined for substrates at a range of concentrations, cell mass being determined from direct weighing of cells and from a dry weight-optical density calibration curve.

Butadiene utilizing organisms were isolated from various water and soil sources using classical enrichment techniques with a mineral salts medium under a butadiene-air atmosphere. Stock cultures were kept at 5°C on Lab Lemco slopes and subcultured regularly.

Preliminary taxonomic studies on the cultures were based on Skerman's Key (Skerman, 1967).

Cultures were grown at 25°C in conical flasks sealed with a silicone bung holding a glass inlet tube fitted with a sterile filter. Butadiene was supplied at 10–15% of the available gas space within the flask. The cultures were shaken on a gyrotary shaker at 150 r.p.m.

Respiration studies were carried out in a conventional Warburg apparatus.

RESULTS

Nocardia sp. 249 was isolated by using classical enrichment techniques to select for growth on butadiene from soil samples.

Respiration of wash cell suspensions of *Nocardia* sp. 249 with butadiene only occurred when the cells were grown on butadiene and was absent in acetate or glucose grown cells. Cell suspensions of butadiene grown *Nocardia* sp. 249 showed a fixed oxygen:butadiene stoichiometry insufficient to account for the complete oxidation of butadiene to carbon dioxide and water (Fig. 1).

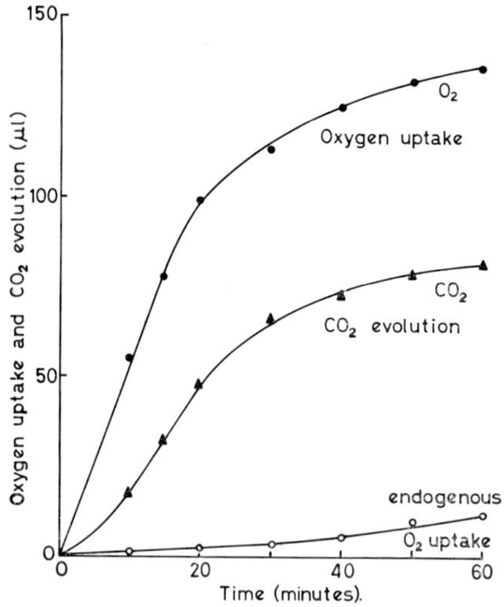

Fig. 1. Gas changes during butadiene metabolism by Nocardia *sp. 249. Protocol: Cells were obtained from a culture grown on butadiene as sole carbon source. Warburg flasks contained in a total of 3 ml:165 μmoles phosphate buffer (pH 7·0), 6·0 mg dry weight cells and 1·8 μmoles butadiene added from the side arm. For oxygen uptakes 0·2 ml 10% KOH was added to the centre well.*

The stoichiometry of oxygen uptake and carbon dioxide evolution was within the range 1 butadiene: 3–3·3 O_2: 2·0 CO_2. Growth of *Nocardia* sp. 249 on butadiene also induced respiration on a variety of four carbon alcohols but at a lower rate than that for butadiene oxidation (Fig. 2). Complete respiration data for *Nocardia* sp. 249 with a range of compounds is presented in Table I.

Table I

RESPIRATION OF *NOCARDIA* SP. 249

Test substrate \ *Growth substrate*	*Butadiene*		*Acetate*	*Glucose*
	% *of butadiene* Q_{O_2}	Q_{O_2}	Q_{O_2}	Q_{O_2}
Butadiene	100	50–120	0	0
Acetate	80	56	148	—
Crotonate	64	36	50	—
Butyrate	71	40	44	—
Acrylate	18	10	10	—
Glyoxylate	0	0	—	—
Glycerate	0	0	—	—
Glycollate	0	0	—	—
Tartrate	5	<5	—	—
Pyruvate	30	25	30	—
Propionate	19	20	—	—
Butadiene monoepoxide	100	70	0	—
Butadiene diepoxide	<5	<5	—	—
Butane-1,3-diol	5	5	—	10
Butane-1,4-diol	10	10	—	25
cis-2-Butene-1,4-diol	12	8	0	0
3-Buten-1,2-diol	10	10	0	—
1-Buten-3-ol	15	10	0	0
1-Buten-4-ol	15	10	0	0
sec-Butanol	<5	<5	—	<5
Methyl ethyl ketone	0	0	0	—
Methyl vinyl ketone	0	0	0	—
Ethyl acetate	75	40	150	—
Ethanol	<5	<5	—	<5
Dihydrofuran	15	10	—	0
Glucose	—	—	—	80

Q_{O_2}—μl O_2 consumed/hr/mg dry weight cells.
— not tested.

Molar growth yields for *Nocardia* sp. 249 were obtained to gain some knowledge of the efficiency of the organism in utilizing butadiene. Two expressions of normalized growth yield, Y_{ave}, and the percentage of substrate carbon assimilated into cell material (Table II), show that significant variations occur between normalized growth yields for different growth substrates.

Isocitrate lyase has served as a useful indicator for the production of acetate units by a catabolic process. Butadiene grown cells contain high levels of isocitrate lyase relative to cells grown on succinate or glucose (Table III). This suggested that butadiene is catabolized via acetate.

Straight chain hydrocarbon catabolism usually provides acetate units by β-oxidation, a process which can be inhibited by acrylate (Thijsse, 1964). Acrylate produced only slight inhibition of butadiene respiration at con-

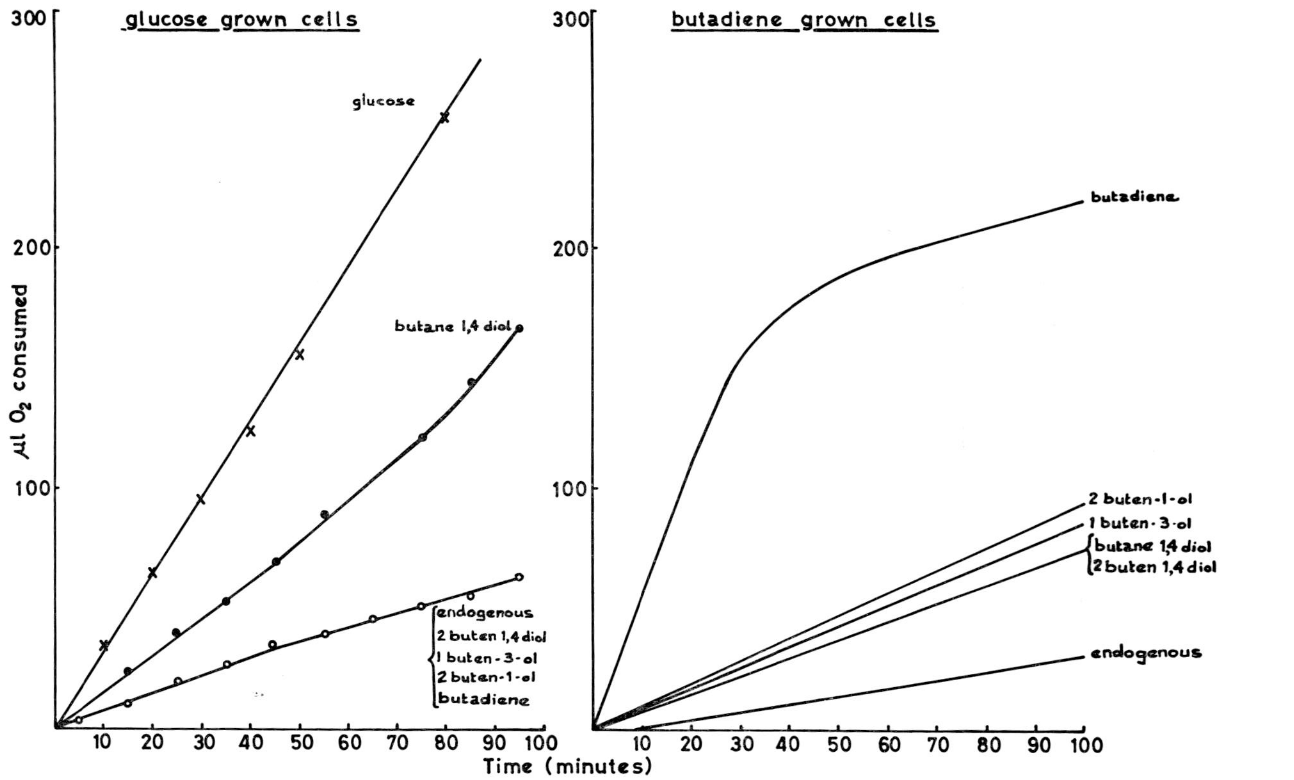

Fig. 2. Respiration of Nocardia *sp. 249 on C4 alcohols. Protocol: Cells were obtained from a culture grown on either butadiene or glucose as sole carbon source. Warburg flasks contained in a total volume of 3 ml:165 μmoles phosphate buffer pH 7·0, 6·0 mg dry weight of cells and 5 μmoles substrate or 2·6 μmoles butadiene, centre well contained 0·2 ml 10% KOH.*

Table II

GROWTH YIELDS FOR *NOCARDIA* SP. 249

Substrate	Y_{sub}*	Y_{ave}†	% *assimilation*‡
Butadiene	37·6	1·71	31
Acetate	10·7	1·34	20
Crotonate	47·3	2·63	40
3-Hydroxybutyrate	45·8	2·54	38
Malate	31·0	2·58	26
Propionate	30·0	2·14	34

* Y_{sub} = dry weight of cells obtained/mole of growth substrate utilized.
† $Y_{ave} = Y_{sub}$: total available electrons per molecule.
‡ % assimilation is calculated on the basis of all the available carbon in the substrate going to cell mass of the composition $(CH_2O)_n$ where n = number of C atoms in the substrate molecule.

Table III

ISOCITRATE LYASE LEVELS FOR *NOCARDIA* SP. 249 AFTER GROWTH ON VARIOUS SUBSTRATES

Growth substrate	*Isocitrate lyase (nmoles/min/mg)*	% *of acetate level*
Glucose	35	8
Succinate	25	6
Propionate	40	9
Lactate	92	22
Butadiene	200	47
Acetate	430	100

Table IV

THE EFFECT OF ACRYLATE ON THE RESPIRATION OF *NOCARDIA* SP. 249

Growth substrate / *Substrate*	*Acetate*			*Butadiene*		
	Q_{O_2}	Q_{O_2}''	% *inhibition*	Q_{O_2}	Q_{O_2}''	% *inhibition*
Endogenous	7	—	—	13	—	—
Butadiene	7	0	—	77	64	—
Acetate	170	163	—	73	60	—
Crotonate	76	69	—	44	31	—
Butyrate	73	66	—	54	41	—
Acrylate	15	8	—	23	10	—
Butadiene + acrylate	15	0+	—	60	37*	42
Acetate + acrylate	105	90+	45	60	37*	39
Crotonate + acrylate	20	5+	93	31	7*	78
Butyrate + acrylate	20	5+	93	31	7*	83

Q_{O_2} = μl O_2 consumed/hr/mg dry weight cells.
Q_{O_2}'' = Q_{O_2}-endogenous Q_{O_2} and -Q_{O_2}'' acrylate where indicated by *.
Protocol: Cells were obtained from cultures grown on butadiene or acetate as sole carbon source. Warburg flasks contained in a total volume of 3 ml 165 μm phosphate buffer (pH 7·0) and 5·2 mg dry weight of cells. 20 μm of substrate (or 2·8 μm butadiene) and 20 μm of acrylate were added from the side arm after a brief equilibration period. 0·2 ml or 10% KOH was added to the centre well.

centrations which completely inhibited the respiration on C-4 β-oxidation substrates, crotonate and butyrate (Table IV). From these observations it appeared that butadiene catabolism did not involve a β-oxidation process.

Fatty acid analyses carried out on cells grown on various growth substrates were provided by Professor J. J. Perry (Table V) and supported the catabolism of butadiene via acetate.

Fluoroacetate prevents the oxidation of compounds via the tricarboxylic cycle. The concentrations of fluoroacetate which completely inhibited respiration on acetate showed a small inhibition of respiration on butadiene but had no effect on the overall stoichiometry of oxygen uptake and carbon dioxide evolution (Fig. 3). Thus, it appears that resting cells do not terminally oxidize

Table V

FATTY ACID COMPOSITION OF *NOCARDIA* SP. 249

Growth substrate	*% Total fatty acids*	
	Even chain	*Odd chain*
Acetate	83·6	16·4
Propionate	11·8	88·2
Butadiene	91·8	8·2

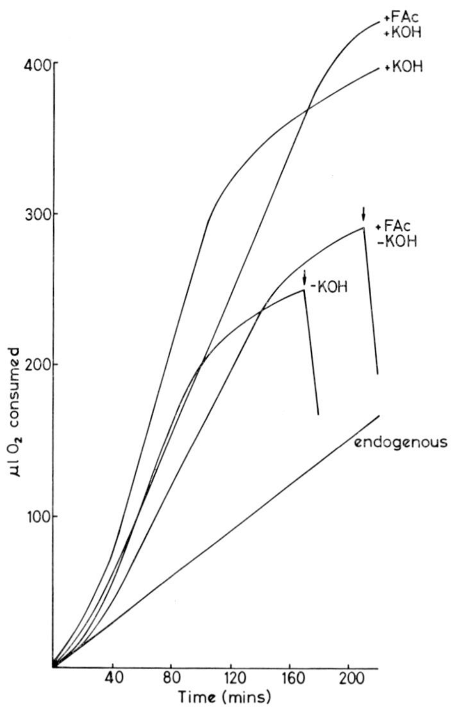

Fig. 3. Effect of fluoroacetate on the stoichiometry of butadiene oxidation by Nocardia *sp. 249. Protocol: Cells were obtained from a butadiene grown culture. Warburg flasks contained in a total volume of 3 ml:165 µmoles phosphate buffer pH 7·0, 5 mg dry weight cells and 2·6 µmoles butadiene. 0·2 ml 10% KOH was added to centre well. Arrows indicate acid tip.*

the acetate produced from butadiene but assimilate it into cell material.

Sodium arsenite (1 mm) partially inhibits respiration rate on butadiene but yields a decreased butadiene:oxygen stoichiometry and provides very little carbon dioxide evolution. Estimation of ketoacids in these reactions shows a small quantity of pyruvate but nothing else. ^{14}C-butadiene oxidation by arsenite-inhibited cells demonstrated an accumulation of an intermediate which had retained all the carbon atoms of butadiene.

The respiration of butadiene grown cells with butadiene monoepoxide provided an epoxide:oxygen stoichiometry of 1:1·25, half of the expected value. Since butadiene monoepoxide is optically active, and a racemic mixture gives the observed stoichiometry, it was found that some epoxide material remained when respiration ceased. ORD measurements showed this epoxide material to be optically active and that the ORD spectrum was modified by thiosulphate, a reaction reasonably specific for epoxide groups.

DISCUSSION

An organism with the ability to utilize butadiene as sole carbon and energy source has been isolated and tentatively identified as a *Nocardia* species. Butadiene oxidation is an inducible system in this organism and simultaneous adaptation studies indicated butadiene monoepoxide as an intermediate but failed to indicate preferentially any one of the possible C4-alkenols.

Growth yields, isocitrate lyase concentrations and lipid fatty acid analyses indicated that butadiene was metabolized via acetate and fluoroacetate inhibition studies demonstrated that only one molecule of acetate is formed per molecule of butadiene and the other two carbon atoms are lost in decarboxylation reactions. Thus, the metabolic sequence indicated is:

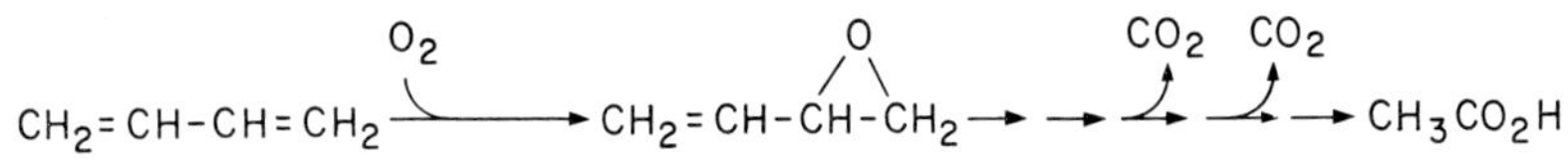

The only scheme consistent with this data involves a $\beta\gamma$-unsaturated-α-keto acid which we believe to be the compound accumulating in the presence of arsenite. We have demonstrated the enzymic reactions necessary for the initial oxygenation of butadiene and the steps leading from aerylate via pyruvate to acetate.

A consequence of the reduced oxygen stoichiometry with racemic butadiene monoepoxide is that the initial 'epoxidation' of butadiene is stereospecific producing only the correct isomer for the subsequent reactions. The overall catabolic pathway is believed to be the following:

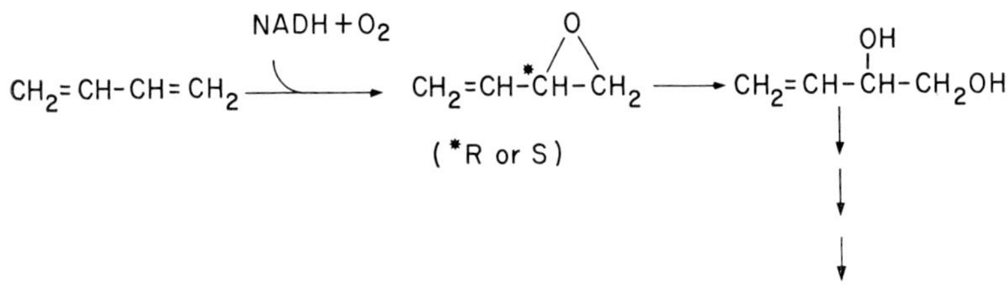

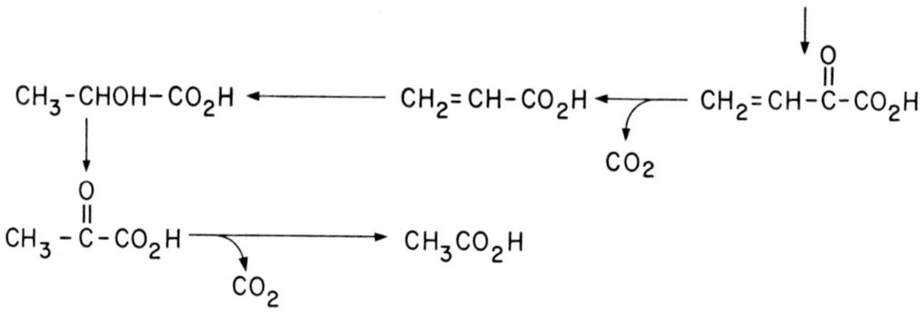

Acknowledgement

The authors would like to thank Professor J. J. Perry for the fatty acid analyses.

REFERENCES

Gibson, D. K., Koch, J. R., and Kallio, R. E. (1968). *Biochemistry*, **7**, 2653.
Ishikura, T., and Foster, J. W. (1961). *Nature*, **192**, 892.
Klug, M. J., and Markovetz, A. J. (1971). *Adv. Microb. Physiol.*, **5**, 1.
Skerman, V. B. D. (1967). *Guide to the Identification of the Genera of Bacteria.* Williams and Wilkins.
Thijsse, G. J. E. (1964). *Biochim. Biophys. Acta*, **84**, 195.

THE MICROBIAL METABOLISM OF CYCLOHEXANE AND RELATED COMPOUNDS

N. A. DONOGHUE, M. GRIFFIN, D. B. NORRIS and P. W. TRUDGILL

Department of Biochemistry and Agricultural Biochemistry, University College of Wales, Aberystwyth, Dyfed, U.K.

Summary

The most abundant natural alicyclic hydrocarbon is cyclohexane. It occurs in crude oil and the ring structure is a common component of many more complex structures. No microorganism has been reliably reported to grow with cyclohexane as the sole source of carbon though its utilization in co-metabolic and commensal situations is well established, with cyclohexanol and cyclohexanone as intermediates. We have isolated strains of *Nocardia* and *Acinetobacter* capable of growth with cyclohexanol. Cleavage of the alcohol involves oxidation to the ketone, followed by a monooxygenase-mediated introduction of a ring oxygen to form a lactone. Lactone hydrolysis and subsequent oxidation yield adipate. The monooxygenases have been purified. They carry FAD as a prosthetic group, have a functional sulphydryl group at the catalytic centre, and an obligatory requirement for NADPH as electron donor. Such enzymes appear to be fundamental to aerobic cleavage of the cyclohexane ring though alternative metabolic routes are discussed.

INTRODUCTION

The microbial degradation of hydrocarbons has been a subject of serious biochemical study only in the last 10–15 years. Although in general somewhat more recalcitrant than their polar analogues and of inevitable low solubility in aqueous media, their study can be justified on the basis of their widespread distribution in Nature and the important and fundamental information on the role and mechanism of oxygenases that has emerged from their study (Gibson *et al.*, 1970a,b; Peterson *et al.*, 1967). Knowledge of the metabolism of paraffins and aromatic hydrocarbons is considerable and increasing rapidly, but in contrast studies of cycloparaffin degradation until recently have been infrequent and limited.

The most abundant natural cycloparaffin is cyclohexane which occurs in low percentage in crude oil and as a structural component of many biologically interesting and important molecules including cyclic terpenes, sterols, and carotenoids. Growth of microorganisms at the expense of cyclohexane was first reported by Imelik (1948) and has been more recently reported by Fredricks (1966) and Jones and Eddington (1968). However, in marked contrast to these observations and in common with several other groups of workers (Ooyama and Foster, 1965; Beam and Perry, 1973; Tonge and Higgins, 1974; de Klerk and van der Linden, 1974), we have been unsuccessful in our attempts to isolate organisms capable of growth with pure cyclohexane as the carbon source.

There have been several reports of metabolic transformations of cyclohexane by organisms typically incapable of growth with it as the sole carbon source but capable of good growth on alkanes. Ooyama and Foster (1965) observed that an isolate grown with 2-methylbutane oxidized cyclohexane to cyclohexanone, which accumulated when the cyclohexane was added to freshly harvested and washed cell suspensions. More recently, Beam and Perry (1974) and de Klerk and van der Linden (1974) have further exploited this observation and achieved complete biodegradation of cyclohexane in a system consisting of two microorganisms, an *n*-alkane oxidizer capable of converting cyclohexane into either cyclohexanol or cyclohexanone during growth on the *n*-alkane and a second organism capable of growth with cyclohexanol or cyclohexanone. Such systems involving co-metabolism and commensalism are perhaps of restricted distribution in the environment but of obvious value in the microbial degradation of such materials as crude oil that contain mixtures of *n*-alkanes, cycloalkanes, and aromatic hydrocarbons.

Because logic and available evidence dictate that microbial attack on cyclohexane be initiated by hydroxylation, we have concentrated our studies on the degradation of cyclohexanol and cyclohexanone. This has the added practical advantage that organisms capable of growth on these compounds as sole sources of carbon are very easily isolated from the biosphere.

MATERIALS AND METHODS

Isolation, maintenance, and growth of organisms

Nocardia globerula CL1 was isolated from woodland soil near Aberystwyth by elective culture in a medium with cyclohexanol as the sole source of carbon. *Acinetobacter* N.C.I.B. 9871 was isolated by Dr P. J. Chapman and deposited in the N.C.I.B. culture collection. Stocks were maintained on nutrient agar, stored at 2°C, and subcultured at monthly intervals.

Culture media and growth of bacteria

Cells were grown at 30°C in the cyclohexanol medium described by Norris and Trudgill (1971). Typically, 25 ml of medium in a 100-ml Erlenmeyer flask were inoculated from a slant and cyclohexanol was supplied from a diffusion tube. After 48 hr of growth in an orbital incubator at 150 rpm, this culture was used to inoculate 1 litre of medium in a 2 litre flask. After 12–14 hr of growth, this culture was harvested for experimental purposes or used to inoculate 10 litre of medium in a fermenter stirred at 250 rpm and supplied with sterile air at the rate of 1 litre/min. Mean generation times were 1·1 hr for *Acinetobacter* N.C.I.B. 9871 and 2·2 hr for *N. globerula* CL1 when grown with forced aeration and stirring on cyclohexanol as the carbon source.

Cells were harvested by conventional continuous or batch centrifugal techniques, washed by resuspension in 21 mM phosphate buffer, pH 7·1, followed by centrifugation.

Preparation of cell-free extracts

Washed cell suspensions were disintegrated by passage through a French pressure cell and clarified by centrifugation as previously described (Norris and Trudgill, 1971).

Gas exchange reactions

Gas exchange catalysed by whole cells or cell extracts was measured by conventional Warburg manometry or with an oxygen electrode.

Enzyme assays

Cyclohexanol dehydrogenase was assayed by following the increase in absorbance at 340 nm when the test extract was added to a 1 cm light-path cuvette that contained in 1 ml, 90 μmol of glycine-NaOH buffer, pH 10, 0·4 μmol of NAD^+, and 1 μmol of cyclohexanol.

Cyclohexanone oxygenase was assayed by following the decrease in absorbance at 340 nm when test extract was added to a cuvette that contained in 1 ml, 90 μmol of suitable buffer, 80 μmol of NADPH, and 0·3 μmol of cyclohexanone. Oxygen consumption, with increased quantities of reaction components, was followed in a 3-ml reaction in either the Warburg or the oxygen monitor.

1-Oxa-2-oxocycloheptane lactonase was assayed by measuring the rate of lactone hydrolysis in reaction mixtures that contained in 2 ml, 8 μmol of substrate, 100 μmol of Tris-HCl buffer, pH 7·1, and test extract. Residual substrate was estimated after a 10-min incubation period by the method of Cain (1961).

6-Hydroxyhexanoate dehydrogenase and 6-oxohexanoate dehydrogenase were assayed by measuring the increase in absorbance at 340 nm when the appropriate substrate was added to a cuvette that contained in 1 ml, 19 μmol of phosphate buffer, pH 7·1, 0·16 μmol of NAD^+ or $NADP^+$, and cell extract.

Preparation of compounds

6-Hydroxyhexanoate was prepared by alkali hydrolysis of 1-oxa-2-oxocycloheptane as previously described (Norris and Trudgill, 1971). 6-Oxohexanoate was prepared from 2-hydroxycyclohexan-1-one by the method of Baer (1942).

Preparation of materials for chromatography

Typically, protein was precipitated by addition of 11 M HCl (1 ml per 4 ml of reaction) and removed by centrifugation. The supernatants were extracted with diethyl ether, the ether extracts dried over anhydrous Na_2SO_4, and evaporated to small volume. Samples were subjected to t.l.c. and g.l.c. directly or after methylation according to the procedure of Metcalf and Schmitz (1961). 2,4-Dinitrophenylhydrazones of neutral compounds were prepared by the procedure of Friedemann and Haugen (1943), harvested by centrifugation, washed with 2 M HCl, washed with water, and dissolved in ethyl acetate.

Chromatographic procedures

T.l.c. of 2,4-dinitrophenylhydrazone derivatives was carried out on 0·25-mm thick layers of Kieselgel G developed with solvent A, benzene:tetrahydrofuran (19:1 v/v) (Byrne, 1965) or with solvent B, hexane:ethyl formate (4:1 v/v) (Dagley and Gibson, 1965) or with solvent C, light petroleum (b.p. 80–100°C): ethyl formate (7:3). T.l.c. of carboxylic acids was carried out on 0·25-mm thick layers of a mixture of Kieselgel G and Kieselguhr G (1:3, w/w) pre-treated with acetic acid vapour and developed with solvent D, the upper phase of a benzene:glacial acetic acid (40:7, v/v) mixture (Miyazaki *et al.*, 1969) or on Kieselgel G developed with solvent E, benzene:dioxan:glacial acetic acid (90:25:1 by vol) or with solvent F, benzene:dioxan:glacial acetic acid (90:15:2 by vol) or solvent G, benzene:methanol:glacial acetic acid (45:8:4 by vol).

2,4-Dinitrophenylhydrazones were detected by direct visual observation, aldehydes were detected by spraying with 0·1% (w/v) 2,4-dinitrophenylhydrazine in 5 M HCl and observing the yellow derivatives thus formed. Carboxylic acids were detected by spraying thoroughly dried plates with 0·1% (w/v) Bromocresol Green in aq. 95% (v/v) ethanol adjusted to pH 6 with NaOH.

Gas–liquid chromatography

Gas–liquid chromatography was performed in a Packard series 7600 instrument on either a 2 mm × 1·9 m column packed with 1·25% (w/w) diglycerol on Chromosorb W or a 4 mm × 1·3 m column packed with 17% DEGS on the same support.

Polyacrylamide gel electrophoresis

Electrophoresis of native enzymes was carried out according to the method of Davis (1964) on 7·5, 10, and 15% (w/v) acrylamide gels. Sodium dodecylsulphate-polyacrylamide gel electrophoresis was performed as described by Weber and Osborn (1969). Gels were stained with either Amido Black 10B or Coomassie Brilliant Blue G-250 and cleared by washing in 7% (w/v) glacial acetic acid or glacial acetic acid-ethanol-water (2:5:13), respectively.

Ultracentrifugation

A Beckman-Spinco Model E analytical centrifuge equipped with a Schlieren phase-plate diaphragm and Rayleigh interference optics was used.

Materials

Sources of most materials were as previously described (Norris and Trudgill, 1971).

RESULTS

Growth spectra

Somewhat surprisingly, both organisms display a strictly limited capability for growth on substituted alicyclic rings. For example, *Acinetobacter* N.C.I.B. 9871 is capable of good growth with cyclohexanol, cyclohexanone, 1-oxa-2-oxocycloheptane, 6-hydroxyhexanoate, *cis*, *trans*-cyclohexan-1,2-diol, and 2-hydroxycyclohexanone. Growth is also good on the straight chain primary alcohols *n*-butanol, *n*-pentanol, *n*-hexanol, and *n*-heptanol but slow with adipate.

Limited substitution of the cyclohexanol ring, other than by hydroxyl or keto substituents at the 2-position, change in ring size and additional complexities render the compounds unacceptable. *Acinetobacter* N.C.I.B. 9871 is incapable of growth with any of the following compounds: cyclobutanol, cyclobutanone, cyclopentanol, cyclopentanone, cycloheptanol, cycloheptanone, cyclooctanol, cyclooctanone, *cis*, *trans*-cyclohexan-1,3-diol, *cis*, *trans*-

cyclohexan-1,4-diol, cyclohexan-1,2-dione, cyclohexan-1,3-dione, 2-methylcyclohexanol, 2-methylcyclohexanone, 3-methylcyclohexanol, 3-methylcyclohexanone, 4-methylcyclohexanol, 4-methylcyclohexanone, 1-oxa-2-oxocyclopentane, 1-oxa-2-oxocyclohexane, menthol, menthone, (+)-pulegone, dihydrocarvone, (+)-camphor, *n*-hexane, *n*-hexadecane, and cyclohexane.

Growth on substituted cyclohexanes is therefore restricted, as we shall see, to those compounds metabolically closely related to cyclohexanol.

Oxidation of substrates by whole cells

Freshly harvested *Acinetobacter* N.C.I.B. 9871 grown with cyclohexanol oxidized the growth substrate at a rate of 230 μl of O_2/hr per mg dry wt with the consumption of 5·6 μmol of O_2/μmol added. Cyclohexanone was oxidized with the consumption of 4·85 μmol of O_2/μmol and a rate of 192 μl of O_2/hr per mg dry wt. Rates of O_2 consumption with other tested compounds were as follows: *cis, trans*-cyclohexan-1,2-diol, 230; 2-hydroxycyclohexanone, 192; cyclohexan-1,2-dione, 24; 1-oxa-2-oxocycloheptane, 44; 6-hydroxyhexanoate, 44; adipate was not oxidized.

With cyclohexanol-grown *N. globerula* CL1, the same general oxidation pattern was observed but with lower oxygen uptake rates and decreased overall consumption (Norris and Trudgill, 1971).

Oxidation of cyclohexanol by cell extracts

Extracts of cyclohexanol-grown *Acinetobacter* N.C.I.B. 9871 catalysed the stoichiometric reduction of NAD^+ in the presence of cyclohexanol at pH 10. The product of the reaction co-chromatographed with authentic cyclohexanone on the diglycerol g.l.c. system and the 2,4-dinitrophenylhydrazone formed from it was indistinguishable from the derivative of the authentic compound in t.l.c. systems A, B, and C. The full specificity range of the enzyme has not been investigated but it is equally active with the cyclohexan-1,2-diols as substrates.

In contrast, attempts to demonstrate the presence of a pyridine nucleotide-dependent cyclohexanol dehydrogenase in extracts of *N. globerula* CL1 have consistently met with failure. Slow oxygen uptake observed with preparations having a significant membrane component leads us to suspect that the enzyme in this organism is membrane bound and feeds electrons directly into the electron transport chain.

Oxidation of cyclohexanone by cell extracts

When extract of either organism grown with cyclohexanol was incubated with cyclohexanone, an NADPH-dependent consumption of oxygen was observed. NADH could not be substituted for NADPH. Investigations of reaction stoichiometry demonstrated that the oxidation of 1 μmol of cyclohexanone was accompanied by the consumption of 1 μmol of NADPH and 0·8–0·93 μmol of oxygen. This stoichiometry is close to that which is theoretically required for a monooxygenase of the external electron donor type, a postulate reinforced by the obligatory requirement for oxygen before substrate stimulated NADPH oxidation can proceed. A possible product of such a monooxygenation would be a hydroxycyclohexanone and the high O_2 uptake rates observed with whole cells and 2-hydroxycyclohexanone would make this compound a possible candidate. However, studies of a number of analogous oxygenations of cyclic and straight chain ketones suggest an alternative possibility. Lactone-forming monooxygenases have been described that attack several cyclic ketones and are involved in the microbial degradation of androstene-3,17-dione (Prairie and Talaley, 1963), (+)-camphor (Trudgill *et al.*, 1966), and cyclopentanone (Griffin and Trudgill, 1972, 1973). This type of monooxygenase is also involved in the microbial cleavage of 2-tridecanone (Forney and Markovetz, 1969) and acetophenone (R. E. Cripps, pers. comm.). The rational advantage of such monooxygenases is that they yield lactones and esters that may be cleaved by esterase action (Scheme 1).

Bulk reactions in which 5–20 μmol of cyclohexanone were incubated with a slight excess of NADPH and dialysed cell extract under aerobic conditions in the Warburg flask were examined by t.l.c. in solvent systems D, E, F, and G and by g.l.c. after methyl esterification. The enzymic product co-chromato-

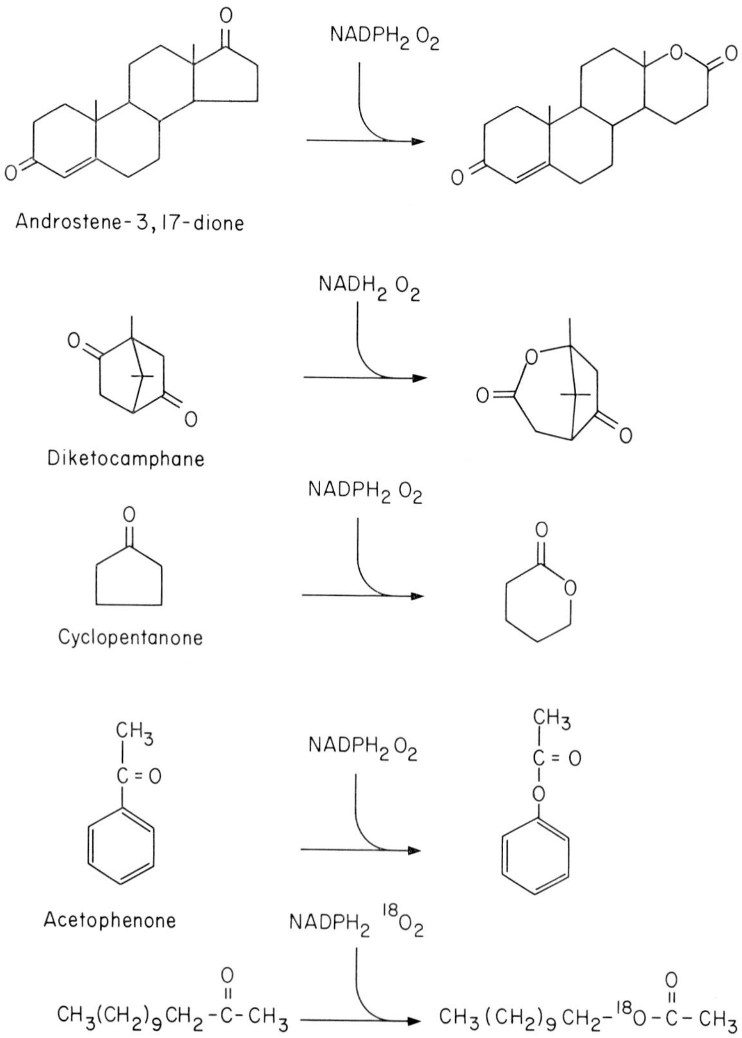

Scheme 1. Some ester-forming monooxygenases.

graphed with 6-hydroxyhexanoate or its methyl ester according to the analytical system used.

A role for 6-hydroxyhexanoate as an intermediate in cyclohexanol metabolism is reinforced by the presence of a rather unstable NAD^+-linked dehydrogenase present in crude cell extracts and of a highly active $NADP^+$-linked 6-oxohexanoate dehydrogenase that yields adipate. Though not so thoroughly investigated, extracts of cyclohexanol-grown *N. globerula* CL1 are capable of converting 6-hydroxyhexanoate to adipate (Norris and Trudgill, 1971).

Intermediates in the conversion of cyclohexanone to 6-hydroxyhexanoate

From the evidence so far presented, it is possible to formulate two alternative routes for the conversion of cyclohexanol to adipate. Scheme 2 depicts

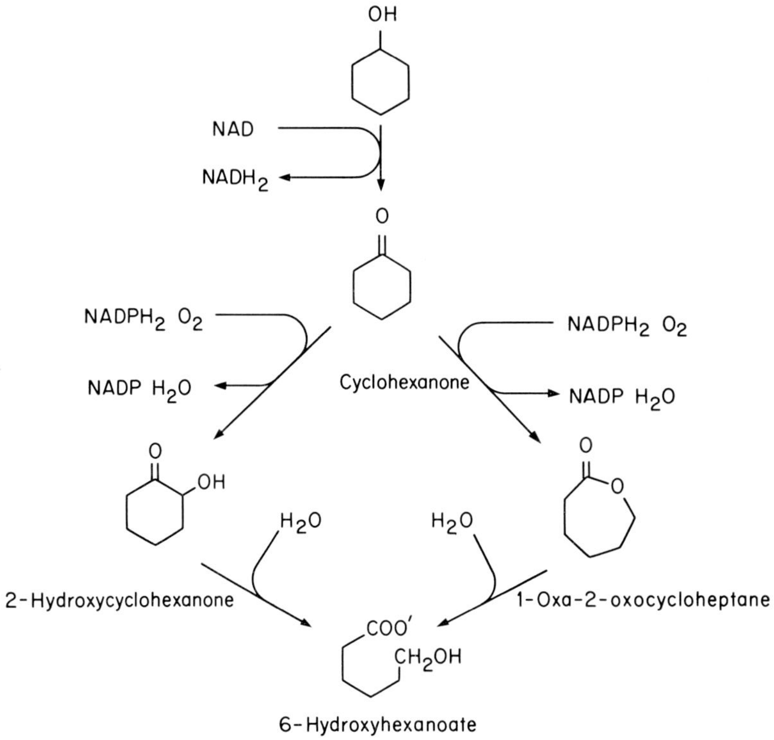

Scheme 2. Alternative routes for the metabolic conversion of cyclohexanol to 6-hydroxyhexanoate.

these and shows clearly how two alternative monooxygenases could be involved. Evidence from whole cell studies is equivocal since both 2-hydroxycyclohexanone and 1-oxa-2-oxocycloheptane are oxidized.

Preliminary evidence for the type of monooxygenase involved came from studies of 1-oxa-2-oxocycloheptane hydrolase activity of crude cell extracts of both organisms. As Table I shows, a highly active hydrolase is induced in

Table I
INDUCTION OF THE ENZYMES OF CYCLOHEXANOL OXIDATION IN *ACINETOBACTER* N.C.I.B. 9871

Enzyme	Activity in extracts of cells grown on: Cyclohexanol (units/mg)	L-Glutamate (units/mg)
Cyclohexanol dehydrogenase	0·28	0·04
Cyclohexanone monooxygenase	0·60	0·001
1-Oxa-2-oxocycloheptane hydrolase	16	0·08
6-Hydroxyhexanoate dehydrogenase	0·04	not detectable
6-Oxohexanoate dehydrogenase	0·60	0·06

Acinetobacter N.C.I.B. 9871 when grown with cyclohexanol; similar data have been obtained for *N. globerula* CL1 (Norris and Trudgill, 1971). It suggests that the monooxygenases are of the lactone-forming type and since the specific activity of the lactone hydrolase in each case exceeds that of the oxygenase by some 30-fold, even a transient accumulation of any intermediate lactone is unlikely.

Confirmation that both monooxygenases are of the lactone-forming type comes from definitive experiments with pure enzymes devoid of lactone hydrolase activity.

Purification of cyclohexanone oxygenase

The oxygenase from each organism was purified by conventional chromatographic and salting out techniques; data for *Acinetobacter* N.C.I.B. 9871 are shown in Table II. Both enzymes were homogeneous in the ultracentrifuge

Table II
PURIFICATION OF CYCLOHEXANONE OXYGENASE FROM *ACINETOBACTER* N.C.I.B. 9871

Step	Total protein (mg)	Total activity (units)	Specific activity (units/mg)	Yield (%)	Purification (-fold)
Crude extract	4 554	2 732	0·60	100	1·0
$(NH_4)_2SO_4$ (50–80% sat.)	1 500	1 920	1·28	70	2·1
DEAE Cellulose (0·21–0·31 M KCl)	169	1 094	6·44	40	10·7
Sephadex G-150	33	492	14·82	18	24·7
Hydroxyapatite (270–350 mM PO_4)	9	189	21·0	6·9	35·0

and sedimented as single symmetrical peaks at all tested protein concentrations. They were pure as judged by polyacrylamide gel electrophoresis; the cyclohexanone oxygenase of *Acinetobacter* N.C.I.B. 9871 migrates as a single band on gels of three pore sizes (Fig. 1), while the enzyme from *N. globerula* CL1 appears to exist as two isoenzymes (Norris and Trudgill, 1974).

Both enzymes are devoid of 1-oxa-2-oxocycloheptane hydrolase activity and the products of bulk oxygenation reactions have been subjected to close scrutiny. Reaction products give strong positive reactions in the hydroxamate test for lactones, give infrared spectra indistinguishable from that of authentic

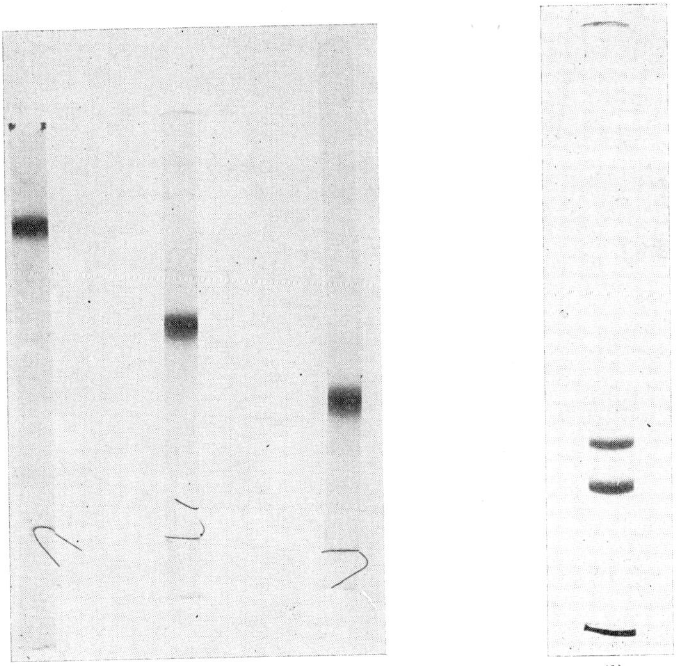

(*a*) (*b*)

Fig. 1. Electrophoresis of purified cyclohexanone oxygenases on polyacrylamide gels. (a) From Acinetobacter *N.C.I.B.9871. Gels were (left) 15%, (middle) 10%, and (right) 7·5% (w/v) with respect to acrylamide. An aliquot of 40 μg of purified protein was applied to each gel. (b) From* N. globerula *CL1. The gel was 10% (w/v) with respect to acrylamide. After electrophoresis gels were stained with Amido Black 10B.*

1-oxa-2-oxocycloheptane, and when subjected to g.l.c., co-chromatograph with the authentic compound (Table III). Since these g.l.c. systems effectively separate 1-oxa-2-oxocycloheptane from the alternative reaction product 2-hydroxycyclohexanone, the enzymes can be unequivocally named as cyclohexanone, NADPH: oxygen oxidoreductases (1,2-lactonizing) (EC

Table III

IDENTIFICATION OF THE PRODUCT OF CYCLOHEXANONE OXYGENATION BY ENZYME PURIFIED FROM *ACINETOBACTER* N.C.I.B. 9871

Sample	*G.l.c. retention times* 1·25% *diglycerol*[a]	17% *DEGS*[b]
	min	min
Cyclohexanone	0·7	1·4
1-Oxa-2-oxocycloheptane	6·2	14·4
2-Hydroxycyclohexanone	2·3	4·9
4-Hydroxycyclohexanone	24·1	26·3
Reaction product	6·1	14·6

[a] 92°C carrier gas at 32 ml/min.
[b] 168°C carrier gas at 60 ml/min.
The reaction system contained in 15 ml of 100 mM Tris-HCl buffer, pH 7·1, 0·4 mg of enzyme, 10 μmol of cyclohexanone. After 10 min, incubation products were extracted and prepared for chromatography.

1.14.13). Only one enzyme of this lactone-forming type had previously been purified to any significant extent from (+)-camphor grown *Pseudomonas putida* C1B, and its properties studied (Trudgill *et al.*, 1966; Yu and Gunsalus, 1969). We have therefore subjected these two cyclohexanone oxygenases to extensive critical study. Some of their more important properties are summarized in Table IV. As indicated, oxygenase specificity is markedly broader

Table IV

PROPERTIES OF CYCLOHEXANONE OXYGENASES

	Source of enzymes	
	Acinetobacter N.C.I.B. 9871	N. globerula CL1
Molecular weight	59 000	53 000
% of soluble cell protein	3	8
Prosthetic group	FAD	FAD
Moles FAD-bound/mole enzyme	1	1
pH optimum	9	8·4
Electron donor	NADPH	NADPH
Substrate specificity	Broad[a]	Broad
Catalytic centre activity	1 390	1 018
K_m for NADPH (μM)	6·9	31
K_m for cyclohexanone (μM)	17·8	1·5
K_d for FAD (nM)	40	not measured
Number of polypeptide chains	1	1
Inhibitors	sulphhydryl group reagents	sulphhydryl group reagents

[a] Good substrates include 2-methylcyclohexanone, 4-methylcyclohexanone, 2-hydroxycyclohexanone, 4-hydroxycyclohexanone, norcamphor, cycloheptanone, cuprizone, and cyclohexanonephenylhydrazone.

than the growth specificity; 2-hydroxycyclohexanone is of interest as an oxygenase substrate because of the equivocal position it occupies in studies of cyclohexanone metabolism.

We have established that the product of oxygenation of this compound, 1-oxa-2-oxo-7-hydroxycycloheptane, which is inherently unstable, rearranges to yield 6-oxohexanoate.

The broad specificity of the cyclohexanol dehydrogenase and cyclohexanone oxygenase of *Acinetobacter* provide a route whereby the cyclohexan-1,2-diols may be metabolized to adipate (Scheme 3, reactions 1, 2, 5, and 6).

Pathway of cyclohexanol oxidation

The reaction sequence whereby cyclohexanol is converted to adipate is shown in Scheme 4. All metabolic steps with the exception of cyclohexanol dehydrogenase of *N. globerula* CL1 have been demonstrated. The further metabolism of adipate has not been investigated in these organisms but may well follow the β-oxidation route already described for strains of *Pseudomonas* by Chapman and Duggleby (1967).

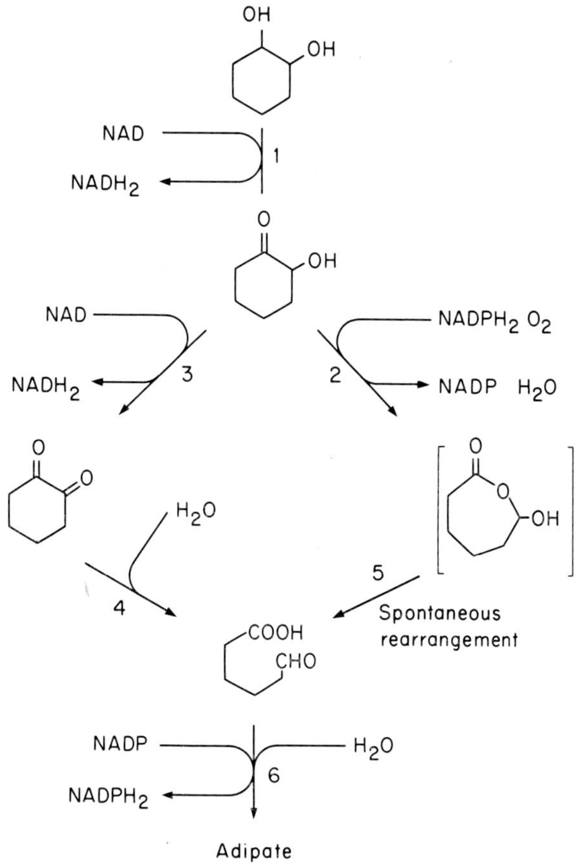

Scheme 3. Alternative routes for the conversion of cyclohexan-1,2-diols to adipate.

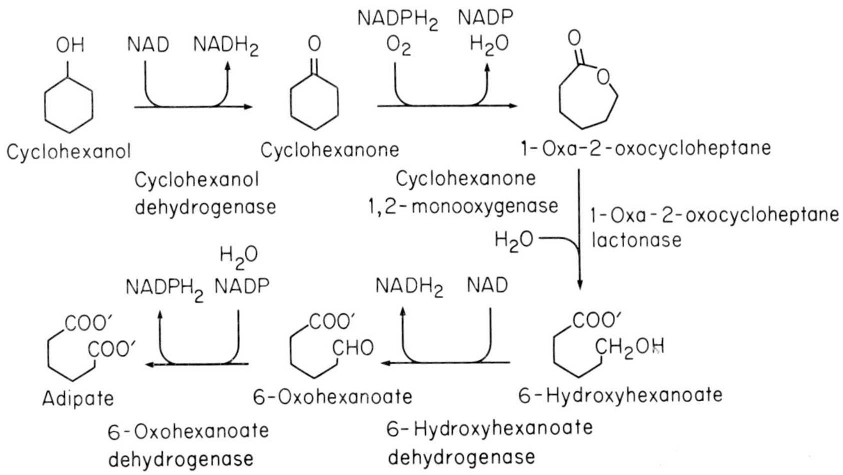

Scheme 4. Reaction sequence for the oxidation of cyclohexanol to adipate.

DISCUSSION

Acinetobacter N.C.I.B. 9871 and *N. globerula* CL1 grow rapidly with cyclohexanol as the sole source of carbon but their capability to use other alicyclic compounds as sole sources of energy and biosynthetic precursors is strictly limited to a few compounds closely related to cyclohexanol. This observation is of interest since cyclohexanol is not of significant natural occurrence. Two possible explanations exist for the ecological advantage that the possession of this inducible capability may give to these organisms. Alicyclic compounds are widely distributed in Nature including terpenes, sterols, and a wide range of more exotic molecules. It is possible that, in natural mixed populations, the degradation of such compounds may yield cyclohexanol or closely related compounds as metabolic intermediates or products of co-metabolism. In addition, Beam and Perry (1973) and de Klerk and van der Linden (1974) have reported the degradation of cyclohexane resulting from the co-metabolism of this compound to cyclohexanol by organisms capable of growth with straight or branched chain hydrocarbons, followed by further degradation of the cyclohexanol by organisms capable of growth with the latter compound. *Acinetobacter* N.C.I.B. 9871 and *N. globerula* CL1 would fulfil the role of the second type of organism in such a commensal situation.

The conflicting reports concerning the ability of microorganisms to grow with cyclohexane as the sole source of carbon are curious but not inexplicable. Imelik (1948) gives few details of his experiments which makes evaluation difficult. Fredricks (1966) used a 2% (v/v) level of substrate, a sufficiently large amount to at least allow the possibility of growth on contaminating material and Jones and Eddington (1968) plated mixed bacterial populations from soil extracts onto agar containing basal salts in the presence of cyclohexanone vapour so that the interaction of more than one bacterial species is a possibility. Current research in a number of laboratories has not substantiated these earlier reports; a variety of enrichment techniques have failed to select an organism convincingly capable of growth with cyclohexane.

Why this should be the case is not clear. Organisms have been isolated, submitted to culture collections, used by several groups of workers, that are well able to grow with *n*-alkanes (Peterson *et al.*, 1967), branched chain alkanes (Pirnik *et al.*, 1974), and aromatic hydrocarbons (Gibson, 1972) which would appear to present similar problems of nonpolarity, lack of functional groups, and negligible water solubility. Initial attack upon these compounds is mediated by monooxygenases that insert one or more hydroxyl groups which opens the way to further oxidation steps. This makes the situation even more strange since several groups of workers (Ooyama and Foster, 1965; Beam and Perry, 1973; de Klerk and van der Linden, 1974) have provided evidence that straight and branched chain hydrocarbon hydroxylases induced when microorganisms are grown on these compounds will gratuitously hydroxylate cyclohexane.

Recently, Tonge and Higgins (1974) have described the slow growth (mean generation time 40 hr) of *Nocardia petroleophila* N.C.I.B. 9438 on methyl cyclohexane and the accumulation of 3-methylcyclohexanol and 3-methylcyclohexanone in the growth medium. Whole cell oxidation of methylcyclohexane is extremely slow (30 nmol of O_2/hr per mg dry wt of bacteria) but it is interesting that initial attack is on the cyclohexane ring rather than the methyl group.

Though we have presented clear evidence that cyclohexanol and cyclohexanone are metabolized through a key lactone-forming monooxygenase step by our strains, Murray *et al.* (1974), working with a strain of *Nocardia* isolated by enrichment culture with cyclohexanone, have suggested on the basis of preliminary oxidation studies that the metabolism of cyclohexanone involves hydroxylation to yield 2-hydroxycyclohexanone (Scheme 2). This presumably is followed by hydrolytic cleavage either of the 2-hydroxycyclohexanone or the product of its dehydrogenation, cyclohexan-1,2-dione.

The suggestion of a direct hydrolytic cleavage of the cyclohexane ring, in contrast to the dependence on monooxygenation, would seem to be theoretically acceptable provided that the ring carries the appropriate substituents. Yugari (1961) reported that a strain of *Pseudomonas* grown with *trans*-cyclohexan-1,2-diol oxidized this to cyclohexan-1,2-dione which was then cleaved hydrolytically to yield 6-oxohexanoate (Scheme 3, reactions 1, 3, 4, and 6). This pathway provides a neat parallel to the proposal of Murray *et al.* (1974) but, unfortunately, the organism has been lost.

It is interesting, though not within the scope of this paper, that alternative substituents on the cyclohexane ring may completely alter the mode of metabolism. For example, introduction of a carboxyl group results in degradation by an aromatization pathway leading to *p*-hydroxybenzoate (Blakley, 1974; Kaneda, 1974).

Acknowledgement

This work was supported by grants from the Science Research Council.

REFERENCES

Baer, E. (1942). Oxidative cleavage of cyclic α-keto alcohols by means of lead tetraacetate. *J. Am. Chem. Soc.*, **64**, 1416–1421.

Beam, H. W., and Perry, J. J. (1973). Co-metabolism as a factor in microbial degradation of cycloparaffinic hydrocarbons. *Arch. Microbiol.*, **91**, 87–90.

Beam, H. W., and Perry, J. J. (1974). Microbial degradation of cycloparaffinic hydrocarbons via co-metabolism and commensalism. *J. Gen. Microbiol.*, **82**, 163–169.

Blakley, E. R. (1974). The microbial degradation of cyclohexanecarboxylic acid; a pathway involving aromatization to form *p*-hydroxybenzoate.

Byrne, G. A. (1965). The separation of 2,4-dinitrophenylhydrazones by thin layer chromatography. *J. Chromatogr.*, **20**, 528–540.

Cain, R. B. (1961). The metabolism of protocatechuic acid by a *Vibrio*. *Biochem. J.*, **79**, 298–312.

Chapman, P. J., and Duggleby, R. G. (1967). Dicarboxylic acid catabolism by bacteria. *Biochem. J.*, **103**, 7C–9C.

Dagley, S., and Gibson, D. T. (1965). The bacterial degradation of catechol. *Biochem. J.*, **95**, 466–474.

Davis, B. J. (1964). Disc electrophoresis II. Method and application of human serum proteins. *Ann. N.Y. Acad. Sci. U.S.*, **121**, 404–427.

Forney, F. W., and Markovetz, A. J. (1969). An enzyme system for aliphatic ketone oxidation. *Biochem. Biophys. Res. Commun.*, **37**, 31–38.

Fredricks, K. M. (1966). Adaptation of bacteria from one type of hydrocarbon to another. *Nature* (*London*), **209**, 1047–1048.

Friedemann, T. E., and Haugen, G. E. (1943). The determination of keto acids in blood and urine. *J. Biol. Chem.*, **147**, 415–442.

Gibson, D. T. (1972). The microbial oxidation of aromatic compounds. *Crit. Rev. Microbiol.*, **1**, 199–223.

Gibson, D. T., Cardini, G. E., Maseles, F. C., and Kallio (1970a). Incorporation of oxygen-18 into benzene by *Pseudomonas putida. Biochemistry*, **9**, 1631–1635.

Gibson, D. T., Hensley, M., Yoshioka, H., and Mabry, T. J. (1970b). Formation of (+)-cis-2,3-Dihydroxy-1-methylcyclohexa-1,4-diene from toluene by *Pseudomonas putida. Biochemistry*, **9**, 1626–1630.

Griffin, M., and Trudgill, P. W. (1972). The metabolism of cyclopentanol by *Pseudomonas* N.C.I.B. 9872. *Biochem. J.*, **129**, 595–603.

Griffin, M., and Trudgill, P. W. (1973). The purification of cyclopentanone oxygenase from *Pseudomonas* N.C.I.B. 9872. *Trans. Biochem. Soc.*, **1**, 1255–1258.

Imelik, B. (1948). Oxydation de Cyclohexan par *Pseudomonas aeruginosa. C.R. Soc. Biol. Paris*, **226**, 2082–2083.

Jones, J. G., and Eddington, M. A. (1968). An ecological survey of hydrocarbon utilizing micro-organisms. *J. Gen. Microbiol.*, **52**, 381–390.

Kaneda, T. (1974). Enzymatic aromatization of 4-ketocyclohexanecarboxylic acid to *p*-hydroxybenzoic acid. *Biochem. Biophys. Res. Commun.*, **58**, 140–144.

de Klerk, H., and van der Linden, A. C. (1974). Bacterial degradation of cyclohexane. Participation of a co-oxidation reaction. *Antonie van Leeuwenhoek*, **40**, 7–15.

Metcalf, L. D., and Schmitz, A. A. (1961). The rapid preparation of fatty acid esters for gas chromatographic analysis. *Anal. Chem.*, **33**, 363–364.

Miyazaki, S., Suhara, Y., and Kobayashi, T. (1969). Separation of aliphatic dibasic acids by thin layer chromatography. *J. Chromatogr.*, **39**, 88–90.

Murray, J. R., Scheikowski, T. A., and MacRae, I. C. (1974). Utilization of cyclohexanone and related substances by a *Nocardia* sp. *Antonie van Leeuwenhoek*, **40**, 17–24.

Norris, D. B., and Trudgill, P. W. (1971). The metabolism of cyclohexanol by *Nocardia globerula* CL1. *Biochem. J.*, **121**, 363–370.

Norris, D. B., and Trudgill, P. W. (1974). Multiple forms of cyclohexanone oxygenase from *Nocardia globerula* CL1. *Trans. Biochem. Soc.*, **1**, 1252–1255.

Ooyama, J., and Foster, J. W. (1965). Bacterial oxidation of cycloparaffinic hydrocarbons. *Antonie van Leeuwenhoek*, **31**, 45–65.

Peterson, A. J., Kusunose, M., Kusunose, E., and Coon, M. J. (1967). Enzymatic ω-oxidation. II Function of rubredoxin as the electron carrier in ω-hydroxylation. *J. Biol. Chem.*, **242**, 4334–4340.

Pirnik, M. P., Atlas, R. M., and Bartha, R. (1974). Hydrocarbon metabolism by *Brevibacterium erythrogenes:* normal and branched alkanes. *J. Bacteriol.*, **119**, 868–878.

Prairie, R. L., and Talaley, P. (1963). Enzymatic formation of testololactone. *Biochemistry*, **2**, 203–208.

Tonge, G. M., and Higgins, I. J. (1974). Microbial metabolism of alicyclic hydrocarbons. Growth of *Nocardia petroleophila* (N.C.I.B. 9438) on methylcyclohexane. *J. Gen. Microbiol.*, **81**, 521–524.

Trudgill, P. W., DuBus, R., and Gunsalus, I. C. (1966). Purification of a tightly coupled electron transport complex in camphor lactonization. *J. Biol. Chem.*, **241**, 4288–4290.

Weber, K., and Osborn, M. (1969). The reliability of molecular weight determinations by dodecylsulfate-polyacrylamide gel electrophoresis. *J. Biol. Chem.*, **244**, 4406–4412.

Yu, C. A., and Gunsalus, I. C. (1969). Monooxygenases. Camphor ketolactonase I and the role of three protein components. *J. Biol. Chem.*, **244**, 6149–6152.

Yugari, Y. (1961). Metabolism of cyclohexane-diol-(1,2)-*trans* by a soil bacterium. *Biken. J.*, **4**, 197–207.

MICROBIAL DEGRADATION OF POLYCYCLIC AROMATIC HYDROCARBONS

DAVID T. GIBSON

Department of Microbiology, University of Texas at Austin,
Austin, Texas 78712, U.S.A.

Summary

A mutant strain of *Beijerinckia*, after growth on succinate in the presence of biphenyl, oxidized benzo[a]pyrene and benzo[a]anthracene to *cis* dihydrodiols. The ultraviolet spectrum of the dihydrodiol formed from benzo[a]pyrene was identical to the spectra given by synthetic *cis*-9,10-dihydroxy-9,10-dihydrobenzo[a]pyrene and 9,10-dihydrobenzo[a]pyrene. Acid-catalysed dehydration of the metabolite gave 9-hydroxybenzo[a]pyrene. These observations together with the information supplied by proton magnetic resonance spectrometry identify the compound as *cis*-9,10-dihydroxy-9,10-dihydrobenzo[a]pyrene.

The compound formed from benzo[a]anthracene gave an identical spectrum to that given by synthetic 1,2-dihydrobenzo[a]anthracene. Acid treatment gave 2-hydroxybenzo[a]anthracene. The proton magnetic resonance spectrum of this product confirmed its structure as *cis*-1,2-dihydroxy-1,2-dihydrobenzo[a]anthracene.

The parent strain of *Beijerinckia* oxidized benzo[a]pyrene to acid products that were not characterized.

INTRODUCTION

Polycyclic hydrocarbons are formed by the thermal diagenesis of organic compounds. The rate of transformation and the composition of the final product is a function of the transformation temperature. Low temperatures favour the formation of alkylated products. Consequently, the polycyclic hydrocarbons found in ancient sediments and fossil fuels are exceedingly complex. Available analytical techniques can only resolve the polycyclic hydrocarbons found in crude petroleum into classes of compounds. Resolution into individual compounds is not yet possible. Pyrolysis at high temperatures (*e.g.* coking of coal, automobile engines, and domestic heating) results in a considerable reduction in the degree of alkylation and favours the formation of unsubstituted polycyclic aromatic hydrocarbons. In urban polluted air, C_1 homologues of the parent hydrocarbons are the predominant alkylated compounds (Lao *et al.*, 1973). In addition, benzo[a]pyrene, benzo[a]anthracene, and alkylated benzanthracenes are found in charcoal-broiled meats (Lijinsky and Shubik, 1964). Recently, Blumer and his associates (Blumer, 1961; Blumer and Youngblood, 1975; Giger and Blumer, 1974) have shown that soils and recent marine sediments contain a complex mixture of polycyclic hydrocarbons in which there is extended alkylhomology although the unsubstituted hydrocarbons are the predominant species. These

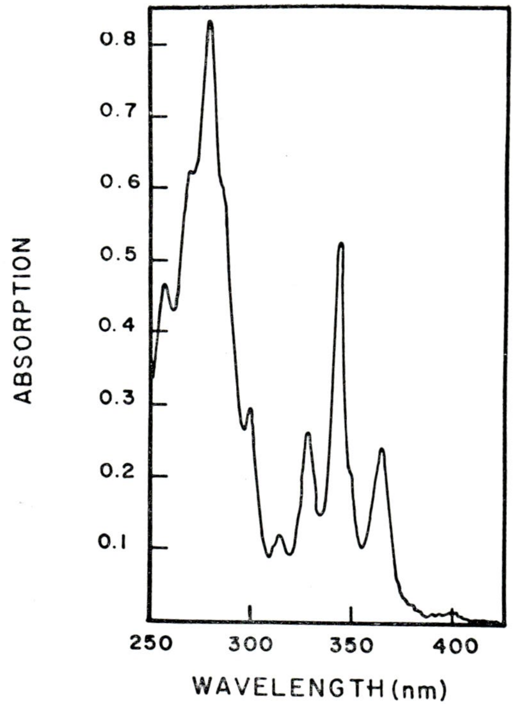

Fig. 1. *Absorption spectrum of the product formed from benzo[a]pyrene by* Beijerinckia *B-836. The solvent was spectral grade methanol.*

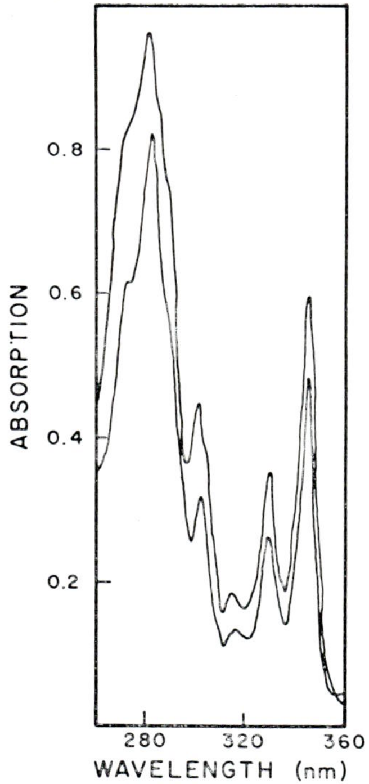

Fig. 2. *Absorption spectra of the benzo[a]pyrene dihydrodiol purified by high pressure liquid chromatography (lower curve) and synthetic 9,10-dihydrobenzo[a]pyrene (upper curve). Samples were dissolved in EtOH;H_2O (75:25) and spectrum recorded on a Cary recording spectrophotometer.*

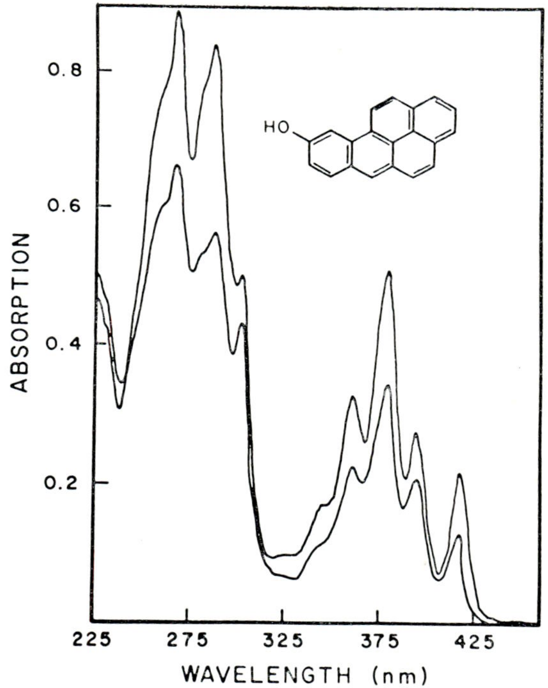

Fig. 3. *Absorption spectra of the phenol formed from the benzo[a]pyrene metabolite by acid treatment (lower curve) and synthetic 9-hydroxybenzo[a]pyrene. Spectra were obtained as described in Fig. 2.*

isolated compound gave a new product that was chromatographically and spectrally (Fig. 3) identical to a synthetic sample of 9-hydroxybenzo[a]pyrene.

The proton magnetic resonance spectrum of the metabolite is given in Fig. 4 and is consistent with the structure shown since a coupling constant of $J_{9,10} = 5{\cdot}2$ Hz requires a *cis* stereochemistry (Jeffrey *et al.*, 1975). Also, the proton magnetic resonance spectrum and the absorption spectrum (Fig. 5) of the bacterial metabolite were identical to a synthetic sample of *cis*-9,10-dihydroxy-9,10-dihydrobenzo[a]pyrene.

Preliminary evidence suggests that the peak at 370 nm (Fig. 1) is part of the absorption spectrum of a second dihydrodiol. Attempts to identify this product are in progress.

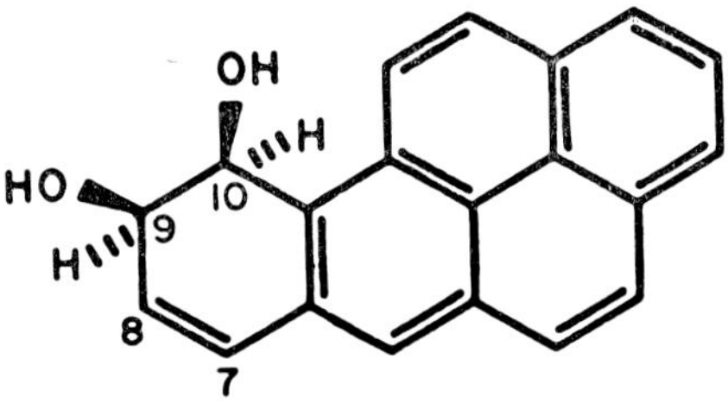

Proton	Chemical shift δ (ppm)	Coupling constants (Hertz)
C7	6·85	$J_{7,8} = 10{\cdot}0$, $J_{7,9} = 2{\cdot}0$
C8	6·11	$J_{8,9} < 1{\cdot}0$
C9	4·77	$J_{9,10} = 5{\cdot}2$
C10	5·63	

[a] Fourier transform 220 MHz spectrum. The sample was dissolved in acetone-d_6 and the spectrum recorded after exchange of the hydroxyl protons with MeOD.

Fig. 4. Analysis of the proton magnetic resonance spectrum of the benzo[a]pyrene dihydrodiol produced by Beijerinckia *B-836.*[a]

Oxidation of benzo[a]anthracene by *Beijerinckia* B836

The product formed from benzo[a]anthracene by *Beijerinckia* B-836 (*see* Materials and Methods) gave the absorption spectrum shown in Fig. 6. Since this spectrum did not resemble any known benzo[a]anthracene derivatives, it was suspected that several compounds were present. High pressure liquid chromatography revealed the presence of a major product and three minor products. The major compound gave an identical absorption spectrum (Fig. 7) to that given by synthetic 1,2-dihydrobenzo[a]anthracene. Acid treatment of the bacterial product gave a compound that was chromatographically and spectrally identical to 2-hydroxybenzo[a]anthracene. The proton magnetic resonance spectrum gave $J_{1,2} = 5{\cdot}4$ Hz which is consistent with a *cis* stereochemistry of the hydroxyl groups. These observations suggest that the product is *cis*-1,2-dihydroxy-1,2-dihydrobenzo[a]anthracene. Studies to characterize this compound further are in progress.

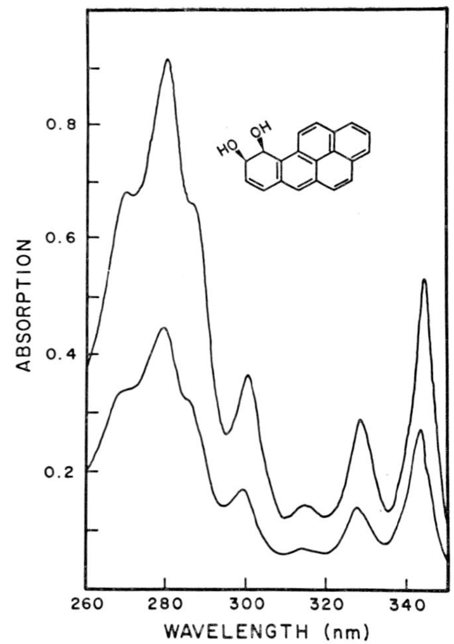

Fig. 5. *Absorption spectrum of the major benzo[a]pyrene dihydrodiol (lower curve) and synthetic* cis-*9,10-dihydroxy-9,10-dihydrobenzo[a]pyrene (upper curve). Spectra were recorded as described in Fig. 2.*

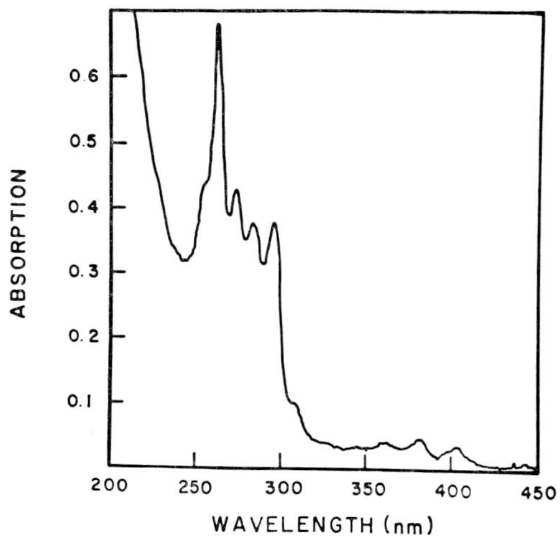

Fig. 6. *Absorption spectrum of the product formed from benzo[a]anthracene by* Beijerinckia *B-836. The sample was dissolved in spectral grade methanol and the spectrum recorded on a Cary recording spectrophotometer. This spectrum was originally thought to be due to* cis-*1,2-dihydroxy-1,2-dihydrobenzo[a]anthracene. It is known to be due to a mixture of four* cis-*benzo[a]anthracene dihydrodiols.*

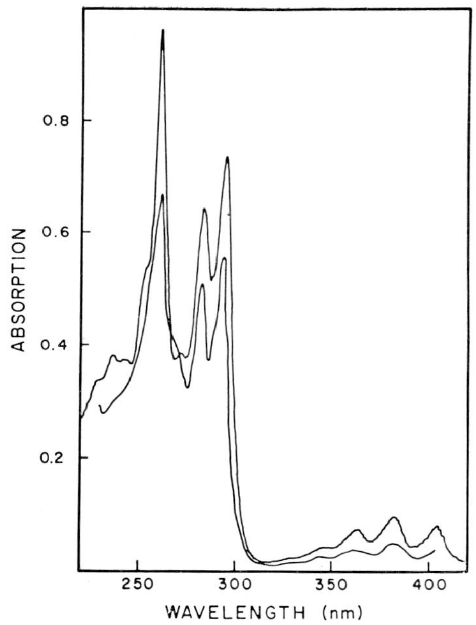

Fig. 7. *Absorption spectra of the major bacterial product and synthetic 1,2-dihydrobenzo[a]anthracene. Spectra were recorded as described in Fig. 2.*

DISCUSSION

Beijerinckia B-836 is a mutant strain that oxidizes biphenyl to *cis*-2,3-dihydroxy-1-phenylcyclohexa-4,6-diene (Gibson *et al.*, 1973). The structural similarities between biphenyl and several polycyclic hydrocarbons (Fig. 8) led us to speculate that the enzyme induced by biphenyl might also have a sufficiently broad specificity that it would also oxidize polycyclic hydrocarbons. This hypothesis is substantiated by the isolation and identification

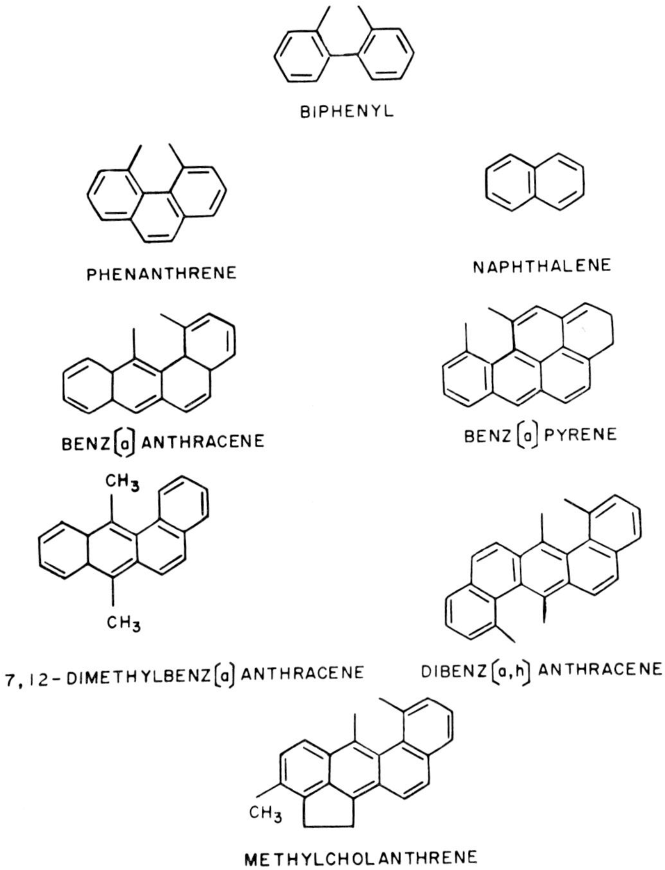

Fig. 8. Structural similarities between biphenyl and polycyclic hydrocarbons. Hydrogen substituents have been omitted for clarity.

of *cis*-9,10-dihydroxy-9,10-dihydrobenzo[a]pyrene (Fig. 9) and *cis*-1,2-dihydroxy-1,2-dihydrobenzo[a]anthracene. The results also suggest that more than one dihydrodiol is produced from each parent hydrocarbon. If this turns out to be correct, it will be the first time that this organism has produced a mixture of positional isomers.

The parent strain of *Beijerinckia* B-836 oxidized benzo[a]pyrene and benzo[a]anthracene to acidic products which were not identified. However, by analogy with the known pathways for the degradation of naphthalene

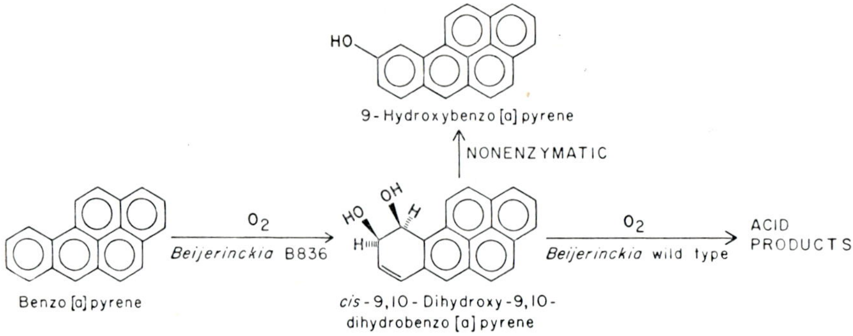

Fig. 9. Major pathway proposed for the oxidation of benzo[*a*]*pyrene by* Beijerinckia.

(Davies and Evans, 1964), phenanthrene, and anthracene (Evans *et al.*, 1965) where salicyclic and hydroxynaphthoic acids are formed, it is conceivable that pyroic and anthroic acids are intermediates in the degradation of benzo[a]pyrene and benzo[a]anthracene, respectively.

The *cis* hydroxylation of the polycyclic hydrocarbons suggests that they are formed by a dioxygenase (Gibson, 1971). This is in contrast to the oxidation of these compounds by mammals where mono-oxygenase activity leads to the formation of arene oxides (Daly *et al.*, 1972). The latter compounds can undergo several reactions, one of which is the enzymatic addition of water to yield *trans* dihydrodiols. At this time *trans* dihydrodiols have not been identified as intermediates in the bacterial degradation of aromatic hydrocarbons. Differences between the mammalian and bacterial oxidation of aromatic hydrocarbons are shown in Fig. 10.

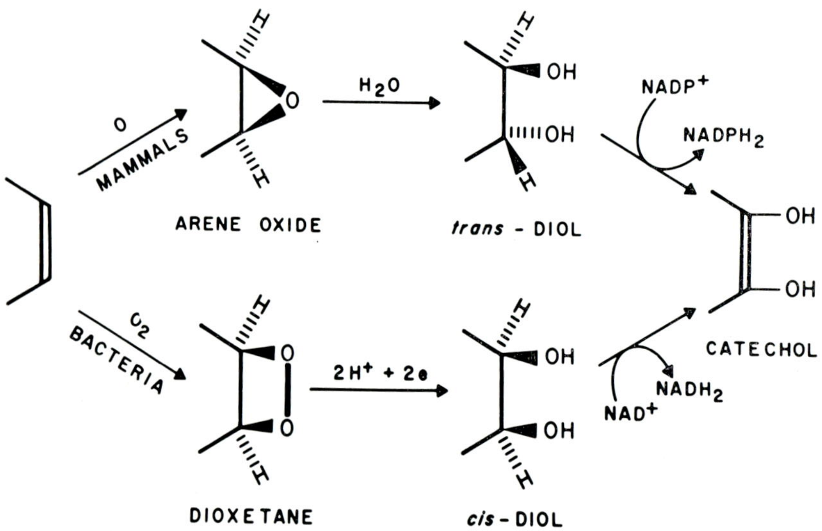

Fig. 10. Pathways utilized by mammals and bacteria for the oxidation of aromatic hydrocarbons.

Acknowledgements

This work was supported by contract NO1 CP 33384 from the National Cancer Institute. The author is a recipient of Public Health Service Career Development Award 1 KO4 ES-70088 from the National Institute of Environmental Health Sciences. The high pressure liquid chromatographic separations and proton magnetic resonance spectra were obtained by Dr D. M. Jerina and his associates at the National Institutes of Health, Bethesda, Md. I thank Zarine Mehta and Judith M. Davis for technical assistance.

REFERENCES

Badger, G. M. (1947). Polycyclic aromatic hydrocarbons. Part XXXII. 2′methoxy- and 2′methoxy-9:10-dimethyl-1:2-benzanthracene. *J. Chem. Soc.*, 940–943.

Blumer, M. (1961). Benzpyrenes in soil. *Science*, **134**, 474–475.

Blumer, M., and Youngblood, W. W. (1975). Polycyclic aromatic hydrocarbons in soils and recent sediments. *Science*, **188**, 53–55.

Borneff, J., Selenka, F., Knute, H., and Maximos, A. (1968). Die Synthese von 3,4-benzpyren und anderen polyzyklischen aromatischen kohlenwasserstoffen in pflanzen. *Arch. Jyg. Bakt.*, **152**, 279–282.

Daly, J. W., Jerina, D. M., and Witkop, B. (1972). Arene oxides and the NIH Shift: the metabolism, toxicity and carcinogenicity of aromatic compounds. *Experientia*, **28**, 1129–1149.

Davies, J. I., and Evans, W. C. (1964). Oxidative metabolism of naphthalene by soil pseudomonads. *Biochem. J.*, **91**, 251–260.

Evans, W. C., Fernley, H. N., and Griffiths, E. (1965). Oxidative metabolism of phenanthrene and anthracene by soil pseudomonads. *Biochem. J.*, **95**, 819–831.

Gibson, D. T. (1971). The microbial oxidation of aromatic hydrocarbons. *Crit. Rev. Microbiol.*, **1**, 199–223.

Gibson, D. T., Roberts, R. L., Wells, M. C., and Kobal, V. M. (1973). Oxidation of biphenyl by a *Beijerinckia* species. *Biochem. Biophys. Res. Commun.*, **50**, 211–219.

Giger, W., and Blumer, M. (1974). Polycyclic aromatic hydrocarbons in the environment: isolation and characterization by chromatography, visible, ultraviolet and mass spectrometry. *Anal. Chem.*, **46**, 1663–1671.

Grimmer, G., and Duvell, D. (1970). Mitt: carcerogene kohlenwasserstoffe in der umgebung des menschen. *Z. Naturforsch.*, **25**, 1171–1175.

Jeffrey, A. M., Yeh, H. J. C., Jerina, D. M., Patel, T. R., Davey, J. F., and Gibson, D. T. (1975). Initial reactions in the oxidation of naphthalene by *Pseudomonas putida*. *Biochemistry*, **14**, 575–584.

Kinoshita, N., Shears, B., and Gelboin, H. V. (1973). K-region and non-K-region metabolism of benzo[a]pyrene by rat liver microsomes. *Cancer Res.*, **33**, 1937–1944.

Lao, R. C., Thomas, R. S., Oja, H., and Dubois, L. (1973). Application of a gas chromatograph-mass spectrometer-data processor combination to the analysis of polycyclic aromatic hydrocarbon content of airborne pollutants. *Anal. Chem.*, **45**, 908–915.

Lijinsky, W., and Shubik, P. (1964). Benzo[a]pyrene and other polynuclear hydrocarbons in charcoal-broiled meat. *Science*, **145**, 53–55.

Poglazova, M. N., Feedoseva, G. E., Khesina, A. J., Meissei, M. N., and Shabad, L. M. (1967a). Destruction of benzo[a]pyrene by soil bacteria. *Life Sciences*, **6**, 1053–1062.

Poglazova, M. N., Feedoseva, G. E., Khesina, A. J., Meissei, M. N., and Shabad, L. M. (1967b). Further investigations of the decomposition of benzo[a]pyrene by soil bacteria. *Dokl. Biol. Sci.*, **176**, 1165–1167.

Poglazova, M. N., Feedoseva, G. E., Khesina, A. J., Meissei, M. N., and Shabad, L. M. (1968). The oxidation of benz[a]pyrene by microorganisms in relation to its concentration in the medium. *Dokl. Biol. Sci.*, **179**, 1460–1462.

Sims, P. (1970). Qualitative and quantitative studies on the metabolism of a series of aromatic hydrocarbons by rat liver preparations. *Biochem. Pharmacol.*, **19**, 795–818.

Sisler, F. D., and Zobell, C. E. (1947). Microbial utilization of carcinogenic hydrocarbons. *Science*, **106**, 521–522.

Stanier, R. Y., Palleroni, N. J., and Doudoroff, M. (1966). The aerobic pseudomonads: a taxanomic study. *J. Gen. Microbiol.*, **43**, 159–271.

THE MICROBIAL DEGRADATION OF DIESEL OIL IN MULTISTAGE CONTINUOUS CULTURE SYSTEMS

P. H. PRITCHARD, R. M. VENTULLO* and J. M. SUFLITA

Department of Biological Sciences, S.U.N.Y. College at Brockport,
Brockport, New York 14420, U.S.A.

Summary

The degradation of diesel oil has been studied in a multistage continuous culture system in which the physical dispersion of an oil layer was not necessary. By using qualitative gas chromatographic analysis, it has been shown that both alkanes and aromatics were significantly removed from the gas chromatographic profile presumedly by bacterial degradation. Some sequential changes in the oil have been detected as the oil passes from one vessel to the next, but the complete removal of gas chromatographable components has not yet been possible. During the degradation, numerous other gas chromatographable components appeared to be synthesized as a result of the bacterial activities. It would appear that our type of system could be quite useful in studying the fate of oil in aquatic environments.

INTRODUCTION

The study of bacterial populations and their degradative potential is a complicated undertaking especially when laboratory simulations are attempted. Since it is well established that the use of batch culture techniques creates a very artificial environment involving numerous 'time-dependent' variables (Hobson, 1969; Jannasch and Mateles, 1974; Tempest, 1970), alternative methods for the study of the activities of bacterial populations have been sought. As a result, continuous culture techniques have been tested for their application to ecological studies and they have in general been found to be very useful (Jannasch and Mateles, 1974). Despite the fact that continuous culture has no exact equivalent in nature, it still encompasses many characteristics which are far more similar to a natural aquatic environment than those offered by other techniques (Gibson, 1968; Hobson, 1969; Kator, 1973).

The goal of our work has been to attempt to determine some of the mechanisms by which bacterial populations completely degrade selected organic pollutants and to assess the environmental factors which affect these degradation processes. A study of the degradation of oil is ideal in this regard for many reasons: it is undoubtedly an important aspect of aquatic pollution (Zeldin, 1971), it can be readily analysed chemically (Blumer *et al.*, 1970; Jobson *et al.*, 1972), it is composed of both recalcitrant and readily degradable hydrocarbons (Zobell, 1970), and it relates directly to the voluminous amount of data already collected on the bacterial oxidation of all types of petroleum hydrocarbons (Gibson, 1968; Klug and Markovetz, 1971).

* Present address: Department of Microbiology, University of Georgia, Athens, Georgia 30602, U.S.A.

To this end, we have attempted to model the degradation of a quiescent oil slick in the laboratory using continuous culture techniques. We have shown that bacteria will readily attack and degrade oil in our model system (Ventullo *et al.*, 1975) and we have shown in this work that multistage continuous culture systems can be used to study the degradation of the oil subsequent to its initial bacterial attack. Evidence is presented to show that both aromatic and aliphatic hydrocarbons are not only degraded but also possibly transformed into other types of products.

MATERIALS AND METHODS

The design of our culture apparatus and its operational details have been described elsewhere (Pritchard *et al.*, 1975). The details of the sequential or multistage continuous culture system used in this study are depicted in Fig. 1. A three stage system (referred to as primary, secondary, and tertiary stages in the text) was employed throughout this study.

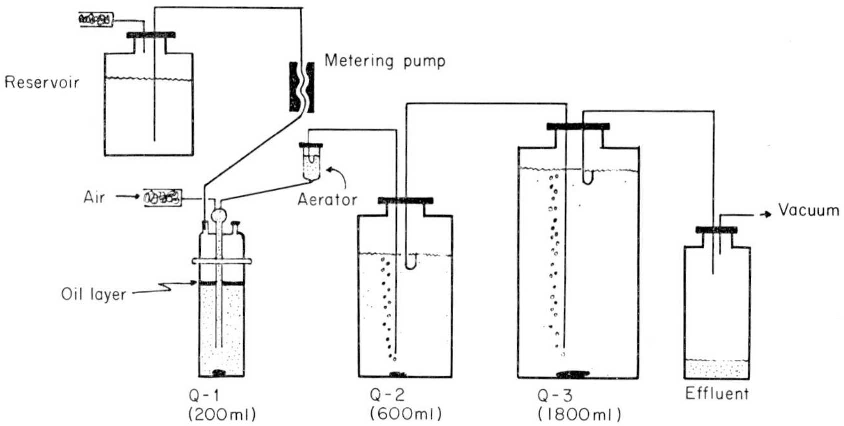

Fig. 1. Multistage continuous culture apparatus used to study the microbial degradation of oil.

To start a continuous culture experiment, an inoculum of 250 ml of Lake Ontario water was placed in the primary culture vessel and amended with 0·7 mM phosphate buffer (pH 7·2), 0·19 mM ammonium chloride, and 0·08 mM magnesium sulphate. Eight millilitres of topped sterile diesel oil or sterile hydrocarbon was added and the system incubated at room temperature for 12–15 hr as a batch culture (no media flow). The flow of reservoir medium was then commenced and continued at a dilution rate of 0·05 hr^{-1}. This dilution rate was used in all experiments. Reservoir media contained the same salts at the same concentration as described above, using a 6:1 mixture of distilled to tap water. The sole source of carbon was the oil or hydrocarbon layer. In the multistage continuous culture systems, the secondary and tertiary vessels were empty at the beginning of the experiment and were eventually filled with the effluent from the primary vessel.

Oil layer samples from the primary growth vessel were taken by touching a disposable capillary pipette to the oil layer. Approximately 50 μl of oil would move into the pipette by capillary action. This oil was then washed out of the pipette with benzene-pentane solvent and subsequently prepared for chemical analysis (*see* below). Samples of oil from the secondary and tertiary stages were obtained by directly extracting 200 ml of culture fluid with pentane-benzene solvent. The 200 ml of culture fluid taken from the vessels was always replaced with 200 ml of sterile distilled water.

Diesel oil used in this study was obtained from the commissary at SUNY Brockport, Brockport, New York. The oil was topped by exposing it to forced draft conditions at 37°C. This process removed volatile materials up to and including normal dodecane, leaving approximately 65% of the weight of the original oil. The topped oil was then filter sterilized (0·45 μmillipore filters).

All oil-containing samples were extracted by shaking with dry pentane-benzene solvent (1:1 by volume).

Gas chromatographic analyses were carried out using OV-1 methyl silicone (Applied Science Laboratories) columns 10 ft in length. A Hewlett-Packard Model 5750 gas chromatograph equipped with a dual flame ionization detector was used in all analyses. The instrument was programmed as follows: linear temperature program 60–670°C at a programming rate of 4°C per minute; injection port temperature, 300°C; flame detector, 300°C; helium carrier flow rate, 20·0 ml/min; hydrogen flow rate, 12·5 ml/min; air flow rate 350 ml/min.

To monitor chemical changes in the oil during the degradation process, samples were analysed by the peak height ratio method of Blumer *et al.* (1970). Peak heights for the *n*-20/envelope ratios were measured from the top of the peaks to a line drawn across the bottom of the peak at the edge of the envelope. The envelope is defined as that area enclosed by the baseline and a line drawn along the bottom of all peaks in the profile.

RESULTS

Gas chromatographic analysis of the samples from a typical multistage continuous culture experiment are shown in Figs. 2–5. All multistage experiments were designated by the symbol Q and the vessels are numbered accordingly. Samples from Q-1 represent material from the oil layer and not the culture fluid below the oil layer. All other samples (Q-2, Q-3, Q-effluent) are taken from the culture fluid. Carbon numbers over the peaks indicate peaks which co-chromatograph with normal alkanes of the indicated lengths and do not infer absolute chemical identification.

The physical and microbiological changes occurring during oil degradation have been reported elsewhere (Walker and Colwell, 1974) but will be briefly reiterated here for continuity. The nondispersed oil layer in our continuous culture systems is initially attacked by hydrocarbon-utilizing organisms eventually leading to the development of a film of bacteria underneath the oil layer. A gradual increase in this bacterial film occurs and the bacteria slowly impregnate the oil layer making it thick and turbid. This promotes a slow flaking process in which small masses of bacteria and oil sluff off the oil layer and wash out with the effluent. With extended incubation (2–3 months for 8 ml of oil spread over a 25 cm^2 area), all of the oil in the oil layer eventually disappeared. The bacterial populations which mediate this degradation process appear to be specifically adapted for attacking a stagnant oil layer and their success in promoting degradation is quite sensitive to different environmental conditions (Walker and Colwell, 1974).

Figure 2 is a composite of gas chromatographs from each of the three vessels (Q-1, Q-2, Q-3) and the effluent from the third vessel (Q-eff) which are representative of the early stages of degradation (about 300 hr incubation). Oil taken from Q-1 and Q-2 showed only minor overall changes in the gas chromatographic profile. The heptadecane (C-17) to pristane ratios and the octadecane (C-18) to phytane ratios typically decreased by 5–10% in Q-1 and Q-2, which indicated early preferential removal of the *n*-alkanes. The C-20 *n*-alkane to envelope ratio decreased by 15–20%. The peaks corresponding to the C-11 and C-12 *n*-alkanes showed the most extensive degree of removal.

Oil taken from Q-3 and Q-3 effluent showed a more extensive alteration in the profiles. The ratios of C-17 *n*-alkane to pristane and C-18 *n*-alkane to phytane were decreased by 10–25% and the C-20 *n*-alkane to envelope ratio had decreased by as much as 20–30%. The peaks corresponding to C-12 and C-13 *n*-alkanes were almost completely missing in these samples.

Also apparent in the profiles from Q-3 and Q-effluent, during this early stage of degradation, was the alteration of the envelope profile. The dotted line indicates the envelope profile of undegraded oil. As indicated by shape and pattern, it was clear that an alteration of the envelope profile shape had occurred on the left hand side (corresponding to the lower boiling branched alkanes and aromatics). This change represents a cutting away of the envelope profile presumedly brought on by the bacterial degradation of the components contained therein.

Another consistent aspect of the sequential degradation process was the appearance of gas chromatographic peaks corresponding to normal alkanes of carbon length C-26–C-34 (Fig. 2—Q-effluent). These peaks were never detected in undegraded oil. To the best of our knowledge they do not appear to result from chemical contamination or from artifacts in our extraction procedure. It is possible they could be metabolic end products resulting from bacterial activities during the degradation process. The makeup of these extra peaks was quite specific in that many of the peaks appeared as a group at about the same relative concentrations.

Along with the apparent synthesis of these extra peaks, there was also an enrichment of certain peaks originally present in undegraded oil. These enrichments were not attributed to sample size or attenuation differences used in the gas chromatographic techniques. In Fig. 2 in the Q-2 and Q-effluent samples, there was a decided enrichment in the peak corresponding to C-25 alkane.

Figure 3 is a composite of gas chromatographic profiles of oil samples extracted from a multistage system after a longer period of incubation (500 hr). The general pattern of degradation is about the same but the particular changes have been intensified. Removal of the *n*-alkanes, especially below C-16, can readily be observed and the components making up the envelope profile have been extensively removed from the lower boiling end. The synthesized peaks (C-25–C-32) are also present but they are now detectable at all three stages. Significant peak enrichment again was apparent especially in Q-2 and specifically in the C-25 peak in Q-effluent.

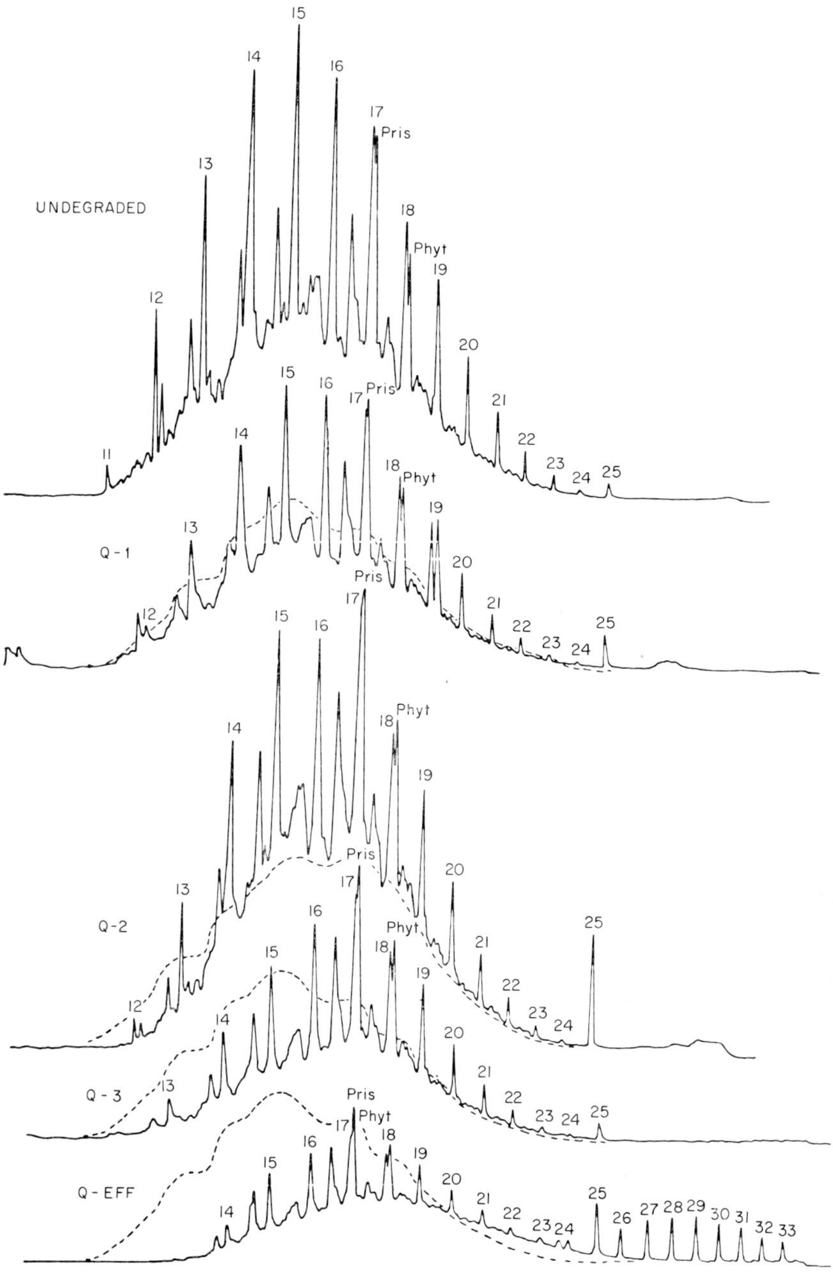

Fig. 2. Gas chromatographic profiles of oil extracted from multistage continuous culture system during the early stages (300 hr incubation) of the degradation of diesel oil.

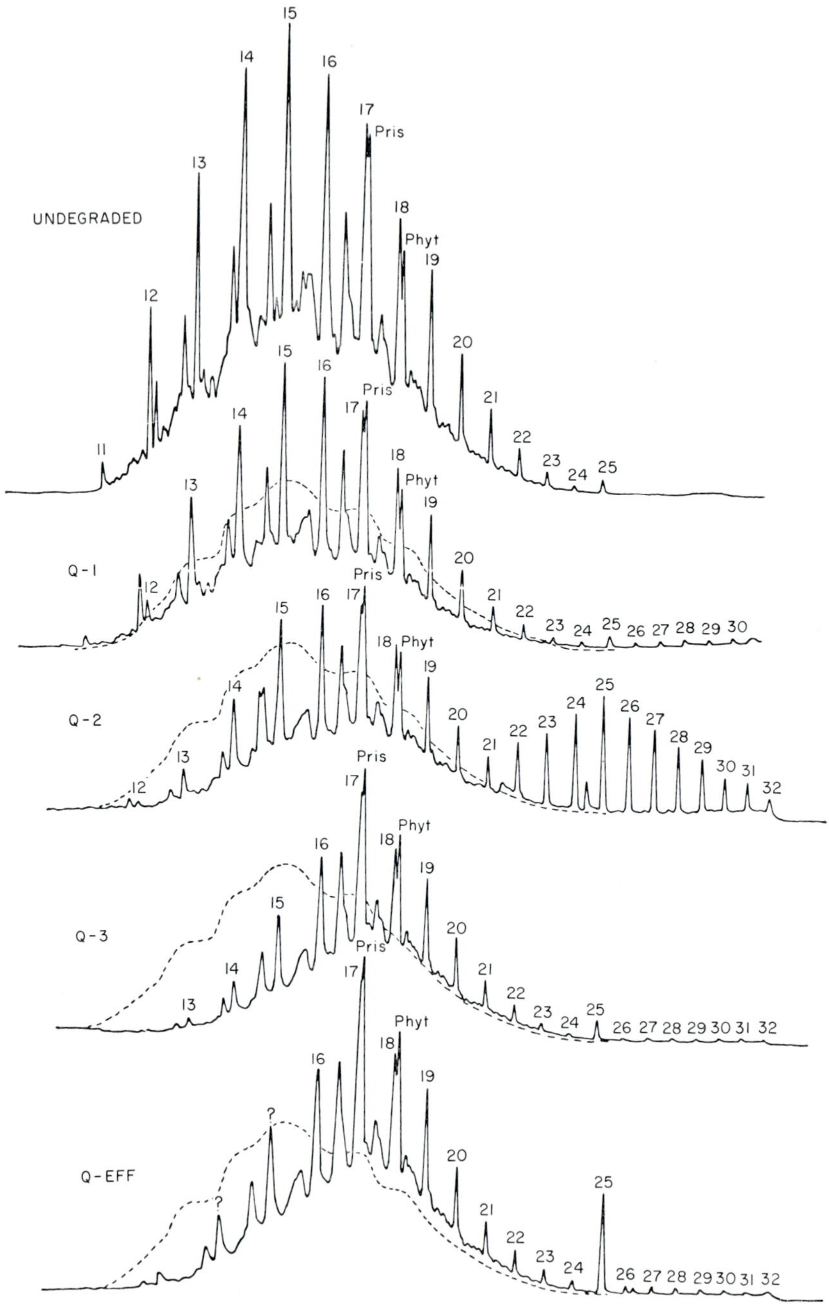

Fig. 3. Gas chromatographic profiles of oil extracted from multistage continuous culture system during the middle stages (500 hr incubation) of the degradation of diesel oil.

As the degradation of the oil layer progressed (1000 hr), greater and greater transformation of the gas chromatographic profiles of the oil was detected in oil extracted from the sequential vessels. Typical gas chromatographic profiles during this stage are shown in Fig. 4. As can be seen, most of the peaks which co-chromatograph with *n*-alkanes of C-12–C-19 were either completely gone or extensively decreased to the point of being barely detectable. The peaks co-chromatographing with pristane and phytane still stand out in most cases but they too were extensively decreased in height.

Almost 75% of the components comprising the envelope profile have been removed or have at least disappeared from the chromatographic profile. This removal was again confined primarily to the low boiling fraction but had now progressed to a point where more and more of the high boiling components had disappeared.

Very obvious in this profile is a large increase in the size of the envelope profile greater than can be accounted for through adjustments in the analytical techniques. A whole new envelope profile, in fact, appears to have been generated which now lies in the higher boiling range. This does not seem to be a shifting of the original envelope profile and could represent metabolic end products. The chemical nature of the components in this new envelope is not known.

At the stage when most of the oil in the oil layer had disappeared (1700 hr) and the remaining oil was a brownish, slimy mass of cells and oil, significant modification of the oil in the oil layer (Q-1) was finally apparent (*see* Fig. 5). This degraded oil reflected the same general pattern of alteration as seen above, *i.e.* removal of *n*-alkanes and envelope components, synthesis and enrichment of unknown peaks, and a shifting of the envelope profile into the high boiling range. The peaks co-chromatographing with the branched alkanes, pristane, and phytane are still quite prominent and have not undergone the extensive removal detected in the next vessel in the chain.

DISCUSSION

The degradation of diesel oil that we have observed in our multistage continuous culture system indicates that this type of system may ultimately prove useful in determining the fate of oil in aquatic environments. In particular, it allows one to follow the degradation of oil after it has been initially attacked and dispersed by bacterial populations. Since the degradation process is continuous, many different environmental and pollution control factors can be tested and evaluated as to the extent and speed of the degradation they promote. Because of the similarities of our system to natural aquatic conditions, much of the information generated can be applied to the design and implementation of oil pollution abatement programmes with considerably more confidence than that derived from other types of experimental systems.

Our use of gas chromatography to follow the oil degradation has revealed a number of interesting aspects, but it should be emphasized that it is a strictly qualitative measure as we have used it. Despite its wide application to oil degradation studies (Blumer *et al.*, 1970; Jobson *et al.*, 1972; Jobson and Westlake, 1974; Walker and Colwell, 1974), there is no way of knowing if the disappearance of a peak actually corresponds to its degradation. The

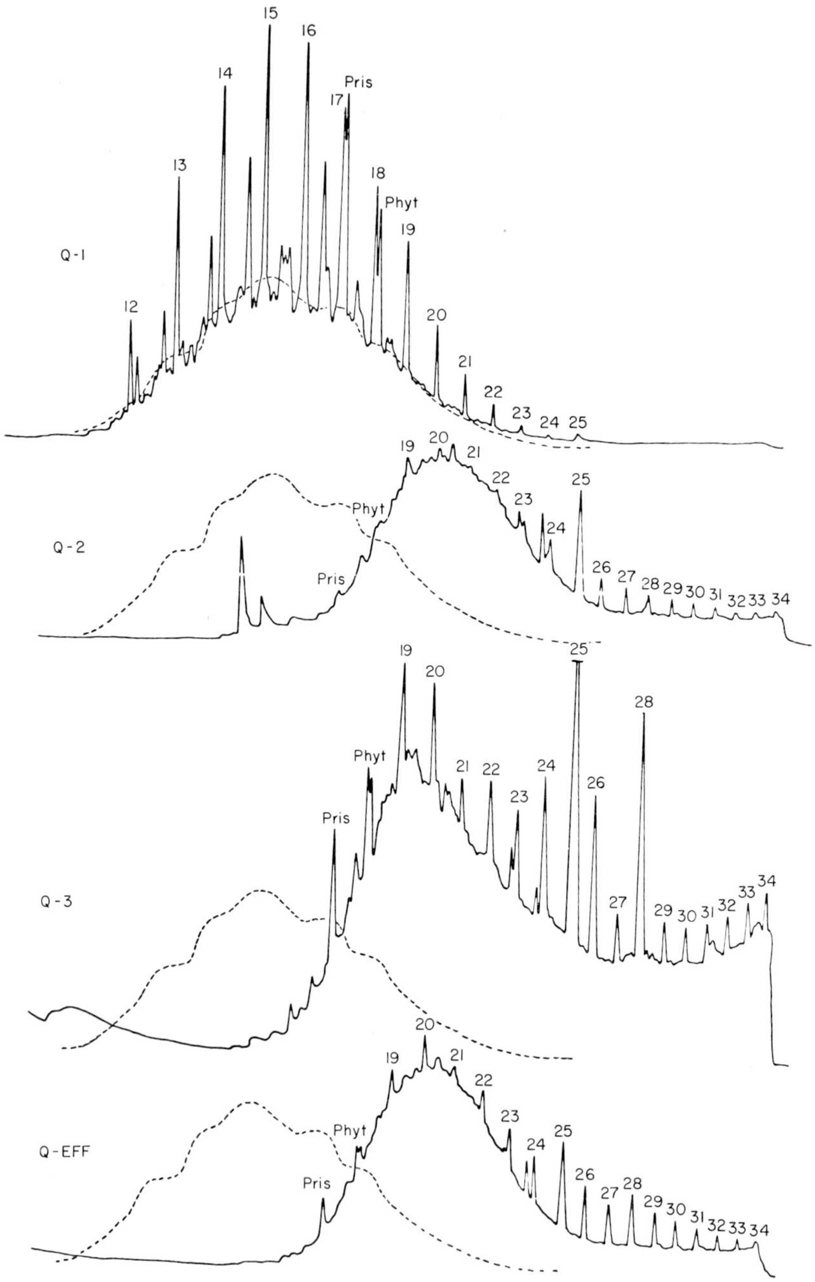

Fig. 4. Gas chromatographic profiles of oil extracted from multistage continuous culture system during the late middle stages (1000 hr incubation) of the degradation of diesel oil.

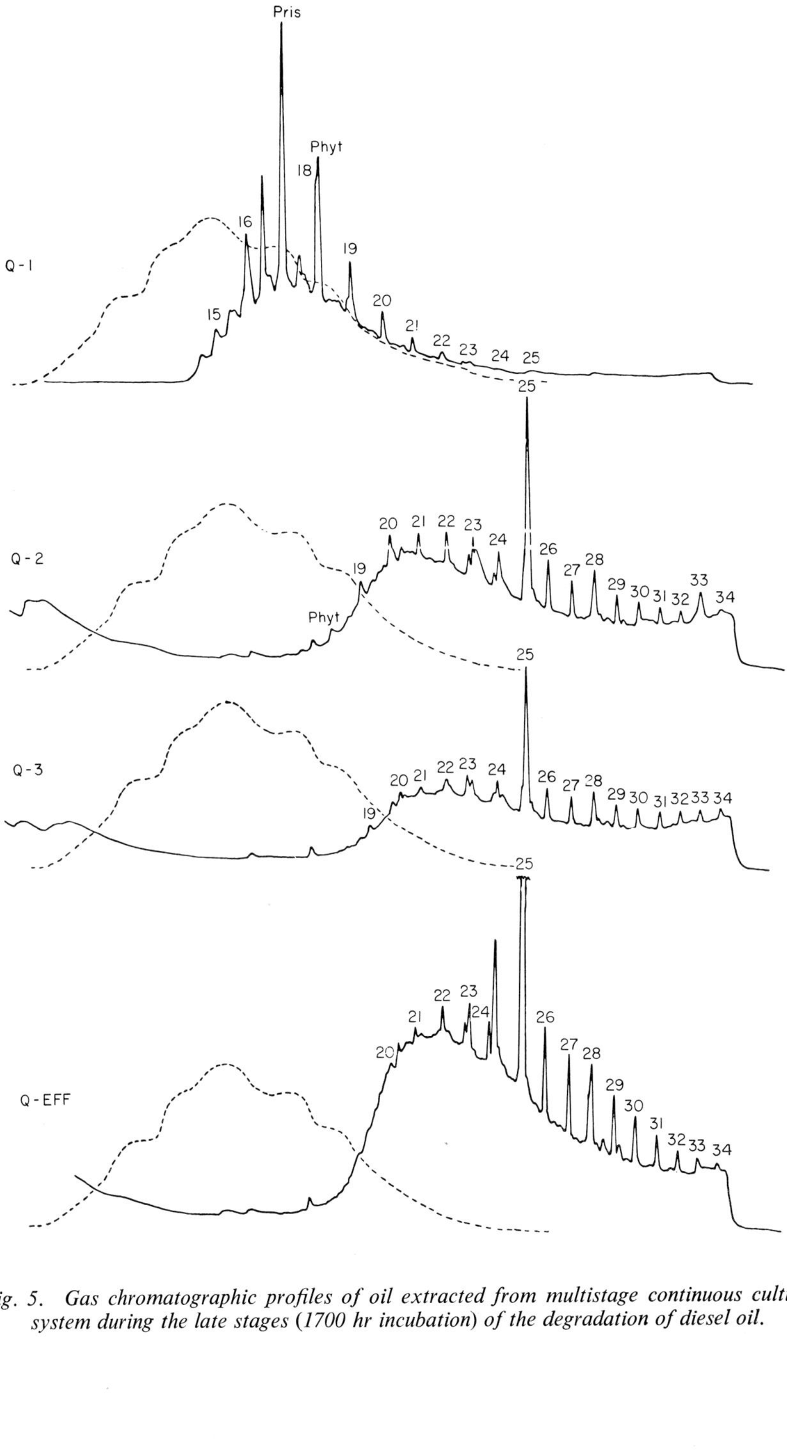

Fig. 5. Gas chromatographic profiles of oil extracted from multistage continuous culture system during the late stages (1700 hr incubation) of the degradation of diesel oil.

removal of peaks from the gas chromatographic profile cannot be equated with complete degradation since a large percentage of the oil may be left undegraded and chromatographically unresolvable.

The relationship of the oil degradation pattern that we have observed in our continuous culture systems to a natural degradation process in an aquatic environment is probably quite close except for the fact that we did not physically disperse our oil. Certainly in protected bays and harbours, spilled oil will remain as a floating slick for long periods (Floodgate, 1973) and will probably undergo much the same bacterial attack that we have observed on the oil layer in our continuous culture systems. Once the oil is dispersed either in the natural situation or in our laboratory model, the oil droplets will carry a significant bacterial population along with them. These populations would then presumably continue their degradation process and thereby release partially degraded oil products into the water column much in the same way that we have observed in our multistage systems. From the observations we have made, we feel that oil which is physically dispersed before bacteria have had a chance to attack it (for example as modelled in a simple batch culture experiment) will suffer an ultimately slower and less extensive degradation. We have seen considerably greater alterations of the gas chromatographic profile than has been seen in many batch culture studies (Jobson *et al.*, 1972).

Blumer *et al.* (1970) have observed a similar degradation of diesel oil in the marine environment. For example, they showed that the *n*-C17 to pristane ratio decreased from 2·0 to 1·5 (25% change) within 2 months (1400 hr). Ratio changes obtained in our systems from oil in the oil layer are comparable to this data. The differences in the availability of nitrogen and phosphorus is difficult to evaluate at this time. Since Blumer obtained his samples directly from the sediments, we would envision this as being equivalent to taking samples directly from our oil layer. Quite possibly, if he had been able to sample the water column above the sediments, degradation products similar to the types we observed in the second and third vessels of our multistage system may have been detected.

Blumer's work also revealed that over a period of a year general degradation of the aromatic fractions of the oil (represented by a decrease in the envelope profile in the gas chromatographs) occurred along with the more rapid alkane degradation. This is in contrast to some reports in the literature (*see*, for example, Jobson *et al.*, 1972; Jobson and Westlake, 1974; Westlake *et al.*, 1974) in which only the degradation of the normal alkane fraction has been observed and very little alteration of the branched and cyclic alkanes or the aromatics has occurred. This unfortunate equating of 'significant oil degradation' with simple *n*-alkane degradation we feel is a limitation of the batch culture method used for studying oil degradation. When a more open type of system is used, such as our continuous culture method, it is quite evident that the aromatic fractions can and will be degraded (as measured by gas chromatography) along with the normal alkane fraction.

The use of the multistage continuous culture system has led to two important observations. First, it is very difficult to get complete removal of the gas chromatographable components of the oil. As was seen in our results, significant amounts of oil and oil degradation products still were detectable in the effluents from the third vessel. In fact, only small amounts of sequential degradation of the oil occurred as it passed down the chain of vessels. Analyses for ammonia, nitrogen and phosphate have shown that these nutrients are still in excess in the effluents and they therefore are not the limiting factor in the degradation. Since experiments in which fresh Lake Ontario water was fed continuously into the growth vessels showed the same extent of degradation, it would appear that bacterial degradation potential was not limiting. Several other types of experiments, such as increasing the size and number of the multistage vessels, will have to be completed before any conclusion can be reached on the factors which predisposed to complete disappearance of the chromatographable components of diesel oil.

The second important observation was the synthesis of gas chromatographable materials which were not originally present in the oil. This has been noted in other oil degradation studies as well (Horowitz *et al.*, 1975). The origin and chemical nature of these components are still unknown at this point and it is quite possible they are artifacts of our analytical techniques. The quantity and extent of these components, however, may indicate that they are metabolic end products generated by the bacteria, perhaps much in the same way as Stewart and Kallio (1959) reported for the synthesis of fatty acid esters. If these extra materials which appear on the gas chromatographic profiles prove not to be artifacts, it is important that their chemical nature be elucidated in light of the fact that they could possibly be more toxic and/or more recalcitrant than the original oil.

Acknowledgements

This work was supported by a grant from the Office of Naval Research #N00014-73-0074. Mr Ventullo and Mr Suflita were gratefully supported by grants from the New York State Assembly Scientific Staff in conjunction with the Office of Intergovernmental Science and Research Utilization of the National Science Foundation and the New York Sea Grant Institute.

REFERENCES

Blumer, M., Sass, J., Souza, G., Sanders, H., Grassle, F., and Hampson, G. (1970). *The West Falmouth Oil Spill.* Woods Hole Oceanographic Institution, Reference Report No. 70-44.

Floodgate, G. D. (1973). A threnody concerning the biodegradation of oil in natural waters. In *The Microbial Degradation of Oil Pollutants* (D. G. Ahearn and S. P. Meyers, eds.), Center for Wet Land Resources, Louisiana State University, Publication No. LSU-SG-73-01.

Gibson, D. T. (1968). Microbial degradation of aromatic compounds. *Science*, **161**, 1093.

Hobson, P. N. (1969). Growth of mixed cultures and their biological control. In *Microbial Growth*, Symposium of Society for General Microbiology, **19**, 43–64.

Horowitz, A., Gutnick, D., and Rosenberg, E. (1975). Sequential growth of bacteria on crude oil. *Appl. Microbiol.*, **30**, 10–19.

Jannasch, H. W., and Mateles, R. I. (1974). Experimental bacterial ecology studied in continuous culture. *Adv. Microb. Physiol.*, **10**, 165–212.

Jobson, A., and Westlake, D. W. S. (1974). Effects of amendments on the microbial utilization of oil applied to soil. *Appl. Microbiol.*, **27**, 166–171.

Jobson, A., Cook, F. D., and Westlake, D. W. S. (1972). Microbial utilization of crude oil. *Appl. Microbiol.*, **23**, 1082–1089.

Kator, H. (1973). Utilization of crude oil hydrocarbons by mixed cultures of marine bacteria. In *The Microbial Degradation of Oil Pollutants*, p. 47 (D. G. Ahearn and S. P. Meyers, eds.), Center for Wetland Resources. Louisiana State University Publication No. LSU-SG-73-01.

Klug, M. J., and Markovetz, A. J. (1971). Utilization of aliphatic hydrocarbons by microorganisms. *Adv. Microb. Physiol.*, **5**, 1–34.

Pritchard, P. H., Kephart, R., Chang, J., and Starr, T. J. (1975). Degradation of octane in continuous culture using non-dispersed conditions. In preparation.

Stewart, J. E., and Kallio, R. E. (1959). Bacterial hydrocarbon oxidation II. Ester formation from alkanes. *J. Bacteriol.*, **78**, 441–448.

Tempest, D. W. (1970). The place of continuous culture in microbial research. *Adv. Microb. Physiol.*, **4**, 223–250.

Ventullo, R. M., Suflita, J. M., and Pritchard, P. H. (1975). The degradation of diesel oil under non-turbulent conditions using continuous culture systems. In preparation.

Walker, J. D., and Colwell, R. R. (1974). Microbial petroleum degradation: use of mixed hydrocarbon substrates. *Appl. Microbiol.*, **27**, 1053.

Westlake, D. W. S., Jobson, A., Phillippe, R., and Cook, F. D. (1974). Biodegradability and crude oil composition. *Can. J. Microbiol.*, **20**, 915–928.

Zeldin, M. (1971). Oil pollution. *Audubon*, **73**, 99–119.

Zobell, C. E. (1970). Fate of petroleum hydrocarbons in the sea. In Proceedings of the Joint Conference on Prevention and Control of Oil Spills, pp. 317–326. American Petroleum Inst., New York.

MICROBIAL DEGRADATION OF PETROLEUM IN THE ARCTIC

R. M. ATLAS and M. BUSDOSH

Department of Biology, University of Louisville, Louisville, Kentucky, U.S.A.

Summary

Microorganisms capable of petroleum utilization were found to be distributed widely in arctic coastal waters with highest concentrations in Prudhoe Bay, Alaska, and lowest in a freshwater pond at Barrow, Alaska. When Prudhoe or Barrow crude oils were incubated with nutrient-enriched water from Prudhoe Bay, about 25% of the oil was lost abiotically and 25% was biodegraded. The weight of the paraffinic and aromatic fractions of the oils decreased during degradation. Oil biodegradation in two coastal ponds at Barrow was nutrient-limited. Addition of an oleophilic fertilizer stimulated biodegradation in the ponds. Seeding with an oil-degrading microorganism also enhanced oil biodegradation in the saline pond but not in the freshwater pond. Direct contact with Prudhoe crude oil was toxic to several invertebrate test organisms. Water-soluble oil components were much less toxic. Enhancing biodegradation by seeding with an oil-degrading microorganism and adding an oleophilic fertilizer did not result in increased toxicity to the invertebrate test organisms.

MATERIALS AND METHODS

Distribution of microorganisms

Surface water samples were collected in sterile bottles at various sites as shown in Fig. 1. Subarctic sites at Valdez and Gambell were included for comparative purposes. Water temperature and salinity were determined with a Yellow Springs Instruments salinity meter. Samples were plated within 12 hr of collection. Heterotrophic bacteria were enumerated by the surface spread technique using trypticase soy agar and marine agar. The highest count on either medium was recorded as the number of heterotrophic microorganisms. Oil-degrading microorganisms were enumerated on oil agar as previously described (Atlas and Bartha, 1973a). Replicate plates were incubated at 25°C for 48 hr and at 5°C for 2 weeks.

Oil biodegradation

Water samples of 100 ml, collected at Prudhoe Bay, were supplemented with 0·5 mM Na_2HPO_4 and 10 mM $NaNO_3$ and incubated in biometer flasks (Bellco Glass Co., Vineland, N.J.) at 5°C with 1 ml of either Prudhoe crude oil (gift of Atlantic Richfield Co.) or Barrow crude oil (gift of Naval Petroleum Reserve #4). Carbon dioxide produced from oil mineralization during 42 days of incubation was determined as previously described (Atlas and Bartha, 1972). Residual oil was also recovered after 42 days. It was then

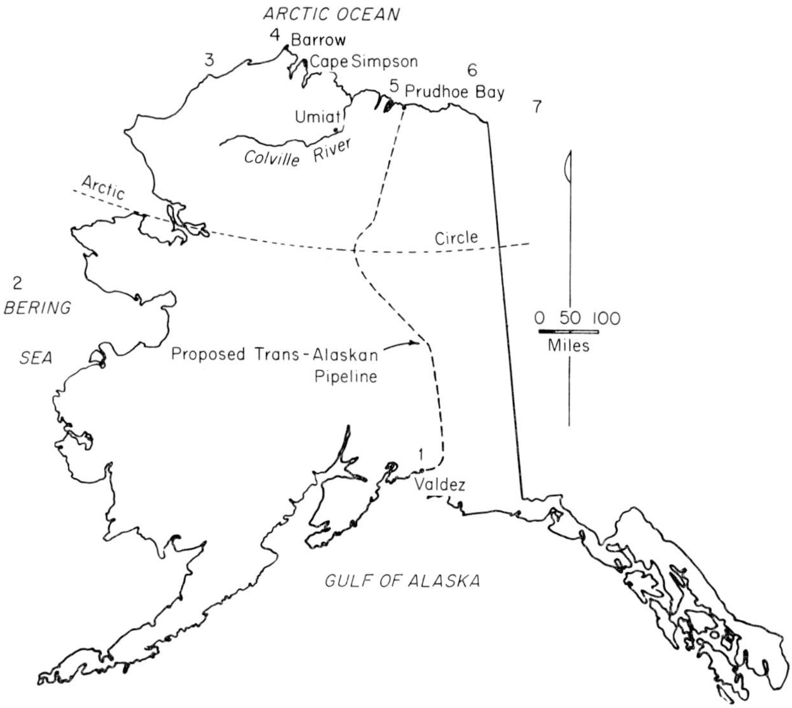

Fig. 1. Sampling sites: 1. Valdez; 2. Gambell, St Lawrence Island; 3. Icy Cape; 4. Barrow; 5. Prudhoe Bay; 6. Barter Island; 7. Inuvik, N.W.T., Canada.

weighed and analysed by gas liquid chromatography (Atlas and Bartha, 1972) and by silica gel chromatography (Atlas, 1975).

In situ biodegradation studies were conducted in freshwater and saline ponds located at Barrow, Alaska. In these studies 1 ml of Prudhoe crude oil was floated in plexiglas cylinders (Atlas and Bartha, 1973b). Some slicks were poisoned by the addition of 1 % $HgCl_2$; some were seeded with a *Pseudomonas* sp. (0·1 mg protein) isolated from Prudhoe Bay (Atlas and Schofield, 1975); some were fertilized with 10 mM $NaNO_3$ and 0·5 mM NaH_2PO_4–Na_2HPO_4 (pH 8); some were fertilized with an oleophilic fertilizer, octyl phosphate 0·5 mM and paraffinized urea 10 mM N (Atlas and Bartha, 1973b); and some were seeded and fertilized. After 21 days of incubation, residual oil was recovered and analysed. Periodically, during the 21 days of incubation, heterotrophic microorganisms in the water underlying the slicks were enumerated. Organisms showing the same colonial appearance as the seed organism were counted.

Bioassay

The effects on several invertebrate organisms of crude oil and oil that had been treated to enhance biodegradation were determined. The organisms tested were amphipods (*Gammarus zaddachi*), Daphnia, rotifers, and fairy shrimp. In vitro experiments were conducted in 500 ml containers, with 2 ml of Prudhoe crude oil added and 20 organisms in each container. In some of the containers a nylon mesh screen was placed below the oil layer to shield the test animals from physically entering the slick. Prudhoe crude oil alone, with and without the nylon mesh screening, oil with seed bacteria and oleophilic fertilizer added, seed bacteria alone, and oleophilic fertilizer alone were tested with the animals. Concentrations of oleophilic fertilizer and seed bacteria were the same in all bioassay experiments as in the biodegradation experiments. The viability of the test organisms was periodically checked.

In situ tests were also conducted with amphipods in a saline pond for 21 days. Amphipods were contained in 1 ft^3 plexiglas boxes with screened openings on the bottoms. There were 50 organisms in each box to which 5 ml of oil was added. Daphnia, fairy shrimp, and rotifers were tested *in situ* in a freshwater pond for 14 days. They were contained in 1 ft plexiglas cylinders, 2 inches in diameter, with dialysis membranes over the bottoms of the cylinders. To each cylinder, 1 ml oil was added. Crude oil, oil fertilized with oleophilic N and P and seeded with bacteria, seed bacteria alone, and oleophilic fertilizer alone were used in the *in situ* tests.

RESULTS

Distribution of microorganisms

Oil-degrading microorganisms were found in concentrations of the order of 100/litre in most locations (Table I). Prudhoe Bay had the highest

Table I

DISTRIBUTION OF MICROORGANISMS IN SURFACE ARCTIC COASTAL WATERS

	Total heterotrophs (hundreds/ml) (25°C)	(5°C)	*Oil degraders* ♯/1	*Salinity* 0/00	*Temp* (°C)	*Date*
Valdez (1)	50	10	130	13	11	8/18
Gambell (2)	2	4	140	28	8	8/4
Icy Cape (3)	100	70	120	27	4	7/20
Barrow (4)	30	5	90	3	5	8/3
Barrow saline pond (4)	2	2	80	3	9	7/9
Barrow freshwater pond (4)	2	2	2	0	5	7/10
Prudhoe Bay (5)	20	50	500	2	7	7/17
Barter Island (6)	50	5	90	5	7	8/3
Inuvik (7)	40	3	100	4	7	7/28

concentration of oil-degrading microorganisms, 500/litre, of any location sampled. A freshwater pond at Barrow had a significantly lower concentration of oil-degrading microorganisms, 2/litre.

Oil biodegradation

When Prudhoe crude oil or Barrow crude oil was incubated for 42 days at 5°C with water from Prudhoe Bay, about 25% was lost abiotically and an additional 22% was biodegraded (Table II). During this period, approximately 10% of the oil was converted to CO_2 (Table II). During degradation of these crude oils, there was a decrease in the weight of paraffinic and

Table VII

21 DAY *IN SITU* BIOASSAY WITH A BARROW SALINE POND INVERTEBRATE *GAMMARUS ZADDACHI*, AN AMPHIPOD

Treatment	*% Mortality*
Control	0
Prudhoe crude oil	6
Oil—oleophilic fertilizer	10
Oil—fertilized, seeded	6

Table VIII

10 DAY IN VITRO BIOASSAY WITH BARROW FRESHWATER POND INVERTEBRATES AT 5°C

Treatment		*% Mortality*	
	Rotifer	*Daphnia*	*Fairy Shrimp*
Control	0	35	35
Prudhoe crude oil	80	100	100
Oil—shielded	0	50	80
Oleophilic fertilizer	0	90	75
Oil—fertilized, seeded, shielded	0	100	100
Seed bacteria	0	50	50

seed bacteria alone did not result in rotifer mortality (Table VIII). Daphnia and fairy shrimp, however, showed higher rates of mortality (Table VIII). There was a significant death rate even in the control populations of these organisms. Addition of fertilizer, seed bacteria, or oil alone or in combination resulted in higher mortality with these organisms. Even organisms shielded from direct contact with oil showed high mortality.

In situ tests with the freshwater invertebrates similarly showed that fairy shrimp and Daphnia were most sensitive to oil and were also adversely affected by fertilizer and bacteria. Here again, controls also showed high mortality rates (Table IX). Rotifers in the *in situ* tests were not killed by

Table IX

14 DAY *IN SITU* BIOASSAY WITH BARROW FRESHWATER POND INVERTEBRATES

Treatment		*% Mortality*	
	Rotifer	*Daphnia*	*Fairy Shrimp*
Control	5	40	50
Prudhoe crude oil	70	90	100
Oleophilic fertilizer	0	70	70
Oil—fertilized, seeded	80	100	100
Seed bacteria	0	70	80

addition of fertilizer or bacteria alone but were killed when oil was added. During the 14 day *in situ* test of freshwater invertebrates, 25% of the originally added oil was lost from the nonfertilized cylinders and 28% was lost from the oil that had been seeded and fertilized.

DISCUSSION

Microorganisms capable of growth on crude oil were found to be widely distributed in arctic coastal waters during summer months. Only a freshwater coastal pond had limiting numbers of oil-degrading microorganisms. The concentration of oil-degrading microorganisms was higher in Prudhoe Bay than in subarctic regions. There was no correlation between the concentrations of oil-degrading microorganisms and total numbers of heterotrophic microorganisms. Likewise, there was no correlation between numbers of microorganisms in a water sample and the salinity or temperature at the times of collection.

Both Prudhoe crude oil and Barrow crude oil were subject to biodegradation at 5°C, with about 25% of the added oil being lost abiotically and 25% being biodegraded during 42 days of incubation. The losses from these oils occurred in the paraffinic and aromatic fractions. Jobson *et al.* (1972) have reported similar findings with other northern crude oils. As had been found previously in other ecosystems (Atlas and Bartha, 1973b; Atlas and Schofield, 1975), oil biodegradation in arctic coastal ponds was nutrient-limited. Addition of an oleophilic nitrogen and phosphorus fertilizer enhanced biodegradation. Water-soluble nitrate and phosphate had no such enhancement effect. Seeding with an oil-degrading *Pseudomonas* sp. coupled with fertilizer addition enhanced biodegradation in a saline pond but seeding did not increase biodegradative losses in a freshwater pond. The seed organism was apparently unable to survive in the freshwater ecosystem, indicating that if seeding is contemplated for stimulating biodegradation, a variety of microorganisms suited for different environments will be needed.

In the bioassay experiments the greatest mortality of invertebrates occurred when the test organisms could physically enter an oil slick. Seeding oil with oil-degrading bacteria and fertilizing with oleophilic nitrogen and phosphorus in order to enhance biodegradation did not result in mortality of amphipods or rotifers. Fairy shrimp and Daphnia had a high natural mortality rate and appeared to be affected easily by any environmental modification, including adding bacteria or fertilizer. In no case did stimulating biodegradation by bacterial seeding and fertilizing result in significantly greater invertebrate mortality than occurred with crude oil alone.

REFERENCES

Atlas, R. M. (1975). Effects of temperature and crude oil composition on petroleum biodegradation. *Appl. Microbiol.* (in press).

Atlas, R. M., and Bartha, R. (1972). Biodegradation and mineralization of petroleum by two bacteria isolated from coastal waters. *Biotechnol. Bioeng.*, **14**, 297–308.

Atlas, R. M., and Bartha, R. (1973a). Abundance, distribution and oil biodegradation potential of microorganisms in Raritan Bay. *Environ. Pollut.*, **4**, 291–300.

Atlas, R. M., and Bartha, R. (1973b). Stimulated biodegradation of oil slicks using hydrophobic fertilizers. *Environ. Sci. Technol.*, **7**, 538–541.

Atlas, R. M., and Schofield, E. A. (1975). Petroleum biodegradation in the Arctic. In *Impact of the Use of Microorganisms on the Aquatic Environment*, pp. 183–198, Ecological Research Series, EPA-660/3-75-001, Environmental Protection Agency, Corvallis, Oregon.

Jobson, A., Cook, F. D., and Westlake, D. W. S. (1972). Microbial utilization of crude oil. *Appl. Microbiol.*, **23**, 1082–1112.

OIL BIODEGRADATION IN THE OCEANS

G. D. FLOODGATE

Department of Marine Biology, Marine Science Laboratories,
Menai Bridge, Anglesey LL59 5EH, U.K.

The research undertaken in many laboratories during the last few years appears to have established which factors govern the fate of oil in the marine environment. These include the physical properties of the oil itself such as its spreading pressure, and the physical properties of the sea such as wave action and tidal flushing. Similarly the chemical characteristics of both oil and sea greatly influence the changes which take place. The oil consists of very many substances in several chemical configurations; the chemistry of the sea varies from place to place. The influence of various biological entities also varies spatially and with time. All these factors interact. Thus oil spread thinly on a calm sunlit sea is liable to photo-oxidation which alters the substrate for microbial attack. Added to this, there are complications in understanding the processes involved due to a wide variety of ways in which the problems of biodegradation have been tackled. Not surprisingly therefore a wide variety of degradation rates have been quoted in the literature (Floodgate, 1972). This means that in spite of all the effort devoted to oil pollution studies there is still very little evidence on which to build a predictive model that will be of value in determining the input that a particular sea area will be able to withstand before serious damage is caused to the environment. From the microbiologist's point of view the most important missing information concerns those processes and factors which limit the rate of decomposition. One of the most important of these limits in biodegradation is the concentration and rate of supply of inorganic nutrients, especially nitrogen and phosphorus (Atlas and Bartha, 1972).

There are two aspects of this problem. Firstly there is the variation in the rate of degradation associated with different concentrations and rates of addition of inorganic nitrogen and phosphorus at levels below those supporting maximum rates, and secondly, there is the biochemical role of the inorganic nutrients within the hydrocarbonoclastic organism. The first of these problems has been tackled in the laboratory by Gibbs (1975) who deals with some of the aspects of his work in a contributed paper to this symposium.

As a matter of policy, it was decided that the work done on our programme should mimic the natural environment as closely as possible so that the maximum rate beyond which biodegradation does not increase was not determined. The evidence from the literature would suggest that the value is much higher than is found in coastal waters around the United Kingdom, so that it is reasonable to assume that biodegradation is always proceeding at rates below the maximum. Again the oils chosen for study were deliberately selected to represent the major crude oil inputs to British refineries. Those selected were Kuwait crude to represent a Middle East oil and a Forties crude to represent the oil now beginning to flow from the North Sea fields. The crudes were 'topped' usually to 350°C to simulate a weathered crude. The oil used in the seawater experiments was spread onto a glass filter so that the thickness of the film could be calculated to be around 2·0 × 10^{-2} mm, so that the greater part of the oil was immediately accessible to microbial attack. Oxygen concentrations varied, but never became low enough to be limiting, particularly when it is remembered that the K_m of the oxygenase involved in the first step in *n*-alkane breakdown indicates a high affinity for the element. The temperatures at which the experiments were run, that is 14°C and 4°C, were chosen to represent summer and winter temperatures of coastal waters. The natural seawater populations were used without any specific enrichment process, and aged seawater taken from the same place whenever required was employed in all these experiments. From the data obtained (Gibbs, 1975; Gibbs *et al.*, 1975), it was noticed that nitrogen added to actively degrading cultures almost entirely disappeared from the seawater within a few days, though the oxygen uptake, which was taken as a measure of the rate of decomposition of oil, slackened only slightly. Moreover when nitrogen was added slowly but continuously the rate of oxygen uptake was correlated to a significant degree with the rate of nitrogen addition and hence uptake. These observations raise the problem of the nitrogen budget of degrading oil, and as already mentioned above, its biochemical role. Some of the nitrogen will become part of the bacterial biomass and incorporated into the cell structure. It would be interesting to know how the nutrient nitrogen partitions between cellular structures and enzymic molecules needed for oil decomposition. It seems very unlikely that the bacterial biomass would simply go on increasing over the months of the experiment. It was impossible to count the number of bacteria at the water oil interface, but the numbers in the seawater remained remarkably constant throughout, and the numbers in the experimental flasks were very close to those in the control (Gibbs *et al.*, 1975). Possibly one destination of the nitrogen once inside the organism is the amino acid pool. Stanley and Brown (1974) have shown that some marine bacteria accumulate glutamic acid. Some nitrogen may be lost as soluble organic nitrogen which would flow out into the surrounding seawater. It was also found that when oxidation stops, usually after a period of up to nine months, then ammonia and phosphate are released into the seawater presumably because some hydrocarbonoclastic cells die due to the lack of utilizable carbon.

It is also interesting to note that the oxidative state of the inorganic nitrogen is immaterial; nitrate or ammonia served equally well. Since the rate of oxygen uptake is more or less lineally related to the rate of nitrogen uptake and supply at natural levels of nutrients, then it should be possible to compute the nitrogen demand, that is the amount of nitrogen involved in the destruction unit quantity of oil, always assuming that oxygen is in excess and that the oil presents a very large surface to the aqueous phase, and that all the components which degrade within several months have all disappeared. The figure found by Gibbs (1975) is around 4 μg—at N mg^{-1} oil. This value is calculated on oxygen uptake, and an assumed TOD of 3·5. If the nitrogen demand is calculated on the basis of oil recovered then the figure comes closer to 2·5. Decreasing the temperature to 4°C from 14°C increased the demand to around 10 μg—at N mg^{-1} oil. This was because nitrogen uptake, and presumably utilization, continued at the same rate as at 14°C, but the oxygen uptake was much slower. These values are of the same order of magnitude as can be calculated approximately using the data of Atlas and Bartha (1972) but very different from that of Robertson *et al.* (1973) who give values of 10 mg N mg^{-1} oil that is 7 × 10^2 μg—at N mg^{-1} oil.

The values for nitrogen demand can be translated into volumes of water containing the required amount of nitrogen, and if the nitrogen turnover rate is known for a given body of water, the rates of degradation to be expected can be calculated for summer and winter temperatures. Making some plausible assumptions this was found to be 30 g m^{-3} yr^{-1} in summer and around 11 g m^{-3} yr^{-1} in winter for the Irish Sea.

Phosphorus, though essential, appears to be less critical than nitrogen as a limit on oil degradation; when added to active oil-degrading cultures the concentration did not decrease to the same extent as did the nitrogen. Nor was there so clear a relationship between phosphorus uptake and oxygen. As the P/N ratio increases, so the rate of phosphate utilization per milligramme of oxygen utilized also tends to increase, possibly because the organisms are indulging in some form of luxury consumption. Again virtually nothing is known about the partitioning of phosphorus between cellular structures and the biochemical processes of oil biodegradation.

In those marine situations in which phosphorus and nitrogen are in plentiful supply, oxidation of oil might be expected to be comparatively rapid. Such a situation may be found on the beaches we have studied (Pugh *et al.*, 1974) and also in the model beaches constructed by Pugh (1975) to mimic beaches of this type. In the case of the natural beach, the ground water contained a mean value of about 15·0 μg—at $litre^{-1}$ nitrate nitrogen and about 6·0 μg—at $litre^{-1}$ of ammonium nitrogen. Phosphorus was present in the natural beach at 1·6 μg—at $litre^{-1}$ and at a very similar value in the models. Both the natural and the model beaches tended to concentrate the nitrogen and phosphorus so that the level of these nutrients in the sand was around three times higher than in the seawater. Since the sand potentially receives fresh supplies of nutrients with each tide it is reasonable to suppose that in this environment the concentration of inorganic nutrients would not be the dominant limiting factor. In view of the fact that the model beaches closely mimic the natural sandy beaches, it was reasonable to expect that the rate of oil degradation on the models would be close to that of a natural beach, and that compared with the water surface the rate would be fast. In fact when either Kuwait or Forties crudes were studied at a constant temperature the rate of change at 4°C was very slow. Alkanes only began to disappear after several months. Either the oxygen tension, the temperature or the state of division of the oil in the beach is the most likely limiting factor. Of these the first is unlikely to be responsible since monitoring the ground water indicates that the oxygen tension is always sufficiently high to allow oxidation to continue, particularly as the K_m of oxygenase indicates a high affinity for the substrate. The temperature effect will become evident shortly as an experiment is now running at 15°C. It is important to notice that temperature effects are dependent on the length of time the experiment runs. Short term experiments will reflect the response of a given population to the change of temperature, whereas in the long term, a change of temperature may produce a different population adapted to the new conditions (Nedwell and Floodgate, 1971). Some fast rates of degradation have been observed at Arctic temperatures (Traxler, 1972; Zobell, 1972; Robertson *et al.*, 1973). The degree of division of the oil within the beach is decided by the way the oil arrives on the beach and its subsequent history. On high energy beaches much of the oil will be simply washed away by wave action while in places where wave-energy is low the oil becomes buried, the conditions anaerobic and so digestion eases. Storms may re-expose the buried oil in some instances.

In our thinking at the moment, the governing factor in beach degradation rates is the way the oil arrives and its degree of division. Large mousse-like lumps exposed in Langston harbour, Portsmouth, were found to be very recalcitrant though the nutrient conditions were excellent (Davis and Gibbs, 1975).

One important gap in our knowledge at present is the variation in degradation rate due to the different origin of the oil. A little is known about the degradation of some fuel oils but the comparative rates at which different types of oil degrade under standard conditions remain unknown. An attempt to compare a series of oils from several sources failed probably due to an unknown factor in the aged seawater. This batch was collected in winter and was either lacking in hydrocarbonoclastic bacteria, which is improbable, or contained an inhibitor of some sort. It is at least possible that once an oil-water interface is colonized by a bacterial population then other species tend to be overgrown and not establish themselves on the surface in any great numbers. If the first colonizers to arrive are a matter of chance, then the subsequent rate of decomposition could also be to some degree a matter of chance.

But not all microbiota associated with decomposing oil are active in the decomposition process. In an earlier paper it was pointed out (Floodgate, 1972) that green sulphur bacteria could be associated with degrading oil. Again lumps of oil which had been exposed to seawater for 3½ years were found to be colonized on the surface by a green mat of common diatoms including colonial and filamentous forms. In addition there were several species of ciliates and flagellates as well as nematodes, these latter occasionally had oil droplets inside the gut. Protozoa are known to take up oil (Andrews and Floodgate, 1974) apparently while feeding. This work has been extended to include fresh water as well as marine protozoa. When a sample of water from a mountain lake was cultured a few protozoans appeared which readily took up a crude oil residue, a vorticellid being particularly active. In another experiment microscope slides were placed in a stream of seawater which was pumped directly from the Menai Straits. The usual ecosystem of algae bacteria and protozoa developed. When crude oil residue was added to the water, it was found that many protozoans contained oil droplets. Similarly when the sand of the model beaches was examined, it was found that many of the protozoans contained oil. These microorganisms appear to be catholic in their taste since a number of different crude oils and 350° residues were all readily ingested.

The ingested oil is released after a period, at least in the case of a *Uronema* spp. A culture of this organism was exposed to finely divided Kuwait crude, and the organisms then separated from the main body of the medium using the method described earlier (Andrews and Floodgate, 1974). The number of cells containing oil declined from 87% to 10% after 22 hr. Presumably this final percentage represents reingestion or possibly dead protozoans.

In general crude oils and residues do not appear to be very toxic to protozoa. *Noctiluca,* which tends to be on the surface in culture flasks, is readily smothered by an oil film, though it also ingests the oil; but generally few adverse effects were seen. However it has been observed that when *Uronema* spp. was exposed to Kuwait crude at a concentration of 10 mg per 100 ml medium, it became less motile and the cells shortened by about 30%. When 350°C Kuwait residue was used the effect was less marked. After 3 or 4 days the cells returned to normal activity and size with no apparent ill effects. Similarly cells killed with 2·5% formalin or mercuric chloride (0·5 mg ml^{-1}) returned to their original length within an hour. These observations were interpreted as being the effect of a slightly toxic material in the crude which evaporated off after a few days. However, soluble extracts of Kuwait crude prepared by the method of Boyan and Tripp (1971) did not show the expected effect.

REFERENCES

Andrews, A. R., and Floodgate, G. D. (1974). Some observations on the interactions of marine Protozoa and crude oil residues. *Marine Biology*, **25**, 7–12.

Atlas, R. M., and Bartha, R. (1972). Degradation and mineralisation of petroleum in sea water: Limitation by nitrogen and phosphorus. *Biotechnology and Bioengineering*, **14**, 308–318.

Boyan, D. B., and Tripp, B. W. (1971). Determination of hydrocarbons in seawater extracts of crude oil and crude oil fractions. *Nature, London*, **230**, 44–47.

Davis, S. J., and Gibbs, C. F. (1975). The effect of weathering on a crude oil residue exposed at sea. *Water Research*, **9**, 275–288.

Floodgate, G. D. (1972). A threnody concerning the biodegradation of oil in natural waters. In *The Microbial Degradation of Oil Pollutants*, pp. 17–24 (D. G. Ahearn and S. P. Meyers, eds.), Louisiana State University. Publication No. LSU-SG. 73-01.

Gibbs, C. F. (1975). Quantitative studies on marine biodegradation of oil. I. Nutrient limitation at 14°C. *Proc. Roy. Soc.* **B188**, 61–82.

Gibbs, C. F., Pugh, K. B., and Andrews, A. R. (1975). Quantitative studies in marine biodegradation of oil. II. Effect of temperature. *Proc. Roy. Soc.*, **B188**, 83–94.

Nedwell, D. B., and Floodgate, G. D. (1971). The seasonal selection by temperature of heterotrophic bacteria in an intertidal sediment. *Marine Biology*, **II**, 306–310.

Pugh, K. B. (1975). The construction and use of an artificial model beach system. *J. Experim. Marine Biol. Ecol.*, **18**, 197–213.

Pugh, K. B., Andrews, A. R., Gibbs, C. F., Davis, S. J., and Floodgate, G. D. (1974). Some physical chemical and microbiological characteristics of two beaches of Anglesey. *J. Experim. Marine Biol. Ecol.*, **15**, 305–333.

Robertson, B. R., Arhelger, S. D., Kirney, P. J., and Button, D. K. (1972). Hydrocarbon biodegradation in Alaskan waters. In *The Microbial Degradation of Oil Pollutants*, pp. 171–184 (D. G. Ahearn and S. P. Meyers, eds.), Louisiana University. Publications No. LSU-SG-73-01.

Robertson, B. R., Arhelger, S. D., Law, R. A. T., and Button, D. K. (1973). Hydrocarbon biodegradation. In *Environmental Studies of Port Valdez*, pp. 449–479 (D. W. Hood, W. E. Shiels, and E. J. Kelley, eds.), Institute of Marine Sciences, University of Alaska, Fairbanks.

Stanley, S. O., and Brown, C. H. (1974). Influence of temperature and salinity on the amino acid pools of some marine pseudomonads. In The Effect of the Ocean Environment on Microbial Activities, Proceedings of 2nd US–Japan Conference on Marine Microbiology, University of Maryland 25–30 Aug. 1972 (R. R. Colwell and R. Y. Morita, eds.), Baltimore University Press.

Traxler, R. W. (1972). Bacterial degradation of petroleum materials in low temperature marine environments. In *The Microbial Degradation of Oil Pollutants*, pp. 163–170 (D. G. Ahearn and S. P. Meyers, eds.), Louisiana University. Publications No. LSU-SG-73-01.

Zobell, C. E. (1972). Bacterial degradation of mineral oils at low temperature. In *The Microbial Degradation of Oil Pollutants*, pp. 153–162 (D. G. Ahearn and S. P. Meyers, eds.), Louisiana University. Publications No. LSU-SG-73-01.

MICROBIAL POPULATIONS IN COASTAL SURFACE SLICKS*†

S. A. CROW, W. L. COOK, D. G. AHEARN

Department of Biology, Georgia State University,
Atlanta, Georgia 30303, U.S.A.

and

A. W. BOURQUIN

U.S. Environmental Protection Agency, Gulf Breeze Environmental
Research Laboratory, Sabine Island, Gulf Breeze, Florida 32561, U.S.A.

Summary

Samples of the upper 10 μm of inshore surface films obtained by adsorption to membranes yielded microbial populations up to 10^8 ml^{-1} or 10^5 cm^{-2}. These populations were typically 10–100 times greater than those in underlying waters at a depth of 10 cm. Predominant bacteria in surface films were motile, nonpigmented, gram-negative rods. Colony-forming units of yeasts and moulds were found in concentrations to 10^4 ml^{-1} or 28 cm^{-2}. The predominant species in surface films were proteolytic and amylolytic but exhibited only weak to negligible hydrocarbonoclastic and lipolytic activities. A greater proportion of the surface film bacteria, as compared to those at 10 cm depth, were capable of growth on fresh-water media.

INTRODUCTION

Surface films are unique microbial habitats occurring at the air-water interface of aquatic systems. They are a common phenomenon in coastal waters and probably occur in ocean waters at most times, depending on the state of the seas. The production of most surface slicks appears to be related to the decay of naturally occurring aquatic organisms (Babkov, 1965). Numerous investigators (Baier, 1970, 1972; Ewing, 1950; Garret, 1965; Sutcliffe *et al.*, 1963) have examined the physical properties of surface slicks but their chemical nature is unclear. Surface films have been shown to contain high concentrations of organic carbon, nitrogen, phosphorus (Williams, 1967), alkanes (Ledet and Laseter, 1974), and chlorinated hydrocarbons (Seba and Corcoran, 1969).

The few studies of microbial populations in surface films have indicated that they contain high densities of bacteria relative to underlying waters. Sieburth (1965) reported bacterial populations up to 4×10^4 ml^{-1} in surface films. The predominant bacteria were pseudomonads which expressed lipolytic activity. Harvey (1966) found that bacteria, small algae, and colourless flagellates were concentrated in the upper 60 μm of surface water. In a preliminary study, Crow *et al.* (1975) found concentrations of total

* Supported in part by Office of Naval Research contract ONR N000-14-71-C-0145 and Environmental Protection Agency contract R 803141-01-0.

† Gulf Breeze Environmental Research Laboratory Contribution No. 254.

heterotrophs up to 10^8 ml^{-1} in coastal surface films. This report examines the microbial populations of surface films occurring in coastal waters of the Gulf of Mexico.

MATERIALS AND METHODS

Surface slick samples were collected by a membrane adsorption technique (Crow *et al.*, 1975). Sterile polycarbonate membranes (Nuclepore®) 47 mm in diameter with a porosity of 0·4 μm were floated on surface waters near Barataria Bay, Louisiana, and along the western coast of Florida near Sabine Island. The membranes with adhering surface film and underlying water were collected with sterile plastic dishes. The membrane was removed from the dish with sterile forceps and either implanted directly on suitable media or placed in 100 ml of sterile seawater. The seawater was agitated vigorously, serially diluted, and various dilutions plated on suitable media. Numbers of microorganisms were calculated on a per-ml basis using a sampling volume of 5·9 μl per membrane and for surface area using 17·3 cm^2 per membrane (Crow *et al.*, 1975). Subsurface samples were collected from a depth of 10 cm with sterile 30 ml disposable syringes fitted with extension tubes.

Total heterotrophic bacterial populations were enumerated using Marine agar 2216 (Difco). Fungal populations were determined with Mycological agar (Difco) prepared with natural seawater and adjusted to pH 4·5 with 0·1 N HCl. For select samples, populations of proteolytic bacteria were enumerated on yeast extract (0·01 %)-skim milk (2·0 %) agar prepared in artificial seawater of 20 ‰ salinity. Hydrocarbonoclastic bacteria were enumerated according to the method of Gunkel (1973). In this method a basal medium containing NH_4Cl 0·005 %, K_2HPO_4 0·005 %, Na_2HPO_4 0·01 %, and 2 % Louisiana crude oil is inoculated with tenfold dilutions of seawater. Utilization of the hydrocarbon is based on the occurrence of turbidity following acidification for the dissolution of inorganic precipitate. All cultures were incubated at 22–25°C; broth cultures were incubated with agitation.

Representative microorganisms, unless otherwise stated, were characterized physiologically with media prepared according to the formula of Colwell and Wiebe (1970). Proteolysis was determined using 2·0 % skim milk and 0·1 % yeast extract in 1·7 % agar and with Thioglycollate gelatin medium (Difco) prepared with artificial seawater. Oxidative or fermentative carbohydrate metabolism was determined with MOF medium (Difco). Lipase and urease activities were determined on Spirit Blue agar (Difco) and Urea agar (Difco), respectively. Both were prepared with artificial seawater at 20 ‰ salinity.

Microorganisms were tested for pristane, decane, tetradecane, ethanol, and crude oil utilization according to previously described methods (Ahearn *et al.*, 1971). Studies of pesticide interference with ethanol and hexadecane metabolism were conducted using selected ethanol-positive bacteria, yeasts, and filamentous fungi. Methoxychlor dissolved in ethanol was added to tubes of yeast extract broth (0·01 %) to obtain a final ethanol concentration of 2 %

® Registered Trademark. Mention of commercial products or trade names does not constitute endorsement by the U.S. Environmental Protection Agency.

and a concentration of 100 μg methoxychlor ml^{-1} of medium. Heptachlor was dissolved in hexadecane to give a final concentration of about 0·1 μg ml^{-1}. Tubes were incubated on roller drums at 20–25°C for 5 days. Cultures were examined at 24-hr intervals and turbidity was compared to controls to determine any interference with ethanol metabolism.

To ascertain the presence of pesticides in surface slicks, large membranes (293 mm in diameter with a 0·4 μm porosity) were used to adsorb surface slick material. These membranes were then placed in 250-ml Erlenmeyer flasks and extracted with 100 ml of pesticide-grade petroleum ether; 10 ml of this material was concentrated 10 times and used for gas chromatographic (GC) analysis. Analyses were performed on a Varian Aerograph 2100 fitted with an electron-capture detector on a 0·64 cm × 1·8 m glass column containing 2% OV-101 on Gas Chrom Q 100/120.

RESULTS

The densities of microorganisms in coastal and estuarine surface slicks ranged as high as $2{\cdot}5 \times 10^5$ cm^{-2}. Typically, underlying waters contained substantially lower populations (Table I). The predominant bacteria were

Table I
CONCENTRATION RANGE OF MICROORGANISMS IN SURFACE SLICKS AND UNDERLYING WATERS

Organisms	*Surface slick* (ml^{-1})	(cm^{-2})	*Subsurface* (ml^{-1})	*No. samples*
Bacteria	10^5–10^8	28–$2{\cdot}5 \times 10^5$	10^2–10^6	24
Yeasts	10^2–10^4	0–5	10^2	16
Moulds	10^3–10^4	0–28	10^2	16

motile, nonpigmented, gram-negative rods, presumptively identified as pseudomonads. Samples from surface films and from 10 cm depths appeared to contain similar proportions of chromogenic bacteria. The most common fungi were isolates of the genera *Aureobasidium*, *Candida*, *Cephalosporium*, and *Cladosporium*. The proportions of hydrocarbon-utilizing and proteolytic bacteria in the total populations at two geographically adjacent stations are presented in Table II. Neither of these stations were subjected to heavy

Table II
AVERAGE NUMBER OF PROTEOLYTIC AND HYDROCARBONOCLASTIC BACTERIA IN SURFACE SLICKS IN FOUR SAMPLES FROM FLORIDA COASTAL WATERS

Bacteria	*Colony-forming units/cm²* *Station* 1	*Station* 2
Proteolytic	$2{\cdot}0 \times 10^3$	$1{\cdot}0 \times 10$
Hydrocarbonoclastic	$3{\cdot}0 \times 10^2$	$2{\cdot}5 \times 10^2$
Total heterotrophic	$2{\cdot}5 \times 10^5$	$1{\cdot}0 \times 10^4$

hydrocarbon pollution; however, station 2, an enclosed saline pond, received laboratory wastes containing pesticides including methoxychlor and heptachlor. GC analyses of extracts of membranes retrieved from surface slicks at both of these stations indicated substances compatible with chlorinated pesticides. Surface slicks from the enclosed pond yielded fewer total heterotrophic bacteria and markedly reduced numbers of proteolytic bacteria. Hydrocarbonoclastic bacteria were present at both stations, but fewer organisms from station 2 exhibited intense growth and none significantly emulsified the oil. Only five isolates of 20 selected cultures from both stations, giving an indication of growth on crude oil, grew well in yeast extract broth, with decane and tetradecane as sole carbon sources. Of these five isolates, four also utilized pristane. Virtually all the yeast isolates representative of *Candida* and *Rhodotorula* grew on the hydrocarbons.

Twenty-one bacteria, representative of the predominant morphological types found at all collection sites, were physiologically characterized (Table III). All organisms were initially isolated on and produced good growth on

Table III

PHYSIOLOGICAL CHARACTERISTICS OF 21 MORPHOLOGICALLY DIFFERENT BACTERIA FROM SURFACE SLICKS

Character	*No. bacteria* (+)	(−)	(±)
Proteolysis			
Skim milk	10	8	3
Gelatin	21	0	0
H_2S production	0	13	8
Amylolysis	11	8	1
Lipolysis	0	21	0
Citrate utilization	3	16	2
Urease activity	2	18	1
Ethanol utilization	15	6	0
Crude oil utilization	1	20	0
MOF reactions			
oxidative	14	4	0
fermentative	5	13	0
no change	3	0	0
Gram reaction	8	11	2
Motility	15	6	0
Growth on freshwater medium	13	8	0
Morphology	rods		

seawater medium (20$^0/_{00}$ salinity); nevertheless, 13 of the 21 strains grew well in yeast extract broth prepared with fresh water. None of the isolates showed lipolysis and only a single isolate grew on crude oil.

Neither methoxychlor nor heptachlor were found to alter the metabolism of ethanol or hexadecane by representative surface slick microorganisms. The concentrations of these pesticides which could be reclaimed by hexane extraction from several yeast and bacterial cultures were reduced in several instances in comparison with uninoculated control flasks by about 25%. Preliminary experiments with C^{14}-labelled heptachlor indicated the radiolabel was bound to the cells and not extractable with the procedure employed. Known biodegradation products of both pesticides were not demonstrated in the culture systems.

DISCUSSION

Microbial populations associated with surface films on coastal waters of the Gulf of Mexico were found to be higher than those reported for films at other oceanic sites collected with either a wire screen or a hydrophilic drum (Sieburth, 1965; Harvey, 1966; Gunkel, 1973). The reports of these investigators that microorganisms were concentrated in the surface films as contrasted to underlying waters were substantiated. Fungi, moulds in particular, occurred mainly in the surface film. In slicks from the eastern Gulf of Mexico coastal waters, hydrocarbonoclastic and lipolytic bacteria appeared uncommon. Sieburth (1965) reported that 95% of surface slick isolates, mainly pseudomonads, from the Pacific Ocean were lipolytic. This anomaly is most likely a reflection of the diverse nature of surface films (MacIntyre, 1974). The ability of most of our isolates to grow on fresh-water media suggests they are of terrestrial origin. Preliminary analysis with culture systems has not demonstrated an effect of heptachlor or methoxychlor on the metabolism of hexadecane or ethanol by representative slick microorganisms. In other studies (Smith *et al.*, 1976), heptachlor was shown to enhance or inhibit hexadecane utilization by *Candida maltosa* from a fresh-water oil slick dependent upon aeration and pesticide concentration. The binding of pesticides to cells in culture systems suggests that the presence of chlorinated hydrocarbons in surface films (Seba and Corcoran, 1969) may be related to the microbial densities of these slicks.

REFERENCES

Ahearn, D. G., Meyers, S. P., and Standard, P. G. (1971). The role of yeasts in the decomposition of oils in marine environments. *Dev. Ind. Microbiol.*, **12**, 126–134.

Babkov, A. I. (1965). The causes of slicks on the surface. *Oceanologiya*, **5**, 102–104.

Baier, R. E. (1970). Surface quality assessment of natural bodies of water. *Proc. Great Lakes Res.* 13th, pp. 114–127.

Baier, R. E. (1972). Organic films on natural waters: their retrieval, identification, and modes of elimination. *J. Geophys. Res.*, **77**, 5062–5–75.

Colwell, R. R., and Wiebe, W. J. (1970). 'Core' characteristics for use in classifying aerobic, heterotrophic bacteria by numerical taxonomy. *Bull. Georgia Acad. Sci.*, **28**, 165–185.

Crow, S. A., Ahearn, D. G., Cook, W. L., and Bourquin, A. W. (1975). Densities of bacteria and fungi in coastal surface films as determined by a membrane-adsorption procedure. *Limnol. Oceanogr.*, **20**, 644–645.

Ewing, G. (1950). Slicks, surface films and internal waves. *J. Mar. Res.*, **9**, 161–187.

Garret, W. D. (1965). Collection of slick-forming materials from the sea surface. *Limnol. Oceanogr.*, **10**, 602–605.

Gunkel, W. (1968). Bacteriological investigations of oil-polluted sediments from the Cornish coast following the 'Torrey Canyon' disaster. In *The Biological Effects of Oil Pollution on Littoral Communities*, pp. 151–158 (J. D. Carthy and D. R. Arthur, eds.), Field Studies Council, London.

Gunkel, W. (1973). Distribution and abundance of oil-oxidizing bacteria in the North Sea. In *The Microbial Degradation of Oil Pollutants*, pp. 127–139 (D. G. Ahearn and S. P. Meyers, eds.), Center Wetland Resources, Louisiana State University, Publication LSU-SG-73-01.

Harvey, G. W. (1966). Microlayer collection from the sea surface: a new method and initial results. *Limnol. Oceanogr.*, **11**, 608–613.

Ledet, E. J., and Laseter, J. L. (1974). Alkanes at the air-sea interface from offshore Louisiana and Florida. *Science*, **186**, 261–263.

MacIntyre, F. (1974). The top millimeter of the ocean. *Sci. Am.*, **230**, 62–77.

Seba, D. B., and Corcoran, E. F. (1969). Surface slicks as concentrators of pesticides in the marine environment. *Pestic. Monit. J.*, **3**, 190–193.

Sieburth, J. McN. (1965). Bacteriological samplers for air-water and water-sediment interfaces. Trans. Jt. Conf. Ocean Sci. Ocean Eng. Mar. Technol. Soc. Am. Soc. Limnol. Oceanogr., pp. 1064–1067.

Smith, G. N., Bourguin, A. W., Crow, S. A., and Ahearn, D. G. (1976). The effect of heptachlor on hexadecane utilization by selected fungi. *Dev. Ind. Microbiol.*, **17**, 331–336.

Sutcliffe, W. H. Jr., Baylor, E. R., and Menzel, D. W. (1963). Sea surface chemistry and Langmuir circulation. *Deep-Sea Res.*, **10**, 233–243.

Williams, P. M. (1967). Sea surface chemistry: organic carbon and organic and inorganic nitrogen and phosphorus in surface films and subsurface waters. *Deep-Sea Res.*, **14**, 791–800.

MODELLING OF THE BREAKDOWN AND THE MOBILIZATION OF HYDROCARBONS IN UNSATURATED SOIL LAYERS

W. VERSTRAETE and R. VANLOOCKE

General and Industrial Microbiology Laboratory,
University of Ghent, Coupure 533, 9000 Ghent, Belgium

R. DE BORGER

Chemical Research Institute, Ministry of Agriculture,
Molenstraat 5, 1980 Tervuren, Belgium

and

A. VERLINDE

Pharmacology, Microbiology and Hygiene Laboratory,
University of Ghent, Apotheekstraat 1, 9000 Ghent, Belgium

Summary

The biotransformation of gasoil in the surface soil layer was monitored by direct chemical analyses of the hydrocarbon compounds and by respirometric methods. The results indicate that in the acidic podzol studied, the pH and the C/N/P ratio are the most prominent factors responsible for the persistence of gasoil. At neutral pH values and upon the addition of nitrogenous fertilizers, up to 60 % biodegradation could be obtained in 42 days for topped Bachaquero gasoil at 20°C. Topped Kuwait turned out to be less recalcitrant than topped Nigeria gasoil, and the latter in turn was more susceptible to microbial breakdown than topped Bachaquero gasoil. The biodegradation phenomena of gasoil in nonadapted soil can be simulated mathematically by asymmetric sigmoidal kinetics. The biodegradation in acclimatized soil can conveniently be fit to zero order kinetics.

The breakdown of hydrocarbons in the water-unsaturated subsurface soil horizons was simulated in lysimeter columns and monitored by chemical analyses. The results obtained from these experiments indicated that upon irrigation of these columns with aqueous solutions of inorganic nitrogenous and phosphorus fertilizers, a substantial amount of the hydrocarbons entrapped in the soil matrix is mobilized and subsequently leached out as a result of the bio-emulgating activity of soil microbiota.

INTRODUCTION

When oil is spilled on the soil surface, it spreads on the soil surface and in the underlaying soil layers. Depending on the permeability of the soil and on the physico-chemical nature of the oil, the pollutant migrates more or less rapidly into the underlaying soil horizon and eventually forms an oil lens on top of the soil-water table (Fig. 1). Three different zones can be distinguished. The surface soil layer is usually aerobic, water-unsaturated, rich in nutrients, and colonized by a diverse microbial community. This layer can be supplemented with various chemicals to combat the pollutant or it can be removed and disposed of on a sanitary landfill. The second zone comprises the water-unsaturated subhorizon through which the oil has migrated. This horizon is usually poor in nutrients, reduced in redox potential, and sparsely populated by microorganisms. The oil absorbed in these layers onto the mineral and the organic soil complex leaches out slowly with the percolating rain water. This process is known to extend over periods of years (Somers, 1966; Working group water and petroleum, 1970). For practical and economical reasons, mechanical removal of these layers is not feasible. The third zone is the oil lens or pancake which lays on top of the groundwater table and forms the bulk of the oil. When the groundwater table descends or rises, the oil lens descends or rises too. In that way the oil can be smeared out over a large volume of

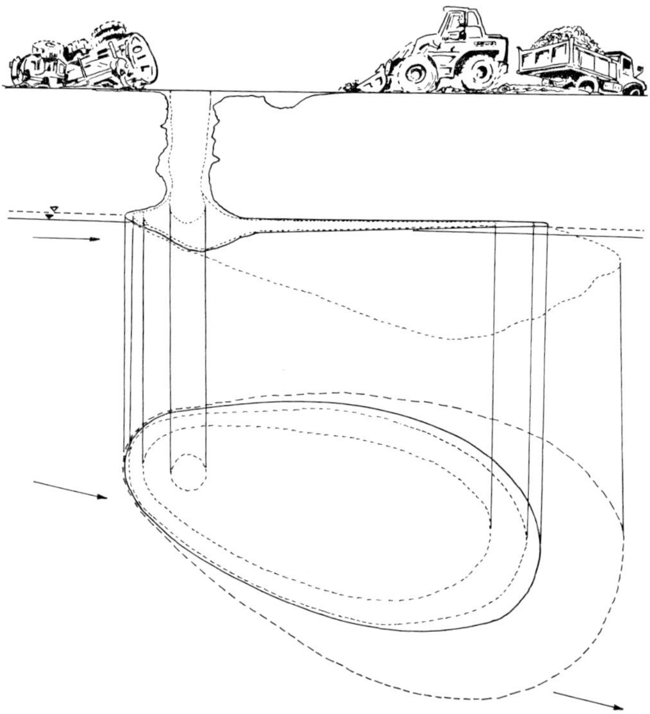

Fig. 1. Groundwater contamination by oil.

soil. This oil lens reaches its maximum horizontal extension a few weeks after the pollution takes place. The bulk of the hydrocarbons entrapped in the oil lens has to be removed by pumping (Fig. 2). It is clear that the oil lens, even when the bulk of the oil phase has been removed by pumping, constitutes the major threat to the groundwaters. Indeed, the hydrocarbons as well as their partly biodegraded derivatives are leached out from this zone by the descending rain water and by the aquifer water as well. In view of the low solubility of the hydrocarbons and of their derivatives in water and considering the slow rates of the oil biotransformation and of the elution processes, this leaching continues for extremely long periods after the pollution has occurred.

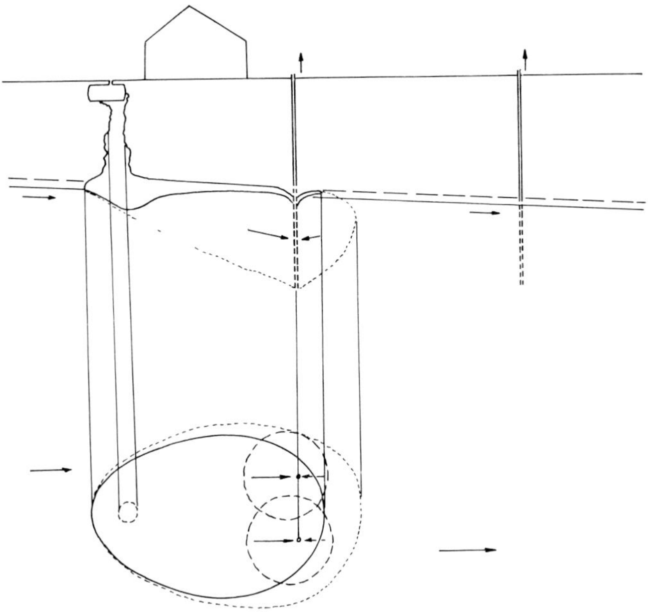

Fig. 2. Removal of the oil lens by pumping.

This paper deals with experiments in the laboratory which simulate oil degradation and oil mobilization processes in the soil surface and the soil subsurface horizons, respectively.

MATERIALS AND METHODS

To study the aerobic breakdown in the soil-surface horizon, an acidic black podzol is used. The principal characteristics of this soil are summarized in Table I. The soil, moistened to about its field capacity, is homogenized and filled in 10 kg aliquots in stainless steel containers with a volume of 20 litres. The containers are covered with aluminium foil to reduce evaporation losses. The soil moisture content is regularly checked and adjusted. After the addition of amendments to the soil (oil, nutrients, . . .) and before sampling, the contents of the containers are mixed carefully.

The principal characteristics of the topped gasoils used in this study are given in Table II. The volatile components are stripped from the gasoil by stirring 5·0 litre batches of the gasoil open to the air over a period of 7 days.

The number of bacteria is counted on nutrient agar (18 g agar/litre) after incubation at 22°C during 7 days. The number of fungi is counted on Martin agar (Martin, 1950) after 7 days at 22°C. All enumerations are done twice. The soil respiration is measured by the Sapromat apparatus (Voets *et al.* 1974).

Five grammes of air-dried soil are extracted twice with 50 ml CCl_4. The combined extracts are evaporated at room temperature under a constant nitrogen flow. The residue is quantified gravimetrically and is referred to as total fatty matter.

Table I

CHARACTERISTICS OF THE ACIDIC BLACK PODZOL

Characteristic	*A Horizon*
pH H_2O	4·50
pH KCl	3·90
% C	3·29
CEC (meq./100 g dry weight)	22·41
BS (% of the CEC)	69·2
Texture	
0–2 μ	0·93
2–53 μ	13·21
53–200 μ	71·14
>200 μ	11·78
total	97·06
water content at	
—saturation	39·0
—field capacity	12·8
—wilting point	8·2
(% of dry weight)	

The CCl_4 extracts of the soil are also analysed by infrared spectrometry at 3·4 μ for their content of aliphatic hydrocarbons (Concawe, 1972).

The aromatic hydrocarbons are extracted by percolating *n*-hexane through 100 g of air-dried soil for a period of 8 hr in a Soxhlett apparatus. The extracts are analysed by uv spectrometry. The absorption peaks covering the 210–330 nm part of the spectrum are integrated and related to a standard curve. The standard curve is established by treating soil samples with known quantities of the gasoil.

For the lysimeter studies, glass columns, 1·0 m in height and 0·2 m in diameter, are used. These columns were set up according to Cassel *et al.* (1974). Thirty kilograms of air-dried soil are packed into the column. Subsequently, the column is irrigated with water. After 2·0 litres of water have percolated through the column, 2·0 litres of topped gasoil are applied. The oil phase is eluted from the soil column by applying 0·5 litre of an aqueous solution three times a week. The water eluted from the column is collected and analysed analogous to the soil samples.

Table II

CHARACTERISTICS OF THE TOPPED GASOILS

Characteristics	*Gasoil*		
	Bachaquero	*Kuwait*	*Nigeria*
Volatile fraction	1·2	11·3	18·1
Nonvolatile fraction	98·8	88·7	81·9
Nonvolatile fraction			
Saturates	59·2	66·1	61·7
Aromatics	32·9	26·0	23·0
Soluble NSO fraction	3·0	3·9	2·9
Insoluble NSO fraction	4·8	4·0	11·8

RESULTS

In the first series of experiments, the aerobic breakdown of the topped gasoil in the soil-surface horizon was examined. Four factors which could influence the breakdown of hydrocarbons in the soil were subsequently investigated: the nutrient status of the soil, the soil microbiota, the soil pH, and the composition of the gasoil. The results of these experiments are summarized in Table III. To facilitate comparison, all results have been converted to percentages of hydrocarbon compounds biodegraded at the end of the incubation period.

Table III
AEROBIC BREAKDOWN OF TOPPED GASOILS IN THE SOIL SURFACE HORIZON

Object	*% Degraded after 42 days of incubation*		
	Alkanes	*Aromatics*	*Total fatty matter*
5000 ppm Bachaquero pH 4·5	22·8	11·5	16·5
5000 ppm Bachaquero N and P fertilizers pH 4·5	24·0	14·1	18·8
5000 ppm Bachaquero Inoculum pH 4·5	20·4	11·5	17·4
5000 ppm Bachaquero N and P fertilizers Inoculum pH 4·5	14·8	12·0	9·9
5000 ppm Bachaquero pH 7·4	37·8	35·5	16·5
5000 ppm Bachaquero N and P fertilizers pH 7·4	60·0	55·7	40·6
5000 ppm Bachaquero N and P fertilizers pH 8·5	30·6	42·5	28·2
5000 ppm Nigeria N and P fertilizers pH 7·4	80·6	62·7	64·3
5000 ppm Kuwait N and P fertilizers pH 7·4	90	67·5	76·6

From these results, it can be seen that the breakdown of gasoil proceeds slowly in the acidic podzol and that the alkanes are more susceptible to microbial degradation than the aromatic compounds. The experiments also indicate that the pH is the primary factor limiting breakdown. Indeed, adjustment to a neutral pH value of 7·4 about doubles the rate of breakdown of alkanes and aromatics. The secondary factor limiting the biodegradation processes is the C/N/P ratio. In effect, adjustment of the C/N/P ratio to 100/10/1 by the addition of NH_3NO_3 and K_2HPO_4 fertilizers to the soil sharply increases the breakdown of gasoil, provided these changes are accompanied by an adjustment of the soil pH. The addition of an inoculum composed of soil preadapted to gasoil had little effect. Finally, the results obtained for the different gasoils show that the Kuwait gasoil is much more biodegradable than the Nigeria gasoil, and the latter, in turn, is considerably more susceptible to microbial breakdown than the Bachaquero gasoil.

In order to evaluate the kinetics of the biodegradation processes, the respiration of the soil treated with the different gasoils was monitored (Fig. 3). The curves relating to the impact of soil pH to the addition of fertilizers and to the influence of the different gasoils correspond to the findings reported in Table III. However, the respiration curves showing the effect of the soil

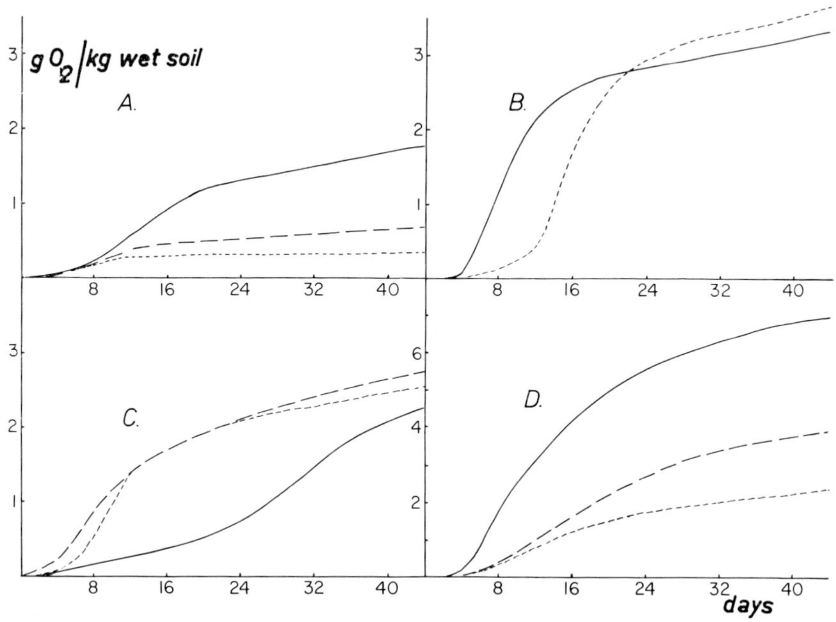

Fig. 3. Respiration curves of soil treated with gasoil. (A) ---- black podzol without treatment; —— black podzol + 5000 ppm Bachaquero; ——— black podzol + 5000 ppm Bachaquero + N and P. (B) ---- black podzol + 5000 ppm Bachaquero + N and P; ——— black podzol + 5000 ppm Bachaquero + N and P + Inoculum. (C) ---- black podzol + 5000 ppm Bachaquero + N and P, pH 4·5; —— black podzol + 5000 ppm Bachaquero + N and P, pH 7·4; ——— black podzol + 5000 ppm Bachaquero + N and P, pH 8·5. (D) ---- black podzol + 5000 ppm Bachaquero + N and P, pH 7·4; —— black podzol + 5000 ppm Nigeria + N and P, pH 7·4; ——— black podzol + 5000 ppm Kuwait + N and P, pH 7·4.

inoculum suggest that there is indeed a stimulation of the soil microbial activity upon the addition of a certain amount of preadapted soil. Unfortunately, this effect is of a transitory nature and does not result in an overall increase in gasoil removal.

The respiration curve permits us to evaluate the kinetics of the breakdown of the gasoil in the soil ecosystem. The first part of the curve has a typical sigmoidal shape. The following equation represents this acclimation phase (Verstraete *et al.*, 1975):

$$\frac{dy^{1/a}}{dt} = k(y^{1/a} + b)(c_o - y^{1/a})$$

$$y = \left(c_o - \frac{b + c_o}{(b/c_o)\exp[k(c_o + b)t] + 1}\right)^a$$

y = amount of oxygen consumed; mg O_2/kg wet soil; t = days; $c_o{}^a$ = oxygen uptake at the end of the acclimation phase; b^a = initial hydrocarbonoclastic biomass; a = coefficient of asymmetry; k = rate constant; k' = slope at the point of inflection.

The parameters a, b, $c_o{}^a$, and k' are computerized by means of the least sum of squares of surface method and represented in Table IV. The high

Table IV
OXYGEN UPTAKE OF A BLACK PODZOL TREATED WITH TOPPED GASOIL

Gasoil	*Fertilizer*	*pH*	*Acclimation phase**				*Stationary phase†*
			b^a	$c_o{}^a$	k'	*ID*	*Respiration rate* (*mg kg*$^{-1}$ *day*$^{-1}$)
a. None	None	4·5	4·53	291	38	0·969	16·7
b. 5000 ppm Bachaquero	None	4·5	9·10	443	50	0·990	21·7
c. 5000 ppm Bachaquero	N and P	4·5	10·98	1 356	111	0·997	26·8
d. 5000 ppm Bachaquero	N and P	7·4	13·55	1 787	155	0·986	36·3
e. 5000 ppm Nigeria	N and P	7·4	13·77	3 152	237	0·981	38·3
f. 5000 ppm Kuwait	N and P	7·4	37·62	5 972	540	0·959	42·8

* Modelled according to an asymmetric S-curve, for symbols *see* text; ID = index of determination.
† Average oxygen uptake of the soils (20°C) in the period between the 20th and the 40th day after the addition of the amendments.

ID values leave no doubt about the appropriateness of the mathematical model. From the values of the parameter b^a and k', it can also be seen that the initial hydrocarbonoclastic biomass as well as the rate of biodegradation increases going from *a* to *f*. A more simple and straightforward way to interpret the respiration curves is to restrict the analysis to the stationary phases. After the initial acclimation periods, the rates of oxygen uptake become more or less constant and the overall processes in the stationary phases approach zero order kinetics. The results given in Table IV clearly show the impact of the various treatments and the gradient of biodegradation rates going from *a* to *f*.

The impact of the various treatments on the number of bacteria is schematized in Fig. 4. The application of gasoil alone does not induce a significant increase of the number of bacteria. However, the addition of fertilizers together with gasoil results in a pronounced increase of the bacterial biomass. Inoculation with preadapted bacteria initially accelerates the increase of the bacterial biomass but does not give rise to the establishment of a significantly higher bacterial population. The effect of the pH is striking: the number of bacteria increases sharply with increasing pH values. The impact on the soil bacteria appears to be quite similar for the different gasoils, although it must be noted that the Kuwait induces important fluctuations.

As to the number of hydrocarbon-degrading microorganisms, an increase

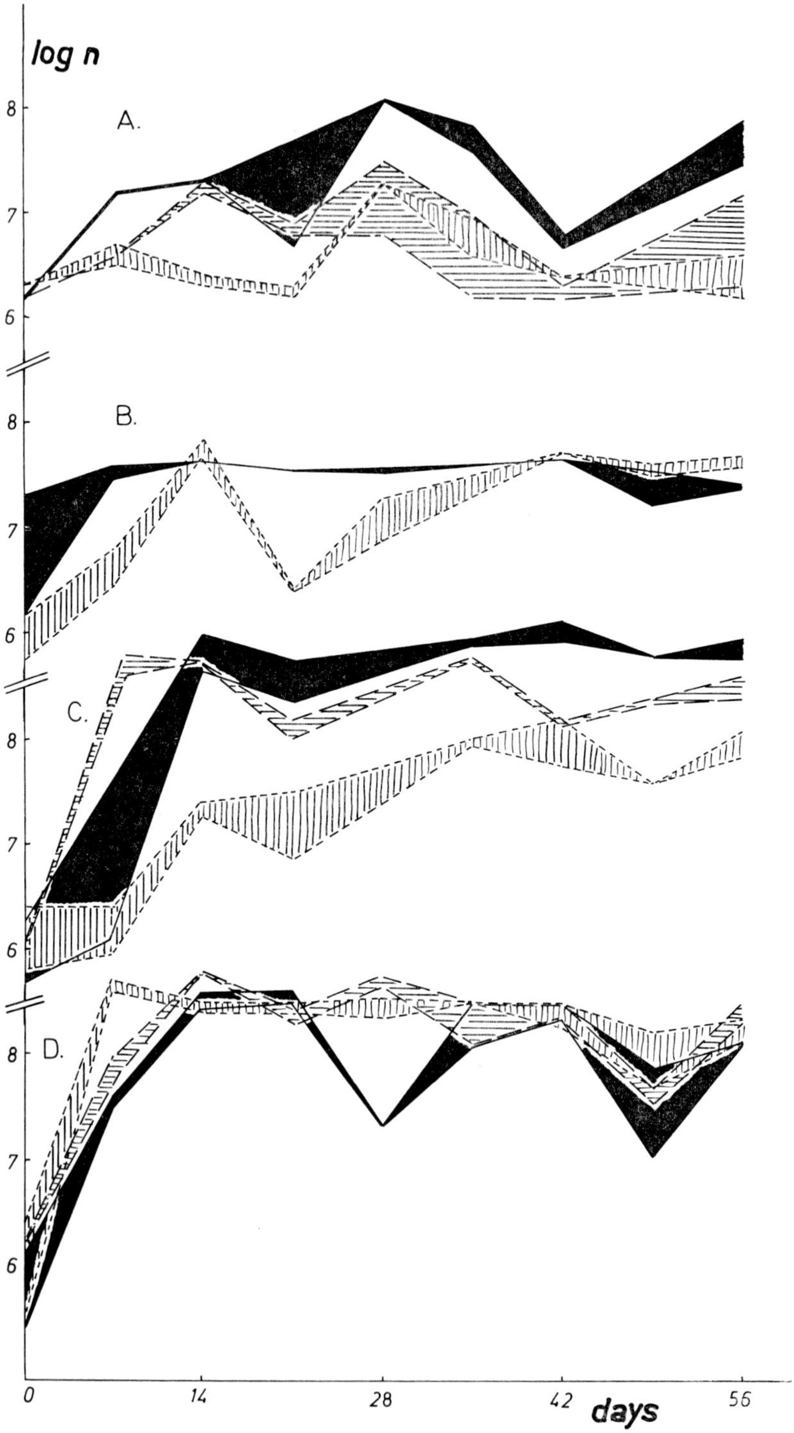

Fig. 4. Impact of gasoil on the number of viable bacteria in soil (for the symbols, see *Fig. 3).*

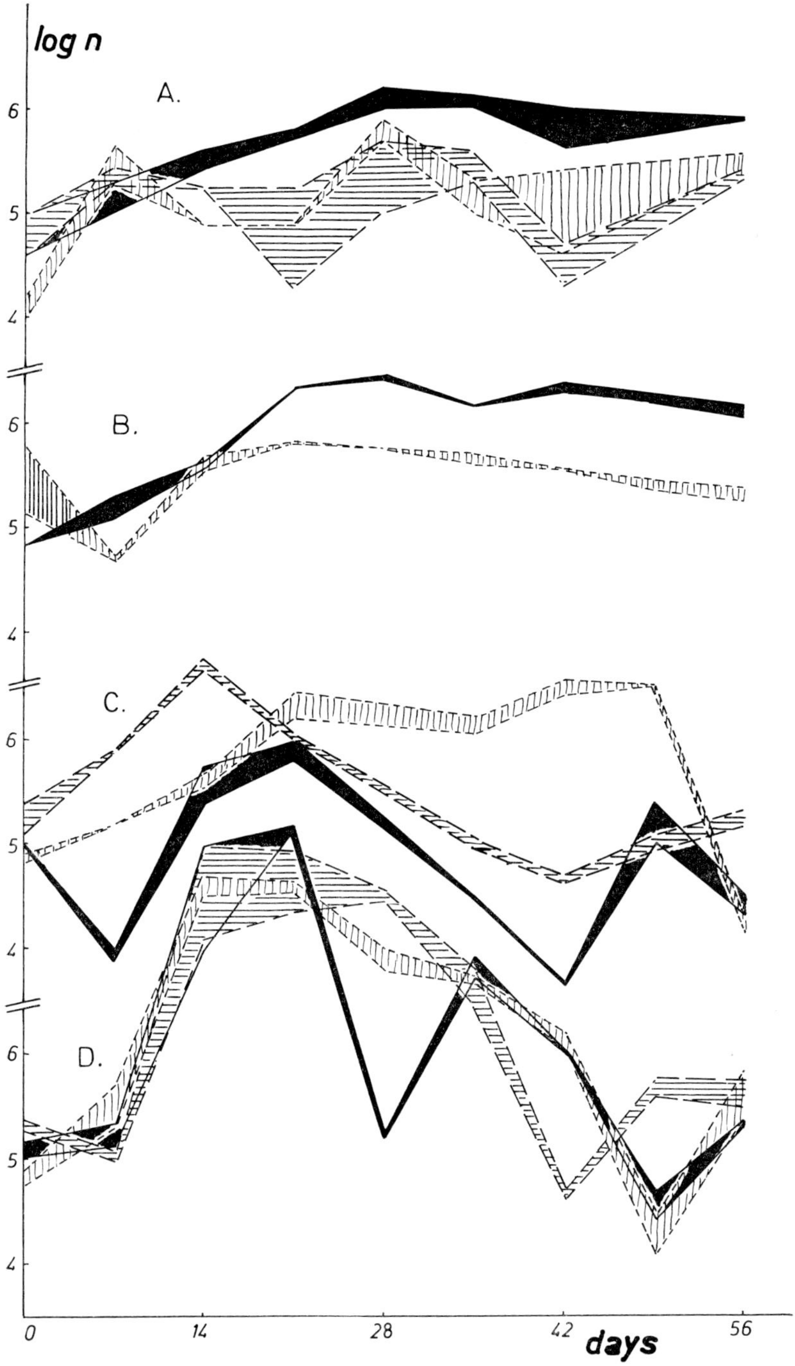

Fig. 5. Impact of gasoil on the number of fungal propagules in soil (*for the symbols*, see *Fig. 3*).

from 10^4 microorganisms to 10^5 per gramme wet soil is noticed upon the application of Bachaquero gasoil to the soil. Addition of N, P, or inoculum does not significantly increase this specific group of microorganisms. However, liming of the soil concomitant with the addition of gasoil results in a 100-fold increase of the MPN of hydrocarbon-degrading bacteria.

The influence of gasoil on the fungal population is, just as for the bacteria population, rather small (Fig. 5). Gasoil supplemented with fertilizers induces a pronounced increase in the number of fungi, however. The inoculation results in the permanent establishment of a higher fungal plate count. The effect of the pH is inverse to that noted for bacteria: the higher the pH value

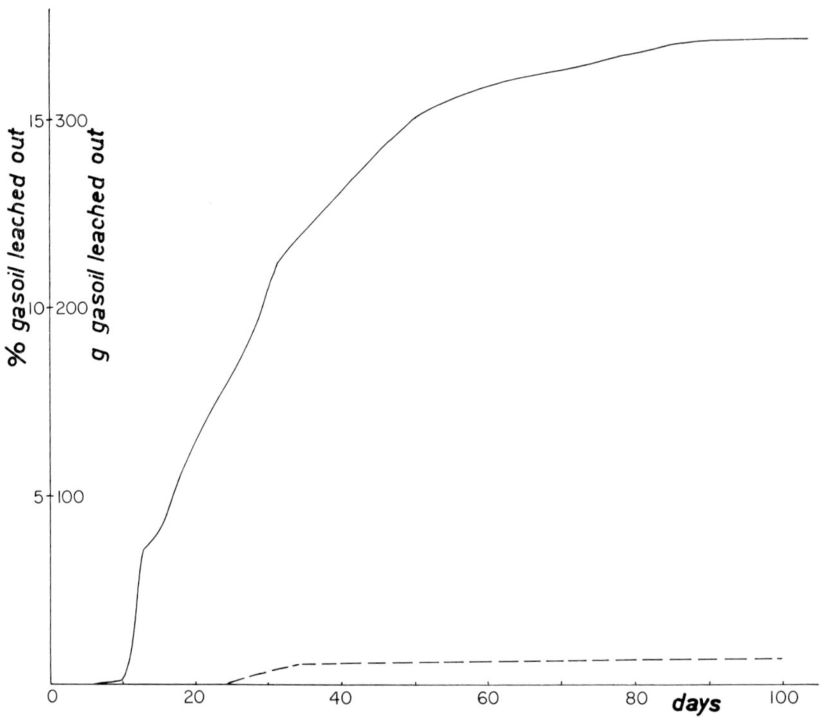

Fig. 6. Oil leaching out of soil columns. ---- irrigation with a N and P solution at a tension of 200 mm Hg; ——— irrigation with a N and P solution at a tension of 350 mm Hg.

the lower the number of fungi. With regard to the comparative impact of the different gasoils, one can see that the patterns of the time-courses of the numbers of fungi are quite similar for the different treatments. One should note, however, that in this experiment the climax of the fungal populations observed after 2 weeks is not stable and is followed by a rapid decline.

In the second series of experiments, the mobilization and breakdown of gasoil entrapped in soil columns were examined. To the columns polluted with oil, water was added at regular intervals. The rate of elution corresponds to about three times the local rate of precipitation. In view of the results obtained in the breakdown experiments described above, two irrigation procedures were tested. To one column distilled water was added, to a second column was added a solution containing 12·14 g NH_4NO_3 and 2·38 g K_2HPO_4 per litre. The results obtained are schematized in Fig. 6. In the soil leached with distilled water, little or no oil migrated with the aqueous phase out of the column. However, in the columns enriched with nitrogen and phosphorus, a remarkable amount of oil was mobilized and leached out together with the percolating water. The amount of oil removed increased with increasing soil tension. To verify if the effect was of a biological nature, control experiments were set up in which the soil columns were regularly fumigated with methylbromide (Fig. 7). In these experiments, no oil was

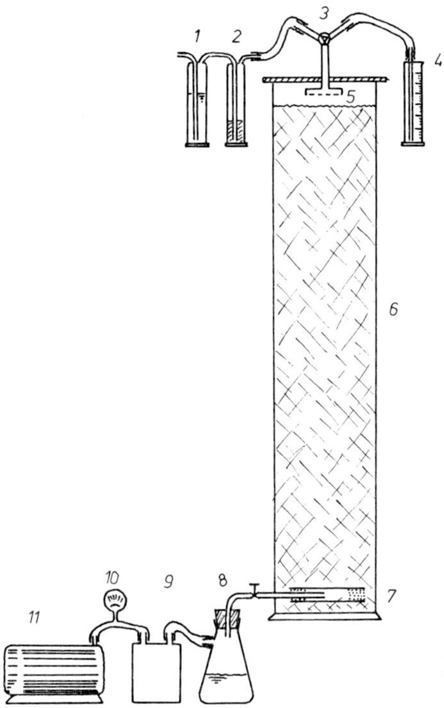

Fig. 7. Lysimeter apparatus for leaching experiments with sterilized soil. (1) air scrubbing with water, (2) air scrubbing with methylbromide, (3) threeway valve, (4) fertilizer solution, (5) sprinkler, (6) lysimeter column, (7) drainage system, (8) effluent collector, (9) buffering bottle, (10) manometer, and (11) vacuum pump.

leached out of the soil column, not even after repeated irrigations with solutions of fertilizers or after prolonged periods of suction at high tension. From this, it was concluded that the mobilization of the oil components from their physico-chemical soil entrapment must be due to the degradative activity of the soil microbiota.

DISCUSSION

The experiments simulating the breakdown of oil in the A-horizon indicate that hydrocarbons in general and aromatics in particular degrade only slowly in soil ecosystems, even if sufficient oxygen is provided. Optimization of environmental parameters such as the pH and the C/N/P ratio increases remarkably the rate of breakdown. In this way, a degradation of up to 90% of the amount of gasoil initially applied could be achieved in a period of 42 days. The applications of oil-adapted bacteria resulted only in a transitory acceleration of the metabolism of the soil biota. Similar studies on the acceleration of the biodegradation of oil by Atlas and Bartha (1973), Fusey and Oudot (1973), and Westlake *et al.* (1974) also indicate that the absence of adequate environmental conditions and not the absence of an adequate microbiota is responsible for the persistence of oil in the environment.

The breakdown of hydrocarbons in soil can be monitored by a variety of methods (Vanloocke *et al.*, 1975). The results obtained indicate that the soil respiration reflects the actual hydrocarbon breakdown phenomena quite well. Furthermore, the respiration can be measured conveniently and continuously by the Sapromat technique. As a consequence, the respiration data can be used to monitor and to model mathematically the actual biodegradation processes. It remains advisable to back up the respiration data by the results of the soil analyses with regard to the overall evolution of the gasoil components. Indeed, as shown by the addition of soil inoculum, activation of the microbiota does not necessarily result in an increase of the breakdown of the pollutant.

Antoniewski and Schaeffer (1972) proposed first order kinetics for the microbial breakdown of hydrocarbons in soil. It is evident from the respiration curves and from the mathematical computations that the phenomena can be better simulated by asymmetric sigmoidal kinetics. The soil biota apparently requires a certain period of time to adapt its enzymatic constitution to the metabolism of the pollutants. Once this adaptation has taken place, the biodegradation proceeds more or less linearly with time. The latter was also reported by Shelton and Hunter (1974) concerning the aerobic decomposition of oil pollutants in sediments.

The changes in microbial numbers upon the applications of the topped gasoils do not suggest a toxicity of these products toward the microbiota. Toxicity of the oil products appears to be due mainly to the volatile fraction (Voets *et al.*, 1974) which is, of course, removed by topping. The subsequent increases in the number of bacteria and fungi indicate that both groups of organisms take part in the biodegradation processes. Shifts in microbial populations have not been investigated specifically but a relative increase of the number of *Penicillium* species was noted in soils treated with gasoil and fertilizers.

The results of the lysimeter experiments indicate that the soil microbiota, while partly degrading the hydrocarbon molecules, simultaneously liberate these molecules from their physico-chemical entrapments. The bio-emulgation of oil components under anaerobic conditions in soil layers has been reported previously by La Rivière (1955). More recently, several studies have reported the fact that in oil-water emulsions, bacteria can enhance the dispersion of the oil into the aqueous phase by the formation of emulsifying agents (Reisfield *et al.*, 1972; Zajic *et al.*, 1974). Our findings indicate that the microbiota exerts a considerable effect on these emulsifying activities in soils, provided these soils are governed by favourable environmental conditions.

Preliminary investigations on the emulsification phenomenon suggest that fatty acids formed by the microbiota from the gasoil might constitute an important fraction of the emulsifying agents. Indeed, the amount of fatty acids present in the column effluent (x, mg) correlates significantly with the amount of gasoil leached out of the soil column, (y, g) ($y = 1{\cdot}02$; $x = 2{\cdot}49$; $r = 0{\cdot}94^{***}$). These experiments also suggest that in soil horizons containing large amounts of absorbed oil, the clean-up process can probably be accelerated considerably by irrigating the soil surface with water enriched with nitrogen and phosphorus. Indeed, such an irrigation procedure will enhance considerably both the biodegradation of the oil within the soil and the leaching of the oil components out of the soil matrix. The aquifer water will, of course, be very heavily polluted during the treatment procedure. It will have to be pumped up and treated appropriately. However, it is felt that, in cases of oil spills on land, such a treatment procedure might considerably reduce the period of pumping required to regain an adequate and a constant quality of the groundwater.

Acknowledgements

This work was supported by the Belgian National R & D Program on the Environment, Project 'Groundwater', Ministry of Science Policy Programming. The authors would like to express their appreciation to the Fina Oil Company for supplying the different gasoils. They also thank L. J. Mercy for his help with the mathematical simulations.

REFERENCES

Antoniewski, J., and Schaeffer, R. (1972). Recherches sur les réactions de coenoses microbiennes des sols imprégnés par des hydrocarbures. Modification de l'activité respiratoire. *Ann. Inst. Pasteur*, **123**, 805–819.

Atlas, R. M., and Bartha, R. (1973). Stimulated biodegradation of oil slicks using oleophilic fertilizer. *Environ. Sci. Technol.*, **7**, 538–541.

Cassel, D. K., Kreuger, T. H., Schroer, F. W., and Norum, E. B. (1974). Solute movement through disturbed and undisturbed soil cores. *Soil Sci. Soc. Am. Proc.*, **38**, 38–40.

Concawe secretariaat, Den Haag (1972). Methods for the analyses of oil in water and soil.

Fusey, P., and Oudot, J. (1973). Note sur l'accélération de la biodégradation d'un pétrole brut par les bactéries. *Mater. Organismen*, **8**, 157–164.

La Rivière, J. W. M. (1955). The production of surface active compounds by micro-organisms and its possible significance in oil recovery. *Antonie van Leeuwenhoek*, **21**, 9–27.

Martin, J. P. (1950). Use of acid rose bengal and streptomycin in the plate method for estimating soil fungi. *Soil Sci.*, **69**, 215–232.

Reisfield, A., Rosenberg, E., and Gutnick, D. (1972). Microbiol degradation of crude oil: factors affecting the dispersion in sea water by mixed and pure culture. *Appl. Microbiol.*, **24**, 363–368.

Shelton, T. B., and Hunter, J. V. (1974). Aerobic decomposition of oil pollutants in sediments. *J. Water Pollut. Control Fed.*, **46**, 2172–2182.

Somers, J. (1966). Pollution of soil and groundwater by mineral oil. *TNO-Nieuws*, **21**, 246–251.

Vanloocke, R., De Borger, R., Voets, J. P., and Verstraete, W. (1975). Soil and groundwater contamination by oil spills; Problems and remedies. *Int. J. Environ. Studies*, in press.

Verstraete, W., Vanloocke, R., and Mercy, J. (1975). Mathematical modeling of biodegradation processes in waters and soils. In *Modeling and Simulation of Water Resources Systems*, pp. 521–534 (G. C. Vansteenkiste, ed.), North-Holland Publishing Co., Amsterdam.

Voets, J. P., Vanloocke, R., and Verstraete, W. (1974). Measurement of the biodegradation of hydrocarbons in soil samples by combined measurement of the O_2-consumption and the CO_2-evolution. *Mémoires du Congrès Internationale des Hydrogéologues*, Montpellier Tome **X**, 159–166.

Westlake, D. W. S., Jobson, A., Philippe, R., and Cook, F. D. (1974). Biodegradability and crude oil composition. *Can. J. Microbiol.*, **20**, 915–928.

Working group water and petroleum (1970). Evaluation and treatment of oil spill accidents on land with a view to the protection of water resources, 2nd edn, Federal Ministry of the Interior, Federal Republic of Germany.

Zajic, J. E., Supplisson, B., and Volesky, B. (1974). Bacterial degradation and emulsification of no. 6 fuel oil. *Environ. Sci. Technol.*, **8**, 664–668.

RESEARCH ON THE FOULING BIOCENOSIS COMPOSITION IN RECYCLING WATER SYSTEMS OF PETROLEUM PLANTS AND MEASURES TO COMBAT IT

R. K. DZHALILOV, N. I. KARAYEVA and N. R. MAGERRAMOVA

Institute of Water Problems, Baku, Azerbaijan SSR, U.S.S.R.

The high growth rate of all branches of industry causes a continuing increase in water consumption along with problems of final disposal of waste waters into the environment. Therefore, the recycling systems of industrial enterprises recently have received more and more attention in the Soviet Union.

Recycling systems offer some advantages: (a) limitations of disposal to the ponds; (b) considerable decrease of fresh-water consumption; and (c) establishment of recycling water treatment, the extent depending on the technological process. It is noteworthy that refineries and petrochemical plants are among the largest water users.

The shortage of fresh water in some southern regions of the country along with the highly developed petroleum industry (in Azerbaijan SSR, Turkmen SSR, Kazakh SSR) results in utilization of seawater in recycling systems as well as in projections of systems employing effluents of rather high mineralization (5–10 mg/litre).

Nevertheless, it involves the intensification of scaling, corrosion, and fouling processes. For this reason the main object of our investigation was to study the species composition of biocenosis of heat-exchanging surfaces of the petroleum and gas-processing plants, and the conditions causing fouling growth. Another object of this study was working out effective preventive methods.

In the present investigations, experiments were conducted both in the laboratory and field (in two plants—petroleum and gas-processing). In the field experiments the Caspian Sea was used as a source of water supply. In the laboratory the mixture of municipal and processing waters went through combined treatment and a ratio of 1:1 was used. Physico-chemical analysis data of used waters are shown in Table I.

During stationary investigations, it was found that the basic mass of fouling biocenosis from the gas-processing plant consisted essentially of microphytes. The species of microfauna and their level were not numerous.

Among bacterial flora, the physiological groups of particular interest are those which take part in metal corrosion processes: nitrifying, sulphate-reducing, and iron-reducing bacteria. Fouling biomass amounts to 0·0624 g/cm^2. Investigations of the fouling biomass formed on open-type, heat-exchanger battery surfaces have established 233 species of algae.

The largest number of species are diatoms and blue-greens—143 and 74 species and varieties, respectively. Algae of these divisions comprise the basic mass of fouling. Other divisions are not as numerous qualitatively or quantitatively. They represent 11 species of green, 2 of red algae, and 3 dinoflagellates.

All algae can be placed in five groups:

Table I

PHYSICO-CHEMICAL ANALYSIS DATA OF STUDIED WATERS

No.	*Sampling*	*t°C*	*pH*	*$Na^+ + K^+$ (mg/litre)*	*Ca^{+2} (mg/litre)*	*Mg^{+2} (mg/litre)*	*Total hardness (me/litre)*	*Cl′ (mg/litre)*
1.	Gas-processing plant	21–44	8·22–8·55	3180–3747	367–480	846–963	88–99	6168–6876
2.	Petroleum plant	40–55	7·67–8·30	2930–5931	240–320	617–697	62–73	4434–9364
3.	Laboratory plant	40	8·15–9·60	1534–3397	189–302	420–703	44–73	2950–4800

No.	*Sampling*	*HCO'_3 (mg/litre)*	*CO''_3 (mg/litre)*	*SO''_4 (mg/litre)*	*O_2 (mg/litre)*	*Oil products*
1.	Gas-processing plant	147–208	9–38	2018–4520	5·8–8·5	0·5–1·7
2.	Petroleum plant	208–353	0–7	2017–3362	4·5–8·6	25–33
3.	Laboratory plant	32–228	6–70	1700–2625	7–15	not examined

I. Dominant, represented by *Nitszchia laevis* Kar., *Lyngbya limnetica* Lemm f. *limnetica, Lyngbya martensiana* Menegh. f. *martensiana, Phormidium valderiae* (Delp.) Geitl. f. *valderiae*.

II. Subdominant, *Spirulina subtilissima* Kütz., *Navicula subagnita* Pr-Lavr., *Synedra tenera* W. Sm., *Amphipleura rutilans* (Trentep) Cl.

III. Algae, developing in small quantities in all seasons. Regular are only diatoms.

IV. Algae, developing sporadically. They consist of 14 species.

V. Group of occasional species dropped into systems from the water-supply source mechanically and evidently not playing a role in biocenosis. Represented by the remainder of 208 species.

Seasonal exchanging of species took place in the two first groups, thus some of them changed their role in biocenosis.

By species composition of Bacillariophyta (diatoms), one can consider the Caspian Sea the only source of fouling since all species discovered during our investigations are Caspian inhabitants.

Seasonal changes of systematic composition observed in algae growth have been shown in some spring and summer reduction of species numbers in diatoms and their increase in autumn and winter, whereas blue-green algae species have displayed the opposite tendency. This phenomenon appears to be evaluated as the antagonism in growth of these two groups. Seasonal changes have occurred also in composition and volume changes of dominant and subdominant complexes.

Microfauna in the investigated fouling was represented by 38 species of infusoria. In all seasons the growth of *Coleps* sp., *Cyclidium citrullus* Cohn, *Euplotes* sp., *Diophrys scutum* Dujardjm was registered. Other species of infusoria were seen sporadically. Their weak growth can apparently be due to elevated temperature. In addition to infusoria, cyclopes, turbellaria, nematoda, and rotatoria were detected in the foulings.

Nitrifying, sulphate- and iron-reducing bacteria were identified when sown on the elective media. Sporadical algae representatives, blue-green and diatoms, were detected at the petroleum-processing plant.

When sown on the elective media, sulphate- and iron-reducing bacteria were found. Weak algae growth and absence of zoo-representatives at the petroleum-processing plant can apparently be explained by the limiting action of temperature and increased content of dissolved oil products in the recycled water.

During laboratory researches at set-ups simulating working conditions, 44 species and varieties related to four divisions were detected at the indicator plates. There were 23 blue-greens, 15 diatoms, 5 green, and 1 Euglenophyte.

The dominant group is herein represented by blue-green algae: *Phormidium valderiae* (Delp.) Geitl f. *valderiae*, *Synechocystis salina* Wisl.; green algae: *Oocystis submarina* var. *variabilis* Skuia.

The group of subdominants consists of blue-green algae: *Lyngbya limnetica* Lemm, f. *limnetica*, *Lyngbya martensiana* Menegh. f. *martensiana*, and diatoms: *Amphora coffeaiformis* (Ag) Cl. var. *coffeaiformis*. The group of rarely seen species consists of seven: 4 blue-green, 2 diatoms, and 1 green. The group of sporadical species is represented by 21 species of algae.

Infusoria of the fouling were kept under observation in laboratory conditions only in autumn. They were represented by only six species. On sowing in the elective media, iron- and sulphate-reducing bacteria were identified also.

It is common knowledge that in recycling systems the processes of scaling, fouling, and corrosion are affected mainly by water conditions, its mineralization and temperature changes. In systems using biochemically treated effluents, it is affected by the presence in water of residual activated sludge.

Table I shows that fouling growth in the investigated systems caused by microorganisms getting into them is affected by the temperature, increased water mineralization, nutrient content, and dissolved oxygen.

Surface fouling of the cooling and condensing systems can have a disastrous effect on the efficiency of heat transfer and system capacity. It involves biocorrosion of structures and secondary contamination of the recycling water with dead organisms. The problem of fouling prevention in the recycling systems is poorly reported in the literature.

Toxins are supposed to be used for this purpose. However, as the fouling embodies the totality of different microorganisms with their individual properties of perception and protection against unfavourable conditions, the availability of a universal toxic remedy is questionable. Again, toxin usage is limited by strict requirements to the blowing water disposal as affecting the steady surface water pollution.

Widely practised chlorination methods as well as combined chlorine and copper vitriol treatments are rather complex and expensive (besides a chlorine shortage in some regions). Therefore chemical treatment is not suitable in certain cases.

The authors have carried out a two-month laboratory experiment for establishing the effect of the residual active chlorine on the biological fouling growth intensity. The investigations were carried out at two laboratory plants both with identical conditions. Plant No. 1 was used as a control, and plant No. 2 was treated by the biocide which was derived from seawater subjected to electrolysis in the electrolytic unit with magnetite medium. The system was

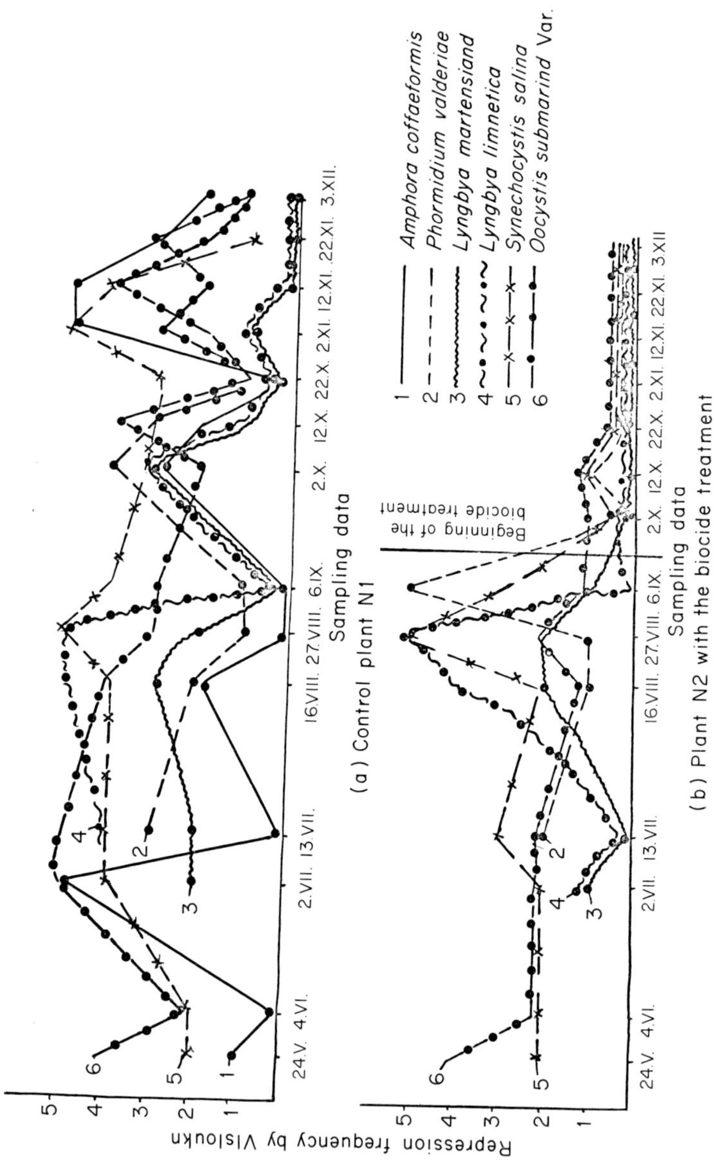

Fig. 1. Growth dynamics of dominant algae species.

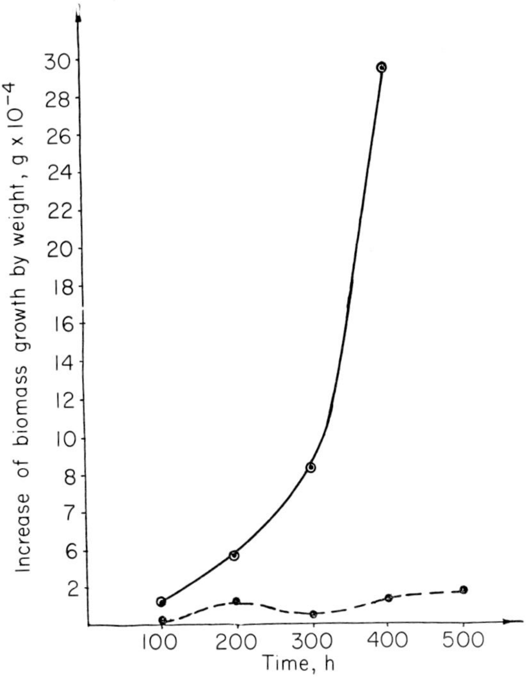

Fig. 2. *Dynamics of biomass increase. —— control plant; – – – electrolysis product treatment (3 mg/litre residual active chlorine concentration).*

fed with a dose of biocide to maintain the residual active chlorine concentration of 2·5–3 mg/litre.

The absence of biological fouling on the indicators observed under microscopic processing was considered as a biocide performance criterion.

During the investigations, samples were analysed qualitatively and quantitatively—14 and 10, respectively. The experimental results are shown in Figs. 1 and 2. The comparison of analytic data from the two laboratory plants—control (No. 1) and with the biocide treatment (No. 2)—has demonstrated the effectiveness of the biocide.

The treatment in the laboratory plant with the biocide began on 20 September. On 2 October, a drastic decrease was observed in algae growth and by 22 October it had stopped completely. The residual active chlorine concentration of 2·5–3 mg/litre was found to be quite effective.

A cost estimation has also been calculated for the proposed method.

BIODEGRADATION OF PETROLEUM HYDROCARBONS BY MARINE MICROBES

RICHARD F. LEE and CHARLOTTE RYAN

Skidaway Institute of Oceanography, P.O. Box 13687,
Savannah, Georgia 31406, U.S.A.

Summary

Different paraffinic and aromatic ^{14}C-labelled hydrocarbons were added to estuarine and offshore water samples followed by the measurement of $^{14}CO_2$ produced after incubation periods of 6–96 hr. The degradation rates ranged from 4·4 μg/litre/day (turnover time of 30 days) for naphthalene at 130 ppb to 0·002 μg/litre/day (turnover time of 3500 days) for benzpyrene at 5 ppb. Benzene, toluene, naphthalene, methylnaphthalene, heptadecane, hexadecane, and octadecane had high degradation rates relative to fluorene and benzpyrene. These rates were affected by season, tide, concentration of hydrocarbon, and incubation time. Benzene and toluene, but not the other hydrocarbons tested, obeyed the kinetic model of degradation, *i.e.* a linear increase in turnover time with an increase in hydrocarbon concentration.

The results support the conclusion that microbes are important in degrading many petroleum hydrocarbons in ocean waters. However, based on the low degradation rates of fluorene and benzpyrene, we hypothesize that microbial breakdown of high molecular weight aromatics is not an important process in ocean water.

INTRODUCTION

It is commonly assumed that microbes are the major degraders of hydrocarbons in the sea since hydrocarbon-degrading bacteria and fungi have been isolated from both polluted and pristine waters (Finnerty *et al.*, 1973; Floodgate, 1972; Gunkel, 1973; Mulkins-Phillips and Stewart, 1974; Robertson *et al.*, 1973; Walker and Colwell, 1974; Zobell, 1971). The hydrocarbon-degrading microbes are generally more abundant in areas of chronic oil pollution than in petroleum-free areas (Atlas and Bartha, 1973). The microbial degradation of soluble organic matter, primarily glucose and amino acids, in marine and fresh waters has been investigated using ^{14}C-labelled compounds (Parsons and Strickland, 1962; Vaccaro and Jannash, 1966; Williams and Askew, 1966; Wright and Hobbie, 1966). This work has provided considerable insight into heterotrophic processes occurring in natural waters. The methods of the above workers have been used in our studies to determine the degradation rate of organic compounds with low water solubility. Concentrations from 3 to 100 ppb of various radiolabelled paraffinic and aromatic hydrocarbons were added to water samples collected from an estuarine and offshore station and their degradation rates studied. The feasibility of the approach used has been shown by Robertson *et al.*

(1973) where ^{14}C-dodecane was added to water from Alaska's Cook Inlet and Port Valdez. Based on the rate of $^{14}CO_2$ released, they calculated a degradation rate of 1 μg/litre/day.

MATERIALS AND METHODS

Surface samples were collected by plastic bucket from a station in the Skidaway River and a location 3 miles off the Georgia coast. The Skidaway River at our collection point was 3 m deep. Water temperatures averaged 13°C in winter and 24°C in the summer, while salinity averaged 20‰ (range of 17–24‰) depending on rainfall and tidal influx. The offshore station was 10 m deep with a salinity of 35‰. Experiments from Skidaway River were started within 10 min of collection while 1 hr was needed before experiments with offshore water could be started.

Bacterial degradation of hydrocarbons was measured by adding ^{14}C-labelled hydrocarbons to 100 ml water samples in a 250 ml flask capped with a silicone stopper. After incubation at *in situ* temperatures, the respired $^{14}CO_2$ was collected and counted with a liquid scintillation counter (Packard Model 3320). The procedure for collecting released $^{14}CO_2$ is described by Hodson *et al.* (unpubl. data). Controls were water samples containing either 2 ml of 2 N H_2SO_4 or mercuric chloride at 1 ppm. All samples were run in triplicate for each concentration and time interval. The turnover times for various hydrocarbons at several different concentrations were calculated by the formula

$$\text{turnover time} = t/f$$

where t is the incubation time and f is the fraction of the labelled hydrocarbon degraded to $^{14}CO_2$ during the incubation period.

The natural concentration of nonvolatile hydrocarbons in the Skidaway River was determined by infrared procedures similar to those described by Brown *et al.* (1973) and Vaughan (1973). Analysis was at 2930 cm^{-1}, using a Perkin-Elmer infrared spectrophotometer (model 467). We also used gas–liquid chromatography to determine these concentrations by extracting 6 litres of water with 300 ml of hexane. The extracts were concentrated to a small volume on a rotary evaporator and applied to activated silicic acid thin-layer plates (Merck). One band corresponding to aliphatics and another containing aromatics were scrapped and eluted with benzene. The fractions were run on a Packard gas chromatograph with an 8 ft column of 10% SP-2100 on Suplecoport (Supelco Co.) and temperature programmed from 70°C to 220°C. Hydrocarbon standards were used for reference identifications. Internal standards were used for quantification.

The radioactive hydrocarbons or hydrocarbon metabolites used were 2-methylnaphthalene-8-^{14}C (7·98 mCi/mmol—California Bionuclear Corp.), 3,4-(benz, 3, 6-^{14}C) pyrene (21 mCi/mmol—Amersham/Searle), ^{14}C-1-naphthalene (5·1 mCi/mmol—Amersham/Searle), ^{14}C-9-fluorene (2·57 mCi/mmol—California Bionuclear Corp.), ^{14}C-(methyl)-benzene (17·5 mCi/mmol—Amersham/Searle), ^{14}C-1-hexadecane (54·4 mCi/mmol—Amersham/Searle), ^{14}C-1-heptadecane (13·5 mCi/mmol—ICN), ^{14}C-1-octadecane (25 mCi/mmol—Amersham/Searle), ^{14}C(U)-phenol (34 mCi/mmol—Amersham/Searle), ^{14}C-1-palmitic acid (57·9 mCi/mmol—

Amersham/Searle), and 1-naphthol-1-^{14}C (20·8 mCi/mmol—Amersham-Searle). All compounds, except benzene and toluene, were purified on silicic acid thin-layer chromatograms before use.

RESULTS

Degradation rates and turnover times (Table I) suggest the presence of an

Table I
MICROBIAL HYDROCARBON DEGRADATION IN WATER SAMPLES FROM COASTAL GEORGIA

Compound and concentration	*Date and locality*[a]	*Incubation time (hr)*	*Degradation rate*[b] *(μg/litre/day)* × 10^2	*Turnover time (days)*
Benzpyrene (5 ppb)	Feb—Sk	96	0	0
Benzpyrene (5 ppb)	May—Sk	96	0·2 ± 0·1	3 500
Benzpyrene (5 ppb)	May—O	96	0	0
Fluorene (30 ppb)	Feb—Sk	72	0	0
Fluorene (30 ppb)	Feb—O	72	0	0
Fluorene (30 ppb)	May—Sk	72	0	0
Fluorene (30 ppb)	June—Sk	48	3 ± 2	1 000
Heptadecane (8 ppb)	May—Sk	24	9 ± 3	80
Heptadecane (15 ppb)	May—Sk	24	18 ± 5	60
Heptadecane (30 ppb)	May—Sk	24	24 ± 4	54
Heptadecane (30 ppb)	May—Sk	12	18 ± 6	170
Hexadecane (25 ppb)	Feb—Sk	48	5 ± 2	500
Hexadecane (25 ppb)	April—Sk	24	12 ± 3	210
Hexadecane (25 ppb)	April—O	48	3 ± 1	1 000
Naphthalene (40 ppb)	Feb—Sk	48	5 ± 1	500
Naphthalene (40 ppb)	May—Sk	24	90 ± 6	46
Naphthalene (40 ppb)	May—Sk	8	56 ± 5	79
Naphthalene (130 ppb)	May—Sk	24	440 ± 12	30
Naphthalene (130 ppb)	May—O	24	4 ± 1	330
Methylnaphthalene (40 ppb)	Feb—Sk	24	2 ± 1	500
Methylnaphthalene (40 ppb)	May—Sk	24	80 ± 5	50
Octadecane (16 ppb)	May—Sk	16	15 ± 6	100
Toluene (20 ppb)	May—Sk	12	12 ± 2	17
Toluene (20 ppb)	May—Sk	18	12 ± 2	17
Toluene (20 ppb)	May—O	24	5 ± 1	40

[a] Sk—Skidaway River, O—offshore water.
[b] ± Standard error.
^{14}C-hydrocarbons were added to surface water samples and the $^{14}CO_2$ produced was collected at the end of the incubation period. Further details of the experiment are given in the Methods section.

active population of heterotrophic microorganisms in our estuarine waters which were capable of relatively rapid degradation of paraffinic and some aromatic hydrocarbons. Naphthalene had the highest degradation rate of the nonvolatile hydrocarbons tested (up to 4 μg/litre/day), followed in decreasing order by methylnaphthalene, heptadecane, hexadecane, octadecane, fluorene, and benzpyrene. The degradation rates of fluorene and benzpyrene, both polynuclear aromatic hydrocarbons, were extremely low (less than 3×10^{-2}

μg/litre/day). During the 5 month study, the degradation rate of all hydrocarbons was consistently higher in the estuarine area than in water from offshore. Incubation times of at least 24 and often 48 hr were needed to obtain sufficient $^{14}CO_2$ for counting from offshore waters.

Calculated turnover times (time required to convert all hydrocarbon to CO_2) in estuarine waters varied from 46 days for naphthalene at 40 ppb to 1000 days for fluorene at 30 ppb. Benzene and toluene showed a linear increase in turnover time with increases in concentration (Fig. 1). Extrapolation of the line to zero shows 50 days as the turnover time for the 'natural' benzene concentration. This line fits the kinetic model for organic compounds, as described by Wright and Hobbie (1965). However, for other hydrocarbons such as hexadecane or naphthalene, there was little, if any, increase in turnover time with increasing concentration.

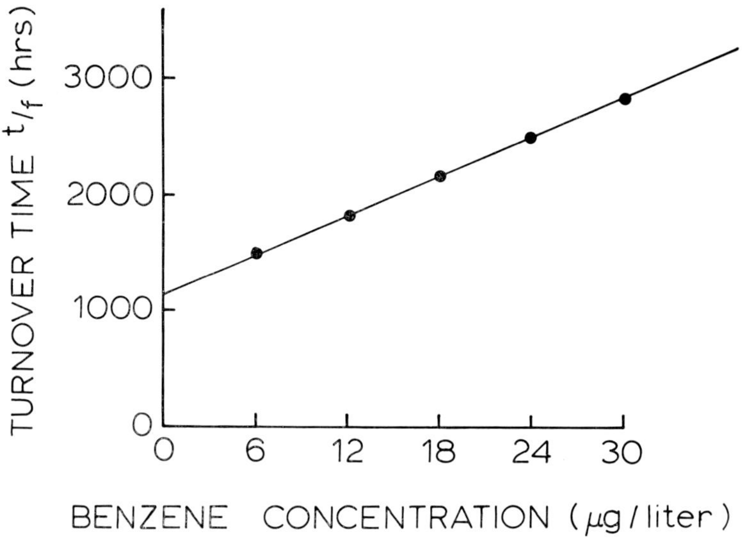

Fig. 1. Turnover times for different concentrations of benzene. ^{14}C-Benzene was added to water samples from Skidaway River. Turnover times were calculated from the amount of $^{14}CO_2$ produced after a 16 hr incubation period.

Total hydrocarbon concentrations in the Skidaway River varied from 10 to 60 ppb. Naphthalene, 1-methylnaphthalene, and 2-methylnaphthalene were not detected (less than 0·5 ppb). Hexadecane varied from 0·5 to 3 ppb. Benzene and toluene varied from nondetectable to 3 ppb (L. Atkinson, unpubl. data). Thus in most of our experiments the added hydrocarbon was at a much higher concentration than normal in the water. In general, there was a large variation in the hydrocarbon degradation rate for different water samples collected at the same time resulting in high standard errors. The exceptions were toluene and benzene which have relatively high water solubilities (benzene, 1780 ppm; toluene, 515 ppm; McAuliffe, 1966). The higher molecular weight hydrocarbons with low water solubility absorb to particles and are presumably degraded there by microbes. The particles vary in size and number between different water samples. When water was passed through a 1 μ filter, allowing passage of free bacteria, there was a low standard error for the hydrocarbon degradation rates of the filtrate. However, the rates measured were only about one-half that of unfiltered water, presumably because particles with attached hydrocarbon-degrading microbes were removed.

Additional variables which affected the degradation rates were season, incubation time, and time of sampling. Degradation rates were very low in the winter (January and February) and high in the spring (April and May). Rates during the remainder of the year are presently under study. Benzene and toluene were degraded at a linear rate for up to 48 hr at all concentrations studied (3 to 30 ppb). For the other hydrocarbons the degradation rate had two phases. First, there was a low rate for 12 hr, followed by a more rapid degradation rate during the next 48 hr. Incubation times of over 12 hr were often necessary to obtain sufficient $^{14}CO_2$ for counting. The effect of time of sampling is still under investigation but preliminary evidence suggests that higher rates occur when water is collected from high tide, possibly because of changes in the amount of suspended sediment.

The results discussed above are concerned with specific hydrocarbons. However, petroleum is composed of thousands of compounds and interactions between these compounds, including competition or inhibition, might be expected during microbial degradation of petroleum. To study this effect, water extracts of fuel oil were added to water samples containing ^{14}C-hexadecane or ^{14}C-naphthalene. The fuel oil hydrocarbon concentration in the water was approximately 50 ppb which was composed of 0·5 ppb hexadecane and 5 ppb naphthalene. The addition of fuel oil water extracts did not decrease the degradation rate of either ^{14}C-hexadecane or ^{14}C-naphthalene. Instead, the degradation rate of ^{14}C-naphthalene showed a slight increase with fuel oil addition.

DISCUSSION

Our results support the conclusion that microbes are important in degrading many petroleum hydrocarbons in ocean waters. The degradation rates of fluorene and benzpyrene, however, were extremely low. We therefore hypothesize that microbial breakdown of high molecular weight aromatics is not an important process in ocean waters. Microbes in marine sediments may play a role in breaking down these hydrocarbons, and larger organisms, such as zooplankton, may be important in their degradation in the water column (Lee, 1975).

Highest degradation rates were observed in estuarine water samples collected in the spring. Much lower rates were seen in the winter in both estuary and offshore samples. Such factors as temperature and nutrients are probably important in explaining these differences (Floodgate, 1972; Gibbs *et al.*, 1975; Gunkel, 1967; Mulkins-Phillips and Stewart, 1974). The estuarine area is characterized by much higher bacteria cell numbers than are offshore areas, presumably because of the higher levels of various soluble organic compounds.

We assume that the hydrocarbon degradation rates presented (Table I) are indicative of 'natural' rates in the waters used for the study. However, several problems require consideration when the method is used for hydrophobic compounds such as hydrocarbons. These include: (1) large water samples needed to observe degradation; (2) adsorption of the compound to the sides of the glass; (3) long incubation times needed to observe appreciable degradation; (4) the small percentage of the microorganisms in the water sample which can degrade the compound in question. Increasing the amount of water used for degradation studies did not affect the degradation rate. Approximately 10% of the radioactive hydrocarbon adsorbed to the sides of the glass. Long incubation times probably introduce some error since rates were very low for the first 8 to 12 hr, with a linear increase after that period. Other studies have shown that adsorption of hydrocarbons to detrital particles increases to a maximum value in 6 hr (Pearl and Lee, unpubl. data). Thus, if degradation takes place on the particles, the degradation rates would be slow until the particles adsorb the hydrocarbons. A second factor could be that hydrocarbon-degrading microbes need some period of time to produce hydrocarbon-degrading enzymes. Problem 4 (above) is perhaps the most difficult to evaluate since some selection may occur for organisms able to degrade hydrocarbons during the incubation time. An addition of water extracts of fuel oil #2 to a large controlled ecosystem enclosure (2 m diameter and 15 m deep—60 000 litres) produced a large increase in the hydrocarbon degradation rates. Three days after petroleum addition, the naphthalene degradation rate increased from 0·2 to 2·5 μg/litre/day (Lee and Takahashi, 1975).

Acknowledgement

This work was supported by NSF/IDOE grant GX-42582.

REFERENCES

Atlas, R. M., and Bartha, R. (1973). Abundance, distribution and oil biodegradation potential of microorganisms in Raritan Bay. *Environ. Pollut.*, **4**, 291–300.

Brown, R. A., Searle, T. D., Elliott, J. J., Phillips, B. G., Brandon, D. E., and Monaghan, P. H. (1973). Distribution of heavy hydrocarbons in some Atlantic Ocean waters. Proc. 1973 Jt. Conf. on Prevention and Control of Oil Spills, American Petroleum Institute, Washington, D.C. pp. 505–519.

Finnerty, W. R., Kennedy, R. S., Lockwood, P., Spurlock, B. O., and Young, R. A. (1973). Microbes and petroleum: perspectives and implications. In *The Microbial Degradation of Oil Pollutants*, pp. 105–125 (D. G. Ahearn and S. P. Meyers, eds.), Louisiana State University, Baton Rouge, Louisiana. Publ. No. LSU-DG-73-01.

Floodgate, G. D. (1972). Biodegradation of hydrocarbons in the sea, pp. 153–171. In *Water Pollution Microbiology* (R. Mitchell, ed.), John Wiley, New York.

Gibbs, C. F., Pugh, K. B., and Andrews, A. R. (1975). Quantitative studies on marine biodegradation of oil II. Effect of temperature. *Proc. Roy. Soc. London*, Series B, **188**, 83–94.

Gunkel, W. (1967). Experimental-okologische untersuchungen uber die limitierenden factoren des mikrobiellen olabbaues. *Helgolander Wiss. Meeresunters*, **15**, 210–225.

Gunkel, W. (1973). Distribution and abundance of oil-oxidizing bacteria in the North Sea. In *The Microbial Degradation of Oil Pollutants*, pp. 127–139 (D. G. Ahearn and S. P. Meyers, eds.), Louisiana State University, Baton Rouge, Louisiana. Publ. No. LSU-DG-73-01.

Lee, R. F. (1975). Fate of petroleum hydrocarbons in marine zooplankton. pp. 549–553. Proc. 1975 Conf. on Prevention and Control of Oil Pollution, American Petroleum Institute, Washington, D.C.

Lee, R. F., and Takahashi, M. (1975). The fate and effect of petroleum in controlled ecosystem enclosures. *Workshop: Petroleum Hydrocarbons in the Marine Environment*, International Council for the Exploration of the Sea. Aberdeen, Scotland, in press.

Mulkins-Phillips, G. J., and Stewart, J. E. (1974). Effect of environmental parameters on bacterial degradation of bunker C oil, crude oils and hydrocarbons. *Appl. Microbiol.*, **28**, 915–922.

McAuliffe, C. (1966). Solubility in water of paraffin, cycloparaffin, olefin, acetylene, cycloolefin, and aromatic hydrocarbons. *J. Phys. Chem.*, **70**, 1267–1275.

Parsons, T. R., and Strickland, J. D. H. (1962). On the production of particulate organic carbon by heterotrophic processes in seawater. *Deep-Sea Res.*, **8**, 211–222.

Robertson, B., Arhelger, S., Kinney, P. J., and Button, D. K. (1973). Hydrocarbon biodegradation in Alaskan waters. In *The Microbial Degradation of Oil Pollutants*, pp. 171–184 (D. G. Ahearn and S. P. Meyers, eds.), Louisiana State University, Baton Rouge, Louisiana. Publ. No. LSU-SG-73-01.

Traxler, R. W. (1973). Bacterial degradation of petroleum materials in low temperature marine environments. In *The Microbial Degradation of Oil Pollutants*, pp. 163–170 (D. G. Ahearn and S. P. Meyers, eds.), Louisiana State University, Baton Rouge, Louisiana. Publ. No. LSU-DG-73-01.

Vaccaro, R. F., and Jannasch, H. W. (1966). Studies on heterotrophic activity in seawater based on glucose assimilation. *Limnol. Oceanogr.*, **11**, 596–607.

Vaughan, B. E. (1973). Effect of oil and chemically dispersed oil on selected biota—a laboratory study. Battelle Pacific Northwest Laboratories, Richland, Wash. American Petroleum Institute Publication No. 4191, 67 pp.

Walker, J. D., and Colwell, R. R. (1974). Microbial petroleum degradation: use of mixed hydrocarbon substrate. *Appl. Microbiol.*, **27**, 1053–1060.

Williams, P. J. LeB, and Askew, C. (1966). A method of measuring the mineralization by microorganisms of organic compounds in seawater. *Deep-Sea Res.*, **15**, 365–375.

Wright, R. T., and Hobbie, J. E. (1965). The uptake of organic solutes in lake water. *Limnol. Oceanogr.*, **10**, 22–28.

Wright, R. T., and Hobbie, J. E. (1966). Use of glucose and acetate by bacteria and algae in aquatic ecosystems. *Ecology*, **47**, 447–464.

Zobell, C. E. (1971). Microbial modification of crude oil in the sea, pp. 317–326. Proc. 1971 Jt. Conf. Prevention and Control of Oil Spills, American Petroleum Institute, Washington, D.C.

METHODS AND INTERPRETATION IN MEASUREMENT OF OIL BIODEGRADATION RATE

C. F. GIBBS

Marine Science Laboratories, University College of North Wales,
Menai Bridge, Anglesey, U.K.

Summary

The interpretation and relevance of various measurements of oil biodegradation rate are discussed, *e.g.* oxygen uptake, recovery of oil, etc. The concepts of degradative capacity of a given body of water c.z. the degradation rate of a small body of oil in unlimited water are considered. Two new respirometric methods are described in which quantitative relationships between oxidation rate and the supply of nutrient nitrogen were obtained at several natural temperatures. Phosphate supply was less important. Unnaturally high nutrient concentrations were avoided by providing a continuous slow supply. Recovery of oil showed a greater destruction than oxygen uptake suggested. The point at which biodegradation ceased was noted, and changes in various properties of the crude oil during biodegradation were measured. Specific gravity increased to above that of seawater. Different crude oils are also contrasted. Kuwait residues exposed on a raft at sea formed a 'chocolate mousse' with subsequent slow changes in physical and chemical properties, apparently without net loss of material. This is contrasted with more rapid changes of the same oil in other experimental situations.

PROBLEMS OF INTERPRETATION AND DIMENSIONS

The many difficulties involved in oil biodegradation measurements have been reviewed before (Floodgate, 1973). One of these, in particular, affects the quantitative interpretation of virtually all experiments, that is the two-phase nature of an oil-seawater system. As well as causing practical problems in obtaining results (*see* later), this prevents the application of normal kinetic considerations for homogeneous solutions, or even those applicable to solids of known and constant surface area. The dimensions of any numerical results obtained are uncertain unless the rate-controlling factor in the experiment concerned is known. For example, is the measured rate of oil destruction in a particular experiment expressed as a rate per quantity of oil present (as in a first order chemical reaction) or per surface area of oil, or per volume of water present, or in relation to the concentration or supply of mineral nutrients, the number of bacteria present, or a combination of these factors? This consideration is critical in drawing conclusions relevant to the natural environment.

In the environmental context, the knowledge required of biodegradation rate studies can be summed up as follows:

(A) How much oil can a given volume of sea cope with per year?

(B) How long will any individual spill last before destruction?
(C) What proportion of any given oil will degrade?
(D) What happens to the remainder?

After reviewing the experimental approaches available, the following sections describe some attempts in this laboratory to measure oil biodegradation rates with some understanding of the rate-limiting factors, under conditions as near natural as possible while still allowing measurements.

EXPERIMENTAL APPROACHES TO BIODEGRADATION RATE MEASUREMENT

There appears to be an environmental equivalent of Heisenburg's uncertainty principle, whereby the constraints required to allow measurements in an experimental model are incompatible with maintaining natural conditions. Oil biodegradation poses particular practical problems, for example:

(a) Since oil is a complex mixture, a short-term measurement may apply only to a minor but particularly labile component, and not represent the process as a whole. Long-term measurements are therefore necessary, preferably continuing until biodegradation ceases. Problems of accumulation of toxic products and depletion of nutrients can then arise.

(b) The two-phase nature of an oil/water system renders 'aliquot' sampling very difficult.

(c) The rate of degradation obtained depends on the characteristic measured experimentally. Table I summarizes a number of experimental approaches, with their direct environmental relevance and limitations.

Disappearance of the surface oil film may be significant, but provides no knowledge of the fate of the oil, such as emulsification as opposed to complete destruction to carbon dioxide and water (mineralization).

Measurement of oil loss by solvent extraction of oleophilic material is attractive at first sight. However, either the problem of representative sampling must be overcome or the whole container must be extracted, thus severely limiting the number of measurements possible per experiment. Conversion of oil to water-soluble products is included in the measurement of oil loss.

Loss of total organic carbon measures complete mineralization and does not respond to chemical modification to soluble products or biomass production.

Carbon dioxide production similarly measures complete mineralization, with the advantage over the above that continuous measurements are possible, without the problem of sampling or destroying the whole experimental unit. This has been used by Atlas and Bartha (1972). It is not free of problems, however, since the aeration by CO_2 free air will affect the carbonate-bicarbonate buffering character of seawater, and, as in any closed system, soluble products can build up and affect the process. The authors considered the latter problem was responsible for the cessation of biodegradation while degradable material and mineral nutrients still remained.

Oxygen uptake is a very attractive method since it is sensitive and non-destructive, allowing long-term measurements. Continuous manometric methods have been used (Bridie and Bos, 1971), but suffer two disadvantages

Table I

SOME PRACTICAL APPROACHES TO OIL BIODEGRADATION RATE MEASUREMENTS

Experimental characteristic	*Relevance*	*Limitations*
Disappearance of visible film	Short-term relevance to effects on birds and surface film chemistry	No knowledge of actual fate (emulsification, chemical modification, mineralization)
Loss of oleophilic (solvent extractable) material	More relevant in long term as 'oily' characteristic destroyed	Sampling destructive (aliquot sampling usually impossible). Is oil mineralized or solubilized?
Loss of total organic carbon	Measures extent of complete mineralization	Conversion of oil to biomass and modification not measured. Sampling destructive.
CO_2 production	Measures extent of complete mineralization	Conversion of oil to biomass and modification not measured. Sampling nondestructive.
Oxygen uptake	Includes both mineralization and partial oxidation	Quick and sensitive, non-destructive. Does not differentiate between partial and complete oxidation, or conversion to biomass.
Properties of oil	Properties affect pollution hazard	Quantitative estimates of loss less reliable.
Use of radioactive substances	Allows examination of single component in oil mixture	Limited to single substrates.
Bacterial numbers	Qualitative indication of organisms concerned	No quantitative measure of rate or extent of degradation.

in seawater. Unless alkali is replaced by a 'carbon dioxide buffer' (Krebs, 1951; Pardee, 1949; Warburg and Krippahl, 1960), the carbonate-bicarbonate buffering character of seawater is destroyed and as well as any effect on pH, this may affect metabolism (Dixon, 1951; Laser and Rothschild, 1939). The closed system again suffers build-up of soluble products, and unless mineral nutrients are present in unnaturally high concentrations initially, their depletion would soon cause a drastic fall in biodegradation rate. However, the author has favoured oxygen uptake measurements and attempts to overcome these problems are summarized here.

1. 'Semi-closed' respirometry (crude oil/seawater)

A method was sought to provide long-term records of oil biodegradation rate, as measured by oxygen uptake, under conditions of temperature and mineral nutrient concentrations as near natural as possible. It was hoped that a slow throughput of seawater would remove soluble products and maintain reasonably natural levels of mineral nutrients. No microbial additions were made to the natural population of the offshore sampled water. First attempts (Gibbs, 1972) to avoid the above problems identified further difficulties to be overcome: (a) the need to maintain a small amount of oil with a large surface area in an apparatus, while allowing a throughput of water; (b) the very dramatic depletion of available nitrogen in the presence of oil.

After various modifications, six respirometers as in Fig. 1 were used. Some

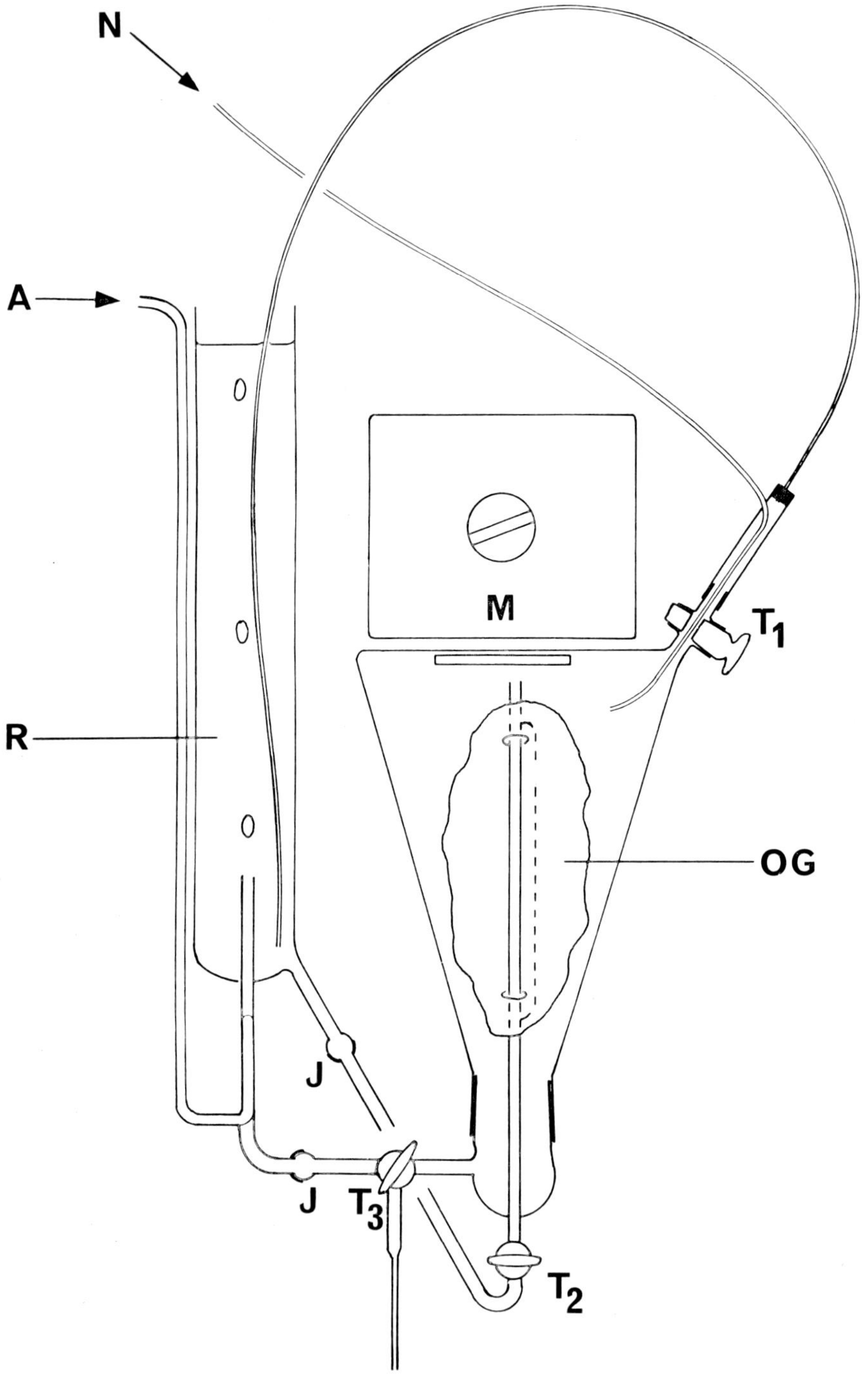

Fig. 1. *'Semi-closed' respirometer vessel. A, air flow; J, ball in socket joints; M, magnetic stirrer; N, nutrient injection; OG, oil on glass fibre filter paper; R, reservoir with air lift pump; T_1, T_2, stopcocks; T_3, 'T bore' stopcock. (Reproduced by courtesy of the Royal Society.)*

of this work has already been described in detail (Gibbs, 1975a; Gibbs *et al.*, 1975). Nitrogen and phosphorus nutrients were supplied continuously via the capillary tubing at a variety of rates and in various ratios. A current of aerated water was maintained through the apparatus by the recycling air lift pump combined in the reservoir, except when oxidation rates were being measured. Duplicate samples for Winkler titration were taken via stopcock T_3 after which the flask was isolated by closing the stopcocks (T_2 and T_3). Further samples were taken after one and two days, and samples were also kept for nutrient analyses. The drop in oxygen concentration multiplied by vessel volume gave the rate of oxidation, and the known addition and measured concentrations of nutrients allowed calculation of their uptake. Temperature was initially held at 14°C. After 12 weeks baseline measurement without oil, 200 mg of Kuwait crude residues absorbed on glass fibre filter paper were added to all but one vessel, this 'control' bearing an unoiled filter.

The processes occurring in the respirometer with maximum nutrient injection are shown in Fig. 2. In the absence of oil, nutrient concentrations built up, but on the addition of the oiled glass fibre they decreased rapidly, nitrate to near zero. Nitrate continued to be utilized as rapidly as it was added until a distinct point where oxidation slowed and nitrate absorption slowed then ceased. Nitrogen was then released as ammonia (not shown in Fig. 2) and phosphate was released (negative values in Fig. 2). This last phase probably occurs when degradable oil components are exhausted, and the oil-degrading organisms die and are decomposed.

When oil components were not limiting, analysis of all results showed that the rate-limiting factor was the rate of supply of nitrate (or ammonia). Figure 3 (upper line) relates the oxidation rate to the uptake of nitrogen (effectively equal to supply rate in this experiment). Phosphate supply did not limit oxidation rate when the P/N supply ratio was reduced to one-fifth of the 'typical' natural concentration ratio. The computed equation for oxygen uptake was:

$$O_2 \text{ uptake, mg per week} = 0{\cdot}72 + 0{\cdot}83 \text{ (nitrogen uptake, } \mu\text{g-atm per week)}$$

$$\text{Correlation coefficient } r = 0{\cdot}89$$

When the temperature was reduced to 4°C, the oxidation rate dropped in relation to nutrient uptake which was little affected (Fig. 3, lower line). The Q_{10} value of 2·7 agreed with the range quoted by ZoBell (1964) for enriched culture conditions.

Assuming a theoretical oxygen demand (t.o.d.) of 3·5 g O_2 per g oil, these results suggest a nitrogen requirement of 4 μg-atm N per mg oil oxidized at 14°C and 11 μg-atm N per mg oil at 4°C in local waters. Extrapolation to natural conditions (always a dubious process) suggested that the degradative capacity of local seawater depends on the (unknown) rate of recycling of available nitrogen. Assuming a one month recycling time (Harvey, 1966) and a total N concentration of 10 μg-atm litre^{-1}, figures in the order of 30 g m^{-3} year^{-1} at 14°C and 11 g m^{-3} year^{-1} at 4°C are tentatively obtained.

However, when these experiments were repeated using different crude oils—Brega (Libyan) and Forties (North Sea)—the seawater used was collected in the winter rather than in summer as in the previous work. The differences noted also applied to one sample of Kuwait oil with this 'winter' water, suggesting that the water and associated biota were responsible for the different behaviour. In particular, nitrate was not reduced to such a low

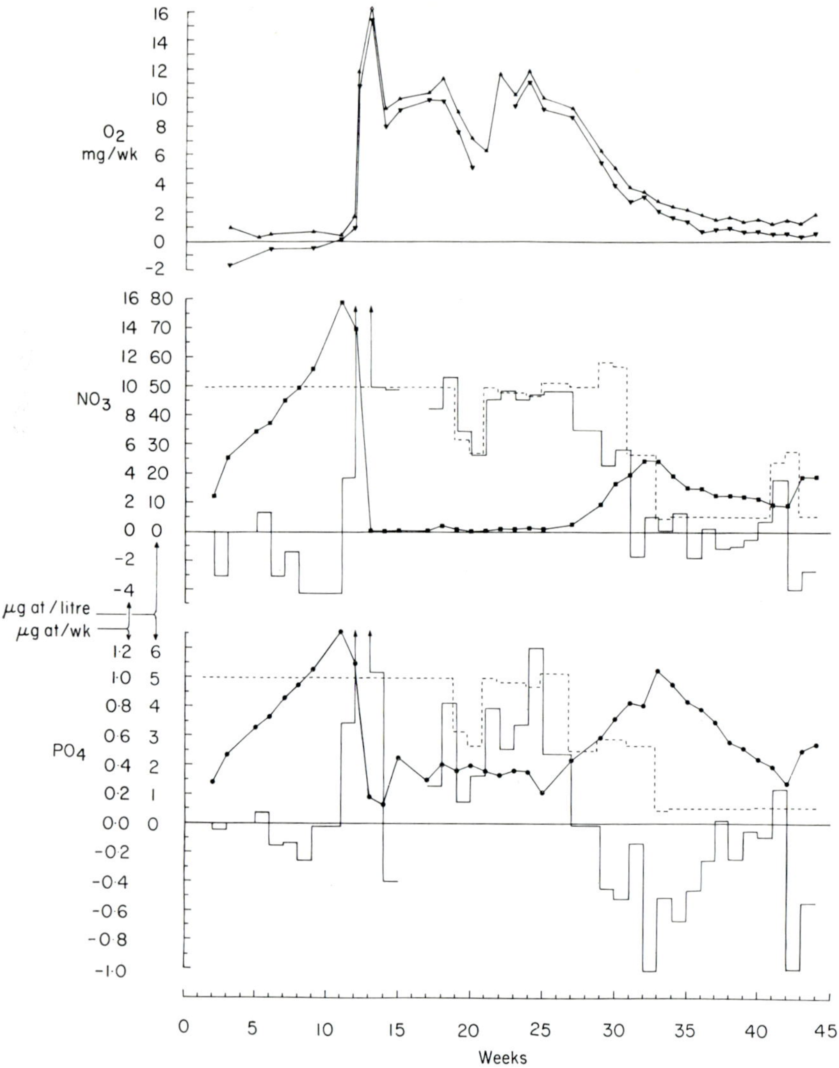

Fig. 2. Activity against time in respirometer with most rapid nutrient addition. ▲, *Rate of oxygen uptake (gross);* ▼, *rate of oxygen uptake less control;* – – – –, *nutrient added;* ———, *nutrient utilized;* ■, *nitrate concentration;* ●, *phosphate concentration.* (*Reproduced by courtesy of The Royal Society.*)

concentration, and at higher temperatures the oxidation rate was not greatly increased in relation to nitrate uptake. This illustrates some of the complexities and dangers involved in work with natural populations and mixed substrates.

In all these experiments, integrated oxygen uptakes suggested about 25% of the oil was oxidized, while recovery of oil indicated a 50% loss, before the process became exceedingly slow. This difference is thought to arise mainly by washing out of soluble derivatives, although the biomass remaining (unmeasured) may also be significant. The undegraded remainders from Kuwait 350°C residue and Brega 350°C residue were dense enough to sink in seawater,

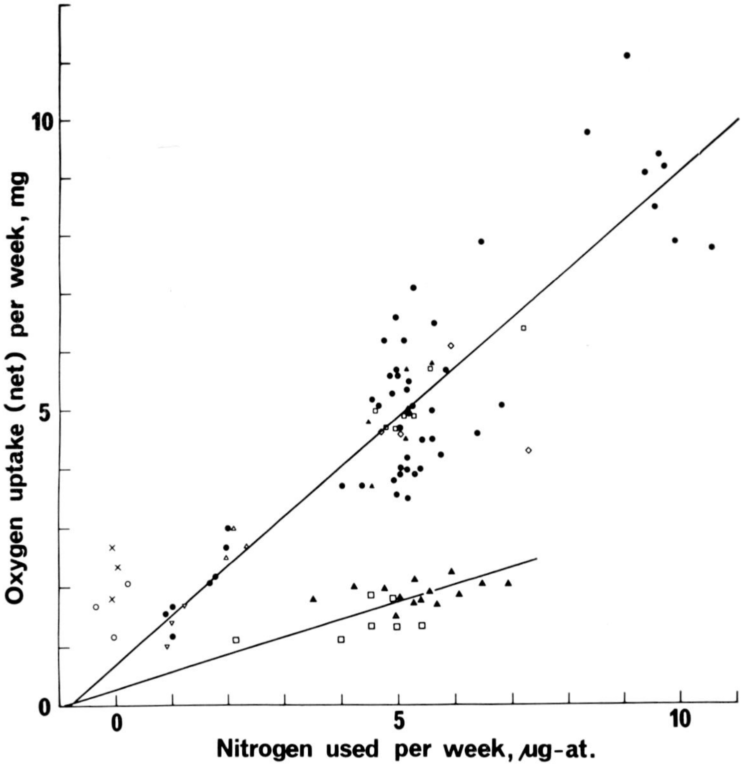

Fig. 3. Rate of oxygen uptake (net) against nitrogen utilized per week at 14 and 4°C. P/N ratios: □, *0·02;* ◇, *0·06;* ●, *0·10;* Δ, *0·25;* ∇, *0·50;* ×, *0·5/0;* ○, *0/0;* ▲, *N as ammonium.*

but that from less severely topped Forties crude (estimated 250°C residue) was not.

This experimental situation thus gives a biodegradation rate per available nitrogen supply which is tentatively interpretable as a degradative capacity of a given volume of seawater (Question A, p. 127).

2. Continuous flow respirometry

(*a*) *Oil/seawater*

The question arose as to whether the situation of an unlimited amount of water acting on a limited amount of oil could be modelled experimentally. This was attempted by pumping seawater through a coiled glass tube (40 ft long) coated internally with oil. The oxygen in the water entering and leaving the coil (and an untreated control coil) was measured by a continuously recording polarographic respirometer (Gibbs, 1975b) (Fig. 4).

The components surrounded by the cross-hatching were thermostatted (initially at 14°C) in a water bath. Water from the reservoir (R) was pumped via a biological filter (gravel column G) to the aeration cylinder (A). Three identical seawater supplies were pumped from (A), direct to the switching system (S), or via the experimental (oiled) coil (E), or control coil (C), respectively. The switching system (S) directed one of these three to the oxygen electrode (O) and flow meter (F), and the other two to waste or the beaker (B). Each of the three seawater lines was sampled in turn, automatic switching being at 30 min intervals. The chart record gave oxygen concentrations as a series of three 'steps', representing 'saturated' input water, and 'control' and 'experimental' effluents; also flow rate through the electrode was recorded. When required, nutrients could be injected via the capillary bore 'h' piece (shown blanked off in Fig. 4). In the system shown, the effluent water was collected in the beaker (B) and returned to the reservoir. A continuous throughput of new seawater is of course also possible.

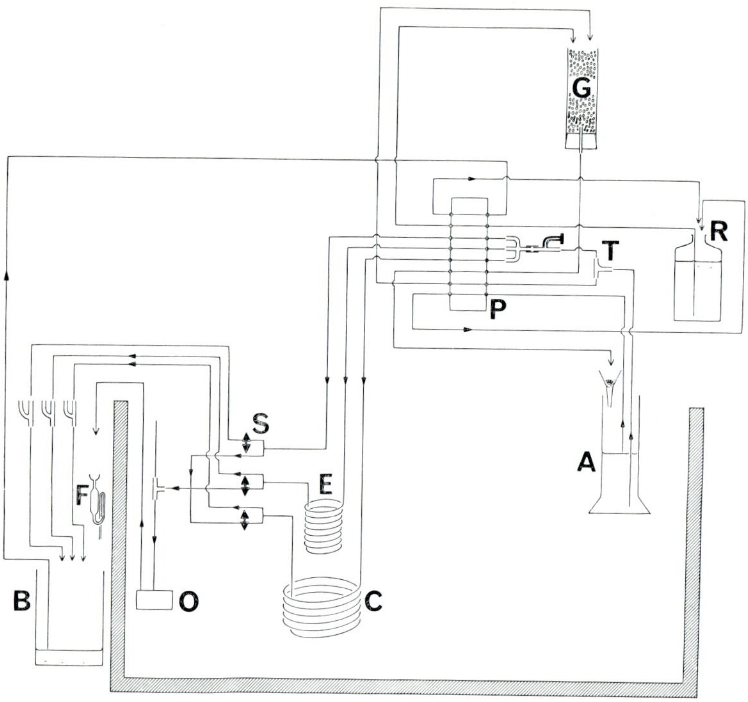

Fig. 4. Continuous flow respirometer. A, aeration cylinder; B, beaker for collection of waste; C, control coil; E, experimental coil; F, flow meter; G, gravel column, O, oxygen electrode; P, peristaltic pump; R, reservoir; S, flow switching system; T, 'T' piece.

At first, oxygen uptake occurred in the oiled coil during the day, but not at night, owing to a purely photochemical process. This probably involved the conversion, discovered by Burwood and Speers (1974), of thiacyclanes to their soluble l-oxides. When this had almost ceased, biodegradation eventually began. A variety of water flow rates, with and without nutrient injection, was employed. This system probably did not achieve the aim of an 'unlimited' supply of seawater to the oil, since oxidation rate was again strongly correlated with rate of nitrate supply:

Oxygen uptake, mg h^{-1} = 0·005 + 0·33 (N supply, μg-atm h^{-1})
Correlation coefficient r = 0·91

However, with unsupplemented seawater at the highest experimental flow rate, the bio-oxidation rate was such as would (by estimation) degrade the labile half of the crude oil present in just over a year. One year may thus be very tentatively suggested as an upper limit for the time required to degrade (as far as possible) an oil of thickness about 7 μm from water interface to centre.

(*b*) *Oil/sand/seawater*

The pioneering work by Johnston (1970) used unenriched seawater percolating through columns of medium sand, contaminated lightly or heavily with oil. Heavy oiling produced anaerobic conditions. Because of the abundance of available nitrogen and the limited access of oxygen in natural sandy beaches, it is probable that oxygen rather than nutrients will limit oil biodegradation. However, with increasing particle size, porosity, and water percolation rates, available nitrogen could again become the most important factor. The following experiment was thus performed with coarse beach material.

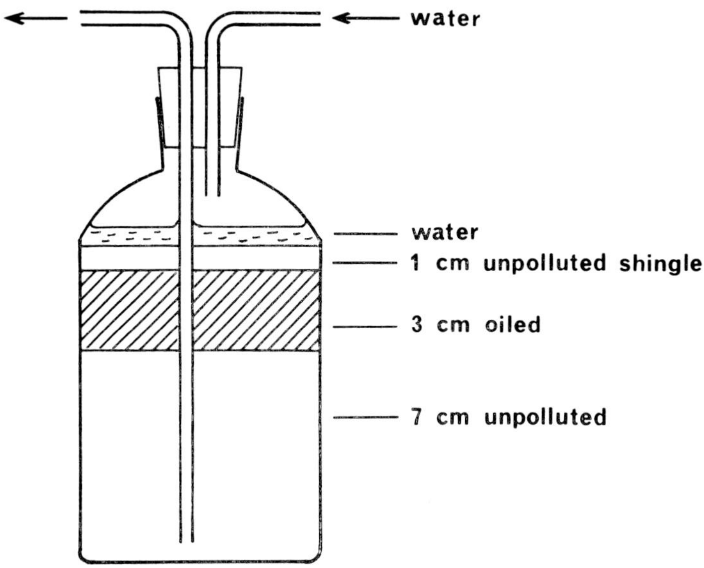

Fig. 5. Column of shingle in 500 ml bottle.

The continuous flow respirometer was set up with two columns of shingle (wide particle size range, median diameter 1·6 mm) replacing the two glass coils of Fig. 4. One column was contaminated with 2·7 g of Kuwait crude residue as in Fig. 5. Biodegradation was slow to start, but once underway the baseline rate (without nutrient enrichment) increased during the first two months. Nutrient supplementation (sodium nitrate and sodium dihydrogen phosphate) within the environmental range (*e.g.*, 10 μg-atm N litre^{-1}) markedly increased oxidation rate in the oiled shingle. The effluent from the oiled column was lower in nitrate than the control. The relationship between oxidation and nutrients, calculated either on the basis of known addition rates or of depletion in the oiled c.z. control column, was complicated by the changing 'baseline' rate. However, the proportionality was similar to the results in other systems, but with a larger 'constant' term arising from nitrogen regeneration in the shingle.

Oxygen uptake, mg h^{-1} = 0·12 + 0·23 (nitrate uptake, μg-atm h^{-1})
Correlation coefficient r = 0·82

The effect of temperature is shown in Fig. 6. One series was recorded during temperature decrease and the other during temperature increase. Q_{10} values in the range 6–21°C were close to the 2·7 found previously.

This work was complementary to studies using model beaches which mimic more closely the natural environment (Pugh, 1975) mentioned below.

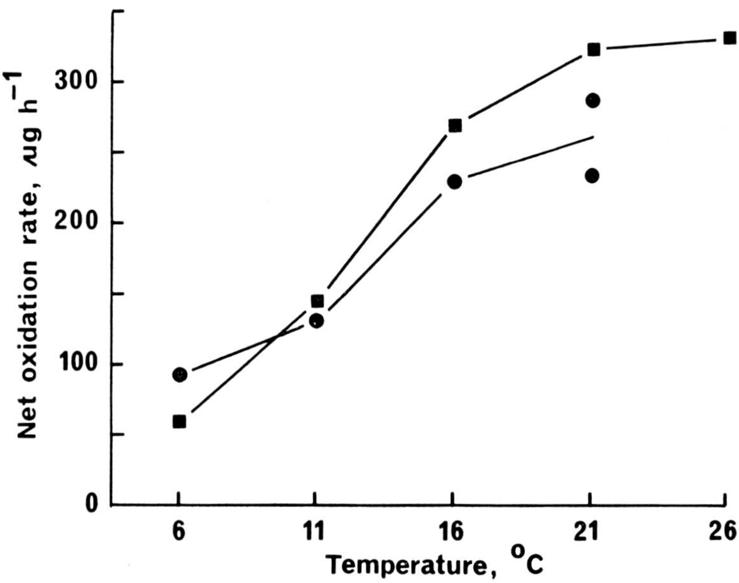

Fig. 6. Effect of temperature on oil oxidation rate in shingle column.

3. Inferences from oil analysis

In some simulated or natural environmental situations, it is not possible to use such methods as quantitative oxygen uptake or CO_2 evolution, and the heterogeneous nature of the system renders estimation of oil content impossible. One is then forced to rely on conclusions drawn from chemical and physical properties of sampled oil. Disappearance of *n*-alkanes has been almost universally used as an indicator of biodegradation. It is very useful qualitatively, but since *n*-alkanes represent a fairly small proportion of involatile oil residues, it does not measure quantitatively the destruction of oil. Vanadium and nickel determination may be the best quantitative measures of oil degradation. Since they are believed to be immune to weathering (Duckworth, 1971; Brunnock *et al.*, 1968), any net loss of oil should be reflected in an increase in their concentration in the residue.

(*a*) *Oil degradation on model beaches*

Of three model beaches as described by Pugh (1975), two have been polluted by crude oils and degradation is being studied by changes in oil properties (Pugh *et al.*, in preparation). Indications are that the process is slow at least at 5°C.

(*b*) *Weathering of 'chocolate mousse' at sea*

Fairly thick layers (7 mm) of Kuwait crude residue were exposed in two tanks on a raft at sea, one tank being open below the waterline to allow tidal

Fig. 7. Oil exposure tanks. (Reproduced by courtesy of the Ministry of Defence.)

flushing and the other being closed. Both were open to the atmosphere (Fig. 7). This oil soon formed a viscous 'chocolate mousse'. Here again, changes in oil properties are the only evidence for degradation. During two years, vanadium and nickel concentrations did not change, suggesting there was no net loss of oil, but substantial chemical modification occurred, probably by

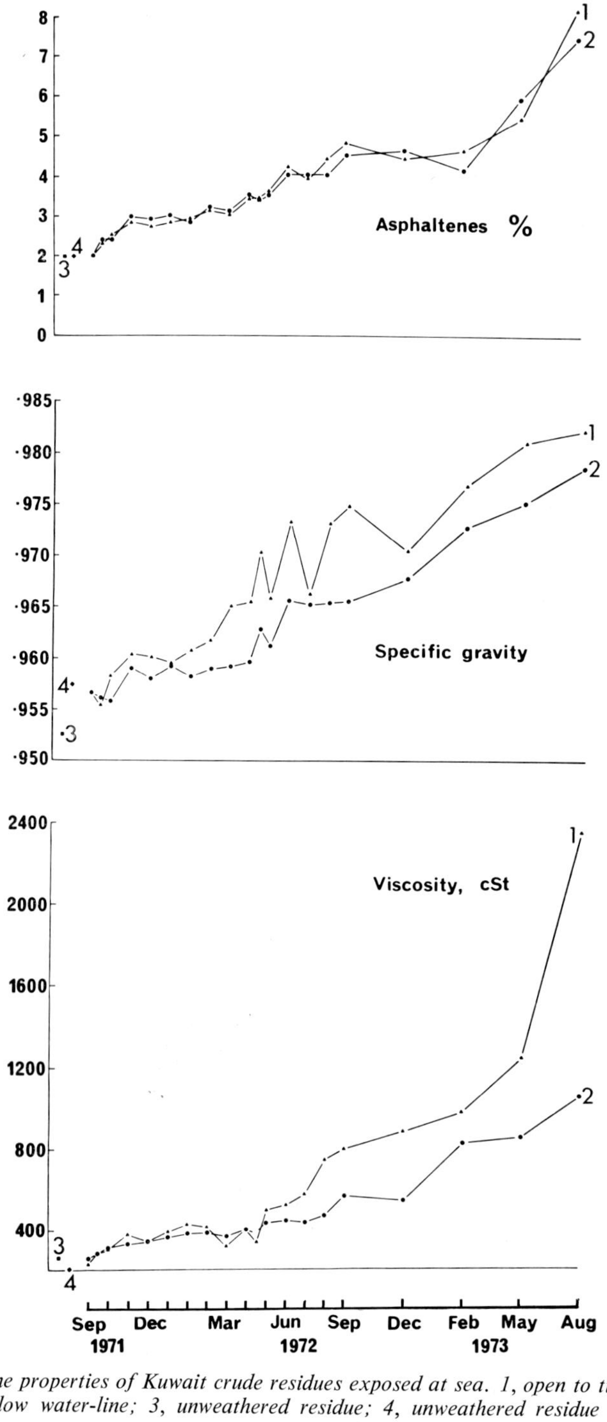

Fig. 8. *Some properties of Kuwait crude residues exposed at sea. 1, open to tidal flushing; 2, closed below water-line; 3, unweathered residue; 4, unweathered residue subjected to recovery procedure.*

both biological and abiological agencies. Some of the property changes are shown in Fig. 8, also *n*-alkane content dropped by about 50% in the open tank (Davis and Gibbs, 1975).

Despite the large oil/water interface in 'mousse', the discontinuity of the trapped water from the external seawater means that biodegradation is limited by the rate of diffusion of oxygen and nutrients into the interior of the mass. It was not possible to give any predictive description of degradation rates in 'chocolate mousse', but it is doubtless very dependent on the degree of physical fragmentation, being very slow in large lumps.

CONCLUSIONS AND SUMMARY

Many of the problems besetting respirometric measurement of oil degradation rate were overcome, and oxygen uptake was applied to rate measurements under comparatively natural conditions. However, supplementary methods are very desirable to indicate the processes occurring. Referring back to the four questions on pp. 127–8:

(A) The quantitative relationships between oil oxidation and nutrient uptake go some way to predicting more reliably the degradative capacity of seawater. Better understanding of nitrogen recycling is called for.

(B) It is more difficult to quantify the lifetime of an individual oil spill. However, for a well-dispersed oil, one year is suggested as the upper limit to degrade the labile 50% of the crudes studied. Thick 'mousse' is resistant unless physically fragmented.

(C) With three crude oil residues, about half appears very resistant to degradation, about one-quarter being mineralized and one-quarter lost by modification, probably as dissolved organic matter.

(D) The recalcitrant remainder from Kuwait and Brega 350°C residues will sink in seawater. Less severely topped Forties crude did not become dense enough in a biodegradation experiment, although extended weathering including photo-oxidation, etc., might cause it to sink.

The conclusions reached from local seawater and these crudes are not necessarily applicable worldwide. The possible contributions from oil should not be ignored in isotopic measurements of the 'age' of organic matter in the sea.

Acknowledgements

For contributions to the work summarized here, thanks are due to Dr K. B. Pugh, Mr S. J. Davis, and Mr A. R. Andrews; also to Dr G. D. Floodgate for encouragement, the Ministry of Defence (Procurement Executive) in connection with the exposure raft, and the B. P. Research Centre for vanadium and nickel determinations.

REFERENCES

Atlas, R. M., and Bartha, R. (1972). Degradation and mineralization of petroleum by two bacteria isolated from coastal waters. *Biotech. Bioeng.*, **14**, 297–308.

Bridie, A. L., and Bos, J. (1971). Biological degradation of mineral oil in sea water. *J. Inst. Petrol.*, **57**, 270–277.

Brunnock, J. V., Duckworth, D. F., and Stephens, G. G. (1968). Analysis of beach pollutants. *J. Inst. Petrol.*, **54**, 310–325.

Burwood, R., and Speers, G. C. (1974). Photo-oxidation as a factor in the environmental dispersal of crude oil. *Estuar. Coastal Mar. Sci.*, **2**, 117–135.

Davis, S. J., and Gibbs, C. F. (1975). The effect of weathering on a crude oil residue exposed at sea. *Water Res.*, **9**, 275–285.

Dixon, M. (1951). *Manometric Methods*. Cambridge Univ. Press, Cambridge. 75 pp.

Duckworth, D. F. (1971). Aspects of petroleum pollutant analysis. In *Water Pollution by Oil*, pp. 165–179 (P. Hepple, ed.), Applied Science Publishers, London.

Floodgate, G. D. (1973). A threnody concerning the biodegradation of oil in natural waters. In *The Microbial Degradation of Oil Pollutants*, pp. 17–24 (D. G. Ahearn and S. P. Meyers, eds.), Workshop held at Atlanta, Georgia, 1972. Publ. no. LSU-SG-01. Centre for Wetland Resources, Louisiana State University, Baton Rouge, U.S.A.

Gibbs, C. F. (1972). A new approach to the measurement of rate of oxidation of crude oil in sea water systems. *Chemosphere*, **3**, 119–124.

Gibbs, C. F. (1975a). Quantitative studies on marine biodegradation of oil. I. Nutrient limitation at 14°C. *Proc. Roy. Soc. Lond. B*, **188**, 61–82.

Gibbs, C. F. (1975b). A continuously recording polarographic respirometer and its use in oil biodegradation studies. Submitted to *Mar. Chem.*

Gibbs, C. F., Pugh, K. B., and Andrews, A. R. (1975). Quantitative studies on marine biodegradation of oil. II. Effect of temperature. *Proc. Roy. Soc. Lond. B*, **188**, 83–94.

Harvey, H. W. (1966). *The Chemistry and Fertility of Sea Waters*, pp. 75–76. Cambridge Univ. Press, Cambridge.

Johnston, R. (1970). The decomposition of crude oil residues in sand columns. *J. Mar. Biol. Assoc. U.K.*, **50**, 925–937.

Krebs, H. A. (1951). The use of 'CO_2 buffers' in manometric measurements of cell metabolism. *Biochem. J.*, **48**, 349–359.

Laser, H., and Lord Rothschild (1939). The metabolism of the eggs of *Psammechinus miliaris* during the fertilization reaction. *Proc. Roy. Soc. Lond. B*, **126**, 539–557.

Pardee, A. B. (1949). Measurement of oxygen uptake under controlled pressure of carbon dioxide. *J. Biol. Chem.*, **179**, 1085–1092.

Pugh, K. B. (1975). A model beach system. *J. Exp. Mar. Biol. Ecol.*, **18**, 197–213.

Warburg, O., and Krippahl, G. (1960). Weiterentwicklung der Manometrischen Methoden (Carbonatgemishe). *Z. Naturf.*, **15b**, 364–367.

ZoBell, C. (1964). The occurrence, effects and fate of oil polluting the sea. *Adv. Water Pollution Res.*, **3**, 85–118.

HYDROCARBON-USING MICROORGANISMS IN THREE FRESH-WATER ECOSYSTEMS

J. J. COONEY and R. J. SUMMERS*

Department of Biology, University of Dayton, Dayton, Ohio 45469, U.S.A.

Summary

Hydrocarbons and hydrocarbon-using microorganisms were studied in three bodies of fresh water over a 2-year period. Each ecosystem had a different history of hydrocarbon pollution. Hydrocarbons did not accumulate in waters but were associated with sediments which may be the site of greatest microbial activity. Petroleum products likely to gain access to these ecosystems contain paraffins as the largest class of hydrocarbons, whereas sediments yielded large fractions of olefins and aromatics. The amount of hydrocarbon in sediments did not appear to promote or hinder the size of the aerobic heterotrophic microbial population. The numbers of total heterotrophic bacteria were at least 100 times as great as the numbers of yeast and filamentous fungi, but the relative biomasses of the two groups may be more nearly equal. Only 0·1% of the total heterotrophic population of bacteria or of yeast and fungi grew with hydrocarbons as the sole carbon source. The numbers of hydrocarbon-using bacteria correlated positively with the numbers of hydrocarbon-using yeast and fungi, suggesting that both groups are important in hydrocarbon-polluted ecosystems. Representative colonies, picked at each sampling, included 54% bacteria, 33% yeasts, and 13% filamentous fungi. About 30% of each group gave evidence of emulsifying kerosene in vitro.

INTRODUCTION

In the last 20 years, the increasing use of petroleum products has been accompanied by a growing concern over the impact of petroleum and its products on the environment. In dealing with this problem, microbiologists have begun by using simplified laboratory systems. This approach has been fruitful, but often it has involved study of the easiest organisms to work with, growing them under laboratory conditions on single, purified substrates.

Over 100 species of microorganisms in some 60 genera can oxidize one or more hydrocarbons. The organisms include bacteria, yeasts, and filamentous fungi. Although the filamentous fungi comprise a significant fraction of the organisms, they have received relatively little attention in comparison to the bacteria and yeasts. Moreover, a number of studies have dealt with total microbial activity without regard to the types of organisms involved.

In addition, efforts have been expended mainly on those fractions of oil

* Present address: Department of Microbiology, Louisiana State University, Baton Rouge, Louisiana 70803, U.S.A.

which are least troublesome ecologically: alkanes and simple alkenes, while cycloparaffins, aromatic hydrocarbons, and asphaltic compounds and alkylated derivatives, which together comprise 40–85% of crude oils, have received much less attention. These fractions persist in fresh water, marsh and marine waters (Horn *et al.*, 1970; Blumer and Sass, 1972; Hites and Biemann, 1972) where they can aggregate as petroleum lumps or 'tarballs'. Tarballs can support growth of microbes and marine organisms (Horn *et al.*, 1970). They can sink and rise under conditions which are poorly understood (Davis, 1967; Hartung and Klingler, 1968; Nelson-Smith, 1970).

Last, more attention has been devoted to marine, estuarine, and marshland systems than to fresh water systems. Few studies have dealt with shallow, fresh water systems. In the ocean or in deep lakes more microbial activity takes place in the upper 75 to 100 m of the water column than in deeper parts of the water column. Paerl (1973) reported that in the upper 75 m of Lake Tahoe, fungi and bacteria played a role in aggregation of detrital particles and in their decomposition. Microbial activity decreased in deeper waters but was noted in bottom sediments. In shallow estuaries and in marine marshes (ZoBell and Prokop, 1966) and in shallow lakes, the mud-water interface is the scene of greater microbial activity than the water column. Suspended particulates, both detritus (Leshniowsky *et al.*, 1970; Paerl, 1973) and particles of clay (Tsernoglou and Anthony, 1971), can enhance microbial activity, presumably by providing adsorptive reaction sites. Microbial activity in sediments and on particulates should have a greater impact on the ecosystem in shallow lakes and rivers than in deep waters, because bottom-sediments can be mixed throughout the entire water column more easily than in deep waters.

The present work had as its objectives:

(1) to determine the prevalence of hydrocarbon-using bacteria, yeasts and filamentous fungi in hydrocarbon-contaminated and noncontaminated water, and to determine the relative numbers of each group;

(2) to determine the effect of hydrocarbon pollution on total microbial populations: hydrocarbon-users as well as those which do not use hydrocarbons;

(3) to assemble a collection of microorganisms which would be representative of hydrocarbon-users found in fresh waters.

MATERIALS AND METHODS

Sampling sites

Three bodies of water and sampling sites are shown in Figs. 1 through 3. Acton Lake (Fig. 1) in Heuston Woods State Park, Ohio, is a man-made recreational lake of about 600 surface acres. It receives limited hydrocarbons since recreational boats are limited to 10-h.p. motors. The main course of water is from a stream at site 4 through site 1, continuing along the north side of the lake and over the spillway. Site 2 is near the south shore and is not in the main water course. Site 3 is at a marina where boats with outboard motors are docked from spring through fall. The water level in Acton Lake is lowered approximately 4 ft from late fall until early spring. Holiday Lake

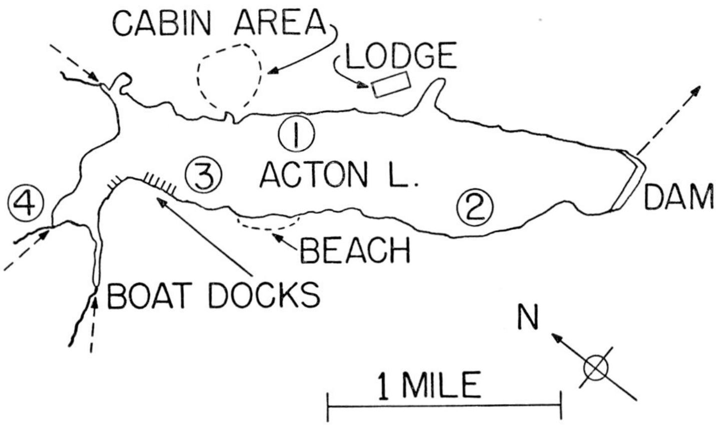

Fig. 1. *Acton Lake and sampling sites. 1–4 indicate sampling sites. Dashed arrows indicate the direction of water movement into or out of the lake.*

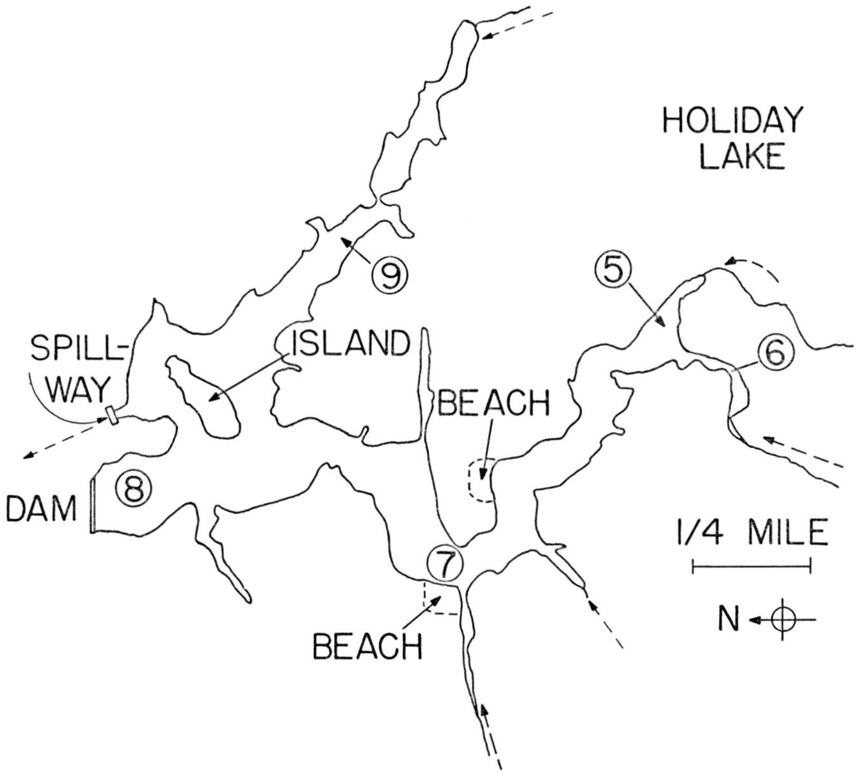

Fig. 2. *Holiday Lake and sampling sites. Dashed arrows indicate the direction of water movement into or out of the lake. The diesel oil spill of October 1, 1970 entered the lake via the stream which empties close to site 5.*

(Fig. 2) near Willard, Ohio, is a man-made recreational lake of about 225 surface acres, which received a heavy influx of diesel oil through one inlet. The lake may also receive chronic hydrocarbon pollution from a sewage disposal plant and a railroad repair shop via the stream near site 5, and from recreational boats. The morphology of the lake permitted sampling near where oil originally gained access, at several points along the main watercourse through the lake, and at other points which may not have been affected as seriously by the spill. Site 5 is close to where a stream bearing oily wastes enters the lake. On October 1, 1970, over 4000 gallons of diesel oil entered the lake through this stream. An oil slick covered much of the lake from site 5 past site 7, which is in the main watercourse downstream. Lesser amounts of oil were evident on the surface near the dam at site 8. Little or no oil was observed near the spillway. As a result of wind action, a thick surface layer of oil was present on much of the north shoreline along this route and tarry

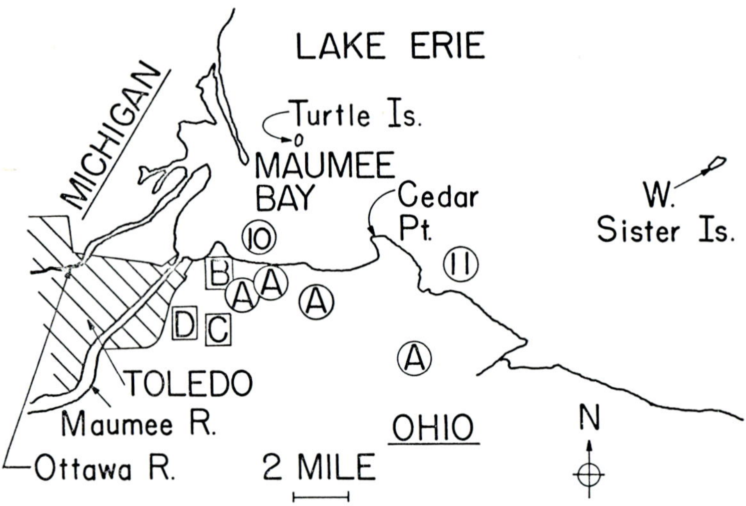

Fig. 3. Lake Erie near Toledo, Ohio. 10 and 11 indicate sampling sites; A designates radio and television masts used for orientation; B, C, and D indicate oil storage, pipeline terminal and refinery facilities, respectively.

residues accumulated on docks and on the shore. No evidence of that accumulation was noted when the present study began in July 1972. But bottom sediments taken at sites 5 and 7 throughout the study appeared black, viscous, and oily and smelled strongly of oil. Site 6 is on a different bay of Holiday Lake than site 5, and the flow of water is from site 6 toward site 7. No oil slick was observed at site 6 when the spill occurred, or at site 9 which is on a separate arm of the lake and did not appear to receive oil as a result of the major spill in 1970. The water level of the lake is not lowered during the winter. Two sites on Lake Erie (Fig. 3) were chosen for their proximity to oil refineries and storage areas. Site 10 is one-half mile offshore and close to a pipeline terminal and to an oil refinery, oil storage, and coal-powered electric generating facilities. The water flow is eastward from site 10. Site 11 is approximately 5 miles east of site 10 and is also one-half mile from shore.

Each site was sampled monthly during April through September and bimonthly during October through March. At each site, the bottom was sampled using an Eckman Dredge and the water column was sampled using a Kemmerer Sampler. Where the depth was greater than 8 ft, two water samples were taken, a surface sample and a sample at the midpoint of the water column.

Microbiological methods

Each sample was iced and returned to the laboratory. On several occasions duplicate microbiological counts were made—in the field and in the laboratory—to establish that counts did not change during transport.

The population of total viable bacteria was estimated by plating appropriate dilutions on the surface of Plate Count Agar (Difco) supplemented with 0·1 % (v/v) amphotericin B (E. R. Squibb, Inc.) to inhibit fungi. The population of yeast and filamentous fungi was estimated by plating on the surface of Littman Oxgall Agar containing crystal violet (Difco). The medium was supplemented with 30 mg streptomycin sulphate per litre to inhibit bacteria. Plates for total viable bacteria or total viable yeast and fungi were incubated for 48 hr at room temperature prior to counting them.

Hydrocarbon-using populations were estimated by the most probable numbers (MPN) technique as described by Cochran (1949), which proved to be better than plate counts on any of several solid media which were tested. Media for estimation of hydrocarbon-using microorganisms were prepared using the salts solution of Bushnell and Haas (1941) (BHS). For bacteria the pH of the salts solution was adjusted to 7·0 and the solution was supplemented with 0·1 % (v/v) amphotericin B. For yeasts and filamentous fungi the pH was adjusted to 5·5 and the solution was supplemented with 1·43 mg crystal violet and 30 mg streptomycin sulphate per litre. Test tubes, each containing 10 ml of sterile BHS, were overlaid with six drops of a 1:1 (v/v) mixture of filter-sterilized kerosene and nondetergent SAE 10W motor oil which served as the carbon source. After inoculation, tubes were incubated at room temperature for 14 days. A tube was scored as positive for growth if it became turbid, if it contained a pellicle at the oil-water interface, or if there was evidence of an emulsion when the tube was compared to a sterile control.

For each sample, several tubes which showed growth were selected and a loopful was streaked on Tryptic Soy Agar (TSA, Difco) or Sabouraud Dextrose Agar (SDA, Difco) to isolate bacteria, yeasts, or filamentous fungi, respectively. Representative colonies were picked and streaked on slants of TSA or SDA and the cultures were added to the collection of hydrocarbon-using organisms. The Gram reaction and cellular morphology of each bacterium was recorded. Fungi were separated into yeasts and filamentous fungi on the basis of colonial morphology and microscopic examination.

In addition, each isolate in the collection was cultured in BHS with kerosene as the carbon source. Cultures were scored for rate of growth, for amount of growth, and for ability to emulsify the mixture of hydrocarbons.

Chemical methods

Each sample was extracted to determine how much carbon it contained. In early experiments, chloroform was used, but samples which contained plant debris also yielded tannins and lignins to chloroform. Therefore, samples were extracted with methylene chloride. For sediment samples, 250 ml of a thick slurry were diluted to 1 litre with distilled water prior to extraction. For water samples, 1-litre quantities were extracted. The sample was extracted with three 50-ml quantities of methylene chloride. The combined methylene chloride extracts were filtered through Whatman No. 2 filter paper to remove any suspended solids. The filtered extract was allowed to stand in a separatory funnel until the water phase separated. The methylene chloride layer was separated and dried over anhydrous sodium sulphate for 8 to 12 hr and then filtered through Whatman No. 2 paper to remove the solid. The filtered extract was reduced to a volume of about 10 ml in a rotary evaporator and it was then transferred quantitatively to a pre-weighed aluminum cup. The solvent was allowed to evaporate to dryness overnight and the amount of hydrocarbon was determined after weighing the cup. Hydrocarbons were fractionated into paraffins, olefins, and aromatics by column chromatography on silica gel (ASTM, 1967).

RESULTS

For the first year of the study, water and sediment samples were taken at every sampling site in the four ecosystems. Regardless of their source, water samples did not differ significantly in hydrocarbon content or in microbial populations. During the second year, only sediment samples were taken.

Comparison of the three ecosystems

(a) Hydrocarbons

When the data for hydrocarbon content of sediments are grouped (all sites combined) by ecosystem (Table I), it is evident that bottom samples from

Table I
COMPARISON OF THE AMOUNT OF HYDROCARBONS IN SEDIMENTS FROM FOUR ECOSYSTEMS

Ecosystem	*No. of sites*	*No. of samples*	*mg Hydrocarbon per litre of sediment (Mean ± SD)*	*Probability of significant difference between ecosystem and:*		
				Acton Lake	*Holiday Lake*	*Lake Erie*
Acton Lake	4	27	17·46 ± 20·30		N.S.[a]	>0·95
Holiday Lake	5	53	30·70 ± 36·78	N.S.		>0·95
Lake Erie	2	14	5·39 ± 6·36	>0·95	>0·95	

[a] The notation N.S. indicates not significant at the 95% confidence interval.

Holiday Lake yielded the largest amount of hydrocarbon, and bottom samples from Lake Erie yielded the least hydrocarbon. Application of analysis of variance techniques indicated that differences between Acton Lake and Holiday Lake were not significant at the 95% confidence interval. The large variation around the mean for each ecosystem reflects the variation inherent in grab samples. It may also reflect the tendency of persistent hydrocarbons to aggregate into tarballs which sink to the bottom (Davis, 1967; Horn *et al.*, 1970).

Kerosene and light motor oil were used as examples of the kinds of hydrocarbons introduced into these systems by man. Paraffins were the major class

Table II
COMPOSITION OF HYDROCARBONS

Hydrocarbons from:	*No. of sites*	*No. of samples*	*Volume %*		
			Aromatics (Mean ± SD)	*Olefins (Mean ± SD)*	*Paraffins (Mean ± SD)*
Kerosene		2	15·0 ± 5·0	28·8 ± 1·0	56·1 ± 4·2
Motor oil		2	10·5 ± 0·7	27·5 ± 1·1	61·6 ± 1·3
Hydrocarbon medium after 14 days growth		1	8·5	82·3	9·2
Acton Lake	4	24	14·8 ± 5·8	64·9 ± 12·1	20·3 ± 12·7
Holiday Lake	5	41	19·4 ± 10·6	61·3 ± 16·1	18·8 ± 11·5
Lake Erie	2	10	24·0 ± 23·0	58·7 ± 25·4	16·9 ± 16·6

of compounds in each case (Table II). In one experiment the residual kerosene and motor oil in the MPN tubes were analysed after 14 days. Paraffins and aromatics had decreased (Table II). Olefins constituted the major fraction of hydrocarbons extracted from sediments from the three lakes (Table II).

(*b*) *Microbial populations*

The grouped values for total viable bacterial populations of the three ecosystems are compared in Table III. Sediments from Lake Erie had a significantly lower bacterial population than the other systems examined, and sediments from Acton Lake yielded significantly fewer viable bacteria than sediments from Holiday Lake.

Table IV compares the grouped data for total numbers of yeasts and fungi in sediments. In each ecosystem, averages were two logarithms lower than for total viable bacteria. The numbers of yeasts and fungi are significantly lower in sediments from Lake Erie than in sediments from the other bodies of water, but other differences are not significant.

Grouped data for hydrocarbon-using bacteria in sediments from the four ecosystems is provided in Table V. In each case, hydrocarbon-users represented only 0·1% of the total bacterial population (Table III). Sediments from Lake Erie yielded the fewest and sediments from Holiday Lake the greatest numbers of hydrocarbon-using bacteria.

Grouped data for hydrocarbon-using yeasts and filamentous fungi are compared in Table VI. Lake Erie sediments yielded the smallest number (less

Table III
COMPARISON OF THE TOTAL VIABLE BACTERIAL POPULATION IN SEDIMENTS FROM FOUR ECOSYSTEMS

Ecosystem	*No. of sites*	*No. of samples*	*Log of bacterial count/ml sediment (Mean ± SD)*	*Probability of significant difference between ecosystem and:*		
				Acton Lake	*Holiday Lake*	*Lake Erie*
Acton Lake	4	39	6·5 ± 1·2		>0·95	>0·99
Holiday Lake	5	61	7·1 ± 1·1	>0·95		>0·99
Lake Erie	2	18	5·3 ± 1·3	>0·99	>0·99	

Table IV

COMPARISON OF THE TOTAL VIABLE YEAST AND FILAMENTOUS FUNGAL POPULATION IN SEDIMENTS FROM FOUR ECOSYSTEMS

Ecosystem	*No. of sites*	*No. of samples*	*Log of yeast and fungal count/ml sediment (Mean ± SD)*	*Probability of significant difference between ecosystem and:*		
				Acton Lake	*Holiday Lake*	*Lake Erie*
Acton Lake	4	39	4·8 ± 1·2		N.S.[a]	>0·99
Holiday Lake	5	60	4·4 ± 1·2	N.S.		>0·99
Lake Erie	2	18	3·0 ± 1·4	>0·99	>0·99	

[a] The notation N.S. indicates not significant at the 95% confidence interval.

Table V

COMPARISON OF THE HYDROCARBON-UTILIZING BACTERIAL POPULATION IN SEDIMENTS FROM FOUR ECOSYSTEMS

Ecosystem	*No. of sites*	*No. of samples*	*Log of bacterial count/ml sediment (Mean ± SD)*	*Probability of significant difference between ecosystem and:*		
				Acton Lake	*Holiday Lake*	*Lake Erie*
Acton Lake	4	23	3·0 ± 0·6		>0·95	N.S.[a]
Holiday Lake	5	43	3·6 ± 1·1	>0·95		>0·99
Lake Erie	2	12	2·0 ± 2·4	N.S.	>0·99	

[a] The notation N.S. indicates not significant at the 95% confidence interval.

Table VI

COMPARISON OF THE HYDROCARBON-UTILIZING YEAST AND FUNGAL POPULATION IN SEDIMENTS FROM FOUR ECOSYSTEMS

Ecosystem	*No. of sites*	*No. of samples*	*Log of yeast and fungal count/ml sediment (Mean ± SD)*	*Probability of significant difference between ecosystem and:*		
				Acton Lake	*Holiday Lake*	*Lake Erie*
Acton Lake	4	24	0·6 ± 1·2		>0·95	>0·99
Holiday Lake	5	43	1·1 ± 0·8	>0·95		>0·99
Lake Erie	2	12	−1·2 ± 0·9	>0·99	>0·99	

than 1 propagule per 10 ml) and Holiday Lake yielded the largest number. In each case hydrocarbon-using fungi comprised a small fraction of the total fungal population (Table IV). Counts were lower than for hydrocarbon-using bacteria (Table V). Holiday Lake yielded significantly larger numbers of hydrocarbon-using fungi than any of the other ecosystems (Table VI). Values for Acton Lake are significantly different from values for Lake Erie.

Comparison within ecosystems

Data for individual sites in each ecosystem are presented in Table VII and the results of applying analysis of variance techniques within each ecosystem are presented in Table VIII.

(a) Hydrocarbons

Sediments from the marina (site 3) at Acton Lake yielded the greatest amounts of hydrocarbon, and sediments from the inlet stream (site 4) yielded the least hydrocarbon. Analysis of variance indicated that all values combined could have been drawn from the same population (Table VIII). However, sediments at the marina yielded significantly more hydrocarbon than a mid-lake site and differences between the marina and the main inlet stream were close to significance ($p = 0{\cdot}91$).

In Holiday Lake, the variance in data for all sites combined indicates that samples were not drawn from the same population. Sediments at site 5 yielded significantly more hydrocarbon than any other site in the lake. Site 6 was close to significance ($p = 0{\cdot}94$), but sediments did not appear oily, nor did they have an oily odour. Differences between sites 7 and 8 are close to significance ($p = 0{\cdot}91$).

Table VII

AMOUNT OF HYDROCARBONS AND THE MICROBIAL POPULATIONS OF SITES WITHIN EACH ECOSYSTEM

Sites	*mg Hydrocarbons per litre of sediment (Mean±SD)*	*Log of* Total viable bacteria/ml (Mean±SD)	*Log of* Total viable yeast and fungi/ml (Mean±SD)	*Log of* (Hydrocarbon-utilizing bacteria/ml (Mean±SD)	*Log of* Hydrocarbon-utilizing yeast and fungi/ml (Mean±SD)
Acton Lake:					
1.	15·4 ± 18·7	6·5 ± 1·2	4·6 ± 1·1	3·1 ± 0·9	0·7 ± 1·2
2.	19·9 ± 26·9	6·7 ± 1·0	4·0 ± 1·1	2·6 ± 0·4	0·6 ± 1·1
3.	31·2 ± 22·0	6·6 ± 1·3	4·2 ± 1·3	3·4 ± 0·8	0·6 ± 1·8
4.	11·7 ± 16·6	6·6 ± 1·3	4·6 ± 1·0	2·9 ± 0·2	−0·4 ± 1·2
Holiday Lake:					
5.	69·9 ± 53·0	7·2 ± 1·4	5·5 ± 1·3	4·5 ± 0·7	1·7 ± 0·8
6.	35·5 ± 38·9	7·1 ± 1·0	4·6 ± 1·0	3·6 ± 0·9	1·0 ± 0·6
7.	27·8 ± 30·0	7·2 ± 0·9	3·8 ± 1·1	3·6 ± 1·2	1·1 ± 0·6
8.	10·2 ± 7·9	6·8 ± 1·1	3·6 ± 0·7	3·2 ± 0·9	0·7 ± 1·0
9.	16·2 ± 15·9	6·7 ± 0·9	4·3 ± 1·0	3·3 ± 1·2	0·7 ± 0·4
Lake Erie:					
10.	6·4 ± 7·2	5·8 ± 0·9	3·7 ± 1·0	3·3 ± 1·6	−1·2 ± 1·2
11.	4·4 ± 5·7	4·8 ± 1·4	2·4 ± 1·6	0·7 ± 2·6	−1·2 ± 0·9

The amount of hydrocarbons in sediments from Lake Erie were not greater close to oil and coal facilities (site 10) than at a point 5 miles away (site 11).

(b) Microbial populations

For Acton Lake the differences in microbial populations were not significant, except that sediments from the stream (site 4) yielded significantly fewer hydrocarbon-using yeasts and fungi (less than 1 ml) than sediments from sites 1 and 2 in the lake (Tables VII and VIII).

Sediments from Holiday Lake did not differ significantly in their total viable bacterial population regardless of the site. For total viable yeasts and fungi, analysis of all sites combined (Table VIII) suggested that more than one population was sampled. Sites 5 and 6 yielded significantly larger numbers of yeasts and filamentous fungi than other sites. Site 5 contained more

Table VIII

PROBABILITY OF SIGNIFICANT DIFFERENCE IN THE AMOUNT OF HYDROCARBONS AND THE MICROBIAL POPULATIONS IN SEDIMENTS BETWEEN SITES WITHIN EACH ECOSYSTEM

Site vs. site	*Probability of significant difference*				
	Hydro-carbons	*Total viable bacteria*	*Total viable yeast and fungi*	*Hydro-carbon-utilizing bacteria*	*Hydro-carbon-utilizing yeast and fungi*
Acton Lake					
1 vs. 2	N.S.[a]	N.S.	N.S.	N.S.	N.S.
1 vs. 3	>0·98	N.S.	N.S.	N.S.	N.S.
1 vs. 4	N.S.	N.S.	N.S.	N.S.	>0·98
2 vs. 3	N.S.	N.S.	N.S.	N.S.	N.S.
2 vs. 4	N.S.	N.S.	N.S.	N.S.	>0·95
3 vs. 4	>0·91	N.S.	N.S.	N.S.	N.S.
All sites combined	N.S.	N.S.	N.S.	N.S.	N.S.
Holiday Lake					
5 vs. 6	>0·96	N.S.	>0·94	>0·97	>0·93
5 vs. 7	>0·99	N.S.	>0·99	>0·92	N.S.
5 vs. 8	>0·99	N.S.	>0·99	>0·99	>0·95
5 vs. 9	>0·99	>0·92	>0·98	>0·98	>0·98
6 vs. 7	N.S.	N.S.	>0·94	N.S.	N.S.
6 vs. 8	>0·94	N.S.	>0·98	N.S.	N.S.
6 vs. 9	N.S.	N.S.	N.S.	N.S.	N.S.
7 vs. 8	>0·91	N.S.	N.S.	N.S.	N.S.
7 vs. 9	N.S.	N.S.	N.S.	N.S.	>0·98
8 vs. 9	N.S.	N.S.	>0·91	N.S.	N.S.
All sites combined	>0·99	N.S.	>0·99	>0·90	>0·96
Lake Erie					
10 vs. 11	N.S.	>0·93	>0·95	>0·94	N.S.

[a] The notation N.S. indicates not significant at the 90% confidence interval.

hydrocarbon-using organisms of both types than other sites, and site 7 yielded significantly more hydrocarbon-using yeasts and fungi than site 9.

Sediments from site 10 in Lake Erie contained more yeasts and filamentous fungi than site 11, but other differences were not significant at the 95% confidence interval.

Kinds of organisms isolated

Of 353 representative organisms isolated, approximately 55% were bacteria, 33% were yeasts, and 13% were filamentous fungi (Table IX). A solid medium used in early experiments yielded more filamentous fungi than the liquid medium used for the bulk of the study. Therefore, the proportion of filamentous fungi may be larger. Gram-negative rods comprised 77% of the bacteria examined. Of the bacteria which grew on kerosene, 30% were capable of emulsifying the kerosene. Similar percentages of yeasts (31%) and filamentous fungi (28%) produced emulsions. It is possible that additional organisms produced micro-emulsions which would not have been detected by visual inspection. In each group of organisms the majority of the isolates were capable of growth on kerosene in a period of 2 weeks or less.

Ten of the gram-negative rods which grew rapidly with kerosene as their

Table IX
CHARACTERIZATION OF MICROORGANISMS ISOLATED

Type of organism	*No.*	*Growth noted in kerosene-salts medium after:*					*No. emulsifying kerosene*
		1 *wk*	2 *wk*	3 *wk*	4 *wk*	*Not at* 4 *wk*[a]	
Bacteria							
gram-positive rods	20	6	5	3	3	3	9
gram-positive cocci	6	0	4	1	0	1	1
gram-negative rods	127	39	41	29	13	5	30
gram-negative cocci	12	0	4	3	0	5	5
unexamined bacteria	28						
Yeasts	114	41	36	18	10	9	33
Filamentous fungi	46	12	14	8	5	7	11
Totals	353	98	104	62	31	30	89

[a] Includes 10 bacteria and 7 yeasts which were lost upon transfer.

carbon source were the subjects of a taxonomic study. Three were identified as strains of *Pseudomonas aeruginosa*, three others as *P. oralis*, *P. dachunae*, and *P. striata*, respectively, two as *Cellulomonas acidula* strains, one as a *Flavobacterium* sp. and one as an *Achromobacter* sp.

DISCUSSION

Differences among these ecosystems were found in the sediments rather than in the water column. Sediments may enhance microbial activity by contributing nutrients and by serving as surfaces for attachment of microorganisms. The effect of particulate matter on microbial activity is not clear but it depends in part on the nature of the particles. Weaver and Dugan (1972) found that several clay materials enhanced bacterial methane oxidation while some nonclay inorganic particulates inhibited oxidation. Wirsen and Jannasch (1974) reported that microbial transformations of various substrates in sea water were stimulated initially by addition of bottom sediments, but the stimulation was followed by a slight depression of activity.

When all the data for sediments from an ecosystem are grouped, the grouping tends to decrease differences between ecosystems. For example, values for the marina at Acton Lake (site 3) increase the mean values for the entire lake and values for sites 6 and 9 at Holiday Lake decrease the mean values for the entire lake. In spite of this tendency of groupings to smooth the data, Lake Erie contained significantly less hydrocarbon than either of the recreational lakes. Site 5 in Holiday Lake differed significantly from grouped data for Acton Lake.

Although paraffins constitute the major class of hydrocarbons likely to enter each ecosystem, olefins are the major class of persistent compounds and future efforts should be directed at olefins.

Based on analyses of grouped data, little correlation is evident between hydrocarbon content of sediments and total microbial populations, suggesting that oil pollution does not inhibit or enhance the numbers of heterotrophic microorganisms. However, a trend is suggested wherein ecosystems which contain more hydrocarbons also yield larger numbers of hydrocarbon-using microorganisms even though the hydrocarbon-users constitute only a small fraction of the total microbial population. Walker and Colwell (1974) reported that the numbers of petroleum-degrading bacteria in sediments from Chesapeake Bay were related to the concentration of oil in each sample. Sediments from an oil-polluted area of Chesapeake Bay were more effective than sediments from a nonpolluted area as inoculum for hydrocarbon degradation (Walker and Colwell, 1975). This suggests that chronic pollution results in enrichment for hydrocarbon-users. Similarly, Berner *et al.* (1975) observed selection for hydrocarbon-using yeasts in a polluted estuary.

In the present work, the numbers of bacteria—total populations or hydrocarbon-users—far exceed the numbers of yeast and filamentous fungi. But cells of yeast and filamentous fungi are larger than bacterial cells, and filamentous fungi often exist in large masses of cells. Therefore, one colony-forming unit for yeasts or fungi represents a larger biomass than a colony-forming unit for bacteria, and the relative biomasses are undoubtedly nearly equal, rather than the hundredfold difference indicated by these counts. Bartha and Atlas (1973) reported that hydrocarbon-using bacteria were abundant in sea water, but that the range of substrates attached by individual organisms was restricted. Thus, degradation of the great variety of hydrocarbons present in crude or refined petroleum probably involves a number of organisms. Beam and Perry (1973) found that cyclohexane was degraded when incubated with fertile soil, but no single organism isolated from the soil could degrade cyclohexane. They suggested that degradation of cycloalkanes in soil proceeds via co-metabolism. Soli and Bens (1973) reported that combined strains of bacteria did not degrade hydrocarbons in flask cultures better than single strains. But Cerniglia and Perry (1973) observed that fungi were more effective than bacteria at decomposing crude oil. They concluded that no single organism can degrade all fractions of crude oil. Schwartz *et al.* (1974) observed use of *n*-tetradecane by mixed cultures but not by pure cultures isolated from the mixture, which suggested a cooperative action. Moreover, Fliegler *et al.* (1974) presented data which suggest that fungi inhibit growth of bacteria in aerobic lake sediments. These reports agree with the results of the present study which suggest that bacteria, yeasts, and filamentous fungi are all active participants in degradation and mineralization of hydrocarbons in aquatic ecosystems.

Comparison of sediments from the marina at Acton Lake (site 4) with other sites at the lake indicates that operation of a number of small outboard engines in the relatively confined area of the marina contributes oil to the ecosystem. However, hydrocarbons from sediments at the marina had the same composition of paraffins, olefins, and aromatics as hydrocarbons from other sites, suggesting that the same general groups of compounds persist whether they are derived from petroleum as at the marina or from plant and animal material as in the stream.

Considerable hydrocarbons entered Holiday Lake through the stream which enters near site 5, and it appears that additional hydrocarbons entered during the course of this investigation. During the 2 years of this study, the water depth at site 5 decreased from 2 ft to about 6 in due to silting in. But as silting progressed, the hydrocarbon content of the sediment-water samples did not decrease, leading to the conclusion that hydrocarbon pollution of Holiday Lake is continuous. Potential sources are a sewage treatment plant and a railroad repair and maintenance shop which contribute water to the stream. Effluent water from the railroad shop enters a holding pond. The shores of the pond are saturated with oil and on a number of occasions the surface of the pond was covered with oil. A mechanical skimmer is intended to remove oil before water leaves the pond and a weir is located downstream from the pond. The weir is intended to retain oil which might escape from the pond, but water leaving the holding pond probably contains near-saturation levels of oil. On several occasions the effluent from the sewage treatment plant made the water turbid in the stream before it entered the lake. Therefore, either or both of these potential sources could contribute hydrocarbons to Holiday Lake. Two other sources of hydrocarbons are the motors of pleasure boats and run-off from farms. It is estimated that 1·0 to $1{\cdot}6 \times 10^6$ gallons of fuel per year are discharged into American waters by two-cycle outboard engines. At concentrations of 5 ppm or less, raw fuel or crankcase drainage from outboard engines inhibits CO_2 fixation by algae (R.P.I., 1972). But boating should not have contributed to the difference between site 5 and other sites because the shallow depth at site 5 decreased boat traffic and the water flow is away from site 5 into the main body of the lake. Agricultural run-off is considered to be the major source of hydrocarbons at site 6, but not at site 5 because of the visible differences in sediment samples.

Hydrocarbons in sediments from Lake Erie were not greater close to oil and coal facilities than at a site 5 miles away from those facilities.

In an attempt to determine if the sizes of the microbial populations in an ecosystem correlated with some key parameters, the data for Acton and Holiday Lakes were analysed using a step-wise multiple regression technique. Data for Lake Erie were not analysed by this technique because only two sites were involved. For each of the four microbial populations, the following variables were entered in the regression analysis: total hydrocarbon content of the sediment; individual amounts of aromatic, olefinic, and paraffinic hydrocarbons; individual percentages of aromatic, olefinic, and paraffinic hydrocarbons; each of the other three microbial populations; and each individual sampling site in the ecosystem.

In both ecosystems, the numbers of total viable bacteria showed a strong positive correlation with the total number of yeasts and filamentous fungi. Thus, the total population of heterotrophic microorganisms appears to vary together. There were some negative correlations between total counts and amount or percent of aromatics and olefins, suggesting that these fractions of petroleum products may be inhibitory or toxic to the microbial flora. However, these correlations were not consistent enough to support a firm conclusion. In Acton Lake the total bacterial population correlated positively with the amount of paraffins and the percent of olefins and correlated negatively with the amount of aromatics in the sediments. The total fungal population correlated positively with the population of hydrocarbon-using bacteria, but the reason is not evident. Hydrocarbon-using yeasts and fungi showed a negative correlation with the inlet stream. In Holiday Lake, which showed the greatest variation in hydrocarbon contents within the ecosystems, the numbers of hydrocarbon-using bacteria showed a strong positive correlation with the numbers of hydrocarbon-using yeasts and filamentous fungi, supporting the suggestion that both groups are important in hydrocarbon-polluted systems.

The microbial populations detected in this study are at best an estimate of the total population present in the ecosystems. Cultures were incubated at 28° ± 2°C for 2 days (total populations) or 14 days (hydrocarbon-users). Thus, psychrophilic or psychrotrophic organisms and slow-growing organisms may not have been tabulated. No attempt was made to detect microaerophiles or anaerobes. Each of these types of organisms plus microscopic algae should be measured in order to have a complete picture of the microbial flora. Moreover, the data obtained provide no information as to whether these organisms are active *in situ* where a variety of substrates is present and where there is competition from and predation by other organisms, nor has any information been gathered about variations in rates of hydrocarbon metabolism from season to season. Atlas and Bartha (1972) reported that in sea water biodegradation of oil was slower at 5°C than at 20°C. In contrast, Wirsen and Jannasch (1974) noted that microbial activity on nonhydrocarbon substrates in sea water was higher in winter than in summer. Low levels of nitrogen and phosphorus may limit microbial activity severely (Bartha and Atlas, 1973; Cerniglia and Perry, 1973; Gibbs, 1975). If so, the hydrocarbonoclastic activity of microorganisms in these ecosystems might be enhanced by judicious seeding with chemicals. In spite of these limitations, the collection of 350 aerobic, mesophilic hydrocarbon-using microorganisms isolated from these four ecosystems represents a library of representative fresh water organisms on which a number of continuing studies can be based, including not only the questions raised above but also such aspects as co-oxidation of recalcitrant substrates (Raymond *et al.*, 1971; Horvath, 1972) and cooperative action among microorganisms (de Klerk and van der Linden, 1974).

REFERENCES

ASTM. (1967). *Hydrocarbon types in liquid petroleum products by fluorescent indicator adsorption.* ASTM Designation D 1319-IP156. Am. Soc. Testing and Materials, 1916 Race St., Philadelphia.

Atlas, R. M., and Bartha, R. (1972). Biodegradation of petroleum in seawater at low temperatures. *Can. J. Microbiol.*, **18**, 1851–1855.

Bartha, R., and Atlas, R. M. (1973). Biodegradation of oil in seawater: limiting factors and artificial stimulation. In *The Microbiol Degradation of Oil Pollutants*, pp. 147–152 (D. G. Ahearn and S. P. Meyers, eds.), Louisiana State University, Baton Rouge, Louisiana. Publ. No. LSU-SG-73-01.

Beam, H. W., and Perry, J. J. (1973). Co-metabolism as a factor in microbial degradation of cycloparaffinic hydrocarbons. *Arch. Mikrobiol.*, **91**, 87–90.

Berner, N. H., Ahearn, D. G., and Cook, W. L. (1975). Effects of hydrocarbonoclastic yeasts on pollutant oil and the environment. In *Impact of the Use of Microorganisms on the Aquatic Environment*, pp. 199–219 (A. W. Bourquin, D. G. Ahearn, and S. P. Meyers, eds.), Nat'l. Environ. Res. Cent., Corvallis, Oregon.

Blumer, M., and Sass, J. (1972). Oil pollution persistence and degradation of spilled fuel oil. *Science*, **176**, 1120–1122.

Bushnell, L. D., and Haas, H. F. (1941). The utilization of certain hydrocarbons by microorganisms. *J. Bacteriol*, **41**, 653–673.

Cerniglia, C. E., and Perry, J. J. (1973). Crude oil degradation by microorganisms isolated from the marine sediment. *Z. Allg. Mikrobiol*, **13**, 299–306.

Cochran, W. G. (1949). *Estimation of bacterial densities by means of the 'most probable number'*, pp. 105–115. Joint Session Eng. Lab. and Statistics Section of A.P.H.A., New York.

Davis, J. B. (1967). *Petroleum Microbiology*. Elsevier Pub. Co. New York, 604 pp.
deKlerk, H., and van der Linden, A. C. (1974). Bacterial degradation of cyclohexane: participation of a co-oxidation reaction. Antonie van Leeuwenhoek, *J. Microbiol. Serol.*, **40**, 7–15.
Fliegler, S. L., McNabb, C. D., and Fields, W. G. (1974). Antibiotic treatment of lake sediments to determine the effect of fungi on decomposition. *Water Res.*, **8**, 307.
Gibbs, C. F. (1975). Quantitative studies on marine biodegradation of oil. I. Nutrient limitation at 14°C. *Proc. Roy. Soc. London B*, **188**, 61–82.
Hartung, R., and Klingler, G. W. (1968). Fisheries and wildlife sedimentation of floating oils. *Papers Mich. Acad. Sci., Arts, Letters*, **53**, 23–27.
Hites, R. A., and Biemann, K. (1972). Water pollution: organic compounds in the Charles River, Boston. *Science*, **178**, 158–160.
Horn, M. H., Teal, J. M., and Backus, R. M. (1970). Petroleum lumps on the surface of the sea. *Science*, **168**, 245–246.
Horvath, R. S. (1972). Microbial co-metabolism and the degradation of organic compounds in nature. *Bacteriol. Rev.*, **36**, 146–155.
Leshniowsky, W. O., Dugan, P. R., Pfister, R. M., Frea, J. I., and Randles, C. I. (1970). Aldrin: removal from lake water by flocculent bacteria. *Science*, **169**, 993–995.
Nelson-Smith, A. (1970). The problem of oil pollution of the sea. *Adv. Mar. Biol.*, **8**, 215–306.
Paerl, H. W. (1973). Detritus in Lake Tahoe: structural modification by attached microflora. *Science*, **180**, 496–498.
Raymond, R. L., Jamison, V. W., and Hudson, J. O. (1971). Hydrocarbon co-oxidation in microbial systems. *Lipids*, **6**, 453–457.
R.P.I. (1972). Rensselaer Fresh Water Inst. at Lake George, *Newsletter*, **2**, 1, Rensselaer Polytech. Inst., Troy, New York.
Schwartz, J. R., Walker, J. D., and Colwell, R. R. (1974). Growth of deep-sea bacteria on hydrocarbons at ambient and *in situ* pressure. *Dev. Ind. Microbiol*, **15**, 239–249.
Soli, G., and Bens, E. M. (1973). Selective substrate utilization by marine hydrocarbonoclastic bacteria. *Biotechnol. Bioeng.*, **15**, 285–297.
Tsernoglou, D., and Anthony, E. H. (1971). Particle size, water-stable aggregates, and bacterial populations in lake sediments. *Can. J. Microbiol*, **17**, 217–227.
Walker, J. D., and Colwell, R. R. (1974). Microbial degradation of model petroleum at low temperatures. *Microbial Ecol.*, **1**, 63–95.
Walker, J. D., and Colwell, R. R. (1975). Some effects of petroleum on estuarine and marine microorganisms. *Can. J. Microbiol*, **21**, 305–313.
Weaver, T. L., and Dugan, P. R. (1972). The eutrophication implications of interactions between naturally occurring particulates and methane oxidizing bacteria. *Water Res.*, **6**, 817–828.
Wirsen, C. O., and Jannasch, H. W. (1974). Microbial transformations of some ^{14}C-labeled substrates in coastal water and sediment. *Microbial Ecol.*, **1**, 25–37.
ZoBell, C. E., and Prokop, J. F. (1966). Micrcbial oxidation of mineral oils in Barataria Bay bottom deposits. *Z. Allg. Mikrobiol*, **6**, 143–162.

MICROBIOLOGICAL ASPECTS OF SUPERSONIC AIRCRAFT FUEL

E. C. HILL
Microbiology Department, University College, Cardiff, U.K.
and
ANGELA R. THOMAS
Department of Plant Pathology, University of Wisconsin-Madison, U.S.A.

Summary

Reasons for the dominance of *Cladosporium resinae* in infections in aircraft fuel are discussed.

The fuel tank lay-outs and fuel temperature cycles in a supersonic transport are given and the properties of a strain of *Aspergillus fumigatus* found to flourish in simulated conditions are described.

The effects of temperature on current fuel biocides and a new biocide are discussed, and results of biocide tests in a simulated environment are given.

MICROORGANISMS IN SUBSONIC AIRCRAFT

Since Bakanauskas (1958) described bacterial sludge in JP4, and Hazzard (1961) and Prince (1961) also noted fungi, there has been a voluminous literature on the types of organisms to be found, their frequency of occurrence, their detection, growth characteristics, inhibition, and of course on the practical consequences of fuel infection.

There is no doubt that over the years a consensus of opinion has grown up which considers that *Cladosporium resinae* is the prime cause of microbial fuel problems. Yet very many other microorganisms can grow in aviation kerosine and from time to time are isolated from aircraft fuel and from ground and ship installations. Some of these can be shown to be faster growing than *C. resinae* and yet are found less frequently.

It is reasonable to suppose that there are therefore pressures involved which select for *C. resinae* and against other organisms. Some of these are fairly obvious; for example because active growth is confined to the water phase associated with kerosine, but infection is largely spread by transferring contaminated kerosine from one location to another, the infectivity of a species depends on its ability to pass from the water phase into the fuel phase and survive there. *Pseudomonas* spp. grow well in fuel water condensate and, with agitation, pass into the fuel phase. (If they are grown separately on laboratory media and then added to fuel they are difficult to disperse.) Figure 1 shows typical survival curves for a *Pseudomonas* sp. in fuel and from these we may infer that storage for a matter of hours will ensure that the fuel is no longer infective. Figure 2 shows survival curves for *C. resinae* spores in fuel and indicates that fuel thus contaminated will be infective for a matter of weeks or months. An additional factor is that *C. resinae* will sporulate freely

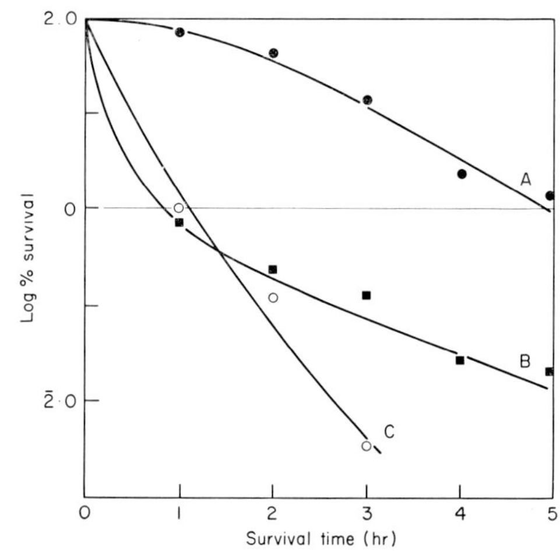

Fig. 1. Survival of Pseudomonas *sp. in kerosine at 25°C. A—100% Relative humidity kerosine; B—94% Relative humidity kerosine; C—Dry kerosine.*

in a fuel/water mixture whilst other fungi which grow in these mixtures do not necessarily sporulate freely in them.

Whilst free water is usually considered to be a pre-requisite for active microbial growth it can be shown that mycelium of *C. resinae* will continue to respire when suspended in kerosine in the absence of visible free water (*see* Fig. 3). We have observed slow growth of *C. resinae* when incubated at 25°C in kerosine containing initially only 80 ppm free water. By determining the wet weight and the dry weight of the mycelium produced after four weeks it was possible to say that the metabolism of the fungus had produced at least 0·94 g of water per litre of fuel.

Selective pressures operating in favour of *C. resinae* are therefore its ability to sporulate and survive in fuel, and to grow in the presence of minimal

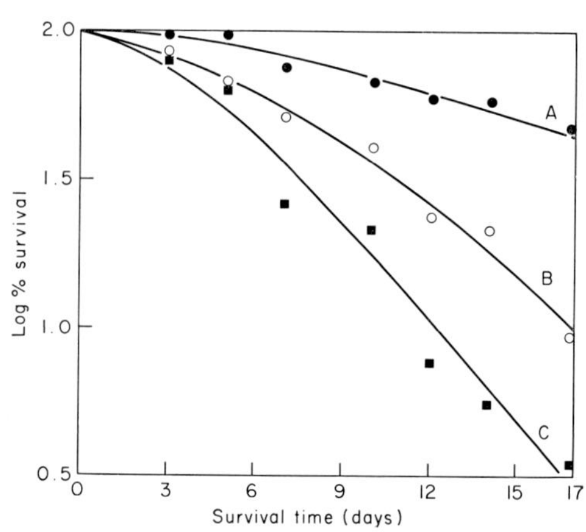

Fig. 2. Survival of Cladosporium resinae *spores in kerosine at 25°C. A—Dry kerosine; B—50% Relative humidity kerosine; C—100% Relative humidity kerosine.*

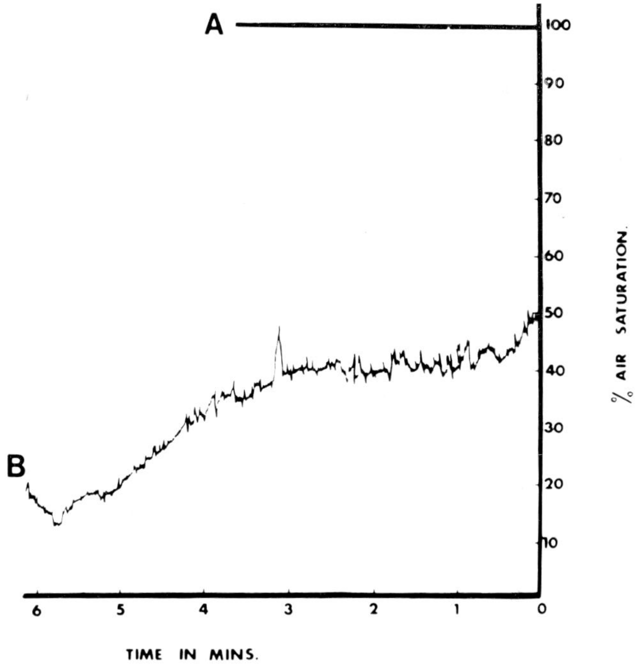

Fig. 3. Oxygen uptake in kerosine (free of visible water) by mycelium of Cladosporium resinae. *A—clean fuel; B—fuel containing* Cladosporium.

free water. There is in fact another factor which is important in conventional aircraft and that is the ability of an organism to survive the repeated cycles of freezing and thawing which occur in the water in the wing during the normal flight pattern of the aircraft. Figure 4 illustrates the results of an experiment

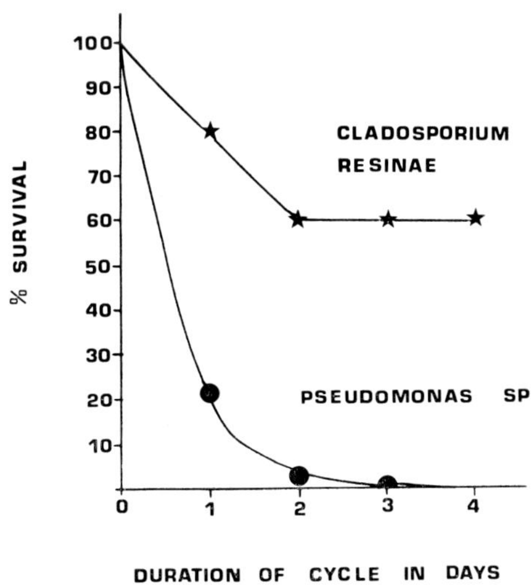

Fig. 4. Survival of Pseudomonas *sp. and* Cladosporium resinae *spores during freezing-thawing cycles of a 1:4 water/kerosine mixture.*

in which *Pseudomonas* sp. and *C. resinae* spores were each suspended in 5 ml of mineral salt medium (Bushnell and Haas, 1941) overlaid with 20 ml of kerosine and frozen to $-22°C$ for 3 hr, thawed at 23°C for 1 hr, frozen and thawed twice more and then left at $-22°C$ overnight (15 hr). The cycle was repeated for four days and samples were taken at 24 hr intervals for plate counts using Tryptose Blood Agar Base and Malt Extract Agar (both Oxoid Ltd) respectively. The data is after Reece and Hill (1971). The ability of *C. resinae* to survive under these conditions must be an important factor in its dominance in the subsonic aircraft environment, but will have no survival value in the fuel environment of supersonic aircraft.

THE FUEL ENVIRONMENT IN SUPERSONIC AIRCRAFT

A publication by Hill (1971) first drew attention to the unusual ecological features of the fuel systems of aircraft designed to operate primarily in the supersonic mode, and Scott and Hill (1971) compared subsonic and supersonic aircraft and gave data for microbial growth in apparatus which simulated supersonic flight. The important ecological difference is that there is normally no freezing cycle in the fuel and in fact kinetic heating raises the fuel temperature. In some parts of the aircraft, this is to levels which might be conducive to microbial growth, whilst in other parts of the aircraft the temperatures are such that the fuel will be self-sterilizing. Figure 5 is a plan of the fuel tank layout of a production Concorde aircraft. This aircraft is designed to cruise at Mach 2·2 at 65 000 ft and Fig. 6 illustrates the predicted fuel tank temperatures during a typical flight. The fuel has a triple role as it also acts as a heat sink, extracting heat from the various aircraft systems

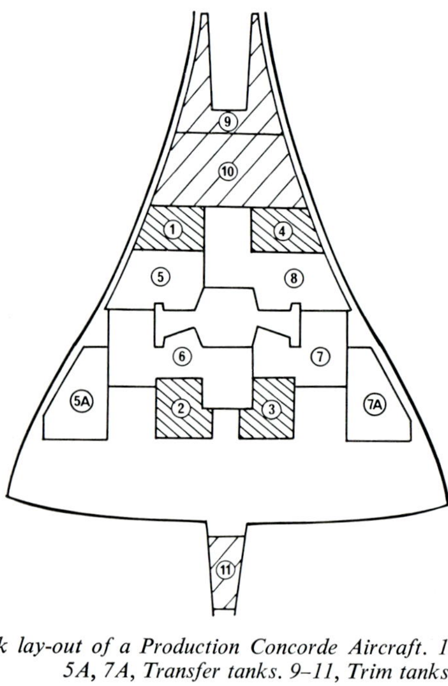

Fig. 5. Fuel tank lay-out of a Production Concorde Aircraft. 1–4, Collector tanks. 5–8, 5A, 7A, Transfer tanks. 9–11, Trim tanks.

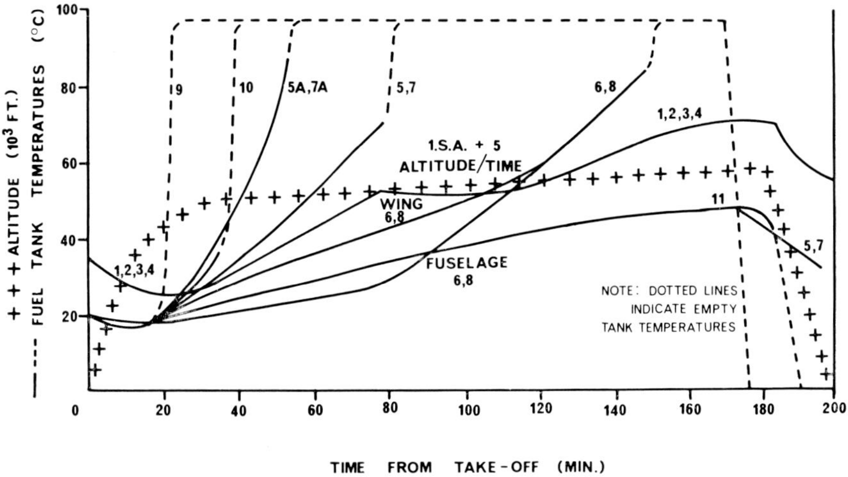

Fig. 6. Predicted fuel temperatures in the fuel tanks of a Production Concorde Aircraft, during supersonic flight.

before it is combusted, and its disposition, particularly between tanks 9, 10, and 11, trims the aircraft in flight.

It should be emphasized that growth conducive temperatures do not guarantee growth, as the critical factor is, as always, the presence of free water. The aircraft is generously equipped with drain points, and void fillers have been extensively used to avoid water trapping. Obviously too, the fuel will not go through such low temperature cycles which condense out water in subsonic aircraft, although some condensation may still occur.

In preliminary experiments, a flask containing 200 ml each of a fuel/mineral salts medium was heated and cooled in a repetitive cycle to simulate possible conditions in tank 11 in an aircraft in continuous service. The temperature cycle used is shown in Fig. 7. Inocula from various sources where thermophilic growth in the presence of oil products was known or suspected were used to infect the aqueous phase, as well as known strains of thermophilic organisms. Every week the flask contents were changed except for 1 ml of the Bushnell and Haas medium which became the inoculum for the next week's growth.

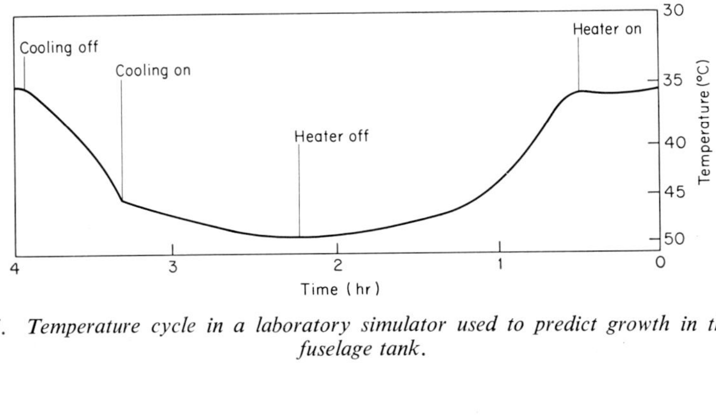

Fig. 7. Temperature cycle in a laboratory simulator used to predict growth in the rear fuselage tank.

Over several months the population settled down to being a mixture of *Pseudomonas* spp., *Nocardia* spp., and *Aspergillus fumigatus*, but these were not necessary thermophils as they may have been growing during the lower temperature of the cycle and merely surviving through the higher temperatures. A typical growth curve of this mixture is shown in Fig. 8 (after Scott and Hill, 1971).

At this point it was considered that there was sufficient justification for a detailed study of microbial growth in this unusual environment and it was decided to pay particular attention to the fungi.

An investigation was undertaken to ascertain whether any hydrocarbon assimilating fungi would be able to survive both the high and low temperatures known to occur during supersonic flight as well as being able to grow in this new environment, and to study the consequences and control of such growth (Thomas, 1973).

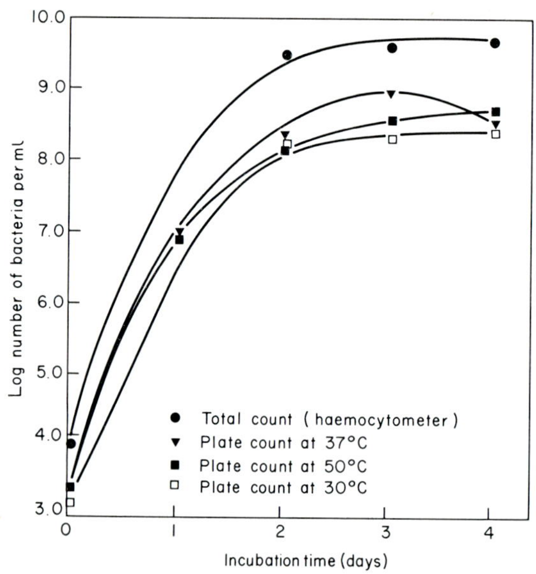

Fig. 8. Growth of bacteria in simulator under conditions as indicated in Fig. 7.

A wide range of fungal isolates obtained from various sources, including thermophilic and thermotolerant fungi and isolates from hydrocarbon environments, were screened for hydrocarbon utilization especially at higher temperatures. Four fungi were selected for further studies; *Cladosporium resinae* and *Paecilomyces varioti*, two mesophilic fungi which commonly occur as contaminants in storage and subsonic aircraft fuel tanks, and two thermotolerant fungi, *Penicillium argillaceum* and an albino mutant strain of *Aspergillus fumigatus.*

The temperature related growth curves of the four selected fungi, when grown in aviation kerosine and mineral salts medium, were established (Figs. 9 and 10). The *C. resinae* isolate grew well over the temperature range 10°C to 37°C, with its optimum at 30°C. A similar growth curve was obtained for *P. varioti*, except growth was barely macroscopically visible at 10°C and 37°C, and the optimum was 25°C. The two thermotolerant fungi, as would be

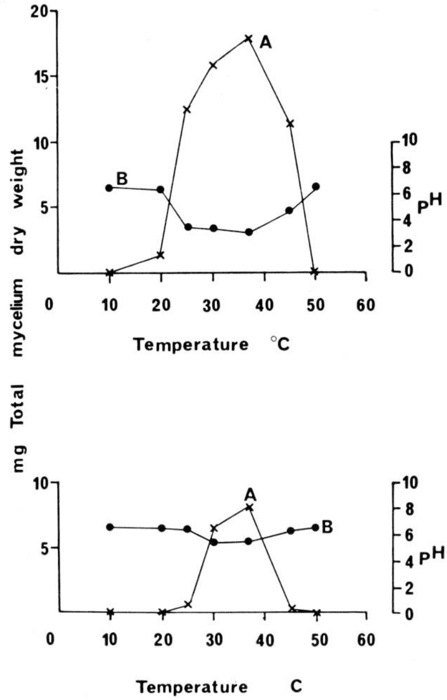

Fig. 9. Temperature related growth curves of Aspergillus fumigatus *(top) and* Penicillium argillaceum *(bottom) in Bushnell and Haas broth and aviation kerosine. A—Mycelial dry weight. B—pH of culture media aqueous phase at time of 'harvesting'.*

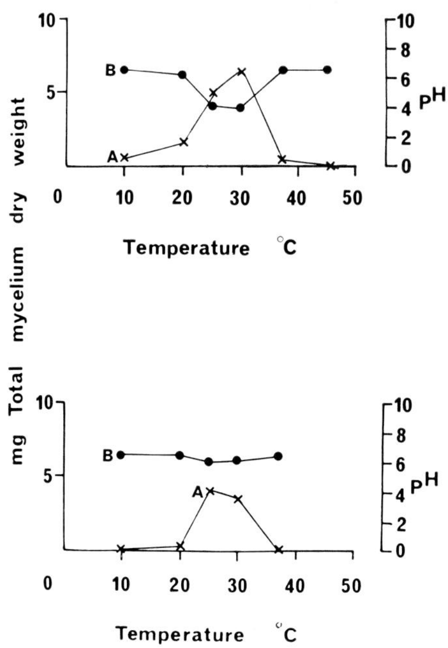

Fig. 10. Temperature related growth curves of Cladosporium resinae *(top) and* Paecilomyces varioti *(bottom) in Bushnell and Haas broth and aviation kerosine. A—Mycelial dry weight. B—pH of culture media aqueous phase at time of 'harvesting'.*

expected, both exhibited better potential for growth in the warm environment of supersonic aircraft fuel tanks. *P. argillaceum* grew within the temperature limits of 20°C and 45°C with an optimum of 37°C. However, although the optimum of the mutant *A. fumigatus* was also 37°C, this isolate grew well at 45°C and had a wide temperature related growth curve between the limits of 10°C and 50°C. This fungus also produced unusual pigments under some conditions. After 4 days, cultures at 37°C produced a yellow pigment soluble in the fuel phase, followed a few days later by red pigment spots in the fungal mycelium itself. The latter, on microscopic examination, were revealed to consist of groups of reddy-brown pigmented 'spicules' ranging from 1·5–4·0 μm in width, which also could be branched and septate. They were the same width as normal hyphae, but had thicker walls and lacked cytoplasmic contents. At 20°C, the cultures of this fungus produced a very pale pink-orange pigmentation in the aqueous phase, after 7 days growth.

Wherever growth occurred, all the four fungi sporulated in the aviation kerosine and mineral salts medium. The conidiospores of *C. resinae* were the same as those produced by the fungus when grown on agar media. Both *P. agrillaceum* and *P. varioti* produced more rounded conidiospores in this environment, compared with the more elongate spores they normally produce on agar media; and, in the latter case, the spores were also larger. The mutant, *A. fumigatus*, produced mainly chlamydospores in the fuel and mineral salts medium. Production of conidiospores by this fungus seemed to be retarded and only developed in older cultures after several weeks, possibly due to an aeration factor.

In all the cultures of the four fungi, where growth occurred, there was an accompanying decrease in pH of the aqueous phase. Further studies were carried out on the effect of the initial pH of the aqueous phase on the growth of the mutant *A. fumigatus* and *C. resinae* in aviation kerosine and mineral salts medium. Both fungi grew in a wide range of initial pH values of the aqueous phase from 3·5 to 8·0, but the mutant *A. fumigatus* grew better when the initial pH values were between 4·5 and 8·0, whereas *C. resinae* grew better at values between 3·5 and 6·0 (Fig. 11).

It was apparent from the temperature related growth curves that both *C. resinae* and the mutant *A. fumigatus* grew far more quickly in the aviation kerosine and mineral salts medium than the other two fungi, since the latter took 6 weeks to produce comparable growth to that produced by the former two in 2 weeks, even though the spore inocula in the latter two cases were known to be slightly greater. Further growth studies, using similar spore inocula, showed that the mutant *A. fumigatus* grew even faster than *C. resinae* in aviation kerosine and mineral salts medium; the former at 30°C and 37°C produced respectively twice and thrice the amount of growth that *C. resinae* produced at 30°C (Fig. 12). By screening for growth on pure alkanes and by gas liquid chromatographic analysis, the mutant *A. fumigatus* was shown to be able to assimilate all the alkanes C_7 to C_{15} inclusively, which comprise the aviation kerosine fraction, and, in fact, produced varying growth responses on pure alkanes ranging from C_6 to C_{36} growing particularly well on the alkanes C_8 to C_{19} inclusively.

The four selected fungal isolates were also screened for growth in aviation kerosine and mineral salts medium when exposed to simulated cyclical temperature conditions of supersonic aircraft fuel tanks. The simulator produced a 4-hr cycle consisting of a 2-hr period of heating up to a set temperature, followed by a half hour of cooling down to an ambient tempera-

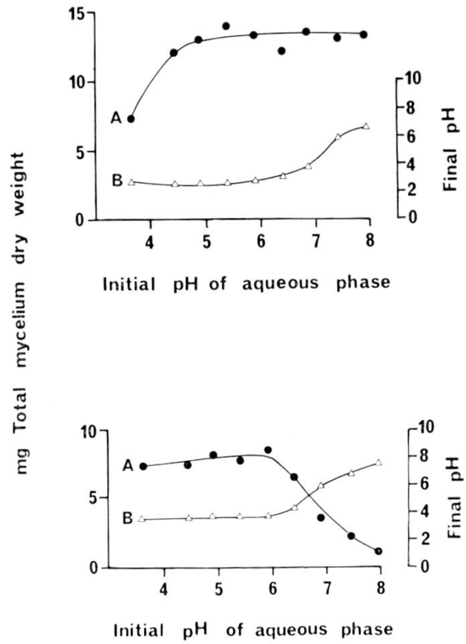

Fig. 11. Growth of Aspergillus fumigatus (*top*) *and* Cladosporium resinae (*bottom*) *in BH broth and aviation kerosine with different initial pH values of the aqueous phase. A—Mycelial dry weight of fungus. B—Final pH of aqueous phase.*

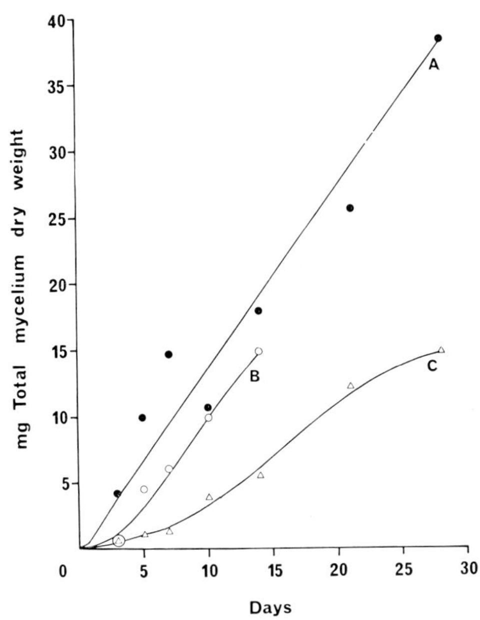

Fig. 12. Growth curves of Aspergillus fumigatus *and* Cladosporium resinae *grown in BH broth and aviation kerosine using equivalent spore concentration inocula. A—Cultures of* A. fumigatus *incubated at 37°C. B—Cultures of* A. fumigatus *incubated at 30°C. C—Cultures of* C. resinae *incubated at 30°C.*

ture of 20°C, at which temperature the simulator remained for a further one and a half hours. The cycles were chosen to represent the temperature conditions experienced in the fuel tanks of a Production Concorde aircraft carrying out 2-hr supersonic flights within every 4-hr period. Three cycles were used, peaking at 40°C, 45°C and 50°C, respectively (Fig. 13). After 3 weeks exposure to a cycle where heating occurred up to 40°C, the mutant *A. fumigatus* was the only one out of the four isolates to produce any significant growth. The others only produced very scant macroscopically visible growth, although control cultures of the three fungi grew well at the respective optimum growth temperature of each of the fungi. The mycelial growth produced by the mutant *A. fumigatus* was equivalent to 62% of its growth produced by controls of the fungus incubated continuously at its optimum growth temperature of 37°C, whereas other controls of the fungus incubated throughout at the ambient temperature of 20°C produced only 9%

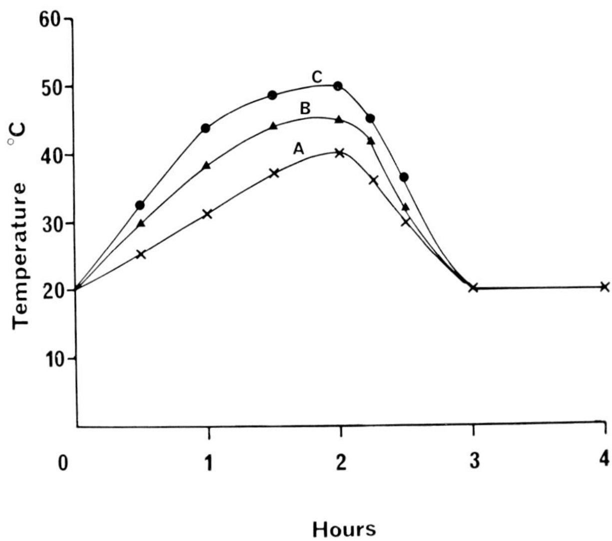

Fig. 13. Cyclical temperature curves in a laboratory simulator with temperature peaks of 40°C (A), 45°C (B) and 50°C (C) respectively.

of the growth produced under optimal temperature conditions. However, when the cycle was set to heat up to 45°C, cultures of the mutant *A. fumigatus* in the simulator only produced 1·4% of the growth produced by control cultures of the fungus grown at the constant optimum temperature. Under the same cyclical conditions, the spore inocula of the other three fungi did not even germinate. Similarly, the spore inocula failed to germinate in cultures of the mutant *A. fumigatus* when exposed to the conditions of a cycle set to heat up to 50°C. In all the above mentioned cases, where the spore inocula failed to germinate, the ungerminated spores did, in fact, germinate normally when incubated at the respective optimum temperature of each fungus.

The cycle heating up to 40°C simulates the temperature conditions produced in tank eleven of a Concorde Production aircraft during a 2-hr flight and those conditions produced in other fuel tanks of the aircraft during flights of shorter duration. Thus, it has been shown that the mutant *A. fumigatus*, under certain conditions, is capable of growing at some of the fuel tank temperatures of Concorde. However, during a 2-hr supersonic flight, the majority of the fuel tanks heat up to temperatures above 50°C under which conditions it has been shown the mutant *A. fumigatus* would not grow, although spores of the fungus may survive at such temperatures, so that growth may occur if more favourable temperature conditions are restored.

Survival studies on the spores of the fungus were carried out to establish their ability to remain viable after exposure to high temperatures produced during supersonic flight and also to low temperatures which might be experienced during prolonged subsonic flight by supersonic aircraft. When examining the survival of fungal spores in the fuel tank environment, their survival has to be considered in relation to both the fuel and aqueous phases, since survival rates may vary considerably in the two media. Even at temperatures within the normal growth limits of the fungus, the survival rates

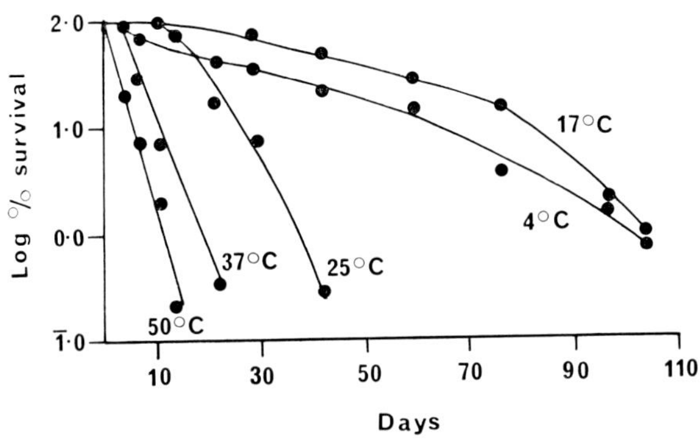

Fig. 14. The survival rates of spores of A. fumigatus *from the same stock culture, in aviation kerosine at different temperatures from 4°C to 50°C.*

of its spores in the fuel phase alone is of importance, since it was shown that the spores will not germinate and grow in the absence of free water. Thus, in fuel tanks at ambient temperatures, containing fuel with no free water, spores would have to be able to remain viable in the fuel until water was available for their further development.

The spore survival rates of the mutant *A. fumigatus*, under the various conditions tested, are only approximate, since some variation in rates was obtained from replicate test samples, which may have been due to differences in aeration of the samples. The survival rates of the spores of the mutant *A. fumigatus* in fuel alone at temperatures within its normal growth limits tended to decrease with increase in temperature, so that 10% of the spores at 4°C were still viable after 70 days, whereas 10% of the spores at 50°C were viable after 6 days and less than 1% of the spores after 12 days; spores exposed to temperatures within those limits exhibited rates between those at 4°C and 50°C (Fig. 14). Similar rates were obtained for spores of the fungus in the aqueous phase alone at 4°C and 50°C; however, at 50°C, less than 1% of the spores were viable after 5 days, whereas at 4°C approximately 50% of the spores were still viable after 50 days. Similar trends of decreased survival with increased temperature, and an ability to survive better in fuel than in water were revealed when spore suspensions of the fungus were heated up to higher temperatures than 50°C. In the fuel phase, 10% of the spores were still viable after 2 days at 60°C, whereas a similar percentage survived only 8 hr at 80°C. Comparatively, in the aqueous phase, 10% of the spores were viable after 4 hr and 1% after 24 hr at 60°C; whereas at 80°C, less than 1% were viable after 1 hr, although, in the latter case, most of the spores were viable after half an hour, but at the time, the test suspension was only just at 70°C and the spores rapidly lost viability as the temperature of the suspension rose to 80°C (Figs. 15, 16, 17, and 18). Thus, a good percentage of the spores of the mutant *A. fumigatus* would be able to survive in the fuel phase of fuel tanks of supersonic aircraft during the average flight period, even when the fuel reached temperatures of 80°C; however, they would be able to survive far less well in the aqueous phase under similar conditions. The water content of the fuel itself, however, would probably not affect spore survival rate, since spores of the fungus survived equally well in water-saturated and dry aviation kerosine.

The spores of the mutant *A. fumigatus* were also shown to be capable of surviving any low temperatures which might be experienced during subsonic flight. Exposure of spores of the fungus to −32°C appeared to have little effect upon them, especially in the fuel phase. Whether the spores were removed after half an hour's exposure or after 4 hr made no difference. After these exposure times, approximately 67% of the spores in the aqueous phase were viable, while about 90% of those in the fuel phase were still viable. In samples which were exposed to cycles of half an hour at −32°C followed by an hour of warming up to room temperature again, the cycling appeared to have little effect on the spores in the aqueous phase. After three such cycles, the spores in water still exhibited 58% viability. However, in the fuel phase, after the second cycle, there was a sharp decrease in spore viability, only 3% still remaining viable. The third cycle, though, had little further effect as 1·5% of the spores were then still viable. After the first cycle, though, about 90% of the spores were viable, as after the prolonged 4-hr exposure at −32°C. Thus, at low temperature, the spores of the fungus exhibited better survival in the aqueous phase than in the fuel phase, whereas at high temperatures the opposite was found.

Although it was shown that *C. resinae* would probably not grow in supersonic aircraft fuel tanks, its spores, however, might be able to survive and be able to germinate under favourable conditions for their growth, such as in a grounded aircraft. Somewhat surprisingly, the spores of *C. resinae* survived better in fuel even at 80°C, than did the spores of the mutant *A. fumigatus*. However, in the aqueous phase, the opposite was found; so that in a test suspension heated up to 60°C, none of the 50% of spores, which had been viable when the suspension reached 60°C, were viable after an hour at that temperature (Figs. 15, 16, 17, and 18).

Thus, the mutant *A. fumigatus* would not only be able to grow at the temperatures of some of the fuel tanks of supersonic aircraft, but, also, its spores might be able to survive in those tanks which are heated to relatively high temperatures during supersonic flight. The consequences of such fungal growth could cause the same problems as occur in subsonic aircraft as a result of fungal growth, such as blockage of fuel filters and fuel pumps, and

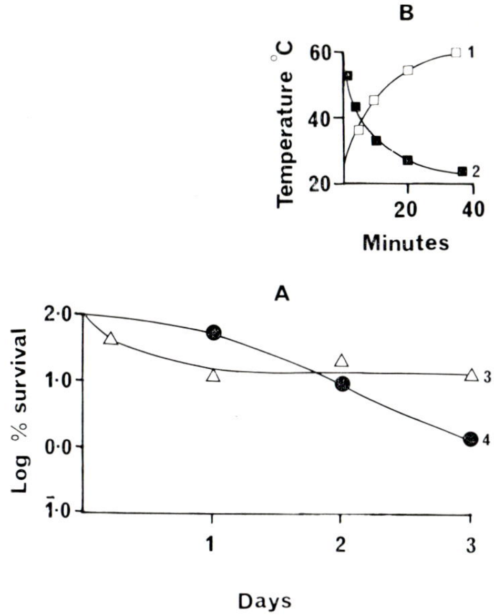

Fig. 15. A—The survival rates of spores of A. fumigatus *and* C. resinae *in aviation kerosine at 60°C. B—Graph inset of the rates of heating and cooling. 1—Rate of heating. 2—Rate of cooling. 3—*C. resinae. *4—*A. fumigatus.

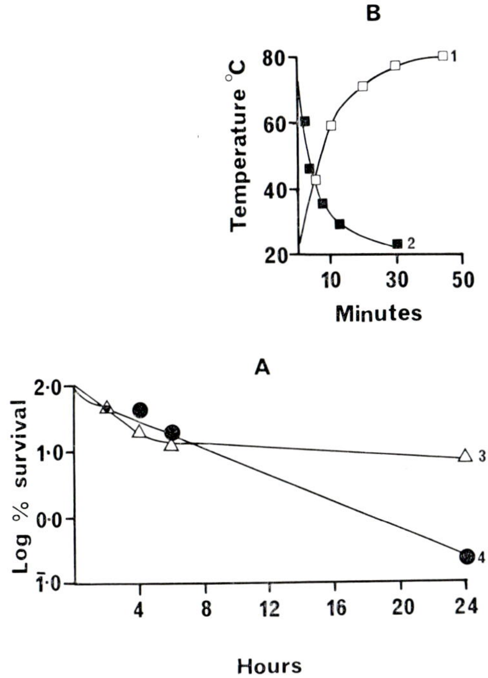

Fig. 16. A—The survival rates of spores of A. fumigatus *and* C. resinae *in aviation kerosine at 80°C. B—Graph inset of the rates of heating and cooling. 1—Rate of heating. 2—Rate of cooling. 3—*C. resinae. *4—*A. fumigatus.

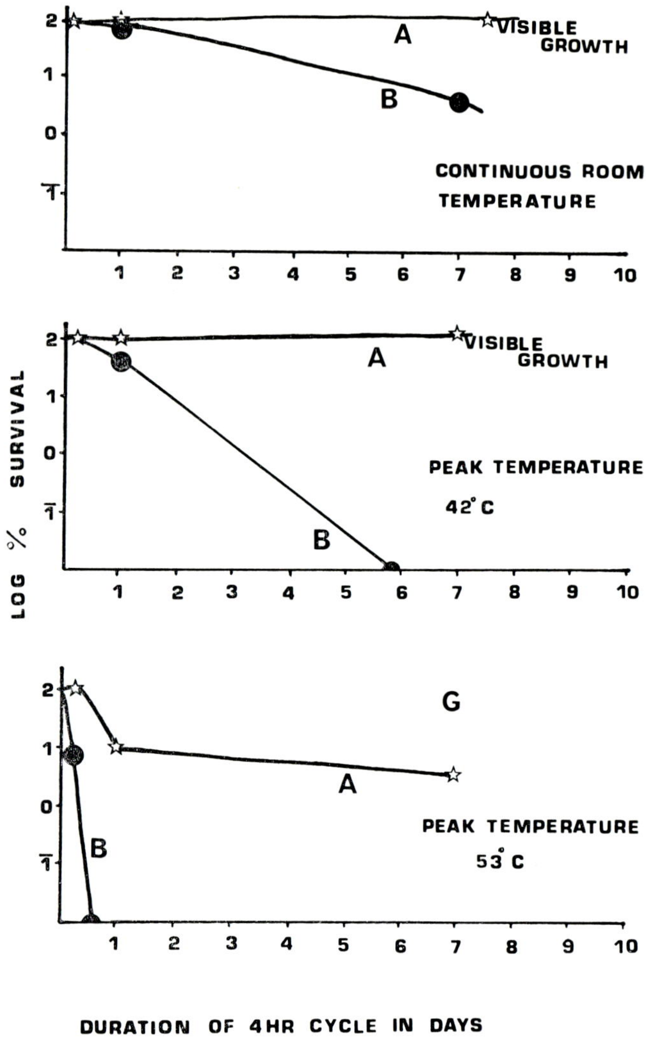

Fig. 19. Survival of A. fumigatus *spores in simulator with 4 hr temperature cycle peaking at 42°C, 53°C compared to continuous room temperature. A—Control no biocide. B—0·15% EGME.*

DISCUSSION

An albino mutant strain of *A. fumigatus* (isolated from an hydraulic oil sample from an aircraft test rig) has been shown capable of growth in some of the temperature conditions experienced in the fuel tanks of supersonic aircraft, and as a consequence of its growth causing corrosion of the fuel tank aluminium alloy. Its spores, also, appear to be able to remain viable even in fuel which is heated to temperatures as high as 80°C.

Type strains of *A. fumigatus* tested for their ability to assimilate aviation kerosine gave varied responses, but even those which gave positive results

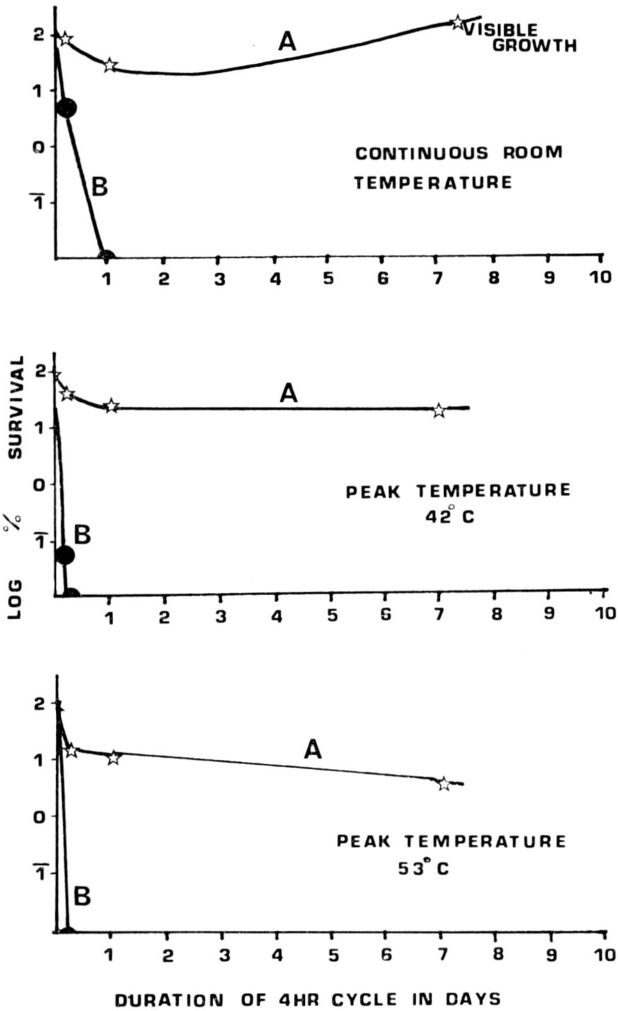

Fig. 20. Survival of A. fumigatus *spores in simulator with 4 hr temperature cycle peaking at 42°C, 53°C compared to continuous room temperature. A—Control no biocide. B—0·00027% of 10% solution of RH886 in EGME.*

produced only very little mycelial growth after prolonged incubation, and then it was not interfacial between the fuel and aqueous phases. One of these strains, however, when tested on pure alkanes grew well especially on the alkanes C_{10} to C_{19} inclusively. Nyns *et al.* (1968) found that different strains of a species had widely differing hydrocarbon assimilating abilities, including strains of *A. fumigatus*. These same workers noted that petroleum fractions appear to be more toxic for fungi than individual alkanes, as some strains were able to assimilate alkanes only when these were extracted from gas oil.

Further isolates of the hydrocarbon assimilating albino mutant strain were subsequently obtained from crude oil samples, and, also, a pale green pigmented strain of *A. fumigatus* was isolated from an hydraulic oil sample.

The latter assimilated hydrocarbons and produced a similar temperature related growth curve and similar morphological growth characteristics in aviation kerosine and mineral salts medium, as those of the original albino mutant strain. Thus, hydrocarbon assimilating strains of *A. fumigatus* appear to occur frequently. The nonhydrocarbon assimilating strains of this fungus might possibly adapt to hydrocarbon assimilation under certain conditions. Also, other hydrocarbon assimilating strains of thermophilic fungi may exist even though those tested during the course of this work failed to assimilate hydrocarbons. On the other hand, mesophilic fungal strains may adapt or even already exist, that are capable of growth in supersonic aircraft fuel tanks. Although none of the *C. resinae* strains examined during this research grew at 45°C, isolates of this species have been reported to be capable of growth at temperatures up to 50°C (Parbery, 1971). Certainly, one strain of *C. resinae* used in this research study, although unable to grow at temperatures above 37°C, proved capable of survival in fuel heating up to 80°C.

From a biological point of view there would thus appear to be an increased potential hazard from microorganisms in the fuel of supersonic transports, as unlike their subsonic counterparts, temperatures are growth stimulating and not growth inhibiting in flight. Fortunately there is likely to be less condensate water in the fuel and biocides, if they are ever needed, will be very much more effective than in subsonic aircraft.

Should biological contamination occur the selective pressures which have operated in subsonic aircraft in favour of *C. resinae* will be less relevant, and new selective pressures will favour growth of strains of *A. fumigatus*.

REFERENCES

Bakanauskas, S. (1958). *Bacterial activity in JP-4 fuel.* WADC Technical Report 58-32.

Bushnell, L. D., and Haas, H. F. (1941). The utilization of certain hydrocarbons by microorganisms. *J. Bact.*, **41**, 653.

Hazzard, G. F. (1961). *Fungal growths in aviation fuel systems. Part I.* Report 252. Australian Defence Standards Laboratories.

Hill, E. C. (1971). Microbiological aspects of Concorde and other supersonic transports. *Proc. G.I.A.M.*, **III**, 203–204. Examiner Press, Bombay.

Nyns, E. J., Auquiere, J. P., and Wiaux, A. L. (1968). Taxonomic value of property of fungi to assimilate hydrocarbons. *Antonie van Leeuwenhoek*, **34**, 441.

Parbery, D. G. (1971). Physical factors influencing growth of Amorphotheca resinae in culture. *Int. Biodetn. Bull.*, **7**, 5.

Prince, A. E. (1961). Microbiological sludge in jet aircraft fuel. *Develop. Ind. Microbiol*, **2**, 197–203. Plenum Press, New York.

Reece, A. P., and Hill, E. C. (Unpubl. data). Univ. Coll., Cardiff.

Scott, J. A., and Hill, E. C. (1971). *Microbiology.* Applied Science Publishers Ltd., London, pp. 25–41.

Thomas, A. R. (1973). Fungal growths in petroleum products. Ph.D. Thesis. Univ. of Wales, Cardiff. 150 pp.

Thomas, A. R., John, D., and Hill, E. C. (1973–5). Reports for M.o.D. Contract: AT/2072/021 MAT. Univ. Coll., Cardiff.

BIODEGRADATION OF BUNKER 6C FUEL OIL

D. LIU and P. T. S. WONG

Environmental Protection Service, Environment Canada,
Canada Centre for Inland Waters, Burlington, Ontario, Canada

Summary

A bacterium isolated from the soil of a local refinery could degrade Bunker 6C fuel oil rapidly. Infrared and gas chromatographic analyses indicated that the major components of Gulf Bunker 6C fuel oil contained medium and high molecular weight *n*-alkanes with a relatively small portion of aromatics. The gas chromatographic spectrum also pointed out that the *n*-alkane fractions were the primary target for degradation by the bacterial isolate. Manometric experiments demonstrated that the bacterial culture could oxidize the Bunker 6C oil over a wide range of concentrations.

INTRODUCTION

Crude and Bunker fuel oils do not persist for a long period of time in the natural environment even when relatively large quantities of oil have been spilled. For example, 14 months after the grounding of the tanker 'Arrow' off the Nova Scotia coast, the average concentration of Bunker C fuel oil in the waters of Chedabucto Bay was found to drop from 90 ppb to 1·5 ppb (1·5 μg/litre), a level of petroleum residue normally found in the marine waters off the Atlantic coast of Canada (Gordon and Michalik, 1971).

The disappearance of the oil in the sea is probably due in large part to the initial degradation by the action of sunlight followed by microbial attack (Davies and Hughes, 1968) or due to the naturally occurring processes such as wave agitation and chemical and biological degradation (Levy, 1971). While microbial degradation of crude oil has received much attention (Atlas and Bartha, 1972; Bridié and Bos, 1971; Soli and Bens, 1972), little work has been reported in the literature with regard to microbial degradation of Bunker C fuel oil. Most investigations were either related to the determination of Bunker oil in the water phase (Forrester, 1971; Jeltes and den Tonkelaar, 1972; Levy, 1972) or to the identification of the original oil source (Ehrhardt and Blumer, 1972).

In recent studies, we have reported that a bacterial culture isolated from a refinery soil could degrade Bunker C fuel oil (Liu and Dutka, 1972). In this paper, we present data on the rate and extent of the degradation of Bunker 6C fuel oil by this bacterium and on the effect of oil concentrations on cell yield and respiration rate.

MATERIALS AND METHODS

Organism

The bacterium, BMO1, used throughout these studies was isolated from soil samples taken from a local refinery (Liu and Dutka, 1972). Preliminary

Fig. 1.

identification of the isolated bacterium indicated it was a *Pseudomonas* species, according to the scheme of *Bergey's Manual* (Breed *et al.*, 1957). It grew rapidly on Bunker 6C oil medium (Fig. 1).

Medium

The mineral salt medium used throughout the studies, unless otherwise specified, contained the following (expressed in grams): K_2HPO_4 anhydrous, 0·66; KH_2PO_4, 0·41; $MgCl_2 \cdot 6H_2O$, 0·10; $FeCl_2 \cdot 4H_2O$, 0·05; $MnCl_2 \cdot 4H_2O$, 0·002; NaCl, 0·10; $(NH_4)_2SO_4$, 1·0; into 1 litre of distilled water. The final pH of the medium was 6·9. The medium was sterilized at 121°C for 15 min, Gulf Bunker 6C fuel oil was sterilized separatedly under the same conditions.

Measurement of cell dry weight

Even after the bacterial growth, small amounts of oil residue still could be found in the culture suspension which would influence the conventional dry weight determination. We found that glass wool could absorb the oil residue and thus would facilitate the dry weight determination.

Fifty millilitres of bacterial suspension was filtered through loosely packed 2-inches thick glass wool and 30 ml of the filtrate was centrifuged at 15 000 × *g* for 20 min. The cell pellet in the centrifuge tube was washed twice in distilled water and was resuspended in 5 ml of distilled water. The washed cell suspension was dried for 20 hr at 105°C.

Respiration studies

In these experiments, BMO1 was grown in a New Brunswick Microferm fermentor for 48 hr before harvesting for use in the respiration studies. Oxygen uptake by the cells was followed at 20°C with the conventional method, using the Gilson Differential Respirometer. After 5 min equilibration, the substrate was added from the side arm into the main chamber of the Warburg flask. The CO_2 was absorbed in 0·2 ml of 20% KOH placed in the centre well. The total fluid volume in each flask was 3·2 ml and the temperature was 20°C.

Gas chromatographic analysis of Bunker 6C fuel oil

Bunker 6C oil was analysed with a gas chromatograph (Beckman GC-65) equipped with a dual flame ionization detector. The dual 1·8 m × 0·63 cm O.D. stainless steel columns contained 2% OV-1 liquid phase on 80–100 mesh chromosorb W-AW-DMCS. Then 0·1 min after sample injection, the oven temperature was programmed to increase from 60°C to 300°C at a rate of 5°C/min and was kept at 300°C isothermally for another 30 min. The temperature at injector and detector was maintained at 250°C and 325°C, respectively. The carrier gas was helium at a flow rate of 60 ml/min. The hydrogen and air were kept at a flow rate of 45 ml/min and 300 ml/min, respectively.

Recovery of Bunker 6C oil from culture suspension

A 200 ml unfiltered culture suspension was centrifuged at 15 000 × *g* for 20 min at 40°C and the supernatant fluid was extracted twice with 5 ml of *n*-hexane. To help break the emulsion, the top layer containing Bunker C oil was centrifuged at 5000 × *g* for 10 min at 4°C. The clear top hexane extract could be easily transferred to a 10 ml volumetric flask with a pasteur pipette and made to the volume with *n*-hexane. This *n*-hexane extract was suitable for immediate gas chromatographic analysis. The extractory procedure covers the major fractions of Bunker C fuel oil (Zajic and Supplisson, 1972).

Infrared analysis of Bunker 6C fuel oil

The infrared spectrum was made with a Pye Unicam SP 1000 IR spectrophotometer, using the potassium bromide (KBr) plate technique. Three to four drops of 50% Bunker 6C fuel oil were evenly spread over two KBr plates to form a thin film of oil. The *n*-hexane solvent was allowed to evaporate in about 15 min, whereupon the KBr plates were pressed together gently to obtain a uniform oil film thickness.

RESULTS

The culture BMO1 was selected for this study from seven other bacterial isolates on the basis of its rapid growth on Bunker 6C fuel oil as the sole carbon and energy source. Figure 2 shows the infrared spectrum of Gulf Bunker 6C fuel oil. This was prepared to give more information on the

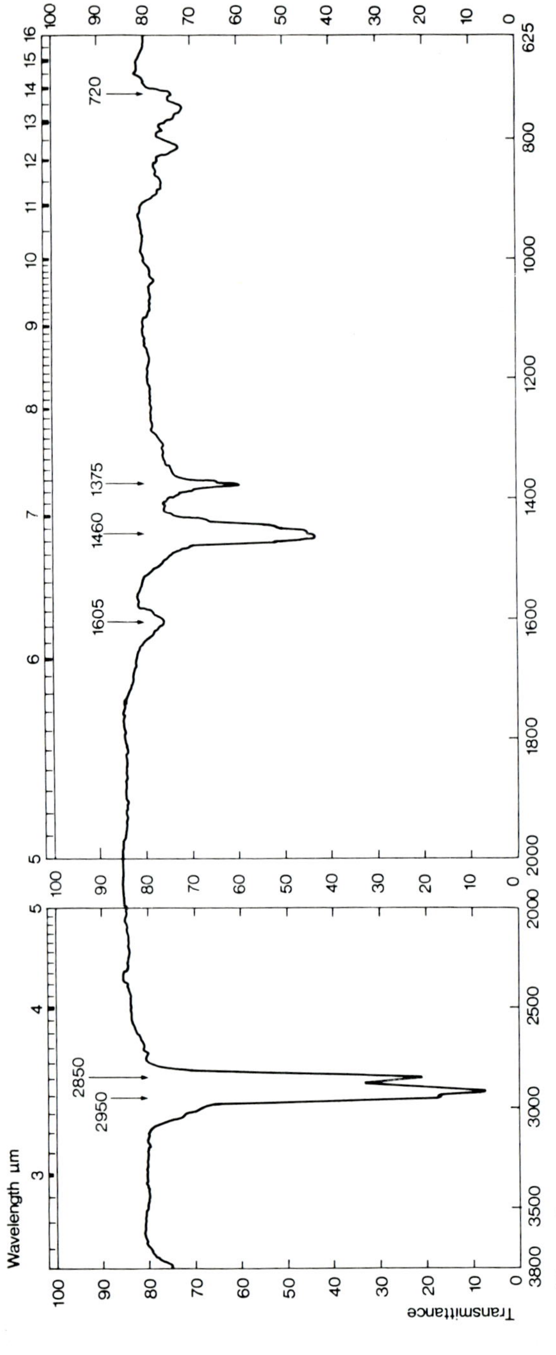

Fig. 2. Infrared spectrum of Gulf Bunker 6C fuel oil.

structure of Bunker 6C fuel oil and to obtain some insight about its major components. The strong absorption bands at 2850 and 2950 cm^{-1} were due to the C–H stretching of alkyl-CH_3, suggesting the presence of high concentrations of alkanes in the Bunker 6C fuel oil. The occurrence of 1375 cm^{-1} and 1460 cm^{-1} bands was due, respectively, to the symmetrical and asymmetrical bending vibration of CH_3 in the alkane molecule (Cross and Jones, 1967). The weak band at 1605 cm^{-1} suggested the presence of low concentration of aromatics in this fuel oil. The gas chromatographic study (Fig. 3) further confirms the findings of infrared analysis that the major components of Bunker 6C fuel oil were *n*-alkane in nature, ranging from *n*-nonane C_9 to *n*-hexatriacotane C_{36}.

To study microbial degradation of oil, it is essential to sterilize the medium. Since membrane sterilization techniques cannot be applied to Bunker 6C fuel oil due to its inherent viscosity, experiments were conducted to investigate the feasibility of using stem autoclaving. The results in Table I showed that

Table I

EFFECT OF STERILIZATION ON THE LOSS OF HYDROCARBON

Hydrocarbon	*% lost*[a] *after autoclaving*
n-heptane (C7)	28·20
n-octane (C8)	1·57
n-nonane (C9)	1·08
n-decane (C10)	1·72
n-dodecane (C12)	0·81
n-tetradecane (C14)	0·18
n-hexadecane (C16)	0·01
Kerosene	0·37
Bunker 6C	0·23

[a] Results are averages of duplicate samples which have been corrected from the control tube containing no hydrocarbon.
Gas chromatographic grade *n*-alkane (between 2–3 g) was weighed to an accuracy of $\pm 0{\cdot}1$ mg in a loosely screw capped culture tube. After autoclaving at 121°C for 15 min, the loss in weight was considered due to autoclaving.

only the light fraction of the reference (C_7) *n*-heptane had considerable evaporation by autoclaving, whereas those hydrocarbons that contained eight carbons or more, including Bunker 6C oil, could be sterilized by autoclaving as described in Materials and Methods. Figure 4 and Table II demonstrated the ability of culture BMO1 to metabolize Bunker 6C oil. After 4 days of incubation, most of the major components of the Bunker 6C fuel oil had been utilized by the bacteria for growth. The gas chromatographic analysis also disclosed that *n*-alkane fractions in Bunker 6C oil were the primary target for degradation by the bacterial culture.

Growth yield study is another convenient way to check whether a bacterium can utilize a particular compound. This is especially true in the study of hydrocarbon biodegradation because Bunker fuel oil contains a mixture of many different kinds of hydrocarbon components. Under carefully controlled experiments, the cell yield would be a reliable indicator that some of the carbon moieties of the hydrocarbon molecule have been degraded and

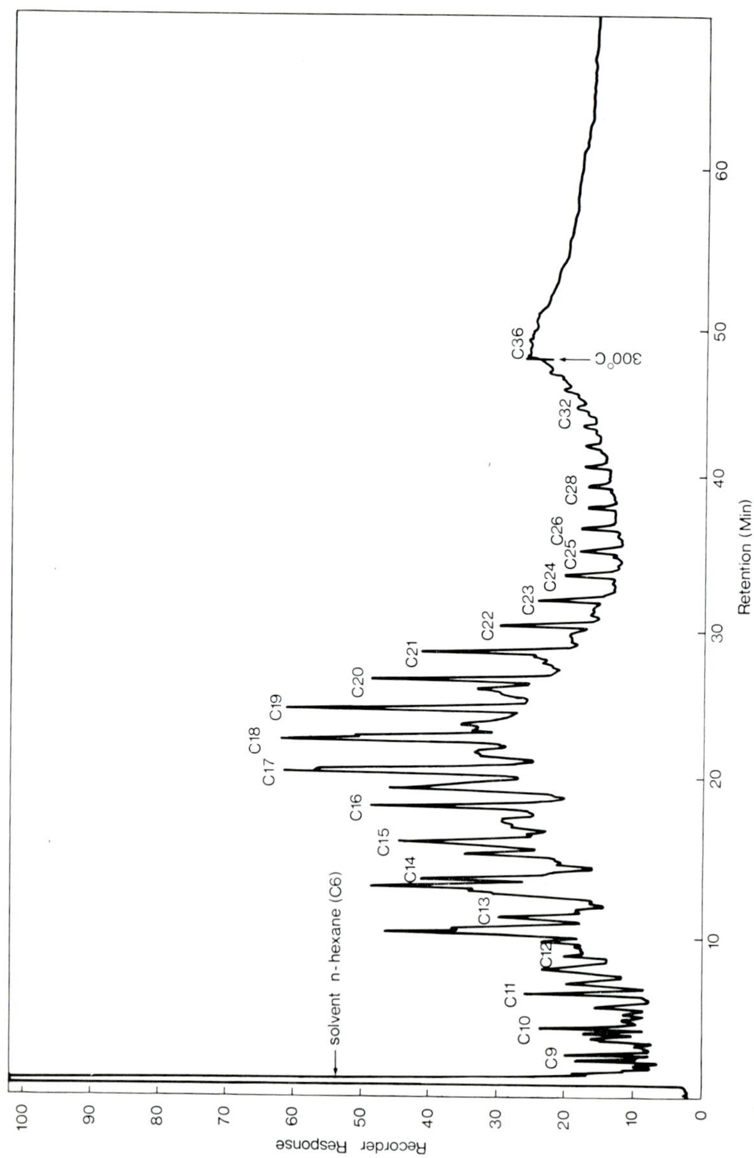

Fig. 3. Gas chromatogram of Gulf Bunker 6C fuel oil.

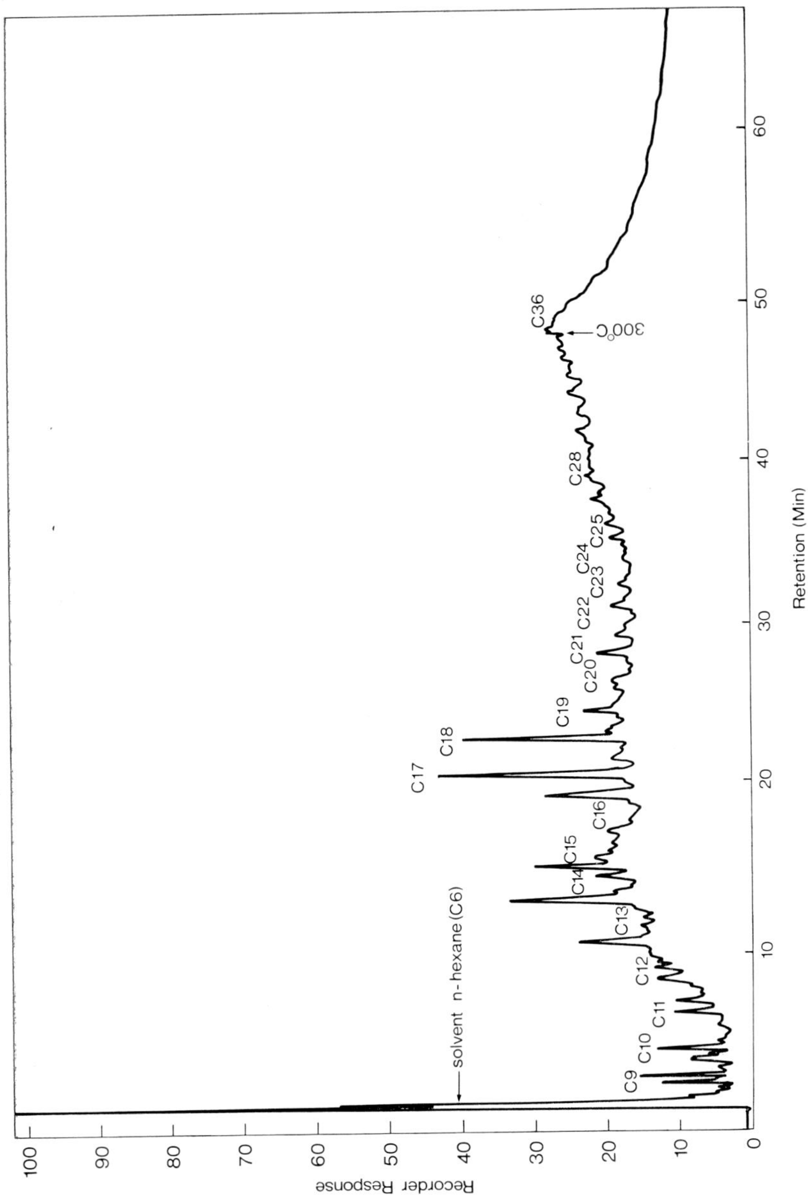

Fig. 4. Gas chromatogram of Gulf Bunker 6C fuel oil after the growth of culture BMO1 for 4 days.

Table II
MEASURED AND CALCULATED AMOUNTS OF *n*-ALKANES DEGRADED IN BUNKER 6C FUEL OIL BY CULTURE BMO1 AFTER 4 DAYS INCUBATION

n-Alkane carbon no.	*% Degraded*[a]
9	80
10	73
11	81
12	67
13	74
14	77
15	74
16	83
17	65
18	68
19	81
20	81
21	75
22	72
23	63
24	57
25	49
26	47
28	37
32	34
36	23
Total Bunker 6C degraded	69

[a] Calculated from the area under the peak of gas chromatograms as shown in Figs. 3 and 4.

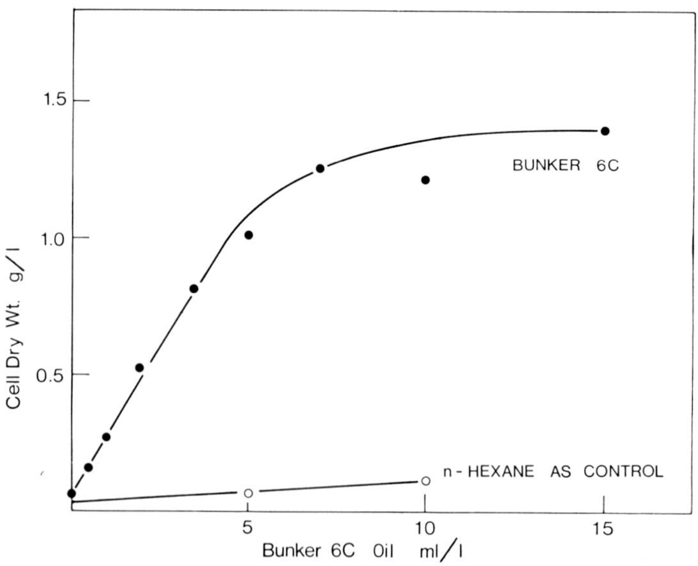

Fig. 5. Effect of Bunker 6C fuel oil on the cell yield of culture BMO1. Each Erlenmeyer flask contained 100 ml of basal medium, 2 ml of cell inoculum, and various amounts of Bunker 6C fuel oil (50% Bunker C in n*-hexane w/w). The experiments were carried out at 20°C on a rotary shaker at 300 strokes/min for 4 days.*

incorporated into the bacterial biomass. Figure 5 shows the relationship between carbon concentration and cell growth. In this experiment, various concentrations of Bunker 6C fuel oil were added to the mineral salt medium. As shown in Fig. 5, it is apparent that the addition of increasing concentrations of Bunker 6C oil to the culture BMO1 in the mineral salt medium resulted in a linear increase of cell yield up to the level of 7 ml of Bunker 6C oil/litre. No significant increase in cell yield was noted above this concentration of oil. The results in Fig. 5 also showed that *n*-hexane supported little or no growth of BMO1. Therefore, *n*-hexane was used to dilute the Bunker 6C to 50% (w/w) to overcome the problem of viscosity in the growth yield study.

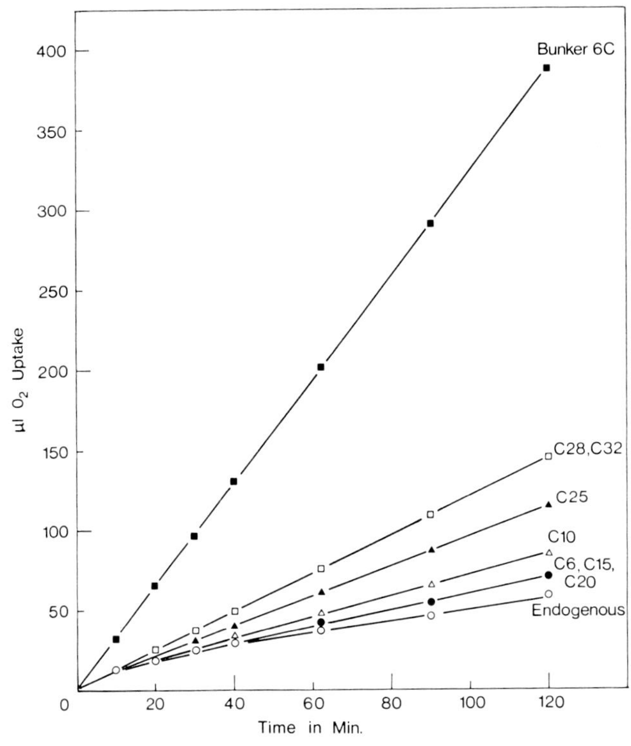

Fig. 6. Effect of various substrates on the rate of oxygen consumption by culture BMO1. Each Warburg flask contained 1 ml of 0·05 M *phosphate buffer (pH 7·0), 1 ml of cell suspension (8·1 mg dry wt), 0·9 ml of H_2O, and 100 μl or 100 mg of substrate.*

The ability of the bacterial culture to degrade Bunker 6C oil was further substantiated by the manometric studies. As shown in Fig. 6, the culture BMO1 had a greater affinity for Bunker 6C oil (Qo_2 = 24·1) than any other kind of *n*-alkanes tested. It is of interest to note that high molecular weight *n*-alkanes, such as *n*-octacosane, C_{28} (Qo_2 = 9·25), were oxidized at a rate faster than the lower ones, such as *n*-hexane (Qo_2 = 4·9).

The success of using hydrocarbon-degrading bacteria to treat oil spills depends on the ability of the bacterial culture to degrade a wide range of oil concentrations in the natural environment. To test the effect of the oil concentration on the rate of oil degradation by the bacterial culture BMO1, various amounts of Bunker 6C oil were added to the Warburg flask. The manometric studies showed the ability of the culture BMO1 to oxidize Bunker 6C oil over a wide range of concentrations (Fig. 7), and thus further substantiated our previous findings that culture BMO1 could degrade Bunker 6C oil rapidly. Figure 8 is the replot of Fig. 7 using the net oxygen uptake values at 60 min against the oil concentration in the Warburg flask. This was intended to show the saturation curve of Bunker 6C oil for the culture BMO1. From the graph, it is obvious that there is a linear relationship between oxygen uptake and the Bunker 6C oil at concentrations below 25 μl of oil per flask (8·3 ml/litre). Increase in oil concentration from 25 to 200 μl

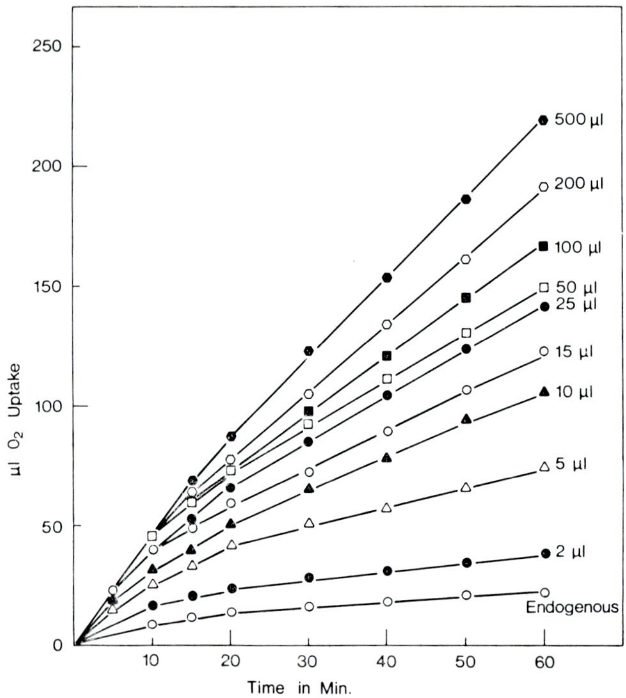

Fig. 7. Effect of varying amounts of Bunker 6C fuel oil on the rate of oxygen uptake by culture BMO1. Each Warburg flask contained 1 ml of 0·05 M *phosphate buffer (pH 7·0), 1 ml of cell suspension (7·9 mg dry wt), and various amounts of Bunker 6C fuel oil (50% Bunker 6C in* n-*hexane w/w).*

stimulated some more oxygen consumption. Further increase to 500 μl did not cause any more oxygen uptake. The rate of hydrocarbon oxidation by bacteria has been demonstrated to be controlled by the oil-water interfacial area (Liu and Townsley, 1970; Wang and Ochoa, 1972).

DISCUSSION

Infrared and gas chromatographic analyses show that the major components of Gulf Bunker 6C fuel oil are of medium and high molecular weight *n*-alkanes with a relative low concentration of aromatics. Due to its high boiling point, Bunker 6C fuel oil can be safely autoclaved without any significant loss of its components. This is probably related to the boiling point of the individual components. For example, 28% of *n*-heptane (b.p. 98·4°C) was lost after autoclaving at 121°C for 15 min; however, its eight-carbon homolog, *n*-octane (b.p. 125·7°C) lost only 1·6% under the same condition.

It is interesting to note that 69% of Bunker 6C oil was degraded by culture BMO1 after 4 days' incubation. It appears that there is no preference for culture BMO1 to degrade odd or even carbon substrate. The growth yield study would further strengthen our belief that BMO1 could use Bunker 6C fuel oil as a sole carbon and energy source as shown by the linear relationship between cell yield and oil concentrations.

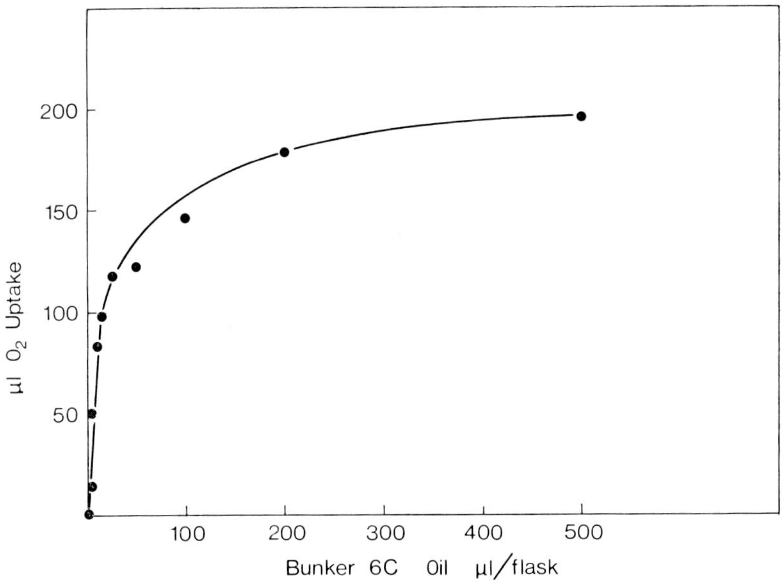

Fig. 8. Concentration of Bunker 6C fuel oil as a function of oxygen uptake by culture BMO1. The experimental conditions were the same as shown in Fig. 7.

While microbial degradation of oil is demonstrable by the gas chromatographic technique, further refinements of this technique are necessary before accurate quantitative estimates of oil degradation can be made. It appears that the manometric technique still holds a considerable promise for future studies on microbial oil degradation, because the amount of oxygen uptake would reflect the amount of oil being degraded under carefully controlled conditions. The manometric experiments had clearly reflected the preference of Bunker 6C fuel oil by culture BMO1 to any other *n*-alkanes tested. The question as to whether the culture will degrade the fuel oil in a natural environment must await further research.

REFERENCES

Atlas, R. M., and Bartha, R. (1972). Degradation and mineralization of petroleum by two bacteria isolated from coast waters. *Biotechnol. Bioeng.*, **14**, 297–308.

Breed, R. S., Murray, F. G. D., and Smith, N. R. (1957). *Bergey's Manual of Determinative Bacteriology*, 7th Ed. Williams and Wilkins, Baltimore.

Bridié, A. L., and Bos, J. (1971). Biological degradation of mineral oil in sea water. *J. Inst. Petrol.*, **57**, 270–277.

Cross, A. D., and Jones, R. A. (1969). *An Introduction to Practical Infrared Spectroscopy*, 3rd Ed. Butterworth, London, pp. 73–80.

Davies, J. A., and Hughes, D. E. (1968). The biochemistry and microbiology of crude oil degradation, pp. 139–144. In *The Biological Effects of Oil Pollution on Littoral Communities* (J. D. Carthy and D. R. Arthur, eds.). Field Studies Council, London.

Ehrhardt, M., and Blumer, M. (1972). The source identification of marine hydrocarbons by gas chromatography. *Environ. Pollut.*, **3**, 179–194.

Forrester, W. D. (1971). Distribution of suspended oil particles following the grounding of the Tanker 'Arrow'. *J. Mar. Res.*, **29**, 151–170.

Gordon, D. C., and Michalik, P. A. (1971). Concentration of Bunker C fuel oil in the waters of Chedabucto Bay, April, 1971. *J. Fish. Res. Bd. Can.*, **28**, 1912–1914.

Jeltes, R., and den Tonkelaar, W. A. M. (1972). Gas chromatography versus infrared spectrometry for determination of mineral oil dissolved in water. *Water Res.*, **6**, 271–278.

Levy, E. M. (1971). The presence of petroleum residues off the east coast of Nova Scotia, in the Gulf of St Lawrence and the St Lawrence River. *Water Res.*, **5**, 723–733.

Levy, E. M. (1972). The identification of petroleum products in the marine environment by absorption spectrophotometry. *Water Res.*, **6**, 57–69.

Liu, D., and Dutka, B. J. (1972). Bacterial seeding techniques: novel approach to oil spill problems. *Can. Res. Develop.*, **5** (4), 17–20.

Liu, D., and Townsley, P. M. (1970). Lignosulfonates in petroleum fermentation. *J. Water Pollut. Control Fed.*, **42**, 531–537.

Soli, G., and Bens, E. M. (1972). Bacteria which attack petroleum hydrocarbons in a saline medium. *Biotechnol. Bioeng.*, **14**, 319–330.

Wang, D. I. C., and Ochoa, A. (1972). Measurement on the interfacial areas of hydrocarbon in yeast fermentations and relationships to specific growth rates. *Biotechnol. Bioeng.*, **14**, 345–360.

Zajic, J. E., and Supplisson, B. (1972). Emulsification and degradation of 'Bunker C' fuel oil by microorganisms. *Biotechnol. Bioeng.*, **14**, 331–343.

BIODEGRADATION OF HIGH-OCTANE GASOLINE

V. W. JAMISON, R. L. RAYMOND and J. O. HUDSON

Sun Ventures, Inc., Marcus Hook, Pennsylvania 19061, U.S.A.

Summary

Thirty-two microorganisms isolated from groundwater that had been contaminated with high-octane gasoline were used to study the biodegradation of individual components of the gasoline. The paraffins and aromatics were sources of carbon for the *Nocardia* and *Pseudomonas* cultures, respectively. However, very few of the remaining components supported growth of the bacteria. Co-oxidation experiments were carried out to determine if co-oxidation could be a possible mechanism for removal of the nongrowth components.

INTRODUCTION

There is relatively little available literature on the utilization of gasoline as either an oxidizable substrate or as the sole source of carbon for microbes. Litchfield and Clark (1973) and McKee *et al.* (1972) referred to the existence of several genera of bacteria present in groundwaters that have been contaminated with gasoline. They tentatively classified the predominant bacteria as belonging to the genera *Pseudomonas* and *Arthrobacter*.

In July 1971, Sun Pipe Line Co. experienced a break in a gasoline pipeline in the area of Ambler, Pa. This spill contaminated the groundwater that was used as the municipal water supply. After physical removal methods for recovery of the contaminating gasoline became nonproductive, biological stimulation of the indigenous hydrocarbon-utilizing microbial flora was carried out. The groundwater was injected with ammonium sulphate, monosodium phosphate, and disodium phosphate while air was sparged into injection wells. After one and one-half years, there was no odour of gasoline or any detectable gasoline when monitoring the producing water wells with an ultraviolet detector.

During the stimulation programme, 32 cultures were isolated from groundwater that was contaminated with gasoline. These cultures that were isolated from the dolomite reservoir used gasoline as a source of carbon. Seventeen of the cultures were gram-negative and the other 15 were gram-positive. Eleven of the gram-negative cultures were motile. Of these, eight had polar flagella and were assumed to be pseudomonads. Ten of the gram-positive cultures were filamentous rods that fragmented after 20 hr and appeared to be actinomycetes. Two of the gram-positive cultures were cocci and were placed in the genus *Micrococcus*. Four of the gram-negative cultures fit the description of *Acinetobacter*. One of the gram-negative cultures has been classified as *Flavobacterium devorans*. The remaining gram-positive culture as well as the remaining six gram-negative cultures have not been classified.

The aim of this preliminary study was to determine the mode of the degradation of the gasoline that had contaminated the groundwater. Growth on the individual components of gasoline as well as co-oxidation experiments will be discussed.

MATERIALS AND METHODS

Microbes

The microorganisms used in this work were those isolated from the groundwater system in the dolomite reservoir. All of the pure cultures could utilize gasoline as the only source of carbon.

Culture conditions

The microorganisms were maintained on nutrient agar slants (Difco). All growth studies were done using a basal mineral salts medium (MM) of the following composition: KH_2PO_4, 0·4 g; Na_2HPO_4, 0·6 g; $(NH_4)NO_3$, 1·0 g; $MgSO_4 \cdot 7H_2O$, 0·2 g; Na_2CO_3, 0·1 g; $CaCl_2 \cdot 2H_2O$, 0·01 g; $MnSO_4 \cdot H_2O$, 0·02 g; $FeSO_4 \cdot 7H_2O$, 0·005 g, and 1·5% washed agar in 1000 ml distilled water. The pH was adjusted to 6·8–7·0 before autoclaving. The hydrocarbons were purchased from numerous commercial sources and were all 95+% purity (most have a purity of 99+%). Desiccators were used as incubators, with the hydrocarbon in a 25 ml filtering flask attached to the desiccators which allowed a mixture of hydrocarbon vapour and air to be in contact with the microorganisms. The plates were streaked with inoculum (six cultures/plate) and incubated at 13°C for three weeks.

Utilization of the components of gasoline

Pint bottles (narrow mouth, metal-lined screw cap) were inoculated with 50 ml Ambler water (well 10, 11/26/73). One-half of the systems were autoclaved and used as control samples. Sunoco 260 (5 μl) was added to all bottles which were then incubated at 13°C for 24–192 hr.

A control and a sample were extracted every 24 hr for four days, followed by two additional samples at 168 and 192 hr. The procedure for extraction was as follows: Freon 113 (5 ml) was added to each bottle and teflon tape was wrapped around the mouth of the bottle to assure a tight seal. After being shaken on a mechanical shaker for 30 min, the contents of the bottle were centrifuged for 15 min to permit recovery of the Freon 113 layer.

Two separate Perkin-Elmer Model 900 Gas Chromatographs equipped with FID detectors (hydrogen—30 ml/min and air—350 ml/min) were used for the capillary GC analyses of the Freon 113 extracts. Prior to each GC analysis, 2 μg of n-C_{13} were added to each extract to serve as an internal standard. Most of the quantitative data were obtained using a 150 ft × 0·01 inch capillary column coated with squalane which was temperature programmed from 25°C to 120°C @ 4°/min after a 16 min initial delay. A small amount of data for the aromatic separations was obtained on the other PE 900 gas chromatograph which was fitted with a 150 ft × 0·01 inch capillary column coated with Ucon HB 550 and programmed from 50°C to 120°C @ 15°/min after a 4 min initial delay.

All samples were injected into the GC at a constant volume of 5 μl using a 10 μl syringe fitted with a Chaney adaptor. The splitter on the injector was closed. Peak areas were integrated using the IBM 1800 computer. Unnormalized peak areas were converted to absolute amounts (microgram) of hydrocarbon by comparison with the internal standard.

Co-oxidation studies

Co-oxidation experiments were run using two components of gasoline (one for growth and the other as a co-oxidizable substrate). Studies were made inoculating MM resin plates (Raymond *et al.*, 1969) with a pure culture and incubating the plates in a desiccator for 10 days at 28°C. The products were recovered from the resin as previously described (Raymond *et al.*, 1969).

Co-oxidation studies were also carried out using quart bottles (narrow mouth, metal-lined screw cap) containing 100 ml MM broth and inoculated

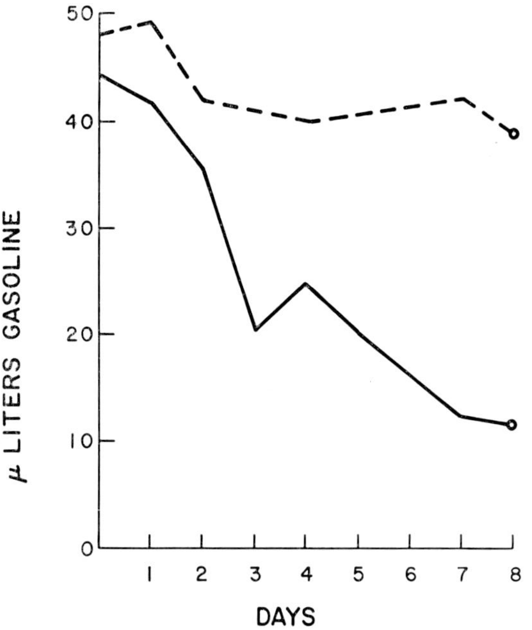

Fig. 1. Rate of degradation of gasoline by mixed culture.

with a pure culture and shaken on a New Brunswick Gyrotory Shaker (Model G10) at 28°C for 10 to 15 days. The broth was acidified to a pH of 2 with concentrated HCl and the products were extracted from the acidified broth with ethyl ether.

Oxidation product analysis

Thin-layer chromatographic (TLC) plates of Silica Gel G (Analtech, Inc., Wilmington, Del.) were employed. The composition of the solvent system for the development of the plates was benzene-acetic acid, 88:12. Acids were detected with a neutral bromocresol green (0·2% in *n*-butanol) spray. Ketones and aldehydes were detected with a 2,4-dinitrophenylhydrazine (0·4 g/75 ml acidic ethanol) spray. Ultraviolet spectra were obtained using a Beckman DK-2A and NMR spectra using a Varian Model A60.

Table I

GROWTH SUBSTRATES FOR GASOLINE—OXIDIZING CULTURES

Unknown genera							Flavo-bacterium	Micro-coccus *sp.*		Acinetobacter *sp.*				Pseudo-monas *sp.*		
5–5	3–10a	3–6b	3–6a	2–20a	2–8	1–11a	1–18	1–13a	1–13	3–10	3–6	2–20	1–11	3–8	3–5	3–4
0	0	0	0	0	0	0	0	0	0	0	0	0	0	0	0	0
2	0	0	0	0	0	0	0	0	1	0	0	0	0	0	0	0
0	0	0	0	0	0	0	0	0	1	0	0	0	2	0	0	0
1	0	0	0	0	0	0	0	0	0	0	0	0	0	0	0	0
3	0	0	0	0	0	0	0	0	0	0	0	0	0	0	0	0
0	0	0	0	0	0	0	0	0	0	0	0	0	0	0	0	0
0	0	0	0	0	0	0	0	0	0	0	0	0	0	0	0	0
0	0	0	0	0	0	0	0	0	0	0	0	0	0	0	0	0
0	0	0	0	0	0	0	0	0	0	0	0	0	0	0	0	0
0	0	0	0	0	0	0	0	0	0	0	0	0	0	0	0	0
0	0	0	0	0	0	0	0	0	0	0	0	0	0	0	0	0
0	0	0	0	0	0	0	0	0	0	0	0	0	0	0	0	0
0	0	0	2	0	0	0	0	1	1	0	0	0	2	0	0	0
0	0	0	0	0	0	0	0	0	0	0	0	0	0	0	0	0
0	0	0	0	0	0	0	0	0	0	0	0	0	0	0	0	0
0	0	0	0	0	0	0	0	0	0	0	0	0	0	0	0	0
0	0	0	0	0	0	0	0	0	0	0	0	0	0	0	0	0
0	0	0	0	0	0	0	0	0	0	0	0	0	0	0	0	0
0	0	0	0	0	0	0	0	0	0	0	0	0	0	0	0	0
0	0	0	0	0	0	0	0	0	0	0	0	0	0	0	0	0
0	0	0	0	0	0	0	0	0	0	0	0	0	0	0	0	0
0	0	0	0	0	0	0	0	0	0	0	0	0	0	0	0	0
0	0	0	0	2	0	0	1	0	1	1	0	0	1	0	0	0
0	0	0	0	0	0	0	0	0	0	0	0	0	0	0	0	0
1	4	0	2	4	4	0	0	0	0	4	4	2	3	0	1	1
0	0	0	0	0	0	0	0	0	0	0	0	0	0	0	0	0
0	0	0	0	0	0	0	0	0	0	0	0	0	0	0	0	0
0	0	0	0	0	0	0	0	0	0	0	0	0	0	0	0	0
0	0	0	3	4	0	0	0	0	0	4	4	2	4	0	0	0
0	1	0	0	2	2	0	0	1	0	3	4	3	3	0	0	0
0	0	0	0	0	4	0	0	0	3	4	4	4	4	4	3	2
0	0	0	0	0	0	0	0	0	0	0	0	0	0	0	0	0
0	0	0	0	0	0	0	0	0	0	0	0	0	0	2	2	1
0	0	0	0	0	0	0	0	0	0	0	0	0	0	1	3	3
1	4	±	±	4	4	±	±	1	1	4	4	4	4	4	3	4

Table I—*contd.*

Pseudomonas *sp.*					Nocardia *sp.*										
3–3	3–2	2–22	2–17	2–16	3–9	3–7	3–1	2–29	2–28	2–21	2–10	2–9	2–7	2–6	
0	0	0	0	0	0	0	0	0	0	0	0	0	0	0	Normal Butane
0	4	0	2	0	4	4	2	0	2	2	3	4	2	4	Normal Pentane
0	4	0	2	0	4	4	3	1	2	2	3	4	2	2	Normal Hexane
0	4	0	3	0	4	4	4	0	0	2	4	4	4	4	Normal Heptane
1	4	0	3	0	4	4	4	0	2	2	4	4	3	4	Normal Octane
0	0	0	0	0	0	0	0	0	0	0	0	0	0	0	Normal *cis*-Butene-2
0	0	0	0	0	0	0	0	0	0	0	0	0	0	0	Normal Pentane-2
0	0	0	0	0	0	0	0	0	0	0	0	0	0	0	2,3-Dimethylbutene-1
0	0	0	0	0	0	0	0	0	0	0	0	0	0	0	Cyclopentane
0	0	0	0	0	0	0	0	0	0	0	0	0	0	0	Cyclohexane
0	0	0	0	0	0	0	0	0	0	0	0	0	0	0	Methylcyclopentane
0	4	0	2	0	3	3	0	0	0	0	1	0	0	4	2-Methylbutane
0	0	0	0	0	3	3	0	0	0	0	2	0	0	3	2-Methylpentane
0	0	0	0	0	0	0	0	0	0	0	0	0	0	0	3-Methylpentane
0	0	0	0	0	0	0	0	0	0	0	0	0	0	0	3-Methylhexane
0	0	0	0	0	0	0	0	0	0	0	0	0	0	0	2-Methylheptane
0	2	0	0	0	4	4	3	1	1	0	3	4	2	4	3-Methylheptane
0	0	0	0	0	0	0	0	0	0	0	0	0	0	0	2,2-Dimethylbutane
0	0	0	0	0	0	0	0	0	0	0	0	0	0	0	2,3-Dimethylbutane
0	0	0	0	0	0	0	0	0	0	0	0	0	0	0	2,2-Dimethylpentane
0	0	0	0	0	0	0	0	0	0	0	0	0	0	0	2,4-Dimethylpentane
0	0	0	0	0	0	0	0	0	0	0	0	0	0	0	2,3-Dimethylpentane
0	0	0	0	0	0	0	0	0	0	0	0	0	0	0	2,3-Dimethylhexane
0	0	0	0	0	0	0	0	0	0	0	0	0	0	0	1,2-Dimethylcyclohexane (*cis-trans* 50/50)
0	0	0	1	0	2	3	4	0	0	2	2	2	0	2	2,2,4-Trimethylpentane (isooctane)
0	0	0	0	0	0	0	0	0	0	0	0	0	0	0	2,3,4-Trimethylpentane (isooctane)
0	0	0	0	0	0	0	0	0	0	0	0	0	0	0	2,3,3-Trimethylpentane (isooctane)
0	0	0	0	0	0	0	0	0	0	0	0	0	0	0	Ethylcyclohexane
0	0	0	0	0	0	0	4	0	0	0	0	0	0	0	Benzene
0	0	0	0	0	1	0	4	0	1	1	1	1	0	1	Ethylbenzene
3	4	3	3	4	0	0	2	1	0	0	0	0	0	0	Toluene
0	0	0	0	0	0	0	0	0	0	0	0	0	0	0	*o*-xylene
2	3	2	0	2	0	0	0	0	0	0	0	0	0	0	*m*-xylene
0	0	0	0	1	0	0	0	0	0	0	0	0	0	0	*p*-xylene
4	4	4	1	4	4	4	4	1	2	2	4	4	2	4	Gasoline

Growth response of a single streak of inoculum on agar plates incubated with the various hydrocarbons is indicated by the following scale:
4 = heavy growth, 3 = moderate, 2 = light, 1 = questionable, 0 = none.

RESULTS

The growth data, using the 32 cultures isolated from the groundwater, employing pure hydrocarbons (major components of the gasoline) are shown in Table I. The ability of the mixed population of microorganisms (the natural flora in the groundwater) to biodegrade the gasoline is shown in Fig. 1. Biodegradation of 54 identifiable components of Sunoco 260 was studied in these systems. Analyses by GC of the extracts from these systems demonstrate biodegradation of many of the individual components (Table II).

Co-oxidation experiments were made using hexane as the growth substrate for *Pseudomonas* sp. 803-2 and hexadecane for the *Nocardia* sp. 802-6, 802-29, and 803-9. Substrates that would not support growth for any of the cultures but were biodegraded in the reservoir were used as the oxidizable substrates. The results using resin plates as the fermentation system are shown in Table III. The only fermentation system that yielded oxidation products of a sufficient quantity for identification were *o*-xylene, *p*-xylene, ethylcyclohexane, and 2-methylheptane.

When fermentations were carried out using an aqueous system, the results obtained are shown in Table IV. Again, as in the case of the resin plates, the significant quantities of material came from the fermentations containing *o*-xylene, ethylcyclohexane, and 2-methylheptane as co-substrates. Identification of products resulting from these co-oxidation studies has been made.

Both the *Nocardia* and *Pseudomonas* cultures were able to oxidize *o*-xylene to an acidic product which was shown to be *o*-toluic acid by UV and TLC methods. The oxidation of ethylcyclohexane by a *Pseudomonas* sp. was confirmed to be cyclohexane acid by NMR. The *Nocardia* sp. studied could oxidize *p*-xylene to *p*-toluic acid under co-oxidation conditions.

The oxidation by the *Pseudomonas* sp. of 2-methylheptane yielded a mixture of several products. The major portion was shown to be positive to dinitrophenylhydrazine (ketones and aldehydes). Two of the spots bleached the dye when sprayed with bromocresol green. No identification has been attempted on these products.

DISCUSSION

The present state of knowledge of the cultures with regard to their ability to utilize hydrocarbons is summarized in Table V. These data are somewhat at odds with the study of the degradation of gasoline by the mixed population, which indicates that most of the components of the gasoline are degraded. It is quite apparent that the *Nocardia* cultures are probably responsible for the major *n*-paraffinic hydrocarbon degradation and the *Pseudomonas* cultures are responsible for the degradation of most of the aromatics.

Since a significant amount of nongrowth substrates, which were major components of the gasoline, were biodegraded, it was natural to look at co-oxidation as the means of the biotransformation for these hydrocarbons. The fermentation chemistry of nongrowth substrate molecules is now commonplace and co-oxidation or co-metabolism is an accepted explanation for the oxygenation of nongrowth substrates.

Davis and Raymond (1961) reported the carboxylation of the side chain when *Nocardia* cells were grown on *n*-alkanes in the presence of a mono-

Table II
BIODEGRADATION OF THE COMPONENTS OF GASOLINE

Components of Sunoco 260	*Initial concen- tration μlitre*	*Concentration at 192 hr μlitre*		*Percent biodegraded over and above control*
		Control	*Sample*	
n-Propane	Trace	Trace	Trace	0
n-Butane	0·63	0·37	0·37	0
n-Pentane	0·55	0·25	0·06	70
n-Hexane	1·36	0·78	0·15	46
n-Heptane	0·37	0·20	Trace	49
n-Octane	0·34	0·18	Trace	54
Olefins—C_4	0·11	0·18	0·12	0
Olefins—C_5	1·04	0·30	0·14	16
Olefins—C_6	0·51	0·16	0·07	18
Isobutane	0·11	0·13	0·12	0
Cyclopentane	0·17	0·05	0·04	0
Cyclohexane	0·12	0·06	Trace	45
Methylcyclopentane	0·41	0·18	0·14	10
Methylcyclohexane	0·05	0·04	Trace	75
2-Methylbutane	3·29	1·34	1·31	0
2-Methylpentane	1·72	0·83	0·73	6
3-Methylpentane	1·30	0·56	0·47	7
2-Methylhexane	0·74	0·53	0·36	23
3-Methylhexane	0·66	0·37	0·48	0
2-Methylheptane	0·35	0·15	Trace	38
3-Methylheptane	0·46	0·31	0·10	45
4-Methylheptane	0·15	0·08	Trace	48
2,2-Dimethylbutane	0·28	0·16	0·09	25
2,3-Dimethylbutane	0·86	0·37	0·36	0
2,2-Dimethylpentane	0·42	0·19	0·15	9
2,4-Dimethylpentane	0·53	0·31	0·25	11
3,3-Dimethylpentane	0·04	0·02	Trace	45
2,3-Dimethylpentane	0·48	0·24	0·21	0
2,5-Dimethylhexane	0·53	0·33	0·22	20
2,4-Dimethylhexane	0·46	0·22	0·20	0
2,3-Dimethylhexane	0·54	0·29	0·19	19
3,4-Dimethylhexane	0·09	0·08	Trace	84
2,2-Dimethylhexane	0·05	0·04	Trace	75
2,2-Dimethylheptane	0·09	0·06	Trace	62
1,1-Dimethylcyclopentane	0·12	0·04	Trace	25
1,2 and 1,3-Dimethylcyclopentane	0·12	0·10	Trace	78
1,3 and 1,4-Dimethylcyclohexane	0·02	Trace	Trace	0
1,2-Dimethylcyclohexane	0·16	0·05	Trace	26
2,2,3-Trimethylbutane	0·03	0·02	Trace	62
2,2,4-Trimethylpentane	3·47	2·40	1·95	13
2,2,3-Trimethylpentane	0·17	0·10	Trace	54
2,3,4-Trimethylpentane	1·89	1·22	0·97	13
2,3,3-Trimethylpentane	1·97	1·35	1·02	16
2,2,5-Trimethylpentane	0·51	0·35	0·23	23
1,2,4-Trimethylcyclopentane	0·03	Trace	Trace	0
Ethylpentane	0·08	0·05	0·04	0
Ethylcyclopentane	0·11	0·04	Trace	31
Ethylcyclohexane	0·06	0·06	Trace	95
Benzene	0·41	0·45	0	100
Ethylbenzene	1·36	1·61	0	100
Toluene	2·22	2·67	0	100
o-Xylene	1·62	2·18	0	100
m-Xylene	3·28	4·29	0	100
p-Xylene	1·03	1·31	0	100
Heavy Ends	8·97	11·80	1·13	87

Table III
CO-OXIDATION RESULTS FROM RESIN PLATES

Microorganism	*Substrate*	*Mg extracted material*
Nocardia sp.		
802-6	methylcyclopentane	1·4
802-29	methylcyclopentane	1·2
803-9	methylcyclopentane	1·3
	ethylcyclohexane	1·6
	cyclohexane	1·5
	2-methylpentane	0·7
	3-methylpentane	1·3
	2,2-dimethylpentane	0·9
	2,3,3-trimethylpentane	0·5
	2,3,4-trimethylpentane	1·4
	o-xylene	2·9
	p-xylene	12·4
Pseudomonas sp.		
803-2	2-methylheptane	3·1
	o-xylene	4·2
	ethylcyclohexane	7·4

alkylcyclohexane. In contrast, Van Ravenswaay Claasen and Van der Linden (1971) reported the hydroxylation of the cyclic ring of monoalkylcyclohexanes when using *Pseudomonas aeruginosa* yielding only 4-alkylcyclohexanols. There was no evidence of the hydroxylation of the ring by either *Nocardia* or *Pseudomonas* cells when ethylcyclohexane was the co-oxidizable substrate, only carboxylation of the side chain took place.

Raymond *et al.* (1967) and Jamison *et al.* (1969) reported several oxidation

Table IV
CO-OXIDATION RESULTS FROM AQUEOUS FERMENTATIONS

Microorganisms	*Substrate*	*Mg extracted material*
Nocardia sp.		
	methylcyclopentane	1·8
802-6	*o*-xylene	63·4
803-9	cyclohexane	4·7
	methylcyclopentane	7·3
	ethylcyclohexane	5·9
	2-methylheptane	1·2
	3-methylpentane	2·6
	2,3,3-trimethylpentane	3·9
	2,3,4-trimethylpentane	5·0
	o-xylene	4·2
Pseudomonas sp.		
803-2	cyclohexane	1·0
	methylcyclopentane	1·2
	ethylcyclohexane	5·0
	3-methylpentane	1·1
	2-methylheptane	35·4
	2,3,3-trimethylpentane	1·6
	2,3,4-trimethylpentane	3·4
	o-xylene	3·0

Table V
SUMMARY OF GROWTH DATA

	Nocardia	Pseudo-monas	Acineto-bacter	Micro-coccus	Flavo-bacterium	*Unclassified*
n-Alkanes	+	−	−	+	−	−
Cyclic alkanes	−	−	−	−	−	−
Alkyl-substituted cyclicalkane	−	−	−	−	−	−
Monomethylalkanes	+	−	−	+	+	−
Dimethylalkanes	−	−	−	−	+	−
Trimethylalkanes	+	+	+	−	−	+
Aromatics	−	+	+	+	−	+

products from *p*-xylene, in both cases *p*-toluic acid is formed as well as one or two other acids. When using the cultures in this study, only *p*-toluic acid was obtained, leading to the speculation that these microorganisms could not hydroxylate the aromatic ring just as there was no evidence for the hydroxylation of the cyclic alkane ring.

Klein *et al.* (1968) reported the co-oxidation of *n*-alkanes to ketones by *Arthrobacter*, so it is not unexpected to see evidence in the fermentation of 2-methylheptane of ketones and aldehydes.

The inability of this preliminary study to show co-oxidation of many of the nongrowth components of gasoline is in part due to very poor growth obtained in the liquid fermentation systems used. Raymond and Jamison (1971) pointed out that an important factor in transformation hydrocarbons is the proper concentration of substrate as well as a high concentration of cells.

SUMMARY

A large number of bacterial species was isolated which could grow on gasoline. Within the genera *Nocardia* and *Pseudomonas*, there was considerable diversity in ability to utilize individual components of gasoline for growth.

The biodegradation of the gasoline by the mixed population is not in agreement with the lack of oxidation of the individual components by pure cultures which suggests that co-oxidation plays an important part during the stimulation or the microorganisms that could grow on branched paraffins, olefins, and cyclic alkanes were not isolated.

The lack of product formation in the co-oxidation studies that have been completed indicates that more work must still be done to establish that co-oxidation is a major pathway in the biodegradation of the nongrowth components of gasoline.

Acknowledgements

This work was supported in part by an API grant. We thank A. J. Raymond and Ruth Jenkins for the gas chromatography studies on the gasoline-utilization and Sun Ventures, Inc., for allowing us to publish this work.

REFERENCES

Davis, J. B., and Raymond, R. L. (1961). Oxidation of alkyl-substituted cyclic hydrocarbons by a *Nocardia* during growth on *n*-alkanes. *Appl. Microbiol.*, **9**, 383–388.

Jamison, V. W., Raymond, R. L., and Hudson, J. O. (1969). Microbial hydrocarbon co-oxidation. III. Isolation and characterization of an α',α'-dimethyl-*cis,cis*-muconic acid-producing strain of *Nocardia corallina*. *Appl. Microbiol.*, **17**, 855–856.

Klein, D. A., Davis, J. A., and Casida, L. E. Jr. (1968). Oxidation of *n*-alkanes to ketones by an *Arthrobacter* species. *Antonie van Leeuwenhoek*, **34**, 495–503.

Litchfield, J. H., and Clark, L. C. (1973). *Bacterial activity in ground waters containing petroleum products.* American Petroleum Institute Publication No. 4211.

McKee, J. E., Laventy, F. B., and Hertel, R. N. (1972). Gasoline in groundwater. *J. Water Pollut. Control Fed.*, **44**, 293–302.

Raymond, R. L., and Jamison, V. W. (1971). Biochemical activities of *Nocardia*. *Adv. Appl. Microbiol.*, **14**, 93–122.

Raymond, R. L., Jamison, V. W., and Hudson, J. O. (1967). Microbial hydrocarbon co-oxidation. I. Oxidation of mono- and dicyclic hydrocarbons by soil isolates of the genus *Nocardia*. *Appl. Microbiol.*, **15**, 857–865.

Raymond, R. L., Jamison, V. W., and Hudson, J. O. (1969). Microbial hydrocarbon co-oxidation. II. Use of ion-exchange resins. *Appl. Microbiol.*, **17**, 512–515.

Van Ravenswaay Claasen, J. C., and Van der Linden, A. C. (1971). Substrate specificity of the paraffin hydroxylase of *Pseudomonas aeruginosa*. *Antonie van Leeuwenhoek*, **37**, 339–352.

PETROLEUM: DEGRADATION BY ESTUARINE MICROORGANISMS

J. D. WALKER* and R. R. COLWELL

Department of Microbiology, University of Maryland,
College Park, Maryland 20742, U.S.A.

Summary

Two environments in Chesapeake Bay, Colgate Creek and Eastern Bay, were examined for petroleum-degrading microorganisms (bacteria, yeast, and fungi) and the capability of these organisms to degrade petroleum. The presence of contaminating oil in Colgate Creek and its absence in Eastern Bay was confirmed by low-resolution mass spectrometry. Seasonal effects on the number and type of petroleum-degrading microorganisms present in water and sediments of the two environments, as well as their degrading capability, were observed. The microbial potential for petroleum degradation in water and sediment varied, depending upon the season, and was not found to be associated consistently with water or sediment. From September 1972 to September 1973, the maximum number of heterotrophic and petroleum-degrading microorganisms present in samples collected at both sites occurred during late winter and early spring. Numbers of petroleum-degrading microorganisms were consistently higher at Colgate Creek compared with Eastern Bay and the amount and type of microbial petroleum degradation were different in the two locations. This study represents the first attempt to measure numbers and activity of petroleum-degrading microorganisms from oil-contaminated and oil-free environments on a seasonal basis. The results also provide information on the susceptibility of petroleum to microbial degradation by indigenous microorganisms of Chesapeake Bay.

INTRODUCTION

A study has been underway in our laboratory since 1972 concerning the degradation of petroleum by estuarine and marine microorganisms. Much of the work has been focused on Chesapeake Bay, the major Atlantic coastal estuary of the United States. Two areas of Chesapeake Bay have been most intensively studied, Colgate Creek, located in Baltimore Harbour and representative of a large industrial harbour receiving continuous contamination with oil and oil products. The other area of interest is Eastern Bay, a relatively unpolluted area where a commercial shellfishing industry is active. Samples collected from Eastern Bay have shown no detectable concentrations of petroleum.

This report will be restricted to considerations of the impact of oil on microorganisms present in the water and sediment at these two areas of

* Present address: Environmental Technology Center, Martin Marietta Corporation, 1450 South Rolling Road, Baltimore, Md. 21227, U.S.A.

Chesapeake Bay. The data presented represent the portion of work accomplished since 1972 pertinent to this aspect of microbial degradation of petroleum.

MATERIALS AND METHODS

Details of the petroleum and petroleum components, media, reagents, and procedures followed in the analysis have been given in several papers (Walker and Colwell, 1974a,b, 1975a,b, 1976a,b; Walker *et al.*, 1975b; Walker *et al.*, 1975a). Benzene extracts of sediment from Colgate Creek and Eastern Bay were analysed by low-resolution computerized mass spectrometry (Walker *et al.*, 1975a).

RESULTS

Hydrocarbons extracted from Colgate Creek sediment samples were very similar in composition to crude oil (Table I). Extracts prepared from Eastern

Table I
COMPARISON OF PERCENT HYDROCARBON COMPOSITION OF EXTRACT OF COLGATE CREEK SEDIMENT WITH CRUDE OIL

Hydrocarbon	*Sediment extract*	*Crude oil*
Paraffins	20·0	41·1
Monocycloparaffins	26·0	19·6
Dicycloparaffins	8·5	4·3
Tricycloparaffins	7·2	0·8
Tetracycloparaffins	2·4	0·2
Pentacycloparaffins	0·5	$<$0·1
Alkylbenzenes	28·8	23·8
Benzcycloparaffins	0·6	5·2
Benzdicycloparaffins	0·2	2·4
Alkylnaphthalenes	3·1	1·3
Alkylacenaphthalenes	1·0	0·5
Fluorenes	1·0	0·4
Phenanthrenes	0·5	0·2
Cyclopentaphenalenes	0·4	0·1

Bay sediment samples, when analysed, revealed several nonhydrocarbon peaks, *i.e.* at m/e 45, 46, 47, 48, 60, 73, 74, and 76, indicating the presence of high molecular weight compounds containing nitrogen, oxygen, and sulphur.

Microbial analyses of the sediment samples included measurement of the ability of the microbial populations in Colgate Creek and Eastern Bay sediment samples to degrade South Louisiana crude oil (Table II). Eastern Bay microorganisms were twice as efficient in degrading the oil than weathering (evaporation) alone. Colgate Creek microorganisms, on the other hand, degraded ca. seven times greater the amount of total oil than was lost via weathering. Alkanes, both normal and branched, were the most susceptible of the components of crude oil to weathering and microbial degradation.

Total numbers of the aerobic, heterotrophic microorganisms, *i.e.* total

Table II

PERCENT COMPONENT OF SOUTH LOUISIANA CRUDE OIL REMOVED BY WEATHERING AND DEGRADATION

Component	*Weathered sample*	*Degraded*	
		Eastern Bay culture	*Colgate Creek culture*
Saturates	25	42	62
Alkanes	26	40	69
1-Ring cycloalkanes	16	19	45
Aromatics	0	5	18
Monoaromatics	0	9	30
Tetraaromatics	0	0	0
Sulphur aromatics	0	0	17
Resins	0	0	61
Asphaltenes	0	0	0
Total residue	7	16	48

viable counts (TVC), and of the petroleum-degrading bacteria (PD) were determined every 4 to 6 weeks from October 1972 to September 1973 for samples collected in Colgate Creek (Fig. 1) and Eastern Bay (Fig. 2). Significantly more petroleum-degrading bacteria were found during the winter months in both Colgate Creek and Eastern Bay samples. To determine if this observation was a result of the petroleum-degrading potential at low temperatures being greater in the winter months than in the summer, degradation of South Louisiana crude oil by bacteria in Colgate Creek sediment samples collected in January and in September 1973 was studied at 0°C and the results were compared (Table III). The amount of total oil

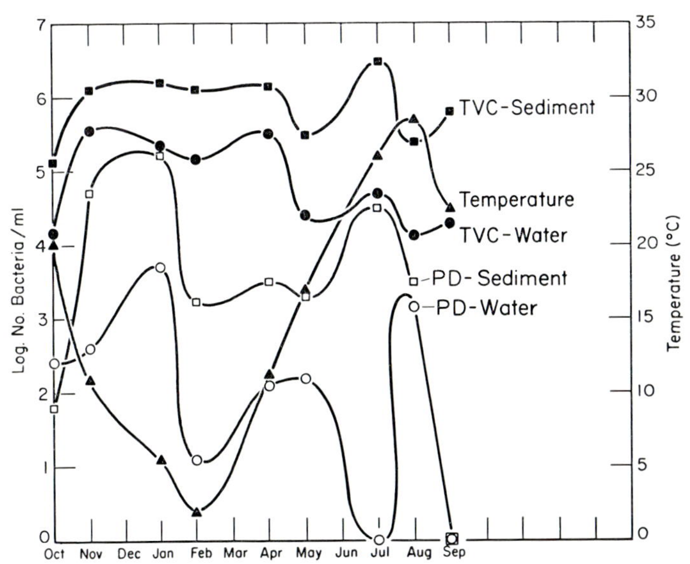

Fig. 1. Seasonal distribution of heterotrophic and petroleum-degrading bacteria for Colgate Creek.

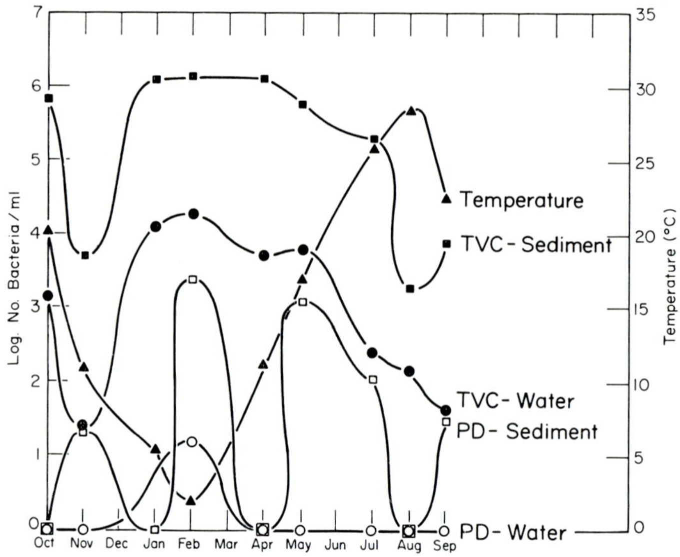

Fig. 2. Seasonal distribution of heterotrophic and petroleum-degrading bacteria for Eastern Bay.

degraded (corrected for weathering), within the 7-week period of degradation, was greater at 0°C for the mixed culture from the sediment sample collected in January. However, except for the 4-ring cycloalkanes and monoaromatics, the pattern of degradation with respect to the components of the crude oil was identical.

The metabolic activity of the petroleum-degrading bacteria in sediment samples collected from Colgate Creek and Eastern Bay was estimated using ^{14}C-1-*n*-hexadecane (Figs. 3 and 4). Conversion of the hexadecane to CO_2 by Colgate Creek water bacteria was greater than that demonstrated by Eastern

Table III

DEGRADATION OF A SOUTH LOUISIANA CRUDE OIL AT 0°C BY MIXED CULTURES OF COLGATE CREEK SEDIMENT BACTERIA COLLECTED IN JANUARY AND SEPTEMBER

Component	*Percent component degraded by culture collected in*	
	January	*September*
Saturates	37	37
Alkanes	36	33
1-Ring cycloalkanes	39	39
4-Ring cycloalkanes	17	10
Aromatics	0	0
Monoaromatics	13	1
Tetraaromatics	0	0
Sulphur aromatics	0	0
Resins	0	0
Asphaltenes	0	0
Total oil	13	2

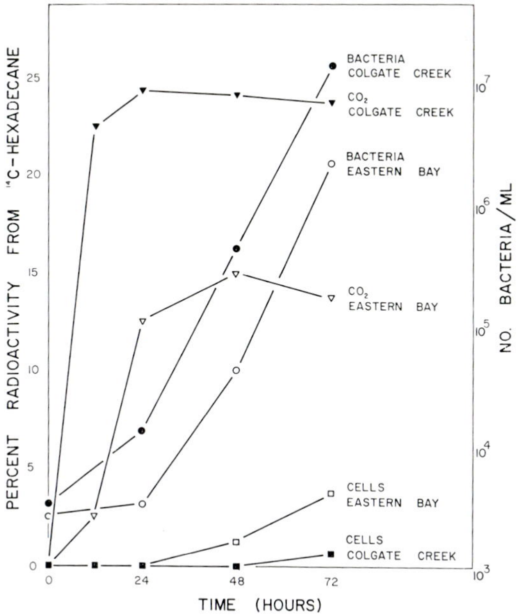

Fig. 3. Comparison of ${}^{14}C$*-1-*n*-hexadecane utilization by Colgate Creek and Eastern Bay bacteria.*

Bay water bacteria collected in July 1974 (Fig. 3). A more pronounced lag phase was observed for growth of Eastern Bay water bacteria on hexadecane than for Colgate Creek water bacteria. Turnover time for hexadecane was calculated to be 15 min for Colgate Creek water bacteria compared with 1 hr for the Eastern Bay water bacteria sampled in March 1974 (Fig. 4).

Growth of Colgate Creek and Eastern Bay sediment bacteria in a medium

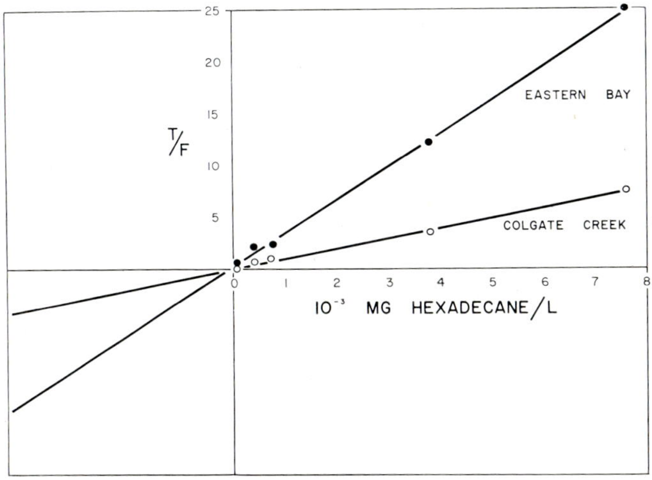

Fig. 4. Plot of mineralization of ${}^{14}C$*-1-*n*-hexadecane by bacteria from water of Colgate Creek and Eastern Bay.*

prepared with a mixed hydrocarbon substrate was compared (Walker and Colwell, 1974a). When water collected from these two areas of Chesapeake Bay was used as the diluent for the medium, variations in growth were observed (Figs. 5 and 6). Colgate Creek sediment bacteria cultured in Colgate Creek water with mixed hydrocarbon substrate yielded higher counts than when grown in Eastern Bay water (Fig. 5). Growth of Eastern Bay sediment bacteria in Eastern Bay water, conversely, was limited by the mixed hydrocarbon substrate (Fig. 6). Colgate Creek water also had a slightly inhibitory effect on the growth of Eastern Bay bacteria.

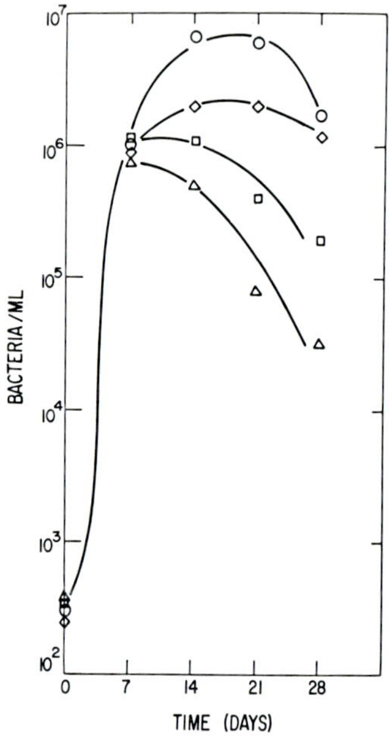

Fig. 5. Growth of Colgate Creek sediment bacteria in Colgate Creek water (○) and Eastern Bay water (□) containing mixed hydrocarbon substrate and in Colgate Creek water (◇) and Eastern Bay water (Δ) without mixed hydrocarbon substrate added.

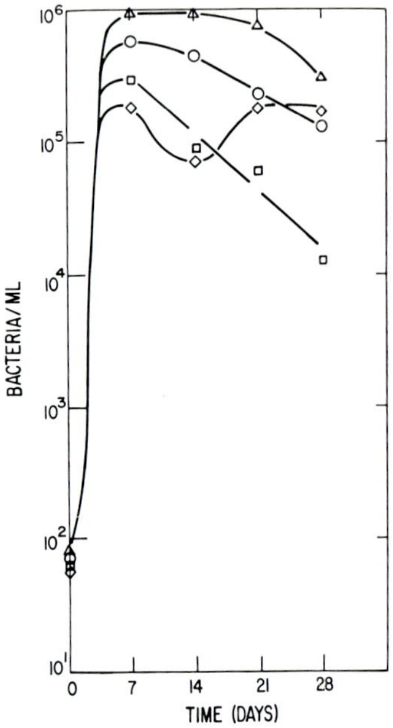

Fig. 6. Growth of Eastern Bay sediment bacteria in Colgate Creek water (○) and Eastern Bay water (□) containing mixed hydrocarbon substrate and in Colgate Creek water (◇) and Eastern Bay water (Δ) without mixed hydrocarbon substrate added.

DISCUSSION

Microbial populations in an estuary will not be uniform, either in numbers or in growth potential. Many factors, including influx of nutrients and/or pollutants, will affect the microorganisms. To establish parameters for growth of Chesapeake Bay microorganisms exposed to oil, we chose two sites for study, an oil-contaminated site and a comparable area which received no detectable oil input. That Eastern Bay water and sediment were, indeed, free of oil was verified by mass spectrometric analyses. Similarly, the presence of oil in Colgate Creek was confirmed (Table I). The slight differences between the hydrocarbon composition of the Colgate Creek extracts and extracts prepared from the crude oil samples can be explained by the fact that substantial amounts of used motor oil enter Colgate Creek. Furthermore, degradation of petroleum that enters the creek does take place, either before or after entry.

As was hypothesized, the populations of microorganisms in the oil-contaminated sediment of Colgate Creek produced significantly greater degradation of South Louisiana crude oil than the Eastern Bay sediment microorganisms. Nearly all of the components of South Louisiana crude oil used in the study were degraded to a greater degree than demonstrated by the Eastern Bay sediment microorganisms. However, the Eastern Bay sediment microorganisms were capable of degrading the oil as evidenced by the pattern of components remaining after degradation, compared with weathering alone. As has been shown repeatedly in studies of microbial degradation of petroleum and petroleum products, the saturated hydrocarbons were most susceptible to degradation (Table II).

The heterotrophic bacterial counts (TVC) were relatively constant in the case of the Colgate Creek sediment bacteria over the seasonal periods studied (Fig. 1). These data were essentially as described by Kaneko and Colwell (1973) for Rhode River (in Chesapeake Bay) sediment bacteria populations over a full annual cycle. However, the number of petroleum-degrading bacteria in sediment and water samples collected in January and July–August did show increases for the Colgate Creek samples. These were periods when the water temperatures were at the minimum and maximum, respectively. Temperature may have a selective effect for the petroleum-degrading bacterial populations in Colgate Creek sediment.

Eastern Bay water and sediment samples revealed heterotrophic counts with largest numbers of heterotrophic bacteria (TVC) occurring during January through May. The petroleum-degrading microorganisms in sediment were at the highest population levels in February–March and in May–July. Very few petroleum-degrading microorganisms were found in the water column in Eastern Bay. Clearly, resident in both Colgate Creek and Eastern Bay sediment is a petroleum-degrading bacterial population that is active in the winter months. Interestingly, the increases in petroleum-degrading bacteria in the water column paralleled the levels in the sediment. The water-sediment interactions, in this case, should be further investigated. The January and September sample comparisons for degradation of South Louisiana crude oil at 0°C, as given in Table III, suggest that a selection for the psychrotrophic forms capable of degrading 4-ring cycloalkanes and mono-aromatics, viz. pseudomonads and *Arthrobacter* spp., may occur in the winter months when the water temperature is reduced. This point remains to be clarified. In any case, the petroleum-degrading microorganism populations are eurythermic rather than strictly psychrophilic or psychrotrophic.

The Colgate Creek and Eastern Bay sediment bacteria grown on ^{14}C-hexadecane showed that the rates of activity for both populations were approximately the same, except for ca. 24 hr lag in growth shown by the Eastern Bay sediment bacteria. However, CO_2 production by the Colgate Creek sediment bacteria was greater. On the other hand, cellular incorporation was slightly greater for the Eastern Bay sediment bacteria (Fig. 3). *In situ* ^{14}C-hexadecane uptake, using Michaelis-Menton kinetics, showed that the turnover time for hexadecane by the Colgate Creek microorganisms was four times faster than the Eastern Bay microorganisms.

Growth studies of microbial populations from oil-polluted habitats transferred to waters of nonoil-polluted environments have not been reported. Since there is much discussion in the literature of seeding such areas if an oil spill occurs, this kind of information is vital to the success of such an endeavour. When Colgate Creek water was inoculated with Colgate Creek sediment bacteria, higher cell yields were obtained than when the same populations were cultured in Eastern Bay water, most likely because of the increased nitrogen content of the Colgate Creek water (unpubl. data, Chesapeake Bay Institute). Addition of mixed hydrocarbon substrate stimulated growth of Colgate Creek bacteria inoculated into a medium made up with Eastern Bay water. Addition of mixed hydrocarbon substrate to Eastern Bay water was found to limit growth of the Eastern Bay sediment bacteria. Obviously, the components of petroleum can have a limiting effect on the growth of microbial populations never before exposed to the oil.

In conclusion, microorganisms in the two areas of Chesapeake Bay that differed in history of previous exposure to oil reacted quite differently in their growth on and degradation of oil and hydrocarbon mixtures. Exposure to oil for 7 weeks, in the case of microorganisms in samples of the nonoil-contaminated sediment, did not result in significant enrichment for microorganisms capable of degrading oil. The microorganisms present in oil-contaminated sediment degrade oil more extensively and at a significantly greater rate. The data clearly indicate that hydrocarbons can limit the growth of bacteria in water and sediment of a nonoil-contaminated environment.

Acknowledgements

Work was supported by Contract N000 14-67-A-0239-0027-02 between the Office of Naval Research and the University of Maryland and by National Science Foundation Grant No. BMS 72-02227-A-02.

REFERENCES

Kaneko, T., and Colwell, R. R. (1973). Ecology of *Vibrio parahaemolyticus* in Chesapeake Bay. *J. Bacteriol.*, **113**, 24–32.

Walker, J. D., and Colwell, R. R. (1974a). Microbial degradation of model petroleum at low temperatures. *Microbial Ecol.*, **1**, 63–95.

Walker, J. D., and Colwell, R. R. (1974b). Microbial petroleum degradation: Use of mixed hydrocarbon substrates. *Appl. Microbiol.*, **27**, 1053–1060.

Walker, J. D., and Colwell, R. R. (1975a). Some effects of petroleum on estuarine and marine microorganisms. *Can. J. Microbiol.*, **21**, 305–313.

Walker, J. D., and Colwell, R. R. (1975b). *Degradation of hydrocarbons and mixed hydrocarbon substrate by microorganisms from Chesapeake Bay.* Seventh Int. Assoc. on Water Pollut. Res. Paris, France. Progs. Water Technol. In Press.

Walker, J. D., and Colwell, R. R. (1976a). Measuring the activity of hydrocarbon-degrading microorganisms. *Appl. Microbiol.* In Press.

Walker, J. D., and Colwell, R. R. (1976b). Enumeration of petroleum-degrading microorganisms. *Appl. Microbiol.* In Press.

Walker, J. D., Colwell, R. R., Hamming, M. C., and Ford, H. T. (1975a). Extraction of petroleum hydrocarbons from oil-contaminated sediments. *Bull. Environ. Contam. Toxicol.*, **13**, 245–248.

Walker, J. D., Colwell, R. R., and Petrakis, L. (1975b). Microbial petroleum degradation: Application of computerized mass spectrometry. *Can. J. Microbiol.* In Press.

USE OF GENETICALLY ENGINEERED MULTI-PLASMID MICROORGANISMS FOR RAPID DEGRADATION OF FUEL HYDROCARBONS

D. A. FRIELLO, J. R. MYLROIE and A. M. CHAKRABARTY

General Electric Corporate Research and Development,
Schenectady, New York 12301, U.S.A.

Summary

In *Pseudomonas*, the genes specifying enzymes governing the biodegradation of a variety of hydrocarbons are clustered together and are borne on transmissible or nontransmissible plasmids. The transfer of such hydrocarbon degradative plasmids from a number of individual *Pseudomonas* species to a single culture produces a multi-plasmid strain which can oxidize a variety of hydrocarbons including aliphatic, aromatic, terpenic, and polynuclear aromatic hydrocarbons. In the presence of suitable inducers, all these plasmid-specified pathways are induced and the hydrocarbon substrates oxidized simultaneously. The multi-plasmid strain is shown to grow much faster with crude oil than any of the individual parents, and may, therefore, be suitable for use in oil spill clean up or in the production of protein from petroleum.

INTRODUCTION

A large variety of microorganisms are known to degrade hydrocarbons. It has been reported that one or more species of 70 genera, including 28 bacteria, 30 filamentous fungi, and 12 yeasts can degrade one or more components of crude oil (ZoBell, 1973; Berner *et al.*, 1975).

There are two important areas in which hydrocarbon-degrading microorganisms can be put to effective use. One, which has recently been investigated, is their use in the production of protein from petroleum (Wang, 1968). The second area involves the use of such microorganisms as agents in the removal of the bulk of the oil from oil slicks (Miget, 1973). While most of the important parameters in the production of single-cell proteins from petroleum have been evaluated rather well (Evans, 1968; Wang, 1968), the use of such microorganisms for the clean up of oil spills is still largely a matter of conjecture due to the unknown ecological disturbances that may result from the use of large quantities of microorganisms in an open environment.

Even though there is a diverse number of microorganisms that can be used to rapidly degrade a variety of hydrocarbons, inherent problems exist when they are used in the areas mentioned above. This has been shown to be the case when yeasts and bacterial strains, such as *Pseudomonas*, have been used for the production of protein from petroleum (Yamada *et al.*, 1968). Yeasts can grow only with *n*-alkanes of limited chain lengths and are incapable of attacking cyclo-paraffinic, aromatic, or polynuclear aromatic hydrocarbons. Because crude oil is a mixture of different kinds of hydrocarbons, a large fraction comprising cyclo-paraffinic, aromatic, and polynuclear aromatic hydrocarbons therefore remains unchanged during growth of yeasts with crude oil. Since many polynuclear aromatic hydrocarbons are potent carcinogens, they along with other nondegradable components must be removed from crude oil either by molecular sieving or by extraction from the yeast cells after growth. Not only does this lead to a decrease in the net amount of cell mass produced but such separation processes substantially contribute to the cost of the protein product.

The use of bacterial strains, like yeasts, also presents problems. Although different strains can degrade different components, such as aliphatics, aromatics, cyclo-paraffins, or polynuclear aromatic hydrocarbons, a single strain can usually attack only a limited number of hydrocarbons (Soli and Bens, 1973). Thus, most strains that can oxidize aliphatic hydrocarbons are incapable of degrading aromatic or polynuclear aromatic hydrocarbons, while strains capable of degrading aromatic hydrocarbons can seldom degrade aliphatic or polynuclear aromatic hydrocarbons (Soli and Bens, 1973). Hence, different bacterial strains, while being more nutritionally versatile than yeasts, exhibit considerable lack of competence in utilizing a large number of hydrocarbon components from crude oil.

In an effort to achieve a greater degree of utilization of all the components of crude oil, Soli and Bens (1973) employed the use of a mixed culture in which each strain had the ability to degrade either aliphatic, aromatic, or polynuclear aromatic hydrocarbons. This approach to solving the problem was unsuccessful in two respects. First, it was shown that the use of a mixed culture system did not lead to any greater loss of crude oil than did the single culture system. Second, it was found that during the growth of the mixed culture, interactions such as the production by one strain of compounds toxic or inhibitory in very low concentrations to others and the difference in growth rates among the strains limited the number of surviving strains.

The limiting factor affecting the rate and range of hydrocarbon degradation in crude oil by microorganisms appears to be the lack of ability of most microbial strains to utilize different components of crude oil. The use of a mixed culture is of little value because of the problems of interaction among different microorganisms leading to the survival of only a few strains (Zajic and Daugulis, 1975). Still another problem with the use of a mixed culture is the lack of quality control: even one strain in a mixture of a large number of strains, if found to be pathogenic or a toxin producer, renders the cell mass unfit for consumption even for lower aquatic animals.

An ideal solution for the rapid microbial consumption of crude oil, either for the production of single-cell protein or for the rapid removal of oil from oil slicks, would be to construct novel strains which would have the genetic potentialities to degrade, simultaneously, a variety of oil components. The use of a single strain would eliminate the interaction among different microorganisms and yet would lead to a loss of all the major hydrocarbon fractions of crude oil. In this report we present some data which suggest that transfer of plasmids, each specifying a complete hydrocarbon degradative pathway, from different naturally occurring cultures to a single strain, produces novel multi-plasmid strains that can degrade a variety of the hydrocarbon components of crude oil. Consequently, such strains might be useful for the rapid clean up of oil slicks or for the enhanced production of protein from petroleum.

MATERIALS AND METHODS

Bacterial strains and their propagation

A list of bacterial strains harbouring specific hydrocarbon-degradative plasmids and their source or origin is given in Table I. The media and growth conditions have been described previously (Chakrabarty, 1972).

Table I

LIST OF BACTERIAL STRAINS

Strain	*Plasmid*	*Degradative pathway*	*References*
P. putida PpG1	CAM	Camphor	Rheinwald *et al.* (1973)
P. putida PpG6	OCT	*n*-Octane	Chakrabarty *et al.* (1973)
P. putida R1	SAL	Salicylate	Chakrabarty (1972)
P. putida PpG7	NAH	Naphthalene	Dunn and Gunsalus (1973)
P. putida arvilla (mt-2)	TOL	*m*-Toluate	Murray and Williams (1974)
	(XAL)	(*m*- or *p*-Xylene)	Mylroie and Chakrabarty (1975)
Pseudomonas Pxy	XYL	*p*- or *m*-Xylene	Friello and Chakrabarty (1975)
P. putida PRS1	MDL	Mandelate	This paper
P. putida PRS1	QUI	Quinate	This paper

Curing and transfer of hydrocarbon-degradative plasmids

The detailed methods describing the conditions of curing and conjugal transfer of plasmids have been enumerated elsewhere (Rheinwald *et al.*, 1973).

Growth with crude oil

For determining the extent of growth with crude oil, approximately 5×10^8 cells of an overnight grown culture of *Pseudomonas aeruginosa*, harbouring different hydrocarbon-degradative plasmids, were inoculated into 50 ml of synthetic mineral medium containing 1 ml of crude oil (Canadian Sweet Crude). Growth was usually continued for five days at 37°C with shaking. At intervals, aliquots of 5 ml were withdrawn. The cell density was determined turbidimetrically in a Bausch and Lomb Colorimeter at 660 nm and viable counts were taken by diluting the samples and plating aliquots on nutrient agar plates.

RESULTS

Construction of a multi-plasmid strain

Genetic analysis conducted recently on a number of strains of *Pseudomonas putida* to determine the linkage relationships in several hydrocarbon-degradative pathways has given rise to some unique and interesting results. In several instances, the genes specifying the enzymes of an entire hydrocarbon-degradative pathway were found to be clustered not on the chromosome, but on transmissible plasmids. For example, the camphor-degradative pathway has been shown to be specified by genes borne on a transmissible plasmid in *P. putida* strain PpGl (Rheinwald *et al.*, 1973). Similar is the case for the salicylate, the octane, and the naphthalene-degradative pathways (Chakrabarty, 1972; Chakrabarty *et al.*, 1973; Dunn and Gunsalus, 1973). A list of bacterial strains, harbouring various hydrocarbon-degradative pathways in the form of plasmids, is shown in Table 1.

The genetic characterization of hydrocarbon-degradative pathways as plasmids has relied mainly on two criteria for designating plasmids. The first criterion is the spontaneous or induced loss (curing) of such pathways from the cells without the resultant death of the cell. The results in Table II clearly show that most degradative plasmids (Chakrabarty, 1973) can be cured at an appreciable frequency when the cells are grown in the presence of mitomycin C. The second genetic basis used for designating plasmids is the conjugal transfer of such genes from one cell to another even in the absence of chromosomal gene transfer. Such naturally occurring hydrocarbon-degradative plasmids as CAM, SAL, XAL, TOL (Williams and Murray, 1974), or NAH can be transferred from one cell to another in either the presence or the absence of any sex factor or other transfer factor at high frequencies: in other words, they appear to be self-transmissible. Further, these self-transmissible plasmids are incapable of initiating mobilization of chromosomal genes to any significant extent. Plasmids such as OCT and XYL are not self-transmissible and require the presence of other transfer plasmids such as factor K for their mobilization (Chakrabarty, 1974; Chakrabarty and Friello, 1974). Another characteristic of the degradative plasmids is their compatibility, *i.e.* they can co-exist in the same cell. Except CAM and OCT (Chakrabarty, 1973) and SAL and XAL (a transmissible xylene-degradative plasmid, Mylroie and Chakrabarty, 1975), all degradative plasmids appear to be compatible. Thus, it is possible to transfer many of these degradative plasmids from their natural hosts to a single strain to construct a multi-plasmid super-strain (Table III) that has the capability to degrade a variety of hydrocarbons including aliphatic (hexane, octane, or decane), aromatic (xylene, toluene), bicyclic monoterpene (camphor), or polynuclear aromatic (naphthalene) hydrocarbons. It must be stressed at this point that although in their mono-plasmid parents, the plasmids appear to be stably retained, the plasmids in the multi-plasmid exconjugants are prone to spontaneous loss, so that selective measures must be adopted to retain these plasmids.

Table II

CURING AND TRANSMISSIBILITY OF DIFFERENT HYDROCARBON-DEGRADATIVE PLASMIDS IN *PSEUDOMONAS*

Plasmid	*Curing by Mitomycin C*	*Frequency of transfer to* P. putida	P. aeruginosa
CAM	+	10^{-2}	10^{-4}
OCT	+	10^{-5}	$<10^{-8}$
SAL	+	10^{-2}	10^{-7}
NAH	+	10^{-3}	10^{-8}
XYL	+	10^{-2}	$<10^{-8}$
XAL	+	10^{-2}	10^{-7}
MDL	+	10^{-6}	$<10^{-8}$
QUI	+	10^{-5}	$<10^{-8}$

Expression of individual plasmids in the multi-plasmid strain

The enzymes involved in the biodegradation of hydrocarbons such as camphor, naphthalene, octane, salicylate, xylene, etc., as they occur in their

Table III

CONSTRUCTION OF A MULTI-PLASMID STRAIN FROM DIFFERENT NATURALLY OCCURRING MONO-PLASMID PARENTS

Donor	*Plasmid*	*Recipient*	*Transfer frequency*	*Exconjugant Phenotype*[a]
P. putida AC 59	CAM-OCT	*P. aeruginosa*	10^{-4}	$Cam^{+}Oct^{+}$
P. putida AC 137	XAL	$Cam^{+}Oct^{+}$ *P. aeruginosa*	10^{-7}	$Cam^{+}Oct^{+}Xyl^{+}$
P. putida PpG7	NAH	$Cam^{+}Oct^{+}Xyl^{+}$ *P. aeruginosa*	10^{-8}	$Cam^{+}Oct^{+}Xyl^{+}Nah^{+}$

[a] Abbreviations used: CAM, Camphor; OCT, Octane; Xyl, *p*- or *m*-xylene; NAH, Naphthalene.

natural hosts, are all inducible. The transfer of the plasmids specifying these enzymes to a number of other hosts produces exconjugants in which these pathways still remain inducible. Since the presence of readily utilizable carbon sources often leads to an inhibition of induced enzyme synthesis (Magasanik, 1961), it was of interest to us to determine if the presence of a mixture of substrate hydrocarbons would inhibit the utilization of other hydrocarbon substrates by the multi-plasmid strain, *i.e.* whether we would observe diauxic lag in the utilization of hydrocarbons like camphor or naphthalene when more readily metabolizable hydrocarbons like mandelate or salicylate were also present. The results in Fig. 1 suggest that the multi-plasmid strain can grow much faster with a mixture of suboptimal concentrations of different hydrocarbons than when individual hydrocarbons alone are provided. The slow growth observed with xylene is presumably due to a rapid loss of the XAL plasmid, since examination of the cells after three days of growth showed the loss of the XAL plasmid from about 40% of the cells. About 10% of the cells appeared to have lost the NAH plasmid. Nevertheless, the majority of the cells retained all the plasmids and could degrade each of the substrates simultaneously, thereby growing at an increased rate.

The ability of the multi-plasmid strain to utilize mixtures of terpenic,

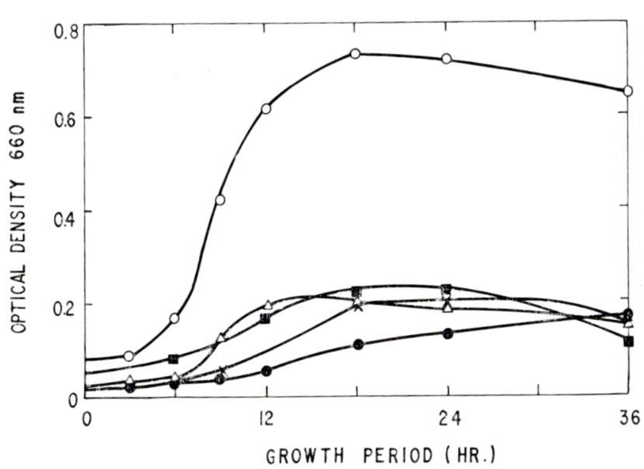

Fig. 1. Growth of multi-plasmid P. aeruginosa *with suboptimal concentration (2·5 mM) of different hydrocarbons*, viz., ●—●, *xylene;* Δ— Δ, *camphor;* ■—■, *octane;* ×——×, *naphthalene, and* ○—○, *mixture (2·5 mM) of each of the above hydrocarbons.*

aliphatic, aromatic, and polynuclear aromatic hydrocarbons at a rapid rate seemed to suggest that such strains might be useful for growth with crude oil since such components are present in abundant quantities in the crude oil. Strains harbouring different kinds of degradative plasmids were, therefore, inoculated into synthetic mineral medium with crude oil as the sole source of carbon and energy and then portions were analysed at intervals to determine the extent of growth. The results in Fig. 2 show that in contrast to most naturally occurring mono-plasmid parents, the multi-plasmid strain grows much faster with crude oil. The extent of growth appears to be a function of the number and nature of the degradative plasmids, which therefore determines the oil-degrading potential of the strain.

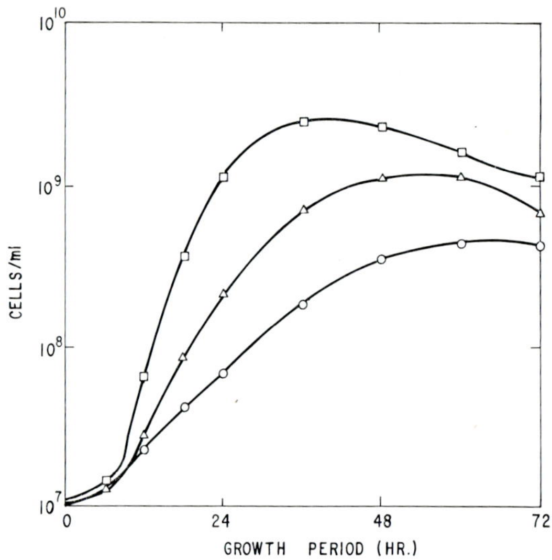

Fig. 2. Changes in the rate of growth with crude oil of P. aeruginosa *harbouring none (○—○), two (Δ— Δ), and four (□—□) known hydrocarbon-degradative plasmids.*

Transfer of nontransmissible hydrocarbon-degradative plasmids

It is recognized that in order to construct a truly superior strain that would have the capability to degrade most of the major components of oil, it would presumably have to harbour a large number of hydrocarbon-degradative plasmids. Aside from the problems associated with the stability of such plasmids in a single strain and the fact that many of the hydrocarbons normally present in crude oil might be recalcitrant to microbial attack and, hence, no known degradative pathway available for them, it is likely that many of the hydrocarbon oxidative pathways known to be possessed by bacterial strains may be specified by genes borne either on the chromosome or on nontransmissible plasmids, making them difficult to be transferred to the multi-plasmid strain. It would, therefore, be useful to have a system whereby nontransmissible degradative plasmids, when occurring in hydrocarbon-oxidizing strains, could be mobilized from one strain to another.

We have previously demonstrated that the plasmid OCT, which specifies a complete *n*-octane or *n*-decane degradative pathway, undergoes dissociation into three independent replicons, OCT, MER, and factor K, when transferred from *Pseudomonas oleovorans* to *P. putida* strain PpG1 (Chakrabarty, 1974; Chakrabarty and Friello, 1974). In strain PpG1, OCT appears to be a non-transmissible plasmid and is rendered transmissible through the presence of the transfer plasmid, factor K. Davey and Gibson (1974) have recently characterized a xylene degradative pathway in a strain of *Pseudomonas* Pxy. Treatment of this strain with mitomycin C produced segregants which had lost their ability to utilize xylene or its metabolites. The segregants appeared to have lost the entire xylene-degradative pathway, since they lacked the activity of several of the enzymes involved in xylene degradation (D. T. Gibson, pers. comm.). However, strain Pxy was unable to transfer the xylene-oxidizing capability to other *Pseudomonas* strains such as *P. putida* strain PpG1 or even to the Pxy segregants (Table IV). Only when a Leu^-

Table IV
TRANSFER OF XYL PLASMID IN PRESENCE OF FACTOR K

Donor	*Recipient*	*Select*	*Transfer frequency*	*Exconjugant phenotype*
AC 138 (Leu^-Pxy)	AC 30 (PpG1 WT)	XYL^+	$<10^{-8}$	
AC 138 (Leu^-Pxy) + AC 4 (Trp^-K^+PpG1)	AC 30	XYL^+	2×10^{-7}	Xyl^+PpG1 WT
AC 138	AC 139 (Xyl^-Pxy Seg)	XYL^+	$<10^{-8}$	
AC 138 + AC 4	AC 139	XYL^+	5×10^{-8}	Xyl^+Pxy

mutant of strain Pxy was grown in association with AC 4, Trp^- PpG1 derivative harbouring factor K, was it able to transfer the xylene-degradative capability to PpG1 strains or to xylene-negative Pxy segregants (Table IV). The XYL^+ PpG1 exconjugants are capable of transferring XYL to Pxy segregants and to other strains provided they also harbour factor K (Friello and Chakrabarty, 1975). Thus, it is shown that factor K is essential for transfer of both the OCT and XYL plasmids. In this regard, it appears that factor K would be very useful in characterizing other nontransmissible hydrocarbon degradative plasmids in *Pseudomonas*.

Formation of mandelate and quinate plasmids

Several aromatic hydrocarbon-degradative pathways have been studied both enzymatically and genetically in a number of *Pseudomonas* species (Ornston, 1971; Wheelis and Stanier, 1970). Apart from hydrocarbon-degradative pathways that occur naturally as transmissible or nontransmissible plasmids, several of the aromatic hydrocarbon-degradative pathways, such as those found in *P. putida* PRS1, have been shown to be specified by genes that appear to be clustered on a small portion of the chromosome (Leidigh and Wheelis, 1973). In order to initiate the transfer of the genes governing these pathways, an efficient chromosomal gene transfer system had to be developed for such strains. Since the transfer plasmid, factor K, had been found to be capable of mobilizing chromosomal genes in *P. putida* strain PpG1 (Chakrabarty and Friello, 1974), we introduced factor K into strain PRS1 and studied the ability of K^+ PRS1 cells to donate chromosomal genes to other PRS1 or PpG1 derivatives. Indeed, a number of chromosomal genes, including those for the degradative pathways for mandelate and quinate, could be transferred from K^+ PRS1 strain to cells of strain PpG1. However, in contrast to other chromosomal markers, the genes specifying the mandelate and quinate pathways appeared to replicate as plasmids in PpG1 exconjugants. They could be cured with mitomycin C, and were transmissible to other PpG1 or PRS1 cells, but not to *P. aeruginosa* (Table V). The transmissibility of the MDL or QUI plasmids is dependent on the presence of factor K. When transduced to other cells by transducing phage pf16, both the MDL and the QUI plasmids become nontransmissible due to the lack of co-transfer of factor K along with these plasmids. It is not entirely clear, however, if the genes for the mandelate or quinate pathways might also exist as part of nontransmissible plasmids in strain PRS1 and are mobilized to strain PpG1 through factor K. The recent use of factor K in determining the time of entry of a number of chromosomal genes by interrupted mating technique as well as physical characterization of these plasmids will be very useful in answering this question.

Table V
FORMATION OF MDL AND QUI PLASMIDS FROM THE CHROMOSOMAL GENES

Plasmid	*Curing by Mitomycin C*	*Transfer frequency to* P. putida *PpG1*	P. aeruginosa
MDL	+	10^{-5}	$<10^{-8}$
QUI	+	3×10^{-6}	$<10^{-8}$

DISCUSSION

The construction of multi-plasmid strains having the capability to degrade aliphatic, aromatic, polynuclear aromatic, and other hydrocarbons appears to be useful in combating environmental pollution caused by oil spills and in the production of protein from petroleum. Several advantages can be cited for using a single hydrocarbon-degrading strain as compared to a mixed culture; the major one, of course, is the ease of documentation of pathogenicity in the experimental organism. When large masses of cells are being used for either cattle feeding or for artificial seeding to clean up oil slicks, it is absolutely essential to determine the presence of toxic components that may be elaborated by any one of the cultures and retained within the cellular environment. It becomes increasingly difficult to study pathogenicity in each of several cultures in a mixed population. It is relatively less laborious and time-consuming to use a laboratory strain which has been extensively studied with regard to toxicity, cellular protein composition, simple nutritional requirements, and other properties. Another advantage of the use of a single multi-plasmid culture is the homogeneity of the protein. Thus, multi-plasmid strains can be constructed with starting cultures that will have a high level of protein of good biological value. In contrast, mixed cultures normally produce proteins whose composition varies depending upon the survival of selective strains and would, therefore, be of limited value as commercial feed.

The technique of hydrocarbon-degradative plasmid transfer to construct superior petroleum-utilizing strains may be one of the ways to solve the nutritional problems associated with oil spill clean up. It is recognized that one reason for the low rate of oil removal by microorganisms is the absence of phosphorus- and nitrogen-containing nutrients in the navigable lake or seawater (Miget, 1973). Although occasional application of oleophilic nitrogen and phosphorus fertilizers has been recommended (Bartha and Atlas, 1973), the large quantities of nitrogen needed by the seeding bacteria would make such a task extremely difficult and costly. However, the use of genetically engineered strains having, in addition to multiple hydrocarbon-degradative plasmids, the capability to fix atmospheric N_2 would solve much of the problem of supplying nitrogenous nutrients. It recently has been possible to transfer nitrogen fixation genes (*nif* genes) from *Klebsiella pneumoniae* to several bacteria, including *E. coli* and *Pseudomonas*, through recombination of the *nif* genes with a drug-resistant plasmid RP4 (Dixon *et al.*, 1975). Since RP4 is transmissible in *Pseudomonas* and is compatible with all of the degradative plasmids, it would be interesting to find out if transfer of RP4 plasmid containing the nitrogen fixation genes to *Pseudomonas* would allow such cells to express the *nif* genes under aerobic conditions. Similarly, transfer of such hydrocarbon-degradative plasmids and the RP4 plasmid harbouring the *nif* genes to psychrophilic *Pseudomonas* strains that have been isolated from miniature oil slicks in the Alaskan Arctic waters from Prudhoe Bay (Atlas and Schofield, 1975) might be one of the ways to control environmental pollution resulting from an accidental oil spill in this region.

Acknowledgement

A portion of the on-going studies presented in this paper is being sponsored by a grant from the National Science Foundation (BMS75-10978).

REFERENCES

Atlas, R. M., and Schofield, E. A. (1975). Petroleum biodegradation in the Arctic. In *Impact of the Use of Microorganisms on the Aquatic Environment*, pp. 183–198. (A. W. Bourquin, D. G. Ahearn and S. P. Meyers, eds.). National Environmental Research Center, Corvallis, Oregon.

Bartha, R., and Atlas, R. M. (1973). Biodegradation of oil in seawater: limiting factors and artificial stimulation. In *The Microbial Degradation of Oil Pollutants* (D. G. Ahearn and S. P. Meyers, eds.). Louisiana State Univ., Center for Wetland Resources. Publ. No. LSU-SG-73-01.

Berner, N. H., Ahearn, D. G., and Cook, W. L. (1975). Effects of hydrocarbonoclastic yeasts on pollutant oil and the environment, In *Impact of the Use of Microorganisms on the Aquatic Environment*, pp. 199–219 (A. W. Bourquin, D. G. Ahearn, and S. P. Meyers, eds.). National Environmental Research Center, Corvallis, Oregon.

Chakrabarty, A. M. (1972). Genetic basis of the biodegradation of salicylate in *Pseudomonas*. *J. Bacteriol.*, **112**, 815–823.

Chakrabarty, A. M. (1973). Genetic fusion of incompatible plasmids in *Pseudomonas*. *Proc. Natl. Acad. Sci. USA*, **70**, 1641–1644.

Chakrabarty, A. M. (1974). Dissociation of a degradative plasmid aggregate in *Pseudomonas*. *J. Bacteriol.*, **118**, 815–820.

Chakrabarty, A. M., and Friello, D. A. (1974). Dissociation and interaction of individual components of a degradative plasmid aggregate in *Pseudomonas*. *Proc. Natl. Acad. Sci. USA*, **71**, 3410–3414.

Chakrabarty, A. M., Chou, G., and Gunsalus, I. C. (1973). Genetic regulation of octane dissimilation plasmid in *Pseudomonas*. *Proc. Natl. Acad. Sci. USA*, **70**, 1137–1140.

Davey, J. F., and Gibson, D. T. (1974). Bacterial metabolism of *para*- and *meta*-xylene: oxidation of a methyl substituent. *J. Bacteriol.*, **119**, 923–929.

Dixon, R. A., Cannon, F. C., and Kondorosi, A. (1975). Derivation of a P-type R factor carrying the nitrogen-fixation genes from *Klebsiella Pseumoniae: nif* expression in unrelated bacteria. *Proc. Soc. Gen. Microbiol.*, **2**, 43.

Dunn, N. W., and Gunsalus, I. C. (1973). Transmissible plasmid coding early enzymes of naphthalene oxidation in *Pseudomonia putida*. *J. Bacteriol.*, **114**, 974–979.

Evans, G. H. (1968). Industrial production of single-cell protein from hydrocarbons. In *Single-Cell Protein* (R. I. Mateles and S. R. Tannenbaum, eds.). The MIT Press, Cambridge, Mass. 243 pp.

Friello, D. A., and Chakrabarty, A. M. (1975). XYL, a non-conjugative xylene degradative plasmid in *Pseudomonas* Pxy. *Abstr. Annu. Meeting of ASM*, 1975, 112.

Leidigh, B. J., and Wheelis, M. L. (1973). The clustering on the *Pseudomonas putida* chromosome of genes specifying dissimilatory functions. *J. Molec. Evolution*, **2**. 235–242.

Magasanik, B. (1961). Catabolite repression. Cold Spring Harbor Symp. *Quant, Biol.*, **26**, 249–256.

Miget, R. (1973). Bacterial seeding to enhance biodegradation of oil slicks. In *The Microbial Degradation of Oil Pollutants* (D. G. Ahearn and S. P. Meyers, eds.), Louisiana State University, Center for Wetland Resources, Publ. No. LSU-SG-73-01.

Mylroie, J. R., and Chakrabarty, A. M. (1975). XAL, a conjugative xylene degradative plasmid in *Pseudomonas putida* (*arvilla*) mt-2. *Abstr. Annu. Meeting of ASM*, 1975, 112.

Ornston, L. N. (1971). Regulation of catabolic pathways in *Pseudomonas*. *Bacteriol. Rev.*, **35**, 87–116.

Rheinwald, J. G., Chakrabarty, A. M., and Gunsalus, I. C. (1973). A transmissible plasmid controlling camphor oxidation in *Pseudomonas putida*. *Proc. Natl. Acad. Sci. USA*, **70**, 885–889.

Soli, G., and Bens, E. M. (1973). Selective substrate utilization by marine hydrocarbonoclastic bacteria. *Biotech. Bioeng.*, **15**, 285–297.

Wang, D. I. C. (1968). Proteins from petroleum. *Chem. Eng.*, **75** (18), 99–108.

Wheelis, M. L., and Stanier, R. Y. (1970). The genetic control of dissimilatory pathways, in *Pseudomonas putida*. *Genetics*, **66**, 245–266.

Williams, P. A., and Murray, K. (1974). Metabolism of benzoate and the methylbenzoates by *Pseudomonas putida* (*arvilla*) mt-2: evidence for the existence of a TOL plasmid. *J. Bacteriol.*, **120**, 416–423.

Yamada, K., Takahashi, J., Kawabata, Y., Okada, T., and Onihara, T. (1968). SCP from yeast and bacteria grown on hydrocarbons. In *Single-Cell Protein* (R. I. Mateles and S. R. Tannenbaum, eds.). The MIT Press, Cambridge. 192 pp.

Zajic, J. E., and Daugulis, A. J. (1975). Selective enrichment processes in resolving hydrocarbon pollution problems. In *Impact of the Use of Microorganisms on the Aquatic Environment*, pp. 169–182 (A. W. Bourquin, D. G. Ahearn, and S. P. Meyers, eds.), National Environmental Research Center, Corvallis, Oregon.

ZoBell, C. E. (1973). Microbial degradation of oil: present status, problems and perspectives. In *The Microbial Degradation of Oil Pollutants*. (D. G. Ahearn and S. P. Meyers, eds.), Louisiana State Univ., Center for Wetland Resources Publ. No. LSU-SG-73-01.

THE BIOCIDAL REQUIREMENTS OF CUTTING FLUIDS

D. E. BOSWELL

Cincinnati Milacron, Cincinnati, Ohio 45209, U.S.A.

Summary

This paper presents a summary of the factors that must be considered in selecting a biocide for a cutting fluid. Special emphasis is placed on biocide efficacy and human safety.

DISCUSSION

Many papers have appeared over the years in which specific biocides have been studied in specific cutting fluids, but there has not appeared a discussion of the total requirements that must be taken into account in selecting biocides for cutting fluids. Biocidal efficacy alone is but one of many requirements. Indeed, from the viewpoint of a representative of a firm that manufactures cutting fluids, frequent contacts with biocide suppliers have shown that there is a broad lack of understanding on the part of these suppliers as to what are the important factors in matching a biocide to a cutting fluid. The need for understanding in this area is apparent when one considers such factors as human safety, cost, and pollution control in the use of cutting fluids today. A fair number of new biocides has been introduced in recent years but a discouragingly small percentage of them have proved to be useful in cutting fluids. One of the problems, again, is a lack of understanding on the part of suppliers about the use of biocides in cutting fluids. Many persons seem to think that the selection of a biocide for a cutting fluid somehow relates in a critical fashion to the ingredients of the cutting fluid. Except for the obvious need to avoid cationic biocides in anionic cutting fluids, and the like, this presumption is not true.

Before discussing biocidal requirements for cutting fluids, it will be useful to define the types of cutting fluids used in industry today and to describe the basic differences among them. Although attempts have been made to standardize cutting fluid nomenclature (Springborn, 1967), we find today a lack of a common nomenclature being used in the industry. This is a lamentable situation since it often leads to misunderstandings in discussions between suppliers and consumers.

Cutting fluids which are liquids at room temperature can be simply classified into four types. These are: (1) cutting oils; (2) soluble oils; (3) preformed chemical emulsions; and (4) solution-type products. Differentiation among these four types can be made easily by their mineral oil content. Category (1), the cutting oils, is essentially 100% mineral oil. They often contain additives to promote lubricity. These additives most often are the well-known EP or 'extreme pressure' additives where the active ingredient lubrication-wise is sulphur, chlorine, or phosphorus or some combination of these. The cutting oils are not mixed with water and therefore do not require biocides. They will not be further considered in this paper. Categories (2), (3), and (4) are all sold as concentrated materials which the user dilutes with water to form the actual cutting fluid. The dilution ranges from 0·5% to about 20% in water depending on the product and the metal working application. However, these three categories can still be readily differentiated by their mineral oil content. Category (2), the soluble oils, contains from 70–90% mineral oil. The remainder of the product is emulsifiers and perhaps other materials such as EP agents and a biocide. The phrase 'soluble oil' is really a misnomer since this type of cutting fluid forms an emulsion and not a solution when added to water. Category (3), the preformed chemical emulsions, contains a small amount of mineral oil of the order of 5–20%. The balance will include emulsifiers, biocides, and often other materials designed to provide other properties such as extra lubricity and corrosion control. Category (4), the solution-type products, contains no mineral oil at all. These types of cutting fluids are, in general, true chemical solutions like sugar in water. As with the other categories, the solution-type products contain a variety of materials which provide, for example, wetting action and microbial and corrosion control. Thus, we can easily categorize fluids by their oil content which ranges from about 100% to 0%. Table I gives the nomenclature used in this paper in several languages.

Table I
NOMENCLATURE OF CUTTING FLUIDS

English	Oil	Soluble oil	Preformed emulsion	Solution
German	Schneid oel	Emulsion bohroel	Halb-synthetische kuhlmittel	Voll synthetische kuhlmittel
French	Huile	Liquide du type emulsion	Semi-synthetique	Synthetique
Italian	Olio	Fluids emulsionabile	Refrigerante semi-syntetico	Refrigerante sintetico
Dutch	Suijolie	Emulsi	Half-synthetisch	Synthetisch
Swedish	Olje	Typ emulsion	Halv-syntetish	Syntetish
Japanese	Abura	Nyukazai tenka	Shyoryo-no kagakubushitsu oyobi abura ganyu	Kagakubushitsu ganyu

Table I shows that the names used for cutting fluids in other countries are suggestive of synonyms often used in the United States. For example, the preformed emulsions are often referred to as semi-synthetics, chemical emulsions, or semi-chemical fluids. The solutions are often referred to as synthetics or chemical fluids and soluble oils, as emulsions or emulsifiable cutting fluids.

Biocides should be used in all cutting fluids which are mixed with water. Water makes up from 80–99·5% of the cutting fluid mix and the mix thus provides an excellent medium in which bacteria and fungi can grow. Some materials used in cutting fluids such as the common sulphonate emulsifiers used in soluble oils are good food sources for certain bacteria. In the absence of any biocide, a cutting fluid will soon become rancid or otherwise suffer from the growth of microorganisms. This is not because the microorganisms are present in the product at the beginning but because bacteria and fungi are everywhere—in the air, on the hands of the people working in the fluid, on the part being cut or ground, occasionally in the water used to make the mix, and often on the surfaces of the cutting fluid tank and delivery system due to inadequate maintenance of the system. These organisms quickly get into the fluid and begin to grow. The most common sign of excessive microbial growth is a foul odour. In extreme situations, microbial growth can lead to a serious breakdown of the product in a chemical sense as the organisms feed on the product and thus destroy its lubricity or rust control or other performance characteristics. In addition to their use in preventing the attack of organisms on the product itself, biocides are needed to combat the effect of the contamination encountered when a cutting fluid is in a machine. By far the most serious contaminants are the oils that leak into the cutting fluids and the metal fines. It is not uncommon to find 3–10% of contaminant oil, usually called 'tramp oil', in a cutting fluid. There are literally hundreds of brands of various oils on the market and they have a very wide ranging chemistry. Often, they contain additives based on sulphur and phosphorus. The sulphur or phosphorus compounds are extracted into the water phase of the product and give the bacteria and fungi an additional food source (Yust, pers. comm.). Metal fines have an enormous surface area and can physically absorb biocides and other product components when the fines are not adequately removed from the cutting fluid. Fines can accumulate in cutting fluid reservoirs and create a local anaerobic condition which can lead to rancid odours. Some biocides are known to be deactivated by metal cations (Rossmoore, pers. comm.).

Thus, all this contamination from so many sources coupled with the presence of a lot of water makes bacterial and fungal control one of the more important properties of a cutting fluid. There are two ways to achieve this control. One is to match a biocide to a product and manufacture the product with a built-in control system. This is by far the more desirable of the two approaches. The other approach is to add biocides to the product as it is being used. This is a cumbersome approach since it requires in-plant handling of the biocides and careful control of the amount and frequency of their addition to the cutting fluid. It is a more expensive approach than using a product which contains a biocide carefully matched to the product itself. Further, biocide additions should be made only on the basis of tests that are run to determine the microbial population of the system. This requires an in-plant capability to do plate counts or some other reliable method of estimating the microbial condition of the system. All this is not to say that biocides should never be added directly to a cutting fluid during its life cycle. The many contaminants that can enter the cutting fluid and lapses in proper control of the cutting fluid such as filter inefficiency or running the fluid too lean can lead to the need to use a biocide to kill the microorganisms while the problem that led to their growth is being corrected. The matching of a biocide to a cutting fluid is a very complicated matter and biocides should never be added to a cutting fluid without prior approval of the cutting fluid supplier.

Again, the most desirable form of microbial control is by the use of cutting fluids that contain an effective built-in biocide system. How does a biocide get matched to a cutting fluid? It is here that the users of cutting fluids can benefit by using products made by competent manufacturers who have the microbiological skills and the know-how necessary to accomplish this critical task. Not only is the capability of the cutting fluid supplier of paramount importance but also important are the understanding and cooperation of the biocide manufacturers who develop and market the biocides in the first place. There are several types of information needed by a cutting fluid manufacturer before he can undertake the expensive and time-consuming evaluation of a new biocide in a cutting fluid. These requirements are given in Table II.

These requirements, if met, will allow the purposeful evaluation of the biocide in a cutting fluid. Most of the requirements speak for themselves, but special emphasis must be placed on efficacy and toxicology. The efficacy of a biocide for use in cutting fluids cannot be, in fact, determined until the biocide is incorporated into a specific fluid and tested microbiologically. With the exception of a few materials such as raw formaldehyde, a given biocide will not work in every cutting fluid. A biocide which will kill all the organisms in a simple water solution may be completely ineffective in some cutting fluids but quite effective in others. The reasons for this phenomenon are probably as numerous as the number of cutting fluids on the market today but this does point up the importance of thorough testing. The testing work done by a cutting fluid microbiologist is expensive and very time-consuming and should not be undertaken until the biocide supplier has provided information on the minimum requirements listed in Table II.

Table II
REQUIREMENTS FOR A CUTTING FLUID BIOCIDE

Solubility. Sufficient water solubility for good biocidal activity.
Chemical stability. Stable in alkaline systems. Unaffected by hard water cations.
Efficacy. Preferably controls both bacteria and fungi.
Environmental Protection Agency (*EPA*) *Registration*. Should be registered or near registration at time of user evaluation.
Toxicology. Should be nontoxic at use dilutions. Should not be a skin sensitizer.
Chemical restrictions. Should be nonphenolic and free of toxic metals.

The determination of the efficacy of a biocide for use in cutting fluids is not a matter of running some straightforward, well-recognized tests. Efficacy determinations and the methods used vary considerably from one manufacturer of biocides to another. However, the basic information should at least include the minimum inhibitory concentration of the biocide in a simple medium against the more common organisms found in cutting fluids such as *Pseudomonas*, *Alcaligenes*, *Desulfovibrio*, *Fusarium*, and *Candida*. It is encouraging that at the present time there is a very active group in the E35.15 Committee of the American Society for Testing and Materials which is well along in developing a standard test procedure for measuring biocide efficacy for cutting fluids (Sharpell, pers. comm.). Their results will be a great help to all biocide suppliers and users since new biocides can be compared to each other and to old familiar biocides to help determine their potential uses. In evaluating finished cutting fluids which contain biocides, there is no alternative to standard microbiological techniques using the organisms mentioned and good counting techniques. Tests involving irregular substrates such as cornmeal are of no value in judging biocide efficacy for cutting fluid applications. This fact was well supported in a report of the Committee on the Bioresistance of Cutting Fluids which is jointly sponsored by the American Society for Testing and Materials and the American Society of Lubrication Engineers (ASTM/ASLE, 1973). Standard test procedures are being developed by the latter committee and are important since they will expedite registration of promising biocides with the Environmental Protection Agency. Even under the latest EPA regulations issued in July of this year, the registration requirements do not include any instructions for efficacy data to support the registration of cutting fluid biocides. Thus, each manufacturer is left to his own devices in developing data that will prove efficacy. It is expected that specific guidelines will be laid down in future years perhaps as a result of the work of the committees mentioned.

Equally important is the toxicological profile of the biocide. The acute toxicity tests required by the Federal Environmental Pesticide Control Act are of great importance in helping judge the applicability of a given biocide to a cutting fluid. Of particular concern is skin safety. While any reputable manufacturer of cutting fluids will avoid using any chemicals that present a skin hazard in his product, it is possible for skin problems to develop from biocides that are not skin irritants but skin sensitizers. The constant exposure of a machinist to the cutting fluid sets up an ideal situation for a skin sensitizer to cause skin disorders. Unfortunately, it usually takes a fairly long time for a sensitizing biocide to reveal its true nature under use conditions. The time may vary from a few months to a year depending upon the concentration of the biocide and the actual extent to which the machine operators are exposed. By the time the problem appears and is identified, the biocide may be widespread and involve a large number of workers.

How can skin sensitization be avoided? First, the biocide itself should be tested by the supplier or by an independent laboratory on animals for range-finding purposes. It may also be helpful to test on humans. At this point, the close cooperation of the biocide supplier and the potential biocide user becomes very important. It is well known that a chemical sensitizer is not active at all concentration levels in causing skin sensitization. Studies on humans may show the material to be completely safe at, say, 250 ppm, but a definite sensitizer at, say, 600 ppm. It would usually not be a skin irritant at either level. The level at which sensitization occurs in an aqueous solution of the biocide should not be taken as an upper limit for the use of the biocide in a cutting fluid. In most cases, the level at which sensitization occurs will vary enormously as one changes the medium in which the biocide is used. For example, a biocide which, in a simple water solution, is a sensitizer at 600 ppm but not at 250 ppm may indeed be a sensitizer at 150 ppm in a medium which has been designed to provide many other performance requirements such as tool life and corrosion control. How can this hidden problem be avoided?

One way is to test the biocide in the cutting fluid in which it is intended to be used on a suitable number of human subjects. Another way is to use animals such as guinea pigs or monkeys. Then, and only then, can the suitability of the biocide be fully substantiated.

The use of biocides in cutting fluids is not simply a matter of preventing microbial growth and prolonging the life and thus lowering the cost of using cutting fluids. While these are very important reasons, biocide use embraces a complex range of considerations. We have already referred to the worker who will be exposed to the biocide. Every effort must be made to make sure that the product is safe for exposure to human beings. In addition to skin sensitization studies, the product should be qualified in the general toxicological sense and checked for skin and eye irritation potential as well as oral and inhalation toxicity. Also of major importance are considerations regarding water pollution. Biocides should be designed with water pollution control in mind. This is why biocides should be nonphenolic and should not contain toxic metals such as mercury or arsenic. Substances of these types have been severely restricted by law from entering municipal sewers, rivers, and streams. It is obviously to everyone's advantage, but especially the cutting fluid user, if substances like these are not used in cutting fluid products in the first place. For many years, for example, phenolic compounds have been a mainstay as biocides in cutting fluids. Most cutting fluid suppliers have long ago removed phenols from their products to avoid the pollution control problems presented by these highly water soluble compounds. Replacements for the phenols in the form of new biocides have been in general difficult to find and this presents a new and major challenge to biocide manufacturers. It appears that the manufacturers are lagging behind in meeting the requirements for nonpolluting biocides. It is hoped that they will accelerate their research work in this area and work closely with the biocide users in order to avoid a crisis in which a sufficient array of biocides is not available for the wide variety of cutting fluids available today. With the biocide being such a critical component of a cutting fluid, and with so many factors influencing our choice of biocides, this close cooperation is mandatory. This is the only way we are going to be able to have the new biocides needed for modern cutting fluids and the only way we are going to be able to continue, at the same time, the admirable human safety record that has existed in the industry for so many years.

REFERENCES

American Society for Testing and Materials/American Society of Lubrication Engineers (1973). D-2, Section L minutes.

Springborn, R. K. (ed.) (1967). *Cutting and Grinding Fluids: Selection and Application*, pp. 31–42. American Society of Tool and Manufacturing Engineers, Dearborn, Michigan.

MICROBIAL ECOLOGY WITH A CUTTING EDGE

H. W. ROSSMOORE, G. H. HOLTZMAN and L. KONDEK

Wayne State University, Department of Biology,
Detroit, Michigan 48202, U.S.A.

Summary

In this presentation two aspects of the microbial ecology associated with metalworking fluids are treated. One study was conducted on-site on the fluid and on the air surrounding the fluid of eight different metalworking operations, using an impinger air sampler and membrane filtration. Both total and coliform counts were done on ambient air during machine operation and during down periods.

Results indicate that there is good correlation between the findings in the fluid and the findings in air with respect to coliforms and the predominant coliform survival in air belongs to the genus *Klebsiella*.

In the second study involving microbial succession, an instrument called an ecologen was used to separate bacterial from fungal growth by different porosity membranes permitting the metabolic products of both to pass freely between chambers. A repeated finding indicated that dialysable molecules from bacterial growth greatly enhance fungal growth in oil-water emulsions, whereas the actual presence of bacteria has a depressing effect.

INTRODUCTION

The microbial ecology of water-soluble metalworking fluids is a result of a number of controllable and uncontrollable interacting factors. These include the type of fluid, the size of system sump, its replacement rate, the degree of external contamination, the type of metal worked, the quality of the water, and the nature and effectiveness of antimicrobial agents if used. Many of these have been documented previously (Bennett, 1972; Rossmoore, 1974). The major groups of organisms associated with water-miscible metalworking fluids are all commonly found in fresh water as well as being gram-negative and mostly belonging to that ubiquitous and omnivorous genus, *Pseudomonas* (Sabina and Pivnick, 1956; Duffett *et al.*, 1943). Other organisms, especially from human contamination, have also been isolated from these fluids (Rossmoore and Williams, 1967; Tant and Bennett, 1956). However, these contaminants tend to be transient and coincidental and usually do not compete successfully with *Pseudomonas* sp. Although gram-positive pathogens, *e.g. Staphylococcus aureus*, survive poorly in cutting fluid milieux (Rossmoore and Williams, 1967) because of the physical and biological environment (DiaGiacinto and Frazier, 1966), the same is not true for certain gram-negative species which can survive alongside the dominant pseudomonads (Maharajah *et al.*, 1956).

The coliforms, a major gram-negative group of public health significance, have been used as a presumptive biodeterioration index (Brandeberry and

Meyers, 1960). However, this was disputed by Rossmoore (1962) who was not able to relate the presence of this group or anaerobic sulphate reduction with deterioration. *Desulfovibrio desulfuricans*, whose metabolic activities result in anaerobic sulphate reduction, has a notoriety far in excess of the damage that it causes, and is an example of microbial succession in cutting fluids (Guynes and Bennett, 1959). Apparently, prior growth of the more active and less demanding pseudomonads is necessary for growth of *D. desulfuricans*, an obligative anaerobe which reproduces during quiescent periods after the oxidation potential has been lowered to -250 millivolts. Even under optimal conditions its generation time is relatively large (ca. 4 hr) (Rossmoore *et al.*, 1964) and counts rarely exceed 10^6/ml in contrast to the aerobic pseudomonads which often reach levels of 10^9/ml. Small numbers do not prevent a kind of microbial exhibitionism, in which fluids become blackened and surrounding air reeks with the pungency of rotten eggs.

An obvious method of control would involve continual aeration to prevent anaerobiosis. But as a reward for this all too simple solution, the perpetrator is literally hoisted on his own petard since the net result of maximal aeration is to increase the levels of the highly oxidative pseudomonads. Bachenheimer and Bennett (1961) pointed out the pitfalls in the utilization of two putatively synergistic inhibitors each aimed at two major bacterial groups, the pseudomonads and the sulphate reducers. Unfortunately, the pseudomonad inhibitor was inactivated by the sulphate reducers whose own special inhibitor was oxidized by the pseudomonads. These pseudomonad-sulphate reducers interactions suggest two ecological principles for the microbial population of cutting fluids: the displacement of one species from its niche may result in equally obnoxious replacements and mixed populations may collectively neutralize one or more microbiocides that single species could not accomplish alone.

This lesson could be applied aptly to fungal-pseudomonad interactions as well. In both a casual (Rossmoore *et al.*, 1972) and a contrived experiment (Rossmoore and Holtzman, 1974) optimal growth of fungi in metalworking fluids appeared dependent on prior growth of bacteria and, furthermore, the survival of the former was enhanced by the reduction of viable bacterial levels. Also, the injudicious use of certain antimicrobial agents (Rossmoore *et al.*, 1971) resulted in the appearance of dominant fungal populations, primarily belonging to the genus *Fusarium*. The reasons for the differential response of fungi to these agents as well as the dependence of fungal populations on bacteria are subjects currently under investigation in our laboratory. Preliminary data from this latter study will be treated in this presentation.

The previously discussed examples of species interactions involving bacteria-bacteria and bacteria-fungi are examples of temporal ecological relationships in which one group is time dependent on the other. Another equally intriguing area involves some problems in spatial ecology in which sites extrinsic to the fluid become populated with organisms from the fluid. These would include the ambient air surrounding the machines and the personnel operating the machines.

There is considerable evidence (Taplin *et al.*, 1965; Ehrencranz *et al.*, 1966; Marples and Kligman, 1969) that the normal flora of human skin which is predominately gram-positive cocci and diphtheroids can be readily displaced by gram-negative species, especially *Pseudomonas*, following selective antimicrobial treatment. Colonization of operator skin by *Pseudomonas* sp. after prolonged contact with cutting fluid is certainly a possibility since these fluids are extremely unfriendly to gram-positive species. Despite its ubiquity, *Pseudomonas* rarely is involved in primary infections and it may be of interest to determine its degree of colonization of the skin in the absence of frank infection.

Although the microbial flora of aerosols in a number of occupational activities has been examined (Micik, 1974; Spendlove, 1974), no such study has ever been conducted on machine tool aerosols. There are indications from epidemiological data that these aerosols have not been an occupational health problem. This may be due to the biology of the organisms in the aerosols as well as the size of the droplets. We have begun an investigation on the microbial flora of the air around machine tool operations and the preliminary findings are presented here.

MATERIALS AND METHODS

Aerobiology

All of the samplings reported here were made on-site in a large multi-functional metalworking plant. Air samples were taken with an air sampler impinger, Millipore Corporation (MC)—with a limiting orifice for 10 litre/min flow which was connected to a miniature vacuum-pressure pump (MC) (Fig. 1). Air was collected in 20 ml sterile impinger fluid, a formula in 1 litre of distilled H_2O consisting of: 2 gr powdered gelatin; 4 gr disodium phosphate,

Fig. 1. Impinger air sampler with limiting orifice and vacuum pressure pump.

anhydrous; 37 gr brain heart infusion; and 0·1 ml octyl alcohol, which was subsequently filtered through 0·45 μm membrane filters (MC ♯HAWG 047 SO). The used filters were transferred aseptically either to pads moistened with coliform medium or total count medium with indicator and incubated in 50 × 12 mm locklid dishes (Falcon Plastics) at 35°C for 24 hr.

At the same time samples of fluid were taken from the machine sumps and used for total aerobic bacterial plate counts and qualitative determination of the presence of members of the coliform group. Trypticase soy agar (TSA, Baltimore Biological Laboratories) was used for the total plate counts and Eosinmethylene blue agar (EMB, Difco Laboratories) was used for the isolation and detection of coliform species. Both were incubated for 48 hr, at a temperature of 35°C. Most typical coliform colonies were picked from the selective air sample plates and speciated by the API system (Analytab Products, Inc.).

Description of sampling sites

(*A*) *Camshaft grinding*

This operation consisted of the grinding of each individual cam on the shaft one at a time, providing an almost continuous extensive mist which was somewhat retarded during normal operations by a safety shield which was removed during the time the samples were taken. As each cam was finished, the machine would cycle, bringing the next cam in line with the grinding wheel. As this happened approximately every 45 sec, the impinger was held so as to keep it in line with the actual grinding operation as the cutting fluid was administered continually at this spot, from a pipe approximately 8 cm above the grinding operation.

(*B*) *Piston rod grinding*

This operation consisted of the grinding down of one side of each piston in a series of four or five steps set up in a circle, thereby providing an extensive continuous mist during normal operation of the machine. Impinger was held to one side of the operator approximately 60 cm from the last operation and 120 cm from the first. Fluid emitted from within each operation drained into the centre. The mist originated evenly from the centre and entrance and exits of the operations.

(*C*) *Grinding of the timing gear*

This operation consisted of the grinding down of a round metal blank down to a gear. The fluid was administered through three pipes opening up about 2·5 cm above the operation, emitting a gentle stream. The mist present was slight and was contained primarily to the machining operation itself, with only regular drops being exuded from the machine. The impinger was held approximately 30 cm away from the actual operation for optimal results.

(*D*) *Rockerarm shafts*

This operation consisted of the drilling of holes through the shaft with the fluid flowing from pipes approximately 5 cm above each drill bit. The impinger was held approximately 60 cm away from the operation due to the complexity of the machine.

(*E*) *Oil pumps*

Actual operation was not functioning during sampling and served as control. The impinger was held approximately 30 cm away from where the operation was performed; however, the air was free of mist.

(*F*) *Cylinder head cam line*

This operation consisted of the drilling and surface grinding of the cylinder heads. Fluid was administered within the operation and any mist produced was confined almost extensively to the operation. The impinger was held as close as possible with difficulty due to the size and complexity of the machine itself.

(*G*) *Flywheels*

This operation consisted of the drilling of holes and grinding of the flywheel in a series of steps set up in a circle. Safety shields constructed partly around the operation allowed the mist to escape only through the top of the shields. The mist originated from the actions of the individual operations and was extensive. The impinger was held above the shields so as to be in an area between two operations where the mist was most extensive, approximately 60 cm from each operation.

(*H*) *Piston grinding*

This operation consisted of the grinding of the upper periphery of the piston to a certain diameter. Fluid was administered approximately 3 cm above the operation and due to the rapid spinning of the operation and flow rate of the fluid, a moderate mist was present. The impinger was held approximately 15 cm away from the operation at eye level.

Fungal-bacterial interaction

The apparatus used in this portion of our study was the Ecologen (Model E-40, New Brunswick Scientific Corp., NBS) housed in Environmental Incubator Shaker (Model G-24, NBS). It consisted of a stainless steel central reservoir (Fig. 2) in a cubical mode, surrounded by four pyrex growth chambers. The growth chambers and reservoir have a working capacity of 100–300 ml and 500 ml, respectively. Membrane filters 60 mm in diameter with varying pore sizes were installed between the central reservoir and growth chambers, allowing differential filtration to occur. Membranes were supplied by The Nucleopore Corp. in the following porosities: 0·0 μm, 0·03 μm, 0·2 μm, and 0·6 μm.

For filtering molecules smaller than 0·03 μm, dialysis tubing (Visking Corp.) with a mean porosity of 0·0048 μm was used. After assembly, the whole unit was sterilized by autoclaving. The fluid environment for this study was a 5% o/w emulsion. Growth chambers contained 200 ml of fluid, while the diffusion reservoir contained 400 ml. The environmental conditions during growth were as follows: incubation temperature, 30°C; shaking rate, 110 rpm.

Inoculum

Bacterial inocula for the central reservoir were *Pseudomonas aeruginosa* ATCC 15442. The cultures were maintained on trypticase soy agar slants (TSA, Baltimore Biological Laboratory), incubated at 30°C, and transferred air levels in operating systems between 10^4/ml and 10^9/ml. Even though the air plates are all TNTC, we have learned that at the comparatively low level of 10^4/ml in the fluid, sufficient organisms survive in the mist to overgrow the plate.

Fungal-bacterial interaction

In the first ecologen experiment both the bacterial and fungal inocula were lower than was intended and as a result the loss in viability on inoculation resulted in very low starting levels of fungi (Table II). In addition, bacteria

Table II
EFFECT OF BACTERIA ON VIABILITY AND GROWTH OF FUNGI IN 5% o/w EMULSION—SMALL FUNGAL INOCULUM

Ecologen mode		*Days after inoculation*					
		2	6	12	20	24	26
Central reservoir	B:	$1{\cdot}4\times10^6$	3×10^5	$1{\cdot}7\times10^6$	5×10^5	2×10^6	3×10^6
	F:	0	0	0	0	$4{\cdot}6\times10^2$	5×10^2
Growth chambers *Membrane pore size*							
0·0 μm	B:	0	0	0	$7{\cdot}3\times10^3$	$6{\cdot}2\times10^4$	4×10^4
	F:	40	2×10^3	7×10^2	$1{\cdot}3\times10^4$	6×10^3	3×10^3
0·2 μm[a]	B:	0	9×10^4	>300	8×10^5	$1{\cdot}4\times10^5$	3×10^6
	F:	50	30	80	0	$3{\cdot}5\times10^2$	4×10^2
0·6 μm	B:	0	9×10^4	>300	5×10^5	7×10^5	$1{\cdot}7\times10^6$
	F:	30	30	30	0	10	50
0·0048 μm	B:	0	0	$1{\cdot}9\times10^2$	9×10^3	$1{\cdot}1\times10^4$	$4{\cdot}7\times10^4$
	F:	20	2×10^2	$4{\cdot}2\times10^3$	8×10^4	7×10^4	6×10^4

[a] Leak detected in this system.
B = Bacteria/ml.
F = Fungal colony forming units/ml.
Inoculum
Pseudomonas aeruginosa—10^7/ml.
Fusarium sp.—270/ml.

appeared in chambers where the pore size of the membranes used should have prohibited their passage. In order to check for leaks without disturbing the apparatus, we added 0·5% aqueous methylene blue to the central reservoir. Throughout the rest of the experiment we detected methylene blue only in the 0·2 μm chamber.

After dismantling the apparatus, we noticed that fungi were growing on the face of all the membranes in the four growth chambers and they were all stained blue. This explains the lack of diffusion of the methylene blue in two of the chambers with pore size sufficiently small to permit free diffusion. We also noticed a tear around the rim of the 0·2 μm membrane. Unfortunately, there is a design shortcoming in this area which we have attempted to overcome in the next experiment. The results with the methylene blue diffusion did not explain to us the reasons for the bacterial presence in the no-pore and in the dialysis chambers. At the conclusion of the 26-day period, the chambers with the highest bacterial counts had the lowest fungal counts, and those with the lowest bacterial counts had the highest fungal counts. In the second experiment (Table III), the size of the inoculum was increased considerably

Table III

EFFECT OF BACTERIA ON VIABILITY AND GROWTH OF FUNGI IN 5% o/w EMULSION—LARGE FUNGAL INOCULUM

Ecologen mode		*Days after inoculation*				
		4	6	10	12	18
Central reservoir	B:	$1{\cdot}1 \times 10^6$	4×10^6	1×10^6	6×10^5	6×10^6
	F:	0	2×10^2	9×10^2	—	$1{\cdot}2 \times 10^3$
Growth chambers						
Membrane pore size						
0·0 μm	B:	0	0	0	0	0
	F:	$1{\cdot}4 \times 10^3$	7×10^4	$1{\cdot}1 \times 10^5$	$1{\cdot}1 \times 10^5$	8×10^4
0·03 μm	B:	0	0	0	0	0
	F:	$7{\cdot}1 \times 10^4$	$1{\cdot}1 \times 10^4$	9×10^4	2×10^4	9×10^4
0·6 μm	B:	TNTC @ 10^2	$1{\cdot}1 \times 10^6$	$1{\cdot}8 \times 10^6$	—	$1{\cdot}6 \times 10^6$
	F:	$1{\cdot}8 \times 10^3$	8×10^3	8×10^3	8×10^3	1×10^4
0·0048 μm	B:	0	0	0	0	0
	F:	$1{\cdot}2 \times 10^5$	$1{\cdot}7 \times 10^5$	4×10^5	3×10^5	3×10^5

B = Bacteria/ml.
F = Fungal colony forming units/ml.
Inoculum
Pseudomonas aeruginosa—5×10^7/ml.
Fusarium sp.—9×10^3/ml.

and, in addition, the following changes were made in filter configuration. The 0·0 μm pore and the 0·0048 μm membranes were used in double thickness and 0·03 μm filters were used in place of the 0·2 as a selective membrane for large molecules. We experienced no leaks in any of the systems, *i.e.* no bacteria were detected in any of the systems where the membrane pore size would preclude their passage. However, fungi did appear in the central reservoir presumably due to propagule passage through the 0·6 μm filter. Again, those systems where there were bacteria present had the lowest fungal levels and the sample with the 0·0048 membrane had the highest.

This experiment is still in progress and therefore we have had no opportunity to examine the membranes.

DISCUSSION

The preliminary investigation of the bacterial flora in mists generated by metalworking operations showed that total counts are of the same order of magnitude as previously reported for activated sludge (Randall and Ledbetter, 1966). The greatly reduced bacterial level in the shielded system is a strong recommendation for the routine shielding of all aerosol-producing operations. The nonoperating control system had levels acceptable even for some non-sterile hospital environments (Gohr, 1970). The *K. pneumoniae* isolations compared with those in activated sludge (Randall and Ledbetter, 1966) which was not surprising, since encapsulated forms are resistant to dessication. Even from this limited study there is an obvious question. What danger do these organisms pose to occupational health?

Randall, C. W., and Ledbetter, J. O. (1966). Bacterial air pollution from activated sludge units. *Am. Ind. Hyg. Assoc. J.*, **27** (6), 509–519.

Rossmoore, H. W. (1962). Correlation of coliform activity and anaerobic sulfate reduction with deterioration of cutting fluids. *Lubr. Eng.*, **18**, 226–229.

Rossmoore, H. W. (1974). Microbiological causes of cutting fluid deterioration. *Soc. Manufact. Eng. Tech. Paper* # MR 74-169.

Rossmoore, H. W., and Holtzman, G. H. (1974). Growth of fungi in cutting fluids. *Dev. Ind. Microbiol.*, **15**, 273–280.

Rossmoore, H. W., and Treusch, P. J. (1975). Comparison of several media for enumeration of cutting fluid fungi. *Dev. Ind. Microbiol.*, **16**, 475–482.

Rossmoore, H. W., and Williams, B. W. (1967). Survival of coagulase-positive staphylococci in soluble cutting oils. *Health Lab. Sci.*, **4** (3), 160–165.

Rossmoore, H. W., DeMare, J., and Smith, T. H. F. (1971). Anti-pro-microbial activity of hexahydro-1,3,5-tris-(2-hydroxyethyl-)-s-triazine in cutting fluid emulsion. In *Biodeterioration of Materials*, pp. 286–293 (A. Harry Walters and E. H. Hueck-Van Der Plas, eds.), Appl. Sci. Publishers, London.

Rossmoore, H. W., Holtzman, G. H., and Rossmoore, L. A. (1972). Microbial succession in cutting fluid milieu. *Abst. Ann. Meeting, Am. Soc. Microbiol., 72nd Abstr.* No. E 15.

Rossmoore, H. W., Shearer, M., and Shearer, C. (1964). Growth studies on *Desulfovibrio desulfuricans*. *Dev. Ind. Microbiol.*, **5**, 334–342.

Sabina, L. R., and Pivnick, H. (1956). Oxidation of soluble oil emulsions and emulsifiers by *Pseudomonas oleovorans* and *Pseudomonas formicans*. *Appl. Microbiol.*, **4**, 171–175.

Schultz, J. S., and Gerhardt, P. (1969). Dialysis culture of microorganisms: design, theory and results. *Bacteriol. Rev.*, **33**, 1–47.

Spendlove, J. C. (1974). Industrial, agricultural and municipal microbial aerosol problems. *Dev. Ind. Microbiol.*, **15**, 20–27.

Steinhauer, B. W., Eickhoff, T. C., Kislak, J. W., and Finland, M. (1966). The *Klebsiella-Enterobacter Serratia* division. Clinical and epidemiological significance. *Ann. Inter. Med.*, 1180–1194.

Tant, C. O., and Bennett, E. O. (1956). The isolation of pathogenic bacteria from used emulsion oils. *Appl. Microbiol.*, **4**, 332–338.

Taplin, D., Zaias, N., and Rebell, G. (1965). Environmental influences on the microbiology of the skin. *Arch. Environ. Health*, **11**, 546–550.

A NEW APPROACH TO THE CONTROL OF MICROBIAL DEGRADATION OF METAL-WORKING COOLANTS AND LUBRICANTS

W. R. SYMES and D. COWAP

Imperial Chemical Industries Limited, Mond Division,
Runcorn, Cheshire, U.K.

Summary

A novel filter based on zirconia fibres is described and laboratory results are reported which demonstrate that it can be used to control microbial contamination in aqueous metal-working coolants and lubricants. The filter is particularly suitable for self-contained machine tools and field trials are described which show that it improves the condition of the cutting fluid and extends coolant life.

INTRODUCTION

Aqueous cutting fluids can provide an ideal environment for the growth of microbes. Depending on conditions, bacteria, fungi, and yeasts may all be found. At alkaline pH and ambient temperature, bacteria generally predominate and many species have been identified in aqueous cutting fluids. In well-aerated emulsions aerobic bacteria such as *Pseudomonas*, *Proteus*, and *Aerobacter* generally predominate. Under favourable conditions these can rise to a concentration of 10^9 organisms/ml.

At such a high bacteria concentration, the cutting fluid is rapidly degraded resulting in unsatisfactory finish of the work-piece, reduced cutting head life, and corrosion of both the work-piece and the machine. Less well-aerated systems will also contain anaerobic bacteria such as *Desulfovibrio* which produce the foul odours associated with badly degenerated and stagnant fluids.

Aqueous fluids are far more prone to bacterial degradation than neat oils. Because of ever rising oil costs, they are now being used for the majority of machining operations and bacterial degradation is a major industrial problem which severely limits cutting fluid life. Renewing the fluids provides only a temporary solution as the bacteria rapidly grow again to a similarly high level. In particularly bad cases cutting fluid life can be reduced to 1–2 weeks. This increases coolant costs, effluent disposal costs, labour costs, and machine down-time.

Attempts have been made to prevent microbes from becoming established by formulating biologically hard fluids. No great success has so far been achieved with this approach and such fluids, being difficult to biodegrade, can cause effluent disposal problems. In an alternative approach, preservatives are added to cutting fluids to prevent microbial growth. To be successful, this requires a suitable method of monitoring the condition of the fluid and adding further quantities of preservative as necessary. While acceptable in very large centralized systems, this technique is less well suited to the self-contained machine tool.

There is a real need for a simple in-line method of limiting microbial degradation of aqueous cutting fluids, particularly in individual machine tools, so that this factor ceases to determine fluid life.

Against this background, ICI has developed the use of 'Saffil' zirconia inorganic fibres for continuously controlling microbial contamination of aqueous cutting fluids by filtration. A compact 'Rembac' filter has been designed for this purpose. ('Saffil' and 'Rembac' are trade marks of Imperial Chemical Industries Limited, Millbank, London, U.K.) This paper reports the results of the initial laboratory and field trials of the filter for the control of microbial degradation of metal-working coolants and lubricants. This work has been concerned mainly with the use of the filter to control bacterial degradation. It is also effective against fungi and yeasts and these contaminants are mentioned where they are important.

THE FILTER SYSTEM

The key advantage in controlling the level of microbial contamination by filtration is that control is in-line and continuous. Maintenance is simple if the filter element is renewed at the same time as the fluid.

Attempts have been made in the past to use proprietary cartridge filters to remove bacteria from cutting fluids. Typically, those with the required efficiency have been rejected because of high pressure loss and rapid blockage. The development of zirconia fibres which have unique advantages over traditional filter media such as glass fibres or cellulose has led us to reconsider the filtration of cutting fluids.

'Saffil' zirconia fibres have closely controlled physical characteristics as shown in Table I. Being available as a well-defined fibre blanket, they are

Table I
SOME PHYSICAL PROPERTIES OF 'SAFFIL' ZIRCONIA FIBRES

Fine fibre diameter	3 μ
Controlled diameter distribution	1 μ < 95% (by no.) < 5 μ
Staple length	2–10 cm

particularly suitable for fabrication into an element which offers depth filtration without excessive pressure loss. Unlike most other fibres (except asbestos which is well known for sterilizing filtration), zirconia fibres are neutral or positively charged at pH 8 or less. This makes them more efficient at collecting bacteria which, because of the proteinaceous nature of their cell walls, develop a negative zeta potential under most conditions (Fig. 1).

To exploit zirconia fibres for this duty, the compact 'Rembac' filter element has been developed. This is shown together with the filter system which houses it in Fig. 2. It is suitable for use with self-contained machine tools and is shown fitted to a production machine in Fig. 3.

Because cutting fluid is continuously recirculated and continuously refiltered, it is not necessary to filter all the fluid at each pass. The normal cartridge filter design has been modified by the inclusion of a small hole in the end-plate as shown in Fig. 4. This allows a partial flow of fluid (*ca.* 30–35%

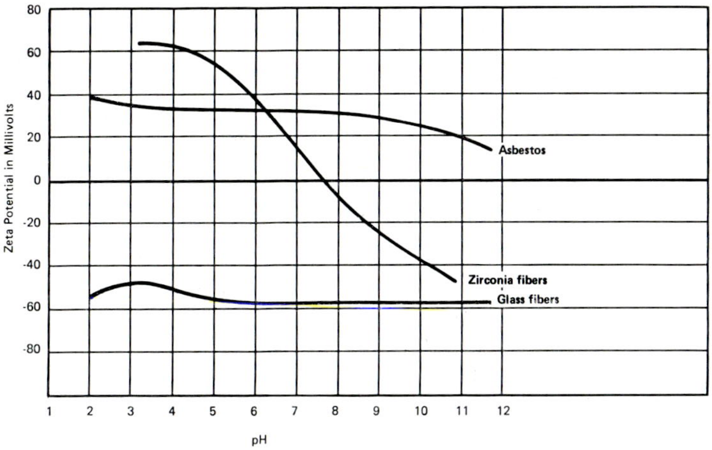

Fig. 1. Zeta potential of zirconia and other inorganic fibres.

on a clean element) parallel to the face of the element. By helping to keep the surface clean and free of swarf and tramp oil, this feature improves filter life. Typical flow-rate/pressure drop characteristics for this filter system are shown in Fig. 5.

Essentially, this new approach aims to establish a steady bacteria concentration in contaminated cutting fluids sufficiently low for bacterial degradation to cease to limit fluid life.

FILTER PERFORMANCE

A wide range of conditions on the shop floor contribute to cutting fluid spoilage. These conditions are difficult to reproduce and control in the laboratory. We have not attempted to copy a machine tool environment but

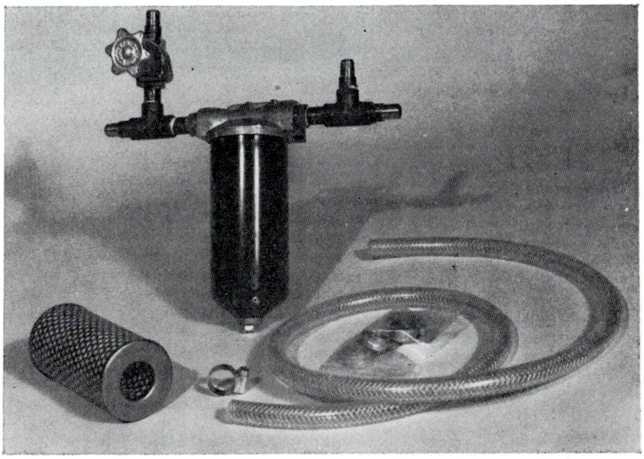

Fig. 2. 'Rembac' filter elements with housing and fittings as used in the laboratory and field trials.

Fig. 3. Filter trial at Brockhouse Transmission Ltd.

have used a laboratory rig (Fig. 6) to compare the level of microbial contamination in filtered and unfiltered cutting fluids. The object of this work was to demonstrate that the filter improves the condition of badly contaminated fluids and to establish the level to which it limits bacteria concentration in freshly made up fluids. Field trials have also been carried out to show that these effects can be achieved under machine shop conditions.

Most of our work has been with three widely used coolants: two mineral oil based emulsions (A and B) and one synthetic coolant (C). Manufacturers closely guard their formulations, but synthetic coolants are often based on polyalkylene glycols and can also contain a limited amount of emulsified mineral oil. Preliminary investigations of 11 other proprietary coolants have suggested that these behave similarly.

Experiments have been carried out using 11 litres of fluid. In the case of the emulsions (fluids A and B), this was made up of 10 litres of fresh emulsion and a 1-litre inoculum of contaminated fluid from a working machine tool sump. It was not necessary to inoculate the synthetic coolant since 11 litres, freshly made up, had a total bacteria concentration of 10^4 organisms/ml.

In each experiment the fluid was circulated at 3 litres/min for up to 8 weeks through a 'Rembac' filter element 140 mm long and 70 mm o.d. An unfiltered control was run for each fluid examined.

At weekly intervals, 1 litre was withdrawn and replaced with 1 litre of freshly made up fluid. This was to ensure that an acceptable nutrient balance was maintained. Samples were taken weekly and the total bacteria count

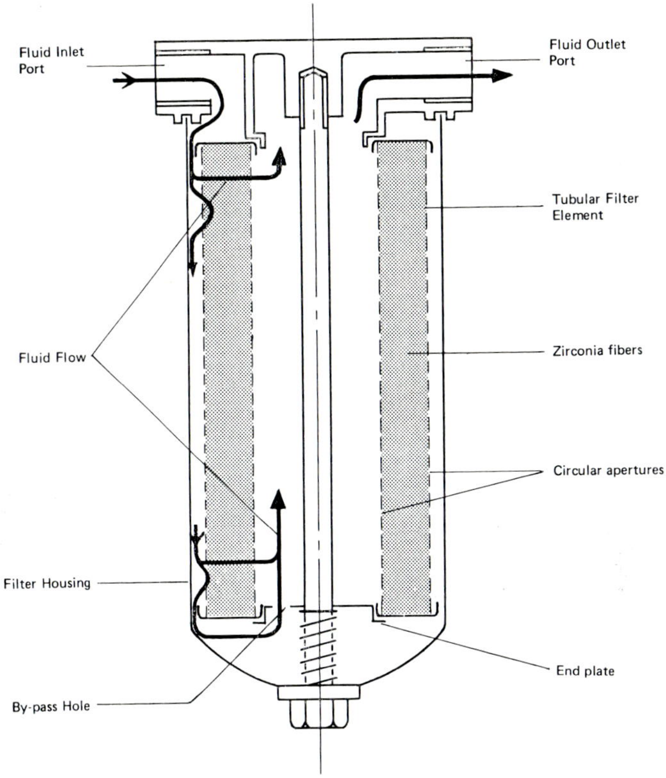

Fig. 4. Bypass filter design.

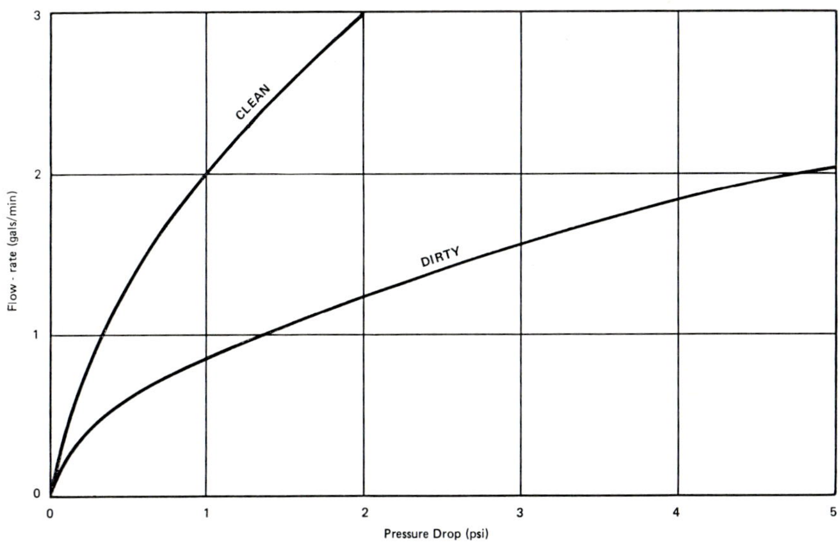

Fig. 5. Flow-rate/pressure drop characteristics of the bypass filter.

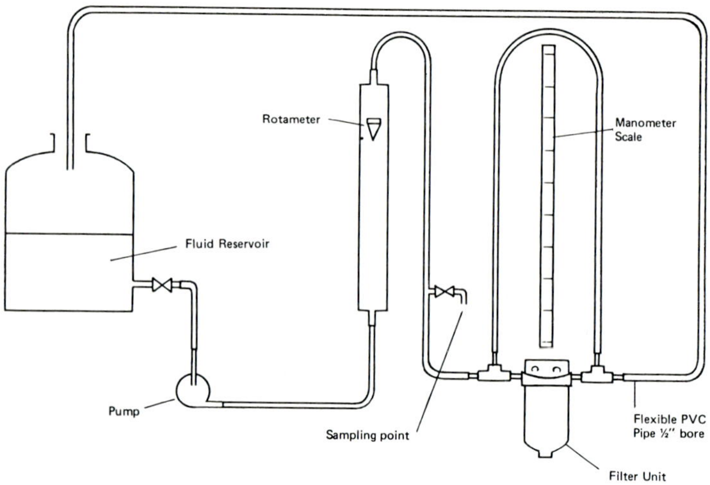

Fig. 6. Laboratory rig layout.

estimated by the viable counting technique of serial dilution using spread plates prepared with tryptose blood agar base and 2,3,5 triphenyltetrazolium chloride. The condition of the fluid and its emulsion stability (where appropriate) were visually assessed. Coolant concentration was monitored using a hand refractometer. From time to time, the levels of fungus and yeast were estimated using proprietary malt agar dip-slides.

The total bacteria counts for the two emulsions (fluids A and B), using the bypass filter design operating continuously at ambient temperature, are plotted in Figs. 7 and 8.

These results show that while the bacteria count in the unfiltered fluid in each case rose to 10^7–10^9 organisms/ml within 2–3 weeks, the level in the filtered fluids was maintained between 10^4 and 10^6 organisms/ml. The

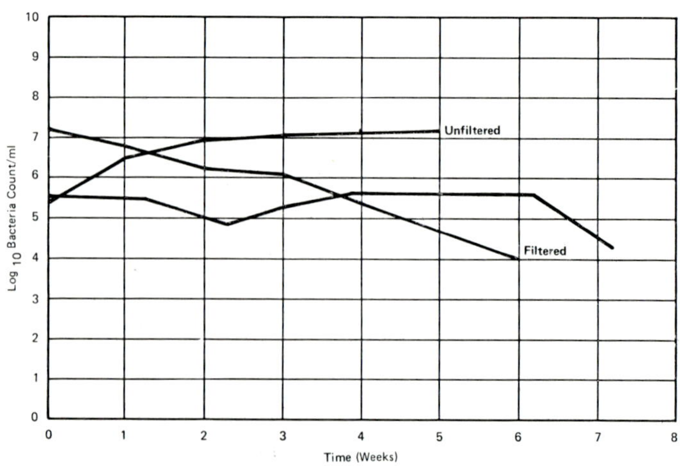

Fig. 7. Emulsion A circulated continuously through a bypass filter at ambient temperature.

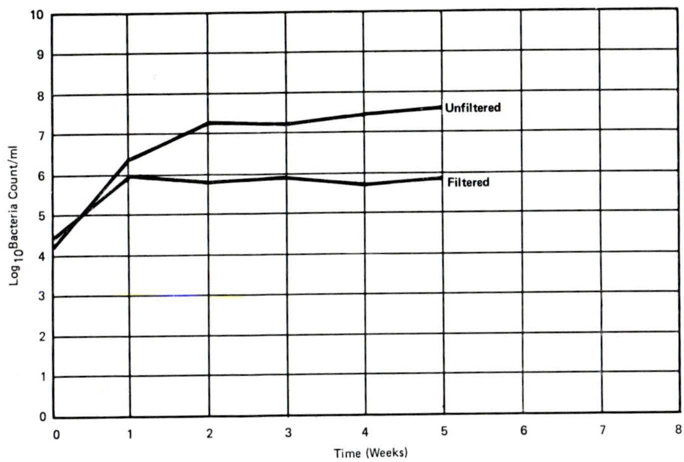

Fig. 8. Emulsion B circulated continuously through a bypass filter at ambient temperature.

filtered emulsions retained their milky appearance and good stability. Free oil was not formed in the reservoir. In contrast, the unfiltered fluids both darkened and, after about 4 weeks, were splitting badly. Figure 7 also shows that where fluid A was allowed to develop a high bacteria concentration initially, filtration reduced the count from about 10^7 organisms/ml to about 10^4 organisms/ml.

Similar results were obtained with the synthetic coolant (fluid C, Fig. 9) under similar conditions. Filtered, this largely retained its clarity and was free of yeasts although, after 4 weeks, signs of fungal growth were observed. The unfiltered control became completely cloudy within a week and showed heavy fungal contamination.

To determine the effect of discontinuous filtration, fluid A was circulated

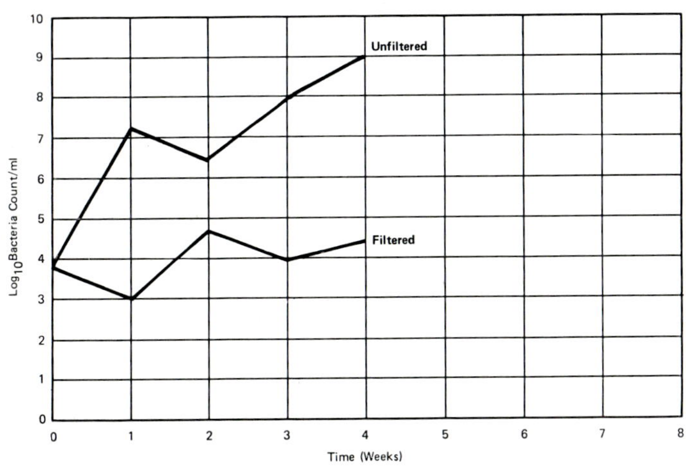

Fig. 9. Synthetic coolant C circulated continuously through a bypass filter at ambient temperature.

for 8 hr each day, 5 days each week, through a bypass filter. The bacteria concentration after 4 weeks was 1×10^6 organisms/ml which is little different to that obtained with continuous fluid circulation (8×10^5 organisms/ml).

These results are summarized in Table II which also shows that when the filter was used without a by-pass (that is with all fluid passing through the element on each pass), fluid circulating continuously, the bacteria concentration after 4 weeks was 4×10^4/ml.

Table II

TOTAL BACTERIA CONCENTRATION AFTER 4 WEEKS FOR EMULSION A UNDER VARIOUS CONDITIONS

Conditions	*Bacteria count/ml*
No filtration	3×10^7
Continuous by-pass filtration	8×10^5
Discontinuous (40 h/w) by-pass filtration	1×10^6
Continuous throughflow filtration	4×10^4

It is concluded that under laboratory conditions the 'Rembac' filter has limited the bacteria concentration to 10^6 organisms/ml or less and maintained the fluid in good condition for up to 7 weeks. Unfiltered, the bacterial level rose to a concentration in excess of 10^7 organisms/ml and rapidly caused degradation of the fluid.

FIELD TRIALS

Two field trials using the bypass design of filter unit have been carried out in the United Kingdom at Rolls Royce (1971) Limited, Derby Engine Division (detailed results to be published in a paper in preparation) and at Brockhouse Transmission Limited. Past experience at these locations has shown that where the bacteria concentration is maintained at 10^6 organisms/ml or less, bacterial degradation does not limit fluid life. At both sites there are machine tools in which bacterial contamination can, in some circumstances, limit fluid tank life.

The object of the field trials was to determine whether filtration of the cutting fluid would extend its life and improve operating conditions. The layout used for the on-machine trials is shown in Fig. 10. The external bypass valve, normally closed, was used occasionally to measure the pressure drop across the filter.

Details of the trials at Rolls Royce (1971) Ltd are shown in Tables III and IV. The semi-synthetic coolant used contained a biocide. At the time of the trial coolant, life on the majority of machine tools was 12 weeks or more. The specific boring machine chosen for this trial was particularly susceptible to microbial spoilage of the cutting fluid and untreated, the fluid had to be renewed every 5–7 days because of obnoxious smells. With the fluid filtered, the bacteria count was maintained at 10^5–10^6 organisms/ml or less for 4 weeks. During this period, the fluid showed little sign of yeast or fungal contamination and functioned satisfactorily without developing a foul odour. After 4 weeks, the fluid was renewed because of buildup of free tramp oil, its life having been extended fourfold by filtration.

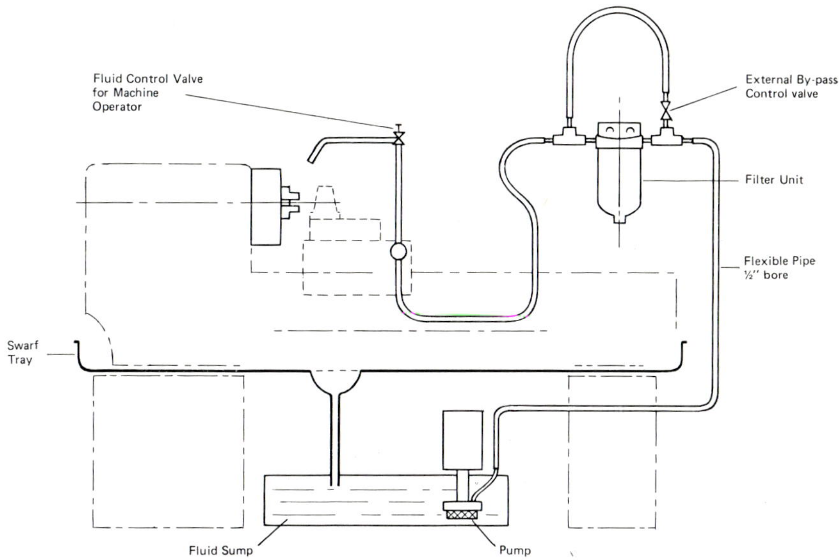

Fig. 10. On-machine filter trial layout.

The conditions and results of the Brockhouse trial are shown in Tables V and VI. The lathe chosen was typical of those in operation at this factory. The unfiltered emulsion was typically renewed at about 6 weekly intervals due to smells and corrosion. The filtered emulsion showed little signs of splitting after 15 weeks and did not smell or cause corrosion. For most of this time the bacteria concentration was maintained at 10^4–10^5 organisms/ml and only rose to 10^6 organisms/ml in weeks 14 and 15. This is perhaps a

Table III

MACHINE TRIAL CONDITIONS AT ROLLS ROYCE (1971) LTD, DERBY ENGINE DIVISION

Machine tool	Webster-Bennet vertical boring machine
Sump capacity	50 gallons
Metals machined	Special steels
Operating conditions	24 hr/day for 6 d/week
Cutting fluid	Semi-synthetic emulsion (50:1)
Flow-rate	2·0 gals/min
Filter element	254 mm × 89 mm

Table IV

MACHINE TRIAL RESULTS AT ROLLS ROYCE (1971) LTD, DERBY ENGINE DIVISION

	Unfiltered	*Filtered*
Coolant life (days)	7	28
Bacteria (count/ml)	10^8 (5 days)	10^6 (26 days)
Coolant condition at change	Smelling/splitting	Slight smell
Filter life	—	>4 weeks

Table V

MACHINE TRIAL CONDITIONS AT BROCKHOUSE TRANSMISSION LTD, WEST BROMWICH, U.K.

Machine tool	Jones & Lampson turret lathe
Sump capacity	15 gallons
Metals machined	Mainly steel also aluminium and cast iron
Operating conditions	2 shift (16 hr/day) for $5\frac{1}{2}$ d/week
Cutting fluid	Emulsion (30:1)
Flow-rate	1 gal/min
Filter element	140 mm × 70 mm

Table VI

MACHINE TRIAL RESULTS AT BROCKHOUSE TRANSMISSION LTD

	Unfiltered	*Filtered*
Coolant life (weeks)	6	15
Bacteria (count/ml)	10^8 (6 weeks)	10^6 (15 weeks)
Coolant condition at change	Smelling/corrosive	Good condition
Filter life	—	>12 weeks

more typical machine tool and the trial indicates that filtration has reduced fluid costs and improved working conditions.

CONCLUSIONS

The laboratory evaluation and field trials of the 'Rembac' filter, reported above, have demonstrated that it was effective in limiting the total bacteria count at about 10^5–10^6 organisms/ml in a number of aqueous cutting fluids. The filtered fluid was maintained in good condition and its life significantly extended in both of the field trials.

This shows that the 'Rembac' filter provides an additional means of controlling microbial degradation of metal-working coolants and lubricants which complements existing methods and is especially suitable for self-contained machine tools. The potential cost savings available from increased fluid life, decreased effluent disposal costs, and decreased labour and machining costs are considerable.

Further work is planned to cover a wider range of cutting fluids and to establish the actual cost savings attainable from the filtration of metal-working coolants and lubricants.

Acknowledgements

We should like to thank the Research and Development Director of ICI Mond Division for permission to present this paper and the management of Rolls Royce (1971) Ltd, Derby Engine Division, and Brockhouse Transmission Ltd for permission to use the results of the field trials carried out at their factories. Thanks are also due to our many colleagues who have contributed to the work discussed here.

SOME ASPECTS OF MICROBIAL DEGRADATION OF ALUMINIUM ROLLING COOLANTS

E. C. HILL

Microbiology Department, University College, Cardiff, U.K.

Summary

The consequences of microbial infection of aluminium rolling coolants are summarized. Recent work on determining the stability of infected coolants using a Coulter Counter and P.64 Analyser is described. Evidence for the deposition of microorganisms on aluminium sheet is presented based on work with the Cambridge Stereoscan microscope. Simple methods for detecting microorganisms in coolants are described and the results of experiments to determine compatibility of biocides are given.

INTRODUCTION

Over the last 10 years there has developed a universal recognition in the aluminium industry of the important role of microorganisms in the deterioration and loss of performance of aluminium rolling coolants. An early paper by Hill (1967) outlined the infection process as follows.

Microorganisms, principally bacteria, can live in rolling coolants, feeding on the ingredients of the formulation and any contaminating matter, and thus bring about chemical and physical changes which are reflected in rolling performance and product quality. The initial infection is supplied by the carry-over from the previous coolant charge, microbial slimes and deposits in the system, infected water used for coolant make-up, and leakages from other contaminated systems—particularly hydraulic systems. Within days or weeks populations of bacteria of up to 10^9 are often achieved.

THE MICROORGANISMS

Large numbers of samples from aluminium mill coolant systems have been examined in this laboratory and when time or available labour permitted, some identification of the dominant species has been attempted. This is always considered a tentative opinion, as this environment has not been subject to detailed study by classical microbiologists of the last generation and undoubtedly many of the strains isolated have not been previously described. In this laboratory we have noted *Pseudomonas aeruginosa*, *Ps. olevorans*, *Ps. ovalis*, *Ps. desmolytica*, *Ps. putrefasciens*, *Ps. rathonis*, *Alcaligenes faecalis*, *A. metalcaligenes*, *Flavobacterium ferrugineum*, *F. peregrinum*, *Achromobacter parvulus*, *Proteus rettgeri*, as well as unidentified strains of *Pseudomonas*, *Achromobacter*, *Alcaligenes*, *Nocardia*, *Desulfovibrio*, and yeasts. In recent years *Pseudomonas* spp. and *Alcaligenes* spp. have been the dominant organisms and occasionally massive infections by *Candida* spp.

have followed injudicious use of antibacterial chemicals. In the industry there has been in some mills a progressive upward shift in the temperature at which the coolants are maintained and coolant temperatures of up to 65°C are not uncommon. In some cases there has been a considerable decrease in biodeterioration but occasionally populations of thermophilic gram-negative bacteria have appeared, mainly *Pseudomonas* spp., and these have proved very difficult to deal with. These organisms when cultured in the laboratory grow well between 45°C and 50°C; few grow above this temperature range but they merely survive, which suggests that in the mill there are cooler pockets where most growth takes place. There has been a noticeable tendency for mills to keep coolants circulating or even aerated and there has consequently been a decrease in *Desulfovibrio* infections. The trend to using higher coolant temperatures on the one hand has improved the antimicrobial activity of biocides used, but it has also resulted in biocide loss from systems due to thermal instability of the biocides.

A phenomenon noted on several occasions has been the disastrous consequences of using water supplies in which there have been algal 'blooms'. The polysaccharides secreted into the water promote thermophilic growth and biodegradation and complete emulsion failure may occur within 3 days.

Infections also occur in other fluid systems in the mill, particularly the hydraulic systems, and these may leak into and re-infect the rolling coolant.

EMULSION INSTABILITY

The biodegradation of emulsifiers and coupling agents and the consequent onset of emulsion instability have always been the most noticeable and undesirable feature of coolant infections.

Degradation is easily demonstrated by inoculating bacteria previously isolated from infected emulsions into Bushnell and Haas (1941) mineral salt medium containing 1% w/v of the emulsifier or coupling agent and establishing a substantial increase in numbers after a few days incubation. Figure 1 gives the growth curve of an unidentified bacterium in mineral salt/triethanolamine medium, a *Pseudomonas* sp. in mineral salt/oleic acid medium, and a *Pseudomonas* sp. in mineral salt/diethylene glycol medium, in all cases incubation being at 30°C in a rotary shaker. The data are from Martin (unpubl. data). This degradation was confirmed by Martin using manometric techniques.

This degradation is reflected in the droplet size of the emulsion, as the overall droplet size must increase to reduce the interfacial area when emulsifiers are degraded. The change in droplet size can be measured optically or can be followed by determining the tendency of an emulsion to change in capacitance over a 15 min period. Figure 2 illustrates the changes in capacitance occurring in rolling emulsions as measured by a Wayne Kerr B641 Autobalance Bridge. This change in capacitance is apparently due to the differing rates of migration of oil droplets of different size. A stable emulsion has a stable capacitance. Unfortunately, this phenomenon is not easy to correlate with other parameters.

The preferred instrument for evaluating emulsion stability is the Coulter Counter (Coulter Electronics Ltd) or its equivalent. Although theoretically an almost complete 'fingerprint' of the oil droplet size distribution can be

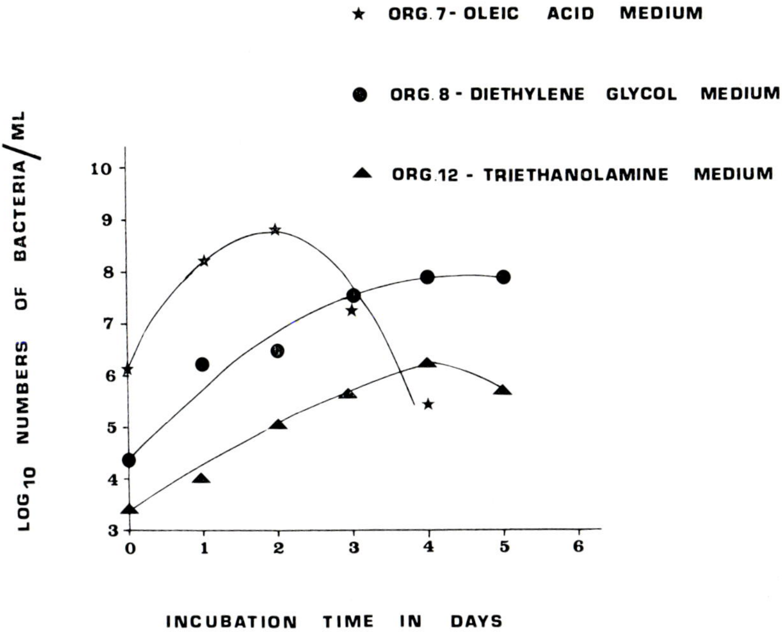

Fig. 1. Growth of bacteria at 30°C in emulsifiers and coupling agents. Pseudomonas *sp. (No. 7) growing in mineral salt/oleic acid medium; Isolate 8 growing in mineral salt/diethylene glycol medium;* Pseudomonas *sp. (No. 12) growing in mineral salt/triethanolamine medium.*

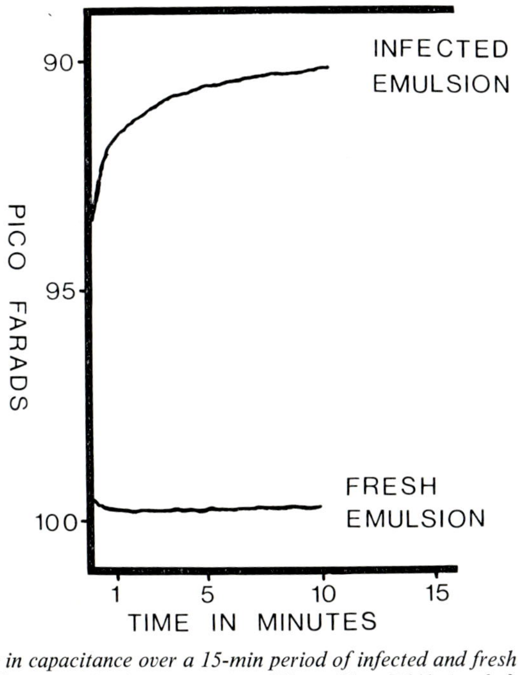

Fig. 2. Changes in capacitance over a 15-min period of infected and fresh aluminium rolling coolants. The measuring instrument is a Wayne Kerr B641 Autobalance Bridge.

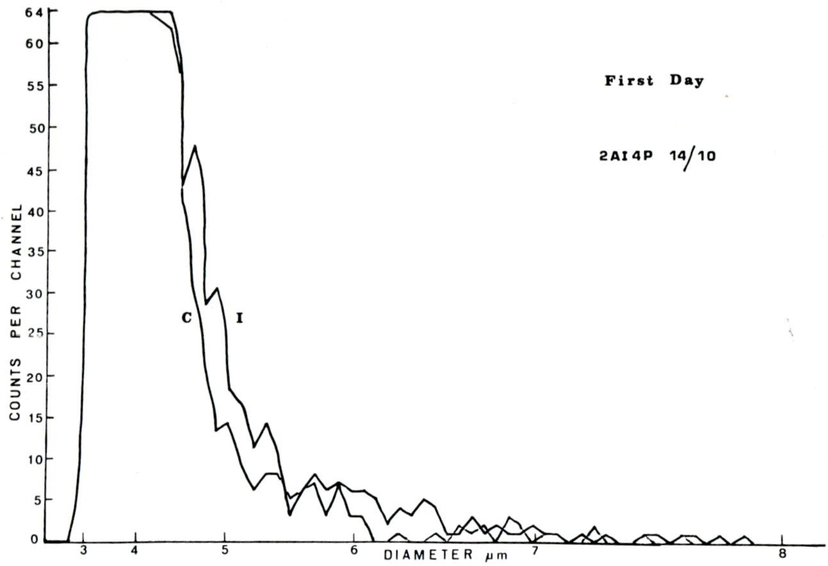

(a)

Fig. 3. *Oil droplet size analyses of infected (I) and uninfected (C) aluminium rolling emulsions after (a) 0, (b) 2, (c) 3, and (d) 10 days. The emulsions were stirred at 46°C and a Coulter Counter ZB fitted with a 70 μm orifice used to produce the data. The data were analysed by a P.64 analyser but the first 3 of the 64 channels were blacked out electrically to eliminate noise. To obtain droplet count per ml, multiply droplet size per channel by 4000.*

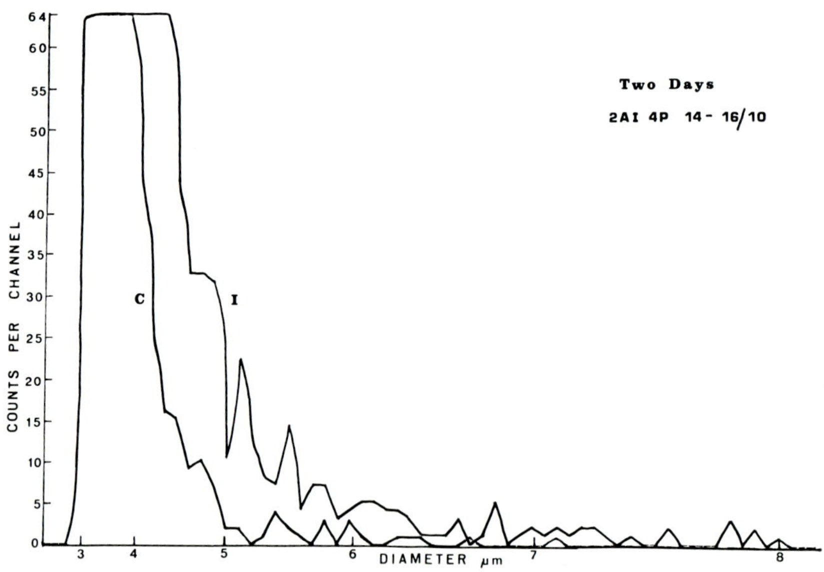

(b)

Fig. 3.—contd.

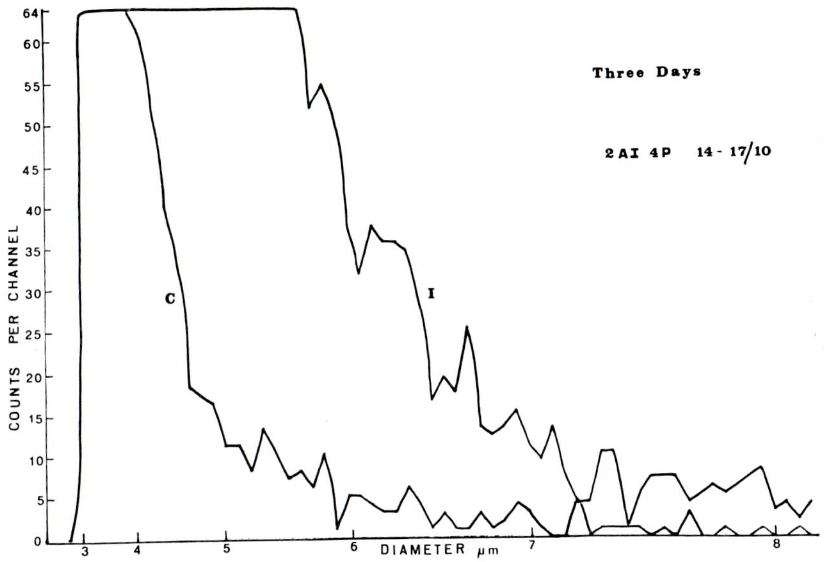

(*c*)

Fig. 3.—contd.

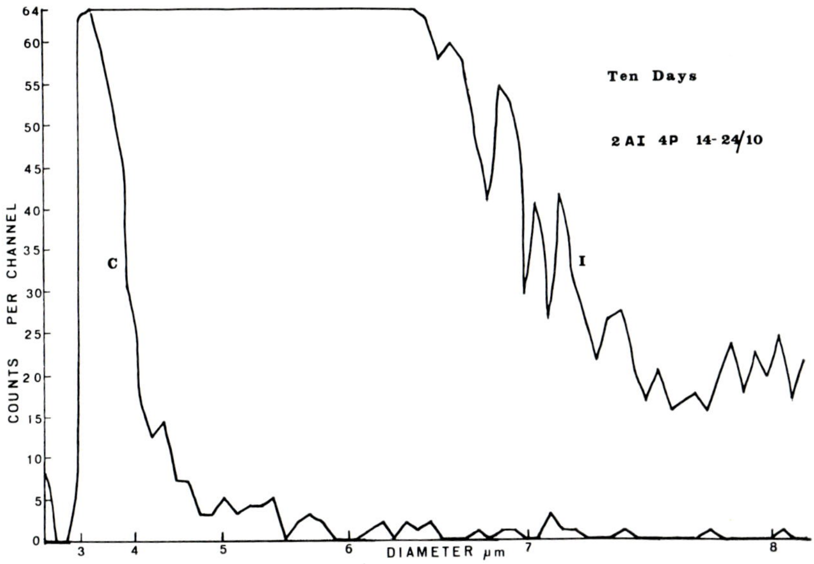

(*d*)

Fig. 3.—contd.

obtained by diluting an emulsion with a conducting, particle-free medium and passing it through both a 30 μm and a 70 μm orifice, the use of the 30 μm orifice demands conditions of cleanliness of the emulsion and freedom from electrical noise which are difficult to achieve. In practice, the use of the 70 μm orifice gives adequate information, as it will detect 'over-size' droplets which arise due to coalescence of smaller droplets. In due course these 'oversize' droplets will disappear as they too coalesce to give rise to free oil.

Figure 3 illustrates typical data obtained from a Coulter Counter Model ZB coupled to a P.64 Analyser and Plotter. The emulsion samples tested were diluted 1:2000 with Steriflex Saline No. 1 (Allen and Hanburys Ltd), and stirred continuously, while 0·5 ml was passed through a 70 μm orifice. In this case two flasks of 200 ml of 4% emulsion in distilled water were stirred at 46°C, one flask (I) being inoculated with bacteria isolated from a similar used coolant and the other not (C). In fact the uninoculated flask eventually became infected but until that occurred, both flasks were tested at intervals of a few days. The shift toward a larger droplet size is obvious in the infected flask. The ancillary equipment is not essential but it has proved valuable to have the facility to vary the full scale deflections of the X and Y axes at will and, of course, the data are obtained within seconds. Without the ancillary equipment, arbitrary counts can be made such as a count of all droplets over 3 μm diameter, or counts can be made in broad diameter ranges (Hill, 1972a,b). In most mills there is a procedure for correlating oil droplet size distribution with coolant performance.

DEGRADATION OF ADDITIVES

The degradation of anticorrosive additives is well known. If extra-pressure additives, such as tri-tolyl phosphate, are present, they may also be biodegraded. Figure 4 gives more data from Martin (unpubl. data) demonstrating

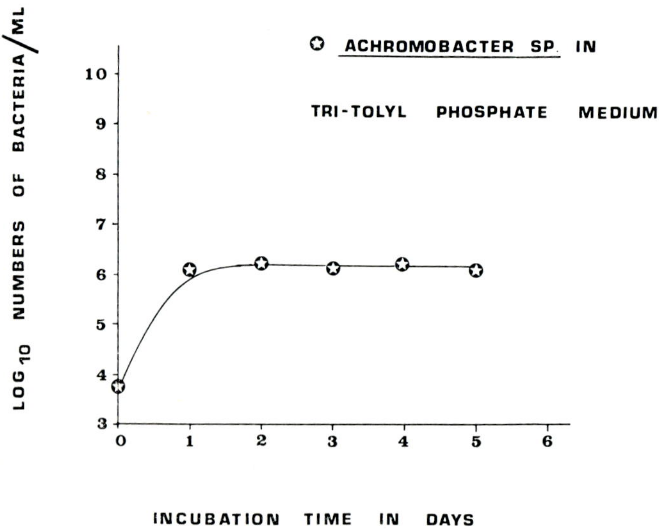

Fig. 4. Growth of Achromobacter *sp. at 30°C in mineral salts medium + 1% tri-tolyl phosphate (after Martin, unpubl. data).*

the growth of an *Achromobacter* sp. on this substance in a mineral salt medium. In this case the tri-tolyl phosphate is probably the carbon and energy source for the organism, but Martin also demonstrated that it could act as the phosphorus source.

CORRELATION WITH PERFORMANCE

It is difficult to prove conclusively that a change in engineering performance of a coolant is directly related to microbial infection but Bartholomew and Crew (unpubl. data) were able to demonstrate increased 'pick-up' when pieces of aluminium–magnesium alloy sheet, 300 mm × 150 mm × 3 mm, were passed through a small 'two-high' electrically driven mill until a 50% reduction had been achieved. The increase in 'pick-up' was noted visually on the rolls when infected ($4{\cdot}8 \times 10^6$ bacteria/ml) emulsion was used as a coolant; the appearance of the rolls when uninfected coolant was used was the basis for comparison. The infected emulsion showed the characteristic increase in the number of oil droplets above 4 μm and had a fatty acid content of 12·3% compared to 11·8% in the uninfected emulsion. They also noted that the surface-finish of the metal rolled was inferior when the infected emulsion was used.

Bartholomew and Crew also submitted infected and noninfected emulsions for Falex testing. The results of duplicate determinations are given in Table I and show a marginal increase in lubricity for the infected coolant.

Table I

FALEX MACHINE RESULTS WHEN THE PIN AND VEE BLOCK OF THE MACHINE ARE IMMERSED IN INFECTED AND NONINFECTED ALUMINIUM ROLLING COOLANT

Jaw load (Newtons)	*Torque (Newton-metres)*	
	Infected emulsion	*Noninfected emulsion*
1 100	1·1	1·1
3 300	2·0	2·1
5 600	2·8	3·2
7 800	4·1	4·2
10 000	5·0	5·3
12 200	6·2	Seizure
14 500	Seizure	—

Thus as well as the more general correlation of engineering problems with microbial infection, which has been derived from mill experience, there is now a growing body of supporting data from laboratory experiments.

DEPOSITION OF ORGANISMS ON THE ROLLED SHEET

It is reasonable to suppose that if the rolling coolant is heavily infected, microorganisms will be deposited on the rolled sheet and may adhere there. Hill and Penberthy (1968) have already shown that this can happen during cold rolling of steel sheet. The size and shape of typical infecting organisms are illustrated by the electron-micrograph of a *Pseudomonas* sp. (Fig. 5) and the photomicrograph of a yeast (Fig. 6), both isolated from aluminium rolling coolants. However, their appearance is distorted when they are dried and heated after deposition on the rolled sheet and they are not obviously recognizable. Figures 7(a) and (b) are photographs taken with a Mk 2A Stereoscan Microscope (Cambridge Scientific Instruments Ltd) of aluminium sheet rolled with coolant known to be very heavily infected (a) and moderately infected (b). The infection was predominantly by yeasts. The defects in the

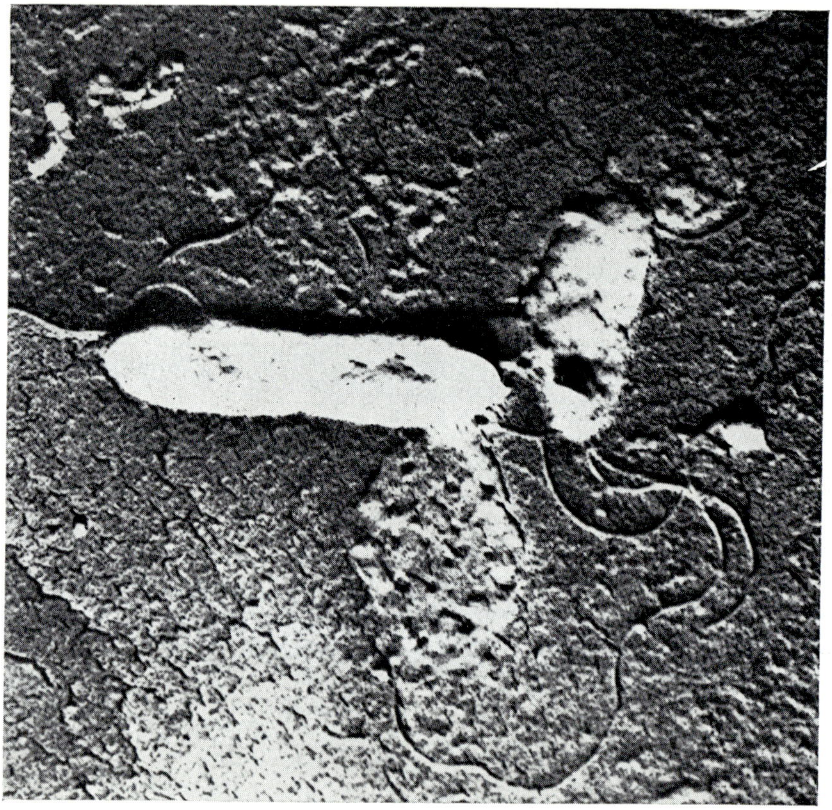

Fig. 5. Electron-micrograph of Pseudomonas *sp. isolated from an aluminium rolling emulsion. Scale: 21 mm equals 1 μm.*

sheet are obvious and clumps of unidentifiable matter can be seen. Fortunately, microorganisms concentrate certain elements into their bodies and one of these, potassium, is easily detected and can be used to specifically indicate which deposits are of biological origin. Figures 8(a) and 8(b) are the same samples as in Figs. 7(a) and 7(b) but a 0·5 μm band has been scanned across the line AB and a trace drawn of the corresponding potassium level. It will be seen that high levels of potassium correspond to the deposits on the metal surface and the inference is that these deposits are microbiological in origin. The significance of this is not yet known but Hill and Penberthy (1968)

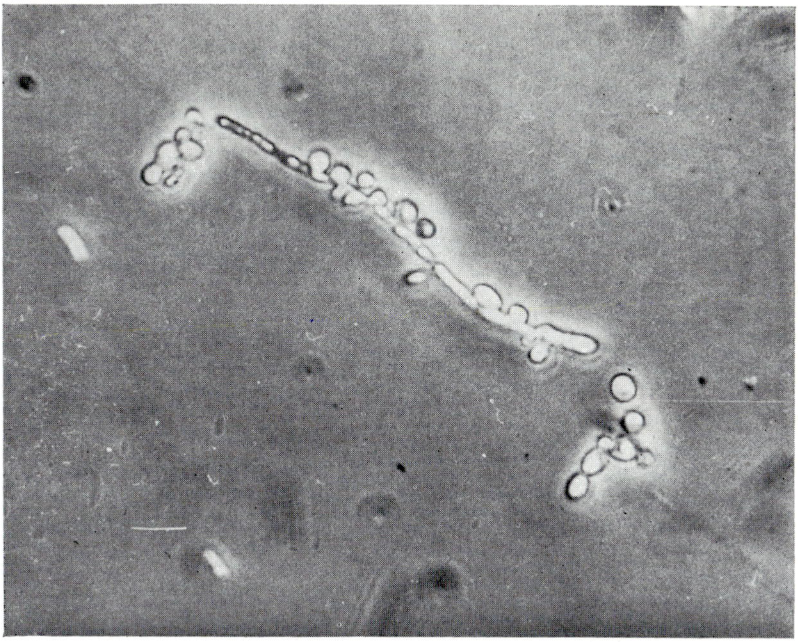

Fig. 6. Photomicrograph of yeast isolated from an aluminium rolling oil emulsion. Scale: 14 mm equals 20 μm.

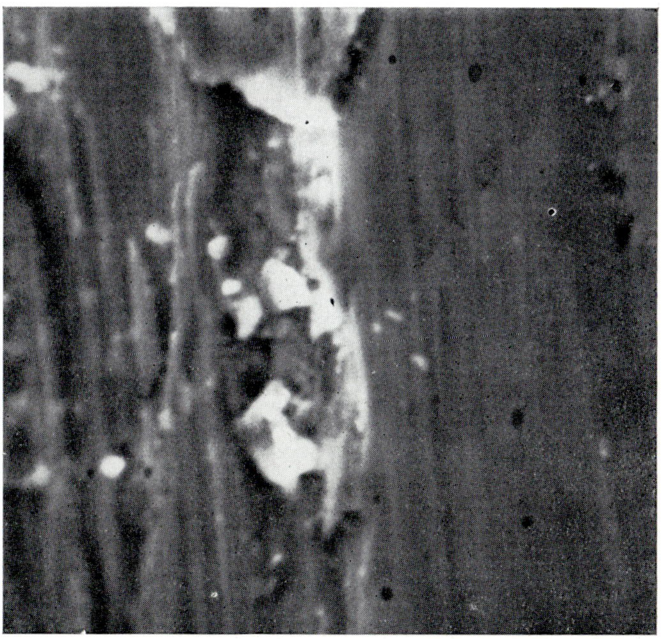

(*a*)

Fig. 7. Electron micrographs (Mk 2A Stereoscan Microscope) of aluminium sheet rolled with heavily infected (a) and moderately infected (b) emulsions. Scale: (a) 10 mm equals 10 μm; (b) 5 mm equals 10 μm.

(*b*)

Fig. 7.—contd.

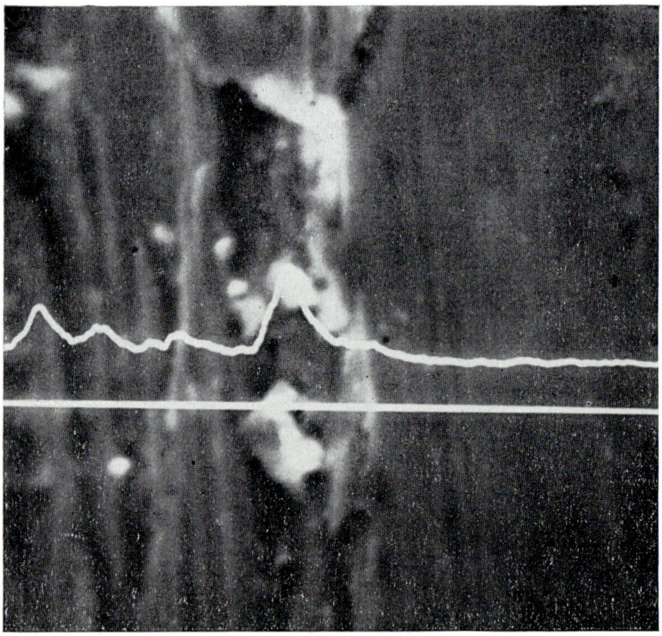

(*a*)

Fig. 8. Same areas of sheet as Fig. 7 but 0·5 μm wide band A–B scanned for potassium level and recorder trace of potassium superimposed. Scale: (*a*) *10 mm equals 10 μm;* (*b*) *5 mm equals 10 μm.*

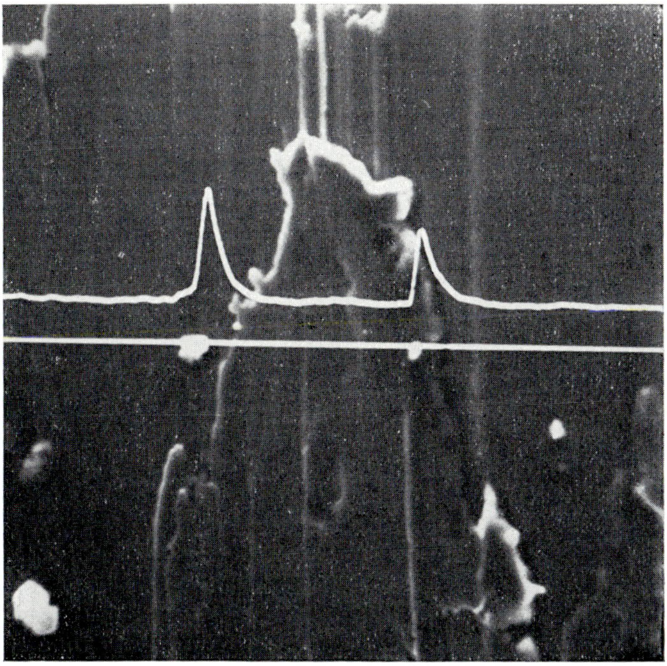

(*b*)

Fig. 8.—contd.

considered that the chemical concentration gradient thus formed could contribute to corrosion of steel sheet.

CHOICE OF ANTIMICROBIAL CHEMICALS

The principles governing the selection of antimicrobial chemicals have been described by Hill (1972a,b). For any one application in the aluminium industry, the biocide selected must be carefully matched to the job it has to do.

It is of prime importance that if it is used in a rolling emulsion, its effect on the oil droplet size distribution is known. Figures 9(a), (b), (c), and (d) show data from the Coulter equipment described previously, when an emulsion containing no biocide (C) and emulsions containing 0·1 % of biocides M, X, and Z were tested. The emulsions were stirred at 46°C and tested after 1 and 11 days. It will be apparent that biocide Z 'tightened' the emulsion, biocide X 'loosened' the emulsion considerably, and biocide M had little effect. Effect on pH is also of importance and it was found that the initial pHs in C, M, X, and Z were 7·3, 7·2, 7·4, and 7·9, respectively. The requirement for thermal stability of the biocide has already been mentioned.

SEMI-QUANTITATIVE MICROBIOLOGICAL TESTS

Close control of a potentially degradable emulsion system can best be achieved by on-the-spot semi-quantitative microbiological testing, using the

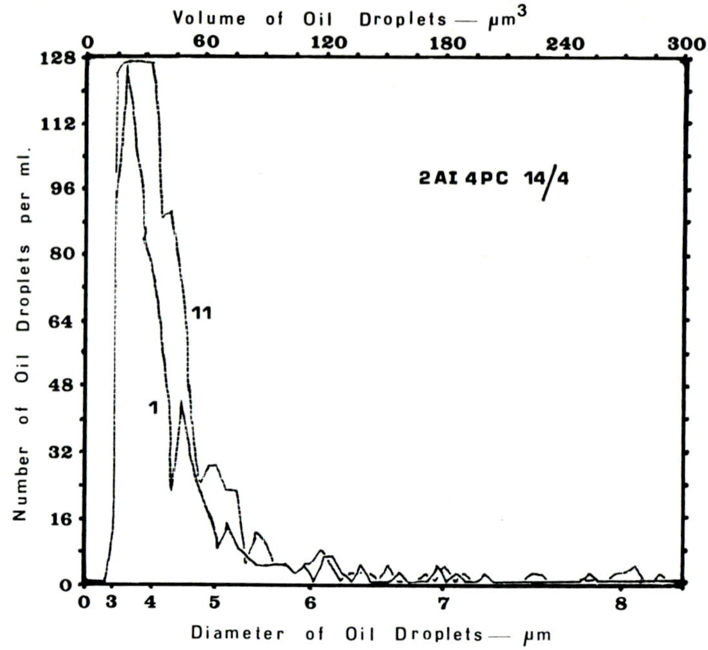

(a)

Fig. 9. Oil droplet size analyses (obtained as for Fig. 3) of emulsions treated with 0·1% of biocides and tested after 1 and 11 days. (a) Untreated; (b) 0·1% Biocide M; (c) 0·1% Biocide X; (d) 0·1% Biocide Z. To obtain droplet count per ml of original emulsion multiply count by 2000.

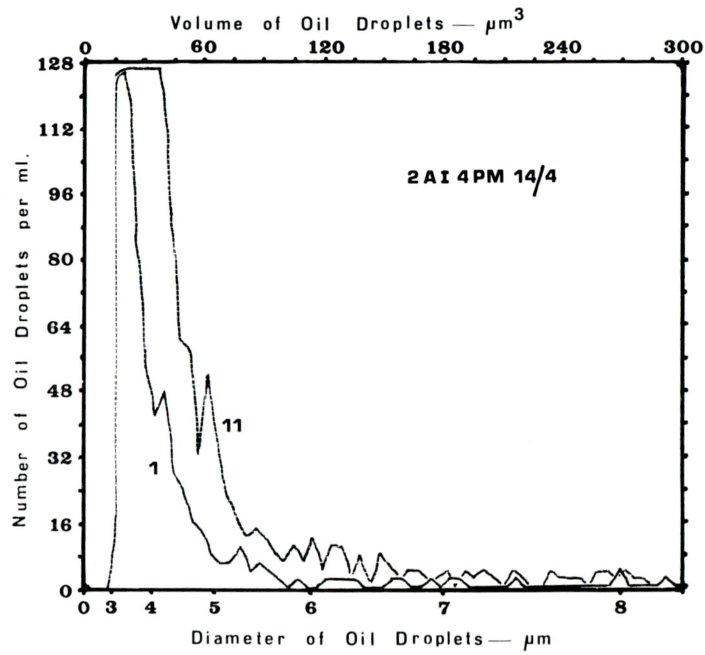

(b)

Fig. 9.—contd.

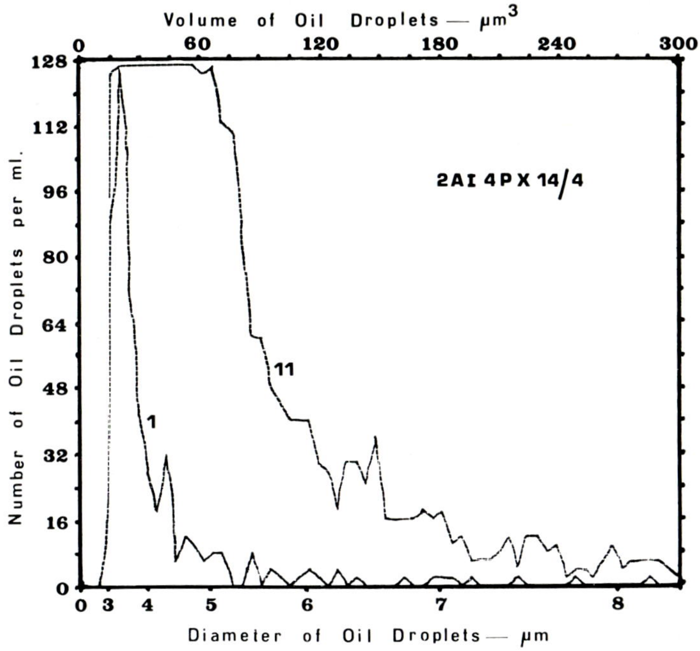

(*c*)

Fig. 9.—contd.

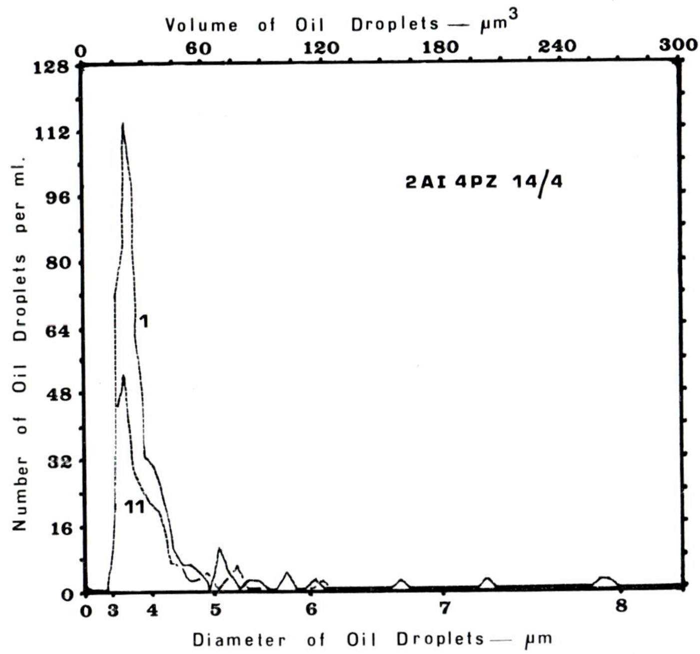

(*d*)

Fig. 9.—contd.

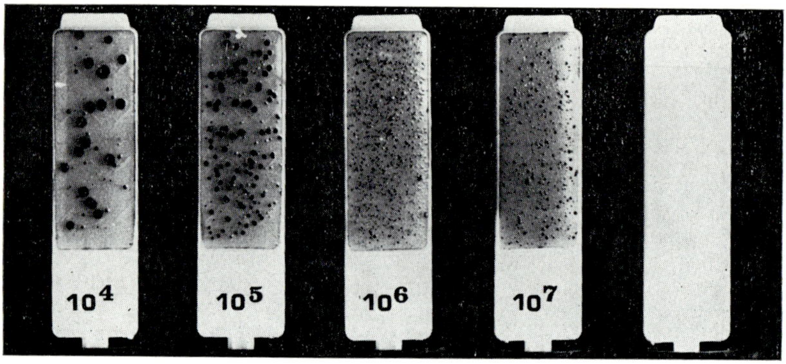

Fig. 10(a). Easicult-TTC slides before use and after dipping in emulsion, draining and incubation. A range of positive results have been selected to correspond approximately to the calibration chart.

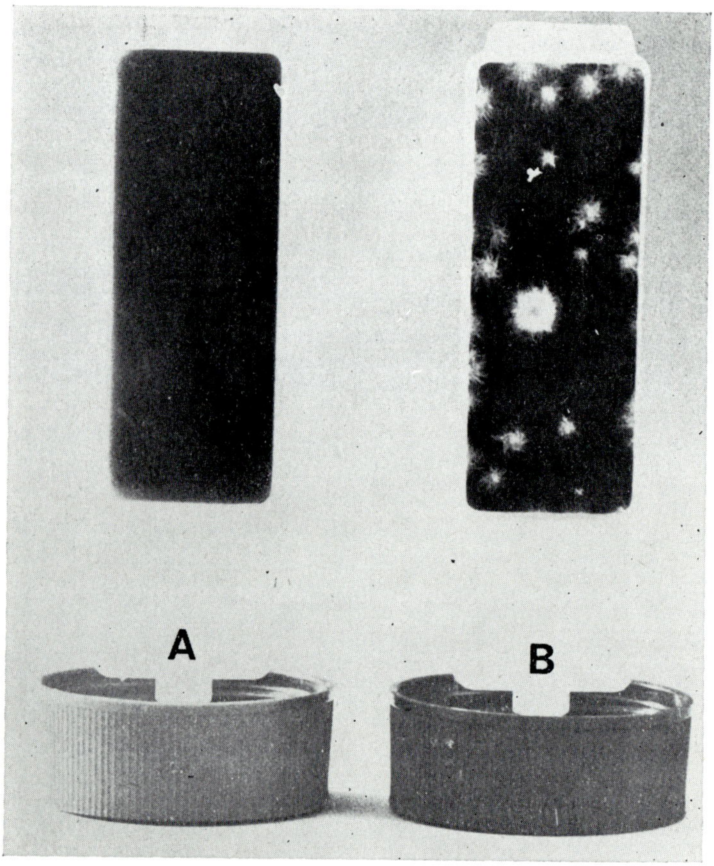

Fig. 10(b). Easicult-M slide before use (A) and after dipping in emulsion, draining, and incubating (B). A light fungal infection is revealed.

results to guide biocide additions and sometimes biocide choice. Hill *et al.* (1967) described the 'Red-spot' test which has been widely used and goes some way to providing microbiological data quickly. There are obvious deficiencies, however, and these have been overcome by the Easicult tests (Orion Diagnostica Ltd, Helsinki). These are a range of tests which make use of a variety of agar-coated slides. The slides are dipped into the coolant, drained, and incubated. The density of colonies produced is matched to a calibration chart. Selective slides are available for fungi/yeasts (M-slide) and bacteria (TBE-slide) and for detecting sulphide generation (Easicult-S). In our hands, the most convenient slide is the Easicult-TTC since the result can be read after overnight incubation; incubation can be achieved by merely suspending the slide and container in the coolant. Figure 10 shows some slides and typical positive results.

The exact calibration of the slides for any system is controlled by the average retention of the coolant on the slides, and it may be found to be slightly different from the standard calibration if the fluid tested is more or less viscous than is normally the case. In practice, it is sufficient to test for bacteria with TTC-slides on a routine basis, with occasional checks for fungi and yeasts with the M-slide. The results are used to indicate the need for an addition of biocide or possibly a change of biocide. Field trials have indicated the importance of testing on-site because transporting samples containing thermophilic bacteria for laboratory testing elsewhere usually leads to substantial population changes.

CONCLUSIONS

Over the last 10 years, there has developed in the aluminium industry an increasing awareness of the nature and consequences of microbial infections of rolling coolants. Most of these consequences are now known in general and specific terms, but there is still a need to investigate more fully the effect on rolling performance, including energy consumption and sheet quality. Undoubtedly, there are more and better antimicrobial chemicals now available for industrial use, and these can be carefully selected, particularly with regard to the effect on emulsion stability, pH, and corrosiveness, so that the success of an application can almost be guaranteed. Coupled with new chemicals and new laboratory compatibility checks, simplified microbiological tests such as those described enable a competent works laboratory to satisfactorily resolve problems associated with microbial infection of rolling coolants.

REFERENCES

Bushnell, L. D., and Haas, H. F. (1941). The utilisation of certain hydrocarbons by microorganisms. *J. Bacteriol.*, **41**, 653–673.

Hill, E. C. (1967). The significance and control of microorganisms in rolling mill oils and emulsions. *Metals and Materials*, **1**, 294–297.

Hill, E. C. (1972a). Biocides for petroleum products. *J. Inst. Petroleum*, **58** (563), 248–253.

Hill, E. C. (1972b). Biodeterioration of metal-working fluids and its significance. Paper MR 72.214, Soc. Man. Eng., U.S.A.

Hill, E. C., and Penberthy, I. O. (1968). The microbiology of steel mills and its relation to sheet corrosion. *Metals and Materials*, **2**, 359–366.

Hill, E. C., Davies, I., Pritchard, J. A. V., and Byrom, D. (1967). The estimation of microorganisms in petroleum products. *J. Inst. Petroleum*, **53**, 275–279.

PHYSICAL, CHEMICAL AND BIOLOGICAL FACTORS INFLUENCING THE DAMAGE OF WOOD AND OTHER MATERIALS BY TERMITES

GÜNTHER BECKER

Bundesanstalt für Materialprüfung, Berlin-Dahlem,
German Federal Republic

Summary

Termites are one of the main insects that destroy materials. Physical, chemical, and biological factors influencing the damage to wood and other materials by termites are discussed. Within the physical factors, environmental influences of temperature, humidity, and electromagnetic field conditions and material properties of hardness, surface, and internal structure and thickness of foils and fibres should be observed. The chemical factors involved are sensorial influences such as repellency, attraction, and stimulation and toxic influences such as toxic efficiency, effect on symbionts, and aging of effective substances. Biological factors include termite behaviour connected with gnawing activity, influence of substrates, and choice or compulsory tests and various other biological factors such as species and race of the termites, number and composition of groups as well as vitality and developmental stage of the colony.

INTRODUCTION

This paper deals with the influences of physical, chemical, and biological factors on the damage to wood and other materials by termites, mainly with respect to designing and evaluating laboratory and field testing.

The deterioration of materials due to termites is of high economic importance in most tropical and subtropical regions as well as in cooler areas of the world. Therefore, knowledge of resistance of materials or of their protection and the respective testing is of wide interest.

The biology and behaviour of termites, which are social insects with a highly developed caste system, are more complex than those of nonsocial insects. The effects of physical and chemical factors concerned are the same for termites and other insects that may destroy wood and other materials. The biological conditions and the reactions of the insects, however, differ greatly.

PHYSICAL FACTORS

Physical factors involved are environmental influences on the termites and properties of the materials. Both groups of factors are important under natural conditions and for testing.

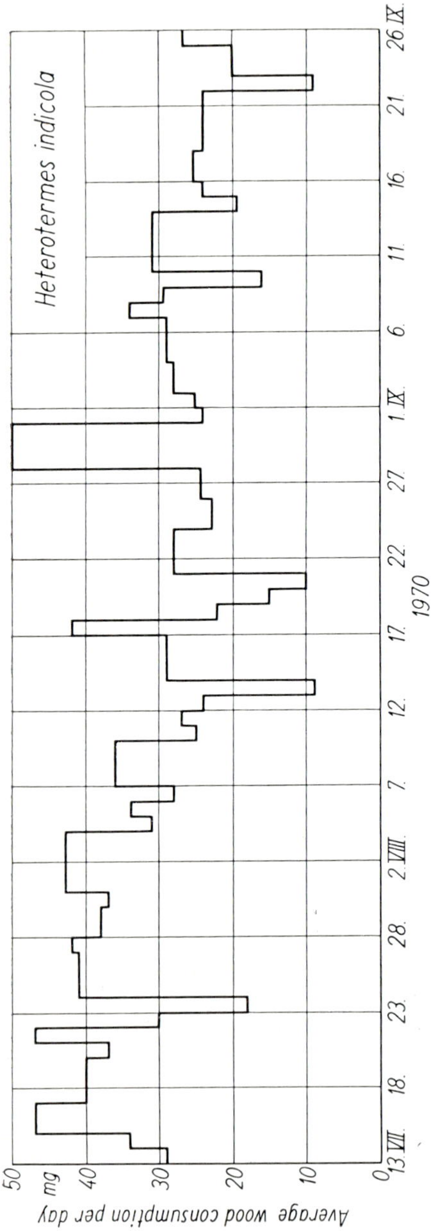
Heterotermes indicola
Average wood consumption per day
mg
50
40
30
20
10
0
13.VII.
18.
23.
28.
2.VIII.
7.
12.
17.
22.
27.
1.IX.
6.
11.
16.
21.
26.IX.
1970

Fig. 1(a)

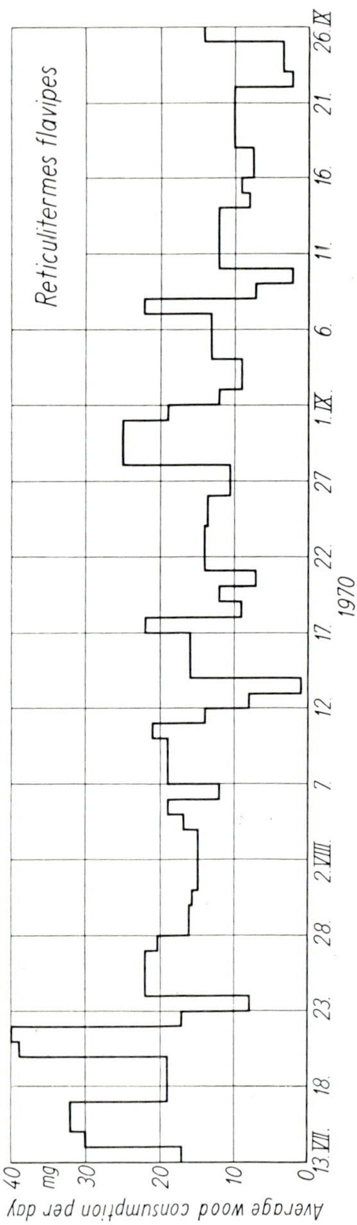
Reticulitermes flavipes
Average wood consumption per day
mg
40
30
20
10
0
13.VII.
18.
23.
28.
2.VIII.
7.
12.
17.
22.
27.
1.IX.
6.
11.
16.
21.
26.IX.
1970

Fig. 1(b)

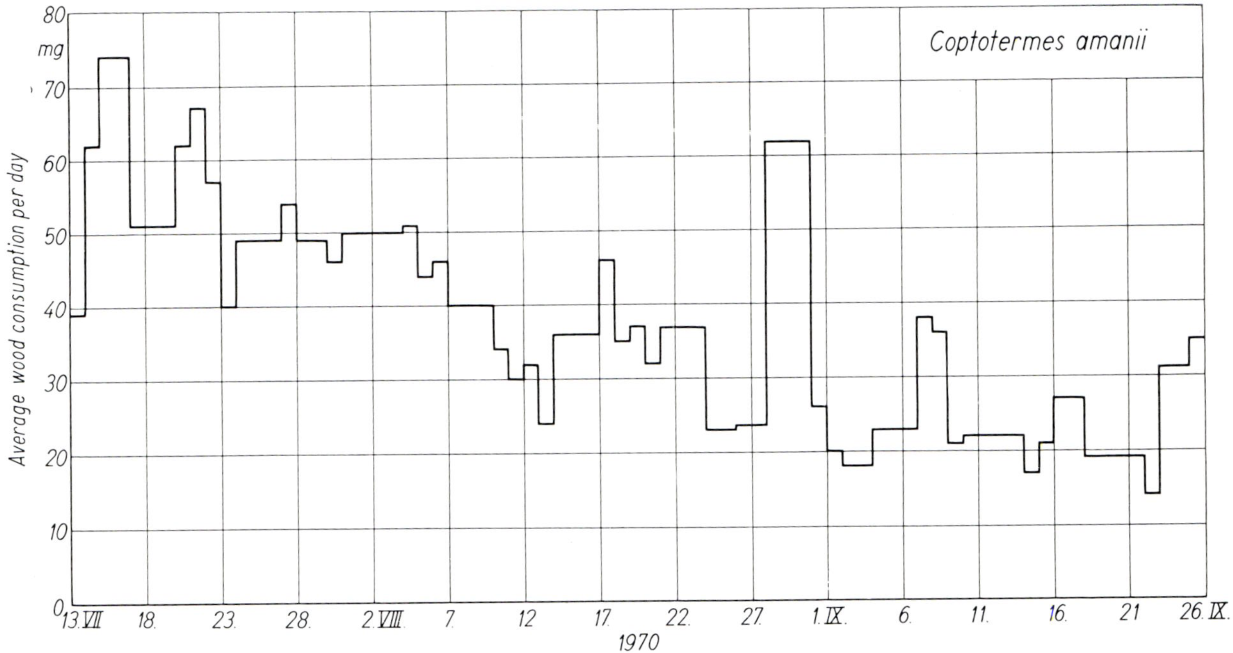

Fig. 1(c)

Fig. 1. Conformable variations of the average food consumption by (a) Heterotermes indicola *(Wasmann), (b)* Reticulitermes flavipes *(Kollar), and (c)* Coptotermes amanii *(Sjöstedt) within the same period.*

Environmental influences

Most activities of insects depend on *temperature*. This, therefore, has been observed in studies on the distribution of termites and in laboratory testing for a long time. However, experimental data are available for only a rather limited number of termite species. Many tropical species have their optimum for feeding activity and development at around 30°C. For *Coptotermes* species, 28°C seems to be more favourable. Some *Reticulitermes* species may prefer about 26°C, and the lowest optimum has been found with *Zootermopsis* species (Becker, 1967). Continuous optimal conditions are advisable if good feeding activity is desired. If prolonged survival is more important, the temperature should be about 2°C lower than the optimum for activity and development. Temperature may also influence the properties of the material or the aging of preservatives. This will be discussed later.

Sufficient *humidity* is another prerequisite for survival and activity of termites (Ernst, 1957, and others). In nature many termites need access to a water supply or to wet wood. For these species at least, saturated air humidity should be provided in the laboratory. Only the so-called dry-wood species of the family Kalotermitidae can develop in a relative air humidity as low as about 90% (Collins, 1969; Williams, 1975). Under natural conditions, materials are more endangered by termite attack in the rainy season than during dry periods. In testing methods, the substrate has to provide the high humidity which the termites need during the testing time (Becker, 1965a). Sand has a very low water-holding capacity of a few percent only and the small quantity of water that can be added evaporates in relatively short periods. A very high water-holding capacity is found with vermiculite, that is, up to more than 300% of its dry weight; its optimal range of water content for termites is rather broad so that re-wetting during the test is not necessary (Becker and Lenz, 1970). In field tests, shading can improve humidity conditions for termites and thus accelerate an attack (Harris, 1961; Williams, 1973). Humidity, too, influences properties of materials and will also be discussed later.

Electromagnetic field conditions including weather factors are another complex of environmental influences which, however, has been long overlooked. It is well known that experiments with organisms, when repeated under the same so-called constant conditions of substrate and food, temperature and humidity, and even air pressure, do not always lead to the same results as previously. In most cases unknown variations of the organisms were thought to be the reason for different results. These can, of course, be involved, but the main influence is of another type.

If groups of termites are kept under constant conditions and food consumption is determined in 24-hr intervals, the quantity of wood eaten may change from one day to the next by 1 : 10 or even more. These changes in frass activity are almost equally followed by replicate groups of the same termite species set up at different times and by groups of different termite species (Fig. 1). The effects are not group- or species-specific and not internally but externally caused (Becker, 1973). Statistical evaluation has shown that results of experiments based on average food consumption are comparable only if the test period lasted for at least about 4 weeks. If the result of an experiment depends on termite activity of the first few days of a test series when it may be extremely high or low, the results of repetitions are sure to differ. This fact should be considered for bioassays and their evaluation. Use of reference materials together with the actual test specimens at the same time should become a standard procedure in order to measure changing physico-environmental factors.

Feeding activity is influenced by weather conditions, including atmospherics. A correlation could be stated between food consumption and disturbances of the magnetic field of the earth in a 27-day rhythm of the rotation of the sun. It is not proven that changes of the magnetic field determine termite activity, but the influences coming from sunspots are one parameter involved in termite reaction (Becker and Gerisch, 1973).

Material properties

Hardness is one of the important properties of wood as well as other materials for an attack by termites. The early wood in the annual rings with comparatively thin cell walls is preferred by termites. They eat the harder summer wood later, and the more resistant lamellae of that may give a typical pattern of destruction by termites. If there is a choice, termites prefer lighter wood species. They are, however, able also to destroy very hard woods, if these are not protected by repellent or toxic substances.

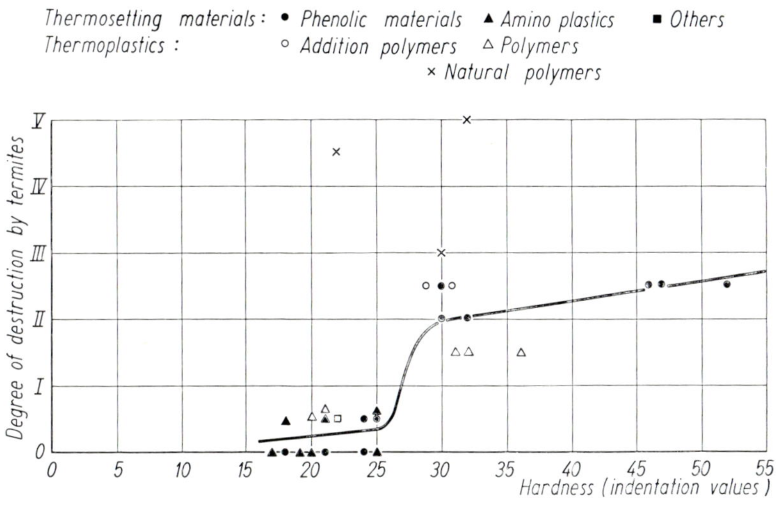

Fig. 2. Degree of deterioration by termites, especially Kalotermes flavicollis (*Fabr.*), *of various plastics in relation to the hardness of the material. Values determined with a micro indentation apparatus.*

The influence of hardness on the attack by termites is also evident with plastic materials (Becker, 1963, 1964; Gay and Wetherly, 1962, 1969; Hrdý, 1964; Southwell *et al.*, 1972; Beal *et al.*, 1973). Comparison of termite attack and an indentation test showed a correlation between hardness and degree of attack and a threshold of resistance against termites (Fig. 2). In the case of thermoplastic materials, the hardness depends on temperature; in the case of wood and fibres, on humidity. This may influence the results of tests which were carried out under different temperature and humidity, but are comparable. Standardization of testing and of storage conditions is necessary for rendering tests repeatable.

Surface and internal structure can also be of importance. Termites cannot attack materials with a very smooth surface such as special plastics or coatings. They start to gnaw at rough spots and edges. This must be considered with respect to test methods. In an anisotropic material such as wood, the cell wall structure influences the mode of attack by termites. The wood fibres of coniferous wood are gnawed in a direction perpendicular to the axis of the tracheids (Schultze-Dewitz, 1960).

Thickness of foils and fibres is important for attacks by termites (Gösswald, 1942; Gay and Wetherly, 1962; Becker, 1963; Becker and v. Seydlitz-Kurzbach, 1972). Foils with a smooth surface can be destroyed only if the foils are thinner than the dimensions of the mandibles. Therefore, the threshold of resistance depends on the size of the termite species used (Fig. 3). Under practical conditions foils may be folded or otherwise deformed so that termites can find a spot to start their attack.

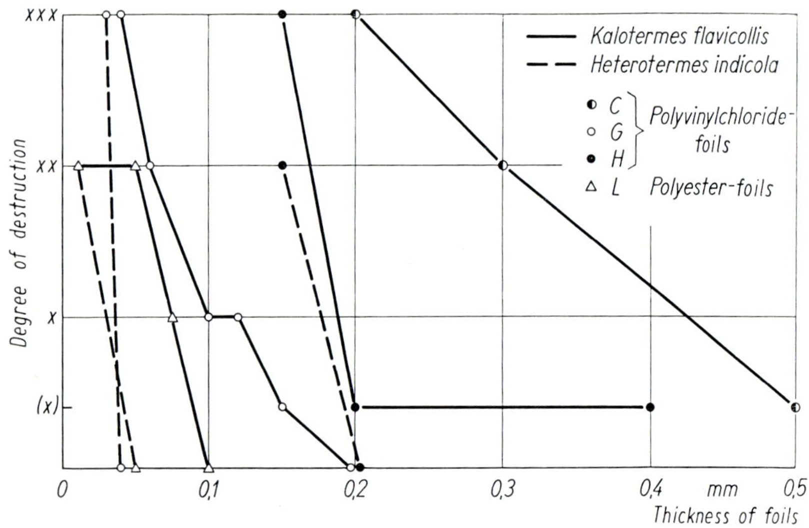

Fig. 3. Degree of damage to plastic foils by two termite species of different size in relation to the thickness of the foils.

CHEMICAL FACTORS

Chemical factors that attract or repel can influence the termites' behaviour and thus results of bioassays in nature or in a laboratory. Toxicity of the material may prevent attack.

Sensorial influences

Attraction of a material may have a considerable influence by inducing attack. The best-known example is the attraction of termites by wood-destroying fungi and fungus-infested wood. The compounds that attract and in some cases even cause trail-following are already known. The attraction of termites by fungus-infested wood is used practically in the United States for killing the insects in special areas, using decayed bait blocks together with insecticides (Esenther and Coppel, 1964; Esenther and Beal, 1974). The consumption of susceptible wood species decreases after their aging (Becker, 1971, 1975a). It is not yet known whether this is due to a loss of attraction or other changes.

Stimulation of feeding and gallery building are provoked by the mycelia of certain *Basidiomycete* species (Becker, 1965b; Becker and Lenz, 1972). Termites eat more uninfested wood in the presence of even the smell of the fungi than they do under chemically uninfluenced conditions. This effect in combination with attraction certainly plays a role under natural conditions and should be considered in laboratory tests. The addition of fungus mycelium may activate the termites in tests. A very special, but informative example is that the oppressive influence by aluminium shielding on vertical gallery building, which has been mentioned, may be compensated for by the presence of fungi in the substrate (Becker, 1975a).

Repellency is a negative result of chemical influences. It prevents the termites from gnawing on a material or even from touching it. Many tropical wood species contain repellent substances in their heartwood. Some of these compounds have been analysed and tested. Termite species may react differently to repellent substances, and especially Kalotermitidae may show individual reactions that are different from those of Rhinotermitidae and Termitidae (Becker, 1971, 1975a). From results with one species only, no general conclusions can be made. An interesting example of a repellent substance in a susceptible European wood species is the efficiency of Furfural in pine sapwood (Fig. 4), which can be leached from the wood by water or evaporated

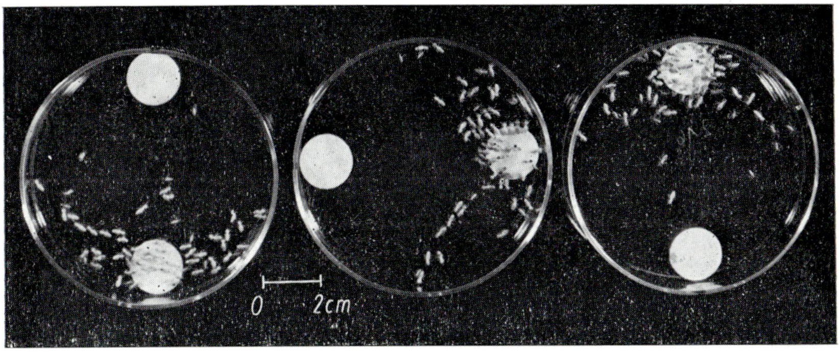

Fig. 4. Avoidance of the repellent compound by Nasutitermes nigriceps (*Haldemann*).

from it (Becker *et al.*, 1971). Repellent wood preservatives either keep the termites away from the treated wood or only feeding on the wood is prevented, but galleries are nevertheless built on its surface. The latter is the case with pentachlorophenol.

Toxic influence

The *toxic efficiency* of a substance may be based on respiratory, contact, or stomach poison effects. Testing methods should consider the three types of poisoning and provide the possibility to distinguish among them. Respiratory effect is mostly connected with repellency and thus acts in two ways. Some water-soluble stomach poisons do not prevent gnawing at treated material.

Slight nibbling at the surface may not be harmful. However, if a large number of termites nibble continuously, the attack may become serious, especially if treated layers of susceptible wood species or other material are penetrated, the termites could destroy the unprotected interior. The European Standards Organization, now dealing with testing methods for wood preservatives against termites, is considering standardizing two different methods for fully impregnated and for nonpressure treated wood with a protected outer zone only.

Toxic substances influence not only the termites but their symbionts as well. When the intestinal flagellates are killed by a poison, the termites suffer from starvation and finally die. Toxic substances may either poison the termites or the symbionts or both (Lenz and Becker, 1972; DaCosta *et al.*, 1975). With termite species containing protozoa as symbionts, all flagellate species may die at the same time, as was found with arsenic acid, boric acid in high concentrations, and—in the case of some termite species—with pentachlorophenol, or the flagellate species disappear successively, as is the case with low concentrations of boric acid and with some toxic wood extracts. In the case of sodium fluoride and—with some termite species—of pentachlorophenol, some of the symbionts survive, but the termites themselves are killed by the poisons. Contact insecticides are examples of a toxic influence on termites but not on the flagellates. If the influence on the flagellates is decisive, termite species without these symbionts, containing only bacteria in the gut, may react differently to a toxic substance.

Aging of effective substances is of great importance with regard to the performance of a material and its service time under practical conditions. Compounds causing natural durability of wood species and preservatives injected into wood or applied to other material may change their chemical properties through oxidation or a decrease in quantity and change in composition through leaching and evaporation (Becker, 1974). Therefore, besides the initial toxicity of preservatives and durability of materials, the influence of aging factors on these properties should be tested before giving recommendations for practical purposes.

BIOLOGICAL FACTORS

Among biological factors influencing the results of laboratory and field tests are various behavioural patterns and differences in vitality of termite species and races, as well as the number of individuals and caste composition of groups. Biology and reaction of termites are also of importance for the performance of materials under service conditions.

Behavioural influences

The continuous *gnawing activity* of termites also leads to damage of non-digestible material. Wood and other plant tissues constitute the normal food of termites. But they may attack rubber, synthetic textiles, plastic materials, paints, and may perforate even soft metals. Holes in power cables, which can lead to short circuits and thus the interruption of operation in important factories, are not a result of termite hunger but of instinctive gnawing when they come across a material on their way. The gnawing activity has to be induced and facilitated in laboratory testing procedures in order to be able to consider potential cases of damage in practice.

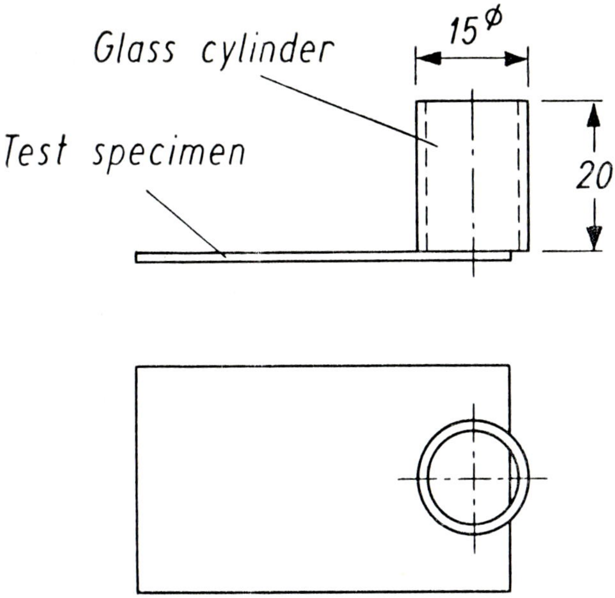

Fig. 5. Testing method (according to G. Becker) for investigating the resistance of materials against termites using the gnawing instinct with narrow openings and the tendency to escape from the glass ring.

A method that the author has used for a long time (Fig. 5) is to keep groups of termites in glass rings on the surface of the material to be tested, leaving a small segmental opening on one side that allows termites to gnaw and finally to escape from the ring (Becker, 1962). A modification for inducing the gnawing instinct is to put a dense plastic foil with holes of the head-capsula diameter of the respective termites on the material to be tested (Gösswald and Kloft, 1959). These holes induce gnawing at their bottom. Another testing method for laboratory and field tests is to put the material to be investigated between blocks or sheets of a susceptible wood species. Also, the settlement of imagines of Kalotermitids can be favoured by the testing method (Butterworth *et al.*, 1966).

Matrix and surrounding conditions are important in a test. While Kalotermitidae and part of the Termopsidae do not need soil—provided that the air humidity is high enough—it is essential for Rhinotermitidae and Termitidae not only for the establishment of favourable moisture conditions but also for gallery building, which is part of the normal behaviour of these termites. These termites can be kept also without soil, but they then live under unnatural environmental conditions. The matrix can be free of nutritional material, as is the case with sand, vermiculite (Becker and Lenz, 1970), or polyurethane foam (Hinterberger, 1975), or it can contain humus or particles of wood. Agar and sawdust have been used, too. The composition of the matrix can influence a laboratory test result remarkably. Also, in field tests consideration of overall conditions at the testing areas is essential (Butterworth and McNulty, 1966; Fougerousse, 1969; Krogh, 1969). In order to attract termites and increase their number at the test plot, feeder blocks or strips are employed (Gay *et al.*, 1957; Williams, 1973). Another factor in testing is the size of testing specimens, especially in the case of wood, for laboratory tests (Becker, 1962, 1969) as well as for field tests (Usher and Ocloo, 1974).

Choice tests and compulsory tests are two ways in which laboratory tests can be implemented. Compulsory conditions may lead to faster results, but the termites live under starvation if the material to be tested has no nutritional value for them. Therefore the first days of the testing time with full activity of the insects constitute the most important period. If during choice tests with food available for the termites they attack an undigestible material or treated wood, this would definitely be endangered under practical conditions.

Various biological factors

Species, even from the same genus, *and geographical races* of termites may react differently to materials and toxic substances, as recent comparative investigations have shown. The size of mandibles, activity and foraging behaviour, dependence on environmental factors, composition of the symbiotic flora of the gut, resistance to toxic agents, and many other properties differ from one genus to the other. The differences are greater with Kalotermitidae and Termitidae than with Rhinotermitidae, but within the latter family, species of the genera *Heterotermes*, *Reticulitermes*, and *Coptotermes* reacted differently to wood species and toxic compounds (Gay *et al.*, 1955; Schultze-Dewitz, 1961; Gay and Schulz, 1965; Becker, 1965c, 1966, 1975b; Becker *et al.*, 1972).

Geographical races can show variations in behaviour, vitality, activity, and tolerance to toxic agents. Within the North American species *Reticulitermes flavipes*, which probably should be divided into two or three subspecies (Becker, 1970a), it was stated that the tendency to gallery building, reaction to sensorial and environmental influences, quantity of wood consumption, and mortality under the influence of toxic compounds can differ a great deal (Becker, 1970b). Racial differences also exist within the European species *Reticulitermes lucifugus* (Rossi). *Reticulitermes speratus* (Kolbe) of East Asia can be divided into three subspecies (Morimoto, 1968). *Kalotermes flavicollis* (Fabr.) from Italy differs from the same species in southern France. Probably variations in reaction and tolerance could be observed with all species. Therefore, the collections or cultures used for testing should be defined sufficiently.

Number and caste composition of groups are important in laboratory tests. Kalotermitidae and Termopsidae can be used in smaller groups than Rhinotermitidae and Termitidae. With *Kalotermes flavicollis* (Fabr.), a species widely used in Europe, small groups of 10 specimens give the same result on the resistance of a material against attack as 100 specimens do (Becker, 1962, 1969); when testing the natural durability of wood species, however, the size of the *Kalotermes* groups may influence the result (Becker, 1975b). *Reticulitermes* species can be used in groups of several hundreds. The number of individuals is not correlated in a linear function with the consumption of wood; larger groups consume relatively less than smaller ones (Schultze-Dewitz, 1962; Becker, 1962, 1969). Mortality becomes smaller with an increasing number of individuals. *Heterotermes* should be used for tests in groups from 500 to 1000 or more individuals, *Coptotermes* and *Nasutitermes* species, in groups of several thousand specimens.

As to the composition of groups, mainly workers or in the case of Kalotermitidae pseudoworkers (=older nymphs) are needed. It is advisable to add a few soldiers or an adequately large number in the case of *Coptotermes* and *Nasutitermes*, *i.e.* genera with a high percentage of soldiers. For tests that are to run for more than a few weeks, some nymphs with wing stumps which can develop easily to neotenics should be added in the case of genera with neotenical development. The testing time in general may influence the result (Schultze-Dewitz, 1965; Becker, 1969).

Vitality and developmental stage of the colony from which the testing specimens are taken may influence the physiological reactions of the termites and thus the results. Again, as a basis for comparison, the use of reference material with well-known properties can help evaluate variations in behaviour and resistance.

CONCLUSIONS

Various physical, chemical, and biological factors influence the damage of wood and other materials under practical conditions and should be observed in testing procedures in order to provide comparability, repeatability, and correct evaluation. The complexity of influences concerned together with the necessity to deal with sufficient numbers of different termite species require further research on the behaviour and reaction of various termites.

REFERENCES

Beal, R. H., Bultman, J. D., and Southwell, C. R. (1973). Resistance of polymeric cable coverings to subterranean termite attack. *Int. Biodeterior. Bull.*, **9**, 28–34.

Becker, G. (1962). Allgemeines über die Laboratoriumsprüfung der Beständigkeit von Werkstoffen und der Wirksamkeit von Schutzmitteln gegen Termiten. *Materialprüf.*, **4**, 215–221.

Becker, G. (1963). Die Widerstandsfähigkeit von Kunststoffen gegen Termiten. *Materialprüf.*, **5**, 218–232.

Becker, G. (1964). Testing results on termite resistance of plastics. RILEM Symposium, Abidjan, 23–30 Nov., 1963, Bull. RILEM n.s. No. 25, pp. 93–97.

Becker, G. (1965a). Feuchtigkeitseinfluss auf Nahrungswahl und -verbrauch einiger Termiten-Arten. *Insectes sociaux*, **12**, 151–184.

Becker, G. (1965b). Versuche über den Einfluss von Braunfäulepilzen auf Wahl und Ausnutzung der Holznahrung durch Termiten. *Material u. Organismen*, **1**, 95–156.

Becker, G. (1965c). Prüfung der Wirksamkeit synthetischer Kontaktinsektizide auf vier Termiten-Arten. *Holz als Roh- u. Werkstoff*, **23**, 469–478.

Becker, G. (1966). Prüfung der Wirksamkeit wasserlöslicher Holzschutzmittel auf drei Termiten-Arten. *Materialprüf.*, **8**, 445–454.

Becker, G. (1967). Die Temperatur-Abhängigkeit der Frasstätigkeit einiger Termitenarten. *Z. angew. Entomol.*, **60**, 97–123.

Becker, G. (1969). Rearing of termites and testing methods used in the laboratory. In *Biology of Termites*, pp. 351–385. Vol. I (K. Krishna and F. M. Weesner, eds.), Academic Press, New York.

Becker, G. (1970a). Vergleichende Untersuchungen zur Ökologie einiger Reticulitermes-Arten (Ins. Isopt.). *Z. angew. Entomol.*, **65**, 183–216.

Becker, G. (1970b). Reticulitermes (Ins., Isopt.) in Mittel- und West-Europa. *Z. angew. Entomol.*, **65**, 268–278.

MATERIALS, CONSTRUCTIONS AND TREATMENTS FOR PROTECTING PACKAGES FROM DETERIORATION BY INSECTS*

HENRY A. HIGHLAND

Stored-Product Insects Research and Development Laboratory,
Agricultural Research Service, USDA, Savannah, Ga. 31403, U.S.A.

Summary

Nearly all flexible packaging materials in use today can be penetrated by one or more species of stored-product insects. Also, the construction of much of the packaging for products susceptible to infestation is not designed to prevent ingress of insects. Materials such as polycarbonate film, aluminium foil laminates, and polyester films are somewhat resistant to boring insects, but the degree of resistance is also dependent upon the thickness of the film. Both invasion and penetration can be reduced by heat-shrinking certain film overwraps on cartons. In addition, treatment with chemicals such as synergized pyrethrins prevents penetration when the insecticide is applied to shipping bags or incorporated in the adhesive of laminated film pouches. Other chemicals have been successful in laboratory tests.

INTRODUCTION

The insects that attack packaged, insect-susceptible food, feed, and fibre in the United States today are generally the same species that attack these stored products around the world. World trade has disseminated the few dozen species responsible for most damage to all areas with similar climates. Wherever these species are found they destroy and contaminate large quantities of stored products. The problem these insects present is twofold: some species (penetrators) such as *Rhyzopertha dominica* (F.), *Tenebroides mauritanicus* (L.), *Lasioderma serricorne* (F.), *Trogoderma variabile* Ballion, *Dinoderus minutus* (F.), and *Corcyra cephalonica* (Stainton) can bore through nearly all flexible packaging materials in use today. Other common species (invaders) usually cannot enter packages unless there is an existing opening. These latter species, however, can easily find the small openings in most packages as currently designed and manufactured.

These categories of penetrators and invaders are somewhat artificial but useful. The apparent capability or failure of any species to penetrate, however, may depend greatly on the testing techniques used.

PACKAGING MATERIALS

Paper is probably the least resistant to penetrating insects of all the flexible

* Mention of a pesticide or a commercial or proprietary product in this paper does not constitute a recommendation or an endorsement of this product by the U.S. Department of Agriculture.

packaging materials in use today. Depending on conditions and species, some insect species can penetrate kraft paper in less than a day, and multi-ply construction adds little to the resistance.

In general, almost all flexible polymer films or combinations of polymer films can be penetrated by one or more species of insects. Polycarbonate is the only film that is notably resistant to insect penetration (Highland and Jay, 1965), but it is not adaptable to food packaging. Likewise, aluminium foil can be penetrated though it is more resistant to penetration than is film or paper.

Rao *et al.* (1972), using small pouches in a beaker, found that insects, both penetrators and some invaders, could penetrate one or more of 18 films and laminates including polyethylene, cellophane/polyethylene, cellophane, polyethylene/paper, saran/cellophane, polyethylene/jute, paper/foil/polyethylene, and aluminium/vinyl, but no penetrations were found in polyester/polyethylene, polyethylene/canvas, foil/polyethylene, or paper/foil/polyethylene. This latter foil was twice as thick as the foil in the penetrated paper/foil/polyethylene. Gerhardt and Lindgren (1954, 1955) found that insects easily penetrated single thicknesses of polyethylene, cellophane and saran, cellophane laminates, and saran/pliofilm, but laminates of saran plus polyester, with the polyester exposed to the insects, were not penetrated. Thus films vary in susceptibility to penetration depending on the basic resin from which the film is made and the combination of materials.

Yerington (1975) showed that when tested as pouches against mostly penetrators, polypropylene-polyethylene combination films were generally more resistant than were other combination films. Also, coextruded polypropylene-ethylene vinyl acetate and coextruded polypropylene-polyethylene films showed more resistance to penetration than other saran-coated polypropylene and polyethlene-coated cellophane films, and oriented polypropylene films were more resistant than nonoriented polypropylene films. Highland *et al.* (1968b) showed that biaxially oriented polypropylene film provided excellent insect-resistant properties when used as an overwrap on paperboard cartons, but was ineffective as a pouch.

Other tests at the Stored-Product Insects Research and Development Laboratory, Savannah, Georgia, have indicated that polyethylene terephthalate (the polyester Mylar) and polyester urethane (Tuftane) films are more resistant to penetration than are most other films. Saran, ionomer (Surlyn), cellophane, and polyvinyl chloride films exhibited very low levels of resistance to penetration by *Rhyzopertha dominica.*

Tests have clearly demonstrated that aluminium foil can also be penetrated, although foil packages are generally more resistant to penetration than are film or paper packages. Indeed, Batth (1970) reported that dried soup mix pouches of 4-mil foil were penetrated within 4 days by *Tenebroides mauritanicus* (L.) and within 2 weeks by *Trogoderma inclusum* LeConte, but not by *Oryzaephilus surinamensis* (L.). Also, at Savannah (unpub. data), paper/foil/polyethylene pouches of cocoa powder and dried soup mix were easily penetrated.

Thicker films are more resistant to penetration than are thinner films made of the same polymer resin. Although polyethylene film is not notably resistant to penetration, unpublished tests conducted at Savannah have shown that some resistance is exhibited by films of 5-mil or greater thickness. Thus, films vary in susceptibility to penetration, depending on thickness, on the basic resin from which the film is made, on the combination of materials, and on the package structure.

PACKAGE STRUCTURES

Shipping containers

Although packaging materials can be carefully chosen and chemically treated to make them resistant to insect penetration, no treatment is effective if the package is not insect tight. For example, bags treated with methoxychlor that had tight, permanent seals remained insect-free during 24 months of exposure to heavy insect populations, but similarly treated bags with sewn, untaped closures were infested within 3 months (Highland *et al.*, 1964).

Multiwall paper shipping bags are made insect-resistant by using a folded-over, stepped-end closure which is sealed to the opposite face of the bag with a heat-activated adhesive (Laudani *et al.*, 1966). This type of bag is used to carry millions of pounds of processed cereals to overseas destinations in the U.S. foreign aid programs. Another insect-resistant multiwall paper shipping bag has synergized pyrethrins-treated paper tape heat-sealed over the stitching on both ends of the bag. Both of these bags are treated with synergized pyrethrins for insect repellency.

The corrugated paper case is another major type of shipping container used in the United States and in most developed countries. However, it is difficult to make these cases resistant to infestation. Well-sealed tape can be placed over all six flap junctures, or the case can be overwrapped with a heat-shrunk, polyethylene film overwrap. However, neither method protects from penetrating insects because both the film and the paper can be penetrated by insects.

Likewise, heat-shrunk polyethylene film overwraps can be applied to pallet loads of bags or cases to provide protection from invading insects. Overwrapped pallets can be protected further from infestation by placing a polyethylene deck sheet between the load and the pallet. During the heat-shrink process, this sheet is partially sealed to the film overwrap. Use of a black deck sheet and an infrared heat source can provide an insect-tight seal with the overwrap.

Consumer-sized packages

Although it is more economical to make the outer shipping container insect-resistant, it is often more desirable to insect-proof the consumer-sized package. Doing so provides protection from infestation throughout the distribution channel, including retail outlets, until the package is opened. The paperboard carton is widely used for consumer packaging in developed countries and in developing countries where food is imported. These cartons can be made insect-resistant by the use of tight overwraps such as those provided by polypropylene film (Highland *et al.*, 1968b). Also, laboratory tests and reports from food-processing companies show that fully adhered paper overwraps are effective, and Collins (1963) showed that fully adhered paper-foil laminated overwraps are quite resistant to insect invasion.

When tight seals are applied to the ends of overwraps, they provide resistance. In the absence of an insect-tight overwrap, the flaps of cartons must be sealed so they are insect-proof, a difficult procedure in modern high-speed operations. Beads of adhesive along the folds and edges of the end flaps and extended folded-down tabs at the sides of the end flaps (the 'Van Buren ear') do provide some insect resistance. Yerington (1971) described a polyethylene-coated carton having heat-sealed flaps that protect raisins from infestation.

More recent tests (Highland, 1975) have shown that composite cans are extremely resistant to both invading and penetrating insects. These cans have rigid paperboard walls that can be covered by laminations of paper, polymer films, and foil. The ends are metal and can be applied in high-speed operations. These cans resisted insect penetration during 29 months of exposure to heavy populations of both penetrating and invading insects.

Insecticide treatments

Insect-repellent treatments are only applied to packaging materials for food and other infestible items in very restricted areas with carefully controlled procedures. Thus, these procedures have a favourable cost-benefit ratio in terms of their impact on the environment when compared with the procedures used to control such insects as mosquitoes, field-crop insects, and forest insects. Nevertheless, the spatial and temporal proximity of chemicals to products destined for human consumption makes this use of pesticides potentially hazardous. These factors must be weighed against the benefits accrued by protecting the commodity from infestation when the commodity has incurred the full cost of growing, harvesting, shipping, storing, and processing. To this we must also then add the costs of packing, shipping, storing, and distributing the processed, packaged commodities. It becomes obvious that to achieve the desired protection without objectionable residues, package treatments must be used judiciously, at minimal dosages, and with all available expertise.

Obviously, the insecticides or repellents used on packages must be prevented from migrating from the treated surface through the package walls and into the commodity. This migration can be reduced or prevented by the use of barrier plies positioned between the treated ply and the commodity. Highland *et al.* (1966) found that four plies of kraft paper prevented unacceptable levels of piperonyl butoxide in cornmeal, flour, rice, and beans stored in synergized pyrethrins-treated bags. Also, saran-coated paper and greaseproof paper were effective barriers to the migration of this chemical (Highland *et al.*, 1968c, 1970), and the copolymer ethylene vinyl acetate in a coating containing synergized pyrethrins reduced the movement of piperonyl butoxide into the packaged food (Highland *et al.*, 1968a).

In the United States and in England much of the early work on insect-resistant packaging was done with the so-called safe insecticides and synergists—methoxychlor, pyrethrins, allethrin, piperonyl butoxide, sulphoxide, and similar materials. More recently, other compounds have been used successfully in laboratory tests, including synthetic pyrethroids (Highland and Merritt, 1973), an experimental antifeeding compound (Loschiavo, 1970), an inert silica gel (Watters, 1966), and carbaryl (Highland, 1967).

Recently, Highland (unpubl. data) found that synergized pyrethrins-treated paper-foil-polyolefin laminated pouches protected cocoa powder from infestation by *Trogoderma variabile* and by mixed populations of other common stored-product insects. Also, a cellophane-polyethylene laminate treated with synergized pyrethrins has been approved in the United States for protecting dried fruit from infestation. The pyrethrins and piperonyl butoxide are applied in the laminating adhesive at the rate of 3·5 and 35 mg/ft^2, respectively (Anon., 1974). Foil-wrapped meat cubes have been protected from infestations by *Ptinus ocellus* Brown (=*P. tectus* Boieldieu), *Stegobium paniceum* (L.), *Dermestes lardarius* L., and *Tribolium confusum* Jacquelin duVal by packaging them in synergized pyrethrins-treated cartons (Dennis, 1962). Likewise, Mallis *et al.* (1961) recommended a mixture of 1,5a,6,9,9a,9b-hexahydro-4a(4*H*)-dibenzofurancarboxaldehyde (MGK® Repellent 11) and *N*-(2-ethylhexyl)-5-norbornene-2,3-dicarboximide (MGK® 264) sprayed on corrugated shipping cases to prevent German cockroaches, *Blattella germanica* (L.), from infesting beer cases. These shipping cases are not sealed tightly enough to prevent the entrance of cockroaches, so short-term prevention of infestation is entirely dependent upon the repellent activity of the chemicals.

Other insecticides such as malathion, DDT, dieldrin, trichlorfon, dichlorvos, lindane, and aldrin have been evaluated as treatments for food and feed packages (Joshi and Kaul, 1965; Lal *et al.*, 1961; Langbridge, 1970). However, many of the attempts had only limited success because of the failure to utilize insect-tight constructions. Also, some of the chemicals migrated from the treated package into the contents and produced residues that were unacceptable (Butterfield *et al.*, 1949).

The development of pesticides and pesticide formulations for use in processing and packaging is inherently slow and difficult. Possible health hazards and effects on the environment must be considered. Also, manufacturers and formulators of pesticides often cannot profitably assume the task of developing materials for such uses because of the relatively small potential sales volume. However, if we are to conserve food, reduce costs, and maintain or raise sanitation standards, we must have the proper tools at hand. Among these tools are pesticides for use on food packaging and the technology for developing and using insect-resistant packaging.

REFERENCES

Anonymous (1974). Piperonyl butoxide and pyrethrins. *Federal Register*, **39**, 38224–38225.

Batth, S. S. (1970). Insect penetration of aluminum foil packages. *J. Econ. Entomol.*, **63**, 653–655.

Butterfield, D. E., Parkin, E. A., and Gale, M. M. (1949). The transfer of DDT to foodstuffs from impregnated sacking. *J. Soc. Chem. Ind.*, **68**, 310–313.

Collins, H. E. (1963). How food packaging affects insect invasion. *Pest Control*, **31**, 26–29.

Dennis, P. O. (1962). Insect infestation. The protection of foodstuffs by synergized Pybuthrin. *Food Proc. Packag.*, **31**, 131–133.

Gerhardt, P. D., and Lindgren, D. L. (1954). Penetration of various packaging films by common stored-product insects. *J. Econ. Entomol.*, **47**, 282–287.

Gerhardt, P. D., and Lindgren, D. L. (1955). Penetration of additional packaging films by common stored-product insects. *J. Econ. Entomol.*, **48**, 108–109.

Highland, H. A. (1967). Resistance to insect penetration of carbaryl-coated kraft bags. *J. Econ. Entomol.*, **60**, 451–452.

Highland, H. A. (1975). Insect resistance of composite cans. ARS-S-74. ARS, U.S. Department of Agriculture.

Highland, H. A., and Jay, E. G. (1965). An insect-resistant film. *Modern Packag.*, **38**, 205–206, 282.

Highland, H. A., and Merritt, P. H. (1973). Synthetic pyrethroids as package treatments to prevent insect penetration. *J. Econ. Entomol.*, **66**, 540–541.

Highland, H. A., Davis, D. F., and Marzke, F. O. (1964). Insectproofing multiwall bags. *Modern Packag.*, **37**, 133–138, 195.

Highland, H. A., Jay, E. G., Phillips, M., and Davis, D. F. (1966). The migration of piperonyl butoxide from treated multiwall bags into four commodities. *J. Econ. Entomol.*, **59**, 543–545.

Highland, H. A., Byrd, R. V., and Secreast, M. (1968a). Effect of ethylene vinyl acetate on the migration of piperonyl butoxide from coatings of synergized pyrethrins on kraft paper. ARS 51-28. ARS, U.S. Department of Agriculture.

Highland, H. A., Guy, R. H., and Laudani, H. (1968b). Polypropylene vs. insect infestation. *Modern Packag.*, **41**, 113–115.

Highland, H. A., Secreast, M., and Merritt, P. H. (1968c). Polyvinylidene-coated kraft paper as an insecticide barrier in insect-resistant packages for food. *J. Econ. Entomol.*, **61**, 1459–1460.

Highland, H. A., Secreast, M., and Merritt, P. H. (1970). Packaging materials as barriers to piperonyl butoxide migration. *J. Econ. Entomol.*, **63**, 7–10.

Joshi, H. C., and Kaul, C. L. (1965). Studies on the protectivity of jute bags impregnated with organic insecticides against red flour beetles and cigarette beetles. *Indian J. Entomol.*, **27**, 491–493.

Lal, R., Srivastava, P. D., and Dhar, P. (1961). Efficacy of insecticide impregnated jute bags against *Sitophilus oryzae* Linnaeus (Curculionidae: *Coleoptera*). *Indian J. Entomol.*, **22**, 204–210.

Langbridge, D. M. (1970). Treatment of paper to protect packaged food from insect attack. *Appita*, **24**, 45–51.

Laudani, H., Highland, H. A., and Jay, E. G. (1966). Treated bags keep cornmeal insect-free during overseas shipment. *American Miller and Processor*, **94**, 14–19, 33.

Loschiavo, S. R. (1970). 4′(3,3-dimethyl-1-triazeno) acetanilide to protect packaged cereals against stored products insects. *Food Technol.*, **23**, 181–185.

Mallis, A., Esterin, W. E., and Miller, A. C. (1961). Keeping German cockroaches out of beer cases. *Pest Control*, **29**, 32–35.

Rao, K. M., Jacob, S. A., and Mohan, M. S. (1972). Resistance of flexible packaging materials to some important pests of stored products. *Indian J. Entomol.*, **34**, 94–101.

Watters, F. L. (1966). Protection of packaged food from insect infestation by the use of silica gel. *J. Econ. Entomol.*, **59**, 146–149.

Yerington, A. P. (1971). Insect resistance research. *Good Packag.*, **32**, 12–13.

Yerington, A. P. (1975). Insect resistance of polypropylene pouches. *Modern Packag.*, **48** 41–42.

INTERNATIONAL DEVELOPMENT AND PRESENT SITUATION OF THE THEORY OF METHODS OF MOTH- AND BEETLE-PROOFING

OSKAR WÄLCHLI

Eidgenössische Materialprüfungs- und Versuchsanstalt,
St Gallen, Switzerland

Summary

A survey of the existing test methods of moth- and beetle-proofing of woollen textiles is given. The aims important for the development of test methods are treated and the question of the application of biological test methods is discussed. The existing test methods are analysed with regard to the choice of test insects, test conditions, and criteria of assessment. The reasons for the necessity to carry out the tests with various wool-attacking insects are given. In addition, various factors which are important for the suitability of a test method are discussed, *e.g.* the influence of the visual and gravimetric determination of the attack of the test insects on the wool specimens, the use of a bait, and of treated and nontreated control materials. Finally, the most important requirements to be met by test methods for moth- and beetle-proofing are summarized.

INTRODUCTION

The moth-proofing treatment of wool is a well-recognized finishing process of the wool fibre. Outstanding moth-proofing agents are available for this. The moth- and beetle-proofing means an effective improvement of the competitiveness of the wool. For a successful application of moth-proofing compounds, we do need suitable test methods, allowing an adequate evaluation of the biological effect against wool-attacking insects. These methods are based mainly on biological tests, using larvae of the most important parasites. On the basis of the results, it must be possible to decide whether a moth-proofing treatment prevents damage by wool-digesting insects and also kills them, and whether a sufficient quantity of the compound is present on the wool. In other cases it is important to know whether the protection is fast against washing, dry cleaning, or other conditions of ordinary service.

Wool-attacking insects and the requirements to be met for testing purposes

All real wool parasites are keratin-digesting insects belonging to the classes of butterflies and beetles. The most important species are compiled in Table I.

For technical and biological reasons, it is not possible to test moth-proofing treatments with all types of wool-attacking insects. A selection based primarily on the frequency and the importance of the respective parasites must be made. In addition, it must be taken into consideration whether

Table I
KERATIN-DIGESTING INSECTS

Lepidopterae	
Tineola bisselliella Hum.	Clothes moth (Fig. 1)
Tinea pellionella L.	Case-bearing clothes moth
Trichophaga tapetzella L.	Tapestry clothes moth
Tinea palescentella Staint.	Large pale clothes moth
Hofmannophila pseudospretella Stt.	False clothes moth
Endrosis sarcitrella Staint.	White shouldered house moth
Coleopterae	
Anthrenus scrophulariae L.	Common carpet beetle
Anthrenus verbasci L.	Varied carpet beetle
Anthrenus pimpinella Fabr.	Woolly bear
Anthrenus museorum L.	Lesser museum beetle
Anthrenus vorax Wat. (*A. flavipes* Lec.)	Furniture carpet beetle (Fig. 2)
Attagenus piceus OL.	Black carpet beetle (Fig. 3)
Attagenus pellio L.	Common furrier's beetle

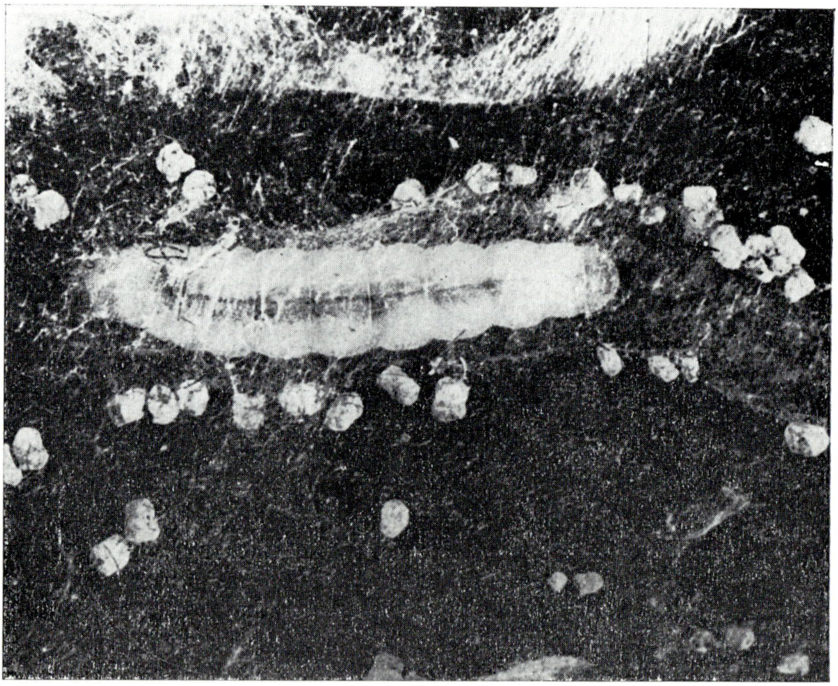

Fig. 1. Clothes moth larva in its silk tube with excrements.

a parasite may be bred under laboratory conditions and be manipulated without difficulties.

Tineola bisselliella can be bred easily on wool baited with yeast. *Attagenus piceus* develops more rapidly on ground dog biscuits (IWTO, 1964) than on wool. Of the various *Anthrenus* species, *Anthrenus verbasci*, living in the moderate climate, and the thermophilic type *Anthrenus vorax*, originating from subtropical regions, can be bred without difficulties on wool baited with dry yeast powder. The low intensity of voracity (Fig. 4) is a disadvantage of *Anthrenus verbasci* (Wälchli, 1955).

A very suitable test insect is *Anthrenus vorax* (Wälchli and Zinkernagel, 1952). Its voracity (Fig. 4) makes it particularly suitable for test purposes. Its development from eggs to larvae and pupae to beetles takes 2–3 months only.

According to ecological properties, *Tineola bisselliella*, *Attagenus piceus*, and *Anthrenus vorax* are three well-suited insects for test purposes which are of economic importance and also easy to breed (Figs. 1, 2, and 3).

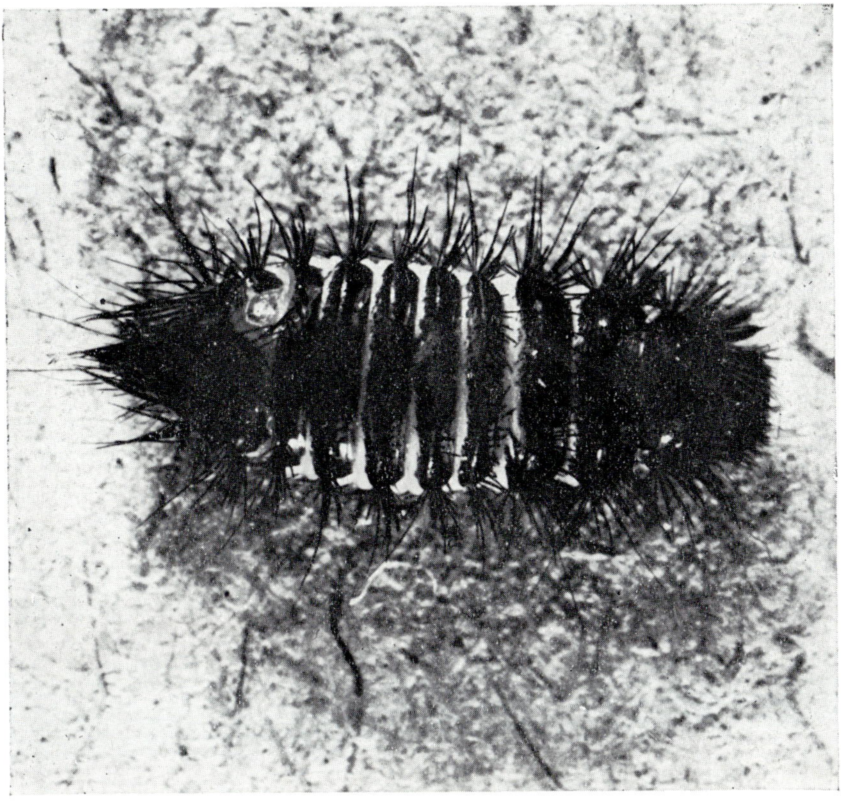

Fig. 2. Larva of the furniture carpet beetle (Anthrenus vorax *Wat.*).

INTERNATIONAL EVOLUTION OF THE THEORY OF TEST METHODS

The first moth-proofing test was an attempt employing bulk cultures of clothes moths in boxes, carried out by Meckbach (1921). This method was proposed later on in the United States as the Sealed Chamber Test (Walton and Shepard, 1960).

The method applied by Hase (1937) consisted of an alternative feeding test, with an untreated and a treated specimen in the same test cage and visual evaluation only. In the test disposition proposed by Stitteler (1938), the specimen with 10 moth larvae is covered with an inverted Gooch crucible (Fig. 5). This method often fails with thick, long pile, wool blankets or carpets as the larvae may escape. Yarns and loose fibres cannot be tested. This method had also been used by the Wool Industries Research Association (WIRA) in England (Hartley *et al.*, 1943).

Fig. 3. Larva of the black carpet beetle (Attagenus piceus *OL.*).

After extensive tests, the American Association of Textile Chemists and Colourists (AATCC) and the National Bureau of Standards established a test method in 1940, prescribing the use of *Attagenus piceus* as the testing insect. Four specimens, each with 10 larvae, were placed into flat glass jars or tins and kept there for 28 days, at 80°F (27°C) and a relative humidity of 50–60% (Fig. 6). In addition to the visual assessment and the determination of the weight of the excrements, the number of surviving larvae was also included as a measure for the toxicity.

Based on the work done by Heal (1942) as well as by Fletcher and Kenaga

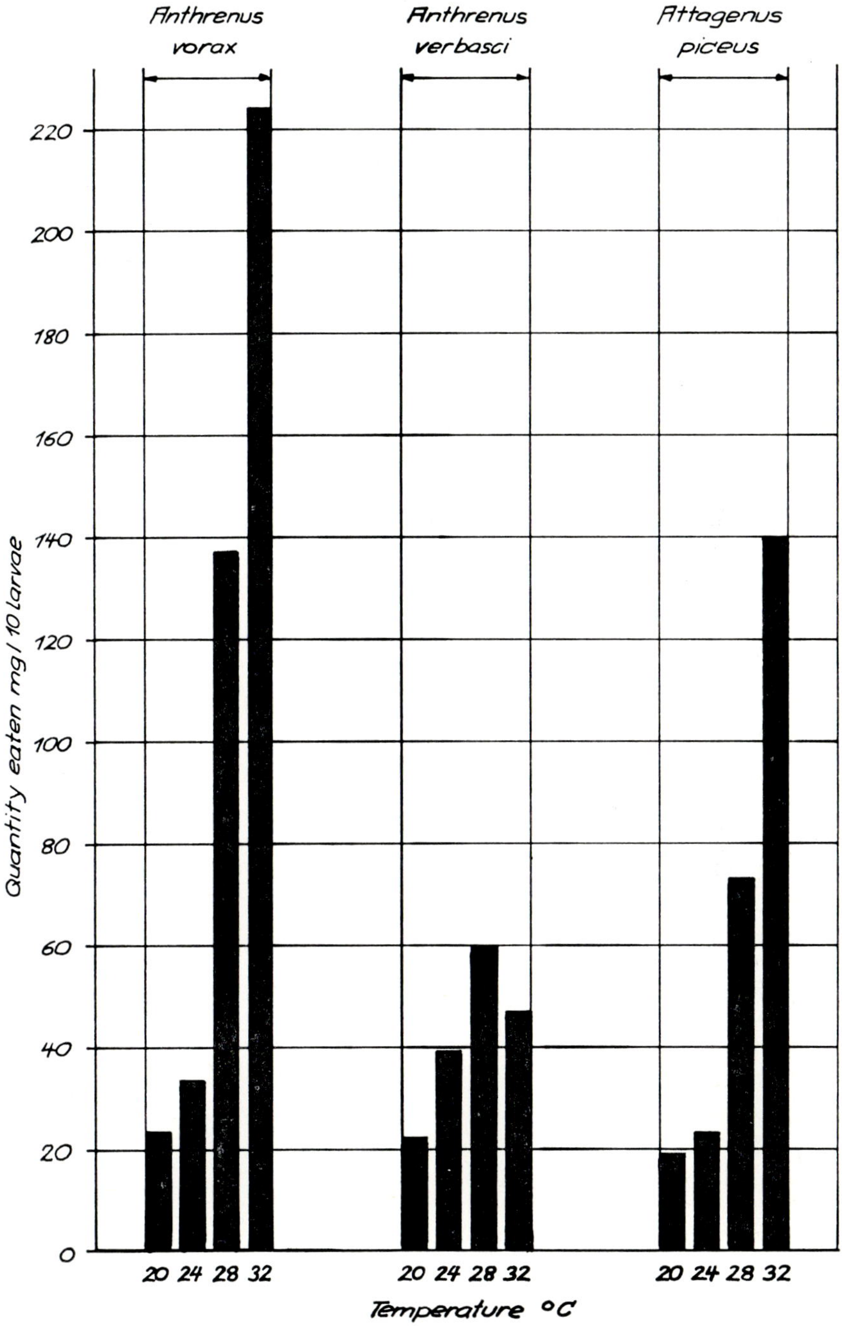

Fig. 4. *Voracity of wool-eating beetle species depending on temperature (Wälchli and Zinkernagel, 1952).*

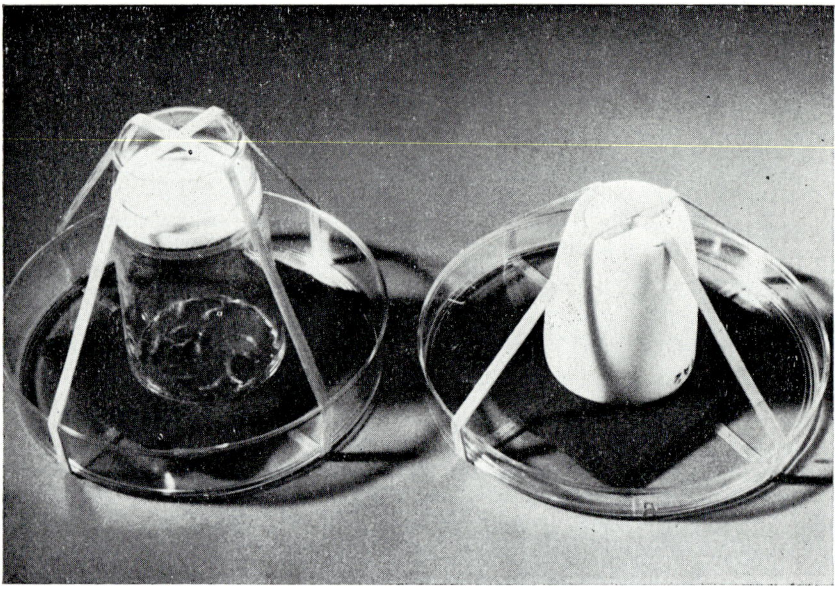

Fig. 5. Stitteler's method using inverted Gooch crucible as test cage.

Fig. 6. AATCC, Swiss, and IWTO methods: Specimens are placed in flat-bottomed glass, metal, or plastic test cages.

(1942), the average weight loss of the specimens was introduced as an improvement. Later on, larvae of the clothes moth, in addition to the *Attagenus* larvae, were proposed as test insects by AATCC (1943). The method published by CSMA (1952) proposed also the use of larvae of *Anthrenus vorax*. As a basis for the evaluation in the tests with beetles, the absolute weight losses of the specimens and excrement weights were used and, for those with moth larvae, the weight losses of the specimens only. These well-proved evaluation criteria have been retained in the revised CSMA specifications (CSMA, 1961) up to the latest method of AATCC (1974). The method published by ASTM (1957) stated for evaluation the absolute as well as the relative weight loss and visual assessment. Later on, visual evaluation was stated as the only criterion by Frey (1958).

In 1949 a method for testing woollen textiles with moth larvae and in 1952 another one for beetle-proofing, using *Anthrenus vorax* as a test insect, were published in Switzerland (SNV, 1949, 1952; Zinkernagel, 1949; Wälchli, 1949; Wälchli and Zinkernagel, 1952) (Fig. 6). In order to increase accuracy, the number of larvae per specimen has been increased from 10 in existing methods to 25 for *Tineola bisselliella* and to 20 for *Anthrenus vorax*. The testing time has been reduced to 14 days. The visual evaluation has been dropped. The most striking characteristic of the Swiss methods is the use of test specimens baited with an aqueous yeast extract. The evaluation of moth- and beetle-proofing is based on the relative weight loss of the test specimens related to the weight loss of the most similar one of six untreated control materials. The above-mentioned principle has been kept up until today. The Swiss methods were revised in 1971 (SNV, 1971). As a further test insect, *Attagenus piceus* has been included. A French method, published in 1963 by AFNOR (1963), follows the Swiss specifications. The different kinds of exposure of the specimens to the test insects in the methods treated are demonstrated in Fig. 7 (Anon., 1949).

Methods used in Europe about 1950 occasionally led to some difficulties when comparing the results. To overcome these difficulties, a subcommittee for larval testing was constituted in 1950, under the auspices of the Technical Committee of the International Wool Textile Organisation (IWTO). As a first result, the method 'Larval testing of moth-proofed wool serge' was published (Fig. 6) (IWTO, 1956). For the first time two treated serge fabrics in addition to an untreated one were used as standard materials. As treatments, two concentrations of 2,4-dinitro-α-naphthol (DAN) were applied. Later on, this method was enlarged and completed in such a way that a wide range of woollen textiles could be tested (IWTO, 1962). Finally, a specification for testing beetle-proofing with *Attagenus piceus* larvae was published (IWTO, 1964). By means of a mathematical transformation of weight losses of the treated controls and test specimens, it is possible to determine an equivalent concentration of DAN for the sample to be tested and consequently to decide whether it can be evaluated as being protected or not (Hueck, 1958). Due to the new principles involved and the precise description of the execution of the tests, these methods yield, without any doubt, the best reproducible results. But the expenditure of work is relatively high, so that this procedure is often used only as an exception, *e.g.* in the case of questions of controversy. Another test method with moth larvae has been published by the South African Bureau of Standards (SABS, 1969), using an untreated flannel and

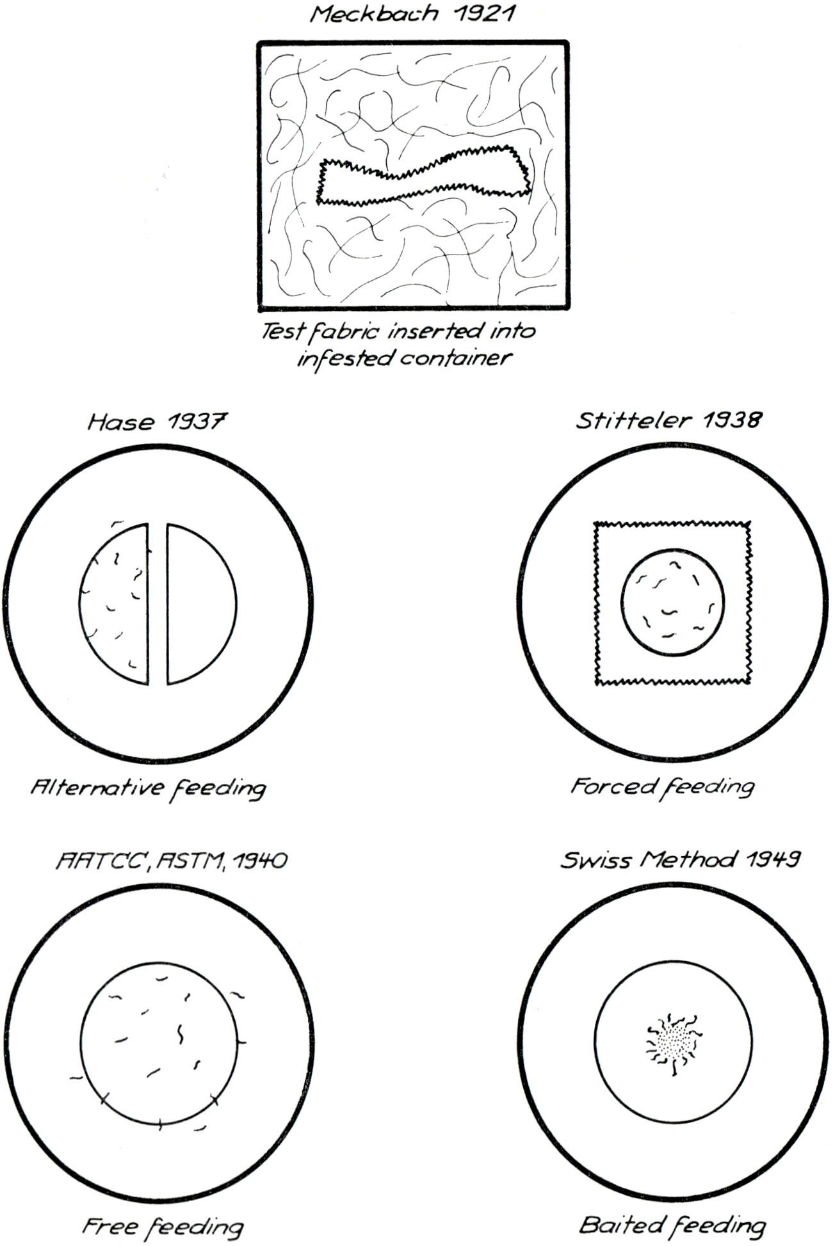

Fig. 7. Different methods of exposing test specimens to the larvae of test insects (Anon., 1949).

two fabrics treated with DAN as standard materials, as in the IWTO method (1962).

Recently, an ISO Working Committee (ISO, 1972) endeavoured to develop a procedure for moth-proofing tests on the basis of gravimetric and visual assessment.

In all methods using gravimetric assessment, weight changes of the specimens due to changes in humidity regain are corrected with the help of humidity control specimens.

REMARKS ON THE TEST CONDITIONS

A test method may be considered suitable if it produces within a laboratory and between different laboratories reproducible results. As a basis for this, test conditions must be defined as exactly as possible. But even then inevitable variations can occur in biological tests between different laboratories. Hueck (1958) has shown, for the IWTO-test method, that it has been possible to keep the variations within relatively narrow limits and that higher numbers of larvae per specimen lead to an increase of the statistical security. According to the U.S. methods, 10 larvae per specimen are used. In the first edition the Swiss methods prescribed 25 moth larvae and 20 *Anthrenus vorax* larvae, respectively. Various investigations within the IWTO-Committee and by Whitfield *et al.* (1958) showed that 15 larvae are more favourable on baited specimens. The voracity weight of various numbers of larvae shows linearity to the point where all baited food is consumed. But when the baited area of the specimen is completely consumed, the feeding activity slows down, which causes a break in the voracity weight-curve and subsequently an undesirable shifting of the relative weight losses by feeding. For this reason the number of larvae has been fixed in the IWTO methods, as well as in the revised Swiss method (SNV, 1971), to 15 larvae for each specimen.

The age and size, respectively, of the larvae are also of importance. For reproducible results it is of utmost importance always to use larvae of the same size and state of health. The feeding activity of very young and full-grown larvae immediately before the pupation is short. The most intensive activity can be observed with middle-aged larvae. For test purposes larvae must be used which are at the beginning of the stage of highest feeding activity, so that they are in the most active stage and do not pupate during the testing time. Normally, the larvae are selected not only according to their age but also according to respective weight. Moth larvae, for example, are collected by putting the wool with larvae on wire netting. The larvae turn away from the light by spinning silk fibres from which they can be collected easily, and then weighed (Fig. 8).

The development of wool-damaging insects is very strongly influenced by temperature. The test temperature should be near the vital optimum, 25°C for moth larvae and about 28°C for larvae of *Anthrenus vorax* and *Attagenus piceus*. Relatively small variations of temperature can produce large differences in weight loss (Fig. 4) (Wälchli and Zinkernagel, 1952). Furthermore, it has been shown that the testing temperature can also influence the protecting effect of the moth-proofing agent (Wälchli, 1955). DDT-treated fabrics, for example, show a decrease of moth- and beetle-proofing with increasing testing temperature.

Fig. 8. Collecting of clothes moth larvae from spun silk fibres.

Another question is whether the test specimens should be baited with yeast extract or not. Both methods, 'with bait' and 'without bait', show advantages and disadvantages. Test procedures without baiting the specimens are quite successful. The use of a bait, as it is prescribed in the Swiss methods, causes a strong increase of feeding activity (Fig. 9) (Wälchli and Zinkernagel, 1952). With baited specimens a better differentiation between moth-proofing treatments of various concentration is obtained, in particular, within the range of the borderline concentrations (Fig. 10). Of course, the quantity of bait to be

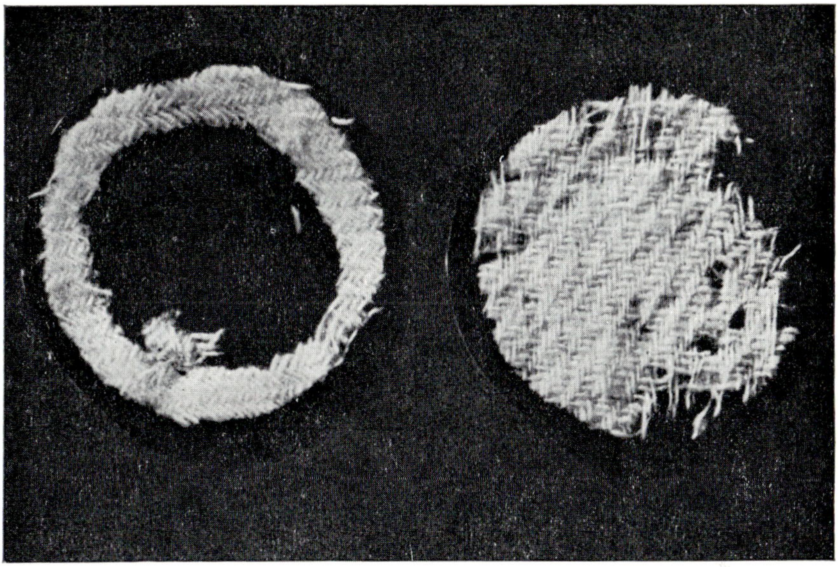

Fig. 9. Attack by clothes moth larvae on baited (left) and unbaited (right) specimens.

applied is difficult to define exactly; therefore we have to put up with certain variations between specimens. However, the baiting has proved so successful after repeated re-examinations that it has been retained in the revised Swiss methods (SNV, 1971).

The evaluation of the moth-proofing effect of textiles of varying texture whether protected or not by the help of the subjective visual method is difficult. The high density of population of larvae in the very small area of the specimens makes recognizable damage hardly avoidable. Powerful contact insecticides may be an exception. Even with the best and most permanent moth-proofing agents, which are stomach poisons, very slight traces of ravage must occur before the insect should be killed by poisoning. But the determination of weight loss, too, is not without difficulties. During pattern-cleaning at the end of the test, one has to remove the insects, the silk spun by the larvae, the excrements, and loose fibres, thus leaving plenty of possibilities

Table II

MOTH-PROOFING TREATMENTS WITH DIFFERENT ACTION
(15 clothes moth larvae on each specimen)

Moth-proofing agent	*Surviving larvae*	*Weight losses*		*Effect*
		mg	%	
A	0/0/0/0	6·7	4·1	Strong ravage inhibition and very strong toxic effect
B	8/10/10/11	5·2	3·2	Strong ravage inhibition and moderate toxic effect
Control untreated	14/15/15/15	164	100	No effect

moth-proofing effect (Table II). In the American methods, an absolute weight loss of 8 mg and a weight of excrements of 5 mg, respectively, are defined as tolerance limits for all materials. In general, the determination of the weight loss of specimens, applicable for moth as well as for beetle tests, seems to be better suited than the weight of excrements. But there are cases, as in the testing of feathers and furs, where the determination of the excrement weight is easier to apply. The general use of one absolute tolerance limit for all types of textiles may lead to differences in evaluation. Depending on the material and on changes in temperature, the attack by wool-damaging insects may vary considerably. In this respect, the determination of the relative weight loss, either in reference to untreated standard materials as defined in the Swiss method or to treated ones according to the IWTO methods, appears to be a more suitable way. The examination of different textile materials according to the two methods yields good corresponding results, as can be shown by some examples (Fig. 11).

Tests with moth and carpet beetle larvae, weakened by starvation, have shown that prolonged periods of starvation lead to a sensible weakening and decreasing of the viability of the insects, which is demonstrated by a decreasing feeding activity (Wälchli, 1970). The results of moth-proofing treatments in absolute and relative weight losses received also show that the determination of the results as relative values gives a better basis for the evaluation of the moth-proofing effect than the absolute weight losses. Naturally, it must be said that for biological tests healthy insects only are to be used.

PERMANENCY REQUIREMENTS OF MOTH-PROOFING TREATMENTS

Moth- and beetle-proofing as a quality of the wool is a real advance only when it is permanent under conditions of ordinary service and the usual cleaning operations. The majority of household moth-proofing products do not give a resistant protection. But there are agents, such as Eulan and Mitin, which provide a durable, long-term moth-proofing effect. Hase (1951) and Wälchli (1968) reported good results in long-term storing tests with Eulan and Mitin, respectively.

In order to improve the possibilities of evaluation and classification, various test proposals have been made. Specifications for such tests have been elaborated already in the years 1940–41 in the United States by AATCC, as well as by ASTM. The latest edition of the AATCC method dates from 1974 (AATCC, 1974) and gives specifications regarding the testing of washing-, dry cleaning-, hot pressing-, water-, perspiration-, light-, and abrasion-fastness. In Europe the Swiss Standard Association published in 1952 specifications for the evaluation and classification of the washing and dry cleaning fastness of moth-proofing finishes (SNV, 1952). In the French standard for moth- and beetle-proofing (AFNOR, 1963), the testing of the washing and dry cleaning-fastness has also been included. Finally, in 1969 a specification for permanent moth-proofing was published in South Africa, according to which the dry-cleaning and the washing durability are to be tested (SABS, 1969).

CONCLUSION

The methods in use for testing moth- and beetle-proofing are generally based on good and sound fundamentals, even if these are not identical for the various methods. In the course of the last 30–40 years, they have been continuously developed. Larvae of *Tineola bisselliella*, *Anthrenus vorax*, and *Attagenus piceus* are in use as test insects. These methods expose textile specimens in separate containers to the attack of a defined number of healthy and active larvae during 14 days. The temperature and humidity conditions are exactly defined.

For the evaluation of results, which can be based either on visually determined damage or weight losses, different methods have been used. There is one method based exclusively on the visual assessment of the damage which must be evaluated as critical. But doubts have also been expressed regarding the use of absolute weight losses only as a criterion for the evaluation. The determination of a relative value, such as the equivalent concentration of treated standards or the relative weight losses related to a most similar, untreated, standard fabric, appears probably to be a better suited criterion for the evaluation. The mortality of the insects, which is not decisive for the moth-proofing effect, is generally mentioned.

It has also been pointed out that, depending on the use of a woollen textile product, further requirements must be met. Particularly important are the durability of a moth-proofing treatment, the resistance to washing, dry cleaning, hot pressing, the fastness against water, perspiration, light, and bleaching. In order to test the permanency, special test methods have been elaborated.

The worldwide experiences which have resulted from the various test methods have shown that it is possible to work with most of the published methods in a satisfactory way. Fortunately, it not only depends on the test procedures but also on the persons who apply them and who are able to interpret them correctly and wisely.

REFERENCES

AATCC Test Method (1943). Resistance of textiles to insect pests. *Technical Manual and Year Book*, American Association of Textile Chemists and Colourists.

AATCC Test Method 24 (1974). Resistance of textiles to insects. *Technical Manual*, pp. 263–267, American Association of Textile Chemists and Colourists.

AATCC Test Method 28 (1974). Insect pest deterrents on textiles. *Technical Manual*, pp. 261–262, American Association of Textile Chemists and Colourists.

AFNOR (1963). Protection des Textiles. Méthode d'essai de résistance des textiles, pelleteries et fourrures, aux insectes kératinophages. Norme française NF X 41–516.

Anon. (1949). The mothproofing of wool. *Wool Sci. Rev.*, No. 3, 16–25.

ASTM Designation D 582-54 (1957). Standard methods of test for resistance of textile fabrics and yarns to insect pests. ASTM Standards on Textile Material, pp. 124–129.

CSMA (1952, 1961). Textile resistance test. Tentative methods of test for resistance of textiles to insect pests. *Soap Blue Book*. Chemical Specialties Manufacturers Association.

Fletcher, F. W., and Kenaga, E. E. (1942). Mothproofing tests. *Soap Sanit. Chem.*, **18** (8), 95–99.

Frey, W. (1958). Die Prüfung von Wollschutz-Imprägnierungsmitteln gegen Larvenfrass der Kleidermotte (Tineola bisselliella Hum.) mit Hilfe der visuellen Beurteilungsmethode. *Nachrichtenblatt des deutschen Pflanzenschutzdienstes*, **10** (12), 181–188.

Hartley, R. S., Elsworth, E., and Barritt, J. (1943). The moth-proofing of wool, *J. Soc. Dyers Col.*, **59**, 266–271.

Hase, A. (1937). Richtlinien zur biologischen Prüfung von Wollschutzmitteln. *Mitt. Biol. Reichsanst. Land- und Forstwirtschaft*. Berlin-Dahlem, Heft 55.

Hase, A. (1951). Wollschäden und Dauerschutz der Wolle durch Eulan-Behandlung. *Mitt. Biolog. Zentralanstalt für Land- und Forstwirtschaft*. Berlin-Dahlem, Heft 72.

Heal, R. E. (1942). Evaluating protection of fabrics from clothes moth and carpet beetle attack. *J. Econ. Entomol.*, **35**, 249–255.

Hueck, H. J. (1958). The development of a comparative method for the evaluation of the moth-proofness of wool serge. *Ann. Appl. Biol.*, **46** (4), 511–528.

ISO/TC 38/SC 8 (1972). Proposal for a draft international standard method of test for proofness of textiles against the common clothes moth. International Organisation for Standardisation.

IWTO (1956). Tentative specification: Larval testing of mothproofed wool serge. IWTO Technical Committee.

IWTO (1962). Method of test and assessment for proofness of wool fabrics against the common (webbing) clothes moth *Tineola bisselliella Hum.* IWTO Technical Committee.

IWTO (1964). Method of test and assessment for proofness of wool fabrics against the black carpet beetle *Attagenus piceus* (Oliv.). IWTO Technical Committee.

Meckbach, E. (1921). Mottenechte Wolle mittels Eulan. *Mell. Textilber.*, **2** (18), 350, (19), 373.

SABS Method 979 (1969). Standard specification for permanent moth-proofing. Council of the South African Bureau of Standards.

SABS Method 580 (1969). Pesticides: Biological evaluation of mothproof textiles containing wool. Standard methods. Council of the South African Bureau of Standards.

SNV 95901 (1949). Methode zur Prüfung der Mottenechtheit von wollenen Textilien. Schweizerische Normenvereinigung.

SNV 95902 (1952). Bestimmung der Käferechtheit von Wolle und andern keratinhaltigen Materialien. Schweizerische Normenvereinigung.

SNV 95910 (1952). Textiles: Valuation of the mothproof and beetleproof effect of wool and other materials containing keratin. Swiss Association for Standardisation.

SNV 95911 (1952). Textiles: Determination of the mothproof and beetleproof effect of wool and other materials containing keratin, after dry-cleaning and washing. Swiss Association for standardisation.

SNV 195901 (1971). Textilien: Bestimmung der Motten- und Käferechtheit von Wolle und andern keratinhaltigen Materialien. Schweizerische Normenvereinigung.

Stitteler, C. H. (1938). A larvae test for moth-resistance of mohair piled fabrics and woollens in general. *Am. Dyest. Rep.*, **27**, 729.

Wälchli, O. (1949). Ueber die Prüfung der Mottenechtheit von wollenen Textilien. *Textil-Rundschau*, **4**, 274–280.

Wälchli, O. (1955). Die Anwendung von DDT-Insektiziden im Textilschutz. In *DDT und seine Bedeutung*, pp. 145–195 (P. Müller, ed.), Vol. I, Verlag Birkhäuser, Basel.

Wälchli, O. (1968). Die Dauerhaftigkeit der Motten- und Käferechtheit durch Mitin. *Textilveredlung*, **3** (7), 345–349.

Wälchli, O. (1970). Einfluss verschiedener Hungerzeiten auf Ergebnisse von Frassversuchen mit keratinfressenden Insektenlarven. *Material und Organismen*, **5** (4), 293–306.

Wälchli, O., and Zinkernagel, R. (1952). Bemerkungen zur Bestimmung der Käferechtheit von wollenen Textilien nach der Prüfvorschrift SNV 95902. *Textil-Rundschau*, **7**, 105–113.

Walton, M. M., and Shepard, H. (1960). *Methods of Testing Chemicals on Insects*, Vol. 2, p. 73. Burgess Publ. Co., Minneapolis.

Whitfield, Sarel, F. G., Cole, J. H., and Whitney, G. F. (1958). The bionomics of *Tineola bisselliella*, Hum. under laboratory culture and its behaviour in biological assay. *Lab. Practice*, **7** (4, 5, 6, 7), 210–217, 275–284, 339–343, 408–413.

Zinkernagel, R. (1949). Mottenschutz heute. *Textil-Rundschau*, **4**, No. 5/6.

AN EVALUATION OF SOME PHYSICAL PARAMETERS WHICH INFLUENCE THE SUSCEPTIBILITY OF PACKAGES TO RAT DAMAGE*

G. K. LAVOIE and J. F. GLAHN

Denver Wildlife Research Center, Denver, Colorado 80225, U.S.A.

Cylindrical packages, 3, 6, and 10 inches in diameter, were made of three plastics (a vinyl, an acetate, and a polycarbonate) in two thicknesses (0·01–0·03 inch) and filled with grain. Each of the 18 package types were exposed for 48 hr to 10 individually caged Norway rats (*Rattus norvegicus*), and the area penetrated on each package was used as a measure of its susceptibility to rat damage. The data were analysed by step-wise multiple regression to determine the contribution of each parameter (curvature, hardness, and thickness) to the package's resistance to rat damage. Indentation hardness had the highest correlation with the area damaged, and only hardness and diameter reduced the variability of the area damaged enough to be considered in the final prediction equation.

Many species of rodents are involved in damage to stored foods, but commensal rodents, particularly Norway rats (*Rattus norvegicus*) and house mice (*Mus musculus*), are the principal pests. These rodents eat and contaminate vast quantities of food in storage and transit: their damage to packaged food also permits entry of various spoilage organisms. Ling (1961) estimated that 33 million tons (3·55 %) of the world's bread grains and rice in storage are lost to rodents each year. Dykstra (1968) reported that up to 80 % of packaged food in some parts of the world was damaged by rodents. Improved food handling and rodent control procedures could reduce these losses, but in some situations, rodent-resistant packages may be the only practical means of controlling rodent damage.

The object of this study was to evaluate a method of measuring specific package parameters as they interrelate to influence rat damage.

Several methods have been developed to evaluate the resistance of materials to rodent damage. Stolurow (1948a) constructed and tested an apparatus, known as the Pittsburgh Obstruction Unit, to evaluate packaging materials and chemicals as rodent deterrents. This device used a photocell activated timer which recorded the time required for a Norway rat to penetrate a test barrier. The test barrier was exposed at a 90° angle to each rat. Extensive conditioning and fasting were used to ensure high, uniform motivation in the rats. Stolurow (1948b) tested numerous chemical repellents on standard barriers, but tested only a few materials which were not chemically treated.

Tarzwell *et al.* (1953) developed a test and evaluated 35 different construction materials in numerous thicknesses and compositions for their resistance

* This study was funded under a project order from the U.S. Army Natick Laboratories, Natick, Massachusetts 01760, U.S.A.

to damage by black rats (*Rattus rattus*). They employed a training and fasting regime for motivation. The test material was exposed as a flat vertical barrier between the rat and a source of food and water. They found that rats required a gnawing edge to penetrate hard materials, but that they could develop a gnawing edge in soft materials. Hard materials with a gnawing edge, including asbestos cement and hard-tempered aluminium alloys, required 34–126 nights of exposure before penetration. Soft materials without a gnawing edge, including asbestos, plywood, sheetrock and asphalt, were penetrated in 1–8 nights.

Hueck and LaBrijn (1959) described a method for testing the resistance of materials to damage by rats and mice. Sex attraction, rather than hunger, was used for motivation. Pairs of male and female rats were separated by a flat vertical barrier of the test material for periods of at least 24 hr. The material's resistance was measured by weighing the barrier before and after exposure. When these investigators tested an unbroken polyvinyl chloride base plastic barrier, they found no damage after several days of exposure. They then repeated the test but drilled a small hole in each barrier to serve as a gnawing edge. The barriers were then chewed and perforated within 24 hr.

These studies indicated that the presence or absence of a gnawing edge is important in the resistance of materials to rodent damage. However, they all evaluated resistance by presenting test materials as a flat or 90°-angled upright barrier. Our studies have concentrated instead on evaluating the factors that contribute to the resistance of materials presented as a three-dimensional food package. In preliminary work, we developed a test chamber with a closed-circuit video camera and recorder that yielded very accurate measurements of the resistance of some packages and materials. However, only relatively soft materials could be tested, and variation in penetration time within and between animals required more replications than were practical in this chamber. As in the studies with barriers, we noted during our evaluation of the chamber that the presence of a gnawing edge made packages easier for a rat to penetrate. We then used the chamber to test fibreboard packages constructed with three different angles (90°, 120°, and 150°) to determine which afforded the least gnawing edge. These tests indicated that the time required for penetration increased only fractionally with the two angles over 90°. Neuhaus (1957) (as cited by Hueck and LaBrijn, 1959) found that cylindrical electrical cables with a diameter of 22 mm or more did not afford a gnawing edge and were therefore much less susceptible to rat damage. Similarly, Landstrom and Pryor (1970) found that pocket gophers (*Geomys bursarius*) were unable to damage electrical cables when the diameter was greater than 2·9 inches. We therefore concluded that rounded, rather than angled, surfaces would offer the least gnawing edge or purchase of any package shape, providing that the diameter exceeded the widest opening of the rat's mouth, and so the method described here tested the materials as cylinders.

In other studies, we had noted also that, as the hardness and thickness of the test material increased, there was a general decrease in the area penetrated by the rat and a general increase in the time required for penetration. Using the area penetrated within a given time as a quantitative measure of a material's resistance, we therefore designed a test to measure the influence of three parameters—curvature (diameter), hardness, and thickness—on the susceptibility of packages to rat damage.

MATERIALS AND METHODS

Three plastics were selected for preliminary evaluation of the test method—a vinyl, an acetate, and a polycarbonate. Indentation hardness, determined with a Shore-D durometer, was 28 for the vinyl, 77 for the acetate, and 79 for the polycarbonate. Two locally available thicknesses, ranging from 0·01 inch to 0·03 inch, of each material were tested.

We made these materials into 5-inch-long cylindrical packages with metal end caps having a 1-inch collar and thus exposing a 3-inch band of plastic to the rat. The metal caps eliminated gnawing edges at the ends of cylinders. Three diameters—3, 6, and 10 inches—were tested for each material. Preliminary tests indicated that cylinders of soft materials, such as the vinyl, would collapse under the weight of a rat, so a 0·025-inch pressboard backing cylinder was used with all materials. Plastic and backing cylinder edges were seamed with glue.

Norway rats were used in this evaluation because they are a principal destroyer of package goods. Rats were contained individually in cages 10 inches wide, 21 inches long, and 15 inches high. Test packages were filled with oats and bolted to the side of the cage with the glued seam positioned on the cage floor to prevent access to it by the rat. Each package type was exposed to each of 10 individually caged rats for 48 hr. During testing periods, water was available ad libitum, but no other food except that in the packages was available. The same rats were used throughout the test period except in one instance where an animal died and was replaced.

Data were analysed by step-wise multiple regression.

RESULTS

Table I summarizes the mean area penetrated for each of the three parameters

Table I
MEAN AREA PENETRATED (cm^2) WHEN CYLINDRICAL PLASTIC FOOD PACKAGES OF DIFFERENT DIAMETERS, HARDNESSES, AND THICKNESSES WERE EXPOSED TO NORWAY RATS FOR 48 hr
(10 rats per package type)

Hardness[a]	28		77		79	
Thickness (in)	0·01	0·02	0·01	0·03	0·02	0·03
Diameter (in)						
3	33·6	34·4	4·1	1·9	2·8	2·2
6	16·3	16·6	4·0	0·3	0·2	0·0
10	16·3	16·6	10·5	1·7	0·1	0·1

[a] Shore-D hardness value.

studied. Hardness vs. area penetrated had a correlation coefficient (r) of $-0{\cdot}630$. Hardness accounted for 39·65% of the variation in area penetrated. Diameter vs. area penetrated had an initial correlation coefficient of $-0{\cdot}145$, but the coefficient increased to $-0{\cdot}194$ when hardness was considered. Diameter and hardness contributed 41·92% to the variation in area penetrated, while thickness did not reduce the variation in area penetrated enough to be considered.

Hardness was the most important parameter of those tested in reducing the area penetrated. The regression equation relating area penetrated in cm^2 ($\hat{Y}$) to hardness (X_1) takes the form:

$$\hat{Y} = 33{\cdot}08 - 0{\cdot}395(X_1) \qquad \text{(S.E. 0·04)} \tag{1}$$

The final equation considering area penetrated ($\hat{Y}$), hardness (X_1), and package diameter in inches (X_2) is:

$$\hat{Y} = 38{\cdot}08 - 0{\cdot}39(X_1) - 0{\cdot}78(X_2) \qquad \text{(S.E. 0·03)} \quad \text{(S.E. 0·29)} \tag{2}$$

DISCUSSION

The test method and analysis appeared to be acceptable for determining the influence of these package and material parameters, providing the values for a particular parameter were sufficiently different. Using the area penetrated in 48 hr as a measure of resistance was not sensitive enough to reflect the 0·01-inch difference between the two thicknesses of vinyl and polycarbonate. However, the difference in thickness of the acetate tested was 0·02 inch and, with this span in values, the differences in areas penetrated became more apparent. Similarly, this measure did not reflect clear differences between the 0·03-inch acetate and polycarbonate which differed only slightly in indentation hardness. The inclusion of midrange hardness value (between 28 and 77 Shore-D) could have established the linearity of the relationship between area penetrated and hardness and would have given greater confidence in predicting resistance.

Equation (1), which considers only hardness regardless of thickness or curvature, predicts that a material of Shore-D hardness 79, such as polycarbonate, will have an actual mean hole size of 1·88 cm^2, or 91% less damage than a material of hardness 28. If this same relationship holds true for even harder materials, damage would approach zero for materials of a Shore-D hardness of 83.

Only hardness and diameter reduced the variability in hole size enough to be considered in the final prediction eq. (2). Apparently, for the range of thicknesses tested, either the measure of resistance (area penetrated) was not sensitive enough or thickness is truly not an important parameter of resistance over an extended time period such as 48 hr. Diameter or curvature did show an interesting relationship with area penetrated over all hardnesses and thicknesses tested. Increasing the diameter from 3 inches to 6 inches reduced the mean area penetrated by more than half, but increasing the diameter from 6 inches to 10 inches increased the mean area penetrated slightly and also increased the variation associated with the mean. It would appear that the curvature associated with about a 6-inch-diameter cylinder is the limit beyond which damage could not be further reduced. Considering this equation with a diameter set at 10 inches, the hardness of the material required to resist penetration by a rat ($\hat{Y} = 0$) in 48 hr would be approximately 79 on the Shore-D scale. This approximates the same results we obtained in the laboratory with thicknesses of 0·02 and 0·03 inch material. Theoretically, we could reduce the thickness to 0·01 inch and obtain similar results.

REFERENCES

Dykstra, W. W. (1968). The economic importance of rodents. In Proc. Asia-Pacific Interchange, *Rodents as Factors in Disease and Economic Loss*, pp. 47–52, Inst. Tech. Interchange, Honolulu, Hawaii.

Hueck, J., and LaBrijn, J. (1959). A simple method of determining the resistance of technical materials to gnawing rodents. *Anz. Schadlingskunde*, **32** (1), 9–10.

Landstrom, R. E., and Pryor, J. (1970). Unpublished annual progress report, Work Unit DF-103.2, p. 10. Denver Wildlife Research Center, U.S. Fish and Wildlife Service, Colorado.

Ling, L. (1961). Man loses a fifth of the crops he grows. *Atlantis*, Special Issue for FAO for freedom from hunger, pp. 1–68.

Neuhaus, W. (1957). Cause and limits of cable damage by rats. *Anz. Schalingskunde*, **30**, 81–82.

Stolurow, L. M. (1948a). Rodent behavior in the presence of barriers. *J. Comp. Physiol. Psychol.*, **41** (3), 219–231.

Stolurow, L. M. (1948b). A standard laboratory method for the evaluation of proposed rodent deterrents. Unpublished report to the U.S. Department of the Army, 82 pp.

Tarzwell, C. M., Stenburg, R. L., Nicholson, H. P., and Lynn, W. D. (1953). The resistance of construction materials to penetration by rats. Publ. Health Monogr. No. 11. Washington, D.C. Public Health Service, 16 pp.

ANTICOAGULANT RESISTANCE AND RODENT CONTROL

WILLIAM B. JACKSON and DALE E. KAUKEINEN

Environmental Studies Centre, Bowling Green State University,
Bowling Green, Ohio 43403, U.S.A.

Summary

Genetic resistance to anticoagulant rodenticides, known in Europe since 1958, was identified in 1971 in a rural, North Carolina population of Norway rats. Since then, in addition to this rural area, the phenomenon is known from 18 cities and 4 agro-business sites. Roof rats and house mice also have been identified with the phenomenon. In the United States we can expect the incidence of resistance to increase as populations in additional areas are examined and with the continued general use of anticoagulants. Except at agro-business sites and in urban areas with a history of intensive use of anticoagulants, the prevalence of resistance in Norway rat populations can be expected to be low ($<10\%$). Resistance in house mouse populations probably will be found to be more prevalent. In some areas of Europe, because of widespread resistance, especially in house mouse populations, effective rodent control with most anticoagulants is not possible. Until really satisfactory alternatives to the present anticoagulants are available in the United States, we can expect increasing problems with commensal rodent control and corresponding environmental deterioration. However, effective management of rodent populations requires both environmental improvement (which reduces the carrying capacity) and the supplemental use of toxicants and devices.

INTRODUCTION

Rodent resistance to anticoagulant rodenticides is not a new phenomenon, having been identified in Europe in 1958 and in the United States in 1971. Numerous discussions of the situation appear in the technological and research literature. The most recent review of the U.S. configuration is Jackson *et al.* (1975).

'Resistance' is variously defined to include bait shyness, physiological tolerance, and even operator incompetence. However, for this discussion only genetic or heritable resistance will be considered as determined from criteria established by the World Health Organization (WHO, 1970). For the Norway rat (*Rattus norvegicus*), surviving a 6-day, no-choice feeding test on food containing 0·005% warfarin, a minimum dose of 12 mg/kg is required. For other commensal rodent species, the criteria are slightly modified to reflect inherent susceptibilities (Jackson *et al.*, 1975).

A resistant rat, because of the mutant allele(s), is able to metabolize vitamin K and produce blood-clotting proteins (prothrombin) even in the presence of warfarin. In the susceptible rat, this biochemical pathway is blocked by warfarin, and the animal dies from internal haemorrhages after several consecutive days of feeding on the toxicant baits.

Resistance to warfarin appears to confer resistance to other anticoagulants (at least those commercially available in the United States), so resistant populations cannot be controlled effectively with this group of rodenticides. Alternative rodenticides (red squill or zinc phosphide) are not wholly suitable substitutes in many situations for safety or efficacy reasons.

With only ineffective or inefficient rodenticides available and in use and without improvements in environmental sanitation, rat populations in affected areas can be expected to increase. With more rats, the incidence of rat bite and rodent-borne diseases and rodent consumption and contamination of foods can be expected to increase. However, such statistics are difficult to develop.

Estimates of rodent numbers and rodent damage are almost always without a valid base, especially when projected on a national base. The one rat per person myth is traced back to Boelter (1909), but the only actual censuses of rats in urban environments were conducted by Davis in 1949 (Davis, 1950; Davis and Fales, 1950), who determined ratios of about 25 persons per rat. The review of Grish *et al.* (1974) exemplifies the difficulty of obtaining actual damage determinations rather than estimates of damages to stored food. Even so, the loss of food stores to rodents is significant, especially in developing countries.

RESULTS

As a result of cooperative studies by our Bowling Green Laboratory and the New York Rat Control Laboratory under the federal (DHEW-PHS) urban rat control programme, resistant Norway rats (*Rattus norvegicus*) populations in the United States have been identified in more than 20 sites (Tables I and II). We may focus attention on four environments where rodent-caused biodeterioration may occur: urban areas, warehouses, agro-business sites, and farm storage facilities.

In the United States, where resistance has been largely an urban phenomenon, inability to limit local rat populations has been apparent to control operators. Initial responses by the PCO are to blame the serviceman for poor technique, to increase the amount of anticoagulant used, or to change the specific anticoagulant employed. None of these improves the situation. Further environmental deterioration occurs.

In the near west side (Lawndale) area of Chicago, where about half the rats tested were designated resistant, active rat sign continued at high levels despite regular control efforts over several years. The area, one of transient human populations, has degraded buildings, and reclamation is difficult.

The number of rat bites reported to the Chicago Health Department was 190 in 1973 and 195 in 1974 and in the first half of 1975 was 116. While these statistics are for the entire city, most reports have come from the Lawndale area. Even so, probably less than half the total bites are ever reported. The current year's increase (the 6-month rate is 20% ahead of last year) may reflect an expansion of the area infested with resistant rats and the inability to control them with anticoagulants.

At several agro-business sites (grain elevators, stockyards), rats were so abundant that they were readily seen moving about day and night. Control efforts with anticoagulants were totally ineffective. Large quantities (actual

Table I
SUMMARY OF 6-DAY FEEDING TESTS WITH WILD NORWAY RATS FROM FEDERALLY FINANCED RODENT CONTROL PROJECT CITIES, 1972–JULY 1975 (NEW YORK STATE DEPARTMENT OF HEALTH AND BOWLING GREEN UNIVERSITY LABORATORIES) WHERE PREVALENCE OF RESISTANCE IS 3% OR MORE

City	*Sample size*	*No. survivors*	*% resistant*
Akron, Oh.	15	4	27
Atlanta, Ga.	48	3	6
Binghamton, N.Y.	114	8	7
Charlotte, N.C.	19	5	26
Chester, Pa.	5	2	40
Chicago, Ill.	86	51	57
Cohoes, N.Y.	165	13	8
East Orange, N.J.	13	2	15
Hoboken, N.J.	73	2	3
Houston, Tx.	61	9	15
Indianapolis, Ind.	48	4	8
Jersey City, N.J.	50	3	6
Newark, N.J.	53	2	4
Norfolk, Va.	149	13	9
North Carolina (Johnson Co.)[a]	114	59	52
Pittsburg, Pa.	94	11	12
Poughkeepsie & Dutchess Co.	309	105	34
Puerto Rico	44	4	9
San Francisco, Ca.	30	3	10
Sioux City, Ia.[b]	15	10	70
Trenton, N.J.	67	2	3
Washington, D.C.	48	4	8
York, Pa.	38	3	8

[a] Original resistance site.
[b] From control operator.

amounts not even estimated) of feed grains were consumed by the rats. Testing of rats from one of these populations indicated 70% of the individuals to be resistant.

Warehouses, where the most concern about damage to goods exists, have not yet been identified with resistance problems. In well-maintained structures resident populations are not allowed to build up, and the rodents (usually mice) found are transients imported with goods. Since control efforts usually are supplemented with a variety of mechanical devices, identification of resistant populations in this environment is unlikely.

The initial U.S. resistance site was on farms in Johnson County, North Carolina. Several other farm areas, characteristically storing grains and/or having animal feeding operations, have since been identified as resistant sites in other parts of the country. This limited occurrence is in contrast to the European experience in which resistance has been found primarily in rural areas and is related to the intensive government control programmes for this environment (Jackson, 1969).

What of the future? Doubtless additional sites of resistance will be revealed. The mutant allele(s) for resistance appear widely scattered among rat populations. Too often the failure to achieve control is attributed to poor materials or techniques, and continued use of anticoagulants intensifies the problem. Probably intensive and general use of anticoagulants in an area for several

Table II

SUMMARY OF 6-DAY FEEDING TESTS WITH WILD NORWAY RATS FROM FEDERALLY FINANCED RODENT CONTROL PROJECT CITIES, 1972–JULY 1975 (NEW YORK STATE DEPARTMENT OF HEALTH AND BOWLING GREEN UNIVERSITY LABORATORIES) WHERE PREVALENCE OF RESISTANCE IS LESS THAN 3%

City	*Sample size*	*No. survivors*
Baltimore, Md.	16	0
Buffalo, N.Y.	30	1
Camden, N.J.	63	1
Cleveland, Oh.	30	0
Duluth, Minn.	4	0
Ft. Wayne, Ind.	9	0
Kansas City, Ka.	18	0
Kodiak, Ak.[a]	19	0
Little Rock, Ark.	4	0
Louisville, Ky.	9	1
Memphis, Tenn.	17	0
Milwaukee, Wis.	34	1
Mobile, Ala.	17	0
Nashville, Tenn.	31	0
New Orleans, La.	23	0
New York, N.Y.	566	12
N.E. Pa. (Scranton, Hazelton, Wilkes-Barre)	46	0
Passiac, N.J.	25	0
Paterson, N.J.	35	0
Philadelphia, Pa.	72	1
Plainfield, N.J.	68	0
Rochester, N.Y.	63	2
Seattle, Wash.	7	0
St. Louis, Mo.	38	0
Syracuse, N.Y.	53	2
Tulsa, Ok.	12	1
Worchester, Mass.	41	0

[a] Non-Federal Programme.

years would provide sufficient selection pressure to achieve a population with a significant proportion of resistant animals; David Drummond of the United Kingdom calculates that 10 generations subject to intensive selection is all that is needed (Drummond, pers. comm.). Because of their large gene pools and persistent control efforts, agro-business sites may be the more likely to be involved. Thus we can expect to find resistance to anticoagulants in other sites in the United States and in other parts of the world, too.

Despite the publicity and concern over resistance, the phenomenon seems limited at the present time; and anticoagulants can continue to be used responsibly in most places without decreased efficacy. In Europe, where the problem has existed for almost two decades but affects less than 5% of the land area, anticoagulants retain their place in control programmes for rats. For most householders and PCO's, fears about resistant rats will not be realized, and they certainly should not be referred to as 'super' rats.

However, alternatives to rodenticides need to be re-emphasized. Sanitation and rodent-proof construction are fundamental yet often ignored. If the environmental supports (food, water, harbourage) for pest populations can be removed or eliminated, the pests themselves disappear. The threat of resistant rats ought to be used as a lever to incorporate environmental management into a city's programme or a PCO's services.

In theory, repellents should provide an effective barrier against rat entry or damage. However, older chemical repellents (*e.g.* Actidione) or more recent market products (*e.g.* Rotran) have not been successful in protecting packaged goods. Treating gunny sacks with malathion provides some repellency but is short-lived (several weeks) (Majumder *et al.*, 1964). At this time repellents cannot be considered a suitable alternative, except for highly specialized uses (*e.g.* protection of buried cables). Demonstrating that physical repellents (*i.e.* ultrasonics) are efficacious under field conditions has been difficult.

While currently registered acute rodenticides are not wholly satisfactory, several new rodenticides offer good portent. Difenacoum and Sorexa CR, neither registered for use in the United States, currently are being used with great effectiveness against resistant populations in Europe. The former is an anticoagulant; the latter, an anticoagulant plus vitamin D_2 (Calciferol)—which is effective in its own right. Only preliminary testing with these compounds has been completed in the United States. An acute carbamate (urea) toxicant, Vacor (R & H 787), has been introduced recently and has been demonstrated to be effective against resistant rodents. A toxicant specific to rats, Raticate (Norbormide), is still under development.

But a very significant part of the problem, like the submerged part of an iceberg, has been avoided. House mouse (*M. musculus*) populations in Europe commonly are resistant (Rowe and Redfern, 1965), and presumptive evidence indicates that the same may be true in the United States. The mouse is prolific and because of its small size and limited movements, more difficult to control. When the environmental pressure is reduced by removal of rats, mice tend to expand their populations and sphere of activity. Resistance in mice is complex, probably polygenic (Rowe and Redfern, 1965). Even without the resistance problem, control of mice is difficult because of their lower susceptibility to the anticoagulants. As in Europe, we are likely to find widespread resistant populations and will quickly need alternatives for their control. Resistant or not, we can expect increasing problems with mice.

Even with immaculate premise sanitation, we will have to use various mechanical and chemical tools and rigidly monitor incoming shipments to warehouses and other food storage sites if the rodent problem is to be contained. Both Sorexa CR (or Calciferol) (not yet registered for use in the United States) and Vacor are effective against resistant mice. DDT as a tracking powder was perhaps the most efficient tool available before its removal by EPA.

In some developing countries, where 80% of the grain never leaves the local harvest area, rodent infestations do levy heavy losses. Major grain storage facilities (godowns) frequently are poorly constructed and heavily infested with rats. Perhaps fortunately, anticoagulants are not too popular in these areas, and the likelihood of their selecting resistant populations is thus reduced.

At this time evidence for anticoagulant resistance outside of Europe and the United States is lacking or involves insufficient data; but in all countries anticoagulants should supplement environmental improvements, be used judiciously, and alternated with acute rodenticides, if possible, annually. The full utilization of the world food supply requires our use of the best possible environmental management measures.

REFERENCES

Boelter, W. R. (1909). *The Rat Problem.* John Bale, Sons and Danielsson, Ltd, London, 165 pp.

Davis, D. E. (1950). The rat population of New York, 1949. *Am. J. Hyg.*, **52**, 147–152.

Davis, D. E., and Fales, W. T. (1950). The rat population of Baltimore, 1949. *Am. J. Hyg.*, **52**, 143–146.

Grish, G. K., Arora, K. K., and Krishnamurthy, K. (1974). Studies on rodents and their control. *Bull. Grain Technol.*, **12**, 139–148.

Jackson, W. B. (1969). Anticoagulant resistance in Europe I, II. *Pest Control*, **37**, 51–54; **37**, 40–43.

Jackson, W. B., Brooks, J. E., Bowerman, A. M., and Kaukeinen, D. E. (1975). Anticoagulant resistance in Norway rats as found in U.S. cities. *Pest Control*, **43**, 12–16; **43**, 14–24.

Majumder, S. K., Krishakumari, M. K., and Krishna Rao, J. K. (1964). Malathion as a repellent for rats. *Curr. Sci.*, **33**, 212–213.

Rowe, F. P., and Redfern, R. (1965). Toxicity tests on suspected warfarin resistant house-mice (*Mus musculus*). *Am. J. Hyg.*, **63**, 389–393.

WHO (World Health Organization—U.N.) (1970). Provisional instructions for determining the susceptibility of rodents to anticoagulant rodenticides. WHO Tech. Rept. Ser. No. 443, 140–147.

BIODETERIORATION BY BIRDS

J. W. CASLICK

New York State College of Agriculture and Life Sciences,
Cornell University, Ithaca, New York 14853, U.S.A.

Summary

An overview of bird damage problems includes brief coverage of their damage to crops, stored foods, forest trees, utility poles, and buildings. Bird problems related to aircraft operation and bird-disease relationships are discussed. Bird damage control techniques and research needs are outlined.

INTRODUCTION

Wild birds are not widely considered to be agents of biodeterioration. Quite generally placed at the other end of the scale, birds are accepted as unique environmental assets, benefiting us economically through their ingestion of insect pests and aesthetically through their attractive appearance, interesting habits, and pleasing vocalizations.

But some of us know that birds can be pests under some circumstances, and it is this aspect which is of interest to us today. Since no previous papers on this topic have been presented at these Symposia, I shall briefly categorize some situations in which birds may serve as agents of biodeterioration, then summarize the general approaches to control. In so doing, I am appreciative of the opportunity to introduce this new topic and invite your suggestions for possible new approaches to bird damage control from your perspectives as microbiologists, biochemists, toxicologists, and other fellow scientists. It is axiomatic that science demands specialization and intensification within narrow disciplines, yet some of our most notable scientific achievements have resulted from such interdisciplinary efforts. We certainly need help from many disciplines in solving bird damage problems.

Public sympathy for birds greatly, and rightly, restricts the acceptable methods of controlling their damage. Imagine for a moment how you microbiologists would approach your research if there were many large organizations actively opposed to killing microorganisms and, in fact, so interested in the continuing welfare of microbes that they were willing to expend large amounts of money to protect them and to produce more. Imagine, too, the further complication of dealing with organisms having a very high level of intelligence and defensive response and perhaps having more highly developed sensory receptors than their human contemporaries! These factors weigh heavily in favour of the birds. Obviously, the development of bird damage control methods must take these factors into account.

To put this session of the Symposium into perspective, I would point out that, among vertebrate animals, the economic losses from birds probably are much less than those from the rodents discussed by previous speakers, although figures or even firm estimates on losses are not available for any large geographic areas. It is also noteworthy that many of the damaging bird species are not native species, but, as with domestic pigeons and house sparrows, have become established in areas where human activities have brought accumulations of those foodstuffs also acceptable to birds. Such damage may occur at two principal times—during the growth period of the basic agricultural crop and during the storage period of raw materials or finished food products.

DAMAGE TO CROPS

In North America, Africa, and elsewhere, wild birds cause heavy damage to maturing corn, rice, wheat, and other grain crops. The destruction of crops by blackbirds in the United States, by waterfowl in Canada, and by *Quelea* in Africa has received wide coverage in the popular press. Where these birds congregate seasonally in flocks of millions, localized crop damage is readily apparent and may be of significant economic importance. For example, in some parts of Africa, bird damage probably makes cereal cultivation impossible (Crook and Ward, 1968). In other areas, smaller flocks of birds damaging crops over long periods of time may inflict sizable losses, in total, without spectacular damage in any particular location. An example of this low-intensity but widespread and significant damage occurred in New York State last year, where it is reported that blackbird damage to corn probably exceeded the damage inflicted by insects and diseases combined (Pardee, 1974).

In North America, scare-devices such as gas exploders, firecrackers, and bird distress cries have been used with varying results. No completely satisfactory chemicals have been developed as repellents, toxic baits, or contact poisons for use in ripening corn or in bird-roosting areas (Caslick, 1968). Large-scale aerial treatment of bird roosts with insecticidal chemical compounds has killed large numbers of birds. However, such treatments are generally considered to be undesirable because of hazardous side effects and doubt about the long-term effectiveness of the partial destruction of populations. At least some of this doubt is based upon experience. Between 1956 and 1960 in South Africa, 400 million *Queleas* were destroyed by spraying, bringing some relief to the problem. But in 1966/67, more than 6000 acres were sprayed, and a record 112 million birds killed, following massive invasions of the farmlands (Crook and Ward, 1968). Newer roost treatment procedures in the United States, which are usually intended to alleviate problems caused by the location of the roosts rather than to reduce crop damage, involve aerial application of surfactant materials in conjunction with rainfall and temperatures below 45°F. Such applications have resulted in a 90% reduction of starlings in a communal night roost (Miller, 1975). The biodegradation of the surfactant materials in bird roosts is of current research interest and is an example of an area in which further cooperation among us might be productive.

Decoy traps for blackbirds have given some relief from crop depredations. A new blackbird bait containing the active ingredient 4-amino pyridine is now commercially available for use in protecting corn from blackbirds. Birds which have ingested the compound exhibit stress symptoms including loss of muscular coordination, often frightening the flock from the treated area.

Bird damage to fruit crops is known throughout the world. Quantitative information on the severity of damage is meagre for large areas, although some intensive studies have been conducted for localized areas such as the Niagara Peninsula of Canada (Brown, 1974), for example. Viewed on a worldwide basis, the control methods employed, if any, are highly dependent upon the current social and economic conditions of that area. These factors determine the true value of the crop to be protected and hence the level of expenditure allowable for labour or its substituted mechanical or chemical counterparts. Control measures vary from employment of human patrols which shout and beat home-made scare devices to automated electronic sound-producing machines which may be programmed to operate at the desired times and frequencies. A considerable amount of research has been conducted on the use of recorded distress calls for frightening birds, including comparisons of behavioural responses of birds to calls of the same species recorded in different countries (Bremond *et al.*, 1968). Decoy traps have been used successfully to protect cherry orchards from attack by starlings (*Sturnus vulgaris*). Sound-producing machines are in wide use in the major grape production areas of the world. Several types of netting are marketed for use in protecting fruit trees and shrubs from birds. Plastic-impregnated paper, nylon, cotton, and polyethylene nettings are sold for this purpose (Caslick, 1968).

DAMAGE TO STORED FOODS

Of all the types of bird damage, that inflicted to stored foods is the most difficult to quantify. Owners of storage structures which allow birds to enter are not likely to wish that this becomes general knowledge. My search of government agency reports has revealed few figures on this, although farmers and operators of feed mills and grain storages frequently mention birds as pests which foul grain with droppings, feathers, and nesting materials. The U.S. Department of Health, Education, and Welfare reported that in 1961 about 4% of all U.S. wheat was contaminated by bird droppings (Scott, 1961). Presumably, much of this occurs in storage. In New York State, Food and Drug Administration officials do not recall having ever condemned a carload of grain because of contamination by birds. They do handle isolated cases of bird contamination in warehouses, and occasionally these may be serious. For example, in a bakery warehouse, birds were picking wrapped loaves of bread on conveyor lines and in storage boxes. These were reported to be sparrows and blackbirds; I presume that the blackbirds may have been starlings. The problem was solved by repairing broken skylights. Such structural modifications usually are the most permanent control methods, although electrical fencing or fumigation with cyanide are sometimes used (Fitzwater, 1971). Bird problems are generally not as serious as rodent problems in food storage areas.

STRUCTURAL DAMAGE AND AESTHETIC CONSIDERATIONS

Structural damage from bird droppings tends to be largely aesthetic, but may be costly in terms of cleanup. I know of no tests which give quantitative information on the effects of droppings from wild birds upon wood, metal, or concrete. Domestic poultry establishments are well aware of the corrosive effects of poultry manure upon metal. Colleagues at Cornell have found chicken manure to be nearly neutral in pH and highly buffering in reaction with acids. The ammonia present in fresh manure may be the corrosive agent; poultry manure, when dry, is relatively noncorrosive.

Bird droppings may accumulate rapidly. Bridge painters attempting to paint beams and gusset plates on New York City bridges have had to use cold chisels and hammers to remove 18 inches of bird guano, mostly deposited by starlings which roost there overnight (Miller, 1975). Several cities around the world have undertaken pigeon control programmes to reduce problems of pigeon excrement upon buildings, statues, and pedestrians. Control methods have included shooting, trapping, poisoning, electric fencing, and the use of sticky repellents and contact avicides. Chemical compounds are available which reduce pigeon reproductive success when fed upon. Urban roosts of starlings have been moved by playing starling distress calls (Pearson *et al.*, 1967). Trapping is effective and selective (Stiles, 1966).

DAMAGE AT LIVESTOCK FEEDLOTS

The annual cost of cattle feed consumed by birds at Denver, Colorado, feedlots has been estimated as $84 per thousand starlings and $2 per thousand redwings (*Agelaius phoeniceus*) (Besser *et al.*, 1968). Feedlot owners are now able to purchase an effective pelletized bait for use in these situations. This bait, commonly known as DRC-1339 (3-chloro-p-toluidine hydrochloride), has been developed as a starling toxicant by the U.S. government (De Cino *et al.*, 1966). When tested on a wintering population of about a quarter-million starlings that occupied a single roost and fed at more than 250 cattle feedlots in the vicinity of Denver, Colorado, the starling population was reduced substantially, probably by more than 60%. Similarly, a flock of 1800 starlings, frequenting a turkey farm in Utah, was reduced 93% in four days, following baiting with DRC-1339 (Royall *et al.*, 1967).

DAMAGE TO FOREST TREES AND UTILITY POLES

A recent five-year study of the feeding habits of the Yellow-bellied Sapsucker (*Sphyrapicus varius varius* L.) quantifies damage done to New England forest trees (Rushmore, 1969). Through drilling holes to consume sap, the bird damages 51 tree species; intensive drilling can kill a small tree in a single season, although two or more years of drilling are usually needed to kill trees larger than 8 inches in diameter at breast height. Earlier studies are cited in which sapsucker damage was estimated as at least 10% for trees within sapsucker winter ranges. In some areas, more than 90% of the trees sustained some damage; 174 tree species were involved. In timber trees, the resulting damage to wood products is known as 'bird peck'. Another type of damage is wood discoloration resulting from sapsucker wounds. Total economic losses are not known but may be considerable, since sapsuckers occur seasonally throughout all the tree-producing areas of this continent.

A more apparent problem exists when woodpeckers (*Picidae*) damage utility poles. Some utility companies have found this damage to be more costly than their losses caused by lightning, wind, and ice (Anon., 1969). The Pileated Woodpecker (*Dryocopus pileatus*), which is about 15 inches long, makes nest cavities in poles which may leave only a one-inch shell of wood around the cavity. Several smaller kinds of woodpeckers do similar damage. They seem to prefer newly creosoted poles over weathered poles and often select utility poles for nesting sites even though the surrounding area offers numerous dead trees. These newly treated poles serve as death traps for the young birds, however, as few eggs hatch and all observed young have died within three days of hatching. In controlled tests, embryos in eggs incubated on creosoted chips usually died within a week. Chemical repellents, bright paints, and mechanical scaring devices have been tried, but the birds became accustomed to them. Hard rubber coatings on poles have shown some promise, but a satisfactory bonding material has not been developed. The best preventative is to wrap one-half-inch mesh hardware cloth around a pole before it is put into service. The 1969 cost for wire was $35 per pole.

BIRDS AND AIRCRAFT

In a most general sense, the problems of bird interaction with aircraft may be considered as biodeterioration, using the definition given in an earlier Symposium (Van Der Kerk, 1972). Bird-aircraft collisions or near-misses are of such widespread concern as to prompt a recent World Conference on Bird Hazards to Aircraft, sponsored by the National Research Council of Canada. Representatives from 21 countries participated. The proceedings of the conference have been published and provide a most complete coverage of the subject (National Research Council of Canada, 1969). The U.S. government has published a concise summary of this problem and remedial measures (Aldrich *et al.*, 1961).

BIRDS AND DISEASE

I wish only to mention briefly the topic of wild birds as vectors of diseases affecting humans and our domesticated animals and plants. An excellent summary of available knowledge about infectious and parasitic diseases of wild birds is available (Davis *et al.*, 1971). Since a number of viral, bacterial, rickettsial, and mycotic diseases as well as parasitic infections of birds may be sufficiently damaging, from our point of view, they might well be considered biodeterioration factors. For example, of current interest in the United States are attempts to prevent further outbreaks of a strain of sleeping sickness that killed 11 persons here during July of this year. There is no known cure for the virus, which is carried from birds to man by mosquitoes. The disease has reached epidemic proportions in Mississippi, with more than 100 cases of the St. Louis strain of the disease either confirmed or suspected. Control efforts presently emphasize mosquito control through drainage and insecticide application. The common pigeon (*Columba livia*) has long been considered a bird having significant importance in public health problems and numerous carefully documented accounts of histoplasmosis epidemics associated with pigeons have been published (Scott, 1961). Examples of the relationships between wild birds and domestic animals include instances in which sparrows have been implicated in transmission of hog cholera (Alboiu, 1965), ravens in transmission of bovine fibrosis (Dennis, 1967), starlings in transmission of histoplasmosis, and wild birds, generally, in transmission of *Salmonella* (Boev and Stamatov, 1967). The relationship of birds to disease transmission in plants is a relatively unexplored area of science; considering the highly mobile and gregarious nature of some species which have adapted to living in close association with man, this might prove to be a productive area for intensified research.

PROBLEMS IN COMBINATION

Thus far, we have considered discrete bird problems, but often they occur in combinations. A recent example is provided by the activities of five million blackbirds which shared a common roost in 30 acres of pine trees in Christian County, Kentucky. In February of this year, *Time* magazine quoted one of the mayors of the area as saying 'This is a pestilence and a scourge. Farmers are in the fields with shotguns, cattle and hogs are driven from feedlots, children's slides are covered with bird droppings.' Bird damage to the area was then estimated at $2·6 million. Since bird flights affected the U.S. Army's flight schedule, the Army expended $20 000 to complete an environmental impact statement and obtained a court ruling to allow treatment of the roost with a surfactant solution, as previously described. This problem was of interest to local citizens, farmers, politicians, military personnel, the Federal Council on Environmental Quality, the Humane Society of the United States, the Society for Animal Rights, Citizens for Animals, and the United States Courts, as well as attracting national news coverage.

CONCLUSION

From this brief presentation, I hope you have gained a general overview of some current bird damage problems and will agree that, in some situations, birds may be agents of biodeterioration. The development of new control measures must take into account the fact that birds may also be highly beneficial. They are of inestimable positive value, economically and aesthetically. They are vital components of fragile ecosystems, and, as such, demand our best efforts in protection. In the future, the balancing of all interests will require increasingly careful assessments of damage, rational approaches to the solution of problems, and the cooperation of scientists of many disciplines in developing more highly selective control techniques.

REFERENCES

Alboiu, M. (1965). Studies of the possibilities of transmitting of hog cholera virus by sparrows. *Rum. Inst. Cercet. Vet. Bioprep. 'Pasteur' Lucrarile*, **2** (1), 105–109.

Aldrich, J. W., Robbins, C. S., and Dykstra, W. W. (1961). Bird hazard to aircraft. U.S.D.I. Wildl. Leafl. 429. Washington, D.C., 10 pp.

Anon. (1969). Stink won't stop 'em—they like it. Forest Research News for the Midsouth, South. For. Exp. Stn. for Serv., USDA, pp. 1–2, 4.

Besser, J. F., DeGrazio, J. W., and Guarino, J. L. (1968). Costs of wintering starlings and red-winged blackbirds at feedlots. *J. Wildl. Manage.*, **32** (1), 179–180.

Boev, B., and Stamatov, T. (1967). Wild birds as Salmonella carriers. *Vet. Med. Nauk.*, **4** (1), 19–24.

Bremond, J. C., Gramet, P.H., Brough, T., and Wright, E. N. (1968). A comparison of some broadcasting equipments and recorded distress calls for scaring birds. *J. Appl. Ecol.*, **5**, 521–529.

Brown, R. G. B. (1974). Bird damage to the fruit crops in the Niagara Peninsula. Can. Wildl. Serv. Rep., Ser. No. 27, Ottawa, 57 pp.

Caslick, J. W. (1968). Bird damage control. *Cons. Circ.*, **6** (3), 1–3, Cornell University, Ithaca, N.Y.

Crook, J. H., and Ward, P. (1968). The *Quelea* problem in Africa. In *The Problems of Birds as Pests*, pp. 211–229 (R. K. Murton and E. N. Wright, eds.), Academic Press, London.

Davis, J. W., Anderson, R. C., Karstad, L., and Trainer, D. O. (eds.) (1971). *Infectious and Parasitic Diseases of Wild Birds*. Iowa State University Press, Ames, 344 pp.

De Cino, T. J., Cunningham, D. J., and Schafer, E. W. (1966). Toxicity of DRC-1339 to starlings. *J. Wildl. Manage.*, **30** (2), 249–253.

Dennis, S. M. (1967). The possible role of the raven in the transmission of bovine fibrosis. *Austral. Vet. J.*, **43** (2), 45–48.

Fitzwater, W. D. (1971). Pigeon control. Univ. Calif. Agr. Ext. Anim. Brief No. 15, 2 pp.

Miller, J. W. (1975). Much ado about starlings. *Nat. Hist.*, **84** (7), 38–45.

National Research Council of Canada (1969). Proceedings of the World Conference on Bird Hazards to Aircraft. Ottawa, Ontario, 542 pp.

Pardee, W. (1974). Lessons from '74. *Am. Agr.*, December, 4.

Pearson, E. W., Skon, E. R., and Corner, G. W. (1967). Dispersal of urban roosts with recorded starling distress calls. *J. Wildl. Manage.*, **31** (3), 502–506.

Royall, W. C., Jr., DeCino, T. J., and Besser, J. F. (1967). Reduction of a starling population at a turkey farm. *Poultry Sci.*, **46** (6), 1494–1495.

Rushmore, F. M. (1969). Sapsucker damage varies with tree species and seasons. USDA For. Serv. Res. Pap. NE-136. Northeast, For-Exp. Stn., Upper Darby, Pa., 19 pp.

Scott, H. G. (1961). Pigeons, public health importance and control. U.S. Dept. of Health, Education & Welfare, Public Health Ser., Atlanta, Ga., 18 pp.

Stiles, D. (1966). Traps reduce starlings 33 percent in Syracuse. *Pest Control*, February, 16–19.

Van Der Kerk, G. J. M. (1972). The chemical approach to biodeterioration prevention: retrospects and prospects. In *Biodeterioration of Materials*, Vol. 2 (A. H. Walters and E. H. Hueck-van der Plas, eds.), Applied Science Publishers, London, 740 pp.

NEW PERSPECTIVES IN RODENT AND MAMMAL CONTROL

REX E. MARSH and WALTER E. HOWARD

Division of Wildlife and Fisheries Biology, University of California, Davis, California 95616, U.S.A.

Summary

Recent and anticipated improvements in the control of rodents and other mammals are reviewed, including genetic control of rats, population genetic studies of rats, chemosterilants, aversive conditioning of rodents and coyotes to avoidance of their normal diets, audible and ultrasonic sound to repel rodents, a new acute rodenticide (RH-787, Vacor®) for rats and mice, encapsulated rodenticides, the anticoagulant difenacoum for resistant rats, and vitamin D_2 (calciferol) as a rodenticide and, along with the chloro analogue of vitamin K_1, as an additive to warfarin for anticoagulant-resistant rats. A brief analysis is given of inherent obstacles to research and development of chemicals and the methodology of controlling vertebrates due to improper federal structuring of this responsibility and the costly registration problems.

INTRODUCTION

Control of population densities of field rodents, moles, bats, rabbits, coyotes, deer, and all other mammalian species except commensal rats and mice is often an emotionally charged subject. The reason is that (1) mammals are warm-blooded and express some emotions and behavioural traits that people can identify with; (2) their control, in contrast to insect pests, involves a methodology that is not very sophisticated and hence not always highly selective; (3) most solutions to vertebrate pest problems are based more on trial and error than on extensive research; (4) the administrative base of field vertebrate pest control and research is improperly structured at the federal level for maximum and responsible achievements; and (5) few good economic studies have been made that document the magnitude of the losses caused by mammalian pests and thus justify reasonable control measures.

RECENT ADVANCEMENTS IN MAMMAL CONTROL

Since rodents cause more serious economic losses to man's crops and possessions than do any other group of vertebrates, their control is emphasized herein. Many recent or potential advances in mammal control are aimed at more effective and safer methods of reducing populations of our three introduced commensal species: the Norway rat (*Rattus norvegicus*), roof rat (*R. rattus*), and house mouse (*Mus musculus*). Fewer advances have been made in the control of our native pest rodents, so there is less to report for that group. In any case, no more than the highlights of some of the most significant developments from all the recent and proposed future advances in mammal control can be presented here.

To simplify information, certain products or trade names of products are specified herein, but no endorsements are implied. All pesticides should be used as directed on their labels and in accordance with the law, so no comments made or implied herein should be construed as circumventing such requirements.

We thank the World Health Organization, Geneva, Switzerland, the National Pest Control Association, and various chemical and pharmaceutical companies for their contributions to our own mammal-control research, which is one of the sources drawn on in assembling this paper.

Rodenticides

Early in 1972, we began laboratory evaluations of compounds belonging to a new series of relatively selective acute rodenticides being developed by the Rohm and Haas Co. Our earlier studies of the experimental compounds coded RH-945 and RH-908 were later expanded to include RH-787. Little research has yet been completed on RH-999, the newest analogue received in our laboratory.

Some early compounds of the series were carbamates, but the one selected for initial marketing is a urea compound, *N*-3-pyridylmethyl-*N'*-*p*-nitrophenyl urea, originally coded as RH-787 and assigned the trade name Vacor® as a ready-to-use rodent bait. Marketed in the United States by Whitmoyer Laboratories as Vacor RatKiller, the product is distributed in small packets for placement to control Norway rats (*Rattus norvegicus*), roof rats (*R. rattus*), and house mice (*Mus musculus*). At 2% on good-quality baits, this unique compound gives remarkable results without prebaiting. Reports indicate effectiveness against anticoagulant-resistant rats. Prepared baits have a relatively high margin of safety for domestic animals, and secondary poisoning seems ruled out as a hazard. A tracking powder containing 10% RH-787, presently registered by the Environmental Protection Agency in the United States, will presumably be marketed later.

RH-787 probably represents one of the most significant breakthroughs in the development of toxic rodenticides since the introduction of warfarin and other anticoagulant rodenticides that followed. RH-787, as indicated, is only one of a series of promising rodenticides. Its relative selectivity and proven efficacy for the three species of common commensal rodents led to its choice over related compounds for marketing to the general public.

Susceptibility to RH-787 appears to vary widely between individual rodents, but that may largely reflect differences in rate of eating the bait, *i.e.* rate of intake. The total amount of bait consumed by rats is nearly always quite low. A 2% bait may be taken in such small amounts (less than 0·1 g) that, although lethal, it is sometimes immeasurable.

RH-787 acts relatively slowly, usually taking a minimum of several hours to kill rats, with some deaths delayed for several days. It reportedly kills by inhibiting niacinamide metabolism, and the rats eventually die from paralysis and pulmonary arrest (Chemical and Engineering News, 1975). We know of no data published by Rohm and Haas researchers on the complexities of the mode of action. Our laboratory studies do not indicate that RH-787 is cumulative.

RH-787 offers a relatively wide margin of safety between target and nontarget animals. The LD_{50} was found to be 4·75 mg/kg for Norway rats, 710 mg/kg for chickens, >1780 mg/kg for pigeons, >500 mg/kg for dogs, and 2000–4000 mg/kg for rhesus monkeys (Peardon, 1974a,b).

Tests on the rhesus monkey and the baboon (Peardon, pers. comm.) showed them to be quite resistant to RH-787. Such results on primates would normally be indicative of man's resistance to a compound; however, there are recent implications of RH-787 in several human deaths in Asia. Certain mycotoxins from contaminated rice are believed implicated along with RH-787 in the human fatalities. Until specific information is available, RH-787 (Vacor) should be considered hazardous.

We are also very encouraged by laboratory and field results with RH-908, of the same series of compounds. Its range of toxicity and efficacy includes many rodents which are not susceptible to RH-787. On a mg/kg basis it is considerably more toxic to rats and mice than is RH-787, but it also appears more toxic to some nontarget species. Although more toxic, it shows a substantial degree of selectivity. If used at low concentrations in appropriate baits, the relative margin of safety to nontarget species should then make this experimental compound environmentally most acceptable. We view this compound as having great potential for several of the most important field rodents in California, particularly pocket gophers (*Thomomys bottae*) and ground squirrels (*Spermophilus beecheyi*), and possibly a suitable replacement for such rodenticides as strychnine and sodium fluoroacetate (1080).

The species-specificity of norbormide, 5-(α-hydroxy-α-2-pyridylbenzyl)-7-(α-2-pyridylbenzylidene)-5-norbornene-2,3-dicarboximide, once marketed by Tavolek Laboratories, exceeds that of all other rodenticides, but this greatly limits the marketing potential. Norbormide, a toxicant specific to the genus *Rattus*, was originally marketed under the trade names Raticate® and Shoxin® for the control of Norway and roof rats. The compound was considerably more toxic to Norway rats than to roof rats, and almost innocuous to other species. Unfortunately, norbormide was not well accepted in baits by rats, which reduced its efficacy. Additionally, sublethal doses brought about a tolerance to the compound.

In England, alphachloralose (Alphakil®) has gained a substantial place in recent years as a rodenticide against house mice. Alphachloralose is a narcotic type of drug which is quite lethal to house mice at ambient temperatures below 60°F (Brooks, 1973). It has not been registered as a rodenticide in the United States and, because temperature influences efficacy, it is most suitable in colder climates.

The rodenticide red squill is probably one of the oldest rodenticides still in use, having been recommended for rodent control as early as the 13th century. The red squill we knew 30 years ago has been drastically improved, and the new stabilized scilliroside glycoside, sometimes referred to as stabilized red squill (SRS), has been under evaluation in this country for a number of years. McLaughlin Gormley King Co. is presently pursuing registration in this country.

We have found this new stabilized scilliroside glycoside to be considerably more toxic than the older fortified red squill, and in fact sufficiently toxic to make an effective tracking powder for house-mouse control, though house mice are less susceptible than are rats to the compound. Results have been very encouraging in Japan, where it is used quite extensively for Norway rat control.

Gophacide®, developed by Chemagro Corporation, is named for its excellence in controlling pocket gophers. This highly toxic organic phosphate compound, 0-0-bis-(*p*-chlorophenyl)-acetimidoyl-phosphoramidothiolate, presents little hazard to nontarget species when formulated at low concentrations in bait and applied underground for pocket gophers. Although effective on other pest species, its toxic characteristics provide little advantage over other rodenticides. Recent reports indicate that Gophacide will be phased out by the marketing firm. That is most unfortunate, for it could effectively replace strychnine, currently used extensively to control pocket gophers.

There is a new anticoagulant with the proposed common name of difenacoum, 3-(3-*p*-diphenylyl-1,2,3,4-tetrahydronaphth-1-yl)-4-hydroxycoumarin, developed in England and manufactured there by the Ward Blenkinsop Chemical Manufacturers and marketed by Sorex Ltd. of London. This new anticoagulant, sold under the trade name of Neosorexa®, has proved effective against anticoagulant-resistant Norway rats, and research is under way to determine its efficacy on roof rats and house mice. We have hopes that this compound will be registered and marketed in the United States.

This same firm earlier developed a product called Sorexa CR®, which also was toxic to anticoagulant-resistant rats (Rennison, 1974) and to house mice (Rowe *et al.*, 1974). That product, a solution of calciferol (vitamin D_2) in corn oil, is used to formulate a bait containing 0.1% calciferol and 0·025% warfarin. Calciferol promotes intestinal absorption of calcium and resorption of bone calcium. Toxic doses produce hypercalcemia, osteomalacia, and metastatic calcification of the blood vessels. Excessive doses of calciferol alone can be toxic to rats (Greaves *et al.*, 1974).

In this country Bell Laboratories has embarked on efficacy studies of vitamin D_3, which holds the same potential as a rodenticide as does calciferol. In some respects it is superior because it is less expensive and more stable. Tests in our laboratory of vitamin D_3 as a tracking powder for house mice have been most promising to date.

The chloro analogue of vitamin K_1, used either alone or in combination with warfarin, has also been found to be an effective rodenticide for anticoagulant-resistant rats. This vitamin K antagonist (2-chloro-3-phytyl-1,4-naphthoquinone) and its anticoagulant action were reported by Suttie (1973) of the University of Wisconsin. We have no information on the current status of the chloro analogue of vitamin K_1.

The anticoagulant chlorophacinone, the accepted common name for 2-[(*p*-chlorophenyl)phenylacetyl]-1,3-indandione), sold under the trade name of RoZol®, is the most recent anticoagulant to appear on the market in this country, although it has been available in Europe for some time. This compound is closely related to diphacinone. Chlorophacinone is available as an oil-soluble concentrate and as a dry concentrate for bait formulation. Considerable credit goes to Chempar Chemical Co. for developing chlorophacinone for the control of several mammal pests of agriculture and forestry (deer mice, *Peromyscus* spp.; pine mice, *Pitymys pinetorum*; and moles, *Scaponus* spp.).

In recent years anticoagulants have been used more and more for the control of field rodents, a trend that will probably continue. Most encouraging is the recent interest of Velsicol Chemical Co. in marketing a ready-to-use anticoagulant bait called Ramik® (active ingredient diphacinone) for agricultural rodents such as ground squirrels and meadow mice.

Researchers in Japan have been exploring the synergistic toxic action between coumarin and indandione anticoagulant derivatives and thallium salts (Kusano, 1969). They reported increased action of the combinations over either material used alone. Whether the effect is truly synergistic or merely additive seems unclear. More recently, they have synthesized a new anticoagulant, 2,4-dioxo-3-(α-acetonylbenzl)-dihydrobenzopyran TI-salt, which is chemically a combination of thallium and warfarin (Kusano and Masubuchi, 1973). Although their research is academically very interesting, it is of no practical value in the United States, where thallium is no longer legal as a rodenticide. Japanese researchers are actively pursuing new rodenticides, possibly more so than in this country. For example, methylene-bis(1-thiosemicarbazide), a new potential rodenticide, was discovered when many synthesized thiourea derivatives were screened (Tokumitsu *et al.*, 1973).

Genetic resistance to rodenticides

It was in 1958 that warfarin resistance was first found, in Scotland (Boyle, 1960). According to Drummond (1971), resistance arises as a result of genetic mutation or recombination. Rats that are resistant to warfarin show a cross resistance, at least to some degree, to most but not all other anticoagulants. A partial exception is coumatertalyl, which is effective against some warfarin-resistant rats (Drummond, 1971). Resistance has appeared in both Norway and roof rats, as well as in house mice. Warfarin anticoagulant resistance in Norway rats and house mice is becoming more widespread in Europe, India, and the United States.

Much effort is being expended on resistant rodents by two U.S. laboratories —one operated by the New York State Department of Health (Brooks and Bowerman, 1974) and the other at Bowling Green State University in Ohio (Jackson *et al.*, 1975).

Webb and Horsfall (1967) indicated evidence of inherited resistance to endrin by the pine mouse (*Pitymys pinetorum*). DDT resistance by cotton rats (*Sigmodon h. hispidus*) has been reported (Ferguson *et al.*, 1965). Ozburn and Morrison (1964) produced a 1·4-fold DDT-tolerant strain of laboratory mice in nine generations, yet Guthrie *et al.* (1971) found little change in the acute toxicity of four insecticides to white laboratory mice after 12–14 generations of selection.

Anticoagulant and multiple-feeding rodenticides not yet registered in the United States that show promise against anticoagulant-resistant rodents include 'novel' 4-hydroxycoumarins (Hadler and Shadbolt, 1975) and, already mentioned Sorexa CR® (Calciferol mixed with warfarin), chloro analogue of vitamin K_1, and difenacoum (Neosorexa®).

Genetic resistance is not as great for the acute rodenticides as for some of the anticoagulant rodenticides. Even so, acute rodenticides may lose efficacy at a much lower level of genetic resistance or acquired tolerance. To illustrate the propensity rodents have for developing genetic resistance with repeated use of acute rodenticides, a test was conducted with Long-Evans laboratory rats and sodium monofluoroacetate (1080). After five generations of non-intensive selective breeding, the approximate LD_{50} changed from 2 mg/kg of 1080 for the parents to 3·5 mg/kg for the F_4 generation (Howard *et al.*, 1973).

Encapsulated rodenticides

Bait aversion or shyness to anticoagulants, although apparent for some time, has recently received more attention. Bowerman and Brooks (1972) published an article on contaminants in anticoagulants and gave examples of how impurities can influence bait acceptance. Krishnamurthy *et al.* (1971) also reported on the effects of warfarin impurities on acceptance. To overcome problems of bait acceptance (toxicant acceptance) and rapid occurrence of bait and/or toxicant shyness, numerous attempts have been made to microencapsulate rodenticides to delay the onset of toxic symptoms and mask unfavourable tastes and/or odours. A research group in England was among the first to explore encapsulation of rodenticides (Greaves *et al.*, 1968).

Rodenticide bait formulations containing encapsulated technical warfarin were found to be 3–10 times as acceptable to rats in laboratory tests as identical baits containing unencapsulated warfarin (Abrams and Hinkes, 1974). This encapsulated warfarin, presently available under the trade name TOX-HID®, is a welcome contribution to improving bait acceptance. Other rodenticide encapsulation studies, with varying degrees of success, have been reported: Chyun (1973), who encapsulated zinc phosphide; Cornwell (1971), who encapsulated warfarin, zinc phosphide, norbormide, and alphachloralose; and Ericsson *et al.* (1971), who reported on encapsulating the male rat sterilant alphachlorohydrin. Fairly recent research endeavours indicate that encapsulating norbormide may substantially improve its acceptance, and hence its efficacy.

Other methods of increasing the acceptance of certain toxicants have been explored. Zinc phosphide can be washed in water to remove some of the odour-emitting impurities, thereby increasing acceptance by rats while reducing the odour of phosphine, objectionable to many people (Hilton and Robison, 1974). Improving methods of warfarin production and maintaining high standards of quality control have been the approaches taken by the largest bait formulator in the United States to assure that only the most acceptable lots of warfarin are used. Quality control involves frequent bioassays for acceptance by rodents.

Lethal tracking powders

Tracking powders containing lethal rodenticides are formulated for dusting lightly in areas where mice or rats will travel over the dusted spots and pick up the powder on their feet and fur. The toxicant is ingested during normal grooming (*i.e.* licking of the feet and fur), a natural behavioural trait with most rodents. Tracking powders can be particularly useful in controlling house mice because their sporadic feeding and nibbling patterns make them difficult to control even with the best of baits. The tracking-powder approach is effective for rats also, especially where other food makes bait acceptance difficult to achieve.

Prior to its ban, DDT (10–50%) tracking powder was used quite extensively by professional pest-control operators (PCO's) for the control of house mice in structures. House mice, unlike rats, are quite susceptible to DDT; with its ban, PCO's were almost immediately faced with uncontrollable house-mouse infestations. At that time the only other tracking powder registered with EPA for house mice was a 2·18% PMP (Valone).

With support from the National Pest Control Association, we began research toward finding tracking powders to replace DDT. We have been involved, to various degrees, in testing almost all the newer tracking powders appearing on the market. We have assisted in developing data on chlorophacinone, diphacinone, Pival®, PMP, and warfarin tracking powders. Most recently, we have found that high concentrations of sulphaquinoxaline added to warfarin tracking powder will significantly increase the control of house mice. The proportion of warfarin to sulphaquinoxaline is far greater than what is commonly found in the rodenticide Prolin®. We theorize that because of this higher sulphaquinoxaline concentration, the intestinal bacterial flora which is responsible for producing vitamin K is more rapidly decreased and hence increases mortality in house mice (unpublished data). Unfortunately, EPA is insisting on extensive field efficacy evaluations, and the expenses of such tests cannot be justified by the market potential. A need for such field efficacy studies seems questionable in view of existing knowledge of the two compounds, the present marketing of warfarin as a tracking powder, and the long-established use of warfarin and sulphaquinoxaline in combination in rodent baits.

RH-787 at a 10% concentration has received EPA registration as a tracking powder. The new stabilized red squill has proved reasonably effective as a house-mouse tracking powder at relatively low concentrations (0·1%). Both 10% zinc phosphide (Marsh, 1972) and 50% lindane tracking powders have proved highly effective for house-mouse control under laboratory conditions, but neither of these are currently registered.

Repellents and attractants

Repellents detected by taste are sometimes practical if they can be coated onto, or impregnated into, materials to be protected. These may be particularly beneficial if they create some physiological ill feeling which the animal then associates with its initial contact with the repellent. Such aversive conditioners, with taste and/or odour serving as the cues used for future avoidance, have worked well in repelling deer mice (*Peromyscus* spp.) from conifer seed. Mestranol is currently being evaluated for this purpose by the Fish and Wildlife Service, Forest-Animal Unit in Olympia, Washington (Lindsey *et al.*, 1974). We have also shown that alpha-naphthylthiourea averts deer mice against eating conifer seeds (Marsh *et al.*, 1974).

Potential chemical rodent repellents have been studied for a long time, though relatively few have been marketed. Probably the most recent repellents developed for rodents are R-55® (tert-butyl-dimethyltrithioperoxycarbamate) and BioMet 12®, which is tri-*n*-butyltin chloride. Both can be used to coat electric cables to protect them against rodent damage, but R-55 is also registered as a temporary barrier and space repellent for commensal rodents.

The most recent achievements in deer repellents are putrified fish fractions, which have proved highly effective in preventing damage to forest trees. For ease in manufacture, a chicken-egg-based repellent (EV) is now replacing the putrified fish fractions with equal repelling qualities (Rochelle *et al.*, 1974).

Current research on both rodent and coyote pheromones in order to make odour lures (attractants) may hold promise in the area of attractants. To our knowledge, no rodent attractants are presently registered.

Frightening sounds (ultrasonic devices)

Many people still hope that commensal rodents might be repelled by units which generate ultrasonics (*e.g.* 20–25 kHz) at high intensities (*e.g.* 100–115 dB at 1 m), with such pulsed sound being intermittent and containing pure sine waves with various harmonics. However, our research with these and other sound-producing devices has always produced essentially negative results (unpublished; Sprock *et al.*, 1967). A pronounced disadvantage to ultrasonics is that they are highly directional, failing to penetrate behind stored items, and attenuate rapidly, failing to carry far. Additionally, rodents quickly become accustomed even to sounds which were frightening when first heard (Marsh *et al.*, 1962). We taped the sounds made when a predator (striped skunk, *Mephitis mephitis*) attacked, killed, and then ate a rat, but the sounds did not deter rats from voluntarily and promptly entering a sound chamber where the tape was being played.

Major breakthroughs will be needed to make this approach to rodent control fruitful, and we have some hope that this may some day come about.

Chemosterilants

Despite considerable research, chemosterilants for rodent control are still in the experimental category in this country, and none has been registered (Marsh and Howard, 1973; Marsh, 1973).

Over the past 8 years we have evaluated some 16 potential rodent chemosterilants on Norway rats. These represent both nonsteroid and steroid compounds. Of the compounds tested in our laboratory, as well as by others, oestrogens as a group hold the greatest promise. Their major drawback is one of acceptance of sufficient amounts of the oestrogen in bait to produce the desired effect on reproduction. Generally speaking, oestrogens have a greater physiological effect on females than males, and the degree of efficacy (*i.e.* duration of inhibited reproduction) is dose-dependent.

Mestranol (17α-ethynylestradiol-3-methyl ether) is the first potential rodent chemosterilant to receive much attention in this country (Marsh and Howard, 1969; Howard and Marsh, 1969). A number of other potent oestrogens of potential value has been reported on since then, and of these the 3-cyclopentyl ether of 17α-hexa-1′,3′-diynyl oestra-1,3,5(10)-trien-17β-ol (BDH 10131) is without a doubt the most potent of any we have evaluated, with some female rats remaining infertile for 11 months after treatment. Comprehensive studies of this compound have been carried out in England (Kendle *et al.*, 1973).

A highly effective male sterilant for Norway rats, studied originally by Upjohn Co., is U-5897 (3-chloro-1,2-propanediol). In mature Norway rats this alpha-chlorohydrin compound causes epididymal lesions, creating permanent sterility in a single feeding (Ericsson, 1970). Baits prepared with 1% alpha-chlorohydrin are lethal to a high percentage of Norway rats under some field conditions and thus have the advantages of both a lethal rodenticide and a male chemosterilant, depending on the dose consumed. This compound has received renewed interest as a lethal rodenticide in the past few years (Andrews *et al.*, 1974).

Another chemosterilant, (1-[2-[*p*-[α-(*p*-methoxyphenyl)-β-nitrostyryl]-phenoxy]ethyl] pyrrolidine monocitrate), which we are currently investigating in great depth, appears to show considerable promise, even though it must be offered on a nearly continuous basis. This experimental compound, coded CI-628 by Parke, Davis, and Co., is a nonsteroidal oestrogen antagonist which is accepted in baits much better than any other chemosterilant previously tested (Gwynn, 1972). We envision it as an additive to be incorporated into anticoagulant baits to prevent anticoagulant resistance from developing in rodent populations, or from such genetic resistance being further spread in populations where it is already developed.

Genetic control

A proportion of zygotes in all animal populations carries genetic factors which render them unable to survive the range of environmental conditions they are likely to encounter, and recessive lethality probably accounts for the bulk of genetic death in natural populations (Whitten and Foster, 1975). The sterile-insect release method, initially so successful in eradicating the screw-worm (*Cochliomyia hominivorax*) from Curacao in 1954 and from Florida in 1959, is not considered a highly practical method of controlling mammals because the released sterile individuals would add to the pest problem and sterile mammals, unlike insects, live much too long.

A number of genetic syndromes has been recognized which affect reproduction in rat and mouse colonies in the laboratory. A colony of genetically developed rats possessing various genetic syndromes which affect fertility of Norway rats is or was maintained by Introgene Foundation at the University of Oklahoma. They investigated the feasibility of introducing such genetic factors into wild populations so as to set in motion a continuing negative effect on fertility. According to Glass (1974), it was demonstrated that genetically sterile male Norway rats introduced into a wild population reduced the reproductive performance of the wild females, but he further points out that 'it seems that the Introgene programme has proceeded as far as it can without massive capitalization'. It will be unfortunate, indeed, if this foundation must give up its research because of inadequate funding.

Since genetic control, even when developed to the practical stage, will not produce an immediate or complete reduction in rodent or other mammal populations, its potential value—like that of chemosterilants—will be limited to particular pest situations. The main difference would be that genetic control would be self-perpetuating to some degree (Marsh and Howard, 1973).

Control of predators

With reference to the control of mammals, the most controversial species today is the coyote (*Canis latrans*) (Howard, 1974a). Actually, the considerable research on coyote control is aimed largely at finding new control materials and improving methodology of reducing livestock losses (Balser, 1974). New innovations include sheep collars containing a cyanide preparation designed to poison depredating coyotes when they bite down on the neck of a sheep. Although this method appears very selective, designed to take only the offending predator, its shortcomings are numerous. In the western part of the country, aversive conditioning of coyotes so they will avoid sheep is receiving much attention (Gustavson *et al.*, 1974), but its practical value may be limited. Lithium chloride is one of the aversive agents being studied, and more recently we have been exploring the possibilities of barium carbonate for the same purpose. Also being studied in great depth are chemosterilants for reducing population densities, and coyote pheromones for attracting coyotes to baits or traps.

No new methodology breakthroughs have occurred in recent years that we are aware of to control other mammalian predators: bobcat, cougar, bear, skunk, fox, raccoon, etc.

If only the ratio of research funds to monetary losses was the same for rodent control as for predator research, rather substantial strides could be made in solving some of our rodent problems.

IMPROVEMENTS NEEDED

Research on more effective, safer, and more efficient ways of controlling rodents is grossly underfinanced, and the costs in developing and registering new control compounds are almost prohibitive. Technology for control of pest mammals needs sophistication. House mice are becoming a worse problem (Marsh and Howard, 1975), and in California, for example, roof rats are continuing to become a more serious pest—living in dense landscaping of freeways, homes, and landscaped industrial sites.

More research is needed on problem mammals as to their general biology, ecology, population dynamics, social structure, genetic resistance to pesticides, and behaviour related to smell, taste, vision, and hearing that might be utilized in control. Almost no good economic studies have been made of losses caused by various mammalian pests.

GOVERNMENT OBSTACLES TO BETTER CONTROL

Most basic research on mammal control would logically be expected to come from government agencies, agricultural colleges in land-grant universities, and other institutions of higher learning. Since the economic incentives are relatively minor, private industry may develop new trap designs, new bait formulations, and, rarely, a new rodenticide, but private industry certainly cannot afford to undertake basic ecological and biological studies on the principles and concepts of pest mammal control.

Vertebrate pest control (*i.e.* rodents, birds, predators, etc.) lacks sophistication today primarily because the U.S. Fish and Wildlife Service (F & WS) has, for all practical purposes, monopolized all but the public-health aspects of this responsibility at the federal level since 1939, when the Biological Survey of the U.S. Department of Agriculture was transferred to the Department of the Interior to become the Fish and Wildlife Service (Howard, 1974b). When animal control was transferred from USDA to USDI, it drastically reduced the possibility that land-grant universities would become involved in basic and applied research and extension of this information to those who desperately need it. For example, in California the state and county agricultural departments spend more than $1 500 000 per year on agricultural rodent control, yet the University of California Agricultural Experiment Station, which receives support from USDA (land-grant university support) and not USDI, in 1974–75 had no line-item funds devoted specifically to basic research on rodent control. Is it any wonder, then, that information on nonlethal methods of relieving damage and on how to reduce environmental hazards from rodenticides, avicides, and predacides is so meagre compared with that on insect control, weed control, animal health, and the like? In 1971 the National Science Foundation (NSF) informed the junior author that USDI would not permit NSF to fund research oriented to animal control, which they claimed could be done only by Interior's Wildlife

Research Laboratories. These wildlife laboratories have never been adequately financed for needed basic research on animal control. Many specialists in vertebrate pest problems believe that the federal responsibility for animal control should be returned to USDA, to allow a healthful check-and-balance arrangement between federal departments of Agriculture and Interior regarding the species in question and the most appropriate methods for relieving the damage they cause.

Another government obstacle to the development of safer and better methods of controlling mammals is the staggering time and expense in bringing new rodenticides and predacides on the market. This is primarily because of the very rigid requirements of the federal Environmental Protection Agency (EPA), some of which appear of questionable merit.

REFERENCES

Abrams, J., and Hinkes, T. M. (1974). Acceptability and performance of encapsulated warfarin. *Pest Control*, **42** (5), 14–16.

Andrews, R. V., Belknap, R. W., and Keenan, E. J. (1974). Demographic and endocrine responses of Norway rats to antifertility control. *J. Wildl. Manage.*, **38** (4), 868–874.

Balser, D. S. (1974). *A review of coyote control research*, Proc. Sixth Vertebrate Pest Conf., pp. 171–177 (W. V. Johnson, ed.), University of California, Davis.

Bowerman, A. M., and Brooks, J. E. (1972). What makes a palatable warfarin? *Pest Control*, **40** (2), 22, 28.

Boyle, C. M. (1960). Case of apparent resistance of *Rattus norvegicus* Berkenhout to anticoagulant poisons. *Nature*, **188**, 517.

Brooks, J. E. (1973). A review of commensal rodents and their control. *Critical Rev. in Environ. Control*, **3** (Issue 4), 405–453.

Brooks, J. E., and Bowerman, A. M. (1974). An analysis of the susceptibilities of several populations of *Rattus norvegicus* to warfarin. *J. Hyg., Camb.*, **73**, 401–407.

Chemical and Engineering News (1975). Rat poison does deed in only one feeding. March 24, 8.

Chyun, S. P. (1973). Studies on the ecology and control of *Rattus norvegicus* Berkenhout in Korea. Dong-Cuk University, Seoul. 30 pp.

Cornwell, P. B. (1971). Studies in microencapsulation of rodenticides, Proc. Fourth Vertebrate Pest Conf., pp. 83–97 (R. H. Dana, ed.), University of California, Davis.

Drummond, D. C. (1971). Warfarin-resistant rats—some practical aspects. *PANS*, **17**, 5–8.

Ericsson, R. J. (1970). Male antifertility compound: U-5897 as a rat chemosterilant. *J. Reprod. Fert.*, **22**, 213–222.

Ericsson, R. J., Downing, H. E., Marsh, R. E., and Howard, W. E. (1971). Bait acceptance by rats of microencapsulated male sterilant alpha-chlorohydrin. *J. Wildl. Manage.*, **35** (3), 573–576.

Ferguson, D. E., Callahan, R. L., and Cotton, W. D. (1965). Comparative DDT-tolerance of cotton rats from three areas differing in past exposure to pesticides. *J. Miss. Acad. Sci.*, **11**, 229–233.

Glass, B. P. (1974). The potential value of genetically sterile Norway rats in regulating wild populations, Proc. Sixth Vertebrate Pest Conf., pp. 49–54. (W. V. Johnson, ed.), University of California, Davis.

Greaves, J. H., Rowe, F. P., Redfern, R., and Ayres, P. (1968). Microencapsulation of rodenticides. *Nature*, **219** (5152), 402–403.

Greaves, J. H., Redfern, R., and King, R. E. (1974). Some properties of calciferol as a rodenticide. *J. Hyg., Camb.*, **73**, 341–351.

Gustavson, C. R., Garcia, J., Hankins, W. G., and Rusiniak, K. W. (1974). Coyote predation control by aversive conditioning. *Science*, **184** (4136), 581–583.

Guthrie, F. E., Monroe, R. J., and Abernathy, C. O. (1971). Response of the laboratory mouse to selection for resistance to insecticides. *Toxic. Appl. Pharmac.*, **12**, 229–241.

Gwynn, G. W. (1972). Field trial of a chemosterilant in wild Norway rats. *J. Wildl. Manage.*, **36**, 823–828.

Hadler, M. R., and Shadbolt, R. S. (1975). Novel 4-hydroxycoumarin anticoagulants active against resistant rats. *Nature*, **253** (5489), 275–277.

Hilton, H. W., and Robison, W. H. (1974). Fate of zinc phosphide and phosphine in the soil-water environment. *Agr. Food Chem.*, **20** (6), 1209–1213.

Howard, W. E. (1974a). The biology of predator control. Addison-Wesley Module in Biology No. 11, Cummings Publ. Co., 2727 Sand Hill Road, Menlo Park, California.

Howard, W. E. (1974b). Predator control: whose responsibility? *BioScience*, **24**, 359–363.

Howard, W. E., and Marsh, R. E. (1969). Mestranol as a reproductive inhibitor in rats and voles. *J. Wildl. Manage.*, **33**, 403–408.

Howard, W. E., and Marsh, R. E. (1974). Rodent control manual. *Pest Control*, **42** (8), D-U. Reprinted by authors as Rat Control Manual, 20 pp.

Howard, W. E., Marsh, R. E., and Palmateer, S. D. (1973). Selective breeding of rats for resistance to sodium monofluoroacetate. *J. Appl. Ecol.*, **10**, 731–736.

Jackson, W. B., Brooks, J. E., Bowerman, A. M., and Kaukeinen, D. E. (1975). Anticoagulant resistance in Norway rats as found in U.S. cities. *Pest Control*, **43** (4), 12–16.

Kendle, K. E., Lazarus, A., Rowe, F. P., Telford, J. M., and Vallance, D. K. (1973). Sterilization of rodent and other pests using a synthetic estrogen. *Nature*, **244** (5411), 105–108.

Krishnamurthy, R., Ramasivan, T., and Singh, D. P. (1971). Studies on rodents and their control: VII. Effects of impurities in warfarin on its acceptability and mortality to black rats (*Rattus rattus*). *Bull. Grain Technol.* (*India*), **9** (4), 252–256.

Kusano, T. (1969). Studies on the improved effectiveness of anticoagulative rodenticides on rodents. I. Synergistic toxic action between coumarin or indandione derivatives and thallium salts on mice. *J. Faculty Agr., Tottori Univ., Japan*, **5** (2), 15–52.

Kusano, T., and Masubuchi, S. (1973). Rodenticidal activity of new indirect anticoagulant, 2,4-dioxo-3-(α-acetonylbenzyl)-dihydrobenzopyran TI-salt. *J. Faculty Agr., Tottori Univ., Japan*, **8**, 43–57.

Lindsey, G. D., Anthony, R. M., and Evans, J. (1974). Mestranol as a repellent to protect Douglas-fir seed from deer mice. Proc. Sixth Vertebrate Pest Conf., pp. 272–279 (W. V. Johnson, ed.), University of California, Davis.

Marsh, R. E. (1972). Recent developments in tracking dusts. Proc. Rodent Control Conf., pp. 60–62 (J. E. Brooks, ed.), N.Y. State Dept. of Health, Glenn Falls, New York.

Marsh, R. E. (1973). Potential chemosterilants for controlling rats. *Beihefte Zeitschrift fur angewandte Zoologie*, **3**, 191–198.

Marsh, R. E., and Howard, W. E. (1969). Evaluation of mestranol as a reproductive inhibitor of Norway rats in garbage dumps. *J. Wildl. Manage.*, **33**, 133–138.

Marsh, R. E., and Howard, W. E. (1973). Prospects of chemosterilant and genetic control of rodents. *Bull. WHO*, **48**, 309–316.

Marsh, R. E., and Howard, W. E. (1975). House-mouse control manual. *Pest Control*, **43**. In press.

Marsh, B. T., Jackson, W. B., and Beck, J. R. (1962). Use of ultrasonics in elevator rat control. *Grain Age*, **3** (11), 27–31.

Marsh, R. E., Passof, P. C., and Howard, W. E. (1974). Anticoagulants and alpha-naphthylthiourea to protect conifer seeds. In *Wildlife and Forest Management in the Pacific Northwest*, pp. 75–83 (H. C. Black, ed.), School of Forestry, University of Oregon, Corvallis.
Ozburn, G. W., and Morrison, F. O. (1964). The selection of a DDT-tolerant strain of mice and some characteristics of that strain. *Can. J. Zool.*, **42**, 519–526.
Peardon, D. L. (1974a). A new series of selective rodenticides, Proc. Sixth Vertebrate Pest Conf., pp. 58–62 (W. V. Johnson, ed.), University of California, Davis.
Peardon, D. L. (1974b). RH-787: a new selective rodenticide. *Pest Control*, **42** (9), 14, 16, 18, and 27.
Rennison, B. D. (1974). Field trials of calciferol against warfarin resistant infestations of the Norway rat (*Rattus norvegicus* Berk.). *J. Hyg., Camb.*, **73**, 361–367.
Rochelle, J. A., Gauditz, I., Oita, K., and Oh, J. H. K. (1974). New developments in big-game repellents. In *Wildlife and Forest Management in the Pacific Northwest*, pp. 103–112 (H. C. Black, ed.), School of Forestry, University of Oregon, Corvallis.
Rowe, F. P., Smith, F. J., and Swinney, T. (1974). Field trials of calciferol combined with warfarin against wild house-mice (*Mus musculus* L.). *J. Hyg., Camb.*, **73**, 353–360.
Sprock, C. M., Howard, W. E., and Jacob, F. C. (1967). Sound as a deterrent to rats and mice. *J. Wildl. Manage.*, **31**, 729–741.
Suttie, J. W. (1973). Anticoagulant-resistant rats: possible control by the use of the chloro analog of vitamin K_1. *Science*, **180** (4087), 741–743.
Tokumitsu, I., Ogushi, K., Yamamoto, H., and Kusano, T. (1973). Studies on the rodenticidal activity of methylene-bis(1-thiosemicarbazide). *Botyu-Kaguku (Scientific Pest Control)*, **38** (4), 202–212.
Webb, R. E., and Horsefall, F. (1967). Endrin resistance in the pine mouse. *Science*, **156**, 1762.
Whitten, M. J., and Foster, G. G. (1975). Genetical methods of pest control. *Ann. Rev. Entomol.*, **20**, 461–476.

PROBLEMS OF MICROBIAL DETERIORATION OF MATERIALS IN EASTERN EUROPE

B. J. ZYSKA

Central Mining Institute, Katowice, Poland

Summary

In the past 30 years, the interest in the knowledge of microbial deterioration of materials has grown rapidly in Eastern Europe. The trends and main achievements in this field are discussed, based on results published in Czechoslovakia, the German Democratic Republic, Hungary, Poland, Romania, USSR, and Yugoslavia. The environment of deep mines is characterized and the progress in protection of materials used in mines is analysed.

INTRODUCTION

During the past 30 years, we have seen the high rate of flow of information on biological deterioration. At first, the valuable sources of information were two American abstract journals, which now are available only in a few libraries of various countries (*Prevention of Det. Abstr.*, *Environmental Effects*). Since 1960, the annual *Developments in Industrial Microbiology* (American Institute of Biological Sciences, and Plenum Press) has been a rich source of information. Since 1965, the *International Biodeterioration Bulletin* with the supplement *Biodeterioration Research Titles* and *Material und Organismen* (Becker, 1965) have also provided valuable information. From among the titles of research works we choose, often unconsciously, the papers to which we have easy access, written in a familiar language. Frequently, the results of interesting investigations escape our notice only because of language barriers.

At the suggestion of the organizers of the Third International Biodegradation Symposium, I shall discuss the problems on microbial deterioration in the countries of Eastern Europe, namely, the problems of the geographical territory with 300 million inhabitants. A distinctive feature of the past period in the countries of Eastern Europe was the dynamic growth in the field of research on the discussed problem, which resulted from the needs of national economies of these countries. The detailed discussion on all developments and the complete list of the references concerned would go beyond the framework of the present paper. I intend to focus my attention on those papers which lose none of their immediate interest and to group them according to the definite materials. This paper concerns the results of research on microbial deterioration obtained in Czechoslovakia, the German Democratic Republic, Hungary, Poland, Romania, USSR, and Yugoslavia.

PAPER, BOOKS, AND ARCHIVAL MATERIALS

Paper is among the materials which particularly easily deteriorate due to microorganisms. Therefore, the microbial processes of deterioration of paper are revealed beginning from the early stages of its production up to the moment of introducing the paper web to the drying cylinder, and then during the whole period of its utilization. Books and archival materials are special items among the final products of the paper industry in which the action of microorganisms results in inestimable cultural losses.

For many years the paper industry has met environmental requirements through the utilization of such scrap materials as waste paper for production of paper and cardboard. Protection of de-inked waste against microbial deterioration is of basic importance for the quality of paper products. The research carried out in Poland showed that the number of microorganisms in 1 g of de-inked waste paper is about 10^4. The dynamics of deterioration of the cellulose increases in the presence of increased microbial populations. However, the cellulose losses do not correlate with the waste paper losses. The growth of microorganisms reduces the whiteness of the waste paper stock, decreases its breaking length, burst, tear, and the number of double folds. *Stachybotrys atra* causes the highest reduction in the quality of the de-inked waste. Temperature of storage of waste paper stock is an important factor determining its technological properties.

It is essential to know the microflora in production lines and the biology of these microorganisms in order to control the biological slime during production of paper. The first information was available from the work of Pavlinova, USSR (Pavlinova, 1933). It is generally known that the most important microorganisms forming biological slime are the bacteria of the genus *Zoogloea*, bacteria-forming capsules and bacteria-secreting slime. From among 67 species of bacteria isolated from biological slime in paper mills in the USSR, 23 species were among the bacteria-forming capsules (Tuleuova, 1969a,b, 1973). *Cytophaga hutsinsonii* and *Sporocytophaga myxococcoides* are found most often among the bacteria-secreting slime (Pavlinova and Tuleuova, 1967). The microorganisms of biological slime in the environment at a temperature within 28° to 30°C are well known. In the newsprint mills where the temperature of paper pulp increases up to 50–60°C, conditions are considerably changed. The thermophilic bacteria of the species of *Bacillus subtilis*, *Bacillus mesentericus*, *B. mycoides*, *Micrococcus caseolyticus*, *Mic. currhiformis*, and *Aerobacter aerogenes* are a slime-producing factor. The optimum growth temperatures for a number of species belonging to the genus *Bacillus* are 45–50°C or 50–55°C. It should be stressed that in pulp and paper mills in the USSR the circulating water is ozonized to control biological slime (Matarucev, 1963; Sergeeva, 1973). The effectiveness of phenylmercuric acetate was tested from this viewpoint in newsprint mills in Romania, with particular regard to the problem of degree of adsorption of this compound on cellulose fibres and environmental pollution (Anton, 1959; Anton *et al.*, 1970).

In the interest of the culture of every nation it is necessary to protect books, archival materials, and historical documents against deterioration in order to enable them to survive as long as possible. During the Second World War, of 22·5 million volumes stored in Polish libraries, only 7·5 million volumes remained. After 1945, the basic difficult task was to prevent the further deterioration of collections of books. The comprehensive research on microorganisms of books and archival materials was carried out in Poland (Kowalik, 1952; Czerwinska *et al.*, 1953; Czerwinska, Kowalik, 1956; Kowalik, 1959; Kowalik *et al.*, 1962), in the USSR (Anchabadze, 1949; Rybakova, 1953; Nyuksha, 1956, 1961), and in Romania (Savulescu and

Lazar, 1970). In the microflora, fungi prevailed. The bacteria were isolated occasionally from archival materials when paper was stored under conditions of 80–90 % relative humidity. The role of Actinomycetes in the deterioration of paper of archival materials cannot be overlooked (Kowalik *et al.*, 1962). Various materials are used for book covers, so their microflora is very different. The growth of fungi on book covers depends also on bookbinding glues (Kowalik and Sadurska, 1956; Nyuksha, 1956, 1961). The microbiology of the air in libraries and archives was the subject of research carried out in Poland by Kowalik *et al.* (1962) in 1953–54. In order to protect books and archival materials against microorganisms, new methods for disinfecting the atmosphere of libraries (Kowalik *et al.*, 1962) and for preserving books and archival materials (Sadurska and Kowalik, 1969) have been developed.

TEXTILE MATERIALS

In 1967, in one of the textile factories of Yugoslavia, the discoloration of 'Malon' acrylonitrile fibres was found, which resulted in many claims. It was found that acrylonitrile fibre was attacked by fungi belonging to the genera *Aspergillus*, *Penicillium*, and *Stachybotrys* and by an unidentified strain. In comprehensive tests, the effect of the mixture of fungi and pure cultures *Chaetomium globosum*, *Myrothecium verrucaria*, *Mucor* sp., *Penicillium cyclopium*, *Trichoderma viride*, *Bacillus subtilis*, *B. mesentericus*, and *B. mycoides* on some properties of 'Malon' acrylonitrile fibre was determined. After 240 days, the fibre swelled under influence of bacteria and fungi. In the case of bacteria, the strength decreased by 12–14 %, and in the case of fungi, by 5–13 %, while the elongation decreased by 18–23 % and 15–32 %, respectively. Swelling of these fibres depended on a type of the microorganism used for testing. Microorganisms change the surface of fibres and their sorptive properties. Adsorption of *p*-nitrophenol with *n*-hexane amounted to 0·375 in polyester fibres. However, for example, under the influence of *Chaetomium globosum*, after 240 days, it decreased to 0·290.

LEATHER AND SHOE MATERIALS

The processes of microbial deterioration in leather and leather products have been known for a long time. However, new observations and research work are being carried out in this field. In Poland the case of black stains occurring on leathers was described, from which *Desulfovibrio desulfuricans* was isolated (Krassowski *et al.*, 1966). The research conducted by Orlita (1968) in Czechoslovakia showed that the distinctive feature of the group of fungi growing on leather products is a great part of fungi belonging to the genera *Aspergillus* and *Penicillium*, which attack the leather already in tanneries during its drying. This fact does not eliminate the problem of pathogenic microorganisms dangerous to man. According to the research carried out in Czechoslovakia (Orlita, 1971), a great number of microorganisms have remained on the bend-leather which is made of high-quality artificial leather of the Texacon 437 type during 16 months. Microbiocides added to shoe materials of Celstelen H, Celstelen I, and Colaten types ensure proper hygiene for shoe use. Orlita and Orlitova (1961) determined the conditions in which the added sodium salt of *p*-nitrophenol, pentachlorophenol, and *o*-phenylophenol protect leather products against moulds. In the USSR the problem of choosing the fungicides for leather was the subject of research carried out by Babakina (1964), while in Romania the same question was considered by Balif *et al.* (1971).

Of special importance are research works on the choice of microbiocides for shoe materials which are toxic for pathogenic microorganisms. The research papers presented during the Second Colloquium on Microbial Deterioration in Poland showed that shoe materials may be protected by 2,4,5-trichlorophenylacetic acid. With regard to the loss of its fungicidal properties, it is used as an aerosol to disinfect sports shoes (Orlita, 1971).

RUBBER AND VULCANIZED RUBBERS

Natural and synthetic rubbers and vulcanized rubbers were the subject of research carried out in many countries from the viewpoint of microbial deterioration. Such research work was also conducted in the countries already mentioned. In addition to Söhngen and Foll (1914) and Spence and van Niel (1936), Kalinenko (1938) is among the group of research workers who showed that natural rubber may be the only carbon source for Actinomycetes and certain bacteria. Within the period of 1952–58, thorough research work was carried out in the USSR (Shaposhnikov *et al.*, 1952a,b; Nette *et al.*, 1959). This showed that pure strains of certain bacteria may cause a 55% mass loss of natural rubber in 70 days, the pure strains of some Actinomycetes may cause a 43·2% mass loss in 45 days, and in the presence of fungi, the mass loss may be as high as 51·7%. The genus *Torulopsis* caused a 41% mass loss of natural rubber sheeting in 67 days.

It is worthwhile to mention the research on the liability of synthetic rubbers to the superficial growth of fungi carried out in the USSR (Dubok *et al.*, 1971), Romania (Lazar, 1973), Czechoslovakia (Blahnik and Zanova, 1963), and in the German Democratic Republic (Schwartz, 1963). It appears that most commercial types of synthetic isoprene rubber are more or less liable to fungal attack. The same is true of butadiene rubbers, butadiene styrene rubbers, and butylene rubbers. The behaviour of butadiene-acrylonitrile rubbers with respect to fungi depends on the type of rubber and the content of the antidegradants. Particular types of ethylene-propylene rubber behave variously with respect to fungi. All research work confirms the resistance of chloroprene rubbers.

Investigations on the microbial resistance of a number of ingredients of rubber mixtures carried out in the USSR (Dubok *et al.*, 1971) and in Romania (Lazar, 1973) describe the behaviour of carbon black, mineral fillers, vulcanizing substances, rubber accelerators, antidegradants, antiozonic wax, softeners, and blowing agents with respect to fungi. The results of these investigations allow us to prepare rubber mixtures more exactly for products in which a high microbial resistance is required. It should be stressed here that the vulcanization process does not improve this resistance but rather reduces it, which was shown in the research conducted by Dubok *et al.* (1971). In some cases, microbial evaluation should be continued for 90 weeks (Dubok *et al.*, 1971).

The choice of microbiocides for rubber used in cables, conductors, and telecommunication lines was the subject of Polish and Romanian research (Pitis, 1972; Pitis *et al.*, 1973; Zyska *et al.*, 1972). In the German Democratic Republic research (Schönborn *et al.*, 1968) was also carried out to find the microbiocides for rubber products which would eliminate the hazard of infecting people with pathogenic organisms, causing mycosis of the legs.

PLASTICS

The microbial deterioration of plastics has been undoubtedly the most important research problem of past years. Certainly, many difficulties must be overcome to learn the mechanism of this deterioration. I have no doubt that the research worker is always ready for the unexpected. As an interesting and original example, I should like to mention the results of tests, conducted by Demmer (1968) in West Germany, which showed that the fungus *Nigrospora sphaerica* caused in polyamide 6,6 deep surface pitting and holes 2 mm in diameter. Opinions on the microbial resistance of plastics may be based only upon tests of plastics of various kinds carried out by a uniform method. Therefore, it is advisable to take into account the results of research carried out in 1953–63 in the German Democratic Republic (Schwartz, 1963) in which more than 300 types of plastics, polymers, aids, and elastomers were tested from the microbiological viewpoint. In this investigation groups of bacteria isolated from plastics were used, and their growth was analysed in the presence of the tested material and compared with the dynamics of growth in media with paraffin and in media without a carbon source. Many series of tests were carried out over a year or even over 3 years. The results of these tests should be called to the attention of every research worker who undertakes the task of evaluation of various plastics as a possible carbon source for microorganisms.

Within the period of 1962–66 in Czechoslovakia, 214 various materials used in electronics and electrotechnics were tested for their resistance to superficial growth of fungi. On the basis of uniform numerical classification, the results of this investigation were compiled together with the results of similar work available till 1967. This is an original source of information providing data on resistances of a great number of plastics, elastomers, and coating materials, and also on the existing scatters of results. In the case of some plastics, the number of considered testing series is more than 30. The collected materials are of great importance for designers of all machines and electrical and electronic equipment which may be exposed to the periodic or continuous attack of fungi (Wasserbauer, 1967a,b).

Lazar, in his research on the resistance of plastics to fungi carried out in Romania in 1960–70, considered in a new way the problem of choosing the species of fungi for tests of plastics by laboratory methods. For a number of fungi used in these investigations, their enzymatic activity was tested according to the species of fungus, its age, media used in the tests, and according to the presence of microbiocides in plastics.

The need to take into account the enzymatic activity of fungi in tests on the resistance of plastics to these organisms is confirmed by the investigation of Kestelman and co-workers (1967, 1971, 1972), carried out in the USSR, the purpose of which was to explain the process of deterioration of plastics under conditions of microbiological synthesis of some organic compounds. The microbial resistance of polyethylene was the subject of the comprehensive research work conducted by Rudakova in the USSR (Bobkova *et al.*, 1971), which showed that it depends on a molecular mass, distribution of particular fractions, and on the kind and amount of the stabilizers used.

The problem of choosing microbiocides for certain plastics is often a subject of research work carried out in various countries. In Romania (Wexler *et al.*, 1971), a fungistatic compound was prepared for PVC used as an electro-insulating material. This compound is a product of synthesis of lead phenolates and lead salts of low-molecular, synthetic fatty acids. It has good processing properties in PVC mixtures while the final product protects plastic against fungi under difficult tropical conditions.

ELECTRO-INSULATING MATERIALS

Microbial processes occurring on the surface of electro-insulating materials are always an undesirable phenomenon and create many problems both for producers and users of electronic and electrical equipment. The research work carried out in Czechoslovakia (Wasserbauer, 1967a, 1969, 1971) shows that the growth of microorganisms on electro-insulating materials is a complex process, which calls for the revision of many opinions and also for the modification of methods for microbiological evaluation of the resistance of these materials. Under conditions of increased air moisture, softened polyvinyl chloride, polyamide, hardened paper, moulding compound 129, and vulcanizate of rubber are attacked in the first place by various bacteria, the phenomenon of succession of various groups of bacteria being observed. In further stages of growth, bacteria may stimulate or inhibit the growth of fungi according to the kind of electro-insulating material. The results of this research are so surprising that the standards for tests of electro-insulating materials will certainly be revised.

During the microbiological tests of electro-insulating materials (Wasserbauer and Blahnik, 1975), the intense growth of microorganisms was observed in the zone of contact of two types of plastics. The mechanism of this phenomenon is not known yet; nevertheless, the Czech standard CSN 03 8484 provides microbiological tests in the zone of contact of fibre glass with hardened paper, and also in the zone of contact of phenol-, urea-, or melamine-formaldehyde moulding compounds with other electro-insulating materials. Studies of the effect of moisture on various electro-insulating materials in the presence of fungi have shown that fungi may change the sorptive properties of these materials (Blahnik and Wasserbauer, 1972).

The knowledge of microbial processes occurring on electro-insulating materials is of particular importance for the evaluation of the reliability of electrical and electronic equipment under tropical conditions. This problem was the subject of research carried out by Czech specialists in the tropical climate of Nigeria, Ghana, Viet-Nam, and China (Rychtera and Niederführova, 1965; Rychtera *et al.*, 1973). The microflora of various materials in the tropics was recognized and the kinetics of growth of microorganisms on these materials was tested and compared with laboratory results. On the basis of the analysis of meteorological data and results of microbiological research, the method for predicting the microbial deterioration of various technical materials was developed for different climatic zones. This method is based upon meteorological hourly data or average monthly data, and it provides that the process of growth of fungi is decelerated or inhibited due to the periodic occurrence of subcritical values of one or more environmental factors.

The processes of microbial deterioration in electrical and electronic equipment occur also during sea transport. For these reasons, research work has been undertaken on Polish ships concerning the microclimate of various spaces on the ship during voyages to Australia and South America (Kaszewski and Woynarowska, 1969; Winiarska and Woynarowska, 1973). The problem of choosing microbiocides for electro-insulating materials is under particular consideration in the USSR, Romania, and Poland (Muskalo *et al.*, 1973; Pitis *et al.*, 1968; Wexler *et al.*, 1971; Jakubowska *et al.*, 1972). However, the use of microbiocides in electro-insulating materials is a complex problem with regard to their effect on metals and the properties of electro-insulating materials, and also on the decrease of microbiocidal properties of these substances during operation of electrical and electronic equipment.

In Poland more comprehensive investigations have been undertaken concerning the efficiency of volatile fungicides with respect to electro-insulating materials (Rytych, 1974). Among others, these investigations determined the ratios of microbiocide mixtures having synergetic properties which do not affect the properties of electro-insulating materials. The stability of volatile fungicides was determined on the basis of a differential thermal analysis. A new manometric method for testing volatile fungicides enabled the double or triple shortening of the time of evaluation of fungicidal properties of these compounds.

DETERIORATION OF MATERIALS IN DEEP MINES

In the process of working coal, iron ores, and nonferrous metal ores, it is necessary to use various materials including those which are liable to microbial deterioration. Below the parallel 60°, the climate of deep mines is featured by high moisture and air temperature above 20°C, and these parameters are not subjected to any periodic fluctuations. In the temperate zone the climate of deep mines approximates humid tropical conditions. (Zyska, 1968).

In the past 90 years, the process of microbial deterioration of mine timber was thoroughly examined and effective methods for timber preservation were developed (Zyska and Pluta, 1957; Zyska *et al.*, 1962). It is estimated that during 1950–60 in the world mining industry 1 million cubic metres of mine timber have been treated annually with the aid of inorganic preservatives. Since 1960, great advances have been made in the knowledge of microflora of other materials used in deep mines. The problem of increasing the microbiological resistance of belts used in belt conveyors was the object of particular concern in Great Britain and Poland (Chamberlain *et al.*, 1965; Zyska, 1964, 1972). Polyamide-cotton fabrics or polyamide-viscose fabrics are used in conveyor belts. The rate of penetration of cellulolytic fungi along the fibres is up to 100 mm during 20 days (Zyska and Elphick, 1965). This process results in the rapid loss of the mechanical properties of fabrics used in the belts (Zyska, 1967). The investigations carried out in Poland found microorganisms growing on electrical rubber cables (Zyska *et al.*, 1972) and on electro-insulating materials used in bituminous coal mines (Zyska *et al.*, 1975).

Mechanization in coal mining calls for the use of new types of supports in which oil emulsions are utilized for fire safety. The investigations conducted in Poland showed the necessity of using microbiocides in these emulsions (Bohdanowicz-Murek and Zyska, 1974; Zyska *et al.*, 1975).

SCIENTIFIC ACTIVITIES

The reviewed investigations in the field of microbial deterioration of materials in a number of countries of Eastern Europe are a part of the scientific achievements of the past 30 years. Therefore, it is advisable to mention the monographs in the field of microbial deterioration of materials published in the past period. In 1960 in Czechoslovakia, the first monograph on tropic proofing of electrical equipment was published (Rychtera and Bartakova, 1960), while in 1963 the monograph concerning the deterioration of all technical materials was edited (Blahnik and Zanova, 1963). In 1971 in the USSR, the comprehensive handbook on the protection of materials against microorganisms was published (Bobkova *et al.*, 1971). The monograph on fungicides for protection of materials (Borecki *et al.*,1965) was published in Poland in 1965 and also the monograph on microbial deterioration of materials (Zyska, 1976) will be published next year. The number of symposia held in Czechoslovakia, Hungary, Poland, Romania, and the USSR gives evidence of the dynamic development of investigations on microorganisms which cause deterioration of materials. This is determined by the scientific progress and needs of national economies of every country.

REFERENCES

American Institute of Biological Sciences (1960). *Developments in Industrial Microbiology*. Vols. 4–13, Washington, D.C.

Anchabadze, T. T. (1949). Materials for investigation of paper-deteriorating fungi. *Vest. Gos. Muzeya Gruzii im. Dzhonashia.*, **14A**, 201.

Anton, I. (1959). Trials of slime control in a paper mill. *Celuloža si Hîrtie*, **8**, 83.

Anton, I., Popescu, Gh., and Giurcanu, C. (1970). The slime control in the newsprint mill. *Celuloža si Hîrtie*, **19**, 206.

Babakina, V. G. (1964). New fungicide for leather. *Kozhevennaja Prom-st*, Seria 2, Inform., **5** (14), TsINTJ Moskva.

Balif, G., Iancu, C., Bratulescu, V., Pitis, I., and Lacatusu, V. (1971). Microbiocidal protection on leather and leather goods. *Microbiologia (Bucuresti)*, **2**, 269.

Becker, G. (ed.) (1965). *Material und Organismen*. Duncker and Hunbolt, West Berlin.

Biodeterioration Research Titles (formerly Reference Index Supplement IBBRIS) (1967) (H. O. W. Eggins, ed.). Biodeterioration Information Centre, Birmingham (U.K.).

Blahnik, R., and Wasserbauer, R. (1972). Effect of microbial attack on moisture absorption of electro-insulating materials. *Int. Biodeterior. Bull.*, **8** (2), 49.

Blahnik, R., and Zanova, V. (1963). *Microbial corrosion*. SNTL, Praha.

Bobkova, T. S., Zlotsevskaya, I. V., Rudakova, A. K., and Tsekunova, L. N. (1971). *Destruction of industrial materials by microorganisms*. Izd. Moskovskogo Universiteta, 147 pp.

Bohdanowicz-Murek, K., and Zyska, B. J. (1974). Microbial deterioration of oil emulsions in coal mines. *Biul. Główn. Inst. Górn.*, **2** (2), 35. In *Przegl. Górn.*, **30** (6), 1974.

Borecki, Z., Czerwinska, E., Eckstein, Z., and Kowalik, R. (1965). *Fungicides.* PWRiL, Warszawa.

Chamberlain, E. A. C., Meek, I. R., Kewell, C. B., and Finlay, A. I. (1965). Microbiological breakdown of conveyor belting in coal mines. *Trans. Inst. Rubber Ind.*, **41** (1), T 24.

Czerwinska, E., and Kowalik, R. (1956). Penicillium species deteriorating manuscripts—remedies. *Acta Microbiol. Pol.*, **5**, 299.

Czerwinska, E., Sadurska, I., and Kozlowska, D. (1953). Actinomycetes deteriorating manuscripts—remedies. *Acta Microbiol. Pol.*, **2**, 160.

Demmer, F. (1968). Contribution to the question of the resistance of fully synthetic plastics to fungal attack. *Material und Organismen*, **3** (1), 19.

Dubok, N. N., Angert, L. G., and Ruban, G. I. (1971). Investigation of rubbers, components and vulcanized rubbers for resistance of moulds. *Kauchuk i Resina*, **30**, 19.

Eggins, H. O. W. (ed.) (1965). *International Biodeterioration Bulletin.* Biodeterioration Information Centre, Birmingham (U.K.).

Environmental Effects on Materials and Environment. Natl. Res. Council, Natl. Acad. Sci., Washington, D.C.

Jakubowska, J., Kusewicz, D., and Szopa, J. S. (1972). Tests on electrical equipment for susceptibility to fungal attack. *Int. Biodeterior. Bull.*, **8** (1), 21.

Jovanovic, R., and Jakševac, J. (1970). Investigation of effects of microorganisms on some synthetic fibers. *Hemijska Industrija* (*Beograd*), **24** (8), 375.

Jovanovic, R., Jakševac, J. R., and Levkova, S. (1970). Some results of investigations of the influence of microorganisms on acrylonitrile fibers. *Hemijska vlakna* (*Beograd*), **10** (3), 11.

Kalinenko, V. O. (1938). The role of Actinomycetes and bacteria in decomposing rubber. *Mikrobiologiya*, **17**, 119.

Kaszewski, B., and Woynarowska, S. (1969). Preliminary tests on exposure of electrical equipment to moulds. *Przegl. Elektrotech.*, **45** (7), 316.

Kestelman, V. N., and Vilnina, G. L. (1971). The influence of polymers on fermentation and other activities of microorganisms. *Int. Biodeterior. Bull.*, **7** (3), 99.

Kestelman, V. N., Gratsheva, I. M., and Tshipizubova, L. I. (1967). The cultivation of fungi in plastics cuvettes. *Prikladnaya biokhimia i mikrobiologia*, No. 3:374.

Kestelman, V. N., Yarovenko, V. L., and Malnikova, E. I. (1972). The corrosion of polymeric materials under the conditions of microbiological synthesis of enzymes. *Int. Biodeterior. Bull.*, **8** (1), 15.

Kowalik, R. (1952). Microorganisms deteriorating manuscripts—remedies. *Prace Min. Przem. Chem.*, No. 2.

Kowalik, R. (1959). Pathogenic microorganisms for man in libraries and archives. *Blok-Notes Muz. im. A. Mickiewicza*, **1**, 175.

Kowalik, R., and Sadurska, I. (1956). Microorganisms deteriorating the leather bookbindings. *Acta Microbiol. Pol.*, **5**, 277.

Kowalik, R., Sadurska, I., and Czerwinska, E. (1962). Microbial deterioration of old book and manuscripts—remedies. *Boll. Ins. Pat. Libro.*, **21**, 116.

Krassowski, B., Sadurska, I., and Kowalik, R. (1966). Contamination of chrome-tanned leather by Desulfovibrio. *Acta Microbiol. Pol.*, **15**, 203.

Lazar, V. (1973). Considerations of the fungus resistance of materials utilized in the rubber industry. *Proc. III Int. Symp. Modelling the Effect of Climate on Electrical and Mechanical Eng. Equip.*, Czech. Acad. Sci. Prague-Libice, p. 261.

Lazar, V. (1975). The study of microbiological corrosion of plastics in Romania. *Int. Biodeterior. Bull.*, **11** (1), 16.

Matarucev, K. W. (1963). Methods of slime prevention. *Bumazhnaya Promyshlennost*, **38** (11), 18.

Muskalo, L. K., Litvinenko, C. N., and Tyltin, A. K. (1973). Fungicides for transformer oil. *Fiziol. asti. vechestva*, No. 5, 126.
Nette, I. T., Pomortseva, N. V., and Kozlova, E. I. (1959). Deterioration of rubber by microorganisms. *Mikrobiologiya*, **28**, 881.
Nyuksha, Yu. P. (1956). Microflora of books and paper. *Botan. Zh.*, **41**, 797.
Nyuksha, Yu. P. (1961). Taxonomic survey of fungi destroying paper, books, and paper pulp. *Botan. Zh.*, **46**, 70.
Orlita, A. (1968). Biodeterioration in the leather industry, In *Biodeterioration of Materials*. Vol. 1, pp. 297–301, Elsevier Publ. Co. Ltd, Amsterdam.
Orlita, A. (1971). Antimicrobial hygienic treatment of synthetic shoe-making materials. *Proc. XIIth Congr. Int. Union Leath. Chem. Soc.*, Praha: III/14, pp. 1–12.
Orlita, A., and Orlitova, I. (1961). Preservation of leather against moulds. *Kožařstvi*, **11**, 201; **11**, 230.
Pavlinova, R. M. (1933). Investigation in Balakhin board mill in connection with the development of fungal growth. *Bumazhnaya Promyshlennost*, No. 20, 51.
Pavlinova, R. M., and Tuleuova, E. T. (1967). Investigations into the microflora of slime on equipment and in the stream of stuff in Balakhin board mill. *Sbornik Trudov TsNIIB*, No. 2, 273.
Pitis, I. (1972). Mycological protection of rubber for industrial products, In *Biodeterioration of Materials*, Vol. 2, p. 294, Appl. Science Publ. Ltd, London.
Pitis, I., Lacatusu, V., and Stan, A. (1968). Contribution to the study of the fungicidal treatment of electrotechnical bakelites. *Lucrarile I.C.P.E.*, No. 19, 97.
Pitis, I., Stanescu, E., Lacatusu, V., Budulan, F., Corici, L., Ciobanica, M., and Dinu, M. (1973). Microbial resistance of rubber blendings for the tropics and the possibility of fungizating them. *Lucrarile I.C.P.E.*, No. 28, 113.
Plenum Press (1960). *Developments in Industrial Microbiology*. Vols. 1–3, New York.
Prevention of Deterioration Abstracts. Natl. Res. Council Natl. Acad. Sci., Washington, D.C.
Rybakova, C. G. (1953). *Sb. Materialov po Sokhrannosti Knizhnykh Fondov*. 2nd ed., Moskva.
Rychtera, M., and Bartakova, B. (1960). *Tropicproofing Electrical Equipment*. SNTL, Praha; Leonard Hill (Books) Ltd, London, 1963.
Rychtera, M., and Niederführova, E. (1965). Relationship between the native microbial corrosion and mould test in laboratory. *Werkst. Korros.*, **16** (2), 116.
Rychtera, M., Genovova, E., and Machova-Nemcova, B. (1973). The atmospheric microbial corrosion of technical materials and its predicting. *Rozp. CSSAV., Rocnik*, **84**, Sešit 3.
Rytych, B. (1974). *Investigations on the Efficiency of Volatile Fungicides for Electrical Insulants Used in Coal Mines*. Centr. Mining Inst. (Katowice).
Sadurska, I., and Kowalik, R. (1969). *Annali della Scuola Speciale per Archivisti e Bibliotecari dell Universita di Roma*, **9** (1–2), 51.
Savulescu, A., and Lazar, V. (1970). Study of saprophytic fungi isolated from archives materials in Romania. *Microbiologia Bucuresti*, **1**, 857.
Schönborn, Ch., Jung, H. D., and Kaben, U. (1968). Antimycotic proofing of rubber and PVC sheeting with paraoxidephenylmethane. *Dtsch. Gesundh. Wesen.*, **23**, 2429.
Schwartz, A. (1963). Microbial corrosion of plastics and their compounds. *Abh. Dt. Akad. Wissen. Berlin, Klasse Chem., Geol. u. Biol.*, No. 5.
Sergeeva, V. V. (1973). Slime formation on equipment of paper mills under high temperature conditions. *Sbornik Trudov TsNIIB*, No. 8: 207.
Shaposhnikov, V. N., Rabotnova, I. L., Yarmola, G. A., and Kuznetsova, V. M. (1952a). On growth of moulds in natural rubber. *Mikrobiologiya*, **21**, 280.
Shaposhnikov, V. N., Rabotnova, I. L., Yarmola, G. A., Kuznetsova, V. M., and Mozokhina-Porshnyakova, N. N. (1952b). On growth of bacteria in natural rubber. *Mikrobiologiya*, **2**, 146.

Söhngen, N. L., and Foll, J. G. (1914). The microbial deterioration of rubber. *Centrbl. Bakteriol. Parasitenk.*, II Abt., **40**, 87.
Spence, D., and van Niel, C. B. (1936). Bacterial decomposition of the rubber in Hevea latex. *Ind. Eng. Chem.*, **28**, 847.
Tuleuova, E. T. (1969a). Microbiological reasons for development of slime on the equipment of board mills. *Sbornik Trudov TsNIIB*, No. 4, 174.
Tuleuova, E. T. (1969b). Microflora of slime on the equipment of board mills. *Prikladnaya Biochimiya i Mikrobiologiya*, No. V, 374.
Tuleuova, E. T. (1973). Microbiological reasons for development of slime on the equipment of board mills. *Sbornik Trudov TsNIIB*, No. 8, 190.
Wasserbauer, R. (1967a). Bacteria on electrical insulants in mould-resistance tests. *Technical Digest SNTL Praha*, **9** (1), 3.
Wasserbauer, R. (1967b). Natural durability of plastics against growth of fungi. *Tropikalisace (Bechovice)*, **8** (2), 26; **8** (3), 16; **8** (4), 22.
Wasserbauer, R. (1969). Ecological relations between microorganisms on the electrical insulating surface and their influence upon electric insulating characteristics. *Tropikalizace (Bechovice)*, **10** (4), 12.
Wasserbauer, R. (1971). Mutual relationship of microorganisms during the colonisation of electrical insulants. *Proc. 2nd Int. Symp. Modelling the Effect of Climate on Electrical and Mechanical Eng. Equip.*, Czech. Acad. Sci. Prague-Liblice, p. 183.
Wasserbauer, R., and Blahnik, R. (1975). Microbial corrosion of electrical equipment. Analysis of causes and consequences: A review. *Int. Biodeterior. Bull.*, **11** (1), 9.
Wexler, T., Ortenberg, E., and Pitis, I. (1971). New fungistatic agents and their behaviour in PVC mixtures. *Chemie et Industrie-Genie Chemique*, **104**, 201.
Winiarska, J., and Woynarowska, S. (1973). Investigations of microclimatic environments on shipboard in relation to the factor of microbial corrosion. *Proc. 3rd Int. Symp. Modelling the Effect of Climate on Electrical and Mechanical Eng. Equip.*, Czech. Acad. Sci. Prague-Liblice, p. 326.
Zyska, B. (1964). Microbial corrosion in conveyor belting. *Biul. Gl. Inst. Górn. (Katowice)*, **14**, 28.
Zyska, B. (1967). Comparative investigations of fabrics for conveyor belting in soil burial tests. *Material und Organismen*, **2** (3), 179.
Zyska, B. (1968). Microbiological deterioration of materials in deep mines. In *Biodeterioration of Materials*, p. 111, Elsevier Publ. Co. Ltd, Amsterdam.
Zyska, B. (1972). Investigations into the protection of PVC conveyor belts against microbiological deterioration. *Prace Gl. Inst. Górn. Kom.*, nr. 543, 1–17.
Zyska, B. (1976). *Microbial Deterioration of Materials.* Wyd. Nauk.-Techn., Warszawa. In press.
Zyska, B. J., and Bohdanowicz-Murek, K. (1975). Studies on the fungal growth on electroinsulating materials. *Proc. 4th Int. Symp. Modelling the Effect of Climate on Electrical and Mechanical Eng. Equip.*, Czech. Acad. Sci. Prague-Liblice. In press.
Zyska, B. J., and Elphick, J. J. (1965). Penetration of coal mine conveyor belting by fungi. *Int. Biodeterior. Bull.*, **1** (2), 56.
Zyska, B., and Pluta, L. (1957). *Mine Timber*. Wyd. Górn.-Hut., Katowice, 299 pp.
Zyska, B., Koszorek, I., and Czech, G. (1962). Efficiency of the preservation of mine timber against fungi. *Prace Gl. Inst. Górn. Katowice*, Kom. nr. 304, 1–23.
Zyska, B. J., Rytych, B. J., Zankowicz, L. P., and Fudaley, D. S. (1972). Microbiological deterioration of rubber cables in deep mines and the evaluation of some fungicides in rubber. In *Biodeterioration of Materials*, Vol. 2, p. 256, Appl. Science Publ., London.
Zyska, B. J., Bohdanowicz-Murek, K., Terecki, S., and Zankowicz, L. P. (1975). Microbial deterioration of oil emulsions used in hydraulic equipment in coal mines. *Proc. 4th Int. Symp. Modelling the Effect of Climate on Electrical and Mechanical Eng. Equip.*, Czech. Acad. Sci. Prague-Liblice. In press.

BIODETERIORATION IN THE UNITED STATES: A REVIEW

EUGENE L. CADMUS

Ventron Corporation, Beverly, Massachusetts, U.S.A.

Faced with the formidable task of describing 'Biodeterioration in the United States,' one gropes for dimensional boundaries to apply. Economic terms were considered: it has been estimated that annual losses due to microbiological deterioration amount to more than $900 000 000. While the magnitude of such an estimate is impressive, the accuracy is questionable, and beyond verification; stark numbers, no matter what their size, are of little use to anyone attempting to understand the complexity of the subject.

A technical frame of reference appeared more promising: starting with a definition of biodeterioration, the subject could be covered in an orderly consideration of the nature of biological attack, its source, its effect, and its control in economically important product areas, such as textiles, coatings, and plastics.

Our first attempts to formulate a working definition of biodeterioration betrayed a bias which confirmed Kaplan's (1968) opinion that one's point of view regarding biodeterioration depends on the position occupied in the producer–user sequence. Thus, as a chemist involved in the production, formulation, and application of biocides, the definition used in an earlier symposium and applied by Van Der Kerk (1971) seems correct: biodeterioration is 'Any undesirable change in the properties of a material caused by the vital activities of organisms'. This definition reflects an interest and concern with biologically induced changes which may or may not interfere with functional integrity.

This becomes important as new synthetic materials are introduced as components of product constructions, replacing more biologically vulnerable materials. For example, textiles composed of blends of synthetic and cotton fibres have greater resistance to the degradative effects of microorganisms than pure cotton, yet their usefulness can be lost because of surface stains of microbiological growth. Such materials, in our point of view, require the additional protection of a biocide with positive antifungal activity, that is, the ability to repel a fungal invasion from whatever source and sustained by whatever nutrients the total construction or its conditions of use provide.

The dynamic technology within the textile industry demands new biocides in a wide variety of formulations to meet the challenge of new fibre blends, complex finishes, and engineering innovations in application techniques. Recently, there has been renewed interest in specialized antibacterial textiles for medical and sanitary applications.

The search for new biologically active materials has been intensified by the threat of legislative action against mercurials, and evaluation of nonmercurial fungicides is the subject of current cooperative studies for both textiles and paint (Edwards and Smith, 1974).

The results of the work in paint indicate that proposed substitutes cannot match the cost/performance effectiveness of the phenyl mercury compounds (Bergfeld *et al.*, 1971). There is also an increasing body of evidence to support the view that the phenyl mercury compounds have been improperly associated with the special, and tragic, toxicity of methyl mercury compounds, and that in paint at levels below well-defined maximum limits the phenyl mercury compounds are not an environmental hazard (Spangler, 1973).

Only recently has the U.S. literature on the microbiological involvement of organic coatings made particular note of what Pauli (1971) described as a 'remote effect' required of an effective paint fungicide. This, as mentioned earlier, is the ability of a fungicide to preserve important surface properties, even though the substrate is not directly affected. Myers (1974) has reported an evolutionary cycle of microorganism growth on an organic coating involving defacement by benign surface growth before more severe attack occurred with the development of *Pullularia*.

The need for this 'remote effect' from the fungicide applies not only to textiles and paints but is especially important in plastics. While almost all synthetic polymers are considered to be resistant to microbiological degradation, classifications of their degree of resistance have been based on the polymers alone, ignoring conditions of use in which polymers become part of a construction involving susceptible materials, such as fabrics, adhesives, wood, and paper; judgements of biological susceptibility are usually made without considering applications which expose the polymer to environments which might alter their biological properties or deposit nutrient surface dirt, such as waxes, oils, etc.

As a practical matter, then, fungicides are often added to plastic formulations, especially in plasticized polyvinylchloride systems (Kaplan *et al.*, 1970; Cadmus, 1974). As described in an earlier symposium (Yeager, 1968), these fungicides should have broad-spectrum activity, compatibility with the vinyl formulation without adding odour or colour, or change any of the physical properties of the material being protected, have the stability to withstand processing temperatures and environmental effects, and low human toxicity. Types of biocides used in plastics include the organometallics, such as organotin and arsenicals, organic sulphur compounds, including captan and folbet compounds originally developed as agricultural fungicides, substituted thiopyridines, chlorinated phenols, and for certain applications, principally in military materials, copper 8-hydroxyquinolinolate.

In my opinion, not all of these materials are effective. Only a biocide which has some mobility within the vinyl system—either derived from its own properties, such as vapour pressure, or resulting from its combination with one of the formulation ingredients, such as solubility in the plasticizer—will be able to exert that 'remote effect' and control externally supported surface growth, or prevent the spread of an actively growing fungus.

Of the organotin compounds, the tributyl salts, such as the acetate or the salicylate, are the usual industrial fungicides. These appear to have some gaps in the spectrum of activity, but, more seriously, they have poor stability, and are inactivated by ultraviolet radiation or through hydrolysis.

The compound, 10,10′-oxybisphenoxarsine, in various formulations involving common plasticizers, enjoys wide use in the plastics industry because of its compatibility and high level of effectiveness; it has good thermal stability and durability in formulations designed for outdoor exposure.

The long history of agricultural use for captan, N-(trichloromethylthio) tetrahydrophthalimide, and folpet, N-(trichloromethylthio) phthalimide, has established their biological activity, but the thermal instability of captan can cause discoloration, and Kaplan *et al.* (1970) have noted an inability to control surface growth on PVC for these organic sulphur compounds.

Copper 8-hydroxyquinolinolate, a well-known, effective fungicide, suffers the disadvantage of its intense yellow-green colour.

The fungal susceptibility of polyurethanes has been reported by Darby and Kaplan (1968). Generally, there is little susceptibility among the polyether urethanes, but the polyester types were found to be quite susceptible. The incorporation of biocides in urethane systems is more difficult than in vinyl formulations. Because of the great variety of components and processing techniques, compatibility of the biocide is a serious problem. More importantly, unlike vinyl polymers, which do not, by themselves, support growth, susceptible polyurethanes involve the actual polymer in microbiological growth. While our experience is limited, we have found that 10,10-oxybisphenoxarsine can be formulated in an unbranched alkane diol, such as 1,4-butane diol, and added to the monomer before reaction with the diisocyanate to provide fungal resistance in the polyurethane without disrupting the critical balance of reactants in the catalysed polymerization.

In conclusion, two further points should be noted: test methods and the search for new biocides.

There are probably more differences of opinion over laboratory evaluation methods than any other subject of concern to those involved with the microbiological properties of materials. Much will depend on the point of view. The biocide producer will favour techniques which demonstrate a high level of activity; the manufacturer will be interested in functional integrity; the user will want a reliable estimate of service life. No single test will provide a complete evaluation. The most useful methods will maintain rigid control of the microbiological aspects of the test by using well-identified test organisms, applied by standardized inoculating techniques, and cultured uniformly. The method should also permit an orderly study of the effects of leaching, weathering, heat, ultraviolet radiation, etc., separately and in combination; and, where possible, should include analytical verification of biocide levels before and after appropriate environmental stress. Each class of materials will require its own special modifications.

In the United States, ASTM methods, where applicable, are generally accepted, but there are few industry-wide standards.

To the high cost and the uncertainty of chemical–biological research directed to the discovery of new biologically active materials, we must now add the expense and difficulties of establishing the potential environmental hazards and possible dangers to public health for any promising biocide candidate. These entirely proper concerns will, I believe, direct research effort to modifications of existing biocides and to adjustments of formulations to meet the demands of new technology.

There is still much to be done; reliable test methods must be developed; a better understanding of the mechanisms of biocide activity and the chemical/physical factors which influence biocide performance for each industrial application is needed to optimize their use. The opportunities are immense and exciting.

REFERENCES

Bergfeld, R. H., Hille, R. H., and Edwards, J. N. (1971). *J. Paint Technol.*, **43**, 80.

Cadmus, E. L. (1974). *Modern Plastics Encyclopedia.*

Darby, R. T., and Kaplan, A. M. (1968). *Appl. Microbiol.*, **16**, 900–905.

Edwards, J. N., and Smith, W. A. (1974). *J. Paint Technol.*, **46**, 37–45.

Kaplan, A. M. (1968). *Biodeterioration of Materials.* pp. 196–204 (A. H. Walters, and J. J. Elphic (eds.), Elsevier, New York.

Kaplan, A. M., Greenberger, M., and Wendt, T. M. (1970). *Polymer Eng. Sci.*, **10**, 241–246.

Myers, R. R. (1974). *Am. Paint and Coatings J.*, Nov., 24.

Pauli, O. (1971). *Biodeterioration of Materials*, Vol. 2, pp. 355–359 (A. H. Walters and E. H. Hueck-Van der Plas, eds.), Applied Science Publishers, London.

Spangler, W. J. (1973). *Science*, **180**, 192.

Van der Kerk, G. J. M. (1971). *Biodeterioration of Materials*, Vol. 2, pp. 1–11 (A. H. Walters and E. H. Hueck-Van der Plas, eds.), Applied Science Publishers, London.

Yeager, C. C. (1968). *Biodeterioration of Materials,* pp. 131–161 (A. H. Walters and J. J. Elphic, eds.), Elsevier, New York.

MECHANISM OF MICROBIAL DETERIORATION OF NATURAL RUBBER VULCANIZATES

P. S. FUDALEJ, B. J. ZYSKA, D. S. FUDALEJ and M. H. KUCZERA

Central Mining Institute, Katowice, Poland

Summary

Structural changes in the natural rubber (NR) vulcanizates caused by soil microorganisms are analysed. The surface deterioration of NR vulcanizates depends on carbon black concentrations. The changes in the NR vulcanizates are followed on the basis of differences in the network chain density of samples, both sterile and deteriorated by soil microorganisms. Using the Flory–Rehner theory, the changes of network chain density in the presence and absence of a reinforcing filler are calculated. The influence of microorganisms on rubber-filler bonding are calculated according to the network model of filled vulcanizates given by Rehner. It is suggested that soil microorganisms or their metabolites are involved in the destruction of the NR vulcanizates network.

INTRODUCTION

The processes of microbial deterioration of elastomers have been the subject of investigations for many years; the first report was dated 1914 (Söhngen and Foll, 1914). In the period after 1945, the main subject of experiments concerned rubbers used in pipe-joint rings (Dickenson, 1965, 1968; Hills, 1967; Leeflang, 1963, 1968; Rook, 1955) and in electrical cables (Blake *et al.*, 1949, 1950, 1955; Zyska *et al.*, 1969, 1972). The mechanism of microbial deterioration of the network structure of NR had not been considered thus far. The aim of this paper is to present a hypothesis of the mechanism of microbial deterioration of elastomers and the experimental verification of the presented hypothesis.

THEORETICAL ASPECTS

The worsening of the mechanical properties of NR vulcanizates because of the influence of microorganisms is undoubtedly connected with changes in the network structure of the elastomer. The actual stage of physical chemistry of elastomers enables the structural properties to be connected with some mechanical properties. It is useful to discuss the hypothetical mechanisms of the network structure of elastomers. All types of rubber are formed, as it is known, of long elastic chains which are cross-linked at random and form a three-dimensional network. There is one crosslink for some tens or hundreds of atoms in the chain.

On a simplified basis it can be assumed that the elastomer network formed by curing compounds is an ideal network, which satisfies the Kuhn–Flory's postulate (Treloar, 1958). The stress–strain relation in a sample of elastomer (Tobolsky, 1964) may be written as

$$f = X/A = \nu RT[L/L_o{}^2 - L_o/L] \tag{1}$$

where X is tensile force, A the area of cross section, ν the number of network chains per volume unit, R the gas constant, T the absolute temperature, L the length of tensioned elastomer sample, and L_o the original length of elastomer sample.

Under a force f at a constant temperature, Young's modulus E is given by the relation

$$E = L(\partial f/\partial L) \tag{2}$$

By applying eqn. (1), Young's modulus E may be written after differentiation as

$$E = \nu RT[2(L/L_o)^2 + L_o/L] \tag{3}$$

but because

$$\nu\,.\,M_c = d_w \tag{3.1}$$

where M_c is the molecular weight of chains between two crosslinks, d_w the elastomer density, and ν the number of effective network chains, then

$$E = d_w RT\,.\,1/M_c[2(L/L_o)^2 + L_o/L] \tag{4}$$

For very small deformations of the sample $L = L_o$ then

$$E = 3d_w RT\,.\,1/M_c \tag{5}$$

From eqns. (4) and (5) results show that for the ideal network, the Young's modulus is in direct relation to the molecular weight of chains between two crosslinks M_c. Since d and R are constant and T is also constant at the isothermal process, the relationship $E(M_c)$ for a given sample is a decreasing function. With the increase in M_c, a significant decrease of the value of Young's modulus is observed. It can be noticed that the number of crosslinks is closely related to the number of effective chains responsible for the stress transmission, hence every change in the number of effective chains results in the change of M_c.

As mentioned before, the influence of microorganisms on the NR vulcanizate results in a decrease in mechanical properties, including Young's modulus. It results from the above considerations that for an ideal elastomer network, the change of the number of effective chains of the elastomer in the process of microbial degradation is to be taken into account for an explanation of the decrease in the strength. The degradation of the chains may be caused directly by microorganisms or indirectly by metabolites, interacting with macromolecules of the NR vulcanizate.

In order to evaluate the decrease in Young's modulus of the NR vulcanizate depending on the period of exposure to microorganisms, the following assumption is established. The changes in ΔM_c are directly proportional to the value of M_c, to the period of microbial degradation t, and to the proportionality factor K. For very small changes of M_c and for very short periods of time, the relationship of this assumption can be written as

$$\mathrm{d}M_c = KM_c\,\mathrm{d}t \tag{6}$$

Under boundary conditions, where $M_c = M_c{}^o$ at $t = 0$, eqn. (6) takes the form

$$M_c = M_c^o \exp(Kt) \tag{7}$$

where M_c is the molecular weight of chains between two crosslinks in a given period of time, M_c^o is the original molecular weight of chains between two crosslinks in the network of the NR vulcanizate, K is the proportionality factor for the system elastomer/microorganism, and t is the exposure time to microorganisms. By substituting eqn. (7) into eqn. (5), we find

$$E = E_o \exp(-Kt) \tag{8}$$

because

$$E = 3dRT/M_c^o$$

where E is Young's modulus after microbial deterioration in period t. The curve of function $E(t)$ is an exponential function linking the duration of microbial deterioration with changes in Young's modulus.

The molecular weight M_c can be determined, among others, by evaluation of the degree of swelling of NR vulcanizate in a solvent, on the basis of the Flory–Rehner equation (Flory and Rehner, 1953; Flory, 1950)

$$M_c = -d_w \,.\, V_0(V_r^{1/3} - 2/f \,.\, V_r)/V_r + \mu V_r^2 + \ln(1 - V_r) \tag{9}$$

where V_o is the molar volume of the solvent in cm^3/mol, V_r is the volume fraction of rubber in the gel, μ is the Huggins solubility parameter, f is the functionality of crosslinks for NR = 4, and d_w is the rubber density.

The degradation of the NR vulcanizate network may result not only in the change of the number of effective chains but also it may lead to the formation of network defects, which considerably reduce the mechanical properties. This statement is based on experimental data (Reszka *et al.*, 1975) which showed that micro- and macrocracks develop in rubber during the attack of microorganisms. On the other hand, it is known that at the top of microcracks a high concentration of stresses appears during the deformation of the elastomer, resulting in a decrease of mechanical properties. The influence of microorganisms is likely to accelerate the ageing of the elastomer.

In practice, most rubber products contain fillers, which increase several times the mechanical properties of the products if compared with unfilled rubber. The effect of microorganisms on the interaction between the filler and rubber is, in addition to the influence on the elastomer network, another main factor determining the final mechanical properties of the NR vulcanizate. The formation of thixotropic spatial structures (Dogadkin *et al.*, 1952; U-Shu-Tsyu *et al.*, 1956; Voyutskii *et al.*, 1951) is one of the substantial conditions for reinforcement of the elastomer by the filler. Model tests on the influence of microorganisms on the regeneration of thixotropic structures showed a considerable decrease in the structure-forming ability of carbon black after microbial degradation (Zyska *et al.*, 1971, 1973). It was found that the higher the filling ratio of the filler, the higher is the microbial deterioration in the model system paraffin oil/carbon black. These results suggest the changes at the phase boundary elastomer/filler caused by microbial metabolites. It may be assumed that microbial deterioration of carbon black-filled vulcanizates has its source not only in changes in the primary network but also on the boundary surface between the filler and macromolecule of the elastomer.

The determination of the influence of microorganisms on the filled elastomer is a question that arises in this connection. The use of eqn. (9) for a sterile filler elastomer and for a deteriorated filled elastomer seems to be an advantageous solution, but the quantitative evaluation must be treated with caution. The introduction of filler particles into the elastomer entails remarkable changes in the structure of the network, while eqn. (9) of Flory–Rehner was established for an ideal network. For evaluation of the changes on the phase boundary elastomer/filler after microbial deterioration, the application of Rehner's theory of the double tetrahedron would be more suitable (Kraus, 1968). This theory enables us to follow changes in the number of bonds between the macromolecule of elastomer and the filler surface. Rehner assumed the existence of two types of tetrahedron forming the spatial network: the volumetric and the superficial. The number of bonds filler/elastomer per 1 cm^2 of filler surface is, according to Rehner,

$$N = 3[3]^{\frac{1}{2}}/10\lambda_s^{\,3} \tag{10}$$

where N is the number of bonds per 1 cm^2 of filler and λ_s is the distance from the base to the top of the superficial tetrahedron. The basic value which must be found in eqn. (10) is λ_s. Between λ and the distance from the base to the top of the superficial tetrahedron, there is a relationship which Rehner expressed as

$$8\lambda_s^{\,3} - 3d_F[1 - R(1 - \phi) - 1/\phi]\lambda_s^{\,2} - 11\lambda^3 = 0 \tag{11}$$

where λ_s is the distance from the base to the top of the superficial tetrahedron, λ is the distance from the base to the top of the volumetric tetrahedron, d_F is the diameter of the filler particle, R is the ratio of the total number of bonds in the filled vulcanizate to the total number in the unfilled vulcanizate, and

$$\lambda^3 = \frac{27[3]^{\frac{1}{2}}}{20}: \text{ number of crosslinks in 1 cm}^3 \text{ of unfilled elastomer} \tag{12}$$

For practical purposes the following substitution may be used in eqn. (11)

$$\lambda_s = x - d_F/8 \,.\, [1 + R(1 - \phi) - 1/\phi] \tag{13}$$

and resolved with respect to x, using Cardan's formulae and limiting oneself to the real roots only. The application of Rehner's equation is sufficient in the determination of the degree of swelling. The results enable changes in the interaction of elastomer macromolecule/filler particle to be followed during the process of microbial deterioration of the vulcanizate.

EXPERIMENTAL

Materials and methods

The NR rubber vulcanizates were made up from the following constituents: white crepe 100 parts per weight, zinc oxide 1 p.p.h.r., thiuram disulphide 0·5 p.p.h.r., sulphur 2·5 p.p.h.r., and progressive loading with carbon black in the range of 10–50 p.p.h.r. Carbon black of Sanex-20 type had a surface area of 22·63 m^2/g, determined in argon by the BET method, and the average particle size, measured under EM, being of 599 × 10^{-8} cm (Fudalej, 1972). Vulcanization was carried out for 2 minutes at 151°C. The samples of NR vulcanizates were exposed to microorganisms in the soil burial test (Zyska *et al.*, 1972) and to sterile soil for periods of 30, 120, 180, and 360 days. After a given time, the tensile strength of the specimens was determined, using the

Polish standard PN-55/C—04206. The vulcanizate density was determined in a pycnometer. For the determination of the degree of swelling, samples weighing approximately 0·3 g were cut out, weighed on an analytical balance, and allowed to swell in 50 ml of chloroform to the equilibrium state at 20°C. The swollen specimens were blotted with filter paper and quickly transferred to tared weighing bottles. The specimens were dried in vacuo at 60°C and 10 mm Hg vacuum to constant weight and then the final weight was determined.

Calculations and results

For the unfilled NR vulcanizate, the value of Young's modulus was calculated depending on the exposure time in the soil burial test and based on eqn. (8) the proportionality factor K was established (Table I). The determined density of the vulcanizates is given in Table II. The degree of specimen swelling was calculated from the formula

$$Q = m_s - m_0/m_0 \tag{14}$$

where m_s is the mass of swollen specimen and m_0 is the mass of the specimen before the swelling (Table III). The part of the elastomer in the swollen gel

Table I

YOUNG'S MODULUS IN UNFILLED VULCANIZATE AFTER EXPOSURE TO SOIL MICROORGANISMS AND TO STERILE SOIL

Duration of the soil burial test (days)	*Young's modulus after exposure to Sterile soil (kg/cm²)*	*Soil microorganisms (kg/cm²)*	*Proportionality factor K (1/day)*
0	4·30	4·30	—
30	4·53	4·27	17·7 . 10^{-5}
120	4·20	4·01	57·5 . 10^{-5}
180	4·19	4·24	7·8 . 10^{-5}
360	4·32	3·64	45·9 . 10^{-5}

Table II

THE EFFECT OF SOIL MICROORGANISMS AND STERILE SOIL ON THE DENSITY OF NR VULCANIZATES OF DIFFERENT CARBON BLACK LOADING

Carbon black loading p.p.h.r.	*NR vulcanizate density in g/cm³ after exposure of* 30 *days*		120 *days*		180 *days*		360 *days*	
	1	2	1	2	1	2	1	2
0	0·932	0·935	0·938	0·934	0·941	0·932	0·932	0·984
10	0·991	0·980	0·985	0·983	0·992	0·997	0·984	1·040
20	1·022	1·010	1·009	1·022	1·020	1·024	1·006	1·072
30	1·043	1·040	1·043	1·046	1·053	1·066	1·043	1·102
40	1·077	1·078	1·079	1·085	1·088	1·096	1·083	1·120
50	1·095	1·089	1·094	1·107	1·105	1·114	1·090	1·148

1 = sterile soil.
2 = soil burial test.

Table III
SWELLING DEGREE OF NR VULCANIZATE IN CHLOROFORM IN RELATION TO THE CARBON BLACK LOADING RATE AND THE EXPOSURE PERIOD TO SOIL MICROORGANISMS AND STERILE SOIL

Duration of the soil burial test (days)	*Carbon black loading rate in p.p.h.r.*											
	0		10		20		30		40		50	
	Swelling degree of NR vulcanizate after exposure to											
	1	2	1	2	1	2	1	2	1	2	1	2
30	7·984	8·183	6·805	6·758	6·067	6·692	5·343	5·612	4·600	4·813	4·701	4·637
120	8·692	8·090	6·360	6·994	6·171	6·524	5·129	5·346	4·339	4·497	4·435	4·366
180	8·909	8·827	6·436	6·967	—	6·185	5·187	5·403	4·015	4·482	4·066	4·431
360	8·299	9·173	6·893	5·844	6·161	4·833	5·495	4·228	4·362	3·845	4·416	3·620

1 = sterile soil.
2 = soil burial test.

was calculated from the formula

$$V_r = \frac{\dfrac{1}{d_w} - \dfrac{G}{104+G}\cdot\dfrac{1}{d_N} - \dfrac{1-(m_p/m_0)}{d_K} - \left[\dfrac{1}{104+G}\left(\dfrac{0{\cdot}5}{d_t} + \dfrac{1}{d_{zn}}\right)\right]}{\dfrac{1}{d_w} + \dfrac{Q}{d_c} - \dfrac{G}{104+G}\cdot\dfrac{1}{d_N} - \dfrac{1-(m_p/m_0)}{d_K} - \left[\dfrac{1}{104+G}\left(\dfrac{0{\cdot}5}{d_t} + \dfrac{1}{d_{zn}}\right)\right]} \qquad (15)$$

where d_w is the NR vulcanizate density, G is the amount of carbon black by weight, d_N is the carbon black density being in this case 1·665 g/cm^3, m_p is the mass of specimen after drying in vacuum, d_K is the rubber density, d is the density of thiuram disulphide, and d_{zn} is the zinc oxide density.

Based on V_r, the molecular weights and the network density were calculated from eqns. (9) and (3.1) and Tables IV and V. Based on eqn. (12), λ for unfilled NR vulcanizate was calculated in relation to the exposure time in the soil burial test (Table VI).

The amount of bonds filler/elastomer in 1 cm^3 of vulcanizate is given in Table VII, which shows the results of calculations according to the formula

$$B = d_N \,.\, \phi \,.\, S \,.\, N \qquad (16)$$

where d_N is the carbon black density, ϕ is the volumetric part of carbon black in the vulcanizate, S is the specific surface of carbon black, and N is the number of bonds in the system filler/elastomer per 1 cm^2 of carbon black surface. The value of N needed for calculation of B, which is the amount of bonds filler/elastomer in 1 cm^3 of vulcanizate, was based on eqns. (10) and (11).

As can be seen from Table I, soil microorganisms bring about a decrease in Young's modulus of the unfilled NR vulcanizate after 360 days of soil burial tests. In a comparable period, no changes were noticed in the sterile soil. For the series of specimens of the soil burial test, eqn. (8) was applied for the calculation of the proportionality factor K. It is obvious that for all tested periods of time the order of factor K remains the same even if some deviation is noticeable. This gives evidence, to some extent, of the correctness of the assumptions made in the deduction of eqn. (8). Nevertheless, it must be stressed that more accurate results could be achieved only after further trials, including a higher number of specimens and longer periods of exposure in the soil burial tests.

The values of the molecular weight M_c for unfilled and for carbon black-filled vulcanizates, given in Table IV, indicate all changes in the elastomer

Table IV
MOLECULAR WEIGHT OF NR VULCANIZATES IN RELATION TO THE CARBON BLACK LOADING RATE AND THE EXPOSURE PERIOD TO SOIL MICROORGANISMS AND STERILE SOIL

Duration of the soil burial test (days)	*Carbon black loading rate in p.p.h.r.*											
	0		10		20		30		40		50	
	Molecular weight of NR vulcanizates after exposure to											
	1	2	1	2	1	2	1	2	1	2	1	2
30	6 948	7 395	7 018	6 483	6 764	7 680	6 409	6 678	5 683	6 072	6 601	6 395
120	8 385	7 093	6 141	7 075	6 798	7 926	5 719	7 516	5 056	5 187	6 084	6 671
180	8 366	8 320	6 553	7 437	—	7 128	6 102	7 068	4 599	6 019	5 236	6 827
360	8 470	12 898	6 946	9 769	6 778	6 120	7 167	5 608	5 154	5 561	6 063	6 558

1 = sterile soil.
2 = soil burial test.

network under the influence of soil microorganisms. In the unfilled NR vulcanizates, the changes in M_c of the elastomer chain after 180 days are insignificant. After 360 days of the soil burial test, the increase in the molecular weight is more than 50% higher, which proves the spatial network degradation by the microorganisms or their metabolites. The decrease in the cross-linked density of the unfilled NR vulcanizates after exposure to soil micro-organisms gives evidence, as can be seen in Table V, of the drop in the number

Table V
NETWORK CHAINS DENSITY IN NR VULCANIZATES IN RELATION TO THE CARBON BLACK LOADING RATE AND THE EXPOSURE PERIOD TO SOIL MICROORGANISMS AND STERILE SOIL

Duration of the soil burial test (days)	*Carbon black loading rate in p.p.h.r.*											
	0		10		20		30		40		50	
	Network chains density $v \times 10^5$ *in NR vulcanizates after exposure to*											
	1	2	1	2	1	2	1	2	1	2	1	2
30	13·4	12·6	14·1	15·1	15·1	13·2	16·3	15·6	19·0	17·8	16·6	17·0
120	11·2	13·2	16·0	13·9	14·8	12·9	18·2	13·9	21·3	20·9	18·0	16·7
180	11·2	11·2	15·1	13·4	—	14·4	17·3	15·1	23·7	18·2	21·1	16·3
360	11·0	7·6	14·2	10·6	14·8	17·5	14·6	19·7	21·0	20·1	18·0	17·5

1 = sterile soil.
2 = soil burial test.

of effective elastomer chains which are responsible for the transmission of stresses in the elongated specimen vulcanizate. The decrease in the volume unit is significant and amounts to approximately 33% of the control samples.

As the unfilled NR vulcanizates have only the primary network, the changes in the structure network concern the changes in the direct interaction elastomer macromolecule/elastomer macromolecule. The addition of fillers to the rubber results in the formation of an additional network, *i.e.* by developing additional bonds filler/elastomer. For these reasons the analysis of the microbial deterioration of carbon black-filled NR vulcanizates must include two different aspects, namely: change in the total number of bonds and changes in the interaction filler/elastomer.

As can be seen in Tables IV and V, the molecular weight M_c increases and the network chain density decreases in the NR vulcanizates with 10, 20, and 30 p.p.h.r. carbon black when exposed to soil microorganisms. In NR vulcanizates with 40 and 50 p.p.h.r. of carbon black, the respective changes in molecular weight and network chain density are negligible. The number of bonds filler/elastomer per 1 cm^3 of tested specimens increases during the soil burial test (Table VII). The increase in bonds filler/elastomer after the

Table VI
THE DISTANCE FROM THE BASE TO THE TOP OF THE VOLUMETRIC TETRAHEDRON IN THE UNFILLED NR VULCANIZATE IN RELATION TO THE EXPOSURE PERIOD TO SOIL MICROORGANISMS AND STERILE SOIL

Duration of the soil burial test (days)	*$\lambda \times 10^7$ in unfilled NR vulcanizates after exposure to*	
	Sterile soil	*Soil microorganisms*
30	3·87	3·95
120	4·11	3·89
180	4·11	4·11
360	4·14	4·68

Table VII
NUMBER OF SURFACE BONDS ELASTOMER/FILLER IN 1 cm^3 OF NR VULCANIZATE IN RELATION TO THE EXPOSURE PERIOD TO SOIL MICROORGANISMS AND STERILE SOIL

Duration of the soil burial test (days)	*Carbon black loading rate in p.p.h.r.*									
	10		20		30		40		50	
	No. of surface bonds elastomer/filler B $\times 10^{-16}$ *after exposure to*									
	1	2	1	2	1	2	1	2	1	2
30	0·153 4	0·083 6	0·652 8	2·414 9	1·208 3	1·617 8	1·185 0	1·626 6	4·529 3	3·840 9
120	0·014 7	0·175 6	0·659 0	3·443 1	0·634 1	4·192 0	0·705 4	0·777 6	2·805 0	4·490 2
180	0·086 6	0·279 0	—	0·990 5	0·853 7	2·192 8	0·478 8	1·529 5	1·388 5	5·361 6
360	0·143 5	0·296 6	0·734 1	0·288 2	2·039 3	0·504 4	0·694 6	1·042 1	2·795 6	3·893 2

1 = sterile soil.
2 = soil burial test.

influence of soil microorganisms is parallel to the carbon black loading rate. Since the structural variables of the primary tetrahedron remained unchanged after microbial deterioration (Table VI), an assumption can be made that in place of strong chemical bonds which exist in the primary network, the microbial factor generates a formation of weak bonds filler/elastomer. Thus the decrease in mechanical properties in NR vulcanizates after the longer exposure to soil microorganisms can be explained by the above-mentioned calculation.

Based on the theoretical aspects and experimental data presented, an assumption can be made that the decrease in the mechanical properties of NR vulcanizates caused by soil microorganisms may be explained by the following phenomena: degradation of the primary structural network by microorganisms or their metabolites, and changes in the number of bonds filler/elastomer by the microbial metabolites.

REFERENCES

Blake, J. T., and Kitchin, D. W. (1949). *Ind. Eng. Chem.*, **41**, 1633.

Blake, J. T., Kitchin, D. W., and Pratt, O. S. (1950). *Trans. Am. Inst. Elect. Engrs*, **72**, 321.

Blake, J. T., Kitchin, D. W., and Pratt, O. S. (1955). *Appl. Microbiol.*, **3**, 35.

Dickenson, P. B. (1965). *Rubber Dev.*, **18**, 85.

Dickenson, P. B. (1968). *Natural Rubber Conference*, Kuala Lumpur, Preprint, pp. 1–10.

Dogadkin, B., Pietskovskaya, K., and Milman, C. (1952). *Koll. Zh.*, **14** (5), 346.

Flory, P. J. (1950). *J. Chem. Phys.*, **18**, 108.

Flory, P. J., and Rehner, J. (1943). *J. Chem. Phys.*, **11**, 512.

Fudalej, P. S. (1972). Ph.D. Thesis. Central Mining Institute, Katowice, Poland.

Hills, D. A. (1967). *Rubber J.*, **149**, 12.

Huggins, M. L. (1958). *Physical Chemistry of High Polymers*, John Wiley and Sons, Inc., New York.

Kraus, G. (ed.) (1968). *Reinforcement of Elastomers*. Izd. 'Chimiya', Moscow.

Leeflang, K. W. H. (1963). *J. Am. Water Works Assoc.*, **55**, 15.

Leeflang, K. W. H. (1968). *J. Am. Water Works Assoc.*, **60**, 1070.

Reszka, J., Zyska, B. J., Fudalej, P. S., and Reszka, K. (1975). *Int. Biodeterior. Bull.*, **11**. In press.

Rook, J. J. (1955). *Appl. Microbiol.*, **3**, 302.

Söhngen, N. L., and Foll, J. G. (1914). *Zbl. Bakt. Parasitenk.*, II Abt., **40**, 87.

Tobolsky, A. W. (1964). *Properties and Structure of Polymers*. Izd. 'Chimiya', Moscow.

Treloar, L. R. G. (1958). *Physics of Rubber Elasticity*. Oxford University Press.

U-Shu-Tsyu, Yampolskii, B. Ya., and Voyutskii, S. S. (1956). *Koll. Zh.*, **18** (6), 751.

Voyutskii, S. S., Zoionchkovski, A. D., Kargin, V. A., and Rubina, S. I. (1951). *Koll. Zh.*, **13** (5), 333.

Zyska, B. J., Rytych, B. J., Koziol, L., and Siekierski, K. (1969). *Biul. Glówn. Inst. Górn.*, **19**, 5.

Zyska, B. J., Fudalej, P. S., and Rytych, B. J. (1971). *Int. Biodeterior. Bull.*, **7** (4), 155.

Zyska, B. J., Rytych, B. J., Zankowicz, L. P., and Fudalej, D. S. (1972). In *Biodeterioration of Materials*, Vol. 2, p. 258. Appl. Science Publ. Ltd, London.

Zyska, B. J., Fudalej, P. S., Rytych, B. J., and Fudalej, D. S. (1973). *Int. Biodeterior. Bull.*, **9** (1–2), 23.

THE HIGHER PLANTS AS DETERIOGENS

D. ALLSOPP and I. D. R. DRAYTON

Biodeterioration Information Centre, University of Aston in Birmingham, Birmingham, U.K.

Summary

The scale of the problem is outlined, with special reference to the factors influencing weed dispersal and the economic significance of the problem. The major deterioration phenomena are described under the headings of physical and visual obstruction, mechanical deterioration, fire risks, changes in micro-climates, changes in soil moisture, shelter, and aesthetic deterioration. The deteriogenic effects of higher plants in individual situations are mentioned, namely, on agricultural land, on buildings, roadways, waterways, railways, and on overhead line routes. Mechanical, biological, and chemical control measures are discussed.

INTRODUCTION

This paper is derived from a much larger and unpublished work (Drayton, 1974) which attempts a review of current literature in the field of higher plants as deteriogens. The purpose of this shorter paper is to summarize the main aspects of the topic and to attempt to crystallize the reader's thoughts in a field where many of the examples are easy to understand but are rarely thought of in a unified way. At a later date, it is hoped to produce more detailed reviews of specific areas within the field.

SCALE OF THE PROBLEM

As man spreads his activities, dwellings, and constructions, greater opportunity is afforded to plants to establish themselves in unwanted places. Not only does man provide the habitat on an increasing scale worldwide but he also provides the means of plant dispersal by his commerce and travel. The problem is therefore worldwide and expands with man's activities and constructions.

Other agencies of weed distribution are the ocean currents, especially when one considers that some weed seeds have been shown to retain their viability after 500 days in seawater, sufficient time for them to drift between most of the world's land masses (King, 1966). Animals and especially birds are also of importance, seeds adhering to plumage or retained in the digestive tract (Holm *et al.*, 1969), viable seeds sometimes remaining for up to 300 hr (Proctor, 1968).

Having reached new territory, the weeds have to propagate and thrive under suitable conditions if they are to become established and spread. One of the general characteristics of weeds is that they exhibit rank and vigorous growth. One contributory factor to this is either the number of seeds produced or the ability to reproduce both sexually and vegetatively. Often, weeds produce vast numbers of seeds, for example, the parasitic weed *Orobanche elatior* may produce up to 270 000 seeds per plant (King, 1966). Rapid vegetative growth is shown by several species, including the now classic deteriogen of waterways the Water Hyacinth, *Eichornia crassipes*. One parent plant is capable of producing 30 plants in 23 days and after 130 days some 1200 plants, forming some 9 m^2 of mat (Evans, 1963).

ECONOMIC CONSIDERATIONS

As with most types of biodeterioration, it is extremely difficult to estimate the cost of damage caused by higher plants, and most of the available statistics are based on the cost of control measures. The figures available are fragmentary but the following give some indication.

On American railways, a conservative estimate (Smith, 1966) put the cost of vegetation control at approximately 110 dollars per mile of treated track, giving a total annual cost of 30 million dollars. On a world scale, this would give a total of the order of 250 million dollars per year.

On roadways in the tropics, one estimate puts roadside vegetation control as costing 10% of the overall maintenance costs of the road (Clarke, 1961).

More specific figures are available for costs of terrestrial weed control along waterways, but for only one specific region of the United States, namely, the 17 Western States (Kirkby and Summers, 1971). The costs of control vary with the methods used. These were estimated in 1971 as being: mowing operations, \$8½ per bank mile; burning, \$10½ per bank mile; chemical spraying, \$11 to 20 per bank mile. The chemical control costs are equivalent to \$75 per acre (Gangstad, 1972).

In controlling aquatic weeds, the costs are high. For example, the Rio Grande irrigation project spends half a million dollars per year (Kirkby and Summers, 1971) and the costs to the 17 Western States, based on the costs in 1966, were then estimated at around \$15 million per year (Westlake, 1960). However, the resulting savings are also high. It is estimated that in these states, the water loss attributable to weeds is some $2{\cdot}4 \times 10^9$ m^3/annum, this being worth some \$30 million (Hagan and Roberts, 1972).

TYPES OF DETERIORATION

We may consider these under a useful set of headings similar to those used by Fisher (1972) in a previous paper. It should be noted that two very important roles of higher plants are excluded from these considerations, namely, plants as agricultural weeds and also plants giving rise to various health risks.

1. Obstruction

Visual

This is of greatest importance on roads and railways (DeWit, 1963; Yazell, 1970), blocking the view of other vehicles, signals, and signs. Power and pipelines may also be obscured and hinder the detection of faults.

Physical

This is a widespread problem of some diversity. In buildings, tree roots may grow into sewers and drains causing a partial blockage (Leonard and Townley, 1971). Overhead power lines may also become fouled in high vegetation (Niering, 1958a,b), and trees near railways may foul equipment or passing trains. Ice falling from branches of trees can also pose a problem with modern high-speed vehicles. Plants may also impair the function of paints or, if actually growing near rails, cause wheel slip in locomotives (Fisher and Ashby, 1965). Excessive plant growth on the track section may also lead to bad drainage and subsequent instability in the ballast (Smith, 1966).

Plant growth in water systems may impair navigation (Guscio *et al.*, 1965; Jernigan *et al.*, 1965; Gangstad, 1972), may slow the flow in irrigation channels (Stephens *et al.*, 1963; Kuowen *et al.*, 1969), and give rise to other problems such as increased silting and increased losses due to evaporation. Weeds which break away in large masses may also block pipes and filters (Timmons, 1966; Holm *et al.*, 1969), foul piers of bridges (Stephens *et al.*, 1963), and even cause the collapse of small structures (Guscio *et al.*, 1965).

2. Mechanical action

Under this heading, we consider actual damage caused directly by plants to materials and constructions. As mentioned previously, roots of plants can be destructive, and in many cases, *e.g.* couch grass (*Argropyron repens*), tough root tips with a high silica content can penetrate into relatively hard materials. A different mechanism is at work in soft-rooted plants such as dandelions (*Taraxacum officinale*), but such plants can still exert pressure on structures. We may consider such plants pushing through asphalt (McCully *et al.*, 1970) (Fig. 1). Such growths provide a starting point for accumulative damage, *i.e.* for further plant growth due to trapping of soil and wind-borne seeds and frost and water damage.

3. Combustion

Terrestrial weeds in waste ground may form up to 2 tons per acre of combustible material each year (Fisher, 1972), and unless this is cleared, it may tend to accumulate (Fig. 2). The threat to buildings and stored products is obvious and any resulting fire may, of course, not be restricted to the immediate area.

4. Microclimate

The presence of vegetation will produce a local or microclimate which may encourage other forms of biodeterioration such as fungal decay. The most important aspects of the microclimate which will be affected will be temperature, humidity, and wind (Old, 1969).

Plant cover will retain heat, and compared with bare ground, will serve to position the maximum temperature further away from ground level.

Plants affect humidity in two ways: first, they add water vapour to the atmosphere due to their transpiration; and second, inhibit removal of water vapour given off by the soil by reducing surface sunlight and wind by the presence of their bodies. The air near the ground thus becomes humid and still, and this combined with the increased temperature will encourage growth of deteriogenic microorganisms.

Water losses from reservoirs may be increased over the normal evaporation levels by the presence of aquatic weeds. Evapotranspiration by aquatic and ditch-bank weeds in the 17 Western States of the United States causes an annual loss of some 2·5 billion cubic metres of irrigation water, enough to irrigate some three-quarters of a million acres of cropland (Hagan and Roberts, 1972).

5. Changes in soil moisture

Extraction of water from soil in a nonuniform manner by plants may lead to uneven settling of structures and the cracking of highway and pavement surfaces. Impairment of the drainage function of railroad track ballast may also lead to problems of rail buckling due to uneven distribution of loads (Yazell, 1970).

Fig. 1. Thistles breaking through newly laid asphalt surfacing of parking lot.

6. Shelter

Rank weed growth provides a habitat for a wide range of insects, rodents, and microorganisms which may act as pests in their own right. Such weed growths may also pose a threat in an agricultural context, where 'agricultural weeds' may infect adjoining 'clean' land.

7. Untidiness

This is a very important but rather subjective property of weeds and depends on the aesthetic sense of the observer. However, when rank weed growth appears around modern buildings and industrial sites, it appears untidy and to most people conveys an impression of inefficiency which may be bad for public relations (Fig. 3).

Fig. 2. Dry plant debris beside parking lot. This is unsightly and a potential fire hazard.

Fig. 3. A variety of plant growth damaging stone work at Ludlow Castle, England.

CONTROL MEASURES

The purpose of this section is to give a mere outline because control measures are well documented elsewhere and also because any detailed account of chemical methods would of necessity be vast and soon out of date. Control methods fall into three categories.

Mechanical control

The pulling or cutting of weeds by hand or with manual implements is a traditional aspect of cultivation of land and tends to be slow and laborious. Burning is also a traditional method, but fire may stimulate the growth of different weed species and merely change the nature of the problem (Kowel, 1966; Old, 1969).

Modern mechanized mowing machines are available for use on edges of roads, railways, and waterways, and heavy duty versions can deal with tree boles up to 10 inches thick. Mechanical techniques are also available for use with aquatic weeds.

A problem of mechanical weed control is that weeds often respond to cutting by vigorous re-growth and permanent control by such methods is almost impossible.

Biological control

Biological control methods often centre upon the use of insects, and as with all cases of biological control, an in-depth picture of possible consequences must be obtained prior to treatment. Insects may spread wider than intended and even if specific to a limited group of plants, such plants may not be considered as weeds in other areas.

Biological control has been successfully used on many types of plants, including the spectacular campaign against the prickly pear (*Opuntia* spp.) in Australia using the Argentine moth borer (*Cactoblastis cactorum*). The list of species being considered for possible biological control is extensive (Holloway, 1964). In the nonagricultural field, the most promising area for biological control of weeds may well be in waterways, where a wide range of organisms have been suggested (Allsopp, 1960; Stott and Robson, 1970; Blackburn *et al.*, 1971).

Chemical control

Chemical control of vegetation is almost an industry in its own right, such is its widespread use. Initially, the development in herbicides was aimed at weeds in agricultural situations, and herbicides were utilized to control weeds in other situations. Now, however, such is the extent of the industry that each aspect has herbicides either formulated or being developed for it.

Toxic herbicides may be classified into three types according to their action.

Contact herbicides

The action of this group of chemicals is caustic, killing the plant by their corrosive action on plant tissue. Many early herbicides were of this pattern.

Systemic herbicides

These enter the plant through the leaves, whereupon they are translocated in the plant body, where their toxic effects cause the ultimate destruction of the plant.

Soil-sterilants

These, when added to the soil, are capable of rendering the soil toxic to plants by entering through the roots, thence being translocated. Thus, these are also systemic herbicides.

A further classification is on the basis of range of action: 'total' herbicides, which control most plant species, and selective herbicides, the latter being further subdivided into those controlling broad-leafed plants and those controlling narrow-leafed plants.

Another group of herbicides comprising the growth regulators or plant hormones (Overbeek, 1968; Bovey, 1971) can be lethal in high doses or in small doses inhibit various aspects of the plants. The major problem in chemical vegetation control is that, in many cases, the long-term fate of the compounds is unknown, and also that many chemical agents have an adverse effect on animal life, especially the lower animals.

CONCLUSION

Man has now a veritable array of weapons with which to combat troublesome growth of vegetation. Over the years, the emphasis of use of these weapons has changed from mechanical to chemical, and is now being directed to more specific and therefore safer herbicides and biological methods. The authors hope that the purpose of this paper in outlining some of the basic aspects of the problems involved in weed control has been served, inasmuch as further interest in this topic may have been generated.

REFERENCES

Allsopp, W. H. L. (1960). The Manatee: Ecology and use for weed control. *Nature*, **188** (4752), 762.

Blackburn, R. D., Sutton, D. L., and Taylor, T. (1971). Biological control of aquatic weeds. *J. Irrigation and Drainage Div. A.S.C.E.*, **97** (IR 3), 421–432. Proc. Paper 8362.

Bovey, R. W. (1971). Hormone-like herbicides in weed control. *Econ. Bot.*, **24** (4), 385–400.

Clarke, K. E. (1961). The use and control of vegetation on roads and airfields overseas. Road Research Laboratory: Technical Paper No. 52, Department of Scientific and Industrial Research.

DeWit, H. C. D. (1963). *Plants of the World: The Higher Plants I.* Thames and Hudson Ltd, London.

Drayton, I. D. R. (1974). The Higher Plants as Deteriogens. Undergraduate dissertation. Department of Biological Sciences, University of Aston, Birmingham, 172 pp, unpublished.

Evans, A. C. (1963). The grip of the Water Hyacinth. *New Scientist*, **19**, 666.

Fisher, G. G. (1972). Weed damage to materials and structures. *Int. Biodeterior. Bull.*, **8** (3), 101–103.

Fisher, G. G., and Ashby, D. G. (1965). Weed control on tropical railways. *Railway Gaz.*, **121** (6), 241–244.

Gangstad, E. O. (1972). Herbicidal control of aquatic plants. *J. Sanitary Eng. Div., A.S.C.E.*, **98** (SA 2), Proc. paper 8851, pp. 397–406.

Guscio, F. J., Bartley, T. R., and Beck, A. N. (1965). Water resources problems generated by obnoxious plants. *J. Waterways and Harbor Div., A.S.C.E.*, **91** (WW 4), Proc. paper 4537, pp. 47–60.

Hagan, R. M., and Roberts, E. B. (1972). Ecological impacts of water projects in California. *J. Irrigation and Drainage Div., A.S.C.E.*, **98** (IR 1), Proc. paper 8780, pp. 25–48.
Holloway, J. K. (1964). Projects in biological control of weeds, In *Biological Control of Insect Pests and Weeds*, pp. 650–670 (P. DeBach, ed.), Chapman and Hall Ltd, London.
Holm, L. G., Weldon, L. W., and Blackburn, R. D. (1969). Aquatic weeds. *Science*, **166** (3906), 699–709.
Jernigan, O. M., Tabita, A., and Wunderlich, W. E. (1965). Control of obnoxious aquatic plants in the South-East. *J. Waterways and Harbor Div., A.S.C.E.*, **91** (WW 4), Proc. paper 4536, pp. 31–46.
King, L. J. (1966). *Weeds of the World: Biology and Control.* Leonard Hill Books, London.
Kirby, J. W., and Summers, A. D. (1971). Weed control in the arid South-West. *J. Irrigation and Drainage Div., A.S.C.E.*, **97** (IR 4), Proc. paper 8570, pp. 571–583.
Kowel, N. E. (1966). Shifting cultivation, fire and pine forest in the Cordillera Central, Luzon, Phillippines. *Ecol. Monogr.*, **36** (4), 389–419.
Kuowen, N., Unny, T. E., and Hill, H. M. (1969). Flow retardance in vegetated channels. *J. Irrigation and Drainage Div., A.S.C.E.*, **95** (IR 2), Proc. paper 6633, pp. 329–342.
Leonard, O. A., and Townley, N. R. (1971). Control of tree roots in sewers and drains. *Calif. Agric.*, **25** (11), 13–15.
McCully, W. G., Bowmer, W. J., and Wiese, A. F. (1970). Control of grasses and weeds growing in asphalt pavements. *P.A.N.S.*, **16** (3), 497–504.
Niering, W. A. (1958a). Principles of sound right of way vegetation management. *Econ. Bot.*, **12**, 140–144.
Niering, W. A. (1958b). A shrub community of *Viburnum lentago*, stable for twenty five years. *Ecology*, **36**, 356–360.
Old, S. M. (1969). Microclimate, fire and plant production in an Illinois Prairie. *Ecol. Monogr.*, **39** (4), 355–384.
Overbeek, J. Van (1968). The control of plant growth. *Sci. Am.*, **219** (1), 75–81.
Proctor, V. (1968). Long distance dispersal of seeds by retention in the digestive tracts of birds. *Science*, **160**, 321.
Smith, R. (1966). Need seen for long range approach in weed control. *Railway Tracks and Structures*, **66** (2), 24–25, 47.
Stephens, J. C., Blackburn, R. D., Seaman, D. E., and Weldon, L. W. (1963). Flow retardance by channel weeds and their control. *J. Irrigation and Drainage Div., A.S.C.E.*, **98** (IR 2), Proc. paper 3550, pp. 31–53.
Stott, B., and Robson, T. O. (1970). Efficiency of Grass Carp (*Ctenopharyngodon idella* Val.) in controlling submerged water weeds. *Nature*, **15** (1), 870.
Timmons, T. L. (1966). Control of weeds harmful to water uses in the West. *J. Waterways and Harbor Div., A.S.C.E.*, **92** (WW 1), Proc. paper 4645, pp. 47–58.
Westlake, D. F. (1960). Waterweed and water management. *J. Inst. Public Health Eng.*, **59**, 148–164.
Yazell, D. H. (1970). Vegetation control by prescription. *Railway Tracks and Structures*, **66** (2), 20–25.

OXYGEN CONSUMPTION, CARBON DIOXIDE EVOLUTION AND ASSIMILATION DURING MICROBIAL METABOLISM OF STARCH, CELLULOSE AND XYLAN

KRISTIINA VOGT and E. E. STAFFELDT

New Mexico State University, Las Cruces, New Mexico 88003, U.S.A.

Bacteria and fungi were isolated from New Mexico and Canal Zone soil samples. Pure cultures of the organisms were grown on starch, cellulose, and xylan as sole carbon sources. A biological oxygen demand (BOD) apparatus was employed during the cultivation of each organism to determine the O_2 utilized and the CO_2 evolved during metabolism. The latter was determined by analysing the KOH used to trap the CO_2 in the BOD chambers. Samples were removed at the termination of the experiment to analyse for reducing power (enzymatic activity), assimilation (increase in cell mass), and accumulation of metabolic products (organic acids). Bacteria were much less effective in degrading two of the test substrates than were the fungi. During starch degradation, it was found that as little as 2·8 % and as much as 39·9 % of the carbon was lost as CO_2, while the carbon utilized in assimilation ranged from 0·0 to 49·6 % and the carbon released as organic acid varied from a low of 0·0 to a high of 33·7 %. The following ranges were observed during cellulose degradation: 3·4–62·7 % lost as CO_2, 0·0–87·7 % assimilated in cell mass, and 0·0–38·8 % released in organic acids. Degradation of xylan yielded ranges but not as wide as those observed for cellulose. These investigations provide a basis for selecting microorganisms for industrial utilization.

INTRODUCTION

Today's world problems of overpopulation, insufficient food, pollution, shortage of energy, and excessive wastes are interrelated. When one portion of the overall problem is considered, generally the remaining portions are set aside or forgotten. The degradative processes of microorganisms on organic wastes and refuse could help satisfy conditions of eliminating excessive wastes and pollution and at the same time supply either needed energy or food and feed substances or both. Insufficient technology has been utilized to convert the wastes being produced so that (1) large volumes of materials can be reduced in a short period of time; and (2) useful products can be derived and harvested during or following the conversion. Microbiological decomposition is the major step in maintaining nature's balance between the opposing synthesis and degradation, and man's utilization of these processes will require additional study of how they proceed in nature.

The selection of microorganisms in this work was based upon the ability of microbes to degrade hemicellulose, cellulose, and starch contained in plant substrates. Microbial metabolism was determined by analysing total O_2 consumed, CO_2 evolved, enzyme activity, organic acid production, and cell mass accumulation.

MATERIALS AND METHODS

Bacteria and fungi were isolated from Panama and the Jornada Experimental Range (25 miles north of Las Cruces, New Mexico) soils. The organisms were identified to genus. *Bacillus* spp. were found to be the most common bacterial genus in both soils, followed by *Pseudomonas* spp. The other bacteria isolated and used in this work were *Vibrio* sp., *Achromobacter* sp., and *Flavobacterium* sp. The fungi isolated and used were *Stachybotrys* sp., *Fusarium* sp., *Chaetomium* sp., *Myrothecium* sp., and *Curvularia* sp. The bacteria and fungi were grown on cellulose, xylan, and starch as the sole sources of carbon.

These organisms were grown in closed flasks, at 22°C, and their O_2 consumption was measured electrolytically (using BOD equipment). KOH boats suspended from the top of the flasks were used to trap the CO_2 released during substrate degradation. A 30% KOH solution (Young *et al.*, 1964) was used to trap the CO_2 released and this was analysed by titration with HCl at the end of a test run (Coleman, 1971).

The dinitrosalicylic acid test (Levinson *et al.*, 1951) was used to determine the amount of reducing sugars (xylose, glucose, maltose) that was produced from the three substrates. The concentration of reducing groups produced correlates with the enzyme activity of the organisms (Bernfeld, 1951).

Gas chromatography was used to analyse short chain organic acids produced during degradation by using a Porapak QS and 2% H_3PO_4 column (Kellogg, 1969). A 10% Carbowax 20 M column was used to determine the presence of Krebs cycle acids.

RESULTS

The seven species of bacteria isolated from the Panama and Jornada soils were tested for their ability in utilizing starch, cellulose, and xylan. This capacity was based on how many milligrams of O_2 were used and how many milligrams of CO_2 were produced by each culture of bacteria in the degradation processes. Enzymatic activity was also examined to determine the ability of the bacteria to convert the substrate to reduced sugars.

Very little enzyme activity was recorded for most of the tested bacteria on the three substrates. Bacteria that did exhibit enzyme activity during starch breakdown were the *Bacillus* sp. isolated from the Jornada which produced 1·42 mg maltose/ml/2 g starch and *Vibrio* sp. (Jornada) which produced 0·10 mg maltose/ml/2 g starch. The bacteria isolated from the Jornada appeared to have greater amylase activity than the bacteria from Panama soils. No measurable enzyme activity was recorded during xylan degradation for any of the bacteria.

The two bacteria utilizing the greatest amount of O_2 and producing the most CO_2 were tested against the fungi on each of the three substrates. Comparisons were based on the above parameters as well as the amount of organic acids produced, the amount of enzyme activity, and when possible, the amount of cell mass of each organism.

The fungi appeared to degrade cellulose to a greater extent than the bacteria when comparing O_2 utilization and CO_2 production (Table I). *Stachybotrys* sp. and *Fusarium* sp., both very active in cellulose degradation,

Table I

MEASURED ACTIVITY OF THE MICROORGANISMS DURING CELLULOSE BREAKDOWN AFTER 22 DAYS

Organism	*O_2 used (mg)*	*CO_2 produced (mg)*	*Glucose/ml (mg)*	*% Increase in biomass*
Achromobacter sp. (Jornada)	240·6	212·5	0	—
Flavobacterium sp. (Panama)	379·0	217·8	0	—
Stachybotrys sp.	914·8	1 393·7	0	682
Fusarium sp.	1 234·8	2 042·9	0	279
Chaetomium sp.	50·2	146·8	0	0
Curvularia sp.	461·6	109·8	0	122
Myrothecium sp.	519·6	538·3	0	0

evolved 6–9 times more CO_2 than the two bacteria and consumed about 4 times more O_2 in the degradation process. Both bacteria exhibited slightly better activity than *Chaetomium* sp.

Fungi degraded more xylan (Table II) than the bacteria tested, except for *Myrothecium* sp. The fungi evolved about 8 times more CO_2 than the most active bacterium, *Vibrio* sp., and consumed 3–6 times more O_2. None of the bacteria exhibited any measurable enzyme activity during xylan degradation.

During starch degradation, *Bacillus* sp. (Jornada) evolved more CO_2 and consumed more O_2 than any of the fungi (Table III). This organism exhibited a low level of measured enzyme production. All of the remaining fungi (except for *Myrothecium* sp.) used more O_2 than *Pseudomonas* sp. (Fig. 1).

In conjunction with the other parameters being determined, organic acid by-products were analysed at the termination of each experiment. Fumaric was the most common acid produced during cellulose degradation. *Achromobacter* sp., *Flavobacterium* sp., and *Myrothecium* sp. produced no detectable acids during cellulose breakdown.

Following xylan breakdown, *Vibrio* sp., *Curvularia* sp., and *Myrothecium* sp. produced no detectable organic acids. *Fusarium* sp. produced a high concentration of fumaric acid which amounted to 1·55 mg/ml/2 g xylan but no other acids were detectable in this work. *Flavobacterium* sp. was the only

Table II

MEASURED ACTIVITY OF THE MICROORGANISMS DURING XYLAN DEGRADATION AFTER 14 DAYS

Organism	*O_2 used (mg)*	*CO_2 produced (mg)*	*Xylose/ml (mg)*	*% Increase in biomass*
Vibrio sp. (Jornada)	89·2	142·1	0·00	—
Flavobacterium sp. (Panama)	109·4	0·0	0·00	—
Stachybotrys sp.	364·0	1 168·9	0·75	196
Fusarium sp.	650·2	1 801·8	0·00	607
Chaetomium sp.	274·0	1 138·7	0·20	245
Curvularia sp.	495·8	1 199·0	0·13	69
Myrothecium sp.	170·0	0·0	0·00	0

Table III

MEASURED ACTIVITY OF THE MICROORGANISMS DURING STARCH DEGRADATION AFTER EIGHT DAYS

Organism	*O_2 used (mg)*	*CO_2 produced (mg)*	*Maltose/ml (mg)*	*% Increase in biomass*
Bacillus sp. (Jornada)	416·2	1 507·0	0·05	—
Pseudomonas sp. (Jornada)	100·4	91·5	0·04	—
Stachybotrys sp.	420·2	1 265·9	0·81	140
Fusarium sp.	77·4	143·0	0·04	32
Chaetomium sp.	186·8	686·4	0·00	135
Curvularia sp.	340·0	1 296·0	0·01	69
Myrothecium sp.	106·4	92·4	0·02	0

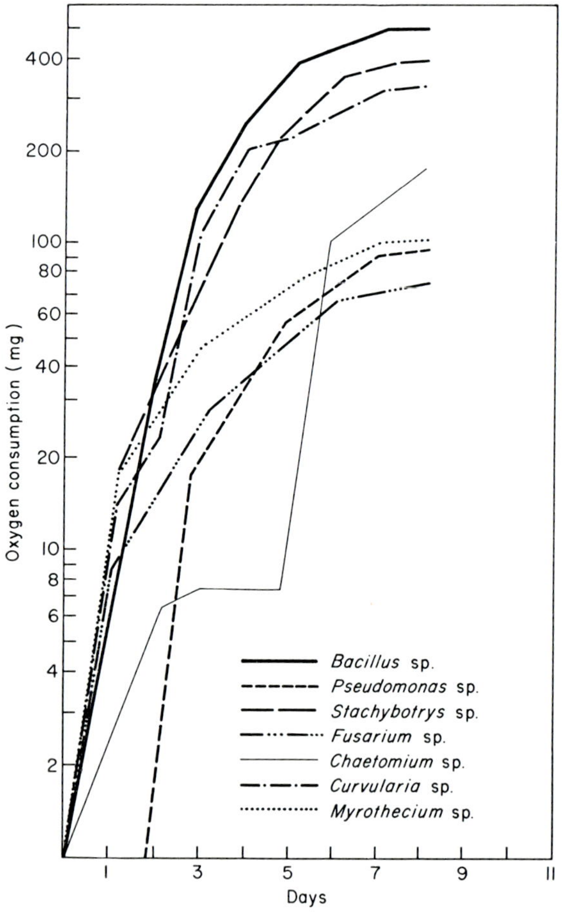

Fig. 1. Oxygen utilization by bacteria and fungi during starch decomposition over an eight day period.

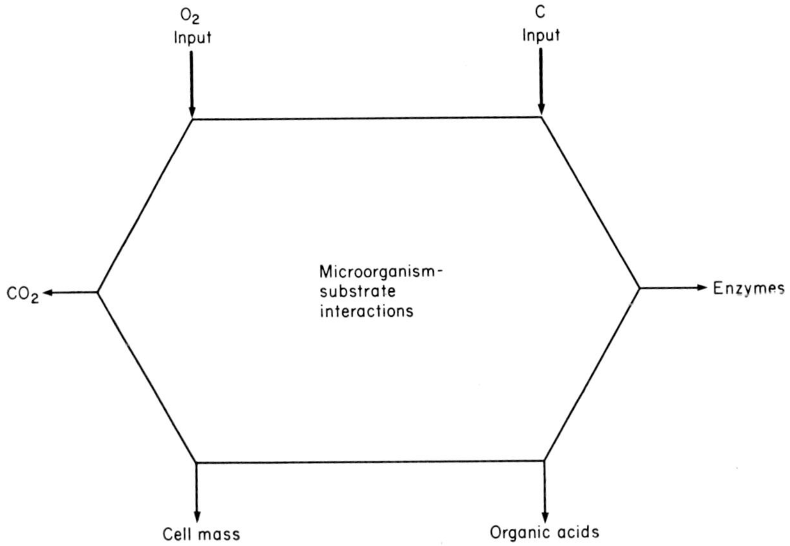

Fig. 2. Diagrammatic representation of inputs to and products of biodegradation.

organism that produced acetic acid, and *Chaetomium* sp. the only one to produce citric acid.

Almost all of the microorganisms produced acetic acid during starch breakdown. *Chaetomium* sp. exhibited the greatest production of acetic acid which amounted to 0·89 mg/ml/2 g starch. It also produced the largest amount of fumaric acid, and was the only organism to produce maleic and gluconic acids. *Myrothecium* sp. produced no detectable acids during starch degradation.

Difficulty arises in fully understanding microbial activity. Activity has generally been measured and evaluated by individual parameters of O_2 utilization, CO_2 evolution, or cell mass accumulation. It becomes necessary in understanding these partial activities to place them into a diagrammatic representation of what occurs during microbial-substrate interaction known as biodegradation, decomposition, or metabolism (Fig. 2).

Table IV

MEASURED ACTIVITY OF THE MICROORGANISMS EXPRESSED AS PERCENT CARBON DURING CELLULOSE BREAKDOWN

Organism	*%C in CO_2*	*%C in enzyme*	*%C as organic acids*	*%C in cell mass*[a]
Achromobacter sp.	6·5	0	0·0	—
Flavobacterium sp.	6·9	0	0·0	—
Stachybotrys sp.	42·7	0	1·0	54·0
Fusarium sp.	62·7	0	16·7	25·2
Chaetomium sp.	4·5	0	38·8	0·0
Curvularia sp.	3·4	0	22·9	87·7
Myrothecium sp.	16·6	0	0·0	0·0

[a] Assumed 40% carbon in mycelia.

In this study individual aspects of these activities are presented which need to be placed in a context of the whole scheme.

Since so many parameters had been analysed, all the measurements were converted to percent carbon in the original substrate. This allowed evaluations of how much carbon from the original substrate was converted to either CO_2, reducing sugars, organic acids, or cell mass. Stoichiometric relationships were used to determine the percent of the substrate that had been converted to each of the parameters being studied and these results are summarized in Tables IV–VI.

Table V

MEASURED ACTIVITY OF THE MICROORGANISMS EXPRESSED AS PERCENT CARBON DURING XYLAN BREAKDOWN

Organism	*%C in CO_2*	*%C in enzyme*	*%C as organic acids*	*%C in cell mass*[a]
Flavobacterium sp.	0·0	0·0	1·9	—
Vibrio sp.	4·2	0·0	0·0	—
Stachybotrys sp.	35·1	15·0	15·5	17·2
Fusarium sp.	54·0	0·0	28·2	43·9
Chaetomium sp.	34·2	4·0	6·9	32·3
Curvularia sp.	36·0	1·4	0·0	48·7
Myrothecium sp.	0·0	0·0	0·0	0·0

[a] Assumed 40% carbon in mycelia.

Table VI

MEASURED ACTIVITY OF THE MICROORGANISMS EXPRESSED AS PERCENT CARBON DURING STARCH BREAKDOWN

Organism	*%C in CO_2*	*%C in enzyme*	*%C as organic acids*	*%C in cell mass*[a]
Bacillus sp.	46·4	1·0	6·6	—
Pseudomonas sp.	2·8	0·8	2·6	—
Stachybotrys sp.	39·0	16·2	8·7	12·6
Fusarium sp.	4·3	0·8	25·8	28·3
Chaetomium sp.	21·2	0·0	33·7	18·2
Curvularia sp.	39·9	0·2	2·7	49·6
Myrothecium sp.	2·8	0·4	0·0	0·0

[a] Assumed 40% carbon in mycelia.

Achromobacter sp., *Flavobacterium* sp., and *Myrothecium* sp. did not appear to be effective in degrading cellulose. *Chaetomium* sp. showed no measurable increase in cell mass but exhibited a high percent substrate C released as organic acids (38·8%) and low evolution of CO_2 (4·5%). The greater substrate conversion into organic acids results in less carbon being converted to the other partial activities or parameters being analysed. Most of the substrate carbon was assimilated into cell mass by *Curvularia* sp. (87·7%) and organic acids (22·9%) and the CO_2 evolved was very low (3·4%). *Fusarium* sp. released the highest percent carbon as CO_2 (62·7%), while the acid (16·7%) and assimilation (25·2%) were moderate.

During xylan degradation, *Flavobacterium* sp., *Vibrio* sp., and *Myrothecium* sp. appeared unable to utilize the substrate very effectively. The rest of the organisms were able to utilize 70–80% of the substrate. *Fusarium* sp. converted the highest percent of the substrate C into organic acids (28·2%), evolved the greatest percent C as CO_2 (54·0%), and assimilated 43·9%. *Curvularia* sp. had the greatest assimilation of substrate C as cell mass (48·7%), no measurable organic acid production, and medium CO_2 evolution (36·0%).

During starch breakdown, *Bacillus* sp. exhibited the greatest percent carbon released as CO_2 (46·4%). The greatest substrate utilization was expressed by *Curvularia* sp., with almost 59% of the substrate carbon going into biomass, 40% into CO_2, and a low organic acid production level (2·7%). *Chaetomium* sp. had a low increase in biomass carbon (18·2%) but had the highest percent carbon (33·7%) released as organic acids, and a low CO_2 production level (21·2%). *Stachybotrys* sp. showed the greatest production of reduced sugars (16·2%), while 12·6% of the substrate C was converted to cell mass and 39·0% into CO_2 production.

DISCUSSION

Microorganisms can be divided into many groups on an interpretive basis of their ecological performance (Moore, 1971). The four considered important in this work were: the decomposers (most rapidly mineralizes the substrate carbon into CO_2 as a major product), producers of cell substances (majority of substrate carbon going into the formation of cell mass), producers of metabolic end-products (majority of substrate carbon going into the formation of organic acids), and the enzyme producers (majority of the carbon going into enzyme production).

Curvularia sp. would be the best organism for cell production during cellulose breakdown, with 87·7% of the substrate carbon converted to cell mass. This organism exhibited low CO_2 evolution (3·4%) but high organic acid production. *Curvularia* sp. completely digested cellulose, facilitating the harvesting of cell mass and organic acids. The greatest organic acid production (38·8%) was exhibited by *Chaetomium* sp.

The highest percent increase in cell mass was exhibited by *Curvularia* sp. during xylan breakdown in which 48·7% of the substrate C was converted to cell production. *Fusarium* sp. would make a good organic acid producer except for a high CO_2 production and high cell mass increase. This organism would have to be classified as the best decomposer of the organisms tested during xylan breakdown because of the high percent substrate carbon released as CO_2 (54·0%). *Stachybotrys* sp. exhibited the highest substrate conversion into enzyme production (15·0%).

During starch breakdown, *Curvularia* sp. exhibited the greatest accumulation of substrate C as cell mass (49·6%). This organism was the most efficient in converting the substrate to cell mass without the production of large amounts of other end-products. *Chaetomium* sp. exhibited the greatest production of organic acids (33·7%). *Stachybotrys* sp. converted the greatest amount of substrate carbon into enzyme production (16·2%) and exhibited very little conversion of starch to cell mass or organic acid production. This organism would be a good enzyme producer. *Bacillus* sp. is the best decomposer during starch breakdown. This organism converted 86% of the substrate C into CO_2.

This work has shown that in understanding microbial activity it is necessary to measure O_2 utilization, CO_2 production, cell mass accumulation, and enzyme and organic acid production during substrate degradation. Since the microorganisms have wide variations in what end-products are predominantly produced, it is necessary to study the individual parameters in context of the total scheme. Microorganisms can be selected for food production when most of the substrate C accumulates as cell mass. Tate and Lyle Ltd (1974) in England have already been working on converting agricultural wastes to microbial protein for use as animal feedstuffs. They can be selected for the formation of fermentation products when enzymes are produced as a major product or for organic acid production.

Acknowledgement

This research was supported in part by the New Mexico Energy Research and Development Program (BEF proposal-66) and US/IBP Desert Biome (Subcontract 0455).

REFERENCES

Bernfeld, P. (1951). Enzymes of starch degradation and synthesis. *Adv. Enzymol.*, **12**, 385.

Coleman, D. (1971). Measurement of total soil respiration at IBP grassland sites. mimeo. IBP Grassland Biome, 4 pp.

Kellog, D. W. (1969). Analysis of rumen fluid volatile fatty acids by chromatography with Porapak QS. *J. Dairy Sci.*, **52**, 1690–1692.

Levinson, H. S., Mandels, G. R., and Reese, E. T. (1951). Products of enzymatic hydrolysis of cellulose and its derivatives. *Arch. Biochem. Biophys.*, **31**, 351–356.

Moore, S. (1971). Enzymatic activity of soil fungi. Unpublished Masters Thesis. New Mexico State University, Las Cruces, New Mexico, 41 pp.

Tate and Lyle, Ltd (1974). *Microbial Protein From Agricultural Wastes.* Reading, Berks., p. 7.

Young, J. C., Clark, J. W., and Garner, W. (1964). The measurement of BOD by electrolysis. *Eng. Expt. Sta. Tech. Rept.*, No. 20, New Mexico State University, University Park, New Mexico, 22 pp.

VARIATION IN CHAIN LENGTH OF DEXTRAN FRAGMENTS PRODUCED BY DEXTRANOLYTIC BACTERIA

JOHN G. EDWARDS and FRED A. ROSENBERG

Department of Biology, Northeastern University,
Boston, Massachusetts 02115, U.S.A.

Summary

Dextrans belong to a family of extracellular bacterial polysaccharides which are polymers of D-glucopyranose with predominantly α-(1 $\rightarrow$ 6) glucosidic linkages. Dextranases (EC 3.2.1.11) are enzymes capable of hydrolysing the α-(1 $\rightarrow$ 6) glucosidic linkages of dextrans if non-α-(1 $\rightarrow$ 6) linkages do not interfere.

Previous studies to elucidate the mechanism of dextranase action by identification of the chain lengths of the hydrolysis products have usually been accomplished by difficult chromatographic techniques that frequently require preparative sample desalting and are limited to separation of short-chained oligosaccharides with degree of polymerization (D.P.) < 10.

We have applied a modification of the gel permeation chromatography (GPC) technique of Granath and Kvist for the molecular weight distribution analysis of dextran following bacterial degradation. The results indicate that *Cellvibrio vulgaris* and *Bacillus subtilis* each show distinctive patterns of dextranase activity.

MATERIALS AND METHODS

Bacteria

The *Bacillus subtilis* culture was obtained from the Northeastern University culture collection; *Cellvibrio vulgaris* was isolated from Fenway soil samples (Back Bay, Boston, Mass.). Pure cultures were obtained by plating diluted mixed culture samples on 2% Dextran T-70 (Pharmacia Fine Chemicals, Piscataway, N.J.) + basal salts (Berg *et al.*, 1972) agar and picking isolated colonies. Identification of *Cellvibrio vulgaris* was accomplished after performing standard morphological and biochemical tests. The test results were submitted to a Control Data Corporation Cyber 70 Model 72-14 computer at the Northeastern University Computation Center, using a computer program to identify bacteria developed by the author (unpubl. data). Pure cultures were maintained on tubed stock culture medium consisting of 2% industrial grade dextran + basal salts (Berg *et al.*, 1972) agar.

Gel permeation chromatography

Ascending solvent flow was employed through a water-cooled Pharmacia K16/100 borosilicate glass column with acrylic jacket (column dimensions = 1·6 × 100 cm) and two A16 adaptors, packed with Sephadex G-100 superfine gel. A jacketed K16/20 glass column (1·6 × 20 cm) was connected in series after the K16/100 column and was packed with Sephadex G-15 gel. The total bed volume (V_t) = 211·0 ml and was calibrated with water using the empty columns. Elution was carried out with 0·3% NaCl and 0·02% NaN_3 as an antimicrobial agent, using a Mariotte flask reservoir. Flow rate was about 6·0 ml/hr. The sample was applied by syringe through a Pharmacia LV three-way valve.

Void volume

The void volume (V_0) was determined before each run of a set of standards using 2·0 ml (2 mg/ml) Sephadex Blue Dextran 2000 ($\bar{M}_n = 2 \times 10^6$). $V_0 = 62{\cdot}0$ ml for the two columns and did not vary over time.

Partition coefficient

The concept of the volume fraction available for a solute in the gel phase was applied according to Laurent and Killander (1964):

$$K_{av} = (V_e - V_0)/(V_t - V_0)$$

where V_e is the elution volume of the species in question, V_t is the total volume, and V_0 the void volume of the gel bed used. All of these parameters can be determined accurately and the resulting K_{av} expresses the penetrability of substances of a certain molecular size into a gel of defined porosity.

Experimental molecular weight determination for standards

After autoclaving and dissolving in sterile basal salts solution (Berg *et al.*, 1972), 0·5 ml (1000 μg/ml) of Pharmacia Dextran T-70, T-40, and T-10, and 0·5 ml (2000 μg/ml) cellobiose and glucose were applied to the columns. Then 2·0 ml fractions were collected in tubes on an automatic fraction collector (GM Chromatography Co., P.O. Box 167, Greenville, Ill.). Molecular weight was correlated with K_{av} following the chromatography of each standard. The procedure was essentially the same as found in Granath and Kvist (1967).

Total reducing sugar

This was determined by the phenol-sulphuric acid method (Dubois *et al.*, 1956), using Dextran T-70 as standard.

Bacterial culture samples for gel permeation chromatography

Pure cultures 24–48 hr old of dextranolytic organisms on dextran agar slants were scraped and suspended in basal salts solution (Berg *et al.*, 1972) prior to inoculation of 0·1 ml into 125 ml metal-capped shaker flasks containing 50·0 ml of 2% Dextran T-70 + basal salts (autoclaved separately after adjusting the pH to 7·2). The flasks were placed in a 30°C water bath (Precision Scientific Co., Chicago, Ill.) with horizontal shaking motion at a rate of 30 cycles per minute. Samples were taken several days after inoculation and were centrifuged at 10 000 × *g* and the supernates were diluted 10^{-1} with distilled water before 0·5 ml was applied to the system. The above conditions were chosen so that the dextran in the collected fractions could be determined without further dilution. Weight average molecular weight ($\bar{M}_w$) and number average molecular weight ($\bar{M}_n$) for culture supernates were calculated from the experimental molecular weight according to the equations (Ingelman *et al.*, 1969): $\bar{M}_w = \Sigma N_i M_i^2 / \Sigma N_i M_i$ and $\bar{M}_n = \Sigma N_i M_i / \Sigma N_i$, where N_i is the number of molecules of kind 'i' present in the mixture and

M_i is their molecular weight. Degree of polymerization was calculated as D.P. = $\bar{M}_w$/molecular weight of glucose.

RESULTS

Correlation of molecular weight and K_{av} for dextran, cellobiose, and glucose on Sephadex

The standards and their elution profiles are presented in Table I and Fig. 1. With larger volume samples, *i.e.* 5·0 and 2·0 ml, band broadening occurred due to longitudinal diffusion. Not shown in Table I is a cellobiose standard, applied at a concentration of 2000 μg/0·5 ml to the single column containing

Table I

PHYSICAL CHEMICAL DATA FOR THE DEXTRAN, CELLOBIOSE, AND GLUCOSE STANDARDS USED IN THE CALIBRATION OF THE GPC DISTRIBUTION ANALYSIS

Standard	$\overline{M_w}$	$\overline{M_n}$	*Sample concentration (μg/ml)*	*Sample volume (ml)*	K_{av}	*Weight (%)*
Dextran T-70	70 000	42 500	1 000	0·5	0·040	48·4
Dextran T-40	39 500	29 500	1 000	0·5	0·121	62·8
Dextran T-10	10 500	6 400	1 000	0·5	0·510	66·3
Cellobiose	360	360	2 000	0·5	0·859	53·8
Glucose	180	180	2 000	0·5	0·872	52·0

Sephadex G-100 superfine gel. Since the elution volume of this low molecular weight standard was very close to the total column volume, the second column containing Sephadex G-75 gel was connected after the first column to increase the eluting power of the system for low molecular weight sugars. The Dextran T-40 and T-10 standards showed a broad molecular weight distribution, while Dextran T-70, cellobiose, and glucose appeared to be homogeneous.

Bacterial degradation of Dextran T-70

Elution profiles for 3-day-old dextran culture supernates of *Cellvibrio vulgaris* and *Bacillus subtilis* are presented in Fig. 2. Reducing sugar and molecular weight distributions for these samples are shown in Tables II and III. *Cellvibrio vulgaris* appeared to split dextran solely into long-chained

Table II

REDUCING SUGAR AND MOLECULAR WEIGHT DISTRIBUTION FOR 72-HR-OLD *CELLVIBRIO VULGARIS* CULTURE SUPERNATE

Peak no.	*Reducing sugar (μg/ml)*	*Percentage under peak*	$\overline{M_w}$	$\overline{M_n}$	*D.P.*
1	164·50	33·43	$>1{\cdot}0 \times 10^5$	$>1{\cdot}0 \times 10^5$	>650
2	92·11	18·72	78 668	78 077	437
3	201·34	40·92	50 100	47 989	278
4	34·11	6·93	27 927	27 474	155

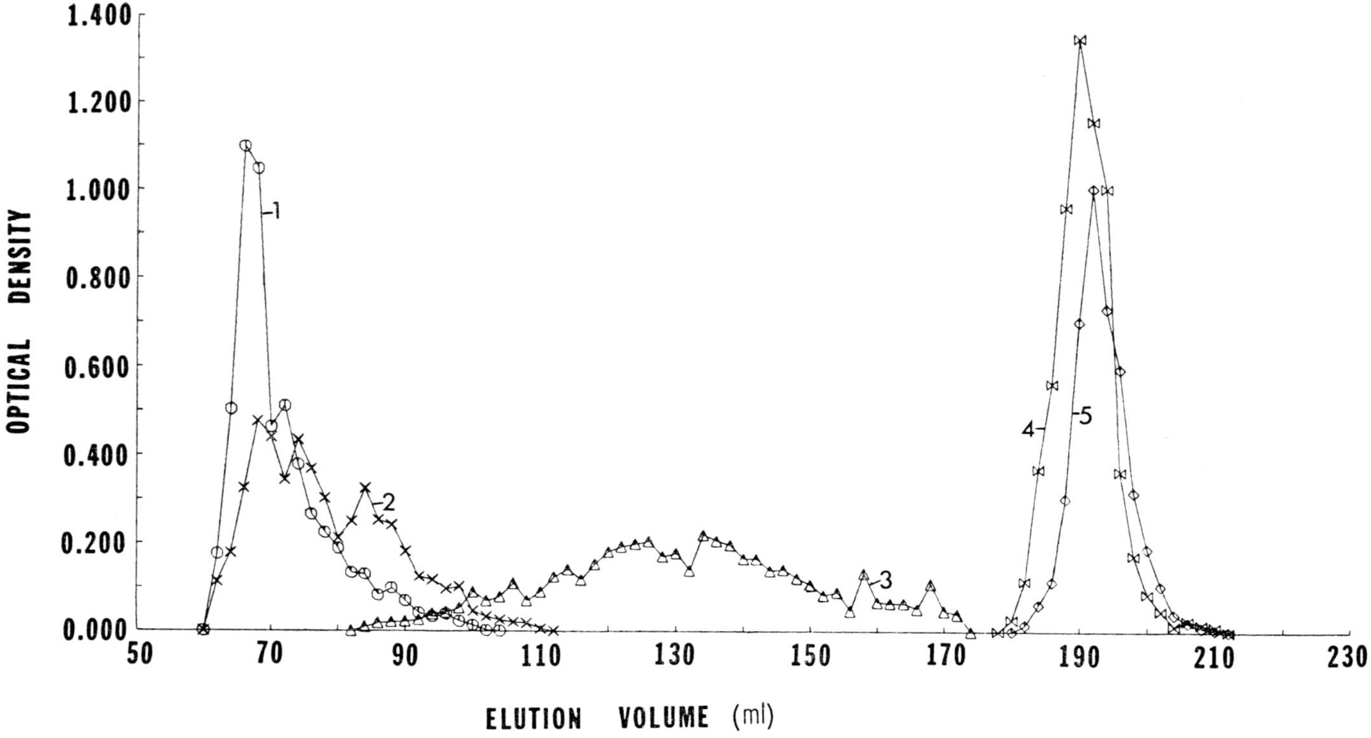

Fig. 1. *Plot of optical density against elution volume for Dextran T-70 (plot 1), T-40 (plot 2), T-10 (plot 3), cellobiose (plot 4), and glucose (plot 5) from Table I.*

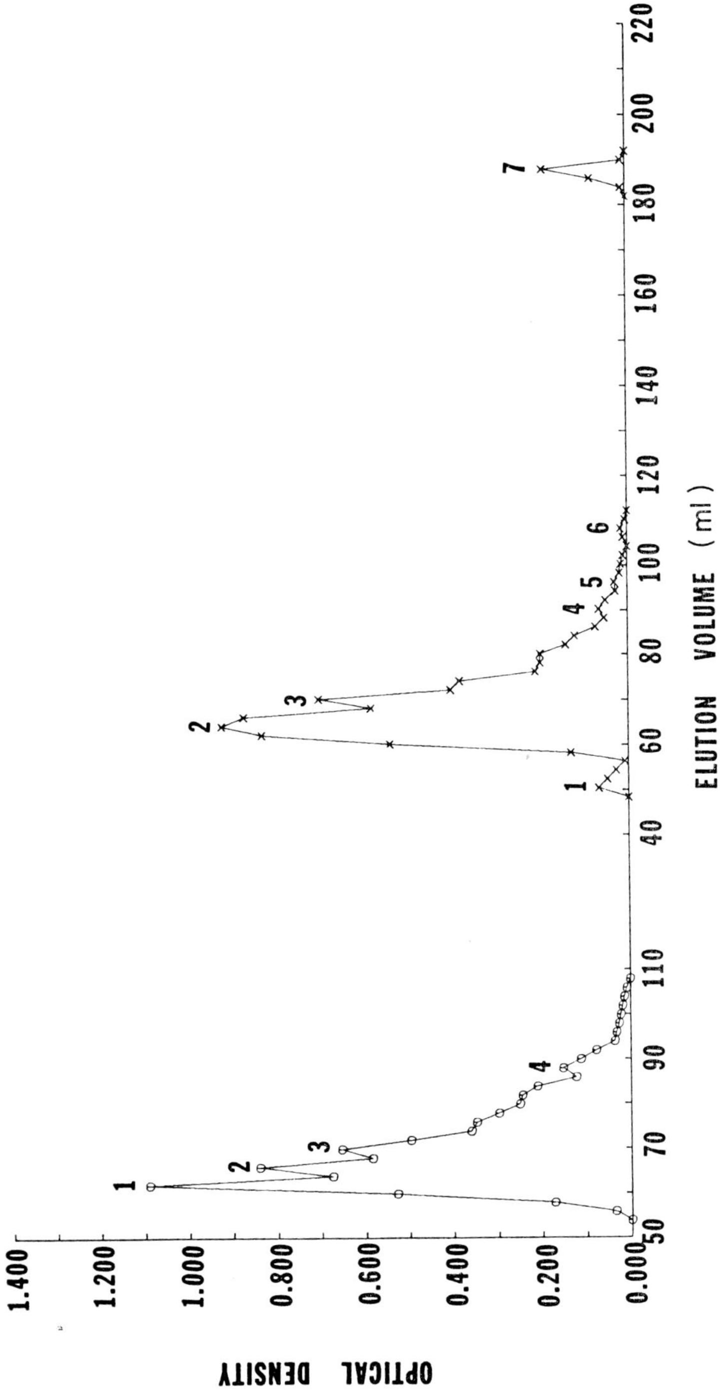

Fig. 2. Elution profiles for 3-day old culture supernates of Cellvibrio vulgaris (○—○), *and* Bacillus subtilis (×—×).

Table III
REDUCING SUGAR AND MOLECULAR WEIGHT DISTRIBUTION FOR 72-HR-OLD *BACILLUS SUBTILIS* CULTURE SUPERNATE

Peak no.	Reducing sugar ($\mu g/ml$)	Percentage under peak	$\overline{M}_w$	$\overline{M}_n$	D.P.
1	9·90	2·36	$>2{\cdot}0 \times 10^5$	$>2{\cdot}0 \times 10^5$	>1 000
2	206·99	49·34	103 888	99 503	577
3	166·12	39·60	49 210	46 873	273
4	9·73	2·32	30 097	30 097	167
5	5·00	1·19	25 752	25 684	143
6	2·05	0·44	20 403	20 392	113
7	10·70	4·70	409	409	2

fragments and apparently no glucose, disaccharides, or short-chained oligosaccharides were formed from the ends of the dextran molecule. The pattern of *B. subtilis* action on Dextran T-70 appeared to be both liquefaction and saccharification, as a significant percentage of disaccharides was formed following the long-chained fragments. This suggests that *B. subtilis* may split off disaccharides from the ends of the dextran chains. Both organisms appeared to synthesize a polysaccharide polymer of $\bar{M}_w > 70\,000$ as shown by the increase in the D.P. of the first peaks.

DISCUSSION

The gel permeation chromatography technique described above provides a powerful tool for the analysis of dextrans of different molecular weights over a broad molecular weight range (7×10^4 to 10^2). Granath and Kvist (1967) obtained excellent results with dextran fractions in the range of $\bar{M}_w =$ 150 000 to 5000, using a single column. In this study, resolution was increased using two columns with two types of Sephadex gel. Dextran T-70, cellobiose, and glucose eluted as nearly Gaussian curves, while the T-40 and T-10 Dextrans appeared to be multipeaked. The reason for this is thought to be due to the heterogeneity of the T-40 and T-10 Dextrans.

The enzymology of bacterial dextranases is still not well understood, but available evidence suggests that possibly there are at least two different types of enzymes with regard to their mode of attack. One mechanism shown by the extracellular dextranases of *Lactobacillus bifidus* (Bailey and Clarke, 1959; Clarke, 1959), *Bacteroides* sp. (Hehre and Sery, 1953), *Cytophaga* sp. (Janson and Porath, 1966), and *Cellvibrio fulvus* (Ingelman, 1948; Ingelman *et al.*, 1969) resembles that of fungal dextranases (Bourne *et al.*, 1962; Hudson and Weigel, 1963; Jeanes *et al.*, 1953; Tsuchiya *et al.*, 1952) and consists of random hydrolysis of interior α-(1 $\rightarrow$ 6) linkages to produce various shorter cello-dextrans. This type of enzyme produces a rapid drop in the degree of polymerization (liquefaction), but only a slow increase in reducing sugar values. The second type is represented by the extracellular dextranases of *Achromobacter* sp. (Sawai *et al.*, 1974), *Bacteroides* sp. (Sery and Hehre, 1956), *Brevibacterium* sp. (Yamaguchi and Gocho, 1973), and the intracellular isomaltodextranase of *Lactobacillus bifidus* (Bailey and Robertson, 1962) which exhibit nonrandom attack on the end of the chain (saccharification).

Except for one report of a cell-bound exodextranase (Zevenhuizen, 1968), bacterial dextranases show liquefying or saccharifying endohydrolase action exclusively.

Previous studies to elucidate the mechanism of dextranase action by identification of the chain lengths of the hydrolysis products were carried out by difficult chromatographic techniques that frequently require preparative sample desalting and are limited to separation of short-chained oligosaccharides with D.P. < 10. The gel permeation chromatography technique described above permits the observation of both liquefying and saccharifying dextranases simultaneously.

Cellvibrio vulgaris appeared to possess only the liquefying type of dextranase, while *B. subtilis* seemed to possess both liquefying and saccharifying capabilities. Glucose had been found in *B. subtilis* cell extracts by Zevenhuizen (1968) as a product of cell-bound exodextranase activity. In the present study, the significant percentage of disaccharide present in the *B. subtilis* culture supernate after 3 days suggests that the saccharifying dextranase activity of this organism may provide short-chained fragments which can pass through the cell wall as substrate for the cell-bound exodextranase.

Both organisms also appeared to synthesize a high molecular weight polysaccharide polymer. It is common in bacterial cultures to find substrate degradation processes occurring concurrently with cellular biosynthetic processes. Degradation processes provide the energy and lower molecular weight substances necessary for synthesis of cell wall and capsular structures. These assimilatory and dissimilatory processes can be demonstrated by sampling the same culture at various times after inoculation and observing the shifts in the chain length distribution during culture growth.

The present investigation shows that different bacteria capable of growth on dextran as the sole carbon source do show distinctive patterns of degradation, and that these patterns can be monitored using GPC. Further studies to isolate and characterize these dextranases are currently under way.

REFERENCES

Bailey, R. W., and Robertson, A. M. (1962). Carbohydrases of a rumen strain of *Lactobacillus bifidus*. *Biochem. J.*, **82**, 272.

Bailey, R. W., and Clarke, R. T. J. (1959). A bacterial dextranase. *Biochem. J.*, **72**, 49.

Berg, V., v. Hofsten, B., and Petterson, G. (1972). Growth and cellulase formation by *Cellvibrio fulvus*. *J. Appl. Bacteriol.*, **35**, 202.

Bourne, E. J., Hudson, D. H., and Weigel, H. (1962). Studies on dextrans and dextranases. *Biochem. J.*, **85**, 158–163.

Clarke, R. T. J. (1959). A dextran-fermenting organism from the rumen closely resembling *Lactobacillus bifidus*. *J. Gen. Microbiol.*, **20**, 549.

Dubois, M., Gilles, K. A., Halmiton, J. K., Rebers, P. A., and Smith, F. (1956). Colorimetric method for determination of sugars and related substances. *Anal. Chem.*, **28**, 350–356.

Granath, K. A., and Kvist, B. E. (1967). Molecular weight distribution analysis by gel chromatography on Sephadex. *J. Chromatog.*, **28**, 69–81.

Hehre, E. J., and Sery, T. W. (1952). Dextran-splitting anaerobic bacteria from the human intestine. *J. Bacteriol.*, **63**, 424.

Hudson, D. H., and Weigel, H. (1963). Studies on dextrans and dextranases. *Biochem. J.*, **88**, 588–591.

Ingelman, B. (1948). Enzymatic breakdown of dextran. *Acta Chem. Scand.*, **2**, 803.

Ingelman, B., Gronwall, A., Gelin, L., and Eliasson, R. (1969). *Properties and Applications of Dextrans*, p. 9, Almquist and Wiksell, Stockholm.

Janson, J. C., and Porath, J. (1966). In *Methods Enzymology*, Vol. VIII, pp. 615–621 (E. F. Neufeld and V. Ginsburg, eds.), Academic Press, New York.

Jeanes, A. R., Wilham, C. A., Jones, R. W., Tsuchiya, H. M., and Rist, C. E. (1953). Isomaltose and isomaltotriose from enzymatic hydrolysates of dextran. *J. Am. Chem. Soc.*, **75**, 5911–5915.

Laurent, T. C., and Killander, J. (1964). A theory of gel filtration and its experimental verification. *J. Chromatog.*, **14**, 317–330.

Sawai, T., Toriyama, K., and Yano, K. (1974). A bacterial dextranase releasing only isomaltose from dextrans. *J. Biochem.*, **75**, 105–112.

Sery, T. W., and Hehre, E. J. (1956). Degradation of dextrans by enzymes of intestinal bacteria. *J. Bacteriol.*, **71**, 373.

Sorenson, W. R., and Campbell, T. W. (1961). *Preparative Methods of Polymer Chemistry*, pp. 32–42. Interscience Publishing Inc., New York.

Tsuchiya, H. M., Jeanes, A. R., Bricker, H. M., and Wilham, C. A. (1952). Dextran-degrading enzymes from molds. *J. Bacteriol.*, **64**, 513.

Yamaguchi, T., and Gocho, S. (1973). Production and properties of alkaline dextranase from a newly isolated *Brevibacterium*. Agr. Biol. Chem., **37** (11), 2527–2533.

Zevenhuizen, L. P. T. M. (1968). Cell-bound exo-dextranase of *Bacillus* species. *Carbohyd. Res.*, **6**, 310.

BIOLOGICAL TREATMENT OF CRESYLIC ACID LADEN WASTE WATER

W. E. OLIVE, Jr., H. D. COBB and R. M. ATHERTON

Trinity University, San Antonio, Texas 78284, U.S.A.

Summary

The effect of temperature (15, 25, 35, and 45°C) and initial pH (6·3, 7·3, 8·3, and 9·3) on the growth rate was determined for recently isolated pseudomonads. Inoculum concentrations were 0·05, 0·1, 0·5, and 1·0 mg/ml. Optimum growth and cresol metabolism occurred at a temperature of 35°C and an initial pH of 8·3. Growth rates at 25°C were slightly less than those at 35°C. At a temperature of 15°C, growth rates were greatly reduced. No growth was observed at a temperature of 45°C; however, if after 8 hr the temperature was decreased to 35°C, the culture did grow and metabolize cresol at a reduced rate. The highest concentration of cresol utilized by organisms in this study was 1450 ppm. A bench scale pilot plant was constructed to determine the practicality of using microorganisms in a biological system for the removal of cresylic acid from waste water. The column was inoculated with a cell suspension containing 1·8 g of organisms. Medium containing a buffered mineral salts solution and cresol was recycled over the column. Removal rate of cresol ranged from 0·75 g/hr to 2·96 g/hr. Factors affecting the rate of removal were: recycling rate, volume of liquid phase, and the concentration of cresol. The column has been operating successfully for 7 months, removing cresol at an average rate of 1·5 g/hr.

INTRODUCTION

The use of phenolic compounds in industrial processes is increasing constantly. The stable and toxic nature of these compounds makes their disposal a particularly important problem. Various military bases encounter this problem as a result of carbon removal and paint-stripping processes involving aircraft maintenance. Solutions containing cresylic acid are used in these processes and are eventually discharged into rivers where they create a potentially serious pollution problem.

Although the physiology of isolated strains of organisms involved in the breakdown of aromatic compounds has been studied in some detail, there have been few investigations into the ecological stability of these organisms when they are cultured simultaneously. In a number of cases organisms which have been demonstrated to be effective in the degradation of phenolic compounds have been found to produce toxic by-products, which tend to reduce the activity of the total ecological system.

The purpose of this study was to isolate microorganisms capable of actively metabolizing cresol and to determine the environmental conditions under which optimal metabolism of this pollutant would occur. After these initial studies, a bench scale pilot plant was to be constructed and inoculated with selected strains of the most promising organisms and the applicability of this design in removal of cresol under simulated field conditions tested.

MATERIALS AND METHODS

Isolation and maintenance of organisms

Samples were taken from a paint-stripping room at locations that were expected to harbour microorganisms capable of degrading cresol. It has been reported that microorganisms capable of degrading hydrocarbons would most likely be found near locations where these compounds were being used (Bushnell and Hass, 1941).

Stock cultures of microorganisms isolated from these samples were maintained on 1·8% agar plates containing basal mineral salts and 900 ppm cresol. The concentrations of the mineral salts used in the medium were (g/litre); $MgSO_4$, 0·20; $CaCl_2$, 0·02; NH_4NO_3, 1·00; $FeCl_3$, 0·005; NaCl, 0·10; K_2HPO_4, 2·00; and KH_2PO_4, 1·00. The 900-ppm cresol was prepared in the proportions of 360 ppm ortho-, 360 ppm meta-, and 180 ppm para-cresol (these concentrations of the isomers are approximately equivalent to those found in technical grade cresylic acid). Stock cultures were exposed to no carbon and energy source other than cresol.

Growth and cresol metabolism

Organisms from 3-day-old stock subcultures were used to inoculate spinner flasks which contained 1 litre of a medium consisting of basal mineral salts and 900 ppm cresol. After the cultures had achieved an adequate cell concentration and were in logarithmic growth, the cells were harvested by centrifugation and resuspended in a small volume of distilled water. Flasks containing 100 ml of medium incubated at the appropriate temperatures were inoculated with a cell suspension sufficient to produce the final required concentration. The contents of these flasks were divided equally between two nephelometer flasks which were placed on a shaker for incubation in an environmental chamber.

Cell concentration during the growth study was determined turbidimetrically with a Klett Somerson colorimeter. One-millilitre samples from each flask were collected at appropriate times and frozen until cresol and pH determinations were made. Cresol concentration was determined with a Barber Coleman Series 5000 gas chromatograph.

During growth and cresol utilization experiments, the following conditions were maintained except when otherwise noted: incubation temperature 35°C, inoculum concentration 0·1 mg/ml, total cresol concentration 900 ppm, and initial pH 8·3.

Pilot plant studies

For pilot plant studies, a glass column 6 inches in diameter and 9 ft tall was constructed (Fig. 1). The column was filled partially with porcelain berel saddles. Fluid in the column containing 900 ppm cresol, mineral salts, and tap water was drawn from the bottom by a centrifugal pump and cycled to the top at a flow rate of 2·3 litre/min where it was distributed by a 3 inch dispersion head onto the saddles. Humidified air was continuously pumped through the column and expelled through a vent in the top at a rate of 0·5 ft^3/min. The column was inoculated with 8 litres of a cell suspension

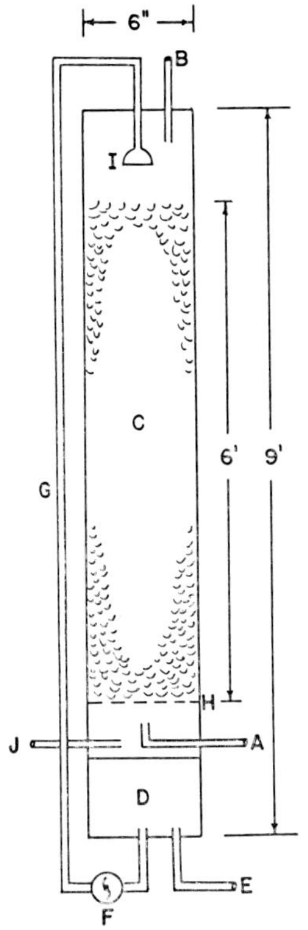

Fig. 1. *Trickling filter pilot plant. A, air inlet; B, vent; C, berel saddles; D, liquid reservoir; E, effluent; F, reflex pump; G, reflux line; H, saddle support; I, dispersion apparatus; J, inlet.*

containing equal portions of the strains tested, at a total weight of 1·8 g.

Prior to each experiment, the column was completely drained and washed by recycling 5 litres of fresh tap water over the saddles for 5 min. The wash water was then removed and an appropriate volume of tap water plus mineral salts and cresol were added. Samples were taken at appropriate times and frozen until the cresol concentration was determined.

RESULTS

Taxonomy

Taxonomic studies demonstrated that all of the organisms could be placed within the fluorescent group of *Pseudomonas* as defined by Stanier *et al.* (1966), which included *P. aeruginosa*, *P. fluorescens*, and *P. putida*. These strains did not correspond exactly to any of those described by Stanier but appeared to be intermediate between those in the fluorescent group (Atherton *et al.*, 1975).

Organisms found to utilize cresol as the sole carbon and energy source were recovered from samples taken from the paint-stripping room and designated J, K, L, O, P, and Q. Nineteen strains were isolated from the samples and coded as follows:

J1, J2, J3, J4
K1, K2, K3

L1, L2, L3, L4
O1, O2, O3
P1, P2
Q1, Q2, Q3

Each sample was obtained from a different location in the paint-stripping room, with each member of a series being isolated from this sample.

Growth and cresol metabolism

Preliminary work indicated that members of the J series were the most promising isolates; therefore, they were selected for this study. The experimental parameters of temperature, pH, cresol, and cell concentrations were examined for their effect on the degradation of cresol.

Growth and cresol utilization rates were determined at different initial pH's of 6·3, 7·3, 8·3, and 9·3. These organisms demonstrated optimal growth and cresol utilization at an initial pH of 8·3, with complete disappearance of cresol occurring in 7 hr. At initial pH 7·3 the growth rate was similar to that at pH 8·3, with complete utilization of cresol occurring in 9 hr. At an initial pH of 6·3, growth was greatly reduced and cresol metabolism was incomplete. Neither growth nor cresol utilization was observed at pH 9·3.

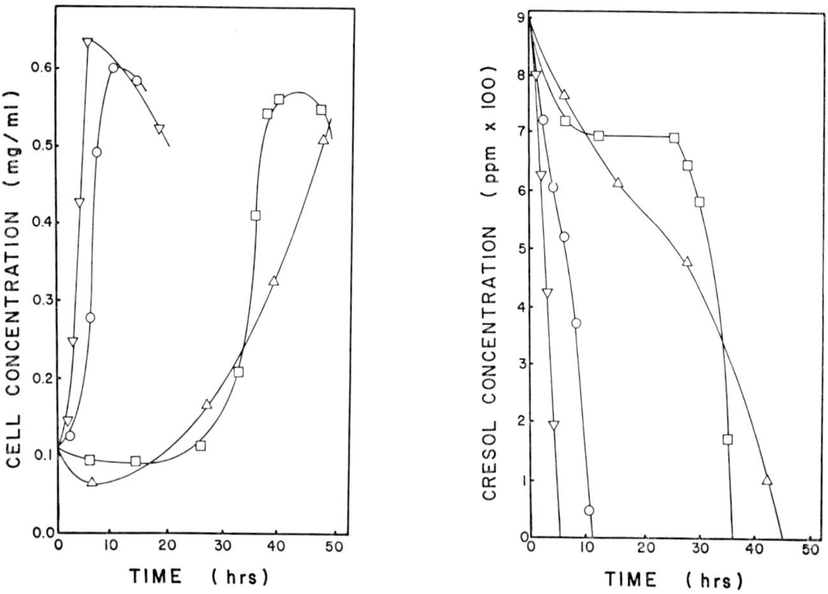

Fig. 2. Growth and cresol metabolism rates at various temperatures: (15°C Δ) (25°C ○) (35°C ∇) (45°C for 8 hr followed by return to 35°C □).

Figure 2 represents growth and cresol metabolism of the organisms at temperatures of 15°C, 25°C, 35°C, and 45°C. All four strains performed optimally at 35°C, with complete utilization of cresol occurring at 7 hr. The organisms did not grow during an 8-hr exposure to 45°C; however, if after 8 hr the incubation temperature was decreased to 35°C, a lag period occurred and growth was re-established with complete utilization of cresol occurring after 37 hr of incubation. While growth at 15°C demonstrated a protracted lag period, the cells adapted and grew at a reduced rate, with disappearance of substrate occurring after 44 hr incubation.

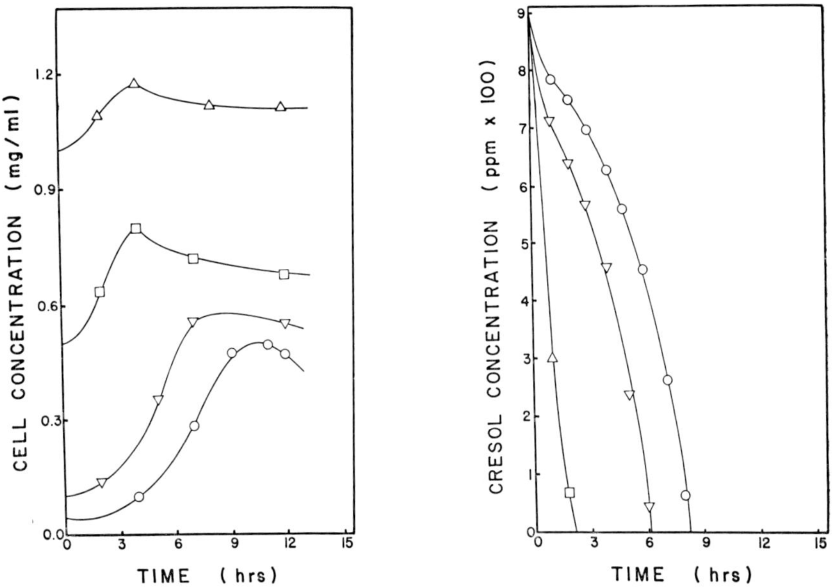

Fig. 3. Growth and cresol metabolism rates at various initial cell concentration: (0·05 mg/ml ○) (0·1 mg/ml ▽) (0·5 mg/ml □) (1·0 mg/ml Δ).

Figure 3 is representative of the growth and cresol metabolism of the organisms with cell concentration as the variable. Cultures were inoculated at 0·05, 0·1, 0·5, and 1·0 mg/ml. Cultures inoculated at 0·05 mg/ml showed a slightly longer lag phase than those inoculated at 0·1 mg/ml; however, the rate of growth was similar under both conditions. The rates of cresol metabolism in cultures with initial inocula of 0·5 and 1·0 mg/ml were identical, with complete utilization occurring in 2 hr. Cresol utilization in cultures with 0·05 and 0·1 mg/ml inocula occurred in 6 and 8 hr, respectively.

To test the effect of intermittent exposure of cresol on culture growth, the cultures were grown under optimal conditions at a temperature of 35°C and at initial cell concentrations of 0·5 and 1·0 mg/ml. Maximum growth and complete utilization of 900 ppm cresol occurred in 3 hr, after which no cresol was added for a period of 21 hr. At this time, the cultures were exposed to an additional 800 ppm cresol. Again, growth and complete utilization of cresol occurred, but at a lower rate, requiring 23 hr.

Cultures were exposed to a series of media containing increasingly higher concentrations of cresol to determine the maximum cresol concentration at which utilization would occur. These organisms were capable of growing and actively metabolizing cresol at concentrations of 900, 1150, and 1450 ppm, with complete utilization occurring at 5, 6, and 7 hr, respectively (Fig. 4). At a cresol concentration of 1500 ppm, growth and cresol utilization rates were dramatically reduced; at a cresol concentration of 1600 ppm, no growth or cresol utilization occurred.

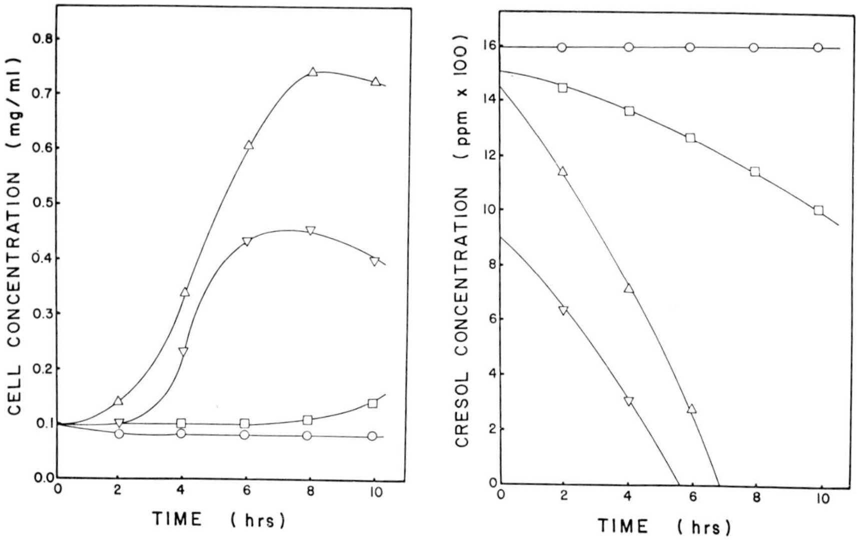

Fig. 4. Growth and cresol metabolism rates at various cresol concentrations: (*900 ppm* ▽) (*1450 ppm* Δ) (*1500 ppm* □) (*1600 ppm* ○).

Pilot plant studies

Ten days after inoculation of the J series on the column a substantial microbial population had been established on the inner surface of the column and on the berel saddles. After 10 days, the column had reached its maximum cresol utilization efficiency, with the entire surface area available for microbial attachment occupied. The column was then tested under a variety of conditions to determine its ability to remove cresol from the medium.

At a volume of 2·5 litres, complete utilization of 900 ppm cresol occurred in 3 hr. At a volume of 5·0 litres, the cresol was completely removed in 4 hr. At a volume of 10·0 litres, 6 hr were required for the complete utilization of the substrate. The rate of utilization of total cresol was observed to increase as the volume of the liquid phase was increased, with the rates of 0·75, 1·12, and 1·50 g/hr corresponding to the volumes of 2·5, 5·0, and 10·0 litres.

Up to this point, the flow rate of fluid through the column in all the experiments had been 2·3 litre/min and at a volume of 5 litres a utilization rate of 1·12 g/hr was obtained. When the flow rate was increased to 4·6 litre/min, the cresol utilization rate increased to 1·63 g/hr. This represents a 31% increase in the utilization rate.

The results of an experiment in which cresol was added to the column at different rates for extended periods of time are illustrated by Fig. 5. The residual cresol, represented by the lower line, is the amount of cresol present in the liquid phase at the time of sampling. The added cresol, represented by the upper line, is the amount of cresol added to the system over a given period of time. The amount of cresol in the liquid phase at zero time was 8 g. Fourteen grammes of cresol had been added by the end of 13 hr and the residual cresol was only 4 g, which indicated that the organisms had not only utilized the 14 g that were added but also 4 of the original 8 g. This yielded a utilization rate of 1·45 g/hr at 13 hr. Over the period from 13 to 24 hr, an additional 21 g of cresol was added which brought the total amount added to

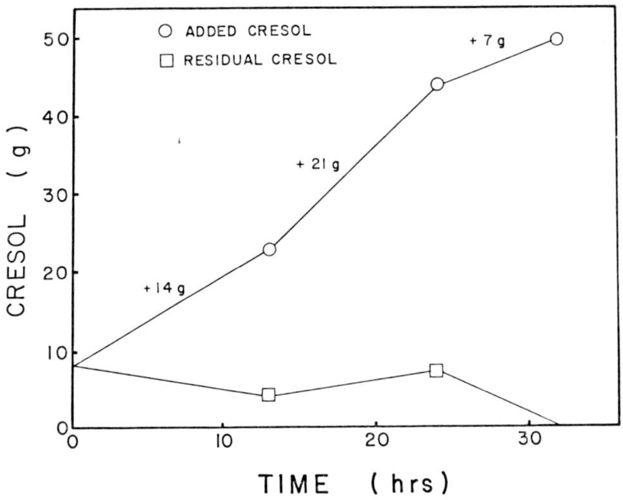

Fig. 5. Total cresol utilization when cresol is added to the liquid phase of the trickling filter pilot plant: (residual cresol □) (added cresol ○).

35 g. The residual cresol at 24 hr was 7 g, which produced a utilization rate of 1·53 g/hr. From 24 hr to 32 hr the rate of cresol addition was decreased, with 7 g of cresol being added. The total amount of cresol added during the 32-hr period was 42 g, and since the residual cresol decreased to zero, a total utilization rate of 1·56 g/hr was obtained.

DISCUSSION

Brebion *et al.* (1963) suggested that for the biological degradation of industrial effluents three criteria should be considered: (1) large numbers of pure strains should be readily available; (2) the behaviour of these strains in different media should be examined, especially with regard to determining the optimum conditions for their development either singly or together; and (3) the choice of these strains should be established with reference to a given problem of purification.

Large numbers of any one of the four strains studied are easily obtained by incubation in 1 litre spinner flasks for approximately 36 hr in a mineral salts medium containing 900 ppm cresol. Physiological studies have established that the optimum conditions for growth and cresol metabolism by the J series was: initial pH range 6·3–8·3 with preference to 8·3, temperature range of 15–25°C with optimum growth at 35°C, 0·5 mg/ml as the minimum cell concentration at which maximum growth occurs, and a substrate concentration not to exceed 1450 ppm cresol. The third suggestion by Brebion *et al.*, namely, selection of the strain on the basis of a specific problem, has surely been established since the J series demonstrated such robust growth and metabolism of cresol.

After determining the optimum conditions for growth and cresol metabolism by the J series, a pilot plant was implemented to determine its capability of removing cresol from simulated waste water under field conditions.

Since all previous study of the J series had been performed on individual strains, it was not known what might occur when they were cultured simultaneously. However, the fact that they were isolated from the same sample site suggested that they were capable of existing in close association with each other. It was hoped that this close association would, in fact, be synergistic, with one strain benefiting the other. For this reason, an equal mixture of all four strains was used to inoculate the column.

The column was tested at liquid phase volumes of 2·5, 5·0, and 10·0 litres, using fresh medium with 900 ppm cresol. As the volume was increased, the total amount of cresol was likewise increased and when the utilization rate was calculated on the basis of grammes of cresol utilized per hour, an increase in volume brought about an increase in the utilization rate.

Any variable that can be altered to increase the efficiency of the column would be of obvious benefit. By increasing the flow rate, a corresponding increase in utilization rate occurred due to increased exposure of the cresol in the medium to the attached microbes.

Under field conditions, the concentration of cresol can be expected to fluctuate. The ability of the column to remove cresol added at several different rates is illustrated in Fig. 5. Manipulation of the flow rate can also be used to compensate for this fluctuation. For example, under conditions of high cresol concentration, the flow rate can be increased to effect a more rapid decrease in concentration. When the cresol concentration is low, a slower flow rate can be used to avoid the total disappearance of cresol, thereby eliminating the need to maintain the column with alternate carbon sources.

For 7 months the four strains in the J series have exhibited the capability of existing in close association within the pilot plant, effectively removing cresol at an average rate of 1·5 g/hr. The stability of the plant for such a long period of time and under a great variety of experimental conditions lends credence to the premise that an ecologically compatible mixture of organisms is desirable over the seeding of a trickling filter of this type with a single strain. The ramifications of such an observation are many and are beyond the scope of this brief paper. It is obvious, however, that the genetic potential of such a mixture would be greatly enhanced, permitting the ecological population to successfully adapt to changing environmental conditions while remaining compatible within itself.

REFERENCES

Atherton, R. W., Cobb, H. D., and Olive, W. E. (1975). An ecological approach to the biological treatment of cresylic acid laden waste water. I. Isolation and characterization of Pseudomonads capable of metabolizing cresols. Texas Reports on Biology and Medicine. In press.

Brebion, G., Prat, S., and Prat, J. (1963). A study of purification by biological means, involving the successive actions of selected and adapted strains of microorganisms for the treatment of residual trade wastes from paper pulp factories. *J. Air Water Pollut.*, 7, 413–425.

Bushnell, L. D., and Haas, H. F. (1941). The utilization of certain hydrocarbons by microorganisms. *J. Bacteriol.*, **41**, 653–675.

Stanier, R. Y., Palleroni, N. J., and Doudoroff, M. (1966). The aerobic pseudomonads: a taxonomic study. *J. Gen. Microbiol.*, **43**, 159–271.

COMPARISON OF VARIOUS *PENICILLIUM FUNICULOSUM* STRAINS AS DEGRADATIVE TEST ORGANISMS

R. E. KLAUSMEIER, E. I. JAMISON and J. L. OSMON

Naval Weapons Support Center, Crane, Indiana, U.S.A.

Summary

Penicillium funiculosum ATCC 9644, long accepted as a standard organism in many fungus resistance tests, develops atypical growth when subcultured on a standard laboratory medium. In searching for a substitute culture, six strains of *Penicillium funiculosum* were tested for their ability to degrade cellulose, plastic formulations and ingredients, and a variety of other substrates. The results indicate that *Penicillium funiculosum* ATCC 9644, whether growing atypically or demonstrating normal growth patterns on a high sucrose medium, is still a satisfactory test organism. In addition, *Penicillium funiculosum* ATCC 11797 has the ability to degrade similar substrates, and could be a satisfactory substitute. Not all strains of *Penicillium funiculosum* show similar degradative capacities, however.

INTRODUCTION

Penicillium funiculosum is widely used as a test organism in a variety of fungus resistance tests. The most commonly used strain is *P. funiculosum* QM 391 (ATCC 9644) although most test methods originating outside the United States do not call for specific strains. It has been noted that *P. funiculosum* QM 391 shows drastic morphologic change, particularly in sporulation characteristics, if repeatedly subcultured on low sugar medium (B. Wiley and E. Simmons, pers. comm.). This characteristic is restored by transfer of the culture back to high sugar media. It has been suggested that another, more stable strain of *P. funiculosum* be substituted in test procedures. This article examines 'normal' and 'deteriorated' cultures of *P. funiculosum*, as well as several stable strains, to determine their suitabilities as test cultures.

MATERIALS AND METHODS

All of the organisms used in this study will be referred to by their QM strain number. These organisms can be obtained from the NLABS culture collection of fungi, Department of Botany, University of Massachusetts, Amherst, Massachusetts, 01002, U.S.A. The six strains of *P. funiculosum* used in this study are described below. QM 391 (ATCC 9644) is the culture which demonstrated morphologic instability (general loss of sporulation and pigmentation) when maintained on a low sugar medium, but 'typical' morphology when cultured on a high sugar medium. QM 9277 is a culture derived from QM 391 by continuous subculture on low sugar medium. This culture also reverts to 'wild type' morphology when returned to high sugar medium. QM 474, QM 475, QM 9417, and QM 9419 are all stable strains of *P. funiculosum*. QM 474 (ATCC 11797) is the culture suggested as a replacement for QM 391.

The standard culture maintenance media for all of these organisms is malt extract agar. However, some of the organisms were maintained on, or transferred to, M40Y medium having the following composition: 400 g sucrose, 20 g malt extract, 5 g yeast extract, and 20 g agar per litre of distilled water. Strains 391 and 9277 were maintained in three different ways as follows: one strain of each organism was maintained on malt extract agar but transferred to M40Y media for preparing inoculum. This strain will be designated MS. The third strain was maintained on 40Y media throughout the study. This strain was designated S. All cultures were maintained on 1-oz prescription bottles containing 6 ml of medium prepared as a slant.

The inoculum was prepared from each of the cultures by introducing 20 ml of sterile distilled water in each culture bottle. The surface of the agar was gently scraped with a nichrome wire loop and the resultant suspension was poured into a sterile, stoppered flask containing 10–15 sterile glass beads and shaken vigorously to break up spore clumps. The resultant suspension was then filtered through glass wool for use in inoculating the various samples. All operations were performed observing aseptic technique. The mineral salts agar used throughout these studies was Crane Mineral Salts (CMS) (Klausmeier *et al.*, 1963) solidified with 2 % agar.

The ability of each culture to degrade filter paper was assessed by placing 1 × 4 cm strips of sterile filter paper on mineral salts agar plates and inoculating each strip with 0·5 ml of a spore suspension. After 21 days incubation at 28°C, the amount of growth was rated as light, moderate, or heavy. The deterioration of the filter paper strips was assessed by removing the paper from the agar plate and attempting to pull the paper apart with the fingers. Deterioration was rated as slight, moderate, or extensive.

The ability of the strains to degrade plasticizers was assessed by preparing homogenized plasticizer agar plates. The various plasticizers individually were blended with molten CMS agar or molten CMS agar to which 1 % yeast extract and 1 % glucose had been added in a Waring blender to effect a final concentration of 1 % of the particular plasticizer. The plates were streaked with a spore suspension and incubated for 21 days at 28°C. Degradation of the plasticizer was indicated by a clear zone or halo around the resultant growth on the agar plate. This is a slight modification of the method used by Klausmeier (1966). The plasticizers examined were butylene glycol polyadipate (BGA), *n*-butyl oleate (*n*BO), dibutyl phthalate (DBP), triethylene glycol dipelargonate (TG-9), dibutyl sebacate (DBS), tributyl citrate (TBC), tricresyl phosphate (TCP), and diisooctyl phthalate (DIOP).

Deterioration of plastic samples was measured by using a weight-loss technique which has been previously described (Osmon and Klausmeier, 1971). Plastic samples were extruded polyvinyl chloride containing 33–40 % plasticizer with approximately 3 % of various pigments, stabilizers, lubricants, etc. The plasticizers used were dibutyl sebacate (DBS) and diisooctyl phthalate (DIOP). Samples were cut using a manually operated die into strips 1 × 100 mm. These samples were tested in groups of five.

The ability of the cultures to degrade cotton twine was assessed by measuring the tensile strength of the string samples after incubation with the various fungal strains and comparing the results with uninoculated controls.

Twine samples were inoculated by dipping the entire string into a spore suspension and placing the inoculated sample on a mineral salts agar plate so that the entire length of the string was in contact with the agar surface. Tensile strength was measured using an Instron TT-C Universal Testing Machine. A gauge length of 7 in and a crosshead speed of 12 in/min were employed for these measurements. Tensile strength was assessed after 0, 10, and 21 days' incubation. Six replicate samples were tested and the average result is reported.

The degree of sporulation of the cultures was assessed as follows: 2 ml of the spore suspension were added to a slant of the appropriate medium. The inoculum was shaken to distribute over the surface of the agar and the excess poured off. Since the surface area of the agar in each slant was approximately equal and the entire surface was covered with fungal growth, this technique resulted in approximately equal amounts of growth being harvested for spore crops. The spores were harvested in the following manner: 15 ml of distilled water containing 0·5 g/litre of wetting agent aerosol OT was introduced into each bottle. The surface of the agar was thoroughly scraped with a nichrome loop. Ten glass beads were then placed on each bottle and the bottles vigorously shaken to break up the spore clumps. The resulting suspensions were filtered through glass wool and the optical density of the suspension read on a Spectronic 20 at 525 nm. To obtain the spore counts, the optical density was compared to standard curves which had previously been prepared by comparing OD to direct counts of spores/ml as determined in the Petroff-Hauser counting chamber.

RESULTS

The various cultures tested showed considerable variety in their ability to degrade various substrates. However, the various cultural conditions for the unstable strains did not have any effect on the degradation capabilities of these cultures.

The abilities of the various cultures to degrade cellulosics are described in Tables I and II. The results of the experiment on the degradation of filter paper are admittedly subjective and nonquantitative. However, the data (Table I) are felt to be sufficient to group the strains into three categories. Strains 391 and 9277 totally degraded the filter paper strips. After 21 days' incubation, the filter paper could not be removed intact from the surface of the agar plate. Strains 474 and 475 degraded the filter paper strips to a lesser

Table I

GROWTH OF CULTURES ON FILTER PAPER

Strain no.	*Growth*	*Deterioration*
QM 391	Heavy	Extensive
QM 474	Light	Moderate
QM 475	Moderate	Moderate
QM 9417	Light	Slight
QM 9419	Light	Slight
QM 9277	Heavy	Extensive

Table II

EFFECT OF VARIOUS CULTURES ON TENSILE STRENGTH OF TWINE

Culture no.	*10 days* Average load (lbs)	*10 days* Percent of control	*21 days* Average load (lbs)	*21 days* Percent of control
Control	52·7	—	42·0	—
QM 474	20·6	39	2·6	6
QM 475	21·4	41	1·6	4
QM 9417	57·4	109	48·2	115
QM 9419	52·0	99	40·5	96
QM 391	13·3	25	0·5	1
QM 391MS	10·4	20	0·5	1
QM 391S	11·1	21	0·4	1
QM 9277	16·0	30	0·5	1
QM 9277MS	13·5	26	0·3	1
QM 9277S	8·6	16	0·5	1

extent. However, the papers were easily torn apart in the fingers. Strains 9417 and 9419 degraded the strips only slightly after 21 days' incubation.

More quantitative data on the cellulolytic abilities of the various strains were derived from studies using cotton twine (Table II). All media variants of strains 391 and 9277 were totally effective in degrading twine. After 21 days' incubation, these cultures were able to reduce the tensile strength of the twine samples to 1% or less of its original value. Strains 474 and 475 were only slightly less effective. Strains 9417 and 9419, however, showed no significant ability to degrade cotton twine.

Some strains of *P. funiculosum* have demonstrated ability to degrade plasticized polyviny formulations. Accordingly, the various strains were evaluated as to their ability to degrade a variety of plasticizers (Table III), as well as two plasticized vinyl formulations (Table IV). In the plasticizer

Table III

DEGRADATION OF PLASTICIZERS BY FUNGI

Medium	474	475	9417	9419	391	391MS	391S	9277	9277MS	9277S
BGA	+	+	+	+	+	+	+	+	+	+
BGA+YEG	+	+	○	○	+	+	+	+	+	+
nBO	+	+	+	+	+	+	+	+	+	+
nBO+YEG	+	+	○	○	+	+	+	+	+	+
DBP	○	+	○	○	○	○	○	○	○	○
DBP+YEG	+	○	○	○	○	○	○	○	○	○
TG-9	+	+	+	○	+	+	+	○	+	○
TG-9+YEG	+	+	+	○	+	+	+	+	+	○
DBS	+	+	+	+	+	+	+	+	+	+
DBS+YEG	+	+	+	+	+	+	+	+	+	+
TBC	○	○	○	○	○	○	○	○	○	○
TBC+YE	+	+	+	○	+	+	+	+	+	○
TCP	○	○	○	○	○	○	○	○	○	○
TCP+YEG	○	○	○	○	○	○	○	○	○	○
DIOP	○	○	○	○	○	○	○	○	○	○
DIOP+YEG	○	○	○	○	○	○	○	○	○	○

+ indicates plasticizer deterioration.
○ indicates no deterioration.

Table IV

WEIGHT LOSS OF PLASTIC SAMPLES AFTER 21 DAYS
PERCENT WEIGHT LOSS

Culture no.	*DBS film*	*DIOP film*
Control	1·0	2·2
QM 474	24·9	2·7
QM 475	23·3	1·2
QM 9417	24·3	1·5
QM 9419	13·4	0·0
QM 391	24·3	0·9
QM 391MS	24·1	2·1
QM 391S	25·0	2·1
QM 9277	24·1	1·6
QM 9277MS	24·2	2·0
QM 9277S	25·1	0·9

Table V

SPORE COUNTS OF CULTURES AFTER 10 DAYS INCUBATION

Strain no.	*Spores/ml* × 10^9	
	Run 1	*Run* 2
474	1·8	1·8
475	2·1	2·1
9417	1·9	1·5
9419	0·8	1·0
391	1·4	1·2
391MS	2·5	2·2
391S	4·3	4·7
9277	1·0	1·0
9277MS	1·2	2·0
9277S	2·6	3·6

degradation studies, strains 474 and 475 were superior in that they degraded the greatest number of plasticizers (Table III). Strain 9419 was significantly inferior to other strains tested in that it was able to affect degradation of only three of the eight plasticizers examined. Interestingly, strain 9277S was significantly inferior to the same strain when grown on other media. The inability of this culture to deteriorate the plasticizers TG-9 and TBC was unexpected and unexplained.

Although all strains examined in this study were able to degrade DBS in agar plate studies, strain 9419 exhibited only approximately half the activity of the other strains when tested against DBS plasticized plastic strips. These results (Table III) indicate that quantitative differences between the various strains may exist even where qualitative differences are absent.

It is interesting to note that although strain 391 and its offshoot 9277 do not sporulate as well on a low sugar medium as on a high sugar one, the spore yield from the low sugar cultures approaches that of some of the stable cultures (Table V). The spore yield from the high sugar medium far exceeds that of other stable cultures.

DISCUSSION

Although *P. funiculosum* QM 391 demonstrates aberrant morphology after several subcultures on low sugar medium, these atypical subcultures are still satisfactory test cultures. The morphological changes are not permanent, and no change in physiology, as related to its abilities as a biodeteriogen, occurs. If, however, a substitute culture showing more stable morphology is desirable, some other strains of *P. funiculosum* are capable of degrading cellulosic and synthetic polymeric formulations. Not all strains are equally satisfactory, however.

REFERENCES

Klausmeier, R. E. (1966). The effect of extraneous nutrients on the biodeterioration of plastics, In S.C.I. Monograph No. 23: Microbial Deterioration in the Tropics, pp. 232–243, Society of the Chemical Industry, London.

Klausmeier, R. E., Kanzig, J. L., Jones, W. A., and Hammersley, V. L. (1963). Mineral salts media for fungus resistance testing. *Dev. Ind. Microbiol.*, **4**, 303–316.

Osmon, J. L., and Klausmeier, R. E. (1971). Candida lipolytica: a nonpathogenic test organism for deterioration of plastics. *Dev. Ind. Microbiol.*, **12**, 197–203.

PROGRESS IN STUDIES OF DETERIOGENIC LICHENS

ARTHUR O. LLOYD

The Hatfield Polytechnic, Hatfield, Hertfordshire, U.K.

Summary

Recently lichens have been seriously recognized as biodeteriogens. Hitherto, *Lecanora dispersa*, a common deteriogen, was the only recorded species in central areas of London and New York. In present studies in Central London, *L. muralis* and *Candelariella aurella* have also been found. First colonizers of asbestos-cement appear to be algae, including minute pustules of *Trebouxia* which appeared to be the source of subsequent lichen development. Algal colonies soon supported a variety of fungi. The role of acari and insects may be significant in the dispersal of propagules. A close affinity between mites, particularly the red spider mite (*Balaustium*), and deteriogenic communities has been observed. Laboratory culture trials are briefly described.

INTRODUCTION

Lichens do not appear to have been seriously regarded as agents of biological deterioration until relatively recently. They were recognized as such by Lloyd (1972, 1974) and Richardson (1973); examples of attack in Venice were shown by Paleni and Curri (1972), and in India by Nair (1972).

Present studies have been conducted in three types of situation: (1) in established lichen communities; (2) on suitable new surfaces exposed to conditions favourable to lichen development; and (3) in the cultivation of deteriogenic species in the laboratory. In the first two situations, work has been toward obtaining detailed knowledge of the process of colonization, in which one of the important factors is the dispersal of propagules. The possibilities of their dispersal by means of air- or water-borne mycobiont spores, algal cells, soredia, or other vegetative particles are well known, but records of actual observations are hard to find. Gerson (1973) stated that their dissemination by arthropods, especially mites, is probable, but reports of actual instances are rare. In the present observations, there was evidence that mites may have a significant role in this respect.

As is the case with dispersal, so in the actual process of lichen development from the propagules, there has been much speculation but few hard facts reported, and some of the apparent facts which have been reported are doubted by many modern workers. Ahmadjian (1967) stated that free-living populations of *Trebouxia* (the most common phycobiont), although reported in the literature, probably do not exist. Present observations indicate that this organism has a limited free-living phase before being invaded by mycobiont—or free-living fungi.

It is only possible here to mention briefly the method and results of attempts at laboratory cultivation of deteriogenic lichens by techniques broadly based on that described by Galun *et al.* (1972) for the culture of Sudanian species.

Fig. 1. Asbestos-cement panels installed in frame on lichen-encrusted concrete parapet of roof at Bayfordbury, Hertfordshire.

Fig. 2. Asbestos-cement panel resting on Lecanora dispersa*-covered asbestos-cement roof of water tank chamber, Press Association Building, London.*

OBSERVATION SITES

Since many of the observations on established communities took place in the same location as where new surfaces were exposed, it is convenient to describe these together.

First, well-washed (pH 7·5), 12-in square panels of $\frac{1}{4}$ in asbestos-cement were installed on the lichen encrusted concrete parapet of the roof of the Polytechnic Science Block, Bayfordbury, Hertford, as shown in Fig. 1. The lichens present on the wall included *Candelariella aurella*, *Lecanora dispersa*, *Physcia* spp., *Caloplaca* spp., *L. muralis*, *Lecidea* spp., *Xanthoria* spp. Similar species were found causing severe fouling and damage to the nearby mansion. Exactly similar panels were set up on the roof of the Polytechnic Chemistry Block at Hatfield, in this case over a bitumenized surface not supporting lichens. In central London three panels were installed, resting directly on the *L. dispersa*-covered, corrugated asbestos-cement roof of the water tank chamber on the roof of the eight-storey Press Association Building in Fleet Street (Fig. 2).

The Polytechnic sites, approximately 20 miles north of London, are in moderately clean-atmosphere districts, whereas the central London site represents a highly polluted atmosphere in which it had been reported (Laundon, 1970) that only one species of lichen, *L. dispersa*, existed. The same has been reported from Central New York (Brodo, 1966).

OBSERVATIONS

When the panels were installed in Central London on 20 November 1973, it was found that although the asbestos-cement roof was covered with *L.*

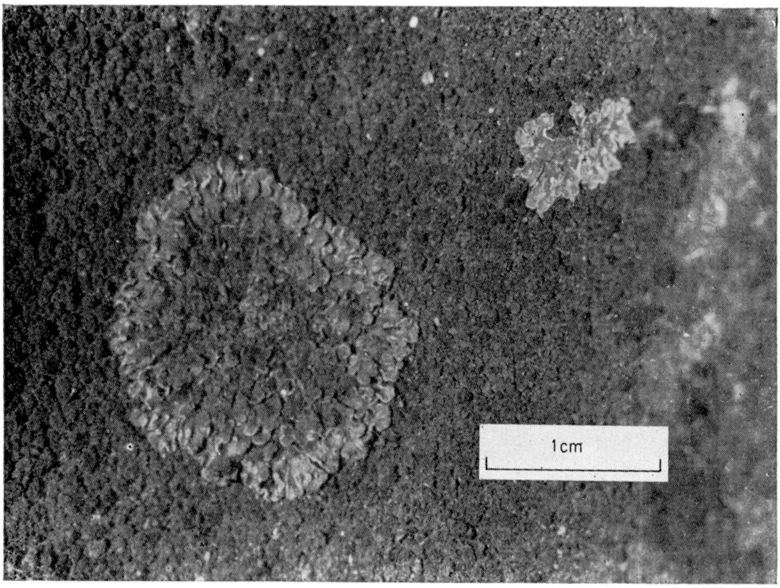

Fig. 3. Lecanora muralis *among old* L. dispersa *incrustation of asbestos-cement on roof of Press Association Building, London.*

dispersa, unexpectedly, there were growths of *L. muralis* present (Fig. 3) and *Candelariella aurella*. The latter organism was also found at ground level on rockery stones in Lincolns Inn Fields in Central London. The Polytechnic test panels were installed during the last week of October 1973.

FLORA

The test panels were examined regularly, both by hand lens and by microscopical examination of Sellotape impressions and surface fragments. Apart from the rich microflora of deposited bird droppings (which on the Bayfordbury roof-wall were very similar in pattern to some of the foliose lichen growths), there was very little organized biological growth to be seen before the autumn of 1974. Until then only individual fungal spores, pollen grains, a few clusters of algal cells, and some small reddish brown stains were found on all the sites, but during September, slight greenish tints were detected and it was obvious that colonization of the surfaces by algae had commenced. Somewhat surprisingly, the most prominent algal development appeared on the Central London site. Perhaps, following rain, drying might have been slower on these panels because they rest directly on the supporting surface, whereas the Polytechnic panels were supported in the air on racks.

The alga showing most rapid development was *Stichococcus bacillaris* in its filamentous stage, but *Chlorella* types and *Pleurococcus*-like cells were also present. Similar algal species were found on all the test-panel sites, but the development of *Stichococcus* was most pronounced in Central London.

On 5 September a 2 in strip was removed from one of the panels and cut into 2 in squares which were placed in petri dishes and incubated at room temperature near a north-facing window, moistening when necessary with sterile distilled water. This promoted more rapid development of the organisms present than occurred outdoors and also prevented destruction or damage of the frail fungal colonies which developed on the algal cover. When the surface of these specimens achieved a strong green coloration, white hyphae developed in the form of tufts and circular colonies on which dark spots developed. These were found to be the picnidia of *Phoma* sp. Other fungi were identified as *Fusarium* sp. and *Fusarium*-like sterile mycelium. Eventually, indoors, the mycelium virtually obliterated the algae. Outdoors, this did not occur but later 'ghost images' of similar whitish growths could be detected on the test panels in London. By 20 November, exactly one year after installation on the Central London site, the greenish tinge had become quite strong and the filaments of the *Stichococcus* had broken up into individual cells and short chains. Also, small yellowish-green raised pustules, about 0·2 mm diameter, were observed within the algal cover. These were found to consist mainly of *Trebouxia* cells.

These *Trebouxia* pustules were seen 2 weeks later at Bayfordbury and Hatfield. At this time no definite fungal growths could be found outdoors but by mid-January (during very mild weather) fungal hyphae were observed, both in the pustules and among the other algae. Apart from fungal associations developing with *Trebouxia* (Fig. 4), associations were also found of other unicellular algae and free-living fungi. *Phoma* was frequently found, also *Cladosporium* spp., *Alternaria* sp., *Stemphylium* types, and unidentified species.

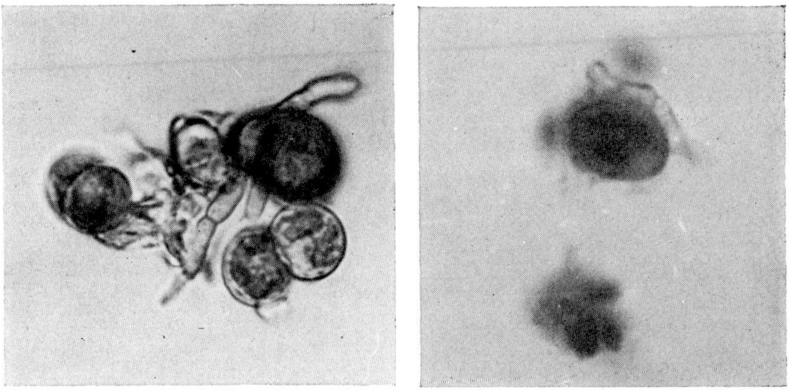

Fig. 4. Fungal hyphae in close association with Trebouxia *cells. Slide preparation of fragment of pustule on surface of test panel in Central London. Scale: 15 mm = 18 μ.*

The first sign of true lichen development was on the Bayfordbury roof site in late April when microscopic particles of bright yellow pigment began to form on some of the pustules, especially those (but not only those) in old bird-dropping areas. These developed until by early June they were just visible with a 10× hand lens, their colour being very similar to nearby *Candelariella aurella* growths. At this time definite organization of algal cells and fungal hyphae into a thallus structure became discernible under the microscope (Fig. 5). In mid-June the second recognizable lichen to appear was in the form of three tiny grey thalli just under one millimetre across and having the general appearance of *Physcia*, two species of which were on the

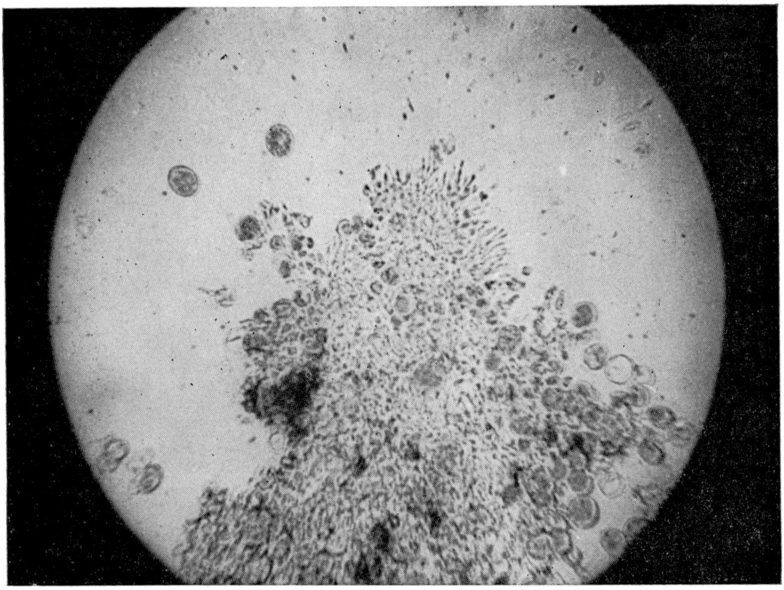

Fig. 5. Squash preparation of fragment of yellow-pigmented pustule, showing organized orientation of hyphae and algae in nascent thallus.

wall nearby. By late July, and after some heavy rain showers, their diameter exceeded 1 mm (Fig. 6).

These three thalli appeared in two old bird-dropping areas of the previous year.

Fig. 6. Minute Physcia-*like thallus on a Bayfordbury test panel. The red spider-mite* (Balaustium) *scurried across the field of view at the instant of flash exposure.*

It should be noted that bacteria are always present, often in great numbers, and that at all stages free-living algae and fungi intermingle with the symbionts. It is from this complex matrix of organisms that the lichen develops, and even mature lichens carry organisms similar to some of the components of this matrix. It would appear that when a lichen association develops and dominates its environment, it does so because of its greater stability to heat, cold, and, in many cases, strong sunlight, and desiccation than is the case with the free-living organisms or associations of organisms within that environment.

FAUNA

In early June 1974, large numbers of red mites were observed on the lichen-encrusted concrete roof-parapet at the Bayfordbury site. By midsummer day, they numbered several thousands over an area of about 100 m^2. During the following week, 14 areas of concrete walls and pavements supporting crustose lichens, ranging over three counties and extending from Cambridge to Princes Risborough, were examined and all, without exception, carried populations of these mites. Following heavy rain in July, when many perished, numbers declined but occasional sightings were made during the following winter, including one in November among lichens on the ruins of Tintern Abbey in South Wales. At this time eggs were found behind partially detached flakes of stonework. (These had been previously found in August in mossy soil deposits protected by the parapet at Bayfordbury.) Also, dark brown and black mites were found at Tintern.

In late May 1975, adult red mites together with a high proportion of larvae and nymphs again appeared on the Bayfordbury walls and on other sites. Their development followed the pattern of the previous year, reaching a peak at midsummer, but this time observation extended as far north as Yorkshire with similar results.

The literature (*e.g.* English and Snetsinger, 1957) cites *Bryobia praetosa*, the clover mite, as the most common red spider mite observed on buildings, but specimens taken from the various sites all differed from *Bryobia* when examined under the microscope. They were identified as *Balaustium* species (Fig. 6). Cornwell (1973) has pointed out that *Balaustium murorum* is in fact often mistaken for *Bryobia praetosa*.

Because all stages of the life cycle of *Balaustium* have been observed on the Bayfordbury roof, high above the ground where there is no other food source, it seems safe to infer that it is a permanent inhabitant of this kind of habitat. The mite has been observed feeding on detached lichen particles, grasping them with its pedipalps and later discarding the residues. It probably feeds on ejected lichen ballistospores too, but these are too small to be seen by direct observation. It can, however, be seen to feed from small clumps of rose pollen grains when these are presented to it. Ascospores and algal cells have been seen embedded among the leg and body hairs when liquid mounts were examined under high power, and their bodies can often be seen covered with loose particles after they emerge from burrowing among the lichens. It seems inevitable that many of these fragments, soredia, spores, and algae must be carried to, and deposited on, adjacent surfaces on which the mites wander, thus effecting the propagation of new colonies.

Other arthropods which were observed and which probably perform similar roles in the dispersal of propagules were the acari, *Phauloppia lucorum* and *Trichoribates trimaculatus*, and the psocid, *Cerobasis guestfalica*, found under the test panels on the roof of the Press Association Building in London.

LABORATORY CULTURE TRIALS

First attempts were aimed at cultivating *L. dispersa* and *L. conizaeoides* after the manner in which Galun *et al.* (1972) cultured Sudanian lichens. These were successful in producing thallus development from vegetative propagules, but the silica gel substrate was found to be a rather difficult support. Second attempts were directed toward finding a more robust substrate. (1) Kieselguhr and (2) well-washed asbestos-cement squares, each impregnated with Bristol's medium and contained in petri dishes, were found to be suitable. Some vegetative growth of *L. dispersa* and *C. aurella* was obtained from fragments deposited on the substrate, using an alternate wet and dry routine at room temperature, under nonsterile conditions, and illuminated by north daylight. The results obtained using rather crude techniques have been sufficiently encouraging to justify proceeding with more elaborate methods in the future.

The present paper is a short progress report on a long-term project. It is not possible to include the colour illustrations, shown at the Symposium and necessary for adequately communicating the results of many of the observations. It is hoped that these will be presented when each section is reported on more fully elsewhere.

Acknowledgements

The writer is indebted to the Press Association for the use of the roof site. Thanks are due for assistance in identifying organisms to Mrs B. W. Parry (acari) and Mr J. R. Laundon (lichens) of the British Museum (Natural History) and Drs J. H. Belcher and E. Swale of N.E.R.C. Culture Centre of Algae and Protozoa, Cambridge, and for the support of Catomance Ltd, Welwyn Garden City, which has made it possible to carry out the present work as a Leverhulme Research Fellow at the Hatfield Polytechnic.

REFERENCES

Ahmadjian, V. (1967). A guide to algae occurring as lichen symbionts. *Phycologia*, **6** (2/3), 153.

Brodo, I. M. (1966). Lichen growth and cities. *Bryologist*, **69**, 427–449.

Cornwell, P. B. (1973). Pest Control in Buildings. Hutchinson, London.

English, L. L., and Snetsinger, R. (1957). The habits and control of the clover mite in buildings. *J. Econ. Entomol.*, **50** (2), 135–141.

Galun, M., Marton, K., and Behr, L. (1972). A method for the culture of lichen thalli under controlled conditions. *Arch. Mikrobiol.*, **83**, 189–192.

Gerson, U. (1973). Lichen-arthropod associations. *Lichenologist*, **5**, 434–443.

Laundon, J. R. (1970). London's lichens. *The London Naturalist*, **49**, 20–69.

Lloyd, A. O. (1972). In *Biodeterioration of Materials*, pp. 185–191 (A. H. Walters and E. H. Hueck-van-der Plas, eds.), Applied Science Publishers, London.

Lloyd, A. O. (1974). Lichen attack on marble at Torcello-Venice. *Petrolio e Ambiente*, 221–224.

Nair, S. M. (1972). *Biodeterioration of Materials*, pp. 401–407 (A. H. Walters and E. H. Hueck-van-der Plas, eds.), Applied Science Publishers, London.

Paleni, A., and Curri, S. (1972). *Biodeterioration of Materials*, pp. 392–400 (A. H. Walters and E. H. Hueck-van-der Plas, eds.), Applied Science Publishers, London.

Richardson, B. A. (1973). Control of biological growths. *Stone Industries*, **8**, 2.

A NEW PRINCIPLE OF BIOCIDE TOXICITY INVESTIGATION FOR PROTECTION OF MATERIALS FROM BIODETERIORATION

D. A. BELENKOV and Y. A. SEROV

The Ural Forest Engineering Institute, Sverdlovsk, U.S.S.R.

Initially, biodeterioration of any material has a local, discrete form. Fungi or bacteria form separate deterioration centres (Fig. 1). The discrete form of deterioration becomes apparent and on resistant materials it remains for a long timc. But there is a rapid transition into complete deterioration on easily damaged materials.

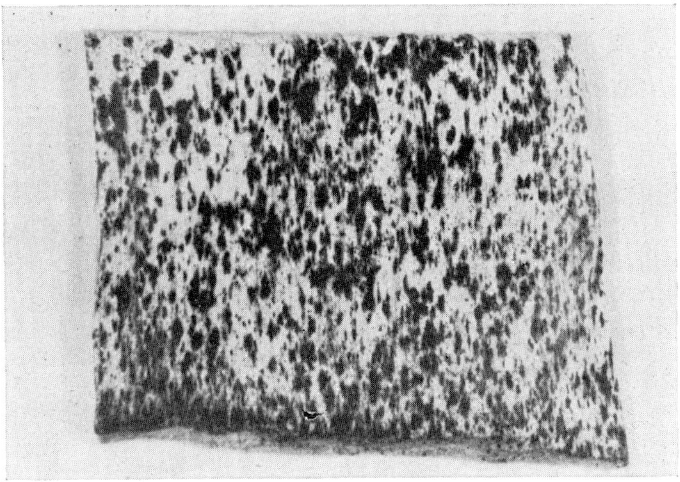

Fig. 1. The initial phase of wood damage with the blue stain fungus. Separate affected locations are distinctly seen.

Working out the method for evaluating toxicity as a protection probability of the material from biodeterioration, a phenomenon of discrete deterioration caused by fungi or bacteria was used as a model for the action of living things upon the material it affected.

Theoretical premises for this method are based on two assumptions:

(1) Quite definite protection of the material affected by living things is guaranteed by every dose of biocide; and

(2) when increasing the biocide dose, the protection probability varies according to the law of normal distribution.

These assumptions were confirmed by testing wood-destroying fungi that cause the initial attack (Belenkov, 1963, 1965). In many cases biodeterioration was caused by spores or spores and hyphaes of different origin simultaneously.

Freshly cut wood and blue stain fungus are suitable materials for biocide toxicity investigation as a protection probability. This kind of wood is easily attacked by blue stain fungus spores and hyphaes.

Ophiostoma coeruleum (Münch) H. et Sud. was used as a tested fungus (Meyer, 1963). This fungus grows quickly on freshly cut sawdust and in agar-mist medium. In a short period of time it attacks the wood, which differs from sound wood in colour. The fungus culture is well preserved in a test tube on an agar-mist slant with inoculation during 1 or 2 months.

The inoculum is prepared from a fungus spore suspension grown in petri dishes on 10-day-old agar-mist. Suspension titre is 10^6 spores per ml.

The most important part of the method is the creation of a uniform inoculation background, where rapid and homogeneous infecting of pine sapwood tested samples without toxics is achieved. Such an inoculation background may be achieved by growing fungus on freshly cut sapwood sawdust with 300% moisture content to which 5% oatmeal is added.

The medium is placed in 1 litre wide-mouthed flasks in 3–4 cm layers. It is sterilized for 1·5 hr at autoclave pressure 1·7 atm.

The medium is aseptically inoculated by even dusting of the spore suspension. Five millilitres of suspension are introduced into every flask. After 10 days, the fungus penetrates the whole volume of the medium and forms even surface mycelium, which is used as an inoculated background. Culture readiness is shown by the appearance of the fungus on the surface (homogeneous growth) and by the inclusion of all the medium.

The tested specimens are 5–6 mm tangential square plates 1 mm thick. Fungus growth may be definitely fixed on such specimens if wood inoculation has taken place. It is possible to test enough specimens simultaneously to get reliable wood protection probability. The specimens are made of freshly cut pine sapwood just before the test.

As a rule, five different biocide concentrations are tested simultaneously. They are taken so as to get a full range of concentration, in which wood protection changes from low to high. For an unknown substance, it is necessary to carry out preliminary tests in a wide range of biocide concentrations.

Specimens are dipped in the biocide solution for half an hour. It is useful to dip control specimens in $KMnO_4$ 0·015% solution for 30 min. This treatment has no influence on fungus attack of the wood, but reduces significantly the emergence of mould. Impregnated specimens are slightly dried in a sterile chamber and placed into flasks on the medium surface which is covered evenly with fungus growth. The scheme of distribution of specimens may be different, but it is necessary to place impregnated specimens of all tested concentrations into every flask. The most convenient and reliable scheme of distribution is shown in Fig. 2. One biocide is tested simultaneously in all 10 flasks. The specimens are incubated on fungus cultures for 10 days.

After that period of time, control specimens and specimens covered with fungus are usually strongly affected, you may see it with the naked eye. Then specimens are taken from the flasks in the sterile chamber and put in sterile petri dishes. These petri dishes are placed in a sterile, humidified container and further incubated for 10 days. This operation is necessary to watch the affected specimens clearly and discern the concealed affected parts.

The tests are conducted at a room temperature of 16–21°C. After the tests specimens are dried, examined carefully, and sorted into infected and uninfected.

It is convenient to paste specimens on white Whatman according to the numbers of the flasks and solution concentrations. This preserves the test

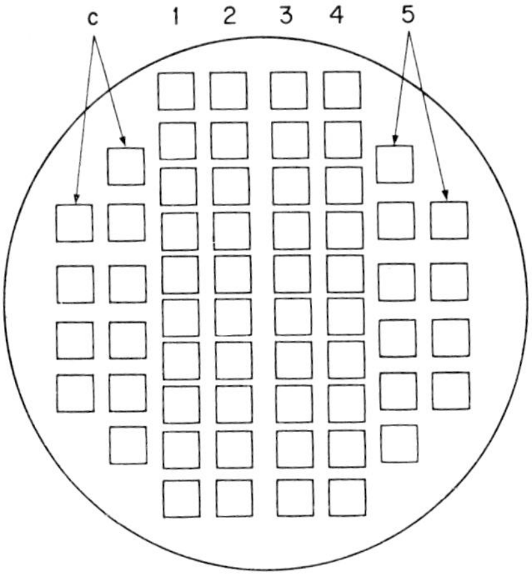

Fig. 2. The distribution of the test specimens over Ophiostoma coeruleum *fungus culture in the tested flask. c, control specimens; 1–5, specimens that contain increasing quantities of biocide.*

results for use in further work. In these tests the degree of fungus in affected specimens was different. The control specimens and specimens treated with a weak biocide concentration were very strongly affected.

The degree of biodeterioration was not considered in the results because it was supposed that if specimens were affected, the degree was a function of time, *i.e.* toxicity in wood did not prevent fungus penetration, but only slowed this process.

Calculation from the experimental data is very simple.

It is necessary only to determine relative quantity of nonaffected specimens for every concentration and change it into probit. According to the probits and biocide concentrations or their logarithms, it is necessary to plot a graph. It should be a straight line, because it has been determined that many living things react to toxicity according to the Laplas-Gauss law (Belenkov, 1963, 1965, 1968; Bessmertny, 1967; Urbah, 1963; Horsfall, 1948; Bliss, 1935; Morrell *et al.*, 1938).

As a criterion of biocide toxicity estimation, we choose that quantity of biocide which protects the wood from being affected with 0·95 probability. The choice of this probability has been confirmed by Belenkov's work (Belenkov, 1966). Biocide doses that protect wood with 0·95 probability are easily determined by 6·65 probit on graphs. Borax, boracic acid, GR-48, and 'Timber' were tested according to this method. The results of the tests are shown in Figs. 3 and 4.

Figures 3 and 4 show that probit-graphs of toxic action coincide well with straight line graphs. Hence, the theoretical premises of the method are correct, and the blue stain fungus reacts to the toxic action according to the same laws as many other organisms in homogeneous totality.

'Timber', GR-48, and boracic acid have the same toxicity for *Ophiostoma coeruleum*. The protecting doses with 0·95 probability are 0·259 %, 0·266 %, and 0·246 % concentration of solution. For the tested fungus, borax is half as toxic. The same doses in two tests were 0·468 % and 0·56 % concentration of solution.

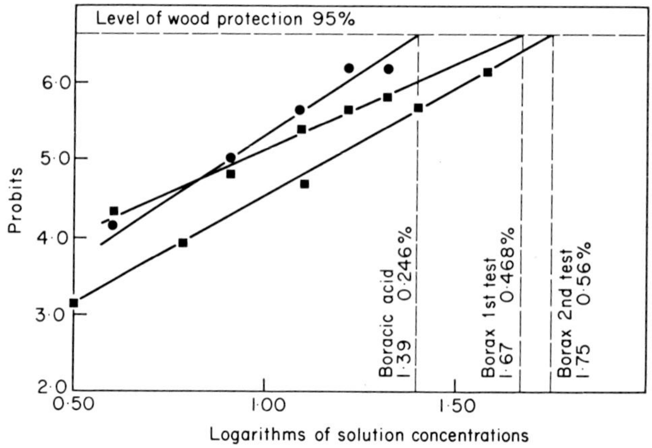

Fig. 3. Toxicity estimation of borax and boracic acid for Ophiostoma coeruleum.

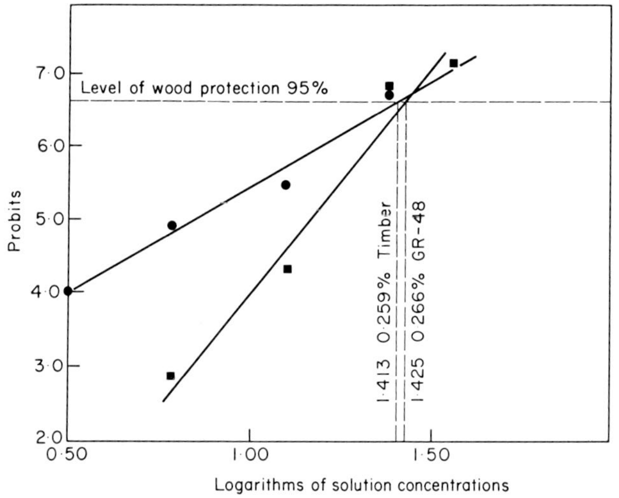

Fig. 4. Toxicity estimation of GR-48 and 'Timber' for Ophiostoma coeruleum.

This way it is possible to solve in the laboratory a number of important problems, using the theorem and laws of the theory of probability and mathematical statistics. The combined biocide toxicity may be evaluated with the help of the theorem of the addition for independent coinciding events (Seletskaya and Belenkov, 1971), to study the process of unpreserving of wood, fixation and washing away of biocide, reliability of wood protection, the quality of impregnation, the influence of post-impregnated treatment of wood upon biocide toxicity, and others. This principle of investigation may be used in working out the evaluation of biocide toxicity for organisms attacking different materials.

REFERENCES

Belenkov, D. A. (1963). Isvestya (News) of Technical Colleges of the U.S.S.R., *Forestry Journal*, No. 6.
Belenkov, D. A. (1965). ITC of the U.S.S.R., *Forestry Journal*, No. 1.
Belenkov, D. A. (1966). ITC of the U.S.S.R., *Forestry Journal*, No. 4.
Belenkov, D. A. (1968). ITC of the U.S.S.R., *Forestry Journal*, No. 2.
Belenky, M. L., and Vitolin, M. A. (1954). Isvestya (News) of Academy of Science of Latvian Soviet Socialist Republic, No. 2.
Belenky, M. L. (1963). Quantity estimation of pharmacology effect.
Bessmertny, B. S. (1967). Mathematical statistics in clinical, prophylactic, experimental medicine.
Bliss, C. I. (1935). *Ann. Appl. Biol.*, **22** (1), 134.
Horsfall (1948). Fungicides and their action. D.G.
Meyer, E. M. (1963). Goslesbumisdat.
Morrell, C. A., Chapman, C. W., and Allemark, M. G. (1938). *J. Pharmacol. Exp. Therap.*, **64**, 14–42.
Seletskaya, V. A., and Belenkov, D. A. (1971). ITC of the U.S.S.R., *Forestry Journal*, No. 2.
Serov, Y. A., and Belenkov, D. A. (1969). Transactions of the Ural Engineering Institute, No. 20, Sverdlovsk.
Urbah, V. I. (1963). Mathematical statistics for biologists and physicians.

SOME FACTORS INVOLVED IN THE SETTLEMENT AND METAMORPHOSIS OF MARINE BIVALVE LARVAE

R. D. TURNER

Museum of Comparative Zoology, Harvard University,
Cambridge, Mass. 02138, U.S.A.

Summary

Teredinid pediveliger larvae settling on white pine metamorphose and penetrate the wood readily, producing a well structured white cone over the entrance of the burrow. Larvae from the same cultures exposed to *Dalbergia*, a naturally resistant wood, are unable to complete metamorphosis, produce the white cones or successfully penetrate the wood. It is shown that the active ingredients in *Dalbergia* are obtusaquinone and obtusastyrene. The pathways by which these chemicals affect shell production are discussed.

INTRODUCTION

The purpose of this paper is to review some of our recent work on the settlement of bivalve larvae which contributes to our understanding of shell formation during metamorphosis and may have a practical application in the control of marine wood borers.

It has been my contention for many years that the effective control of an organism is most readily obtained through a knowledge of its life history and the factors controlling it. A review of the literature on the control of marine boring and fouling organisms shows that in the past more emphasis has been put on the material to be protected than on the organisms doing the damage. Treatments to date have been mainly of two types. One involves the use of highly toxic, nonspecific chemicals which devastate a broad spectrum of organisms. Such treatments are no longer acceptable to the EPA and other environmental groups and should be replaced. The other protects by mechanical or physical means, *e.g.* sheathing with metal, plastic, fibreglass, or jacketing with cement. Published results of experiments and tests report the degree of success or failure but seldom consider the stage in the life cycle of the organism when the treatment is effective nor do they question the reasons for success or failure.

Detailed studies of the life history of several species of marine bivalves, including species in the families Teredinidae and Pholadidae, have shown that the most critical period in their life history is at the time of settlement and metamorphosis. Most larvae can delay metamorphosis if the proper substratum is not available, but once the substrate is found and the larvae become competent to metamorphose, an ordered sequence of events occurs which must follow one another if the organism is to survive. At this time major changes occur in all biological systems, including the method of feeding, respiration, and excretion as well as the musculature of the foot and growth of the shell. Any interruption or alteration in the normal sequence of events affects the ability of the larvae to progress to the next step whether that be the

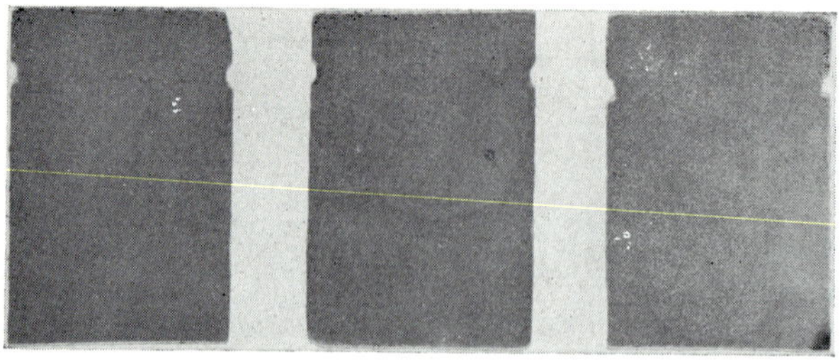

Fig. 1. Dalbergia retusa *heartwood. Small panels cut from larger panels which had been exposed for 51 months at Naos Island, Panama. Exposed for 15 months at Nahant, Mass., and 23 months at Woods Hole. X-ray shows no evidence of borers.*

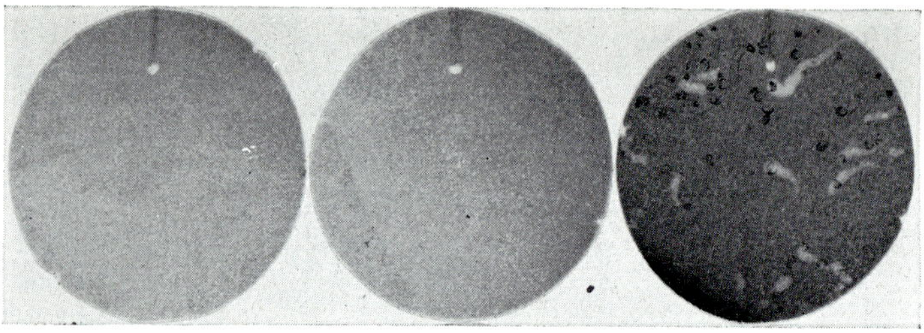

Fig. 2. Dalbergia retusa *sapwood. Exposed for 12 months at Nahant and 3 months at Woods Hole. The evident difference in attack may be attributed to the region of the sapwood from which the discs were cut. This is an important point to take into consideration when evaluating tests or types of natural woods and emphasizes the need for replicating experiments.*

loss of a larval organ or the acquisition of a postlarval one, and such interruptions usually result in the death of the organism. Following settlement, all larvae, in addition to developing siphons and ctenidia to replace the lost velum (the larval swimming, breathing, and feeding organ), must immediately begin adding both proteinaceous and calcareous material to the larval shell. In addition, the Teredinidae must form a protective cone over the entrance of the burrow, line the burrow, and produce pallets (Turner, 1966; Turner and Johnson, 1971), all of which require the deposition of calcium. The calcareous crystals of the growing molluscan shell are deposited on a skeleton or framework of conchiolin (Wilbur, 1964; Hillman, 1961). The abilities to produce both the proteinaceous or conchiolin framework and the calcareous crystals are equally important to the success of the developing mollusk.

Karande and Menon (1972, 1973) used the presence of white calcified cones on the surface of the wood as an indication of the ability of newly settled *Teredo furcifera* to lay down calcareous material during the initial penetration of various timbers. They showed that young *furcifera* were highly successful when placed on *Abies pindrow* but were unsuccessful on the known resistant timber, *Tetrameles nudiflora.* They also reported on the inability of newly settled young to secrete calcium when the fir test pieces into which they were boring were exposed to solutions of seven compounds (all known calcium inhibitors) including 2,4-dinitrophenol. Unfortunately, Karande and Menon did not carry their experiments further and gave no explanation as to why the cones were not produced. Our experiments with 2,4-dinitrophenol, impregnated into white pine, showed it to be effective for a short time, but it is

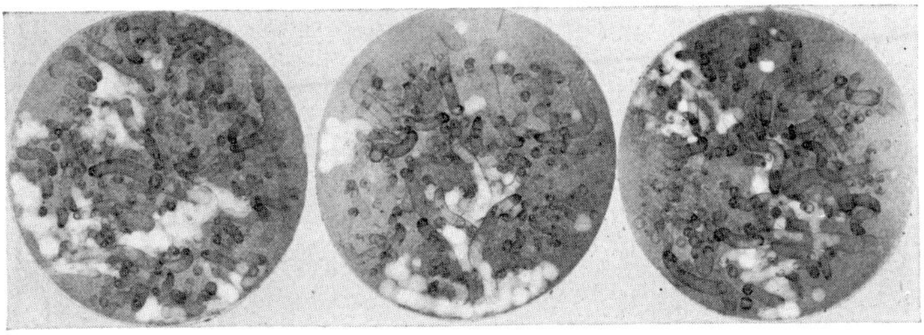

Fig. 3. White pine controls after 6 months exposure. The discs were beginning to fall apart and had to be replaced.

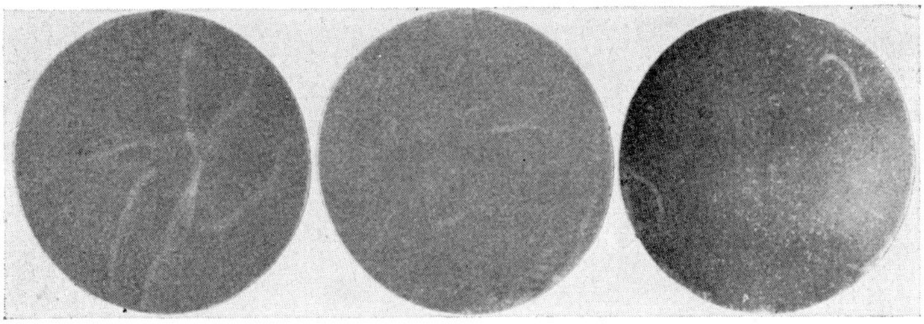

Fig. 4. White pine impregnated with obtusaquinone X-rayed after 8 months exposure at Nahant. The small white specks are pits made by Lyrodus *larvae that could not penetrate deeper. The long white trails are* Limnoria *tunnels.*

highly soluble in water and did not bond to the wood so leached rapidly, and the panels were riddled in less than a year. Consequently, we did not pursue this further.

Our first experiments were conducted to observe and record the normal behaviour of the larvae of *Lyrodus pedicellatus* Quatrefages during settlement and penetration in white pine. Following this, larvae were exposed to a naturally resistant wood in order to detect behavioural and morphological differences and the extent of penetration. Samples of *Dalbergia retusa*, a Panamanian rosewood commonly called cocobolo, were obtained from Dr J. D. Bultman who has done extensive field studies on naturally resistant woods (Southwell *et al.*, 1970, 1971; Bultman and Southwell, 1972; Bultman *et al.*, 1973). His field experiments showed that the heartwood was far more resistant than the sapwood and so it was this that we used. Fifty pediveliger larvae of *L. pedicellatus* known to be competent to metamorphose were placed on the *Dalbergia* and their behaviour observed using a binocular microscope. An equal number of larvae from the same culture were placed on white pine as controls. All larvae crawled quite normally and within a day began to bore into the wood, but none succeeded in penetrating the *Dalbergia* to a depth greater than about 1·5 times the length of their valves (*ca.* 500 μ) and all of the specimens on *Dalbergia* were dead in three days. By this time the control specimens were well into the wood and had well-developed protective cones, siphons, and pallets. This experiment was repeated three times with the same results. The young borers on *Dalbergia* all died leaving empty valves in shallow pits surrounded by shallow debris rings but no protective cones were formed. Obviously, the wood was not too hard for the larvae to excavate, and the pits appeared quite typical when compared with

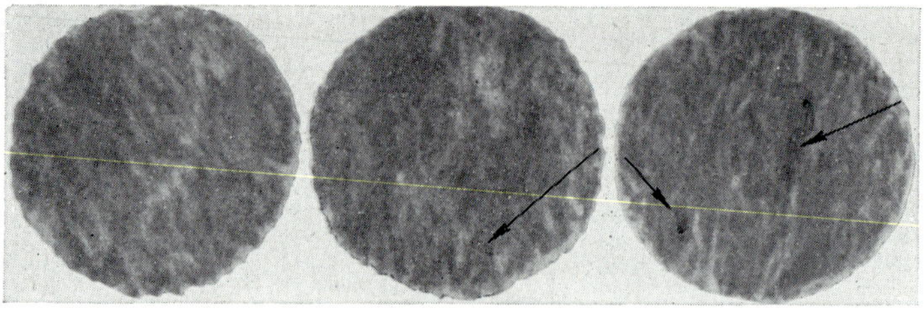

Fig. 5. The same discs after being held dry for a year and then exposed for an additional 7 months at Woods Hole. The Limnoria *attack is heavy but only three* Lyrodus *succeeded in penetrating.*

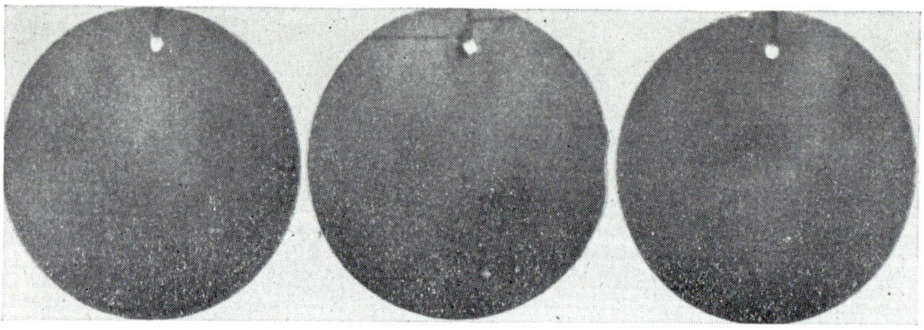

Fig. 6. White pine impregnated with obtusaquinone after 7 months exposure at Woods Hole.

those of the controls except that the debris rings were not consolidated and impregnated with calcium to form a protective cone over the young borer. The white cone not only protects the entrance of the burrow but extends into the wood as a lining to the tiny burrow. It is to this lining that the muscles of the pallets are attached, and it is against the inside of the cone that the pallets fit when closing the burrow. Consequently, the inability of the larvae to produce a cone also means the lack of pallets and the means of protection against predators. Examination of the specimens removed from the *Dalbergia* also showed that the young borers were incapable of adding well formed shelly material to the anterior margin of the larval shell or of producing pallets, though the siphons were well formed.

Bioassays using whole extracts of *Dalbergia* and swimming pediveligers showed no effect, as would be expected, because the larvae at that stage are not involved in laying down calcium. However, whole extracts of *Dalbergia* impregnated into white pine showed that it did indeed prevent pediveliger larvae placed on these test pieces from metamorphosing and producing protective cones, shells, and pallets. It was now essential to know what was in the *Dalbergia* and to test these chemicals individually to ascertain the active ingredient.

Jurd *et al.* (1972) made a detailed chemical examination of the extractives of *Dalbergia retusa* Hemsley and compared them with the extracts of *Dalbergia obtusa* Lecomte. The major constituent of the heartwood of both trees was obtusaquinone. The trees differ in that *D. obtusa* contains the cinnamyl phenol, obtusastyrene, which appears to be lacking in *D. retusa*. This has interesting implications in the light of later findings in our experiments and those of Waite and Wilbur (1974).

Dr Jurd purified crystals of obtusaquinone and obtusastyrene and further

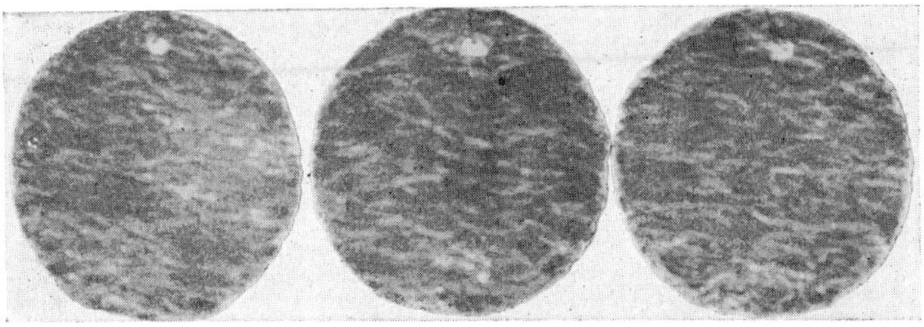

Fig. 7. The same discs 7 months later. Again the Limnoria *attack is heavy, but only one minute* Lyrodus *succeeded in penetrating.*

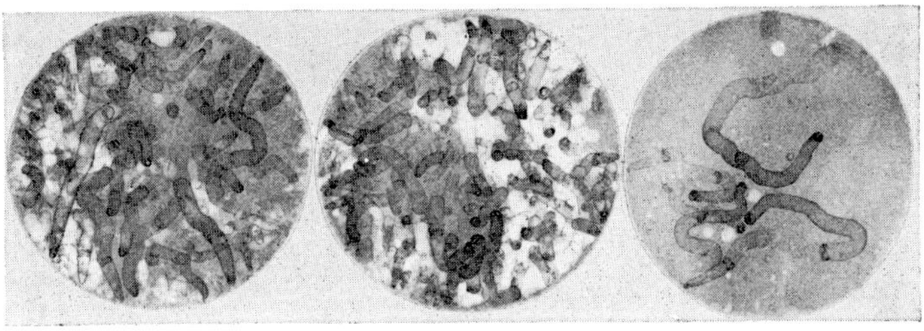

Fig. 8. White pine impregnated with obtusastyrene X-rayed after 7 months exposure at Woods Hole. Note lack of Limnoria *but very heavy* Lyrodus *attack.*

experiments, conducted in cooperation with Bultman and Jurd, and designed to ascertain which substance in the *Dalbergia* prevented the larvae from producing well-formed cones, shells, and pallets, showed that obtusaquinone was the active ingredient. This and related quinones are produced by many tropical trees and there is a considerable body of literature on the natural protection of trees against wood-boring insects. Obtusastyrene, the other chemical extracted from the *Dalbergia*, proved somewhat effective against *Limnoria* but not *Lyrodus*. However, when obtusaquinone and obtusastyrene were both impregnated into white pine, there appeared to be a synergistic effect for the combination was better than either of the chemicals alone.

Experiments to determine the biochemical pathways by which obtusaquinone and obtusastyrene affect the production of shells by *Lyrodus* were designed by Waite and Wilbur at Duke University. The results of their experiments show that obtusaquinone, presumably in a metabolized form, inhibits periostracal or conchiolin phenoloxidase and hence blocks the production of the conchiolin network, the base to which the calcium crystals are attached in the production of the shell (Waite and Wilbur, 1975; Waite, 1975). They have also shown that the settling pediveliger larvae exposed for 48 hr to wood impregnated with obtusaquinone have a calcium-45 uptake of only one-third that of larvae exposed to control wood. In addition, the analysis of the homogenate of pediveliger larvae which were competent to metamorphose showed that they have considerable phenoloxidase activity. This suggests that the larvae were actively producing conchiolin at the same time they were fixed for study. The control larvae settled immediately they were placed on wood. Waite also showed that there is a synergistic effect when small amounts of obtusastyrene are added to obtusaquinone, and this confirmed our laboratory experiments using a double impregnation. Waite

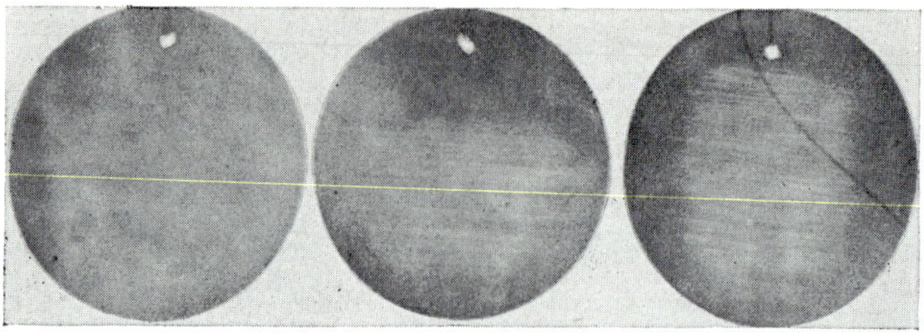

Fig. 9. White pine impregnated with obtusaquinone and obtusastyrene X-rayed after 3 months exposure at Woods Hole.

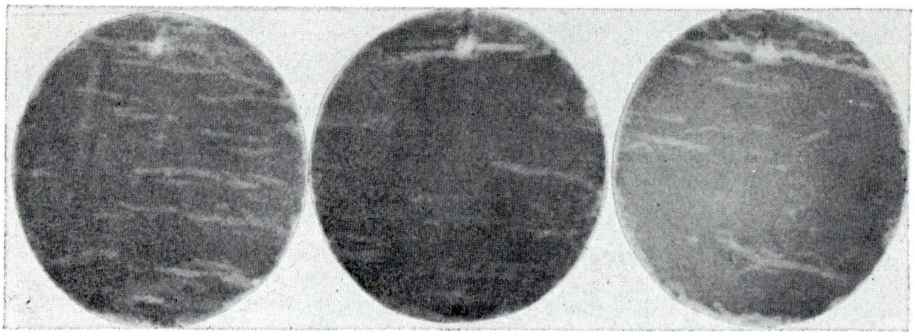

Fig. 10. The same discs after 10 months exposure. Note that the Limnoria *attack is light in comparison with that shown in Figs. 5 and 6 and that there is no evidence of* Lyrodus. *This suggested a synergistic effect when both obtusaquinone and obtusastyrene are used and Waite's experiments proved it.*

has shown that by itself obtusastyrene has no modifying effect on phenoloxidase. However, obtusastyrene is oxidized by phenoloxidase, and the product of this oxidation inhibits the normal action of the phenoloxidase. Further, obtusastyrene conversion to the toxic form is accelerated by the presence of comparable amounts of obtusaquinone, thus adding to the production of enzyme inhibitors.

Mr Waite's experiments have shown how these naturally occurring chemicals prevent the larvae from producing a shell and hence from entering the wood and completing their metamorphosis (Fig. 11). The rosewood cinnamyl derivatives, obtusaquinone and obtusastyrene, appear to impair the formation of the molluscan exoskeleton by inhibiting the enzyme phenoloxidase. Phenoloxidase plays a critical role in the tanning or quinone-crosslinking of the periostracum, a cuticular structure on the outer surface of the shell which must be deposited before mineralization can occur. Although obtusaquinone and obtusastyrene have no direct effect on phenoloxidase, the decomposition and oxidation products of these compounds profoundly inhibit the intended activity of this enzyme. Obtusastyrene is hydroxylated (I) and dehydrogenated by molluscan phenoloxidase. Either product II or the tannin formed through decomposition is responsible for the inhibition of the natural catechol dehydrogenase activity (lower half of diagram). Obtusaquinone, on the other hand, enhances the rate of formation of product I and also inhibits the dehydrogenase activity of the enzyme probably through tannins formed during decomposition. The inhibition of enzyme activity is competitive in nature. The dotted arrow drawn from compound II to the periostracum was included to demonstrate that this quinone is also capable

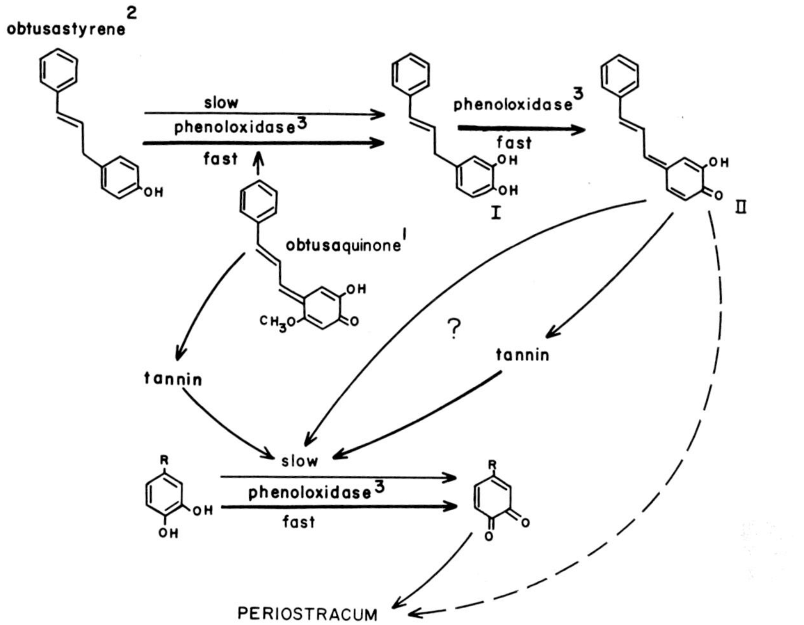

Fig. 11. Obtusaquinone and obtusastyrene impair formation of exoskeleton of Lyrodus. (*By J. Herbert Waite, Department of Zoology, Duke University, Durham, N.C.*)

of tanning the periostracum but probably in a defective manner. The enzyme (3) used here was isolated from the mantle and periostracum of the horse mussel *Modiolus demissus* Dillwyn; obtusaquinone (1) is from *Dalbergia retusa* Hemsley and *D. obtusa* Lecomte, and obtusastyrene is found in *D. obtusa*. The substituent –R of the natural tanning substrate has not been determined. However, from a practical point of view, rosewood is rare and expensive and it is even difficult to get sufficient purified extract for experimental work. Many quinones are general in their effects and obtusaquinone is probably one of these. Assuming this, and knowing the structure of obtusaquinone and obtusastyrene, it is now possible to synthesize similar molecules and test their effects. Several of these experiments are in progress. We are also looking at other naturally occurring and more common chemicals of similar structure and presumably similar function such as humic acids and phenolic compounds.

During the period from July 1971 to January 1973, our experiments were conducted at the Marine Science Institute, Northeastern University, Nahant, Massachusetts. At the time we moved to the Marine Biological Laboratory, Woods Hole, Mass., all discs were removed, allowed to dry, and were stored for about a year before being reexposed. Figures 4 and 5 show that the chemicals were still effective against *Lyrodus* but less so against *Limnoria*.

Duplicate series of discs impregnated at the same time as the original series but stored dry have now been in the tanks for three months. The rate and type of attack on these discs is comparable to that of the original series, again indicating the feasibility of dry storage, an important factor from an economic point of view.

All experimental discs were of white pine, 5 cm in diameter and 5 mm thick, and were evacuated for one hour before treatment. The chemicals to be impregnated were dissolved in ethanol or acetone and impregnated in a pressure cell at 100 lb/in^2 for one hour. After treatment, the discs were washed for a week to leach excess chemicals. They were then suspended on strings (one treatment per string) in running seawater tanks with large pieces of 'bait wood' which contained adult, spawning *Lyrodus pedicellatus*. The tanks were maintained at 15–20°C. Examination with the binocular microscope and X-rays were made at two to eight month intervals as needed, after initial observations of the behaviour of the larvae at the time of settlement.

Acknowledgement

This work is being supported by ONR contract N0014-67-A-0298-0027, NR104-686 with Harvard University and in cooperation with J. D. Bultman, L. Jurd, K. Wilbur, H. Waite, R. Mitchell, and I. Chet.

REFERENCES

Bultman, J. D., and Southwell, C. R. (1972). A preliminary investigation of the Marine Borer resistance of the tropical wood *Dalbergia retusa*. NRL Report 7416, 1–12, Figs. 1–6.

Bultman, J. D., Fetsenstein, D., and Ritchie, D. D. (1973). Inhibitory effects of obtusaquinone on the growth and reproduction of two marine fungi. NRL Report 7653, 1–7, Figs. 1–4.

Hillman, R. E. (1961). Formation of the periostracum in *Mercenaria mercenaria*. *Science*, **134**, 1754–1755.

Jurd, L., Stevens, K., and Manners, G. (1972). Phenolic and quinoidal constituents of *Dalbergia retusa*. *Tetrahedron Letters*, No. 21, 2149–2152.

Karande, A. A., and Menon, K. B. (1972). Inhibition of calcium deposition by chemical inhibitors. *Current Science*, January 5, 1972, **41** (1), 12–13, Figs. 1 and 2.

Karande, A. A., and Menon, K. B. (1973). Inhibition of calcium secretion by chemical inhibitors in shell dwelling organisms. 3rd Int. Congr. on Marine Corrosion and Fouling 1973, pp. 806–810, Figs. 1–4.

Southwell, C. R., and Bultman, J. D. (1971). Marine borer resistance of untreated woods over long periods of immersion in tropical waters. *Biotropica*, **3** (1), 81–107, Figs. 1–4.

Southwell, C. R. *et al.* (1970). Biological Deterioration of Wood in Tropical Environments. Part 2—Marine borer resistance of natural woods over long periods of immersion. NRL Report 7123, 1–42, Figs. 1–10.

Southwell, C. R. *et al.* (1971). Biological Deterioration of Wood in Tropical Environments. Part 3—Chemical wood treatments for long term marine borer protection. NRL Report 7345, 1–12.

Turner, R. D. (1966). A Survey and Illustrated Catalogue of the Teredinidae. Museum of Comparative Zoology, Harvard University, 265 pp., 25 text figs., 64 pls.

Turner, R. D., and Johnson, A. C. (1971). Biology of Marine Wood-Boring Molluscs.

Waite, H. (1975). Obtusaquinone-obtusastyrene inhibition of molluscen phenoloxidase—a natural coincidence (in prep.).

Waite, H., and Wilbur, K. M. (1974). Molluscan phenoloxidase and obtusaquinone inhibition. Paper given at NRL Tropical Woods Workshop.

Waite, H., and Wilbur, K. M. (1975). Phenoloxidase in the periostracum of Modiolus demissus Dillwyn (in press).

Wilbur, K. (1964). Shell formation and regeneration, In *Physiology of the Mollusca*, Vol. 1, pp. 243–282 (K. M. Wilbur and Sir C. M. Yonge, eds.), Academic Press, New York.

ALGAL SETTLEMENT AND SECRETION OF ADHESIVE MATERIALS

A. H. L. CHAMBERLAIN

Department of Biological Sciences, Portsmouth Polytechnic, Portsmouth, U.K.

Summary

Sessile diatoms from a fouling community are shown to possess adhesive systems of varying morphology. These comprise mucilaginous acidic polysaccharides and appear as irregularly meshed fibrous materials when observed by scanning electron microscopy. Motile diatoms seem to attach by the raphe material alone.

The processes of spore settlement and adhesive production and secretion are described as exemplified by the red alga *Ceramium*. Finally, the present knowledge of diatom and algal spore settlement and adhesion with respect to the fouling problem is reviewed briefly.

INTRODUCTION

Any object introduced into a natural aquatic environment is a potential surface for settlement and colonization by aquatic organisms. This is a natural process involving competition and succession but only attains economic importance when the cumulative effects become deleterious to the functioning of the structure. This situation constitutes the problem of aquatic fouling which is of particular importance in the marine milieu.

Fouling by algae has become predominant since the advent of supertankers and container ships and the development of paint compositions effective in reducing animal fouling to a minimum. It was quickly realized that the most critical phase in the development of an algal fouling community was the initial attachment of the spores or diatoms. Although mucilaginous materials had long been implicated in the attachment of many diatoms (Butcher, 1932) and red algae (Chemin, 1937), only after the fine structural observations of *Oedogonium cardiacum* zoospores by Hoffman and Manton (1962) were internally produced adhesives suggested for flagellate spores. Prior to this the secretion of algal spore adhesives was barely considered (Anon., 1952). Subsequently, a number of fouling algae have been examined from this aspect and are reviewed herein.

In addition to observations of the mode of attachment, the factors affecting settlement and adhesion have also received some attention, especially with regard to water movement, surface texture, and the necessity for a preconditioning surface layer. These are also considered in this paper.

MATERIAL AND METHODS

Specimens for transmission electron microscopy (T.E.M.) were processed according to the method previously described (Chamberlain, 1974) and examined in either an A.E.I.Em 6b or a Phillips EM 300. Material for scanning electron microscopy (S.E.M.) was obtained from Langstone Harbour, Hampshire, U.K., and treated as follows: spores or diatoms were allowed to settle on glass coverslips for 24 hr, rinsed in several changes of sterile sea-water, and fixed for 2 hr in a 5% v/v glutaraldehyde solution in 0·1 M sodium cacodylate buffer (pH 7·2) containing 0·25 M sucrose. Following two washes in the same buffer containing 0·33 M sucrose, post-fixation in 1% w/v osmium tetroxide in buffer was carried out for 2 hr. After rinsing in buffer solution, the material was dehydrated through an ethanol series and transferred via ethanol/acetone mixtures to pure acetone in which it was held for 1 hr. Subsequent treatment included critical point drying, using a Polaron E3000 instrument and liquid CO_2, followed by mounting on stubs, gold coating, and final examination in a Cambridge Stereoscan.

A modified version of the sequential alcian blue/alcian yellow staining technique of Parker and Diboll (1966) was used to determine the anionic groups of the diatom extracellular polysaccharides. This was merely a reversal of the dyes, *i.e.* alcian yellow pH 0·5, alcian blue pH 2·5. All light microscopy was carried out using a Wild M20 microscope. Spore detachment studies were carried out according to the method of Christie *et al.* (1970).

RESULTS

Diatom adhesion

Chemically cleaned glass slides placed in Langstone Harbour for varying periods of time (2–16 days) showed a typical colonization sequence. Initially, only bacteria were present but within 2 days, solitary specimens of several diatoms were observed. Diatom numbers increased with immersion time and after 16 days, both solitary and colonial diatoms were present.

The diatoms most commonly found as sedentary members of the algal fouling community exhibited a range of attachment systems, all based on the presence of mucilaginous material which assumed various morphological forms. The following types were recorded:

(1) Attachment by raphe substances alone, *e.g. Navicula* spp., *Pleurosigma balticum.*

(2) A basal or unipolar adhesive pad, *e.g. Synedra fulgens, Asterionella* sp.

(3) An investing mucilage capsule surrounding solitary cells, *e.g. Amphora* sp.

(4) A tubular or flattened mucilage sheath surrounding colonies, *e.g. Navicula Grevillei.*

(5) A mucilaginous stalk which may branch to yield colonies, *e.g. Licmophora flabellata, Achnanthes longipes.*

As noted above, the materials functioning as adhesives were frequently also of importance in colony formation.

The staining reactions of the attachment material of Group 1 species were impossible to detect in the mixed fouling diatom communities due to overall staining of the cell wall materials and some cytoplasmic structures. However, examples of the other classes were examined. The results are given in Table I.

Table I

REACTION OF DIATOM ATTACHMENTS WITH SEQUENTIAL ALCIAN YELLOW-BLUE METHOD

Species	*Group*	*Alcian yellow pH* 0·5	*Alcian blue pH* 2·5
Asterionella sp.	2	−	+
Cocconeis scutellum	2	−	+
Grammatophora marina	2	+	−
Synedra fulgens	2	−	+
Amphora sp.	3	−	+
Navicula Grevillei	4	+	−
Licmophora flagellata	5	±	+
Achnanthes longipes	5	−	+

The predominant staining with alcian blue pH 2·5 indicates that the main components are carboxyl group containing polysaccharides. The reactions of *Grammatophora* and the Group 4 species *Navicula Grevillei* indicate that sulphated polysaccharides predominate. *Licmophora* gave variable results. If a small basal pad subtended one or two cells, the adhesive stained only with alcian blue. However, if the attachment material was elongated to form a stripe, then the mid-region was stained yellow while the most distal and proximal portions were alcian blue positive.

Examination of mixed sedentary diatom communities with the scanning electron microscope confirmed the general attachment structures seen by light microscopy. The flat dorsi-ventral thalli of *Cocconeis scutellum* and *Cocconeis sublittoralis* were firmly adherent due to extracellular material located between the substratum and the edge of the lower valve (Fig. 1).

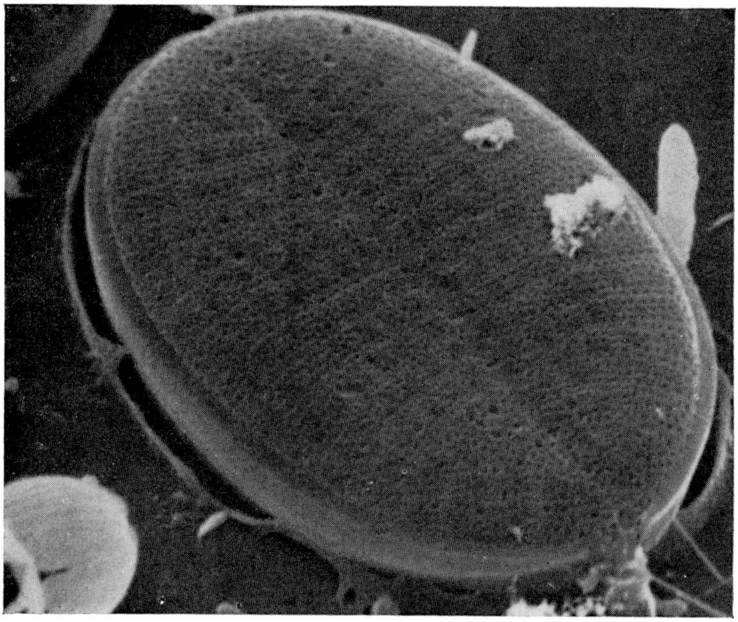

Fig. 1. Cocconeis scutellum *with mucilaginous adhesive around the periphery of the lower valve* (×*15 000*).

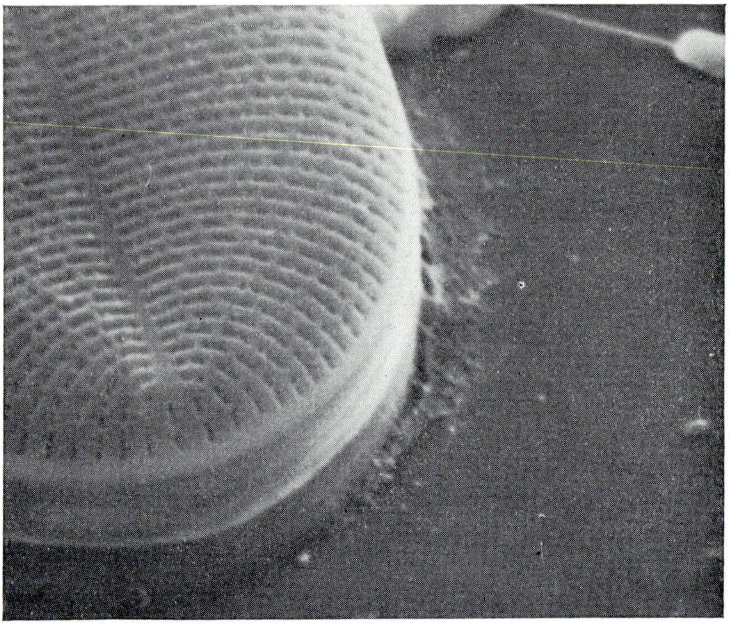

Fig. 2. Cocconeis scutellum *with more abundant adhesive showing the rather fibrous-fibrillar nature* (×*30 150*).

Fig. 3. Licmophora flabellata. *Whole cell with basal mucilage pad* (×*6150*).

Fig. 4. Licmophora flabellata. *The basal region with mucilage pad showing irregular perforations and frustule pores* (*arrow*) (×*17 250*).

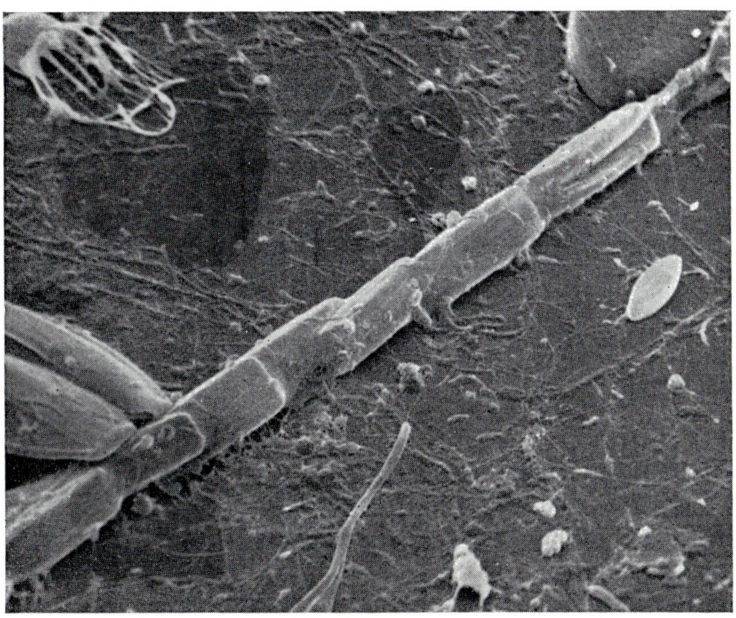

Fig. 5. Navicula Grevillei. *Part of a filamentous colony surrounded by the mucilage tube. The cells are obliquely arranged* (×*5100*).

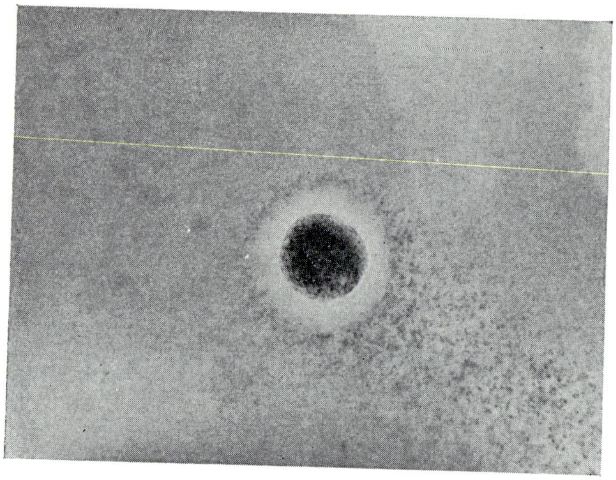

Fig. 8. A released tetraspore mounted in Indian Ink to demonstrate the mucilage envelope which almost doubles the spore diameter. Note the 'tail' of low viscosity material streaming out as the medium flows from top left to bottom right (×1500).

tetraspores were centrifuged for 25 min at 3000 *g*, the cell contents became stratified as follows: basal layer, floridean starch grains, and adhesive vesicles; equatorial layer, plastids with some vacuoles in upper levels; supra-equatorial layer, Golgi dictyosomes and mitochondria; apical polar region, vesicles, e.r. and fibrillar vacuoles. The nucleus occupied a position in the dictyosome/mitochondria region.

The adhesive vesicles come to occupy a peripheral location in the released spore (Fig. 10). Once an initial, tenuous attachment has been effected by the mucilage envelope, this second material is liberated. This process may be observed under the light microscope as the formation of a refractile, hyaline pad beneath the spore (Fig. 9).

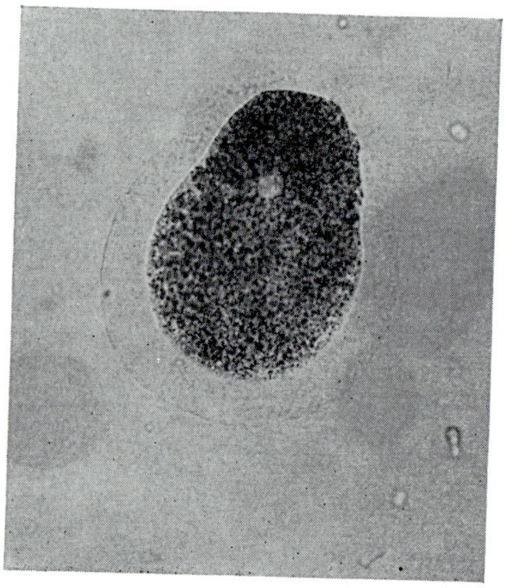

Fig. 9. *A newly settled spore after secretion of the highly refractile adhesive pad (×475).*

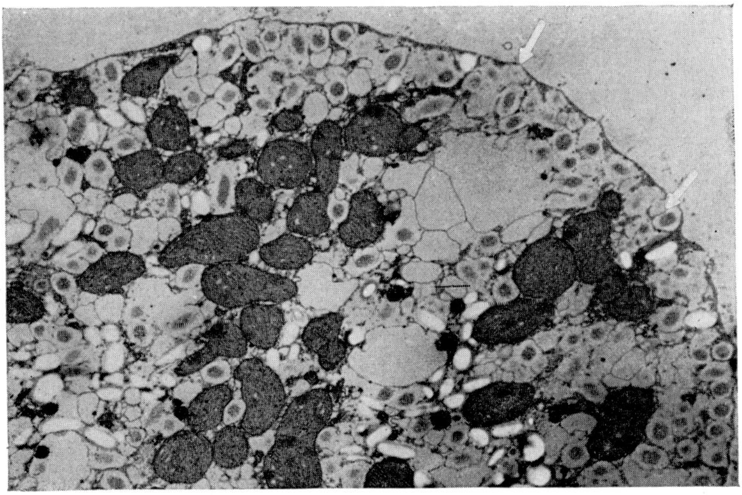

Fig. 10. Part of a tetraspore showing the peripheral distribution of the adhesive vesicles (arrows) during spore settlement prior to adhesion ($\times 3000$).

The strength of practical adhesion is dependent upon surface contact time. Thus using the seawater hosing method of Christie *et al.* (1970), settled tetraspores of *Ceramium* yielded the following results after a gentle pre-rinse to remove unattached spores.

Surface contact time (*h*)	*Percentage spores detached*
3	100
18	30–40
36	0

This indicates that adhesion increases with the length of time of surface contact. Similar results have been reported with other red algal spores by Charters *et al.* (1973).

Adhesive liberation is not light-dependent. Settlement occurs just as well in the dark, presumably the large reserves of floridean starch yield sufficient metabolic energy. Settlement will occur at a wide range of temperatures although at low temperatures, firm attachment takes longer, possibly due to an increase in the viscosity of the adhesive. However, once the adhesive has been liberated, metabolic functioning is not required to maintain adhesion; dead spores remain firmly attached. The same is true of diatoms even when only raphe material is involved in the attachment.

S.E.M. examination of settled spores shows the spore surface is covered by a layer of material which is continuous with the basal adhesive pad. After processing, this pad consists of a circular plate of material which is a continuous sheet in the region nearest to the spore but becomes perforated in the outer regions which ultimately gives way to an open fibrous meshwork at the periphery (Fig. 11). In view of the fibrillar nature of the adhesive (Fig. 12) (Chamberlain and Evans, 1973), this fibrous structure probably represents bundles of irregularly arranged fibrils. That these are not ordered is shown by the optical isotropy when the adhesive pad is viewed by polarized light. The

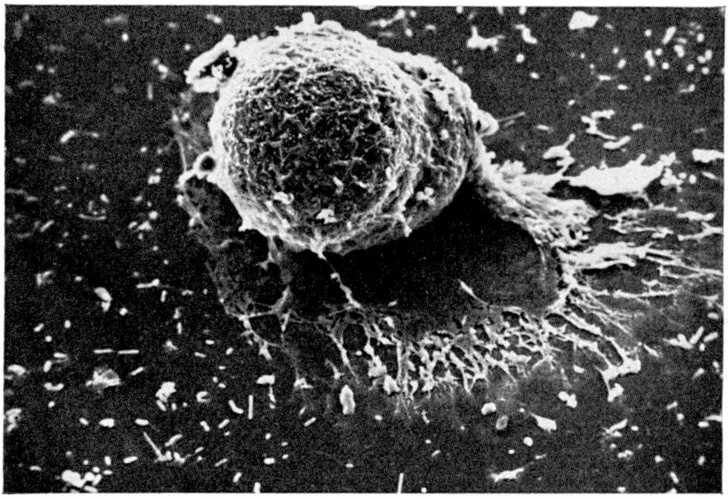

Fig. 11. *A settled spore with secreted adhesive pad. The material becomes more reticulate at the periphery* (×*3420*).

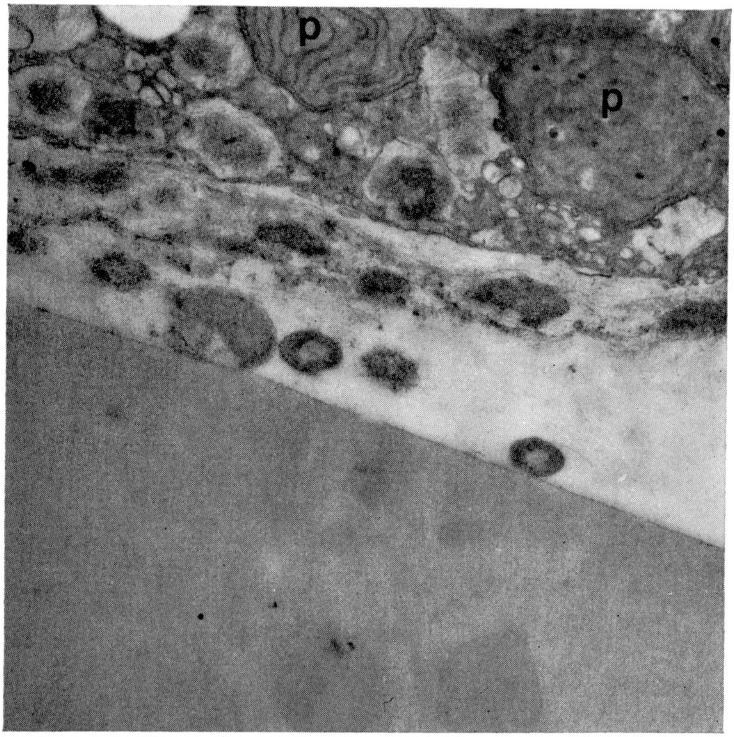

Fig. 12. *Part of the adhesive pad with newly secreted adhesive vesicles outside the plasma membrane. Other adhesive vesicles lie inside in close proximity to the spore surface:* (*p*) *Plastids* (×*15 000*).

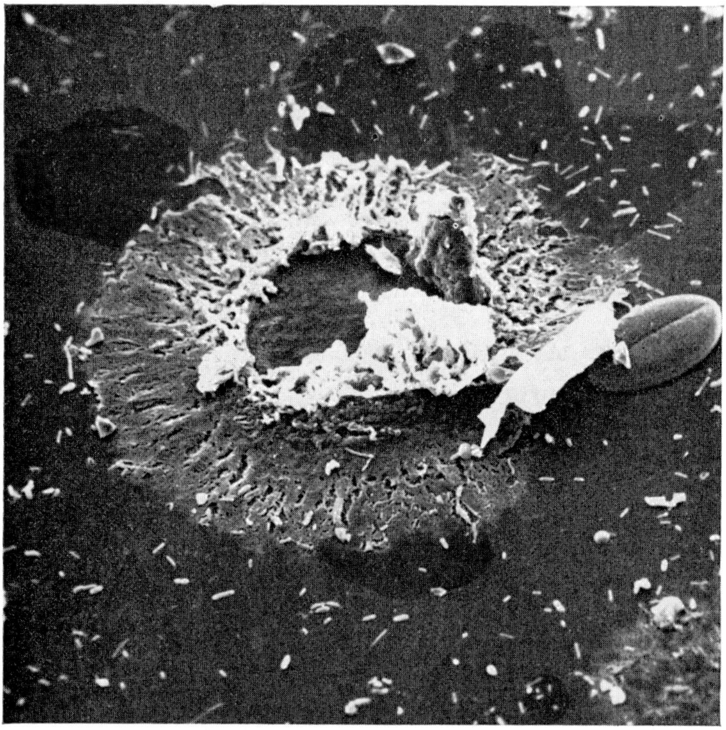

Fig. 13. An adhesive pad after mechanical removal of the spore. The central area retains part of the spore wall/membrane (×3450).

radial variations in structure probably reflect the damage inflicted by dehydration on adhesive layers of diminishing thickness. The forcible removal of adherent spores demonstrates that neither cohesive nor adhesive failure occurs in the adhesive system. Instead, the plasma membrane ruptures (Fig. 13) and may be seen as a smooth structure overlying the centre of the adhesive pad.

Following germination, the rhizoid grows down through the adhesive pad and either enters or grows along the substratum depending upon its penetrability.

DISCUSSION

Diatom adhesion

Diatoms are commonly the primary algal colonists on any newly immersed surface. It has frequently been suggested that their extracellular 'slimes', together with those of bacterial origin, may be of importance in aiding settlement of other organisms, especially on toxic surfaces (Wood, 1967). Lewin (1955) showed that the capsulate *Navicula pelliculosa* produced copious additional mucilage when grown in media depleted of certain essential nutrients. However, it seems unlikely that the mucilaginous materials described in this paper are present in sufficient quantity or physical state to form an all-enveloping film.

Early workers thought that diatom mucilages were pectic in nature (Mangin, 1908; Liebisch, 1929) due to their positive staining by ruthenium red. However, Luft (1971) demonstrated that this stain was not pectin-specific but acted merely as a basic dye. Reimann *et al.* (1965) suggested that the organic wall constituents of all diatoms give a positive stain with ruthenium red.

Gross chemical analyses of mucilages from a number of diatoms have been published and establish that the basic composition may vary considerably from genus to genus. Table II lists these analyses which unfortunately have mainly considered fresh-water species.

Table II
ANALYSES OF DIATOM EXTRACELLULAR MUCILAGE COMPOSITION

Species	*Author*	*Principal Components*
Amphipleura rutilans	Lewin, R., 1958	Xylose, mannose, rhamnose, sulphate, silica, and possibly protein
Gomphonema olivaceum	Huntsman and Sloneker, 1971	D-galactose, D-xylose ester sulphate, uronic acid, calcium, and magnesium
Navicula pelliculosa	Lewin, J., 1955	Glucuronic acid
Phaeodactylum tricornutum	Lewin, J. *et al.*, 1958	Xylose, mannose, fucose, galactose
Thalassiosira fluviatilis	Blackwell *et al.*, 1967	Chitin
Cyclotella cryptica	Blackwell *et al.*, 1967	Chitin
Thalassiosira fluviatilis	Dweltz *et al.*, 1968	Chitan

The presence of sulphated polysaccharide(s) in the mucilage tubes of *Navicula Grevillei* is paralleled by the results of Lewin (1958) for *Amphipleura rutilans*, while the presence of both uronic acid and ester sulphate in *Gomphonema olivaceum* (Huntsman and Sloneker, 1971) correlates with the present results for stipitate colonies of *Licmophora*. Perhaps future studies may show a correlation between anionic groupings and the extracellular adhesive morphology. The cytoplasmic origin of these mucilages has not been demonstrated convincingly although their polysaccharide nature suggests that the Golgi apparatus is involved. Dawson (1973) described several vesicle types in *Gomphonema parvulum* of which two were frequently closely associated with the plasmalemma and, quoting unpublished results of Crawford studying *Melosira*, suggested that the granular content of the one with a Golgi origin was perhaps associated with extracellular mucilage secretion.

Hufford and Collins (1972) showed that stalk material of *Cymbella cistula* was extruded from a terminal pore field, similar to that observed in *Licmophora* in this study. The function of such 'Gallertporen' was discussed by Gerloff (1959) studying *Asterionella formosa* and by Geissler and Gerloff (1963). The stalks of *Gomphonema olivaceum* were stained metachromatically with Azure B and bore linear striations (Drum, 1969) probably caused by convolution of the stalk surface during processing for S.E.M. The stalk material was shown to be fibrillar when viewed by T.E.M., similar to the mucilaginous tube polysaccharides of *Mastogloia* (Drum *et al.*, 1966) and *Amphipleura* (Drum unpublished, quoted Drum, 1969).

The concept that all motile epipsammic diatoms adhere by raphe mucilage trails while sedentary forms have a mucilage system was advanced by Harper and Harper (1967). The present S.E.M. observations appear to extend this theory to fouling diatoms. Further evidence was supplied by Hufford and Collins (1972) who indicated that the raphe of *Cymbella cistula* was inactive when the organism was mounted upon a mucilaginous stalk. These authors further suggested that the origin of the stalk material may be the 'crystalloid bodies' first described by Drum and Hopkins (1966) with reference to locomotion and adhesion of cells with the raphe only system of attachment.

Following the recent brief review by Bishop *et al.* (1974) of the effects of diatom fouling on anti-fouling compositions, it seems that a reappraisal of diatom adhesion systems and the chemical species involved is long overdue and could well supply valuable information in terms of future attacks upon the fouling problem.

Algal spore settlement

Several factors become operative once the spore is released from the parent sporangium. Spore sedimentation rates depend upon size and density, with additional modifying factors such as spore shape, motility, and the frictional drag of enveloping mucilage. Coon *et al.* (1972) have discussed the mathematical basis of these sedimentation rates for a range of red algal spores.

The relatively high water-current velocities found in marine environments suggest that in open water only under exceptional conditions will the swimming velocity of motile zoospores fulfil any transport function. Thus even motile spores are essentially at the mercy of the tide. However, this ceases to be the case when a spore passes into the relatively slow-moving, laminar sublayer of the boundary layer about a surface. Almost all algal spores have dimensions which allow them to occupy a 'safe' position in this region (Houghton, pers. comm.) and here spore motility may be advantageous for surface selection. Positive surface selection has been reported for *Enteromorpha* zoospores (Christie and Shaw, 1968) and *Chorda filum* (South and Burrows, 1967).

The surface properties of potential fouling substrata have not received very much attention although work in other fields of biological adhesion has stressed their importance (Baier, 1970). Good adhesion is related to effective 'wetting' of a surface by the adhesive, the higher the surface free energy, the more likely an adhesive will 'wet' it. However, spores will settle on both high and low energy surfaces, *e.g.* polyethylene—31 dynes/cm, polystyrene—33 dynes/cm, and even Teflon 18 dynes/cm (values from Sharpe and Schonhorn, 1963) as against glass [approximately 700 dynes/cm (Maroudas, 1973)]. A complication in any determination of substrate surface tension is the presence of suspended and/or dissolved organic materials which may be rapidly adsorbed to yield a low energy surface. Even the presence of water will have a considerable effect (Baier, 1970). Thus in the marine environment all potential fouling surfaces may be reduced to a common low surface energy.

The actual process of spore adhesion appears to depend, initially at least, upon the physical forces such as the London forces (Chamberlain and Evans, in press). Chemisorption phenomena cannot occur until molecular contact is established, that is, after Van der Waal's-type forces have become operative. The increase in adhesion strength with respect to surface contact time may be a measure of the development of secondary chemical crosslinking reactions although on theoretical grounds, the dispersion forces alone would be sufficient to prevent any breakdown of the adhesive at an interface. This suggests that detachment may be due to cohesive failures in the bulk adhesive

Lewin, J. C., Lewin, R. A., and Philpott, D. E. (1958). Observations on *Phaeodactylum tricornutum*. *J. Gen. Microbiol.*, **18**, 418–426.
Lewin, R. A. (1958). The mucilage tubes of *Amphipleura rutilans*. *Limnol. Oceanogr.*, **3**, 111–113.
Liebisch, W. (1929). Experimentelle und kritische Untersuchungen uber die Pektinmembran der Diatomeen unter besonderer Berucksichtigung der Auxosporenbildung und der Kratikularzustande. *Z. Bot.*, **22**, 1.
Luft, J. H. (1971). Ruthenium red and violet. I. Chemistry, purification, methods of use for electron microscopy and mechanism of action. *Anat. Rec.*, **171**, 347–368.
Mangin, M. L. (1908). Observations sur les diatomées. *Ann. Sci. Nat.* (*Bot.*), **8**, 177.
Maroudas, N. G. (1973). Chemical and mechanical requirements for fibroblast adhesion. *Nature*, **244**, 353–354.
Massalski, A. (1969). Cytological studies as filamentous Xanthophyceae. Ph.D. Thesis. University of Leeds.
Micalef, H., and Gayral, P. (1972). Quelques aspects de l'infrastructure des cellules végétatives et des cellules reproductrices *d'Ulva lactuca* L. (Chlorophycées). *J. Microscopie*, **13**, 417–428.
Parker, B. C., and Diboll, A. G. (1966). Alcian stains for histochemical localisation of acid and sulphated polysaccharides in algae. *Phycologia*, **6**, 37–46.
Reimann, B. E. F., Lewin, J. C., and Volcani, B. E. (1965). Studies on the biochemistry and fine structure of silica shell formation in diatoms. *J. Cell Biol.*, **24**, 39–54.
Retallack, B., and Butler, R. D. (1972). Reproduction in *Bulbochaete hiloensis* (Nordst.) Tiffany. I. Structure of the zoospore. *Arch. Mikrobiol.*, **86**, 265–280.
Rosowski, J. C., and Kugrens, P. (1973). Observations on the euglenoid *Colacium* with special reference to the formation and morphology of attachment material. *J. Phycol.*, **9**, 370–383.
Sharpe, L. H., and Schonhorn, H. (1963). Surface energetics, adhesion and adhesive joints. *Adv. Chem. Ser.*, **43**, 189–201.
South, G. R., and Burrows, E. M. (1967). Studies on marine algae of the British Isles. 5. *Chorda filum* (L.) Stackh. *Brit. Phycol. Bull.*, **3**, 379–402.
Wood, E. J. F. (1967). *Microbiology of Oceans and Estuaries*. Elsevier Oceanography Series, 3, Elsevier, Amsterdam, London, New York.

SECRETION OF ADHESIVE POLYMERS AND ATTACHMENT OF MARINE BACTERIA TO SURFACES

WILLIAM A. CORPE, LINDA MATSUUCHI and BARBARA ARMBRUSTER

Department of Biological Sciences, Columbia University, New York 10027, U.S.A.

Summary

Various organic particles and glass slides submerged in seawater enriched with 0·005% (w/v) peptone supported populations of irreversibly attached, primary film-forming bacteria. The population, developed within a period of three days, seemed to bear no relationship to the chemical nature of the substrata. Isolated film-forming bacteria displayed a variety of cell surface structures that are probably related to their sorption, including surface 'blebs', 'droplets', fibrils, pili, and other appendages. *Pseudomonas atlantica* mutants lacking the voluminous mucopolysaccharide still retained the propensity for attachment. An acidic glycoprotein produced by mutants and other marine bacteria may be responsible for attachment. Enhancement of aggregation and adhesion of cells of bacteria and algae by polymers produced by primary film formers is described and evaluated in terms of biodeterioration.

The means by which bacteria become attached to surfaces have not been thoroughly explored, but it is generally accepted that irreversible attachment involves cell surface structures or extracellular materials that bridge cells to solid substrata (Corpe, 1970a; Marshall *et al.*, 1971).

In the present paper, the surface morphology of a few periphytic marine bacteria is examined and the importance of bacterial surface polymers in microbial film formation evaluated.

MATERIALS AND METHODS

Cultures

Cultures used in this study were isolated from agar inoculated with seawater samples, or from the surface of glass slides that had been submerged in the sea for from one to three days. The former included gram-negative rods D_113, D_117, D_313, D_511, and D_1E. The latter were *Pseudomonas atlantica* strains 4b and T_6c and *Pseudomonas* sp., strains 6b and Ma 13, all previously described (Corpe, 1973). All of the cultures could attach irreversibly and colonize glass slides immersed in artificial seawater (Instant Ocean, Aquarium Systems, Inc., Wickliffe, Ohio) containing 0·005% (w/v) Bacto Peptone, within 12 hr at 25°C. All cultures were maintained on slants of Bacto Marine agar (Difco Co., Detroit, Michigan).

Isolation of the exopolysaccharide-less mutants

Pseudomonas atlantica, strain T_6c, was grown in flasks containing sterile Bacto peptone (0·5% w/v) and glucose (1% w/v) in seawater (PGSW medium). Flasks were mounted on a rotary shaker revolving at 150 rpm in a 25°C incubator. Logarithmic phase cells were treated with mutagen N-methyl N-nitro N-nitroguanidine (NTG, Aldrich Chemical Co., Milwaukee, Wis.) in 0·2 M tris maleate buffer (pH 6·0) using the procedure described by Adelberg (1965). Treated cultures and untreated controls were diluted and surface plated on Bacto Marine agar. Inoculated plates were wrapped in aluminium foil and incubated at 25°C. Colonies were inspected for lack of gumminess. Stable, nongummy colonies were found in cultures exposed to the mutagen for 1 and 1·5 hr in which 99·99 % of the viable cells had been killed. The mutants were compared with the wild type T_6c as to cell morphology, growth, and culture viscosity in PGSW broth and the ability to hydrolyse agar, a distinctive feature of *Pseudomonas atlantica* (Corpe, 1970b).

Recovery and analysis of extracellular polymers from cultures

Extracellular polymers produced by marine bacteria were recovered from four-day-old PGSW culture, supernatant fluids, by acetone precipitation. The insoluble material was dialysed, lyophilized, and chromatographed on diethylaminoethyl (DEAE) cellulose as described by Corpe (1973). Carbohydrate, uronic acid, and protein content of samples were analysed by methods previously described (Tosteson and Corpe, 1975).

Other cultural methods

Primary films were developed on glass slides immersed in a marine aquarium containing various bacteria, algae, and invertebrate animals. It was illuminated by daylight and the water was not aerated or filtered. Fresh seawater samples collected at the New Jersey shore, in the vicinity of Atlantic City, were dispensed in 150-ml flasks and supplemented with sterile, 5 % (w/v) peptone solution to give a final concentration of 0·005 % (w/v). Pairs of flasks were supplemented with each of the following dry insoluble materials, added so as to provide about 1 mg/ml: crab chitin (powder, prepared by D. M. Reynolds), Bacto 'purified' agar (Difco, Detroit, Mich.), alpha cellulose (Nutritional Biochemicals Co., Cleveland, Ohio). After various periods of incubation at 25°C, cultures were examined with a Leitz microscope, using phase optics or air-dried films fixed and desalted with acetic acid (2 %) and stained with Hucker's crystal violet for 2 min.

Hydrolysis of casein, starch, chitin, cellulose, and agar was examined, using aerobic conditions and seawater nutrient agar supplemented with the appropriate substrate. The methods followed were as described by Rodina (1972).

Suspensions of algal cells

Enrichments of algae containing both unicellular, green algae, and diatoms were prepared using Bristol's sodium nitrate solution (Allen, 1949). One volume of the medium was mixed with one volume of a freshly collected seawater sample and incubated in daylight at room temperature. Subenrichments were prepared and used as a source of algal cells. The mixed algae were centrifuged at 3000 × *g* and washed twice with 0·1 M tris-glycine buffer, pH 7·8, in 50 % (v/v) seawater. Dense, uniform suspensions were prepared in buffered seawater and diluted to give an optical density reading of about 0·9 at 600 nm in a Bausch and Lomb Spectronic '20' Instrument (Bausch & Lomb Inst. Co., Rochester, N.Y.). The suspension was mixed with T_6c polymers, prepared as a distilled water solution containing 2 mg/ml.

Electron microscopy

Half percent peptone broth cultures or seawater suspensions of cells were fixed with 5% (v/v) glutaraldehyde in seawater for 30 min at room temperature. One drop of the suspension was placed on a grid (covered with a carbon coated nitrocellulose film). After 20 sec, the drop was withdrawn with filter paper and a drop of uranyl acetate [2% (w/v) or 1% (w/v) phosphotungstate at pH 7·0] applied for 20 sec, then withdrawn with filter paper. The grid was allowed to dry in air and examined in the electron microscope. Aqueous Alcian Blue 1% (w/v) was also used as a staining reagent for electron microscopy. It was applied in the same manner as with the negative stains. Specimens were examined with an RCA EMU 4B electron microscope operating at 50 kV.

RESULTS

Primary film-forming bacteria

Figure 1 shows photographs of firmly attached bacteria on glass slides which were suspended in a marine aquarium for up to 72 hr. A variety of morphological types of bacteria were seen, including rods, coccid rods, and filamentous forms. Hyphomicrobia and Caulobacters were recognized particularly in Figs. 1B and 1C. The shapes of the cells and their surface morphology was in general quite irregular. This was particularly evident in crystal violet-stained cells. Cells observed in wet mount with phase optics seemed to have a much more discrete cell boundary.

Figure 2 shows photomicrographs of chitin, agar and cellulose particles that had been colonized by bacteria. When particles were washed on polypropylene, filters (10 micron pore size) removed unattached cells but the bacteria-covered particles appeared the same as in Fig. 2. Rods, coccoid, and stalked forms were common. No chitin or cellulose-digesting

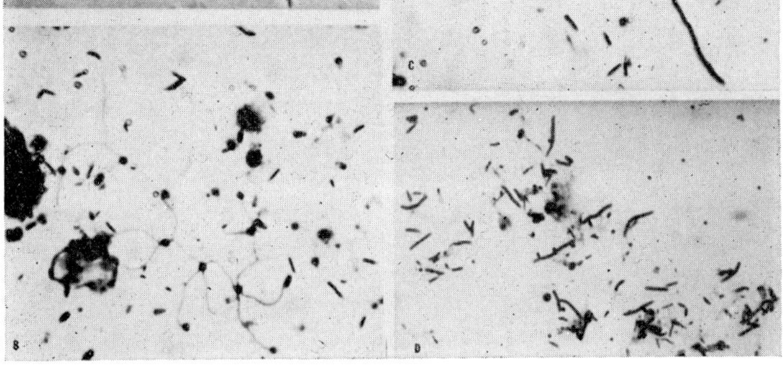

Fig. 1. Firmly attached bacteria on glass slides suspended in a marine aquarium. Crystal violet stained. (A) 12 hr; (B), (C), (D), 72 hr. Marker = 10 μm.

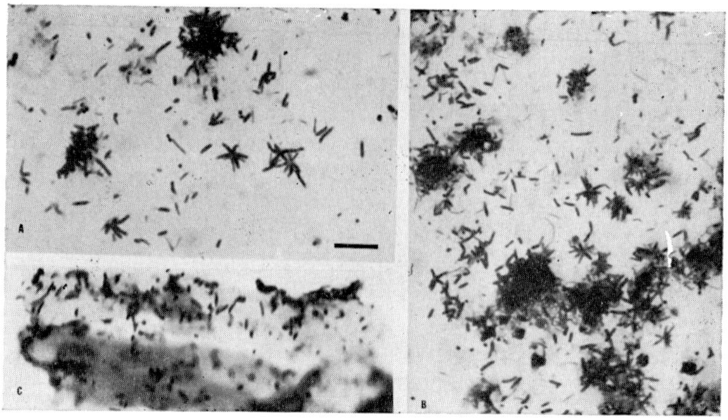

Fig. 2. Marine bacteria attached to Agar particles (A); Chitin particles (B), Cellulose fragments (C). Stained with crystal violet. Marker = 10 μm.

bacteria were found in the chitin or cellulose enrichments for up to 72 hr. A few agar digesters were found after 24 hr when the agar enrichment was streaked onto marine agar. Agar digesters were not observed when the cellulose enrichment was streaked onto agar plates. Slides bearing dried-on smears of starch or protein (albumin) developed a substantial population of bacteria when immersed in seawater that formed an almost continuous film

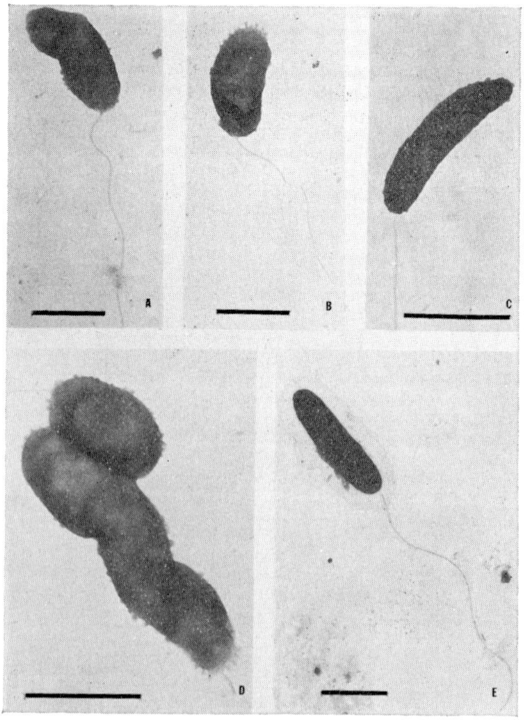

Fig. 3. Negatively stained marine bacteria. Culture 4b (A, B, and D), culture 6b (C) stained with 2% uranyl acetate. Culture D3-13 (E) stained with 1% PTA. Marker = 10μm.

of cells, a large portion of which were starch or protein hydrolysers.

Bacteria that became attached to chitin, cellulose, or agar particles were not compared in quantitative experiments; however, pure cultures of isolated bacteria obtained from streak plates of each enrichment appeared to be similar with regard to cell morphology, staining, pigmentation, and colonial form. There was a greater proportion of protein digesters in the chitin enrichment than in the agar or cellulose enrichments.

Surface morphology of some primary film-forming bacteria

Negatively stained, logarithmic phase cells of various isolates were examined with the electron microscope. Figure 3 (A–D) shows cells with very irregular or convoluted surfaces. 'Blebs' or irregular 'droplets' were visible on some cells, particularly numerous at the poles of the cells. Culture D1-13 (Fig. 3E) showed a uniform capsular material overlaying a smooth surface. Extended, spindle-shaped droplets were seen in preparations of Ma 13 (Fig. 4, A–D), but in the same preparation (4A and 4C) cells with a fuzzy, hair-like quality were observed. The cells were often joined together by firm bridges (Fig. 4A), forming large clumps. As shown in Fig. 5, bacteria with surface appendages were not uncommon. Cells with pili are shown in Fig. 5 (D–G). Figures 5D, E, and F show pili in tufts at the poles of the cells. Figures 5 (A–C) show appendages covering the entire cell surface, displaying a tubular or globular appearance.

Cells shown in Fig. 6 were stained with Alcian Blue, a copper phycocyanin

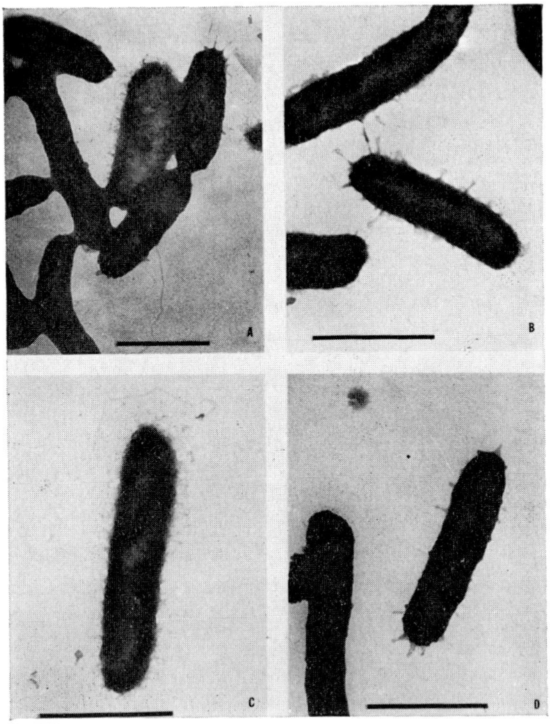

Fig. 4. Negatively stained cells of Ma 13 grown on PGSW agar, suspension fixed in 5% glutaraldehyde, stained with 1% PTA. Marker = 10 μm.

dye. They were prepared as a suspension from a gummy agar colony of *Pseudomonas atlantica*, strain 4b. Young broth cultures of the same organism were shown in Fig. 3. When these cells were stained with Alcian Blue, surface 'blebs' were not observed, rather the cells had an appearance not unlike that of cells shown in Fig. 6, from which the strands of loose polymer had been removed. The dye was complexed and precipitated both by the loose exopolymer and by material retained at the cell surface. The polymeric fibrils of the type associated with *Ps. atlantica* strains were not seen in other cultures.

Exopolysaccharide-less mutants of T_6c

Nongummy mutants or those that could produce only reduced amounts of exopolymer were isolated from cultures treated with NTG. All were agar hydrolysing rods which grew in PGSW broth without production of viscosity nor did they produce gummy or viscous colonies on agar. Some of the properties of two mutants compared with the wild type are shown in Table I. The yield of exopolymer by mutants was greatly reduced in quantity. The percentage of protein in the mutant polymer was much greater than in the wild type polymer. The mutant polymer contained less uronic acid than the wild type polymer.

The rate of attachment of mutants as compared with wild type cells is shown in Table II. The mutants became firmly attached more slowly than the

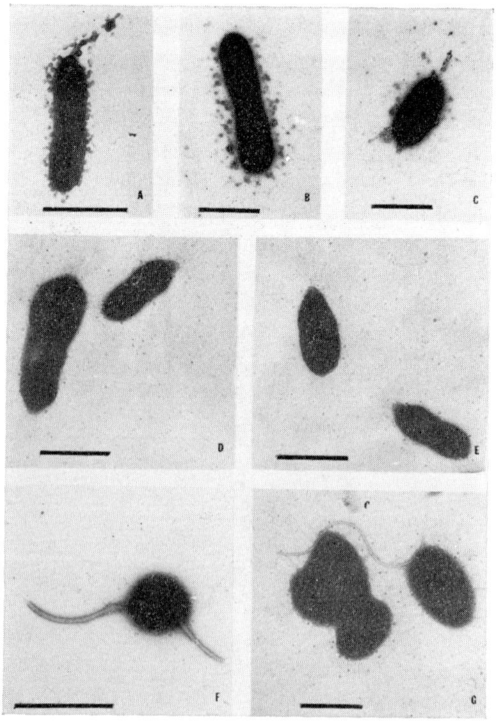

Fig. 5. Negatively stained cells of marine bacteria culture D1-17 (A, B, C), culture D3-13 (D, E), culture D5-1-11 (F) and DIE (G). Fixed with 5% glutaraldehyde and stained with 2% uranyl acetate. Marker = 10 μm.

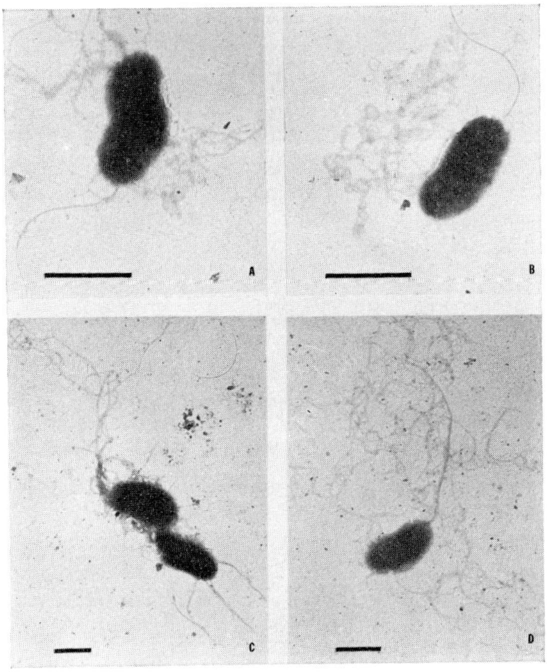

Fig. 6. Suspension of 4b cells prepared from an agar colony. Stained with 1% Alcian Blue. Marker = 10 μm.

Table I
SOME YIELD AND COMPOSITIONAL DIFFERENCES BETWEEN EXOPOLYMERS PRODUCED BY WILD TYPE AND MUTANTS OF T_{6C}

Culture	*Acetone precipitate*	*Exopolymer yield (mg/100 ml broth)*	*Carbohydrate*	*Percentage[a] uronic acid*	*Protein*
Wild type	gel	10·5	54·3	43·8	2·0
Mutant 18	floc.	0·8	54·2	2·8	43·2
Mutant 26	floc.	2·7	30·8	9·3	59·8

[a] Analysis of polymer which had been absorbed onto DEAE cellulose and eluted with 0·05 M NaOH. Percentage calculated on a weight basis after correction for moisture and ash.

Table II
ATTACHMENT OF *PSEUDOMONAS ATLANTICA* TO GLASS SLIDES, COMPARISON OF WILD TYPE AND MUTANTS

Culture	*Average no. of cells per microscopic field[a]* 24 *hr*	48 *hr*	72 *hr*	*Average no. of cells per ml of medium, 24 hr[b]*
Wild type	3·80	6·88	12·72	$5·0 \times 10^6$
Mutant 16	0·68	7·84	10·44	$4·25 \times 10^6$
Mutant 26	0·16	4·84	23·28	$9·0 \times 10^6$

[a] Average no. cells in 25 oil immersion microscopic fields on slides that were rinsed, fixed, and stained. Slides were submerged for the time indicated in 200 ml of seawater containing 0·005% peptone.
[b] Counts made with Petroff-Hauser counting chamber.

wild type, but after 48 hr, the number of cells per field was similar to, or even exceeded, that of the wild type. Cells per millilitre of medium were similar in mutant and wild type cultures.

Enhancement of algal aggregation by T_6c polymer

Washed cells of a unicellular green alga, mainly a *Chlorella* species, were mixed with T_6c polymer in test tubes. Experimental tubes were inspected for clumping of the algal cells by naked eye and with the microscope. Pairs of tubes were incubated in daylight and in the dark at room temperature for 12 hr. Clumping of the algal cells incubated in daylight was evident after 6 hr but no dark-incubated cells had clumped. After 12 hr in the dark, however, some aggregation was noted. The control suspension showed some aggregation; so the effect of the bacterial polymer was to enhance aggregation.

DISCUSSION

Bacterial surface films

The firm attachment of bacteria to solid surfaces occurred only after a period of incubation of the living cells in the presence of a carbon and nitrogen source (Corpe, 1974). In the sea, the amount and quality of nutrient available as well as the physical and chemical character of the substrata and surrounding waters probably determine the rate of primary film formation and which organisms appear first. All representative marine bacteria tested appear to be periphytic but they attach and grow at somewhat different rates (Corpe, 1974). Figures 1 and 2 display the variety of bacterial types commonly observed on various surfaces. Photographs of this sort do not adequately emphasize that once a film has formed, the surface properties have changed and will remain so long after the viable population and even the microscopically demonstrable primary film-forming bacteria have disappeared.

Attachment of primary film-forming bacteria to solid surfaces

Both direct and indirect evidence have been provided to show that irreversible attachment of bacteria involves polymeric bridging of cells to substrata by extracellular polymeric fibrils (Marshall, 1973). The fibrils might be located mainly at the poles of the cells, causing the latter to be oriented at right angles to the surface. Marshall (1973) showed that interference of protein synthesis by chloramphenicol inhibited the attachment of several marine pseudomonads. Streptomycin inhibited growth but not adhesion, suggesting that the bacterium synthesized bridging polymer prior to the addition of antibiotic.

Figures 3–6 show the electron microscopic appearance of various bacterial surfaces. Soft ‘blebs’ or ‘droplets’ and what appear to be pili can be seen. Any or all of these may serve as the site of attachment or sites where polymeric bridging material is concentrated. Young, washed cells of T_6c, similar in appearance to cells in Figs. 3A, 3B and 3D, still required a period of incubation of up to 5–6 hr before a substantial number of firmly attached cells could be demonstrated. Thus the presence of bridging polymers may not be all that is necessary for attachment. If we visualize the adhesive bonds to be of the London or van der Waals type (Baier, 1973), intimate contact must be made at an adequate number of sites to accomplish adhesion. The rate at which quantitatively significant contacts are made may be influenced by a number of metabolic functions, not only the capacity to secrete adhesive polymers.

Figures 4 and 5 show cells that are clumped through pili or other bridging materials. The clumping of cells of *Ps. atlantica* strains by means of a matrix of polymeric material has been reported (Corpe, 1970b). Strains 4b and T_6c produced gummy colonies in which cells were embedded in polymeric fibrils. Figure 6 shows cells of 4b which have been partially disengaged from the embedding polymer.

In the present work, using polysaccharide-less mutants, it is clear that the acid polysaccharide (Table I) is not needed for attachment (Table II). This information generally confirms the view of Fletcher and Floodgate (1973), who studied another acid polysaccharide-producing species and suggested that a layer more intimately associated with the cell wall was the actual site of attachment to the substratum. Like the acid polysaccharide, the cell wall component could bind Alcian Blue but contained protein as a major component. Material of this type was produced by most of the marine bacteria studied (Corpe, 1974) and recovered from culture supernatants as well as from cell surfaces. It may be identical to the loosely held 'blebs' or 'droplets' on cells as shown in Figs. 3 and 4.

Ecological importance of extracellular products of primary film-forming bacteria

The acid polysaccharide fraction of the *Ps. atlantica* exopolymer was previously described as an important agent for adhesion of particulate materials to surfaces (Corpe, 1970b). Even though the present work shows the polysaccharide was not necessary for attachment of the producing cells, because of its surface charge and other properties, it serves as a polymeric bridge for other cells.

The acid polysaccharide greatly enhanced the adhesion of bacteria to glass when smeared slides were suspended in the sea (Corpe, 1970b). T_6c polymer produced aggregation of algal cells (Fig. 7) which generally confirmed the

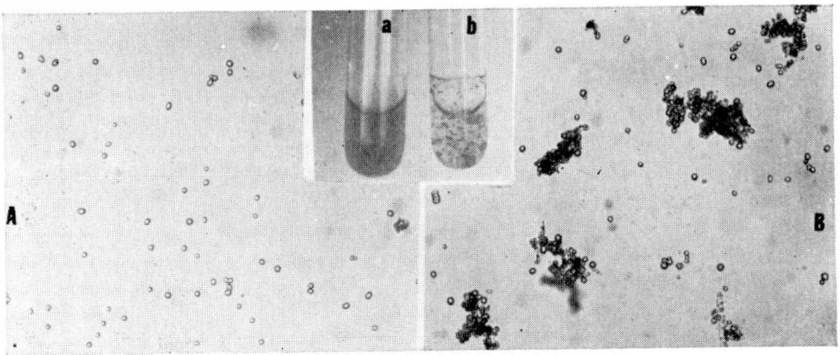

Fig. 7. Algal suspension in buffered seawater mixed with T_6c polymer (B, b) and without polymer (A, a).

action of T_6c polymer and other bacterial products on the marine *Chlorella vulgaris* (Tosteson and Corpe, 1975). The polymer used in the present work contained both a mucopolysaccharide component and glycoprotein which distinguished by reflective infrared spectrophotometry (Baier, pers. comm.) which polymers produced by the mutants were glycoprotein.

Materials similar to those produced by marine microorganisms have been recovered from seawater (Tosteson and Corpe, 1975). Baier (1973) has suggested that solid surfaces in contact with seawater become modified by absorption of glycoprotein which presumably 'conditions' the surface and enhances the adhesion of cells and other particulate materials.

The possibility that primary film-forming bacteria could facilitate adhesion of fungus spores or mycelium fragments, algal cells, and animal larvae should not be overlooked by those who are studying aspects of biodeterioration involving such organisms.

Acknowledgements

The authors wish to thank Dr Robert Baier for his willingness to examine various samples by the MAIR infra-red spectroscopic technique. This work was supported in part by a grant from the Environmental Protection Agency (EPA R803-113).

REFERENCES

Adelberg, E. A., Mandel, M., and Chein Ching Chen, G. (1965). Optimal conditions for mutagenesis by N-methyl-N-nitro-N-nitrosoguanidine in *Escherichia coli* K12. *Biochem. Biophys. Res. Comm.*, **18**, 788.

Allen, O. N. (1949). *Experiments in Soil Microbiology*. Burgess Publ. Co., Minneapolis, Minn.

Baier, R. (1973). Influence of initial surface condition of materials on bio-adhesion. In Proc. 3rd Int. Congr. on Marine Corrosion and Fouling, pp. 633–639, Northwestern University Press, Evanston, Ill.

Corpe, W. A. (1970a). Attachment of marine bacteria to solid surfaces. In *Adhesion in Biological Systems*, pp. 72–87 (C. S. Manly, ed.), Academic Press, Inc., New York.

Corpe, W. A. (1970b). An acid polysaccharide produced by a primary film-forming marine bacterium. *Dev. Ind. Microbiol.*, 402–412.

Corpe, W. A. (1973). Microfouling: The role of primary film forming marine bacteria. In Proc. 3rd Int. Congr. on Marine Corrosion and Fouling, pp. 598–609, Northwestern University Press, Evanston, Ill.

Corpe, W. A. (1974). Periphytic marine bacteria and the formation of microbial films on solid surfaces. In Proc. 2nd U.S.–Japan Conf. Marine Microbial, Effect of the ocean environment on microbial activities, pp. 397–417 (R. R. Colwell and R. Y. Morita, eds.), University Park Press, Baltimore, Md.

Fletcher, M., and Floodgate, G. D. (1973). An electron microscopic demonstration of an acidic polysaccharide involved in the adhesion in a marine bacterium to solid surfaces. *J. Gen. Microbiol*, **74**, 325–334.

Marshall, K. C. (1973). Mechanism of adhesion of marine bacteria to surfaces. In Proc. 3rd Int. Congr. on Marine Corrosion and Fouling, pp. 625–632, Northwestern University Press, Evanston, Ill.

Marshall, K. C., Stout, R., and Mitchell, R. (1971). Selective sorption of bacteria from seawater. *Can. J. Microbiol.*, **17**, 1413–1416.

Rodina, A. G. (1972). *Methods in Aquatic Microbiology*. University Park Press, Baltimore, Md.

Tosteson, T. R., and Corpe, W. A. (1975). Enhancement of adhesion of *Chlorella vulgaris* to glass by polymers from the sea and exudates from microbial cultures. *Can. J. Microbiol.*, **21**, 1025–1031.

POST-GERMINATION ATTACHMENT MECHANISMS IN MARINE FOULING ALGAE

R. L. FLETCHER

The Marine Laboratory, Department of Biological Sciences,
Portsmouth Polytechnic, Hayling Island, PO11 0DG, Hants, U.K.

Summary

Culture studies were carried out on spores of a number of green, brown, and red algae collected from marine-fouled habitats on the south coast of England in order to determine the mechanism of germling development and attachment, using both conventional light and scanning electron microscopy. In all species examined, the basal rhizoidal (attachment) filaments were found to secrete a mucilaginous material from the apical regions which facilitated good contact with the substrate and functioned as an adhesive. This mucilage was produced abundantly in the discoid (holdfast) type of rhizoidal base and provided them with considerable powers of adhesion. All the species studied demonstrated the ability to initiate new erect shoots from the basal rhizoidal system.

INTRODUCTION

Marine algae contribute significantly to the fouling communities on a wide range of immersed substrates. However, it is their relatively recent role as the dominant fouling organisms on the hulls of container ships and oil tankers which has commanded particular attention. These vessels usually have short stays in port and often commute between tropical and temperate waters. Only some marine algae, and in particular species of *Enteromorpha* and *Ectocarpus*, appear to have adapted to survive these arduous conditions, and in addition appear little affected by the conventional antifouling compositions used. The severe economic problems caused by the algal growths have stimulated numerous investigations into the 'biology' of some of the more important species. The purpose of these studies is to understand the underlying mechanisms of growth and development and use this knowledge to develop toxicants which will act more selectively on particular growth stages.

One aspect of this 'biological' approach to the problem which is receiving considerable attention is the initial spore colonization process. A number of papers dealing with reproductive spore formation, liberation, settlement, and adhesion in some of the more important ship foulers have been published (see especially Baker and Evans, 1973a,b; Chamberlain and Evans, 1973; Christie, 1973; Christie *et al.*, 1970; Evans and Christie, 1967, 1970; Houghton *et al.*, 1973; Jones and Babb, 1968). However, after the initial settlement process, the success of an alga as a fouling organism will depend upon a number of additional factors, not least of which will be the ability to form an extensive and efficient anchorage system. Not only will this be required to maintain an expanding erect shoot system but also to combat the effects of wave action and ship movement. Despite the importance of the prostrate system to a fouling alga, very few studies have dealt specifically with this aspect of morphogenesis. Very little is known about the range of secondary attachment systems found in marine fouling algae and the factors affecting their formation and adhesion.

It was with these aims in mind that the present study was initiated. A cultural investigation was carried out on spores of a number of green, brown, and red algal species collected from various fouling ecosystems on the south coast of England, and particular emphasis was placed on an examination of the germination process and early attachment mechanisms using conventional light and scanning electron microscopy (SEM).

MATERIALS AND METHODS

Samples of marine algae were collected from fouling habitats (buoys, marinas, ships, etc.) in Portsmouth and Langstone Harbour during the spring and early summer period, and used to set up unialgal cultures. Fertile portions of green and brown algae were cleaned with cotton wool, partially dried overnight in a refrigerator at 4°C, and then small portions immersed in the compartments of a repli dish containing filtered seawater (F.S.W.). Upon release, the motile spores were concentrated by making use of their particular phototactic responses (for experimental details, *see* Fletcher and Chamberlain, 1975) and when in sufficient concentration, pipetted off in drops, either directly into the culture vessels (plastic petri dishes, 90 mm diam.) or presettled on to glass coverslips (18 mm diam. no. 1½). The culture vessels were then flooded with culture medium and placed in the growth rooms.

For release of red algal spores, portions of fertile thalli were wiped clean and then carefully laid out over the surface of a repli dish which had been flooded with seawater. Each compartment of the repli dish contained a glass coverslip. A constant flow of seawater was then directed over the surface of the repli dish to prevent desiccation. This method usually resulted in the release of large numbers of viable spores which floated down and settled on the surface of the coverslips.

Throughout the experimental period all cultures were grown at 15°C under 16–8 daylength conditions, using Grolux fluorescent tubes at an intensity of approximately 1000 lux. The culture medium used was that of Von Stosch (1963).

For S.E.M. observations, 10 mm square portions of the coverslips and petri dish floor containing the attached germlings were cut out, prefixed in 3% glutaraldehyde (phosphate buffered at pH 7·2) for 3–5 hr at 4°C, washed in 0·2 M solution of phosphate buffered sucrose (pH 7·2) and post-fixed in 1% osmium tetroxide buffered to 7·2 with 0·1 M phosphate. The material was then washed in the phosphate buffer, dehydrated in a graded ethanol series, and critically point dried before coating with gold for examination.

RESULTS

Cultures were set up of the following 20 species of algae:

Chlorophyceae (zoospore cultures)—*Blidingia marginata*, *Enteromorpha intestinalis*, *E. linza*, *E. prolifera*, *Ulothrix flacca*, *Ulva lactuca*, *U. rigida*, and *Urospora penicilliformis*.

Phaeophyceae (plurispore cultures)—*Ectocarpus fasciculatus*, *E. siliculosus*, *Giffordia granulosa*, *Petalonia fascia*, *Pilayella littoralis*, and *Scytosiphon lomentaria*.

Rhodophyceae (carpospore and tetraspore cultures)—*Ceramium rubrum*, *Polysiphonia brodiaei*, *P. elongata*, *P. nigrescens*, *P. spiralis*, and *P. urceolata*. Taxonomic nomenclature follows that of Parke and Dixon (1968).

Light microscope observations

Chlorophyceae

Blidingia marginata. Zoospore germination was apparent within 2–3 days of settlement. In this species no distinct germination tube was formed; the original spore cell appeared to simply extend in one direction and divide into two cells. Further divisions of these 2 cells, at right angles to the first division, produced a small rosette-shaped, light-green coloured disc of four cells. Outward growth of these cells combined with both transverse and oblique divisions produced a slowly expanding disc (Fig. 1), the central region of which gradually became polystromatic (crustlike) and dome-shaped.

The erect fronds, typical of the species, were produced from the surface of the expanding prostrate system at various stages of development, usually when it was dome-shaped and well established (Fig. 2) although occasionally from the initial 10–15-celled discoid stage.

Examination of the margins of the young discs usually indicated the presence of a cuticular-like substance surrounding the peripheral meristematic cells which appeared to be binding them together. However, this outer

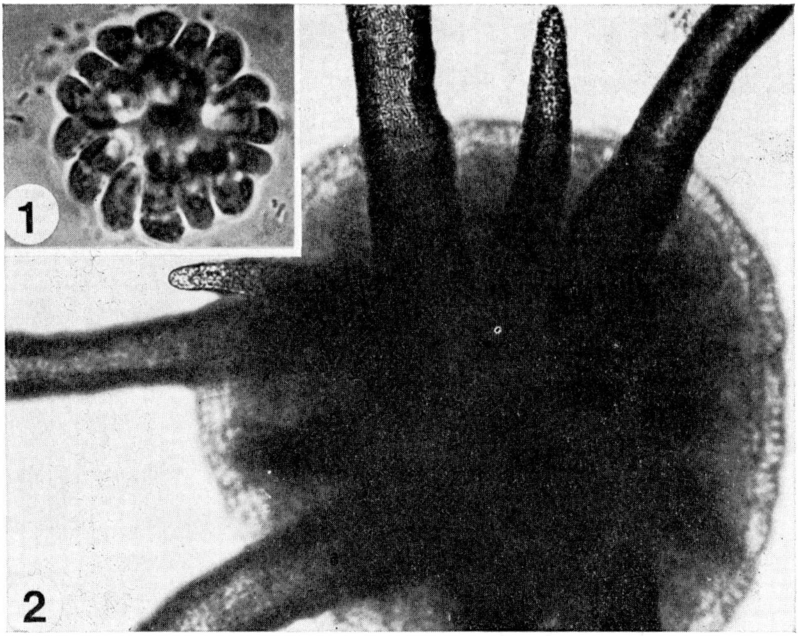

Fig. 1. Blidingia marginata. *2-week-old disc (×900).*

Fig. 2. Blidingia marginata. *3-month-old basal 'cushion' giving rise to erect shoots (×243).*

margin was less evident around the more advanced 'cushions' although some cultures did appear to be surrounded by a halo of material which attracted bacteria. In old cultures individual filaments of cells extended from both the margin and upper surface regions and provided an additional means of surface contact.

Enteromorpha *spp.* In *E. intestinalis* the settled zoospores usually showed signs of growth within 2–3 days in culture. Occasionally, a distinct germination tube was evident which grew out from the side of the zoospore away from the light. In this case a transverse cell wall was laid down dividing the germling into a rhizoid initial cell which spread over the surface and a cell which grew into the erect shoot. In the majority of cultures, however, the zoospore grew vertically into the upright filament without the visible formation of a horizontal germ tube and rhizoidal cell. Only when the erect filament was 3–4 cells long was a prostrate rhizoidal protuberance observed emerging at the base. This basal extension of the rhizoidal cell often appeared from all around the edge and bifurcated to form a small 2–4 cleaved disc. As the erect filament enlarged, the basal cells proximal to the original spore cell also sent down rhizoidal filaments for attachment purposes.

In many of the cultures the rhizoidal filaments extended only a short distance over the substrate and cohered to form a distinct basal disc (Figs. 3 and 4). Close examination of this disc revealed the constituent rhizoidal filaments to be photosynthetic, multicellular, branched, occasionally bifurcated at the tip, and surrounded by a thick mucilaginous cell wall. These discs were apparent in cultures up to 3 weeks old; thereafter, the rhizoidal filaments tended to extend individually across the surface (Fig. 5).

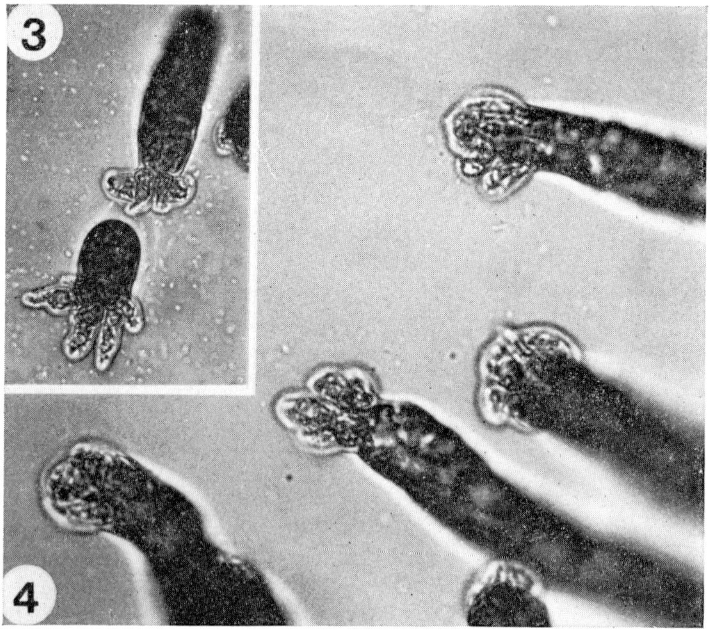

Fig. 3. Enteromorpha intestinalis. *1-week-old germlings* (×*435*).

Fig. 4. Enteromorpha intestinalis. *12-day-old germlings arising from basal discs* (×*500*).

However, this was not always the case; in many of the older cultures the basal system remained disc-like with the constituent filaments enveloped in mucilage (Fig. 6).

As the erect shoot continued to enlarge, greater numbers of rhizoids were produced from the basal shoot cells which spread over the surface. The terminal regions of the filaments were usually white, while the proximal regions were often dark-green due to the presence of photosynthetic cells. Sometimes lateral cohesion of filaments was apparent while some rhizoids appeared to be surrounded by a mucilaginous sheath.

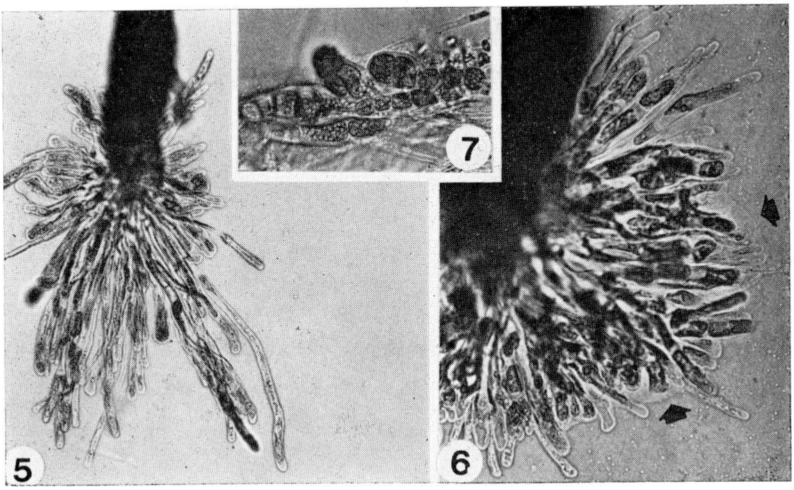

Fig. 5. Enteromorpha intestinalis. *Base of 2-month-old germling showing rhizoidal filaments (×338).*

Fig. 6. Enteromorpha intestinalis. *Base of 2-month-old germling showing the rhizoids enveloped in mucilage (arrowed) (×282).*

Fig. 7. Enteromorpha intestinalis. *Young erect shoots arising from rhizoidal filaments (×278).*

Particularly noteworthy was the large number of young erect shoots which were produced directly from the rhizoidal filaments. Up to 30 small dormant shoots were counted at the base of some mature plants, each produced from the localized activity of a large intermediate cell on the filament (Fig. 7). Very often these young plants became independent of the parental plant and developed freely. The great majority, however, appeared constrained in their growth, probably as a result of inhibitory action by the parent plant.

In *E. linza* and *E. prolifera* the great majority of cultures produced a distinct germ tube from the spore side which spread out into a multicellular rhizoidal filament. This filament later branched and occasionally bifurcated at the tip, while the original spore cell enlarged vertically and developed into the erect shoot. As the basal shoot regions enlarged by vertical divisions, many of the cells sent down rhizoidal filaments. Although at first these tended to cluster together, giving a disc-like appearance to the base; later they always spread out individually across the substrate.

Ulothrix flacca. Although some cultures produced a distinct germ tube from one side of the spore, which later transversely divided into an erect shoot cell and a rhizoidal cell, in most cultures the spores immediately produced an erect filament, this occurring without any apparent indication of an additional attachment mechanism to that of the original spore adhesive. Close examination of the base, however, usually revealed that the lower rhizoidal cell was raised up from the floor with contact being affected by the pointed tip of a downwardly extending germ tube. At the point of contact, a small circle of mucilage appeared to be present (Fig. 8).

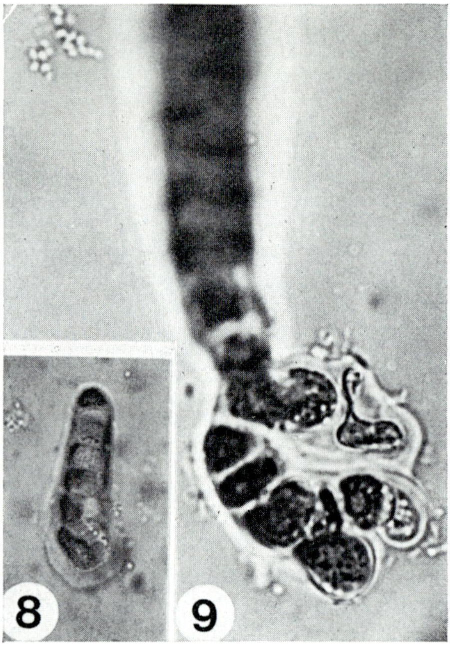

Fig. 8. Ulothrix flacca. *Young erect filament with basal mucilage pad* (×*900*).

Fig. 9. Ulothrix flacca. *Base of 2-week-old germling with cohered rhizoidal filaments* (×*1125*).

As the erect filament developed, this basal germ tube grew out over the substrate and assumed the appearance of an elongated rhizoid cell. However, this cell usually extended only a short distance across the substrate, despite the increased growth of the erect filament, and this was undoubtedly due to the copious secretions of mucilaginous material which spread out from the rhizoidal cell and assisted in adhesion. Later, the rhizoidal cell produced 1–2 small branches, while some of the basal cells of the erect filament also sent down additional rhizoids. All the basal filaments appeared to be enveloped in this mucilaginous matrix which gave a disc-like appearance to the base (Fig. 9).

Ulva *spp.* The early development of the sporelings was similar for both *U. lactuca* and *U. rigida*, with the formation of a germ tube which spread out over the substrate and provided attachment for the developing erect shoot. The elongated rhizoidal cell remained largely unicellular and colourless and tended to remain limited in development. With the ensuing erect shoot growth, additional rhizoidal filaments emerged from the basal cells and cohered to form a small disc-shaped base. Occasional oblique divisions in the apical regions enabled the disc to increase in diameter.

However, these discs, although providing good adhesion for the young erect shoots, gradually became irregular in outline as individual rhizoids spread out over the surface. After 3 months in culture, all plants had an extensive basal mat of colourless, thin, unicellular rhizoids spreading over the surface (Fig. 10).

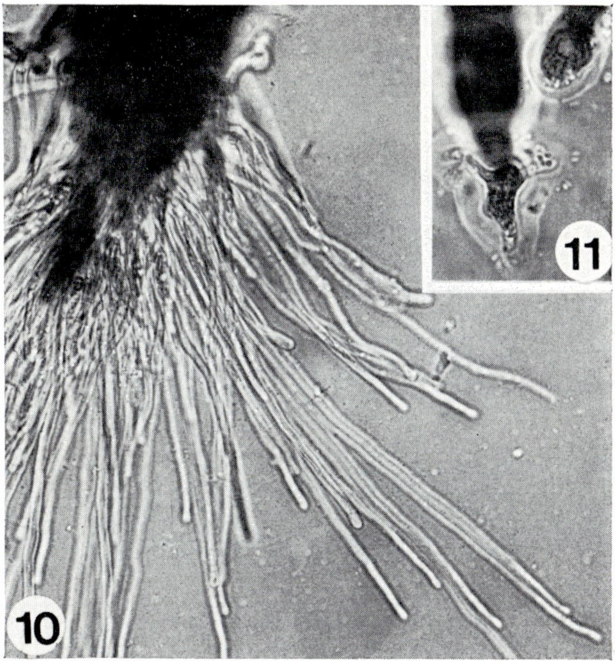

Fig. 10. Ulva rigida. *Base of 3-month-old plant showing extensive rhizoid production* (×250).

Fig. 11. Urospora penicilliformis. *Base of 2-week-old germling with rhizoidal cell enclosed in a mucilaginous matrix* (×700).

Urospora penicilliformis. The settled zoospores, which appeared to lie unattached on the culture vessel floor, formed a narrow germ tube which often became hook-shaped at the end. This germ tube sometimes extended over the floor as a rhizoidal cell, although usually it quickly attached its tip to the floor and secreted a small pad of mucilaginous material. During this process the spore cell grew into the erect filament, support for which was apparently provided by the increased quantities of mucilage observed around the rhizoid apex. Later this apex produced 2–4 finger-like protusions which spread out, each enclosed within a mucilaginous sheath (Fig. 11). By branching and dividing, they increased their surface contact while additional rhizoids, also enclosed in mucilage, were produced from the lower halves of the basal shoot cells.

Phaeophyceae

Ectocarpus. The settled plurispores of both *E. fasciculatus* and *E. siliculosus* usually germinated within 1–3 days. During germination, a small protuberance was produced on the side away from the light source which grew out over the

out a thick (27–30 μm) multicellular rhizoid filament which usually produced a terminal disc (Fig. 22). Thereafter, additional rhizoids produced tended to be slightly thinner (13–14 μm) and these spread out over the substrate (Fig. 23). They were multicellular and branched frequently but did not form discs upon contact.

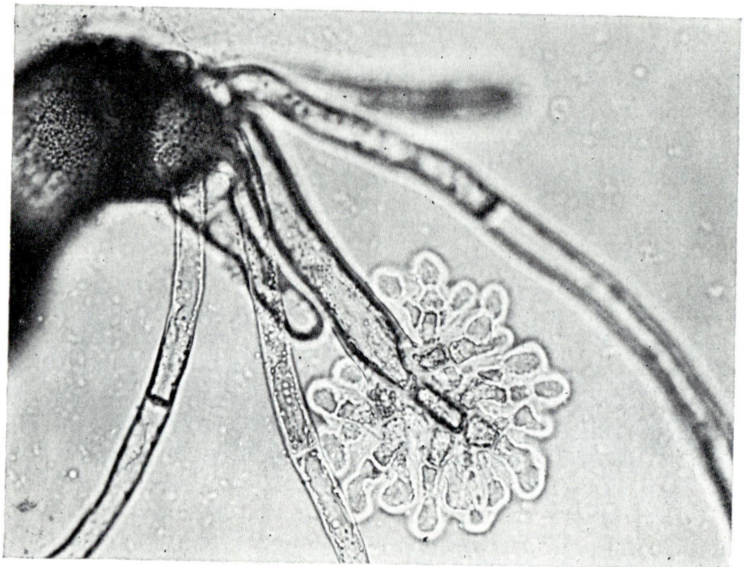

Fig. 23. Ceramium rubrum. *Base of 4-week-old tetraspore germling showing initial attachment disc and several rhizoidal filaments (×375).*

Later, the prostrate discs sent out individual rhizoids from their perimeter and these contributed to the bulk of the filamentous multicellular branch rhizoids which were formed not only from the initial rhizoid cell but also from the cortical cells of the nodal regions on the erect shoot.

Electron microscope observations

Chlorophyceae

Some preliminary S.E.M. observations were made on the attachment processes of 5 green algae: *Blidingia marginata*, *Enteromorpha intestinalis*, *E. linza*, *Ulothrix flacca*, and *Ulva rigida*.

In *Blidingia* the characteristic basal cushion was found to be initiating a number of small erect shoots from the raised central regions (Fig. 24). The cushion itself was composed of large numbers of small intertwined rhizoidal filaments (Fig. 25). All the rhizoids were covered by a fibrous matrix which connected to adjacent filaments and probably helped to adhere them together.

In the early formed discoid holdfast of *E. intestinalis* (Fig. 26), observation of the constituent rhizoidal filaments was prevented by an enveloping mucilaginous sheath which not only embedded the rhizoids but also spread around the marginal area. Only later, when the rhizoids emerged individually, was it possible to examine them in more detail, and it was interesting to find that in both this species and *E. linza* and *Ulva rigida* (Fig. 27) a mucilaginous material was present at their tip.

In agreement with the light microscope observations, a similar mucilage

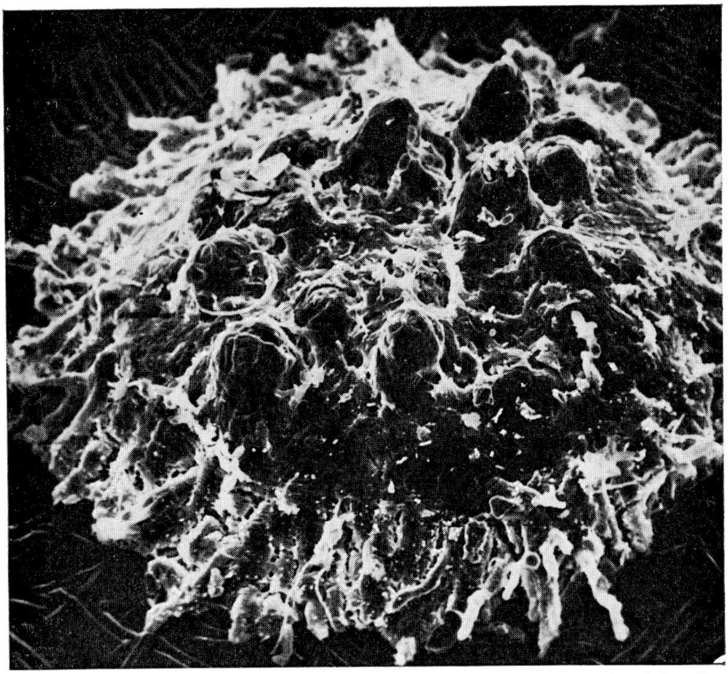

Fig. 24. Blidingia marginata. *Scanning electron micrograph (S.E.M.) of basal cushion showing young shoots arising from the central regions* (×2378).

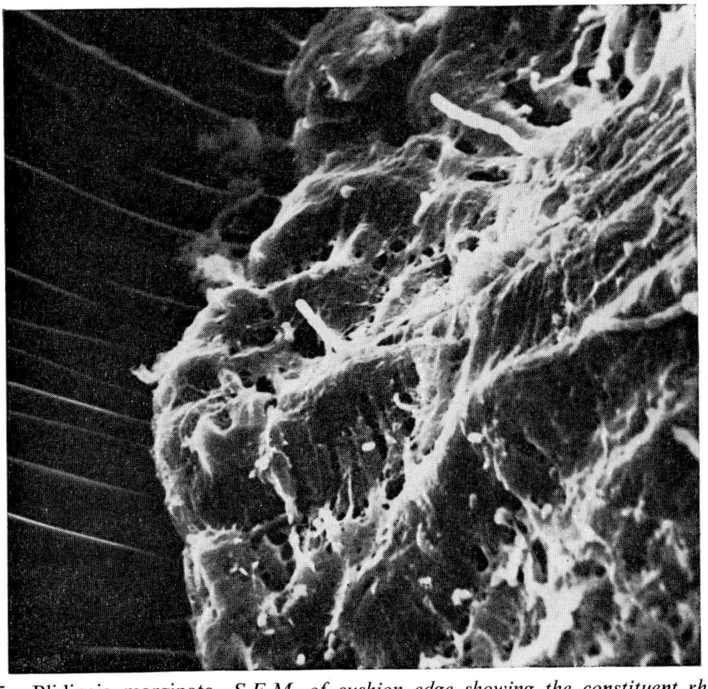

Fig. 25. Blidingia marginata. *S.E.M. of cushion edge showing the constituent rhizoidal filaments enclosed in a fibrillar material* (×5460).

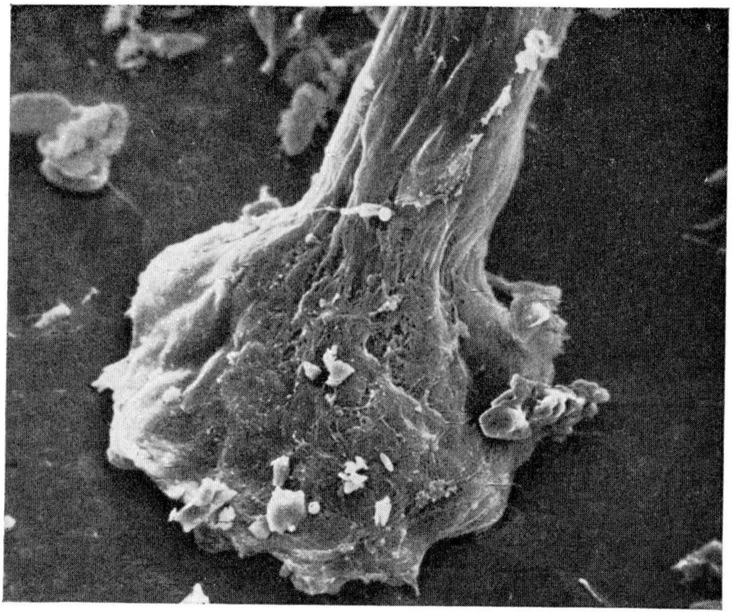

Fig. 26. Enteromorpha intestinalis. *S.E.M. of young basal disc (×9090).*

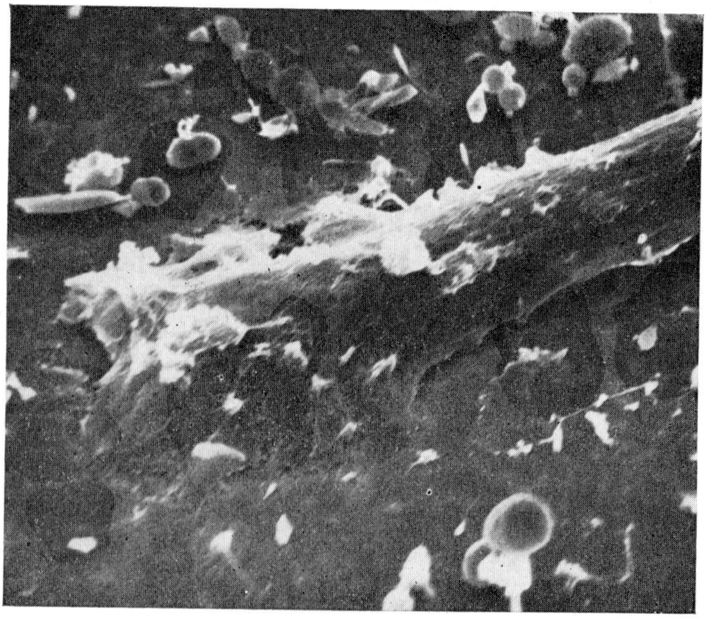

Fig. 27. Ulva rigida. *S.E.M. of rhizoid tip showing the release of exudate onto the substrate (×9555).*

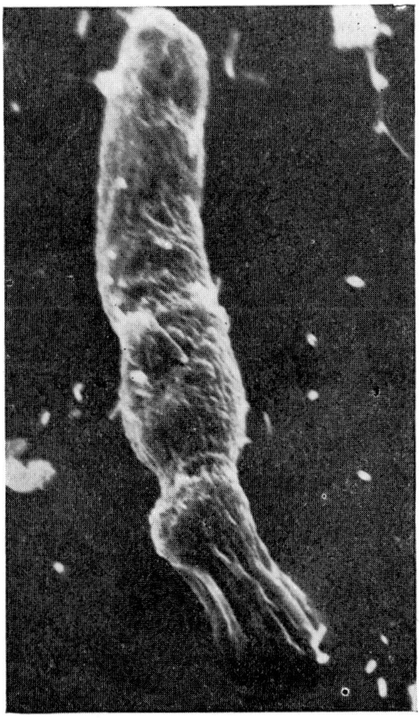

Fig. 28. Ulothrix flacca. *S.E.M. of young filament attached by terminal adhesion of rhizoidal cell (×9566).*

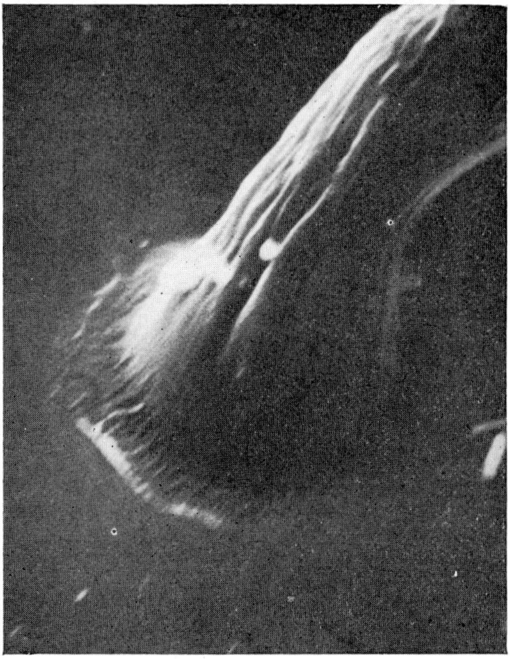

Fig. 29. Ulothrix flacca. *S.E.M. of rhizoidal cell apex showing release of exudate on to the substrate (×22 050).*

appeared to terminate the rhizoidal cell of the young *Ulothrix* plants (Fig. 28). The mucilage, which appeared in considerable quantity from the blunt rhizoid apex, flowed out over the substrate and seemed to be continuous with the surface material of the alga (Fig. 29).

Phaeophyceae

S.E.M. observations were made on only one species, a 2-week-old culture of *E. fasciculatus*. As described for the green algae, the actively growing apices of the filamentous prostrate system appeared to give rise to a mucilaginous material (Fig. 30). This material spread over the substrate both terminally and laterally to give a semi-circular shape to the growing apex.

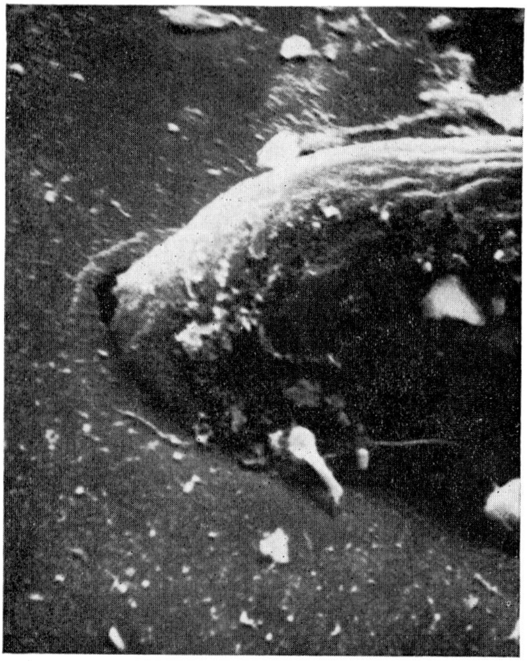

Fig. 30. Ectocarpus fasciculatus. *S.E.M. of basal filament apex showing release of exudate (×20 000).*

Rhodophyceae

Attention was paid to an examination of the discoid rhizoids of *Polysiphonia urceolata* produced on the petri dish floor (Fig. 31). Usually, the characteristic four-lobed structure of the disc was evident (Fig. 32), although this was not always the case, and the disc was sometimes observed to spread uniformly from the central rhizoidal filament (Fig. 33). The thick, mucilaginous covering prevented observation of the constituent outwardly spreading filaments. This mucilage was also produced terminally and spread over the substrate (Fig. 34) much as it did around the green and brown algal rhizoids.

DISCUSSION

The occurrence of a mucilaginous coating around the rhizoidal filaments in

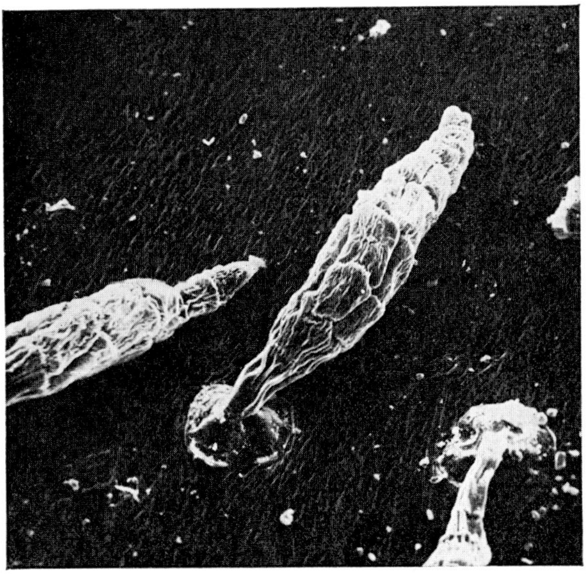

Fig. 31. Polysiphonia urceolata. *S.E.M. of 7-day-old tetraspore germling grown on petri dish floor* ($\times$ *1668*).

all 3 groups of algae investigated is in agreement with the findings of other workers (*see* especially Moss, 1973, 1975). This extracellular mucilage, which was produced at the growing apices, undoubtedly serves as an adhesive and facilitates good contact between the rhizoidal filament and the substrate. Although quite effective for the adhesion of outwardly spreading rhizoidal filaments, it is clearly of even greater importance in the adhesion of the

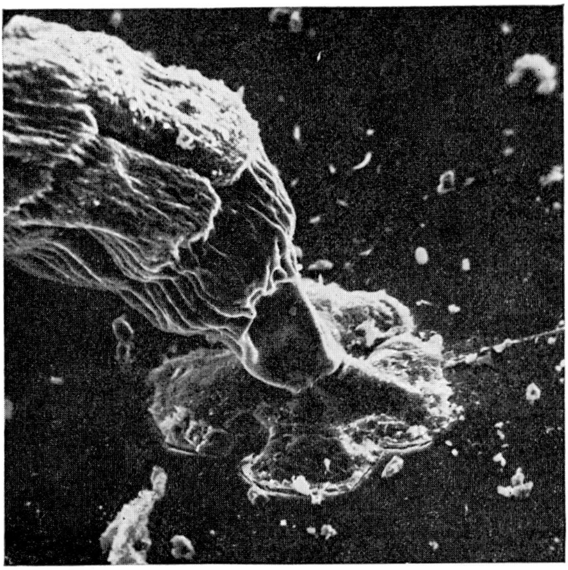

Fig. 32. Polysiphonia urceolata. *S.E.M. of base of tetraspore germling showing the lobed structure of the attachment disc* ($\times$ *4022*).

Fig. 33. Polysiphonia urceolata. *S.E.M. of basal disc* (×*8096*).

discoid stages. Its effectiveness was demonstrated by the fact that quite vigorous brushing of the thalli (especially in *Polysiphonia* spp.) usually failed to dislodge them. The presence of downward projecting, finger-like probes, shown in vertical sections of the discs, would be of no importance in attachment to the relatively smooth surface of the petri-dish floor, and on this substrate it is the cuticle-like secretions between these filaments and under the

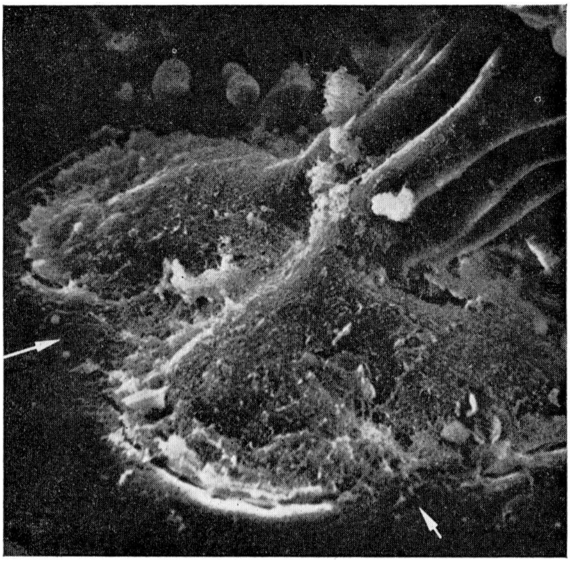

Fig. 34. Polysiphonia urceolata. *S.E.M. of basal disc. Note the extension of disc material onto substrate* (*arrowed*) (×*7838*).

disc which effect adhesion. However, on more natural substrates, these probes would be particularly effective in seeking out and penetrating any surface cracks and crevices that are present.

Despite the widespread occurrence of this rhizoidal mucilage and its importance in algal adhesion, very little is known regarding its origin in the growing apex. Undoubtedly, the Golgi complex plays a key role in its production, and numerous transmission E.M. studies of actively growing rhizoid tips (Chen, 1967; Rawlence and Taylor, 1972; Sievers, 1967; pers. unpubl. data) have usually revealed an accumulation of secretory vesicles. However, it is perhaps noteworthy that the function of these vesicles is generally attributed to that of cell wall formation which would suggest, therefore, that the mucilage might simply be excess gelatinous cell wall material which has exuded from the tip. Certainly this would appear to be a less complicated procedure than that involved in the initial attachment of the reproductive spore stages.

Particularly interesting was the effect of substrate type on the rhizoidal expression in *Polysiphonia*. The occurrence of the discoid holdfasts on the petri dish floor but not on the glass coverslips was probably due to the difference in surface nature of the two substrates; either surface 'roughness' was the causal factor or more likely it was the difference in the surface tension. This surface nature effect could usually be overcome by slight pressure, *e.g.* the weight of a coverslip on top of the germlings, and disc formation on glass coverslips was stimulated in *Petalonia* and *Scytosiphon* by this means.

The importance of the initial spore adhesive in germling attachment during the early stages of development in some of the green and red algae deserves particular mention. We have seen that many of the species investigated very quickly formed an erect filament from the settled spore before any indication of the production of a rhizoidal process, suggesting that attachment was dependent on the spore adhesive pad. However, it may well have been that after or just before the first division into 2 cells the lower rhizoidal germ tube immediately produced mucilage at the base which supplemented that of the spore adhesive. This would certainly account for the 'maturation' process observed for the adhesive material released around the *Polysiphonia* sporelings.

Not only do the rhizoidal systems provide attachment for the erect shoots they probably also serve as both perennating and propagating organs. The occurrence of plastids in the rhizoidal filaments of all the species investigated would undoubtedly enable them to survive independently of the parental material and they were indeed shown to be capable of initiating new shoots directly from parts of the filament. This ability of the probably more hardy rhizoidal filaments to assume a vegetative role probably adds much to the alga's chances of survival, especially perhaps during unfavourable periods as a fouling organism. It is likely that even if the erect shoots were killed off by severe environmental changes or scrubbed off mechanically during docking periods, portions of the prostrate system would remain to initiate further growths.

Acknowledgements

This work was supported by a Ministry of Defence (Navy) Contract No. W/CP 3698/70/H1041 to Dr E. B. Gareth Jones and is published with their approval. I would like to thank Mrs Y. Butler for helpful criticism of the manuscript.

REFERENCES

Baker, J. R. J., and Evans, L. V. (1973a). The ship-fouling alga *Ectocarpus*. I. Ultrastructure and cytochemistry of plurilocular reproductive stages. *Protoplasma*, **77**, 1–13.

Baker, J. R. J., and Evans, L. V. (1973b). The ship fouling alga *Ectocarpus*. II. Ultrastructure of the unilocular reproductive stages. *Protoplasma*, **77**, 181–189.

Chamberlain, A., and Evans, L. V. (1973). Aspects of spore production in the red alga *Ceramium*. *Protoplasma*, **76**, 139–159.

Chen, L. C-M. (1967). Rhizoids in *Polysiphonia urceolata* (Family Rhodomelaceae). M.Sc. thesis, University of New Brunswick, Fredericton, Canada.

Christie, A. O. (1973). Spore settlement in relation to fouling by *Enteromorpha*. In Proc. III Int. Congr. Mar. Corrosion and Fouling, pp. 674–681 (R. F. Acker, B. Floyd Brown, J. R. De Palma and W. P. Iverson, eds.), Northwestern University Press, Evanston, Ill.

Christie, A. O., and Shaw, M. (1968). Settlement experiments with zoospores of *Enteromorpha intestinalis* (L.) Link. *Br. Phycol. Bull.*, **3**, 529–534.

Christie, A. O., Evans, L. V., and Shaw, M. (1970). Studies on the ship-fouling alga *Enteromorpha*. II. The effect of certain enzymes on the adhesion of zoospores. *Ann. Bot.*, **34**, 467–482.

Evans, L. V., and Christie, A. O. (1967). The fouling of ships; an electron microscopial approach. Trans. Br. Assoc. Adv. Sci., Sept. 1967, pp. 116–117.

Evans, L. V., and Christie, A. O. (1970). Studies on the ship-fouling alga *Enteromorpha intestinalis* I. Aspects of the fine-structure and biochemistry of swimming and newly settled spores. *Ann. Bot.*, **34**, 451–466.

Fletcher, R. L., and Chamberlain, A. (1975). Marine fouling algae. In *Methods in Biodeterioration*, pp. 59–81. Ann. Tech. SAB; Academic Press, New York.

Houghton, D. R., Pearman, I., and Tierney, D. (1973). The effect of water velocity on the settlement of swarmers on *Enteromorpha* spp. In Proc. III Int. Congr. Mar. Corrosion and Fouling, pp. 682–690 (R. F. Acker, B. Floyd Brown, J. R. De Palma and W. P. Iverson, eds.), Northwestern University Press, Evanston, Ill.

Jones, W. E., and Babb, M. S. (1968). The motile period of swarmers of *Enteromorpha intestinalis* (L.) Link. *Br. Phycol. Bull.*, **3**, 525–528.

Moss, B. (1973). Observations on the breakdown of paint surfaces by ship fouling algae. In Proc. III Int. Congr. Mar. Corrosion and Fouling, pp. 39–47 (R. F. Acker, B. Floyd Brown, J. R. De Palma and W. P. Iverson, eds.), Northwestern University Press, Evanston, Ill.

Moss, B. (1975). Attachment of zygotes and germlings of *Ascophyllum nodosum* (L.) Le Jol. (Phaeophyceae, Fucales). *Phycologia*, **14**, 75–80.

Parke, M., and Dixon, P. S. (1968). Check-list of British marine algae, 2nd rev., *J. Mar. Biol. Assoc. U.K.*, **48**, 783–832.

Rawlence, D. J., and Taylor, A. R. A. (1972). A light and electron microscopic study of rhizoid development in *Polysiphonia lanosa* (L.) Tandy. *J. Phycol.*, **8**, 15–24.

Sievers, A. (1967). Elektronenmikroskopische Untersuchungen zur geotropischen Reaktion. II. Die polare Organisation des normal wachsenden Rhizoids von *Chara foetida*. *Protoplasma*, **64**, 225–253.

Von Stosch, H. A. (1963). Wirkung von Jod und Arsenit auf Meeresal gen in Kultur. In Proc. IV Int. Seaweed Symp. (D. DeVirville and J. Feldmann, eds.), Pergamon Press, Oxford.

PHYSICAL AND CHEMICAL MECHANISMS OF BARNACLE ATTACHMENT

ELEK LINDNER and CAROL A. DOOLEY

Naval Undersea Center, San Diego, California, U.S.A.

Summary

Besides increasing drag on ships, barnacles can contribute to corrosion and degradation of many materials. Although a dense population of barnacles could act as a protective layer, their unique growth and attachment pattern tend to damage protective coatings and trap constituents, which enhance corrosion. In the attachment process of barnacles, adhesive secretion is spread between shell basis and substratum. Using the adhered basis as an anchor, the sharp-growing edges of the shell are pressed against the substratum by a specialized muscle system and can result in cutting through the soft coatings down to the base material. The barnacle adhesive is a proteinaceous bulk material cross-linked with quinone bridges. Typical absorption bands between 330 and 345 nm were found to be suitable for characterization of quinone cross-linked proteins. Kinetic studies indicate that enzyme-generated *o*-quinones are superior to *p*-quinones as cross-linking agents because (1) in concentrations less than 10^{-3} M, the reaction rate of *o*-quinones is an order of magnitude higher than that of the *p*-quinones and (2) while *p*-quinones react mainly with α-amino groups, *o*-quinones react with α- and ε-amino groups equally fast. Since in the *o*-quinone reaction the initial step producing a monosubstituted derivative has a limiting high rate, the *o*-quinone reaction favours an autocross-linking mechanism where the quinone is formed from peptide-bound phenolic sidechains.

NOTATION

λ	wavelength	Hq	diphenol (hydroquinone)
nm	nanometer (10^{-9} metre)	C	catechol
D	optical density (absorbance)	E	phenolase enzyme
a	reactant concentration	A	amino compound
$-da/dt$	reactant concentration decrease	BuA	n-butylamine
r	resulting product concentration	Gly	glycine
dr/dt	product concentration increase	Val	L-valine
v	initial reaction velocity	Lys	L-lysine
M	molarity (moles per litre)	Tyr	L-tyrosine
h	hours	Ala_4	tetra-L-alanine
o	ortho	Lys_4	tetra-L-lysine
p	para	Lys_n	poly-L-lysine
Q	benzoquinone	P	protein (denatured trypsin)

INTRODUCTION

Barnacles, a major group of fouling organisms, are mainly responsible for increasing the drag of ships by attaching to the hull, resulting in loss of speed and overconsumption of fuel. An obvious question that presents itself is whether the barnacles damage the material to which they are attached.

Some species of barnacles, such as *Alcippe lampas* and the species of the genus *Lithotrya*, do indeed cause damage by burrowing into their substratum, mainly mollusc shells, coral, or limestone. The whale barnacles *Coronula* and *Xenobalanus* probably do not burrow directly into the skin of the cetaceans, but by attachment cause irritation which encourages the epidermis to grow over and into their shells in order to reinforce their attachment. These species are not true parasites since they are only hitching a ride and are feeding on plankton rather than on their host. The true parasitic barnacles, such as the species *Laura* living on echinoderms and corals and *Sacculina* living on crabs, do much more extensive damage to their hosts. However, the rock-boring, the whale, and the parasitic barnacles are not considered fouling organisms.

Species of barnacles commonly associated with fouling do not deliberately attack their substratum. They attach to the surface by secreting an adhesive rather than burrowing into the substratum for mechanical fixation or protection and they feed on plankton rather than on the substrate material. In fact, it is not uncommon to find a dense community of barnacles protecting the substratum from environmental attack. Weiss (1948) reported that a heavy growth of fouling can protect wooden structures from attack by wood borers. When one of two identical wooden blocks exposed near Miami, Florida, was cleaned periodically, it was attacked severely by wood borers, while the other one was covered with a dense population of barnacles and remained almost intact inside. Also, steel plates exposed to a seawater environment often show smooth, uncorroded circular areas under attached barnacle bases, while areas not covered by barnacles corrode severely.

Even if barnacles cannot prevent corrosion, they may provide unexpected benefit. In 1862, the iron hull plates of the H.M.S. TRITON were corroded to such paper thinness that, according to her commander, she was kept from foundering only by her fouling, practically sailing home on her barnacles (Young, 1867).

Corrosion of copper-nickel alloy does occur in association with barnacles, however, but almost exclusively under the bases of dead specimens rather than under those of living ones (Fig. 1) (Friend, 1932; LaQue and Clapp, 1945). It appears that corrosion results from the action of decaying bacteria rather than that of the barnacle or its secretion. Production of acidic conditions and hydrogen sulphide by dying and decaying members of the fouling community may create a condition favourable to corrosion (Ragg, 1935). Of course, the presence of fouling, both alive or dead, may favour the accumulation and growth of microorganisms, particularly sulphate-reducing bacteria, to which Legendre (1938) has attributed the corrosion of iron under marine conditions. These bacteria secure their needed oxygen under anaerobic conditions by reducing sulphate. In this process, hydrogen is consumed and the resulting depolarization of the metallic surface favours corrosion (Von Wolzogen Kuhr, 1938).

Sessile barnacles may cause direct damage to paint coatings (Ragg, 1925;

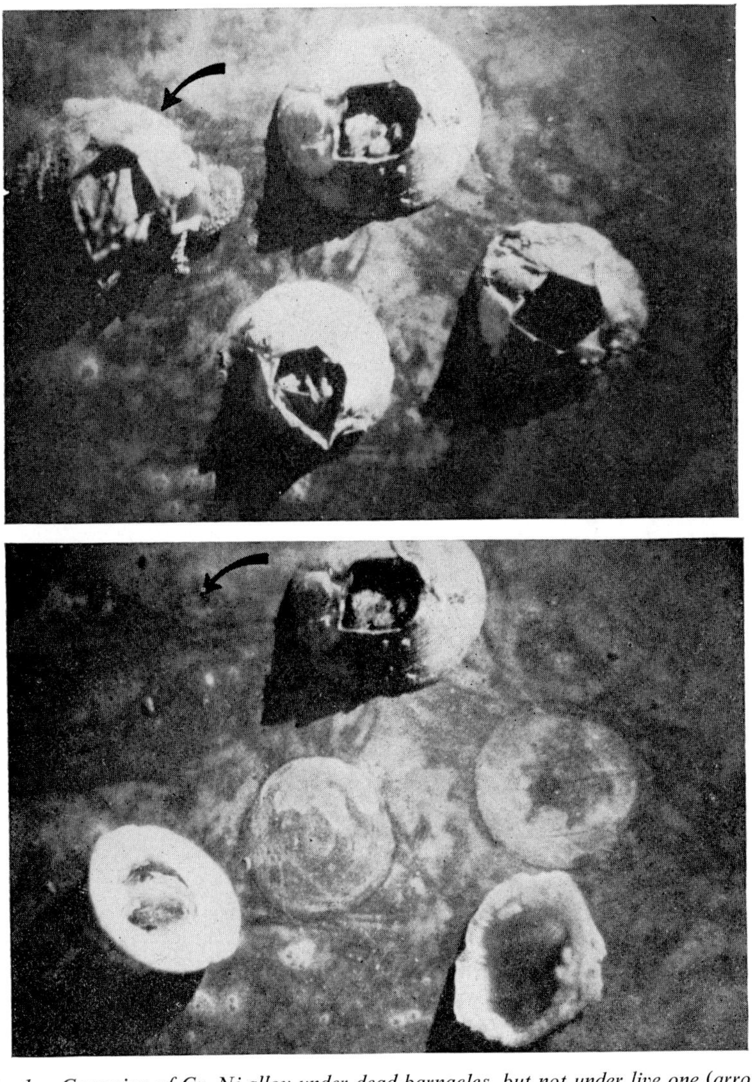

Fig. 1. Corrosion of Cu–Ni alloy under dead barnacles, but not under live one (arrow).

Nue, 1932; Barenfanger, 1939; Harris, 1946) due to their unique way of growth. The shell of the sessile barnacles consists of calcareous plates forming the walls and a basis, which is also calcareous in a large number of species. The plates are separated from each other and from the basis by a thin region of living tissue, where the shell actually grows. The plates and the basis, therefore, are not rigidly connected to each other but are capable of limited movement. As we described previously (Saroyan *et al.*, 1968, 1970), the shell walls and the basis are tightly connected by numerous muscle tissues around the basal perimeter at the inside joints of the leading edges (Fig. 2). The contraction of these muscles pulls the basis upward and, at the same time, the shell walls downward. Since the basis is usually firmly cemented to the substratum, the wall is pulled down and presses the basal edge of the shell

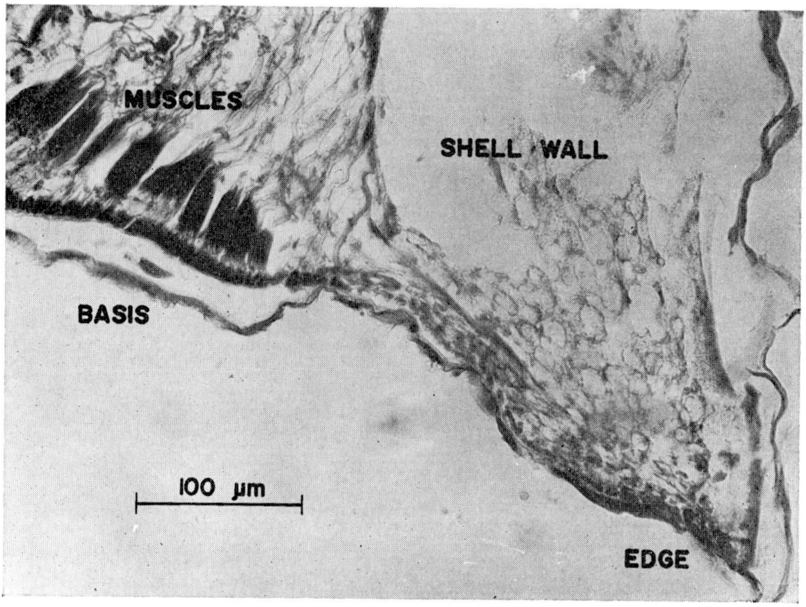

Fig. 2. Muscles connecting shell wall and basis.

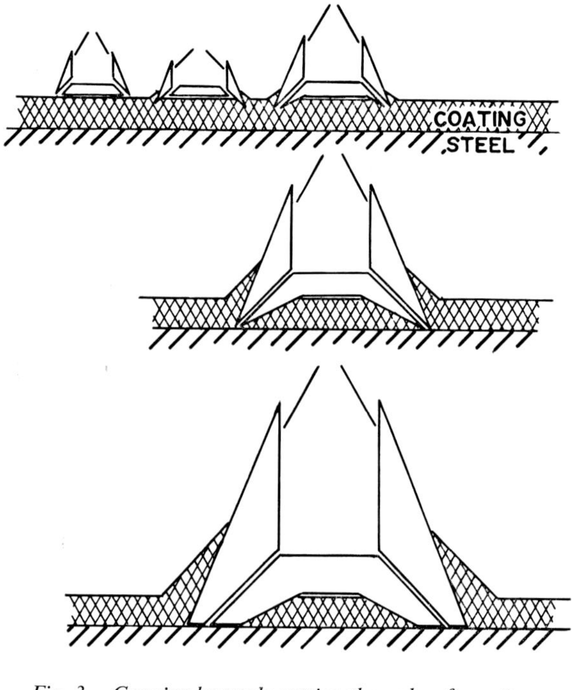

Fig. 3. Growing barnacle cutting through soft coating.

wall tightly down on the substratum. If the substratum, for instance a paint coating, is soft enough, the sharp edges of the shell will indent or even cut into the coating. The muscle contraction is probably repeated a number of times, especially after moulting which coincides with the periodic growth (Saroyan *et al.*, 1968). Thus, with each growing period, the edges of the shell wall may cut deeper and deeper into the coating and eventually slice through to the bare metal. During subsequent growing cycles, the shell will grow on the surface of the metal, ploughing away the protective coating (Fig. 3). This behaviour and anatomical construction ensure that the barnacle attaches to a solid, firm substratum rather than to loose deposits, fouling organisms, or detritus which can easily wash away. Also, the shell wall rests upon the growing edge of the basis, which is a thin, flexible membrane of chitin, regardless of whether or not the species in question has a calcareous basis (Costlow, 1956; Newman *et al.*, 1967). Consequently, this thin, flexible rim of the basis is also pressed very tightly to the substratum, leaving only a very slight gap. Because of this peripheral pressure, the basis may grow into the recessions and over the irregularities of a solid substratum, faithfully duplicating the surface structure (Pilsbry, 1916; Gregg, 1948). This creates a maximum contact area for adhesion and, after the adhesive is spread to fill any gap between the basis and the substratum (Darwin, 1854), presses together the two surfaces to be adhered. For good adhesion it is probably essential to spread the adhesive in a very thin layer between the two surfaces since the hardened adhesive of the sessile barnacle lacks good tensile strength (Lindner and Dooley, 1972; Lindner *et al.*, 1973). Because of peripheral pressure exerted by the barnacle, the thickness of the adhesive layer is normally less than 5 μ (Saroyan *et al.*, 1970a).

The adhesive is secreted through a complex duct network with terminal orifices around the perimeter. Since the ducts become imbedded in the calcareous basis after each moult, that is, in each new growing phase, a complete new duct network develops to supply adhesive to the enlarged basal perimeter. The old ducts embedded in the basis, although usually not functional, can secrete adhesive in certain cases. This is possible because the ducts are not plugged by hardened adhesive, but flushed out after the primary secretion. Thus, when the barnacle is detached or damaged, it makes an attempt to reattach or repair the damage by a secondary secretion of adhesive through the old ducts (Saroyan *et al.*, 1970a; Saroyan *et al.*, 1970b). Based upon this reattachment attempt of the barnacles, a method for increased adhesive production was developed in our laboratory by carefully detaching barnacles and collecting the adhesive droplets from basal areas not in contact with a new substratum. Thus, the adhesive could be collected relatively free of contamination in quantities sufficient for analysis.

The results of the analytical investigation compiled evidence consistent with the general scheme of quinone cross-linking of a proteinaceous material as the hardening mechanism of the barnacle adhesive (Lindner and Dooley, 1972; Lindner *et al.*, 1973) (Fig. 4). The major arguments in support of this theory are: (1) the bulk of the adhesive is protein with an abundance of free amino groups, which are the major functional groups in the quinone cross-linking mechanism; (2) the presence of phenolic precursors and by-products, such as melanin and stable free radical quinhydrone; and (3) especially the presence of the phenolase enzyme in the secretory glands and ducts.

Although the basic principles of quinone cross-linking are relatively well

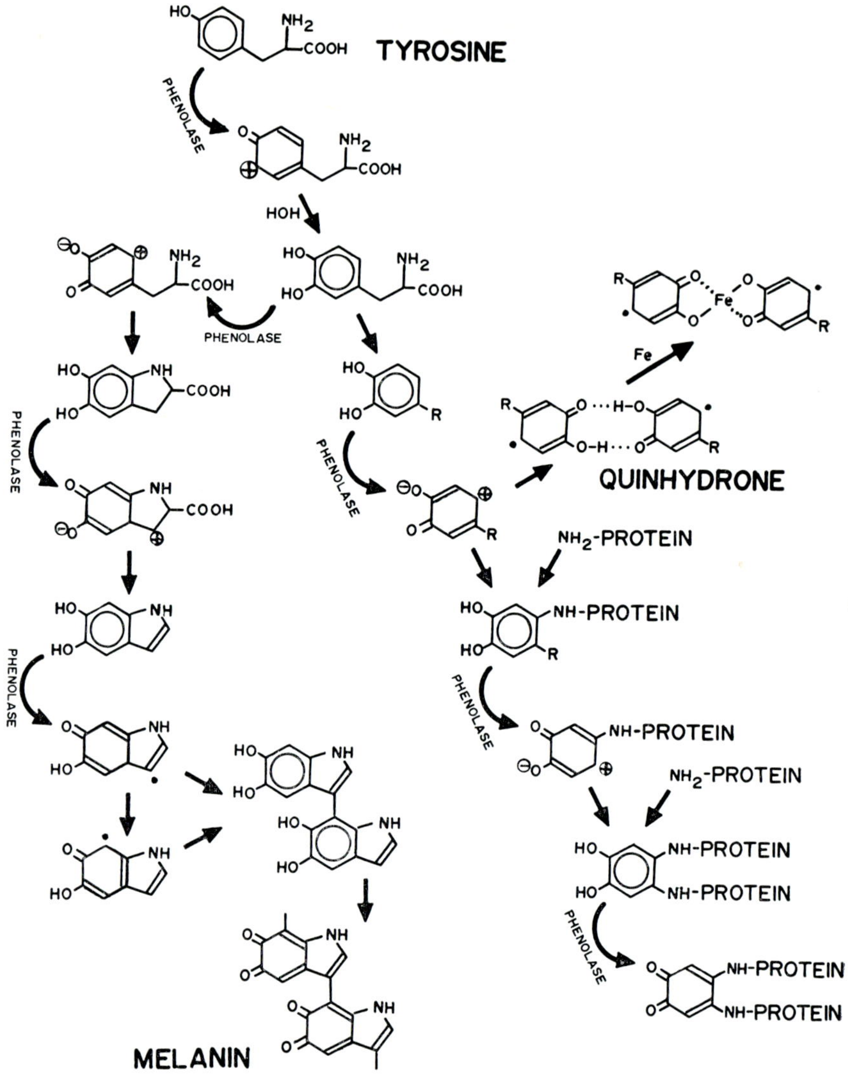

Fig. 4. Quinone cross-linking mechanism.

established, namely, that the free amino groups of the protein are bound into the quinonoid ring, the reaction can take different routes. These variations of the quinone cross-linking mechanism were postulated mainly on the basis of some spectral data and on the presence or absence of the catalyst enzyme and the expected precursors, the phenolics and the protein. A more thorough study of the mechanism, however, is hampered by the extreme insolubility of the product. Because of these difficulties, model reactions were selected for elucidation of the quinone cross-linking mechanism.

In this paper some aspects of the reaction kinetics of the hardening mechanism of the barnacle adhesive will be discussed on the basis of some model reactions.

MATERIALS AND METHODS

Amines such as butylamine and indole, amino acids such as valine and lysine, oligopeptides such as valyltyrosine, tyrosylglycylglycine, and tetra-L-alanine, polypeptides such as polylysine, and well-defined proteins such as crystalline trypsin denatured by heating were reacted with quinones and their phenolic derivatives through the oxidative action of the phenolase enzyme to elucidate the quinone cross-linking mechanism. The reactions were monitored with a Beckman DB-G ultraviolet-visible recording spectrophotometer.

Except where otherwise noted, the concentrations of the model compounds were adjusted to give a final concentration of 0·5 mM with respect to the phenolic component. This concentration produced suitable absorption values when the undiluted reaction mixture was scanned in the u.v. in a 1 mm path cell.

Generally, the reaction volumes were 3·0 ml, consisting of 0·3 ml of phenolase (1·25 mg/5 ml in dist. H_2O) and 2·7 ml containing the remainder of the reactants in Sorenson's phosphate buffer, pH 7.

Throughout this study, concentrations of the test solutions and path-lengths of the cells were selected to give an appropriate absorbance; the details of the experiments and spectral measurements for each set of curves are shown in Appendix I. In most cases our primary interests were the wavelength or the position of the absorption bands; the absolute or relative absorbance had only secondary importance. In these cases the spectral curves were displaced along the ordinate axis (absorbance) to avoid overlapping in the region of interest and to insure clarity. Where absorbance was of importance, the molar absorptivity of the absorbing species was expressed according to the Beer-Lambert Law. The spectra were scanned from 200 to 400 nm, but the absorption bands near 210 nm were not considered because increased light scattering in these shorter wavelength regions makes evaluation of the spectra difficult.

RESULTS

Since aromatic compounds absorb in the u.v. domain, u.v. spectrophotometry was selected to monitor the reaction in the investigation of the quinone cross-linking mechanism.

As reported previously (Lindner and Dooley, 1972; Lindner *et al.*, 1973), solutions of partial hydrolysates of hardened barnacle adhesive and mussel byssus showed absorption bands in the 300–350 nm region (Fig. 5). Since these bands were not assigned to a specific chromophore system, model reactions were designed for the purpose of elucidating the characteristics of the chemical species which show spectra with absorption bands between 300 and 350 nm and of investigating the possibility of the chemical mechanisms which are capable of producing such chemical species.

When amino compounds were incubated with *p*-benzoquinone in the presence of air, a new absorption band developed near 345 nm (Fig. 6), indicating reaction between the amino compounds and quinone. Also, the development of a peak near 290 nm suggested that some of the quinone had been reduced to hydroquinone. These spectral data were in good agreement with the postulated reaction between amino compounds and quinones:

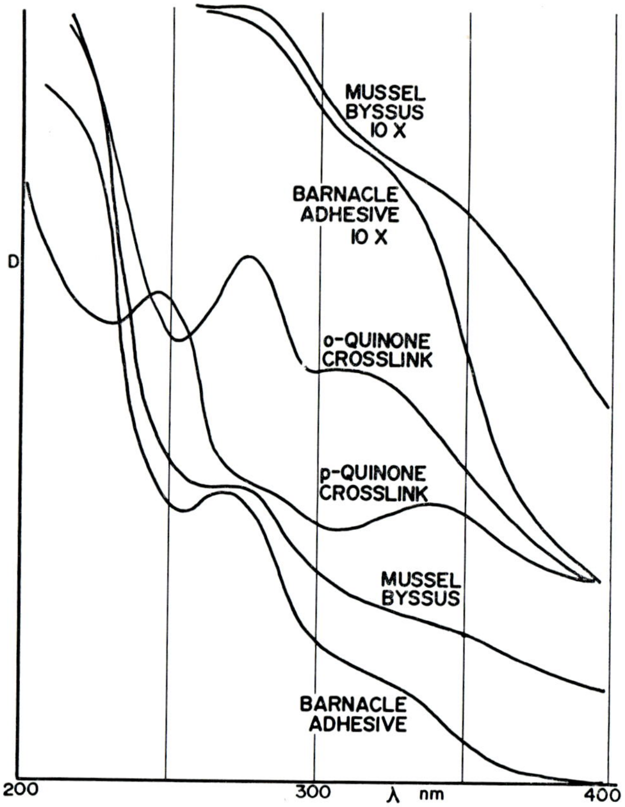

Fig. 5. U.V. spectra of hydrolysates of barnacle adhesive, mussel byssus.

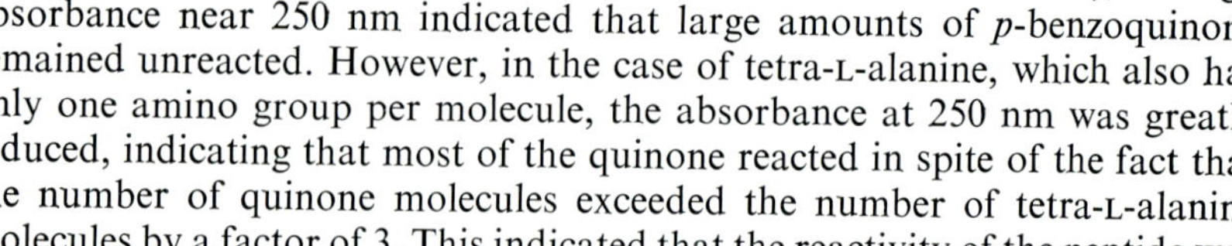

Although the small molecular amino compounds, butylamine, valine, and lysine, were in excess relative to the *p*-benzoquinone (mole ratio 2:1), the high absorbance near 250 nm indicated that large amounts of *p*-benzoquinone remained unreacted. However, in the case of tetra-L-alanine, which also has only one amino group per molecule, the absorbance at 250 nm was greatly reduced, indicating that most of the quinone reacted in spite of the fact that the number of quinone molecules exceeded the number of tetra-L-alanine molecules by a factor of 3. This indicated that the reactivity of the peptide was higher than that of the free amino acid or amine. The complete disappearance of the absorption peak of the *p*-benzoquinone at 250 nm from the spectra of tetra-L-lysine, poly-L-lysine, and protein (crystalline trypsin) further indicated the increased reactivity of the peptides. The increased absorbance at 345 nm of the product with the lysine peptides indicated that the ε-amino groups of the lysyl residues of the peptide chain readily reacted with the *p*-benzoquinone.

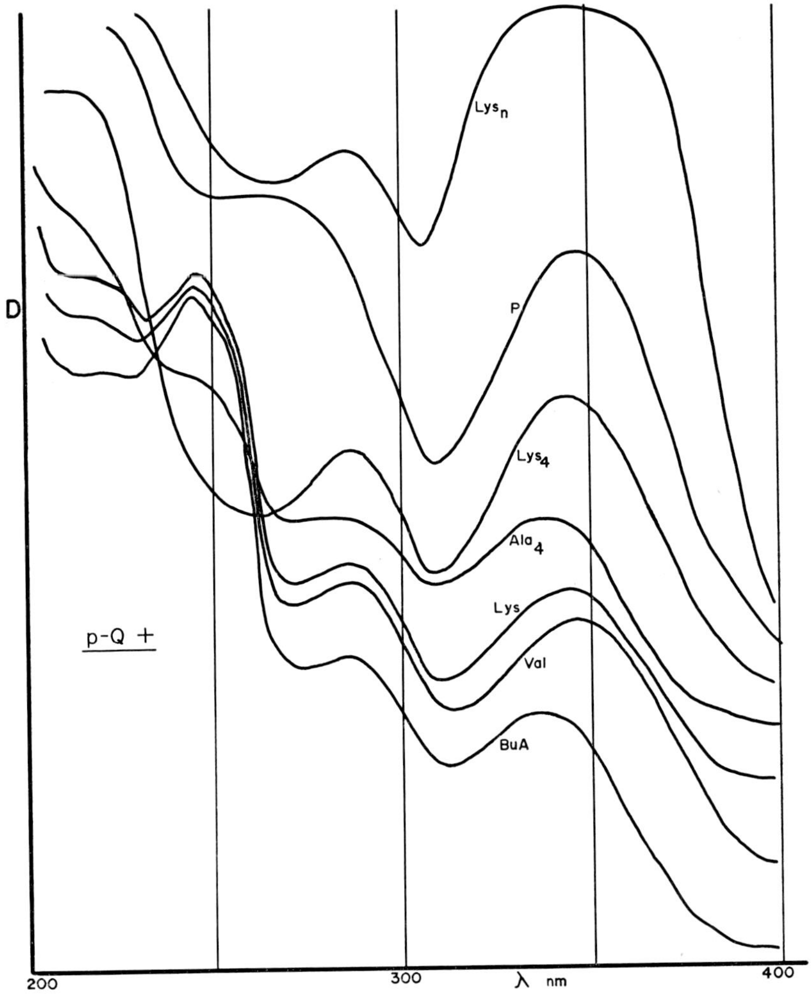

Fig. 6. Spectra of products of p-*quinone and amino compounds.*

Amino compounds also reacted with enzymatically generated *o*-benzoquinone from catechol (Fig. 7) and within a few hours an appreciable amount of the product developed as evidenced by a moderate absorption between 300 and 350 nm, producing a shoulder rather than a pronounced peak. The shoulder appeared to be near 325 nm rather than near the 345 nm observed with *p*-benzoquinone, except with tetra-L-lysine and polylysine, which showed a bathochromic shift, possibly due to the steric effect of numerous cross-links through ε-amino groups.

When protein was incubated with tyrosine and *o*-diphenol derivatives in the presence of phenolase and also with *p*-diphenol and *p*-benzoquinone (Fig. 8), the difference in the absorption of the products with the *o*- and *p*-diphenolic compounds was clearly shown.

In order to determine the actual absorption band of the product developed from the reaction of trypsin and enzyme-generated *o*-benzoquinone, the

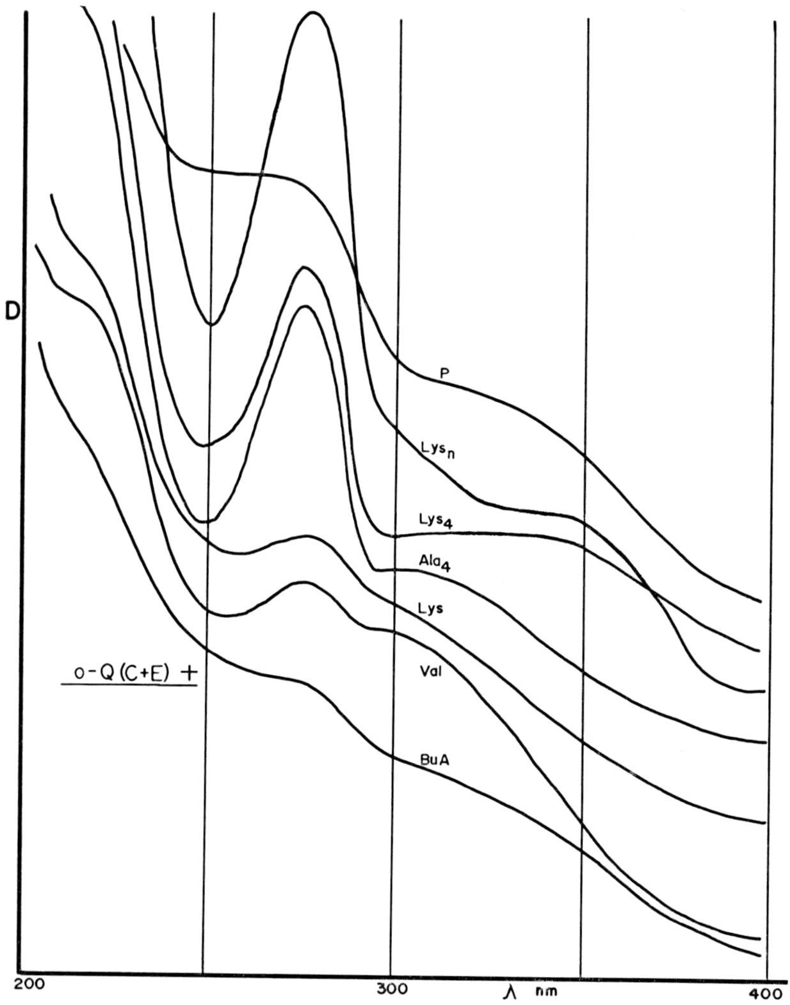

Fig. 7. Spectra of products of enzyme-generated o-*quinone and amino compounds.*

difference spectrum of the trypsin-catechol-phenolase and the catechol-phenolase mixtures was established (Fig. 9). The difference spectrum was resolved into a minimum number of Gaussian curves since a single absorption band of a solution approximates a curve of Gaussian shape. The result of this procedure was a relatively wide and weak band with a maximum at 330 nm, a relatively strong band with a maximum at 250 nm, and several smaller peaks which probably reflect the general absorption observed in aged solutions and which probably arise from polymerized phenolic products. Based upon these results, we found that the product of the reaction between amino compounds and enzyme-generated *o*-benzoquinone had a wide absorption band of moderate strength with a maximum at 330 nm. This maximum was shifted toward shorter wavelengths by about 15 nm as compared to that of the product from *p*-benzoquinone, a difference similar to that between the

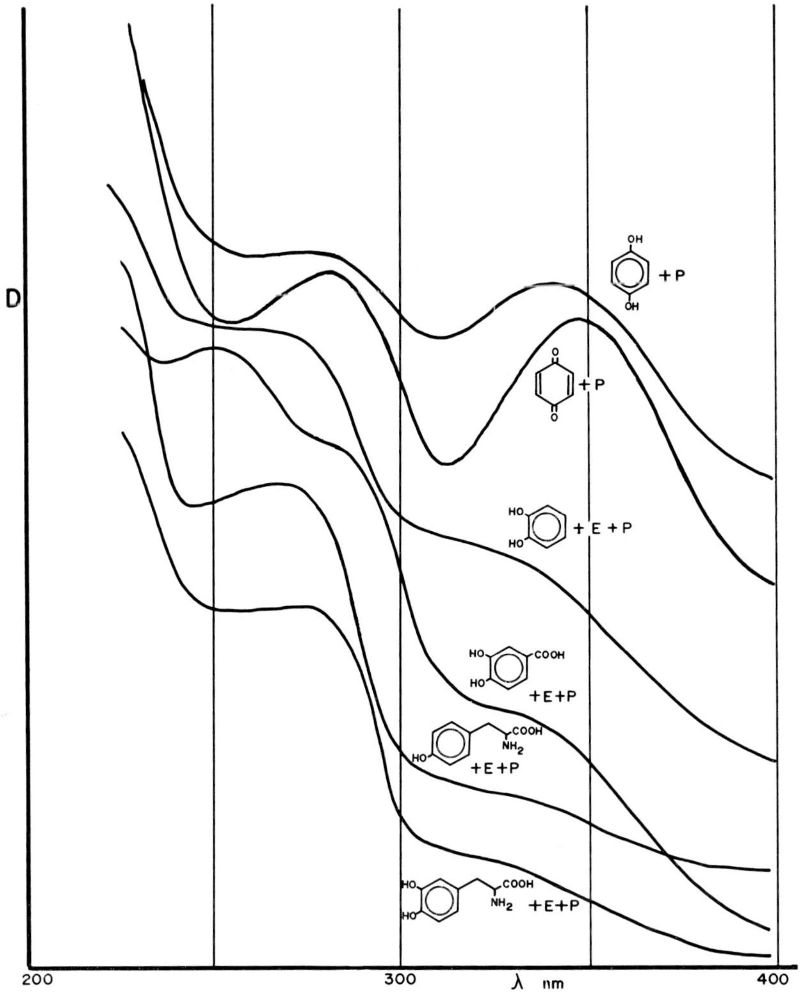

Fig. 8. *Spectra of products of protein and various quinones.*

absorption bands of *o*-diphenol and *p*-diphenol. The absorption at 250 nm was probably due to the resonance form of a *p*-substituted *o*-diphenol.

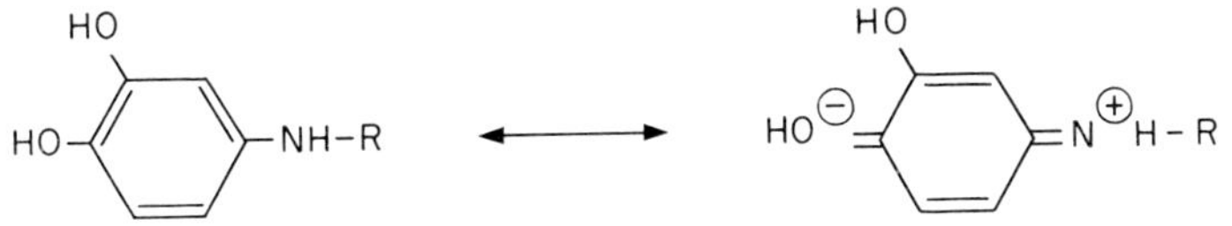

The spectral changes seen during the reaction between tetra-L-alanine and *p*-benzoquinone in a nitrogen atmosphere are shown in Fig. 10. The development rate of the absorption band at 345 nm was gradual and slowed down after 20 hr. The absorption near 290 nm, indicating hydroquinone, develops faster with little change after about 4 hr; the quinone peak, on the other

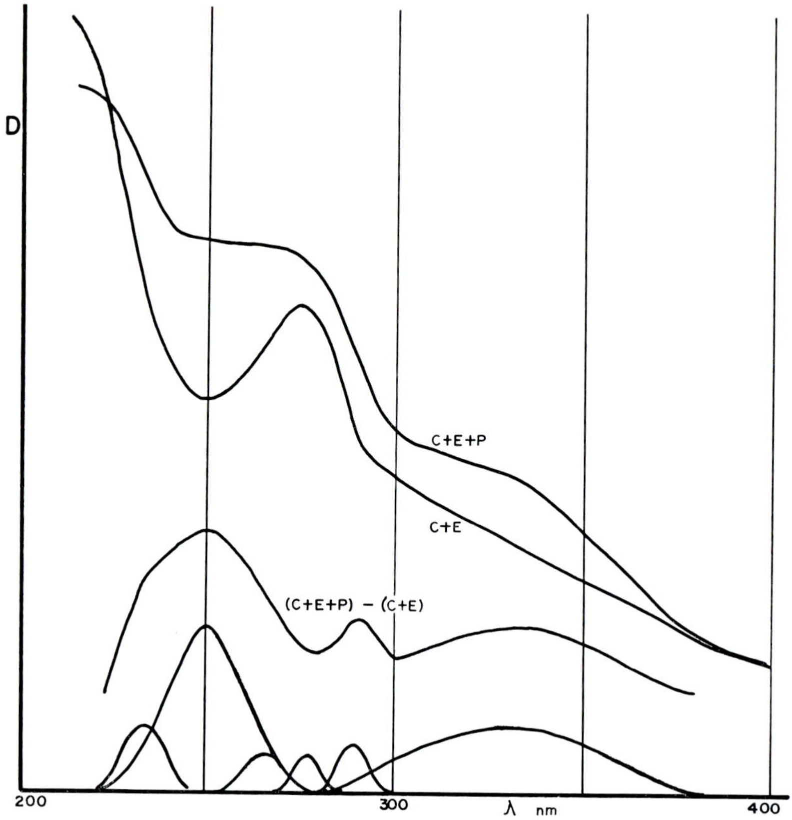

Fig. 9. Resolved difference spectrum of products of o*-quinone and protein.*

hand, decreased rapidly throughout the reaction until it almost disappeared after 36 hr. The isobestic point at 263 nm indicated that during the reaction only two species are present, the quinone and the hydroquinone, which absorb in this region. The selected mole ratio of 3 moles of quinone to 2 moles of tetra-alanine in this reaction theoretically should have produced one quinone cross-link between two amino compounds, with 2 moles of hydroquinone as by-product (Fig. 11). One molecule of quinone results in monosubstituted hydroquinone, two molecules of quinone result in disubstituted hydroquinone, and the third molecule of quinone is used for oxidizing this product to a disubstituted quinone. The molar absorptivity calculated for this diamino-quinone type of product is $\varepsilon = 16\,000$.

The reaction with the enzyme-generated *o*-quinone results in products similar to the *p*-quinone reaction, but here the phenolase enzyme produces the quinone by catalysing the oxidation of the dihydroxyphenol derivatives; therefore, practically no excess quinone is necessary to complete the reaction (Fig. 12). The absorption band of the product of the *o*-quinone developed at 330 nm has a shoulder rather than a peak (Fig. 13). This reaction was much faster than the *p*-quinone reaction and was completed within 2 hr. The calculated molar absorptivity was $\varepsilon = 4700$. The catechol peak at 275 nm

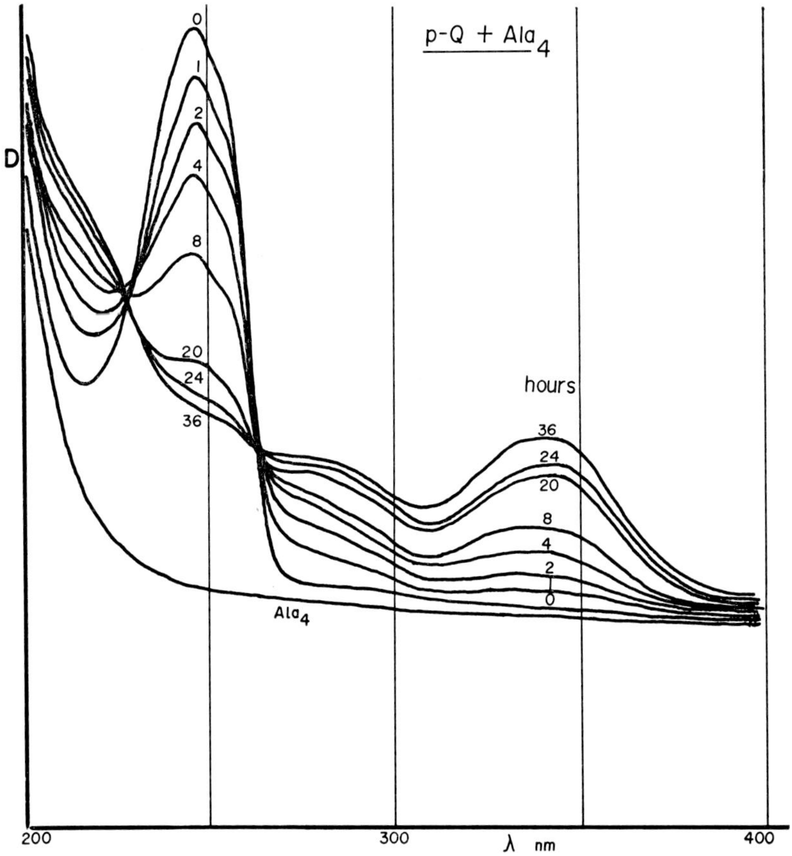

Fig. 10. Spectral changes during reaction of p-*quinone and amino compounds.*

never really decreased, and there was no isobestic point indicating that a number of different types of compounds were probably produced and took part in the reaction.

Comparison of the rate of formation of the products (Fig. 14) calculated from the molar extinction coefficients indicated that the enzyme-generated *o*-quinone reacts extremely quickly, reaching 50 % of completion within 30 min and 90 % completion in about 7 hr, while the reaction with the *p*-quinone showed a detectable amount of product only after 2 hr and proceeded at a steady slow rate over 24 hr.

In order to study the reaction kinetics, the reaction at various concentrations of *o*- or *p*-quinone and valine mixtures was monitored spectrophotometrically. In the case of *o*-quinone, we used the absorbance of the product at 330 nm, but in the case of *p*-quinone the reaction was more conveniently monitored by the reduction of the *p*-quinone peak at 245 nm because the development of the product peak at 345 nm was delayed. Data for the kinetic studies can be found in Appendix II.

Figure 15 shows the plots of the initial reaction velocities versus the reagent concentrations. The curve for the reaction with the enzyme generated

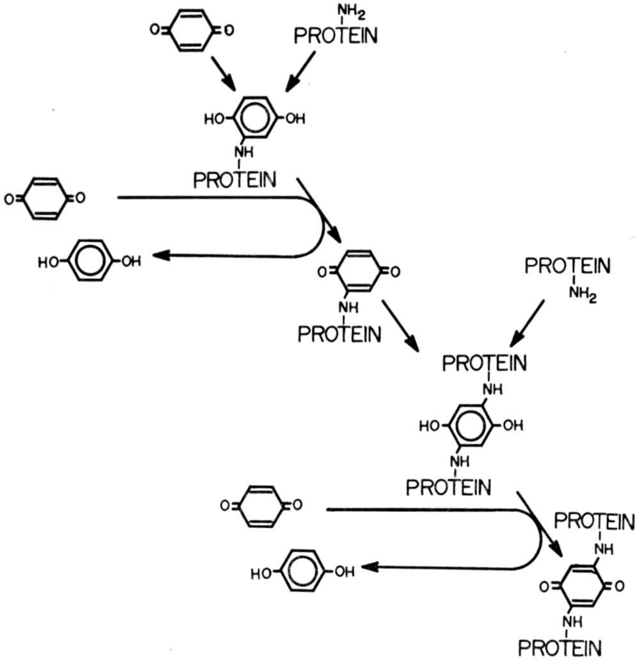

Fig. 11. *Reaction of* p-*quinone with protein.*

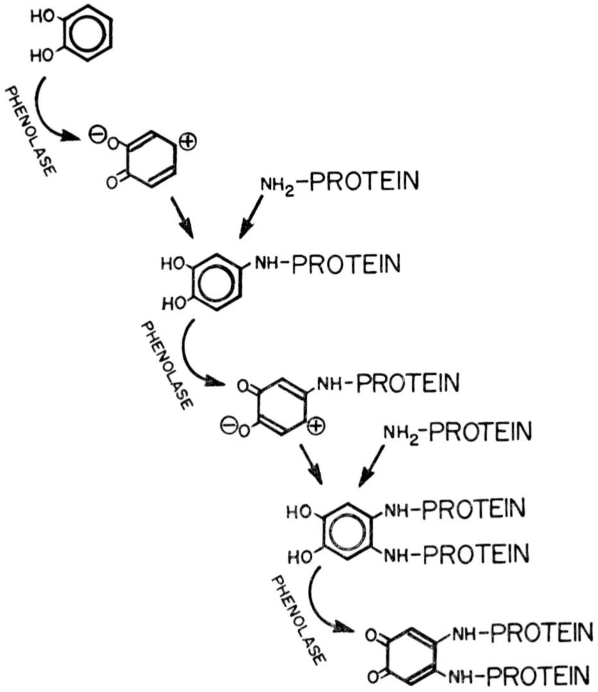

Fig. 12. *Reaction of* o-*quinone with protein.*

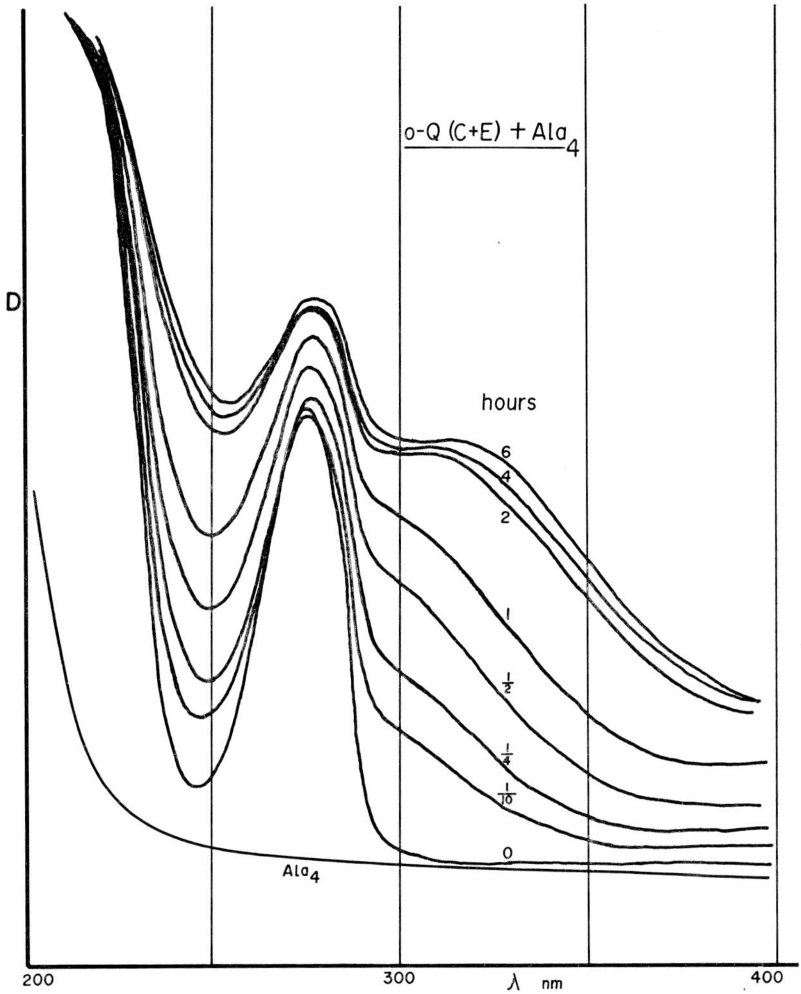

Fig. 13. Spectral changes during reaction of o-*quinone and amino compounds.*

o-quinone [*o* − *Q*(d*r*/d*t*)] is a rectangular hyperbola and the reaction follows the classical Michaelis-Menten (1913) mechanism. This indicates that *o*-quinone reacted practically instantaneously as it was generated by the enzyme. The *p*-quinone reaction [*p* − *Q*(−d*a*/d*t*)] on the other hand, shows sigmoidal kinetics. At low quinone concentrations the initial velocities were low and did not change with the changes of concentration. Above 1 mmol, the velocity increased rapidly, and above 3 mmol, the curve became a rectangular hyperbola. The dotted line is the extrapolation of the hyperbolic section. The plot of velocities calculated from the product of *p*-quinone [*p* − *Q*(d*r*/d*t*)] shows a similar sigmoidal shape, but the velocities are much lower. Although the maximum velocities for the two quinones were comparable, in concentrations less than 1 mmol the velocity of the *o*-quinone reaction was about an order of magnitude higher than that of the *p*-quinone.

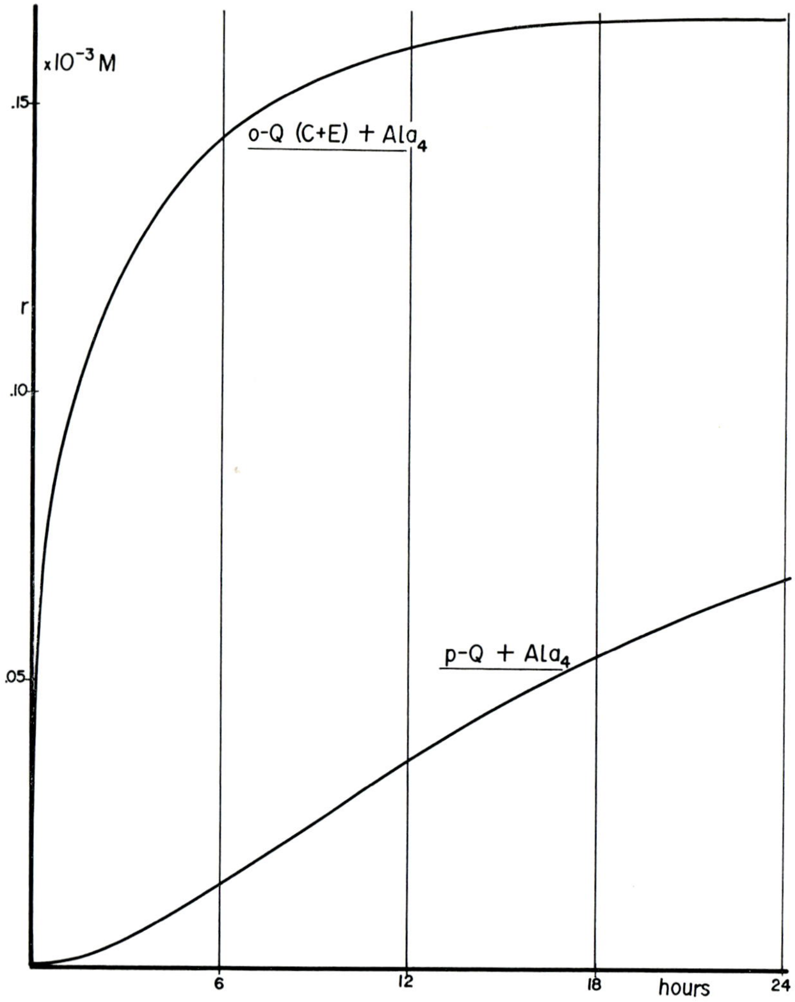

Fig. 14. Reaction rates of quinones and amino compounds.

The maximum velocities of these reactions can be determined by plotting the reciprocal of the velocities versus the reciprocal of the concentrations (Fig. 16). From these Lineweaver-Burk (1934) plots, the maximum velocity for the *o*-quinone reaction is $v = 32{\cdot}8 \times 10^{-6}$ M min^{-1} and that for the *p*-quinone is $v = 77 \times 10^{-6}$ M min.

The sigmoidal kinetics of the *p*-quinone reaction indicated that more than one type of interaction was involved and the order of the reaction, that is, the number of the participating molecules, may vary with the quinone concentration. The order of the reaction was determined from the plot of the logarithms of the velocities versus the logarithms of the concentration (Fig. 17) because

$$v = -\frac{\mathrm{d}a}{\mathrm{d}t} = ka^n$$

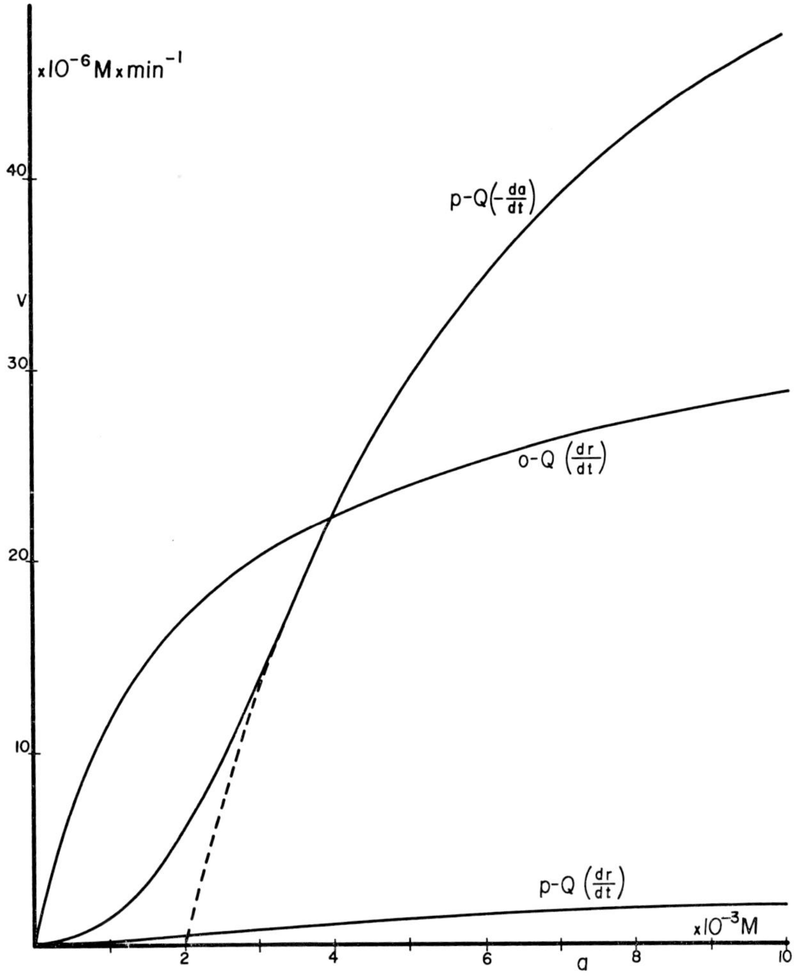

Fig. 15. Reaction velocity as function of concentration.

the velocity (v), that is the rate of reduction of the reactant quinone concentration ($-\mathrm{d}a/\mathrm{d}t$) is proportional to the nth power of the reactant concentration. The 'n' is the number of reactant molecules interacting, and 'k' is the reaction rate constant. The logarithmic equation is

$$\log v = n \log a + \log k$$

and the order of the reaction 'n' is the slope of the plot of log v versus log a.

After an initial rise, the plot of log v versus log a levelled off between 3×10^{-5} and 10^{-4} M, indicating zero-order kinetics. Above this concentration, the curve rose again to reach a steep straight line between 5×10^{-4} and 3×10^{-3} M, showing maximum-order kinetics. At higher concentrations, the plot levelled off again approaching zero-order.

In order to determine the order of the reaction more precisely, the plot of log v versus log a was divided into five sections and equations for each section

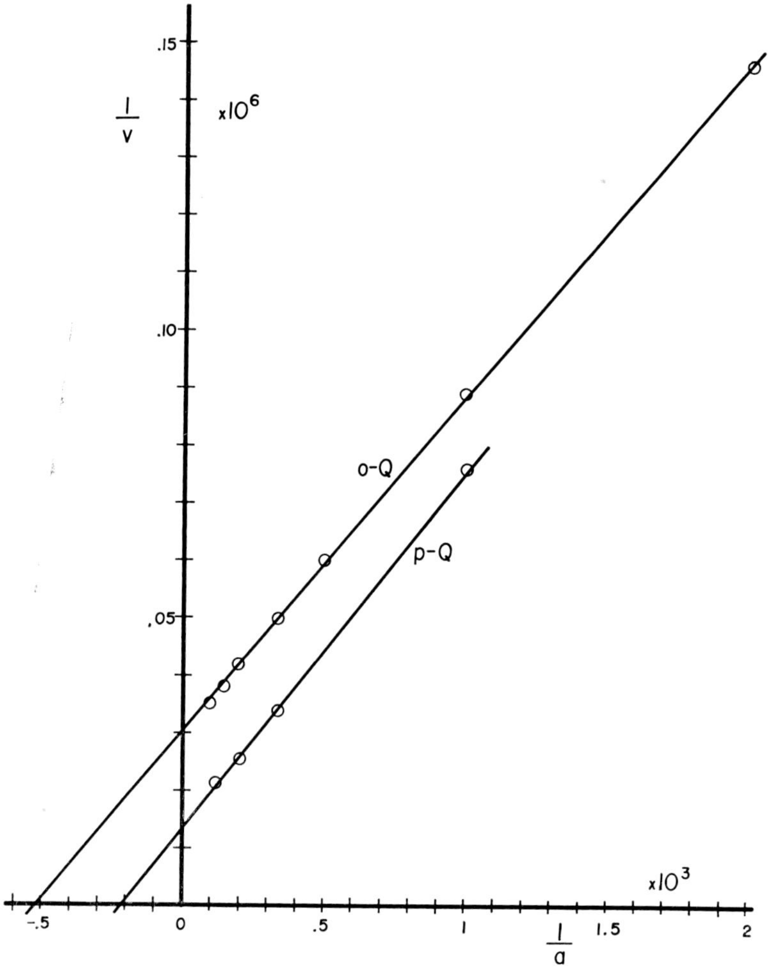

Fig. 16. Lineweaver-Burk plot.

were determined. The first differentials of these equations gave straight lines, representing the order of the reaction at each concentration. From these differentials it was clear that the initially first-order reaction approaches zero-order and remained so between 3×10^{-5} and 10^{-4} M. Above this concentration range, the reaction became a mixed order and reached a maximum of second-order in *p*-quinone at about 5×10^{-4} M. The reaction gradually began to decrease in order as the concentration increased above 3×10^{-3} M, approaching zero-order, where the reaction reached a maximum velocity which is independent of concentration variations. These sigmoidal kinetics of the *p*-quinone and amine reaction indicated that the final oxidation step to produce a diamino-quinonoid derivative had a low reaction rate, since the maximum order of the reaction was two, suggesting that statistically only two molecules of quinone took part in the reaction at any given time. As was noted, with two molecules of quinone the primary product was

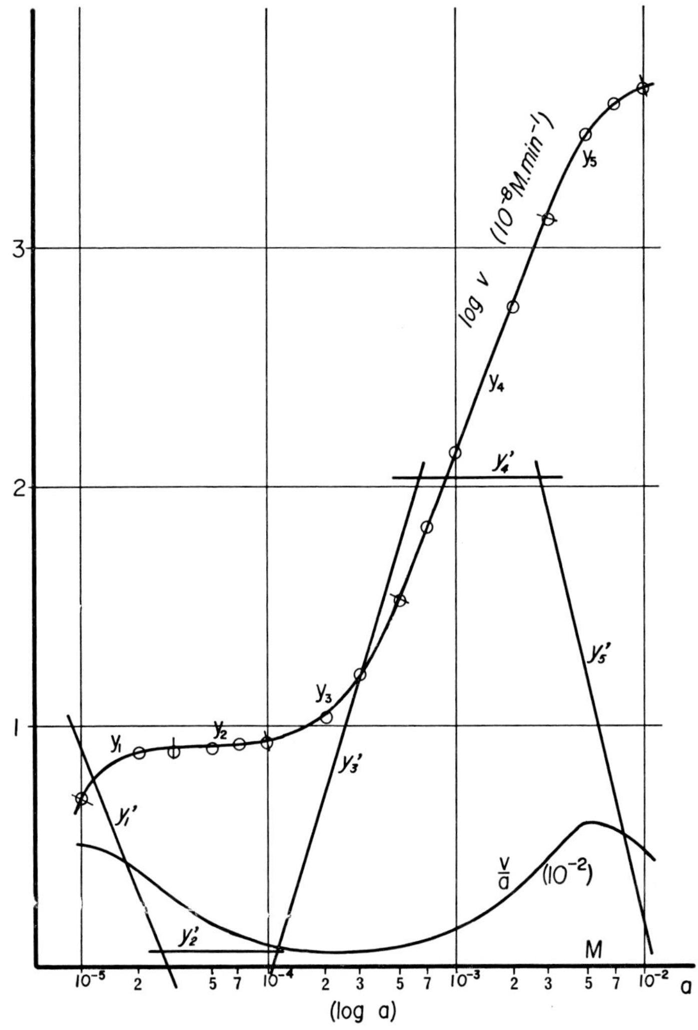

Fig. 17. Reaction kinetic characteristics of p-*quinone.*

disubstituted hydroquinone and the third molecule of quinone oxidized this product to a quinonoid derivative.

The plot of the velocities relative to the concentration (v/a) shows that the relative velocity of the *p*-quinone reaction dropped to a minimum at about 3×10^{-4} M. The significance of this data is revealed by examining the spectra of the reaction mixtures of various concentrations after various incubation periods (Fig. 18).

As is expected, the most concentrated (10 mM) sample reacted quickly and after 4 hr, the peak at 345 nm indicated a quinone-amine product. The quinone peak at 250 nm reduced and the hydroquinone peak at 290 nm developed. Although the 1 mM and the 0·1 mM solutions showed little change, the strong reduction in the quinone peak and the development of the hydroquinone peak indicated that some reaction does proceed in the least

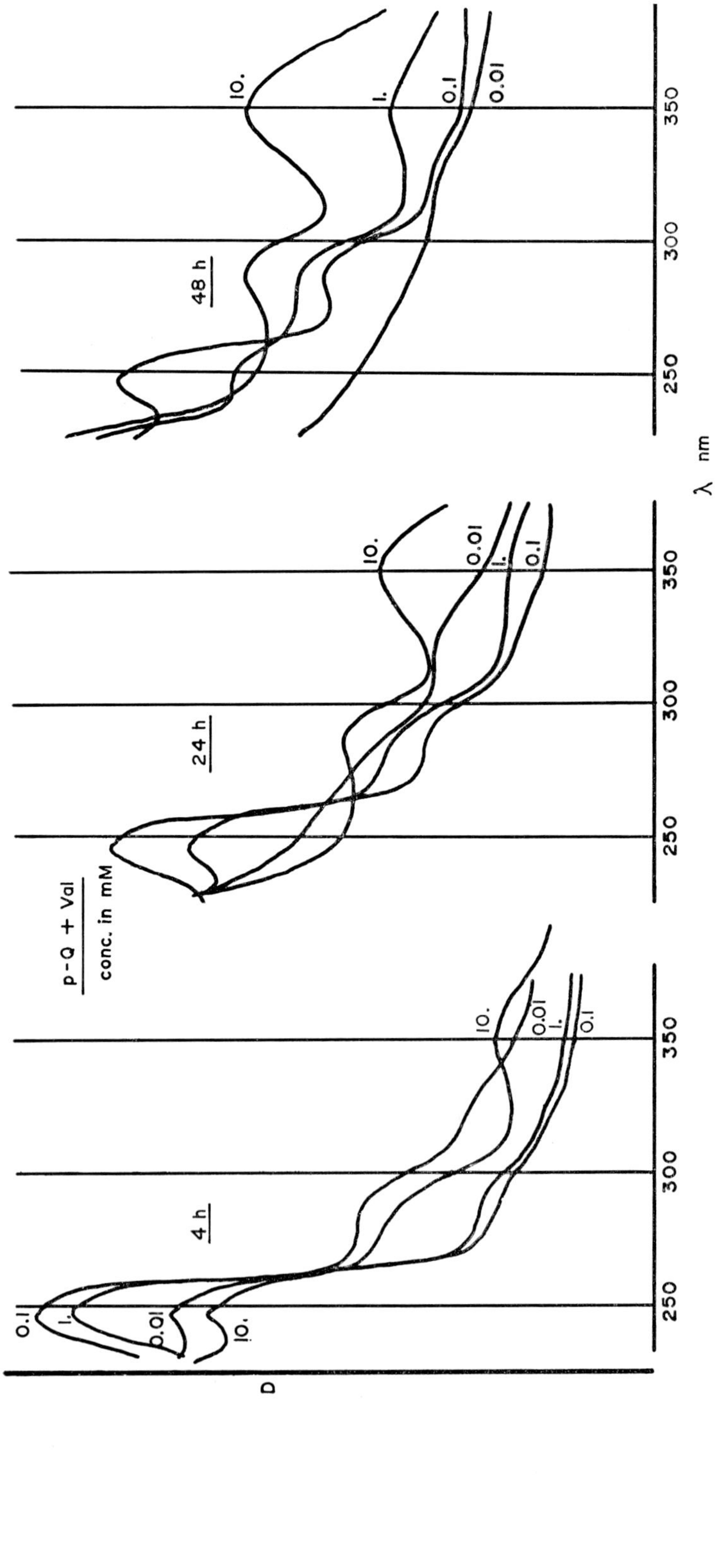

Fig. 18. Spectra of products of p-*quinone and amino compounds in various concentrations.*

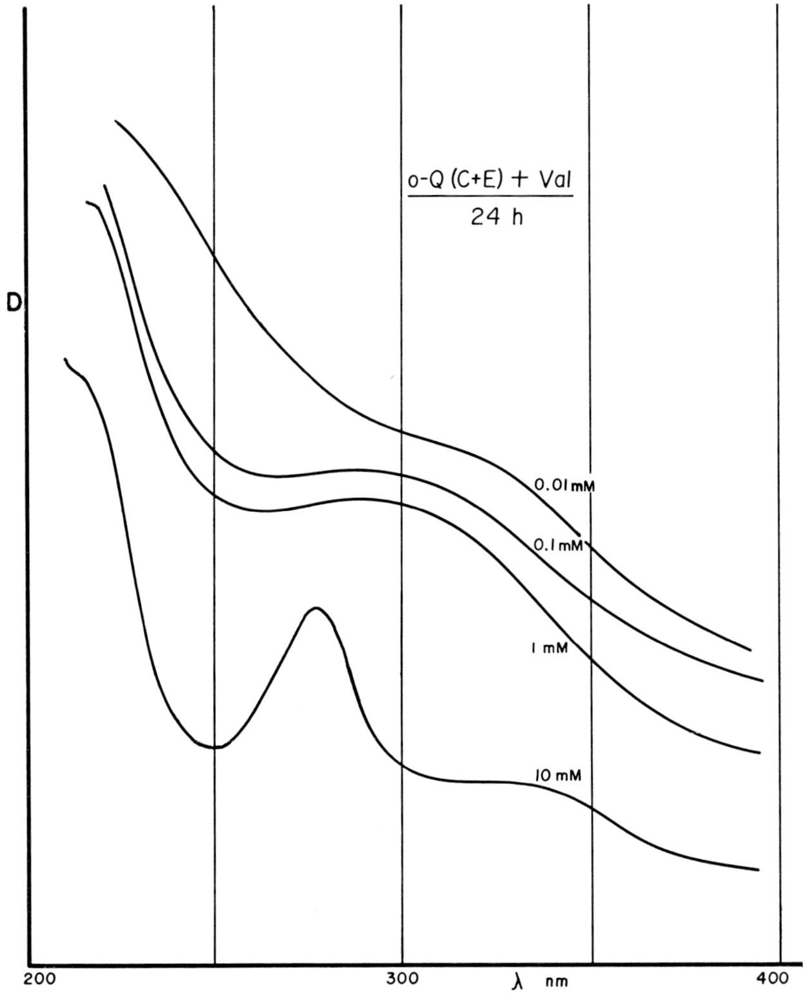

Fig. 19. Spectra of products of o-*quinone and amino compounds in various concentrations.*

concentrated (0·01 mM) reaction mixture. But instead of a peak at 350 nm, a shoulder developed near 330 nm, indicating a product of the *o*- rather than a *p*-quinone type.

After 24 hr, the 1 mM solution showed a spectrum similar to that of the 10 mM solution, but there was little change in the spectrum of the 0·1 mM solution. In the spectrum of the least concentrated mixture, the absorption band near 330 nm was clearly visible.

After 48 hr, the absorption bands of the product with *p*-quinonoid character at 350 nm have increased in the spectra of the two more concentrated mixtures, and absorptions characteristic of the *o*-quinonoid product have developed near 330 nm in the spectra of the two less concentrated mixtures. Based on these results, apparently the character of the reaction product of an amino compound and *p*-quinone was dependent upon the concentration of the reaction mixture. At a very low concentration, the product unexpectedly

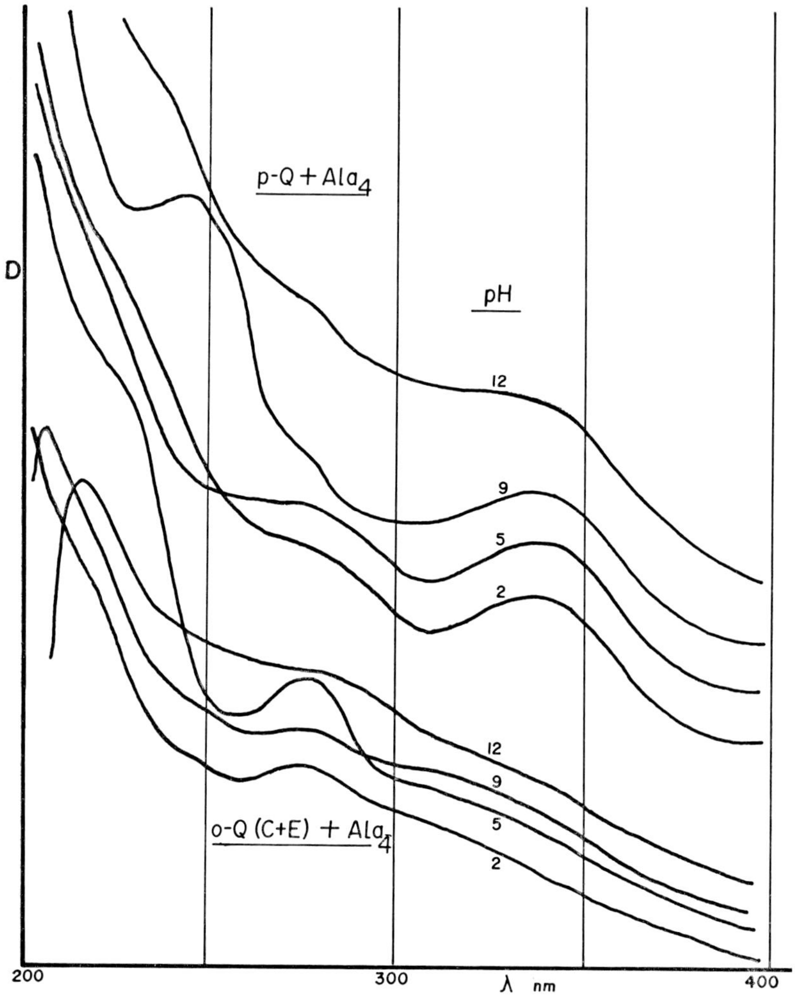

Fig. 20. Spectra of quinone-amino compounds at various pH's.

had an *o*-quinonoid character, while at higher concentrations a *p*-quinonoid character appeared. Further experiments in varying the concentrations showed that the character of the product changed from ortho to para at around 3×10^{-4} M, where the relative velocity has been shown to reach a minimum.

When enzyme-generated *o*-quinone reacts with an amino compound, a similar dependence of product character on concentration could be detected (Fig. 19). After 24 hr, in the mixtures of lower concentration the product showed absorption as a weak shoulder near 330 nm, typical of an ortho derivative, but in the mixture of highest concentration, a stronger absorption developed with a slight bathochromic shift, which may indicate that products with *p*-quinonoid character may be involved.

Variation of pH often causes shifts in the absorption bands of phenolic compounds. In solution of pH values higher than pH 8, phenolic compounds show a bathochromic shift caused by ionization of the phenolic hydroxyl.

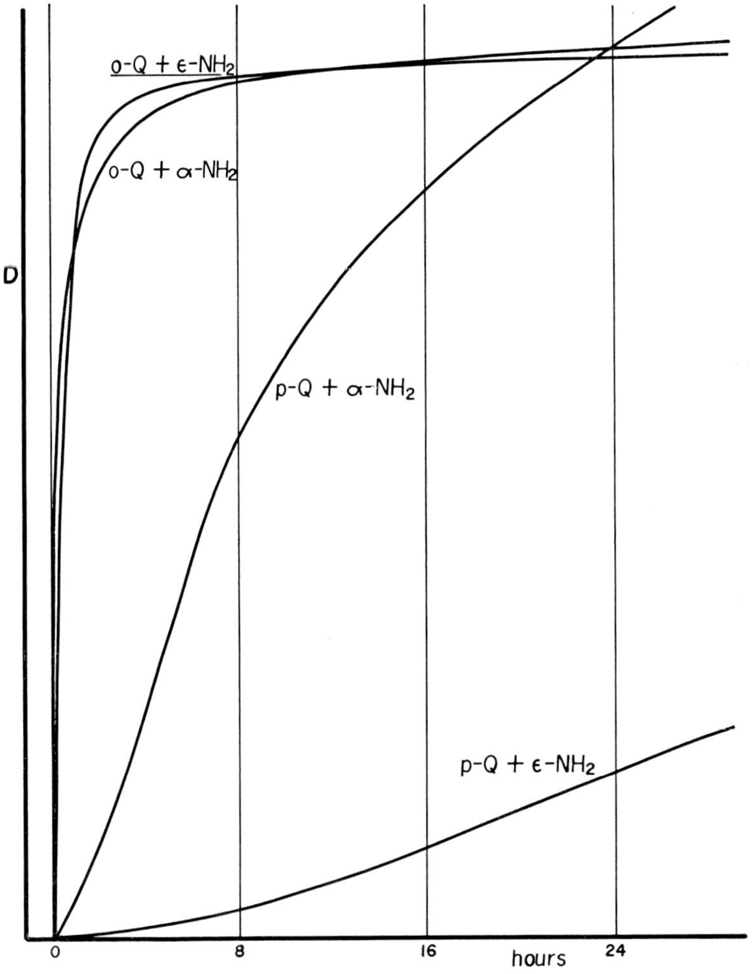

Fig. 21. Reaction rates of quinones with α- and ε-amino groups.

Figure 20 compares the reaction mixtures of tetra-alanine with catechol in the presence of phenolase and with *p*-benzoquinone measured in media of different pH's. Unlike the case with phenolic compounds, a strong alkaline medium (pH 12) caused a hypsochromic shift and a hypochromic effect on the product of the *p*-benzoquinone-tetra-alanine reaction, producing a shoulder near 330 nm that resembled the absorption of an *o*-quinone derivative. This can probably be explained by ionization of both a carbonyl and an amino group, producing an *o*-quinonoid zwitterion.

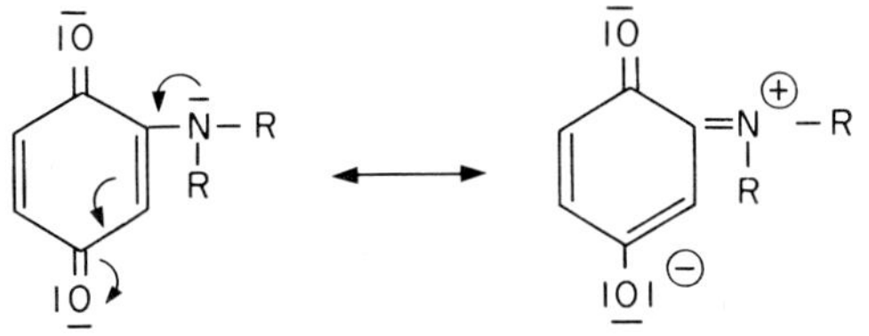

There was no change in wavelength or absorbance near 345 nm at pH values lower than pH 8. Also, no shift was found in the absorption of the product of the *o*-quinone when the pH was changed, although maximum absorbance was obtained in near neutral media.

In order to achieve effective cross-linking by quinones, reaction with other than terminal amino groups in the protein chain is important. If the reaction progresses mainly with terminal α-amino groups, end-to-end linkage of only two peptide chains will occur, producing a linear structure with a molecular weight only a little higher than twice that of the original peptide; this can hardly result in the desired hardening or curing of the liquid precursor mixture. Therefore, it is essential that other reactive groups, such as the ε-amino groups of the lysine residues in the peptide chain, can take part in the reaction with quinone, resulting in a high molecular weight, multiply cross-linked, steric copolymer of peptide and quinone. To investigate the reaction preference between α- and ε-amino groups, *p*-benzoquinone and catechol-phenolase systems were reacted separately with tetra-L-alanine (MW = 302·3), which has only a free α-amino group, and with hippuryl-L-lysine (MW = 307·4), which has only a free ε-amino group. Figure 21 compares the reaction rates of these four reactions monitored at 330 nm for the ortho and at 345 nm for the para derivative. The results showed that with *p*-benzoquinone the reaction rate was higher with the α-amino group, but with phenolase-generated *o*-quinone no difference in reaction rate with either α- or ε-amino groups could be detected.

DISCUSSION

During efforts to elucidate the quinone cross-linking mechanism, typical absorption bands in the near u.v. were encountered which are suitable for detection and characterization of quinone cross-linked proteins. When amino compounds reacted with *p*-quinone, a characteristic absorption band developed near 345 nm with a molar extinction coefficient of ε = 16 000. This value was in good agreement with the molar extinction coefficient of the product of monochlorotrifluoro-*p*-benzoquinone and glycine where ε = 16 600 at 350 nm (Nakaya *et al.*, 1967). When amino compounds reacted with enzyme-generated *o*-quinone, the weak and broad absorption band of the product had a maximum at 330 nm, with a molar extinction coefficient of ε = 4700.

The kinetic study of the quinone cross-linking mechanism strongly indicated that cross-linking with *o*-quinones was more advantageous than with *p*-quinones. The reaction with enzyme-generated *o*-quinones followed the classical Michaelis-Menten (1913) mechanism, suggesting that the *o*-quinones react with amines instantaneously. Although the maximum initial velocities of the reaction with *o*- and *p*-benzoquinones were comparable ($V_0 = 32{\cdot}8 \times 10^{-6}$ M min^{-1}; $V_p = 77 \times 10^{-6}$ M min^{-1}) at saturation concentrations, at less than 10^{-3} M the reaction velocity of the *o*-quinone was about an order of magnitude higher than that of the *p*-benzoquinone. This is in agreement with the studies of Doskočil (1950) who concluded from redox potential measurements of unsubstituted *o*- and *p*-quinone and of quinones with amino acid substitutions that since the loss of free energy is greater in the *o*-quinone substitution (295 mV) than in the *p*-quinone substitution (216 mV), the *o*-quinone appears to be more reactive.

The products of the reactions with both quinones can exist in both *o*- and *p*-quinonoid resonance structures (Wiesner, 1942; Doskočil, 1950). The final character of the product was found to be pH-dependent and, more interestingly, concentration-dependent. This effect was particularly pronounced with *p*-benzoquinone. At initial concentrations lower than 3×10^{-4} M, the product of the reaction of *p*-benzoquinone and an amino compound showed absorptions indicating an *o*-quinonoid character, and above that concentration, a *p*-quinonoid character appeared. Also, the product of *o*-quinone and amino compounds showed somewhat *p*-quinonoid character at high initial concentrations. These results suggested that the product of the reactions with both quinones can exist in both *o*- and *p*-quinonoid resonance structures, depending on the initial concentration of the reaction mixture.

As was shown by previous investigators (Hilpert and Brauns, 1925; Wiesner, 1942; Evans and deHeer, 1950; Akerfeldt, 1953; Morrison *et al.*, 1969), the quinone cross-linking process initially leads to a monoaminohydroquinonoid derivative. This is followed by an oxidation and second substitution to produce a diaminohydroquinonoid derivative, which is oxidized to a diaminoquinonoid form in the final step. Since there is an initial time lag in the reaction of *p*-benzoquinone and amine before the absorption band of the product near 345 nm appears, apparently disubstitution was preferred and the monosubstituted derivative did not absorb in these wavelengths. On the other hand, the sigmoidal kinetics of the *p*-benzoquinone and amine reaction indicated that the final oxidation step to produce a diaminoquinonoid derivative has a low reaction rate since the maximum order of reaction in *p*-benzoquinone was two, suggesting that statistically only two molecules of quinone took part in the reaction at any given time. With two molecules of quinone, the principal reaction will produce only diaminohydroquinone ($2A + 2Q \rightarrow AHq + Q + A \rightarrow AQ + A + Hq \rightarrow AHqA + Hq$). On a similar basis, the initially first-order reaction with *o*-quinone indicated formation of a monoaminocatechol derivative.

Although the reaction rate of *p*-benzoquinone was higher with terminal α-amino groups than with sidechain ε-amino groups, *o*-quinone reacted equally rapidly with both amino groups. This did not support the hypothesis that the terminal α-amino group of a peptide is almost certainly more facile in reaction than the ε-amino groups, as postulated by Mason (1955a,b) on the basis of the ‘peptide effect’, by which increasing the number of peptide bonds increases the reaction rate of the terminal amino group with enzyme-generated *o*-quinones. Higher reactivity with peptides over simple amino compounds was found only with *p*-quinone, not with enzyme-generated *o*-quinone. The equally high reaction velocity of the α- and the ε-amino groups can be an important factor in effective cross-linking, since it increases the probability of the large protein molecules orienting into a position which is favourable for cross-linking. This, in combination with their higher reaction velocity, makes *o*-quinones more suitable than *p*-quinones for cross-linking in nature. The advantage of *o*-quinone as a cross-linking agent readily explains the presence of various *o*-dihydroxyphenol derivatives in natural quinone-cross-linked proteins (Pryor, 1940a,b, 1962; Pryor *et al.*, 1946; Hackman and Todd, 1953; Dennell, 1958a; Karlson *et al.*, 1962; Ravindranath and Ramalingam, 1972) but, on the other hand, does not support the theories promoting cross-linking through *p*-quinones (Dalgleish, 1955; Dennell, 1958b). It is not surprising, either, to find that the absorption bands of the barnacle adhesive hydrolysate indicated an *o*-quinone type of cross-link. But, since there is only one type of gland in the barnacle to secrete the protein, the phenolics, and the enzyme and there is a lack of extractable phenolic precursors (Lindner and Dooley, 1972), the question is raised whether autocross-linking, first hypothesized by Brown (1949) for *Mytilus* byssus, can be involved in the hardening mechanism of the barnacle adhesive. Autocross-linking (Fig. 22) was proposed to occur

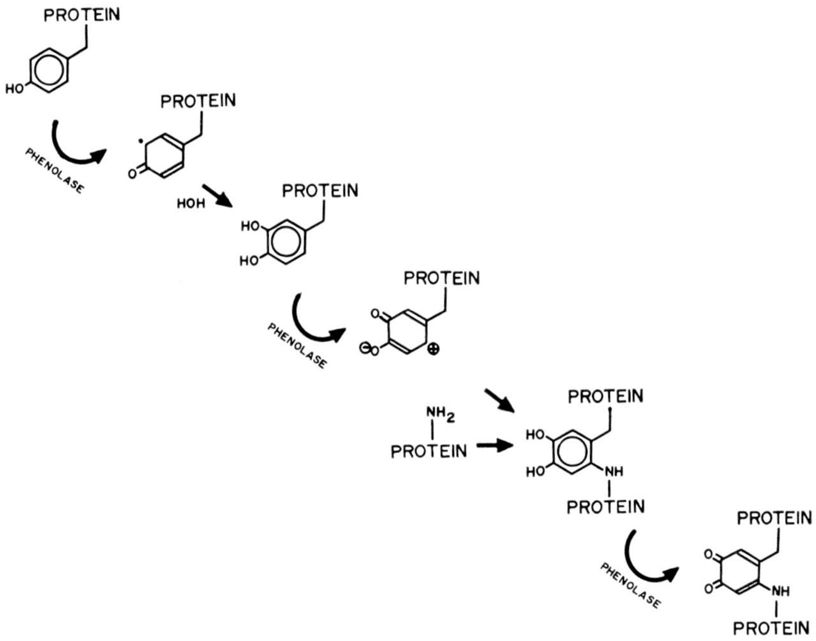

Fig. 22. Autocross-linking mechanism.

between the tyrosyl residues of protein sidechains and free amino or other reactive groups of another protein molecule with the aid of the phenolase enzyme. The fact that in the *o*-quinone reaction the initial step that produced a monosubstituted derivative had a high limiting rate is in favour of the autocross-linking mechanism, since coupling with only one amino group of another molecule is sufficient to form an intermolecular cross-link.

REFERENCES

Akerfeldt, S. (1953). *Arkiv f. Kemi*, **7**, 75.
Barenfanger, C. (1939). *Angew. Chem.*, **52**, 72.
Brown, C. H. (1949). *Exp. Cell Res. Suppl.*, **1**, 351.
Costlow, J. D. (1956). *J. Morphol.*, **99**, 359.

Dalgleish, C. E. (1955). *Arch. Biochem. Biophys.*, **58**, 214.

Darwin, C. (1854). *A Monograph on the Subclass Cirripedia—the Balanidae*. Royal Society, London, 684 pp.

Dennell, R. (1958a). *Proc. Roy. Soc.*, **B148**, 280.

Dennell, R. (1958b). *Proc. Roy. Soc.*, **B149**, 176.

Doskočil, J. (1950). *Coll. Trav. Chem. Tchecosl.*, **15**, 780.

Evans, M. G., and deHeer, J. (1950). *Quart. Rev. London*, **4**, 94.

Friend, J. N. (1932). *Engineering*, March 11, 320.

Gregg, J. H. (1948). *Biol. Bull.*, **94**, 161.

Hackman, R. H., and Todd, A. R. (1953). *Biochem. J.*, **55**, 631.

Harris, J. E. (1946). *J. Iron Steel Inst.*, **154**, 297.

Hilpert, S., and Brauns, F. (1925). *Collegium*, 64.

Karlson, P., Sekeris, C. E., and Sekeri, K. (1962). *Physiol. Chem.*, **327**, 86.

LaQue, F. L., and Clapp, W. F. (1945). *Trans. Electro-Chemical Soc.*, **87**, 165.

Legendre, R. (1938). *Rev. Scientifique, Paris*, **76**, 142.

Lindner, E., and Dooley, C. A. (1972). Proc. 3rd Int. Congr. Marine Corrosion and Fouling, pp. 653–673, Gaithersburg, Md., Northwestern University Press.

Lindner, E., Dooley, C. A., and Clavell, C. (1973). Proc. 4th Inter-Naval Conf. Marine Corrosion, pp. 358–391, Naval Research Laboratory.

Lineweaver, H., and Burk, D. (1934). *J. Am. Chem. Soc.*, **56**, 658.

Mason, H. S. (1955a). *Nature*, **175**, 771.

Mason, H. S. (1955b). *Adv. Enzymol.*, **16**, 105.

Michaelis, L., and Menten, M. L. (1913). *Biochem. Z.*, **49**, 333.

Morrison, M., Steele, W., and Danner, D. J. (1969). *Arch. Biochem. Biophys.*, **134**, 515.

Nakaya, K., Horinishi, H., and Shibata, K. (1967). *J. Biochem. (Tokyo)*, **61**, 331,

Newman, W. A., Zullo, V. A., and Wainwright, S. A. (1967). *Crustaceana*, **12** (2). 167.

Nue, W. (1932). *Int. Rev. Hydrobiol.*, **37**, 104.

Pilsbry, H. A. (1916). The Sessile Barnacles (*Cirripedia*) Contained in the Collection of the U.S. National Museum, Including a Monograph of the American Species. Government Printing Office, Washington, D.C., 365 pp.

Pryor, M. G. M. (1940a). *Proc. Roy. Soc.*, **B128**, 378.

Pryor, M. G. M. (1940b). *Proc. Roy. Soc.*, **B128**, 393.

Pryor, M. G. M. (1962). *Comparative Biochemistry*, Vol. 4, 371 pp. Academic Press, New York.

Pryor, M. G. M., Russell, P. B., and Todd, A. R. (1946). *Biochem. J.*, **40**, 627.

Ragg, M. (1925). *Die Schiffsboden- und Rostschutz-Farben*. Berlin.

Ragg, M. (1935). *Farbenzeitschr.*, **40**, 1167.

Ravindranath, M. H., and Ramalingam, K. (1972). *Acta Histochem.*, **42**, 87.

Saroyan, J. R., Lindner, E., and Dooley, C. A. (1968). Proc. 2nd Int. Congr. Marine Corrosion and Fouling, Athens, pp. 495–512, Tech. Chamb. Greece.

Saroyan, J. R., Lindner, E., and Dooley, C. A. (1970a). *Biol. Bull.*, **139**, 333.

Saroyan, J. R., Lindner, E., Dooley, C. A., and Bleile, H. R. (1970b). *Ind. Eng. Chem., Prcduct R & D.*, **9**, 122.

von Wolzogen Kuhr, A. A. H. (1938). *Water*, **22**, 33 and 45.

Wiesner, K. (1942). *Chem. Listy*, **36**, 313.

Weiss, C. M. (1948). *Ecology*, **29**, 120.

Young, C. F. T. (1867). The Fouling and Corrosion of Iron Ships: their Causes and Means of Prevention with the Mode of Application of Existing Ironclads. London, 212 pp.

APPENDIX I

Fig.	*Reactants*	*Reaction* Conc. (*mM*)	*Reaction* *E*	*Reaction* *Q:A*	*r × n time* (*hr*)	*Spectral measurement* Conc. (*mM*)	*Spectral measurement* Path (*cm*)	*Notes*
6	*p*-Quinone +							
	n-butylamine	45·5	—	1:2	24	0·15	1	
	L-Valine	28·4	—	1:2	24	0·1	1	
	L-Lysine	22·8	—	1:2	24	0·08	1	
	Tetra-L-alanine	1	—	3:1	24	1	0·02	
	Tetra-L-lysine	0·63	—	6:1	24	0·63	0·02	
	Poly-L-lysine	1–5	—	1:1	24	1–5	0·1	
	Protein (trypsin)	0·21	—	12:1	24	0·007	1	
7	Catechol + E +							
	n-Butylamine	0·33	20	3:2	72	0·33	0·1	
	L-Valine	0·33	20	3:2	4	0·33	0·1	
	L-Lysine	0·17	20	3:2	4	0·17	0·1	
	Tetra-L-alanine	3·04	25	3:1	3	3·04	0·02	
	Tetra-L-lysine	3·82	23	3:2	3	3·82	0·02	
	Poly-L-lysine	~2·5	20	~1:1	¼	~2·5	0·1	
	Protein (trypsin)	0·21	20	12:1	24	0·007	1	
8	Protein (trypsin) +							
	Hydroquinone	2·5	—	12:1	24	0·084	1	
	p-Quinone	2·5	—	12:1	24	0·084	1	
	Catechol + E	2·5	20	12:1	24	0·084	1	
	3,4-Dihydroxybenzoic acid + E	2·5	20	12:1	24	0·084	1	
	Tyrosine + E	2·5	20	12:1	24	0·084	1	
	Dopa + E	2·5	20	12:1	24	0·084	1	
9	Catechol + E	2·5	20	—	24	0·084	1	
	Catechol + E +	2·5	20			0·084		
	Protein	0·21		12:1	24	0·007	1	
10	*p*-Quinone +							
	Tetra-L-alanine	0·63	—	3:2	0–36	0·63	0·02	in N_2 atm

Opercular rim cells

Sandwiched between the opercular wall and perculum plate cells at the opercular rim is a small group of columnar cells, on average 23·8 μm high. These may be distinguished from other cell types by their microvillous border of short microvilli and the absence of intracellular vacuoles (Figs. 2 and 3). Below the microvillus border are numerous membrane-bound vesicles which range from 0·1 μm to 0·3 μm in diameter and possess contents of varying density similar to those in peduncular wall epithelial cells. These vesicles appear to originate in Golgi complexes which are situated in the vicinity of the nucleus (Fig. 3). Several lysosome-like bodies are present in this region but no multivesicular bodies have been observed (Fig. 3). A moderate granular endoplasmic reticulum is characteristic of these cells but few mitochondria are evident. The intercellular junctions resemble those already described, although they lack distensions in the intercellular space.

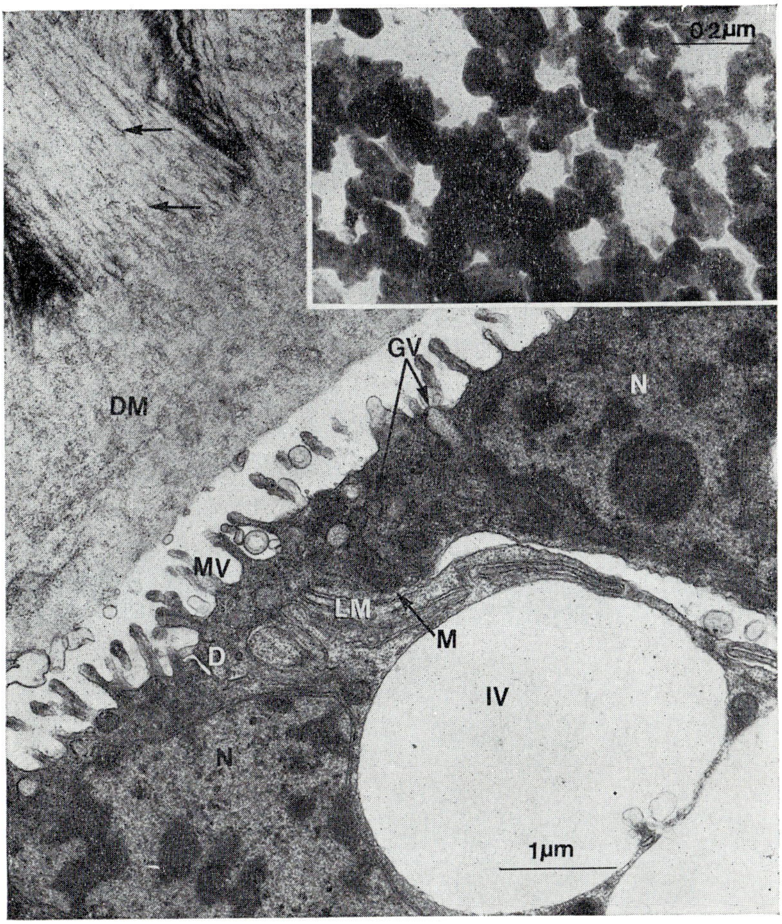

Fig. 5. Electron micrograph showing the relationship of decalcified organic matrix (DM) of the opercular plate reflecting crystal outline (arrows) and the underlying opercular plate cells. Inset: Electron micrograph showing the crystals of the opercular plate. D, desmosome; GV, Golgi vesicles; IV, intracellular vacuoles; LM, lateral membrane; M, mitochondria; MV, microvilli; N, nucleus.

Opercular plate epithelial cells

Adjacent to the operculum rim cells, the cells of the operculum plate epthelium are approximately 9·8 μm high but become progressively shorter towards the centre of the epithelium where they are 7·2 μm high and cuboidal in profile. The epithelial cells possess a microvillous border consisting of short microvilli. Bundles of tonofilaments are prominent in these cells which originate in the core of microvilli and eventually terminate on hemidesmosomes on the basal membrane. The basal membrane is deeply infolded and is seated on a distinct basal lamina. Intercellular junctions resemble those described earlier and the lateral membranes are infolded and interdigitate along their whole length. Distensions are not observed in the intercellular spaces. The cell apex has typically one to three Golgi complexes and several mitochondria, some of which are also evident in other regions of the cell. A number of spherical, membrane-bound vesicles, approximately 0·95 μm in diameter and with a compact fibrous material of varying density, are usually evident beneath the microvillous border. The contents are frequently observed to be released from the cells (Fig. 5). Throughout the cytoplasm, short, granular, endoplasmic reticulum profiles are evident, as are free ribosomes. A feature of these cells is the presence of large intracellular vacuoles, some of which contain a fibrous material. These vacuoles occupy large areas of the cell volume.

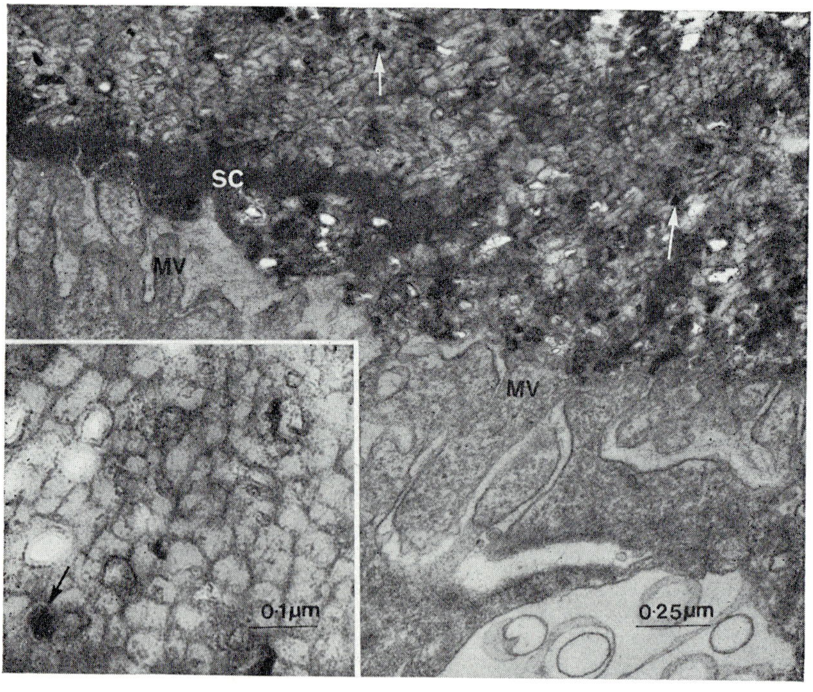

Fig. 6. Electron micrograph showing the organization of the organic matrix in the vicinity of the opercular rim into a mosaic pattern of polygons, some containing crystals (arrows). Note also the layer of secreted cuticular material (SC). Inset: High magnification electron micrograph showing the organization of the organic matrix into regular polygonal profiles. Also note the presence of a crystal (arrows). MV, microvilli.

Opercular plate

The cuticle. The calcareous plate is overlaid by a solid cuticle which is first apparent at the opercular rim (Figs. 2 and 3). In this region it is up to 5·3 μm thick and consists of three distinct layers: an outer layer approximately 0·04 μm thick; a middle layer, comprising the bulk and composed of a dense amorphous matrix in which fibrils and some microvilli profiles are evident; and an inner electron-dense layer (Fig. 3). The cuticle progressively decreases in thickness to 1·6 μm across the top of the opercular plate, mainly due to a reduction in the second layer. In the vicinity of the opercular rim, aggregations of electron-dense material are arranged into layers and added to the inner surface of the cuticle (Fig. 6). The addition of material to the inner surface appears to give rise to a fourth, more electron-lucid, layer.

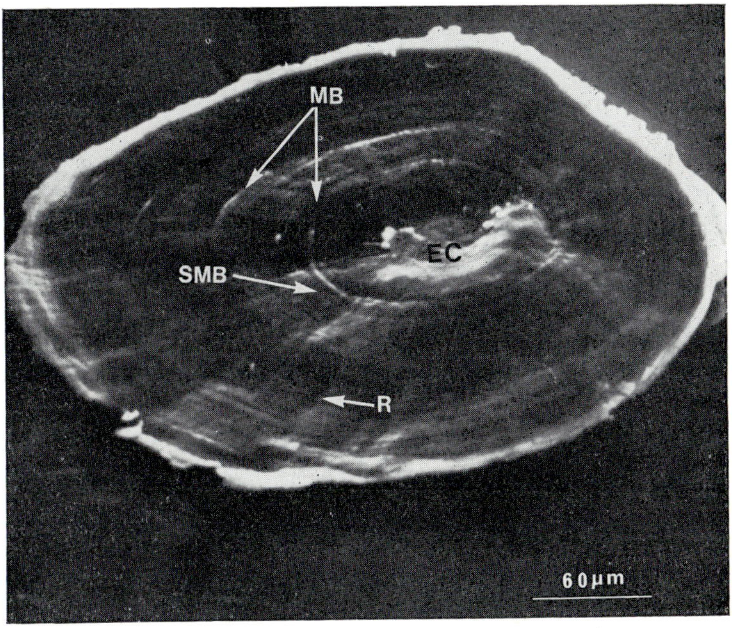

Fig. 7. Scanning electron micrograph showing the dorsal surface of the solid (tanned) cuticle overlying the operculum plate. Note the presence of a eccentric core (EC) and major (MB) and minor (SMB) concentric bands. Also note the radial striations (R).

Scanning electron microscopy reveals that the surface of the cuticle is composed of an eccentric core, approximately 50 μm in diameter, around which are evident a number of major concentric bands, on average 50 μm in width. Each major band is subdivided into a series of minor concentric bands from 4 μm to 8 μm in width (Fig. 7). Also evident on the surface of the cuticle are striations which radiate from the eccentric core like the spokes of a wheel (Fig. 7).

The calcareous plate. At the rim of the operculum, between the opercular plate cells and the solid cuticle, a matrix, exhibiting a mosaic pattern of polygons from 0·03 μm to 0·06 μm in diameter, is evident (Fig. 6). In this region, although most of the matrix profiles are apparently empty, however, some are filled with electron-dense contents indicating calcium carbonate

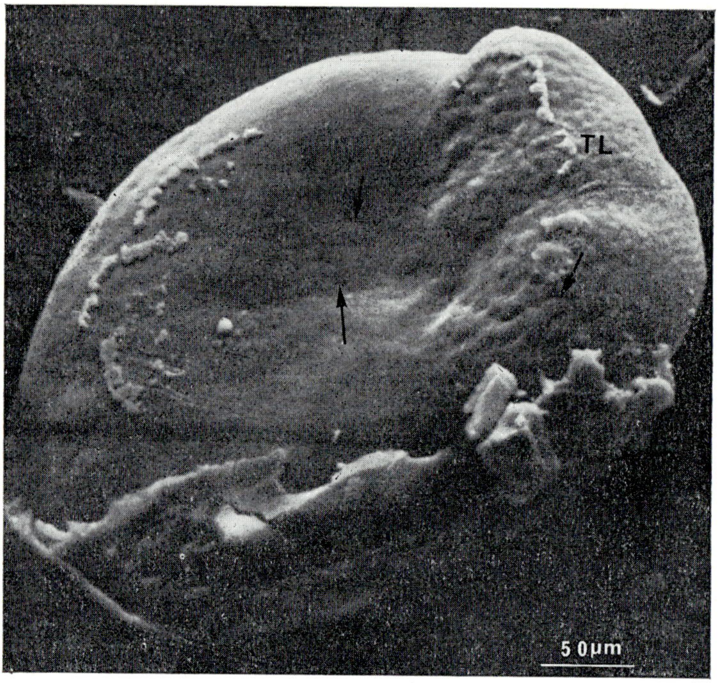

Fig. 8. *Scanning electron micrograph of the ventral surface of the opercular plate showing the bolster-like talon (TL). Note the presence of large angular profiles on the surface (arrows).*

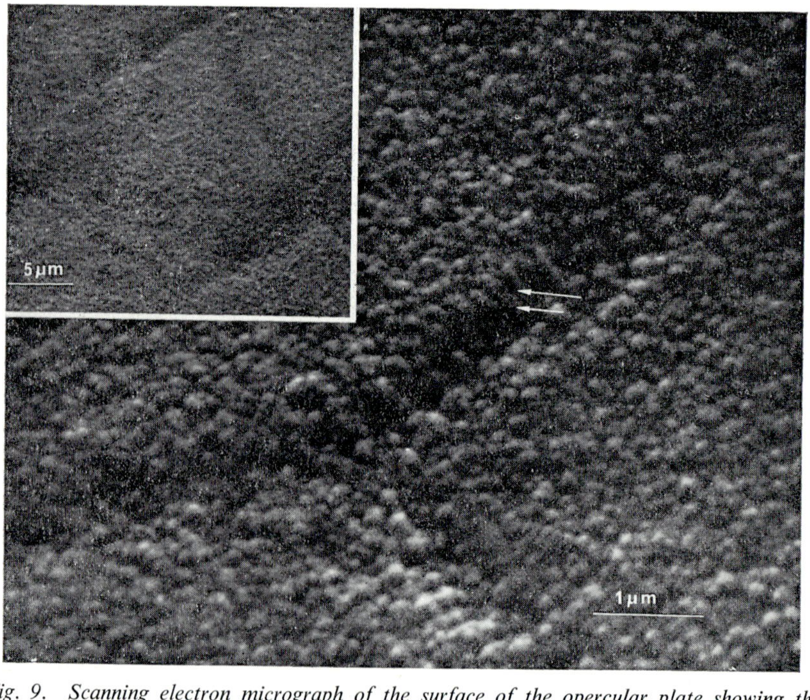

Fig. 9. *Scanning electron micrograph of the surface of the opercular plate showing the arrangement of the crystals (arrows). Inset: Scanning electron micrograph showing angular profile characteristic of the ventral surface.*

crystals (Fig. 6). Away from the rim, the opercular plate is made up of tightly packed polygonal-shaped crystals which average 0·05 μm in diameter (Fig. 5). In among these tightly packed crystals, aggregations of electron-dense material are occasionally evident. Decalcified sections of the opercular plate reveal a matrix which reflects the profiles of the crystals (Fig. 5). When examined by the scanning electron microscope, the ventral surface of the operculum plate is seen to be asymmetric. From one side of the plate a prominent talon is evident (Fig. 8). On the surface of the talon is a mosaic pattern of large angular profiles which are more clearly defined in other parts of the plate (Fig. 8). These angular profiles average 20 μm in width. At higher magnification each polygon is made up of tightly packed crystals, approximately 0·12 μm in diameter (Fig. 9). At the junction of the angular figures, a lateral

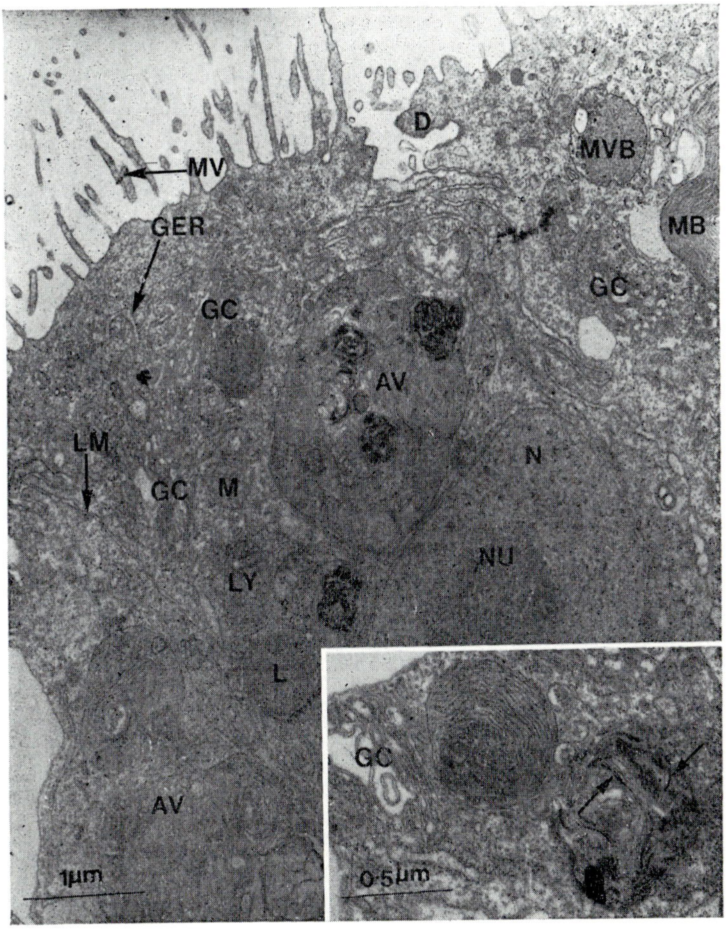

Fig. 10. Electron micrograph showing the apex of cells after 5 days regeneration showing autophagic vacuoles (AV), multivesicular body-like bodies (MVB), myelin bodies (MB), lysosome-like bodies (LY), lipid bodies (L), and Golgi complexes (GC). Inset: Electron micrograph showing lysosome-like bodies with diverse contents. Note the presence of needle-like microcrystals (arrows). D, desmosome; LM, lateral membrane; GER, granular endoplasmic reticulum; M, mitochondrion; MV, microvilli; N, nucleus; NU, nucleolus.

profile of some of the crystals is observed and appears to consist of tightly packed columns of crystals (Fig. 9). Each angular profile appears to reflect the secretory activity of a single cell.

Opercular regeneration

Following the excision of the operculum, the cut surface of the peduncle becomes sealed by a layer of cells by the end of 3 days. After 5 days, a complete epithelial layer is continuous across the peduncle. The early stages of opercular regeneration were not studied during this investigation. The operculum was examined after 5, 9, 12, and 15 days.

At the end of 5 days, the epithelial cells are columnar with a microvillous border consisting of long microvilli, while basally several large intracellular vacuoles are evident (Figs. 10 and 11). Intercellular junctions, comprising a

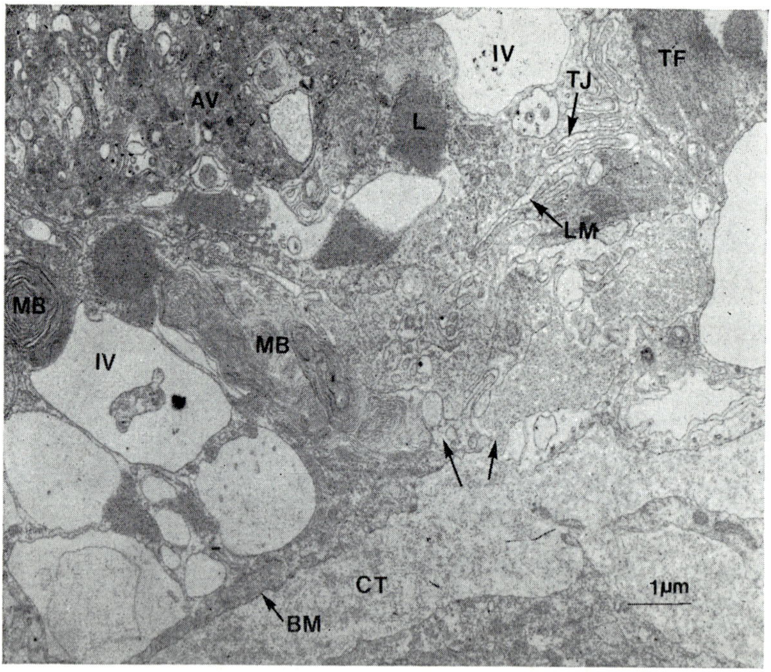

Fig. 11. Electron micrograph showing the basal region of the cells after 5 days regeneration. Note the presence of a myelin body (MB) (arrow) and material resembling connective tissue material (CT) in distensions of the intercellular space (arrows). AV, autophagic vacuole; BM, basal membrane; IV, intercellular vacuole; LM, lateral membrane; L, lipid, TF, tonofilaments; TJ, tight junction.

desmosome, separate the cells. Below this the lateral membranes are slightly infolded. Numerous distensions in the intercellular space are apparent. These are filled with either fibrillar material or myelin bodies (Fig. 11). The intercellular space is obliterated periodically by the formation of tight junctions. The basal membrane is infolded and packed with connective tissue material. In the cell apex, two well-developed Golgi complexes with associated vesicles are usually evident above the nucleus (Fig. 10). Associated with the latter organelle are one to two membrane-bound bodies which possess a homogeneous material and several microvesicles (Fig. 10). Similar microvesicles are associated externally with the bodies and are also observed below the apical plasma membrane. A few vesicles with electron-dense contents are also evident in the vicinity of the apical membrane in some cells. These appear to fuse with one another. Scattered throughout the cytoplasm are a few myelin bodies, autophagic vacuoles, several lysosomes, and lipid bodies (Fig. 10). In most regions of the cell granular endoplasmic reticulum profiles are evident, but are not extensively developed. A few small mitochondria are also observed in the cytoplasm. Bundles of tonofilaments, which terminate on the basal membrane, are prominent in most regions. A rudimentary operculum is apparent by the end of 9 days. It is covered by a prominent cuticle in which three distinct layers are evident. The cuticle is of a uniform width of 1·8 μm along the whole length of the operculum. No evidence of a calcareous plate

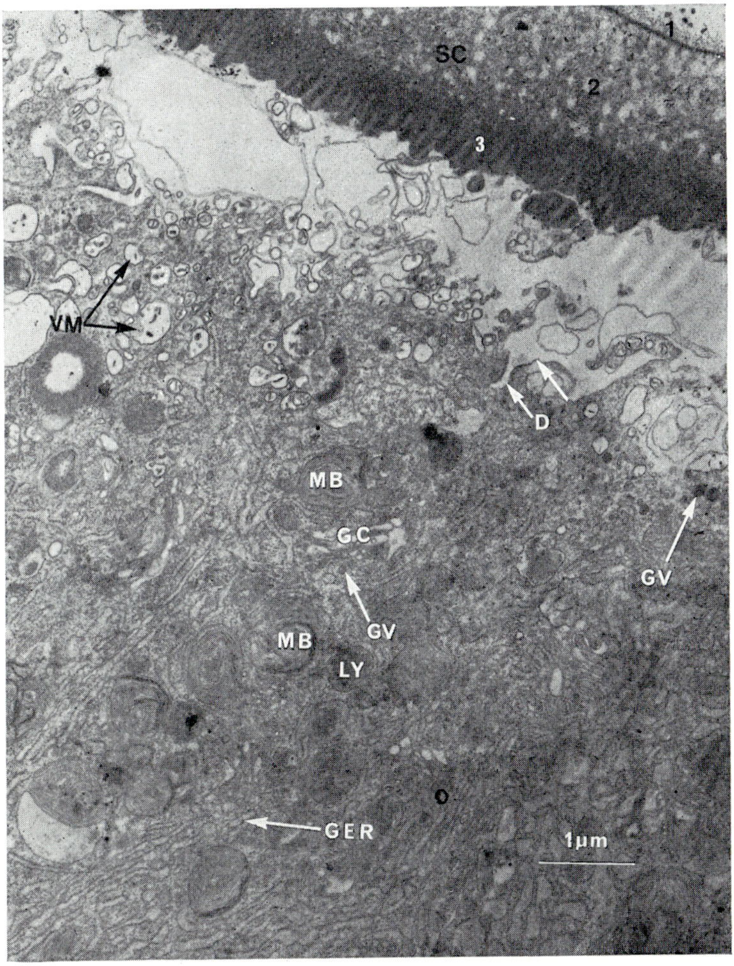

Fig. 12. Electron micrograph showing three-layered cuticle (SC) and cell apex after 12 days regeneration. Note the presence of vesicles containing microcrystals (VM), vesicles (GV) which appear to arise in Golgi complexes (GC), and large numbers of myelin bodies (MB) and lysosomes (LY). D, desmosome; GER, granular endoplasmic reticulum.

is observed at this stage. The underlying cells, while possessing all the features already mentioned, exhibit several differences. At the cell apex large numbers of membrane-bound vesicles are usually present which contain a few needle-like microcrystals in their lumina (Fig. 12). Also evident are smaller vesicles with electron-dense contents (Fig. 12). Larger numbers of Golgi complexes are evident while the granular endoplasmic reticulum is better developed. Large numbers of lysosomes, lipid bodies, autophagic vacuoles, myelin bodies, and intracellular vacuoles are still present throughout the cytoplasm (Fig. 12). After 12 days, an almost normal operculum has developed. At its rim a group of operculum rim cells are evident which lack intracellular vacuoles and, in addition, are characterized by a microvillous border of short microvilli and a well-developed granular endoplasmic reticulum system. They

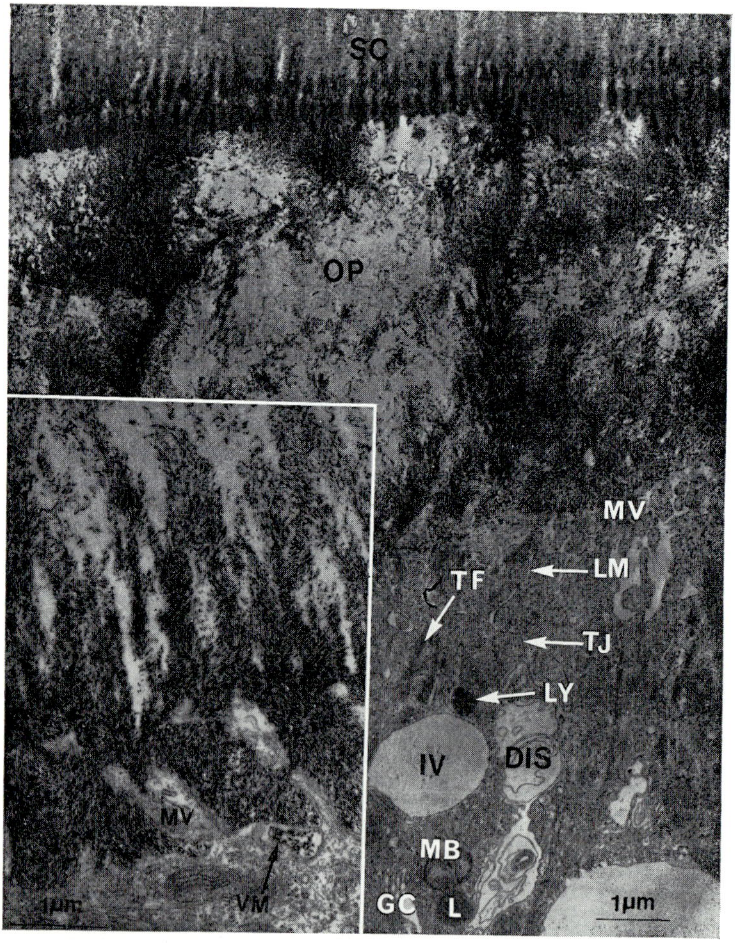

Fig. 13. Electron micrograph showing the opercular plate (OP) after 15 days regeneration comprising loosely arranged microcrystals. Inset: Electron micrograph showing the needle-like microcrystals orientated perpendicular to cell surface. VM, microcrystal containing vesicle. DIS, distended intercellular space; GC, Golgi complex; IV, intercellular vacuole; L, lipid; LM, lateral membrane; LY, lysosome; MB, myelin body; MV, microvilli; TF, tonofilaments; TJ, tight junction; SC, solid (tanned) cuticle.

differ from normal opercular rim cells in possessing myelin bodies and autophagic vacuoles. The remaining prospective opercular plate epithelial cells, however, retain the features of those described for 9 days' regeneration. In the apical region larger numbers of vesicles, possessing needle-like microcrystals, are evident. These microcrystals are released from the cells and become aligned on the inner surface of the overlying cuticle. Needle-like microcrystals are also evident in the lumina of lysosomal-like bodies present in the cells (Fig. 10). The calcareous plate is not evident at this stage; instead, a diffuse fibrous matrix separates the cuticle from the underlying cells.

Specimens examined after 15 days show that the normal opercular profile is evident. Opercular rim cells are prominent and only differ from those in the normal operculum in possessing several myelin bodies and a few autophagic vacuoles. The opercular plate epithelial cells remain columnar in the vicinity of the rim, becoming reduced in height toward the centre of the epithelium. These cells still possess large numbers of myelin bodies, lysosomes, lipid bodies, autophagic vacuoles, and distensions along the intercellular space. Below the three-layered cuticle, which is of uniform thickness, a fibrous matrix containing needle-like microcrystals is first evident in the vicinity of the opercular rims. The microcrystals constitute the opercular plate (Fig. 13). At the surface of the cells the microcrystals are orientated perpendicular to the

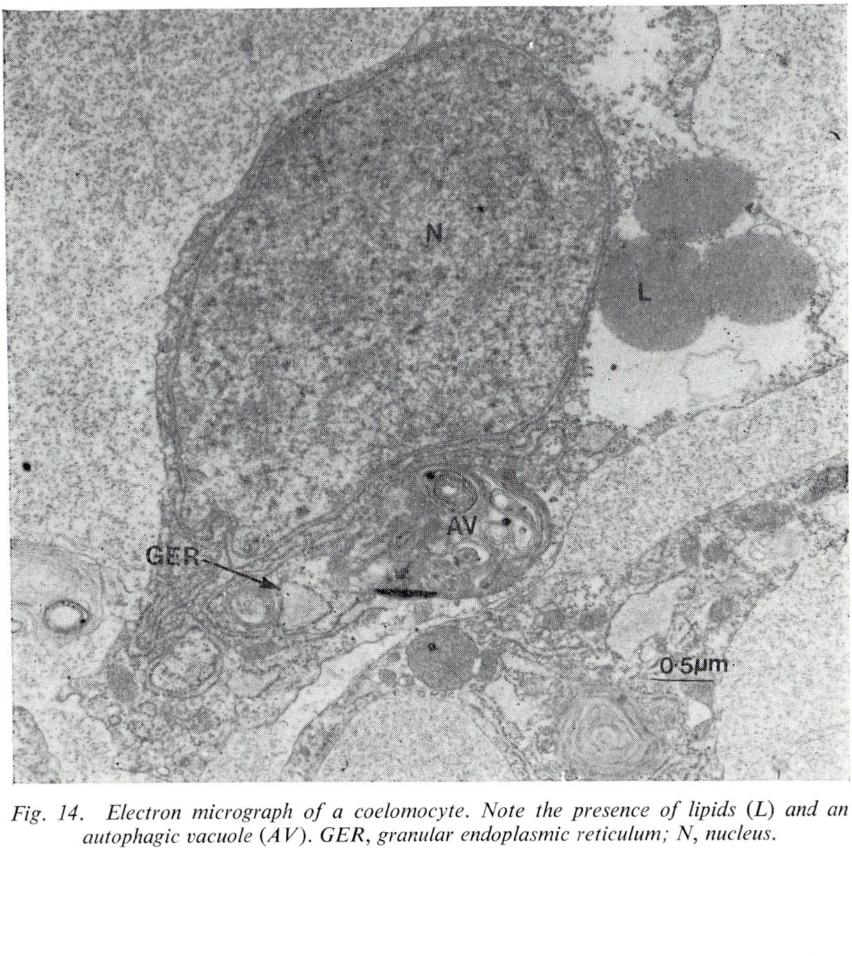

Fig. 14. Electron micrograph of a coelomocyte. Note the presence of lipids (L) and an autophagic vacuole (AV). GER, granular endoplasmic reticulum; N, nucleus.

apical membrane and enclose the microvilli (Fig. 13). In some instances it appears as if some microvilli are nipped off in to the opercular plate (Fig. 13). The opercular plate apparently consists of somewhat loosely arranged parallel vertical columns of compact microcrystals alternating with electron-light spaces of some outline (Fig. 13). Larger, spherical to irregularly outlined electron-light areas are also evident in the opercular plate (Fig. 13). From the underlying cells some myelin bodies and lipid bodies are released into the opercular plate, the latter bodies aggregating among the microcrystals.

Coelomocytes

In the connective tissue below the opercular plate epithelium cells, large numbers of coelomocytes exhibit a variety of shapes, *i.e.* spindle, oval, or spherical. The cytoplasm is moderately dense and contains an oval, often lobed nucleus situated either centrally or eccentrically in which a nucleolus is occasionally evident. Prominent in these cells are elongate granular endoplasmic profiles, with normally distended cisternae also prominent in these cells (Fig. 14). Clusters of ribosomes are evident throughout the cytoplasm and at regular intervals along the nuclear membrane. Also, a small number of mitochondria, with few cristae, are normally evident in the cytoplasm. In most cells one or two Golgi complexes, with accompanying vesicles, are closely associated with the nucleus. In some cells a few large, membrane-bound bodies containing a homogeneously electron-dense material are present, as, also, are one or two multivesicular body-like inclusions. In some cells several lipid, large autophagic, and myelin bodies, which occupy most of the cell volume, are also observed (Fig. 14). Throughout the period of regeneration, large numbers of coelomocytes are evident beneath the epithelial cells of the developing operculum and are frequently closely associated with their basal membranes. There is some evidence to suggest that they enter the epithelial cells.

DISCUSSION

The results of this investigation show that the operculum of *S. spirorbis* comprises three distinct cell types which are responsible for the deposition of a fibrillar (collagenous) cuticle, a solid (tanned) cuticle, and a calcareous plate. An essentially similar situation has been reported in the opercula of other serpulids (Bubel, 1973a). The cuticle of both the operculum and peduncle appears to be delicate and composed of an amorphous matrix within which fibrils lie in planes parallel to the epithelial surface. The organization of this cuticle strongly resembles that described for other annelids (Michel, 1969; Storch and Welsch, 1970), where the fibres of the cuticles have been identified as collagen and the matrix as mainly polysaccharide. The collagen fibres of annelids differ from those of vertebrate collagen in containing a high proportion of bound carbohydrate and lacking any axial periodicity (Rudall, 1968). In annelids the epithelium underlies the cuticle and separates it from the rest of the organism. Accordingly, it has been suggested that epithelial cells are responsible for the synthesis and orientation of the cuticular fibres (Krall, 1968; Goodman and Parrish, 1971; Bubel, 1973a). In the epithelial cells of the annelids already studied and also that of *S. spirorbis*, granular endoplasmic reticulum, many mitochondria and Golgi complexes are near the cell apex which may indicate that epithelial cells synthesize collagenous precursor of the cuticle fibres. Krall (1968) and Bubel (1973a) have presented fine structural but purely circumstantial evidence suggesting that vesicles arising in the Golgi complexes of the epidermal cells and opercular wall cells in the oligochaete, *Dero obtusa*, and serpulids, *S. granulatus* and *S. pusilloides*, may be the source of the cuticular fibres. The situation in the opercular wall cells of *S. spirorbis*, however, appears to be different in that small vesicles, with contents of varying density, are evident below the cell membrane as are a group of larger bodies. The contents of the former frequently appear to be released. In both cases no definite connection between them and the Golgi complexes has been established. Instead, a close spatial relationship exists between Golgi complexes, multivesicular bodies, and heterogeneously dense bodies (lysosomes). A similar situation has been reported in endocrine cells of vertebrates where it has been suggested that they function in the control of the secretory process (Smith and Farquhar, 1966). In *S. spirorbis* cells there is, however, a close relationship and similarity between connective tissue fibres beneath the cells and material found in both the distended intercellular lumina and the intracellular vacuoles. This may indicate that they play a role in cuticle deposition. If this is the case, the proposed collagenous cuticle of *S. spirorbis* would be mesenchymatous in origin (Rudall, 1968). It is obvious that further work and other approaches to the problem are needed to determine the precise origin of the cuticle.

At the operculum rim, the nature of the cuticle changes from a delicate structure to a solid one that has a natural yellowish colour which is refractory to stains. It has been suggested that this cuticle is composed of tanned protein (Thorp, 1961; Thorp and Segrove, 1975). The opercular rim cells, which are characterized by several prominent Golgi complexes with associated vesicles, appear to be responsible for the rhythmic deposition of this cuticle and the presence of regularly arranged concentric bands (Fig. 7) supports this. The factors controlling the rhythmic deposition of the cuticle have still to be elicited. In an earlier investigation of a few serpulid opercula, it was suggested that Golgi vesicles may contain phenol oxidase enzymes which are incorporated into the three-layered solid cuticle to effect its tanning by oxidizing phenolic amino acids to quinone (Bubel, 1973a). This, however, still has to be conclusively proven. It is pertinent to mention that the situation at the opercular rim is analogous to that at the mantle edges of molluscs and barnacles, where specialized cells secrete tanned protein structures, namely, the periostracum and epicuticle, which provide a substratum on which the deposition of the calcareous layers of the shell may occur (Bubel, 1973b, 1975). Results obtained from this investigation show that, initially, at the opercular rim, the opercular plate cells secrete an organic matrix which is organized into a series of orderly polygonal profiles. Crystal formation appears to occur within these organic matrix profiles. A fundamentally similar situation appears to occur in shell deposition in both molluscs and barnacles (Bevelander and Nakahara, 1969; Bubel, 1975). In the absence of experimental evidence, the pathway of calcium through the cells is difficult to assess, especially since it is absent in the form of deposits. It seems likely, therefore, that it moves through the cells in a soluble, ionised form. Recent evidence in crustacea and molluscan cells suggests, however, that the main route of calcium transport occurs through the intercellular spaces (Graf, 1971; Neff, 1972). A similar situation may exist in opercular plate cells but this does not preclude the possibility that the cells may play a part in the generation and control of calcium transport.

After the excision of the operculum, the exposed peduncular surface becomes completely sealed by an epithelial layer within 5 days. In general, repair of wounds is characterized by three processes, namely, cell migration, proliferation, and differentiation. During this investigation, the former two processes were not investigated. Results of the investigation show that a degree of differentiation occurs in the cells as they sequentially deposit a fibrillar (collagenous) cuticle, a solid (tanned) cuticle, and a calcareous opercular plate. A characteristic feature of the cells is the presence of large numbers of lipid bodies, lysosomes, autophagic vacuoles, and myelin bodies. Similar observations have been reported during the regeneration of epithelial cells of other annelids after the amputation of some anterior segments (Coulson and Thouveny, 1974; Burke, 1974). Lipid deposition, myelin figures, and autophagic vacuoles have been recorded in cells under the metabolic stress of differentiation, or as a response to hormonal stimulation or even to injury (Boilly, 1968; Schilling, 1968; Ericsson, 1969; Michaels *et al.*, 1971; Bubel, 1973c). The changes observed in *S. spirorbis* cells may be, initially, a response to hormonal stimulation, since it is well known that, in annelids, a supraoesophageal ganglion produces a hormone enabling regeneration to occur (Clarke, 1965). It would appear that, during the differentiation of the *S. spirorbis* cells, increased protein synthesis occurs which is reflected by a well-developed granular endoplasmic reticulum and several Golgi complexes. The concomitant increase in autophagic vacuoles, lysosomes, and lipid bodies could, therefore, provide chemical constituents which may be reutilized in the cells' economy for both increased secretory activity and the necessary energy.

After 15 days, the newly deposited calcareous plate differs in structure from the normal in that it comprises loosely arranged needle-like microcrystals. The microcrystals appear to be deposited on a fibrous matrix which is initially secreted by the opercular plate cells. A similar situation has also been reported in the regeneration of the shell of *Helix*, where small crystals of calcium carbonate are deposited in an organic matrix which then joined to form larger crystals, eventually giving rise to a calcified layer (Saleuddin and Chan, 1969). The needle-like microcrystals may similarly give rise to the typical polygonal-shaped crystals which normally characterize the operculum plate. Preceding the deposition of the new opercular plate, the underlying cells possess several vesicles which contain needle-like microcrystals in their lumina. Although the origin and formation of these vesicles was not determined, their presence may indicate that calcium in the differentiated cells is transported in a different form from that under normal circumstances. Similarly, the proposed lysosomes in these cells also possess needle-like microcrystals. It is interesting to mention in this connection that, in his study of the calcium-secreting glands in the serpulid *Pomatocerus caerulus*, Neff (1971) suggested that a lysosome-like vacuolar system could be involved in calcium secretion.

Throughout the regeneration of the operculum, large numbers of coelomocytes are evident in the connective tissue, but their precise role in the differentiation of the operculum is not at all obvious. An investigation of the stages of wound healing of the peduncle may provide such information. The results of several investigations of wound healing into other polychaetes have also shown that coelomocytes accumulate near wounds (Clarke, 1965). It is believed that coelomocytes are primarily concerned in wound healing and regeneration by providing considerable quantities of nutritive material to the tissue mainly in the form of lipid granules (Liebmann, 1942). This, however, has yet to be determined during *S. spirorbis* peduncle wound healing and subsequent opercular regeneration.

Acknowledgements

The author wishes to thank Dr C. H. Thorp for helpful criticism of the manuscript. Appreciation is also extended to Prof. H. Mutvei of the Swedish Museum of Natural History, Stockholm, for providing facilities and much help with the SEM work. This work was supported by a Ministry of Defence (Navy) Contract No. AT/2198/010 CDL and published with their permission.

REFERENCES

Bergen, P. (1953). On the anatomy and reproduction biology in *Spirorbis* Daudin. *Nytt Mag. Zool.*, **1**, 1–26.

Bevelander, G., and Nakahara, H. (1969). An electron microscope study of the formation of the nacreous layer of the shell of certain bivalve molluscs. *Calc. Tissue Res.*, **3**, 84–92.

Boilly, B. (1968). Étude ultrastructurale de l'evolution des tissus impliqués dans la régénération céphalique et caudale de *Syllis amica* Q. (Annélide Polychète). *J. Microscopie*, **7**, 865–876.

Bubel, A. (1973a). An electron microscope investigation into the cuticle and associated tissues of the operculum of some marine serpulids. *Mar. Biol.*, **23**, 147–164.

Bubel, A. (1973b). An electron microscope study of periostracum formation in some marine bivalves. II. The cells lining the periostracal groove. *Mar. Biol.*, **20**, 222–234.

Bubel, A. (1973c). An electron microscope study of periostracum repair in *Mytilus edulis*. *Mar. Biol.*, **20**, 235–244.

Bubel, A. (1975). An ultrastructural study of the mantle of the barnacle, *Elminius modestus* Darwin in relation to shell formation. *J. Exp. Mar. Biol. Ecol.*, (in press).

Burke, J. M. (1974). Wound healing in *Eisenia foetida* (Oligochaeta) II. A fine structural study of the role of the epidermis. *Cell Tiss. Res.*, **154**, 61–82.

Clarke, M. E. (1965). Cellular aspects of regeneration in the polychaete *Nephtys*, In Proc. Regeneration in Animals, pp. 240–249 (V. Kiortsis and H. A. L. Trampusch, eds.), North Holland Publ. Co., Amsterdam.

Coulson, J., and Thouveny, Y. (1974). Étude ultrastructurale de la dédifférenciation cellulaire dans la régénération enterieurure de l'annélide *Owenia fusiformis* (Delle Chiaje). I. L'epiderme (phénomène d'autophagie). *Ann. Embrylol. Morphol.*, **7**, 61–79.

Ericsson, J. L. E. (1969). Mechanisms of cellular autophagy. In Lysossomes in Biology and Pathology, Vol. 2, pp. 345–394 (T. J. Dingle and H. B. Fells, eds.), North-Holland Publ. Co.

Goodman, D., and Parrish, W. B. (1971). Ultrastructure of the epidermis in the ice worm, *Mesenchytraeus solifugus*. *J. Morphol.*, **135**, 71–86.

Graf, F. (1971). Dynamique du calcium dans l'epithelium des caecums posterieurs d'*Orchestia cavimana* Heller (Crustace, Amphipode). Role de l'espace intercellulaire. *C.r. hebd. Seanc. Acad. Sci. Paris*, **273**, 1828–1831.

Krall, F. J. (1968). The cuticle and epidermal cells of *Dero obtusa* (Family Naididae). *J. Ultrastruct. Res.*, **25**, 84–93.

Liebmann, E. (1942). The coelomocytes of Lumbricidae. *J. Morphol.*, **71**, 221–250.

Michaels, J., Albright, J., and Patt, D. (1971). Fine structural observations on cell death in the epidermis of the external gills of the larval frog, *Rana pipiens. Am. J. Anat.*, **132**, 301–318.

Michel, C. (1969). Ultrastructure et histochimie de la cuticle pharyngrenne chez *Eulalia viridis* Muller (annelide polychete erante, Phyllodocidae). *Z. Zellforsch mikrosk. Anat.*, **98**, 54–73.

Neff, J. M. (1971). Ultrastructural studies of the secretion of calcium carbonate by the serpulid polychaete worm, *Pomotoceros caeruleus. Z. Zellforsch.*, **120**, 160–186.

Neff, J. M. (1972). Ultrastructure of the outer epithelium of the mantle of the clam, *Mercenaria mercenaria* in relation to calcification of the shell. *Tissue Cell*, **4**, 591–600.

Potswald, H. E. (1968). The biology of fertilization and brood protection in *Spirorbis* (*Laeospira*) *morchi. Biol. Bull. Mar. Biol. Lab.*, Woods Hole, **135**, 208–222.

Rudall, K. M. (1968). Comparative biology and biochemistry of collagen. In *Treatise on Collagen*, Vol. 2, pp. 83–137 (B. S. Gould, ed.), Biology of Collagen Part A, Academic Press, New York.

Saleuddin, A. S. M., and Chan, W. (1969). Shell regeneration in *Helix*: shell matrix composition and crystal formation. *Can. J. Zool*, **47**, 1107–1111.

Schilling, J. (1968). Wound healing. *Phys. Rev.*, **48**, 274–423.

Schochet, J. (1973). Opercular regulation in the polychaete. *Hydroides dianthus* (Verrill, 1873). I. Opercular ontogeny, distribution and flux. *Biol. Bull.*, **144**, 400–420.

Smith, R. E., and Farquhar, M. G. (1966). Lysosome function in the regulation of the secretory process in cells of the anterior pituitary gland. *J. Cell. Biol.*, **31**, 319–347.

Storch, V., and Welsch, U. (1970). Uber die Feinstruktur der Polychaeten-Epidermis (Annelida). *Z. morph. Tiere*, **66**, 310–322.

Thorp, C. H. (1961). The Reproductive cycles of Spirorbinae (Polychaeta: Serepulida) and their synchrony with structural changes in the operculum. Ph.D. Thesis. University of Sheffield.

Thorp, C. H. (1975). The structure of the operculum in *Pileolaria granulata* (L.) (Polychaeta: Serpulidae) and related species. *J. Exp. Mar. Biol. Ecol.*, (in press).

Thorp, C. H., and Segrove, F. (1975). The opercular moult in *Spirorbis spirorbis* (L) and *S. pusilloides* Bush (Polychaeta: Serpulidae). *J. Exp. Mar. Biol. Ecol.* (in press).

The evidence presented here suggests a new approach to controlling marine fouling. The use of nontoxic repellents has the potential to replace the toxic compounds which are commonly used, without their environmental hazards.

Acknowledgement

This research was supported in part by Contract No. N00014-67-A-0298-0237 between Harvard University and the U.S. Office of Naval Research.

REFERENCES

Adler, J. (1973). A method for measuring chemotaxis and use of method to determine optimal conditions for chemotaxis by *Escherichia coli*. *J. Gen. Microbiol.*, **74**, 77–91.

Dyckman, E. J., Montemarano, J. A., and Fischer, E. C. (1973). Environmentally compatible antifouling materials: organometallic polymers. *Naval Eng. J.*, Dec., 33–37.

Liberatore, G. L., Dyckman, E. J., Montemarano, J. A., and Cohn, M. L. (1972). Antislime coating, Part II: Preconditioning of slime for barnacle attachment. *Nav. Ship and cen.*, Aug., 28–233.

Marshall, K. C. (1971). *Sorptive Interaction between Soil Particles and Micro-organisms in Soil Biochemistry,* Vol. 2, Chapt. 14. (A. D. McLaren and J. J. C. Knjins, eds.), Marcel Dekker, New York.

Marshall, K. C., Stout, R., and Mitchell, R. (1971). Mechanism of the initial events in the sorption of marine bacteria to surfaces. *J. Gen. Microbiol.*, **68**, 337–368.

Sieberth, J. M. W., and Conover, S. T. (1965). Sargassum tannin an antibiotic which retards fouling. *Nature*, **208**, 52–53.

Smith, J. L., and Doetsch, R. N. (1969). Studies on negative chemotaxis and the survival value of motility in *Pseudomonas fluorescens*. *J. Gen. Microbiol.*, **55**, 379–391.

Tso, W., and Adler, J. (1974). Negative chemotaxis in *Escherichia coli*. *J. Bacteriol.*, **118**, 560–576.

Wood, E. J. F. (1967). *Microbiology of Oceans and Estuaries*. Elsevier, Amsterdam.

Young, L. Y., and Mitchell, R. (1973a). The role of microorganisms in marine fouling. *Int. Biodeterior. Bull.*, **9**, 105–109.

Young, L. Y., and Mitchell, R. (1973b). Negative chemotaxis of marine bacteria to toxic chemicals. *Appl. Microbiol.*, **25**, 972–975.

MICROBIAL DECOMPOSITION OF ANIMAL MUSCLE BURIED IN DESERT SOIL

MARK CAHENZLI and E. E. STAFFELDT

New Mexico State University, Las Cruces, New Mexico 88003, U.S.A.

Summary

Weighed portions of animal muscle were placed in nylon mesh bags and buried in soils contained in plastic, asbestos tubes. In the laboratory, these tubes contained soils adjusted to field capacity and were then subjected to constant temperatures of 5, 20 and 40°C. Evolving CO_2 and volatilizing NH_3 from decomposing muscle were trapped in 0·6 M KOH and 0·5 N H_2SO_4, respectively. This was accomplished by enclosing the open end of the tube with a layer of plastic wrap followed by a layer of aluminium foil after the vials containing KOH and H_2SO_4 were placed on the soil surface. The vials containing alkali and acid were exposed for a 24-hr period and were exchanged for other solution samples to enable determinations to be made on a daily basis for the 3-week decomposition period. Similar plastic, asbestos tubes were driven into the soil on the desert and determinations were made on an alternate day basis during the summer and fall of 1974. When moisture and temperature were not limiting, 70–90% of the muscle was decomposed within 21 days. Nitrogen from the muscle was detected as volatilized NH_3, salts in the soil and protein assimilated in microbial cells. Substantial amounts of carbon evolved as CO_2.

For centuries, dead animal tissue has been known to decompose and disappear rapidly from the environment. This decaying proteinaceous material is believed to release substantial amounts of gases and minerals. Investigations have been conducted employing substitute protein materials, placing them with soil in a flask under laboratory conditions and ultimately extrapolating these results to the field from which the soil was removed (O'Brien, 1972; Walquist, 1971; Woods, 1970). The natural protein material, while under varying field conditions, was tested during this investigation. These efforts were concentrated on determining how animal muscle was decomposed by evaluating CO_2 evolution, NH_3 volatilization, and weight loss during the natural process of microbial degradation.

MATERIALS AND METHODS

In the laboratory, weighed portions of rabbit muscle (*Lepus californicus*) were buried 2–3 cm in soil adjusted to field capacity. The soil was contained within sealed plastic, asbestos tubes and maintained at 5, 20, and 40°C. The CO_2 evolved and NH_3 volatilized were trapped in 0·6 M potassium hydroxide (KOH) and 0·5 N sulphuric acid (H_2SO_4), respectively. These solutions were contained within plastic, screw-cap vials and were changed on a daily basis. CO_2 evolution determinations generally followed methods outlined by Coleman (1971). The trapped CO_2 was precipitated as carbonate and titrated with 0·6 M hydrochloric acid (HCl). The amount of HCl was then converted to mg CO_2/24 hr. NH_3 volatilization determinations generally followed methods outlined by Vogt *et al.* (1975). The exposed H_2SO_4 was diluted and Nesslers reagent added. This was then read on the spectrophotometer and the reading converted to μg NH_3/24 hr. The open tops of the tubes were sealed with a square of plastic wrap followed by a square of aluminium foil, each retained in place with a rubber band. Muscle tissues were removed after a specified period of time, oven dried, weighed, ashed, and the ash weighed to determine weight loss.

Similar plastic, asbestos tubes were driven into soil in an area representative in moisture, temperature, and soil conditions of that of the Jornada Desert in New Mexico. Weighed portions of rabbit muscle were buried 2–3 cm in these tubes on 7 June, 22 June, 18 July, 16 August, 19 September, and 1 November 1974. CO_2 evolution and NH_3 volatilization determinations were made on an alternate day basis using the same techniques employed in the laboratory. Muscle was removed from the field on 27 July (June burials), 6 August (July burial), 10 September (August burial), 9 October (September burial), and 22 November (November burial) 1974. These were oven dried, weighed, ashed, and the ash weighed to determine weight loss.

Soil moisture and temperature readings, inside and outside the tubes, were obtained from gypsum blocks buried at the 4–5 cm depth. These were the same type as those employed by the Jornada Validation Site Personnel (Whitford *et al.*, 1972). Daily rainfall and maximum and minimum air temperature data were obtained from the Weather Service, New Mexico State University (Malm, 1974).

Tests for nitrate, nitrite, and ammonium were conducted on soil from the 0–3 and 4–10 cm depths employing techniques outlined by Nelson *et al.* (1954), Bray (1945), Bremner (1965), and Bear (1955). Later, when it became apparent that some of the nitrogen introduced from the decomposing muscle may have leached, several tubes were sealed on the bottom to trap any leachates from the soil above. Tests for nitrate, nitrite, and ammonium were conducted on soil from the 0–3 and 17–20 cm depths.

Weighed segments of rabbit were placed in flasks with 1 g of soil and 100 ml of distilled water. Following 7 days of decomposition, the microbial cells were harvested and Kjeldahl's estimations were conducted to determine the protein content of the cells. Similar tests were conducted using small segments of plant stems as a substrate.

RESULTS

The laboratory portion of this experiment was designed to give baseline data for what might occur in the field. Results from these studies showed that the greatest volatilization of NH_3 and evolution of CO_2 occurred from samples at 40°C with lesser amounts at the other temperatures. Weight loss due to decomposition at 40°C amounted to 90·23% in 15 days as compared to 20·53% and 48·83% for the samples at 5 and 20°C, respectively. These results were used as a guide for the field experiment.

Soil temperatures were generally taken between 7.30 and 9.30 a.m. up to 16 August, after which they were taken between midday and 1.00 p.m. The temperature of the soil inside the tubes was essentially the same as that recorded for the soil on the outside. The soil moisture was also essentially the same inside the tubes as that of the soil surrounding the tubes except for 27–28 June and 15–27 July. These differences resulted from the tubes being sealed for an experimental run in which moisture condensed on the internal surfaces on 27 and 28 June and a prolonged drying period which extended from 15 to 27 July 1974. These differences proved to have little effect on the rate of decomposition of the rabbit muscle. The level of moisture tended to remain between −0·1 and −1·0 bars, both inside and outside the tubes, except for the periods mentioned above as a result of the large amount of precipitation experienced during 1974. The total amount (mm) of rain for the months of June to November were 1·52 (June), 82·20 (July), 86·11 (August), 79·25 (September), 49·28 (October), and 7·87 (November).

Maximum and minimum air temperatures both showed increases to about the 20th of June and were followed by a steady decline. The high temperature in September appeared to have a very positive effect on the decomposition of the rabbit muscle. The prolonged period of dryness in June created high suction pressures in the range of −128 to −130 but the rainfall on 4 July reduced the suction pressures.

Carbon dioxide evolution was determined on an alternate day basis and therefore a piecewise linear relationship was employed to determine CO_2 evolution on the intervening days. Fewer total mg CO_2 were evolved from animal muscle buried on 7 June than the other burials (Table I). Similarly

Table I
CO_2 EVOLUTION FROM MUSCLE BURIED IN JUNE, JULY, AUGUST, SEPTEMBER AND NOVEMBER, 1974

Burial	*Date*	*mg CO_2*	*Days buried*	*mg CO_2 evolved/day*
1st	7 June	1 037·65[a]	50	20·75
2nd	22 June	1 236·18[b]	35	35·32
3rd	18 July	1 222·51[c]	19	64·34
4th	16 Aug.	1 328·65[c]	25	53·15
5th	19 Sept.	1 749·22[c]	21	83·30
6th	1 Nov.	1 123·83[c]	21	53·52

[a] Mean of 18 observations.
[b] Mean of 15 observations.
[c] Mean of 12 observations.

the lowest mg CO_2 evolved per day was recorded for this burial. This was due to the lack of soil moisture. Muscle buried during July, August, September, and November yielded from 53·15 to 83·20 mg CO_2/day and was generally not subjected to reduced soil moisture. During the 7 June burial, 125 mg CO_2 or 12% of the total evolved during the first 4 days when the muscle was drying. The 22 June burial was also influenced by the initial lack of soil moisture (Fig. 1). Only the August burial was subjected to a period of reduced soil moisture, but the number of days it was exposed to greater than −30 bars was minimal.

Lower soil temperatures allowed for greater CO_2 evolution in the initial portion of the 7 June burial than the 22 June burial. High soil temperatures between 22 and 27 July reduced the microbial activity and changed the slope of the line. The combination of soil moisture and temperature could not be utilized to explain high CO_2 evolution recorded during September. Lower

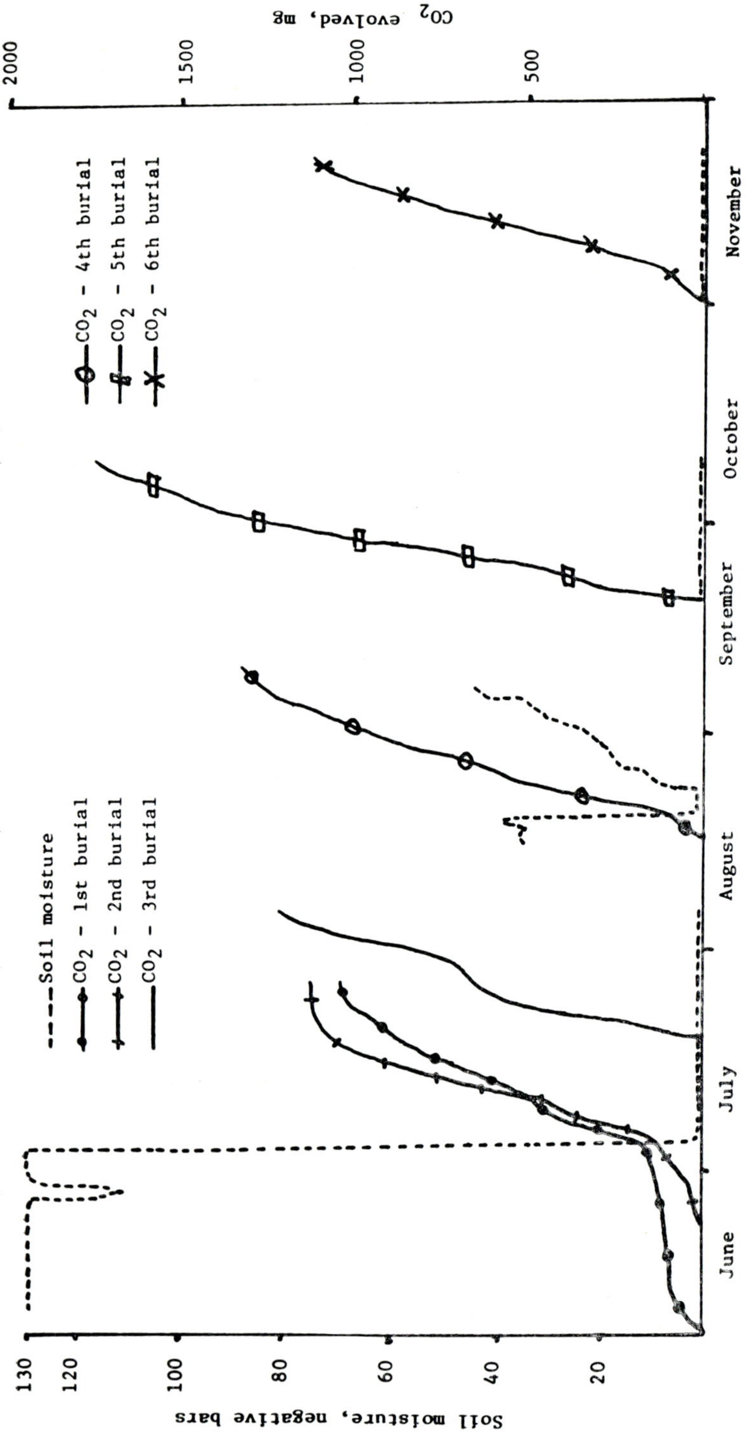

Fig. 1. *Soil moisture and CO_2 evolved from decomposing animal muscle, June–November, 1974.*

soil temperatures reduced CO_2 evolution in the November burial.

Weight loss data was collected and evaluated in conjunction with CO_2 evolution as a measurement of decomposition. Based on weight loss, 63·70–90·63% of the animal muscle was degraded within 19–50 days during the summer and fall of 1974 (Table II). When this data was compared to the CO_2 evolution data (Table III), differences of 3·82–12·05 were found and were

Table II
PERCENT DECOMPOSITION OF MUSCLE AS MEASURED BY WEIGHT LOSS DETERMINATIONS

Burial	*Date*	*% muscle decomposed*	*Days decomposed*	*% decomposition per day*
1st	7 June	78·48[a]	50	1·77
2nd	22 June	84·97[b]	35	2·43
3rd	18 July	74·53[c]	19	3·92
4th	16 Aug.	63·70[c]	25	2·55
5th	19 Sept.	90·63[c]	21	4·32
6th	1 Nov.	72·65[c]	21	3·46

[a] Mean of 18 observations.
[b] Mean of 15 observations.
[c] Mean of 12 observations.

Table III
COMPARISON OF PERCENT MUSCLE DECOMPOSITION USING CO_2 EVOLUTION AND WEIGHT LOSS AS PARAMETERS

Burial	*Date*	*% as CO_2*	*% as weight loss*	*Difference*
1st	7 June	82·30	78·48	3·82
2nd	22 June	95·10	84·97	10·13
3rd	18 July	85·80	74·53	11·27
4th	16 Aug.	105·00	63·70	41·30
5th	19 Sept.	140·00	90·63	49·77
6th	1 Nov.	85·50	72·65	12·85

considered within the realm of the experimentation. The differences of 41·30 and 49·77 in August and September required additional observation. Maggots were found in these burials and these animals appeared to be responsible for the increased CO_2 evolution. Very few small maggots were found in the November burial and did not induce increased CO_2 evolution. Maggots were not observed in the June and July burials.

The largest amount of NH_3 was evolved in the shortest time period from muscle buried during July (Table IV). Volatilization based on mg NH_3/day was lowest in the 7 June burial which was due to the prolonged lack of soil moisture. Tissues buried during July, August, September, and November yielded from 19·06 to 42·07 μg NH_3/day and were generally not subjected to reduced soil moisture. During the first 27 days of the 7 June burial, 50 mg NH_3 of the total volatilized and was trapped. Of the 50 mg NH_3, 45 mg or 90% was volatilized during the first 4 days while the muscle was drying. Volatilization, during the August burial, was not influenced substantially by the minimal period of reduced soil moisture.

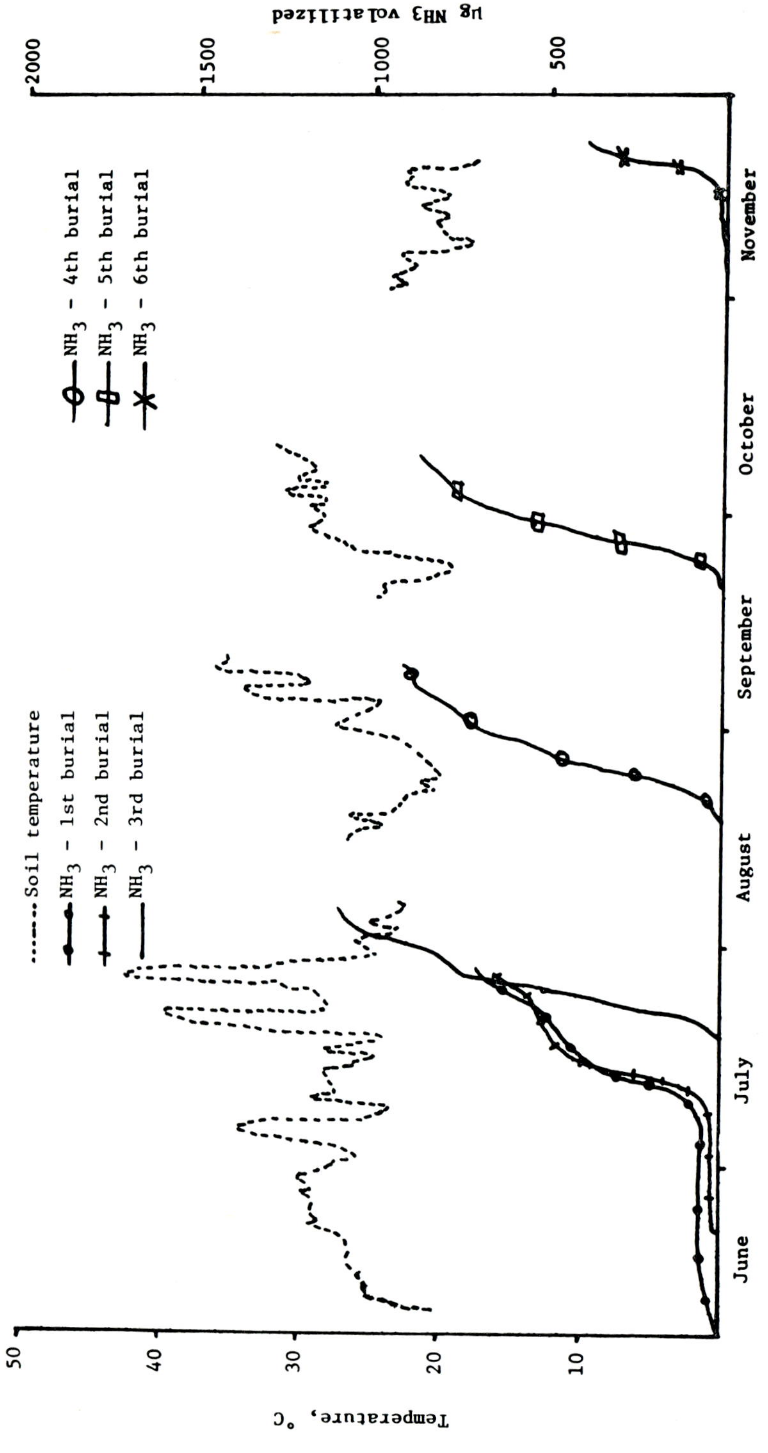

Fig. 2. Soil temperature and NH_3 volatilized from decomposing animal muscle, June–November, 1974.

Table IV
NH_3 VOLATILIZATION FROM MUSCLE BURIED IN JUNE, JULY, AUGUST, SEPTEMBER AND NOVEMBER, 1974

Burial	*Date*	*μg NH_3*	*Days buried*	*μg NH_3 volatized/day*
1st	7 June	688·88[a]	50	13·78
2nd	22 June	641·16[b]	35	18·32
3rd	18 July	1 086·09[c]	19	57·16
4th	16 Aug.	907·69[c]	25	36·31
5th	19 Sept.	883·29[c]	21	42·07
6th	1 Nov.	400·27[c]	21	19·06

[a] Mean of 18 observations.
[b] Mean of 15 observations.
[c] Mean of 12 observations.

Greater NH_3 volatilization was associated with lower soil temperatures in conjunction with low soil moisture in the initial portion of the 7 June burial (Fig. 2). High soil temperatures between 22 and 27 July appeared to influence NH_3 volatilization as it did CO_2 evolution. Lower soil temperatures were associated with reduced NH_3 volatilization and low temperatures in November emphasized this effect. Maggots were observed in the September burial and did not appear to change the NH_3 volatilization as it had the carbon evolution. Lower soil temperatures in November delayed NH_3 volatilization for 12 days, and the quantity evolved was much reduced when compared to other burials.

The NH_3 volatilized during decomposition amounted to 0·62–1·64% of the total nitrogen in the muscle (Table V).

Table V
PERCENT OF TOTAL MUSCLE NITROGEN VOLATILIZED AS NH_3 DURING DECOMPOSITION

Burial	*Date*	*Total μg NH_3 evolved*	*Days buried*	*NH_3 as % of total*
1st	7 June	688·88[a]	50	1·17
2nd	22 June	641·66[b]	35	1·02
3rd	18 July	1 086·09[c]	19	1·64
4th	16 Aug.	907·69[c]	25	1·50
5th	19 Sept.	883·49[c]	21	1·44
6th	1 Nov.	400·27[c]	21	0·62

[a] Mean of 18 observations.
[b] Mean of 15 observations.
[c] Mean of 12 observations.

Soil tests for nitrate, nitrite, and ammonium accounted for 5·6% of the nitrogen lost from muscle during decomposition. Greater production of nitrate, nitrite, and ammonium was observed in the initial portion of the 7 June burial in association with lower soil temperatures. The nitrate and ammonium levels increased again in the fall when lower soil temperatures prevailed while the nitrite levels decreased.

Data revealed that as much as 1% of the total nitrogen from the muscle was leached to lower depths in the form of NO_3. NO_2 levels also varied at the different depths. This aspect was discovered late in the investigation and was carried out only on the November burial.

Accounting for 0·62–1·64% of the muscle nitrogen as NH_3 and 5·6% as nitrate, nitrite, and ammonium salts in the soil, laboratory studies were conducted to locate possible sources of the remainder. Kjeldahl determinations revealed that substantial differences occurred in the % protein contained in the muscle decomposing microbes (75–82%) as compared to those microbes decomposing plant stems (13%). Utilizing the laboratory values, the following determinations were made. Originally 51 mg of nitrogen was contained in the meat. After 7 days decomposition 21·68 mg nitrogen remained in the meat, 20·55 mg nitrogen were found in microbial cells, 0·82 mg nitrogen volatilized as NH_3, and 2·81 mg nitrogen was found as ammonium, nitrite, or nitrate. These quantities amounted to 89·9% of the total nitrogen available.

DISCUSSION

Soil moisture and soil temperature were found to be important variables during the decomposition of rabbit muscle. If soil moisture (−0·1 to −30 bars) and soil temperature (23·9–37·8°C) are suitable decomposition is rapid. If the soil temperature range of 18–41°C or the soil moisture above −30 bars is exceeded, as seen in June or November, the environmental conditions influence the capability of microorganisms to decompose animal muscle. CO_2 evolution exhibited the greatest influence due to lack of soil moisture while NH_3 volatilization increased and decreased as the soil temperature increased and decreased.

Summer and fall of 1974 proved to be very moist in this arid area of New Mexico. During previous years, at least two and occasionally three drying periods (with soil moisture recorded at −128 bars) occurred during the summer and fall (Staffeldt and Vogt, 1974). Since the moisture remained between −0·1 and −30 bars most of the season, the usual influence of wetting and drying was not observed. Under conditions of adequate moisture and temperature, the majority of the animal muscle was usually decomposed in 21 days.

Staffeldt and Vogt (1974) started using CO_2 evolution measurements in conjunction with weight loss determinations. Using the two parameters, they were able to detect losses due to plant leaching in the early stages of decomposition and 'negative weight loss' due to microbial accumulation on moist tissue during drying cycles later in the season. Later, using these two parameters, they were also able to give approximate values for losses due to termites. In the latter case, the low CO_2 values induced by the microorganisms did not correspond to the high weight loss determinations due to the termites removing the plant materials. During this work still another difference was observed. The increased CO_2 evolution measurements did not correspond with the lower weight loss determinations due to the presence and activities of the maggots. In all of the above instances cited, there would have occurred a misinterpretation of data if both measurements were not employed. Either CO_2 evolution or weight loss determinations can be overevaluated or underevaluated by using only one of the measuring techniques. Either one of the above measurements would give an indication of the decomposition process, but based on the data accumulation it should be emphasized that combinations of measurements should be employed whenever possible. Large quantities of NH_3, as much as 80% of the available nitrogen, had been volatilized during decomposition of isoelectric casein (O'Brien, 1972). However, throughout this experiment only 1·64% of the available nitrogen from the muscle was volatilized as NH_3. The smell of NH_3 was never detected which it should have been if the majority of the nitrogen had come off as a gas.

Tests for soil nitrate, nitrite, and ammonium accounted for an additional 5·6% of the nitrogen available from the muscle. Volatilization occurred throughout the test period and could possibly occur throughout the year. Nitrification appeared to be enhanced by lower soil temperatures in the fall and winter and may be adversely influenced by high soil temperatures.

Additional investigation will be necessary to substantiate the conversion of organic residue nitrogen into microbial cell nitrogen. Determinations of approximate quantities immobilized in this process and the length of time this nitrogen remains immobilized will have to be evaluated. In all probability the major nitrogen reservoir in numerous soils is microbial cell mass.

Acknowledgement

This research was supported in part by the New Mexico Energy Research and Development Program (BEF proposal-66).

REFERENCES

Bear, F. E. (1955). *Chemistry of the Soil.* Reinhold Publishing Company, New York, 373 pp.

Bray, R. H. (1945). Determinations of nitrates and nitrites. *Soil Sci.*, **60**, 219.

Bremner, J. M. (1965). Nitrogenous compounds. In *Soil Biochemistry*, pp. 19–61, Marcel Dekker, Inc., New York.

Coleman, D. (1971). Measurements of total soil respiration at IBP Grassland sites. Mimeo., IBP Grassland Biome, 4 pp.

Malm, Norman R. (1974). Personal correspondence. Weather Service, New Mexico State University, Las Cruces, New Mexico.

Nelson, J. L., Kurta, L. T., and Bray, R. H. (1954). Rapid determination of nitrates and nitrites. *Analytical Chem.*, **26**, 1081–1082.

O'Brien, Robert T. (1972). Proteolytic activity of soil microorganisms. US/IBP Desert Biome, Res. Memo 72-42, 8 pp.

Staffeldt, Eugene E., and Vogt, Kristiina Besmond (1974). Measurements of carbon and nitrogen changes in soil. US/IBP Desert Biome, Res. Memo 74-38, 15 pp.

Vogt, Kristiina Besmond, Cahenzli, Mark A., and Staffeldt, Eugene E. (1975). Ammonia volatilization determinations. Unpublished procedure. New Mexico State University, Las Cruces, New Mexico.

Walquist, Brent T. (1971). The effects of soil moisture and soil temperature on the decomposition of chitin and chitin containing complexes in soil. Unpublished doctoral dissertation. New Mexico State University, Las Cruces, New Mexico, 69 pp.

Whitford, W. G. (coordinator) and Collaborators (1972). Jornada Validation Site Report. US/IBP Desert Biome, Res. Memo 72-4.

Woods, Donald Eugene (1970). Nitrogen transformation in a desert grassland soil. Unpublished masters thesis. New Mexico State University, Las Cruces, New Mexico, 50 pp.

Continuous culture

An attempt was made to simulate the dynamic conditions in a housewater tank, using the laboratory-scale 2-stage continuous culture apparatus shown in Fig. 1. A medium reservoir (5-litre) feeds a primary culture vessel (500-ml) fitted with a weir overflow at the 250-ml level. This is connected to a secondary culture vessel fitted with a constant level device maintaining a working volume of 1 litre. By means of a micro-metering pump fitted with different diameters of tubing, fresh medium may also be added to the secondary vessel at a rate four times that of its addition to the primary vessel. Preservative solution can be metered to the secondary vessel at selected rates. Both culture vessels are magnetically stirred and maintained at 30°C by thermostatically controlled hot plates. The medium is not aerated, but a slight positive pressure of sterile air is maintained in the vessels. The medium used was synthetic cane juice which is similar in composition to housewater (Table III).

Table III

COMPOSITION OF SYNTHETIC CANE JUICE

Ingredient	*Concn (g/l)*
Sucrose	165
Glucose	14
KH_2PO_4	1·5
$MgSO_4$	0·5
$CaSO_4$	0·5
Soluble starch	1·0
Tryptone	5
Yeast extract	5
Distilled water to	1 litre

pH 5·5
Autoclave 10 min at 10 p.s.i.

The test organism was an unidentified, asporogenous yeast isolated from spoiled housewater. A pure culture was inoculated into sterile medium in the primary vessel and grown in batch culture for 24 hr; continuous operation then commenced, with overflow into the secondary vessel. Equilibrium in this vessel was attained after about 36 hr, at which stage addition of preservative to the secondary vessel commenced. A quaternary ammonium compound was used, at a final concentration of 25 ppm in the medium. Measurements were made of pH, refractive Brix, and sucrose content (by G.L.C.) in both the fresh medium and spent medium at equilibrium. Viable yeast counts were made by pour plate techniques, using a neutralizer for the preservative.

RESULTS AND DISCUSSION

1. Microbiological survey of sweetwater systems

Only 4 of the 10 selected sites, representing 2 locations fed from 2 sources, exhibited significantly high microbial populations; these were the rotoclones (sites 5 and 6) and housewater tanks (sites 7 and 8) (Table IV). In both these locations the predominant organisms were mesophilic bacteria, mesophilic

Table IV
GEOMETRIC MEAN COUNTS PER ml ON VARIOUS MEDIA OF SWEETWATERS TAKEN FROM 10 SAMPLE POINTS
(each result represents the mean of approximately 10 weekly samples)

Sample Point	*Medium*					
	NA 30°*C*	*NA* 55°*C*	*RCA*	*MEA*	*OSMO*	*ROGOSA*
1	$2{\cdot}9\times10^3$	6	1	35	2	3
2	$1{\cdot}2\times10^3$	1	1	75	2	1
3	$9{\cdot}6\times10^3$	716	5	151	1	72
4	$1{\cdot}3\times10^4$	$2{\cdot}2\times10^3$	13	120	1	3
5	$2{\cdot}5\times10^7$	23	2×10^4	$1{\cdot}1\times10^7$	$2{\cdot}5\times10^6$	$3{\cdot}3\times10^3$
6	$6{\cdot}4\times10^7$	7	$1{\cdot}8\times10^4$	$1{\cdot}9\times10^6$	$8{\cdot}8\times10^3$	$3{\cdot}9\times10^4$
7	$4{\cdot}2\times10^6$	$2{\cdot}8\times10^3$	$1{\cdot}1\times10^7$	$4{\cdot}9\times10^6$	$1{\cdot}1\times10^6$	$1{\cdot}7\times10^6$
8	$2{\cdot}2\times10^6$	795	$1{\cdot}5\times10^6$	$8{\cdot}1\times10^6$	$1{\cdot}2\times10^5$	$2{\cdot}2\times10^4$
9	$5{\cdot}3\times10^3$	87	49	188	9	5
10	$5{\cdot}6\times10^3$	$1{\cdot}3\times10^3$	18	110	5	3

Table V
MEANS AND STANDARD DEVIATIONS FOR pH, TEMPERATURE, BRIX, AND DEXTRAN CONTENT OF SWEETWATERS TAKEN FROM 10 SAMPLE POINTS
(each result represents the mean of approximately 10 weekly samples)

Sample point	*pH*	*Temp.* (°*C*)	°*Brix*	*Dextran* (% *w/v*)	
				Weeks 1–6	*Weeks* 7–12
1	7·0 ± 0·49	75 ± 6·5	0	0·065	0·013
2	6·8 ± 0·46	74 ± 5·5	0	0·022	0·005
3	6·9 ± 0·32	66 ± 8·0	0·3 ± 0·6	0·051	0·005
4	6·6 ± 0·47	66 ± 9·0	0·25 ± 0·3	0·014	0·008
5	5·3 ± 1·02	14 ± 4·0	25 ± 7·5	0·032	0·016
6	6·2 ± 1·30	14 ± 4·0	9·5 ± 6	0·041	0·018
7	4·8 ± 0·72	23 ± 1·0	6·5 ± 4	0·112	0·330
8	4·8 ± 0·73	23 ± 2·0	8 ± 6	0·079	0·340
9	6·1 ± 0·68	72 ± 5·0	2·5 ± 2·5	0·057	0·007
10	5·6 ± 0·65	74 ± 4·0	4 ± 4	0·032	0·016

yeasts, and osmotolerant yeasts, while thermophilic bacteria counts were low. Nearly all the anaerobes were facultative, not obligate. An inspection of the physical conditions existing at the various sites (Table V) suggests that temperature is the main parameter influencing counts. Conditions in the rotoclones and housewater tanks are favourable for growth of mesophilic, acid-tolerant organisms. Counts of lactic acid bacteria on Rogosa Agar were not as high as expected; possibly this medium is too selective for sugar refinery strains of *Lactobacillus* and *Leuconostoc*. Dextran contents were generally lower in the second part of the investigation, due to an intensive hygiene campaign; however, this was not the case with housewater, which contained very high levels of dextran. Rotoclones are a type of recirculating dust collector which are operated at ambient temperature. As noted by Hayes (1966), they are especially susceptible to growth of *Leuconostoc mesenteroides* and related organisms. Since this investigation, the rotoclones have been modified to operate at higher temperatures and counts have been reduced. While counts of thermophilic bacteria at other sites were generally low, they

Table VI
NUMBERS OF DIFFERENT BACTERIAL GENERA AT EACH SAMPLING SITE

Genus	*Sampling site* 1	2	3	5	7	10	*Total*
Bacillus	0	3	1	3	6	1	14
Coliforms	0	0	0	0	2	0	2
Lactobacillus	0	1	1	2	2	0	6
Leuconostoc	0	3	0	12	23	2	40
Micrococcus	0	0	0	0	2	0	2
Pseudomonas	0	0	0	2	1	0	3
Streptococcus	0	7	0	2	5	1	15
Total	0	14	2	21	41	4	82

Table VII
NUMBERS OF THE DIFFERENT YEAST GENERA AT EACH SAMPLING SITE

Genus	*Sampling site* 1	2	3	5	7	10	*Total*
Candida	0	0	0	0	1	1	2
Debaryomyces	0	0	0	0	0	1	1
Hansenula	0	1	2	5	13	0	21
Pichia	0	1	0	1	5	1	8
Saccharomyces	0	5	2	1	17	0	25
Schizosaccharomyces	0	1	0	0	3	1	5
Torulopsis	0	6	4	5	19	1	35
Total	0	14	8	12	58	5	97

Table VIII
IDENTIFICATION AND FREQUENCY OF OCCURRENCE OF 97 YEAST CULTURES ISOLATED FROM SWEETWATERS AT A REFINERY

Torulopsis	Saccharomyces	Hansenula	Pichia
8 *aeria*	14 *cerevisiae*	5 *silivicola*	5 *etchellsii*
7 *globosa*	5 *exiguus*	4 *subpelliculosa*	2 sp.
4 *dattila*	2 *fructuum*	4 *anomala*	*Schizosaccharomyces*
3 *candida*	2 *rosei*	3 *schneggii*	5 *pombe*
3 *stellata*	1 *elegans*	1 *membranefaciens*	*Candida*
3 *versatilis*	1 *veronae*	5 sp.	2 *guillermondii*
1 *ernobii*			1 *melinii*
1 *lactis-condensi*			*Debaryomyces*
5 sp.			1 *hansenii*

could affect counts of thermophilic spores in final products, although all refined products are practically sterile (Murphy, 1965; Scarr, 1968).

2. Identification of biodeteriogenic microorganisms

A total of 97 yeast isolates and 82 bacterial isolates was selected for identification. As expected, the predominant genera of bacteria found in sweetwaters are lactic acid bacteria (*Leuconostoc*, *Lactobacillus*, and *Streptococcus*) and *Bacillus* (Table VI). The ability of these organisms to form dextran and levan is well known and their growth must lead to processing problems in other stages of sugar refining (Imrie and Tilbury, 1972). *Torulopsis*, *Saccharomyces*, and *Hansenula* were the predominant yeast genera (Table VII). A wide range of species was found (Table VIII), many of them known to be osmotolerant and frequently occurring in environments of low water activity. The ability of many of the species to rapidly ferment or assimilate sucrose, produce invertase, and lower pH is well known (Devilliers, 1955). Furthermore, their presence in a sugar refinery may lead to problems in storage stability of final products such as liquid sugar.

3. Estimation of sugar loss in sweetwater systems

It is extremely difficult to estimate sugar loss caused by microbial growth in sweetwaters. This is because the circuits are complex in nature, dynamic but discontinuous, and the liquors have a constantly changing chemical composition, temperature, pH, Brix, and retention time. However, it is hoped that the relatively crude estimate summarized in Table IX is of the right order

Table IX

ESTIMATED LOSS OF SUGAR DUE TO GROWTH OF MICRO-ORGANISMS IN SWEETWATER TANKS AT A REFINERY

Sweetwater tank	*Average loss in tons sugar*[a] *per 4 hr retention time* *June* 1974 (*mean of* 9 *samples*)
C 41, 'Square tank', main section	0·081
C 41, 'Square tank', end section	0·039
C 21/1 'No. 9 tank'	0·162
Total loss per 4 hr, tons sugar	0·282
Total loss per week, tons sugar	11·844
Total loss per week as % R.S.O.[b]	0·49
Value of weekly sugar loss at £200/ton	£2369

Data by D. Clark.
[a] Calculated from °Brix change, corrected to 20°C reading and expressed in dry ton/ft^3, multiplied by volume of tank in ft^3.
[b] R.S.O. = Refined solids output.

of magnitude. A sugar loss of 0·5% refined solids output, valued at about £2000 per week, is very significant and accounts for a substantial proportion of the current 'unknown' loss. Repeat tests in October 1974, using 22 samples, gave similar results. Obviously, it is important to introduce suitable control measures to minimize microbial growth in sweetwaters.

4. Evaluation of preservatives

Recent national shortages of raw sugar forced a reduction in the working week from 7 to 5 days at one refinery. This necessitated the storage of quantities of sweetwater over the weekend. Steam was not available to main-

Table X
EFFECT OF VARIOUS PRESERVATIVES ON BACTERIAL AND YEAST POPULATIONS IN REFINERY HOUSEWATER AFTER INCUBATION FOR 48 HR AT 30°C

Preservative	*Concn (ppm)*	*Decimal reduction (−) or increase (+) in count/ml housewater after* 48 *hr incubation at* 30°*C* — *Bacteria*		*Yeasts*	
		Expt. 1	*Expt.* 2	*Expt.* 1	*Expt.* 2
Na benzoate	500	−4·8	+0·7	−4·4	+0·2
	1 000	−4·7	+0·8	−4·3	−0·6
	2 000	−4·3	−0·6	−2·4	−0·9
K sorbate	500	−2·1	−2·2	−2·2	+1·0
	1 000	−2·6	+2·4	−2·5	+0·3
	2 000	−5·3	+0·8	−4·6	−1·0
Na metabisulphite	500	−2·7	−6·3	−2·4	−1·9
	1 000	−5·8	−4·4	−6·4	−2·4
	2 000	−8·0	−3·2	−6·5	−5·2
Formaldehyde	50	−2·8	−5·4	−4·2	−5·2
	200	−7·2	−7·8	−9·6	−5·2
	500	−10·0	−6·3	−9·6	−0·9
Nipaheptyl	12·5	−2·7	+0·9	−2·1	−1·1
	25	−2·7	N.T.	−2·4	N.T.
	50	−4·5	N.T.	−4·5	N.T.
Nipasol	200	N.T.	−1·5	N.T.	−0·9
Nipaheptyl + Nipasol	12·5 + 200	N.T.	−4·5	N.T.	−1·9
Control	—	−0·6	+2·4	0	+2·0
Initial counts/ml juice		1×10^{10}	8×10^{7}	6×10^{9}	2×10^{5}

1st expt: initial °Bx 12·5; initial pH 4·3.
2nd expt: initial °Bx 3·8; initial pH 5·0.

tain elevated temperatures or to evaporate liquors to concentrated solutions, so a search was made for a suitable preservative to minimize microbial growth at ambient temperature. The need to comply with food additive regulations restricted the choice of preservatives to benzoate, sorbate, SO_2, and para-hydroxybenzoate esters. The latter were included following reports of their successful use in similar situations (Badgley, 1970; Brit. Pat. No. 1236036,

Table XI
EFFECT OF CONTINUOUS ADDITION OF A QUATERNARY AMMONIUM COMPOUND AT 25 ppm ON YEAST GROWTH AND SUGAR LOSS IN A CONTINUOUS CULTURE SYSTEM
(mean of two runs)

Before Q.A.C. addition	
% sucrose in fresh medium	17·9
% sucrose in spent medium at equilibrium	10·3
% initial sugar consumed	43
Viable yeast count/ml at equilibrium	3×10^{8}
After Q.A.C. addition	
% sucrose in spent medium at equilibrium	13·9
% initial sugar consumed	23
% improvement in sugar consumed due to treatment	46·5
Viable yeast count/ml at equilibrium	7×10^{5}
% kill due to treatment	97·7

1971). Formaldehyde was also included since its use is accepted in the beet sugar industry (Oldfield *et al.*, 1968). Microbiological results, summarized in Table X, indicated that formaldehyde and sodium metabisulphite were the most effective preservatives. Data for changes in Brix and pH during storage (not given here) largely confirmed these conclusions. On a cost/effectiveness basis, formaldehyde treatment at 50 ppm (0·7 p/m^3 sweetwater) was cheaper than metabisulphite at 500 ppm (6·1 p/m^3 sweetwater). However, since formaldehyde is not a permitted food preservative, the final choice was sodium metabisulphite at 500 ppm. This recommendation has been implemented at the refinery and is reported to be successful in prevention of sugar loss.

For reasons previously outlined, batch-type experiments do not simulate the dynamic conditions found in sweetwater systems during normal refinery operations. Hence, it was decided to use a two-stage continuous culture system to simulate housewater tanks at one refinery. Mean results of two experiments are shown in Table XI. Under steady-state conditions, continuous addition of a Q.A.C. preservative (25 ppm) reduced by nearly 50% the sugar consumed by yeast growth. Extrapolating this result to a typical housewater tank containing 15 tons sweetwater/day at an average of 15° Brix, preservative treatment would result in a saving of 0·23 tons sugar/day or 84 tons/annum. The value of this sugar saving is about £17 000 p.a. for a treatment cost of about £140. Further work is needed to select the optimum treatment.

Acknowledgements

The authors wish to thank the Directors of Tate & Lyle Ltd, Professor A. J. Vlitos (Chief Executive, Tate & Lyle Group R & D), and Professor J. E. Smith (University of Surrey) for permission to publish this paper. The helpful cooperation of refinery staff, especially chemists and microbiologists, made this study possible. Special thanks are due to Dr C. Abram, Mr D. Norman, and Mr D. Clark.

REFERENCES

Badgley, G. R. (1970). Sugar industry use for new preservative. *Food Technol.*, **24** (138), 32.

Beech, F. W., Davenport, R. R., Goswell, R. W., and Burnett, J. K. (1968). Two simplified schemes for identifying yeast cultures. In *Identification Methods for Microbiologists, Part B* (B. M. Gibbs and D. A. Shapton, eds.), Academic Press, London.

Brit. Pat. No. 1236036, F. & M. Schaeffer Brewing Co. and Washine Chemical Corp. Method for preserving dilute sugar solutions.

Devilliers, P. L. (1955). Microbial infection in sugar manufacture. *Sucre France*, **96**, 101–113.

Hayes, F. W. (1966). Microbiology in sugar production. *Process. Biochem.*, **1** (5), 285–289.

Imrie, F. K. E., and Tilbury, R. H. (1972). Polysaccharides in sugar cane and its products. *Sugar Technol. Rev.*, **1**, 291–361.

Jarvis, B. (1973). Comparison of an improved Rose Bengal–chlortetracycline agar with other media for the selective isolation and enumeration of moulds and yeasts in foods. *J. Appl. Bact.*, **36**, 723–737.

Keniry, J. S., Lee, J. B., and Mahoney, V. C. (1969). Improvements in the dextran assay of cane sugar materials. *Int. Sugar J.*, **71**, 230–233.

Lodder, J. W., and Kreger van Rij, N. J. W. (1952). *The Yeasts—a Taxonomic Study*. North-Holland Publishing Co., Amsterdam.

Murphy, R. P. (1965). Thermophilic bacterial contamination in sugar manufacture and the effect of ultraviolet rays. *Int. Sugar J.*, **67**, 82–83.

Oldfield, J. F. T., Dutton, J. V., Grierson, D., Heaney, R. K., and Teague, H. J. (1968). Comparison of sulphur dioxide and formaldehyde as bacteriostats in diffusion. *Int. Sugar J.*, **70** (Pt. I), 296–298; (Pt. II), 330–333.

Scarr, M. P. (1959). Selective media used in the microbiological examination of sugar products. *J. Sci. Food Agric.*, **10** (12), 678–681.

Scarr, M. P. (1968). Thermophiles in sugar. *J. Appl. Bact.*, **31**, 66–74.

BIODEGRADATION OF BREWERS' GRAINS—I

J. R. EDWARDS

Bournemouth College of Technology, Bournemouth, Dorset, U.K.

and

R. N. SMITH

The Hatfield Polytechnic, Hatfield, Hertfordshire, U.K.

Summary

The biodegradation of brewers' grains by pure cultures of six mesophilic and three thermophilic fungi has been investigated. Optimal conditions for the growth of these organisms were established. The fungi were divisible into three groups based upon their mode of colonization and degradation of the substrate. Group I: Comprising *Aspergillus glaucus*, *A. niger*, *Trichoderma viride*, and *Thermoascus aurantiacus* brought about rapid solubilization of the material during a short phase of vegetative growth, the substances released consequently being metabolized during a distinct reproductive phase. The Group II pattern was demonstrated by *Humicola lanuginosa* where vegetative and reproductive phases occurred simultaneously, resulting in a steady degradation of the material. The final grouping comprised: *Chaetomium globosum*, *Ch. thermophile*, *Fusarium culmorum*, and *Papulaspora* sp. and these fungi achieved the maximal degree of degradation of the grains, this being associated with a failure to reproduce themselves either sexually or asexually.

MATERIALS AND METHODS

Brewers' grains are the residue remaining after the production of malt extract from barley. They consist of approximately 20% ethanol and water soluble materials, 20% protein, 30% hemicellulose, and 30% cellulose and are traditionally sold as foodstuff for ruminants (Fagan and Allen, 1909). In this investigation a variety of mesophilic and thermophilic fungi have been examined for their ability to degrade brewers' grains. The definition of thermophily proposed by Cooney and Emerson (1964) has been adopted. Some fungi were selected from references in the literature recommending their degradative ability when grown on similar natural substrates, while others were isolated directly onto brewers' grains from a variety of soil types using a number of different techniques (Edwards, 1974). Each organism selected had, in a previous series of experiments, caused a reduction in excess of 18% in the dry weight of brewers' grains. The species used are given below, together with the optimum temperature of each organism for degradation of brewers' grains.

Nine fungi were used: *Aspergillus glaucus*, *Aspergillus niger*, *Chaetomium globosum*, *Papulaspora* sp., and *Trichoderma viride* were grown at 30°C; *Fusarium culmorum*, at 25°C; and *Chaetomium thermophile*, *Humicola lanuginosa*, and *Thermoascus aurantiacus*, at 50°C.

Samples of brewers' grains (4·0 g) were placed in 9-cm petri dishes, moistened with a standard salts solution (Table I) to give a final water content of 67% and autoclaved at 1·05 K_5/cm^2 for 15 min.

Table I
STANDARD SALT SOLUTION

Salt solution		*Volume (added/litre)*	*Weight (added/litre)*
Solution A			
K_2HPO_4	60 g in 200 ml	2·0 ml	600 mg
KH_2PO_4	40 g		400 mg
Solution B			
$MgSO_47H_2O$	50 g in 100 ml	0·2 ml	100 mg
Solution C			
$CaCl_2$	10 g in 100 ml	0·2 ml	20 mg
Solution D			
$(NH_4)SO_4$	50 g in 100 ml	2·0 ml	1·0 g

The samples of brewers' grains were inoculated by placing a 0·8 cm diameter plug of agar bearing actively growing mycelium in the centre of each dish. The cultures were incubated for 14 days. There were three replicates of each treatment; two cultures were taken at 3- and 4-day intervals for dry weight determination and chemical analysis. Samples were dried to constant weight at 70°C; the dried material was then ground to a fine powder in a Moulinex coffee grinder and subjected to a series of analyses based on the methods used by Chang (1967).

The ethanol soluble fraction was determined by extraction of the material with industrial methylated spirit in a Soxhlet thimble until the extract was free of sugar. This process removed simple sugars, glucosides, amino acids, peptides, and colouring matter. *The water and diastase* soluble fraction was determined by diastasic hydrolyses of the ethanol extracted material (Weinmann, 1947). *The bound protein fraction* was determined by the micro-Kjeldhal method. *The hemicellulose fraction* was determined by alkali extraction of the diastase-treated material using 24% KOH. This treatment removed both protein and hemicellulose and the weight of hemicellulose was determined by subtracting the weight of bound protein from the weight loss following alkali extraction. *The cellulose, lignin, and ash fraction* was determined by adding together the fractions given above and subtracting this from the initial dry weight of the sample.

The preferential sequence of carbon utilization in brewers' grains was studied by enriching brewers' grains with either glucose, starch, casein, xylan, or cellulose (Whatman's Chromomedia). These supplements were mixed with brewers' grains to give a final proportion of 10% dry weight. The mixture was then moistened, placed in vertical tubes, and inoculated at the top with a spore suspension of *Trichoderma viride*. A controlled flow of moist, sterile air was passed through the grain and the exhaust gases were analysed for the presence of ammonia (Edwards, 1974).

RESULTS

Degradation of brewers' grains by pure cultures

All treatments caused a marked reduction in the dry weight of the material

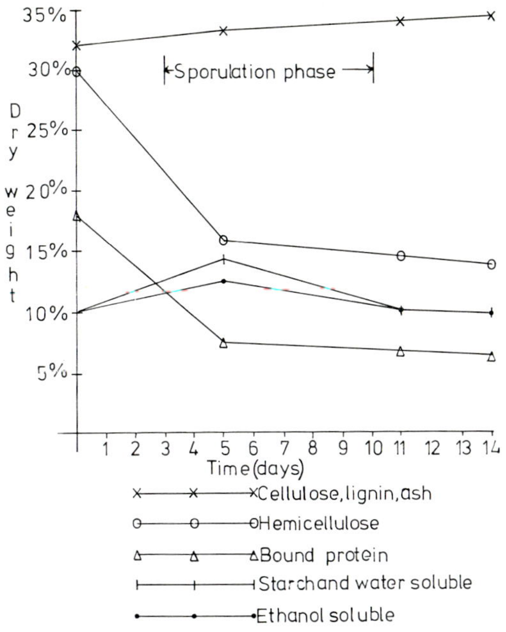

Fig. 1. *Composition of Brewers' grain during the growth of Group I fungi.*

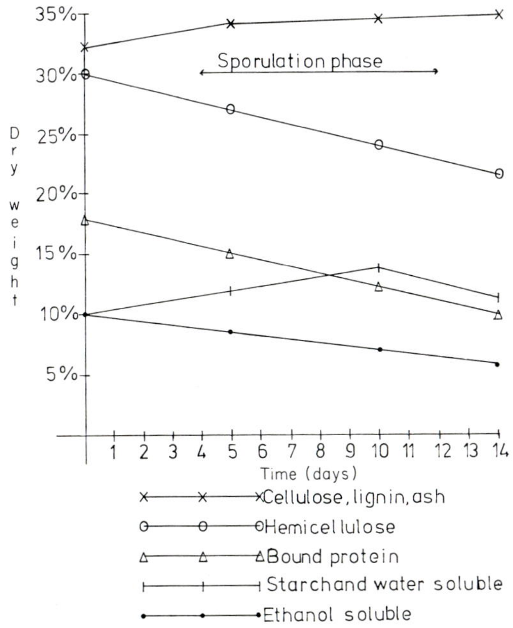

Fig. 2. *Composition of Brewers' grain during the growth of Group II fungi.*

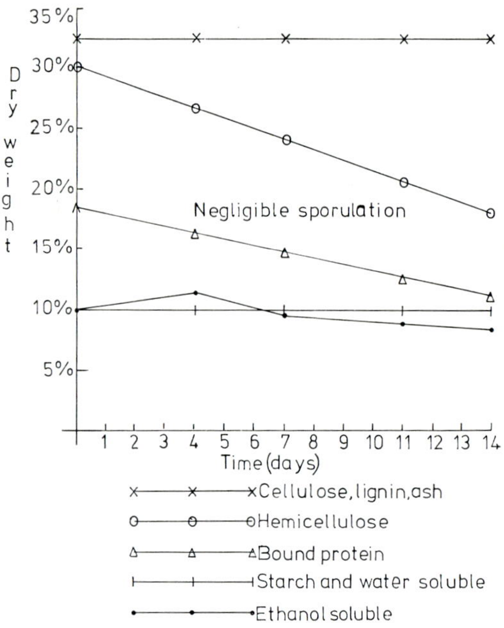

Fig. 3. Composition of Brewers' grain during the growth of Group III fungi.

which was largely due to degradation of the insoluble protein and hemicellulose fractions. There were three discernible patterns of degradation and the behaviour of the nine organisms under investigation is summarized in Figs. 1, 2, and 3.

Degradation of brewers' grain enriched with glucose, starch, casein, xylan, or cellulose by ***Trichoderma viride***

The degradation of enriched brewers' grains and the release of ammonia following deamination of protein are displayed in Tables II and III.

DISCUSSION

There were three discernible patterns of colonization and these were reflected in patterns of substrate utilization. The first (Fig. 1) was characterized by fungi, including *A. glaucus*, *A. niger*, *T. viride*, and *T. aurantiacus* whose vegetative mycelium rapidly invaded the grains, and then sporulated profusely over the surface of the substrate. This behaviour was characteristic of the 'penicillium' mode of colonization also observed when brewers' grains had been buried in soil for the isolation of degradative organisms. One isolate, *T. aurantiacus*, did not conform to the general pattern in showing a typical 'penicillium' mode of substrate utilization although no sporulation occurred.

Examination of substrate utilization by these cultures revealed rapid degradation of both the bound protein and hemicellulose fractions of the grains during the first 4–5 days of incubation. This was accompanied by a rise in the level of soluble material over this period, but these levels gradually fell until day 11 of incubation, after which there was no appreciable change.

Table II

VARIATION IN THE COMPOSITION OF CARBON-ENRICHED BREWERS' GRAINS AFTER 7 DAYS GROWTH OF *TRICHODERMA VIRIDE*

Composition	*Dry wt loss*	*Ethanol soluble*	*Starch and wat. sol.*	*Bound protein*	*Hemicellulose*	*Cellulose lignin*
Control						
Original		10·0%	10·0%	18·0%	30·0%	32·0%
Change	23%	−3·1%	−0·9%	−7·2%	−12·8%	−1·3%
Glucose						
Original		19·0%	9·0%	16·2%	26·8%	29·0%
Change	27%	−10·2%	−1·6%	−3·1%	−12·8%	+0·4%
Starch						
Original		9·0%	19·0%	16·2%	26·2%	29·0%
Change	28%	+0·4%	−9·9%	−4·2%	−10·8%	−0·1%
Casein						
Original		9·0%	9·0%	26·2%	26·9%	29·0%
Change	22%	+1·5%	+1·8%	−14·8%	−10·3%	−0·1%
Xylan						
Original		9·0%	9·0%	16·2%	36·8%	29·0%
Change	28%	−0·4%	−0·1%	−6·0%	−20·3%	−1·2%
Cellulose						
Original		9·0%	9·0%	16·2%	26·8%	39·0%
Change	26%	−1·3%	−0·3%	−6·3%	−11·1%	−7·0%

Table III

AMMONIA EVOLVED FROM 'BLOWN BED' CULTURES OF *T. VIRIDE* GROWN ON CARBON ENRICHED BREWERS' GRAINS

Culture	*Amount of ammonia released*
Control brewers' grains	1·0
Brewers' grains + glucose	0·5
Brewers' grains + starch	0·5
Brewers' grains + casein	3·0
Brewers' grains + xylan	0·9
Brewers' grains + cellulose	0·9

The amount of ammonia released is expressed as a ratio to that evolved from a control culture of Brewers' grains.

The second colonization pattern (Fig. 2) was seen in cultures of *H. lanuginosa* where vegetative growth and sporulation occurred simultaneously during the invasion of the substrate. In these cultures there was a far more gradual and prolonged loss of protein and hemicellulose and ethanol soluble substances, while the amount of starch- and water-soluble substances increased for the first 10 days and then declined. In both these colonization patterns there was a slight increase in the cellulosic fraction, which may have been due to the accumulation of spores and mycelium.

The final colonization pattern (Fig. 3) was typified by those fungi which either did not spore at all, for example, *Papulaspora* sp., or produced only very few reproductive structures, for example, *C. globosum* and *F. culmorum*. The substrate utilization in these cultures closely resembled that observed in cultures of *H. lanuginosa* in that there was a steady decline in protein. The hemicellulose content gradually declined throughout the time of the experiment. The correlation between sporulation and the cessation of active growth of soil fungi was also noted by Chesters (1948). The results suggest that the patterns of substrate utilization and sporulation in pure cultures of fungi grown on brewers' grains are related.

There was little degradation of the brewers' grains after 14 days and with the exception of *C. globosum* (Smith *et al.*, 1975), the cultures may be considered to have entered the stationary phase. The reasons for the cessation of growth are not clear; readily available nutrients were still present since the concentration of ethanol and water soluble materials which might be present in solution never fell below 2·3 and 3·3%, respectively. Similar findings have been reported from the degradation of wheat straw (Chang, 1967). Staling due to the liberation of ammonia may have been a contributing factor since in most cultures the pH rose to 7·0–7·5 by day 14 of incubation. However, in grains supporting *A. niger*, degradation was complete by day 14 even though the pH never rose above 5·6. In no treatment was the cellulose and lignin component of grains degraded.

In all the enriched cultures the supplements were substantially degraded and, except for casein where the increased release of ammonia may have enhanced staling, there was a concomitant rise in the total loss in dry weight. The addition of glucose and starch reduced the release of ammonia to half that of the unsupplemented control. However, xylan and cellulose enrichments had no effect on ammonia release, and therefore on protein degradation and its use as a carbon source even though both of these materials were also attacked by *T. viride*. Xylan is the major product from the hydrolysis of barley hemicellulose (Preece, 1931). These results show that either there is no preference for hemicellulose over protein as a carbon source or else the natural hemicellulose incorporated in brewers' grain is less susceptible to enzyme action. Experiments with cellulytic fungi *in vitro* show that the cellulose enzymes are repressed by relatively low levels of simple carbon sources (Bravery, 1968). Since cellulase enzymes were present in the cellulose-enriched grains, it is also probable they were present in the nonenriched samples. The degradation of cellulose under these conditions may be interpreted as occurring in a highly heterogeneous environment in which localized nutrient deficiencies permit the induction of cellulase enzymes. The failure of the cellulose enzymes to degrade the cellulose component of normal brewers' grains may be due to the tissue resistance of the material and in particular to the protective action of lignin. A similar phenomenon occurred in a composting where the destruction of paper fibre was prevented by the protection afforded by the presence of lignin (Poincelot and Day, 1973; Regan *et al.*, 1973).

In conclusion, the biodegradation of a natural material such as brewers' grains is dependent not upon the degradability of the individual components but rather upon the overall tissue resistance which is in turn dependent upon the manner in which the most recalcitrant component is incorporated and combined with the other components. Thus the biodegradation of lignin would appear to be the key to the complete utilization of brewers' grains.

Acknowledgement

The authors would like to acknowledge the support given by Mr M. H. Baker-Munton of Munton and Fisons Limited, Cedar's Factory, Stowmarket, Suffolk, for this investigation.

REFERENCES

Bravery, A. F. (1968). Microbiological breakdown of cellulose in the presence of alternative carbon sources. *J. Sci. Food Agr.*, **19**, 133–135.

Chang, Y. (1967). The fungi of wheat straw compost. II. Biochemical and physiological studies. *Trans. Brit. Mycol. Soc.*, **50** (4), 667–677.

Chesters, C. G. C. (1948). A contribution to the study of fungi in the soil. *Trans. Brit. Mycol. Soc.*, **30**, 100–117.

Cooney, D. C., and Emerson, R. (1964). *Thermophilic Fungi. Enmycota.* W. H. Freeman Publ. Co., San Francisco.

Edwards, J. R. (1974). A study of the Biodegradation of Brewers' Grains by Fungi. Ph.D. Thesis. The Hatfield Polytechnic, Hatfield, Herts.

Fagan, T. W., and Allan, C. (1909). Composition of brewers' and distillers' grains. *Analyst*, **34**, 218.

Poincelot, R. P., and Day, P. R. (1973). Rates of cellulose decomposition during composting of leaves combined with several municipal wastes and other additives. *Compost Sci.*, **14** (3), 23–35.

Preece, I. A. (1931). Studies on hemicelluloses. I. The hemicelluloses of brewers' grains. *J. Inst. Brewing*, 409–413.

Regan, R., Jevis, J. G., Gasser, R., McCann, K., and Hudek, J. (1973). Cellulose Degradation in Composting. *U.S. Natl. Tech. Inform. Ser.*

Smith, R. N., Odell, D. E., and Edwards, J. R. (1975). Biodegradation of brewers' grains. II. 3rd Int. Biodegrad. Symp.

Weinmann, H. (1947). Determination of total available carbohydrates in plants. *Plant Physiol., Lancaster*, **22**, 279–290.

BIODEGRADATION OF BREWERS' GRAINS—II

R. N. SMITH and D. E. ODELL
The Hatfield Polytechnic, Hatfield, Hertfordshire, U.K.
and
J. R. EDWARDS
Bournemouth College of Technology, Bournemouth, Dorset, U.K.

Summary

The biodegradation of brewers' grains, a waste product from the production of malt extract, has been investigated. Higher rates of degradation than were achieved by pure cultures were produced by certain combinations of fungi, especially those involving either *Thermoascus aurantiacus* or *Chaetomium thermophile*. Prior colonization by *T. aurantiacus* did not prevent subsequent colonization by other fungi and may have rendered the material more susceptible to degradation by them. *Ch. thermophile* was unique in being able to follow most other species on brewers' grains. Biochemical studies showed that the weight loss by brewers' grains was due to the reductions in bound protein and hemicellulose. Five degradation patterns were recognized using mixed cultures: these were (1) synergistic; (2) enhanced; (3) dominant; (4) compatible; and (5) incompatible or inhibitory.

MATERIALS AND METHODS

The biodegradation of any material under natural conditions is accomplished by a mixed microflora (Chesters, 1950), and it was found in isolation experiments with brewers' grains that there was always more than one fungus active in the material. The fungi used in dual and tandem culture experiments were all those previously studied in pure culture (Edwards and Smith, 1975) and, in addition, two thermophilic isolates from brewers' grains *H. grisea* and *P. duponti*, which grew only weakly in pure culture on virgin grains.

The influence of mixed cultures upon the degradation was studied using dual cultures, tandem cultures, and by soil burial. Fungi used in the dual and tandem studies were grown on 4-g samples of brewers' grains in petri dishes as described by Edwards and Smith (1975). The dual cultures were inoculated simultaneously and incubated for 11 days, while the tandem cultures were inoculated first with one organism and then 11 days later with the second. Cultures were not autoclaved before the second inoculation. Studies showed that autoclaving cultures at the end of the first 11 days prior to inoculation with the second organism had no effect on the extent of degradation.

For soil burial experiments, 2-g samples of brewers' grains were moistened, enclosed in nylon net sachets, autoclaved, buried in beakers of soil at 75% water-holding capacity, and incubated at 30 or 50°C to permit the development of mesophilic and thermophilic fungi, respectively. The nylon net was gauge 12 N and offered no obstruction to the entry of soil microorganisms into the samples.

At the end of the incubation period, cultures were examined for loss in dry weight and analysed biochemically to determine (1) the ethanol-soluble fraction containing sugars, glucosides, amino acids, and essential oils; (2) the diastase and water-soluble fraction consisting of starch and possible fungal glycogen; (3) the bound protein fraction; (4) hemicellulose; and (5) cellulose and lignin. Details of the analyses are given by Chang (1967) and Edwards and Smith (1975).

Dual mesophilic cultures were examined visually and patterns of colonization were recorded. Four mesophilic dual cultures were selected for respiratory studies. The cultures were prepared by inoculating 15 g of sterile brewers' grains with inocula consisting of 2 g of grain bearing a pure culture of the test organisms. Measurements of oxygen uptake were made using a Gilson respirometer. Measurements of oxygen uptake were made at 12, 24, 48, 72, 96, and 120 hr after inoculation.

RESULTS

The results for reduction in dry weight and pattern of colonization of dual

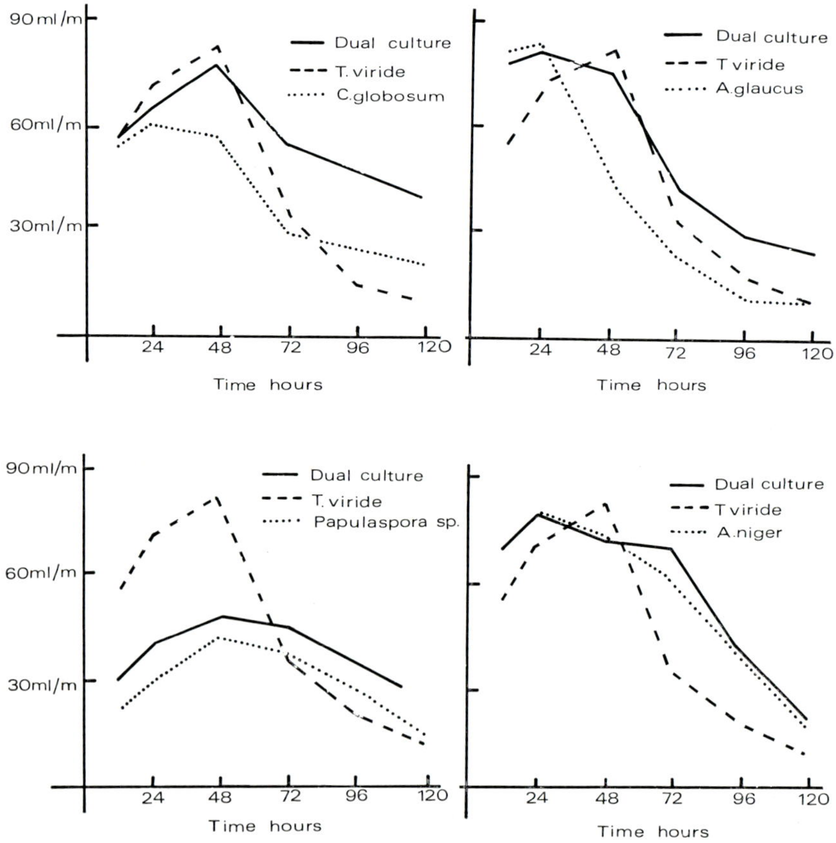

Fig. 1. Oxygen uptake (μlitres/min) of pure and dual cultures of selected mesophilic fungi grown on brewers' grains.

Table I
DEGRADATION OF BREWERS' GRAINS BY *DUAL* CULTURE RECORDED AS % LOSS IN DRY WEIGHT AND % COLONIZATION AFTER 11 DAYS

(a) Mesophilic combinations

No.	*Combination* A	B	*% Dry wt loss*	*% Colonization* A	B
1	*A. glaucus* (25)	*A. niger* (23)	25	75	25
2	*A. niger* (23)	*F. culmorum* (20)	23	100	0
3	*T. viride* (26)	*C. globosum* (19)	26	85	15
4	*T. viride* (26)	*F. culmorum* (20)	26	100	0
5	*Papulaspora* sp. (26)	*F. culmorum* (20)	26	100	0
6	*T. viride* (26)	*Papulaspora* sp. (26)	26	100	0
7	*A. glaucus* (25)	*T. viride* (26)	25	65	35
8	*A. niger* (23)	*T. viride* (26)	23	65	35
9	*A. glaucus* (25)	*C. globosum* (19)	23	65	35
10	*A. glaucus* (25)	*F. culmorum* (20)	23	100	0
11	*A. niger* (23)	*C. globosum* (19)	22	65	35
12	*A. niger* (23)	*Papulaspora* sp. (26)	24	100	0
13	*C. globosum* (19)	*Papulaspora* sp. (26)	24	35	65
14	*Papulaspora* sp.(26)	*A. glaucus* (25)	23	0	100
15	*C. globosum* (19)	*F. culmorum* (20)	17	65	35

(b) Thermophilic combinations

No.	*Combination*		*% Dry wt loss*
16	*H. grisea* (5)	*T. aurantiacus* (24)	30
17	*C. thermophile* (25)	*T. aurantiacus* (24)	30
18	*H. lanuginosa* (19)	*T. aurantiacus* (24)	28
19	*P. duponti* (6)	*T. aurantiacus* (24)	27
20	*H. grisea* (5)	*P. duponti* (6)	10
21	*H. lanuginosa* (19)	*P. duponti* (6)	20
22	*H. lanuginosa* (19)	*H. grisea* (5)	18
23	*C. thermophile* (25)	*H. lanuginosa* (19)	22
24	*C. thermophile* (25)	*H. grisea* (5)	20
25	*C. thermophile* (25)	*P. duponti* (6)	20

Figures in parentheses are the dry weight losses in pure culture.

cultures are given in Table I, which contains 15 mesophilic and 10 thermophilic combinations. Respiratory studies using four mesophilic dual cultures are displayed in Fig. 1.

The effect of degradation by tandem cultures is displayed in Table II and a summary of the biochemical analysis of these and soil burial experiments is presented in Table III.

DISCUSSION

Waksman (1931) found that while the degradation of straw was greater with mixed cultures of *Zygorhynchus* sp. and *Trichoderma* sp., the degradation of other materials such as alfalfa was accomplished as readily by *Trichoderma* sp. alone as by a variety of dual cultures or even by a 'natural' soil microflora. However, Sharp and Eggins (1970) have demonstrated enhanced degradation of Beechwood veneer strips when two cellulolytic fungi were applied together. The results of the current investigation on the degradation of brewers' grains

Table II

DEGRADATION OF BREWERS' GRAINS BY *TANDEM* PERMUTATIONS OF MESOPHILIC AND THERMOPHILIC FUNGI AFTER 22 DAYS

	Primary colonizer									
Secondary colonizer	*A.g.*	*A.n.*	*C.g.*	*P.sp.*	*T.v.*	*C.t.*	*H.l.*	*H.g.*	*P.d.*	*T.a.*
A.g.	(25)	—	—	—	S	31				30
A.n.	—	(23)	—	—	—	—	S	—	—	25
C.g.	—	—	(23)	—	—	S	30	—	—	28
P.sp.	—	—	—	(26)	—	—	—	—	—	31
T.v.	—	—	—	—	(25)	—	—	—	—	25
C.t.	32	26	27	32	31	(28)	24	—	—	30
H.l.	—	—	S	—	S	S	(19)	—	—	24
H.g.	—	—	—	—	S	S	24	(NT)	—	24
P.d.	—	—	28	—	—	S	24	—	(NT)	27
T.a.	—	—	S	—	S	S	24	—	—	(24)

Figures refer to the % loss in dry weight. 'S' represents slight growth by the second organism in which less than 20% of the surface was colonized; dry weight losses and biochemical analyses were not determined for these cultures; (—) represents no growth by secondary colonizer; the weight losses in parentheses are for the pure culture after 22 days, and NT indicates not tested.

A.g. = *A. glaucus*, A.n. = *A. niger*, C.g. = *C. globosum*, P.sp. = *Papulaspora* sp., T.v. = *T. viride*, C.t. = *C. thermophile*, H.l. = *H. lanuginosa*, H.g. = *H. grisea*, P.d. = *P. duponti*, and T.a. = *T. aurantiacus*.

Table III

SUMMARY OF THE COMPOSITION OF BREWERS' GRAINS FOLLOWING *TANDEM* DEGRADATION BY FUNGI GIVEN IN TABLE II AND *SOIL BURIAL* AFTER 22 DAYS

	% Component fractions					
	% Dry wt loss	*Ethanol soluble*	*Diastase soluble*	*Bound protein*	*Hemi-cellulose*	*Cellulose and lignin*
1. Mean composition noncellulose-degrading spp.	27·35	8·1	8·7	7·8	15·2	32·4
2. Mean composition cellulose-degrading spp.	27·9	7·0	10·1	8·1	18·9	28·0
3. Mean 22 day pure culture controls	24·5	9·3	7·9	8·8	17·1	32·4
4. Soil burial						
(a) Mesophilic	39·5	4·8	3·9	7·9	12·9	31·0
(b) Thermophilic	40·0	3·6	1·4	9·5	15·0	30·5
5. Untreated control	0·0	10·0	10·0	18·0	30·0	32·0

suggest that certain *tandem* cultures were more effective than pure cultures and more effective than a natural 'soil microflora' in cellulose and lignin degradation. Porter (1924) classified the reactions of the two species in a dual fungal culture as inhibition at a distance, slight inhibition, intermingling, and overgrowth. In this work on brewers' grains, there was no evidence of inhibition at a distance although the other reactions were observed.

Some thermophilic combinations showed increased degradation of brewers' grains as compared with pure cultures. One combination, *H. grisea* with *T. aurantiacus*, caused slightly more degradation than the combined rates of the two partners and may have demonstrated true synergism. Five combinations (Table I Nos. 17, 18, 19, 20, and 21) showed enhanced degradation which was greater than the more degradative partner in pure culture. *T. aurantiacus* occurred as one partner in four of the six combinations showing synergistic or enhanced degradation and may well have played a special role in the degradation of this material. Synergistic and enhanced degradation may have been due to cross feeding or toxin naturalization, as has been suggested by Veldkamp (1972) in studies on bacterial growth, or to enzyme complementation. Enzyme complementation is based on the fact that complex natural materials such as plant cell walls require a large battery of enzymes to effect their degradation (Selby and Maitland, 1967; Halliwell and Riaz, 1970). Sahgal and Agarwal (1973) have demonstrated that the activities and specific properties of the cellulolytic enzymes produced by cellulolytic fungi vary from species to species. It is therefore possible that the combination of two sets of enzymes from two species could complement each other and have more degradative action than either set alone. However, since the organisms in these experiments were involved in trophic degradation and competing for the same substrate, synergism could be expected only when the rate of degradation of one or both partners was low as in the case of *H. grisea* with *T. aurantiacus*. Eight of the fifteen pairs of mesophilic fungi (Table 1:1–8) caused degradation at the same rate as one of the partners, and in two combinations (Table I:7 and 8), the extent of degradation was the same as the less degradative combinant. In one of these combinations (Table I:6), degradation by the two partners alone and together proceeded at the same rate. Five mesophiles (Table I:9–13) and four thermophilic combinations (Table I:22–25) caused degradation which was midway between the two combinants alone.

Garrett (1970) has emphasized the importance of competitive saprophytic ability in the ecology of fungi. In those combinations in which degradation was the same as one of the combinants or intermediate between the two, degradation may have been determined by the competitive saprophytic ability of the two organisms. Visual estimates in mesophilic cultures of the proportion of colonization of the two partners showed that in those cases where the extent of degradation was the same as that of the more vigorous partner (combinations 1–5) this species was completely dominant or nearly so on the grains. Similarly, co-dominance was recorded in those combinations in which degradation was intermediate between the two partners.

While visual observations on the predominance of one partner over the other may assist in interpreting the role of the two organisms in degradation studies, it is also possible that active nonsporing hyphae of one partner may be present although not recorded. This has been supported by results from the respiratory studies. Thus, the respiratory patterns of *T. viride* and *Papulaspora* sp. (Table I:6; Fig. 1, bottom left) and *T. viride* with *C. globosum* (Table I:3; Fig. 1, top left) indicate that *T. viride* was not enjoying the dominance which results for the extent of degradation and visual observations of dominance would suggest. The respiratory pattern indicated that *Papulaspora* was metabolically dominant in the first case and that in the second, *C. globosum* is having a significant effect. However, the respiratory pattern of the *A. niger*/*T. viride* combination (Fig. 1, bottom right) was virtually identical with that of *A. niger* alone, suggesting that *A. niger* is metabolically dominant. This was in agreement with the percentage weight loss which was also identical with *A. niger* alone (Table I:13). The respiration pattern of grain inoculated with *T. viride* and *A. glaucus* tended again to confirm the earlier suggestions of dominance by *A. glaucus*. Thus the respiratory 'finger print' can be combined usefully with visual and other parameters when determining the role of two organisms growing together on a substrate.

Two combinations, *Papulaspora* sp. with *A. glaucus* (Table I:14) and *C. globosum* with *F. culmorum* (Table 1:15), were incompatible, and degradation was inhibited, being less than that by either partner alone. Tandem cultures in which a second species was introduced onto the grains when the growth of the first colonizer was almost complete (Table II) revealed that comparatively few fungi (21 out of 90 permutations) were able to colonize the substrate successfully. In 14 permutations degradation was enhanced, *i.e.* greater than that which was recorded for either pure culture after 22 days incubation (Table II). However, the synergism which had been observed in dual cultures of *T. aurantiacus* and *H. grisea* did not occur in tandem culture and it may be that synergism can occur only when the hyphae of both partners grow together in a young active condition. In previous work, *P. duponti* and *H. grisea* were able to cause little degradation of virgin brewers' grains, and their performance in dual and tandem culture places them in the role of specialized secondary colonizers along with *C. thermophile*.

Loss in dry weight of brewers' grains in soil at thermophilic and mesophilic temperatures occurred at approximately the same rate, and this was markedly greater than that which occurred in any dual or tandem culture (Table III). This was due to greater utilization of the ethanol- and diastase-soluble fractions since the cellulose and lignin components were not attacked until between days 22 and 29. Thus degradation by the 'full soil microflora' had been more effective in the utilization of the ethanol- and diastase-soluble fractions. Regan *et al.* (1973) reported similar findings in a composting experiment in which little cellulose degradation occurred. However, cellulose degradation did occur in four tandem permutations within 22 days with *H. lanuginosa* followed by *C. globosum* or *H. grisea*, and *T. aurantiacus* followed by *C. globosum* or *C. thermophile*. In these cultures cellulose degradation occurred despite a high background level of ethanol- and diastase-soluble materials, and in preference to hemicellulose; the mean levels of hemicellulose in the cellulose-degraded cultures being 18·9 as compared with 15·1 in the remaining permutations. These four permutations would appear to be examples of enzyme complementation whereby cellulose is degraded sooner than would be otherwise expected.

The results from degradation studies using pairs of fungi on brewers' grains suggest that when a substrate is invaded by more than one fungus, one of five possible degradation patterns occurs:

1. *True synergism* in which degradation is greater than the sum of the two partners in pure culture.
2. *Enhanced degradation* which is greater than that of either pure culture alone but less than the sum of the pure cultures.
3. *Compatible colonization* in which degradation occurs at the same rate as one of the two partners or lies somewhere between the two; and in which growth of both partners occurs.

4. *Dominant degradation* in which one or other partner determines the rate of degradation and is dominant on the material.

5. *Mutual inhibition or incompatibility* in which degradation is less than that caused by either partner in pure culture.

Acknowledgement

This work was part of a study sponsored by Munton and Fisons Limited, Cedars Factory, Stowmarket, Suffolk, and the authors wish to acknowledge the encouragement given by the Managing Director, Mr M. H. Baker-Munton.

REFERENCES

Chang, Y. (1967). The fungi of wheat straw compost, II. Biochemical and physiological studies. *Trans. Brit. Mycol. Soc.*, **50** (4), 667–677.

Chesters, C. G. C. (1950). Presidential Address. *Trans. Brit. Mycol. Soc.*, **32.**

Edwards, J. R., and Smith, R. N. (1975). Biodegradation of brewers' grains, I. 3rd Int. Biodegrad. Symp.

Garrett, S. D. (1970). *Pathogenic Root Infecting Fungi.* Cambridge Univ. Press, London.

Halliwell, G., and Riaz, M. (1970). The formation of short fibres from native cellulose by components of *Trichoderma koningii* cellulose. *Biochem. J.*, **116,** 35–42.

Porter, C. L. (1924). Growth characteristics of mixed cultures, I. *Am. J. Bot.*, **2,** 168–188.

Regan, R., Jeris, J. S., Gasser, R., Melann, K., and Hudek, J. (1973). Cellulose degradation in composting. National Technical Information Service. U.S. Department of Commerce, P.B. 215722.

Sahgal, D. A., and Agarwal, P. N. (1973). Differences in mode of action of cellulolytic enzymes from *Curvularia lunulata* and *Chaetomium globosum. Indian J. Exp. Biol.*, **2** (1), 40–42.

Selby, K., and Maitland, C. C. (1967). The cellulase of *Trichoderma viride*. Separation of the components involved in the solubilization of cotton. *Biochem. J.*, **104,** 716–724.

Sharp, R. F., and Eggins, H. O. W. (1970). The ecology of soft rot fungi 2. Interaction. *Int. Biodeterior. Bull.*, **6** (2), 65–74.

Veldkamp, H. (1972). Mixed culture studies with the chemostat. *J. Appl. Chem. Biotechnol.*, **22,** 105–123.

Waksman, S. A. (1931). Decomposition of the various chemical constituents of complex plant materials. *Arch. Mikrobiol.*, **2,** 136–154.

CHEMICAL CONTROL OF POSTHARVEST DISEASES

J. M. OGAWA, B. T. MANJI and A. H. EL BEHADLI

Department of Plant Pathology, University of California,
Davis, California, U.S.A.

The evolution of chemicals for control of postharvest deterioration of fresh agricultural products came with the desire of the industry to provide quality products for the consumer and to develop new distant markets for their increasing supply. Losses of certain shipment lots were not uncommon a few years ago, and we will continue to see deterioration of such perishable commodities as strawberries, peaches, and lettuce during marketing. Significant advancements in decay control were the parallel sophistication in chemical application, better packaging, and, most important, the cooling of commodities before and during transit, storage, and retailing. The impact of these advancements has resulted in extending the availability of decay-free produce in domestic and international markets.

Before the utilization of refrigeration, chemical control, and processing, man usually consumed commodities within the growing season. The serving of fresh fruits and vegetables with each meal is a cherished custom but with the explosion in population during the last century, processed products became necessary to provide food and fibre for all the people. Chemicals have played a major role in providing fresh fruits and vegetables as well as canned, frozen, and dried products in-season and off-season.

As a preamble to chemical control of postharvest diseases, we recall the adage, 'after the horse has left the barn it is a little late to close the gate' and 'one rotten apple in a basket spoils the whole basket'. Once infection has occurred, protective treatments are a little late and the prevention of pathogen spread is more difficult than one realizes.

The fact sometimes overlooked by growers and packers is that postharvest decay problems are related to field problems. Thus, the first attempt to reduce deterioration before the fruit is harvested is by preventing infection by the reduction of propagules which contaminate crops in the field or cause incipient or latent infections. A few examples to show the importance of preventing early incipient infections in the field are: *Botrytis* blossom blight on cherries is not serious in California, but blossom sprays are needed to prevent incipient infections on developing green fruit. Preharvest sprays during the blossoming period are essential to postharvest control of *Botrytis* on grapes. *Gloesporium* infections on bananas are not visible when harvested at three-quarters maturity, but as the fruit ripens, the fungus quickly rots the fingers.

Another point that needs emphasis in the use of chemicals is the proper identification of pathogens involved in postharvest decay. Too often decay is reported as caused by moulds or mildew, which is meaningless. We emphasize the need to identify certain pathogens by their generic and specific epithets because different strains vary in their pathogenicity. It is not uncommon to find that a chemical is highly effective on *Rhizopus stolonifer* but not on *R. arrhizus*, *R. oryzae*, and *R. circinans*. Sometimes chemically tolerant and sensitive strains of the same fungus are found together and special tests are required to identify them.

From the standpoint of decay by pathogens, injuries are the most common sites for infection. During normal hand or mechanical harvest operations, visible and invisible injuries occur. Also many of the pathogens, including *Monilinia*, *Botrytis*, *Penicillium*, and *Alternaria*, can penetrate healthy areas. The genetic characteristics of a fruit, such as a tendency to crack or have a deep suture, make chemical control more difficult or impossible. On bananas (Wardlaw, 1972) the cut portion of the hand (cushion) and on pineapples the cut stems become the main avenues for infection. An important aspect of decay control is to minimize open wounds and to protect these infection sites. To prevent such open wounds, citrus fruits are carefully harvested by clippers to retain the fruit button and the button kept alive with 2,4-dichlorophenoxyacetic acid (2,4-D) applications to prevent infection from *Diplodia*. One then asks, must these open wounds be protected immediately? The answer is that a certain amount of delay is not adverse because germinated spores are more susceptible to chemicals.

The environment during marketing lends to infection and decay. After harvest, special precautions are taken to remove quickly the field heat from the produce by such methods as hydrocooling, vacuum cooling, and forced air cooling. Packaged products are held at various temperatures (13°C for bananas and sweet potatoes and below 5°C for many other products) to delay decay. Yet, when the package reaches the wholesale market, it is exposed to ambient temperatures optimum for decay. Free water necessary for bacterial multiplication or spore germination and infection is provided by condensation resulting from change in ambient temperature. The wholesaler, retailer, or the consumer must ripen fruit such as peaches, bananas, and tomatoes by holding them at a higher temperature. Decay is most likely to occur during this ripening period. Yet, it is surprising to find relatively few reports of unsaleable lots of produce.

The chemicals currently registered for use (U.S.E.P.A.), the spectrum of pathogens controlled, the methods of application, and chemical residues permitted in the United States are presented in Table I. Listed also in Table I are chemicals used for animal feed, because mycotoxins are associated with deteriorated feed and discussed in this session, and treatments for food containers and processing.

The possible mechanisms of decay control are also discussed because they are important in the contamination of the crops. The total effort is to reduce the pollution by fungi and bacteria which compete with man for food, and possibly expose his life and animals to acute poisoning or chronic diseases.

The mechanisms of action of the 32 chemicals listed as postharvest decay control chemicals in Table I are largely that of *fungistasis* instead of a fungicide. The chemicals inhibit spore germination or reduce the rate of germination and growth. Some compounds, such as benomyl and thiabendazole, have little effect on germination but show specific action on hyphal growth. These chemicals must come in direct contact with the pathogen and remain at a minimum specific concentration for inhibition which is 100 ppm for captan against *Monilinia*, 2·0 ppm for Botran against *Rhizopus*, and 0·1 ppm for benomyl against *Penicillium*.

The most extensive use of CA in this country and throughout the world has been with apples. Originally, CA was used to lengthen the storage life of apples susceptible to cold damage by preventing such physiological diseases as soft scald, brown core, and soggy breakdown. Today, it is used to extend the storage life of many cultivars of apples (Anderson, 1967; Dalrymple, 1967; Dewey *et al.*, 1969). Commercially, fruits are sealed in a room so that their respiration reduces the O_2 and increases CO_2 in the atmosphere. The O_2 content may also be reduced by use of a catalytic oxygen burner. When the desired atmosphere is reached, the proper proportions of O_2 and CO_2 are maintained by ventilation and by operation of scrubbers. Lime may be used to keep the CO_2 level down. In the United States, the desired atmosphere is 2–3% O_2 and 1–5% CO_2, depending on the cultivars. 'Delicious' apples may be stored in CA throughout the year and still be of acceptable quality. CA has been reported to extend the storage life of 'Bartlett' pears and is used commercially for their storage. Early-harvested fruit can be stored up to 5 months at 32°F (0°C) in an atmosphere of 0·5–1% O_2 + 5% CO_2 and still ripen to acceptable quality. Later-harvested fruits are injured by CO_2 and should be stored in 0·5–1% O_2 without CO_2 (Claypool, 1973).

CA also has shown definite promise with peaches and nectarines but, to date, has not been used commercially for their storage. The normal storage life of most cultivars of peaches is about 3 weeks in air at 32°F (0°C) and only about 10 days at 40°F (5°C). However, some cultivars of peaches stored up to 9 weeks at 32°F, in an atmosphere of 1% O_2 and 5% CO_2, were still of acceptable quality (Anderson and Penney, 1975; Anderson *et al.*, 1969) (Fig. 4). Peaches from this atmosphere ripened normally in air at 65°F (18·3°C) and

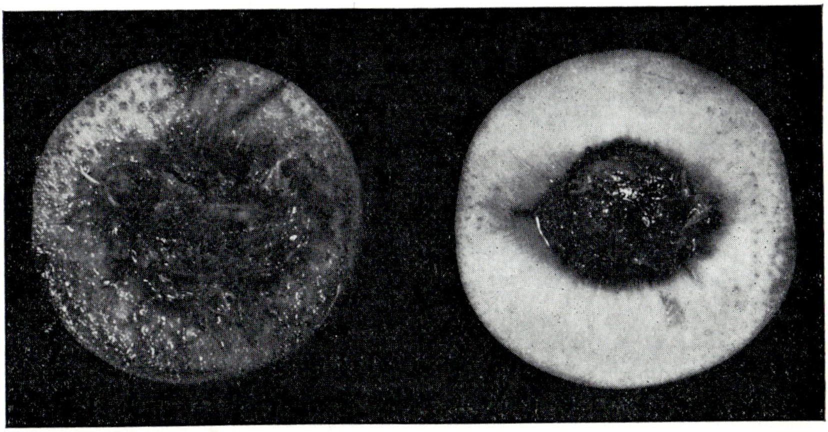

Fig. 4. Nectarines stored 9 weeks at 32°F in air (left) and in CA of 1% O_2 + 5% CO_2 (right).

did not develop the red to brown, dry, grainy flesh so characteristic of fruit held in air at 32°F. The 1% O_2 + 5% CO_2 atmosphere inhibited the development of decay during the storage period, but decay developed when the fruits were ripened in air at 65°F. Studies also have shown that intermittent warming of fruit during CA or air storage will lengthen the expected storage life (Anderson and Penney, 1975).

In another form of CA, very high concentrations of CO_2 (20–30%) have been used to control postharvest decays, particularly of strawberries and cherries (Brooks *et al.*, 1932). Often dry ice is used during transit both to lower temperature and increase CO_2 (Harvey, 1967). As with other forms of CA, this treatment usually has no lasting effect against decay when produce is moved to a normal atmosphere.

The value of CA in preserving the quality of vegetables is not as clearly demonstrated as it is with fruits (Stoll, 1974). Studies show that atmospheres of 3–5% O_2 and 3–5% CO_2 help maintain the green colour of stored cabbage but have little effect on decay (Isenburg and Sayles, 1969). Although its value to lettuce is somewhat controversial, CA has been used extensively in commercial shipments of this commodity. Mechanically refrigerated cars are flushed with nitrogen until the O_2 level is reduced to about 2%, and build-up of CO_2 during transit is partially controlled by the inclusion of bags of lime. The main benefit is reduction of a physiological disease known as 'russet spotting'. Accumulation of CO_2 to 2% and above may cause CO_2 injury of lettuce. The injury has been named 'brown stain' (Lipton, 1968). CA also has a harmful effect on potatoes, since with low O_2 and high CO_2, wound barriers do not develop, and decay-producing organisms gain entrance (Lipton, 1971). Other research studies show that an atmosphere of 3% O_2 with or without 5% CO_2 is beneficial in extending the storage life of mature green tomatoes (Parsons *et al.*, 1964) and that high CO_2 concentrations help extend the storage life of asparagus.

Hypobaric storage

Experimentally, hypobaric storage, or storage at less than normal atmospheric pressures, has been shown to aid in preserving quality of certain fruits and vegetables. Mature green tomatoes stored at 180–190 mm of Hg (1/4 atmosphere) did not ripen to the same extent as those stored at 380 mm Hg and higher (Tolle, 1969). Similar results were obtained with immature sweet cherries. Both the tomatoes and cherries ripened to full colour after removal to normal atmosphere (Tolle, 1972). Studies also indicated that strawberry decay was controlled by 1/4 atmospheric pressure (Tolle, 1972). Reduced atmospheric pressure appears to have a variable effect on growth and survival of fungi and bacteria in culture, depending on the species.

Relative humidity

Regardless of storage or holding temperatures, relative humidity of 80% and above is essential to prevent desiccation of the produce. Studies have shown that shrivelling of apples and pears can be prevented simply by using plastic liners in the storage boxes or bins (Hardenburg, 1974). This practice has been widely adopted commercially. High humidity obtained with polyethylene-lined cartons also reduces weight loss and scarring of bananas during transit in refrigerated ships (Hardenburg, 1974). Most workers formerly believed that very high relative humidity increased decay of produce. Recent studies showed that decay of carrots, parsnips, potatoes, and cabbage was no greater when they were stored at 98–100% than at 90–95% relative humidity (Van den Burg and Lentz, 1973). The produce was stored in rooms with double walls and with cooled air circulated above the ceiling, within the walls, and beneath the floor (Lentz and Van den Burg, 1973) (jacketed storage). Under those conditions, temperature change within the storage area was largely prevented and the probability of moisture condensing on the produce was lessened. Moisture condensed on the produce, not the relative humidity of the air, is generally believed to enhance decay.

Heat treatments

Heat treatments to control decay of fresh produce were first used on citrus and tomatoes in about 1920 (Fawcett, 1922; Rosenbaum, 1920). They were never adopted to any extent commercially until recently. In Hawaii and Australia the main emphasis has been on the use of hot humid air (Akamine and Arisumi, 1952; Tindale, 1961). In the United States and other parts of the world, emphasis has been on hot-water treatments. Interest in hot-water treatments was renewed when they, unlike existing chemical treatments, were found to destroy even those pathogenic organisms that had penetrated beneath the skin of the produce (Smith, 1962). Moreover, pathologists have been under considerable pressure to develop treatments for control of post-harvest decay with the least possible chemical residues.

In the United States, the first recent report of the effectiveness of hot-water treatments was on the control of peach decays (Smith, 1962). The study showed that 2·5 min dips in water at 130°F (54°C) effectively controlled *Monilinia* and *Rhizopus* decays of peaches and nectarines, even though the organisms were beneath the skin of the fruit (Fig. 5). Since that study, the

Fig. 5. Brown rot of treated and nontreated peaches after 6 days at 65°F. (Left): nontreated; (right): treated 125°F (52°C) water for 2½ min.

author has used a lower temperature, 125°F (52°C), as sometimes fruit had been injured by the higher temperature water. The hot-water treatment has been adopted commercially to a limited extent by the stone fruit industry in this country and for fresh market fruit in France. Although the 125°F water treatment caused no visible injury to fresh market peaches, it caused severe injury to peaches stored for extended periods in air or CA.

The reports on hot water for the control of peach decays attracted worldwide interest and resulted in tests conducted on citrus, mangoes, apples, papayas, melons, cherries, tomatoes, bell peppers, and raspberries (Coney and Follstad, 1966; Hardenburg, 1967; Johnson, 1968; Shiffman-Nadel and Cohen, 1965; Sharples, 1967; Smith, 1962, 1965; Smoot and Melvin, 1963; Smoot and Segall, 1963). Usually temperatures of about 125°F (52°C) were found to be the safest and most effective. As a result of these tests, commercial adoption of hot-water treatment is now widespread for mangoes and somewhat less so for cherries and melons. With melons in particular, 30-second exposure to 130°F (54°C) water has been especially effective in controlling stem-end rots.

Previous to the report of hot water on peaches were reports from Hawaii and Australia on the use of hot moist air to control decay of papayas and peaches (Akamine and Arisumi, 1952; Shoji, 1951; Tindale, 1961). Use of hot moist air or hot water has become standard for control of anthracnose on papayas in Hawaii, South Africa, and other areas of the world. For the hot-air treatment, fruits are usually exposed to saturated air of 110°F (43°C) for 8 hr and to air of about 115°F (46°C) for 5 hr; for the hot-water treatment, fruits are placed in water at 115–128°F (49°C) for 20 min. In Australia, peaches are treated with about 104°F (40°C) air for 24 hr before they are stored or ripened for canning (Tindale, 1961). Treatment in hot saturated air at 110°F for $\frac{1}{2}$ to 1 hr also controlled *Monilinia* and *Rhizopus* decays of peaches, and treatment in saturated air of 104°F for 1 hr controlled decay of strawberries, raspberries, and figs (Coney and Follstad, 1966; Smith, 1965).

When compared on the same fruits, hot-water and hot-air treatments are about equally effective in controlling decay. The hot-water treatments are faster and are less apt to injure the fruits than hot-air treatments. Especially with hot-water treatments, the temperature rise is on and just under the skin of the fruit and usually does not necessitate long cooling periods to reduce fruit temperatures afterwards. The hot-water treatments also have distinct possibilities for home or consumer use since so little additional equipment is necessary. With both hot-water and hot-air treatments, careful sanitation and proper packing are usually necessary to prevent recontamination of the produce. Because heat treatments do not protect against recontamination, perhaps their best use would be in combination with chemical treatments. Tests show that when fungicides are added to hot water both the water temperature and concentration of the fungicide can be lowered without affecting decay control.

Sanitation

Unsanitary conditions prevailing during packing and marketing often contribute heavily to postharvest decay. This is illustrated by Table II. In these tests peaches were sampled when they arrived at the packing house and at different places in the packing line. All fruits then were held for six days at 65°F (18·3°C) for ripening and possible decay development. Decay increased due to contamination of the fruit on unsanitary packing lines (Table II). Similar increases in decay also occur on many other fruits and vegetables under similar conditions. Washing or steam cleaning of the packing lines would greatly reduce the population of contaminating organisms and, hence, decay. If the produce becomes contaminated, a subsequent treatment with hot water, for example, as shown in the table, would quite effectively prevent an increase in postharvest decay.

Table II
DECAY OF PEACHES SAMPLED AT DIFFERENT LOCATIONS IN THE PACKING LINE AND THEN HELD 6 DAYS AT 65°F

Location of sample	*Decay* (%)
Field bins	2·0
After water dump	32·0
After defuzzer	38·5
After treatment in 52°C H_2O for 2 min	2·0
After hydrocooling	6·0

SUMMARY

This brief report summarizes the many nonchemical methods that are used to control postharvest deterioration of fruit and vegetables. Each of them effectively controls within its specific limitations. Only the heat treatments destroy organisms which cause postharvest decay. For any of the non-chemical treatments to control postharvest deterioration for long periods, careful sanitation to prevent recontamination must be practised.

REFERENCES

Akamine, E. K., and Arisumi, T. (1952). The control of postharvest storage decay of papaya. Hawaii Agr. Expt. Sta. Progr. Notes, **82**, 6 pp.

Anderson, R. E. (1967). Experimental storage of eastern-grown 'Delicious' apples in various controlled atmospheres. *Proc. Am. Soc. Hort. Sci.*, **91**, 810–820.

Anderson, R. E., and Penney, R. W. (1975). Intermittent warming of peaches and nectarines stored in a controlled atmosphere or air. *J. Am. Soc. Hort. Sci.*, **100**, 151–153.

Anderson, R. E., Parsons, C. S., and Smith, W. L. Jr. (1969). Controlled atmosphere storage of eastern-grown peaches and nectarines. U.S.D.A., ARS, MRR-836, 19 pp.

Artschwager, E. (1927). Wound periderm formation in the potato as affected by temperature and humidity. U.S.D.A., *J. Agr. Res.*, **35**, 995–1000.

Bennett, A. H. (1963). Thermal characteristics of peaches as related to hydrocooling. U.S.D.A., AMS Tech. Bull. 1292, 38 pp.

Bennett, A. H., Soule, J., and Yost, G. E. (1965). A prototype commercial forced-air precooler. U.S.D.A., AMS 52-9, 11 pp.

Brooks, C. *et al.* (1932). Effect of solid and gaseous carbon dioxide upon transit diseases of certain fruits and vegetables. U.S.D.A. Tech. Bull. 318, 60 pp.

Claypool, L. L. (1973). Further studies on controlled atmosphere storage of 'Bartlett' pears. *J. Am. Soc. Hort. Sci.*, **98** (3), 289–293.

Couey, M. H., and Follstad, M. N. (1966). Heat pasteurization for control of postharvest decay in fresh strawberries. *Phytopathology*, **56**, 1345–1347.

Dalrymple, D. G. (1967). The development of controlled atmosphere storage of fruit. U.S.D.A. Div. of Marketing & Utilization Services. Fed. Ext. Serv., 56 pp.

Dewey, D. H., Herner, R. C., and Dilly, D. R. (1969). Controlled atmospheres for the storage and transport of horticultural crops. Mich. State Univ. Hort. Rept. 9, 155 pp.

Fawcett, H. S. (1922). Packing house control of brown rot. *Calif. Citrog.*, **7** (7), 223, 254.

Fidler, J. C., and Coursey, D. G. (1969). Low temperature injury to tropical fruit. Proc. Conf. Tropical and Subtropical Fruits, pp. 103–110, Tropical Products Inst., London, Sept. 1969.

Hardenburg, R. E. (1967). Hot water and chemical treatments to control scald in 'Stayman' apples. *Proc. Am. Soc. Hort. Sci.*, **90**, 484–490.

Hardenburg, R. E. (1974). Use of plastic films in maintaining quality of fresh fruits and vegetables during storage and marketing. Symp. Semi-annual Meeting ASHRAE, Chicago, Ill., ASHRAE Bull. CH-73-7, 19–29.

Harvey, J. M. (1967). Modified atmospheres in transport. United Fresh Fruit and Vegetables Assoc. Yearbook, pp. 193–194, 196.

Hruschka, H. W., Smith, W. L. Jr., and Baker, J. E. (1969). Reducing chilling injury of potatoes by intermittent warming. *Am. Pot. J.*, **46** (2), 38–53.

Isenburg, F. M., and Sayles, R. M. (1969). Modified atmosphere of Danish cabbage. *J. Am. Soc. Hort. Sci.*, **94** (4), 447–449.

Jorgensen, E. G. (1973). Jacketed fruit and vegetable storages in Canada—Design and operating experience. Symp. Semiannual Meeting ASHRAE, Chicago, Ill., ASHRAE Bull. CH-73-7, 35–39.

Johnson, H. B. (1968). Heat and other treatments for cantaloups and peppers. United Fresh Fruit and Vegetables Assoc. Yearbook, pp. 51–52, 54, 56.

Lentz, C. P., and Van den Berg, L. (1973). Factors affecting temperature, relative humidity, and moisture loss in fresh fruits and vegetables. Symp. Semiannual Meeting ASHRAE, Chicago, Ill., ASHRAE Bull. CH-73-7, 5–12.

Lipton, W. J. (1968). Low O_2 atmospheres benefits and dangers. United Fresh Fruit and Vegetable Assoc. Yearbook, pp. 99–100, 103.

Lipton, W. J. (1971). Controlled atmosphere effects on lettuce quality in simulated export shipments. U.S.D.A. ARS 51-45, 14 pp.

Lyons, J. M. (1973). Chilling injury in plants. *Ann. Rev. Plant Physiol.*, **24**, 445–466.

McColloch, L. P., and Worthington, J. T. (1952). Low temperature as a factor in the susceptibility of mature green tomatoes to alternaria rot. *Phytopathology*, **42**, 425–427.

Moline, H. E. (1975). Ultrastructural changes associated with chilling of tomato fruit. *Physiol. Plant Pathol.*

Parsons, C. S., Gates, J. E., and Spalding, D. H. (1964). Quality of some fruits and vegetables after holding in nitrogen atmospheres. *Proc. Am. Soc. Hort. Sci.*, **84**, 549–556.

Rosenbaum, J. (1920). Infection experiments on tomatoes with *Phytophthora terrestria* Serb. and a hot water treatment of the fruit. *Phytopathology*, **10**, 101–105.

Shiffmann-Nadel, M., and Cohen, E. (1965). The influence of time of treatment and of conditions in the grove on the effectiveness of heat in control of brown rot of citrus fruits. Volcani Inst. Agr. Res. Prelim. Rept. No. 485, 8 pp.

Sharples, R. O. (1967). The effect of a post-harvest heat-treatment on the storage behaviour of Cox's orange pippin apple fruits. *Ann. Appl. Biol.*, **59**, 401–406.

Shoji, K. (1951). Storage studies of vapor-heat and ethylene-dibromide treated papaya. Hawaii Agr. Exp. Sta. Progr. Notes 67, 5 pp.

Smith, W. L., Jr. (1962). Reduction of postharvest brown rot and rhizopus decay of eastern peaches with hot water. U.S.D.A. *Plant Dis. Rept.*, **46** (12), 861–865.

Smith, W. L., Jr. (1965). Heat treatments—New techniques for reduction of decay of fresh produce. United Fresh Fruit and Vegetable Assoc. Yearbook, pp. 187–193.

Smith, W. L., Jr., and Smart, H. F. (1956). Relation of soft rot development to protective barriers in Irish Potato slices. *Phytopathology*, **45**, 649–654.

Smoot, J. J., and Melvin, C. F. (1963). Hot water as a control for decay of oranges. *Citrus Ind.*, **45** (2), 5, 7, 13.

Smoot, J. J., and Segall, R. H. (1963). Hot water as a postharvest control of mango anthracnose. U.S.D.A. *Plant Dis. Rept*, **47** (8), 739–742.

Stoll, K. (1974). The storage of vegetables in controlled atmospheres. *Int. Inst. Refrigeration*, **54** (6), 1302–1324.

Tindale, G. B. (1961). Cool storage of desert peaches. *J. Agr. Victoria*, **59**, 258.

Tolle, W. E. (1969). Hypobaric storage of mature-green tomatoes. U.S.D.A. ARS MRR-842, 9 pp.

Tolle, W. E. (1972). Hypobaric storage of fresh produce. United Fresh Fruit and Vegetable Assoc. Yearbook, pp. 27–38, 43.

Van den Berg, L., and Lentz, C. P. (1973). Effects of relative humidity on decay and other quality factors during long term storage of fresh vegetables. Symp. Semi-annual Meeting ASHRAE, Chicago, Ill., ASHRAE Bull. CH-73-7, 12–18.

NATURE AND IMPORTANCE OF MYCOTOXINS IN GRAINS

URBAN L. DIENER, NORMAN D. DAVIS and GARETH MORGAN-JONES

Department of Botany and Microbiology, Auburn University
Agricultural Experiment Station, Auburn, Alabama 36830, U.S.A.

Summary

Mycotoxins are chemical substances produced by fungi that are toxic to animals when ingested. This is particularly important to man because foodstuffs such as rice, corn, wheat, rye, barley, peanuts, and cottonseed and their products may be contaminated by mycotoxins that sometimes result in serious mycotoxicoses in man and animals. Fungi that elaborate the more widely investigated mycotoxins are both plant pathogens and postharvest saprophytes. They are classified principally in the Deuteromycotina Class Hyphomycetes in *Aspergillus*, *Fusarium*, and *Penicillium* genera, except for *Claviceps*, which is an Ascomycete. Mycotoxins can be extracted readily from naturally contaminated foods and feeds with a range of polar and nonpolar solvents, and analysed with such instrumentation as GLC, TLC, IR, UV, NMR, MS, and HPLC. Molecular weights range from 154 to 697, with most falling between 300 and 600. A single toxin may affect a wide range of experimental and domesticated animals (including man) or be relatively specific to only one or two species. Mycotoxins have been found to be carcinogenic, teratogenic, and/or tremorgenic; most are hepatotoxins, nephrotoxins, or neurotoxins. The LD_{50} in rats ranges from 5–7 mg/kg for the common mycotoxin aflatoxin B_1 to over 200 mg/kg for citrinin.

INTRODUCTION

Any discussion of 'the nature and importance of mycotoxins in grains' could start with the pioneering research of Clyde Christensen and his students and the reviews he has prepared over the past 20 years (Christensen, 1957, 1971; Christensen and Kaufman, 1965, 1969, 1974).

According to Christensen and Kaufman (1974), fungi are a major cause of spoilage in stored grains and seeds, and probably rank second only to insects as a cause of deterioration and loss in all kinds of stored products throughout the world. In technologically advanced countries, where insects and rodents are controlled in commercial storage, fungi probably cause more spoilage than any other agent. Due to their inconspicuousness, the fungi that cause discoloration, heating, and spoilage of stored grains often go unrecognized. Christensen (1957) classed such fungi into two ecological groups: (1) field fungi—because they invade seeds before harvest and require a high moisture content in equilibrium with relative humidities of 90 to 100%, and (2) storage fungi—because they invade during storage starting on the farm or at the elevator and have the ability to grow at moisture contents in equilibrium with relative humidities of 70 to 90%. The storage fungi comprise only about 10 to 15 group species of *Aspergillus* and several species of *Penicillium* (Raper and Fennell, 1965). Under environmental conditions favourable for growth, storage fungi cause decreases in germinability, discoloration, heating, mustiness, total decay, and produce mycotoxins.

Mycotoxins are toxic fungal metabolites that have caused mortality in man and animals since the days of 'Holy Fire' (ergotism) in France during the Middle Ages to the epidemics of alimentary toxic aleukia (ATA) in Russia in 1932 and 1942–47. Recent interest in mycotoxins resulted from the discovery of aflatoxin (a toxic metabolite of the *Aspergillus flavus* group), headlined by the disturbing fact that aflatoxin B_1 is the most carcinogenic, naturally occurring substance known (Wogan and Newberne, 1967). Research on aflatoxin has turned the spotlight on the total problem of mycotoxins in human foods and animal feeds. This subject has been thoroughly reviewed recently (Ciegler *et al.*, 1971; Goldblatt, 1969; Kadis *et al.*, 1971, 1972; Lillehoj *et al.*, 1970; Purchase, 1971, 1974). This paper will be limited to the nature and importance of the aflatoxins, ergot, ATA, *Fusarium* toxins in rice, and the yellowed rice syndrome.

A reviewer of scientific literature attempting to evaluate data on the importance and occurrence of mycotoxins in grains is provided with (a) direct evidence from analysis of field samples of mouldy or nonmouldy grains for naturally occurring mycotoxins, and (b) indirect evidence from analysis of sterilized grains or lab media fermented by pure cultures of fungi isolated from grains or feeds suspected to be the cause of a toxicosis of animals or man. The data presented on aflatoxin are primarily analyses of naturally occurring contamination of grains on the farm or in storage. Much of the remaining data is a mixture of direct and indirect evidence, with the emphasis being placed on direct evidence where available.

NATURE AND IMPORTANCE OF THE AFLATOXINS

Of the known mycotoxins, the most important from the viewpoint of possible hazard to human health are the aflatoxins (Scott, 1973). They are certainly the most studied and illustrate several aspects of mycotoxin research that need to be explored with other fungal toxins. The aflatoxins are primarily produced by *Aspergillus flavus* and *A. parasiticus* of the *A. flavus* group (Raper and Fennell, 1965) and may be the only source, despite some data to the contrary (Diener and Davis, 1969). The *A. flavus* group is a constituent of the microflora in air and soil and is found on or in living or dead plants and animals throughout the world. The role of *A. flavus* as one of the storage fungi that contributes to deterioration in stored grains, flour, soybeans, and peanuts was reviewed recently (Christensen and Kaufman, 1974). Aflatoxins have been detected not only in cereal grains, oilseeds, and other major food crops, but in numerous tropical and exotic agricultural commodities, and meat and cheese, which indicates the ubiquitous nature and widespread distribution of this fungus.

Conditions favouring growth of *A. flavus* and *A. parasiticus* and aflatoxin production in grain and grain products are moisture contents in equilibrium

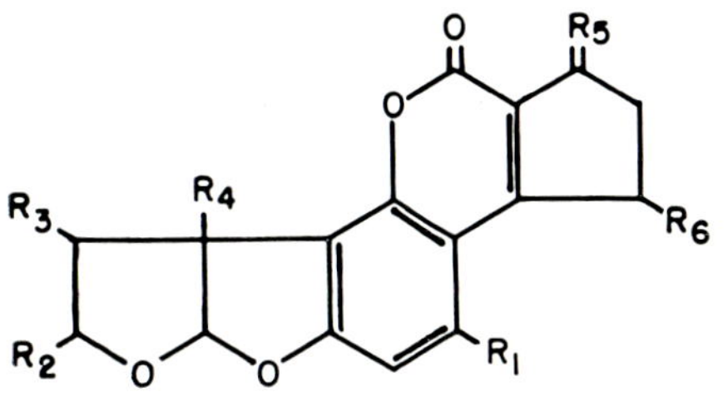

AFLATOXIN B series

Toxin	R_1	R_2	R_3	R_4	R_5	R_6
B_1	OCH_3	H	H	H	O	H_2
B_2	OCH_3	H_2	H_2	H	O	H_2
M_1	OCH_3	H	H	OH	O	H_2
M_2	OCH_3	H_2	H_2	OH	O	H_2
B_{2a}	OCH_3	H,OH	H_2	H	O	H_2
R_o	OCH_3	H	H	H	OH	H_2
1-methoxy-aflatoxin B_2	OCH_3	CH_3O	H_2	H	O	H_2
2-methoxy-aflatoxin B_2	OCH_3	H_2	CH_3O	H	O	H_2
1-ethoxy-aflatoxin B_2	OCH_3	C_2H_5O	H_2	H	O	H_2
1-acetoxyaflatoxin B_2	OCH_3	CH_3CO_2	H_2	H	O	H_2
3-hydroxy-aflatoxin B_1 or Q_1	OCH_3	H	H	H	O	OH
P_1	OH	H	H	H	O	H_2

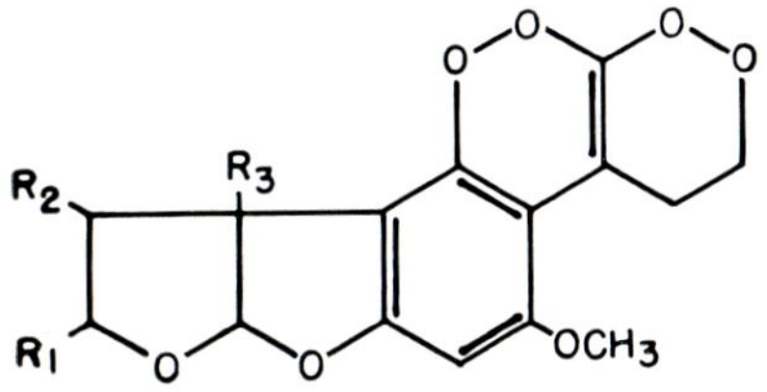

AFLATOXIN G series

Toxin	R_1	R_2	R_3
G_1	H	H	H
G_2	H_2	H_2	H
G_{2a}	OH	H_2	H
GM_1	H	H	OH
1-ethoxy-aflatoxin G_2	C_2H_5O	H_2	H

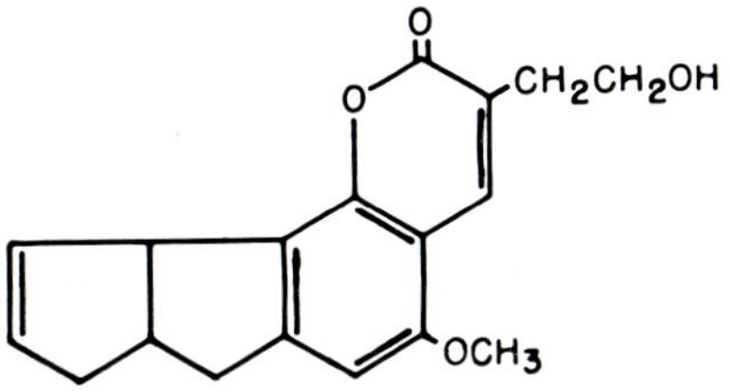

PARASITICOL

(Aflatoxin B_3)

Fig. 1. Structures of the aflatoxins.

with relative humidities (RH) of 85% or higher and temperatures of 25–30°C (Diener and Davis, 1969), with maximal production being attained in 5–9 days, depending on the substrate. Aflatoxin production is highest (1–2 g/kg) on natural substrates such as rice and wheat, while a simple semisynthetic medium of 15–20% sucrose and 2% yeast extract gave yields of 45–63 mg of total aflatoxins per 100 ml of media in stationary culture (Davis *et al.*, 1966).

The structure and chemistry of the aflatoxins have been extensively reviewed (Buchi and Rae, 1969; Detroy *et al.*, 1971). There are 18 aflatoxins described in the literature (Fig. 1). Detroy *et al.* (1971) gave structural formulas for 16, with two others, P_1 and Q_1, being described later by Dalezios *et al.* (1971) and Masri *et al.* (1974). However, Ciegler (1975) indicated 13 current naturally occurring aflatoxins.

Aflatoxin B_1, the most widely known mycotoxin, has an empirical formula of $C_{17}H_{12}O_6$, a molecular weight of 312, and a melting point of 268–269°C. Crystals precipitate from benzene, methanol, or chloroform-ethanol. B_1 fluoresces blue on TLC at R_f 0·50 in chloroform-methanol (97:3).

There is much known about the biological activity, toxicology, and biochemical and carcinogenic effects of the aflatoxins. The reader is referred to numerous reviews, chapters, and books by the authority in this field (Wogan, 1965, 1966a, 1966b, 1969a, 1969b, 1972, 1973; Mateles and Wogan, 1967; Wogan and Mateles, 1968; Wogan and Pong, 1970). In summarizing aflatoxin carcinogenesis, Wogen (1973) pointed out that the aflatoxins are potent toxins in a broad spectrum of biological systems and are carcinogenic to some mammals, fish, and birds. Although direct evidence on human responsiveness to aflatoxin is scanty, most of the direct and indirect evidence linking exposure to disease incidence has been reviewed by Wogan (1973) and most recently by Campbell and Stoloff (1974). Man's response to aflatoxin in the diet is assumed to depend on many factors such as age, nutritional status, concurrent exposure to toxic herbal agents, genetic factors, concurrent illness, as well as the dose and duration of exposure. Campbell and Stoloff (1974) reviewed the published data of aflatoxin research on comparative metabolism, primate studies, inadvertent human incidents, and epidemiology.

Aflatoxin B_1 is the most common and potent toxin of all the aflatoxins. LD_{50} oral in day-old ducklings was 364 μg/kg, ip in guinea pigs was 1·4 mg/kg, and oral in male rats was 5·55 mg/kg. Aflatoxin B_1 was carcinogenic to rats when fed at 15 ppb in the diet. Levels of 4–8 ppb in the daily diet induced tumors in rainbow trout in 12 months; 0·8 ppb caused tumors in 20 months.

A chapter in a recent book on grain storage presents an excellent review of mycotoxins in stored grain, feeds, and grain products (Scott, 1973). A modified version of Scott's table of data on reported natural occurrences of aflatoxins in grains and grain products is presented in Table I. In general, the incidence of aflatoxins in grains and grain products appears low in the United States and most other countries except France. However, Campbell and Stoloff (1974) summarized more recent data of several investigators in the Philippines, South Africa, Thailand, and Uganda relative to aflatoxin B_1 contamination of over 2000 samples of corn, rice, peanuts, millet, legumes, and other foods. These data showed that aflatoxin contamination of selected foods was not uncommon, with levels occasionally reaching 0·5–1·0 ppm.

Corn and other grains are currently undergoing widespread investigation for mycotoxin-producing fungi and the presence of mycotoxins. One of the more detailed, early investigations in the United States of a serious mycotoxin outbreak was that of mouldy corn toxicosis, which occurred in west Florida and Georgia between 1952 and 1957 and resulted in extensive losses of swine and cattle (Forgacs and Carll, 1962; Detroy *et al.*, 1971). Initial studies reproduced the disease in swine by feeding them mouldy corn (Sippel *et al.*, 1953). Later, Burnside *et al.* (1957) isolated 13 fungi (9 of which were *A. flavus*)

Table I

NATURAL OCCURRENCE OF AFLATOXINS IN GRAINS AND GRAIN PRODUCTS

Grain	*Country*	*Samples*	*Afla-ppb*	*Reference*
Wheat	U.S.	2 of 531	B_1–7, G_1–2	Shotwell *et al.*, 1969a
	Ethiopia	1	B_1–42; G_1–21	Coady (*see* Scott: 21)
	France	47%	B_1 up to 225	LaFont & LaFont, 1970
	U.K.		traces	Raymond, 1966
Flour	France	28%	B_1 up to 150	LaFont & LaFont, 1970
	Germany			Frank & Eyrich (Scott: 38)
Spaghetti	Canada	1	B_1–13	Walbeek *et al.* (Scott: 141)
Corn	U.S.	35 of 1 311		Shotwell *et al.*, 1969b
	U.S.	6 of 283	B_1–12–25	Shotwell *et al.*, 1970
	U.S.	8 of 293	B_1–7–25	Shotwell *et al.*, 1971
	U.S.	394 of 1 283	6% < 50	Hesseltine *et al.*, 1975
	U.S.-ALA		B_1 up to 8 733	Diener & Davis, 1969
	U.K.		traces	Raymond, 1966
Rice	U.S.	?	B_1–50	Schroeder *et al.*, 1968
	Taiwan	1	B_1–226	Tung and Ling (Scott: 132)
	Senegal	53%	30–2 000	Toury *et al.* (Scott: 127, 128)
Oats	U.S.	3 of 304	B_1–6	Shotwell *et al.*, 1969a
	France	42%	B_1–>100, G_1>65	LaFont & LaFont, 1970
Sorghum	U.S.	6 of 533	B_1–3–10; G_1–3–19	Shotwell *et al.*, 1969a
	Australia	1	B_1–8 000	Connole & Hill, 1970
	France	25%	B_1>100	LaFont & LaFont, 1970
	Ethiopia	1	B_1 75	Coady (Scott: 21)
	Senegal			Toury *et al.* (Scott: 128)
Rye	France	19%	B_1>100	LaFont & LaFont, 1970
Barley	France	7%	B_1>10	LaFont & LaFont, 1970
	U.K.		traces	Raymond, 1966
Soybeans	U.S.	2 of 866	B_1 7–10	Shotwell *et al.*, 1969b
Feeds	Poland	306	7·3% 4·3%<100 ppb 2·6%<1 ppm	Strzelecki & Gasiorowska, 1974

and found 2 fungi, *A. flavus* and *P. rubrum*, which when cultured on sterilized corn and fed to swine produced a similar haemorrhagic toxicosis. *Penicillium rubrum*-contaminated corn was found to be more toxic than *A. flavus*-contaminated corn (Forgacs and Carll, 1962). Bailey and Groth (1959) concluded that a disease of similar etiology in dogs, which they called hepatitis-X, was identical to mouldy corn toxicosis. Later Wilson *et al.* (1967) demonstrated that the symptoms of hepatitis-X could be simulated by the administration of aflatoxin, thus showing that this disease was produced by *A. flavus*.

Whether aflatoxin B_1 or rubratoxin B or both were involved in mouldy corn toxicosis is not clear, but the animals' symptoms were not typical of those usually associated with aflatoxin.

During 1966, the mycoflora of mouldy corn from 25 storage bins in central Iowa was determined by Richard *et al.* (1969). Of 246 fungal isolates obtained from the mouldy corn, ether extracts of rice cultures of 99 isolates (13 genera) produced death in mice and/or ducklings. Forty-two isolates were *Aspergillus* species, and 15 were *Penicillium* species. Other toxic genera isolated were *Trichothecium*, *Fusarium*, *Chaetomium*, and *Nigrospora*. Aflatoxin production of 2–8 ppm in naturally infected, high quality corn in South Africa was associated with temperatures of 25°C and kernel moisture contents of 19·5–20% when stored for 14–28 days at RH of 80% (van Warmelo *et al.*, 1968). Martin *et al.* (1971) isolated 80 species of fungi from 635 samples of corn, peanuts, sorghum, corn meal and peanut meal, and other legumes collected in rural areas of Eastern Transvaal and Swaziland. Of 531 isolates tested for toxicity to ducklings, 45 of the 52 species tested were toxic to some degree. A high degree of toxicity was noted for *Aspergillus flavus*, *Nigrospora oryzae*, *Fusarium equiseti*, *Penicillium citrinum*, *P. rubrum*, and *P. rugulosum*. Also, 17 other fungi showed moderate toxicity with 37–56% of the isolates being toxic. These studies constitute both direct and indirect evidence of occurrence of aflatoxin and other mycotoxins in corn.

The natural incidence of *A. flavus* and aflatoxin is probably high in the mid to deep South of the United States because of favourable temperatures and moisture. This is a corn deficit area, so that all local corn produced, *e.g.* in Alabama, is fed to animals by the grower or in neighbouring counties. Incidence of aflatoxin in Alabama has varied with environmental conditions. In 1973, severe losses from aflatoxin were experienced in swine. Incidences in which aflatoxin analysis of the feed and study of liver pathology of the swine were done at Auburn University involved the loss of 600 swine on three widely scattered farms in the southwest, mid-west, and northeast areas of the state. Aflatoxin ran as high as 2000 ppb aflatoxin B_1. In all three cases, storage of high-moisture corn at harvest had resulted from dryer breakdown or conflict with other farm operations for a few days. Temperatures are usually favourable in the deep South, and high RH and the seasonal rains of July during harvest periods affords ideal conditions for aflatoxin production in corn unless it is dried adequately.

In the mid-South from Missouri and Tennessee to Virginia and North Carolina, aflatoxin apparently occurs only sporadically, varying with the season and management. This statement is supported by the research of Shotwell *et al.* (1973). They found that of 60 samples of white and yellow corn (grades 2–5) collected from Alabama, North Carolina, South Carolina, Tennessee, and Virginia, 21 samples contained aflatoxin. Aflatoxin B_1 levels varied from 6 to 308 ppb. Incidence in 1970 was 35% as compared to incidences of 1·7–2·3% in 1964, 1965, and 1967. Twelve samples of the 1969 crop collected in the southeast contained no aflatoxin. Later, Hesseltine *et al.* (1975) reported that 394 samples of 1283 truck loads of 1972 white corn in Missouri contained aflatoxin. Of these 394 samples, 235 had aflatoxin levels of 20 ppb or lower, while 159 had levels exceeding 20 ppb, with 77 exceeding 50 ppb. North of this area, the most severe and common mycotoxin problems are generally those caused by metabolites of *Fusarium* spp. and other fungi that produce toxins optimally at lower temperatures. The cold, wet autumn seasons of the past several years have been ideal for growth of *Fusarium* spp. in the mid-western Corn Belt. Because of corn deficit, Alabama and other southern states received the major portion of *Fusarium*-contaminated corn produced in Ohio, Indiana, and Michigan in 1972.

As pointed out by Scott (1973), rice and sorghum samples with over 1000 ppb aflatoxin have been reported from Senegal and Australia, respectively. Seitz *et al.* (1975) reported isolating *Alternaria* metabolites from grain sorghum, and earlier Burroughs and Sauer (1971) showed that the mycotoxin-producing genera *Alternaria, Cladosporium, Fusarium, Penicillium,* and *Trichothecium* as well as *A. flavus* invaded sorghum seed under varying conditions of moisture and temperature.

Since aflatoxin is the most studied mycotoxin, it seems appropriate to comment on progress of control of aflatoxin, particularly in peanuts (Diener, 1973). Processed peanuts and peanut products of the United States are now relatively free of aflatoxins. Control is not absolute but reduction and prevention of contamination of peanuts in the field, during harvest, handling, and storage has been the first step. Once the basic knowledge of the fungus and mycotoxin was obtained, progress in control was rapid. In addition, removal of toxin-containing kernels by separation and chemical inactivation or detoxification of contaminated meals for animal feeds is now possible (Gardner *et al.*, 1971). Besides a potential loss in life or health, an expensive side-effect of a mycotoxin problem is the economic losses incurred due to increased costs in curing, drying, and handling techniques of the agricultural commodity, chemical testing, segregation and elimination of contaminated lots, and detoxification of toxic lots for other than edible uses.

NATURE AND IMPORTANCE OF ERGOTISM

No presentation on mycotoxins would be complete without some consideration of ergot and ergotism. Ergotism is the earliest and best known mycotoxicosis, with recorded effects on man as far back as the Middle Ages. However, it was not until the 1930s that the alkaloids responsible for the biological effects of ergot were isolated and identified. The last major incidence of ergotism in the United States involving humans was in 1825, but serious outbreaks occurred in 1926–27 in Russia and in 1928 in England (Wogan, 1966a). Ergotism was recently reviewed by Gröger (1972).

In man ergotism or ergot poisoning has resulted from the ingestion of bread that has been prepared from grain parasitized by ergot fungi. Cattle, sheep, and swine may also be poisoned from ergot-contaminating feedstuffs or pasture grasses in the field; the latter is a recurring problem in the southern United States.

The five species of ergot-producing fungi that parasitize grain crops and fodder grasses are *Claviceps purpurea, C. paspali, C. fusiformis, C. gigantea,* and *Sphacelia sorghi* (Mantle, 1969). The most important is *C. purpurea,* which parasitizes rye, wheat, barley, and Triticale, an amphidiploid artificial genus resulting from the cross of rye and wheat (Robbers *et al.*, 1975). The ergot fungi may be divided biochemically into three groups: (1) *C. purpurea* and related species that synthesize clavine alkaloids, simple lysergic acid derivatives, and peptide alkaloids; (2) fungi (*C. paspali* type) that produce clavine alkaloids and simple lysergic acid derivatives but not the peptide alkaloids; and (3) fungi that form the ergoline ring system starting from tryptophane and mevalonic acid. These strains are unable to oxidize elymoclavine to lysergic acid. All of the ergot alkaloids and lysergic acid derivatives are produced by ergot fungi in submerged or surface culture on synthetic media (Gröger, 1972).

The structures of the ergot alkaloids have been published by Gröger (1972). Ergot alkaloids are 3,4-substituted indole derivatives with the tetracyclic ring system being designated ergoline. They may be divided into two groups: (1) acid amide type derivatives of lysergic acid and stereoisomeric isolysergic acid consisting of the ergotamine, ergotoxine, ergoxine, ergobasine, and some miscellaneous groups of alkaloids; and (2) the clavine alkaloids, of which more than 20 compounds are known.

The biological effects of natural ergot alkaloids can be divided into three groups, which are separated by their site of action: (1) peripheral effects such as vasoconstriction and contraction of the uterus; (2) neurohumoral effects such as antagonism to adrenalin and serotonins; and (3) effects of the central nervous system in reducing the activity of the vasomotor centre, while the sympathetic structures in the mid-brain, particularly the hypothalamus, are stimulated by ergot alkaloids.

Although the hazard of ergot and ergotism in the human food supply has been essentially eliminated, contamination of feeds or pasture grasses still occurs and is frequently observed by veterinarians.

NATURE AND IMPORTANCE OF ALIMENTARY TOXIC ALEUKIA

Alimentary toxic aleukia (ATA) appears to be the most recent and only mycotoxicosis in grains other than ergotism that has caused many human deaths (Wogan, 1969b). The first outbreak in 1932 involved mortality of 62% of those afflicted in one district in Russia. ATA was especially severe from 1942 to 1947, when in one district in 1944 up to 30% of the people were affected and mortality from 2 to 80% of those affected occurred in the various districts. Incidence of ATA has been restricted to the U.S.S.R. and extensive Russian research and literature exists concerning this toxicosis. There are a number of pertinent reviews (Bilay, 1960; Forgacs and Carll, 1962; Joffe, 1965, 1971; Wogan, 1966a, 1969b; Wogan and Mateles, 1968).

ATA results from the ingestion of food processed from cereal grains, particularly proso millet but also wheat and barley, that have overwintered in the fields under snow. The most frequently occurring and most toxic fungi isolated from the grains were *Fusarium poae*, *F. sporotrichioides*, *Cladosporium epiphyllum*, and *C. fagi*, each of which produces a specific toxin-poaefusarin, sporofusarin (fusariogenin), epicladosporic acid, and fagicladosporic acid, respectively. Toxin production by these fungi, which occurred at or near subzero temperatures, was maximal during sporulation. Field conditions in 1944 (peak year of ATA in Russia) approximated those for optimal toxin formation in the laboratory. The chemical structures for all four compounds have been reported by Joffe (1965, 1971). Poaefusarin is a monoglucoside with an empirical formula of $C_{35}H_{39}O_{12}$ containing xylose and a steroidic aglucone. Sporofusarin is a saponinic material with an empirical formula of $C_{65}H_{95}O_{25}$. Molecular weights of these two compounds are 951 and 1270, respectively.

Grains or laboratory media inoculated with one of these toxin-producing fungi and incubated under favourable conditions caused death in several experimental animals from a single feeding. Among the various bioassay systems, the cat responded similarly to man and represented the only species in which the entire clinical picture of ATA could be experimentally reproduced. ATA toxins are relatively heat stable in contaminated grain; withstanding a temperature of 188°C for 18 hr or longer. Grain samples stored 6 years retained their toxicity. Toxins are extractable with lipid solvents and the standardized biological assay is based on rabbit dermal toxicity of ether extracts of suspected material.

Symptoms of the disease in man include fever, haemorrhagic rash, bleeding from the nose, throat, and gums apparently from direct contact with the toxins, necrotic angina, anaemia, and leukopenia (Forgacs and Carll, 1962; Joffe, 1971). These latter symptoms grow progressively worse in 3 to 4 weeks with continued intake of toxic grains, terminating in severe anaemia and total exhaustion of bone marrow with panleukopenia and agranulocytosis. This toxicosis was observed almost exclusively in rural areas and primary preventive measures consisted of educating the rural populations as to the etiology and clinical symptoms of ATA; such measures have greatly reduced outbreaks of ATA (Joffe, 1971).

NATURE AND IMPORTANCE OF CERTAIN *FUSARIUM* TOXINS IN JAPAN

Rice is another important grain that is invaded by fungi and contaminated by mycotoxins. Rice, like other grains, is susceptible to mould damage if storage environments provide the moisture content and temperature suitable for fungal growth. Several systematic surveys in Japan have been summarized, revealing the extensive contamination of rice and other cereals by species of *Fusarium*, *Penicillium*, *Aspergillus*, and other fungi (Kinosita and Shikata, 1965; Tsunoda, 1970; Udagawa *et al.*, 1970; Saito and Tatsuno, 1971). For many years the Japanese have associated mouldy rice, wheat, barley, and oats with diseases and disorders in man. Widespread serious damage from akakabi-byo, the red mould or scab disease of wheat, barley, rice, corn, and oats first reported in 1901, occurred in almost every district in Japan, and is associated with heavy rain and cold spring seasons. Toxicity in horses, cows, and pigs and with headaches, vomiting, and diarrhoea in man was associated with scabby grain. Inspection of rice revealed that *F. graminearum* occurred in highest frequency, followed by *F. fujikuroi*, *F. nivale*, and *F. kuhnii*. In wheat and other grains, *F. graminearum* was the most frequently occurring fungus followed by *F. nivale*, *F. poae*, and *F. oxysporum*.

Nivalenol and fusarenon-X are two 12,13-epoxytricothecene mycotoxins produced by *F. nivale* grown on rice, wheat, barley, or corn and also on Czapek medium supplemented with 10 g/litre of peptone. Laboratory production of these metabolites on rice is optimal at 27–28°C for 14 days. Maximal yield of fusarenon-X was produced at 27°C in 6–8 days on peptone-supplemented Czapek's medium. Ueno (1971) indicated that fusarenon-X is a major metabolite of *F. nivale*, while nivalenol, a minor metabolite, was presumed to be derived from fusarenon-X.

The chemical structure of nivalenol and fusarenon-X (Fig. 2) indicates these metabolites are scirpenols, scirpenes, 12,13-epoxytricothecenes, and sesquiterpenoids with molecular weights of 312 and 354, respectively. The

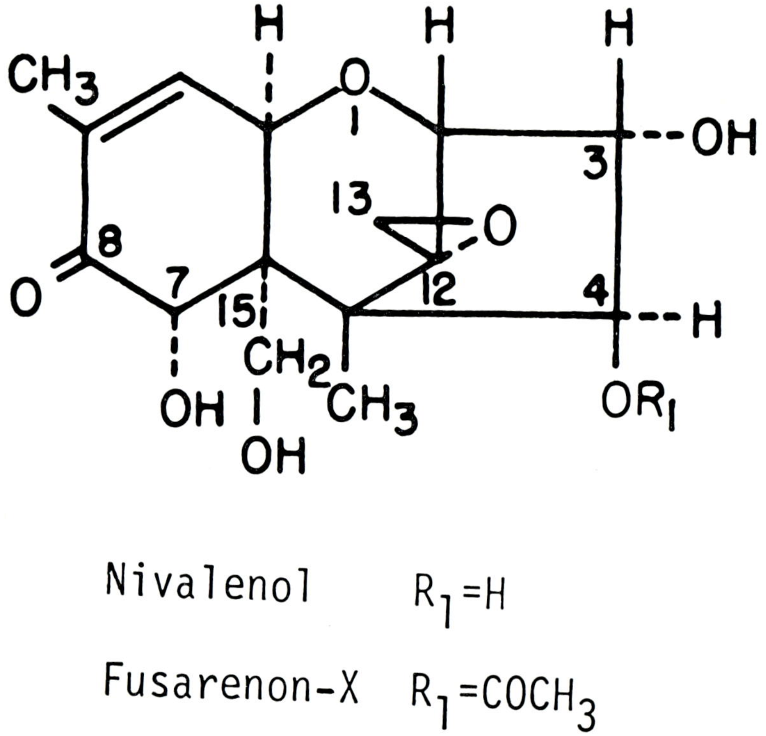

Fig. 2. Structures of nivalenol and fusarenon-X.

empirical formulas for nivalenol and fusarenon-X are $C_{15}H_{20}O_7$ and $C_{17}H_{22}O_8$, respectively. The chemistry of these two toxins was reviewed by Morooka and Tatsuno (1970).

Fusarenon-X is the most potent of these toxins and is toxic to most experimental animals by oral or ip administration. LD_{50} ip was 3·6 mg/kg in adult mice as compared to 4·1 mg/kg for nivalenol. Lesions were primarily localized in rapidly proliferating tissues such as the haemopoeticells of the bone marrow, the lymphatic tissues, and intestinal epithelium. Disturbance in DNA and protein synthesis may be the cause of these radiomimetric effects (Newberne, 1974). Ueno (1971) concludes that toxicological evidence indicates that fusarenon-X is one of the causal factors in *Fusarium* mycotoxicoses in grains in Japan.

NATURE AND IMPORTANCE OF CERTAIN *PENICILLIUM* TOXINS IN RICE

From 1947–54 more than 100 000 tons of rice imported into Japan were declared unfit for human consumption. Rice contains an abundant mycoflora as indicated by surveys of Kinosita and Shikata (1965) and those summarized by Saito *et al.* (1971), with the dominant genus being *Aspergillus*, followed closely by *Penicillium*, with the occurrence of 'toxic fungi' around 10%. Safe storage moisture for rice is 14% or less, the moisture content at which no visible mould is observed for at least 1 year (Saito *et al.*, 1971). Optimal moisture for the growth of most storage fungi is 15 to 17%. A number of *Penicillium* species have been isolated from yellowed rice and many are toxin producers.

Citreoviridin is produced by *Penicillium citreo-viride*, a fungus which was first isolated as *P. toxicarum* from yellowed rice by Miyake *et al.* (1940). It is also produced by *P. pulvillorum* and *P. ochrasalmoneum*. Inoculated polished rice incubated for 1 month at 20–24°C yielded 12–13 g of the toxin per kg of rice. Its highest production in nutrient culture was on Ushinsky's medium. Production of this toxin was extensively reviewed by Ueno (1971), including

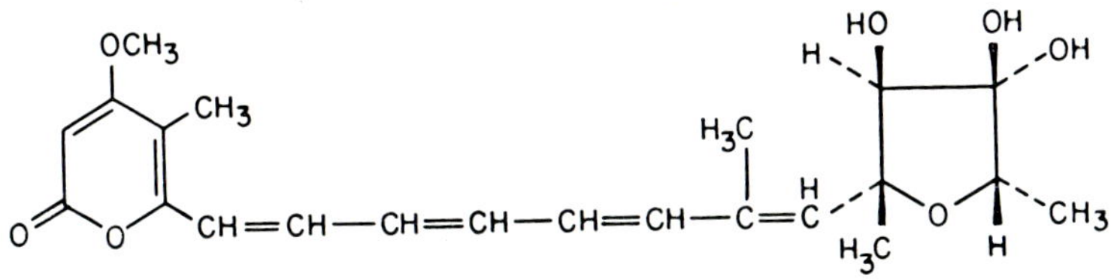

Fig. 3. Structure of citreoviridin.

the structure which is shown in Fig. 3. Citreoviridin has an empirical formula of $C_{23}H_{30}O_6{\cdot}CH_3OH$ and a molecular weight of 434.

Citreoviridin is toxic to seven mammals and five other vertebrates. It acts in three stages, excitation, ascending paralysis, and respiratory paralysis. Nervous symptoms are prominent in dogs, cats, and monkeys. Uraguchi (1971b) has carried out extensive toxicological and epidemiological investigations on mouldy rice in Japan. He demonstrated that the ethanol extract of *P. citreo-viride* was neurotoxic to several animals, causing symptoms identical to acute cardiac beriberi ('Shoshin-kakke'). He noted that rice inspection introduced in 1910 decreased the high incidence of the disease; however, this was before the discovery of vitamins so avitaminosis was not a reasonable theory for the cause of 'Shoshin-kakka' (Uraguchi, 1971a, 1971b).

Penicillium islandicum was isolated by Tsunoda in 1948 and the most systematic and detailed investigations were directed to this organism and its metabolites, luteoskyrin and cyclochlorotine (Saito *et al.*, 1971). These investigations were stimulated by the findings that rice infested by this fungus caused acute and chronic liver damage when fed to mice or rats; this was of interest in view of the high incidence of liver diseases, including primary hepatic carcinoma among Asiatic populations (Wogan, 1969b). There are two excellent reviews of this research (Uraguchi, 1971a; Saito *et al.*, 1971).

The fungus was grown on malt extract-amended Czapek liquid medium at 27°C for 2 weeks. Mycelial mats of 350 g yielded 4 g of luteoskyrin. The structure of luteoskyrin is shown in Fig. 4, and has an empirical formula of $C_{30}H_{22}O_{12}$ with a molecular weight of 574.

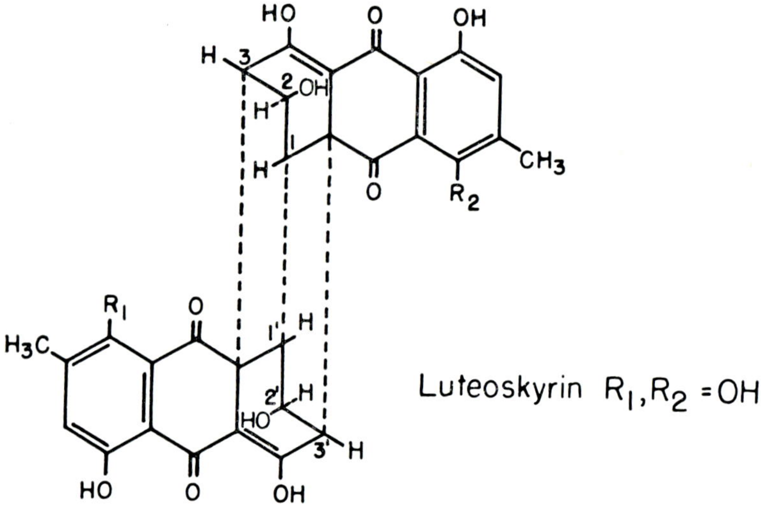

Fig. 4. Structure of luteoskyrin.

Luteoskyrin is a lipid soluble pigment which is relatively slow acting. Microscopic changes in the liver occur after 24 hr; liver turns cloudy yellow, soft, and dotted with red spots. Histologically, changes are centrolobular necrosis and fatty degeneration of the liver cells. Acute liver damage induced by luteoskyrin and revealed by electron microscopy showed swelling of mitochondria, cystic dilatation of the endoplasmic reticulum, and positive fatty droplets. Chronic administration causes periportal fibrosis and a small incidence of hepatoma formation.

Cyclochlorotine, the chlorine-containing peptide, was isolated from the culture filtrate of *P. islandicum* with a yield of 20 to 40 mg/50 litres. It was also obtained from mycelial mats and cultured rice. The structure of cyclochlorotine is shown in Fig. 5, and has an empirical formula of $C_{25}H_{36}N_5O_8Cl_2$.

In rats the toxin causes the disappearance of glycogen granules in the injured liver, the main target organ; fibrosis and cirrhosis are typical of this toxin; no necrosis occurs as with luteoskyrin but it affects the liver more rapidly. The liver becomes pale and anaemic 5 min after administration. Toxin affects blood capillaries, causing necrosis of the epidermis with a low incidence of hepatoma.

Although toxin production by a large number of fungi isolated from rice has been reported, the natural occurrence of the yellowed rice toxins has not yet been demonstrated (Campbell and Stoloff, 1974).

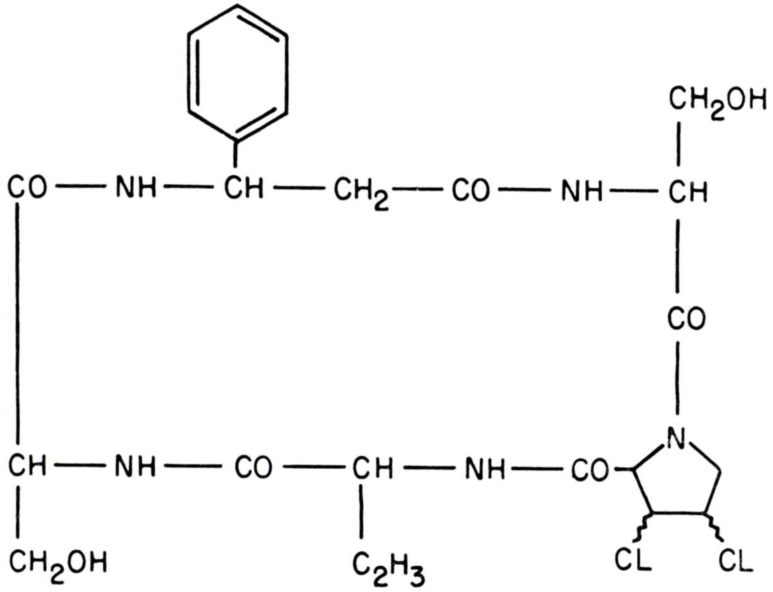

Fig. 5. Structure of cyclochlorotine.

CONCLUSION

There is undoubtedly worldwide contamination of human foodstuffs with a variety of mycotoxin-producing fungi. Surveys of foods and animal feeds for the actual presence of mycotoxins have been limited primarily to aflatoxin. However, in the last 5 years field surveys of the occurrence of other mycotoxins such as citrinin, ochratoxin A, patulin, penicillic acid, sterigmatocystin, and zearalenone have been reported. The point is that presence of mycotoxin-producing fungi in foods demonstrates a hazard, but the proof is the actual presence of the mycotoxin. Certainly, in light of the data summarized by Campbell and Stoloff (1974) and Wogan (1973), there is little doubt that mycotoxins are a probable source of naturally occurring carcinogens in humans.

REFERENCES

Bailey, W. S., and Groth, A. H., Jr. (1959). The relationship of hepatitis-X of dogs and moldy corn poisoning in swine. *J. Am. Vet. Med. Assoc.*, **134**, 514–516.

Bilay, V. I. (1960). Mycotoxicoses of man and agricultural animals. U.S.S.R. (transl). U.S. Joint Pub. Res. Serv., Washington, D.C. 205 pp.

Buchi, G., and Rae, I. D. (1969). Structure and chemistry of the aflatoxins, pp. 55–75. In *Aflatoxin* (L. A. Goldblatt, ed.), Academic Press, New York. 472 pp.

Burnside, J. E., Sippel, W. L., Forgacs, J., Carll, W. T., Atwood, M. B., and Doll, E. R. (1957). A disease of swine and cattle caused by eating moldy corn. *Am. J. Vet. Res.*, **18**, 817–824.

Burroughs, R., and Sauer, D. B. (1971). Growth of fungi in sorghum grain stored at high moisture contents. *Phytopathology*, **61**, 767–772.

Campbell, T. C., and Stoloff, L. (1974). Implications of mycotoxins for human health. *J. Agr. Food Chem.*, **22**, 1006–1015.

Christensen, C. M. (1957). Deterioration of stored grains by fungi. *Bot. Rev.*, **23**, 108–134.

Christensen, C. M. (1971). Mycotoxins. *Crit. Rev. Environ. Control*, **2**, 57–80.

Christensen, C. M., and Kaufman, H. H. (1965). Deterioration of stored grains by fungi. *Annu. Rev. Phytopathol.*, **3**, 69–84.

Christensen, C. M., and Kaufman, H. H. (1969). Mycotoxins and grain quality, pp. 76–93. In *Grain Storage—The Role of Fungi in Quality Loss*. Univ. Minnesota Press, Minneapolis. 153 pp.

Christensen, C. M., and Kaufman, H. H. (1974). Microflora, pp. 158–192. In *Storage of Cereal Grains and Their Products* (C. M. Christensen, ed.), Am. Assoc. Cereal Chemists, St. Paul. 549 pp.

Ciegler, A. (1975). Mycotoxins: Occurrence, chemistry, biological activity. *Lloydia.*, **38**, 21–35.

Ciegler, A., Kadis, S., and Ajl, S. J. (1971). *Microbial Toxins, Vol. VI*. Academic Press, New York. 563 pp.

Connole, M. D., and Hill, M. W. M. (1970). *Aspergillus flavus* contaminated sorghum grain as a possible cause of aflatoxicosis in pigs. *Aust. Vet. J.*, **46**, 503–505.

Dalezios, J., Wogan, G. N., and Weinreb, S. M. (1971). Aflatoxin P_1: A new aflatoxin metabolite in monkeys. *Science*, **171**, 584–585.

Davis, N. D., Diener, U. L., and Eldridge, D. W. (1966). Production of aflatoxin B_1 and G_1 by *Aspergillus flavus* in a semisynthetic medium. *Appl. Microbiol.*, **14**, 378–380.

Detroy, R. W., Lillehoj, E. B., and Ciegler, A. (1971). Aflatoxin and related compounds, pp. 3–178. In *Microbial Toxins, Vol. VI* (A. Ciegler, S. Kadis, and S. J. Ajl, eds.), Academic Press, New York. 563 pp.

Diener, U. L. (1973). Deterioration of peanut quality caused by fungi, pp. 523–557. In *Peanuts—Culture and Uses*. Am. Peanut Res. Educ. Assoc., Stillwater, Oklahoma. 684 pp.

Diener, U. L., and Davis, N. D. (1969). Aflatoxin formation by *Aspergillus flavus*, pp. 13–54. In *Aflatoxin* (L. A. Goldblatt, ed.), Academic Press, New York. 472 pp.

Forgacs, J., and Carll, W. T. (1962). Mycotoxicoses. *Adv. Vet. Sci.*, **7**, 273–382.

Gardner, H. K., Jr., Koltun, S. P., Dollear, F. G., and Rayner, E. T. (1971) Inactivation of aflatoxins in peanut and cottonseed meals by ammoniation. *J. Am. Oil Chem. Soc.*, **48**, 70–73.

Goldblatt, L. A. (1969). *Aflatoxin*. Academic Press, New York. 472 pp.

Gröger, D. (1972). Ergot, pp. 321–373. In *Microbial Toxins, Vol. VIII* (S. Kadis, A. Ciegler, and S. J. Ajl, eds.), Academic Press, New York. 400 pp.

Hesseltine, C. W., Bothast, R. J., and Shotwell, D. L. (1975). Aflatoxin occurrence in some white corn under loan, 1971: IV. *Mold flora. Mycologia.*, **67**, 392–408.

Joffe, A. Z. (1965). Toxin production by cereal fungi causing toxic alimentary aleukia in man, pp. 77–85. In *Mycotoxins in Foodstuffs* (G. N. Wogan, ed.), M.I.T. Press, Cambridge. 291 pp.

Joffe, A. Z. (1971). Alimentary toxic aleukia, pp. 139–189. In *Microbial Toxins, Vol. VII* (S. Kadis, A. Ciegler, and S. J. Ajl, eds.), Academic Press, New York. 401 pp.

Kadis, S., Ciegler, A., and Ajl, S. J. (1971). *Microbial Toxins, Vol. VII*. Academic Press, New York. 401 pp.

Kadis, S., Ciegler, A., and Ajl, S. J. (1972). *Microbial Toxins, Vol. VIII*. Academic

Press, New York. 400 pp.
Kinosita, R., and Shikata, T. (1965). On toxic moldy rice, pp. 111–132. In *Mycotoxins in Foodstuffs* (G. N. Wogan, ed.), M.I.T. Press, Cambridge. 291 pp.
LaFont, P., and LaFont, J. (1970). Contamination de Produits Cerealiers et d'Aliments du Betail par l'Aflatoxine. *Food Cosmet. Toxicol.*, **8**, 403–408.
Lillehoj, E. B., Ciegler, A., and Detroy, R. W. (1970). Fungal toxins, pp. 2–136. In *Essays in Toxicology, Vol.* 2 (F. R. Blood, ed.), Academic Press, New York. 218 pp.
Mantle, P. G. (1969). Role of alkaloids in the poisoning of mammals by sclerotia of *Claviceps* spp. *J. Stored Prod. Res.*, **5**, 237–244.
Martin, P. M. D., Gilman, G. A., and Keen, P. (1971). The incidence of fungi in foodstuffs and their significance, based on a survey in the eastern Transvaal and Swaziland, pp. 281–290. In *Symposium on Mycotoxins in Human Health* (I. F. H. Purchase, ed.), MacMillan, London. 306 pp.
Masri, M. S., Haddon, W. F., Lundin, R. E., and Hsieh, D. P. H. (1974). Aflatoxin Q_1. A newly identified major metabolite of aflatoxin B_1 in monkey liver. *Agr. Food Chem.*, **22**, 512–515.
Mateles, R. I., and Wogan, G. N. (1967). *Biochemistry of Some Foodborne Microbial Toxins.* M.I.T. Press, Cambridge, Mass. 171 pp.
Miyake, I., Naito, H., and Tsunoda, H. (1940). Studies on the production of toxin in the saprophyte-growing rice grains under storage. *Beikoku Riyo Kenkyujo Hokoku*, **1**, 1–30. (In Japanese.)
Morooka, H., and Tatsuno, T. (1970). Toxic substances (Fusarenon and Nivalenol) produced by *Fusarium nivale*, pp. 114–119. In *Toxic Microorganisms* (M. Herzberg, ed.), UJNR Joint Panels and U.S. Dept. Interior, Washington, D.C. 490 pp.
Newberne, P. M. (1974). The new world of mycotoxins—animal and human health. *Clin. Toxicol.*, **7**, 161–177.
Purchase, I. F. H. (1971). *Symposium on Mycotoxins in Human Health.* MacMillan Press, London. 306 pp.
Purchase, I. F. H. (1974). *Mycotoxins.* American Elsevier Scientific Pub. Co., New York. 433 pp.
Raper, K. B., and Fennell, D. I. (1965). *The Genus* Aspergillus. Williams and Wilkins Co., Baltimore. 686 pp.
Raymond, W. D. (1966). Mycotoxin problems in the United Kingdom and the tropics. *Food Technol.*, **20** (7), 54–60.
Richard, J. L., Tiffany, L. H., and Pier, A. C. (1969). Toxigenic fungi associated with stored corn. *Mycopathol. Mycol. Appl.*, **38**, 313–326.
Robbers, J. E., Krupinski, V. M., Sheriat, H. S., and Huber, D. M. (1975). A method for detection of ergot contamination in ground Triticale grain. *Phytopathology*, **65**, 455–457.
Saito, M., Enomoto, M., and Tatsuno, T. (1971). Yellowed rice toxins: luteoskyrin and related compounds, chlorine-containing compounds, and citrinin, pp. 299–366. In *Microbial Toxins, Vol. VI* (A. Ciegler, S. Kadis, and S. J. Ajl, eds.), Academic Press, New York. 563 pp.
Saito, M., and Tatsuno, T. (1971). Toxins of *Fusarium nivale*, pp. 293–316. In *Microbial Toxins, Vol. VII* (S. Kadis, A. Ciegler, and S. J. Ajl, eds.), Academic Press, New York. 401 pp.
Schroeder, H. W. (1970). Aflatoxins in rice in the United States, pp. 56–60. In *Toxic Micro-organisms* (M. Herzberg, ed.), UJNR Joint Panels and U.S. Dept. Interior, Washington, D.C. 490 pp.
Scott, P. M. (1973). Mycotoxins in stored grains, feeds, and other cereal products, pp. 343–365. In *Grain Storage: Part of a System* (R. N. Sinha and W. E. Muir, eds.), Avi. Pub. Co., Westport. 481 pp.
Seitz, L. M., Sauer, D. B., Mohr, H. E., Burroughs, R., and Paukstelis, J. V. (1975). Metabolites of *Alternaria* in grain sorghum. Compounds which could be

mistaken for zearalenone and aflatoxin. *Agr. Food Chem.*, **23**, 1–4.

Shotwell, O. L., Hesseltine, C. W., Burmeister, H. R., Kwolek, W. F., Shannon, G. M., and Hall, H. H. (1969a). Survey of cereal grains and soybeans for the presence of aflatoxin: 1. Wheat, grain sorghum, and oats. *Cereal Chem.*, **46**, 446–454.

Shotwell, O. L., Hesseltine, C. W., Burmeister, H. R., Kwolek, W. F., Shannon, G. M., and Hall, H. H. (1969b). Survey of cereal grains and soybeans for the presence of aflatoxin: 2. Corn and soybeans. *Cereal Chem.*, **46**, 454–463.

Shotwell, O. L., Hesseltine, C. W., and Goulden, M. L. (1973). Incidence of aflatoxin in southern corn, 1969–1970. *Cereal Sci. Today*, **18**, 192–195.

Shotwell, O. L., Hesseltine, C. W., Goulden, M. L., and Vandegraft, E. E. (1970). Survey of corn for aflatoxin, zearalenone, and ochratoxin. *Cereal Chem.*, **47**, 700–707.

Shotwell, O. L., Hesseltine, C. W., Vandegraft, E. E., and Goulden, M. L. (1971). Survey of corn from different regions for aflatoxin, ochratoxin, and zearalenone. *Cereal Sci. Today*, **16**, 266–273.

Sippel, W. L., Burnside, J. E., and Atwood, M. B. (1953). A disease of swine and cattle caused by eating moldy corn. *Proc. Am. Vet. Med. Assoc.*, 174–181.

Strzelecki, E. L., and Gasiorowska, U. W. (1974). Aflatoxin B_1 in feedstuffs. *Zbl. Vet. Med. B*, **21**, 395–400.

Tsunoda, H. (1970). Micro-organisms which deteriorate stored cereals and grains, pp. 143–162. In *Toxic Micro-organisms* (M. Herzberg, ed.), UJNR Joint Panels and U.S. Dept. Interior, Washington, D.C. 490 pp.

Udagawa, S. I., Ichinoe, M., and Kurata, H. (1970). Occurrence and distribution of mycotoxins producers in Japanese foods, pp. 174–184. In *Toxic Micro-organisms* (M. Herzberg, ed.), UJNR Joint Panels and U.S. Dept. Interior. Washington, D.C. 490 pp.

Ueno, Y. (1971). Toxicological and biological properties of Fusarenon-X, a cytotoxic mycotoxin of Fusarium nivale Fn-2B, pp. 163–178. In *Symposium on Mycotoxins in Human Health* (I. F. H. Purchase, ed.), MacMillan Press, London. 306 pp.

Uraguchi, K. (1971a). Pharmacology of mycotoxins, pp. 143–299. In *Pharmacology and Toxicology of Naturally Occurring Toxins* (H. Raskova, ed.), IEPT, Sect. 71, Vol. II. Pergamon Press, Oxford. 316 pp.

Uraguchi, K. (1971b). Citreoviridin, pp. 367–380. In *Microbial Toxins. Vol. VI* (A. Ciegler, S. Kadis, and S. J. Ajl, eds.), Academic Press, New York. 563 pp.

Van Warmelo, K. T., vd Westhuizen, G. C. A., and Minne, J. A. (1968). *The production of aflatoxins in naturally infected high quality maize*. Dept. Agr. Tech. Serv. Tech. Comm. 71. Pretoria. 5 pp.

Wilson, B. J., Teer, P. A., Barney, G. H., and Blood, F. R. (1967). Relationship of aflatoxin to epizootics of hepatitis among animals in southern United States. *Am. J. Vet. Res.*, **28**, 1217–1230.

Wogan, G. N. (1965). *Mycotoxins in Foodstuffs*. M.I.T. Press, Cambridge. 291 pp.

Wogan, G. N. (1966a). Mycotoxin contamination of foodstuffs. *Adv. Chem. Ser.*, **57**, 195–215.

Wogan, G. N. (1966b). Chemical nature and biological effects of the aflatoxins. *Bacteriol. Rev.*, **30**, 460–470.

Wogan, G. N. (1969a). Naturally occurring carcinogens in foods. *Progr. Exp. Tumor Res.*, **11**, 134–162.

Wogan, G. N. (1969b). Alimentary mycotoxicoses, pp. 345–451. In *Foodborne Infections and Intoxications* (H. Riemann, ed.), Academic Press, New York. 698 pp.

Wogan, G. N. (1972). Mycotoxins and liver injury, pp. 161–181. In *The Liver* (E. A. Gall and K. F. K. Mostofi, eds.), Int. Acad. Pathol. Mono. No. 13. Williams and Wilkins Co., Baltimore. 525 pp.

Wogan, G. N. (1973). Aflatoxin carcinogenesis, pp. 309–344. In *Methods in Cancer Research, Vol. VII* (H. Busch, ed.), Academic Press, New York. 418 pp.

Wogan, G. N., and Mateles, R. I. (1968). Mycotoxins, pp. 149–175. In *Progress in Industrial Microbiology, Vol.* 7 (D. J. D. Hockenhull, ed.), CRC Press, Cleveland. 232 pp.

Wogan, G. N., and Newberne, P. M. (1967). Dose and response characteristics of aflatoxin B_1 in carcinogenesis in the rat. *Cancer Res.*, **27**, 2370–2376.

Wogan, G. N., and Pong, R. S. (1970). Aflatoxins. *Ann. N.Y. Acad. Sci.*, **174**, 623–635.

MYCOTOXINS OTHER THAN AFLATOXINS

C. W. HESSELTINE

Northern Regional Research Laboratory, Agricultural Research Service,
U.S. Department of Agriculture, Peoria, Illinois 61604, U.S.A.

Summary

Some fungi growing on food and feed produce toxic products both in the field and after harvest, in addition to the *Aspergillus flavus* group that form aflatoxins. Effects of these other toxins on animals are varied and include some carcinogenic and estrogenic symptoms, loss of weight, liver damage, haemorrhage, and rejection of feed. In some instances, toxins are produced by the host plant in response to the fungus infection or physical injury rather than being produced directly by the fungus. Some of the most important mycotoxins, as well as the aflatoxins, will be described in respect to causal agent, chemistry, natural occurrence, and acute and chronic toxicity. Generally, the toxins are produced by members of the genera *Aspergillus*, *Penicillium*, *Fusarium*, *Alternaria*, and the class Fungi Imperfecti. These compounds are ochratoxin, T-2, zearalenone, penicillic acid, citrinin, patulin, corn rejection factor in swine, alternariol and related compounds, and sterigmatocystin.

INTRODUCTION

Mycotoxins are toxic chemicals produced by fungi. Although toxic to higher animals and sometimes to plants and microorganisms, they are not infectious. During growth, fungi elaborate secondary metabolites, and those that are harmful are referred to as mycotoxins as contrasted to antibiotics (likewise secondary metabolites) which are useful in controlling diseases caused by microorganisms. Some of the currently important mycotoxins, such as citrinin and patulin, were discovered many years ago during the search for new antibiotics. Typically, mycotoxins are produced by fungi when they have reached their maximum growth. The resultant toxin may remain in the food or feed product for years, long after the mould has died. In one instance, we know of corn 12 years old that still contains aflatoxin probably formed shortly after harvest.

Among the numerous reviews available on the subject of mycotoxins in various books and journals, a few are listed under the general references (Wogan, 1965; Hesseltine, 1969, 1974; Herzberg, 1970; Ciegler *et al.*, 1971; Purchase, 1974). It is not my purpose to review the literature because it has been done adequately in the references cited. Rather, I have chosen to summarize the information on the most important mycotoxins. Certainly others now known but not discussed or even those yet to be discovered may become more important in the future. I have purposely omitted ergot from the chapter since the knowledge that ergot poisoning was known to be caused by a fungus goes back a hundred years or more, and the literature is extensive and exhaustive in detail. Also excluded are the aflatoxins produced by *Aspergillus flavus* and *A. parasiticus*, as well as the variety of toxins produced by mushrooms.

In this review, I have tried to follow an outline for each toxin giving first the mould, its occurrence, the chemistry of the toxic compound, assay methods, and toxic effects in animals. Only a few key references to each mycotoxin are cited. The references are arranged separately for each toxin treated, except patulin and penicillic acid are combined.

Four criteria governed my choosing the most important mycotoxins: (1) The mycotoxin is a carcinogen. There are five: aflatoxins, penicillic acid, patulin, rugulosin, and sterigmatocystin. (2) The mycotoxin occurs in agricultural commodities in nature. (3) The mycotoxin is known to cause disease in man and animals. (4) The mycotoxin is produced by fungi commonly found in the farm commodity either in the field or in storage. The genera most often implicated in natural poisoning outbreaks are species of *Penicillium*, *Aspergillus*, and *Fusarium*.

In selecting those mycotoxins of greatest importance to animals and man, besides ergot and aflatoxins, I have not attempted to rank them in importance.

Penicillic acid

Like patulin, penicillic acid is a lactone metabolite synthesized by a number of species of *Aspergillus*, *Penicillium*, and *Paecilomyces ehrlichii*. *Aspergillus* species producing penicillic acid include *A. ochraceus*, *A. sulphureus*, *A. melleus*, *A. sclerotiorum*, *A. alliaceus*, and *A. ostianus*, all members of the *A. ochraceus* series. The *Penicillium*-producing species are *P. lividum*, *P. simplicissimum*, *P. cyclopium*, *P. thomii*, *P. roqueforti*, *P. martensii*, *P. fennelliae*, *P. aurantio-virens*, *P. janthinellum*, *P. viridicatum*, *P. palitans*, *P. baarnense*, and *P. puberulum*. The majority of these species belong in the *Asymmetrica Fasciculata* group and some form both penicillic acid and patulin.

Penicillic acid has been found in corn and beans (Thorpe and Johnson, 1974), and Snow *et al.* (1972) identified it in tobacco. Kurtzman and Ciegler (1970) used *P. martensii* NRRL 3612 to produce penicillic acid on a solid substrate. Fifty grams of corn plus 100 ml of water was incubated at various temperatures for varying periods of time. Yields were highest at temperatures of 20°C or below. They reported yields up to 8·2 mg/g of dry corn. Generally, commodities high in protein, such as soybeans, are poor substrates for toxin formation.

Penicillic acid has a molecular weight of 170 with a melting point of 84–87°C. While slightly soluble in cold water, it is soluble in hot water, alcohol, ether, benzene, chloroform, and ethyl acetate but insoluble in hexane and petroleum ether. Penicillic acid is stable at 100°C at both high and low pH's. It absorbs Br_2, reduces ammoniacal silver nitrate, turns yellow in alkaline solution, and upon exposure to ammonia, turns dark red (Wilson, 1975). Its ultraviolet absorption maximum is about 227 in water. The infrared spectra have bands at 3270, 1728, 1643, 1353, 1223, 909, and 811 in KBr. Penicillic acid, like patulin, interacts with the SH group of enzymes and inhibits DNA, RNA, and protein synthesis. The structure is shown in Fig. 1.

No official assay method exists for penicillic acid, but methods based on bioassay, colorimetric, thin-layer chromatography (TLC), and gas-chromatographic methods have been developed. Thorpe and Johnson (1974) obtained the trifluoroacetate of penicillic acid for gas chromatographic detection in

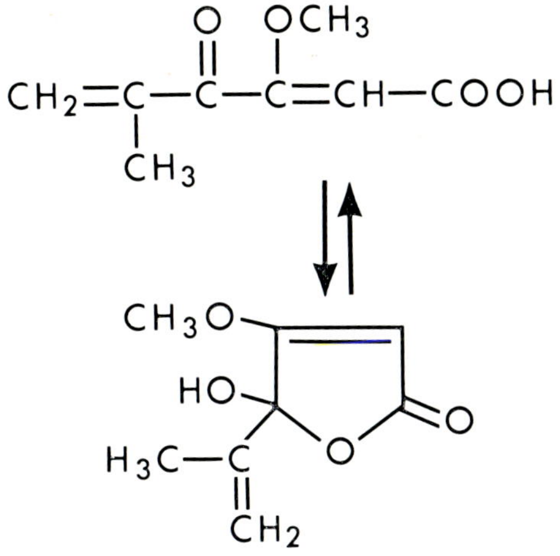

Fig. 1. Penicillic acid.

beans and corn, and they used gas chromatograph-mass spectrometry (gc-ms) for confirmation.

Penicillic acid was discovered by Alsberg and Black (1913) from a strain of *Penicillium* isolates from toxic corn. It is an antibiotic substance with more gram-negative activity than gram-positive activity against bacteria, and it has antiviral activity. The LD_{50} of penicillic acid was 100 mg/kg for mice by subcutaneous injection and 250 when given intravenously. Penicillic acid-treated Hela cells at 10 μg/ml increased mitotic cells and enlarged interphasic cells, which contained large nuclei with dotty chromatin and with large and irregular nucleoli (Wilson, 1975). Both penicillic acid and patulin produced malignant tumors in rats when they were injected subcutaneously (Dickens and Jones, 1961).

Patulin

Patulin, also known as clavicin, clavitin, claviformin, expansin, leucopin, mycoin C, penicidin, and ternicin, is a toxin produced by a number of species of typical storage moulds belonging to the genera *Penicillium*, *Aspergillus*, and *Byssochlamys* [*P. urticae* (*P. patulum*), *P. expansum*, *P. granulatum*, *P. lanosum*, *P. claviforme*, *P. melinii*, *P. divergens*, *P. griseofulvum*, *P. novae-zeelandiae*, *P. cyclopium*, *P. lapidosum*, *P. leucopus*, *P. equinum*, *A. clavatus*, *A. giganteus*, *A. terreus*, and *Byssochlamys nivea*] (Wilson, 1975). The *Penicillium* species, therefore, mostly fall in the *Asymmetric Fasciculata* series (*P. urticae*, *P. granulatum*, *P. expansum*, *P. cyclopium*, and *P. claviforme*). Like aflatoxin, patulin is a carcinogen and a mutagen (producing genetic changes in biological systems). The toxin has been found in commercial apple juice, presumably because mouldy apples were used in making the juice (Scott *et al.*, 1972) and in wheat straw (Norstadt and McCalla, 1963). Isolates of Penicillia have been obtained from malt feeds which produced patulin.

Patulin can be produced on either Czapek-Dox and Raulin Thom media or modifications of these. Yields as high as 2·7 g/litre have been reported in a potato dextrose broth after 14 days' growth at 25°C.

Patulin has a molecular weight of 154 with a melting point of 110–112°C.

Pure patulin is a colourless crystalline product, soluble in water, alcohols, acetone, ethyl acetate, and chloroform, but less soluble in ether and benzene, and insoluble in petroleum ether. It has an absorption maximum at about 276 nm. Both patulin and penicillic acid react with sulphydryl and amino groups, but adducts have little biological activity. Patulin is unstable in alkaline solutions but can be heated to 100°C in 0·1 N HCl for 30 min and still retain its activity. Patulin is unstable in wheat flour, ground sorghum, and wet ground corn, but not in dry corn. The structure for patulin is shown in Fig. 2.

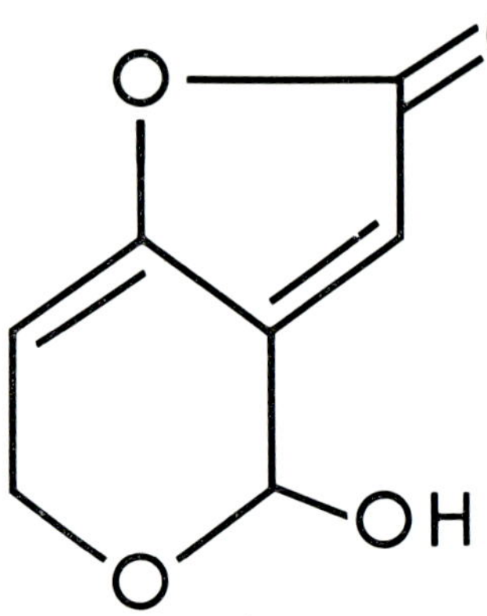

Fig. 2. Patulin.

No simple and rapid method of detection is known although identification is accomplished by TLC and by gas and liquid chromatography. Scott (1974a) described a collaborative study on detection of patulin in apple juice in which column chromatography was followed by TLC with 3-methyl-2-benzothiazolinone hydrazone HCl as the spray reagent. Recoveries in apple juice ranged from 70–98% at three contamination levels. During its investigation as an antibiotic, its inhibitory activity against certain bacteria served as an assay.

Patulin is a widespread biocide affecting microorganisms including bacteria, protozoa, fungi, and some viruses, as well as animals and plants. It has an LD_{50} in mice via the oral route of 35 mg/kg and 8–10 mg/kg via the subcutaneous. Patulin has a marked antidiuretic effect in rats. Animal poisoning results in generalized edema, and capillary damage with death preceded with convulsions. Considerable experimental information is available on its toxicity to man as a result of clinical trials to determine whether patulin could be used as an antibiotic.

Sterigmatocystin

Sterigmatocystin and its closely related forms—5-methoxysterigmatocystin, demethylsterigmatocystin, dihydrosterigmatocystin, and dihydrodemethylsterigmatocystin—are produced by *Aspergillus versicolor*. Sterigmatocystin is an important mycotoxin because it is a known carcinogen as determined by both subcutaneous injection and by oral administration, but its carcinogenic property is much less than aflatoxin. Other species of fungi-producing sterigmatocystin are *Bipolaris* sp. and *A. nidulans*. More recently, O-methylsterigmatocystin (Burkhardt and Forgacs, 1968) and aspertoxin (Rodricks *et al.*, 1968) have been isolated from aflatoxin-producing cultures of *A. flavus*. Sterigmatocystin has been found in grain, coffee beans, and food samples from Mozambique (van der Watt, 1974).

Yields of 12 g of sterigmatocystin per kilogram of maize have been obtained with *A. versicolor*. The culture was grown on a solid medium at 27°C for 21 days (van der Watt, 1974).

Sterigmatocystin is a yellow crystalline compound, which may occur up to 1·3% of the dried mycelium and has an m.p. of 246 (Roberts, 1974). Since it belongs to the group of furofuran compounds, it is related to aflatoxin. The structure of sterigmatocystin is shown in Fig. 3. Sterigmatocystin is insoluble

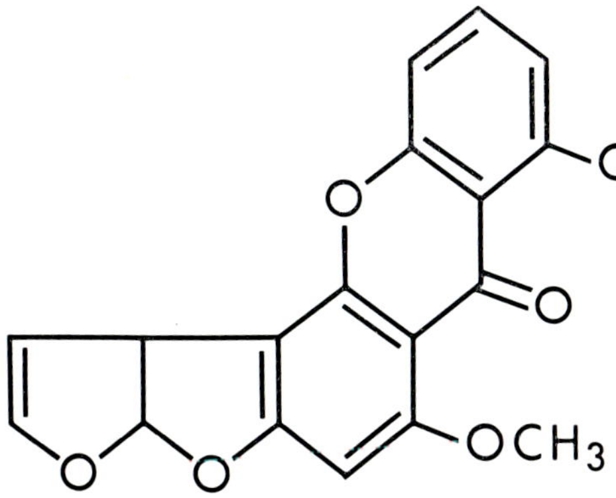

Fig. 3. Sterigmatocystin.

in water, aqueous sodium carbonate, and aqueous sodium hydroxide, and gives a deep yellow colour with the last compound. It is quite soluble in chloroform and pyridine but sparingly soluble in most other organic solvents (Davies *et al.*, 1960).

An official method for sterigmatocystin assay in wheat and barley has been published (AOAC, 1975) involving TLC and column chromatography. When TLC plates are sprayed with $AlCl_3$ solution, the sterigmatocystin spots give a bright yellow fluorescence under short ultraviolet light (UV). Overall recovery is about 90% with detection levels as low as 30 μg/k.

The LD_{50} in rats indicate 166 mg/kg body weight in males and 120 mg/kg in females. The i.p. LD_{50} in rats was between 60 and 65 mg/kg body weight. Postmortem findings are marked by hepatic and renal necrosis. Holzapfel *et al.* (1966) reported the *per os* LD_{50} was 120 mg/kg in albino rats when administered in wheat germ oil. Van der Watt (1974) states that the LD_{50} i.p. was about 32 mg/kg in the male vervet monkey. In this nonhuman primate, at high doses, livers became enlarged, yellow, friable, and haemorrhages were present in the kidney, gastrointestinal tract, epicardium, and on all serosal surfaces.

Alternaria Toxins

Grain exposed to rain and snow often becomes weathered, discoloured, and invaded by typical field fungi. Among the common fungi encountered, not host specific, are *Alternaria*, *Cladosporium*, and *Pullularia*. Recently, a series of compounds produced by *Alternaria tenuis* and *A.* sp. were found.

Because *Alternaria* is so common, this group of toxins should be given more attention, particularly in crops in the field such as sorghum where the fruit is exposed to the weather. Seitz *et al.* (1975) isolated both alternariol and alternariol monomethyl ether from grain sorghum in both the autumn of 1973 and overwintered sorghum.

Various workers have studied the toxicity of crude preparations made from *Alternaria* and report that many isolates produce poisonous substances. Doupnik and Sobers (1968) stated that out of 96 isolates of *Alternaria* tested, 31 were lethal to chicks and 12 depressed growth 20% or more. In another study, Christensen *et al.* (1968) found 90% of their isolates to be lethal to rats when fed on a moulded corn-rice mixture.

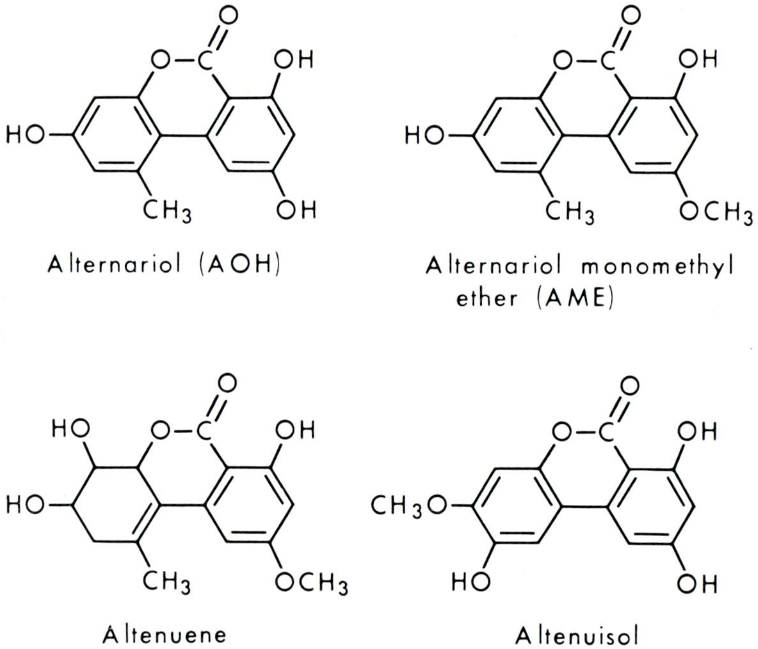

Fig. 4.

In the laboratory, toxin is formed in flasks with a medium consisting of 50 g polished whole rice moistened with 75 ml of Czapek-Dox broth fortified with 2% yeast extract. The flasks, inoculated with NRRL 5255 (M-1 of Pero) Pero *et al.* (1971a), were incubated in stationary culture for 28 days at 25°C in the dark.

Alternariol and its methyl ether are produced in large quantities, reportedly up to 13% of the dry mycelial weight in the better producing strains. Other compounds from *Alternaria* are formed in much smaller amounts but usually occur with alternariol and its methyl ether.

Alternaria produce three types of mycotoxins: dibenzo-α-pyrones, tetramic acids, and a group called altertoxins ($C_{20}H_{16}O_6$ + $C_{20}H_{14}O_6$), the structures of which have not been determined. The dibenzo-α-pyrones and their derivatives are eight in number, and the structure of four are given as part of the scheme (Fig. 4).

Altenuisol can be isolated with 70% acetone and is soluble in ethanol, acetone, and ethyl acetate, but only slightly soluble in chloroform or diethyl ether, and insoluble in benzene, hexane, and water. The UV absorption maxima in ethanol are 216, 256, and 278. Altenuene has a UV spectrum from ethanol at 319, 278, and 240. For further characterization of these compounds, *see* Pero *et al.* (1971a, 1971b; 1973a, 1973b).

Analytical methods for the detection of these compounds are reported by Seitz *et al.* (1975), based on TLC. Pero *et al.* (1971b) report a method for purification of altenuene, alternariol, and alternariol monomethyl ether. A simultaneous quantitative analytical method for these three compounds is given based on gas–liquid chromatography with detection limits of about 0·1 μg for each. TLC analysis separates each compound with a detection limit of 0·1 μg.

Alternariol and its methyl ether are colourless, crystalline compounds that have respective melting points of 350 and 267°C. Since they are phenolic in nature, both exhibit intense purple ferric reactions and fluoresce bright blue under UV irradiation.

The dibenzo-α-pyrones have a high cytotoxicity. The LD_{50} values of alternariol and alternariol monomethyl ether for Hela and lymphoma cells are 6 and 8 μg/ml, respectively. They also have an antibiotic effect on *Bacillus mycoides* ATCC 6462, with alternariol having activity at 60 μg per disc and the methyl ether at 500 μg. If together, they show inhibition at 0·25 μg/disc. The same synergistic effect has been demonstrated in Hela cells. Altertoxin II was the most toxic compound, having a LD_{50} of 0·5 μg ml for Hela cells. Tenuazonic acid is more potent against mice with a LD_{50} of 81 mg/kg, whereas altertoxins I and II have LD_{50} of 150 mg/kg. Alternariol at 100 mg/k body weight administered subcutaneously for 9 to 12 days significantly increased dead and reabsorbed fetus/litter and runts/litter in mice. A mixture of this compound and the methyl ether at the level of 25 mg/k increased both dead and reabsorbed fetuses. Sodium tenuazonate has antiviral activity against enteroviruses, respiratory viruses, vaccina, and herpes simplex at levels of 100–500 μg/ml.

Emetic and rejection factors

Such compounds are produced by a number of species of *Fusarium*, including *F. graminearum* (*Gibberella zeae*, *G. saubinetii*), *F. avenaceum*, *F. poae*, *F. moniliforme*, *F. roseum*, *F. sporotrichoides*, *F. equiseti*, and *F. culmorum* (Prentice *et al.*, 1959). According to some classifications of Fusaria, some of these species are synonyms of *F. graminearum* (*F. roseum*, *G. saubinetii*). *Gibberella* is the sexual stage of many Fusaria. Crops affected appear to be cereals, especially wheat, barley, and corn. In barley and wheat, the disease is known as scab. The above authors reported the production of a toxin in 10 days in some strains on Richard's solution or on a nutrient broth containing beef extract 3 g, peptone 10 g, glucose 10 g, and 1 litre of water as tested in the pigeon. The toxin persists for more than 4 years in infected grain (Prentice and Dickson, 1964) and when the wheat was milled by the conventional roller milling, the emetic principle appeared in the bran and tail-end shorts. Reports of emesis in cereals goes back to 1923 at least (Prentice and Dickson, 1968). Their report concerned wheat flour baked into bread that caused nausea and vomiting in humans in Russia, but other reports on scabby cereals causing poisoning in the Orient go back much further in connection with Akakabibyo disease of wheat and barley.

Emetic material is soluble in water, butanol, methanol, and ethanol and somewhat soluble in ether, amyl alcohol, toluene, pyridine, and slightly soluble in acetone, chloroform, and ethyl acetate. It is stable to heat and drying. At the same time, Vesonder *et al.* (1973) and Yoshizawa and Morooka (1973) isolated deoxynivalenol which has received the trivial name vomitoxin. Earlier, Morooka *et al.* (1972) reported on the isolation of a toxin that they called Rd toxin from cereal grains in 1970; they gave its melting point as 151°C along with an isolation procedure and infrared and mass spectra data. Vesonder *et al.* (1973) isolated the material from corn naturally infected by *Fusarium graminearum* (*Gibberella zeae*). The corn was from the 1972 Ohio crop and was assayed by intubation in swine. From these two papers the compound causing emesis was identified as a trichothecane, 3,7,15-trihydroxy-12,13-epoxy-trichothe-9-en-8-one (Fig. 5). It was named deoxynivalenol by

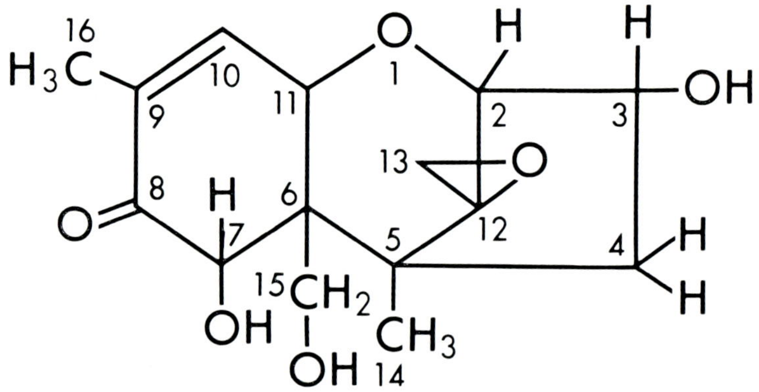

Fig. 5. Vomitoxin (deoxynivalenol).

Yoshizawa and Morooka (1973) who isolated it from a *Fusarium roseum* strain from barley. The compound forms fine crystals when recrystallized from ethyl acetate and petroleum ether and has a melting point at 151–153°C. More chemical data appear in the paper by Yoshizawa and Morooka (1973) and Vesonder *et al.* (1973).

The emetic material especially affects single stomach animals such as swine, dogs, man, and pigeons. Vesonder (pers. comm.) has determined that vomitoxin is also the swine rejection factor in corn. Vomiting may be induced by either the oral or intravenous routes. The LD_{50} of deoxynivalenol injected i.p. in mice was 70 mg/kg body weight in males and 76 for females. The monoacetate also isolated from the same mould was 49 mg/kg for males and 46·9 for females (Yoshizawa and Morooka, 1974). Deoxynivenol inhibited multiplication of cells of *Tetrahymena pyriformis*. The LD_{50}'s, therefore, are less than nivalenol and fusarenon, which have LD_{50}'s of 6·9 and 5·6 mg/kg in male mice. Vomiting can be induced within 25 minutes of intubation in swine according to Vesonder *et al.* (1973). The acute symptoms were marked dilation and haemorrhage of the gastrointestinal tract and engorgement of the testes. In Peking ducklings, the LD_{50} by subcutaneous route was 27 mg/kg with a minimum vomiting dose of about 10 mg/kg. Via the same route, vomiting and diarrhea in dogs were 0·1 and 0·2 mg/kg, respectively. The compound showed marked cytotoxicity to cultured mammalian cells. Ueno *et al.* (1971) reviewed numerous outbreaks of poisoning in Japan, including the consumption of mouldy rice by 25 young men. After eating, they experienced nausea, vomiting, headache, and drowsiness. In a later paper, Ueno *et al.* (1974) investigated fraction IV from corn, supplied by the Northern Regional Research Laboratory, which had caused vomiting in swine. This fraction, presumably deoxynivalenol, produced vomiting in ducklings and cats and was lethal to mice. Histological examination of the poisoned mice and cats revealed extensive cellular destruction and karyorrhexis of the bone marrow, thymus, spleen, and lymph nodes. It also markedly inhibited the uptake of ^{14}C-leucine in rabbit reticulocytes.

Zearalenone (F-2)

The effect of zearalenone in swine has been reported many times by farmers and agriculturalists and pre-dates the discovery of aflatoxin. The usual observation is that female swine consuming mouldy corn and feed develop enlarged vulvas and mammary glands and in extreme cases prolapsed vaginas. Generally, this toxin has been observed in corn, and occasionally in barley. The fungi involved are *Fusarium graminearum* (*F. roseum*), *F. moniliforme*, *F. sporotrichoides*, *F. oxysporium*, and *F. tricinctum*. *Fusarium graminearum* has *Gibberella zeae* (*G. saubinetii*) as its perfect stage. *G. zeae* is a common plant pathogen on cereals in the field. Infected kernels appear bleached and in severe infections may possess reddish discolouration and are chaffy.

Zearalenone appears to be a commonly occurring mycotoxin in corn in the upper Midwest of the United States. Out of 68 samples of feed submitted by farmers and veterinarians from farms reporting reproductive symptoms (Mirocha and Christensen, 1974), 45 % of the samples contained zearalenone up to a level of 2909 ppm. Levels of 1–5 ppm are considered physiologically significant in swine. It has been found occurring naturally in England, Finland, Mexico, France, and Yugoslavia. In surveys in the United States on corn, it occurred at levels up to 800 ppb (Shotwell *et al.*, 1971). In 293 samples of commercial corn in various grades assayed for zearalenone, five positive samples ranged from 450 to 750 ppb.

Isolates of *F. graminearum*, like other fungi, produce zearalenone in varying amounts from none to levels of 38 000–40 000 ppm on solid media, such as rice or corn (Mirocha and Christensen, 1974). Moisture level must be adjusted to about 45 % for corn and about 60 % for rice. Cultures were incubated at 24–27°C for 12 weeks until a mass of mycelium was produced

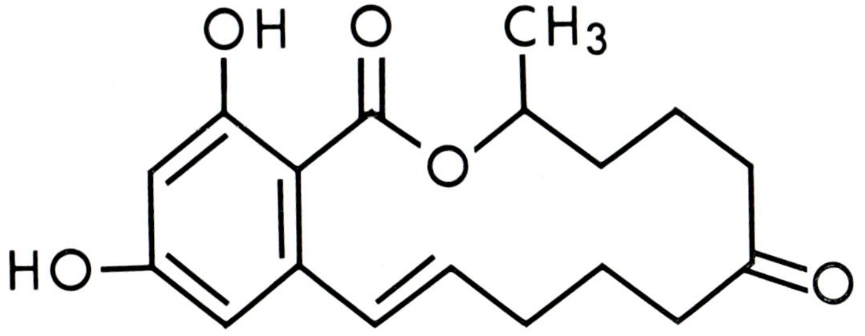

Fig. 6. Zearalenone.

and then incubation was continued at 12–14°C for 4–6 weeks. These are the conditions under which zearalenone is produced in nature.

Zearalenone has 16 derivatives (Fig. 6). The chemistry of zearalenone was given in U.S. Patent 3,196,019 issued to Andrews and Stob (1965), based upon *F. roseum* NRRL 2830. Zearalenone was referred to as the FES factor and is useful in animals because of its anabolic and estrogenic activity. It is a white crystalline product with a molecular weight of 318. Zearalenone is highly soluble in methylene chloride, ethanol, butyl acetate, 1,4-dioxane, and ethyl ether; soluble in methanol, butenol, chloroform, and acetone; slightly soluble in hexane, petroleum ether, and heptane; and insoluble in water, carbon disulfide, carbon tetrachloride, and dimethylformamide. It has a melting point of 164–165°C. UV absorption has maxima at 236, 274, and 313 mμ.

Although this mycotoxin is being studied extensively and a collaborative study is currently underway, an official method for its detection remains to be established (AOAC, 1975). However, the method used widely in surveys for zearalenone in corn is that described by Eppley (1968) which detects ochratoxin, aflatoxin, and zearalenone simultaneously. However, its sensitivity is low (200 ppb). It makes use of the system described below under ochratoxin A.

Zearalenone produces an estrogenic syndrome in swine consuming infected corn, characterized by swollen vulva in females, shrunken testes in young males, and enlarged mammary glands in both sexes. In females, it interrupts oestreus and may cause abortion in pregnant sows. In 6-week-old gilts fed 5 mg/day for 5 days, greatly enlarged uteri and atrophied ovaries resulted. The animals became normal after termination of the treatment. Three general affects of zearalenone in swine are observed (1) hyperoestrogenism, (2) infertility, and (3) abortion. Other animals are affected by zearalenone—mice, guinea pigs, rats, cattle, chickens, and turkeys. In poultry, the effects are less marked but according to Mirocha and Christensen (1974), there is a marked reduction in body weight in turkeys fed 40% *Fusarium*-infected corn for 46 days. In chicks, a ration containing 300–800 ppm of zearalenone stimulated growth. At low levels, zearalenone and some of its derivatives increase weight gain in farm animals. Curiously, zearalenone at low levels 1–100 ppm has a stimulatory effect on the production of perithecia of some Ascomycetes (Mirocha *et al.*, 1968) and appears as a sex regulating hormone for *G. zeae*.

Ochratoxin

Ochratoxin is one of the more important and extensively studied mycotoxins. Ochratoxin is produced by *Aspergillus ochraceus*, *A. sulphureus*, *A. sclerotiorum*, *A. alliaceus* (all closely related species) (Hesseltine *et al.*, 1972) and by *Penicillium viridicatum*, *P. variable*, *P. cyclopium*, *P. commune*, and *P. purpurescens* (Ciegler *et al.*, 1972). Optimum conditions for toxin formation by Aspergilli are 28°C for 5 days' incubation on solid substrates shaken in Fernbach flasks. On the other hand, optimum temperature for production of ochratoxin by *P. viridicatum* under the same conditions is 20°C.

Shotwell *et al.* (1969) detected the first natural occurrence of ochratoxin A in corn, but since then, it has been found in wheat, oats, barley, peanuts, white beans, and mixed feed grain from Canada (Scott *et al.*, 1972) and in coffee (Levi *et al.*, 1974). Ochratoxin was also detected in two surveys of corn from the midwestern United States collected from commercial channels (Hesseltine, 1974).

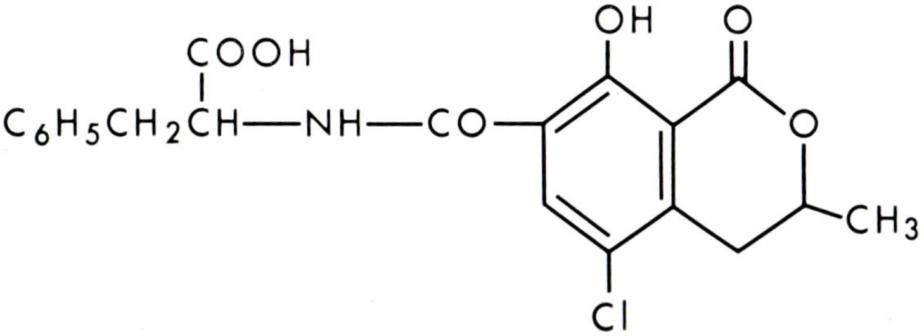

Fig. 7. Ochratoxin A.

The structure of ochratoxin A, the most important of the ochratoxins, is shown in Fig. 7 (Chu, 1974).

Three other ochratoxins have been described: Ochratoxin B (the dechloro analog), C, and D (the respective ethyl and methyl esters of A). Ochratoxin A is a colourless crystalline compound, soluble in benzene, chloroform, alcohol, acetone, and alkaline solutions. Ochratoxin A gives a red ferric chloride test, has a melting point of 94–96°C (with benzene of crystallization) and 169°C (from xylene), UV absorption maxima of 213, 331, and 378 nm (ε 32 700, 4100, and 2050, respectively), and γ maxima (KBr disc) 1730, 1680, and 1535 cm^{-1}. Under UV light (365 nm), ochratoxins A, B, and C appear as bright green fluorescent zones on TLC silica gel plates, but on exposure to ammonia, the colour of the zones changes from green to light blue.

Analytical procedures for ochratoxins have presented difficulties in recoveries and sensitivities. Ochratoxin A has been determined, along with aflatoxin and zearalenone, by placing a portion of a chloroform-water extract of the commodity to be analysed on a silica gel column and eluting the toxins stepwise after appropriate solvent washes of the column (Eppley, 1968). The sensitivity limit of the multitoxin procedure is 25–50 ppb and recoveries of 60% can be expected. A method of determining ochratoxins in barley has been approved for official first action in the AOAC book on Natural Poisons (1975), whereby chloroform-water extracts of barley are placed on an aqueous sodium bicarbonate-diatomaceous earth column. The column is eluted first with hexane-chloroform to remove ochratoxin esters and then with formic acid-chloroform to remove ochratoxins A and B. Ochratoxins can be determined in purified extracts by TLC either by visual or densitometric comparisons with standards on TLC plates. The initial confirmatory test is the change from green to blue in fluorescence on exposure to ammonia. Final confirmation depends on the formation of the ethyl ester. In the same book, a procedure is given for official first action for ochratoxin in green coffee.

According to Steyn (1971), the oral LD_{50} dose in day-old ducklings was 150 μg/duckling and that of ochratoxin B and the ethyl ester of ochratoxin A were less toxic. Peckham *et al.* (1971) reported the oral LD_{50} of ochratoxin A in 7 day-old chicks was 116–135 μg and ochratoxin B was less toxic (1890 μg/chick). Chicks fed ochratoxin showed acute nephrosis, hepatic degeneration, and enteritis. Suppression of haematopoiesis in the bone marrow and a reduction in the lymphoid elements in the spleen and bursa of Fabricins were also seen. The principal gross finding was visceral gout. Unlike earlier studies, these investigators reported that ochratoxin B was about one-tenth as toxic as ochratoxin A.

When rats are administered ochratoxin, enteritis and renal necrosis occur; also, the quantity of glycogen in the liver increases and the phosphorylase enzyme system is inhibited (Steyn, 1971).

Krogh *et al.* (1974) studied the effect of ochratoxin in swine at levels of 200, 1000, and 4000 μg per kilo of feed. These levels correspond to the toxin levels sometimes found naturally in animal feeds in Denmark. The swine were fed crude ochratoxin at these levels for about 4 months. A characteristic symptom was porcine nephropathy, observed in swine at time of slaughter in commercial slaughter houses. The changes in kidney structure consisted of degeneration of the proximal tubules, interstitial formation of connective tissue, and, in later stages, development of atrophied and sclerotized glomerator tufts. Ochratoxin A was detected in the kidney, liver, adipose, and muscular tissue of these experimental animals.

Citrinin

Citrinin was originally discovered by Hetherington and Raistrick (1931), later, in 1941 found to be an antibiotic. It is produced by a number of fungi including *Aspergillus niveus*, *A. terreus*, and *Penicillium citrinum*; the last one is a common storage mould occurring in various cereals. In addition, reportedly citrinin is formed by *P. implicatum*, *P. lividum*, *P. fellutanum*, *P. jensenii*, *P. citreo-viride*, *P. expansum*, *P. notatum*, and *A. candidus*. The toxin occurs naturally in wheat, rye, barley, and oats in both Canada and Denmark (Hesseltine, 1974). Since *P. citrinum* isolates are frequently found in cereals and since most strains produce citrinin, it may be an important mycotoxin when its occurrence can be investigated with a suitable assay.

Davis *et al.* (1975) state that *P. citrinum* NRRL 5927 will produce up to 1·75 g/litre of medium when grown in stationary culture in carboys containing 4 litres of a 4% sucrose-2% yeast extract medium. The carboys, laid on their sides to give maximum surface growth, were incubated at 30°C. Citrinum formed on the eighth day and increased for 21 days.

Although no official method for detection of citrinin is known, Neely *et al.* (1972) reviewed available methods, investigated the electronic spectral properties of the material, and recommended fluorometric determination in $CHCl_3$ as the most sensitive analytical method. As an antibiotic, it was assayed by measuring zones of inhibition against selected bacterial strains.

Crystals of citrinin are yellow, with absorption maxima at 250 and 331 mμ and a melting point of 172°C. It has an UV spectra absorption at 320. It is insoluble in water but soluble in alcohol, dioxane, and dilute alkali. Citrinin fluoresces greenish yellow under UV light. Solutions change colour with pH changes—lemon yellow in acid and cherry red in alkali. Both methyl and dihydro forms are known. The first is soluble in hot alcohol, moderately soluble in chloroform, and insoluble in petroleum ether. The dihydro is soluble in alcohol, acetone, and chloroform, and sparingly soluble in benzene and petroleum ether. Citrinin has the structure shown in Fig. 8.

In animals, citrinin retards growth and causes renal damage. There is an increased urinary output in rats. Kidneys become grey white, enlarged, and turbid and histological changes show degeneration and dilation of the lower nephrons. The LD_{50} via i.p. route was 35 for the mouse and 19 mg/kg body weight for the rabbit.

The most detailed paper on the toxicity of citrinin is that of Ambrose and DeEds (1946). Typical symptoms observed in rabbits given intravenous

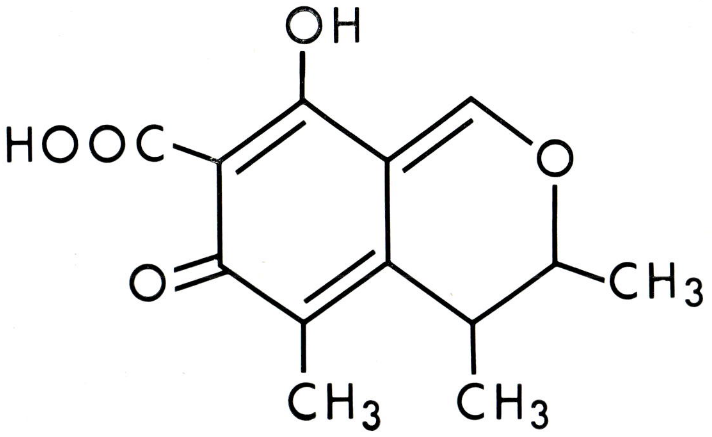

Fig. 8. Citrinin.

injections of citrinin included hyperemia, miosis, salivatin, increased bronchia secretion, and lacrimation. In dogs, an evanescent fall in blood pressure is noted, which is associated with dilation of the blood vessels of the skin, kidney, and extremities. Citrinin acts as an antibiotic against some fungi and gram-positive but not gram-negative bacteria.

T-2 Toxin

T-2 is the only trichothecene mycotoxin isolated from naturally infected feed. The trichothecene toxins, at least 27 are known, are probably responsible, according to Smalley and Strong (1974), for a number of diseases, including low-temperature mouldy corn toxicosis, mouldy cereal emesis, 'akakabibyo' toxicosis, alimentary toxic aleukia, mouldy bean hull intoxication, stachybotryotoxicosis, dendrodochiotoxicosis, and fusariotoxicosis.

T-2 toxin is formed by various species of *Fusarium* but probably especially by *F. tricinctum*. This species has been isolated from field and sweet corns, sorghum, tall fescue, turf grasses, peas, cranberries, and rice. The toxin has been isolated only once from nature, and this was from cattle feed containing 60% mouldy corn in Wisconsin (Hsu *et al.*, 1972).

Burmeister (1971) using *Fusarium tricinctum* NRRL 3299, an isolate from mouldy corn, reported yields of 7·5 g of T-2 per 1 k of white corn grits. The fermentation was carried out on this solid substrate without agitation for 3 weeks at 15°C, with recoveries of 2·3 g of crystalline material per 1200 g of substrate.

T-2 toxin, 3a-hydroxy-4β,15-diacetoxy-8a-(3 methyl-butyryloxy-12,13-epoxy-Δ^9-trichothecene has a melting point of 151–152°C and a molecular formula of $C_{24}H_{34}O_9$. It is a colourless, crystalline, optically active solid which is stable and not destroyed by normal cooking. It is soluble in moderately polar organic solvents, but only very slightly soluble in water. The structure of T-2 toxin is shown in Fig. 9. Bamberg and Strong (1971) discuss the chemistry of all the trichothecenes in detail including T-2.

No assay for the specific detection of T-2 toxin is as yet known, although skin testing has been used. T-2 should be suspected when lesions are produced by application of extracts of possibly contaminated substances and with the

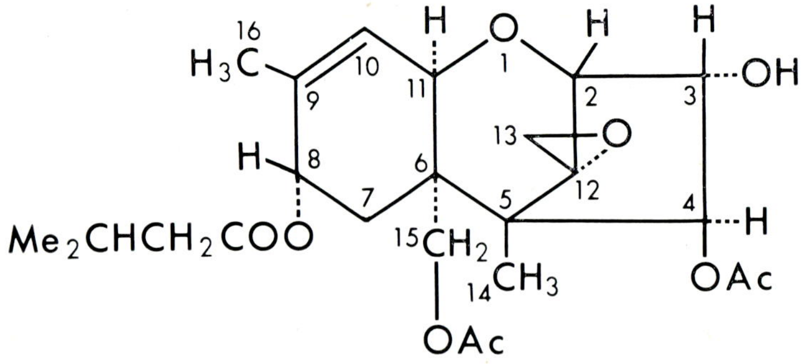

Fig. 9. T-toxin.

mycological detection of *F. tricinctum* from the suspicious product. Besides T-2 toxin, several other trichothecenes possess dermatitic activity. If T-2 is applied to the shaved skin of rats, in a few hours the area becomes swollen, followed by a dry crusty appearance of the skin, which becomes dark red or purple; with a concentrated solution, suppuration and extensive necrosis occur. The material is hazardous to humans handling it, and extreme care must be taken not to come into contact even with dilute solutions. Bamburg and Strong (1971) report on one such accident in the laboratory. Even though the contact was for only 2 minutes, the effect of an extract on the hands was a burning sensation, replaced by numbness and complete loss of sensitivity. The skin layer peeled off, and sensitivity was not restored until new skin had grown back. The LD_{50} (oral) of T-2 was 3·8 mg/kg body weight in the rat and 6·1 mg/kg in trout. The i.p. LD_{50} in the mouse was 3·04 mg/kg.

Smalley (1973) investigated the mouldy corn syndrome caused by T-2 toxin. When T-2 was administered to rats or other animals, the mycotoxin induced necrotic lesions and haemorrhages in the intestines, liver, and kidneys. The toxin can be absorbed through the skin, resulting in local inflammation and in death. He determined the LD_{50} in swine and rats as about 4 mg/kg of body weight. In a natural poisoning in cattle resulting in seven deaths in a herd of cows, T-2 was isolated from the corn being fed as well as the fungus *Fusarium tricinctum*. The level of T-2 toxin in this corn was 2 mg/kg of mouldy corn but recovery was incomplete. Wyatt *et al.* (1973) noted that chickens ingesting T-2 toxin developed raised yellowish-white lesions on mouth parts and that growth rates were reduced. T-2 toxin also inhibits germination of plant seeds, retards growth of plants, and acts as an antibiotic against some microorganisms, such as yeasts.

REFERENCES

General References on Mycotoxins

AOAC. Natural poisons (1975). Chapter 26, Official Methods of Analysis. Assoc. Off. Anal. Chem., Washington, D.C. 24 pp.

Ciegler, A., Kadis, S., and Ajl, S. J. (eds.) (1971). Microbial Toxins. Vol. VI. Academic Press, New York. 563 pp.

Herzberg, M. (ed.) (1970). Proc. First U.S.–Japan Conf. Toxic Microorganisms, Unnumb. Publ., U.S. Dept. Interior and UJNR Panels on Toxic Microorganisms. 490 pp.

Hesseltine, C. W. (1969). Mycotoxins. *Mycopathol. Mycol. Appl.*, **39**, 371–383.
Hesseltine, C. W. (1974). Natural occurrence of mycotoxins in cereals. *Mycopathol. Mycol. Appl.*, **53**, 141–153.
Purchase, I. F. H. (ed.) (1974). Mycotoxins. Elsevier Scientific Pub. Co., New York. 443 pp.
Wogan, G. N. (ed.) (1965). Mycotoxins in Foodstuffs. MIT Press, Cambridge, Mass. 291 pp.

Penicillic Acid and Patulin

Alsberg, C. L., and Black, O. F. (1913). Contributions to the study of maize deterioration: Biochemical and toxicological investigations of *Penicillium puberulum* and *Penicillium stoloniferum*. *U.S. D.A. Bull.*, **270**, 5–48.
Dickens, F., and Jones, H. E. H. (1961). Carcinogenic activity of a series of reactive lactones and related substances. *Brit. J. Cancer*, **15**, 85–100.
Kurtzman, C. P., and Ciegler, A. (1970). Mycotoxin from a blue-eye mold of corn. *Appl. Microbiol.*, **20**, 204–207.
Norstadt, F. A., and McCalla, T. M. (1963). Phytotoxic substance from a species of *Penicillium*. *Science*, **140**, 410–411.
Scott, P. M. (1974a). Collaborative study of a chromatographic method for determination of patulin in apple juice. *J. Assoc. Off. Anal. Chem.*, **57**, 621–625.
Scott, P. M. (1974b). Patulin. In *Mycotoxins*, pp. 383–403 (I. F. H. Purchase, ed.), Elsevier Scientific Pub. Co., New York.
Scott, P. M., Miles, W. F., Toft, P., and Dube, J. G. (1972). Occurrence of patulin in apple juice. *J. Agr. Food Chem.*, **20**, 450–451.
Snow, J. P., Lucas, G. B., Harvan, D., Pero, R. W., and Owens, R. G. (1972). Analysis of tobacco and smoke condensate for penicillic acid. *Appl. Microbiol.*, **24**, 34–36.
Thorpe, C. W., and Johnson, R. L. (1974). Analysis of penicillic acid by gas–liquid chromatography. *J. Assoc. Off. Anal. Chem.*, **57**, 861–865.
Wilson, D. M. (1975). Patulin and penicillic acid. *Adv. Chem.* In press.

Sterigmatocystin

Burkhardt, H. J., and Forgacs, J. (1968). O-Methylsterigmatocystin, a new metabolite from *Aspergillus flavus* Link ex Fries. *Tetrahedron Letters*, **24**, 717–720.
Davies, J. E., Kirkaldy, D., and Roberts, J. C. (1960). Studies in mycological chemistry. Part VII. Sterigmatocystin, a metabolite of *Aspergillus versicolor* (Vuillemin) Tiraboschi. *J. Chem. Soc.*, 2169–2178.
Holzapfel, C. W., Purchase, I. F. H., Steyn, P. S., and Gouws, L. (1966). The toxicity and chemical assay of sterigmatocystin, a carcinogenic mycotoxin and its isolation from two new fungal sources. *South Afr. J.*, **40**, 1100–1101.
Roberts, J. C. (1974). Aflatoxins and Sterigmatocystins. *Fortschr. Chem. Org. Naturst.*, **31**, 119–151.
Rodricks, J. V., Henery-Logan, K. R., Campbell, A. D., Stoloff, L., and Verrett, M. J. (1968). Isolation of a new toxin from cultures of *Aspergillus flavus*. *Nature*, **217**, 668.
van der Watt, J. J. (1974). Sterigmatocystin. In *Mycotoxins*, pp. 369–382 (I. F. H. Purchase, ed.), Elsevier Scientific Pub. Co., New York.

Alternaria Toxins

Christensen, C. M., Nelson, G. H., Mirocha, C. J., and Bates, F. (1968). Toxicity to experimental animals of 943 isolates of fungi. *Cancer Res.*, **28**, 2293–2295.
Doupnik, B., and Sobers, E. K. (1968). Mycotoxicosis: Toxicity to chicks of *Alternaria longipes* isolated from tobacco. *Appl. Microbiol.*, **16**, 1596–1597.
Harvan, D. J., and Pero, R. W. (1975). The structure and toxicity of the alternaria metabolites. *Adv. Chem.* In press.

Pero, R. W., Owens, R. G., Dale, S. W., and Harvan, D. (1971a). Isolation and identification of a new toxin, alternuene, from the fungus *Alternaria tenuis*. *Biochem. Biophys. Acta*, **230**, 170–179.

Pero, R. W., Owens, R. G., and Harvan, D. (1971b). Gas and thin-layer chromatographic methods for analysis of the mycotoxins altenuene, alternariol, and alternariol monomethyl ether. *Anal. Biochem.*, **43**, 80–88.

Pero, R. W., Harvan, D., and Blois, M. C. (1973a). Isolation of the toxin, altenuisol, from the fungus, *Alternaria tenuis* Auct. *Tetrahedron Letters*, **12**, 945–948.

Pero, R. W., Posner, H., Blois, M., Harvan, D., and Spalding, J. W. (1973b). Toxicity of metabolites produced by the '*Alternaria*.' *Environ. Health Perspect.*, (June) pp. 87–94.

Seitz, L. M., Sauer, D. B., Mohr, H. E., Burroughs, R., and Paukstelis, J. V. (1975). Metabolites of *Alternaria* in grain sorghum. Compounds which could be mistaken for zearalenone and aflatoxin. *J. Agr. Food Chem.*, **23**, 1–4.

Emetic and Rejection Factors

Morooka, N., Uratsuji, N., Yoshizawa, T., and Yamamoto, H. (1972). Studies on the toxic substances in barley infected with *Fusarium* spp. *J. Food Hyg. Soc. Japan*, **13**, 368–375.

Prentice, N., and Dickson, A. D. (1964). Emetic material in scabbed wheat. *Cereal Chem.*, **41**, 548–550.

Prentice, N., and Dickson, A. D. (1968). Emetic material associated with *Fusarium* species in cereal grains and artificial media. *Biotechnol. Bioeng.*, **10**, 413–427.

Prentice, N., Dickson, A. D., and Dickson, J. G. (1959). Production of emetic material by species of *Fusarium*. *Nature*, **184**, 1319.

Ueno, Y., Ishikawa, Y., Nakajima, M., Sakai, K., Ishii, K., Tsunoda, H., Saito, M., Enomoto, M., Ohtsubo, K., and Umeda. M. (1971). Toxicological approaches to the metabolites of Fusaria. I. Screening of toxic strains. *Japan J. Exp. Med.*, **41**, 257–272.

Ueno, Y., Ishii, K., Sato, N., and Ohtsubo, K. (1974). Toxicological approaches to the metabolites of Fusaria. VI. Vomiting factor from moldy corn infected with *Fusarium* spp. *Japan J. Exp. Med.*, **44**, 123–127.

Vesonder, R. F., Ciegler, A., and Jensen, A. H. (1973). Isolation of the emetic principle from *Fusarium*-infected corn. *Appl. Microbiol.*, **26**, 1008–1010.

Yoshizawa, T., and Morooka, N. (1973). Deoxynivalenol and its monoacetate: New mycotoxins from *Fusarium roseum* and moldy barley. *Agr. Biol. Chem.*, **37**, 2933–2934.

Yoshizawa, T., and Morooka, N. (1974). Studies on the toxic substances in infected cereal (III). Acute toxicities of new trichothecane mycotoxins: Deoxynivalenol and its monoacetate. *J. Food Hyg. Soc. Japan*, **15**, 261–269.

Zearalenone

Andrews, F. N., and Stob, M. (1965). Anabolic and estrogenic compound and process of making. *U.S. Pat.* 3,196,019.

Eppley, R. M. (1968). Screening method for zearalenone, aflatoxin, and ochratoxin. *J. Assoc. Off. Anal. Chem.*, **51**, 74–78.

Mirocha, C. J., and Christensen, C. M. (1974). Oestrogenic mycotoxins synthesized by *Fusarium*. In *Mycotoxins*, pp. 129–148 (I. F. H. Purchase, ed.), Elsevier Scientific Pub. Co., New York.

Mirocha, C. J., Christensen, C. M., and Nelson, G. N. (1968). Physiologic activity of some fungal estrogens produced by *Fusarium*. *Cancer Res.*, **28**, 2319–2322.

Shotwell, O. L., Hesseltine, C. W., Vandegraft, E. E., and Goulden, M. L. (1971). Survey of corn from different regions for aflatoxin, ochratoxin, and zearalenone. *Cereal Sci. Today*, **16**, 266–273.

Ochratoxin

AOAC. Natural Poisons (1975). National Research Council 1973. Toxicants

could be regarded as weak pathogens, while the other isolates which were unable to cause rot could be considered only as secondary invaders.

Mycelial growth was observed externally on the corms at the sites of wounds and at the points of emergence of rootlets, which were broken off during harvesting. No penetration occurred through the intact skin of the corms. It is thus indicated that the penetration of the fungi takes place through wounds created as a result of mechanical injury during harvesting or otherwise.

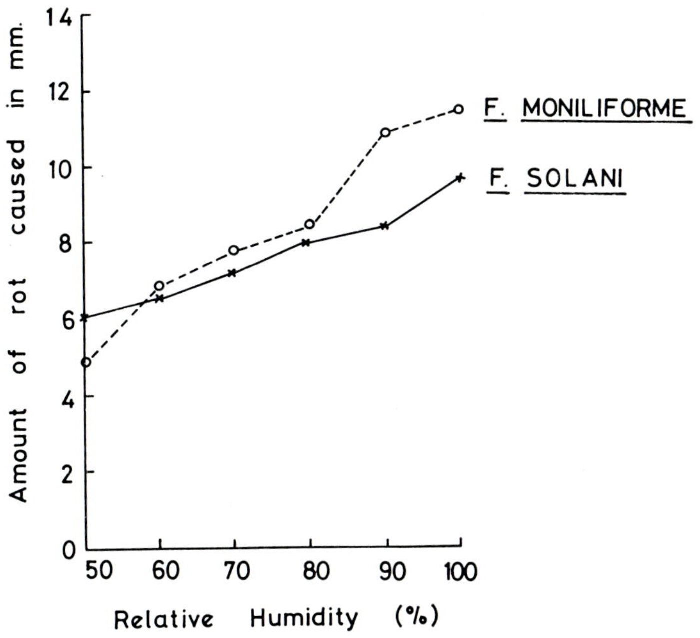

Fig. 2. Effects of relative humidity on rot caused by the pathogens.

The effect of environmental factors on the ability of the fungi to infect cocoyam and cause rot is shown in Figs. 1 and 2. Rot occurred over the range of 20–30°C, with maximum rotting between 26 and 30°C. The results of investigations into the effect of relative humidity (R.H.) on the ability of the pathogens to cause rot as shown in Fig. 2 indicate that no rot occurred at relative humidities less than 50%, but rotting progressively increased with R.H. from 50 to 100%, appreciable rot occurring at R.H. above 80%.

DISCUSSION

Probably the most serious storage losses are caused by a group of rots (Coursey, 1968); losses due to rots have been shown to reach 45–48% after 3–4 months storage in Cameroun (Praquin and Miche, 1971).

Several species of fungi have been reported as associated with root rots of the growing plants (Wright, 1930); of these several species of *Pythium* are thought to be important (Wright, 1930; Posnette, 1945; Plucknett *et al.*, 1970; Trujillo, 1967) and have been shown to cause serious losses of yield in *Colocasia* in Hawaii (Parris, 1941). *Diplodia* spp., *Fusarium solani*, *Sclerotium rolfsii*, and *Bacillus carotoverus* Jones were found to be the cause of storage rots of various economic aroids, including *X. sagittifolium*. Sclerotium rot has also been reported in Hawaii (Parris, 1941) and in the Philippines (Fajardo and Mendoza, 1935).

Of the microorganisms associated with cocoyam rot in this study, *F. solani* has previously been associated with storage rot of aroid root crops (Gollifer and Booth, 1972). *F. moniliforme* is recorded for the first time in this context although this fungus has been associated with soft rot of stored yams in Nigeria (Ogundana *et al.*, 1970). *S. rolfsii*, isolated in this study, behaved only as a weak pathogen though it had been reported to have caused cocoyam rot in Hawaii, while Gollifer and Booth (1972) have shown the same fungus to be pathogenic on *Colocasia* corms in the British Solomon Islands Protectorate.

Since it has been shown that infection normally occurs through the points of physical damage inflicted either during harvesting or subsequently, prevention of storage rot therefore demands that harvesting should be carried out with the greatest care and mechanical damage to the corms avoided.

The extent of rotting caused by each of the pathogens at various temperature levels after 2 weeks as shown in Fig. 1 shows that the optimum range of development of rot was between 26 and 30°C. This is in agreement with the results obtained by Burton (1970) who found the optimum temperature for rot development as being between 22 and 29°C. It is not surprising that optimum rot caused by the pathogens occurred at high relative humidity (80–100%) and 26–30°C since these are the natural storage conditions of Nigeria.

Acknowledgements

The author expresses his appreciation to D. C. Coursey and R. H. Booth of the Tropical Products Institute, London, for their critical appraisal of the manuscript.

REFERENCES

Burton, C. L. (1970). Diseases of tropical vegetables on the Chicago market. *Trop. Agric. Trin.*, **47** (4), 303–313.

Coursey, D. C. (1968). The edible aroids. *World Crops*, **20** (3), 25–30.

Fajardo, T. G., and Mendoza, J. M. (1935). Studies on the *Sclerotium rolfsii* Sacc. *Philipp. J. Agric.*, 387–424.

Gollifer, E. D., and Booth, R. H. (1972). Storage losses of taro corms in the British Solomon Islands Protectorate. *Ann. Appl. Biol.* (in press).

Ogundana, S. K., Naqui, S. H. Z., and Ekundayo, J. A. (1970). Fungi associated with soft rot of yams (*Dioscorea* spp.) in storage in Nigeria. *Trans. Br. Mycol. Soc.*, **54** (3), 445–451.

Parris, G. K. (1941). Diseases of taro in Hawaii and their control, with notes on field production. *Circ.* 18 *Hawaii Agric. Exp. Stn.*
Plucknett, D. L., de la Pena, R. S., and Obrero, F. (1970). Taro, Colocasia esculenta. *Field Crop Abstr.*, **23** (4), 413–426.
Posnette, A. F. (1945). Root-rot of cocoyams (*Xanthosoma sagittifolium* Schott.) *Trop. Agric. Trin.*, **22** (9), 164–170.
Praquin, J. Y., and Miche, J. C. (1971). Essai de conservation de taros et macabos au Cameroun IRAT Rapport preliminaire No. 1, Station de Dachang.
Trujillo, E. E. (1967). Diseases of the genus *Colocasia* in the Pacific area, and their control. Proc. 1st Int. Symp. Trop. Root Crops, Trinidad, Vol. 2 (4), 13–19.
Wright, J. (1930). Root-rot of Cocoyams. *Gold Coast Dept. of Agric. Year-book* 1930. Govt. Printer, Accra.

BIODEGRADATION OF TASTE AND ODOUR COMPOUNDS

WESLEY O. PIPES

Drexel University, Philadelphia, Pennsylvania, U.S.A.

Summary

Organic taste and odour compounds in public water supplies originate either from industrial wastes or from metabolic by-products produced by microorganisms in the water source. This is a report on an investigation of a technique of using bacteria to degrade a typical odour compound in water as it passes through a sand filter. The filter was heavily inoculated with bacteria capable of degrading the odour compound, but it required 1–10 days of continuous filtration before the bacteria became active enough to reduce the odour below the detectable limit.

INTRODUCTION

The two major sources of taste and odour compounds in a public water supply are industrial wastes and microorganisms in the raw water source. Most odorous wastewater components are readily oxidized by chlorine. The group of organic compounds in industrial wastes which cause the greatest problems are phenols. Chlorination of water containing phenols causes the formation of chlorophenols which have even worse odours than the phenols themselves. Complete oxidation of phenols requires massive doses of chlorine.

Only within the last 10 years have some of the naturally occurring odorous compounds been chemically identified. These compounds include geosmin (Gerber and Lechevalier, 1965), mucidione (Dougherty and Morris, 1967), and isoborneol (Medsker *et al.*, 1969), but there are several other such compounds yet to be characterized. The odorous compounds are derived mostly from actinomycetes.

Silvey and Roach (1964) found that some species of *Bacillus* were able to oxidize odorous actinomycete metabolites. Hohen (1965) tested several strains of *Bacillus* for ability to reduce odours in water and found that the most effective was *Bacillus cereus* ATCC 10876. Silvey *et al.* (1970) reported on using a mass culture of *Bacillus cereus* ATCC 10876 to produce enough organisms to inoculate an entire reservoir and thereby achieve removal of the odour from the raw water supply. Narayan and Nunez (1974) reported on laboratory studies in which they investigated the rate of oxidation of geosmin by four different species of *Bacillus* and six other species of bacteria which they isolated from enrichment cultures.

This study was an investigation of a method of using biodegradation for removal of odours from water. The odour compound used as a model was phenol since this compound causes significant odour problems in many public water supplies, it is readily available in pure form, and satisfactory chemical methods for measuring the concentration of phenol in water are available.

METHODS

A model rapid sand filter was constructed from 3-inch inside diameter plexiglas tubing. The filter consisted of 24 in of sand over 6 in of gravel. The sand had an effective size of 0·42 mm, a uniformity coefficient of 1·4, and a porosity of 40 %. The filter was fitted with six manometers so that the pressure loss in the filter could be monitored.

The water used for the experiments was tap water which was dechlorinated by passing through an activated carbon column. After dechlorination, the water flowed into a constant head tank. A sterilized solution with a phenol concentration of 10 mg/litre was mixed into a constant head tank at a rate so that it was diluted 1000 to 1 with dechlorinated tap water to give a phenol concentration of 10 μg/litre. The phenol solution being filtered was essentially free of suspended solids.

The bacteria used for inoculation of the filter were a strain of *Bacillus cereus* which had been isolated from an activated sludge system treating wastewaters containing some phenol. The inoculation culture was grown on a broth containing 500 mg/litre glucose, 100 mg/litre peptone, and 100 mg/litre yeast extract. After one day of growth, the bacterial population density was checked by a direct microscopic count using a Petroff-Hauser counting chamber. The inoculation culture was then diluted with sterile water to give a bacterial density of approximately 10^8 cells/ml. Before each run, the filter was sterilized with a solution containing 40 mg/litre of free chlorine. The filter was then backwashed with sterile, distilled water and the inoculum culture was passed through the culture. The flow of the phenol solution was then adjusted to 370 ml/min. This corresponds to a filtration rate of 2 gal/min/ft^2 (81 litre/m^2/min), a value common in American water works practice. At this rate the contact time of the phenol solution in the filter was 3·76 min.

The phenol in the influent water and filtered water was measured at least twice daily by the aminoantipyrine method (American Public Health Association, 1971). This procedure gives a detection limit of 1 μg/litre. The bacterial count in the filtered water was determined at least twice a day. Both the direct microscopic count using the Petroff-Hauser counting chamber and the plate count using standard pour plates with nutrient agar were used.

During each test run, one sample of the filtered water was collected for a determination of the chlorine dose required to kill the bacteria coming off the filter. At the end of each filter run, a core sample of the sand in the filter was removed using a sterile tube so that the numbers of bacteria per gramme of sand could be determined.

RESULTS

For the entire series of experimental runs, the influent phenol concentration was found to vary between 8 mg/litre and 13 mg/litre. The concentrations of phenol in the filtered water from filter runs 1 and 2 are plotted versus time in Fig. 1. Both these runs were made with a water temperature between 16 and 17°C. The filter was inoculated with *Bacillus cereus* for run 1, but run 2 was an uninoculated control. Run 1 demonstrated that bacteria could grow in the filter in sufficient numbers so that they could completely remove phenol. The removal of phenol in run 2 was ascribed to the development of a natural

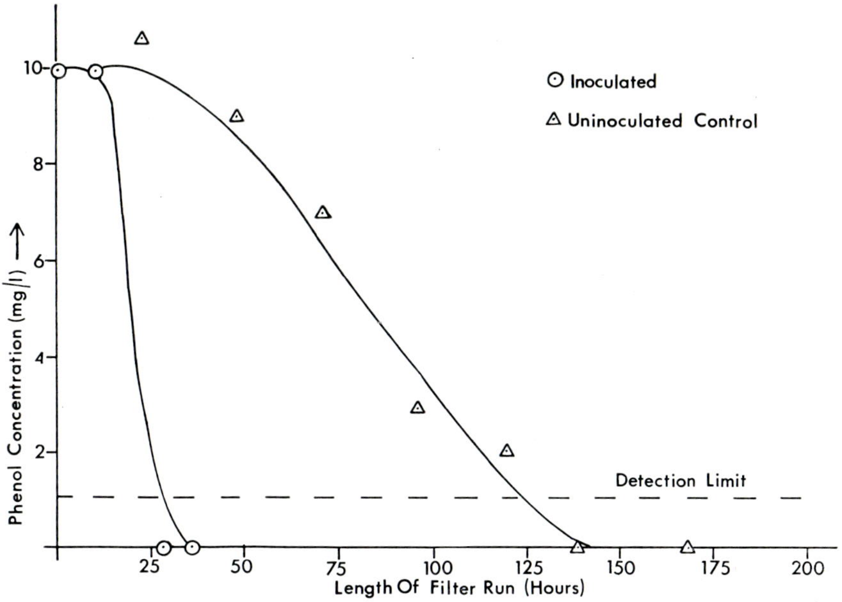

Fig. 1. Filtered water phenol concentration as a function of length of filter run for inoculated and uninoculated filters.

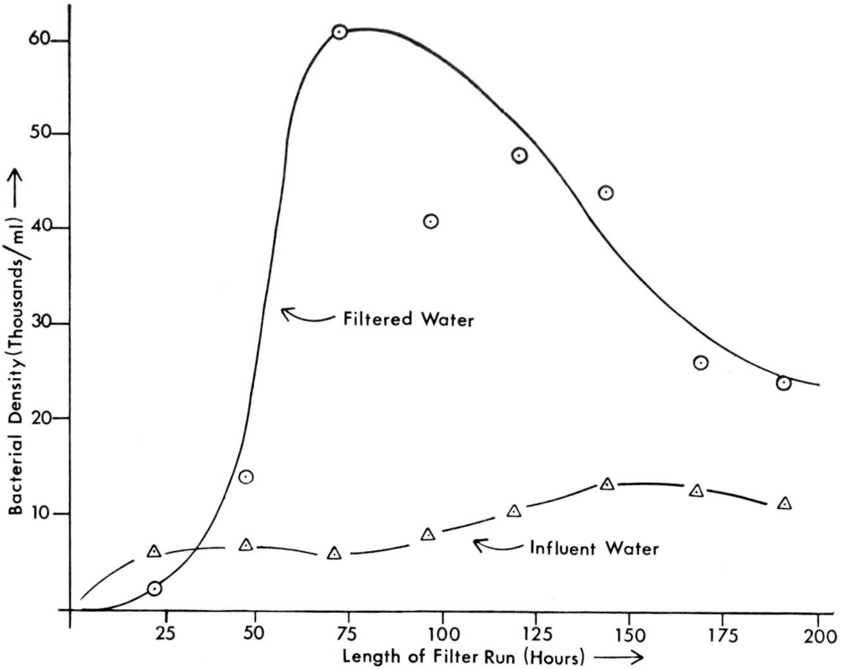

Fig. 2. Bacterial counts on water from uninoculated filter.

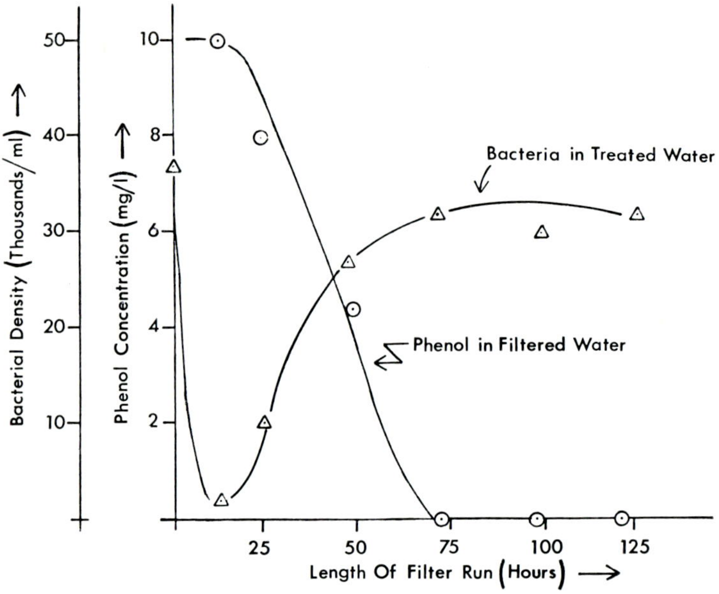

Fig. 3. Phenol and bacteria in filtered water (Run 3).

flora of phenol degrading bacteria in the filter. The water being filtered was not sterilized before it was mixed with the sterile phenol solution and had bacterial plate counts which varied between 5000 and 10 000/ml (Fig. 2). From Fig. 2 it is clear that there was extensive bacterial growth in the filter which could only have been due to degradation of phenol. The bacterial plate counts on the filtered water during run 3 are presented in Fig. 3. This same pattern was found for all the other filter runs which were inoculated. In these runs the bacterial counts on the filtered water dropped off initially to levels similar to that in the influent and then increased to level off at average values between 13 000/ml and 35 000/ml when the phenol was being completely removed. In the uninoculated run, the growth was initially in the water and 4 or 5 days were required before the bacterial growth was retained in the filter.

Table I

BACTERIAL DATA FROM FILTER RUN

Run no.	*Inoculum bacteria count (per ml)*	*Temperature (°C)*	*Time to complete phenol removal (hr)*	*Bacteria in filtered water (per ml)*	*Bacteria in sand (per g)*
1	$1{\cdot}7 \times 10^8$	16–17°C	<24	$3{\cdot}5 \times 10^4$	$8{\cdot}4 \times 10^3$
2	none	16–17°C	144	$2{\cdot}5 \times 10^4$	—
3	$2{\cdot}6 \times 10^8$	13–14°C	68	$2{\cdot}9 \times 10^4$	$9{\cdot}3 \times 10^3$
4	$4{\cdot}8 \times 10^8$	11–12°C	132	$1{\cdot}3 \times 10^4$	$5{\cdot}9 \times 10^4$
5	$3{\cdot}7 \times 10^8$	10–11°C	156	$1{\cdot}9 \times 10^4$	$1{\cdot}6 \times 10^4$
6	$1{\cdot}4 \times 10^8$	8–10°C	264	$2{\cdot}8 \times 10^4$	$1{\cdot}1 \times 10^4$

The bacterial counts in the filtered water and in the sand from the filter for all runs are presented in Table I. The bacterial counts given for the sand were from the top 2 cm in each case. The bacterial numbers decreased at lower levels in the filter so that at the bottom of the sand only about one-quarter of the count of the top layer was found. The variability in bacterial counts between filter runs was surprisingly small and there was no detectable trend either up or down with the temperature change.

The experience with run 2 pointed out the significance of the time required to establish a bacterial population large enough to oxidize the phenol. This time was designated as 'seasoning time'. In the subsequent runs, water temperature and seasoning time were the major variables studied. Since these

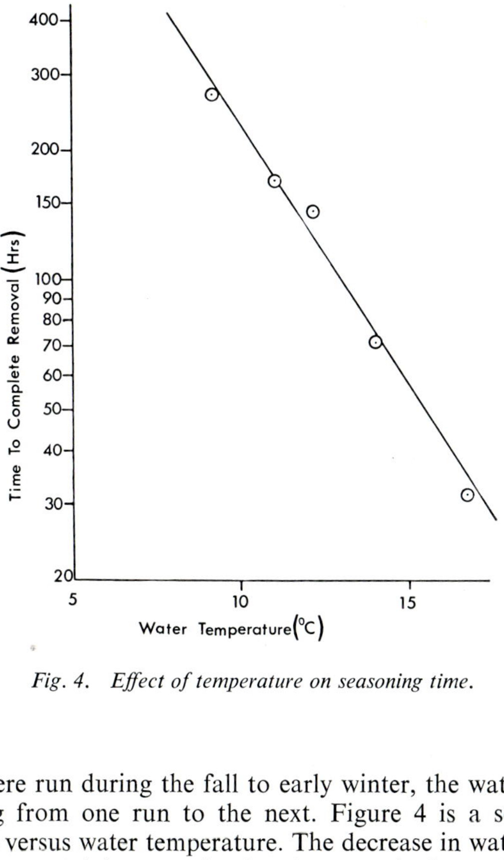

Fig. 4. Effect of temperature on seasoning time.

experiments were run during the fall to early winter, the water temperature was decreasing from one run to the next. Figure 4 is a semilog plot of seasoning time versus water temperature. The decrease in water temperature caused an exponential increase in the time required to obtain complete removal of the phenol.

For all filter runs the pressure loss through the filter varied between 3 and 3·5 ft and did not increase significantly during the runs. The chlorine doses used to kill the bacteria in the filtered water sampled were varied between 1 and 10 mg/litre. A 30-min contact time was used. In all cases it was found that a chlorine dose of 4 mg/litre was adequate to reduce the total plate counts to less than 1 mg/ml.

DISCUSSION

In most water treatment plants, a filter bed is used until the pressure loss reaches some preselected value such as 8 or 10 ft. The filter is then backwashed and the cycle started again. The pressure loss values found in this series of experiments, 3–3·5 ft, represented mostly the frictional resistance of the sand itself and never did reach values high enough that the filter had to be backwashed. The virtual absence of suspended solids in the water being filtered was responsible for the lack of pressure loss buildup. In an operating plant the filter would have been backwashed after 30–100 hr of operation.

When the filter was not inoculated with *Bacillus cereus* (run 2), it required almost six days to build up a bacterial population adequate to remove all of the phenol. In practice, the filter would have been backwashed long before it was seasoned and the effect of the biodegradation of the phenol would not have been noticed. Of the runs when the filter was heavily inoculated with *Bacillus cereus*, runs 4, 5, and 6 required seasoning times longer than the normal operating time of a filter before backwash. Run 3 required three days seasoning time, and it is problematical whether or not the biodegradation of the phenol would have been observed in plant operation before the filter was backwashed. Only in run 1 when the seasoning time was one day would the biodegradation have been likely to have been observed in plant operation. Even in this case the biodegradation would not have been a great deal of help in controlling odours in the finished water because one-half to one-quarter of the filtered water would have detectable concentrations of phenol.

It is clear that the idea of inoculating a rapid sand filter with microorganisms capable of degrading odour compounds does not appear to be suitable for routine plant operation. However, it has been demonstrated that, after adequate seasoning time, it is possible to build up a population of bacteria which are capable of degrading odour compounds. This suggests that it may be possible to develop a process which would be suitable for routine plant operation.

CONCLUSIONS

(1) It has been demonstrated that heavy inoculation of a rapid sand filter with bacteria capable of degrading odour compounds can, after adequate seasoning of the filter, achieve complete removal of the odour compounds from water.

(2) The time required for seasoning the filter is temperature dependent, requiring about 1 day at 16–17°C and about 10 days at 9°C.

(3) The seasoning time requirement is so great that this concept probably will not result in a useful water treatment process. These results do suggest that a useful water treatment process for biodegradation of odour compounds could be developed.

Acknowledgements

The efforts of M. J. A. Cavalier, Mrs C. Wienke, and Mrs B. Katz in operating the filters and collecting the data are gratefully acknowledged. At the time this work was carried out, W. O. Pipes was Professor of Civil Engineering, Northwestern University, Evanston, Illinois.

REFERENCES

American Public Health Association (1971). *Standard Methods for the Examination of Water and Wastewater*. 13th ed., Am. Public Health Assoc., New York, pp. 501–510.

Dougherty, J. D., and Morris, R. L. (1967). Studies on the removal of actinomycete musty tastes and odors in water supplies. *J. Am. Water Works Assoc.*, **59**, 1320–1326.

Gerber, N. N., and Lechevalier, H. A. (1965). Geosmin, an earthy-smelling substance isolated from actinomycetes. *Appl. Microbiol.*, **13**, 935–938.

Hohen, R. C. (1965). Biological methods for the control of tastes and odors. *Southwest Water Works J.*, **47**, 3: 26–35.

Medsker, L. L., Jenkins, D., and Thomas, J. E. (1969). Odorous compounds in natural waters: 2-exo-hydroxy-2-methylborane, the major compound produced by several actinomycetes. *Environ. Sci. Technol.*, **3**, 476–478.

Narayan, L. V., and Nunez, W. J. (1974). Biological control: isolation and bacterial oxidation of the taste-and-odor compound, geosmin. *J. Am. Water Works Assoc.*, **66**, 532–536.

Silvey, J. K. G., and Roach, A. W. (1964). Studies on microbiotic cycles in surface waters. *J. Am. Water Works Assoc.*, **56**, 60–72.

Silvey, J. K. G., Henley, D. E., Nunez, W. J., and Hohen, R. C. (1970). Biological control of naturally occurring tastes and odors by microorganisms. *Proc. Natl. Biol. Congr.*, Detroit, pp. 132–139.

THE BIODETERIORATION OF NATURAL RUBBER PIPE-JOINT RINGS IN SEWER MAINS

A. M. CUNDELL

Plant Pathology–Entomology Department, University of Rhode Island, Kingston, Rhode Island 02881, U.S.A.

and

A. P. MULCOCK

Microbiology Department, Lincoln College, Canterbury, New Zealand

Summary

Natural rubber pipe-joint rings collected from sewer mains in Christchurch, New Zealand, were found to be deteriorated by *Streptomyces* spp. The characteristics of the sewage and the physiological requirements of the rubber-utilizing microorganisms are discussed. The biodeterioration of vulcanized rubber samples was demonstrated in pure culture experiments. Evidence that antioxidants and antiozonants protect vulcanized natural rubber from microbiological deterioration was obtained by thin-layer chromatography of extracts of the pipe-joint rings. Antioxidants were absent or at a significantly reduced level in deteriorated pipe-joint rings, whereas undeteriorated rings contained high levels of two substituted phenylene diamine antioxidants.

Rubber o-rings are used to form flexible water-tight joints in underground pipelines conveying potable water, storm water, and sewage. Since these pipelines have an extended service life, it is imperative that the rubber rings sealing their joints be resistant to microbiological deterioration. A survey of the prevalence of deteriorated natural rubber pipe-joint rings in the Christchurch area showed, on the basis of a sample of 583 rings, that the estimated percentage of rings subjected to microbial attack in domestic sewer sidelines was 63% (Cundell and Mulcock, 1973a). Investigations by the authors indicated that the formulation of vulcanized rubber markedly influences the resistance of the rubber to biodeterioration under laboratory conditions (Cundell and Mulcock, 1973b,c; Cundell *et al.*, 1973). In this publication, the role of rubber-utilizing microorganisms in the deterioration of natural rubber pipe-joint rings from sewer mains is examined and the efficacy of antioxidants in preventing the biodeterioration of the rings under field conditions established.

MATERIALS AND METHODS

Collection of pipe-joint rings

Natural rubber pipe-joint rings were collected from sewer mains within the Christchurch area during alteration or repair of the concrete pipelines. The temperature of the sewage was related to ground and groundwater temperatures. It varied during the year from 10 to 20°C and was usually anaerobic. Sewer mains are jointed, on average, every 6 ft; however, extra joints occur where manholes, flush tanks, rising mains, and pumping stations are present in the disposal system. The number of pipe-joint rings laid between 1946 and 1970 was estimated as 283 772 in 342 miles of sewer mains.

During the collection, the orientation of the rings was recorded, details of the location of rings, date collected, number of years in use after installation, and the extent and nature of the deterioration of the rubber ring were noted.

Scanning electron microscopy

Deteriorated natural rubber pipe-joints were maintained in a mineral salts medium at 24°C for a month to encourage the formation of an aerial mycelium by the *Streptomyces* spp. growing within the deteriorated ring and prepared for examination with a scanning electron microscope (Cundell and Mulcock, 1975).

Identification of antioxidants in the pipe-joint rings

Rubber from the pipe-joint rings was finely divided by buffing on an electric grindstone and 0·25 g samples of each rubber were extracted with 10 ml of chloroform, methanol, and acetone, respectively, for not less than 72 hr. The extracting solvents were combined, evaporated to dryness, the residue dissolved in 0·5 ml of carbon tetrachloride, spotted on to silica gel thin-layer chromatography plates, and developed with a benzene/ethyl acetate/acetone 100/5/2 v/v mixture. The spots were located by spraying the plates with a 0·2% solution of 2,6-dibromo-*p*-benzoquinone-4-chlorimide in ethanol and the compounds identified by comparison with commercially available antioxidants (after Davies and Kam, 1968).

Isolation of rubber-utilizing microorganisms

Potential rubber-utilizing microorganisms were isolated on a natural rubber latex agar (Spence and van Niel, 1936), using the residue from the collar of pipe-joints as a source of inocula. The cultural characteristics and biochemical activities of the *Streptomyces* sp. most commonly isolated were determined.

Strips of vulcanized rubber 2·5 × 0·5 in were cut from tensile sheets approximately 0·07 in thick. A hole was punched at each end of the strip and the rubber was mounted on glass spanners. The mounted strips were placed in 100-ml conical flasks and partially immersed in 30 ml of a mineral salts solution (0·05% K_2HPO_4, 0·05% KH_2PO_4, 0·1% KNO_3, and 0·02% $MgSO_4 \cdot 7H_2O$). The flasks were plugged with nonabsorbent cotton wool, capped with foil, and sterilized for 15 min at 115°C. The flasks were inoculated with the rubber-utilizing microorganisms and incubated on a rotary shaking device in a constant temperature room maintained at 24°C for 18 months. The flasks were topped up with sterile distilled water at intervals to replace the water lost from the medium by evaporation.

At the termination of the pure culture experiments, the presence of nitrite-nitrogen and ammonia-nitrogen in the culture, the pH, and the surface tension of the medium were determined. The extent of growth of the inocula was noted and the loss of weight of the individual rubber strip found by removing the microbial growth from the rubber surface with a wet, soft bottle brush and drying the strip to constant weight in a vacuum drying oven at 45°C.

DEGRADATION OF WHEAT STRAW BY THE FUNGI ISOLATED FROM SYNTHETIC MUSHROOM COMPOST

D. S. CHAHAL,* AMARJIT SEKHON and B. S. DHALIWAL

Punjab Agricultural University, Ludhiana, India

Summary

Achaetomium luteum, *Aspergillus fumigatus*, *Chaetomium* sp., *Mucor pusillus*, and a member of Gymnoascaceae, possibly *Myxotrichum* sp., were isolated from synthetic compost (for *Agaricus bisporus*) during its peak degradation (fermentation). Some of the isolates are a new addition to the already recorded fungi from mushroom compost. The *Chaetomium* isolate seems to be a new species as it does not resemble the known species of this genus. All the isolated fungi were tested for their cellulolytic activity (CMCase) and degradation of wheat straw. The *Chaetomium* sp. showed the highest CMCase activity and maximum degradation of wheat straw under shake flask conditions at 37–40°C. Degradation of wheat straw was almost doubled when it was delignified.

The cultivation of mushrooms (*Agaricus bisporus* [Lange] Imbach) has recently been introduced in the Punjab State, India, where this is made feasible by an abundance of wheat straw and a suitable environment during winter (Chahal, 1974). Synthetic compost was prepared according to the method devised by E. F. K. Mantel (Chahal, 1974). During the degradation (fermentation) of straw, a large number of microorganisms developed which converted the straw into a suitable substrate for the growth of mushrooms. A large number of thermophilic and thermotolerant fungi and actinomycetes have been isolated from mushroom compost by various workers (Fergus, 1964; Fergus and Amelung, 1971; Chang and Hudson, 1967). The thermophilic mycoflora isolated from mushroom compost prepared from wheat straw grown under the Punjab agro-climatic conditions is quite different from that already reported.

MATERIALS AND METHODS

Isolation of fungi

Samples of about 100 g of decomposing straw were taken from a depth of 5–10 cm and immediately placed in sterile, wide-mouthed bottles and sealed. The temperature of the compost heap at this depth varied from 40–50°C, while it was 50–65°C below 10 cm. Fungi and actinomycetes were predominantly recorded in the depth range 5–10 cm, while a few actinomycetes and most of the bacteria observed occurred at depths greater than 10 cm. Almost no microorganisms were recorded from the core of the compost heap where the temperature attained 70°C (Fig. 1).

* Present address: Visiting Scientist, Department of Nutrition and Food Science, Massachusetts Institute of Technology, Cambridge, Massachusetts 02139, U.S.A.

Fungi were isolated on a modified Czapeks (Dox) agar (Ainsworth, 1971) in which sucrose was replaced with 15 g of glucose or 10 g of cellulose powder, and dibasic potassium phosphate was replaced with monobasic potassium phosphate in order to provide a pH of 4·8–5·0 in the medium. Four small pieces of decomposing straw were taken from the sampling bottles and placed aseptically at four equidistant points opposite each other on agar in 10-cm diameter petri-dishes. The medium contained glucose in one series of isolations and cellulose powder in another. Plates were incubated at 40°C. As soon as radiating hyphae were observed, their tips were cut and transferred onto new plates of the appropriate medium.

A dilution plate method was also used to isolate the fungi. In this, 10 g of decomposing straw was mixed with 90 ml of sterile water and shaken in 250-ml Erlenmeyer flasks on a rotary shaker for 1 hr. Various dilutions (10^{-2}, 10^{-3}, and 10^{-4}) were used for both media described above. Individual colonies were transferred onto new plates for further purification and identification.

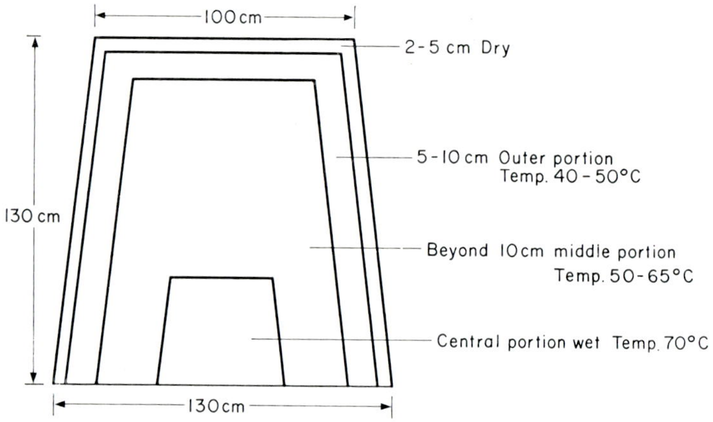

Fig. 1. Transverse section of compost pile.

The cultures of all the isolated fungi were maintained on modified Czapeks (Dox) agar slants without sugar but with a sterile filter paper of each tube prior to inoculation. All the cultures were incubated at 40°C. Within a week, all the cultures showed good sporulation; thereafter they were transferred to a refrigerator to avoid desiccation in the incubator.

Inoculum

The inoculum was prepared as follows: 5 ml of sterile distilled water was used to obtain a suspension of spores or ascospores from the slants with matured perithecia (in the case of *Achaetomium* and *Chaetomium* species). Two millilitre aliquots of spores-ascospores suspension were used to inoculate each flask containing 50 ml of Czapek medium with 15 g glucose. Two- or three-day-old cultures from two flasks were washed with sterile distilled water and then mixed with 100 ml of sterile distilled water in a Waring blender for 2–3 min until a thick suspension of small mycelial fragments was obtained. Two millilitre aliquots of this mycelial suspension were used to inoculate each flask containing Cahal and Gray medium with 500 mg of wheat straw or other cellulosic substrate. All the inoculated flasks were incubated for 5 days at 37–40°C on a rotary shaker with 130 rpm and amplitude of 3·2 cm.

Carboxymethylcellulase (CMCase) activity

Since CMCase is extracellular, the filtrate was used to measure its activity by the procedure described by Miller *et al.* (1960). This method is based on the increase in the reducing power of a solution of cold carboxymethylcellulose (CMC).

CMCase unit

A unit of carboxymethylcellulase (CMCase) activity has been taken as the activity produced by that amount of enzyme present in 1 ml of the filtrate which would produce an amount equivalent to 1 mg of glucose as the reducing sugar under the following conditions: 1·0 ml of 1·0% CMC in 0·1 M citrate buffer at pH 5·0 incubated at 50°C for 20 min.

Estimation of degradation of cellulosic substrates

Degradation of wheat straw and other cellulosic materials was recorded as loss in weight although the mass obtained after filtration contained the unutilized cellulosic materials along with the fungal mass synthesized. It was further correlated with the fungal growth which was estimated in terms of crude protein synthesized.

Estimation of crude protein

Estimation of mycelial crude protein is based on the determination of the total nitrogen content by the micro-Kjeldahl method of McKenzie and Wallace (1954) and multiplied by a factor of 6·25.

Selection of medium

In the preliminary study for the growth of CMCase production by the isolated fungi, the following media were tried: Yeast-Starch medium (Y_pS_s) and Yeast-Glucose medium (YG) used by Cooney and Emerson (1964) for thermophylic fungi; a medium used by Rautela and Cowling (1966) for isolation of cellulolytic fungi; a medium used by Chahal and Gray (1968) for the growth of cellulolytic fungi; and a modified Czapek (Dox) medium—a general-purpose medium (Ainsworth, 1971). The medium used by Chahal and Gray (1968) proved to be the best of all the media tried. This was used throughout the present study with some change in nitrogen source in certain experiments.

Substrates

Wheat straw was washed thoroughly with ordinary water to remove all the dirt. It was ground to a fine powder of 60 mesh with a Wiley mill after drying. Delignification was done by the method described by Myhre and Smith (1960). For all the experiments, 1% of each substrate was used.

RESULTS

A large number of fungi were isolated from wheat straw compost by two methods, *e.g.* bits-transfer and pour-plate method. All the isolated fungi were grouped together according to their common characteristics. With the results the following fungi were recorded as the distinct species:

1. *Achaetomium luteum* Raj & Tewari
2. *Chaetomium* sp. nov.
3. *Aspergillus fumigatus* Fres. (A)
4. *Aspergillus fumigatus* Fres. (B)
5. *Aspergillus fumigatus* Fres. (C)
6. *Aspergillus fumigatus* Fres. (D)
7. *Mucor pusillus* Lindt
8. *Myxotrichum* sp.

During isolation, it was found that most of the time the plates were over-crowded with the species of *Aspergillus* and *Mucor*. Species of *Chaetomium* and *Achaetomium* were found only when isolation was carried out by placing the small, degrading bits of straw onto the agar plates with cellulose. A member of Gymnoascaceae, which was later identified as possibly a species of *Myxotrichum*, was isolated only once.

All the *Aspergillus* isolates were grouped into four distinct types, which could be easily distinguished from each other by their colony characteristics, and these were kept as four different species of *Aspergillus*. A detailed microscopic examination indicated that all four different species closely resembled *Aspergillus fumigatus* Fres. Because they showed distinct colony character, they were named as different strains of *A. fumigatus* (A), (B), (C), and (D).

A. luteum, *Chaetomium* sp., and *Myxotrichum* sp. were recorded for the first time from wheat compost. *Chaetomium* species presented characteristics quite different from those of known species of this genus. This is being described as a new species of genus *Chaetomium* by Chahal and Hawksworth elsewhere.

Weight loss

The weight loss in wheat straw varied from 51–56% in the case of *A. fumigatus* (A), (C), and (D), *Chaetomium* sp., and *A. luteum* (Table I). But when

Table I

DEGRADATION OF WHEAT STRAW, CMCASE AND PROTEIN PRODUCTION BY THE ISOLATED FUNGI

Organism	*Final pH*	*Wt of mycelium and unutilized straw (mg)*	*Weight loss (%)*	*CMCase (units/ml)*	*Protein (%)*	*Total protein per flask (mg)*
Achaetomium luteum	6·4	218	56·4	1·7	22·0	49·1
Chaetomium sp.	6·0	242	51·6	1·2	26·0	62·9
Aspergillus fumigatus (A)	5·9	238	52·4	2·8	21·0	49·9
Aspergillus fumigatus (B)	6·0	252	49·6	0·7	16·1	40·5
Aspergillus fumigatus (C)	6·1	245	51·0	1·2	16·8	41·2
Aspergillus fumigatus (D)	6·3	245	51·0	0·9	14·0	34·3
Mucor pusillus	6·2	282	42·4	1·0	10·7	30·8
Myxotrichum sp.	5·7	485	3·0	—	0·7	3·4

Nitrogen source: KNO_3.

the degradation of straw was measured as mycelial mass in terms of protein produced, then *Chaetomium* sp. ranked the best, followed by *A. fumigatus* (A) and *A. luteum*. *M. pusillus* caused less weight loss and also produced less protein. *Myxotrichum* sp. was rated the poorest of the species in weight loss and protein produced.

Almost the same amount of weight loss and protein production was recorded by these fungi on cellulose powder (Table II). Although these fungi

Table II

DEGRADATION OF CELLULOSE POWDER, CMCASE AND PROTEIN PRODUCTION BY THE ISOLATED FUNGI

Organism	*Final pH*	*Weight of mycelium and unutilized straw (mg)*	*Weight loss (%)*	*CMCase (units/ml)*	*Protein (%)*	*Total protein per flask*
Achaetomium luteum	5·9	327	34·6	1·2	16·9	55·2
Chaetomium sp.	5·4	370	26·0	1·0	17·5	64·7
Aspergillus fumigatus (A)	5·6	352	29·6	0·9	12·6	44·3
Aspergillus fumigatus (B)	5·6	341	31·8	0·8	16·1	54·9
Aspergillus fumigatus (C)	5·6	372	23·6	0·6	14·0	52·1
Aspergillus fumigatus (D)	5·8	422	15·5	0·4	8·7	36·7
Mucor pusillus	5·9	405	19·0	1·1	7·1	28·7
Myxotrichum sp.	5·6	480	4·0		no growth	

Nitrogen source: KNO_3.

produced less weight loss and also a low percentage of protein, the increase in total amount of protein synthesized indicated that they grew better on cellulose than on wheat straw. The high percentage of protein production on straw by *Chaetomium* sp. might be due to the fact that less biomass had been produced and most of the straw had been degraded to CO_2.

The degradation of straw by *Chaetomium* sp. increased to a great extent when it was delignified and when weight loss was compared in terms of protein production (Table III). Microcrystalline cellulose proved to be less susceptible than powdered cellulose when compared in terms of protein production.

Table III

DEGRADATION OF DIFFERENT CELLULOSIC SUBSTRATES BY *CHAETOMIUM* SP.

Substrate	*Final pH*	*Weight of mycelium and unutilized straw (mg)*	*Weight loss (%)*	*CMCase (units/ml)*	*Protein (%)*	*Total protein per flask (mg)*
Wheat straw	6·2	246	49·2	1·2	25·4	62·5
Delignified straw	5·9	210	58·0	1·9	42·3	88·8
Cellulose (powder)	5·8	370	26·0	1·4	17·5	64·7
Cellulose (microcrystalline)	5·6	327	34·6	0·5	16·9	55·3

Nitrogen source: KNO_3.

CMCase activity

A. fumigatus (A) produced the highest CMCase activity on straw and about one-third less on cellulose powder (Tables I and II). The other fungi with high CMCase activity were *A. luteum*, *Chaetomium*, *A. fumigatus* (C), and *M. pusillus*. No activity was recorded in *Myxotrichum* sp. The CMCase activity of *A. fumigatus* (A) was almost double that of *Chaetomium* sp., but it produced comparatively less protein than *Chaetomium* on straw. Nevertheless, both the organisms caused the same amount of weight loss in straw.

Similarly, there was no correlation between the CMCase activity and the weight loss and protein production by all these fungi on cellulose powder. *A. luteum* and *Chaetomium* sp. produced almost the same CMCase activity but *Chaetomium* produced more protein on cellulose powder.

A. luteum produced high CMCase activity and also high weight loss on both the substrates. In general, CMCase activity was high in almost all the fungi when straw was used as a substrate.

DISCUSSION

Eight different fungi were isolated from synthetic wheat compost when it was at its peak degradation (fermentation).

A. fumigatus, a thermotolerant fungus, is omnipresent in degrading organic matter. This has also been reported from wheat straw compost by Fergus (1964) and Chang and Hudson (1967). *M. pusillus* is another common thermophilic fungus which has been recorded in wheat compost by these workers.

A. luteum, *Chaetomium* sp., and *Myxotrichum* sp. have been reported for the first time from wheat straw compost as indicated by a survey of the literature. Different mycoflora of compost might result from different composting procedure, additional organic matter used, and different geographical and ecological locations where wheat was grown.

The first *Chaetomium* species recorded from straw at high temperature was by La Touche (1950), and he named it *C. thermophile*. Later on, the same *Chaetomium* species was reported by Fergus (1964) and Chang and Hudson (1967) from wheat straw compost. Recently, a new thermotolerant species of *Chaetomium*, *C. rectopilium*, has been reported by Fergus and Amelung (1971) from spent wheat compost after mushroom harvest. But the *Chaetomium* species isolated from wheat straw compost this time was quite different from all the known species of the genus *Chaetomium*. This is being reported by Chahal and Hawksworth elsewhere.

CMCase activity and degradation of cellulose by thermophilic fungi have not been studied thoroughly. Fergus (1969) reported that *C. thermophile* var. *coprophile* and *C. thermophile* var. *dissitum* caused a high loss in weight of filter paper when incubated for 25 days at 45°C. He further reported that these fungi produced a good amount of CMCase when grown on CMC for 15–20 days at 45°C. There is another report by Chang (1967) that *C. thermophile* and *A. fumigatus* caused a high loss in weight of filter paper when incubated at 45°C for 3 weeks. The present study reports a great loss in weight of wheat straw within 5 days and very high CMCase activity by *Chaetomium* sp., *A. luteum*, and *A. fumigatus* (A). These thermophilic fungi have shown very high cellulolytic activity as compared to others reported in the literature.

CMCase activity was higher when the substrate was wheat straw as compared to the other cellulosic materials. *Chaetomium* sp. and *A. luteum* seem to be the best organisms to be exploited for the production of single-cell protein (SCP) from cellulosic materials.

Acknowledgements

The authors are very grateful to Drs Hawksworth, Onion, Stockdale, and Lunn of the Commonwealth Mycological Institute, Kew, Surrey, U.K., for their help in confirming the identification of the isolated fungi.

REFERENCES

Ainsworth, G. C. (1971). *Ainsworth & Bisby's Dictionary of Fungi*. 6th ed., C.M.I. Kew, Surrey, 358 pp.

Chahal, D. S. (1974). *Mushroom Growing for Beginners*. Punjab Agricultural Univ., Ludhiana, India, 20 pp.

Chahal, D. S., and Gray, W. D. (1968). Growth of selected cellulolytic fungi on wood pulp. In *Biodeterioration of Materials: Microbiological and Allied Aspects*, pp. 584–593 (A. H. Walters and J. S. Elphick, eds.), Elsevier Publ. Co., Ltd, U.K.

Chang, Y. (1967). The fungi of wheat straw compost. II. Biochemical and physical studies. *Trans. Brit. Mycol. Soc.*, **50**, 667–677.

Chang, Y., and Hudson, H. J. (1967). The fungi of wheat straw. I. Ecological studies. *Trans. Brit. Mycol. Soc.*, **50**, 649–666.

Cooney, D. G., and Emerson, R. (1964). *Thermophilic Fungi*. W. H. Freeman & Co., San Francisco, 188 pp.

Fergus, C. L. (1964). Thermophilic and thermotolerant molds and actinomycetes of mushroom compost during peak heating. *Mycologia*, **56**, 267–284.

Fergus, C. L. (1969). The cellulolytic activity of thermophilic fungi and actinomycetes. *Mycologia*, **61**, 120–129.

Fergus, C. L., and Amelung, R. M. (1971). A new thermotolerant species of *Chaetomium* from mushroom compost. *Mycologia*, **63**, 1212–1217.

La Touche, C. J. (1950). On thermophilic species of *Chaetomium*. *Trans. Brit. Mycol. Soc.*, **33**, 94–104.

McKenzie, H. A., and Wallace, H. S. (1954). The Kjeldahl determination of nitrogen. Critical study of digestion conditions, temperature, catalysts, and oxidizing agents. *Aust. J. Chem.*, **17**, 55–79.

Miller, G. L., Blum, R., Glennon, W. E., and Burton, A. L. (1960). Measurement of carboxymethylcellulase activity. *Anal. Biochem.*, **2**, 127–132.

Myhre, D. V., and Smith, F. (1960). Constitution of the hemicellulose of Alfalfa. *Medicago sativa*. Hydrolysis of hemicellulose and identification of neutral and acidic compounds. *J. Agr. Food Chem.*, **8**, 359–364.

Rautela, G. S., and Cowling, E. B. (1966). Simple cultural tests for relative cellulolytic activity of fungi. *Appl. Microbiol.*, **14**, 892–898.

products from the animals. In nonintensive situations, the animal acts as its own disposal system, distributing its waste about the land on which it grazes or looks for food. The soil then acts as a treatment and recycling system via the food chains. The land does not become overloaded and no pollution problems arise. When the animals are intensified (U.K. Ministry of Agriculture, Fisheries and Food—MAFF—define an intensive unit as one having more than 1 cow, 4 pigs, or 100 chickens per acre of arable land), the amount of waste produced per acre of land increases until the soil flora and fauna cannot cope and all activity ceases. Further complications arise when the animals are placed in an intensive situation away from the land, inside concrete-floored pens. The wastes must be removed daily and disposed of on the land. This procedure is both costly and not possible all of the year, particularly in the United Kingdom where land access is restricted in the winter months. Also, the large amounts of waste require large quantities of land for safe disposal. In the United Kingdom land cannot be sacrificed continually for waste disposal. Due to land shortages, it must be put to grazing or crops for part of the year.

An example will serve to demonstrate the magnitude of the problem on intensive units in the United Kingdom. An intensive pig farm may contain 5000 animals on between 5 and 10 acres of land, and each pig will excrete one gallon (U.K.) of urine plus faeces slurry (MAFF Short term leaflet No. 67) per day on average. This means that 5000 gallons of waste are produced per day. For continuous safe disposal, this requires between 150 and 300 acres of land (Jones, 1969). The cost of transportation of such large quantities of predominantly water (90–95%) may prohibit removal further than a 2 or 3 mile radius from the unit. The waste must thus either be stored, which leads to seepage problems and storage area requirements, or treated to reduce its polluting ability (notably the biochemical oxygen demand BOD), so that the resulting liquid will be acceptable for discharge to a water course or sewerage system. This latter solution may be costly, with no return in terms of reclamation of nutrients for the farmer, especially if he has no land on which to spread any sludge from the process.

Many of the potential nutrients in the waste arise from the feed regimes which the intensive animal receives. Pigs, for example, may be fed 16–20% protein together with fats, carbohydrates, and roughage (Agricultural Research Council, 1967). The efficiency of utilization of these materials varies from nonutilization of the roughage to between 33 and 50% utilization of protein. Thus the waste contains much nitrogen and carbon together with phosphorus- and sulphur-containing compounds. These nutrients are available for microbial utilization under a range of concentrations. Outside this range, in overloading conditions, microbial activity and utilization may cease.

Probably the largest plant waste in the United Kingdom is that from cereal production. The straw stem is either left in the field to be ploughed in or burned in an attempt to control pathogens and put minerals back into the soil. About 3·5 million tons (38·2%) out of a total crop of 9 million tons of straw is a waste product and most of this is burned in the field (National Farmers' Union, 1973). Ploughing in of straw presents mechanical and biological problems (McLean, 1973) in that the straw may need to be chopped beforehand; once in the soil it can affect drainage, and the increased carbon in the soil which the straw contributes may increase the carbon:nitrogen ratio so that microbial breakdown of the straw is reduced. The burning of straw represents a loss of energy, this energy being the sum of captured solar energy and the input from the cultivation of the crop. Approximately 50% of the carbon and energy in a cereal crop is found in the waste straw at harvesting (Phoenix *et al.*, 1974; O'Callaghan, 1975). As discussed above, agriculture is relatively inefficient in its energy balance and any wastage of captured energy should be avoided.

Straw contains appreciable amounts of cellulose and hemicellulose (about 70% of dry matter) (O'Callaghan, 1975) and these are potential carbon and energy sources for many organisms from the microfungi up to the ruminants The availability of the cellulose for utilization is dependent upon the association that cellulose has in nature with lignin. Very little is known about the ligno-cellulose association or bond, but it is known that cellulose is more

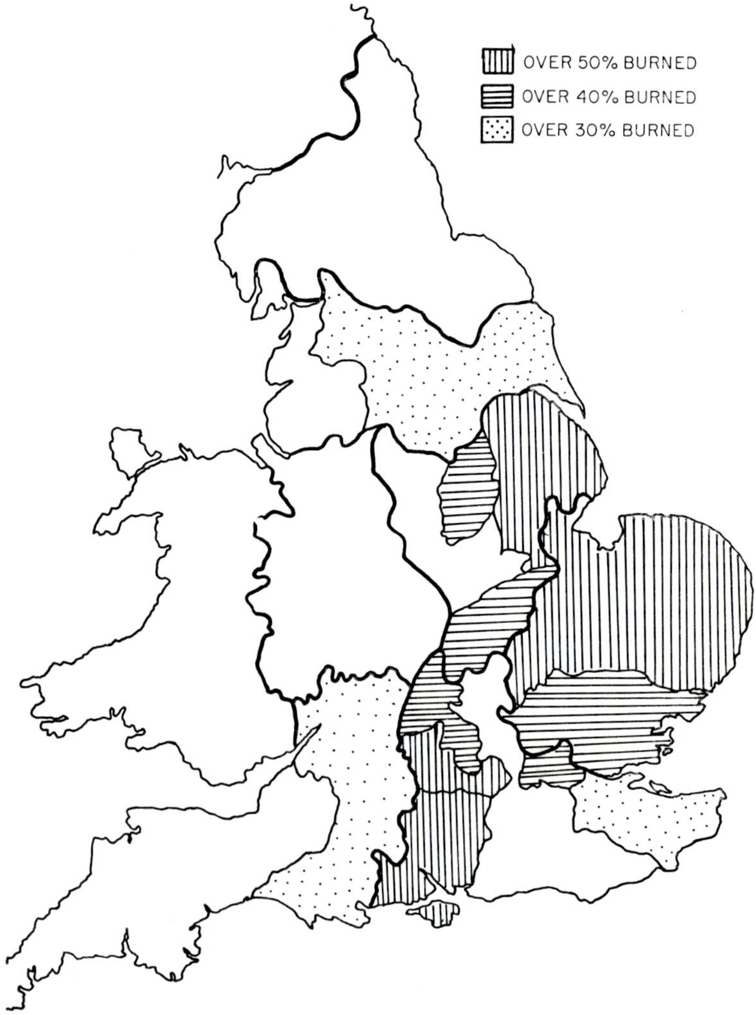

Fig. 1. Map to show straw production and burning areas.

available to cellulolytic organisms when this bond is broken, either chemically or microbiologically. Mature cereal straws contain about 14% lignin (Sarkanen and Ludwig, 1971) and this reduces the availability or digestibility of the straw considerably, making the plant a poor source of cellulose for organisms, particularly ruminants which contain a cellulolytic flora not able to degrade lignin (Van Soest and Mertens, 1974). The improvement of straw as a feed ingredient for ruminants has been studied by many workers (Carmona and Greenhalgh, 1972; Phoenix *et al.*, 1974; Chandra and Jackson, 1971; Pigden and Bender, 1972). Most of the methods to date have concentrated on chemical and physical methods, both of which achieve increases in digestibility, but both of which are energy-consuming and present practical handling problems. As many organisms, particularly microorganisms, are able to utilize cellulose and some to utilize lignin, it should be possible to devise biodegradative processes for the conversion of straw into either a protein or carbohydrate source.

When studying how microorganisms might be involved in the biodegradation of agricultural waste products, mention should be given to the distribution of the wastes in the United Kingdom. As stated in the introduction, the wastes are concentrated geographically in United Kingdom. This is shown in Fig. 1, where we see that 30% of the straw produced is grown in our eastern counties. In contrast, as a result of the availability of grass favoured by a wetter climate, animal production is concentrated in the western counties. In broad terms this means that there is an imbalance of nutrient distribution, there being higher nitrogen-containing wastes in the east (Willets, 1974). A redistribution of these wastes would help alleviate nutrient shortages and reduce the amount of artificial high energy-requiring fertilizers, much of which needs to be imported. In the regions where there is overlap between animal and crop production, biodegradation could play an important role in treating both wastes by using the nutrients from the animal and crop wastes in a combined process to produce recycled nutrients for the animals. Using the nitrogen and phosphorus from the animal waste and the carbon from the cellulose in straw, there is the potential for microbial conversion. The economics of such a process will depend upon straw transport costs and the equipment used in the process. These costs would have to be balanced against the advantages gained by the removal of pollutants and the value of the products.

THE ROLE OF MICROORGANISMS

Indication has already been given above as to the use of microorganisms in the biodegradation of agricultural wastes. The waste nutrients present ample food sources for a wide variety of organisms as shown by the different types of reactions which take place in wastes when they are subjected to varying environmental conditions. As far as pollution and public health are concerned, there are certain reactions involving undesirable microorganisms such as those pathogenic species which may survive when given their optimum habitat. The encouragement of desirable reactions should result in the following conditions:

(1) A reduction in the polluting ability of the waste.

(2) A large reduction in the nuisance aspect such as odour or fly attraction.
(3) An increase in the ease of handling and disposal.
(4) An increase in the value of the product either as a fertilizer or a feedstuff.

All these points will vary in importance according to the type of waste.

There are several types of desirable reactions already known in agricultural waste treatment. These include aerobic oxidation ditch treatment (Gowan, 1972; Robertson, 1972), composting (Poincelot, 1972), anaerobic digestion of animal wastes (Hobson and Shaw, 1973), and the chemical treatment of straw to increase its digestibility (Carmona and Greenhalgh, 1972). All of these methods succeed to varying degrees in achieving the above four points, but they all rely on an external energy source, either for aeration, mixing, or pretreatment. Microorganisms, and in particular the fungi, offer many advantages using very simple control conditions imposed by the fungi themselves. Fungi are able to grow under semi-solid fermentation conditions. This is by virtue of their ability to ramify through a solid substrate, colonizing it very quickly (Hudson, 1972). Bacteria rely upon water in order to distribute themselves through a substrate. If the process needs too much water, it must be contained within a vessel and there may be a problem of wastewater disposal. The ability of fungi to translocate nutrients means that the nutrients in the substrate need not necessarily be uniformly distributed and the ability of fungi to penetrate intact tissue removes the need for pretreatment (chopping or grinding) of a plant waste such as straw (Eggins and Allsopp, 1975).

Microorganisms are very important breakdown agents in the soil and they may be selectively encouraged to grow outside the soil on the waste substrates (Hartley *et al.*, 1974). By studying the parameters for optimum growth rates and conversion of the wastes, induced processes may be developed to work efficiently in aiding the natural microbial conversion which may be in danger of becoming overloaded.

Figure 2 shows the involvement of microorganisms in the biodegradation and recycling of waste nutrients. The flow diagram clearly indicates that there must be an input of energy, but that this comes from the sun via the green plants. Induced processes would neither completely replace the natural cycles nor the use of green plants. However, in situations where waste production is concentrated and the soil is unable to cope, induced microbial activity or biodegradation could be a particularly low energy-consuming process for achieving the four points mentioned earlier in this section.

CONCLUSIONS

The use of biodegradation offers a cheap method for recycling nutrients efficiently and, when optimized, at a faster rate than under natural conditions. It would appear to be low in its energy requirements, but this, together with economic considerations, still needs study. It is able to assist the natural recycling systems in situations where these systems could not cope with the large amounts of waste introduced to them, and where costs prohibit removal of the waste to large distances from the site of production. The recycling of waste nutrients as a feedstuff could help the United Kingdom to decrease the present 50% imported animal feeds and help the balance of payments. It is

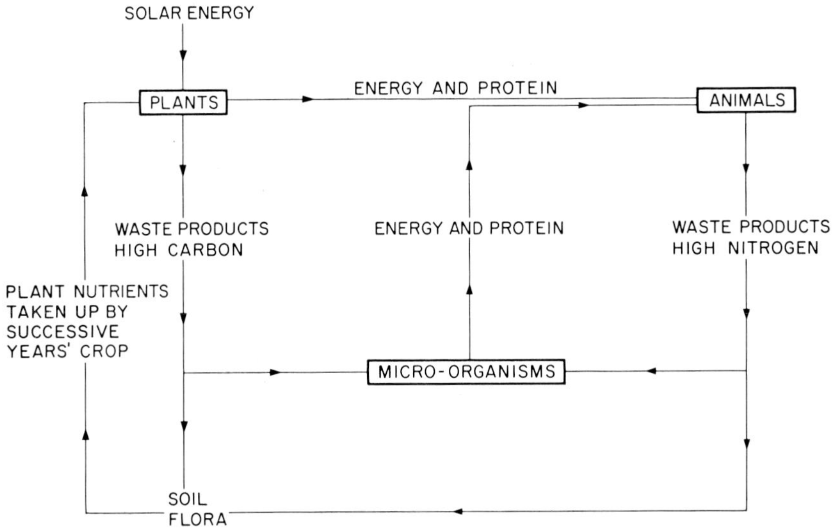

Fig. 2. Flow diagram to show microorganism involvement (natural and induced) in the recycling of animal and plant wastes.

necessary that a deeper look be taken at the management of agricultural wastes and the value of their nutrients. Although not the complete solution, biodegradation could be very important in the extraction of these nutrients at a high rate and then channelling them either directly to animals or, via the soil, into plants.

REFERENCES

Agricultural Research Council (1967). The nutritional requirements of farm livestock, No. 3. Pigs. ARC, London.

Anthony, W. B. (1969). Cattle manure re-use through wastage feeding. From Conference on Animal Waste Management. Cornell University, pp. 105–113.

Carmona, J. F., and Greenhalgh, J. F. D. (1972). The digestibility and acceptability to sheep of chopped or milled barley straw soaked or sprayed with alkali. *J. Agr. Sci. Ca mb.*, **78**, 477–485.

Chandra, S., and Jackson, M. G. (1971). A study of various chemical treatments to remove lignin from course roughages and increase their digestibility. *J. Agr. Sci. Camb.*, **77**, 11–17.

Eggins, H. O. W., and Allsopp, D. (1975). Biodeterioration and biodegradation by fungi. In *The Filamentous Fungi*, Vol. 1, pp. 301–319 (Smith and Berry, eds.), Edward Arnold, London.

Gowan, D. (1972). *Slurry and Farm Waste Disposal*, Farming Press Ltd., Ipswich.

Hartley, R. D., Jones, E. C., King, N. J., and Smith, G. A. (1974). Modified wood waste and straw as potential components of animal feeds. *J. Sci. Food Agr.*, **25**, 433–437.

Heslop-Harrison, J. (1975). Crops, commodities, and energy capture. *Biologist*, **22** (2), 60–67.

Hobson, P. N., and Shaw, B. G. (1973). The anaerobic digestion of waste from an intensive pig unit. *Water Res.*, **7** (3), 437–449.

Hudson, H. J. (1972). *Fungal Saprophytism.* Studies in Biology No. 32, Edward Arnold, London.

Jones, P. H. (1969). Theory and future outlook of animal waste treatment in Canada and the United States. Animal Waste Management, Conference on agricultural waste management, Cornell University, pp. 23–36.

Kent-Jones, D. W., and Amos, A. J. (1967). *Modern Cereal Chemistry.* 6th ed., Ford Trade Press Ltd, London.

MAFF (1961). Making the most of farmyard manure. Advisory Leaflet 435. MAFF Farm Waste Disposal. Short term leaflet No. 67.

McLean, K. A. (1973). Straw Disposal. ADAS Quarterly Review No. 11, pp. 107–112.

National Farmers' Union (1973). Report on the use and disposal of straw. NFU Working Party on Straw Disposal.

O'Callaghan, J. R. (1975). The waste products of agriculture. *J. Royal Soc. Arts*, **123** (5223), 139–149.

Phoenix, S. L., Bilanski, W. K., and Mowat, D. N. (1974). In vitro digestibility of barley straw treated with sodium hydroxide at elevated temperatures. *Trans. Am. Soc. Agr. Eng*, **17** (4), 780–782.

Pigden, W. J., and Bender, F. (1972). Utilisation of ligno cellulose by ruminants. *World Animal Rev.*, No. 4, 7–10.

Poincelot, R. P. (1972). The biochemistry and methodology of composting. Connecticut Agricultural Experiment Station Publication. Bulletin 727.

Robertson, A. M. (1972). Treatment of livestock wastes. *Process Biochem.*, **7** (6), 21–24.

Sarkanen, K. V., and Ludwig, C. H. (1971). Lignin-classification and distribution, In *Lignins*, pp. 43–94 (Sakanen and Ludwig, eds.), Chapter 3, Wiley-Interscience, New York.

Seal, K. J., and Eggins, H. O. W. (1972). The role of microorganisms in the biodegradation of farm animal wastes with particular reference to intensively produced wastes. A review. *Int. Biodeterior. Bull.*, **8** (3), 95–100.

Townshend, A. R., Reichert, K. A., and Nodwell, J. H. (1969). Status report on water pollution control facilities for farm animal wastes in the province of Ontario. *Animal Waste Management*, pp. 131–149, Cornell University.

Van Soest, P. J., and Mertens, D. R. (1974). Composition and nutritive characteristics of low quality cellulosic wastes. *Fed. Proc.*, **33** (8), 1942–1944.

Willets, S. L. (1974). A model for alleviating farm waste pollution in England. Part 1. *Environ. Pollut. Mgmt*, **4** (6), 265–273.

THE USES OF THERMOPHILIC FUNGI IN THE BIODEGRADATION OF PIG WASTES

K. J. SEAL and H. O. W. EGGINS

Biodeterioration Information Centre,
University of Aston in Birmingham, Birmingham, U.K.

Summary

Some concepts are presented on the usefulness of thermophilic fungi in upgrading agricultural wastes. Laboratory studies were carried out on pig waste from an intensive unit as follows: a range of thermophilic fungi was isolated from straw-containing waste at 40°C and 48°C. Studies were carried out on the temperature and pH optima for the growth of these fungi, and the effects of addition of inocula from other natural sources. It was found that a temperature of about 50°C and a slightly alkaline pH gave reduced pathogen counts, increased the rate of biodegradation, and suppressed fungal pathogens such as *Aspergillus fumigatus* and *Mucor pusillus*. A soil inoculum and the addition of 25% of the recycled end-product further enhanced biodegradation. The products were reduced in bulk and odour and easier to handle. It is suggested that the product might be recycled as an animal feedstuff.

INTRODUCTION

This paper is concerned with the application of thermophilic biodegradation to the treatment of pig wastes. The use of biodegradation in the treatment of agricultural wastes has been presented elsewhere in the symposium and need not be discussed here. In the United Kingdom approximately 25% of pigs are intensively housed and, due to the high polluting ability of these wastes compared to human wastes (Table I), problems of disposal can occur (Seal and Eggins, this symposium). An investigation was undertaken to attempt to assess the use of thermophilic fungi as agents of biodegradation in treating intensively produced pig waste. The aim was to determine the parameters for

Table I
POLLUTING ABILITY OF ANIMAL WASTES IN U.K. AS COMPARED WITH HUMAN WASTE

Livestock	*Effluent (million gals)*	*BOD's population equivalents (millions)*
Cattle	9 000	30·7
Pigs	2 300	11·2
Poultry	1 330	12·4
Total	12 630	54·3

From O'Callaghan (1975).

growth of thermophiles in the waste and to encourage an increase in the rate of biodegradation by optimizing these parameters.

Various workers have studied the occurrence and activity of thermophiles in agricultural wastes such as horse manure (Waksman and Cordon, 1939; Crisan, 1959), self-heated retted quayle (Cooney and Emerson, 1964), composts (Mills, 1973; Barnes, 1974; Hayes, 1969), straw (Chang and Hudson, 1967), and palm kernel stacks (Eggins and Coursey, 1964). An excellent review of thermophilic fungi has been given by Cooney and Emerson (1964) and their physiology and biochemistry need not be discussed in this paper.

In this study emphasis was laid on the encouragement of thermophilic fungi in the process for the following reasons:

(1) There are many cellulolytic species able to use the large amounts of cellulose in agricultural wastes as an energy source.

(2) They occur naturally in animal and plant wastes.

(3) They are obligate aerobes and do not produce the unpleasant odours usually associated with anaerobicity.

(4) Their pH optima tend to discourage bacteria which contain pathogenic groups.

(5) They are good sources of protein, being low in nucleic acid levels.

(6) Their mycelial habit gives them a colonizing advantage in that they can spread over solid substrates and penetrate intact tissue without any pre-treatment.

(7) Their optimum growth range suppresses bacterial pathogens.

(8) The number of fungi able to grow is reduced drastically and those that can have had their physiological and biochemical activities well studied.

It was decided to study the indigenous thermophilic fungal flora and the conditions necessary to optimize biodegradation of pig waste. This would then allow predictions for a larger scale process, bearing in mind that it should operate in an 'on farm' situation.

MATERIALS AND METHODS

1. Isolation programme

Two types of pig waste were studied—mixtures of faeces, urine, and straw bedding (60–70% moisture), and slurries of faeces and urine (90–95% moisture). These two types of waste occur under practical conditions in large amounts and can pose disposal problems. The samples were freshly collected in sterile containers for transport to the laboratory. Dilutions in sterile distilled water of 10^{-1}, 10^{-2}, and 10^{-3} were made and 1 ml samples were then pipetted into petri dishes before pouring the agars and mixing thoroughly. Two agars were employed: Eggins and Pugh cellulose agar (Eggins and Pugh, 1962) and an Eggins and Pugh mineral salts agar, substituting for cellulose a mixture of glucose (5 g/litre) and starch (5 g/litre). The agars were buffered to pH 4·0, 5·6, 6·6, 7·6, and 8·6 before autoclaving using McIlvaine's citrate-phosphate buffer. The pH's changed slightly on autoclaving and Table II shows the final pH's. The plates were incubated at 40°C and 48°C under high humidity conditions. Ten replicate plates were prepared for each temperature,

Table II
pH OF AGAR AFTER AUTOCLAVING

Eggins and Pugh		*Glucose/starch*	
Solid waste	*Slurry*	*Solid waste*	*Slurry*
4·0	4·2	4·1	4·5
5·6	5·9	5·6	5·5
6·6	6·4	6·4	6·2
7·4	7·3	7·4	7·0
8·2	8·1	8·2	7·9

pH, and media employed for each sample of waste. Identifications and frequencies of occurrence were determined after 12 days' incubation. Observation and subsequent identification of fungal species were made using the 'Sellotape' technique of Lloyd (1965). This enabled whole fruiting structures to be observed which might otherwise have broken up using normal mounting procedures. Percentage frequency of isolation (PFI) was determined by noting on how many of the replicate plates a particular fungus occurred and expressing this figure as a percentage.

2. Biodegradation of pig waste

In order to study the parameters to encourage biodegradation to take place, laboratory-scale fermenters were constructed (*see* Fig. 1) of a size that

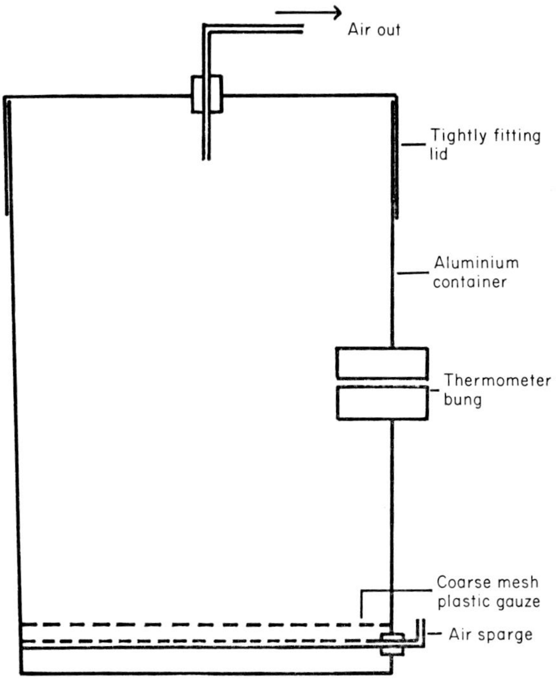

Fig. 1. Laboratory-scale fermenter.

could be placed in an incubator. The fermenters were charged with mixtures of straw, faeces, and urine as found under practical conditions on the farm and a slow sterile air stream of about 0·25 litre/min/kg waste was passed through the fermenter. The following investigations were carried out: (1) the effects of temperature on biodegradation and pathogen removal; (2) the effects of pH; and (3) the effects of inocula additions.

For the first investigation, the following temperatures were employed: 25, 40, and 50°C. Biodegradation was determined as the strength loss of cellulose chromatography paper strips placed in a modified screened substrate tube (*see* Fig. 2) and placed in the fermenters. Coliforms and nonlactose-fermenting bacteria including the *Salmonellae* were determined at 40, 50, and 56°C by means of plate counts on selective media. McConkey's agar number 3

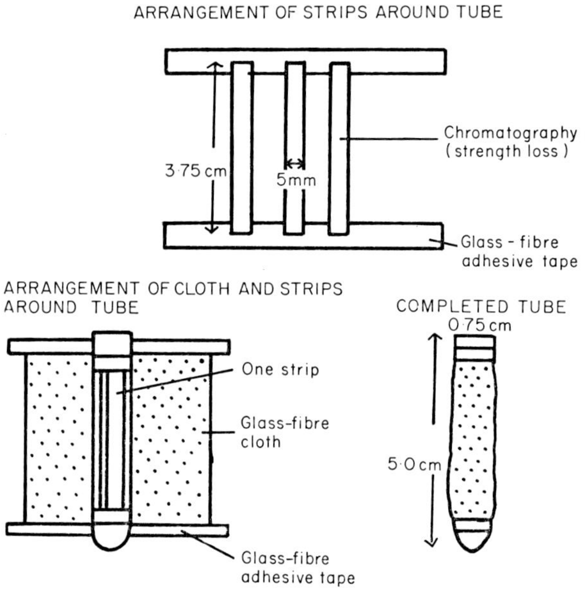

Fig. 2. The modified screened substrate tube used in the fermenters to determine fungal successions.

(Oxoid Ltd) was used for coliforms, and nonlactose-fermenting fermenters and Desoxycholate agar (Hynes) (Oxoid Ltd) were used for the *Salmonellae*. For the second investigation, initial pH's of the waste mixtures were adjusted to 4·0, 5·0, and 6·0 and then incubated at 50°C. In the third investigation, 10% (wet weight) additions of composted town refuse, sewage sludge/town refuse ad-mixture, and soil were made to the waste and the mixtures incubated at 50°C. The value of end-product recycling was also tested by adding 10, 25, and 40% of a product previously treated for 17 days in a laboratory-scale fermenter at 50°C.

Biodegradation was measured by determining the change in carbon, nitrogen, cellulose, and pH in the waste. Note was also made of the handling and aesthetic quality of the product.

RESULTS

1. Isolation programme

These are presented in Tables III and IV. A wide range of thermophiles was found in the straw-based wastes but very few in the liquid waste. Differences in PFI's of the thermophiles can be seen between the two temperatures and the two media showing some form of temperature and substrate selection. Also it can be seen that many of the thermophiles were able to grow well above pH 7 and that some, such as *Mucor pusillus* and *Aspergillus fumigatus*, favoured acid pH's. This selection by pH could be important in controlling dominant fungi present.

Table III

THE FREQUENCY OF OCCURRENCE OF THERMOPHILIC FUNGI ISOLATED FROM SOLID WASTE AT A VARIETY OF pH'S ON

(a) *Eggins and Pugh cellulose agar*

Fungal species	*pH*	*% frequency of isolation* 40°C	48°C
Aspergillus fumigatus	4·0	0	20
	5·6	100	20
	6·6	60	—
	7·4	40	—
	8·2	10	—
Chaetomium thermophile	4·0	—	—
	5·6	60	100
	6·6	—	20
	7·4	—	40
	8·2	—	80
Humicola insolens	4·0	—	—
	5·6	100	20
	6·6	—	—
	7·4	100	—
	8·2	—	—
Humicola lanuginosa	4·0	—	—
	5·6	20	60
	6·6	100	20
	7·4	—	50
	8·2	100	100
Mucor pusillus	4·0	60	—
	5·6	30	20
	6·6	30	50
	7·4	—	—
	8·2	—	—
Stilbella thermophila	4·0	—	—
	5·6	100	—
	6·6	30	—
	7·4	—	—
	8·2	30	—
Torula thermophila	4·0	—	—
	5·6	30	20
	6·6	20	90
	7·4	100	20
	8·2	100	100

Table III—*contd.*

(b) *Glucose starch agar*

Fungal species	*pH*	% *frequency of isolation* 40°*C*	48°*C*
Aspergillus fumigatus	4·1	30	—
	5·6	90	80
	6·4	30	60
	7·4	80	20
	8·2	—	—
Chaetomium thermophile	4·1	—	—
	5·6	—	90
	6·4	10	20
	7·4	10	40
	8·2	—	—
Humicola grisea	4·1	—	—
	5·6	—	—
	6·4	20	20
	7·4	—	—
	8·2	—	—
Humicola insolens	4·1	—	—
	5·6	—	—
	6·4	—	—
	7·4	90	—
	8·2	—	—
Humicola lanuginosa	4·1	—	—
	5·6	—	20
	6·4	70	60
	7·4	—	50
	8·2	—	—
Mucor pusillus	4·1	80	—
	5·6	60	20
	6·4	70	20
	7·4	30	30
	8·2	—	—
Penicillium dupontii	4·1	20	—
	5·6	50	—
	6·4	—	—
	7·4	—	—
	8·2	—	—
Torula thermophila	4·1	—	—
	5·6	—	20
	6·4	10	—
	7·4	60	20
	8·2	—	—

— = no growth.

2. Biodegradation of pig waste

The effects of temperature on biodegradation can be seen in Figs. 3 and 4 and Table V. It can be seen that for all three fermentations there was an initial rise in pH up to between 8·2 (40°C) and 8·5 (25°C) within 2 days, which then levelled out and began to fall. However, at 40°C the pH fell to 7·5 after 3 days and then began to rise over the remainder of the incubation period to pH 8·0. The 25°C and 50°C fermentations both fell slowly to give final pH values of 7·5 and 7·8, respectively. The effect of temperature on the cellulolytic activity is quite distinct; at 25°C the strips lost only 71% strength

Table IV
FREQUENCY OF OCCURRENCE OF THERMOPHILIC FUNGI ISOLATED FROM LIQUID WASTE AT A VARIETY OF pH's ON

(a) *Eggins and Pugh cellulose agar*

Fungal species	*pH*	*% frequency of isolation* 40°C	48°C
Aspergillus fumigatus	4·2	—	10
	5·9	—	—
	6·4	10	—
	7·3	—	—
	8·1	—	—
Humicola lanuginosa	4·2	—	—
	5·9	—	10
	6·4	—	—
	7·3	—	—
	8·1	20	—

(b) *Glucose/Starch agar*

Fungal species	*pH*	*% frequency of isolation* 40°C	48°C
Aspergillus fumigatus	4·5	—	—
	5·5	10	—
	6·2	10	—
	7·0	10	—
	7·9	10	—
Aspergillus niger	4·5	50	—
	5·5	40	—
	6·2	30	—
	7·0	20	—
	7·9	20	—
Humicola lanuginosa	4·5	—	—
	5·5	30	10
	6·2	—	—
	7·0	—	—
	7·9	—	—
Mucor pusillus	4·5	—	—
	5·5	10	—
	6·2	—	—
	7·0	10	—
	7·9	—	—

Table V
CARBON AND NITROGEN CHANGES AT VARYING TEMPERATURES

Time	*Nitrogen %*			*Carbon decrease (%)*		
	25°	40°	50°	25°	40°	50°
0	2·05	1·40	1·85	0	0	0
2	1·45	1·55	1·50	—	—	—
3	1·40	1·70	1·55	1·5	0·5	6
4	1·40	1·65	1·60	—	—	—
6	1·40	1·75	1·65	—	—	—
7	1·35	1·75	—	3	1	7
9	1·40	1·80	1·55	—	—	—
12	1·45	—	1·50	4	1·5	12
15	1·45	1·85	1·50	—	—	—
17	1·40	—	—	5	5	13

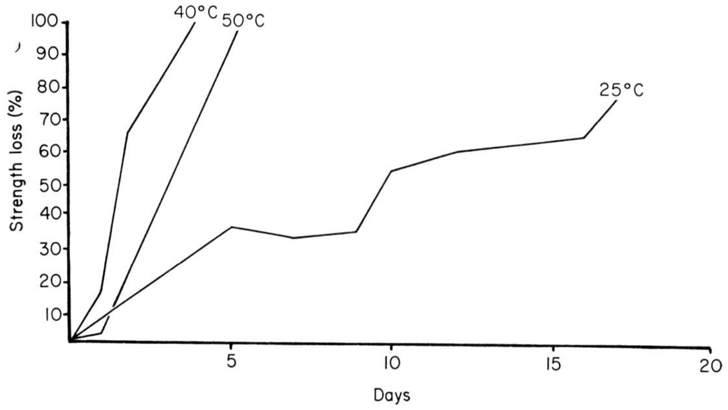

Fig. 3. Effect of temperature on cellulolytic activity.

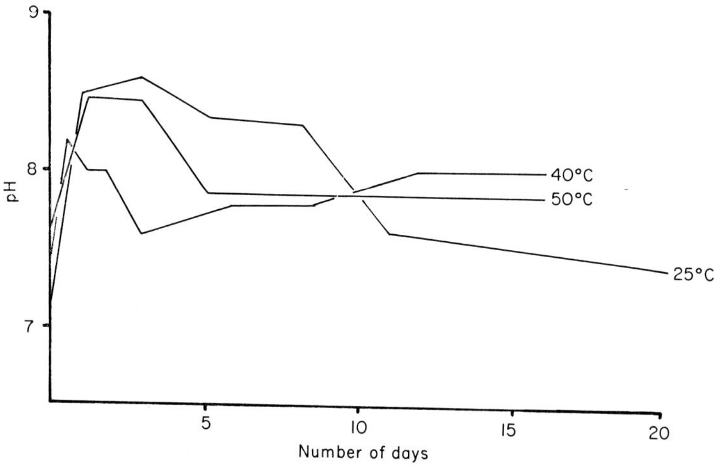

Fig. 4. Effect of temperature on pH.

after 17 days, while at 40°C and 50°C 100% loss was demonstrated after 4 and 5 days, respectively. The lag at 50°C did not seem to affect the rate of cellulolytic activity after the first day. The nitrogen determinations (Table V) showed a considerable drop on the first day of fermentation followed by either a stabilization (25°C) or a slight increase and decrease with a levelling off after 9 days (50°C). However, at 40°C there was an apparent increase in nitrogen and a levelling off after 6 days. The decreases in carbon are shown in Table V. The greatest decrease was given by the 50°C process of 13% after 17 days, whereas at 25°C and 40°C the loss was only 5%.

In the second investigation (Fig. 5) it can be seen that the lower the initial pH the longer was the time that the fermentation remained acidic. At pH 4·0 there was no change for 7 days, followed by an increase to 7·8 at the end of the fermentation period. At an initial pH of 5·0 there was a decrease to 3·8 in the first 3 days, followed by a gradual increase over the next 5 days when the pH reached 8·4. This then stabilized out at 7·3 after 10 days. At pH 6·0

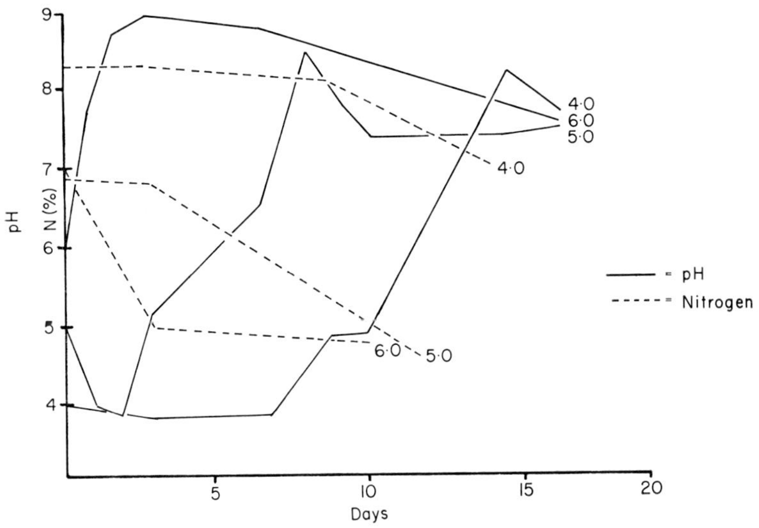

Fig. 5. Effect of initial pH on nitrogen content.

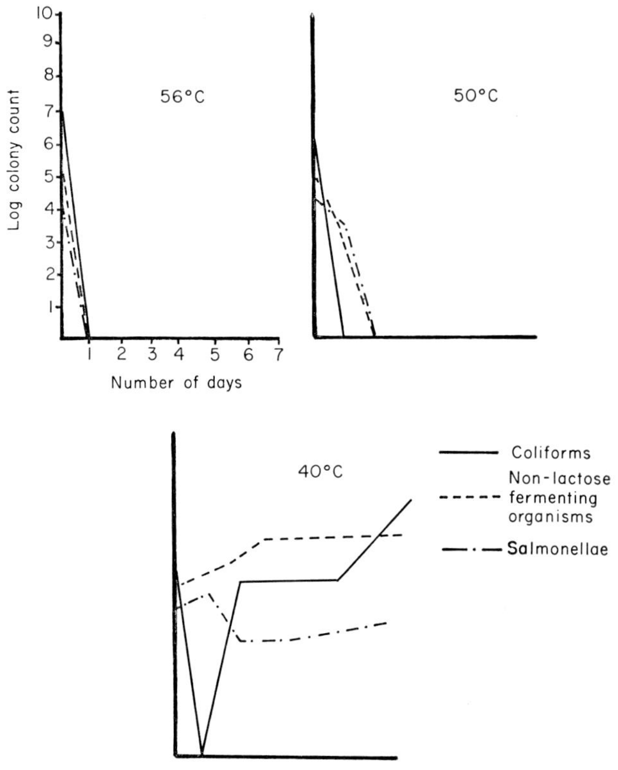

Fig. 6. Graphs to show effects of temperature on pathogenic bacteria present in pig slurry/straw mixtures.

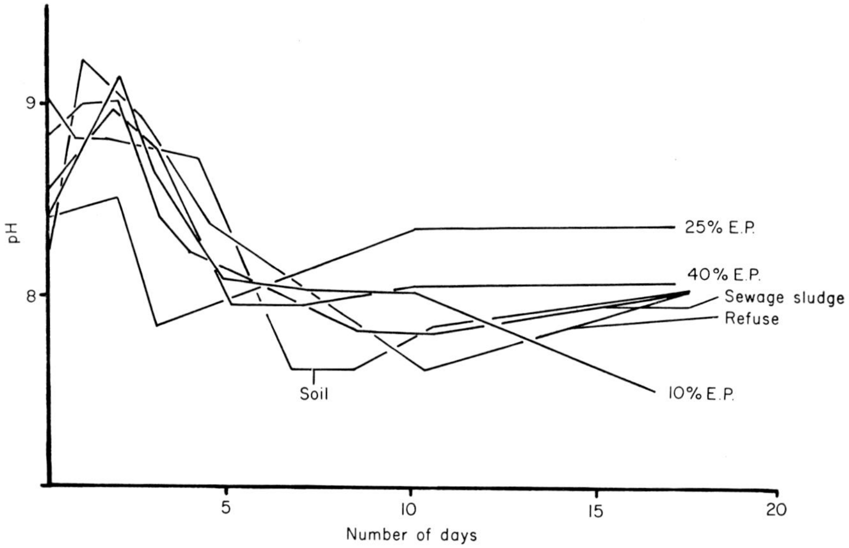

Fig. 7. Effect of inocula additions on pH.

the pH increased to 8·8 after 3 days and then decreased to 7·2 after 14 days. The nitrogen results (Fig. 5) show that nitrogen decreased when the pH entered the alkaline regions. The results from the pathogen counts (Fig. 6) indicated quite clearly that 50°C was a minimum requirement for pathogen suppression. Below this temperature the pathogens, after an initial temperature shock, were able to multiply at 40°C.

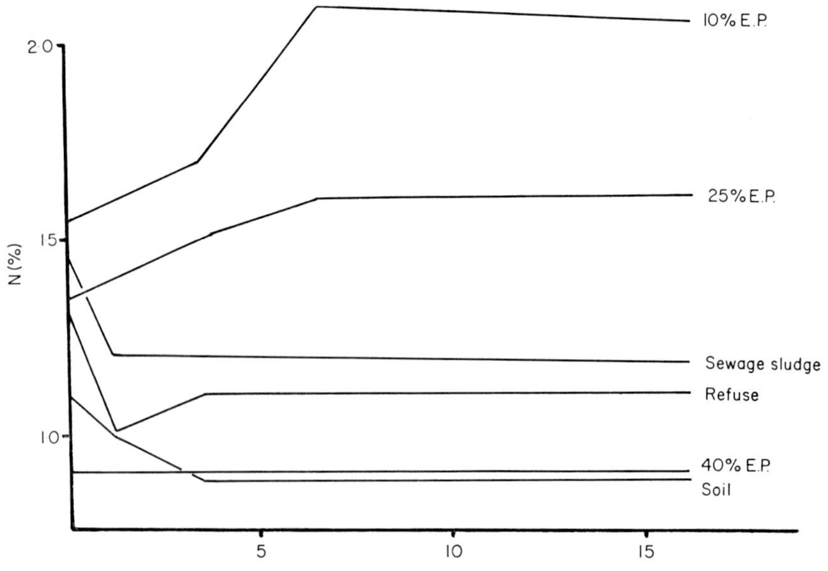

Fig. 8. Effect of inocula on nitrogen.

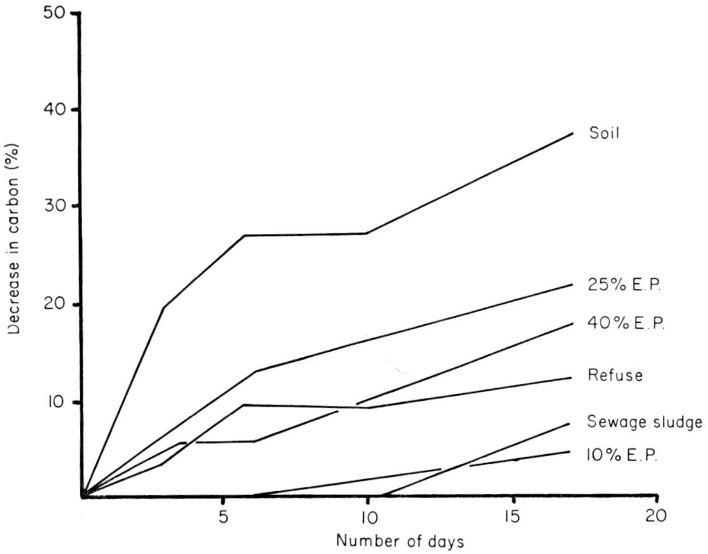

Fig. 9. Effect of inocula additions on carbon reduction.

The results of the inocula additions are shown in Figs. 7–9. The pH curves show that in all but two cases there was an initial appreciable increase in pH (9·3 being the highest recorded for the refuse inoculum). The terminal pH for the additions varied between 8·3 and 7·5. The soil and 25% end-product addition gave, respectively, a fall and very little increase in pH. The fall may have been due to the higher initial pH of 9·05. There were organic nitrogen losses for the sewage sludge, refuse, and soil inocula, while the end-product additions—with the exception of 40%—increased the organic nitrogen content.

The carbon decreases show the difference in the biodegradation processes of the 6 inocula. The soil was found to give the greatest decrease (38%). the 25% end-product gave 22% and the 40% end-product gave 18%.

Cellulose determinations carried out on the inoculated samples are summarized in Table VI. The greatest cellulose reductions were obtained at 50°C from the soil and 25% end-product inoculum. In some investigations there was an apparent increase in cellulose content. This may have been due to reduction of other components in the waste with little or no reduction in

Table VI
CELLULOSE DECREASE IN SAMPLES FROM INOCULATED FERMENTERS

Treatment	*Decrease in cellulose (%)*
Sewage sludge	1·8
Refuse	+3·6
Soil	29·7
10% end-product	16·4
25% end-product	19·3
40% end-product	9·3

+ indicates an apparent increase.

cellulose, resulting in an apparent increase in percentage of the original dry weight.

DISCUSSION

The results indicate that a thermophilic flora, in particular the fungi, is present in animal waste and that it can be encouraged to utilize the waste to produce an upgraded product. The optimum operating temperature for breakdown of cellulose has been shown to be 50°C. However, at 56°C pathogen removal was more complete. This result compares with that of Jalal (1969) who found that the nonspore formers could be removed at this temperature. If the product is to be used as a feed additive, it may be more worthwhile to sterilize it, perhaps combining this with drying. The acidic pH's would appear to be preferable from a nitrogen loss viewpoint. Culp and Culp (1971) have shown that ammonium ions tend to change to ammonia gas as the pH increases above 7. The reduction of proteins via proteolysis to ammonia ions can encourage loss of nitrogen to the atmosphere if the pH rises above 7. Initial control of the pH of the waste can reduce this loss and encourage the thermophilic fungi to grow. Mills (1973) has shown that maximum growth and enzyme activity of the thermophilic fungi occur in the acid region and it may thus be necessary to keep the pH acid to encourage waste utilization, particularly cellulose. However, Mills did find that his refuse-inhabiting thermophiles produced activity peaks in the alkaline regions and this appears to agree with this isolation programme where the *Humicola* spp. and *Torula thermophila* were isolated to greater degrees above pH 7.

The control of pH is also important in discouraging pathogenic species. Bacterial pathogens may be suppressed in the acid ranges. However, two important thermophilous fungi causing animal mycoses are able to grow in these ranges, they are: *Aspergillus fumigatus* and *Mucor pusillus*. The results from the isolation programme indicate control mechanisms. Both organisms favour acid pH's of between 4 and 6 and occur in greater amounts at 40°C than at 50°C. Mills (1973) has found this to be so in refuse heaps. Cooney and Emerson (1964) report that *M. pusillus* has an optimum growth range of between 35 and 45°C. Thus by keeping the temperature at about 50°C and the pH to just above 7, such pathogens may be discouraged and nitrogen loss kept to a minimum.

The investigations on inocula additions show that certain types of inocula are able to enhance the rate of biodegradation. The work of Golueke *et al.* (1954) indicated that inocula additions made no difference to the biodegradation of municipal refuse, and they concluded that an adequate bacterial population was present. The work with farm waste suggests that the addition of more organisms can increase the biodegradation rate. Town refuse, composted sewage, and soil have all been found to be good sources of thermophilic fungi (Mills, 1973; Barnes, 1974; Eggins and Malik, 1969). The use of recycled end-product is quite common in waste treatment where it serves as a starter inoculum. The admixing of fresh with a small amount of the end-product would appear to increase biodegradation over other inocula additions, with the exception of soil. However, the use of soil may be limited if the product is to be used as a feedstuff. Soil is mostly nutritionally inert and as such reduces the overall nutritional value of the product.

A pilot scale 'on-farm' process based on these laboratory results (Seal and Eggins, 1975) has shown that the wastes can be changed from a highly polluting material into an odourless, nutritionally enhanced and handleable material. The use of recycled pig waste has received little attention, especially in the area of microbial upgrading. Harmon *et al.* (1972) have studied only the use of oxidation ditch residue—the dried sludge—as a potential feedstuff. In this process no thermophilic stage was involved and the predominant organisms were probably bacteria. However, they found good essential amino acid levels.

The whole area of microbial waste utilization is still in its very early stages and it would be foolish to implement incomplete processes at this stage. As indicated by this contribution, much more work needs to be done at laboratory and on-site stages together with continuous economic assessment of potential processes. Particularly important is the distribution of the waste in order to spread nutrients, concentrated by man-made efforts, back to the cycles where they can be dealt with efficiently. Where this is not possible, biodegradation could help in maintaining Nature's efficiency.

REFERENCES

Barnes, T. G. (1974). Ecological aspects of the biodegradation of town wastes. Ph.D. Thesis, University of Aston in Birmingham.

Chang, Y., and Hudson, H. J. (1967). The fungi of wheat straw compost. 1. Ecological studies. *Trans. Brit. Mycol. Soc.*, **50** (4), 649–666.

Cooney, D. G., and Emerson, R. (1964). *Thermophilic Fungi, an Account of Their Biology, Activities and Classification.* W. H. Freeman & Co., San Francisco.

Crisan, E. V. (1959). The isolation and identification of thermophilic fungi. M.Sc. Thesis, Purdue University, Lafayette, Ind.

Culp, R. L., and Culp, G. L. (1971). *Advanced Wastewater Treatment.* Van Nostrand Rheinhold Co., New York.

Eggins, H. O. W., and Coursey, D. G. (1964). Thermophilic fungi associated with Nigerian oil palm produce. *Nature*, **203** (4949), 1083–1084.

Eggins, H. O. W., and Malik, K. A. (1969). The occurrence of thermophilic fungi in a pasture-land soil. *Antonie van Leeuwenhoek*, **35** (2), 178–184.

Eggins, H. O. W., and Pugh, G. J. F. (1962). Isolation of cellulose-decomposing fungi from the soil. *Nature*, **193** (4810), 94–95.

Golueke, C. G., Card, B. J., and McGauhey, P. H. (1954). A critical evaluation of inoculums (sic) in composting. *Appl. Microbiol.*, **22**, 45–53.

Harmon, B. G., Day, D. L., Jensen, A. H., and Baker, D. H. (1972). Nutritive value of aerobically sustained swine excrement. *J. An. Sci.*, **34** (3), 403–407.

Hayes, W. A. (1969). Microbial changes in composting wheat straw/horse manure mixtures. *Mushroom Sci.*, **8**, 173–178.

Jalal, K. F. (1969). A technological evaluation of composting for community waste disposal in Asia. *Compost Sci.*, **10** (1–2), 20–25.

Lloyd, A. O. (1965). An adhesive tape technique for the microscopical examination of surfaces supporting, or suspected of supporting, mould growth. *Int. Biodeterior. Bull.*, **1** (1), 10–11.

Mills, J. (1973). The biodegradation of certain components of town waste by thermophilic fungi. Ph.D. Thesis. University of Aston in Birmingham.

O'Callaghan, J. R. (1975). The waste products of agriculture. *J. Royal Soc. Arts,* **123** (5223), 139–149.

Seal, K. J., and Eggins, H. O. W. (1975). The upgrading of agricultural wastes by thermophilic fungi. In Proc. Symp. Food from Waste, held at Weybridge College of Food Technology, April 7–9, 1975 (in press).

Waksman, S. A., and Cordon, T. C. (1939). Thermophilic decomposition of plant residues in composts by pure and mixed cultures of micro-organisms. *Soil Sci.*, **47**, 217–225.

limits of detectability. Urease producers were present throughout the term of the experiment but undoubtedly utilized other substrates.

Many previous studies (Chang and Hudson, 1967; Eastwood, 1952; Poincelot, 1972) have indicated total numbers of microorganisms in composting material. Such data generally show only the abundance of the microbial flora and do not indicate if the organisms can grow and produce degradative enzymes at the prevailing pile temperature. Samples of our compost incubated at the prevailing pile temperature always showed more bacteria than fungi (about 2 log units greater) (Figs. 2 and 3). The numbers of

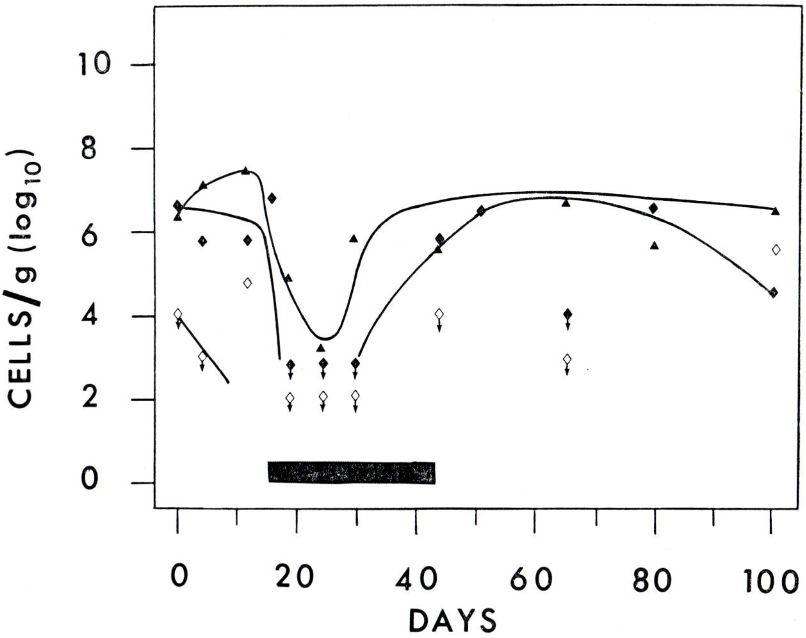

Fig. 3. Total fungal count (▲), numbers of fungi able to produce proteases at the prevailing pile temperature (◆), and numbers of fungi able to grow and excrete proteases under thermophilic conditions (◇) in compost sampled over a 100-day test period. Points with arrows indicate count below this level. Horizontal bar indicates thermophilic stage.

bacteria able to grow as mesophiles fluctuated during the test period, whereas the population of fungi able to grow as mesophiles remained fairly stable. The opposite was true for thermophilic bacteria and fungi. The former remained stable, while the latter decreased during the thermophilic stage. This does not mean that the population of bacterial thermophiles *per se* remained constant but probably indicates a change in overall population, *i.e.* different species or groups predominating at different periods.

As the pile began to heat up (thermophilic stage), there was a decline in the total bacterial population (Fig. 2), but bacteria able to produce proteases remained stable (Fig. 2) while those that produced lipases declined. This may indicate a progressive change in bacterial flora as the test period progressed, or a change in the ability of the various bacteria to produce these extracellular enzymes at the incubation temperature used.

The numbers of fungi able to grow and produce proteases and lipases at the prevailing pile temperature fluctuated considerably and declined substantially during the thermophilic stage (Figs. 3 and 4). The lipid content of the pile declined only after the thermophilic stage (Fig. 4). Fungal thermophiles able to produce proteases were always at a low level (Fig. 3).

Bacteria able to degrade pectin, starch, alkanes, and those able to produce acids were also sought. Samples incubated at the prevailing pile temperature showed that the number of bacteria able to degrade these substrates or excrete acids were either killed during the thermophilic stage or were not able to grow. The latter is more reasonable since the same samples incubated at a mesophilic temperature provided numbers of bacteria as high as the total bacterial count.

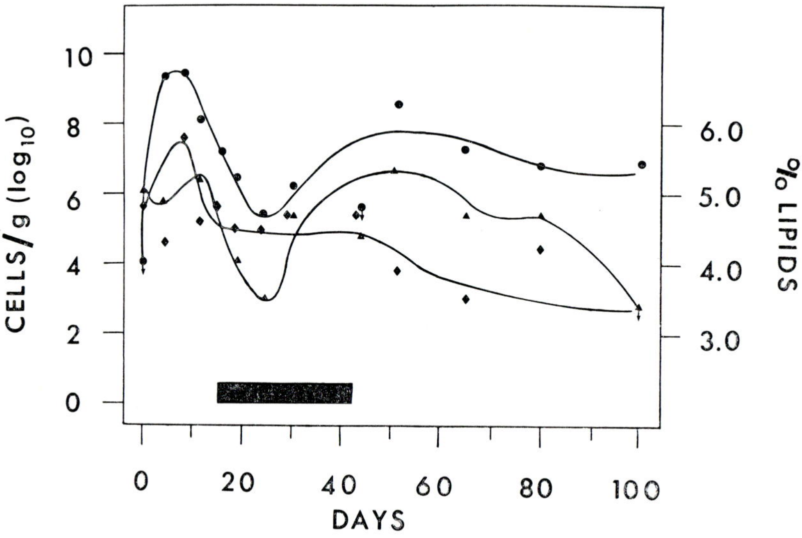

Fig. 4. Percentage lipids (◆), total number of bacteria (●), and fungi (▲) able to excrete lipases at the prevailing pile temperature in compost sampled over a 100-day test period. Horizontal bar indicates thermophilic stage.

The number of proteolytic fungi declined substantially during the thermophilic stage, while amylolytic fungi declined only slightly during this stage. As with the bacteria, the same samples incubated at 30°C showed a stable population for the amylase producers, but the proteolytic fungi which are mesophiles appeared to be killed when the pile was hot. The numbers of pectolytic fungi, both those able to produce pectate transeliminase and polygalacturonase, declined during the thermophilic stage; they increased after the thermophilic stage and again declined at the end of the test period.

Compost is usually considered to be usable when the C/N ratio drops below 20. However, the use of the C/N ratio alone to assess finished compost can be misleading. In our leaf pile the C/N ratio dropped below 20 on the

23rd day and visual inspection indicated a finished compost. Yet, the compost was probably neither finished nor usable because the temperature of the pile was 54°C and it still contained large numbers of microorganisms which were still producing degradative enzymes. Our pile started in the fall was completely finished by the following spring, however, in time for use in the growing season.

In 1974 a second leaf compost pile was constructed. As a substitute for urea, grass clippings and chopped apples were added to one-half of the pile to adjust the C/N ratio; the other half contained only leaves. Additionally, perforated pipes were placed within the pile to induce better air flow. The pile that contained only leaves did not provide a useful compost, even after 5 months, while the pile that contained grass and apples was useful much sooner. The pile containing leaves only heated sporadically, and this was probably the single most significant change.

DISCUSSION

The task of duplicating exactly the composition of successive compost piles in repetitive experiments to control substrate composition and amount and numbers of microorganisms in the initial inoculum is probably impossible. Thus, each experiment reflects only the conditions at the time of the experiment. Data obtained may be useful as long as the operating parameters are defined and different studies may be compared. Our leaf compost pile appears to be unique in many respects, both in overall composition as well as in the ability of the microflora to excrete degradative enzymes. Overall, we have noted no large differences in total numbers of either bacteria or fungi as reported in other studies. Further, the pattern for the decline and fall of organisms during the thermophilic stage follows patterns as shown by others. Our experiment differs in that we attempted to demonstrate declines or increases in numbers of organisms able to excrete specific enzymes related to temperature changes in the pile.

The numbers of bacteria and fungi able to excrete enzymes change, especially in the thermophilic stage, even though the total count remains fairly stable. Further, our experiment shows that some organisms probably are not killed by the high temperature but are not able to excrete specific enzymes.

Since the numbers of fungi declined during the thermophilic stage of our pile, the indication is that bacterial degraders were probably more active in our pile than fungal degraders. This is opposite to results found by others, but attests to the difficulties of comparing different kinds of compost piles. However, Kane and Mullins (1973), working with municipal wastes, reached the same conclusions as we did. The use of urea in our 1973 pile did not appear to be the best choice for supplemental nitrogen since it was degraded almost immediately. However, a leaf pile constructed in 1974 without the addition of any supplemental nitrogen failed to react normally in that the leaves did not decompose properly. In this case the leaves for the second pile came from the same trees as for the 1973 pile and shows the need for a long-lasting nitrogen amendment. However, the slight, short-term reduction in the

C/N ratio afforded by the urea appears important since without it the pile probably would not have gone to completion.

Finally, our experiment shows the complex heterogeneity of micro-organisms in a successful compost pile. This attests to the need for a variety of microorganisms able to excrete different types of enzymes, under a variety of temperatures and conditions, to bring about a usable composted material. Studies of the interactions of microorganisms within a compost pile are certainly in order, yet the intricacies of such experiments almost baffle the imagination.

Acknowledgements

We thank Margaret Staba and Reneé Shapiro for excellent technical help.

REFERENCES

Chang, Y., and Hudson, H. J. (1967). The fungi of wheat straw compost. I. Ecological Studies. *Trans. Brit. Mycol. Soc.*, **50**, 649–666.

Dickinson, C. H., and Pugh, G. J. F. (Eds.) (1974). *Biology of Plant Litter Decomposition.* Vols. I and II, Academic Press, New York.

Eastwood, D. J. (1952). The fungus flora of composts. *Trans. Brit. Mycol. Soc.*, **35**, 215–228.

Hankin, L., and Anagnostakis, S. L. (1975). The use of solid media for detection of enzyme production by fungi. *Mycologia*, **67**, 597–607.

Hankin, L., and Sands, D. C. (1974). Bacterial production of enzymes in activated sludge systems. *J. Water Pollut. Control Fed.*, **46**, 2015–2025.

Hankin, L., Zucker, M., and Sands, D. C. (1970). Improved solid medium for the detection and enumeration of pectolytic bacteria. *Appl. Microbiol.*, **22**, 205–209.

Hankin, L., Sands, D. C., and Hill, D. E. (1974). Relation of land use to some degradative enzymatic activities of soil bacteria. *Soil Sci.*, **118**, 38–44.

Horwitz, W. (Ed.) (1965). *Official Methods of Analysis of the Association of Official Agricultural Chemists*, 10th Ed. Association of Official Agricultural Chemists, Washington, D. C., 957 pp.

Hudson, H. J. (1968). The ecology of fungi on plant remains above the soil. *New Phytol.*, **67**, 837–874.

Kane, B. E., and Mullins, J. T. (1973). Thermophilic fungi in a municipal waste compost system. *Mycologia*, **65**, 1087–1100.

Kochitizky, O. W., Seaman, W. K., and Wiley, J. S. (1969). Municipal composting research at Johnson City, Tennessee. *Compost Sci.*, **9**, 5–16.

Lossin, R. D. (1971). Compost studies. *Compost Sci.*, **12**, 12–13.

Martley, F. G., Jayashankar, S. R., and Lawrence, R. C. (1970). An improved agar medium for the detection of proteolytic organisms in total bacterial counts. *J. Appl. Bacteriol.*, **33**, 363–370.

Poincelot, R. P. (1974). A scientific examination of the principles and practice of composting. *Compost Sci.*, **15**, 24–31.

Poincelot, R. P. (1972). The biochemistry and methodology of composting. Bull. 727. Conn. Agr. Expt. Sta., New Haven, Conn., 38 pp.

Poincelot, R. P., and Day, P. R. (1972). A simple dye release assay for determining the cellulolytic activity of fungi. *Appl. Microbiol.*, **23**, 875–879.

Regan, R. W., and Jeris, J. S. (1970). A review of the decomposition of cellulose and refuse. *Compost Sci.*, **11**, 17–20.

Satriana, M. J. (1974). *Large Scale Composting.* Noyes Data Corp., Park Ridge, N.J.

Sierra, G. (1957). A simple method for the detection of lipolytic activity of micro-organisms and some observations on the influence of the contact between cells and fatty substrates. *Antonie van Leeuwenhoek*, **23**, 15–22.

Stutzenberger, F. J. (1971). Cellulase production by *Thermospora curvata* isolated from municipal solid waste compost. *Appl. Microbiol.*, **22**, 147–152.

Stutzenberger, F. J., Kaufman, A. J., and Lossin, R. D. (1970). Cellulolytic activity in municipal solid waste composting. *Can. J. Microbiol.*, **16**, 553–560.

University of California at Berkeley (1953). Reclamation of municipal refuse by composting. Tech. Bull. 9, Sanitary Eng. Res. Project.

Werkman, C. H. (1951). In *Manual of Methods for Pure Culture Study of Bacteria.* Soc. Amer. Bacteriologists, McGraw-Hill, New York.

Zucker, M., and Hankin, L. (1970). Regulation of pectate lyase synthesis in Pseudomonas fluorescens and *Erwinia carotovora. J. Bacteriol.*, **104**, 13–18.

ENZYMATIC ACTIVITY OF SOIL FUNGI

STEPHANIE MOORE and E. E. STAFFELDT

New Mexico State University, Las Cruces, New Mexico 88003, U.S.A.

Summary

Fungi were isolated from New Mexico grassland and Canal Zone soil samples to compare the metabolic activities of organisms found in soils of two dissimilar natural conditions. Pure cultures of the isolated fungi were grown in liquid solutions containing cellulose as the sole source of carbon. Growth curves were established for each fungus prior to the primary investigation to ensure extraction of samples at time intervals during the lag and accelerated growth phases. The enzyme activity of these extracted aliquots was determined by incubating the sample with fresh substrate and measuring the changes in reducing power throughout the exponential growth phase. Additional aliquots were removed for metabolic by-product determinations which yielded acetic, propionic, butyric, oxaloacetic, fumaric, malic, glutaric, and citric acids. All fungi tested degraded to some degree the cellulose employed. Comparisons of the enzyme production-cellular assimilation interactions revealed that the tested organisms possessed varying capabilities in the degrading process. This differential between enzyme production and cellular assimilation is suggested for selecting microbes in mass cell culturing for microbial feeds and foods and in high production of metabolic products during fermentation of organic substrates.

Recently, increased emphasis has been placed on the conversion of wastes and refuse to solve problems associated with more adequate means of sewage and refuse disposal, reduced pollution, obtaining additional energy as power, and possibly feed or food. Much of this organic residue is of plant origin. Fungi are important entities in the degradation of major plant components. They are considered to be primary decomposers of cellulose, hemicellulose, and lignin. Alexander (1965) included lists of microbial genera capable of utilizing the above substrates, but difficulty arose in knowing if the same organism possessed the capacity of degrading all the plant constituents or was limited in breadth of its enzyme system. Garrett (1963) stated that two aspects of successful competitive saprophytic ability were rapidity of spore germination (or growth) and broadness of the enzyme system possessed by the microbe. Efforts were then made to test these possibilities.

MATERIALS AND METHODS

Soils were obtained from the Jornada Grassland Biome site located in the southwestern corner of the Jornada Experimental Range, 25 miles north of Las Cruces, New Mexico, and from plots established in the Albrook Forest which adjoins Fort Clayton, Canal Zone.

Sturdy alfalfa stems 5 to 7 cm long devoid of leaves were sterilized by autoclaving. These organic traps were aseptically layered in petri dishes, separated by soil, and incubated at ambient temperature. Five stems were removed from each soil after 1, 2, 4, and 8 days incubation. They were washed in running tap water, rinsed five times in tap water, and finally five times in sterile, distilled water. The traps were placed between sterile paper towels to remove the excess water. Finally, the ends were removed, each was cut into three equal portions and plated onto carrot decoction agar contained in petri dishes. Fungi growing from these organic traps were isolated for subsequent experimentation.

Growth curves were determined for each fungus on cellulose, following the methods described by Calam (1969). The fungus species were then grown in 80 ml of culture medium base, with 1 g of cellulose added as the sole carbon source. This culture medium base consisted of 1 g NH_4Cl, 6 g K_2HPO_4, 3 g KH_2PO_4, 5 g NaCl, 0·1 g $MgSO_4$ in 1000 ml distilled water. Culture incubation occurred on a shaker (108 oscillations/min) maintained at ambient temperature. Samples for enzyme determinations were removed at 6 hr intervals during the first day and 12 hr intervals on subsequent days. The enzyme activity of these extracted aliquots was determined by inoculating the sample with fresh substrate and measuring the reducing power. Amounts of reducing sugar were determined employing the dinitrosalicylic test (Levinson *et al.*, 1951). Bernfeld (1951) reported that concentration of reducing sugar correlates with enzyme activity.

Samples were also taken during the exponential and stationary phases for metabolic by-product analysis. A Varian Aerograph 1600 gas chromatograph was used to analyse for low molecular weight organic acids employing a Porapak Q S plus 2% phosphoric acid column (Kellogg, 1969). Further analysis was conducted to determine certain larger monocarboxylic and dicarboxylic acids. A 5% Carbowax plus 5% isophthalic acid column was used (Harmon and Doelle, 1969).

RESULTS

Representatives of the genus *Fusarium* were the most common fungi observed colonizing buried stems in both soils. *Aspergillus* sp. and *Trichoderma* sp., both from Panama soil, were the next most prevalent fungi. *Curvularia* sp., from Jornada soil, and *Rhizopus* sp., from Canal Zone soil, occurred a moderate number of times. The least common fungi were *Penicillium* sp., from Canal Zone soil, and *Chaetomium* sp., *Myrothecium* sp., and *Stachybotrys* sp., from Jornada soil. Fewer genera of fungi were observed colonizing straws in the Canal Zone soils, but the number of individuals representing each genus were higher.

Fungi representing three genera, *Rhizopus* sp., *Trichoderma* sp., and *Chaetomium* sp., were observed as fast-growing microbes as they advanced from the stems and onto the agar. Representatives of all other genera exhibited moderate growth rates with the exception of *Myrothecium* sp. which appeared relatively slow. Since all of the fungi included in this work were relatively successful colonizers, the enzyme aspect required further elucidation.

Growth curves were determined for the 12 test organisms on cellulose and the average dry weight increase in mycelium was plotted against time. The duration of the lag phase and the time required to double the mycelial weight

Table I
TIME OF LAG PHASE AND TOTAL TIME REQUIRED TO DOUBLE MYCELIAL WEIGHT WHEN FUNGI WERE GROWN ON CELLULOSE

Soil	*Fungus*	*Days*	
		Lag phase	*Total time*
Jornada	*Alternaria* sp.	2	6
	Chaetomium sp.	4	10
	Curvularia sp.	2	4
	Fusarium sp.	1	4
	Myrothecium sp.	3	5
	Rhizopus sp.	1	5
	Stachybotrys sp.	2	5
Panama	*Aspergillus* sp.	1	3
	Fusarium sp.	1	4
	Penicillium sp.	1	5
	Rhizopus sp.	3	4
	Trichoderma sp.	1	3

during the exponential growth phase was extrapolated from these growth curves (Table I). Although *Chaetomium* sp. grew rapidly on a solid medium such as agar, it did not assimilate rapidly in liquid culture and on cellulose as the sole source of carbon. Four or more days were required for most of the test fungi to double their weight.

Chaetomium sp. and both species of *Fusarium* and *Curvularia* sp. produced the largest quantities of measurable cellulase (Table II). Representatives of *Rhizopus* from both Jornada and Canal Zone soil exhibited the lowest cellulase activity. Within the first day, the test fungi had produced between 50–100% of the maximum enzyme levels that were observed in this experimentation. It appeared that values near constant levels were attained in 72 hr. Cellobiase production by *Fusarium* sp. from the Jornada soil was slower than that produced by other organisms (Table III). Other than *Fusarium* sp. and

Table II
ENZYMATIC ACTIVITY IN SAMPLES AS DETERMINED BY THEIR ABILITY TO REDUCE CELLULOSE TO GLUCOSE

Soil	*Fungus*	*Glucose (mg/ml) produced after various incubation periods (hr)*							
		6	12	18	24	36	48	60	70
Jornada	*Alternaria* sp.	0·04	0·06	0·10	0·16	0·19	0·20	0·23	0·25
	Chaetomium sp.	0·09	0·22	0·46	0·62	0·68	0·73	0·79	0·79
	Curvularia sp.	0·11	0·19	0·31	0·49	0·55	0·58	0·60	0·63
	Fusarium sp.	0·15	0·21	0·29	0·35	0·57	0·66	0·66	0·68
	Myrothecium sp.	0·06	0·09	0·16	0·23	0·34	0·40	0·42	0·42
	Rhizopus sp.	0·03	0·05	0·05	0·07	0·07	0·07	0·07	0·07
	Stachybotrys sp.	0·10	0·15	0·21	0·28	0·37	0·45	0·45	0·47
Panama	*Aspergillus* sp.	0·12	0·19	0·28	0·30	0·30	0·31	0·31	0·31
	Fusarium sp.	0·08	0·28	0·49	0·65	0·68	0·70	0·73	0·74
	Penicillium sp.	0·09	0·16	0·21	0·25	0·26	0·28	0·29	0·29
	Rhizopus sp.	0·03	0·06	0·06	0·08	0·09	0·11	0·13	0·13
	Trichoderma sp.	0·04	0·16	0·28	0·42	0·49	0·52	0·52	0·52

possibly *Chaetomium* sp., these fungi appeared to attain the maximum production of cellobiase within 48 hr. Therefore, the measurable cellulase and cellobiase in the solutions had generally reached a constant level before there was a doubling of mycelial weight. Again, the two isolates of *Rhizopus* produced very little measurable cellobiase activity.

All of the fungi tested diffused organic acids into the culturing solutions (Table IV). *Chaetomium* sp. and *Alternaria* sp. produced the highest levels of low molecular acids, while *Fusarium* sp. and *Rhizopus* sp. produced the highest levels of mono- and dicarboxylic acids. All of these fungi were isolated from the Jornada soil. *Stachybotrys* sp. isolated from Jornada soil and *Trichoderma* sp. isolated from Canal Zone soil produced the lowest levels of acid.

Table III

ENZYMATIC ACTIVITY IN SAMPLES AS DETERMINED BY THEIR ABILITY TO REDUCE CELLULOSE TO CELLOBIOSE

Soil	*Fungus*	*Cellobiose (mg/ml) produced after various incubation periods (hr)*					
		6	12	18	24	36	48
Jornada	*Alternaria* sp.	0·11	0·18	0·26	0·33	0·33	0·34
	Chaetomium sp.	0·08	0·19	0·35	0·46	0·61	0·65
	Curvularia sp.	0·05	0·12	0·21	0·38	0·47	0·49
	Fusarium sp.	0·11	0·23	0·31	0·49	0·62	0·71
	Myrothecium sp.	0·10	0·18	0·21	0·37	0·39	0·39
	Rhizopus sp.	0·12	0·15	0·14	0·15	0·15	0·16
	Stachybotrys sp.	0·04	0·11	0·20	0·29	0·30	0·30
Panama	*Aspergillus* sp.	0·11	0·22	0·35	0·36	0·36	0·37
	Fusarium sp.	0·15	0·41	0·58	0·58	0·59	0·59
	Penicillium sp.	0·11	0·19	0·27	0·31	0·31	0·31
	Rhizopus sp.	0·01	0·04	0·08	0·09	0·09	0·09
	Trichoderma sp.	0·10	0·26	0·36	0·45	0·46	0·46

DISCUSSION

Rapidity of enzyme production and the attainment of a constant enzyme level prior to the doubling of the mycelium dry weight by all of the test fungi were not anticipated. Similarly, the ultimate utilization of the converted cellulose was not predictable in most instances. Capacities and capabilities of the fungi tested herein were variable. Variability would also be expected within the same genus and species. Evaluation of this variability was attempted in order to make these findings more useful tools for other investigators. The enzyme production and microbe assimilation interactions were compared and as the comparisons increased, indecision became apparent about the well-known stereotyped terminology that had been utilized. Enzyme levels varied from 0·23 mg/ml and less which were considered little, 0·59–1·12 mg/ml which were considered moderate, to 1·33–1·44 which were considered extensive. This allowed the development of comparisons between the enzyme production

Table IV

BY-PRODUCT FORMATION OF ORGANIC ACIDS DURING CATABOLISM OF CELLULOSE

Soil	Fungus	*mg/ml/g Mycelium dry wt*					*mg/ml*			
		Acetic	*Propionic*	*Butyric*	*Oxaloacetic*	*Malic*	*Fumaric*	*Gluconic*	*Citric*	*Value*
Jornada	*Alternaria* sp.	0·86	0·78	0·51	0·24	0·13	0·16	0·02	0·20	2·95
	Chaetomium sp.	0·91	0·88	0·44	0·21	0·12	0·10	0·18	0·25	3·13
	Curvularia sp.	0·32	0·18	0·23	0·17	0·15	0·17	0·01	0·15	1·28
	Fusarium sp.	0·84	0·78	0·68	0·41	0·38	0·42	0·08	0·43	4·02
	Myrothecium sp.	0·37	0·28	0·15	0·13	0·11	0·10	0·03	0·16	1·33
	Rhizopus sp.	0·68	0·63	0·29	0·28	0·25	0·29	0·05	0·20	2·67
	Stachybotrys sp.	0·18	0·15	0·08	0·06	0·02	0·04	0·05	0·05	0·53
Panama	*Aspergillus* sp.	0·50	0·35	0·19	0·38	0·39	0·35	0·10	0·35	2·61
	Fusarium sp.	0·39	0·31	0·30	0·24	0·26	0·28	0·09	0·28	1·75
	Penicillium sp.	0·39	0·28	0·26	0·19	0·21	0·20	0·04	0·18	2·15
	Rhizopus sp.	0·34	0·25	0·21	0·16	0·09	0·07	0·04	0·15	1·31
	Trichoderma sp.	0·21	0·16	0·13	0·13	0·04	0·04	0·02	0·11	0·84

and conversion of substrate to microbial cell assimilation (Table V). Grades were then assigned to these interacting parameters.

Table V
SCHEMATIC ASSIGNMENT OF GRADE LEVELS BASED UPON ENZYME PRODUCTION AND CELL ACCUMULATION

Relative level of enzyme produced	*Relative amount of cell substance*	*Grade assigned*
None	None	0
Little	Little	1
Little	Moderate	2
Little	Extensive	3
Moderate	Little	4
Moderate	Moderate	5
Moderate	Extensive	6
Extensive	Little	7
Extensive	Moderate	8
Extensive	Extensive	9

It was assumed that an organism could then primarily decompose, be active as a competitor, produce cell substance, produce acid, produce enzyme, or be involved in various combinations of the above. The decomposer evaluation failed because CO_2 information was lacking, and it was necessary to determine mineralization levels by measuring the quantity of carbohydrate that was converted to CO_2. Organisms that would be successful in competition with others were assumed to produce enzymes extensively and be capable of rapid assimilation (Table VI). Remarks were included to indicate relative levels based on the authors' opinions and this aspect would be open to criticism. Measurements of spore production could modify this interpretation. Evaluation of enzyme production-microbial assimilation interactions as they influence cell production and harvest is summarized in Table VII. It was assumed that as an organism more completely utilizes a substrate

Table VI
SUCCESS OF MICROBES IN COMPETITION WITH OTHERS

Grade	*Remarks*	*Grade*	*Remarks*
0	No	5	Fair
1	Poor	6	Good
2	Fair	7	Fair
3	Good	8	Good
4	Poor	9	Excellent

Table VII
SUCCESSFUL CELL PRODUCTION FOR HARVEST

Grade	*Remarks*	*Grade*	*Remarks*
0	No	5	Fair
1	Fair	6	Good
2	Good	7	Poor
3	Excellent	8	Fair
4	Poor	9	Good

in cell mass, efficiency increases and cost decreases. Some of the remarks were assigned to certain numbers reluctantly, but assignments were made as uniformly as possible. Microbes producing substantial quantities of enzyme and little cell mass would convert the substrates into metabolic products such as organic acids (Table VIII). After assigning the remarks to the grades based

Table VIII
SUCCESS FOR PRODUCTION OF END-PRODUCTS IN METABOLISM

Grade	*Remarks*	*Grade*	*Remarks*
0	No	5	Fair
1	Good	6	Poor
2	Fair	7	Excellent
3	Poor	8	Good
4	Good	9	Fair

on ability to compete with others, to produce cell mass, and to produce metabolic products, these grades and remarks were ascribed to the fungi in this experiment (Table IX). Comparisons were made to check the possible

Table IX
CAPABILITIES OF TEST FUNGI AS COMPETITORS, PRODUCERS OF CELL SUBSTANCE AND END-PRODUCTS WHEN UTILIZING CELLULOSE

Organism	*Grade based on cellulose utilization*	*Possible utilization*		
		Saprophytic competitor	*Cell production*	*Fermentation end-product*
Alternaria sp.	4	Poor	Poor	Good
Chaetomium sp.	7	Fair	Poor	Excellent
Curvularia sp.	5	Fair	Fair	Fair
Fusarium sp. (Jornada)	8	Good	Fair	Good
Myrothecium sp.	5	Fair	Fair	Fair
Rhizopus sp. (Jornada)	1	Poor	Fair	Good
Stachybotrys sp.	5	Fair	Fair	Fair
Aspergillus sp.	6	Good	Good	Poor
Fusarium sp.	9	Excellent	Good	Fair
Penicillium sp.	4	Poor	Poor	Good
Rhizopus sp.	3	Good	Excellent	Poor
Trichoderma sp.	6	Good	Good	Poor

utility of these grades and remarks in depicting the fungi as possible organic acid producers or saprophytic competitors. Remarks included under fermentation end-products were compared with values compiled for organic acids (Table IV). The organism listed as excellent produced a value of 3·13. Those listed as good exhibited values of 1·75, 2·67, 2·95, and 4·02, while microbes classified as fair showed values of 0·53, 1·28, 1·33, and 2·15. Those in the poor category produced values of 0·84, 1·31, and 2·61. Of the five organisms listed as excellent or good in producing metabolic end-products, four of them were observed as producing the highest quantities in this experiment. The real misfit within these groupings was *Aspergillus* sp. which was placed in the poor acid-producing category but still yielded a value of 2·61. Based on the evaluation suggested for saprophytic competitors, the initial colonization of stems was re-examined. The organism rated as excellent was observed 59 times, those rated as good were observed 23, 44, 46, and 53 times, those rated as fair were observed 11, 14, 15, and 36 times, and those rated as poor were observed 14, 18, and 18 times. Since the stems contained other compounds such as starch, hemicellulose, lignin, and other compounds, values for these substrates need to be obtained to determine a mean or better still the stem material should be used as the substrate to determine the value for comparison. After all of these aspects were summarized, each test fungus acquired different characteristics and potentials than had been originally anticipated.

Additional data could eliminate some of the discrepancies observed or focus on the biological interpretation of why or how they appear as discrepancies. Also, environmental parameters such as temperature, pH, and nitrogen balances need to be examined to determine how they influence the microbial-enzyme-substrate interactions.

Acknowledgement

This research was supported in part by the New Mexico Energy Research and Development Program (BEF proposal—66).

REFERENCES

Alexander, M. (1965). *Introduction to Soil Microbiology*. John Wiley and Sons, New York. 472 pp.

Bernfeld, P. (1951). Enzymes of starch degradation and synthesis. *Adv. Enzymol.*, **12**, 385.

Calam, C. T. (1969). The evaluation of mycelial growth, In *Methods in Microbiology, Vol. I*, pp. 572–575 (J. R. Norris and D. W. Ribbons, eds.), Academic Press. New York.

Garrett, S. D. (1963). *Soil Fungi and Soil Fertility*. The MacMillan Co., New York. 165 pp.

Harmon, M. A., and Doelle, H. W. (1969). Gas chromatographic separation and determination of microquantities of the esters of tricarboxylic acid cycle acids. *J. Chromatography*, **42** (2), 157–169.

Kellogg, D. W. (1969). Analysis of rumen fluid volatile fatty acids by chromatography with Porapak Q S. *J. Dairy Sci.*, **52**, 1690–1692.

Levinson, H. S., Mandels, G. R., and Reese, E. T. (1951). Products of enzymatic hydrolysis of cellulose and its derivatives. *Arch. Biochem. Biophys.*, **31**, 351–356.

DEGRADATION OF LIGNOSULPHONATES BY *STREPTOMYCES* SP.

T. HARAGUCHI and N. MOROHOSHI

Tokyo University of Agriculture and Technology, Tokyo-Fuchu, Japan

Summary

It was observed that lignosulphonate was utilized by a strain of *Streptomyces* (S5) isolated from soil. The lower molecular weight fraction of lignosulphonate was degraded and *p*-hydroxybenzoic, vanillic, and ferulic acids were detected as degradation products. When crude enzyme prepared from culture filtrate of S5 was used, similar results were obtained. When sulphonated dehydrogenation product was incubated with the crude enzyme, its oligomer fraction decreased. It was shown that both the lignin-related dimer having a phenyl coumaran bond and the lignin-related dimer having a β-aryl ether bond were enzymatically attacked, but pinoresinol was not changed. Sulphonated dimers seemed difficult to degrade because the dimers were polymerized to some extent during sulphonation reaction. Some monomers underwent complete breakdown by the same crude enzyme.

INTRODUCTION

Lignosulphonate (LS) is a major component of sulphite pulp waste liquor and is highly resistant to the action of microorganisms. White-rot fungi, which are capable of degrading protolignin in wood, cannot utilize LS (Shibamoto *et al.*, 1960). Ferm and A. Nilsson (1969), however, recognized that over 15% of LS could be removed from the growth medium through cultivation of the yeast genus *Trichosporon*. On the other hand, a mixed population of microorganisms consisting mainly of *Pseudomonas* and *Geotrichum* was found to utilize primarily the low molecular weight portion of LS (Ferm and R. Nilsson, 1970). Selin and Sundman (1971) demonstrated that only low molecular weight components of LS (MW 1000) could be utilized by a mixed population consisting of bacteria and protozoa and observed the colour increase of the LS medium. Using an enrichment technique, Haraguchi (1970) isolated *Streptomyces* (S5), capable of utilizing LS as the sole carbon source. The purpose of this study was to investigate mainly by means of gel filtration the degradation of LS and the related model compounds by mycelium of S5 and the crude enzyme separated from the culture filtrate of S5.

MATERIALS AND METHODS

Lignosulphonate

The LS solution obtained from Kôjin, Saiki Pulp Mill, Japan, was precipitated with β-naphthyl amine and deionized by passing the alkali solution of the precipitate through an Amberlite IR 120 column. The lignosulphonic acid was neutralized with potassium hydroxide. A residual sugar content determined by the Somogy method was about 2%.

Sulphonated dehydrogenation product

The dehydrogenation product (DHP) was prepared by a modification of the method of Harkin (1966/67). Sulphonation of DHP was carried out in such a way that 10 ml of calcium bisulphite solution containing 5·6% of total SO_2 was added to 100 mg of DHP and the mixture was cooked at 137°C for 3 hr. The product was concentrated under reduced pressure and was poured into a large amount of methanol to obtain the preparation of sulphonated DHP.

Lignin-related dimers and monomers

The three kinds of dimers—guaiacylglycerol-β-coniferyl ether (GCE), dehydrodiconiferyl alcohol (DDA), and pinoresinol (PR)—and the three kinds of monomers—*p*-hydroxycinnamic acid (PHCA), ferulic acid (FA), and sinapic acid (SA)—were prepared by the method of Freudenberg and Hübner (1952).

Growth experiment

LS medium had the following constituents: potassium lignosulphonate, 1·0 g; ammonium nitrate, 1·0 g; potassium dihydrogen phosphate, 1·5 g; crystalline magnesium sulphate, 0·5 g; crystalline ferrous sulphate, 0·01 g; water to 1000 ml, pH 7·0. The medium was sterilized by autoclaving. Inoculum was taken from mycelia of S5 cultured in the above medium for more than 1 month. After incubation in a thermostat at 40°C for 1 month, the cells were removed by centrifugation and the remaining LS was fractionated by means of gel filtration. Metabolic products in the aliquots taken from the supernatant containing the remaining LS were also examined.

Preparation of crude enzyme

The culture filtrate of S5 was saturated with ammonium sulphate to 0·8 saturation. The precipitate obtained was dissolved in potassium phosphate buffer solution (pH 7), dialysed against cold water for 24 hr and the buffer solution for 48 hr, and used as a crude enzyme preparation, the protein content of which was 0·16 mg/ml.

Enzymic degradation experiments

The components of reaction mixtures were as follows: crude enzyme solution, 4 ml; 10^{-3} M substrates, 4 ml; and potassium phosphate buffer solution (pH 7), 2 ml. The reaction mixture was shaken at 28°C for 1 hr (for dimers and monomers) to 9 hr (for LS and sulphonated DHP). After incubation, the mixture was made up to the original volume with distilled water, 2 ml of which was applied to the column of gel filtration. As regards sulphonated DHP and lignin model compounds such as GCE, DDA, and PR (dimers) or PHCA, FA, and SA (monomers), aliquots were taken from the reaction mixture and UV spectra were measured after purifying.

Gel filtration

The fractionation of LS, sulphonated DHP, and model compounds according to their molecular size was carried out by means of gel filtration on Sephadex G-25 and G-10 in columns of various size. Detail of column size and eluting solvent is described in the figures below. The effluent was monitored by measuring the absorbance at 280 nm.

Ultraviolet spectrum

The following solvents were used to dissolve samples for measuring UV spectra by Hitachi spectrophotometer 139: dioxane for the degradation products from lignosulphonates; dioxane:water (1:1 v/v) for GCE and DDA; water for sulphonated DHP, PR, and monomers PHCA, etc.

Thin-layer chromatography

Aliquots from the reaction mixture were acidified with sulphuric acid and extracted with ether. After concentration of the ether extract under reduced pressure, a small amount of the concentrated matter was spotted on silica gel plate (Wakogel B-5), and developed by the ascending method with a mixture of toluene:ethyl formate:formic acid (5:4:1 v/v/v). Diazotized sulphanilic acid and a UV lamp (254 nm) were used for detecting spots on the silica gel plate.

Gas chromatography

The ether extract of the reaction mixture was dried in vacuo, dissolved in a small portion of dioxane, and applied to a gas chromatograph, Hitachi 53. Column used: Golay Z-45 at 180°C; carrier gas: nitrogen at a flow rate of 5 ml/min.

Mass spectrometry

Shimazu LKB-9000 was used to determine the degradation products isolated from culture filtrate of LS medium.

RESULTS

Degradation of LS by *Streptomyces* (S5)

When S5 adapted in LS was incubated in the LS medium, the blackish-brown colour of the medium was found to pale. The gel-filtration curves before and after incubation are shown in Fig. 1. Viscosity-average molecular weight of fraction 1 is 80 600 and those of fractions 2, 3, and 4 are 33 800, 13 200, and 6700, respectively. The change in fraction 2 was remarkable but that in fraction 1 remained almost unchanged. Three compounds (I, II, and III) were isolated from culture filtrate. I and II were identified as *p*-hydroxybenzoic acid and vanillic acid, respectively, by comparing their R_F values of thin-layer chromatogram, UV spectra, gas chromatogram (Fig. 2), and mass spectra with those of the authentic compounds: TLC R_F I: 0·45, II: 0·53; UV λ max nm I: 253, II: 264, 290; Ms m/e I: 166 (M^+), II: 196 (M^+). From the retention times of gas chromatograms (Fig. 2), compound III was presumed to be ferulic acid.

Degradation of LS by the crude enzyme prepared from S5

Figure 3 shows that the crude enzyme solution prepared from culture filtrate of S5 degraded slightly fraction 2 LS, though the remarkable degrada-

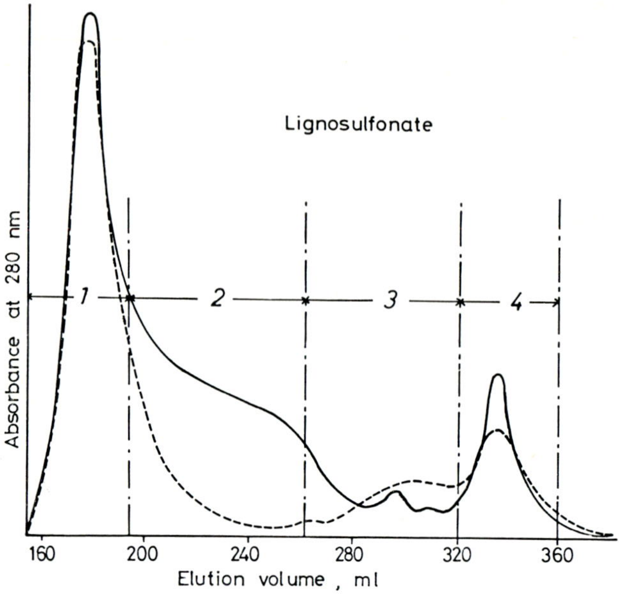

Fig. 1. Gel filtration of lignosulphonate through a Sephadex G-25 column 2·5 cm in diameter and 150 cm long. ——— *control,* – – – – *cell-free supernatant after growth of* Streptomyces *(S5).*

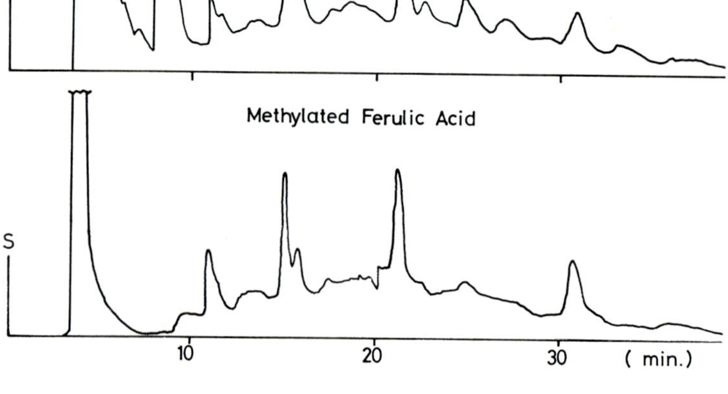

Fig. 2. Gas chromatogram of ether extracted sample of lignosulphonate after growth (above) and ferulic acid (below). Both sample and ferulic acid were methylated prior to use. (I): p-*hydroxybenzoic acid, (II): vanillic acid.*

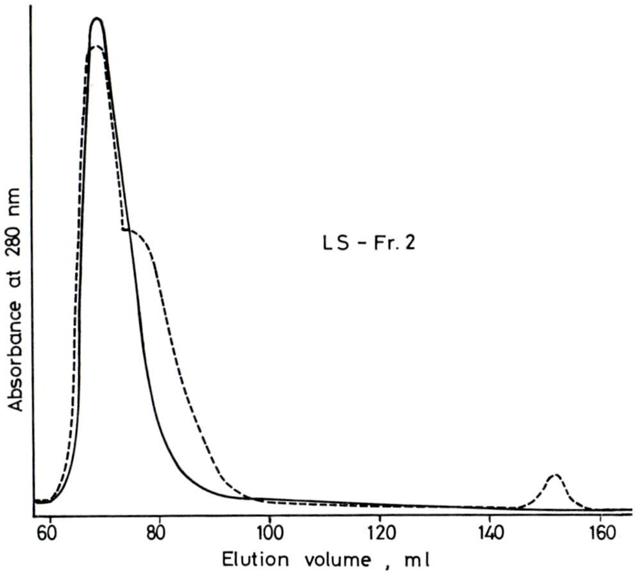

Fig. 3. *Gel filtration of lignosulphonate-fraction 2 through a Sephadex G-25 column 3 cm in diameter and 40 cm long.* ——— *control,* – – – – *after incubation with crude enzyme prepared from culture filtrate of S5.*

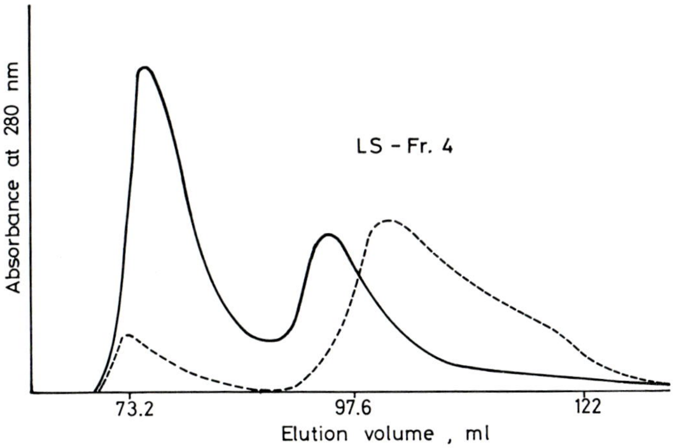

Fig. 4. *Gel filtration of lignosulphonate-fraction 4 through a Sephadex G-10 column 3 cm in diameter and 40 cm long.* ——— *control,* – – – – *after incubation with crude enzyme.*

tion as shown in the living cell was not observed. It can be seen from Fig. 4 that a considerable amount of LS was depolymerized, and compounds I and II were also isolated from the reaction mixture at the end of incubation.

Degradation of sulphonated DHP by the crude enzyme

Sulphonated DHP as a model compound of LS was incubated with the crude enzyme. In this case, as can be seen from Fig. 5, depolymerization as well as a tendency to polymerize was observed.

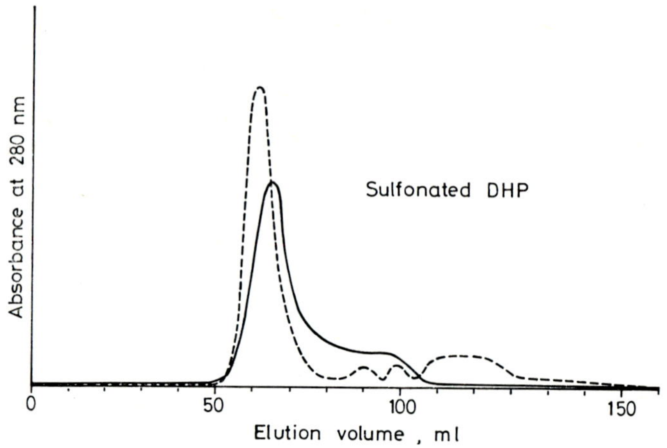

Fig. 5. Gel filtration of sulphonated DHP through a Sephadex G-10 column 1·6 cm in diameter and 76 cm long. Elution with dioxane–water (1:1). ——— control, - - - - after incubation with crude enzyme.

Degradation of lignin model compounds—dimers—by the crude enzyme

Dimers containing β-aryl ether, phenyl coumaran, and pinoresinol bonds, *i.e.* GCE, DDA, and PR, were incubated with crude enzyme respectively in order to make clear which bond is the most susceptible to enzymic attack. The results are summarized in Figs. 6 and 7. GCE and DDA changed their UV spectra during incubation, especially at 280 nm and 340 nm. That some changes in molecular size occurred is seen from the gel filtration curve in Fig. 7. GCE is supposed to change into polymerized compounds and DDA mainly into depolymerized ones, one of which was confirmed to correspond

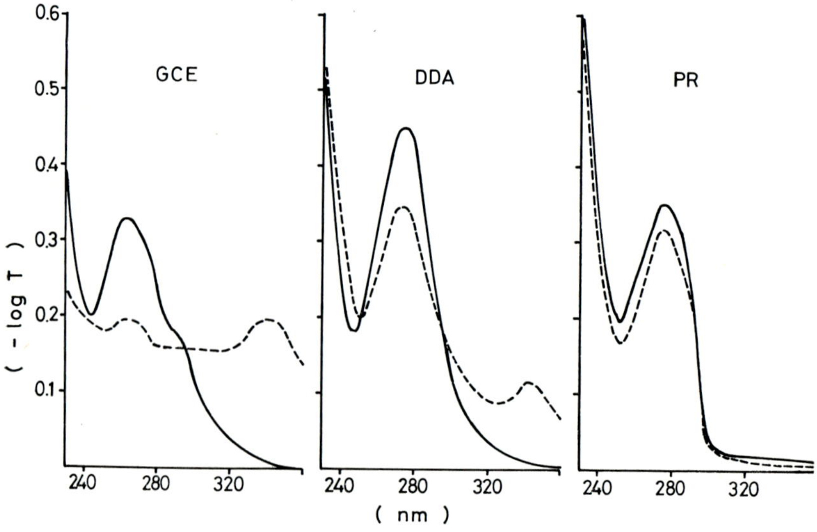

Fig. 6. UV spectra of lignin compounds—dimers. GCE: guaiacylglycerol-β-coniferyl ether, DDA: dehydrodiconiferyl alcohol, PR: pinoresinol. ——— control, - - - - after incubation with crude enzyme.

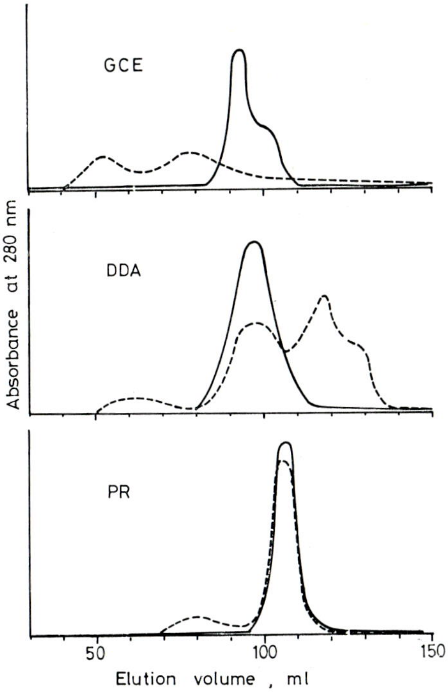

Fig. 7. *Gel filtration of lignin model compounds—dimers. GCE: guaiacylglycerol-β-coniferyl ether, DDA: dehydrodiconiferyl alcohol, elution with dioxane–water (1:1) on a Sephadex G-10 column 1·6 cm in diameter and 76 cm long. PR: pinoresinol, elution with water on a Sephadex G-10 column 2·8 cm in diameter and 32 cm long. ——— control, – – – – after incubation with crude enzyme.*

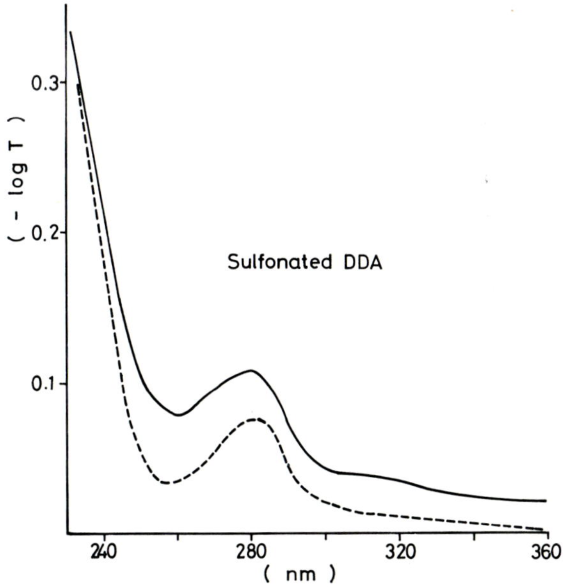

Fig. 8. *UV spectra of sulphonated dehydrodiconiferyl alcohol. ——— control, – – – – after incubation with crude enzyme.*

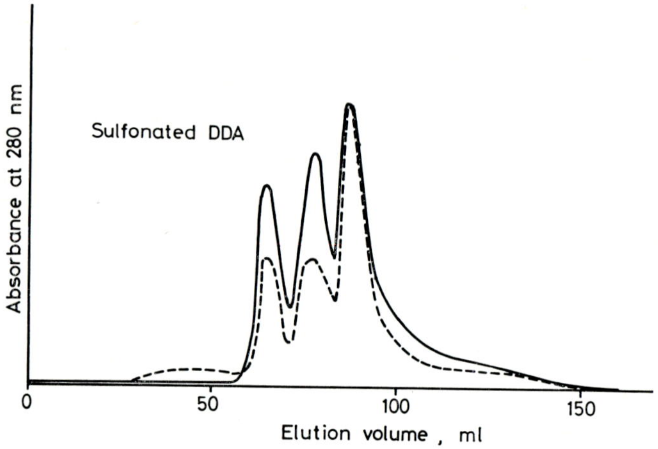

Fig. 9. *Gel filtration of sulphonated dehydrodiconiferyl alcohol, elution with water on a Sephadex G-10 column 2·8 cm in diameter and 39 cm long. ——— control, - - - - after incubation with crude enzyme.*

to a monomer by means of gel filtration. In PR, only a slight polymerization occurred. Although sulphonated samples of these three dimers were also subjected to enzymic degradation, no remarkable change was observed. One example is shown in Figs. 8 and 9.

Degradation of lignin model compounds monomers by crude enzyme

PHCA and FA were almost completely degraded, but SA was not (Figs. 10

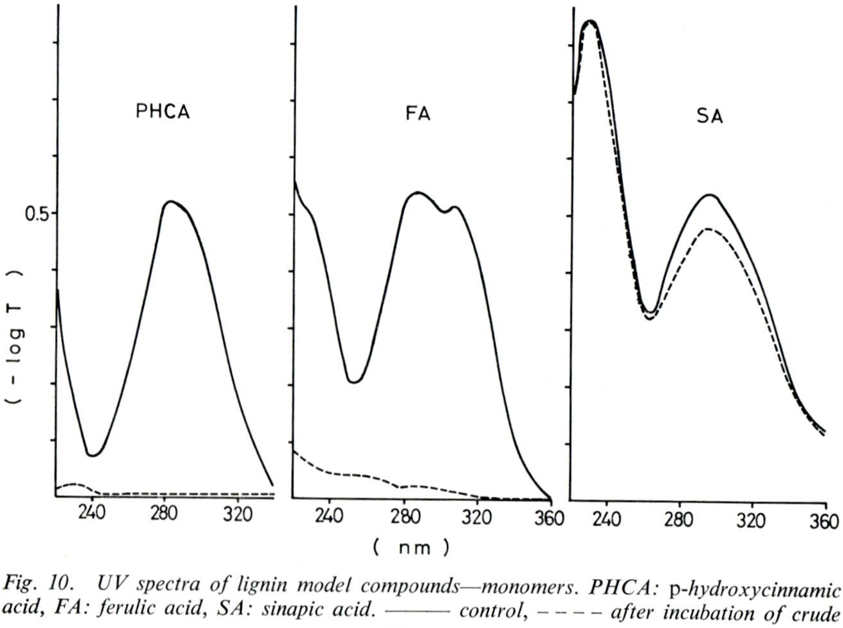

Fig. 10. *UV spectra of lignin model compounds—monomers. PHCA:* p-*hydroxycinnamic acid, FA: ferulic acid, SA: sinapic acid. ——— control, - - - - after incubation of crude enzyme.*

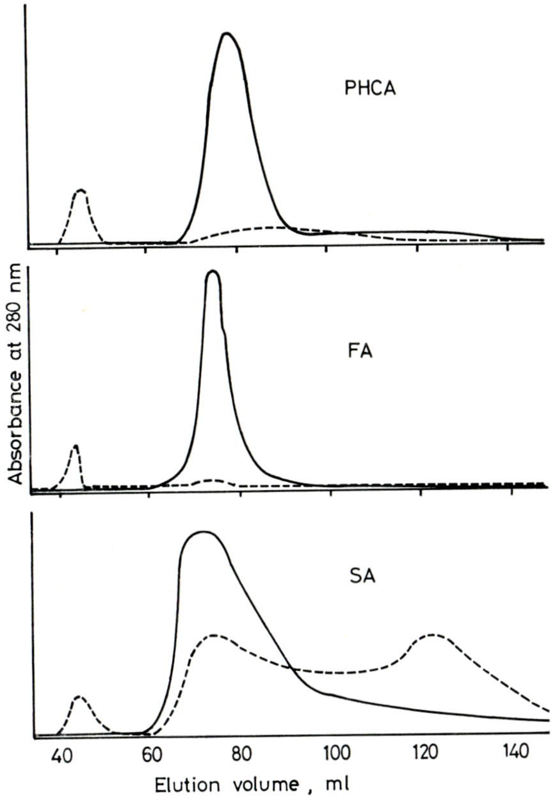

Fig. 11. Gel filtration of lignin model compounds—monomers. PHCA: p-*hydroxycinnamic acid, FA: ferulic acid, SA: sinapic acid, elution with water on a Sephadex G-10 column 3 cm in diameter and 32 cm long.* ——— *control,* – – – – *after incubation with crude enzyme.*

and 11). In each case, a small amount of polymerized fraction was detected. Results mentioned above are summarized in Fig. 12.

DISCUSSION

Compared with a living cell, the crude enzyme showed only a slight degradation activity to LS. It was, however, shown that the activity to low molecular weight substrates such as GCE, DDA, PHCA, and FA was strong.

The polymerization which was observed in the enzymic reaction of these dimers and monomers may be considered to occur as a result of the recombining with unstable or activated intermediates liberated by the enzyme action.

Since a monomer fraction was recognized by gel filtration in the enzymic degradation of DDA, the phenyl coumaran ring is considered to be ruptured.

As Nimz (1966) pointed out, biosynthesis of GCE from coniferyl alcohol is accompanied by production of 1,2-diguaiacyl propane-diol-1,3. Therefore, the GCE preparation used here was a mixture of GCE and diaryl compounds. In such a preparation, the enzyme might break down both β-aryl ether and

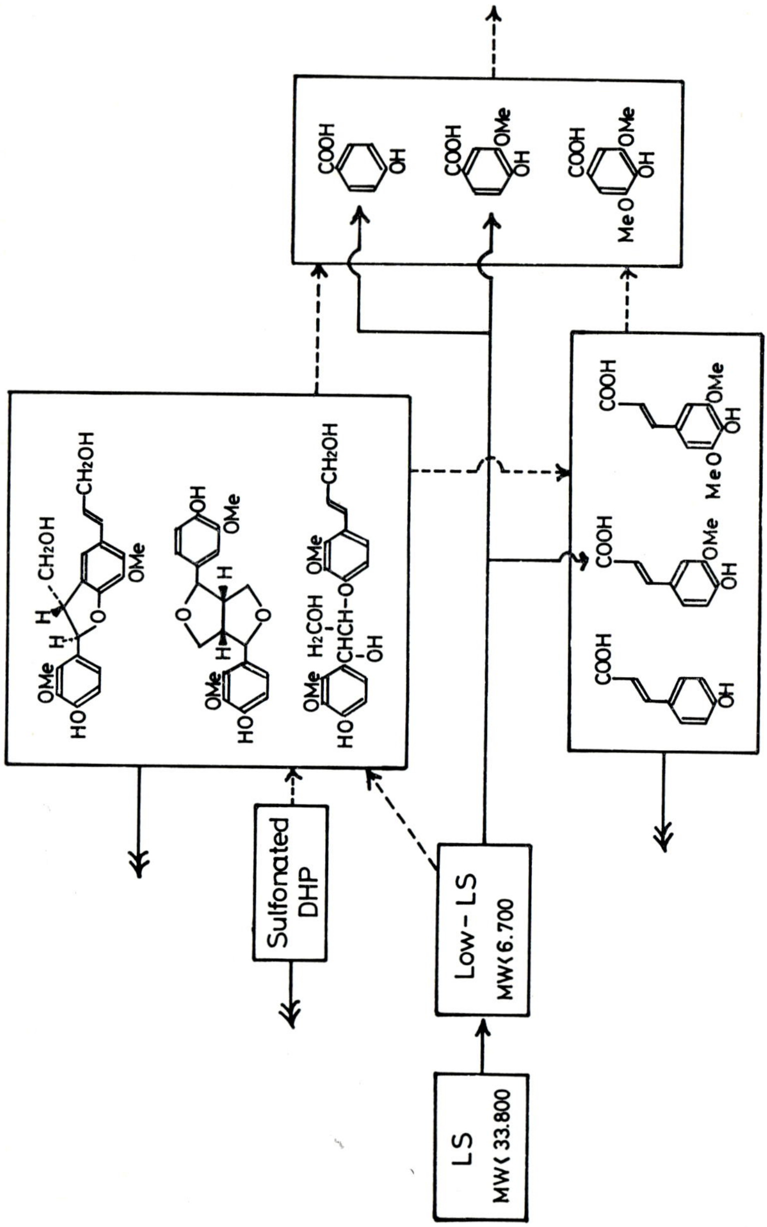

Fig. 12. *Degradation of lignosulphonate by* Streptomyces *(S5).* → *Degradation (demonstrated),* - - → *Degradation (postulated),* ↞ *Polymerization (demonstrated).*

β-1 bonds, and produce various degradation products. Polymerization observed in this case is presumably a result of a complex reaction among the degradation products.

The reason why sulphonated DDA was not easily degraded may be explained from the fact that DHP was apparently polymerized during the sulphonation reaction (cf. Figs. 7 and 9).

REFERENCES

Ferm, R., and Nilsson, A.-C. (1969). *Svensk Papperstidning*, **72**, 531–536.

Ferm, R., and Nilsson, R. (1970). *Svensk Papperstidning*, **73**, 283–286.

Freudenberg, K., and Hübner, H. H. (1952). *Chem. Ber.*, **85**, 1181–1191.

Haraguchi, T. (1970). Proceedings of the 15th Symposium on Lignin Chemistry 8–9 October, Sapporo, Japan. pp. 17–20.

Harkin, J. M. (1966/67). Unpublished work. In *The Constitution and Biosynthesis of Lignin* by K. Freudenberg and A. C. Neish. 1968. Springer-Verlag, Berlin. p. 84.

Nimz, V. H. (1966). *Holzforschung.*, **20**, 105–109.

Selin, J.-F., and Sundman, V. (1971). *Finska Kemists Medd.*, **80**, 11–19.

Shibamoto, T., Fukuzumi, T., Mikawa, H., and Hayashi, M. (1960). *Mokuzai Gakkaishi*, **6**, 135–140.

DESIGN, SYNTHESIS AND DEGRADATION OF POLYMERS SUSCEPTIBLE TO HYDROLYSIS BY PROTEOLYTIC ENZYMES

S. J. HUANG, J. P. BELL, J. R. KNOX, H. ATWOOD, D. BANSLEBEN, M. BITRITTO, W. BORGHARD, T. CHAPIN, K. W. LEONG, K. NATARJAN, J. NEPUMUCENO, M. ROBY, J. SOBOSLAI and N. SHOEMAKER

Departments of Chemistry and Chemical Engineering, Biological Sciences Group and Institute of Materials Science, University of Connecticut, Storrs, Connecticut 06268, U.S.A.

Summary

Phenyl-substituted polyamide, polyester, poly(amide-urethane), and poly(ester-urea) derived from mandelic acid, phenylalanine, and benzylmalonic acid were synthesized and were found to be biodegradable. Modification of biodegradable poly(caprolactone) and gelatin with 1,6-diisocyanatohexane gave new biodegradable materials of improved physical properties.

INTRODUCTION

Recent literature in the field of biodegradability of synthetic polymers is concerned mostly with the problem of preventing or retarding attack on synthetic polymers and plasticizers by microorganisms (Kaempf *et al.*, 1973; Monk, 1972; Osman *et al.*, 1971; Rodriques, 1971; Tirpack, 1970; Darby and Kaplan, 1968; Wassell, 1964; Jen-Hao and Schwartz, 1971; Hueck, 1960). Potts *et al.* (1973) reported in 1973 an extensive Union Carbide Corporation study on the biodegradability of synthetic polymers by microorganisms such as fungi and bacteria. They reported that aliphatic poly(esters) and urethanes were the only class of synthetic high molecular weight polymers found to be biodegradable in a study embracing a large number of synthetic polymers such as poly(esters), poly(amides), poly(ethers), and poly(olefins). Unfortunately, the degradable aliphatic poly(esters) generally lack the desirable physical properties necessary for many applications. The microbial degradation of cellophane and amylose films has been studied by Carr and co-workers (Engler and Carr, 1972; Bradley *et al.*, 1973).

It has been reported by several workers that low molecular weight normal paraffins and low molecular weight poly(ethylene) supported some microbial growth (Potts *et al.*, 1973; Merdinger and Merdinger, 1970; Barna *et al.*, 1970; Miller and Johnson, 1966). Efforts have been directed toward the synthesis of photodegradable polymers so that segments from photodegradation of polymers might become biodegradable (Guillet, 1973; Cooney *et al.*, 1973; Scott, 1973). Attempts have also been made to improve the degradability of polymer composites by mixing biodegradable biopolymers such as starch with nondegradable synthetic polymers (Griffin, 1973).

Currently, there are very few reported studies on truly biodegradable synthetic polymers that have the desirable physical properties necessary for the production of a wide range of biodegradable plastics, films, and fibres. Bailey and co-workers have discovered that a regular alternating co-polymer of glycine and ε-aminocaproic acid (nylon 2/nylon 6) was metabolized by microorganism (Ennis and Kramer, 1975). It has been reported by Gilbert *et al.* (1974) that Borax 210 [a poly(acrylonitrile) based resin] was utilized by the microorganism *Penicillium notation.* Poly(enamines) are also reported to be degradable (Kim, 1974). Biodegradable polymers have been obtained from the modification of cellulose (Kim *et al.*, 1973). Most of these studies involved soil burial tests or culturing on agar-immersed samples. While our degradation testing includes these older methods, we have developed procedures for exposing polymer materials to concentrated enzyme solutions in order to accelerate the degradation reaction and to be able to establish the chemistry and kinetics of a single reaction in the absence of side reactions originating in the microorganism.

It became apparent from the above-mentioned literature that useful biodegradable polymers will have to come from new synthetic approaches designed to incorporate biodegradable structural units into polymer chains having desirable physical properties. Our efforts have been directed toward two areas. First, we are designing, synthesizing, and testing new polymers that might be biodegradable and have the desirable physical properties. New polymers that have been studied are polymers derived from α-aminoacids and α-hydroxyacids as well as substituted nylons. Second, we are modifying biodegradable polymers to improve their physical properties so that useful biodegradable plastics, films, and fibres can be produced. The polymers that have been studied are poly(caprolactone) and gelatin.

BIODEGRADATION METHODS

Two methods were used to test the biodegradability of polymer samples: (1) degradation by enzymes, and (2) degradation by fungi.

1. Degradation by enzymes

A. Exposure of materials to enzymes

After the synthesis and purification of a new material, a degradation study was initiated as follows: into 16 ml corrosion-resistant borosilicate glass vials were added 100–200 mg of powdered polymer (depending on the availability of the polymer). Samples were run in duplicate for each condition. The vials for the samples to be subjected to enzyme attack were filled with 14 ml of 0·2 M buffer solution at the appropriate pH for the enzyme used. Azide was added to prevent microbial contamination. Then 1 ml of enzyme solution at about 1 mg/ml concentration was added. Two controls were used for each sample: the first contained the polymer plus the buffer (15 ml) and was used to determine the amount of polymer leached or hydrolysed by the buffer alone. The second contained buffer (14 ml) and enzyme solution (1 ml) to serve as a blank for the amine end-group determination on the supernatant liquor of the treated sample.

The sample plus controls were placed on a shaker table at ambient temperature for 6–10 days. During this period, fresh enzyme was added to the sample and the enzyme-buffer control every 24 or 48 hr. One millilitre was withdrawn and saved, while 1 ml of the enzyme solution at the original concentration was added.

At the end of the last day of the reaction, the solids (if present) were filtered in the predried, tared, sintered glass or cellulose filters. The filtrate was saved and the remaining solid washed with at least 50 ml of distilled water. The solids were then air dried for 24 hr, then oven dried for 2 hr at 60°C. The filters and solids were then weighed and the weight loss for the sample determined. Solids were kept for viscosity and molecular weight measurements. The filtrates (from both the sample and the material removed during enzyme replenishment) were analysed for amine end groups using either a colorimetric ninhydrin assay or a titrimetric assay with HCl.

B. Enzyme assays

It was necessary to monitor enzyme activity as a function of time and temperature to determine how frequently the enzyme must be replenished during the 10-day runs. Standard methods were used to assay activities using known substrates. We also verified that the enzymes were not inactivated by physical adsorption to our samples.

C. Analytical procedures for measuring degradation

(a) *Weight loss.* By weighing the solid polymer before and after exposure to enzyme, a simple measure of degradation is possible.

(*b*) *Amino end-group analysis.* The NH_2 groups generated by enzyme cleavage of amide HN—CO bonds can be detected in filtrates or even in the remaining solids if a suitable solvent is available. Two procedures have been used: titration with acid and colorimetric analysis with ninhydrin reagent.

(*c*) *Molecular weight changes of polymer samples.* Molecular weight changes of polymer samples are measured by viscosity and vapour pressure osmometry.

2. Degradation by fungi

Four cultures were used: *Aspergillus niger*, *Aspergillus flavus*, *Penicillium funiculosum*, and *Chaetomium globosum*. Fungal growth with polymer as the only carbon source was viewed as an indication of polymer degradation.

A. Agar plates

A standard technique for testing fungal growth on polymers is provided in ASTM method D1924-61T. The procedure requires three plates: polymer, filter paper, and a control, all using non-nutrient agar devoid of utilization carbon. The filter paper control should show heavy growth indicating fungal viability, while the blank control should show no growth on the agar medium itself.

B. Spore suspensions

To confirm further that these moulds are assimilating the polymer and not just the residual sugars of the agar, another experiment was initiated. Polymer strips, cut from polymer plates pressed by ram compression, were inserted half way into a spore suspension. No agar was used, but basal salts were in the suspension. Two controls were used: a piece of sterile filter paper was used for a spore viability control; and a polymer strip in basal salts was used to show whether any degradation occurred due to the salt water. Samples were done in duplicate or triplicate at 30 $\pm$ 2°C.

RESULTS

1. New polymers

A. Benzylated nylons

Our initial testing of nylon-6 and nylon-6,6 showed that these polymers resist enzyme hydrolysis. Since several enzymes (chymotrypsin and pepsin, for examples) are specific in cleaving peptide linkages adjacent to aromatic groups, we decided to synthesize α-benzylated nylons, anticipating that the introduction of the α-benzyl group would make the nylons more susceptible to enzyme cleavage.

(*a*) *Nylon-n,*6 *Bz series.* Nylon-n,6 Bz was synthesized according to Scheme 1.

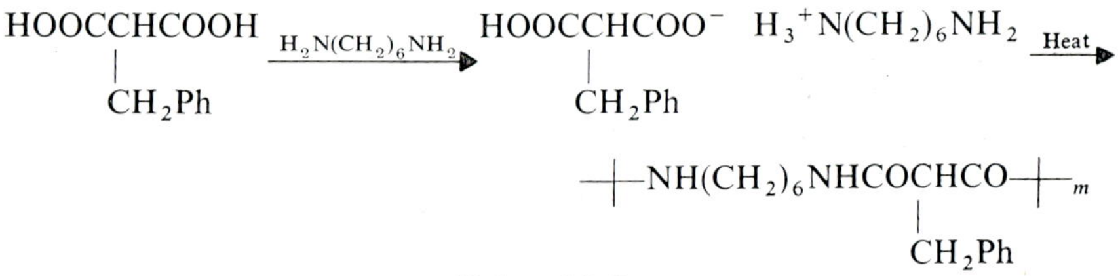

Nylon-6,3 Bz

Scheme 2. Synthesis of Nylon-6,3 Bz

The nylon-n,6 Bz polymers thus synthesized have M_n (by amino end group titrations) of the order of 2×10^3 with softening points: 2,6 Bz, 145°; 4,6 Bz 140°; 6,6 Bz, 120°, and 8,6 Bz, 105°. Enzyme degradation studies using chymotrypsin, subtilisin, thermolysin, pepsin, and elastase were all negative.

(*b*) *Nylon*-6,3 *Bz*. Nylon-6,3 Bz, prepared according to Scheme 2 has M_n ranging from 10^3 to 10^4, dependent upon the polymerization conditions. Since nylon-6,3 Bz contains two 'phenylalanine-like' amide linkages quite different from that of the nylon-n,6 Bz series, it might have different degradability compared to that of the nylon-n,6 Bz series.

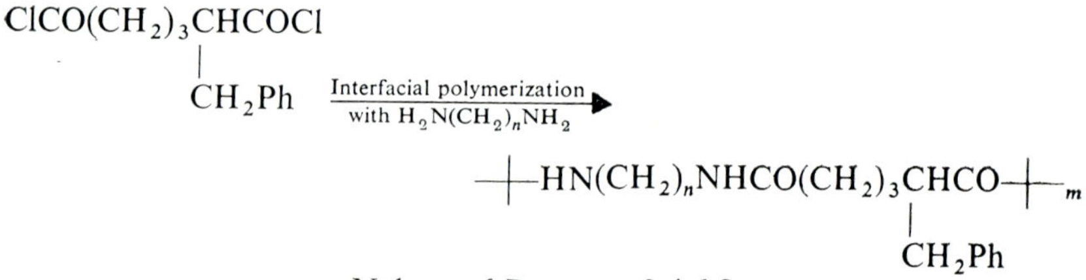

Nylon-n,6 Bz $n = 2,4,6,8$

Scheme 1. Syntheses of Nylon-n, 6 Bz

A sample of nylon-6,3 Bz after exposure to imidazole-buffered chymotrypsin showed relatively high weight loss (42% in buffer and 47% in buffered chymotrypsin) as compared with the negligible weight loss results of that of the nylon-n,6 Bz series.

B. Poly(ester-urea) derived from α-amino acids

(*a*) *Phenylalanine containing poly(ester-urea)*. Since chymotrypsin is known preferentially to cleave proteins at the phenylalanine site and also polyesters generally are found to be utilized by microorganisms (Potts *et al.*, 1973; Fields *et al.*, 1974), we decided to explore the possibility of preparing a poly(ester-urea) containing phenylalanine residues. We first prepared a model di-ester-di-urea, C_1-EsPheUa-C_6-UaPheEs-C_1, according to Scheme 3.

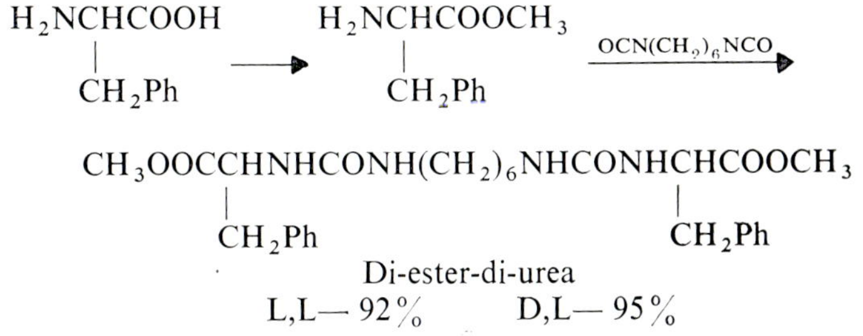

Scheme 3

Degradation studies on the model di-ester-di-urea were then carried out using various enzymes. Chymotrypsin, elastase, papain, and urease were found to degrade the di-ester-di-urea, with chymotrypsin being the most effective enzyme.

Low M_n poly(ester-urea), copoly(L-phenyl-alanine/ethylene glycol/1,6-diisocyanate hexane was prepared from ethylene glycol phenyl-alanine ester according to Scheme 4.

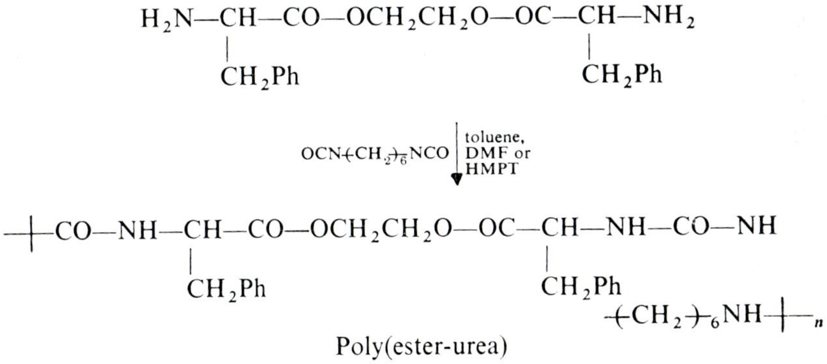

Scheme 4

The polymer thus prepared has M_n between 1930 and 2640 as determined by vapour-phase osmometry with mp 194–198°C. Substantial degradation was observed after the polymer (M_n 1930) was exposed to buffered chymotrypsin at room temperature for 10 days (Table I). A 19·3% weight loss was observed together with the detection of 23·7% of phenylalanine residue in the soluble fragments. The fact that 87% of the added enzyme was found to be absorbed on the remaining insoluble polymer (which can be washed out by saline solution) provided proof that there was interaction between enzyme and polymer. The fact that little amino end-group increase was observed is in

Table I
DEGRADATION OF POLY(UREA-ESTER) BY CHYMOTRYPSIN[a]

Run	1	2	3	*Average*
% weight loss[b]	19·2	19·4	19·2	19·3 ± 0·1
% Enzyme[c] absorbed on solid polymer	85·8	87·0	88·2	87 ± 1·2
% Phe in soluble[d] fragments	23·1	22·2	24·8	23·7 ± 1·5
Increase in —NH_2, 10^{-4} m mol[e]	7·78	7·77	7·80	7·78 ± 0·3

[a] After exposure to buffered chymotrypsin at room temperature for 10 days. Data corrected over controls (polymer and buffer).
[b] Corrected for absorbed enzyme.
[c] Determined by u.v.
[d] Determined by u.v.
[e] This represents less than 1% of —NH_2 group in the polymer.

agreement with the known specific acyl-cleavage activity of chymotrypsin—only ester linkages in the polymer should be hydrolysed, generating no new amino groups.

It is interesting to compare the degradation of the polymer with that of the model di-ester-di-urea (Table II). Under the same conditions, 79·9% weight

Table II
RESULTS OF CHYMOTRYPSIN DEGRADATION OF L,L-DI-ESTER-DI-DIUREA AND L-POLY(ESTER-UREA)

Substrate	M_n	*No. ester linkages*	*Degradation span (days)*	*% weight loss*[b]
Monomer	526[a]	2	10	79·9 ± 0·2%
Polymer	1930	8	10	19·3 ± 0·1%

[a] $\overline{M}_n = \overline{M}_w$.
[b] Weight loss differences from exposure to buffered enzyme solution compared with that of exposure to buffer only.

loss was observed for the model compound whereas 19·3% weight loss was observed for the polymer, which contains four times as many ester linkages as the model compounds. This suggests that the polymer shows identical degradability as the model compounds.

(*b*) *Poly(ester-urea) derived from glycine.* A similar polymer was prepared with glycine in place of phenylalanine. However, degradation studies with enzymes and *Aspergillus niger* to date showed no appreciable biodegradability for the unsubstituted poly(ester urea).

C. Polymers derived from α-hydroxyacids

Aliphatic polyesters are found to be generally degradable by microorganisms (Potts *et al.*, 1973; Fields *et al.*, 1974). Poly(glycolate) is used as sutures (Schmitt *et al.*, 1968; Chirgwin, 1968, 1969). Polyurethanes derived from polyesters are found to be more degradable by microorganisms than those derived from polyethers (Darby and Kaplan, 1968). It was reasoned that homopolymer and co-polymers derived from mandelic acid (phenylglycolic acid) should be degradable by enzymes.

(*a*) *Poly*(*mandelate*). Poly(mandelate), prepared by thermal polymerization, was found to be degradable by subtilisin (Table III).

Table III
RESULTS OF DEGRADATION STUDIES ON POLY(MANDELATE)[a]

Enzyme used (0·1 *mg/ml*)	*Buffer, pH*	*Weight loss* %[b]
Subtilisin	phosphate, 6, 7, 8	3 at pH 8
Axion detergent[c]	phosphate, 8, 10	40
Chymotrypsin	imidazole, 7·5	<1

[a] Data are averages from duplicates or triplicates.
[b] Weight loss difference from exposure to buffered enzyme solution compared with that from exposure to buffer only.
[c] It is believed that Axion contains subtilisin and surface active agents. The large difference between Axion and buffered subtilisin might be due to surface active agents that will loosen up the polymer.

Poly(mandelate) was also found to support the growth of *Aspergillus niger*.

(*b*) *Poly*(*amide-urethane*) *from mandelic acid*. Since polymers containing aliphatic ester main chains are generally low melting and have wax-like physical properties, we decided to prepare co-polymers containing ester as well as amide, urea, and urethane linkages in the main chain. The introduction of H-bond forming linkages should increase the melting points of polyester and perhaps also the fibre-forming properties of the polymer.

We have prepared a poly(amide-urethane) from mandelic acid and 1,6-dicyanatohexane with M_n 7500 and mp 110–115°C.

$$\left[\begin{array}{c}\text{OCCHOOCNH}(\text{CH}_2)_6\text{NHCOOCHCONH}(\text{CH}_2)_6\text{NH}\end{array}\right]_m$$

(with Ph substituents on each CH)

Poly(amide-urethane)

Samples of the polymer became soluble after one day exposure to buffered (pH 8·8) elastase at room temperature. At the end of the 14-day exposure period, 36% amino end-group increase (by ninhydrin analysis) was observed, indicating extensive degradation. Samples of the polymers were converted into viscous gels after 14-day exposure to subtilisin (buffered at pH 8), again indicating degradation. Extensive fungal growth was observed when polymer samples were exposed to *Aspergillus niger*, indicating microbial degradation of the polymer.

(*c*) *Poly*(*amide-urethane*) *derived from glycolic acid*. Similar poly(amide-urethane) prepared from glycolic acid was found to be nonbiodegradable by enzymes and *Aspergillus niger*.

2. Modification of degradable polymers

A. Poly(*caprolactone*)

Poly(caprolactone) was reported to support a mixture of four fungi: (*Aspergillus niger*, *Aspergillus flavus*, *Penicillium funiculosum*, and *Chaetomium globosum*) growth (Potts *et al.*, 1973) and is degradable by *Pullularia pullulans*.

We found that when used separately, *Aspergillus flavus*, *Penicillium funiculosum*, and *Chaetomium globosum* degraded poly(caprolactone) whereas *Aspergillus niger* was not effective. A 32 % viscosity M_w decrease was observed when the polymer was exposed to acid protease from *Rhizopus chinensis* for 10 days. All these results show that poly(caprolactone) is biodegradable. However, its poor physical properties (mp 60°, low tensile strength) limit the use of this polymer in a wide range of applications. We have converted poly(caprolactone) into a new co-polymer containing ester, amide, and urethane linkages—co-poly(caprolactone/ethanolamine/1,6-diisocyanato hexane) according to Scheme 5.

$$H{-}[O(CH_2)_5CO]_n{-}OH$$

Poly(caprolactone)

$$\Big\downarrow \; H_2NCH_2CH_2OH$$

$$H\,[\,O(CH_2)_5CO]_xNHCH_3CH_2OH$$

$$\Big\downarrow \; OCN(CH_2)_6NCO$$

$$-\!\!\!-\!\!\!-[O(CH_2)_5CO]_xNHCH_2CH_2OOCNH(CH_2)_6NHCO]\!\!-\!\!\!-_m$$

Co-poly(caprolactone/ethanolamine/1,6-diisocyanatohexane)

Scheme 5

This procedure increased the nitrogen controgen content of the finished polymer to 1·8 %, indicating that one of each 19 caprolactone units was reacted to the diisocyanate. Only a very slight melting peak at 320–330°K, due to residual crystalline polycaprolactone, was observed.

The polymer from this procedure was extruded into a monofilament, using an Instron capillary rheometer and a variable speed windup. Below 95°C the melt seemed inhomogeneous; the extrusion temperature used was 97°C. The monofilament thus produced could be drawn by hand. The ultimate percentage elongation of the drawn fibre was approximately 350 %, and the undrawn, 1000 %. The strength of the drawn fibre was in excess of 10 000 psi, although exact figures were not obtained due to nonuniformity of the cross-sectional area. The original poly(caprolactone) has tensile strength well below 2000.

The co-polymer was found to support growth of *Aspergillus flavus* and *Penicillium funiculosum* but not *Chaetomium globosum* and *Aspergillus niger*. Ten-day exposure to urease resulted in 25 % degradation (measured by viscosity changes and ninhydrin analysis of amino end-group increase). Renin, an enzyme produced by *Aspergillus* bacteria strains, also degraded the co-polymer but not as effectively as urease.

B. Cross-linked gelatin

Gelatin is an animal protein by-product which is available in quantity at relatively low cost. It has many commercial uses, among which is its use as the main component of photographic emulsions. Gelatin has been cross-linked with many reagents—usually aldehydes (Sinclair, 1973; Veis, 1964; Anderau *et al.*, 1969; Young and Luce, 1967; Courts, 1967; Hafter and

Hoermann, 1970; Taylor, 1971; Nishio *et al.*, 1967; CIBA, 1969; Tabor, 1968; Thies *et al.*, 1968; Coops, 1970). Our goal has been to produce a material which is cross-linked, water insoluble, yet pliable enough to be used as fibres, ribbons, or sheets. The polymer would at the same time be easily biodegradable since the main component is protein. We have succeeded in cross-linking gelatin using different conditions: interfacial cross-linking, bulk cross-linking, and cross-linking of drawn fibres. Cross-linked, drawn fibres were prepared by blend glycols or glycerine with gelatin prior to extrusion and cross-linking. All the cross-linked gelatin samples were found to be degradable. Results are listed in Tables IV and V.

Table IV
DEGRADATION OF MODIFIED GELATINS: UNDRAWN BULK MATERIAL[a]

Gelatin polymer	*Enzyme used*	*Length of soluble fragments (as number of monomer units)*[b] *in enzyme solution*	*in buffer control*
Native uncross-linked	papain	7–9	70–80
	subtilisin	7–11	85
	urease	no degradation	
	trypsin	no degradation	
Interfacially cross-linked	papain	9–11	10–12
	subtilisin	4–6	7–9
	urease	no degradation	
	trypsin	10–12	10–12
Solid-state bulk cross-linked	papain	20–22	140
	subtilisin	5–7	8–10
	urease	no degradation	
	trypsin	17–19	100
Hot-water washed bulk cross-linked	papain	7–9	550
	subtilisin	7–9	1200
	urease	no degradation	
	trypsin	no degradation	

[a] Cross-linked with 1,6-diisocyanato.
[b] Number of monomer units in the original gelatin chain is estimated at 250.

DISCUSSION AND CONCLUSION

Most known biopolymers can be hydrolysed by specific enzymes normally present in living organisms. Proper design of a polymer to more closely resemble the essential molecular architectural ingredients of biopolymers is expected to increase the susceptibility of the polymer to enzyme hydrolysis. Polymers degraded by enzymes would have an added advantage in that the degradation fragments are of a narrow molecular weight distribution and potentially reclaimable. Much of the specificity associated with proteases and esterases results from a preference for recognizable polar or hydrophobic groups adjacent to the amide or ester linkages at which the enzyme attacks. Chymotrypsin and subtilisin enzymes preferentially attack amide and ester linkages adjacent to hydrophobic aromatic substituents. Since the structures and mechanism of the catalytic activities of these easily available enzymes are reasonably well understood, we chose to design, synthesize, and test the degradability of new polymers derived from phenyl group bearing acids (mandelic acid, phenylalanine, and benzyl malonic acid), anticipating that these polymers will be degradable by selected enzymes. Our results show that phenyl substituted polyamide, polyester, poly(ester-urea), and poly(amide-urethane) are biodegradable and the unsubstituted analogues are not degradable. The fact that nylon-6,3 Bz is biodegradable whereas nylons-n,6 Bz are not suggests that biodegradability also might be dependent on the length of the chain (number of atoms) between neighbouring hydrolysable functional groups. Preliminary results also show that condensation polymers with substituents other than phenyl (methyl and hydroxy) are also biodegradable although more work has to be done before any generalized conclusion can be drawn.

Table V
DEGRADATION OF MODIFIED GELATINS: DRAWN FIBRES

Gelatin fibre gel:plast:H_2O[a]	*Enzyme used*	*Length of soluble fragments*[c] *(as number of monomer units)*	
		in enzyme solution	*in buffer control*
Uncross-linked	papain	5–7	6–8
20:10:10	subtilisin	8–10	30
polyethylene glycol	urease	66	85
Cross-linked[b]	papain	7–9	28
20:10:10	subtilisin	8–10	35
polyethylene glycol	urease	48	90
Uncross-linked	papain	5–8	6–9
20:10:10	subtilisin	7–9	30
glycerine	urease	66	72
Cross-linked	papain	5–8	32
20:10:10	subtilisin	7–9	35
glycerine	urease	36	68
Uncross-linked	papain	4–6	5–7
20:8·6:9·6	subtilisin	6–8	30
ethylene glycol	urease	86	90
Cross-linked	papain	4–6	38
20:8·6:9·6	subtilisin	6–8	28
ethylene glycol	urease	50	58

[a] Ratio of gelatin: plasticizer: H_2O.
[b] The cross-linking agent was 3% diisocyanatohexane in toluene, 69°C.
[c] After 7-day exposure.

Modification of poly(caprolactone) and gelatin with 1,6-diisocyanatohexane greatly improved the physical properties of the polymers without substantially changing their biodegradabilities.

Our results show that it is feasible to design and synthesize biodegradable polymers having a variety of structural units and properties from easily available starting materials. The potential use of these materials in manufacturing of slow release drugs, insecticides, and fertilizers should be interesting.

Acknowledgement

Financial supports from U.S. Army Natick Research Laboratories and the University of Connecticut Research Foundation are gratefully acknowledged.

REFERENCES

Anderau, W., Walter, O., Oetiker, A., and Deuschel, W. (1969). Ger. Offen. 1,914,955.

Bailey, W., private communication.

Barna, P. K., *et al.* (1970). *Appl. Microbiol.*, **20**, 657.

Bradley, S. A., Engler, P., and Carr, S. H. (1973). *Appl. Polym. Sym.*, **22**, 269.

Chirgwin, L. D. Jr. (1968). S. African Pat 6801, 144, American Cyanamid Co.

Chirgwin, L. D. Jr. (1969). U.S. Pat. 3,422,181, American Cyanamid Co.

CIBA Ltd (1969). Fr. Pat. 1,560,704.

Cooney, J. D., Colin, G., and Wiles, D. M. (1973). *Am. Soc. Test. Mater. Spec. Tech. Publ.*, **533**, 17.

Coops, H. I. (1970). *J. Polym. Sci.*, Part A-1, **8**, 1793.

Courts, A. (1967). Brit. Pat. 1,191,837.

Darby, R. T., and Kaplan, A. M. (1968). *Appl. Microbiol.*, **16**, 900.

Engler, P., and Carr, S. H. (1973). *J. Polym. Sci. Polym. Phys. Ed.*, **11**, 313.

Ennis, D. M., and Kramer, A. (1975). *J. Food Sci.*, **40**, 181.

Fields, R. D., Rodrigues, F., and Finn, R. F. (1974). *J. Appl. Polym. Sci.*, **18**, 3571.

Gilbert, S. G., Giacin, K. J., Van Gordon, T., Vahidi, A., and Giacin, J. R. (1974). *Coating and Plastics Reprints*, **34**, 114.

Griffin, G. J. L. (1973). Ger. Offen., 2,333,440.

Guillet, J. (1973). Polymers and Ecological Problems, pp. 1–44 (J. Guillet, ed.), Plenum Press, New York and Baum, B., and White, R. A., *ibid.*, pp. 45–60.

Hafter, R., and Hoermann, H. (1970). *Leder*, **21**, 237.

Hueck, H. J. (1960). *Plastics*, 419.

Jen-Hao, L., and Schwartz, A. (1961). *Kunstoffe*, **51**, 317.

Kaempf, G., Papenroth, W., and Holm, R. (1973). *Farbe Lack*, **59**, 9.

Kim, S. L. (1974). *Diss. Abstr. Int. B.*, **35**, 1121.

Kim, S., Stannett, V., and Gilbert, R. D. (1973). *J. Polym. Sci. Polym. Lett.*, **11**, 731.

Merdinger, E., and Merdinger, R. P. (1970). *Appl. Microbiol.*, **20**, 561.

Miller, T. L., and Johnson, M. J. (1966). *Biotechnol. Bioeng.*, **8**, 567.

Monk, D. W. (1972). *Text. Res. J.*, **42**, 741.

Nishio, F., Yamamoto, N., Iwakura, Y., and Nakabayashi, N. (1967). Japan Pat. 12,556.

Osman, J. L., Klausmeier, R. E., and Jamison, E. I. (1971). Mater. Proc. 2nd Int. Biodeterior. Symp., 66.

Potts, J. E., Clendining, R. A., Ackart, W. B., and Niegisch, W. D. (1973). *Polymers and Ecological Problems*, pp. 61–79 (J. Guillet, ed.), Plenum Press, New York.

Rodriques, F. (1971). Chem. Tech. 409.

Schmitt, E. E., Epstein, M., and Polistina, R. A. (1968). S. African Pat. 68 01, 143, American Cyanamid Co.

Scott, G. (1973). *Chem. Brit.*, **9**, 267.

Sinclair, R. G. (1973). *Environ. Sci. Technol.*, **7**, 955.

Tabor, B. E. (1968). *J. Appl. Polym. Sci.*, **12**, 1967.

Taylor, L. D. (1971). U.S. Pat. 3,576,795.

Thies, C., Cuthbertson, S., and Yoshida, N. (1968). *J. Colloid Interface Sci.*, **27**, 673.

Tirpack, G. (1970). *SPE J.*, **26**, 26.

Veis, A. (1964). *The Macromolecular Chemistry of Gelatin.* Academic Press, New York.

Wassell, C. J. (1964). *SPE Trans.*, 193.

Young, H. H., and Luce, S. B. (1967). U.S. Pat. 3,300,470.

BIODEGRADATION OF SYNTHETIC POLYMERS: THE C-14 METHOD APPLIED TO POLYETHYLENES

ANN-CHRISTINE ALBERTSSON and BENGT RÅNBY

Department of Polymer Technology, The Royal Institute of Technology, Stockholm, Sweden

Summary

The C-14 method for studies of enzymatic degradation rate of linear and branched polyethylenes has been applied to a series of microorganisms, fungi, and bacteria. The organisms are grown in a culture containing the synthetic polymer dispersed to different exposed surface areas per unit weight. The CO_2 evolved is absorbed in aqueous KOH and the amount of C-14 is analysed by liquid scintillation counting. The sensitivity is better than 0·001 % degraded polymer. The data obtained so far indicate that both linear and branched polyethylenes without additives or u.v.-light exposure degrade at an initial rate between 0·005 and 0·10 % per month, dependent on conditions. In addition, a very slow release of CO_2 from polyethylene without micro-organisms has been found as a zero reading in the degradation experiments.

INTRODUCTION

A simple way to investigate biodegradation is to bury the material in soil. Natural soil burial tests have been carried out since 1967 by our department. This kind of experiment takes a very long time and conditions such as temperature and humidity are not controlled. Another way to investigate biodegradation is the use of laboratory tests. Since 1971, we have tested polymers with the agar plate method and in liquid cultures with oxygen. When the rate of degradation of the polymers is low, it is necessary to use special techniques for the measurements. During the conference 'Degradability of polymers and plastics' in 1973, Dr Nykvist read a paper 'Biodegradation of low density polyethylene'. He gave the first result from the C-14 method for studies of the degradation of polymers. Polyethylene is mixed with organic substances that decompose more readily. To distinguish between the CO_2 formed by decomposition of the polymer and that formed from the readily decomposable substances, polyethylene containing the radioactive isotope ^{14}C is prepared and used. We have continued the studies reported by Nykvist (Nykvist, 1974). The experiments started in 1972 in the College of Forestry have been continued in our laboratories since January 1974, *i.e.* for about three years. The C-14 method used now is modified and improved. It has been applied to series of experiments with both low density and high density polyethylene.

MATERIAL AND METHOD

Material

The C-14 method needs a specially prepared polymer containing the radioactive isotope ^{14}C. It is possible to synthesize this polymer in different ways. The C-14-labelled low density polyethylene (LDPE) is prepared by Imperial Chemical Industries. First, a polymer is made from ethylene with a rather high concentration of ^{14}C, using the high pressure free radical process. Second, this polymer is blended with another batch of a similar commercial polymer to get a polymer with a rather low concentration of ^{14}C.

The C-14-labelled high density polyethylene (HDPE) is prepared in cooperation with Unifos Kemi AB, Sweden. In this case already the C-14-labelled monomer is blended with the usual monomer. The resulting ethylene gas mixture with a low concentration of ^{14}C is polymerized.

The two methods give different kinds of distribution of the ^{14}C atoms. In LDPE all ^{14}C is located in a few molecules, while in HDPE the ^{14}C randomly occurs in all polymer molecules. A polymer containing ^{14}C is expected to be less stable than a usual polymer due to the radiation from ^{14}C. The degradation caused by the radioactive isotope ^{14}C can be measured, using polymers containing different amounts of ^{14}C atoms.

The amount of ^{14}C atoms is determined by combustion of the polymer with oxygen in an elementary analysis apparatus. All CO_2 evolved is absorbed in aqueous potassium hydroxide and the isotope ^{14}C is analysed in a liquid scintillation spectrometer. General information about the ^{14}C isotope is given by Catch (1961).

Equipment

The apparatus is shown schematically in Fig. 1. It comprises five connected vessels with different functions. In the middle vessel (No. 3) the polymer is charged under the conditions to be studied, *e.g.* together with soil, nutrient solution, or in a sterile system. Vessels No. 1 and 5 contain aqueous potassium hydroxide. Vessels No. 2 and 4 are empty just to stop back-flow. Air is pumped through the system. In vessel No. 1 all CO_2 is absorbed. In No. 3 the CO_2 is formed by the processes and is carried by the air stream through vessel 4 and absorbed in vessel 5. The CO_2 is measured by titration with hydrochloric acid and the amount of isotope ^{14}C is analysed in a liquid scintillation spectrometer (Packard Tri Carb, Model 3375).

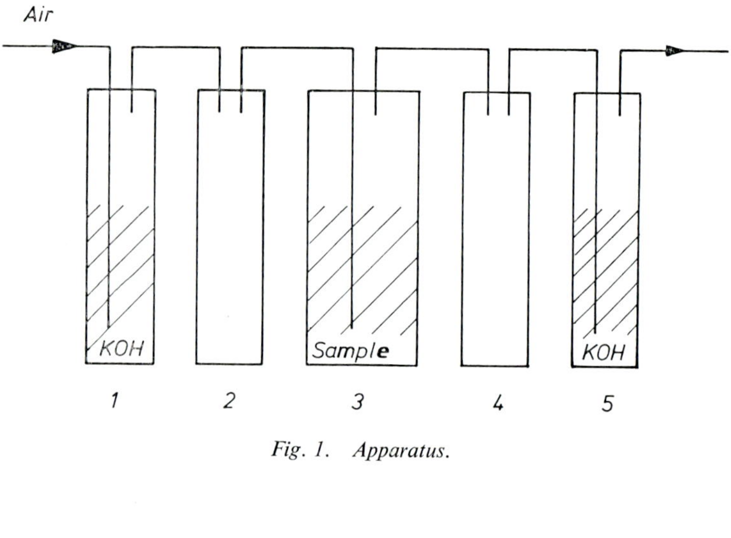

Fig. 1. Apparatus.

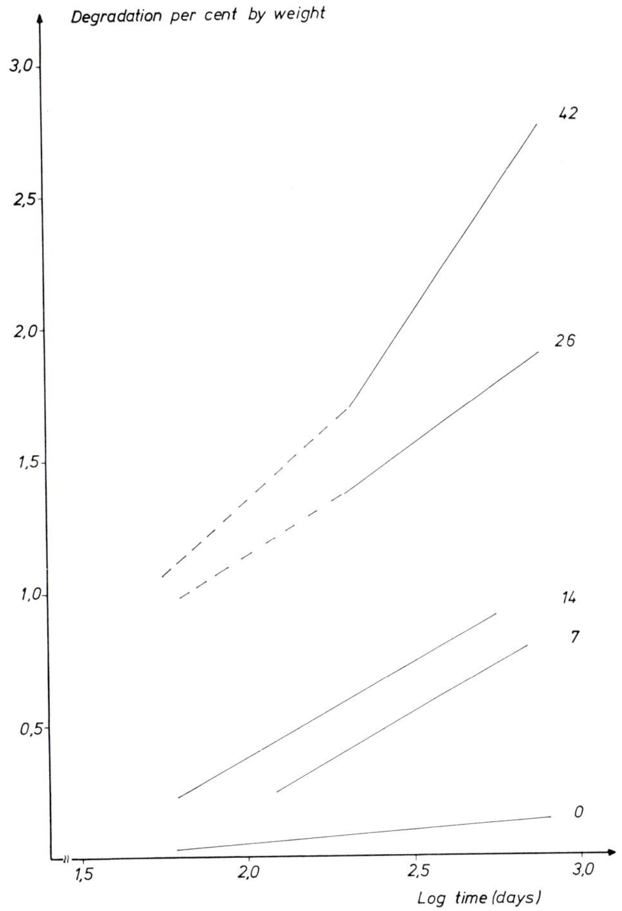

Fig. 7. *The degradation of NDPE exposed to u.v. light 0, 7, 14, 26, and 42 days as a function of logarithmic time.*

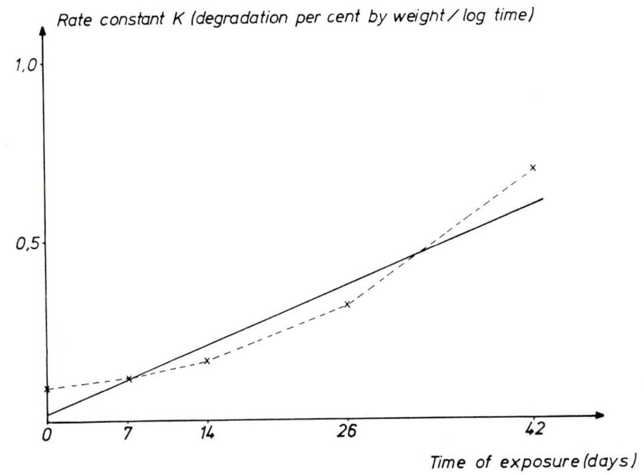

Fig. 8. *The rate constants for O-PE exposed to u.v. light as a function of time for exposure.*

packaging materials (Hocking, 1973). But even with this exposure the total degradation after $2\frac{1}{2}$ years for NDPE is less than 2% (Fig. 4). If the degradation continues according to the assumed logarithmic rate equation (eqn. 1), it is possible to postulate the degradation for longer times. From extrapolation of Figs. 6 and 7, it follows that O-PE and NDPE with 26 days of u.v. irradiation after 100 years is expected to degrade 1% and 3·5%, respectively. Without u.v. exposure, the degradation of LDPE would be less than 0·5%.

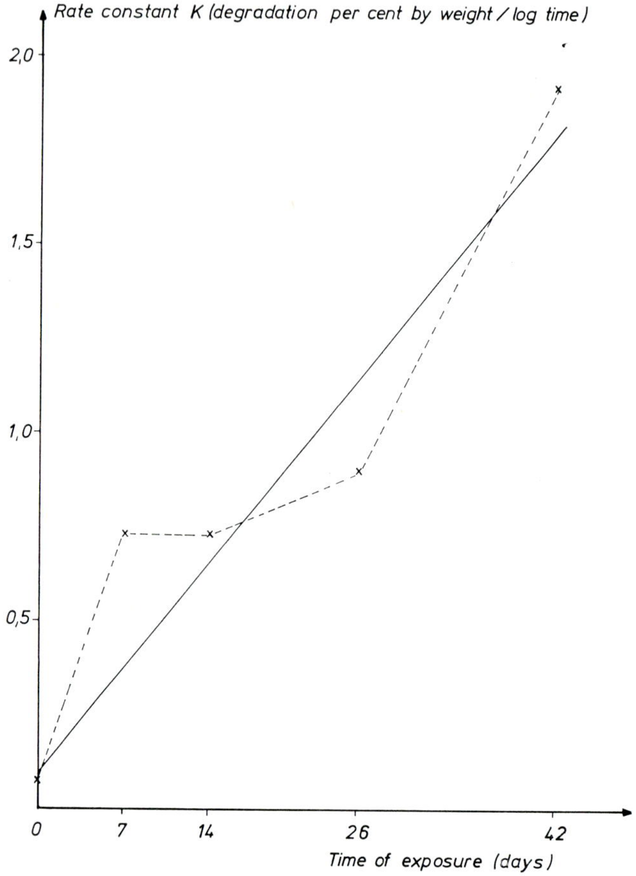

Fig. 9. The rate constants for NDPE exposed to u.v. light as a function of time for exposure.

The straight lines in Figs. 6 and 7 seem to justify the use of the assumed kinetics. In three cases the degradation lines are broken (O-PE u.v. irradiated 42 days and for NDPE irradiated 26 days and 42 days). The reason may be trivial. At the time when the lines change direction, the equipment was moved to our laboratories and the scintillation spectrometer changed from an old Beckman LS 100 to a new Packard Tri Carb, Model 3375. Before the move, the temperature was allowed to fluctuate, but after the move it was constant at 25%.

According to Figs. 8 and 9, the rate constants increase with increased u.v.

exposure time but presumably not linearly. U.V. irradiation changes the material and gives higher carbonyl content, higher polarity, and decreased molecular weight (Ranby and Råbek, 1975). All these changes are expected to increase the rate of biological degradation. Important for the biological degradation is also the change in the physical state of the samples from film to powder. Powder means a larger contact area between polymer substrate and the enzymers, *i.e.* higher accessibility of the polymer. The biological degradation rates in these experiments are very low. We have found that it is possible to accelerate the degradation. In the experiments reported here, as much as 10 g plastics per 40 g soil were used. Higher degradation rates have been observed when 1–2 g plastics are used under similar conditions. Special microorganisms in selected nutrient solution and at an optimal temperature can also increase the degradation. The conditions in these experiments were chosen to be similar to outdoor conditions and not to give as high a degradation rate as possible.

The C-14 method is suitable for studies of enzymatic degradation of any labelled polymer. The problem is to have a good biological system which can be operated under controlled conditions for very long times.

REFERENCES

Catch, J. R. (1961). *Carbon*-14 *compounds*. Butterworths, London.

Hocking, C. S. (1973). The relative rates of degradation of paper based and synthetic packaging materials. *Degradability of polymers and plastics*. Preprints. Plastics Institute, London, 27 Nov.

Nykvist, N. B. (1974). Biodegradation of low density polyethylene. *Plastics Polymers*, **42**, 195–199.

Rånby, B., and Rabek, J. F. (1975). *Photodegradation, Photo-oxidation and Photostabilization of Polymers*. John Wiley & Sons, London, p. 121 ff.

Table IV
EVIDENCE OF GROWTH OF BACTERIA ON POLYMERS, MIXED INOCULUM

Media	*Maximal growth, viable units/0·1 ml* × 10^4	*Fold increase over MM*	*% growth compared to glucose*
MM only	190	1	0·07
PE1	250 000	1 315	89
PP1	34 000	178	12
PE1 + PP1	170 000	895	61
Glucose	280 000	1 500	100

previous experiments was due to the presence of carbon sources other than the polymeric additive, the polymers were washed in MM solution. They were added at 5 % concentration to MM in flasks. The flasks were shaken vigorously for two days under sterile conditions to dissolve any adventitious carbon source. The polymer slurry was then filtered twice through a 0·22 μ Millipore filter and all products were then checked for sterility.

A series of flasks with 0·5% polymer in MM (or the extract from the equivalent weight of polymer) were prepared as shown in Table V. The

Table V
POLYMER SYSTEMS USED WITH BACTERIA INOCULUM

Flask	*Additive*
1	MM only
2	MM + 0·5% washed PE1
3	MM + 0·5% washed PP1
4	MM + washings of PE1 (equivalent to 0·5%)
5	MM + washings of PP1 (equivalent to 0·5%)
6	MM + 0·5% unwashed PE1 + PP1
7	MM + 0·2% glucose

inoculum was a mixture of genera isolated from the previous experiments. This was diluted to give a low inoculum of 500 cells/ml. The growth of bacteria for a period of 10–15 days was monitored by taking viable counts from each flask. The results are given in Table VI.

Table VI
BACTERIAL GROWTH: WASHED POLYMERS

Media	*Maximal viable units/0·1 ml culture* × 10^4	*Fold increase over MM*
MM only	90	
PE1 washed	70 000	780
PP1 washed	7 000	80
PE1 washings	500	6
PP1 washings	100	1
PE1 + PP1 unwashed	110 000	1 220
Glucose	140 000	1 550

It is evident from Table VI that the washings from PP1 and PE1 did not support growth significantly more than the minimal medium. On the other hand, both washed and unwashed polymers supported a significant amount of growth. It is clear that for polypropylene the rate was slower than for polyethylene.

In the above experiments the inocula used consisted of a mixture of the 10 different genera selected from previous experiments. Growth was assayed by viable counts on nutrient agar plates and usually all of the genera were judged to be present during most of the growth cycle. This could be accomplished since most of the organisms give a distinct morphological appearance on nutrient agar plates. The mixed inocula were used so as to simulate the type of mixed flora one would find in the soil which would normally be present to degrade the polymers.

However, in order to determine whether each isolate was capable of growth individually on the polymers (PP1 + PE1), experiments were carried out using inocula from a single isolate. Growth was monitored by viable counts on nutrient agar every 80 hr for a 240-hr period. A summary of the results appears in Table VII. If the growth with polymer is 10 times greater than

Table VII
GROWTH OF BACTERIA ON POLYMER: INDIVIDUAL ISOLATES

Organism no.	*Maximal growth viable counts/0·1 ml* × 10^4	*Relative to glucose*[a] %	*Fold increase over MM*[a]	*Capability of growth on polymer*[b]
1-1	<1	0	0	−
1-2	1 000	5	50	+
3-1	340	0·5	50	+
3-2	19 000	38	2 800	+
3-3	11 000	20	1 500	+
4-1	<1	0	0	−
4-2	9 000	12	750	+
5-1	33 000	44	2 750	+
5-2	12 000	16	1 000	+
6-1	7 000	9	580	+
6-2	10 000	13	850	+
6-3	400	0·6	2·5	−
6-4	4 000	6·4	25	+
7-1	3 100	5	20	+
7-2	800	1·3	5	−
8-1	2 000	3·2	13	+
8-2	5 000	8	45	+
9-1	4 000	6·4	25	+
9-2	25 000	40	150	+
Mixed inoculum in glucose	45 000	100	5 000	
Mixed inoculum in MM	1	0·02	1	

[a] These values are relative to the growth of the same organism in glucose or MM only.
[b] All organisms which yield a low or negative result were rechecked.

that with MM, the test is considered positive, *i.e.* the organism utilizes the polymer as a carbon source. Most figures greatly exceed this figure. It can be observed that different organisms appear to utilize the polymer to different extents. The range of value for the increase over growth in MM only is

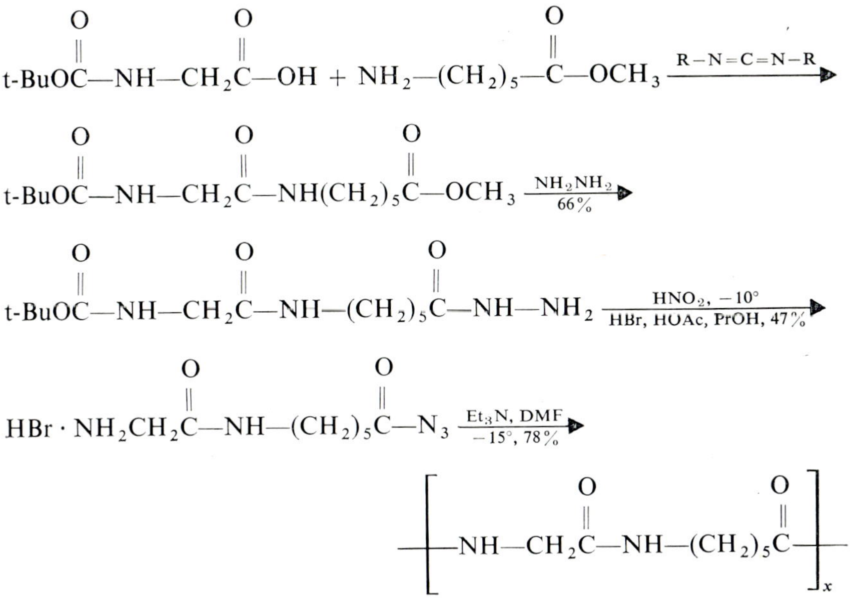

The aminoacid azide hydrobromide was polymerized in DMF at $-40°$ by the addition of triethylamine over a period of 50 min. A 78% yield of the regularly alternating copolyamide of nylon 2 and 6 was obtained with a melting point of 268–270° and an intrinsic viscosity (*m*-cresol) at 25° of 0·46 dl/g.

This copolyamide was highly crystalline as evidenced by an X-ray determination and a relatively high heat of fusion in a differential thermal analysis. Fibres pulled from the molten polymer and oriented had a tenacity of about 6 g/denier. The melting point, 268–270°, is intermediate between that of nylon 6, 220°, and polyglycine, which decomposes at about 400°. It is hydrophilic, absorbing about twice as much water as nylon 6. It has the same average concentration of amide groups as nylon 4 but is much more thermally stable, less flammable and lower melting.

Since the copolyamide contained two different amide bonds, both of which were similar but not identical to a peptide bond, it was hoped that the extracellular enzymes of the bacteria or fungi would cleave the polymer to small fragments that could be utilized by the microorganisms as a food. In order to make fast progress on this research project, it was considered essential to have a rapid screening test available. A group in the Department of Horticulture at the University of Maryland cooperated on this project and did indeed develop such a fast-screening procedure (Ennis and Kramer, 1974, 1975). In their method, a culture of bacteria and/or fungi was fed a standard amount of hydrolysed casein in a modified Warburg apparatus and the amount and rate of carbon dioxide liberated was monitored by a gas chromatograph. In order to determine whether a material was biodegradable on the time scale used, the material was added to a similar flask and any increase in the rate or amount of carbon dioxide over the control was noted.

It was not surprising that glycine and ε-aminocaproic acid were readily consumed by the microorganisms as food, since the former is a common constituent of proteins and the latter is used in massive doses (25 g per day) to treat certain nerve problems in humans. In this test nylon 6 is inert as is polyglycine, presumably because of its very low solubility and relatively unswellability in water. A large series of amides and amide derivatives was screened in an attempt to develop a correlation between structure and biodegradability. When a dimer of ε-aminocaproic acid was tested, it was degraded but at a slower rate than the monomer. A trimer showed only very slow degradation. When the regularly alternating copolyamide (nylon 2/nylon 6) was tested, it was degraded by both fungi and bacteria. The fungus *Aspergillus niger* completely degraded the copolymer in about 3 weeks and the bacteria, *Flavobacterium*, *Flavobacterium* spp., and *Alcaligenes*, all degraded the material but did not give a quantitative yield of carbon dioxide. Examination of the photomicrographs of the polymer during degradation indicates that certain portions, presumably the amorphous regions that are swollen in water, are attacked first and other regions, presumably the more crystalline parts, are attacked more slowly.

In order to see if hydrophilic character and water solubility was an important factor in the rate of biodegradation, a regularly alternating copolyamide of serine and ε-aminocaproic acid was prepared by the following set of reactions:

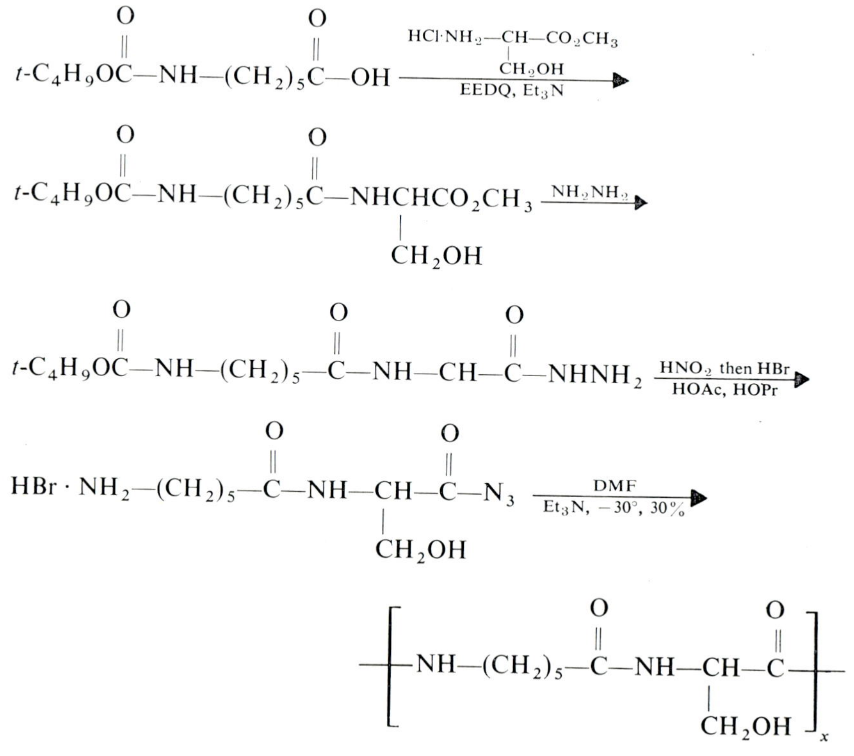

This water-soluble polymer was readily degraded in the screening test described.

Thus a new class of biodegradable polymers has been discovered which have very attractive physical properties and apparently have the versatility in structure so that a large variety of materials with a wide range of physical and chemical properties can be prepared.

EXPERIMENTAL

t-Butyloxycarbonylglycyl-ε-aminocaproyl hydrazide

Into a 500-ml Erlenmeyer flask cooled to 0–5°C were placed 25·0 g (120 mmoles) of freshly distilled dicyclohexylcarbodiimide, 17·5 g (100 mmoles) of *t*-butyloxycarbonylglycine, mp 88–89°C, 21·7 g (120 mmoles) of methyl ε-aminocaproate hydrochloride and 7·9 g (70 mmoles) of 1,4-diazobicyclo-[2.2.2]-octane in 400 ml of freshly distilled methylene chloride. After the mixture had been allowed to stand at room temperature for 1 day, a small amount (about 5 ml) of acetic acid was added. The precipitated urea was removed by filtration after the solution had been allowed to stand 0·5 hr at 0°. The filtrate was washed successively with 40 ml of cooled 1 N hydrochloric acid, 40 ml of water, 40 ml of a saturated sodium bicarbonate solution, 40 ml of water, and then dried over anhydrous calcium sulphate. After the solvent was removed under reduced pressure on a rotary evaporator and the residue was dissolved in 200 ml of acetone, the solution was kept at 0°C overnight. A further quantity of urea was removed by filtration and the filtrate was concentrated by evaporation under reduced pressure. The residue containing the blocked ester was dissolved in 80 ml of methanol, and 19 ml of 95% hydrazine was added dropwise with stirring. The solution was stirred for 3 days at 45–50°C. The solvent was removed on a rotary evaporator under reduced pressure to give an oily product, which was kept standing at −20°C overnight to produce a white solid. The solid was washed with cold ether and dried over concentrated sulphuric acid to give 20·0 g (66%) of *t*-butyloxy-carbonylglycyl-ε-aminocaproyl hydrazide, mp 60–62°C.

Anal.

Calcd. for $C_{13}H_{26}N_4O_4$: C, 51·64; H, 8·67; N, 18·53. Found: C, 51·94; H, 8·66; N, 18·30.

The infrared spectrum of this blocked hydrazide showed absorptions at 3300 (N—H stretching), 2930 (C—H stretching), 1695 (C=O acid hydrazide), 1650 (C=O urethane and amide), 1530 (CNH), and 1170 cm^{-1} (NH_2). The nmr spectrum ($CDCl_3$) showed resonance peaks at δ values 1·42 (singlet, 9, $(CH_3)_3C$), 1·15–1·9 (multiplet, 6, C—(CH_2)3—C), 2·1 (multiplet, 2, N—CH_2—C), 3·25 (multiplet, 2, C—CH_2—CO), 3·79 (doublet 2, N—CH_2—CO), 3·80 (singlet, 2, NH_2), 5·75 (broad, 1, NH), 6·80 (broad, 1, NH), and 7·96 ppm (broad, 1, NH).

Glycyl-ε-aminocaproyl azide hydrobromide

Into a 200-ml, three-necked flask fitted with a thermometer, a stirrer, and a dropping funnel were added 3·66 g (12 mmoles) of *t*-butyloxycarbonylglycyl-ε-aminocaproyl hydrazide in a mixture of 30 ml of 1 N hydrochloric acid and 6 ml of acetic acid previously cooled to −10°C. A 50-ml layer of cold ether was poured into the solution. With vigorous stirring, 1·05 g (15 mmoles) of sodium nitrite in 10 ml of water were added slowly to the solution at −10–0°C. After the addition was complete, the mixture was stirred for 10 min. The blocked azide was extracted into the ether layer as it formed. After the ether layer was separated and dried over anhydrous calcium sulphate for a few minutes, this ether layer was placed in the dropping funnel which was cooled with a Dry Ice-methanol bath. This solution of the blocked azide was added dropwise to a mixture of 25 ml of acetic acid, 25 ml of propionic acid, and 5 ml of acetic anhydride previously saturated with anhydrous hydrogen bromide at −20° to −30°C. When the addition was complete, the mixture was stirred for 2 hr at −20° to −30°C. After the mixture had been subjected to partial vacuum for 1 hr to remove excess hydrogen bromide, the mixture was poured into 1 litre of previously cooled anhydrous ether (−30°C) to give a yellowish solid. These crystals were collected by suction filtration, washed with cooled ether, and then recrystallized from methanol-ether to give 1·65 g (47%) of glycyl-ε-aminocaproyl azide hydrobromide. This compound was unstable at room temperature. The infrared spectrum of this compound showed absorptions at 3270 (N—H stretching), 2145 (N_3 asymmetric stretching), 1720 (C=O stretching of acid azide), 1650 (C=O stretching of amide), and 1560 cm^{-1} (CNH).

Poly(glycyl-ε-aminocaproic acid)

Into a 15-ml flask was added a suspension of 1·61 g (5·5 mmoles) of glycyl-ε-aminocaproyl azide hydrobromide in 4 ml of DMF at −40°C. Triethylamine, 1·0 ml, was added to the mixture with a syringe over a period of 50 min. A heavy white precipitate was gradually formed. After the mixture had been stirred for 30 hr at −15°, the resulting mixture was filtered, and the collected white solid was washed with hot water, then reprecipitated by the addition of a hexafluoroisopropanol solution to water to give 0·73 g (78%) of poly(glycyl-ε-aminocaproic acid). This polymer was dried over phosphorous pentoxide *in vacuo* at 140°C for 3 weeks.

Anal.

Calcd. for $(C_8H_{14}N_2O_2)_n$: C, 56·45; H, 8·29; N, 16·46. Calcd. for $(C_8H_{14}N_2O_2 \cdot \frac{1}{4}H_2O)_n$: C, 55·00; H, 8·36; N, 16·03. Found: C, 54·74; H, 8·06; N, 16·17.

The infrared spectrum of this polymer showed absorptions at 3290 (N—H stretching), 3090 (combination band of the C=O stretching and N—H bending band), 2930 and 2850 (C—H stretching), 1640 (amide I), 1555 (amide II), and 690 cm^{-1} (N—H wagging). The nmr spectrum (CF_3COOH) showed resonance peaks at δ values 1·12 (broad singlet, 6, C—$(CH_2)_3$—C), 2·09 (broad multiplet, 2, C—CH_2—N), 3·00 (broad singlet, 2, C—CH_2—CO), 3·85 (broad singlet, 2, N—CH_2—CO), 7·29 (broad singlet, 1, NH), and 7·69 ppm (broad singlet, 1, NH). The intrinsic viscosity of the polymer in freshly distilled *m*-cresol at 25° was 0·46 dl/g, and the polymer melted at 268–270°C.

Acknowledgement

This work was supported in part from a grant from the Material Science Center of the University of Maryland through a block grant from the National Science Foundation.

REFERENCES

Anon. (1974a). *Chem. Eng. News*, **52** (28), 16.

Table III
POLY(TETRAMETHYLENE ADIPATE) PREPARATIONS

Identification	*Intrinsic viscosity* $[\eta]$ *dl/g (chloroform, 25°C)*	*Molecular weight,* M_v *(see eqn below)*	*Method of polymerization*
3A	0·076	$0{\cdot}11 \times 10^4$	Solution, no catalyst
3B	0·23	$0{\cdot}54 \times 10^4$	Melt, Ti(i-Pr)$_4$
3C	0·74	$3{\cdot}1 \times 10^4$	Melt, Ti(i-Pr)$_4$

$[\eta] = 6{\cdot}6 \times 10^{-4} M_v^{0{\cdot}68}$ (Darby and Kaplan, 1968).

The reduced viscosities, a measure of the molecular weight, were measured for all the polymers, except polycaprolactone, in chloroform solutions at 25°C. All of the solutions had relative viscosities less than 2. Tables I, II, and III list the intrinsic viscosities of the polyesters synthesized as calculated from solution viscosity.

The melting temperatures and heats of fusion of the polyesters were measured using a Perkin-Elmer differential scanning calorimeter. Table IV lists the heats of fusion of the polyesters along with their melting temperatures. All of the polyesters showed crystalline melting points. The low, 50°C, melting point of the copolymer as compared with the other two 1,4-butanediol polyesters indicates that it was probably less crystalline than the other two.

Table IV
MELTING TEMPERATURE, T_m, AND HEAT OF FUSION, H_f

Polyester	$[\eta]$, *dl/cm*[a]	T_m, °C	H_f, *cal/g*
Polycaprolactone	0·47	60	16·6
Polycaprolactone	0·29	60	18·0
Poly(tetramethylene adipate)	0·74	60	9·12
Poly(tetramethylene sebacate)	0·64	68	20·5
Poly(tetramethylene adipate-sebacate) 50:50 molar ratio of adipic:sebacic acid	0·41	50	13·7

[a] Chloroform, 25°C, except polycaprolactone, benzene, 30°C.

Microorganisms

A variety of microorganisms was investigated in the course of this work. However, the results reported in this paper are confined to the five described here.

P. pullulans was obtained from the U.S. Army Quartermaster Corps. *Pseudomonas fluorescens* ATCC 12842, an organism that utilizes caproic acid, was obtained from the American Type Culture Collection.

A strain of *Penicillium* was obtained in the following manner: an agar medium of mineral salts, 0·5% yeast extract, and 0·5% glucose was prepared. While the medium was hot, 0·5% polycaprolactone (1250 MW) was added and suspended by mixing in a Waring blender. This dispersed the insoluble polymer in the agar. This culture medium was inoculated with the supernatant from a soil suspension. Those organisms which had clear zones around them were ones that were hydrolysing the polymer (Klausmeier, 1966). Presumably, exoenzymes excreted by the microorganisms caused polymer hydrolysis.

Using this clear zone method, soil bacteria and fungi, actinomycetes, and a *Penicillium* were isolated that produced polyester-depolymerizing enzymes.

Soil bacteria cultures were isolated on a polymer-Acti-dione medium. Acti-dione (The Upjohn Co.) is a fungistat that inhibits mould growth but does not affect bacteria (Phillips and Hanel, 1950). The medium contained 0·0001% Acti-dione, 0·4% yeast extract, 0·5% suspended 2000 molecular weight polycaprolactone, and mineral salts. The medium was sterilized, then inoculated with a soil suspension, and incubated at 30°C. Bacteria colonies that cleared the medium (evidence of de-polymerase activity) were streaked out on new plates to obtain cultures of depolymerizing bacteria. Also, an enriched soil bacterial culture was obtained with a perfusion method (Lees, 1944).

Degradation

The primary measure of degradation was weight loss under controlled conditions. As previously described (Fields *et al.*, 1974), the method used a layer of polymer covered by a nutrient agar medium. The plates were inoculated with a suspension of the microorganisms being used and incubated at 30°C. Sterile plates were wetted with a 1% mercuric chloride solution to insure that they remained sterile. Since low molecular weight polyesters were slowly hydrolysed by water, these sterile plates distinguished between enzymatic hydrolysis and hydrolytic instability of the polymer. Typically, both inoculated and sterile samples were incubated for 3 weeks.

RESULTS

One primary parameter investigated here is polymer structure. The significant properties of a series of polyesters is summarized in Table V along with the

Table V
SUSCEPTIBILITY OF POLYESTERS TO ATTACK BY *P. PULLULANS*

Polyester	*Intrinsic viscosity, (dl/g)*[a]	*Weight loss (mg/cm²)*			
		Sterile	*Inoculated*	*Enzymatic*	T_m *(°C)*[b]
Poly(decanediol dodecanedioate)	0·084	0·089	0·943	0·854	73
Poly(hexamethylene sebacate)	0·091	2·25	16·5	14·2	59
Poly(tetramethylene adipate)	0·12	2·28	23·7	21·4	50
Poly(tetramethylene succinate)	0·14	3·18	7·65	4·47	109
Poly(tetramethylene oxalate)	low	33·0	31·2	—	109
Poly(β-propiolactone)	0·12	3·67	11·3	8·7	87
Polycaprolactone, PCP-0230	0·064	2·73	8·60	5·87	47
Polycaprolactone, PCP-0240	0·086	1·18	3·11	1·93	53

[a] Measured in chloroform at 25°C except for PCP-0230 and PCP-0240 in benzene at 30°C.
[b] Measured in polarized light on hot-stage microscope.

susceptibility to attack by *P. pullulans*. As indicated, the weight loss from the sterile sample, assumed to be due to hydrolysis, was subtracted from the loss for the inoculated sample in order to arrive at an estimate of the loss due to enzymatic attack.

Various organisms were compared as degrading agents for a low-molecular weight polycaprolactone (Table VI). Three variations in culture medium were

Table VI
COMPARISON OF DEPOLYMERASE ACTIVITY OF VARIOUS ORGANISMS
(weight loss of 2000 MW polycaprolactone)

Organism	*Culture media*		
	Sodium succinate yeast extract mineral salts	*Suspended polymer sodium succinate yeast extract mineral salts*	*Suspended polymer mineral salts*
P. pullulans	7·05	4·87	4·91
Penicillium	12·1	6·13	4·63
Soil bacteria	6·82	4·57	2·28
Enriched soil bacteria	5·36	5·50	3·92
Ps. fluorescens	4·96	4·02	3·4
Sterile	0·95	—	—

included in a test which ran 3 weeks at 30°C. In a separate test, the degradation of the same polycaprolactone was compared with a copolymer using two pure cultures and a mixed soil bacterial colony (Table VII). The copolymer is described more completely in Table II.

Table VII
DEGRADATION BY SOIL MICROORGANISMS

Organism	*Weight loss after 21 days (mg/cm^2)*	
	Polycaprolactone PCP-240	*Copolymer 2F (Table II)*
Sterile	1·83	1·67
P. pullulans	7·86	18·7
Soil bacteria	11·9	21·9
Penicillium	9·45	14·9

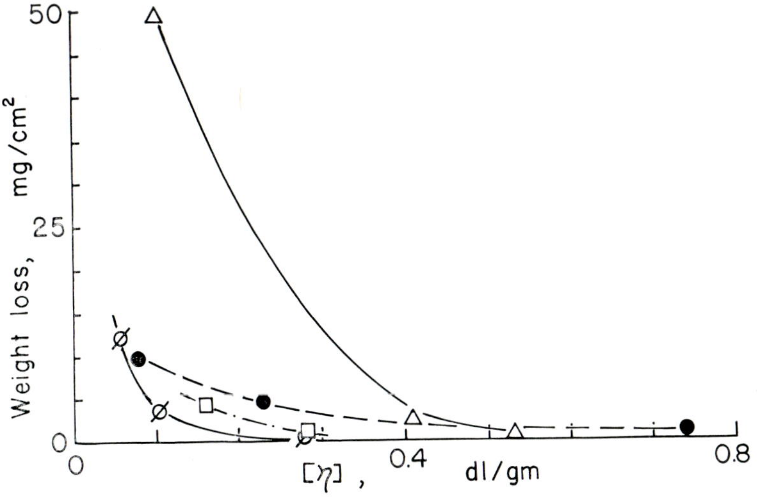

Fig. 1. Degradation of poly(tetramethylene adipate-sebacate), 50:50 mole ratio of acids (Δ) compared with adipate (●) and sebacate (□) homopolymers and polycaprolactone (ϕ). Weight loss measured after 21 days at 30°C with P. pullulans.

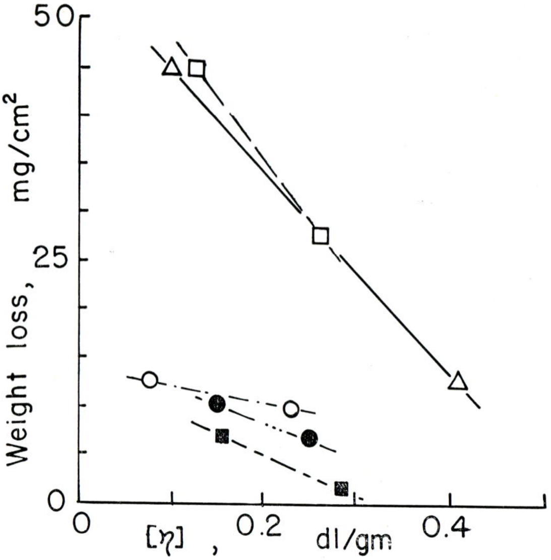

Fig. 2. *Degradation of polymers made from 1,4-butanediol and a mixture of adipic and sebacic acids where the mole percent adipic acid is 100 (○), 75 (□), 50 (Δ), 25 (●), and 0 (■). Weight loss measured after 21 days at 30°C with* P. pullulans.

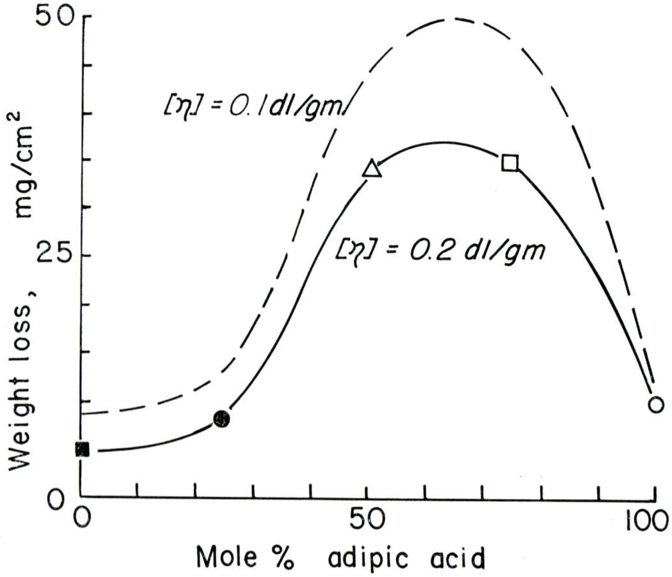

Fig. 3. *Dependence of weight loss on polymer composition for copolymers made from 1,4-butanediol and a mixture of adipic and sebacic acids. Data interpolated from Fig. 2 for two values of intrinsic viscosity.*

Attempts to test the effect of polymer morphology on degradation rate took the form of taking the more promising candidates from the first structure tests and making copolymers. The rate of degradation for a 50-mole % adipic-sebacic copolymer is compared in Fig. 1 with the rates for homopolymers at various molecular sizes. The intrinsic viscosity is used here as a measure of molecular weight. In a subsequent run (Fig. 2), the range of copolymer compositions was extended. A cross-plot (Fig. 3) indicates the influence of copolymer composition on loss by biodegradation due to *P. pullulans* after 21 days at 30°C.

DISCUSSION

A previous quantitative study of the biodegradation of polycaprolactone had shown a marked increase in resistance to attack as molecular weight increased (Fields *et al.*, 1974). In that work, there are many literature references to degradation of polyesters by microorganisms (*see also* Rodriguez, 1971). In these, most workers have reported qualitative results in terms such as 'no growth, growth, and luxuriant growth'. In a field where there are so few data points, such results have their place. However, as more data accumulate, it is desirable that quantitative measurements be made in order to permit extrapolation to compositions with desired properties. The weight-loss method from a surface of known area yields such results.

Effect of structure on rates of polyester degradation

Polycaprolactone has certain advantages over other polyesters. It is commercially available in high molecular weights and it is compatible with a wide variety of other polymers. This presumably expands its range of application. Nonetheless, generalizations about molecular structural influences on biodegradability can be arrived at only through the study of other systems. One lactone analogue, poly(β-propiolactone), was included in the present work. Because they are convenient to prepare, several dyadic polyesters based on diacids and dialcohols were used. For each composition, several molecular weights were tested so that results could be interpolated to a common base.

Knowing the template-like nature of enzymatic catalysis, it is to be expected that mere changes in the distance between ester groups can have an influence on degradability. Therefore, representative polymers with C_2–C_{12} diacids and C_4–C_{12} dialcohols were studied.

Initial work, using the clear zone method to detect degradation, showed that polyesters synthesized from short chain dicarboxylic acids, such as oxalic acid, were not as rapidly broken down as the ones made from longer dicarboxylic acids, such as adipic acid. To describe this effect quantitatively, several low molecular weight (approximately 2000) polyesters made from different glycols and acids were degraded using the quantitative method of measuring the weight loss of degraded samples.

The sterile and inoculated samples lost weight because the low molecular weight polyesters were slowly hydrolysed, but the inoculated ones also lost weight because of enzyme-catalysed hydrolysis. The difference in the weight loss between a sterile and inoculated sample would give the weight loss attributable to biological degradation (Table V).

The polyester based on the very long glycol (10-carbon decanediol) and the long acid (12-carbon, dodecanedioic acid) was not degraded as rapidly as those based on shorter acids and glycols. Poly(tetramethylene oxalate), made from a 2-carbon acid, oxalic acid, showed no significant biological degradation. Apparently, the polyesters based on very long or short moieties were not as easily degraded as the ones of intermediate length such as poly-(tetramethylene adipate).

In a study of plasticizer utilization by fungi (Berk *et al.*, 1957), it was concluded that the *n*-hexyl ester of adipic acid showed greater growth than higher or lower alcohols. Also, it was seen that, in the homologous series of diacids, the 4-carbon succinic acid was utilized more effectively than any other acid in the 2–9 carbon range. However, sebacic acid (10 carbons) gave even greater growth than succinic acid. These results were obtained with simple esters using colony growth as a criterion. Among the two dozen fungi tested by Berk *et al.*, *P. pullulans* gave the least growth. In the present work, it has been found more effective than some others.

It does appear in the present work that certain diols and diacids favour growth more than others. Maximum enzymatic degradation occurs with diols of 4 or 6 carbons and diacids of 6–10 carbons (Table V). Smaller units show degradation due to simple hydrolysis. As already noted, larger units seem to approach the inertness of polyethylene. Although polyethylene in the molecular weight range of 2000 can be degraded, it is practically resistant under the conditions employed here.

Several workers in the past have reported biodeterioration of polyesters such as poly(tetramethylene adipate). However, either the molecular weight was not included (Klausmeier, 1966; Berk *et al.*, 1957) or the growth was reported in qualitative terms (Darby and Kaplan, 1968; Potts *et al.*, 1973).

It would be interesting to pursue the study of polylactones. The polymer of β-propiolactone seemed more susceptible to attack than that of caprolactone. Unfortunately, the intermediate lactones were not readily available except for 4-butyrolactone, which is a carcinogen.

Activity of various organisms

The fungi recommended by the ASTM had been tested on low molecular weight polyesters for their ability to clear a polymer suspension. This qualitative test of which organism created the largest clear zones showed that *P. pullulans* was the most active. There is no reason to think the ASTM fungi should be the most effective for degrading polyesters since their primary recognition was due to their effect on cellulose. The other microorganisms, both fungi and bacteria, that had been isolated from soil were tested along with *P. pullulans* by measuring the weight loss of a low molecular weight, 2000 polycaprolactone, to see which ones had the most depolymerase activity.

Three different culture media (Table VI) were used to determine nutritional requirements for rapid degradation. In the first, sodium succinate is a carbon source that does not repress adaptive enzyme formation as glucose can. Suspended polymer was excluded from the medium to see if intimate contact between the microorganisms and the polymer was required to produce depolymerizing enzymes. In the second, 0·5% of polycaprolactone (MW 1250) was suspended in agar with the previous formula. This medium would show whether or not suspended polymer enhanced enzyme production. The third was a formula with mineral salts and 0·5% suspended polymer. This would show if the microorganisms could utilize the polyester without an additional carbon source.

In a quantitative test (Table VI), the pattern which emerges is quite consistent. Suspended polymer decreases the attack on the underlying solid sheet of polymer. Only the weight loss by the sheet is reported, not that by the suspended polymer. On the other hand, the presence of yeast extract and sodium succinate certainly favours degradation.

When the clear zone method was used earlier to see which fungi-polyester combinations resulted in the utilization of the polyester, it had been observed that polymer-utilizing fungi and the degradable polyesters formed a coherent set. In a quantitative test (Table VII), a copolymer which had been found to be more degraded than polycaprolactone by *P. pullulans* proved to be attacked also by soil bacteria and a *Penicillium* in like manner. One concludes that the order of biodegradability of polyesters found with one aggressive microorganism is not likely to be changed greatly when a different microorganism is substituted. It is rather obvious considering the enormous number of possibilities for polyester-microorganism combinations that such a generalization can only be tentative.

Effect of morphology on degradation rate

The polyesters dealt with here have glass transition temperatures, T_g, well below room temperature and melting temperatures, T_m, above room temperature. Degree of crystallinity cannot be controlled by quenching under these circumstances. In order to explore the effect of amorphous-crystal ratio on degradability, several polymers were included which were based on mixtures of diacids.

To see if amorphous polymers were more rapidly degraded than crystalline ones, a series of 1,4-butanediol polyesters with different degrees of crystallinity was degraded. Poly(tetramethylene adipate) and poly(tetramethylene sebacate) were degraded by the method previously described. A copolymer made from 1,4-butanediol and a 50:50 molar ratio of adipic and sebacic acids was also used. This polymer was crystalline, but showed a lower T_m than the homopolymers. This would indicate that there was less order and more amorphous material in the copolymer.

If the degree of crystallinity did not effect the rate of enzymatic depolymerization, the rate of degradation for the copolymer would be expected to fall between, or near, the rates for the two homopolymers. If amorphous polymers were more easily attacked, then the rate should be faster for the copolymer. To eliminate molecular weight as a variable, three molecular weights of each polyester were used.

The results, a graph of weight loss versus reduced viscosity, are shown in Fig. 1. The adipate and sebacate polyesters showed similar rates of degradation. At low molecular weights, the copolymer degraded much more rapidly than the more crystalline ones. This effect was not as pronounced at the higher molecular weights.

Other copolymers, with molar ratios of adipic to sebacic acids of 25:75, 50:50, and 75:25, were degraded to see if there was an optimum ratio of the two acids. Two polymers, a low and an intermediate molecular weight of each composition, were degraded for 21 days by *P. pullulans*. Figure 2 shows the weight loss versus reduced viscosity for these copolymers. The weight loss at a reduced viscosity of 0·25 was then plotted against polymer composition (Fig. 3) to show the dependence of weight loss on composition independent of molecular weight.

CONCLUSIONS

There are organisms in the soil that have the ability to utilize polyesters. This property was not unique to any one organism, but several, both fungi and bacteria, were found.

Esterases, the enzymes that hydrolyse polyesters, are known to be low in specificity. They will hydrolyse a large variety of substrates, although not all at the same rate. The polyesters most rapidly hydrolysed were those made from moieties containing six carbon atoms. Increasing or decreasing the spacing between ester groups made the polymers less susceptible to enzymatic attack.

Polymer crystallinity did affect the rate at which polyesters were hydrolysed. Figures 1 and 2 show how the rate of degradation decreases as the reduced viscosity increases. If degradation occurred at chain ends, the rate of degradation would be expected to decrease as the molecular weight increased. The copolymers of adipic and sebacic acids with 1,4-butanediol were more rapidly degraded than the homopolymers. This may be because their amorphous structure provides more available chain ends for enzymatic attack and the diffusivities of water and enzyme may be higher in amorphous polyesters than in crystalline polyesters.

Acknowledgement

The authors are grateful to Mr Alex Tannahill for his advice and assistance.

REFERENCES

ASTM Method E-123-64 (1972). Annual Book of ASTM Standards, Part 30. Am. Soc. Testing Mats., Philadelphia, Pa., 225 pp.

Berk, S., Ebert, H., and Teitell, L. (1957). *Ind. Eng. Chem.*, **49**, 115.

Darby, R. T., and Kaplan, A. M. (1968). *Appl. Microbiol.*, **16**, 900.

Fields, R. D., Rodriguez, F., and Finn, R. K. (1974). *J. Appl. Polym. Sci.*, **18**, 3571.

Gresham, T. L., Jansen, J. E., and Shaver, F. W. (1948). *J. Am. Chem. Soc.*, **70**, 998.

Klausmeier, R. E. (1966). *Soc. Chem. Ind. (London) Monogr.*, **23**, 232.

Lees, H. (1944). *Chem. Ind.*, **22**, 238.

Phillips, G. B., and Hanel, E., Jr. (1950). *J. Bacteriol.*, **60**, 104.

Potts, J. E., Clendinning, R. A., and Ackart, W. B. (1973). In *Degrability of Polymers and Plastics*. Conference Preprint Booklet. The Plastics Institute, London.

Rodriguez, F. (1971). *Chem. Technol.*, **1**, 409.

Zavaglia, E. A., Billmeyer, F. W., and Mosher, W. A. (1965). *J. Paint Technol.*, **33**, 229.

THE ROLE OF MICROORGANISMS IN THE TRANSPORT OF CHLORINATED INSECTICIDES

HAROLD P. VIND

Civil Engineering Laboratory, Naval Construction Battalion Center,
Port Hueneme, California 93043, U.S.A.

Summary

Based on the results of previous investigations of the resistance of certain insecticides to biodegradation and leaching in seawater, it was hypothesized that the water-insoluble chlorinated insecticides are transported in seawater primarily by being absorbed in the fatty tissues of microorganisms suspended in the water. The hypothesis predicts that less fat-soluble insecticides than DDT will not enter the ocean food chain as readily as DDT. Therefore, they should not be as hazardous to the environment as DDT and yet might be just as toxic and persistent. Several experiments are underway at the Civil Engineering Laboratory to determine the validity of the hypothesis.

INTRODUCTION

Studies on the biodegradation and leaching of chlorinated insecticides in seawater were carried on at the Civil Engineering Laboratory (CEL) for several years (Vind *et al.*, 1971–73). The studies were sponsored by the Office of Naval Research (ONR).

As a part of these studies, a preliminary test was conducted to ascertain if DDT-tolerant microorganisms are prevalent in seawater. When serial dilution plates were made of the microorganisms that accumulated on the surfaces of small sticks of wood exposed in the ocean, there were no significant differences in the numbers of colonies that developed on seawater agar containing no DDT and 10 ppm of DDT. Also, there were no significant differences in the numbers of microorganisms which accumulated on the surfaces of unimpregnated and DDT-impregnated wood. It was concluded that at least most of those marine microorganisms that accumulate on wood are tolerant to DDT.

In the leach-resistance tests, insecticide-impregnated wood was preleached either in the ocean or in hot seawater and then exposed in aquaria to attack by wood-borers of the species *Limnoria tripunctata*. Sixteen different chlorinated insecticides were tested. The apparent decline in toxicity of the wood, caused by preleaching it, was greater in the ocean than in hot seawater. The decline was greater for wood impregnated with the most fat-soluble insecticide, lindane, than for the least fat-soluble one, kepone. The decline in toxicity was intermediate for DDT and other insecticides of intermediate fat-solubility.

In the biodegradation tests, which are summarized in Table I, eight different chlorine-containing compounds were deposited on diatomaceous earth and cultured with mixed wild strains of marine microorganisms under

Table I
TOTAL WEIGHT AND CHLORINE CONTENT OF RESIDUES OF ORGANIC COMPOUNDS EXPOSED FOR ONE YEAR TO MARINE MICROORGANISMS IN SEAWATER

Organic chlorine compound	*Hypothetical* 0-*Month* *Total* (*g*)	*Cl* (*mg*)	*Experimental* 0-*Month* *Total* (*g*)	*Cl* (*mg*)	*Aerobic* 12-*Month* *Total* (*g*)	*Cl* (*mg*)	*Anaerobic* 12-*Month* *Total* (*g*)	*Cl* (*mg*)
Aldrin	0·20	117	0·19	114	0·18	105	0·19	102
Chlorinated paraffin	0·20	N/A	0·20	90	0·19	80	0·21	82
DDD	0·20	89	0·20	88	0·20	87	0·20	87
DDE	0·20	88	0·19	88	0·21	87	0·20	87
DDT	0·20	100	0·20	98	0·21	98	0·20	98
HCB	0·20	145	0·18	143	0·17	123	0·16	113
Kepone	0·20	145	0·20	114	0·19	108	0·19	120
Lindane	0·20	146	0·20	146	0·19	136	0·19	138

both aerobic and anaerobic conditions. None of the compounds was degraded appreciably in one year. In view of these results, it is unlikely that biodegradation of insecticide had caused the decline in the toxicity of the wood in the leach-resistance tests.

To explain the seemingly contradictory results of the previous experiments, it was hypothesized that the chlorinated insecticides are transported in seawater primarily by being absorbed in the fatty cell membranes, or other fatty tissues, of heat-sensitive microorganisms suspended in the water. The hypothesis predicts that less fat-soluble insecticides than DDT will enter the oceanic food chain less readily than DDT and will at the same time be more persistent. Chlorinated insecticides are very insoluble in water and if they cannot be absorbed by microorganisms, they have no apparent way to be transported in water from their site of application. Several experiments are now underway at CEL to determine the validity of the hypothesis.

EXPERIMENTAL METHODS

In all of the experiments now underway at CEL, small sticks of wood of the kind used for making matches are employed as test specimens. The matchsticks were impregnated with solutions of various chlorinated insecticides at a concentration of 20 mg/ml. The solvent, which in most instances was xylene, was then permitted to evaporate. The insecticide-impregnated matchsticks are being exposed to seawater under a variety of conditions.

In two of the three experiments now underway, gas chromatography is being employed to determine residual insecticide in the matchsticks after various periods of exposure. A research model gas chromatograph with a hydrogen flame ionization detector and a glass column packed with silicone-coated diatomaceous earth is being used. The chromatograms are developed isothermally at a column temperature of 242°C. Terphenyl is employed as an internal standard. The analyses are performed on solvent extracts of the matchsticks.

In the first experiment, DDT-impregnated matchsticks are being leached with seawater of different microbial populations. Some are being leached in hot seawater, some in cold, some in sediment-laden seawater, some in clear, some in cultures of microorganisms, and some in the ocean. The water is changed at least twice weekly in most of the containers; but in four of them, the water is not changed. One pair of the latter containers is maintained under aerobic conditions; the other pair, anaerobic. Bacterial counts on the seawater in the containers and chemical analyses of the residual insecticide in the matchsticks are made periodically.

In the second experiment, the relative rates of biodegradation of the chlorine-containing insecticides listed in Table II are being determined.

Table II

COMMON NAMES OF THE CHLORINATED ORGANIC INSECTICIDES, OR RELATED CHLORINATED COMPOUNDS, THAT ARE BEING SUBMITTED TO BIODEGRADATION TESTS

Aldrin	Hexachlorobenzene
Bulan	Isobenzan
Chlorinated paraffin	Kepone
Chlordane	Lindane
m,p′-DDD	Methoxychlor
o,p′-DDD	Mirex
p,p′-DDD	Perthane
o,p′-DDT	Pentachlorophenol
p,p′-DDT	Prolan
Dicofol	Strobane
Dieldrin	TDE
Endrin	Tetradifon
Heptachlor	Toxaphene

Matchsticks impregnated with the compounds are being exposed to marine bacteria in containers of seawater, some of which are being maintained under aerobic and some under anaerobic conditions. In this experiment, the seawater is not being changed. Volume changes caused by evaporation are adjusted by addition of distilled water; but these changes are seldom required, because a high humidity is maintained in both the aerobic and anaerobic chambers. After periods of one and two years, residual insecticide in the matchsticks will be estimated by gas chromatography.

In the third experiment now underway at CEL, 10 matchsticks each, impregnated with each of the 26 insecticides listed in Table II, are being exposed in a seawater aquarium. Each week the aquarium is stocked with 100 or so wood-borers of the species *Limnoria tripunctata*. The aquarium is supplied with running seawater. The times required for the wood-borer to sever the matchsticks are employed as indices of the rates of leaching of the various insecticides from the matchsticks. The rates will be compared to the known physical properties of the compounds.

RESULTS AND DISCUSSION

The experiments now underway at CEL do not differ too much from some of those that were performed previously, but the techniques have been improved. In most instances gas chromatography rather than 'wet-chemistry' is being employed to estimate the relative quantities of residual insecticide.

An insufficient supply of test animals was available for the bioassays in the previous experiments, whereas the supply is now abundant. Formerly, the bioassays were conducted in an aquarium in which the seawater was changed periodically. Now they are conducted in an aquarium supplied with running seawater. The tendency for the aquarium water to become toxic from the leached insecticides has thereby been reduced.

The only experimental data that have been obtained from the current experiments are summarized in Table III. The results of the analyses for

Table III
DDT CONTENT OF DDT-IMPREGNATED MATCHSTICKS WHICH WERE LEACHED IN VARIOUS MEDIA FOR 1, 12, AND 18 MONTH PERIODS

Leaching medium	*Leaching temperature (°C)*	*Weekly changes of water*	*Microbial population (per ml)*	*DDT-content per matchstick* 1 Month (mg)	12 Month (mg)	18 Month (mg)
Seawater	25	3	10^5	0	0	0
Seawater	50	3	10^0	0	0	0
Air	25	0	N/A	7	6	7
Ocean	20	0	10^3	7	6	5
Deionized water	25	3	10^4	6	6	6
Deionized water	50	3	10^0	7	6	6
Seawater	25	3	10^5	7	6	6
Seawater	50	3	10^0	7	5	5
Seawater and sediment	25	3	10^5	5	6	7
Seawater and sediment	50	3	10^0	5	5	6
Seawater in light	25	3	10^5	5	6	6
Seawater in light	50	3	10^1	5	6	6
Microbial culture	25	3	10^6	6	6	6
Microbial culture	50	3	10^3	7	5	6
Aerobic seawater	25	0	10^0	no data	6	6
Anaerobic seawater	25	0	10^5	no data	6	6

residual insecticides in the matchsticks after 1, 12, and 18 months are presented. It is impossible to ascertain relative leaching rates from the data because too little change has occurred in the quantities of residual insecticides in the matchsticks.

The results attest to the very slow rate at which DDT is leached from wood and to its great resistance to biodegradation. Repeated changes of seawater assures a repetitive introduction of marine microorganisms. Yet the DDT does not disappear. Also, the extracted insecticide, which was originally a highly purified grade of p,p′-DDT, continues to be eluted from the chromatographic column as a single peak shortly after the peak for the terphenyl standard.

The data of Table III are consistent with the data of Table I from the previous experiments at CEL. In Table I, the chlorine content of the aged samples of kepone and hexachlorobenzene (HCB) were 25% or so lower than the theoretical initial chlorine content. However, the lower value for the chlorine content for these two compounds was traced to incomplete extraction and reaction in the analytical procedure rather than to an actual loss of chlorine. Overall, the analyses showed that no extensive loss of chlorine occurred during the experimental period.

From the experimental data that have been obtained thus far, it is not possible to compare leaching rates with bacterial counts of the seawater or with the fat solubilities of the insecticides. However, there are literature references that are consistent with the hypothesis upon which these experiments are based. For example, investigators (Leshniowsky *et al.*, 1970) at Ohio State University found that when floc-forming Lake Erie bacteria settle, they remove all of the aldrin from the water columns. Other investigators (Odum *et al.*, 1969) studied the distribution of DDT in estuaries. Concentrations in particulates ranged up to 78 ppm and were 10 000 to 100 000 times higher than the amounts in the filtrates.

One of the basic principles of pharmocology, which was formulated more than 40 years ago (Collander and Barlund, 1933), is also consistent with the hypothesis. The principle states that the permeability of a compound through biological membranes is directly proportional to the oil water partition coefficient of the compound, that is, compounds that are soluble in fats but insoluble in water are readily taken up by cells. Only a few specific compounds that are soluble in water but insoluble in fats can penetrate cell walls. This stems from the fact that most cell membranes are made up of layers of lipoproteins. Many of the chlorinated hydrocarbon insecticides are no doubt soluble in the membranes covering bacterial cells and they are usually very insoluble in water. Hence, bacterial cells could conceivably increase the apparent solubility of insecticides in water in much the same way that hemoglobin increases the apparent solubility of oxygen in blood.

The practical significance of the new concept is that it predicts that insecticides only slightly soluble in fats might be just as stable and perhaps even more persistent than DDT and yet not necessarily constitute an environmental hazard. Insecticides that remain at their site of application for a long time might not be as damaging to the environment as less persistent ones. They would not be transported as readily in the waterways and would not as readily enter the ocean food chain.

REFERENCES

Collander, R., and Barlund, H. (1933). Permeabilitatsstudien on chara ceratophylla. II. Die Permeabilitat fur Nichteliktrolyte. *Acta Bot. Fenn.*, **11**, 1. Reviewed in: *Principles of Drug Action*, by A. Goldstein, L. Aronow, and S. M. Kalman. Harper and Row, New York, 1969.

Leshniowsky, W. O., Dugan, P. R., Pfister, R. M., Frea, J. I., and Randles, C. I. (1970). Aldrin: Removal from lake water by flocculent bacteria. *Science*, **169**, 993–995.

Odum, W. E., Woodwell, G. M., and Wurster, C. F. (1969). DDT residues absorbed from organic detritus by Fiddler Crabs. *Science*, **164**, 576–577.

Vind, H. P., Muraoka, J. S., and Mathews, C. W. (1971–73). Biodeterioration of Navy insecticides in the ocean. ONR-Annual Reports 1, 2, and 3.

THE ENZYMOLOGY OF TRINITROTOLUENE REDUCTION

R. E. KLAUSMEIER, J. A. APPLETON, E. S. DuPRE and K. TENBARGE

Naval Weapons Support Center, Crane, Indiana, U.S.A.

Summary

Bacillus subtilis, whose growth is inhibited by 2,4,6-trinitrotoluene (TNT), possesses the ability to metabolically reduce TNT to a variety of amino- and hydroxylaminotoluene derivatives. Cell-free extracts of this culture are capable of reducing TNT, utilizing either nicotinamide-adenine dinucleotide (NAD) or nicrotinamide-adenine dinucleotide phosphate (NADP) as co-factors. The NAD-linked system has a narrower pH range and is more sensitive to heat than the NADP-mediated system. Both react with essentially the same nitroaromatic substrates, elute from Sephadex columns at the same time, and show the same solubilities in ammonium sulphate solutions. The cell-free extract from *Pseudomonas aeruginosa*, whose growth is not inhibited by TNT, also has NAD- and NADP-linked TNT reductases with almost identical characteristics as those from *B. subtilis*.

INTRODUCTION

We have previously reported that the explosive 2,4,6-trinitrotoluene (TNT) is almost universally metabolized by microorganisms during growth (Klauss meier *et al.*, 1974). This is true, at subinhibiting concentrations, for those cultures whose growth is prevented by TNT as well as for those culture-resistant to TNT. TNT interferes with oxygen uptake of washed-cell suspensions of sensitive cultures with a variety of substrates, including endogenous respiration (Klausmeier and Adang, unpubl. data). The metabolic products formed are all reduced forms of TNT, reduction occurring on the nitro-groups. These studies describe initial enzymatic studies undertaken to determine whether the enzymatic mechanisms of TNT reduction differ between susceptible and nonsusceptible strains, and whether any such differences can account for the differences in susceptibilities.

METHODS

Two cultures were utilized during these studies: *Pseudomonas aeruginosa* ATCC 13388 and a *Bacillus subtilis* of unknown origin carried in our collection as B13. The organisms were grown in a mineral salts medium (Gary and Bard, 1952) for 24 hr at 28°C after inoculation from a 24 hr seed grown in the same medium. The *B. subtilis* culture will impart a pink pigmentation in the medium. An occasional nonpink culture was found not to be satisfactory.

Cells were harvested by centrifugation, washed twice with physiological saline, and resuspended in distilled water at a concentration of 40 to 50 times the growth concentration. The suspension was then subjected to four 30-

second intervals of ultrasonic vibration in a Biosonik disintegrator at full power. The suspension was chilled in an ice bath at that time. Eighty to 90% of the cells were broken during this process. The suspension was then centrifuged at 15 000 *g* at 4°C for 15 minutes. The precipitate was discarded.

Ammonium sulphate fractionation was performed by mixing the proper volume of saturated aqueous ammonium sulphate solution with the cell extract, allowing to incubate at 4°C for approximately 15 minutes, then centrifuging at 15 000 *g*. The pellets were resuspended in water, but were not desalted since preliminary experiments indicate that ammonium sulphate did not interfere with activity at low concentrations. High-salt-containing supernatants were desalted with Sephadex G-10. Other manipulations of the cell extract will be described with the results.

Oxidation-reduction reactions were measured by following changes of absorbancy at 340 nm, utilizing either nicotinamide-adenine dinucleotide (NAD) or nicotinamide-adenine dinucleotide phosphate (NADP) co-factor. Typical reaction mixtures were as follows: 1·5 ml Tris buffer, pH 7·8; 0·5 ml diluted extract; 0·1 ml aqueous $MgCl_2$ (0·1 M); 0·1 ml aqueous reduced NAD or NADP (0·0027 M); 0·25 ml aqueous TNT ($5{\cdot}7 \times 10^{-4}$ M).

Reaction products in sufficient quantity to attempt identification were generated by coupling the TNT-reducing enzymes with commercial glucose-6-phosphate dehydrogenase and an excess of glucose-6-phosphate to maintain a constant source of reduced NADP, or with yeast ethanol dehydrogenase and ethanol to provide a supply of reduced NAD. All commercial preparations were from Sigma Chemical Co.

Thin-layer chromatography was performed on either benzene or acetone extracts of reaction mixtures. The aqueous phase was saturated with $CaCl_2$ prior to extraction. The solvent extracts were evaporated to dryness and the residue taken up in the minimal amount of solvent for application to the plates. After spotting on silica gel plates, the plates were developed with the solvent systems described by either Hoffsommer (1970) or Chandler *et al.* (1972). After the solvent had been evaporated, the plates were examined either under ultraviolet light or were sprayed with a 1:5 mixture of ethylenediamine:dimethylsulphoxide.

Spot tests for nitroso compounds were as described by Feigl (1960).

RESULTS

Preliminary experiments were conducted with *B. subtilis*. Various methods of cell disruption were utilized—ultrasonic disintegration, alternate freezing and thawing, extraction from acetone- or freeze-dried cells, and mechanical grinding with abrasive. The most satisfactory method was sonic disintegration.

Inasmuch as all biological reactions with TNT involve reduction of the nitro-groups to hydroxylamino or amino groupings, the first reactions measured were the involvement of the oxidation-reduction coenzymes nicotinamide adenine dinucleotide (NAD) and nicotinamide adenine dinucleotide phosphate (NADP). Spectrophotometric observation indicated that the reduced forms of both co-factors (NAD · H and NADP · H) were oxidized in the presence of cell extract and TNT. (There is an active system in this preparation which reoxidizes the reduced NADP in the absence of added substrate. By dilution of the cell extract this activity was so reduced as to be not detectable under experimental conditions.) That the reaction involved was truly oxidation of NADP · H was demonstrated by allowing the decrease in absorbancy mediated by addition of TNT to continue to completion. At that time glucose-6-phosphate (G-6-P)—the preparation contains an active glucose-6-phosphate dehydrogenase—was added. The absorbancy immediately began increasing, indicating reduction of the oxidized co-factor (Fig. 1). No NAD-linked dehydrogenases were readily discernible in the cell extract. Corresponding experiments, which demonstrated conclusively that the NAD · H–TNT-extract mediated decrease in absorbancy was indeed oxidation of the co-factor, were not performed.

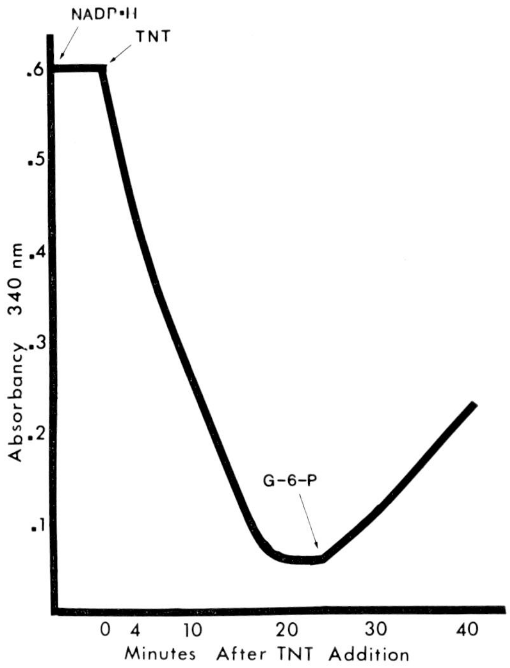

Fig. 1. An NADP-linked TNT reductase.

Incubation in a boiling water bath for 10 min destroyed the activity of both the NAD- and NADP-linked reactions.

The reaction rates are a function of the TNT concentration or the amount of enzyme present (Fig. 2). Apparently, the enzyme is saturated at TNT concentrations approximating 4×10^{-5} M (approximately 10 ppm).

The enzyme preparations showed a broad pH range at which the NADP-linked reaction was equally active (Fig. 3). The NAD-linked system was most active over a narrower pH range.

The NAD- and NADP-linked systems also showed differences in heat lability, the NAD-linked system being completely destroyed (based on the spectrophotometric assay procedure) after 11 min at 55°C, whereas the NADP-linked activity was reduced only by half.

Attempts to separate the two enzyme systems (NAD- and NADP-linked) by ammonium sulphate fractionation or by use of Sephadex columns were

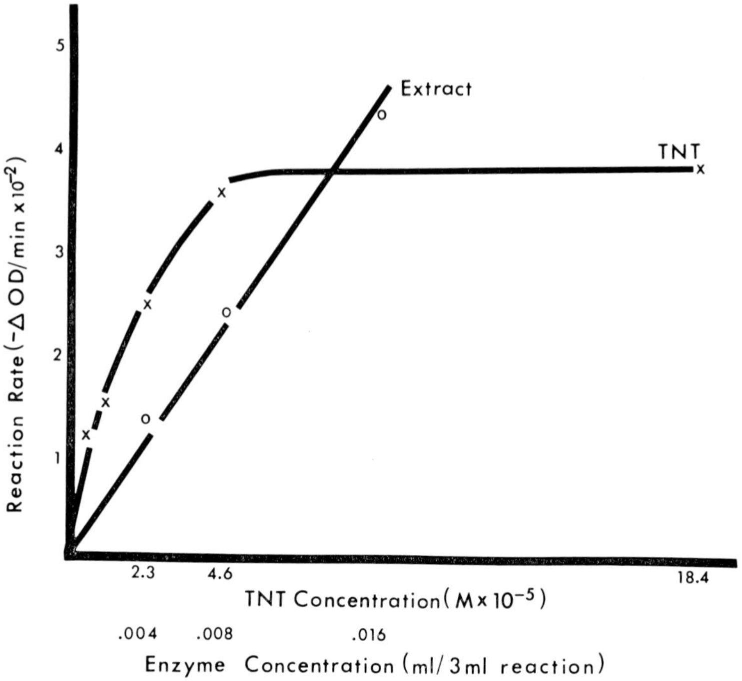

Fig. 2. The effect of substrate or enzyme concentration on NADP-linked TNT reductase.

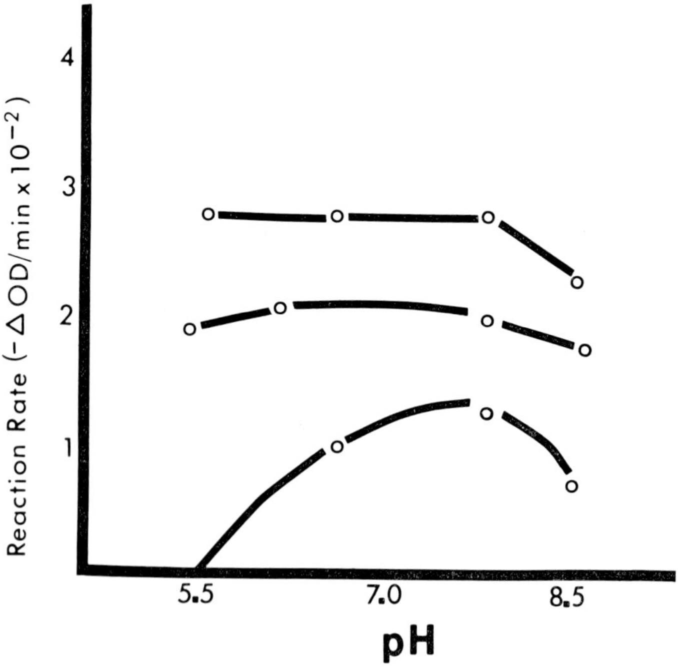

Fig. 3. The effect of pH on NAD- and NADP-linked TNT reductase.

not fruitful. Both systems remained soluble in 50% saturated ammonium sulphate, and were precipitated in 80% saturated solutions. On the Sephadex G-100 column at a bed height of 39 cm, at a flow rate of 25 ml Tris buffer, pH 6·0, per hour, the NAD-linked activity began to elute after 50 ml of buffer had passed through the column, with the most activity in fractions collected at 105 and 110 ml. The NADP-linked activity was confined almost exclusively to fractions collected at 100 and 105 ml.

For comparison, extracts prepared by ultrasonic disintegration of *P. aeruginosa* were examined for TNT-reductase activity. These extracts also contained NAD- and NADP-linked systems. The NADP-linked enzyme was approximately 10 times as active as the NAD-linked system. The bulk of both systems was soluble at 50%, but precipitated by 80% saturated ammonium sulphate. Further, the NAD-linked enzyme is very heat labile (all activity is lost in 3 min at 55°C) whereas the NADP-linked system retains over 70% of its activity after 11 min at the same temperature.

The extract from *B. subtilis* was examined to determine its ability to oxidize the two reduced co-factors in the presence of a variety of organic substrates. All substrates were first dissolved in a small amount of ethanol, then diluted with water to yield a concentration which, in the reaction cuvette, approximated 8×10^{-4} M. Almost all of the substrates most active with NADP were also the most actively metabolized substrates with NAD. The sole exception is 2,4-dinitrotoluene, which reacts rapidly with NADP but at only 40% of the TNT rate with NAD. Further, none of the substrates which reacted slowly, or not at all, with NADP reacted differently with NAD (Table I).

Table I
THE USE OF VARIOUS SUBSTRATES BY *B. SUBTILIS* EXTRACT WITH NAD OR NADP

Substrate	*Reaction rate* (−Δ*OD*/*min*) *NADP*	*NAD*
Trinitrotoluene	0·021 8	0·019
m-Dinitrobenzene	0·025 0	0·011
p-Dinitrobenzene	0·019 4	—
2,4-Dinitrotoluene	0·192	0·008
1,3,5-Trinitrobenzene	0·019 0	0·026
4-Hydroxylamino-2,6-dinitrotoluene	0·018 0	—
0-Dinitrobenzene	0·016 5	0·010
2,6-Dinitrotoluene	0·001 5	0·004
Nitrobenzene	0·002 5	—
Toluene	0	0
Hexahydro-1,3,5-trinitro-5-triazine	0	0
Ammonium picrate	0	0
4-Amino-2,6-dinitrotoluene	0·006	0

It is interesting to note that only compounds containing two or more nitro groups react. However, not all such compounds react. Similar studies were not carried out with the *P. aeruginosa* extract.

Some preliminary studies of reaction products were carried out with *P. aeruginosa* extracts. After the reaction had proceeded for an adequate time (15 min to 3 hr), the reaction mixture was extracted with benzene or acetone. The extract was concentrated and examined by thin-layer chromatography.

The reaction products appeared to be the same for both systems. Other than residual TNT, the reaction products included 2,2′,6,6′-tetranitro-4,4′-azoxytoluene, a dimer formed by nonenzymatic condensation of two 4-hydroxylamino-2,6-dinitrotoluene molecules. Two other observed products included one having an Rf of 0·49 in Chandler's solution and coloured red-brown when sprayed with ethylene diamine, and a second, having an Rf of 0·21 and an orange colour after spraying. This latter spot moves at the same rate as 4-hydroxylamino-2,6-dinitrotoluene in both Chandler's and Hoffsommer's developing solvents. The colour, however, differed from that of a known sample of hydroxylaminodinitrotoluene. This spot also reacted with sodium pentacyanoamine ferroate to give a tan colour which darkens to a greenish brown. A green or purple colour is characteristic of nitroso compounds. Lack of sufficient sample precluded further identification of the reaction products.

Thin-layer chromatography with extracts from the *B. subtilis* culture demonstrated reaction products having identical Rf values and whose reaction colours after spraying are identical with the reaction products of *P. aeruginosa* extracts.

DISCUSSION

Both TNT-susceptible (*B. subtilis*) and nonsusceptible (*P. aeruginosa*) cultures enzymatically reduce TNT. The cultures show extremely similar enzymatic characteristics. These include: use of either NAD or NADP as coenzyme; faster reaction rates with NADP than with NAD; heat lability of the NAD-linked system; relative solubilities in ammonium sulphate solutions; and similar products of reaction. The differences in susceptibilities of these two cultures, therefore, cannot be explained by differences in the initial enzymatic reactions of the two with TNT. We have named this enzyme system(s) TNT reductase.

There is also an amazing similarity between the activities of the NAD- and the NADP-linked TNT reductase systems. The two move at the same rate on Sephadex columns, have the same ammonium sulphate solubilities, react with the same substrates, and produce the same products after reaction with TNT. They do show somewhat different reactions to pH changes, and the NAD system is much more heat labile. However, one is inclined to speculate that there is only one enzyme, but that it reacts either with NAD and NADP. The pH and heat-related differences may merely reflect the site of reaction of the coenzymes.

From examination of the various nitroaromatic compounds which react with this system, it seems probable that the substrate must contain at least two nitro groups. Neither mononitrobenzene or any of the mononitrotoluenes react. These groupings need not necessarily be separated at two-carbon intervals, since all three dinitrobenzene isomers react equally well. Further, groupings between the nitro groups (*i.e.*, 2,6-dinitrotoluene) interfere with the reaction. However, addition of polar groups to the ring seem to prevent utilization of the compound. None of the nitrosubstituted phenols (2,4-dinitrophenol or ammonium picrate) reacted. Any speculation on actual mechanism of reaction based on these data, however, would be premature.

BIODEGRADATION OF ETHYLENE/VINYLACETATE CO-POLYMERS

G. J. L. GRIFFIN and H. MIVETCHI

Brunel University, Middlesex, U.K.

Summary

The common synthetic plastics have not generally been regarded as biodegradable, apart from the special case of plasticised PVC. Because of the conflicting technological objectives in, on the one hand, maintaining the stability of plastics used in exposed conditions as structural materials and, on the other hand, in achieving some control of the interaction between throw-away plastics packaging and the environment, any observed exceptions to the biological inertness merit special study. This work reports observations on the growth of fungi in association with EVA co-polymers. The extent of the growth is shown to increase with the vinyl acetate content and also the presence of starch as a filler. A simple optical technique is described for recording the extent of fungal invasion, and the actual penetration of hyphae into the body of the samples is confirmed by microscopy. The results of experiments with pure cultures and burial in soil are listed. Gross changes in the physical properties of some samples are reported with, in the extreme case of pure polyvinyl acetate, the appearance of partial water solubility.

INTRODUCTION

If we limit our definition of biodegradation to 'a direct utilization of a major component of a material in the metabolic processes of microorganisms', this, in practice, implies a situation in which the normal enzyme secretions of that microorganism may be regarded as corrosive agents toward the material and also that the corrosion products diffuse back into the microorganisms with nutritionally beneficial results. This happens in the case of plasticised PVC and it is generally accepted (Booth and Robb, 1968) that the mechanism involves a diffusion of the low molecular weight plasticiser out of the body of the plastics material to the surface region which is being depleted by the corrosion process. Let us call this Type I. In this particular case, the physical consequences would be a loss in weight accompanied by an increase in hardness and tensile strength. There is, as yet, no evidence for attack on the vinyl chloride polymer chain itself. If we restrict the definition still further to 'utilization of part of the polymer chain structure proper', then the ability of the microorganism to cause a real collapse of the structure is implied and there are very few authenticated examples of synthetic polymers which can be quoted as demonstrations of this type of process, which we can call Type II. The most celebrated are fungal corrosion of certain of the polyester polyurethanes and the effect was originally identified by Levisohn and Evans, (1966) in these materials. It is very clear from microscopy that, in this case, 'chemical tunnelling' is going on with the implication that a limited region of the growing hyphal tip is active. The ASTM fungal resistance test procedure does not call for any attempt to discriminate between overgrowth and penetration, *i.e.*, Type I and Type II, so that the useful survey of microbiological sensitivity given by Potts *et al.* (1973) is not so subclassified. The absence of plasticiser from a polymer formulation does not of itself prove that only a Type II mechanism is prevailing because there are usually other minor ingredients present with sufficient mobility or, simply, the low molecular weight fraction occurring in the usual polymer chain-length distribution may be mobile enough. There is the further possibility that a Type II biocorrosion might occur without hyphal penetration.

One polymer listed by Potts *et al.* (1973) as giving a positive ASTM test response is PVA, while LDPE appears in his table as a nil response. In view of the fact that a whole range of co-polymers of vinyl acetate and ethylene are available, it would appear interesting to examine any progressive change in biological sensitivity with a change in vinyl acetate content of such co-polymers and such an effect has been claimed by Worne (1970). The effect of the introduction of an auxiliary nutrient in the form of whole-grain starch filler could also be examined.

EXPERIMENTAL

Test environments

Two exposure environments are reported here. The first was soil burial in John Innes No. 3 potting soil which had been used for its proper purpose for two seasons so as to ensure a rich content of microorganisms. Disc samples 50 mm in diameter were set out horizontally in the centre plane of normal seed trays filled with soil so that the samples were covered by 20–30 mm of soil. The trays were stored under the benches in an unheated greenhouse and they were watered regularly as part of the greenhouse routine.

The second environment was essentially that of the ASTM D 1924 test with sample discs 20 mm in diameter incubated on *Trichoderma viride* cultures (CMI 45553).

Preparation of samples

Polymers. A variety of commercial EVA co-polymers were selected to cover a wide range of vinyl acetate contents and molecular weights. Typical data, as provided by the manufacturers, are listed in Table I, some of it on a batch analysis basis.

Two series of sample discs were prepared, one series from the polymers alone, the other from polymer blended with 40 phr (0·197 volume fraction or 28 % ww) of rice starch which had been oven pre-dried at 80°C to a moisture content less than 2 %, by mixing on a 27 × 15 cm 2-roll mill steam heated to the minimum temperature necessary to flux the resin in order to avoid the compound sticking to the rolls. In each case the rough hides were hot pressed into sheets about 0·5 mm thick between polished steel plates using 'Melinex' polyester foil as a facing material, enabling the rather sticky co-polymers to be released easily from the press and handled without fear of contamination until required for testing.

Table I

Supplier	*Code*	*Melt flow index*	*Density g cm^{-3}*	*Vinyl acetate content %*
ICI	Q1388	2	0·916	0
ICI	EVA 49	10	0·937	18
USI Chemicals	UE 631	1·4	0·941	18–21
USI Chemicals	UE 633	20	0·941	17–19
USI Chemicals	UE 634	3	0·950	26–28
USI Chemicals	UE 636	22	0·948	26–28
USI Chemicals	UE 646	16	0·949	26–28
USI Chemicals	UE 643	9	0·941	18–20
USI Chemicals	EY 902	34·2	0·967	41·7
USI Chemicals	20150	7·3	0·973	
Cairn Chem. Co.	S7	Mw 86 000	—	100

Assessment methods

Optical. Gross changes could be noted visually and recorded photographically. Less dramatic effects needed a more refined approach. In an attempt to quantify the extent of the overall change in the samples, they were scanned by a standard microdensitometer and the change in optical density between the exposed sample and a stored control sample recorded as an index of attack. Figure 1 shows a typical scan across a control and a soil

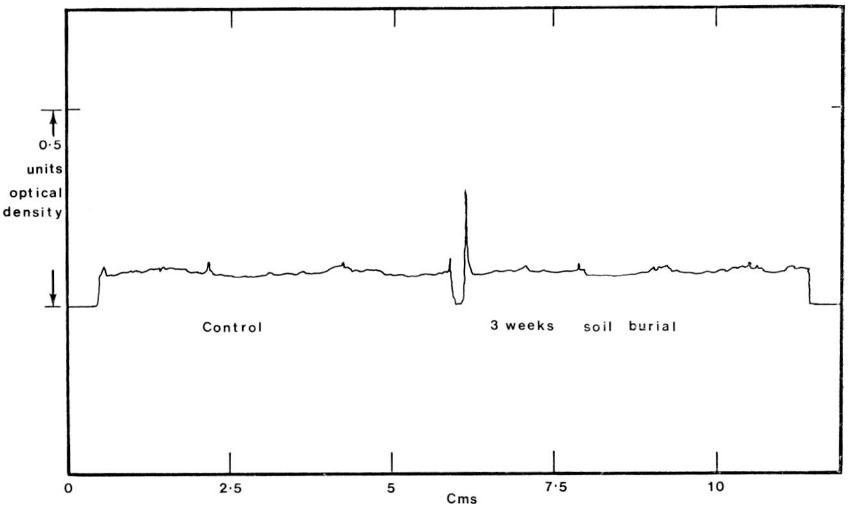

Fig. 1. Change in optical density of a typical EVA sample sheet after exposure to soil burial for 3 weeks. Reproduced from a single microdensitometer trace. Sample was a proprietary partly hydrolysed EVA moulded sheet.

burial sample with almost no effects, while Fig. 2 shows the effect of starch addition to an otherwise identical pair. Data extracted from six sets of samples for a typical EVA/starch blend are shown in Fig. 3 plotted against time of burial. Even when the growth was irregular and the traces correspondingly broadened, it was easily possible to draw in a mid-line for the purposes of extracting a single density change figure. The method was sensitive to lack of parallelism between the faces of the samples and, if present,

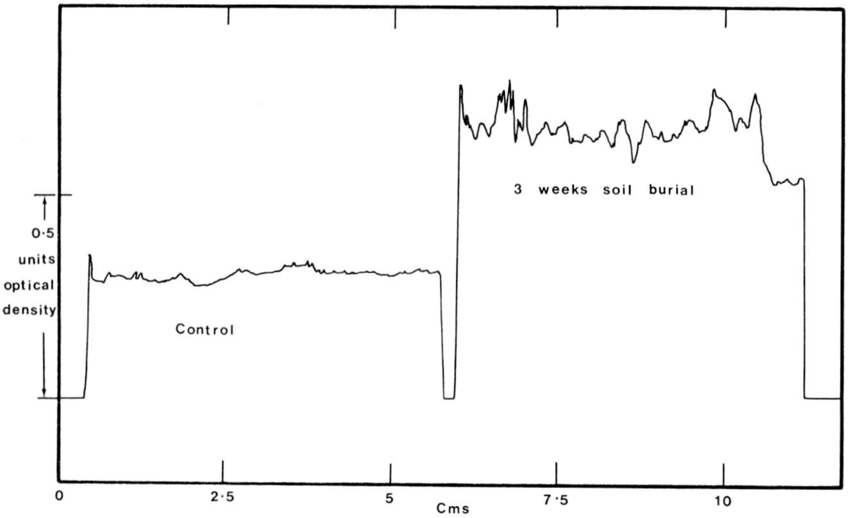

Fig. 2. Change in optical density of a typical EVA/starch sample after exposure to soil burial for 3 weeks. Reproduced from a single microdensitometer trace. Sample was a proprietary partially hydrolysed EVA with 0·197 vol. fraction rice starch.

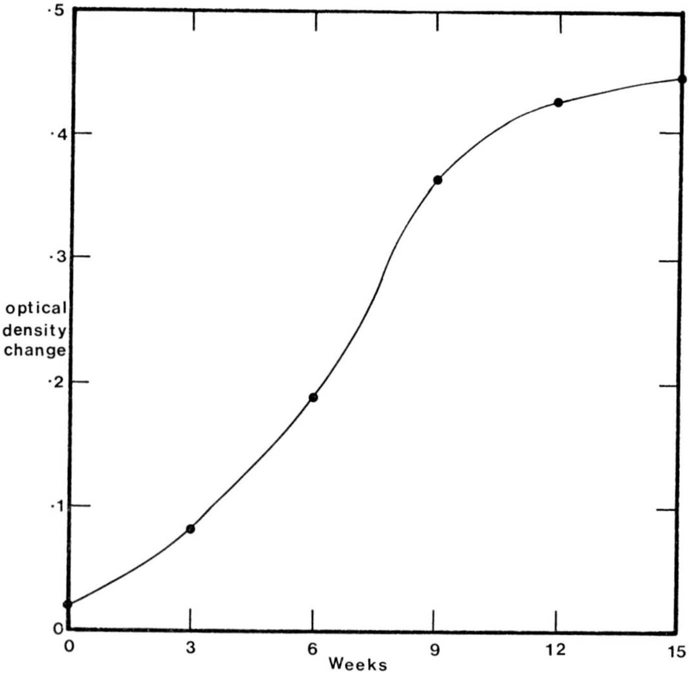

Fig. 3. Increase in optical density over control sample plotted against time of soil burial for an EVA/starch blend. Sample was UE633 of 20% VA content, MFI 20 density 0·941 g . cm^{-3} and containing 0·197 vol. fraction of rice starch.

this became evident in the form of an imposed gradient on the trace which could be accommodated by, again, drawing in the mid-line and taking its centre as the measurement point.

The samples were also examined by a high power optical stereo-microscope and, in certain cases, sections cut, stained in lactophenol cotton blue, and mounted in lactophenol for photomicroscopy.

Mechanical. Because of the extreme difficulty of handling the quantity of material that would have been necessitated by a decision to use conventional tensile testing procedures, hardness was adopted as an alternative index of mechanical change. Because the specimens were thin (ca. 0·5 mm), it was necessary to use a very small indentor and this was available in the ICI pattern microhardness tester.

Metabolic. Because evidence for major breakdown of the material was being sought, this aspect of the work was confined to following weight loss of the samples after careful conditioning to constant weight at 35°C.

RESULTS

Because the test period was limited to 15 weeks and the testing environments were mild, as compared to the more chemically aggressive nature of a composting garbage environment as reported by Wallhäuser (1973) and Griffin (1975), polyethylene itself showed very little reaction in any of these tests. At the other end of the scale, however, polyvinyl acetate (Mw 86 000 visc. 26–32 cps 86 g/1000 ml benzene) developed a high degree of water sensitivity very quickly, especially if starch filled, and could not then be separated from the soil. On drying the soil a 'biscuit' of impregnated soil could be recovered.

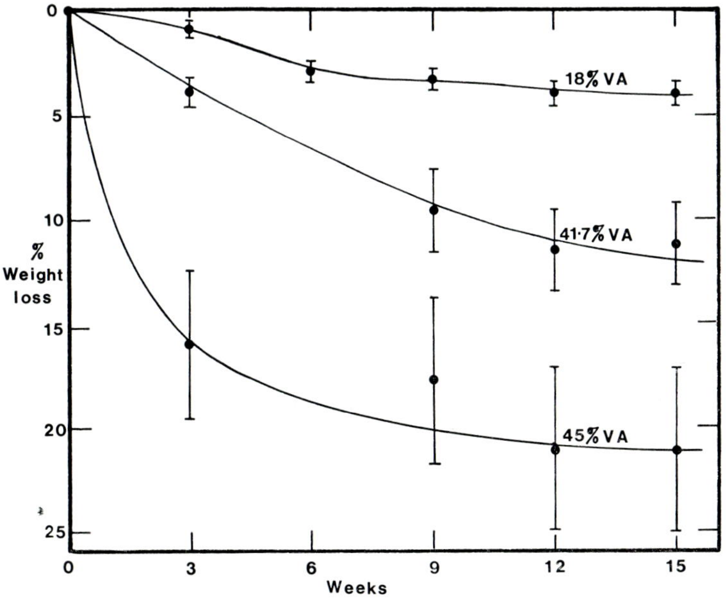

Fig. 4. Variation in percentage weight loss with time on exposure to soil burial conditions of 28% ww maize starch loaded EVA with three different vinyl acetate contents.

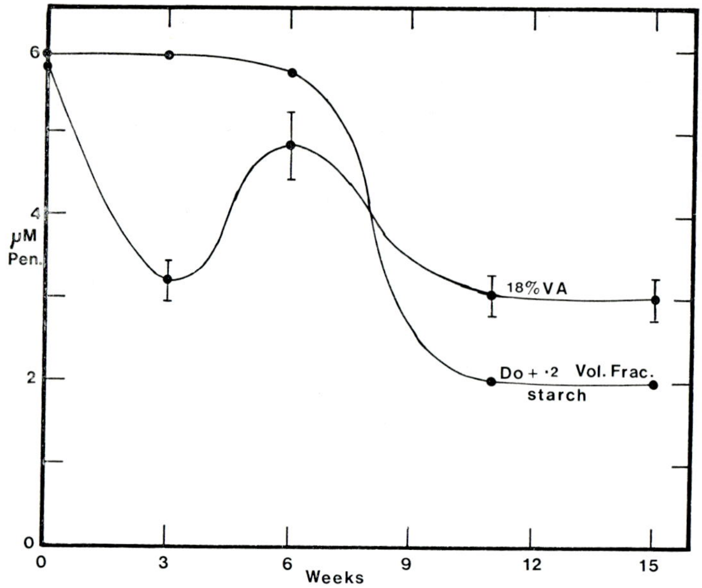

Fig. 5. Variation in microhardness with time for EVA and EVA/starch mixture exposed to Trichoderma viride *culture.*

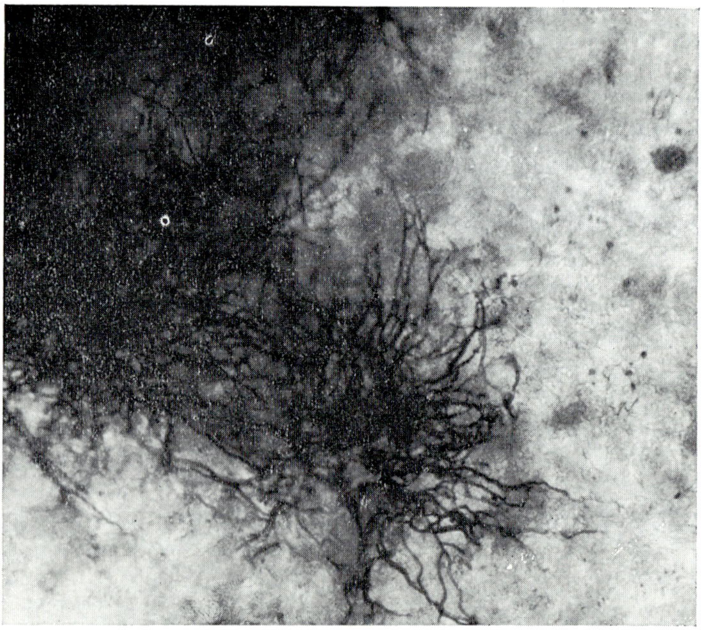

Fig. 6. Photomicrograph of the surface of an EVA sheet with 41·7% vinyl acetate content and containing 0·197 vol. fraction of rice starch after 9 weeks soil burial. The surface has been stained with lactophenol cotton blue. The texture in the background is due to the swollen starch grains and the parts of the hyphal growth below the surface can be fairly readily identified. Scale 3 mm equals 2 µm.

POLYVINYL ALCOHOL BIODEGRADATION BY OXYGEN-ACTIVATED SLUDGE

J. P. CASEY and D. G. MANLY

Air Products and Chemicals, Inc., Allentown, Pennsylvania 18105, U.S.A.

Summary

VINOL 325 Polyvinyl alcohol (PVOH) may be totally biodegraded as illustrated by controlled respirometric assays. Organisms which consume PVOH are viable in mixed culture activated sludges as determined by both synthetic and actual domestic wastewater studies in which PVOH was added to influent to a small scale multi-stage OASES correlation unit. Mixed culture growth rates analysed by appreciation of oxic solids residence time and clarifier damping factors permit sequential analyses of *Nitrosomonas*, *Nitrobacter*, and PVOH-degrading organism proliferation in activated sludge operated at successively lower dilution rates. A step increase in PVOH concentration allowed an initial calculation of PVOH organism growth rate within the oxygen activated sludge system of $\mu = 0{\cdot}34$ day^{-1} at 20 ± 1°C. A second scratch startup and examination of PVOH organism proliferation at a lower dilution rate confirmed this growth rate. Efficiencies of mixed culture PVOH degradation reached 13 mg PVOH/hr/g VSS with 3 hr influent residence time for a domestic wastewater primary clarifier effluent. Organisms degrading PVOH in mixed culture remained viable under H_2O_2 treatment conditions which were toxic to filamentous and nitrifying organisms. Systems removing >90% of 50 mg/litre PVOH, >95% of 40 mg/litre NH_3 and >90% of BOD_5 were operated with excellent macroscopic sludge properties.

DISCUSSION

In light of early documentation of microbial metabolism of PVOH to CO_2 and H_2O (Nord, 1936) and later reference thereto (Pritchard, 1970), it is surprising that aqueous solutions of PVOH were long accepted as minimally biodegradable. This conclusion has been affirmed by reports of 0·2% (Clemson, 1971) −1% (Burr, 1971) BOD/weight ratios determined by *Standard Methods* (APHA, 1971). The weakness of BOD testing to determine 'latent' biodegradability has been appreciated (Burr, 1971), yet other more rigorous methods have been employed to document the refractory nature of specific commercial PVOH products (Wirick, 1974).

Because of the extensive use of PVOH as a textile sizing agent and the need for nonpolluting textile wastewater treatment systems, considerable recent work on PVOH has appeared. A study reporting variable long-term BOD results for PVOH (Porter and Snider, 1974) protested the difficulty in relating controlled biodegradability tests to wastewater treatment operations. An integrated study of textile wastewater treatment technology dealt with activated sludge acclimatization to, and removal of, PVOH on a 2 litre scale (Bryan, 1972). A 20 litre and pilot scale study (Baines, 1973) documented

PVOH removal by activated sludge. Further work has generated dimensionless data proposing a loading ratio of 0·1 COD/MLSS and a solids aeration time (Sat) of 8 (g MLSS/L)(days) for efficient treatment (Hahn *et al.*, 1975).

Extensive work by Nippon Synthetic Chemical Industry Co., Ltd (Nishikawa and Fujita, 1975) has resulted in proprietary wastewater treatment technology claims involving additive and process control tactics, yet some textile plants have performed satisfactorily on an independent basis (Anon., 1974).

Clearer understanding of the fundamentals of PVOH microbial metabolism has been published (Suzuki *et al.*, 1973), but no satisfactory pure culture or mixed culture growth rate data are presently available. Such information is critical for application of bioengineering principles in wastewater treatment. The present study was undertaken to provide such information and develop

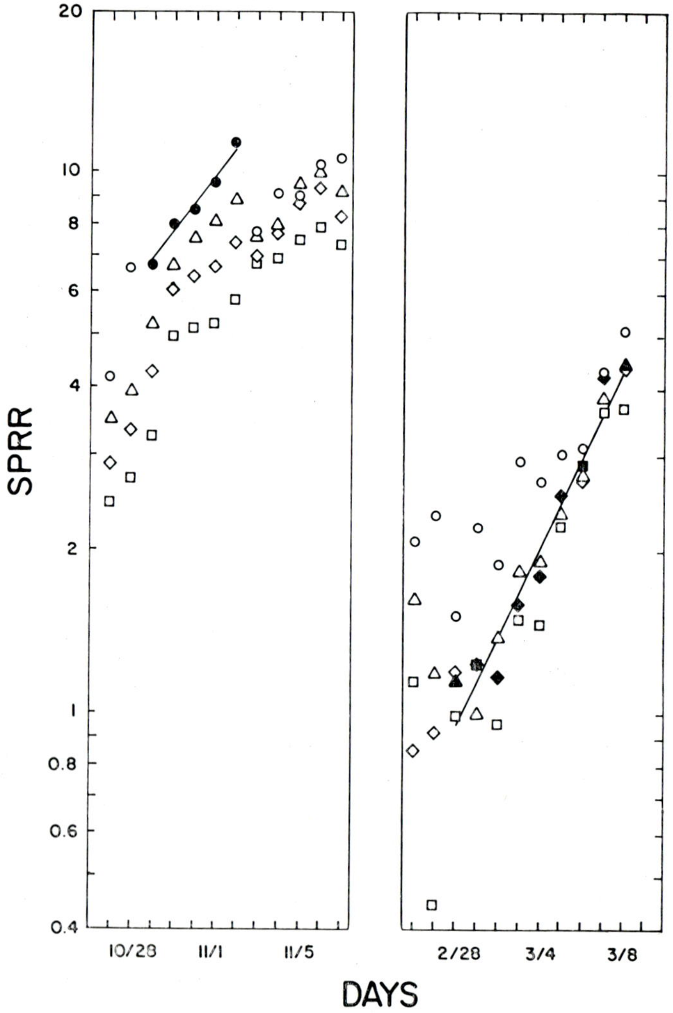

Fig. 1. Analyses of observed growth rate. ○ SPRR Stages 1–2; Δ SPRR Stages 1–3; ◇ SPRR Stages 1–4; □ SPRR Stages 1–5. Filled symbols represent data for least squares line fit. Period I, slope 0·120 day^{-1}. Period II, slope 0·188 day^{-1}.

design criteria for treatment of wastewater containing polyvinyl alcohol.

Multiple-stage OASES correlation unit operation permitted stage by stage substrate determination on a daily grab sample basis. The interstage change in VINOL 325 concentration normalized to MLVSS level and stage residence time permitted calculation of a Specific PVOH Removal Rate (SPRR) expressed in mg PVOH/hr/g VSS. With the viable assumption that active mass degrading PVOH constituted a small fraction of the total MLVSS, the change in SPRR could be used to assay increased or decreased presence of PVOH-degrading bacteria. Given the assumption that SPRR is proportional to PVOH-active biomass (X), the expression $\mu = \int dX/dt/X$ to obtain the growth rate of active mass could be solved by obtaining the slope of *ln* SPRR plotted *vs.* days of operation.

Two summary periods of analysis are presented in Fig. 1. The volume-weighted averages of daily values of SPRR for multiple stages inclusive of the first were used for a least square line fit. Analysis of response to a step increase of VINOL 325 from 50 to 100 mg/litre during period I at constant flow rate results in an observed growth rate of 0·120 day^{-1}. Analysis of the increased rate of VINOL 325 degradation following trace seeding with supplemental active organisms during period II results in calculation of observed μ of 0·188 day^{-1}.

In each case the actual organism growth rate may be calculated after correction for the unit dilution rate, for the observed rates are reduced at higher system dilution rates. Furthermore, a factor which expresses the damping effect of clarifier inventory on observed system response is required. Given rigorous daily and summary average period data, this damping factor is easily calculable from reduction of a set of full system rate equations. Table I presents the specific data utilized to calculate actual growth rates $\mu = \mu_{diln} +$

Table I

Period *Dates*	*I* 10/29–11/2	*II* 2/28–3/8
# days	5	9
μ_{obs} (day^{-1})	0·120	0·188
Correlation coefficient	0·989	0·973
MLVSS (mg/litre)	4 458 ± 290	4 482 ± 174
Aeration volume (L)	16	40
VSS produced/day (g)	12·4 ± 3·7	10·9 ± 5·3
SRT (day)	5·90	16·51
μ_{diln} = 1/SRT (day^{-1})	0·170	0·060
f	1·353	1·468
$\mu_{diln} + f\mu_{obs}$ (day^{-1})	0·333	0·338
T stage #1 (°C)	22·1 ± 0·3	20·3 ± 1·2
T last stage (°C)	21·0 ± 0·4	19·0 ± 1·4

$f\mu_{obs}$. These are 0·33 day^{-1} in response to the step increase in loading and 0·34 day^{-1} for response during the period of higher SRT. In mixed culture with activated sludge in the range of 19–22°C, the organism doubling time is therefore of the order of 2 days. Depending on the solids residence time of the chosen activated sludge system and the damping factor provided by ineffectual clarifier inventory, the time for doubling of observable organism presence may be substantially higher than the 5·8 or 3·7 days noted in the periods analysed above.

This work demonstrates that definitive mixed culture growth rate data can be generated for organisms which biodegrade PVOH in mixed industrial-domestic wastewaters. Furthermore, the current scope of practicable bio-engineering technology (Landis *et al.*, 1974) can be utilized for the design and operation of activated sludge systems for PVOH removal.

MATERIALS AND METHODS

VINOL 325, degree of polymerization 1725 ± 25, degree of hydrolysis 98–99% and VINOL 540, degree of polymerization 2400 ± 50, degree of hydrolysis 87–89% were obtained from commercial lots produced at Air Products and Chemicals' Calvert City, Kentucky facility. PVOH-3 was obtained from a commercial lot imported for use as a textile sizing agent, and is a crotonic acid-vinyl alcohol co-polymer of medium viscosity and 94% degree of hydrolysis.

Suspended solids, BOD and COD measurements were performed following 'Standard Methods' (APHA, 1971). Nitrification of ammonia in the BOD test was supressed by addition of $1{\cdot}6 \times 10^{-4}$ M allylthiourea. Filtered BOD and COD data refer to filtrates obtained from aspirator vacuum filtration using 2·4 cm Reeve Angel glass fibre filter paper. BOD dissolved oxygen (DO) measurements, mixed liquor DO measurements, and dynamic oxygen uptake measurements were performed using a Yellow Springs Instrument stirred BOD probe model #54.

Ammonia, nitrite, nitrate plus nitrite, and hydrolysable phosphate determinations were performed on filtered samples using Technicon Autoanalyser II methods. VINOL concentration determinations using a boric acid-I_2/I_3^- procedure (Finley, 1961) were initially performed using a Perkin-Elmer Model 202 spectrophotometer, then automated for unit operations by adaptation to the Technicon Autoanalyzer II at 80 samples per hour using a 660 nm colourimetric filter for a full-scale range of 0–50 mg/litre VINOL 325.

Unit operations were performed using plexiglas aeration chambers in series. Submerged candle spargers supplied O_2/air/N_2 mixtures to the open-staged mixed liquor in which the gas pressures to each stage were separately reduced and gas flows were separately regulated and monitored by flow tubes. This gas system permitted simultaneous variation of desired DO levels and gas flows providing intrastage mixing at approximately 0·15–0·20 hp/1000 gal. Dynamic oxygen uptake measurements taken serially beginning with the final stage were facilitated by the independent gas flow controls, allowing dilution corrections of the form $dO_2/dt = (DO_t - DO_{t+1})/dt + (DO_{N-1} - (DO_t + DO_{t+1})/2)\,(Q/V)$ for the Nth stage upon terminating gas flow to that stage and providing mechanical stage mixing while recording the DO.

Domestic wastewater was obtained twice a week in epoxy-lined 55 gallon drums from the primary clarifier overflow at the Easton, Pa. sewage treatment plant. Desired concentrations of PVOH were obtained by addition of 5% PVOH concentrate to the barrels. Barrel contents were stirred mechanically to maintain influent solids suspension and were removed, using a Cole-Parmer positive displacement pump capable of drawing influent from the barrel bottom through a $\frac{1}{8}$ inch mesh wire screen protecting the tip of a $\frac{1}{4}$ inch ID plexiglas tube connected to tygon tubing on the pump suction side. Activated sludge recycled from the inclined cylindrical bottom of an 18 litre,

24 inch tall, 6 inch OD clarifier equipped with a 10 inch deep centrewell of 2 inches OD and sludge rake was introduced, via a second peristaltic pump, alongside the influent directly into the first aeration stage.

Daily data included determinations of DO, pH, temperature, TSS, VSS, $NH_3 - N$, $NO_2^- - N$, $NO_3^- + NO_2^- - N$, and PVOH for influent, effluent, and each aeration stage. The oxygen uptake rate (OUR), sludge volume index (SVI), and O_2/air/N_2 gas flows were recorded for each stage. A zone settling velocity (ZSV) and 30 minute stirred SVI (SSVI) were determined on the last stage mixed liquor in a 1 litre graduate using a 6 rph rake. Influent and effluent SS, BOD, and COD were determined at the beginning of the data day on grab samples and their filtrates obtained after overnight steady-stage operation. The Calcomp figures herein present 3-day running average data for daily values except for initial data days at new process conditions, which are represented by a 1-day and 2-day value, respectively.

RESULTS

VINOL 325 was first demonstrated to be nontoxic and noninhibitory upon short-term testing by addition to activated sludge organisms degrading a glucose-based synthetic wastewater. Response of 2500 mg/litre MLVSS from a state of endogenous oxygen uptake to a state of maximum oxygen uptake following addition of 1000 mg/litre glucose was unchanged upon addition of 400 mg/litre VINOL 325. A second indication of nontoxicity was obtained in seeded BOD tests using unacclimatized biomass. In tests of 5–20-day duration, the change in DO of the seeded BOD blanks was not substantially exceeded, but was equalled by, the change in DO (up to 4–5 mg/litre in some heavily seeded tests) of the BOD tests containing 3–7 mg/litre VINOL 325.

A survey of textile wastewater treatment plants was undertaken in parallel with initial acclimatization tests, and consisted of analytical work documenting PVOH removal by select plants, then collection of mixed liquor sludges capable of removing PVOH. A culture from one textile wastewater treatment plant in the Southeast was enriched in fill and draw tests to provide seed organisms to document PVOH biodegradation. Die-away analyses of VINOL 325 confirmed by COD analyses (COD/VINYL 325 = 1·64 g/g, theory ~1·82 g/g) indicated removal rates of 9–15 mg/hr/g VSS.

Supernatant from an endogenous mass of sludge acclimatized to PVOH was used as seed in a series of BOD tests performed without and with 3×10^{-5} M ATU to inhibit nitrification. In addition to replicate seed controls, each substrate was tested at 10, 15, and 30 × dilution of a 100 mg/litre stock solution. DO levels were monitored at 5-day increments through 30 days or until DO depletion, at which points PVOH, $NH_3 - N$, $NO_2^- - N$, and $NO_3^- + NO_2^- - N$ levels were monitored by Autoanalyzer II test methods.

Figure 2 presents summary plots of BOD data indicating levels of precision where multiple samples remained undepleted. Table II presents, based on stoichiometric oxygen consumptions for $NH_3 - N$ to $NO_2^- - N$; $NO_2^- - N$ to $NO_3^- - N$; and PVOH to CO_2 and H_2O, calculated and observed charges in DO levels. The agreement in nondepleted samples strongly suggests total oxidation of VINOL 325 and, after an initial lag, of the less fully hydrolysed VINOL 540. PVOH-3 proved substantially less degradable by the seed organisms acclimatized to VINOL 325.

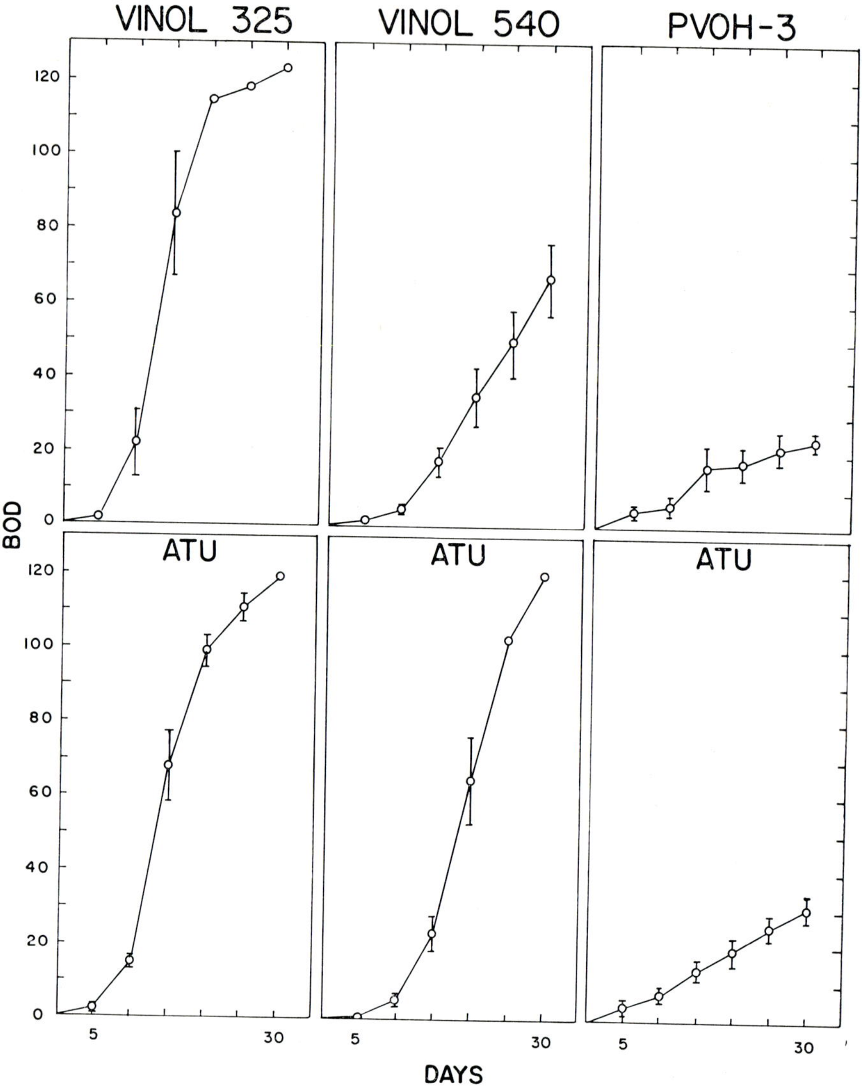

Fig. 2. *Thirty day BOD tests without and with ATU inhibition of nitrification, corrected* vs. *corresponding blanks. I indicates Standard Deviation of nondepleted samples of varying dilutions of 100 mg/litre VINOL 325.*

Table II
DISSOLVED OXYGEN BALANCES IN 30 DAY BIODEGRADATION OF 3·33 mg/litre PVOH

	$\Delta PVOH$	ΔDO $PVOH$	ΔNOs_x	ΔDO NOs_x	*Calculated sum* ΔDO	*Actual sum* ΔDO
VINOL 325	3·33	5·46	0·38	1·72	7·18	7·10
VINOL 325 with ATU	2·98	4·88	0·08	0·34	5·22	4·90
VINOL 540	1·58	2·61	0·46	2·08	4·69	5·25
VINOL 540 with ATU	3·33	5·49	0·06	0·26	5·75	4·90

Table I

REACTIONS OF SPECIFIC PESTICIDE LINKAGES IN SOILS AND/OR MIXED FUNCTION OXIDASE SYSTEMS

Linkage	*Reaction*	*Relative*	*Pesticide*
Ring compounds	ring epoxidation	slow	aldrin heptachlor
	ring hydroxylation	variable	2,4-D MCPA
	reductive dechlorination	rapid	lindane
	oxidative dechlorination (probably chemical)	slow	atrazine
Functional group reactions			
Dealkylations			
R—N(CH_3)$_2$	*N*-dealkylation	slow	many phenylureas
R—O—CH_3	*O*-dealkylation	slow	dicamba linuron methoxychlor
R—S—CH_3	*S*-dealkylation	rapid	prometryne
Dehalogenations			
R—Cl_2—COOH	aliphatic dehalogenation	rapid	dalapon
R—Cl_3	aliphatic dehalogenation	rapid	DDT
ϕ—CF_3	oxidative dehalogenation	slow	trifluralin
Hydrolysis			
R—NH—COOR	ester or amide hydrolysis	rapid	chlorpropham propham
R—NH—COR	amide hydrolysis	rapid	propanil
R—COOR	ester hydrolysis	rapid	esters of phenoxy herbicides
Oxidation			
RO, S, R, RO, OR (RO and S above R, RO and OR below R)	oxidation (probably chemical)	rapid	parathion
R—C N	oxidation	rapid	cyanazine
Reductions			
ϕ—NO_2	reduction	rapid	trifluralin dinitramine

2,2-*bis*(*p*-chlorophenyl)ethylene] are resistant to rapid metabolism in most aerobic environments. Synthesis of a large number of analogs of DDT, primarily by substitution of various ring groups in the *p*- position, has provided valuable information on biodegradability. Detoxification was measured in insects (*Musca domestica* and *Phormia regina*) in which mixed cation for simple *p,p'*-substituents were Cl < H < CH_3 < CH_3O < CH_3S.

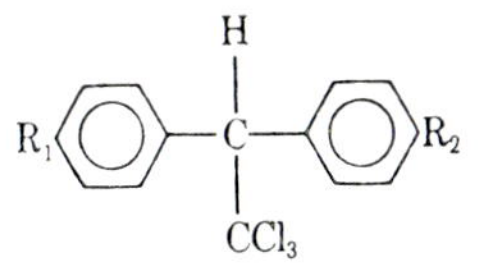

For various *p,p'*-dialkoxy substituted analogs, the rate of detoxification was ranked as follows: C_2H_5O < C_3H_7O < CH_3O < *iso*-C_3H_7O. Asymmetrical compounds, where the *p*-substituents were different, were generally more degradable than the symmetrical analogs.

The biodegradability of three asymmetrical DDT analogs was compared in a model ecosystem and their metabolic pathways studied in mice and insects (Kapoor *et al.*, 1973). The *p,p'*-substituents were CH_3S—, CH_3O— (methoxy-methiochlor), CH_3—, C_2H_5O— (methyl-ethoxychlor) and Cl—, CH_3 (chloro-methylchlor). Based on a 'biodegradability index' (ratio of polar products detected/nonpolar metabolites detected in the ecosystem) the following values were reported in fish: chlor-methylchlor 3·43, methoxy-methiochlor 2·75, methyl-ethoxychlor 1·20, and DDT 0·015. Metabolic studies revealed that methoxy-methiochlor was completely detoxified by oxidation to the sulphoxide and sulphone analogs. Some dehydrochlorination of methoxy-methiochlor occurred, since the unsaturated analogs similar in structure to DDE were also detected. Methoxy-methiochlor, when compared to the other two DDT analogs, was the most readily metabolized. The methylthio group, as suggested in Table I, is one of the most liable substituents that can be incorporated in a pesticide molecule. Due to its rapid oxidation, methoxy-methiochlor exhibits a low tendency to bioaccumulate in fish. The overall pathway for the metabolism of methoxy-methiochlor is shown in Fig. 1.

Methyl-ethoxychlor was metabolized by *O*-dealkylation and side chain oxidation. Oxidation of the methyl group led to the formation of the carboxylic acid derivative of methyl-ethoxychlor, whereas oxidation of the ethoxy group led to the hydroxylated derivative. The pathway for methyl-ethoxychlor is shown in Fig. 2.

Chloro-methylchlor was altered by side chain oxidation and dehydrochlorination. Chloro-methylchlor, with only one biodegradable group, *i.e.* the methyl group, and with an aryl chlorine atom was bound to behave more like DDT when compared to the other analogs. It was concentrated 1400-fold in fish. The parent compound represented about 25% of the recovered ^{14}C, the remainder being almost entirely polar metabolites. The overall pathway for metabolism of cloro-methylchlor is found in Fig. 3.

A DDT analogue lacking any chlorine atoms, dianisylneopentane, or 1,1-*bis*(*p*-methoxyphenyl)-2,2-dimethylpropane was metabolized by *O*-demethylation to form mono- and diphenol- derivatives (Coats *et al.*, 1973). The neopentyl group proved to be about as stable in biological systems as the isosterically equivalent trichloromethyl group of DDT.

A series of silicon-containing analogs of DDT have been prepared and examined for effectiveness as insecticides and for biodegradability (Fahmy *et al.*, 1973). The rationale behind the silicon DDT analogs as biodegradable insecticides was the greater instability of the C—Si bond compared to the C—C bond in ionic and radical reactions. Unfortunately, virtually all of the approximately 20 silicon analogs were ineffective against houseflies, showing little or no toxicity at 500 $\mu g/g$ alone or in combination with the synergist piperonyl butoxide.

Most of the effort toward modifying the persistence of DDT has involved substitution of a biodegradable group on the ring portion of the molecules. Oxidation of these groups leads to the formation of a more polar molecule that exhibits a lower tendency to accumulate in the fatty tissue of aquatic or terrestrial organisms. These substituents may also ultimately lead to a more accelerated rate of fragmentation of the intact molecule and concomitant loss of insecticidal activity. It is important to recognize, however, that only certain molecular parameters can be altered and still retain activity in the

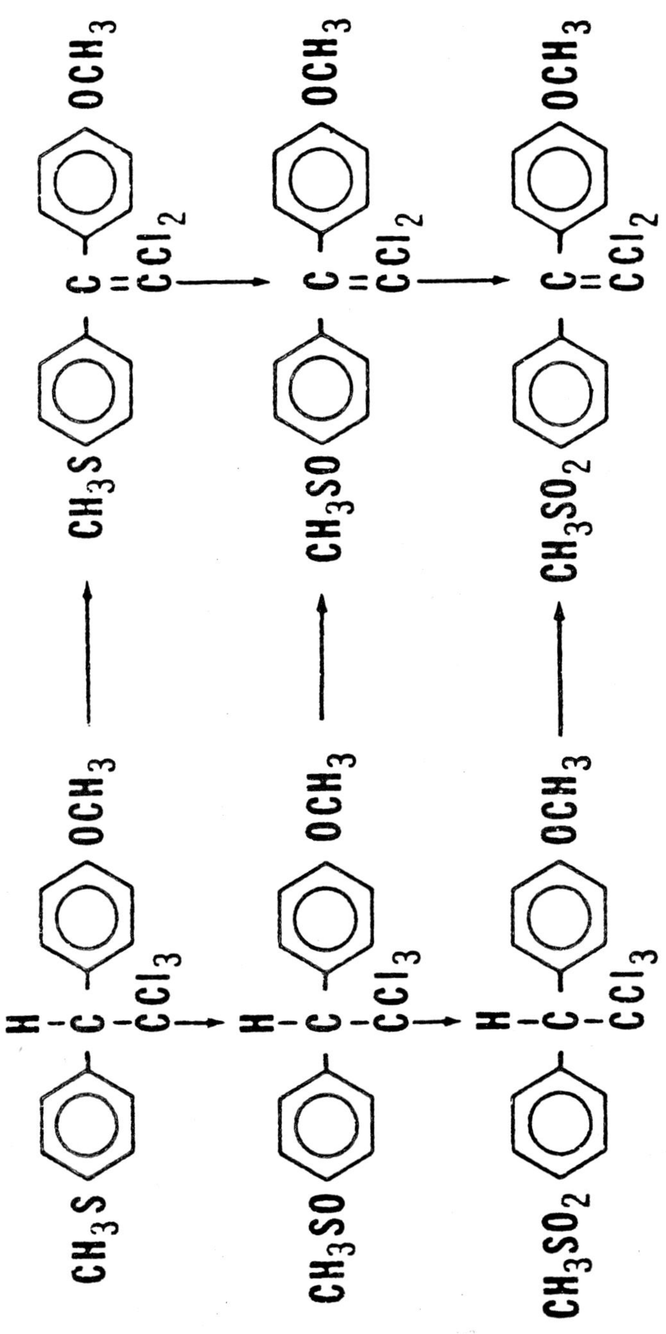

Fig. 1. *Pathways of methoxy-methiochlor metabolism in mouse, salt marsh caterpillar, and in a model ecosystem (After Kapoor* et al.*, 1973).*

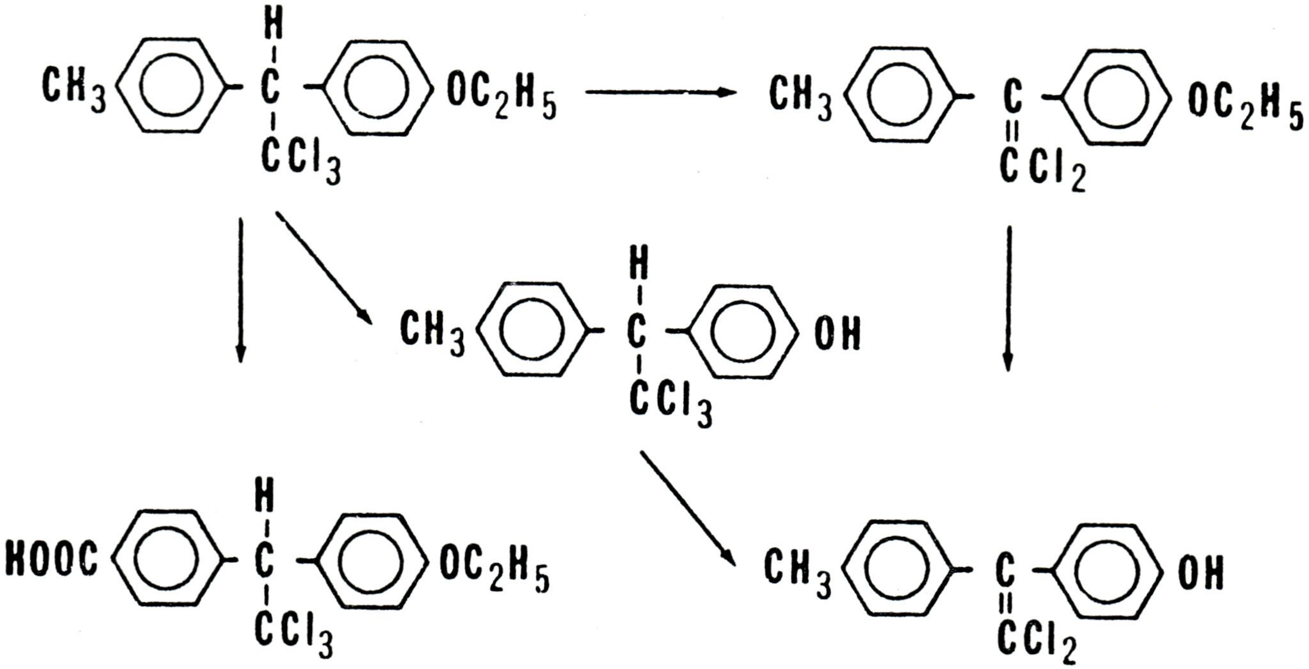

Fig. 2. Pathways of methyl-ethoxychlor metabolism in mouse, salt marsh caterpillar, and in a model ecosystem (After Kapoor et al., *1973).*

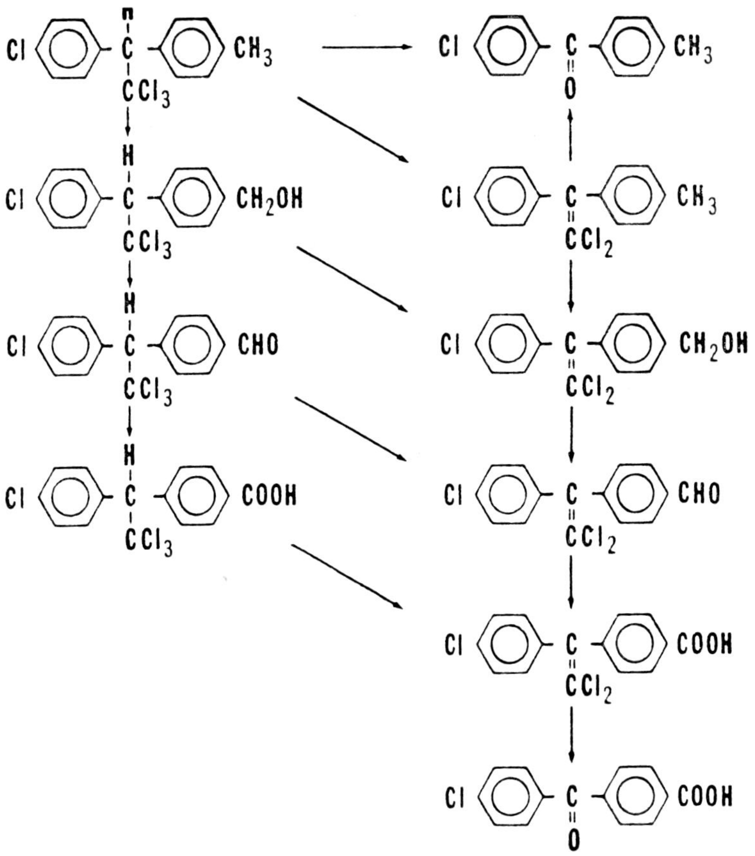

Fig. 3. Pathways of chloro-methylchlor metabolism in mouse, salt marsh caterpillar, and in a model ecosystem (After Kapoor et al., *1973).*

parent molecule. Within these molecular restraints, it is possible to put a biodegradable 'handle' on an insecticide that may, when oxidized, generate a more polar molecule that is less likely to bioaccumulate.

HERBICIDES

s-Triazines

There are several recent examples of structural modifications of effective herbicidal molecules that have increased their biodegradability. The largest usage of a pesticide in the United States today is of atrazine (2-chloro-4-(ethylamino)-6-(isopropylamino)-*s*-triazine). The United States production of atrazine in 1971 was estimated to be 95 million pounds (von Rumker *et al.*, 1974). Under certain climatic and soil conditions, it is considered one of the more persistent herbicides (Kearney *et al.*, 1966). Atrazine degradation in soils proceeds by two mechanisms (Kaufman and Kearney, 1970). The major reaction is oxidative dechlorination to yield hydroxyatrazine [2-hydroxy-4-(ethylamino)-6-(isopropylamino)-*s*-trazine]. The reaction proceeds by nucleophilic substitution of the chlorine at the 2 position, which is facilitated by delocalization of the ring electrons when the molecule is adsorbed onto clay particles (Armstrong and Chesters, 1968). A second reaction mediated by several soil microorganisms involves *N*-dealkylation to remove the ethyl and isopropyl groups in the 4- and 6-positions, respectively. Soil fungi responsible for dealkylating atrazine include *Aspergillus fumigatus*, *A. flavipes*, *A. ustus*, *Rhizopus stolonifer*, *Fusarium moniliforme*, *F. roseum*, *F. oxysporum*, *Penicillium decumbens*, *P. janthinellum*, *P. rugulosum*, *P. luteum*, and *Trichoderma viride*. As indicated in Table I, these are relatively slow reactions.

A structural modification of one of the alkyl substituents on the *s*-triazine molecule has resulted in a novel series of herbicides which are considerably more biodegradable in soils. The most important of these is cyanazine (2-[4-chloro-6-(ethylamino)-*s*-triazine-2-yl amino]-2-methyl-propionitrile). Substitution of a nitrile group for a methyl group in the isopropyl side chain results in a herbicide that is considerably less persistent in soils (Chapman *et al.*, 1968). Metabolism of cyanazine in soils proceeds by conversion of the nitrile group to an amide and then to an acid (Beynon *et al.*, 1972). The pathway is illustrated below:

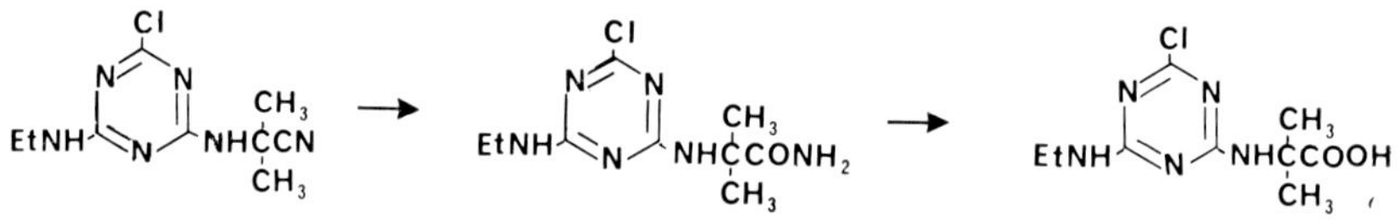

The persistence of atrazine and cyanazine has been compared in a Perth clay loam soil under actual field conditions (Sirons *et al.*, 1973). Cyanazine was found to disappear rapidly from soils, with the formation of a derivative lacking a propyl group (2-chloro-4-amino-6-ethyl-amino-*s*-triazine) as a major phytotoxic metabolite. Cyanazine amide [2-chloro-4(1-carbamoyl-1-methylethyl-amino)-6-ethylamino-*s*-triazine] was detected. Atrazine was degraded more slowly than cyanazine. Atrazine was converted to the dealkylated derivative minus the ethyl group [2-chloro-4-amino-6-isopropylamino-*s*-triazine] as a major phytotoxic metabolite. Thus substitution of an easily degradable substituent in an otherwise relative persistent herbicide has altered its longevity in soils.

Another analog of the *s*-triazine herbicides that was synthesized originally with 2-chloro-*s*-triazines was the 2-methylthio series. Although these 2-methylthio-*s*-triazines have a more limited commercial market than the 2-chloro-derivatives, they illustrate the effect of biodegradable groups on structurally related molecules. One 2-methylthio-*s*-triazine that has been studied to some extent in soils is prometryne [2,4-*bis*-(isopropylamino)-6-(methylthio)-*s*-triazine]. Prometryne is metabolized to the sulphoxide and sulphone by soil microorganisms (Kearney and Plimmer, 1969). The sulphone is unstable in water and hydrolyses to hydroxy propazine.

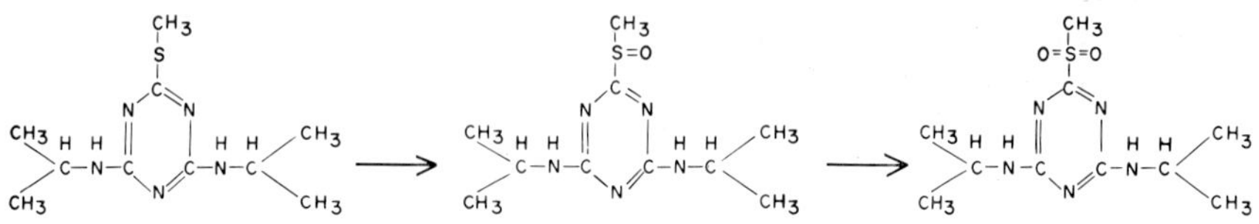

Major modifications of the triazine herbicides have been on various groups bonded to the heterocycling ring structure. Recent introduction of asymmetrical triazines, such as metribuzin [4-amino-6-*tert*-butyl-3-(methio)-*as*-triazine-5(4H)one], has provided information on the biodegradation of an altered heterocyclic ring. The persistence of metribuzin and two related compounds, the isopropyl analog [4-amino-6-isopropyl-3-(methylthio)-*as*-triazin-5(4H)-one] and the cyclohexyl analog [4-amino-6-cyclohexyl-3-(methylthio)-*as*-triazin-5(4H)-one] in soil was measured under laboratory and field conditions (Hyzak and Zimdahl, 1974). At 5 and 20°C, metribuzin and the isopropyl analog degraded more rapidly than the cyclohexyl analog. At 35°C, the rates of degradation were similar. The calculated half-lives were 44 and 16 days for the three herbicides at 20 and 35°C, respectively. The pattern of herbicide degradation in the field resembled the laboratory studies at 20°C. Although structural comparisons are not entirely justified because of the differences in ring substituents, the *as*-triazines would appear to be less persistent than the symmetrical triazine herbicides.

Only a few examples can be cited where biodegradable alternatives to persistent pesticides have been marketed. There is evidence that existing pesticides, such as DDT and atrazine, can be modified to reduce persistence. Primary emphasis in current screening tests for new pesticides is aimed at selectivity and efficacy. As awareness of residue problems increases, it is entirely possible that less effective but more biodegradable candidates in a biologically active class of compounds may receive more attention in the future.

REFERENCES

Armstrong, D. E., and Chesters, G. (1968). Adsorption catalyzed chemical hydrolysis of atrazine. *Environ. Sci. Technol.*, **2**, 683–689.

Beynon, K. I., Stoyden, G., and Wright, A. N. (1972). The breakdown of the triazine herbicide cyanazine in soils and maize. *Pestic. Sci.*, **3**, 293–305.

Chapman, T., Jordan, D., and Payne, D. H. (1968). WL19805—*A new triazine herbicide*. Proc. 9th Brit. Weed Conf. pp. 1018–1025.

Coats, J. R., Metcalf, R. L., and Kapoor, I. P. (1973). Metabolism of methoxychlor isostere, dianisylneopentane, in mouse, insects and a model ecosystem. *Pestic. Biochem. Physiol.*, **4**, 201–211.

Fahmy, M. A. H., Fukuto, T. R., Metcalf, R. L., and Holmstead, R. L. (1973). Structure-activity correlations in DDT analogs. *J. Agr. Food Chem.*, **21**, 585–591.

Hyzak, D. L., and Zimdahl, R. L. (1974). Rate of degradation of metribuzin and two analogs in soil. *Weed Sci.*, **22**, 75–79.

Kapoor, I. P., Metcalf, R. L., Hirwe, A. S., Coats, J. R., and Khalsa, M. S. (1973). Structure activity correlations of biodegradability of DDT analogs. *J. Agr. Food Chem.*, **21**, 310–315.

Kaufman, D. D., and Kearney, P. C. (1970). Microbial degradation of *s*-triazine herbicides. *Residue Rev.*, **32**, 443–446.

Kaufman, D. D., and Plimmer, J. R. (1972). In *Water Pollution Microbiology*, pp. 173–203 (R. Mitchell, ed.), John Wiley & Sons, Inc., New York.

Kearney, P. C., and Plimmer, J. R. (1969). Prometryne degradation in soil. *Abstr. 158th Meet. Am. Chem. Soc., Div. Pestic. Chem. No.* 30.

Kearney, P. C., and Plimmer, J. R. (1970). Relation of structure to pesticide decomposition. In *Pesticides in the Soil: Ecology, Degradation and Movement*, pp. 65–72. Int. Symp. Pestic. Soil, Michigan State Univ. East Lansing.

Kearney, P. C., Nash, R. G., and Isensee, A. R. (1966). In *Chemical Fallout: Current Research on Persistent Pesticides*, pp. 54–67 (M. W. Miller and G. G. Berg, eds.), Charles C. Thomas, Springfield, Illinois.

Metcalf, R. L., Kapoor, I. P., and Hirwe, A. S. (1971). Biodegradable analogues of DDT. *Bull. World Health Org.*, **44**, 363–374.

Sirons, G. T., Frank, R., and Sawyer, T. (1973). Residues of atrazine, cyanazine and their phytotoxic metabolites in a clay loam soil. *J. Agr. Food Chem.*, **21**, 1016–1020.

von Rumker, R., Lawless, E. W., and Meiners, A. F. (1974). *Production, distribution, use and environmental impact potential of selected pesticides*. 439 pp. EPA 540/1-74-001 Office of Pestic. Prog. Envir. Protection Agency.

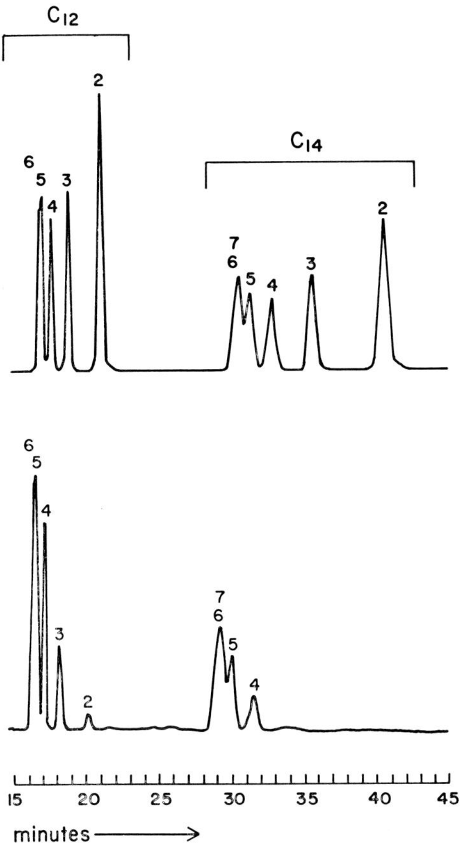

Fig. 1. Biodegradation of C_{12} + C_{14} LAS mixture in river water, detected by desulphonation-gas chromatography (Swisher, 1963a). Upper: before biodegradation, day zero, 5 ppm. Lower: after 50 % biodegradation, day 15, 2·5 ppm.

biodegradation. This seems to be quite general among the alkylbenzene sulphonate surfactants. It undoubtedly reflects relations between hydrophobe conformation and enzyme topology, made more complex by charge interactions between the sulphonate group and the enzyme.

BRANCHED ALKYLBENZENE SULPHONATES

Minimal branching of the chain has little effect on the speed of primary biodegradation. Figure 2 shows three compounds, all with effective chain length of 11 carbons. No. 1 is plain. No. 2 has an extra methyl group at the near end of the chain (actually this is simply the linear 2-phenyl C_{12} LAS). In No. 3 the methyl group is at the far end of the chain, making a fork. The upper chromatogram shows the composition of a mixture of the sulphonates in river water at time zero, and the lower one 4 days later when about 70 % had disappeared. The degradation rates of the three sulphonates were almost

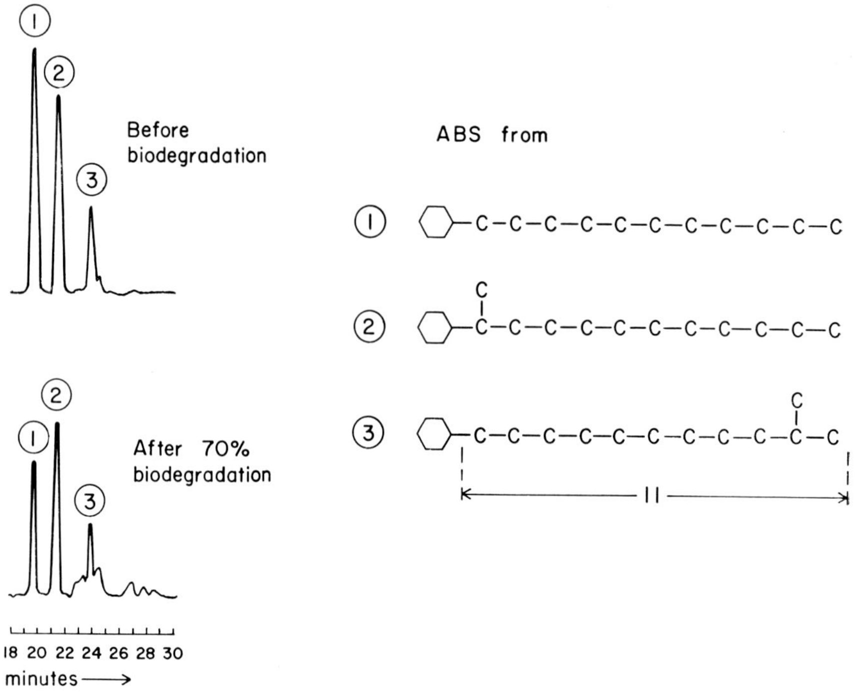

Fig. 2. Biodegradation of ABS mixture, effective chain length C_{11}, detected by desulphonation-gas chromatography (Swisher, 1963a). Upper: day zero, 7·4 ppm. Lower: day 4, 2·2 ppm.

identical. The plain chain, No. 1, was slightly faster, 77 % degraded compared to 64 % each for the other two. Thus a single methyl branch has only a very slight retarding effect, whether it is at the near end of the chain or at the far end. Again, the degradation rate appears to be mainly a function of the distance between the sulphonate group and the far chain end; with equal distances, equal rates.

On the other hand, another methyl group at the end of the chain, forming a terminal quaternary structure, markedly retards the degradation. Figure 3 shows two such compounds. In this case the lower chromatogram shows the composition when it was half gone, 32 days after mixing in river water. (It was almost all gone seven days later.) At the halfgone point about three-fourths of the longer one had disappeared, but only one-fourth of the shorter. So here, too, with these much more resistant structures, we find more rapid degradation with increased distance of the sulphonate group from the remote chain end.

But the terminal quaternary group does not necessarily retard the primary biodegradation. Compound A in Fig. 4 does contain such a group, but a linear open end chain is also present in the molecule. In this run it was compared with compound B, 6-phenyldodecane LAS. The bottom chromatogram was taken from the 10 day sample, when it was a little over half degraded. At that time the linear isomer was all gone and the branched one was still all

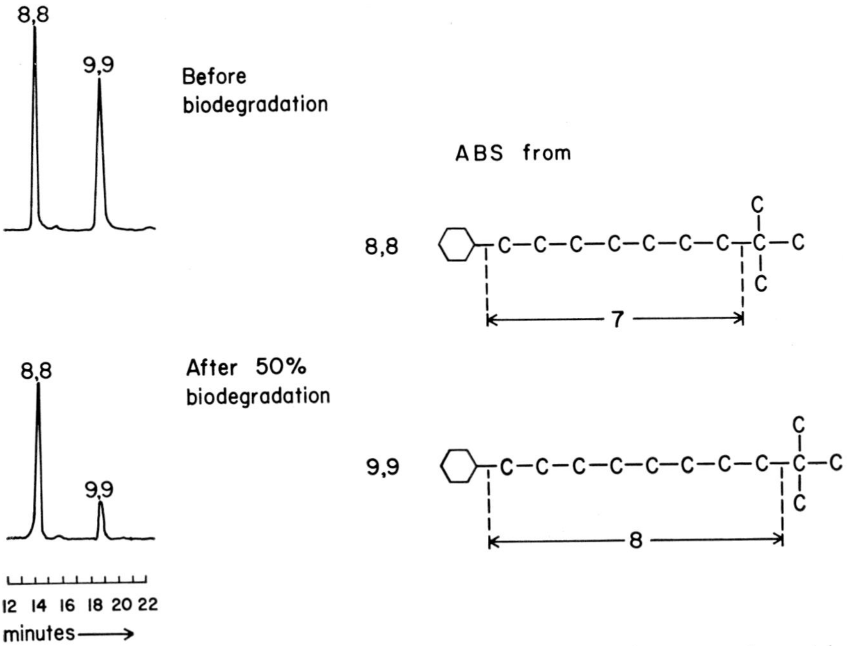

Fig. 3. *Biodegradation of terminal quaternary ABS mixture in river water, detected by desulphonation–gas chromatography (Swisher, 1963a). Upper: day zero, 5 ppm. Lower: day 32, 2·6 ppm.*

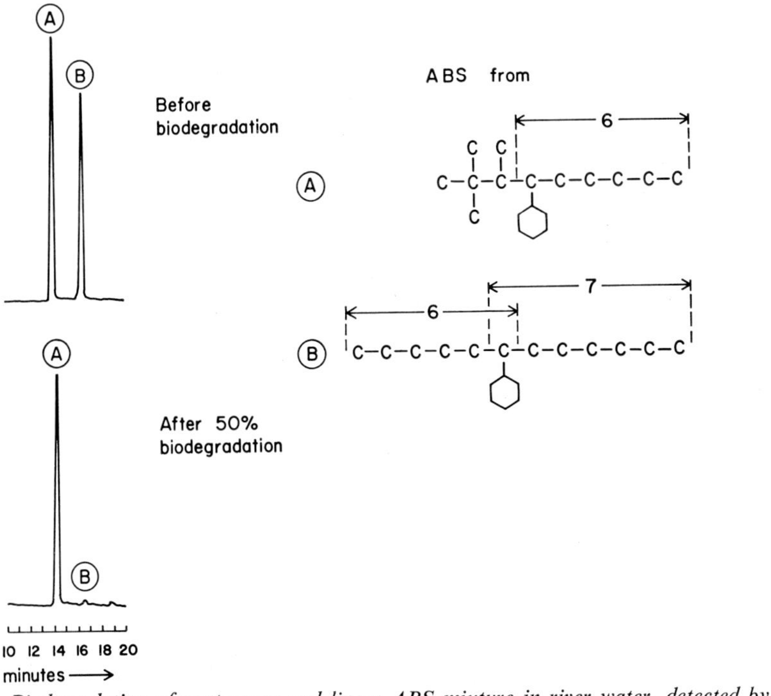

Fig. 4. *Biodegradation of quaternary and linear ABS mixture in river water, detected by desulphonation-gas chromatography (Swisher, 1963a). Upper: day zero, 5·5 ppm. Lower: day 10, 2·25 ppm.*

there. Three days later the branched one had disappeared also. This shows that the terminal quaternary group does not necessarily by its mere presence impart recalcitrance. With an open end chain also present in the molecule, compound A degraded only a little slower than B, which has two open ends of comparable length.

The recalcitrance of the tetrapropylene hydrophobe, as exemplified by TBS, can reasonably be related to its branching. Spectroscopic evidence suggests that it averages four to five branches per molecule, and that terminal quaternary groups are present. Five branches would make the effective chain length seven carbons at most, when the phenyl group is attached at the end. If the average phenyl attachment is in a more central location, the effective chain length would be even less. This alone might well be sufficient to impart the observed recalcitrance. A further contribution is probably made by the terminal quaternary groups.

LAS BIODEGRADATION PATHWAYS

Information on primary biodegradation such as that reviewed above says nothing about the chemical mechanisms involved—what metabolic pathways have been followed; nor does it reveal how far those pathways might extend, whether all the way to ultimate biodegradation or perhaps to reach an earlier dead end in the form of some more or less recalcitrant structure.

Several lines of evidence show that the LAS biodegradation pathways are open all the way. In activated sludge biodegradation systems, the dissolved organic carbon and benzene rings of LAS disappear to the extent of 80–90% or more in a 6 hr exposure (Swisher, unpubl. results). Under less vigorous conditions in shake flask systems, carbon dioxide evolution, both from the benzene ring and from the rest of the molecule, reaches 50% of theoretical within 10 days and continues on to 60% or more (Gledhill, 1975).

The biodegradation pathways involved in these transformations quite clearly are the ones commonly used for other compounds—ω-oxidation at the end of the alkyl chain followed by rapid β-oxidation of the chain followed by oxidation of the ring (Huddleston and Allred, 1963; Swisher, 1963b). Using enzymological methods, this sequence has been eloquently confirmed by Cain and his group (Cain *et al.*, 1971; Willetts and Cain, 1972a, 1972b; Bird and Cain, 1974), showing that ring degradation proceeds as usual via the catechol derivative. Either ortho or meta cleavage of the catechol may occur, depending on the exact structure of the substrate and the microbial species involved.

Less familiar than those well-known metabolic reactions is the desulphonation step, which Cain also investigated enzymologically. This cleavage occurred prior to ring degradation, and depending on the circumstances gave either (i) sulphate, having a hydrogen at the vacated position on the ring, or (ii) sulphite and the corresponding phenolic derivative. In some cases desulphonation preceded the ω, β-oxidation of the chain (mainly when the chain was short) and in other cases it occurred after the chain oxidation.

Most of Cain's work was done with the linear primary alkylbenzene sulphonates, in which the ring is attached to the end carbon of the chain. Little if any of these isomers are present in commercial LAS, which has the phenyl linkage mainly at nonterminal carbons. This structural difference may introduce minor differences into the pathway.

EFFECT OF FOUR HEAVY METALS (Mn, Ni, Cu and Co) ON SOME BACTERIA FROM THE DEEP SEA

SHON HUA YANG and HENRY L. EHRLICH

Department of Biology, Rensselaer Polytechnic Institute, Troy, New York 12181, U.S.A.

Summary

The individual effects of Mn, Ni, Cu, and Co on growing cells of cultures 138 and BIII 88 depended on the type of bacterial strains, the peptone concentration in the medium, the metal ion concentration, and temperature. Mg ions reversed the toxic effects of Mn, Cu, Co, and Ni for culture BIII 88 growing in a medium made up in 3% NaCl. Some metal ion combinations were more or less toxic to culture 138 than the individual metal ions. A hydrostatic pressure of 340 atm did not increase significantly the susceptibility of resting cells of cultures 138 and BIII 88 in seawater to added Ni and Co. It caused a slight increase in the case of Cu. Ni showed a slight protective effect for culture 138 at 340 atm.

INTRODUCTION

Ferromanganese nodules and sediments harbour bacteria, some are able to catalyse Mn(II) oxidation and others are able to catalyse Mn(IV) reduction (Ehrlich, 1975). The Mn(II)-oxidizing bacteria, by generating Mn(IV) oxides on nodules, can promote the scavenging of cations of such elements as Mn, Ni, Cu, and Co, while Mn(IV)-reducing bacteria can cause the release of these elements from ferromanganese (Ehrlich *et al.*, 1973). The following is a study of the tolerance of some of these bacteria for Mn, Ni, Cu, and Co under different conditions.

MATERIALS AND METHODS

Cultures. Two different bacterial isolates were employed in these studies, strains 138 and BIII 88. Strain 138, isolated from a Pacific ferromanganese nodule, is a gram-negative rod which requires seawater for growth and is capable of oxidizing Mn(III). Strain BIII 88 is a gram-negative rod isolated from Pacific deep-sea sediment. It is capable of reducing Mn(IV). It can grow in media made up in 3% NaCl as well as in seawater but not in fresh water. Neither culture ferments glucose, lactose, or sucrose, nor do they reduce nitrate or hydrolyse starch. For most experiments in this report, inocula were prepared from cells grown overnight at 18°C on Roux slants of nutrient agar (Difco) prepared with seawater. The cells were harvested and washed by centrifugation (2400 × *g*) in 20-ml volumes of sterile seawater.

Measurement of bacterial growth. Bacterial growth was followed turbidimetrically at 450 nm with a Coleman Junior spectrophotometer in a series of optically matched culture tubes (18 × 150 mm) containing 10 ml of seawater broth with concentrations of glucose, peptone, and, when appropriate, one or more of the metal salts, as specified for each experiment. The following reagent grade salts of the metals were used: $MnSO_4 . H_2O$; $Ni(NO_3)_2 . 6H_2O$; $CuSO_4 . 5H_2O$; and $Co(NO_3)_2 . 6H_2O$. The cell inoculum for each growth tube consisted of 0·01 ml of cell suspension, containing about 10^8 cells determined turbidimetrically. Before measuring turbidity in each growth tube at appropriate times, the content was thoroughly mixed on a vortex mixer.

Effect of metals on resting cells at hydrostatic pressures of 1 *and* 340 *atm.* One-day-old cells of strains 138 and BIII 88 from test tube slants of seawater nutrient agar were harvested and washed three times by centrifugation (2400 × *g*) in 5-ml portions of sterile seawater. The washed cells were resuspended in sterile seawater and diluted to a concentration which was usually of the order of 10^5 cells/ml. Six millilitres of diluted cell suspension were then added to each of two solutions, one consisting of 54·06 ml of sterile seawater and the other of 54 ml sterile seawater and 0·06 ml of an appropriate metal solution. One set of four tubes (18 × 65 mm) was then completely filled with the first inoculated solution and a second set of four such tubes, with the second inoculated solution. Each tube was stoppered with a sterile No. 1 neoprene stopper, and two tubes of each set were incubated at 15°C and 1 atm and the remaining two tubes of each set, at 15°C and 340 atm, using the method of ZoBell and Oppenheimer (1950). Initial cell numbers and surviving cells after 2 days of incubation were determined by plate counts.

RESULTS

Effect of individual heavy metals on bacterial growth at 18° *and* 5°*C.* The growth of cultures 138 and BIII 88 responded differently to the presence of the salts of Mn, Ni, Cu, or Co at 18° and 5°C in a seawater medium containing 0·1% glucose and 0·025% peptone, as shown in Fig. 1. It is seen that at 18°C both strains were least sensitive to Mn. Strain 138 was most sensitive to Cu while strain BIII 88 was most sensitive to Co. Growth of strain BIII 88 was significantly stimulated by up to 5 μg of Ni per ml. At 5°C in the same medium, the sensitivity of strain 138 to all four metals was greater than that of strain BIII 88. Ni lost its stimulatory effect on strain BIII 88 at this temperature. However, 1 μg of Mn per ml became stimulatory to the growth of strain 138. The growth rate of both cultures, even in the absence of added metals, was slower at 5°C than at 18°C.

Effect of medium composition on metal toxicity. Peptone has the ability to complex heavy metal ions and may thus lessen their toxicity for bacteria. To determine to what extent metal toxicity is affected by different peptone concentrations, growth at 18°C of strains 138 and BIII 88 in the presence of Mn, Ni, Cu, and Co was individually tested in seawater containing 0·1% glucose and either 0·025% or 0·005% peptone. The metals were tested at

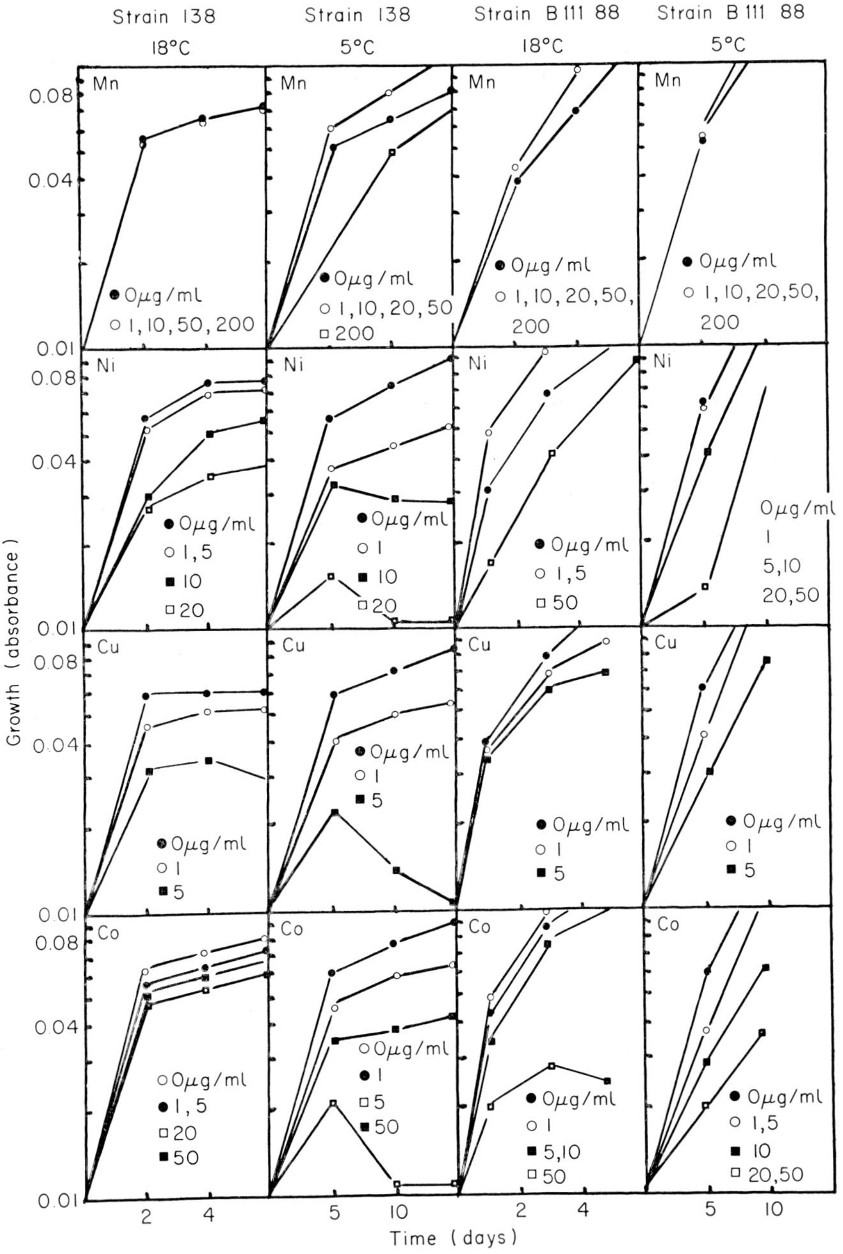

Fig. 1. Effect of different concentrations of metal ions on the growth of cultures 138 and BIII 88 at 18° and 5°C in a seawater broth containing 0·1% glucose and 0·025% peptone. Absorbance was measured at 450 nm.

concentrations of 0, 1, 5, 10, 20, 50, and 200 μg/ml, as appropriate. The results are shown in Table I. It is seen that strain 138 was significantly more

Table I
EFFECT OF VARYING PEPTONE CONCENTRATION ON THE GROWTH OF CULTURES 138 AND BIII 88

Peptone conc. (%)	*Minimum inhibition metal concentration (μg/ml)*[a] *Culture* 138				*Culture BIII* 88			
	Mn	*Ni*	*Cu*	*Co*	*Mn*	*Ni*	*Cu*	*Co*
0·025	>200	10	5	50	>200	20	5	5
0·005	50	1	1	5	>200	20	1	5

All cultures were incubated at 18°C in seawater containing 0·1% glucose. Progress of growth was followed turbidimetrically.
[a] Minimum metal concentration refers to smallest metal concentration tested which showed but did not necessarily stop growth.

sensitive to the metals at 0·005% peptone than at 0·025% peptone. Strain BIII 88, on the other hand, showed very similar sensitivity to Mn, Ni, and Co at the two peptone concentrations but increased sensitivity to Cu at the lower peptone concentration.

Effect of Mg ions on heavy metal toxicity for strain BIII 88. Strain BIII 88, unlike strain 138, was found capable of growth in medium containing 0·5% glucose and 0·05% peptone and prepared in 3% NaCl solution. In this medium, the addition, singly, of Mn (50 μg/ml), Cu (1 μg/ml), Ni (1 μg/ml), or Co (1 μg/ml) inhibited growth to varying degrees (Fig. 2a). The addition of Mg as $MgSO_4 \cdot 6H_2O$ to a final concentration of 1·3 mg/ml of medium,

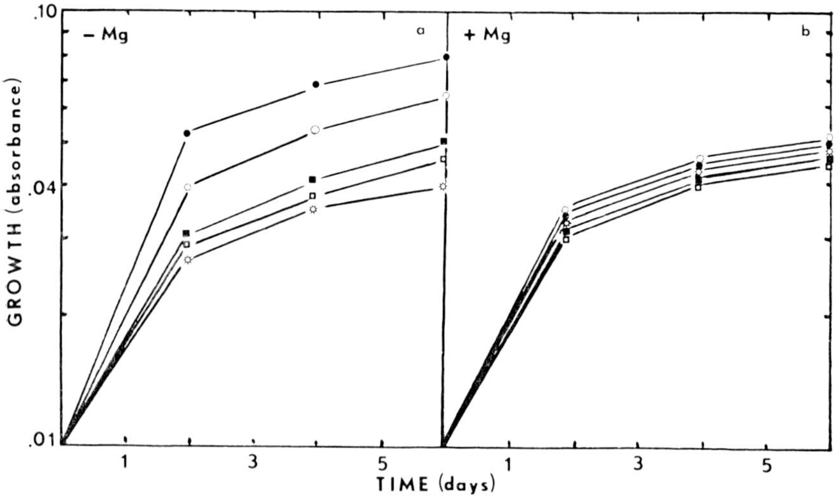

Fig. 2. Effect of Mg ions on the toxicity of heavy metal ions for the growth of culture BIII 88 in a 3% NaCl solution containing 0·5% glucose and 0·05% peptone incubated at 18°C; (a) without added Mg ions; (b) with added Mg ions to a final concentration of 1·3 mg/ml. Explanation of symbols: ● *no added heavy metals;* ○ *50 μg Mn/ml;* ☼ *1 μg Ni/ml;* ■ *1 μg Cu/ml;* □ *1 μg Co/ml.*

while depressing growth in general (compare controls without added metal salts), eliminated the toxicity of Mn and nearly eliminated it for Ni, Cu, and Co (Fig. 2b).

Comparison of toxicities of different metal combinations for strain 138. The effect of different combinations of metal ions (10 μg/ml, each) in seawater enriched with 1% glucose and 0·05% peptone on the growth at 18°C of strain 138 was tested. The results are shown in Fig. 3. While Mn and Co by

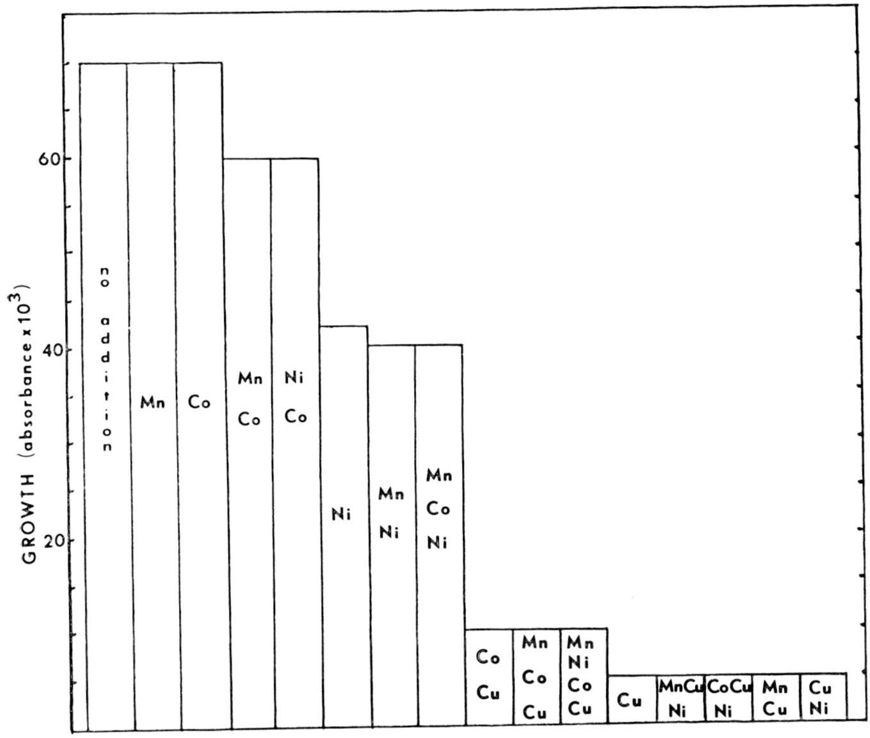

Fig. 3. Toxicity effects of various combinations of heavy metal ions on the growth of culture 138. The concentration of each metal ion was 10 μg/ml in a seawater medium containing 1% glucose and 0·05% peptone. Incubation was at 18°C for 7 days.

themselves had no effect on growth in this medium, their combination did. The combination of Co and Ni was more toxic than Co alone but less toxic than Ni alone. Cu was most toxic of all. Its toxicity was slightly relieved by Co or by the combination of Co and Mn, or Co, Mn, and Ni, but by no other metal combination tried. The mixtures of Mn and Ni and of Mn, Co, and Ni were more toxic than their individual components.

The effect of hydrostatic pressure on the toxicity of heavy metals for resting cells of strains 138 *and BIII* 88. Although no nutrient had been added to the seawater in these experiments, strain BIII 88 at an initial cell concentration of about 10^5 per ml passed through 3–4 cell generations during incubation (Fig. 4). As shown in Fig. 4, neither Co nor Ni, each tested at a concentration

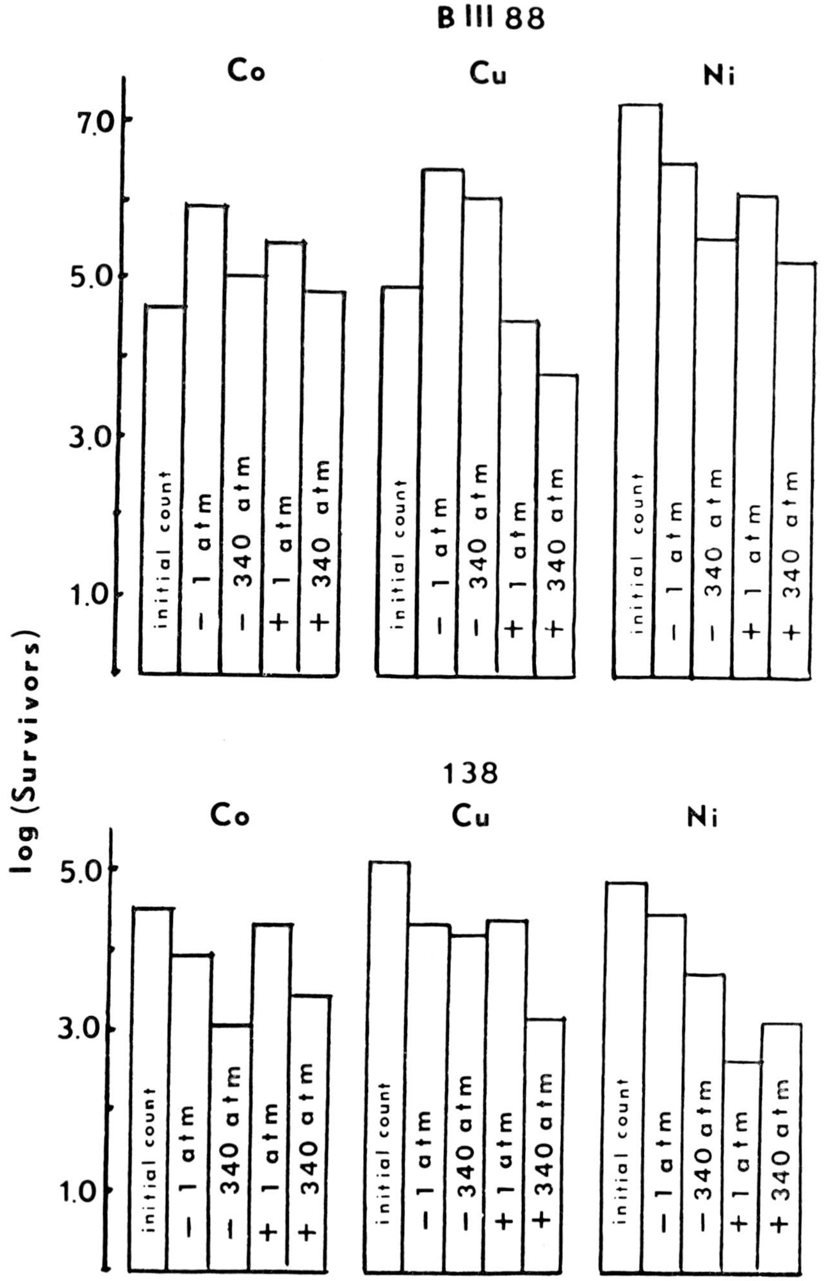

Fig. 4. Effect of 1 and 340 atm of hydrostatic pressure on the survival of 'resting cells' of cultures BIII 88 and 138 in seawater in the presence of heavy metal ions. Incubation was for 2 days at 15°C. The concentrations of Co and Ni ions were 1 μg/ml for both cultures. The concentration of Cu ions was 0·1 μg/ml for culture BIII 88 and 0·01 μg/ml for culture 138. + signs signify presence of added metal ions; − signs signify absence of added metal ions.

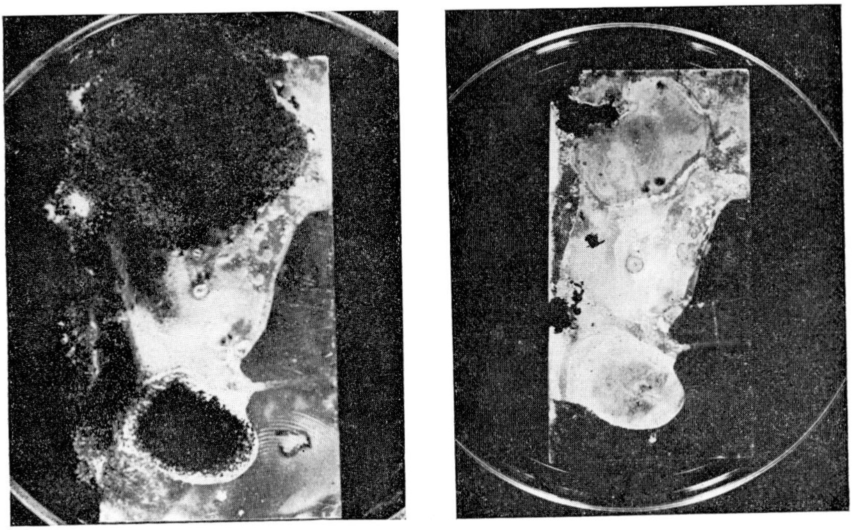

Fig. 1. *Aluminium 1100 after incubation in 95–100% humidity at 28–29°C for one month. The medium was bouillon-Tang® and faeces. (Left) the coupon with growth; (right) the metal after removal of growth. Note the areas where the aluminium has been completely consumed.*

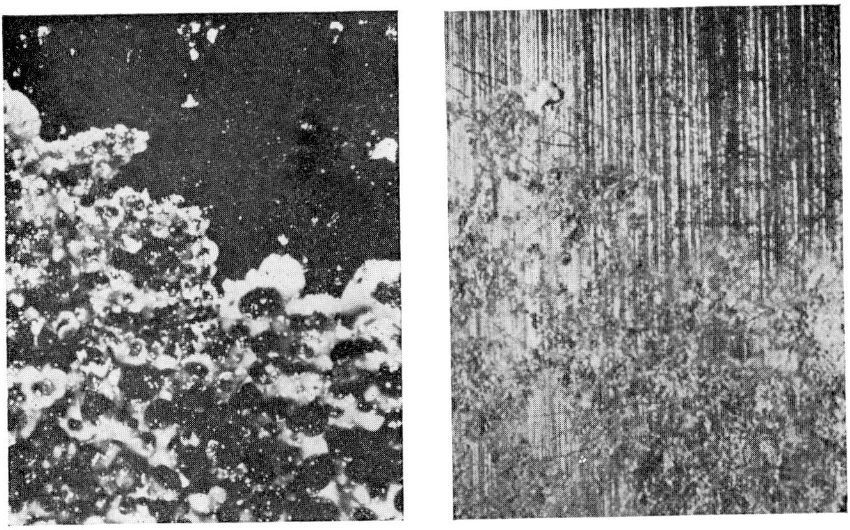

Fig. 2. *Metal 2014-T6 after 2 months incubation in beef bouillon at 25°C with an inoculum of 0·1% faeces. (Left) the rich bacterial growth; (right) the same metal after removal of growth. Note the etching and erosion of the metal in area previously covered by the growth.*

RESULTS

Preliminary experiments involved incubating the metal strips in solutions of Tang® and faeces, and bouillon and faeces. Aluminium 1100 (Fig. 1 (left and right)) and aluminium alloy 2014-T6 (Fig. 2 (left and right)) were degraded in the presence of microbial growth, whereas aluminium alloy 6061-T6 and stainless steel 321 were not visibly affected in these initial studies. Closer examination of the centre of Fig. 1 (left) reveals formation of small bubbles. This fact prompted a subsequent study on the gaseous atmosphere after growth on bouillon and Tang® using faeces as an inoculum. Gas chromatographic analysis of atmospheres showed that methane, carbon monoxide, and hydrogen had been formed during growth on metals.

A large number of microbiological species was identified after being isolated from bathwater or faeces. Of these, only those organisms listed in Tables II and III were employed.

Table I

EFFECT OF OXYGEN CONCENTRATION ON THE CORROSION OF SELECTED METALS BY 25 PURE CULTURES OF MICROORGANISMS

Metal	*Oxygen tension* 20%		50%		75%		*Total*	
	♯ *Coupons tested*	% *Corroded*	♯ *Coupons tested*	% *Corroded*	♯ *Coupons tested*	% *Corroded*	*Total ♯ Coupons tested*	% *Showing corrosion*
6061-T6	38	18·4	34	29·4	39	23·1	111	23·4
2014-T6	38	26·3	32	53·1	40	62·5	110	47·3
321	31	3·2	29	0·0	26	0·0	86	1·2
1 100	36	2·8	31	0·0	32	3·1	99	2·0
Total	143	13·3	126	21·4	137	25·5	406	20·0

Conditions: These studies were conducted in appropriate atmospheres under ambient temperature, pressure and humidity for 2 months. The medium used was either Tang® or bouillon at pH 7·0.

It was clearly established that SS 321 and aluminium alloys 6061-T6 and 2014-T6 were corroded by human faeces and by microorganisms derived from human faeces or bathwater when incubated with Tang® and bouillon. Corrosion was evident even when a light (0·1%) inoculum was used. While corrosion of these test metals has been demonstrated to occur even under sterile conditions (Tennyson and Brown, 1975), all of the corrosion noted in these studies has been in the presence of microbiological growth. Increased oxygen tension and increased humidity enhanced the growth of these microorganisms.

In tests carried out with 25 isolates, representing four fungal and three bacterial genera, there was no apparent increase in growth when the oxygen concentration was increased from 20% to 50% or 75%. On the contrary, there was a substantial increase in corrosion with increases in oxygen concentration from 20% to the higher values (50–75%) (Table I). Also, the metals inoculated with fungi exhibited considerably more corrosion than did those inoculated with bacteria (Tables II and III). Aluminium alloys 6061-T6 and 2114-T6 were more susceptible to corrosion than was SS 321 and aluminium 1100 (Table I). When the aluminium alloys were tested, there was a marked increase in corrosion with an increase in oxygen concentration from 20% to 50% or 75%, with alloy 2014-T6 exhibiting greater corrosion than alloy 6061-T6. Tests using Tang® as the substrate showed a 91·3% increase in corrosion over comparable tests with bouillon as the substrate (Table II).

Experiments conducted with a 50% oxygen atmosphere at ambient pressure, 1 psi above ambient, and 5 psi above ambient showed essentially no difference in growth with the 10 cultures employed in this study. There was a significant decrease in corrosion from 32% at ambient pressure to 23% at 1 psi above and 11% at 5 psi above.

Table II

EFFECT OF SUBSTRATES AND PURE CULTURES ON CORROSION OF VARIOUS METALS

Cultures	# *Isolates*	*Bouillon* # *Coupons tested*	*Bouillon* % *Corroded*	*Tang®* # *Coupons tested*	*Tang®* % *Corroded*	*Total* # *Coupons tested*	*Total* % *Corroded*
Achromobacter	8	71	8·5	36	25·0	107	14·0
Alcaligenes	4	33	15·2	19	15·8	52	15·4
Aspergillus	4	43	9·3	38	39·5	81	23·5
Trichoderm	1	11	18·2	10	20·0	21	19·1
Rhizopus	1	10	30·0	9	66·7	19	47·4
Mycobacterium	1	11	0·0	4	0·0	15	0·0
Unknown	6	48	29·2	55	25·5	103	27·2
Total	25	227	15·0	171	28·7	398	20·8

Conditions: These studies were conducted on the appropriate medium in air under ambient temperature, pressure and humidity for 2 months. The pH of the medium was 7·0. The metals used in this study were: aluminium 1100; aluminium alloys, 6061-T6 and 2014-T6; and stainless steel 321.

The effect of humidity on growth and the corrosive ability of 10 selected microbial cultures showed an observable increase in growth with increased humidity with Tang® as the substrate, but very little effect when bouillon was the substrate. Probably these results are a reflection of the experimental design which enabled considerable growth to take place in the system prior to reaching equilibrium throughout the system for a given humidity. Contrarywise, there was a marked increase in corrosion of aluminium 1100 when the humidity was increased from 10% to 50% or 90% (Table IV). Virtually all (95%) of the aluminium coupons showed evidence of corrosion at the higher humidity levels (50–90%).

Acknowledgement

The authors wish to acknowledge the support received from NASA Grant NASA-NGL 25-001-032 from the NASA George C. Marshall Space Flight Center.

Table III

EFFECT OF OXYGEN CONCENTRATION ON CORROSION OF VARIOUS METALS BY PURE CULTURES OF MICROORGANISMS

Culture	# *Isolates*	*Oxygen tension* 20% # *Coupons tested*	20% % *Corroded*	50% # *Coupons tested*	50% % *Corroded*	75% # *Coupons tested*	75% % *Corroded*	*Total* # *Tested*	*Total* % *Corroded*
Achromobacter	8	47	8·7	28	14·3	32	21·9	107	14·2
Alcaligenes	4	19	5·3	14	7·1	19	26·3	52	13·5
Aspergillus	4	27	18·5	27	29·6	28	25·0	82	24·4
Trichoderm	1	7	14·3	8	0·0	6	50·0	21	19·0
Rhizopus	1	8	62·5	6	0·0	6	66·7	20	45·0
Mycobacterium	1	5	0·0	4	0·0	6	0·0	15	0·0
Unknown	6	31	9·7	40	35·0	39	23·1	110	23·6
Total	25	144	13·3	127	21·3	136	25·7	407	20·0

Conditions: These studies were conducted in appropriate atmospheres under ambient temperature, pressure, and humidity for 2 months. The medium used was either Tang® or bouillon at pH 7·0. The metals used in this study were: aluminium 1100; aluminium alloys, 6061-T6 and 2014-T6; and stainless steel 321.

Table IV

THE CORROSION OF STAINLESS STEEL 321 AND ALUMINIUM 1100 BY PURE CULTURES UNDER AN ATMOSPHERE OF 70% OXYGEN/30% AIR AND VARIOUS HUMIDITIES

Metal	*Tang®*						*Bouillon*						*Total*					
	% Humidity						*% Humidity*						*% Humidity*					
	10		50		90		10		50		90		10		50		90	
	# *tested*	% *corroded*	# *tested*	% *corroded*	# *tested*	% *corroded*	# *tested*	% *corroded*	# *tested*	% *corroded*	# *tested*	% *corroded*	# *tested*	% *corroded*	# *tested*	% *corroded*	# *tested*	% *corroded*
321	8	0·0	10	0·0	10	10	10	10·0	10	10·0	10	0·0	18	5·5	20	5·0	20	5·0
1100	10	0·0	10	100·0	10	100·0	10	0·0	10	100·0	10	90·0	20	0·0	20	100·0	20	95·0
Total	18	0·0	20	50·0	20	55·0	20	5·0	20	55·0	20	45·0	38	2·6	40	52·5	40	50·0

Conditions: These studies were conducted in appropriate media and humidities at ambient temperature and pressure for 2 months.

REFERENCES

Breed, R. S., Murray, E. G. C., and Smith, N. R. (1957). *Bergey's Manual of Determinative Bacteriology*, 7th ed., Williams and Wilkins Co., Baltimore, Md.

Hodgman, C. D., Weast, R. C., and Selby, S. M. (eds.) (1955). *Handbook of Chemistry and Physics*, 37th Ed. Chemical Rubber Publ. Co., Cleveland, Ohio.

Society of American Bacteriologists (1957). *Manual of Microbiological Methods.* McGraw-Hill, New York.

Society for Industrial Microbiology (1966). *Proposed Procedures for the Screening of Microbial Inhibitors in Hydrocarbon/Water Systems by Committee on Deterioration of Fuels: Standard Methods*. SIM Special Publication No. 2.

Tennyson, J. M., and Brown, L. R. (1975). The microbial degradation of aluminium 1100. Proc. 3rd Int. Biodegrad. Symp. (this volume, p. 883).

THE MICROBIAL DEGRADATION OF ALUMINIUM 1100

JOHN M. TENNYSON* and LEWIS R. BROWN

Mississippi State University, Mississippi State, Mississippi 39762, U.S.A.

Summary

Under the experimental conditions employed, pressure and oxygen concentration did not affect the corrosion of Aluminium 1100. Greater corrosion was observed when Tang® served as the substrate, rather than bouillon. Corrosion was evident at humidities of 50% and 90% but not at 10%. Fungi promoted corrosion to a greater extent than did bacteria. Control metal strips exhibited slight etching, while strips in the presence of microorganisms were pitted or had holes. Attempts to quantitate Al losses failed. Micatex®, a coating material, did not inhibit microbial growth. Teflon®-coated aluminium appeared to resist corrosion better than did other coating materials (Micatex® with colour coating), although #37886 (Light Beige) and alodined 1200 of the aluminium 1100 showed some promise.

INTRODUCTION

It was clearly demonstrated (Brown and Pabst, 1975) that microorganisms make a significant contribution to the biodeterioration of aluminium 1100 in the presence of a suitable substrate. Because of the utilization of this metal in the Manned Orbital Laboratory, it was of interest to determine the extent to which microorganisms contributed to the process as well as to determine if coated or treated aluminium would be more resistant to corrosion.

MATERIALS AND METHODS

Materials

The test materials were: (1) aluminium 1100 (99% pure); (2) an 1100 series foil designated as an alodined 1200; (3) aluminium 1100 coated with Micatex® with different colour coatings—#26314, Medium Grey; #36231, Grey; #37886, Light Beige; #37855, Dark Beige; #25102, Blue; #35231, Blue; #30226, Yellow-Green; #34552, Green; #33538, Yellow; #37038, Black; #21105, Red; (4) Teflon®-coated 1100.

The metal coupons (0·076 mm × 9·5 mm × 38·1 mm) and 25·4 mm discs were cleaned by placing them in acetone, then *n*-hexane and subjecting them to ultrasonic cleaning for 10 min.

Human faeces, fingerprints, sputum, and urine were used as naturally occurring mixed cultures. The pure cultures were a *Rhizopus* species and *Scopulariopsis* species. The culture media were: (1) Tang® Instant Breakfast Drink, 134·8 g/litre of distilled water; (2) Armour Star®—Beef Bouillon Cubes, four cubes dissolved in 474 ml of distilled water; (3) Micatex®—

* Present address: Petroleum Products Division, Mississippi State Chemical Laboratory, Mississippi State, Mississippi 39762, U.S.A.

mineral salts medium [as used by Brown *et al.* (1964) except that NH_4NO_3 was used instead of KNO_3]; and (4) Difco Nutrient Agar and Nutrient Broth.

Effect of environmental conditions on corrosion

Procedures employed in this investigation are described in the preceding paper by Brown and Pabst (1975).

Aluminium loss due to biodeterioration

Each metal test strip was placed in an individual container together with 2 ml Tang® and the test organism and incubated at reduced pressure (6 psi). The 95% humidity was maintained as stated by Lange (1956). The gas mixtures were 20% O_2–80% N_2, 50% O_2–50% N_2, and 70% O_2–30% N_2.

A *Rhizopus* species, a *Scopulariopsis* species, and normal faeces were employed as inocula. Three individual containers were removed periodically, the Al strips removed and the liquid analysed for Al using a Perkin-Elmer 290 Atomic Absorption Spectrophotometer in accordance with the method described in *Analytical Methods for Atomic Absorption Spectrophotometry* (1966).

Effect of coatings on the corrosion of aluminium

To determine if Micatex® would serve as a substrate for microbial growth, the following tests were conducted. Prescription bottles each containing 100 ml nutrient broth and 0·5 ml Micatex® were inoculated with a 1 g faeces. After one week incubation at room temperature on a rotary shaker, the samples were examined for growth by the spread-plate technique with nutrient agar as a substrate. Pure cultures were obtained using the conventional streak-plate technique but were not identified. Each culture was tested for its ability to grow in nutrient broth and Micatex®. These isolates, along with cultures of *Rhizopus* and *Scopulariopsis*, were transferred from nutrient agar slants to prescription bottles containing 100 ml of Micatex®-mineral salts medium, incubated at room temperature and observed for growth.

To determine the effect of an array of suspected corrosive agents on coated and treated samples, another experiment was undertaken. Tang® was employed as the substrate on these samples, in an atmosphere of 70% O_2 and 30% N_2, static conditions, a pressure of 5 psi, and a humidity of 95%. Inocula were *Rhizopus* and *Scopulariopsis* and undiluted human sputum, urine, and a 0·1% dilution of faeces. In addition, human fingerprints were used as an inoculum, without a substrate, but under similar incubation conditions. The controls, set up in a separate chamber, were sterilized with ethylene oxide before gassing.

RESULTS

Effect of environmental conditions on corrosion

It has been demonstrated (Brown and Pabst, 1975) that microbiological activity contributed substantially to the corrosion of aluminium 1100. Further, it was noted that not only did the microflora of naturally occurring materials (*e.g.* faeces and bathwater) produce this effect (Fig. 1) but also some pure culture isolates produced the same effect.

The effect of pressure and substrate upon the degradation of aluminium 1100 by 45 microbial isolates was tested under an atmosphere of 50% oxygen and 50% air. An appreciable amount of corrosion was observed in most of the test samples, but no correlation between pressure and corrosion was observed. Corrosion was more extensive when Tang® was the growth substrate rather than bouillon. Growth and corrosion were not affected by oxygen concentrations when tested at 20%, 50%, and 70% oxygen in the atmosphere. Although there was no observable difference in growth at humidities of 10%, 50%, and 90%, no corrosion was evident at 10% humidity but it was present at 50% and 90%.

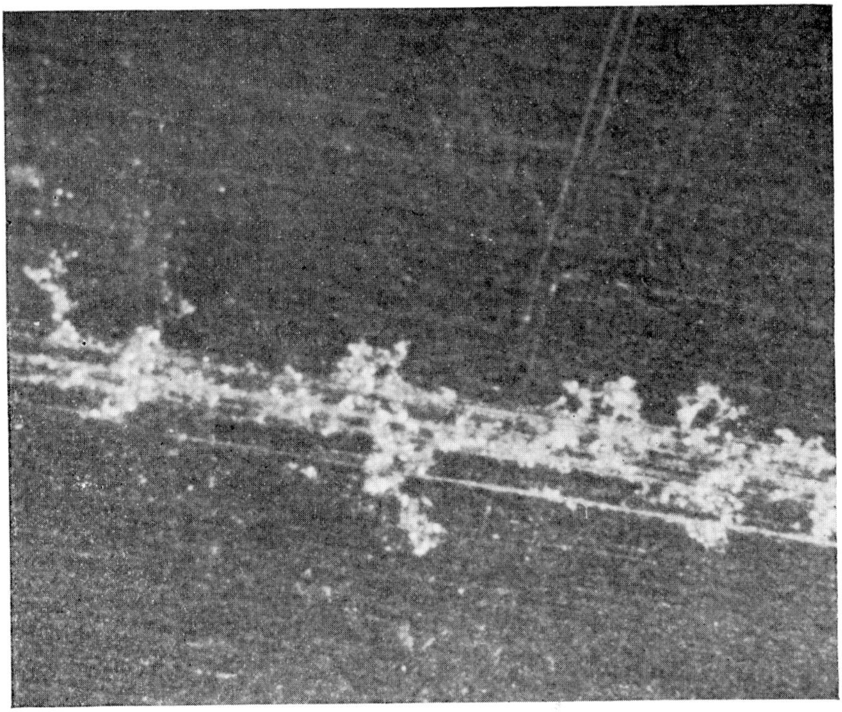

Fig. 1. The characteristic type of corrosion of aluminium caused by faecal matter (approximately 15×). Note the degradation along the 'rolling marks' on the aluminium strip.

It was observed in the preceding tests that fungi produced a greater amount of corrosion than did the bacteria under the test conditions employed. A *Scopulariopsis* species and a *Rhizopus* species caused more extensive corrosion than did the other isolates. Characteristic corrosion exhibited by these cultures is shown in Fig. 2. It was noted that a majority of the corrosion occurred along the 'rolling marks'. Pin-holes began to appear in the aluminium which was placed in Tang® inoculated with the *Scopulariopsis* species after 32 days, and in the test systems inoculated with *Rhizopus* after 53 days. The control samples which were placed in sterile media exhibited only slight etching, but not pits or holes.

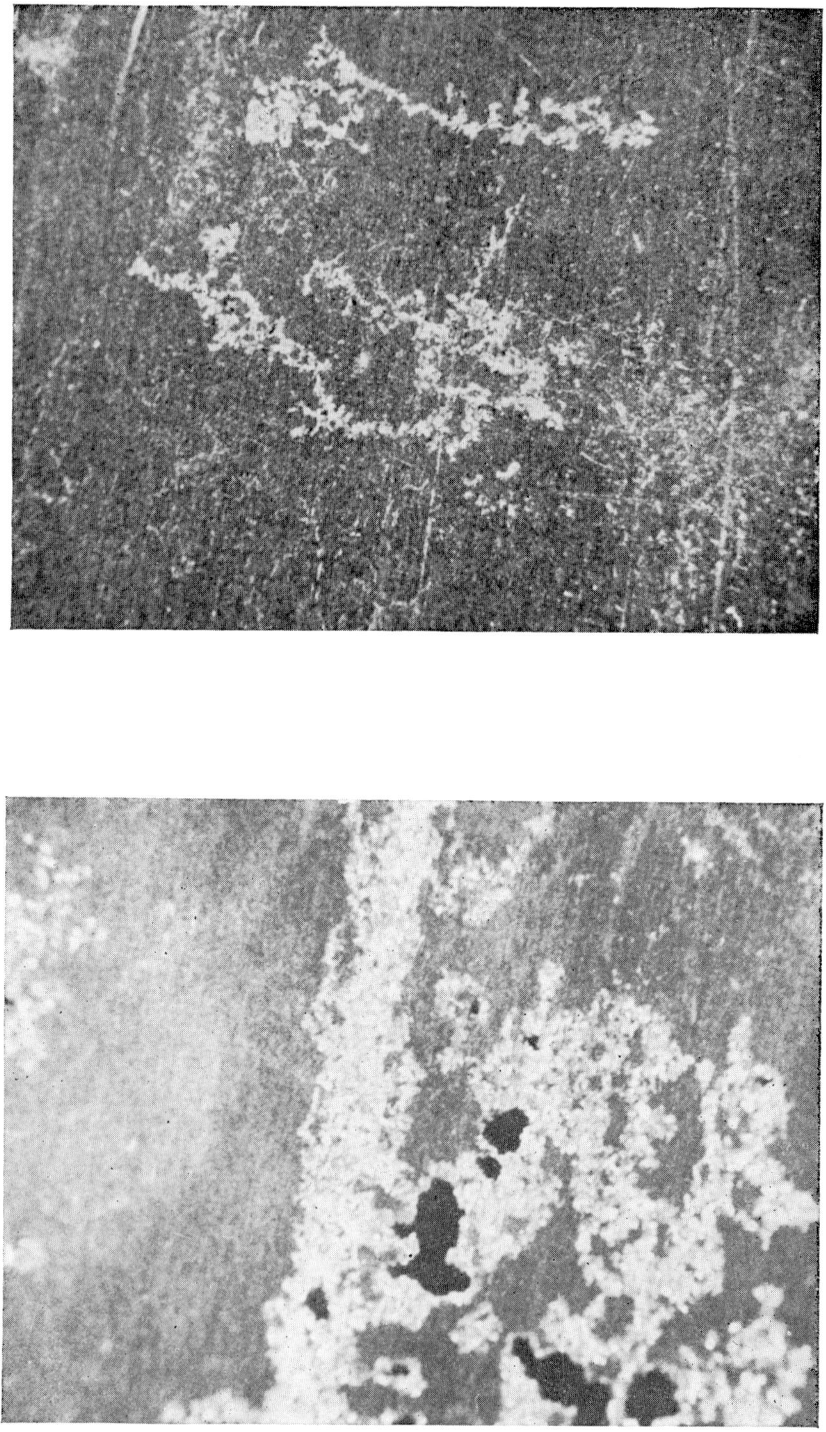

Fig. 2. Characteristic corrosion of aluminium (approximately 15×). (Top) Rhizopus; *(bottom)* Scopulariopsis. *(Note: the dark areas in the bottom photograph are holes.)*

Aluminium loss during biodeterioration

The preceding experiments clearly indicated that microorganisms contributed substantially to the corrosion of aluminium 1100, but all of the tests were qualitative rather than quantitative. Therefore, it seemed desirable to quantitate the corrosion process. Monitoring weight losses during the corrosion was inadequate due to the limited sensitivity of the balance, hence attempts were made to quantitate the process by monitoring the dissolution of aluminium. Samples were removed periodically and the amounts of aluminium remaining and in solution were monitored using atomic absorption spectrophotometry. The aluminium concentrations in all solutions, including the control, increased with time; but the overall increases were too small to make any meaningful interpretations. In all cases, the aluminium losses after 59 days of incubation were less than 1 %. Microscopic examination of the metal strips in the controls revealed only a slight etching, while those in the test systems revealed either pitting or holes. Under the circumstances, no further attempts were made to quantitate the aluminium losses.

Effect of coating and treatment of aluminium on corrosion

The question arose as to whether coatings or treatment of the aluminium would prevent or retard corrosion. The first experiment was designed to determine if Micatex® (a coating material) would serve as a substrate for microbial growth in a mineral salts medium. It was found that all of the test organisms grew well in the Micatex®-mineral salts medium. The growth of *Rhizopus* and *Scopulariopsis* is shown in Fig. 3.

Because of the severity of corrosion noted previously, and since Micatex® did not exert an inhibitory effect, an experiment was conducted to determine

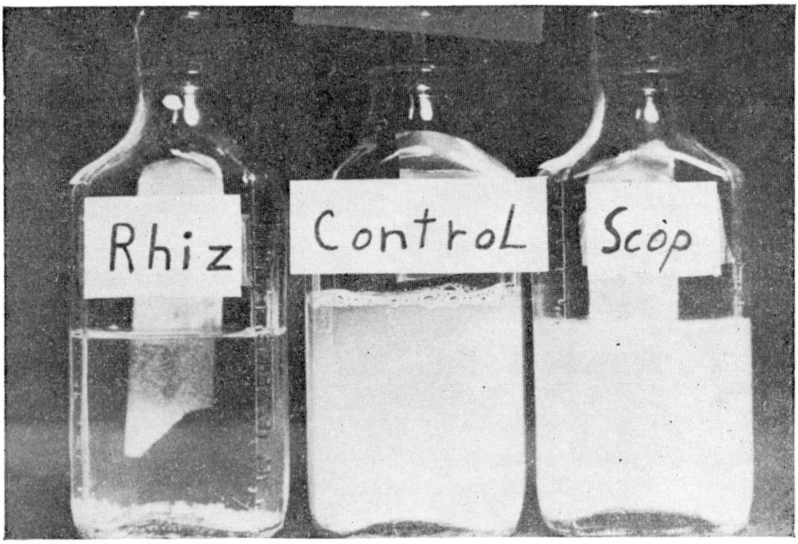

Fig. 3. Growth of Rhizopus (*Rhiz*) *and* Scopulariopsis (*Scop*) *in mineral salts medium containing Micatex®. Note the pellets of* Rhizopus (*Rhiz*) *growth present on the bottom of the bottle and the completely clear medium and also the* Scopulariopsis (*Scop*) *growth evident on the bottom of the bottle and the partial clearing of the medium.*

the effectiveness of a variety of pigmented coatings, of Teflon coating, and of treatment in inhibiting corrosion.

The Teflon® coating appeared to be the best corrosion inhibitor, although #37886 (Light Beige) and alodined 1200 of the aluminium 1100 showed some promise in this respect (Table I).

Table I

CORROSION OF COATED ALUMINIUM DISKS AND ALODINED ALUMINIUM 1200 STRIPS

Metals	*Inocula*						
	Control	*Faeces*	*Sputum*	*Urine*	Rhizopus	Scopul-ariopsis	*Finger-prints*
26314 (Medium Grey)	—	+	+	+	+	+	—
36231 (Grey)	—	+	+	+	+	+	—
37886 (Light Beige)	—	—	—	+	—	—	—
37855 (Dark Beige)	—	+	+	+	+	—	—
25102 (Blue)	+	+	+	+	+	+	—
35231 (Blue)	+	+	+	+	+	+	—
30226 (Yellow green)	+	+	+	+	+	+	—
34552 (Green)	+	+	+	—	+	+	—
33538 (Yellow)	+	+	+	+	+	+	—
21105 (Red)	+	+	+	+	+	+	—
37038 (Black)	+	+	+	+	+	+	—
Teflon cylinders	—	—	—	—	—	—	—
Alodined 1200	—	—	⊕	⊕	—	—	—

+ = corrosion of coating and/or metal, — = no observable effect on the sample, ⊕ = hole.

Note: Corrosion of controls, especially in the absence of corrosion of fingerprint samples, is indicative of chemical corrosion; this is borne out in that the degree of corrosion noted on inoculated samples was far greater than on controls.

Acknowledgement

The authors wish to acknowledge the support received from NASA Grant NASA-NGL-25-001-032 from the NASA George C. Marshall Space Flight Center.

REFERENCES

Analytical Methods for Atomic Absorption Spectrophotometry (1966). Perkin-Elmer, Norwalk, Conn.

Brown, L. R., and Pabst, G. S. (1975). Biodeterioration of stainless steel and aluminium alloys. Proc. 3rd Int. Biodegrad. Symp. (this volume).

Brown, L. R., Strawinski, R. J., and McCleskey, C. S. (1964). The isolation and characterization of *Methanomonas methanooxidans* (Brown and Strawinski). *Can. J. Microbiol.*, **10**, 792–799.

Lange, N. A. (ed.) (1956). *Handbook of Chemistry*, 9th Ed. Handbook Publishers, Inc., Sandusky, Ohio.

THE ASSIMILATION OF TRACE METAL IONS BY THE MARINE BACTERIA, *ARTHROBACTER MARINUS* AND *PSEUDOMONAS CUPRODURANS*

GALEN E. JONES

Department of Microbiology and Jackson Estuarine Laboratory, University of New Hampshire, Durham, New Hampshire 03824, U.S.A.

LESLIE G. ROYLE* and LINDSAY MURRAY†

Department of Oceanography, The University of Liverpool, Liverpool, U.K.

Summary

During growth on a basal seawater medium containing 0·1% yeast extract and tryptone in a synthetic seawater adjusted to a salinity of 26 g/kg, two marine bacterial cultures, *Arthrobacter marinus* and *Pseudomonas cuprodurans*, assimilated cations against the gradient as follows: Fe > Cd > Cu > Zn > Pb ≏ Ni ≏ Mn. These trace elements were all concentrated by 10^2 to more than 10^3-fold. The major cations in seawater (Na, Mg, Ca, K) were not concentrated by the cells by more than 2·6-fold, and in *P. cuprodurans* all major cations were at lower concentrations in the cells than in seawater. Some of these major cations in seawater may have been leaked from the cells during washing with 0·5 M ammonium formate.

Zinc assimilation by *A. marinus* at graded concentrations of $ZnCl_2$ to toxicity indicated a gradual increase in Zn at lower concentrations in the basal seawater medium. As the chelating capacity of the nutrients in the medium was almost exceeded, the Zn assimilation was accelerated. Where 3×10^{-4} M $ZnCl_2$ was added to the medium which had the highest concentration of Zn permitting growth, the cells contained 0·37% Zn. The concentration factor for Zn in the cells of *A. marinus* was greatest at 3×10^{-4} M Zn and where no Zn was supplemented to the basal seawater medium.

Marine bacteria assimilated trace elements from seawater medium to similar concentrations as trace metals are concentrated from seawater by marine plants and animals.

INTRODUCTION

The assimilation of trace metal ions against the gradient by marine plants and animals by hundreds to thousands of times has been well established (Webb, 1937; Noddack and Noddack, 1940; Vinogradov, 1953; Bowen, 1966; Riley and Segar, 1970; Segar *et al.*, 1971; Riley and Roth, 1971; Martin and Knauer, 1973; Jones, 1973). The trace element analyses of bacteria are generally limited (Porter, 1946; Bowen, 1966). The mechanism by which trace elements are concentrated and the manner in which they are held in bacteria is unknown or incompletely understood (Perlman, 1965). The abundance of trace elements in marine bacteria is unknown.

Some trace elements are essential for organisms and others do not have a

* Present address: York Water Authority, York, U.K.
† Present address: Ministry of Agriculture and Fisheries, Burnham-on-Crouch, Essex, U.K.

firmly established physiological function. For example, Harvey (1939, 1947) indicated suboptimal amounts of Fe and Mn in seawater which may affect adversely the growth of marine phytoplankton. Few other trace elements in seawater have been implicated as limiting to the growth of phytoplankton. Nevertheless, significant aspects of the chemistry of the sea are regulated by marine life (Redfield *et al.*, 1963).

The function, if any, of many of the trace elements in metabolism is fragmentary or unknown. Certain cations such as Mg, Mn, Ca, K, Fe, and Co activate enzyme reactions (Bowen, 1966). Other trace elements such as Ba as sulphate in the rhizopod *Xenophyophora* serve structural functions (Bowen, 1966). Porphyrins such as chlorophyll-containing Mg, haem-containing Fe, vitamin B_{12}-containing Co as well as Cu- and Zn-porphyrins are well known. The functional significance of many other trace elements is obscure. There may be no functional significance for many trace elements, their presence in cells or tissues merely reflecting their high biological reactivity.

The purpose of this investigation was to determine whether two marine bacteria, *Arthrobacter marinus* and *Pseudomonas cuprodurans*, have the ability to concentrate trace elements against the gradient similarly to marine plants and animals. The addition of Zn to a basal medium until Zn toxicity was expressed for *A. marinus* was used to determine a specific metal ion effect on elemental composition of the cells.

MATERIALS AND METHODS

Organisms and growth medium

Two cultures of marine bacteria, *Arthrobacter marinus* (ATCC 25374) and *Pseudomonas cuprodurans*, both isolated from littoral seawater samples from Woods Hole, Massachusetts, were utilized for their ability to assimilate inorganic cations. *A. marinus* is noteworthy for its formation of plasmolysed spheroplasts (megalomorphs) as the concentration of trace metal cations approaches toxic concentrations in the medium (Cobet, 1968; Cobet *et al.*, 1970). *P. cuprodurans* is characterized by relatively high Cu ion tolerance (McCarthy, 1971). Stock cultures of both bacteria were maintained on basal seawater agar slants at 20 $\pm$ 2°C.

The basal seawater medium contained Bacto-yeast extract 1·0 g (Difco Laboratory, Detroit, Michigan); Oxoid-tryptone 1·0 g (Oxoid Ltd, London); and 75% synthetic seawater (Table I), 1000 ml. The synthetic seawater (Lyman and Fleming, 1940) was prepared with reagent grade chemicals (International Chemicals Inc., Ltd, England). The final pH of the medium was 7·5–7·8. $ZnCl_2$ converted from ZnO was prepared as aqueous solutions and the desired dilutions were added to the basal medium prior to sterilization by autoclaving at 118°C for at least 15 min where trace metals were added.

Culturing and harvesting of cells

The basal seawater medium was dispensed in 500 ml aliquots into 1-litre, acid-cleaned Erlenmeyer flasks, autoclaved at 118°C for 15 min, cooled to ambient temperature, and inoculated aseptically with a culture of the test bacterium grown on basal seawater medium at 20 $\pm$ 2°C to early stationary growth. The flasks were placed on a Griffin Orbital Shaking Machine (Model S36-706) at 185 rpm at 20 $\pm$ 2°C. The optical density of the culture was determined by removing subsamples and reading in an E.E.L. Absorptiometer using a 601 filter (3800–4700 Å), with a 1 cm light path.

Table I
COMPOSITION OF SYNTHETIC SEAWATER HAVING A SALINITY OF THE OPEN OCEAN AND 75% SYNTHETIC SEAWATER USED IN BASAL SEAWATER MEDIUM, COMPUTED FROM LYMAN AND FLEMING (1940)

	Concentrated in g/kg of distilled water	
Constituent formula	*L. and F. synthetic seawater (full strength)*	*L. and F. synthetic seawater (75%)*
$NaCl$	23·476	17·607
$MgCl_2 \cdot 6H_2O$	10·635	7·977
Na_2SO_4	3·917	2·937
$CaCl_2$	1·102	0·827
KCl	0·664	0·498
$NaHCO_3$	0·192	0·144
KBr	0·096	0·072
H_3BO_3	0·026	0·020
$SrCl_2 \cdot 6H_2O$	0·040	0·030
NaF	0·003	0·002 3
Salinity, g/kg	34·481	25·861

At the end of stationary growth, the cells were harvested in an M.S.E. (model Super Medium) centrifuge at 3500 × *g* for 20 min in the $ZnCl_2$ experiment. In the general cation composition experiments, a M.S.E. (model High Speed 25) refrigerated (4°C) centrifuge at 8000 × *g* for 10 min was employed.

The cells from less than a litre of medium due to subsampling and evaporation were washed three times with 150–180 ml each of 0·5 ammonium formate chilled to 4°C. The cell pellet was resuspended in the 0·5 M ammonium formate solution after each wash by gently agitating with an acid-cleaned, glass, stirring rod. The final cell pellet was transferred from the glass, centrifuge cups into acid-cleaned, dry, tared, teflon beakers using a few millilitres of double distilled water. The beakers were dried in an oven at 105–110°C for several days to constant weight and reweighed. The ammonium formate was volatilized during drying (Riley and Roth, 1971).

Ashing the cells

Twenty-five millilitres of concentrated nitric acid which had been redistilled from a silica still were added to the cells in the beakers. The cells were digested by covering and placing the beakers on a hot plate at 90°C for 18 hr in a ventilated hood. Subsequently, the temperature of the hot plate was reduced to 30°C and the lids were removed from the beakers. When the solution had evaporated to about 10 ml, the beakers were placed under an u.v. lamp for 1 hr to digest residual organic matter in solution. The beakers were returned to the hot plate at 30°C and their contents evaporated to dryness. Care was taken to remove the beakers from the hot plate before any charring occurred.

A millilitre of 2·0 N redistilled concentrated nitric acid and a few millilitres of double distilled water were added. The beakers were warmed gently. The resulting solutions were adjusted to exactly 10 ml with double distilled water and stored in closed polythene bottles. Reagent blanks were prepared by placing 25 ml aliquots of the redistilled concentrated nitric acid in clean, teflon beakers and treating with the procedure described.

Ashing and preparation of basal seawater medium constituents

Since the organic nutrients in the basal seawater medium complex with the trace elements in the organic nutrients and the trace elements as impurities in the reagent grade inorganic chemicals, the Bacto-yeast extract, Oxoid tryptone, and synthetic seawater were analysed separately. The trace elements in synthetic seawater were determined by concentrating the elements using the anion exchange resin, Chelex 100 (Bio-Rad Laboratories, Richmond, Calif.) in the Na form (Riley and Roth, 1971). Ten litres of the synthetic seawater were passed through the Chelex 100 column, and then a second Chelex 100 resin to determine the efficiency of the first column for removal of trace elements.

Ten litres of double distilled water were passed through a third Chelex 100 column to serve as a column and reagent blank. The columns were eluted with 50 ml 2·0 N redistilled nitric acid. The acid eluates were collected in the teflon beakers and evaporated to dryness. The residues were redissolved in 1 ml of 2·0 N nitric acid and a few millilitres of double distilled water. The solution was adjusted to 10 ml with double distilled water and stored in closed polythene bottles for analyses.

The organic nutrients of the basal seawater medium were prepared separately by adding approximately 1 g in a tared teflon beaker and reweighed accurately. The only differences in the digestion procedure cited above were the use of 40 ml of 2·0 N redistilled nitric acid in each beaker and an u.v. lamp exposure of 1·75 hr.

Determination of trace element and major cation concentrations

Zn, Fe, Cu, Pb, Ni, Mn, Cd, Mg, and Ca were determined by atomic absorption spectrophotometry, and Na and K by flame photometry (Riley and Roth, 1971). The atomic absorption spectrophotometer was a Techtron AA 5 instrument, with a grating spectrophotometer connected to a Techtron D1-20 digital indicator unit.

A matrix effect was detected only for the Ca determination which was probably due to the presence of very stable Ca compounds such as $Ca_2P_2O_7$ which did not dissociate in the flame. The matrix effect was overcome by using the hotter nitrous oxide/acetylene flame rather than the air/acetylene flame used for the other elements.

The flame photometry, a form of emission spectroscopy in which the spectral lines of the element are excited by spraying a solution of the element into a flame, was performed on an E.E.L. flame photometer with a coal gas flame and appropriate Na and K filters. Mixed standards contained 1, 2, 5, and 10 ppm of Na and K in 0·4% NH_4SO_4 solution.

RESULTS

In order to investigate the assimilation of cations from solution by bacteria, the exact composition of the cations in the uninoculated medium was determined. The cationic composition of the basal seawater medium for the elements investigated was determined separately for the Bacto-yeast extract, Oxoid-tryptone, and 75% synthetic seawater (Table II). The summation of the three determinations for each element represented the calculated elemental composition of the basal medium. The concentrations of the major ions in seawater (Na, Mg, Ca, K) were calculated from the salts added to 75% synthetic seawater (Table I). When these concentrations of cations in seawater were compared with those of Goldberg (1965) for open ocean seawater (Table II), a reduction in the concentration of Na and Mg was apparent which was not made up completely by the Na and Mg impurities in the organic nutrients. The disparity was not considered serious due to the abundance of these cations in seawater. In fact, 75% seawater is more effective for growing marine bacteria than full strength seawater (ZoBell, 1946). The Ca and K concentrations in the medium are very close to those expected from the reduction of the salinity to 25·861 g/kg. Due to the impurities in reagent grade chemicals used to prepare synthetic seawater, and in the Bacto-yeast extract and Oxoid tryptone, the trace cations investigated were concentrated from 6- in Mn to 600-fold in Pb compared to those estimated in natural seawater (Table II). The concentration of these cations in the basal seawater medium constitutes the basis of concentration factors by bacterial cultures grown in the medium.

Table II
ELEMENTAL COMPOSITION OF SELECTED CATIONS IN NATURAL AND SYNTHETIC SEAWATER, MEDIA CONSTITUENTS, AND CALCULATED IN BASAL SEAWATER MEDIUM

Element	*Concentration in µg/litre*				
	Natural seawater[a]	*Synthetic seawater*[b] (75%)	*Bacto yeast extract*	*Oxoid tryptone*	*Basal seawater medium*
Na	10 500 000	7 443 828	7 988	7 439	7 459 255
Mg	1 350 000	954 117	1 262	47	955 426
Ca	400 000	298 641	1 531	606	300 778
K	380 000	284 847	12 870	146	297 863
Zn	10	13·5	140	42	196
Fe	10	1·0	96	69	166
Cu	3	1·7	15·5	2·6	20
Pb	0·03	6·7	9·1	3·5	19
Ni	2	14·3	2·9	0·3	16
Mn	2	4·8	2·9	5·7	13
Cd	0·11	1·2	0·3	0·4	2

[a] From Goldberg (1965).
[b] Data calculated for major ions from Lyman and Fleming (1940).

A comparison of the cationic concentration by *A. marinus* grown under standard conditions on eight separate occasions and *P. cuprodurans* on two separate occasions is presented in Table III. When the mean concentration of the major cations in seawater (Na, Mg, Ca, and K) was examined for both bacteria, they appeared to be among the most abundant cations in the cells. Nevertheless, an examination of the concentration factors which were determined by dividing the cation concentrations in the basal seawater medium (Table II) with the ions in the bacterial cells (Table III) indicated that these major ions were not concentrated more than three-fold in either organism. In many instances, the major cations were not concentrated at all, particularly in *P. cuprodurans*. The washout and leaking of these major cations during the ammonium formate washes were a partial explanation.

Nevertheless, it is evident that the major ions of seawater were not concentrated, particularly, by the marine bacterial cells.

The trace elements, in contrast, were concentrated from 100- to 1000-fold in most instances. The Fe and Cd were concentrated greatest by both bacterial cultures, as evidenced by the concentration factors. The Zn and Cu were both concentrated between 500- and 1000-fold by both bacteria and Pb, Ni, and Mn were concentrated from about 100- to 500-fold. The minor cations of seawater were concentrated to a much greater extent by both marine bacterial cultures from basal seawater medium than the major ions of seawater.

Table III
METAL CONTENT OF *A. MARINUS* and *P. CUPRODURANS* CELLS GROWN ON BASAL SEAWATER MEDIUM AT 20 ± 2°C, HARVESTED AT STATIONARY GROWTH, AND WASHED THREE TIMES WITH 150–180 ml OF COLD 0·5 M AMMONIUM FORMATE. CONCENTRATION FACTORS ARE CALCULATED FROM THE METAL CONTENTS OF THE BASAL SEAWATER MEDIUM IN TABLE II

	Concentration in ppm/dry cell weight of cells			
Metal	*A. marinus*		*P. cuprodurans*	
	Concentration	*Concentration factor*	*Concentration*	*Concentration factor*
Na	5 600	0·7	255	0·03
Mg	1 580	1·6	292	0·3
Ca	790	2·6	272	0·9
K	700	2·3	36	0·1
Zn	122	622	112	571
Fe	257	1 548	422	2 542
Cu	14·4	720	19	950
Pb	9·7	510	3	158
Ni	1·7	106	4·5	281
Mn	1·7	131	0·9	69
Cd	2·8	1 400	4·6	2 300

An experiment was designed to determine the maximum concentration of $ZnCl_2$ for the growth of *A. marinus* in basal seawater medium and its effect on elemental cellular composition. Eighteen flasks of basal seawater medium in 500 ml aliquots were autoclaved and supplemented with neutralized $ZnCl_2$ solutions prepared from ZnO in concentrations ranging from 1×10^{-5} to 5×10^{-4} M $ZnCl_2$ in duplicate. The flasks were inoculated with 1·0 ml of a 28 hr culture of *A. marinus* grown on a gyrotary shaking machine at 185 rpm at 20 ± 2°C. The development of the cells in the presence of graded Zn (Fig. 1) was somewhat similar to that of *A. marinus* in $NiCl_2$ solutions (Cobet *et al.*, 1970). There was a slight stimulation of growth from the $ZnCl_2$ addition at 1×10^{-5} M. However, increasing toxicity developed at increasing concentrations of Zn as indicated by longer lag periods. Little toxicity was evident at 5×10^{-5} and 1×10^{-4} M $ZnCl_2$. A lag period of 20 hr characterized the curve where 2×10^{-4} M $ZnCl_2$ was present in the medium. At 3×10^{-4} M $ZnCl_2$ a lag period occurred which lasted more than 50 hr. No growth of *A. marinus* took place at concentrations of 4 and 5×10^{-4} M $ZnCl_2$.

The balances of Zn in the media and in the cells of *A. marinus* are compared in Table IV. Analyses of Zn in the sterile medium before inoculation and Zn in the spent medium after growth permitted the calculation by difference as

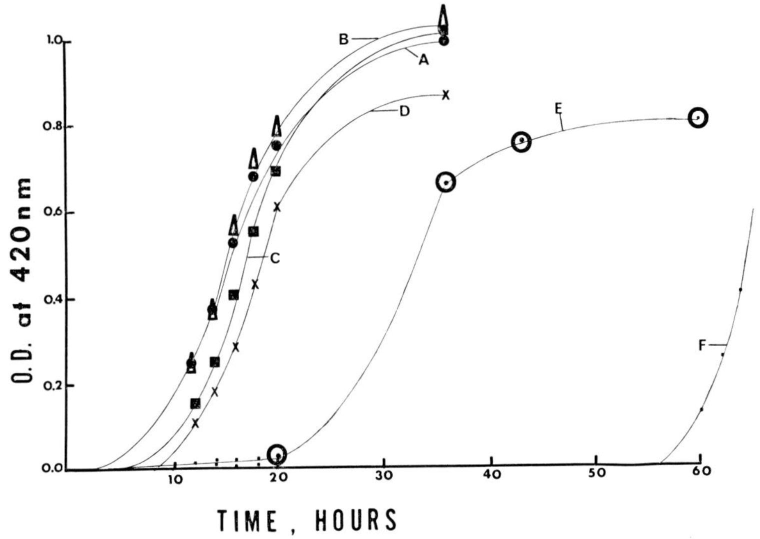

Fig. 1. Growth of A. marinus *as optical density at 420 nm in basal seawater medium supplemented with (A) 0, (B) 1 × 10⁻⁵, (C) 5 × 10⁻⁵, (D) 1 × 10⁻⁴, (E) 2 × 10⁻⁴, and (F) 3 × 10⁻⁴* M *$ZnCl_2$ at 20 ± 2°C on a gyrotary shaking machine at 185 ppm.*

the Zn removed from solution during growth. The Zn-uptake equivalent is the amount of Zn in the cells of the organism (Table V) corrected for the volume of the medium when harvested. The Zn balance between the cells and the medium is the difference between the Zn removed from solution and that accounted for in the cells. The cells made a heavy demand on the

Table IV

ASSIMILATION OF Zn FROM BASAL SEAWATER MEDIUM BY *A. MARINUS* AT 20 ± 2°C WITH ADDITIONS OF $ZnCl_2$ SHOWING Zn BALANCES IN MEDIA AND CELLS

$ZnCl_2$ added (M)	*Zn^{++} added (ppb)*	*Zn^{++} detected in sterile medium (ppb)*	*Zn^{++} detected in spent medium (ppb)*	*Zn^{++} removed from solution during growth (ppb)*	*Zn^{++} uptake equivalent (ppb)*	*Zn^{++} balance between cells and medium (ppb)*	*Volume of medium harvested (ml)*
0	0	350	350	0	31·7	−31·7	870
1×10^{-5}	654	1 050	1 020	30	40·9	−10·9	910
5×10^{-5}	3 269	3 650	3 480	170	62·2	+107·8	870
1×10^{-4}	6 538	6 960	6 430	530	90·9	+439·1	880
2×10^{-4}	13 076	13 390	12 350	1 040	286·4	+753·6	850
3×10^{-4}	19 614	18 350	13 040	5 310	480·8	+4 829	845
4×10^{-4}	26 152	—	—	—	—	—	—

available Zn when no Zn was added, or very little (1×10^{-5} M Zn) was added. As increasing amounts of Zn were available in the medium, more of it became undetectable. Much of this undetected Zn may be accounted for as a visible precipitate, which was more noticeable after autoclaving as the Zn concentration increased.

As indicated in Fig. 1 and Table V, the cells with the highest concentration

Table V
GROWTH PERIOD, pH CHANGE, AND CELL RECOVERY DURING Zn ASSIMILATION BY *A. MARINUS* FROM BASAL SEAWATER MEDIUM AT 20 ± 2°C

$ZnCl_2$ added (M)	Period of incubation (hr)	pH Initial	pH Final	Dry weight cells (mg)	Analytical weight Zn in cells (μg)
0	36	7·5	7·7	393·97	27·7
1×10^{-5}	36	7·5	7·7	366·45	37·2
5×10^{-5}	36	7·4	7·6	268·80	54·1
1×10^{-4}	36	7·4	7·3	240·04	80·0
2×10^{-4}	64	7·2	7·8	252·38	243·4
3×10^{-4}	64	7·2	7·7	110·09	408·7

of Zn where growth appeared were allowed to grow for 64 as opposed to 36 hr before harvest. Flasks containing 4 and 5×10^{-4} M Zn were incubated for 12 days before final removal from the gyrotary shaking machine, at which time no growth had occurred.

The metal content and concentration factor for Zn as Zn was added to the basal seawater medium are shown in Table VI. Increased amounts of Zn

Table VI
METAL CONTENT AND CONCENTRATION FACTOR DURING Zn ASSIMILATION BY *A. MARINUS* FROM BASAL SEAWATER MEDIUM AT 20 ± 2°C

$ZnCl_2$ added (M)	Concentration in ppm/dry weights of cells						
	Zn	Zn concentration factor	Fe	Cu	Mn	Mg	Ca
0	70·1	200	176	14·2	2·3	1 927	937
1×10^{-5}	101·5	97	784	7·6	3·0	1 714	742
5×10^{-5}	201·3	55	249	7·1	3·0	1 641	536
1×10^{-4}	333·3	48	227	12·1	1·8	1 208	275
2×10^{-4}	964·4	72	167	4·4	2·0	1 740	313
3×10^{-4}	3 712·4	202	250	7·3	1·8	1 272	1 326

Ni and Cd less than 0·1 ppm in all samples.

were concentrated by the cells up to 3×10^{-4} M Zn in basal seawater medium, at which point the cells contained almost 0·4% Zn. Interestingly, the concentration factor for Zn where no Zn was added to the medium was almost the same as it was for the most highly tolerated concentration of Zn, 200 times. A falling off at the intermediate concentrations of added Zn indicated that cells at low concentration of a metal have a fairly high concentration factor which diminished until the cells approached high metal stress conditions. The Zn concentrations in this experiment were not as high as those reported for *A. marinus* in Table III. Possible explanations for this exist in different analyses, culturing, handling, and washing of the cells by individual investigators. No particularly outstanding observations may be made on the other elements analysed other than the high concentration of Fe in the presence of 1×10^{-5} M $ZnCl_2$ and the markedly increased concentration of Ca in the presence of 3×10^{-4} M $ZnCl_2$ (Table VI). The experiments suggest that heavy metal ions were concentrated against the gradient by bacteria but that the highest concentration factors were achieved at very low and high concentrations of the cation.

DISCUSSION

Although most of the elements in the periodic table are present in the electrolyte solution seawater, 99·9 % of the dissolved solids in the seawater consist of 10 ionic species, namely Cl, Na, SO_4, Mg, Ca, K, HCO_3, Br, Sr, and B (Garrels and Thompson, 1962). These major ions in seawater are the relatively conservative or unreactive elements, which is why they have tended to be retained by the ocean (Goldberg, 1965). The trace or minor elements in the ocean, which comprise less than 0·1 % of dissolved solids, are involved in some of the more important inorganic and biochemical reactions of the marine environment.

Marine bacteria assimilated trace elements against the gradient in a manner similar to that of marine plants and animals. Many of the elements in seawater are chelated by the small amounts of organic matter in the sea (Jones, 1973). The interrelationship between organic matter and trace elements is a governing factor in the availability of trace metals to microorganisms. The chelating potential of a culture medium regulates the toxicity of trace metal ions (Cobet, 1968; McCarthy, 1971). Trace metal ions in seawater are a principal factor in the death of freshwater bacteria in seawater (Jones and Cobet, 1975). The oligodynamic effect of metals in which small amounts are beneficial and large amounts are detrimental has been established for several bacteria (Cobet *et al.*, 1970; Sadler and Trudinger, 1967).

Probably the two greatest sources of error in determining the elemental composition of bacteria are the handling of the cells during washing and the nature of the washing solution. Leaking and lysis of cells are common. Other sources of error are the care of glassware cleaning, standardization of synthetic seawater and media preparation, exact conditions of autoclaving, care in digestion of cells, and analytical procedures. We are concentrating our efforts on improved washing and handling procedures.

Marine bacteria concentrate minor ions from a basal seawater medium and probably from the sea itself. The function of many of these ions in cells and the effect of bacteria on marine geochemistry is still a matter of speculation.

Acknowledgements

This work was supported in part by a grant GA-34101 from the National Science Foundation, Washington, D.C. The senior author would like to express his gratitude to Professor J. P. Riley, Department of Oceanography, The University of Liverpool, Liverpool, England, in whose laboratory this work was conducted. His generosity in equipment, encouragement, and ideas has been appreciated by all of us.

REFERENCES

Bowen, H. J. M. (1966). *Trace Elements in Biochemistry*, pp. 1–241, Academic Press, London.

Cobet, A. B. (1968). The effect of nickel ions on *Arthrobacter marinus*, a new species. Ph.D. dissertation, pp. 1–212, University of New Hampshire, Durham, N.H.

Cobet, A. B., Wirsen, C., Jr., and Jones, G. E. (1970). The effect of nickel on a marine bacterium, *Arthrobacter marinus* sp. nov. *J. Gen. Microbiol.*, **62**, 159–169.

Garrels, R. M., and Thompson, M. E. (1962). A chemical model for sea water at 25°C and one atmosphere total pressure. *Am. J. Sci.*, **260**, 57–66.

Goldberg, E. D. (1965). Minor elements in sea water. In *Chemical Oceanography*, Vol. 1, pp. 163–196 (J. P. Riley and G. Skirrow, eds.), Academic Press, London.

Harvey, H. W. (1939). Substances controlling the growth of a diatom. *J. Mar. Biol. Assoc. U.K.*, **23**, 499–520.

Harvey, H. W. (1947). Manganese and growth of phytoplankton. *J. Mar. Biol. Assoc. U.K.*, **26**, 562–579.

Jones, G. E. (1973). An ecological survey of open ocean and estuarine microbial populations. I. The importance of trace metal ions to microorganisms in the sea. In *Estuarine Microbial Ecology*, pp. 233–241 (L. H. Stevenson and R. R. Colwell, eds.), University of South Carolina Press, Columbia, S.C.

Jones, G. E., and Cobet, A. B. (1975). Heavy metal ions as the principal bactericidal agent in Caribbean seawater, In *Int. Sym. on Discharge of Sewage from Sea Outfalls*, pp. 199–208 (A. L. H. Gameson, ed.), Pergamon Press, Oxford.

Lyman, J., and Fleming, R. H. (1940). *J. Marine Res.*, **3**, 135.

Martin, J. H., and Knauer, G. A. (1973). The elemental composition of plankton. *Geochim. Cosmochim. Acta*, **37**, 1639–1653.

McCarthy, L. R. (1971). The effect of copper on *Pseudomonas cuprodurans*, sp. nov. Ph.D. dissertation, pp. 1–263, University of New Hampshire, Durham, N.H.

Noddack, I., and Noddack, W. (1940). Die Haufigkeiten der Schwermetalle in Meerestieren. *Ark. Zool.*, **32A**, 1–52.

Perlman, D. (1965). Microbial production of metal-organic compounds and complexes. In *Advances in Applied Microbiology,* Vol. 1, pp. 103–138 (W. W. Umbreit, ed.), Academic Press, New York.

Porter, J. R. (1946). *Bacterial Chemistry and Physiology*, pp. 1–1073. Wiley, New York.

Redfield, A. C., Ketchum, B. H., and Richards, F. W. (1963). The influence of organisms on the composition of seawater. In *The Sea*, Vol. 2, pp. 26–77 (M. N. Hill, ed.), Interscience, New York.

Riley, J. P., and Roth, I. (1971). The distribution of trace elements in some species of phytoplankton grown in culture. *J. Mar. Biol. Assoc. U.K.*, **51**, 63–72.

Riley, J. P., and Segar, D. A. (1970). The distribution of the major and some minor elements in marine animals. I. Echinoderms and Coelenterates. *J. Mar. Biol. Assoc. U.K.*, **50**, 721–730.

Sadler, W. R., and Trudinger, P. A. (1967). The inhibition of microorganisms by heavy metals. *Mineralium Deposita*, **2**, 158–168.

Segar, D. A., Collins, J. D., and Riley, J. P. (1971). The distribution of the major and some minor elements in marine animals. II. Mollusca. *J. Mar. Biol. Assoc. U.K.*, **51**, 131–136.

Vinogradov, A. F. (1953). The elementary composition of marine organisms, pp. 1–647, Trans. from Russian. Sears Found. Mar. Res., Mem. II. Yale University, New Haven, Conn.

Webb, D. A. (1937). Studies on the ultimate composition of biological materials. Part II. Spectrographic analysis of marine invertebrates with special reference to the chemical composition of their environment. *Scient. Proc. R. Dubl. Soc.*, **21**, 505–539.

ZoBell, C. E. (1946). *Marine Microbiology: A Monograph on Hydrobacteriology,* pp. 1–240, Chronica Botanica Co., Waltham, Mass.

minant not additive with *cadA*, and *mad* a locus marginally increasing cadmium sensitivity. The *cadB* and *mad* loci have been mapped by Smith and Novick (1972). Studies of their physiological or biochemical bases have simply not started.

Recombination studies between point mutants in different genes yield an independent (but precisely the same) genetic map as derived from deletion analysis (Hayes, 1968, Smith and Novick, 1972). Genetic analysis is thus a very powerful tool in studying heavy metal resistances as in so many other aspects of physiology and biochemistry.

Plasmid transfer

In nature, not all members of a bacterial species harbour plasmids and plasmids may be either acquired or lost (Fig. 1). There is a balance depending on the selective advantages of the plasmid functions and on the metabolic burden of synthesizing and maintaining the extra DNA and determined proteins. Experiments largely confined to the conjugative plasmid F have shown the occasional loss of the plasmid during growth and cell division as the multiplication of the plasmid falls behind the pace of the multiplication of the host cell. Loss during growth can be favoured by changes in growth conditions such as temperature, pH, and the presence of some deleterious agents. Plasmids may be transferred directly from one cell to another by a process requiring physical contact between the cells (bacterial conjugation or cell mating). They may also be transferred from cell to cell by means of viral intermediates (transduction) or, apparently more rarely, a cell may acquire a loose plasmid DNA molecule floating in its milieu (transformation).

DNA is replicated (doubled) during conjugation and, therefore, the donor cell retains all of the genes and attributes of the plasmid. The newly acquired molecule replicates in its new host cell and the process can be repeated again and again very rapidly. This potential for rapid proliferation and exponential spread within a population of bacteria is one of the key characteristics and a serious practical danger of plasmids. Because of this, bacteria have an 'open-ended' genome which allows for quick adaptation to new environmental situations without going through the slow processes of random mutation and natural selection at a time when the environment has become hostile. The first report of antibiotic-resistant plasmids appeared in 1959 from a dysentery case in Japan, and in a few years most dysentery bacteria had acquired these genetic determinants. Since then, antibiotic-resistant bacteria have been found all over the world even in countries where antibiotics would not seem to be an important factor in the bacterial environment.

Although selective pressures have not been studied with the determinants of heavy metal resistances in the detail that antibiotic-resistance factors have been studied, one would expect them to exist. Cadmium resistance has thus far been found only on plasmids of *S. aureus* where it is the most common of heavy metal resistances; it is either absent, nonfunctional, or rare on enteric plasmids. Cadmium resistance occurs in clinical isolates where cadmium resistance would not seem to have any survival advantage. Mercury resistance is found on 25–75% of plasmids in many species. Mercury resistance was found on 25% of the over 800 enteric plasmids tested that had been originally isolated on the basis of antibiotic resistance (Schottel *et al.*, 1974a). Organo-mercurial resistance never occurs without concomitant resistance to Hg^{2+} (as will be discussed below) but has a more narrow plasmid range than mercuric ion resistance. Plasmids conferring mercury resistance but none toward tested antibiotics also have been found (Dyke *et al.*, 1970; Laub-Kupersztejn *et al.*, 1974). Since the proportion of estuary pseudomonads resistant to mercurials shows variation from season to season and from place to place depending on the degree of industrial waste present (Nelson *et al.*, 1973; Walker and Colwell, 1974; Colwell and Nelson, 1974), we expect that the introduction and rapid spread of plasmids for heavy metal resistances will be a frequent aspect of microbial degradation cycles in nature.

Mercury and cadmium are the two 'clean' resistances associated with plasmids, where the difference in tolerated level between the sensitive and the resistant bacteria is about 100-fold in concentration (Fig. 7). This is in contrast to the other heavy metal resistances reported on some bacterial plasmids including arsenate, arsenite, antimony, bismuth, cobalt, lead, nickel, and silver. These have been two- to threefold resistances.

With this brief introduction to bacterial genetics and plasmids, we can turn to studies of microbial resistance to heavy metals and the mechanisms involved. The symposium on plasmid structure and function in Schlessinger (1975) and the classic on bacterial genetics by Hayes (1968) are recommended to people wishing a more thorough treatment of the genetic topics that we have just passed over.

MERCURY RESISTANCE

The toxicological problems arising from mercury in food chains and even more so from industrially produced organomercurials (Bakir *et al.*, 1973) have caused a great deal of attention to be centred on the mercury cycle in nature. The current status of this problem was presented by Wood (1974) and Jernelöv and Martin (1975). Although chemical disproportionation reactions cause two mercurous ions ($Hg_2{}^{2+}$ in solution) to produce one mercuric and one metallic mercury molecule, all of the other reactions shown in Fig. 4 involve either bacteria or bacteria-synthesized enzymes (Wood, 1974). Bacteria can convert ionic Hg^{2+} to organomercurials, such as methyl and dimethyl mercury (Wood *et al.*, 1968; Yamada and Tonomura, 1972), and other bacteria can reduce these and other organomercurials to metallic Hg°. Our laboratory has been primarily studying this latter reductive detoxification mechanism. Microbial transformations of mercury compounds can be approached using either pure cultures in laboratory settings or with mixed

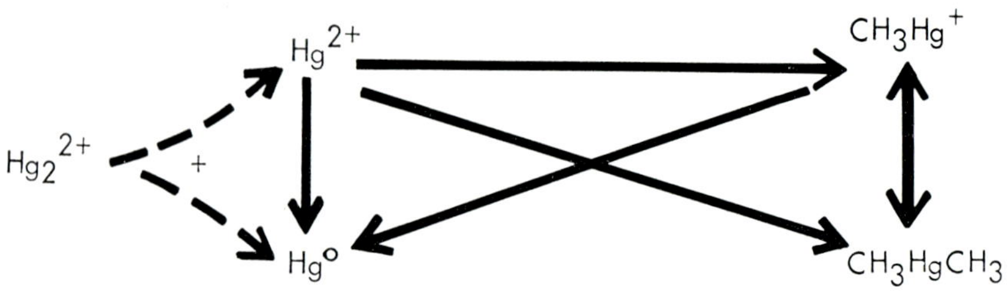

Fig. 4. A simplified mercury cycle in the biosphere (after Wood, 1974). The dashed lines represent a chemical disproportion reaction. All the solid lines represent reactions which involve bacteria or bacterial enzymes.

microbial populations in something approaching a natural ecosystem (Jensen and Jernelöv, 1969; Spangler, *et al.*, 1973). Both must be done and the information acquired from each complements the other.

The interest in microbial mercury methylation was generated by the finding that essentially all of the mercury in fish (both fresh and salt water) was methylmercury (Wood, 1974; Jernelöv and Martin, 1975). Since methylmercury is more toxic to man than inorganic mercury, and since mercury from fish represents a large portion of our total mercury consumption, the question has been: what bacteria are responsible for conversion of the normally inorganic mercury found in fresh and salt water to the methylmercury found in fish?

When seeking methylating organisms, one may look either for rare unusual organisms or for the ability of more common bacteria to methylate mercury. Wood *et al.* (1968), using an anaerobic methanogenic bacterium originally isolated from canal mud, found an enzyme that could methylate Hg^{2+} using methylcobalamine as a methyl donor. Jensen and Jernelöv (1969) incubated sediments from fresh water lakes with Hg^{2+} and found conversion of mercury to methyl and dimethylmercury, caused by anaerobic bacteria that could be isolated from these sediments. They also found bacteria on dead fish that could methylate Hg^{2+}. These results have been confirmed in other laboratories and led to the general consensus shown in Wood's (1974) mercury cycle: that methylation of mercury is carried out by anaerobic bacteria residing in the sediments below bodies of water.

But fish usually do not swim in bottom sediments, and especially in oceanic waters it is difficult to account for the methylmercury in, for example, tuna on the basis of bottom sediment bacteria. This dilemma was pointed out by Donn Kushner at last year's NRC Conference in Ottawa (Schottel *et al.*, 1974b); and the possible solution was suggested in the work of Vonk and Sijpesteijn (1973) who found methylation of mercury by common aerobic microorganisms including *Pseudomonas fluorescens*, *Bacillus megaterium*, *Aerobacter aerogenes*, and *E. coli*! Perhaps we have been looking in the wrong places for exotic anaerobes as responsible agents for mercury methylation, while in normal ecosystems methylation is due to common and ubiquitous aerobic organisms.

With this hypothesis in mind, we sent Dr Iverson bacterial strains that we have been studying as organomercurial decomposers (*see* below). The results from his apparatus (Blair *et al.*, 1974) for measuring microbial mercury methylation are given in Table I. Both *E. coli* strains, B which was used also by Vonk and Sijpesteijn (1973), and K-12, which is the classical organism of bacterial geneticists, produced volatile Hg° and methylmercury. This experiment was under anaerobic conditions and Vonk and Sijpesteijn (1973) reported that *E. coli* B produces methylmercury at about half the rate anaerobically as aerobically. No dimethylmercury was detected. With a culture of *E. coli* harbouring the plasmid R831 that controls decomposition of methylmercury and Hg^{2+} to Hg° (Schottel *et al.*, 1974a,b), no detectable Hg° or methylmercury was found in the 'head space' gas. The Hg° that was produced early in the 9-day incubation period must have escaped from the system. Any CH_3Hg^+ that was synthesized would be rapidly degraded because the rate of functioning of the organomercurial reduction system is over a range of minutes, whereas days are needed for appreciable mercury methylation. With *P. aeruginosa* the pattern is less clear but similar to that in *E. coli*. The plasmid-

Table I
METHYLATION OF MERCURY(II) BY BACTERIA WITH AND WITHOUT PLASMIDS

Bacterial strain		Hg°	Methyl-Hg
E. coli			
	B	++	++
	K-12	+++	+++
	K-12(R831)	—	—
P. aeruginosa			
	PU21	+	+
	PU21(*p*MG1)	++	++
	PU21(*p*MG2)	*	+
	PU21(FP2 +)	?	?
	PU21(R3108)	trace	trace

Experiment by Warren Iverson. Bacteria were transferred to slants of Nelson *et al.* (1973)'s medium containing 1 ppm Hg as $HgCl_2$. The slants were capped with teflon mininert valves and incubated for 9 days at 22 ± 1°C. 1 ml of the head atmosphere above the slants was introduced into the gas chromatography-AAS unit and retention times and peak heights were noted. The results are expressed in relative peak heights: —, no detectable product above background; ?, questionable peak above background; trace, small but positive peak; +, ++, and +++, well defined peaks of product. *, represents a peak at 1 min 18 sec, which was not typical for Hg°.

less *Pseudomonas* produced volatile Hg° and CH_3Hg^+. Cells with the plasmids $FP2^+$ and R3108, both of which determine the ability to decompose CH_3Hg^+ (Schottel *et al.*, 1974b), show only questionable or trace CH_3Hg^+ in the gas phase, while cells with plasmids pMG1 and pMG2, which can reduce Hg^{2+} but not methylmercury, show measurable production of methylmercury during the 9 days' incubation (Table I).

Since the same *E. coli* bacteria can synthesize and degrade methylmercury, what will be found in nature will be an equilibrium between inorganic and organic mercury depending upon the duration of exposure to mercury(ials) and the presence or absence of plasmids in the normally abundant bacteria. This is a major conclusion—and I phrase it as such to stimulate a response from Dr Wood: normally abundant bacteria can be responsible for both sides of the methylmercury synthesis-degradation cycle. We should look for the normal gut and gill flora of fish and invertebrates for candidates for methylating bacteria. We might hope that normal mammalian intestinal flora (with plasmids) would alter the fraction of ingested methylmercury that is accumulated by the body as well.

A great range of studies of the genetics, physiology, and biochemistry of mercury and mercurial resistance has occurred in the last few years in our own (Summers *et al.*, 1972, 1973, 1974, 1975; Schottel *et al.*, 1974a,b) and other laboratories (Furukawa and Tonomura, 1972a,b; Izaki *et al.*, 1974; Tonomura *et al.*, 1972; Tonomura and Kanzaki, 1969). For this reason we will present only a few titbits of previously unpublished data to demonstrate points of interest. Mercury volatilization can be measured both by gas–chromatography–mass spectroscopy [as utilized by our session Chairman (Blair *et al.*, 1974) and ourselves (Summers and Silver, 1972)] and also by using radioactive mercurials (Summers *et al.*, 1972, 1973, 1974, 1975; Schottel *et al.*, 1974a,b).

We have studied the plasmid-determined decomposition of a wide range of organomercurials (Schottel *et al.*, 1974a,b). The two major classes of mercury-resistant plasmids are those with and those without resistance to organomercurials, but the organomercurial-resistant plasmids also vary in the range of compounds which their plasmid-determined enzymes can decompose. This topic is still inadequately studied, but we have measured decomposition of methylmercury (Schottel *et al.*, 1974b), phenylmercury (Schottel *et al.*, 1974a,b; Summers *et al.*, 1975), *p*-hydroxymercuribenzoate (*p*HMB), merbromin, and thimerosal (Fig. 5), and fluoromercuriacetate. Figure 6 shows the chemical structures of these compounds. We will happily look at any other mercurial that you consider of particular interest. An adaptation of the standard flameless atomic absorption spectroscopy (AAS) assay for

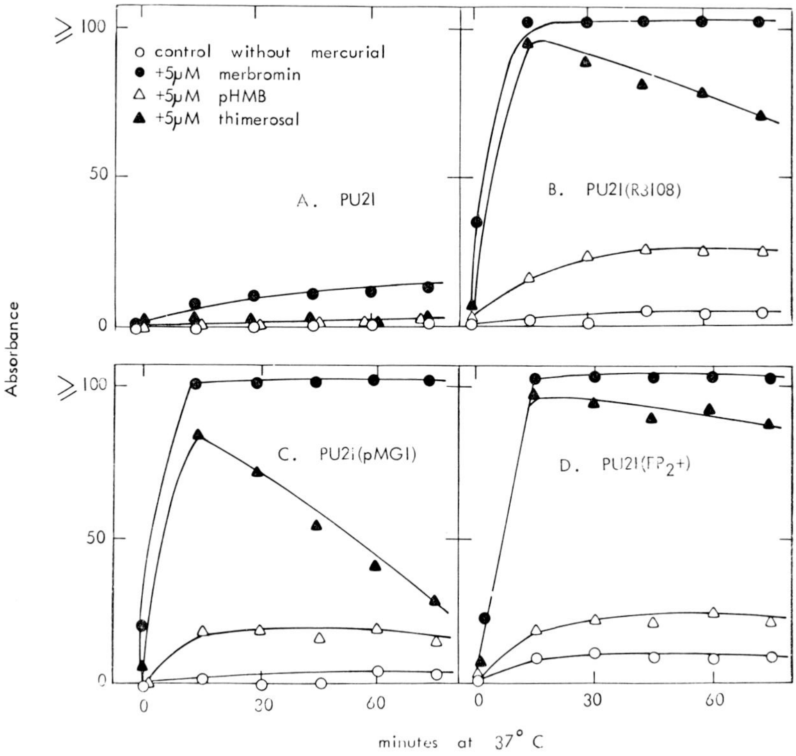

Fig. 5. Plasmid determined decomposition of organomercurials. Absorption at 254 nm (*Hg*°) *was measured as described in the text and in Schottel* et al. (*1974b*).

mercury volatilization that we described at the 1974 NRC Symposium (Schottel *et al.*, 1974b) makes this easy for anyone possessing a flameless mercury AAS apparatus. In the experiment in Fig. 5, bacterial suspensions were incubated in buffer in standard BOD bottles in a 37°C water bath. Every 15 min the bottles were attached to the standard flameless Hg AAS apparatus and the Hg° that had been generated during the incubation was aerated into the AAS cell and recorded. The plasmidless strain of *P. aeruginosa*, PU21, was unable to reduce merbromin, *p*HMB, or thimerosal to volatile mercury forms. The low level volatilization with merbromin is chemical disproportionation that occurs also in the absence of any bacteria.

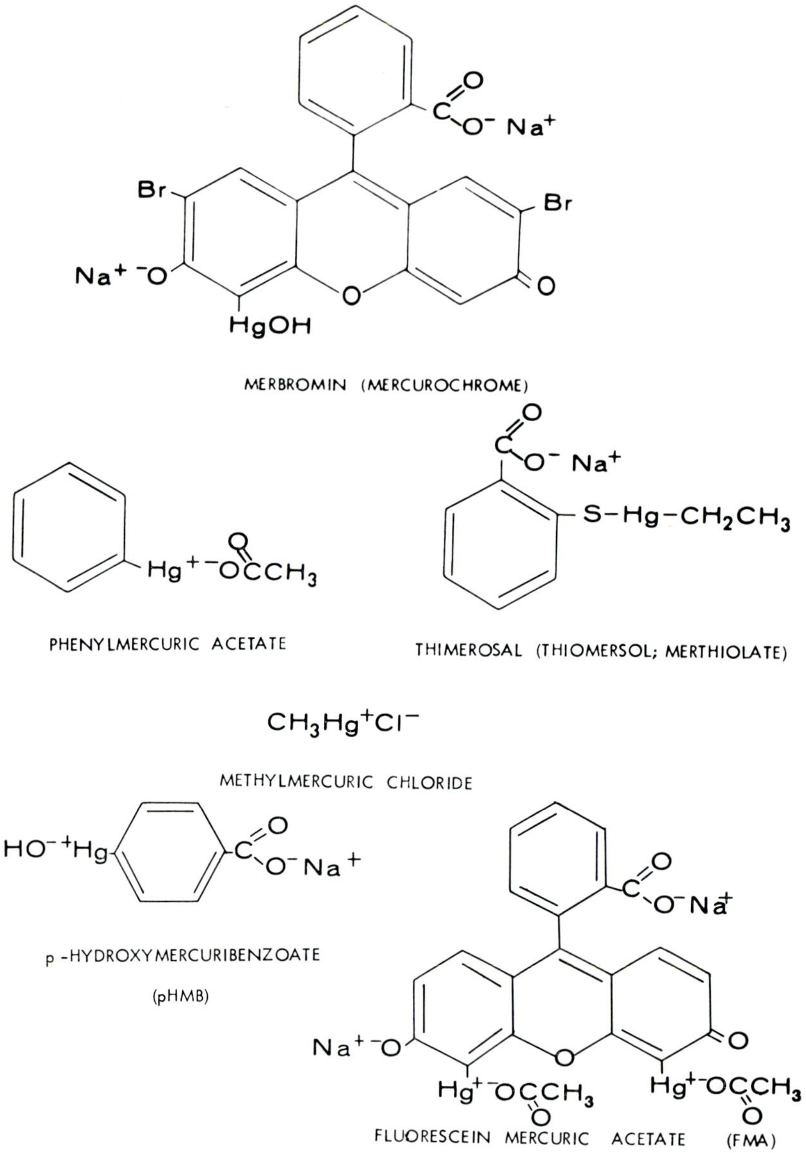

Fig. 6.

Plasmid R3108 controls decomposition of merbromin, thimerosal, and at a slower rate *p*HMB. Plasmid pMG1 controls volatilization of these three mercurials at rates very similar to those with cells with plasmid R3108, even though pMG1 does not determine the ability to volatilize methylmercury (Schottel *et al.*, 1974b). Here we have an example of a difference in substrate range among the organomercurial-resistant plasmids. Plasmid $PF2^+$, which is the original pseudomonad plasmid controlling conjugation (Holloway *et al.*, 1971), also determines decomposition of all three organomercurials. I would happily expand on our studies of mercurial resistance, but have been asked by the chairman to resist this temptation.

CADMIUM RESISTANCE

Cadmium is frequently a by-product of metal processing, particularly in the refining of zinc. Thousands of tons of cadmium are released into the air and water annually. The presence of cadmium in fertilizers and pesticides (McCaull, 1971), in polluted air and water, and in microorganisms which

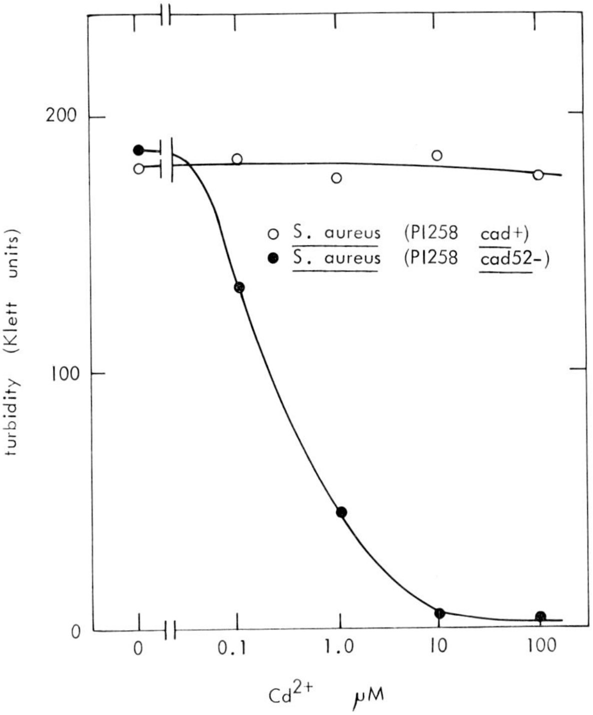

Fig. 7. Cadmium resistance in S. aureus. *Growth (turbidity) after 5 hr incubation with aeration at 37°C in nutrient broth with or without added cadmium.* ○, *with a derivative of plasmid PI258 with* cad+ *(resistance);* ●, *with a derivative of plasmid PI258 with the* cad52 *(sensitivity) mutation in the* cadA *gene.*

accumulate large quantities of cadmium (Doyle *et al.*, 1975) results in a recycling of cadmium through food chains, including our own.

The mechanism of cadmium resistance (Fig. 7) is totally different from that of mercury resistance and has been studied by Chopra (1971, 1975) and others, including ourselves. The two current alternatives for the mechanisms of cadmium resistance are (1) the existence of a trans-membrane 'active transport' system for bringing cadmium into the cells, that is inactive or altered in resistant cells (Fig. 8A); and (2) a change in membrane composition of the resistant cells, lowering the surface binding and permeability to cadmium. We tentatively favour the first hypothesis. There is no evidence of cadmium-

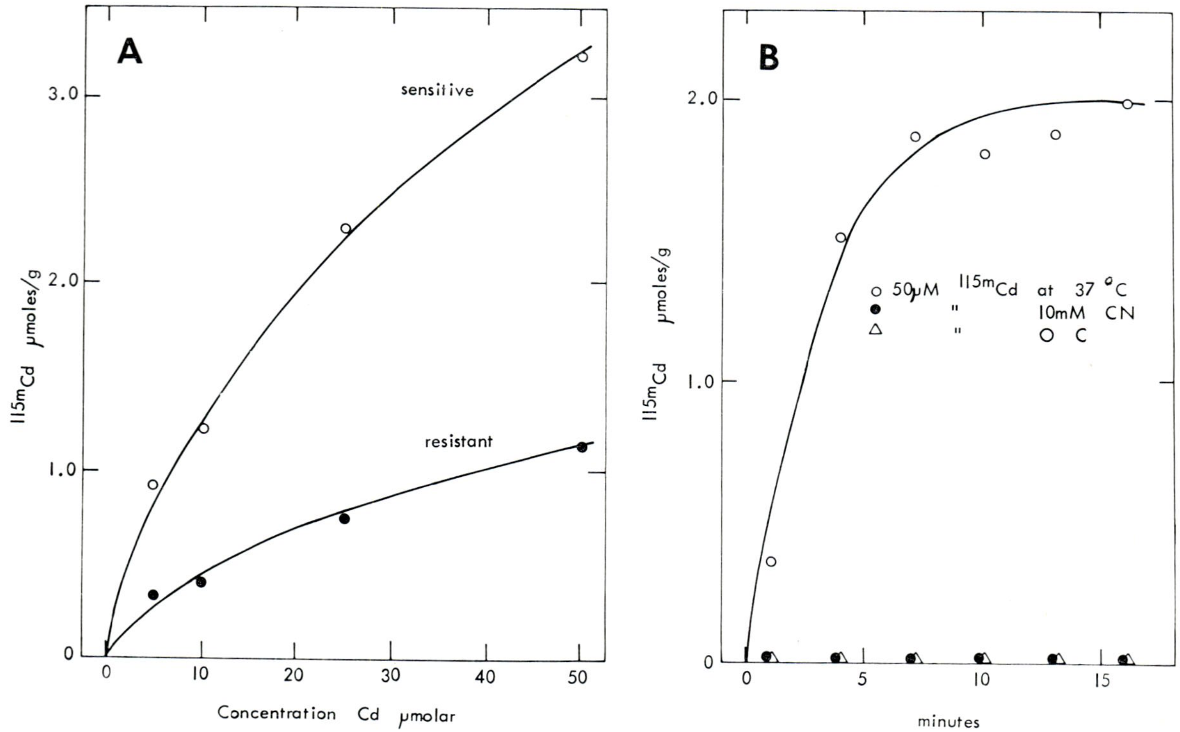

Fig. 8. *Cadmium uptake by sensitive (A and B) and resistant (A only)* S. aureus *and energy requirements for cadmium uptake. Cells were grown in tryptone broth to a turbidity of 50 Klett units, centrifuged and resuspended in glycylglycine buffer (pH 7·3) plus 1% NaCl and 0·8% glucose.* ^{115m}*Cd was added and uptake was measured after 18 min (A) or at various times (B) by withdrawing samples, filtering through millipore filters, and counting in a planchet counter.*

binding metallothionen-like proteins, such as are found in mammalian systems (Cherian, 1974).

Cadmium uptake by sensitive cells is temperature dependent (Chopra, 1975; Kondo *et al.*, 1974; Tynecka *et al.*, 1975; Fig. 8B) and inhibited by inhibitors of energy-dependent metabolism, cyanide, for example, to an extent similar to the inhibition of other active transport processes (data not shown). Plasmid-determined changes in an active transport system are not unique to cadmium, as the same mechanism has been proposed for plasmid-determined resistance to the antibiotic tetracycline (Levy and McMurry, 1974). With tetracycline resistance a new membrane protein is found in the resistant cells, but only after exposure to tetracycline (Levy and McMurry, 1974; *i.e.* inducible). We now know that cadmium resistance does not depend on prior exposure to cadmium but is always expressed by the cells (*i.e.* constitutive).

If a toxic material such as cadmium is accumulated by an active transport process, one can ask what the *normal* physiologically desirable substrate for this transport system might be, assuming that the cells have not evolved a system specific for cadmium. Preliminary evidence from this laboratory shows a highly specific relationship between cadmium resistance (and uptake) and the normal manganese transport system (Fig. 9). Manganese transport systems occur in all bacterial species at which we have looked (Silver *et al.*, 1970; Eisenstadt *et al.*, 1973; Jasper, 1975). *S. aureus* also has a highly specific manganese transport system that discriminates against other divalent cations such as magnesium and calcium (unpublished). With the cadmium-sensitive cells, 50 μM cadmium rapidly inhibits manganese transport (Fig. 9A), whereas under the same conditions manganese transport by the resistant cells is unaffected (data not shown). This effect of cadmium is specific for the manganese transport system, since 50 μM cadmium does not have a similar effect on ^{86}Rb (a substrate for the potassium transport system) uptake (Fig. 9B). The results in Fig. 9 clearly require a thorough kinetic analysis of the relative affinities of cadmium and manganese for the cellular transport system, as we have done with other organisms (Silver *et al.*, 1979; Eisenstadt *et al.*, 1973), but we thought it more useful to give you our current working model than to simply leave the question of mechanism of cadmium resistance hanging.

The alternative hypothesis of a difference in membrane components binding cadmium is still tenable. Chopra (1975) has shown that spheroplasts from resistant cells retain the resistance to cadmium that is characteristic of the intact cells. Since the spheroplasts are produced by removing the cell walls, this experiment demonstrates that the cell membrane and not the cell wall contains the component distinguishing sensitive from resistant bacteria. Since toluene treatment, which alters membrane permeability to small molecules, caused cadmium-resistant cells to become cadmium-sensitive, as measured by inhibition of protein synthesis, Chopra (1975) considered it likely that the membrane was functioning as a permeability barrier keeping out the cadmium that could otherwise flow nonspecifically to intracellular sites of inhibition. Thus Chopra (1975) was led to favour the permeability rather than the transport explanation for the difference between cadmium-sensitive and cadmium-resistant bacteria, even though he was unable to find a difference in either thiol groups or lipid moieties—the two classes of membrane components tested.

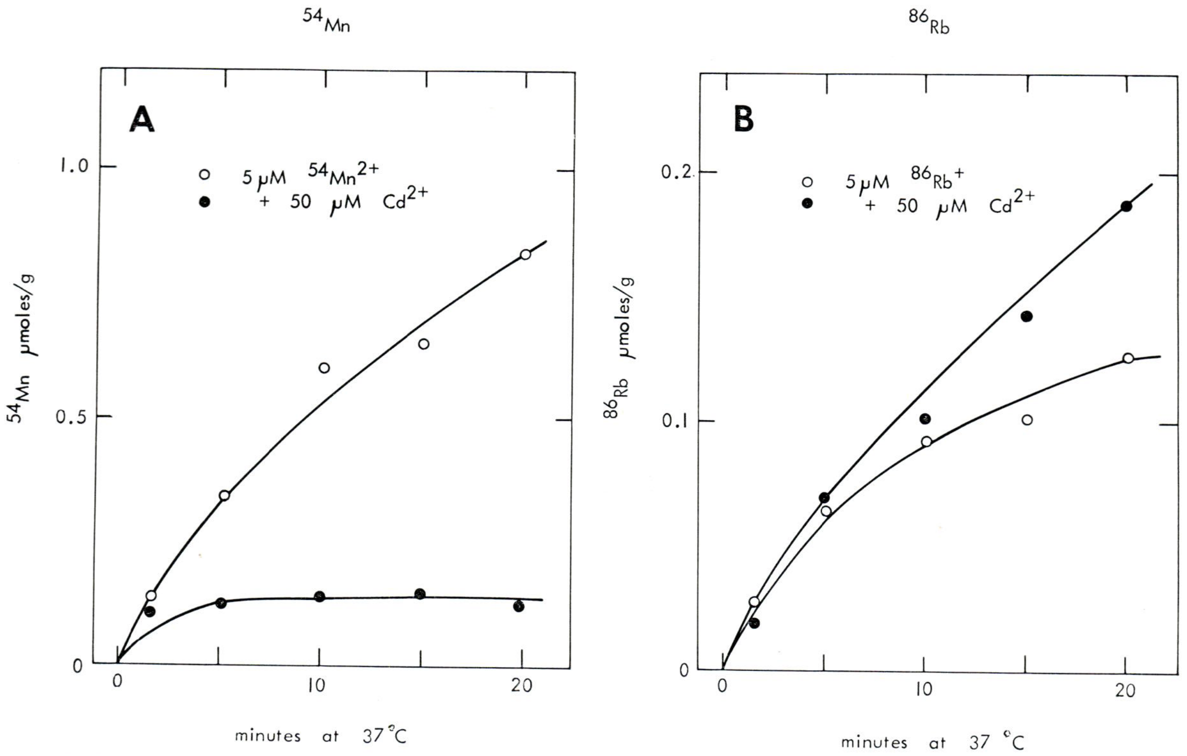

Fig. 9. *Cadmium inhibition of manganese, but not rubidium transport. The cadmium sensitive* S. aureus *strain was grown and centrifuged as described in Fig. 8, resuspended in broth with* ^{54}*Mn or* ^{86}*Rb and sampled as before.*

Pertinent to the mode of inhibition by cadmium is the distribution of cadmium that is taken up by the cells. The data in Table II show a difference in distribution between the sensitive and the resistant cells. Resistant cells took up less cadmium and proportionately less of the cadmium was found in the cytoplasm (high speed supernatant). Eighty percent of the cadmium in the sensitive cells was found in the cell cytoplasmic fraction.

Permeability barriers are not the only way that microorganisms can cope with toxic levels of cadmium. Mitra *et al.* (1975) found that continued exposure to cadmium resulted in an 'accommodation' so that the cells that were still genetically cadmium-sensitive were physiologically cadmium-resistant. During this process of inhibition by cadmium, loss of the ability to form colonies and large-scale intracellular vacuolization occurred. This was followed by 'accommodation', regaining of colony-forming ability *without*

Table II

CELLULAR DISTRIBUTION OF ^{115m}Cd

	Sensitive cells	*Resistant cells*
Intact cells before disruption	5·5 μmole/g (100%)	1·7 μmole/g (100%)
After disruption		
low speed supernatant	96%	68%
low speed pellet	10%	23%
high speed supernatant	78%	38%
high speed pellet	8%	11%

$^{115m}CdCl_2$ was added at 100 μM to sensitive (RN1008) or resistant (RN23) cells grown aerobically to a density of 0·16 mg/ml dry weight. After 30 min additional incubation at 37°C, the bacteria were harvested by centrifugation at 8000 × *g* for 20 min and washed twice with 100 ml 0·9% saline (each time). Total cellular cadmium was determined from the turbidity (dry weight) and radioactivity (cadmium) present at this stage. The cells were disrupted in a Braun cell disintegrator for 6 min. The broken cell suspensions were filtered and washed with 100 ml saline to remove the glass beads, centrifuged at 10 000 × *g* for 10 min (low speed to remove residual intact bacteria), and the supernatant fluid from the low speed centrifugation was recentrifuged at 25 000 × *g* for 20 min (high speed to remove cell envelope fragments).

growth and clearing of the vacuoles from the cytoplasm. The cadmium was largely displaced from the cytoplasm to the cell wall and membrane during accommodation.

In summary of the current state of knowledge of plasmid-determined cadmium resistance, this is probably the best studied heavy metal resistance from the genetic point of view. Surprisingly, plasmid-determined cadmium resistance has only been found in *S. aureus*, unlike mercury resistance which is found associated with plasmids in many different bacterial species. The mechanism of cadmium resistance is not a chemical transformation—neither methylation nor reduction occurs—but rather a permeability change takes place, so that the plasmid-containing resistant cells do not accumulate the toxic levels of cadmium accumulated by the sensitive cells. The resistant plasmids seem to block an energy-dependent transport system functional in most bacterial cells.

RESISTANCE TO OTHER HEAVY METALS

Arsenate, arsenite, antimony, bismuth, cobalt, lead, nickel, and silver are the other heavy metal ions for which plasmid-determined resistance has been reported. Several of these resistances have been mapped as separate loci on the genetic maps of staph plasmids such as PI258 (Fig. 3). In each case the resistance was so marginal that it discourages attempts at physiological experiments. In each case there are no data to our knowledge relating to a possible mechanism for plasmid-determined resistances. Unlike mercury resistance and cadmium resistance which are found only on plasmids, there are a few cases of chromosomal mutations affecting some of these ion sensitivities and of explanations for the resistances. These may give us some idea of how plasmid-regulated systems could function.

Arsenate is an alternative substrate for one of two phosphate transport systems in several organisms. Arsenate-resistant mutants have been isolated that lack the *ipt* (*i*norganic *p*hosphate *t*ransport) system but are able to accumulate adequate levels of phosphate for growth by means of the *pst* (*p*hosphate *s*pecific *t*ransport) system (Bennett and Malamy, 1970; Medveczky and Rosenberg, 1971). Similarly, mutants of *Salmonella typhimurium* (and other organisms) selected for chromate resistance are defective in the sulphate transport system for which chromate is an analogue substrate (Ohta *et al.*, 1971). Cobalt and manganese are alternative substrates for magnesium transport system(s) in *E. coli* (Nelson and Kennedy, 1972; Silver *et al.*, 1972) and in other organisms. Manganese-resistant mutants have been mapped on the primary chromosome, and cells with such a mutation showed decreased affinity for manganese on the magnesium transport system. The cobalt-resistant mutants have been used to demonstrate that there are two separable magnesium systems in *E. coli*, one of which is missing in cobalt-resistant mutants (Nelson and Kennedy, 1972). It is not known whether the two classes of mutants affect the same transport system but cobalt-resistant mutants are not manganese-resistant and manganese-resistant mutants are not cobalt-resistant. So here we have three cases of chromosomally determined resistance that are associated with changes in the specificity of transport systems, as we propose to be the case for the plasmid-determined cadmium resistance.

Plasmid-determined silver resistance has not been studied. The plasmid-specified mercury reducing systems described above also have silver as a substrate (Ag^+ is reduced to Ag°; Summers *et al.*, 1975). Yet this ability to reduce silver cations does not confer any resistance to silver on the plasmid-bearing strains. We believe this is because silver is a noneffective 'inducer' of the mercury-reducing system, which is only found in cells grown with mercury or mercurials (Furukawa and Tonomura, 1972b; Summers and Silver, 1972). Therefore, cells exposed to silver cannot produce the enzyme(s) needed for reduction of silver. That leaves five additional heavy metal ions for which plasmid-determined resistance has been reported. Although there are occasional studies of accumulation and toxicity of these materials with bacteria or other microbes, the information available is so sparse as not to justify further discussion.

We hope that we have persuaded you of the utility of plasmids in providing heavy metal resistances for the bacteria and of the utility of genetic and laboratory-type physiological and biochemical studies in addressing the problems of microbial resistances and transformations with which we are increasingly concerned.

Acknowledgement

Research in our laboratory has been sponsored by National Institutes of Health grant AI08062. Anne Summers and Amal Mandal carried out some of the mercury studies. Dan Clark and Kathleen Farrelly provided technical help. We thank W. P. Iverson for the mercury methylation data in Table I and John Langridge and R. P. Novick for the photographs in Fig. 2.

REFERENCES

Bakir, F., Damluji, S. F., Amin-Zaki, L., Murtadha, M., Khalidi, A., Al-Rawi, N. Y., Tikriti, S., Dhahir, H. I., Clarkson, T. W., Smith, J. C., and Doherty, R. A. (1973). Methylmercury poisoning in Iraq. *Science*, **181**, 230–241.

Bennett, R. L., and Malamy, M. H. (1970). Arsenate resistant mutants of *Escherichia coli* and phosphate transport. *Biochem. Biophys. Res. Commun.*, **40**, 496–503.

Blair, W., Iverson, W. P., and Brinckman, F. E. (1974). Application of a gas chromatograph-atomic absorption detection system to a survey of mercury transformations by Chesapeake Bay microorganisms. *Chemosphere*, **3**, 167–174.

Cherian, M. G. (1974). Isolation and purification of cadmium binding proteins from rat liver. *Biochem. Biophys. Res. Commun.*, **61**, 920–926.

Chopra, I. (1971). Decreased uptake of cadmium by a resistant strain of *Staphylococcus aureus*. *J. Gen. Microbiol.*, **63**, 265–267.

Chopra, I. (1975). Mechanism of plasmid-mediated resistance to cadmium in *Staphylococcus aureus*. *Antimicrob. Agents Chemother.*, **7**, 8–14.

Colwell, R. R., and Nelson, J. D., Jr. (1974). Bacterial mobilization of mercury in Chesapeake Bay. Proceedings of the International Conference on Transport of Persistant Chemicals in Aquatic Ecosystems. National Research Council, Ottawa, pp. III-1–III-10.

Doyle, J. J., Marshall, R. T., and Pfander, W. H. (1975). Effects of cadmium on the growth and uptake of cadmium by microorganisms. *Appl. Microbiol.*, **29**, 562–564.

Dyke, K. G. H., Parker, M. T., and Richmond, M. H. (1970). Penicillinase production and metal-ion resistance *Staphylococcus aureus* cultures isolated from hospital patients. *J. Med. Microbiol.*, **3**, 125–136.

Eisenstadt, E., Fisher, S., Der, C.-L., and Silver, S. (1973). Manganese transport during growth and sporulation in *Bacillus subtilis*. *J. Bacteriol.*, **113**, 1363–1372.

Furukawa, K., and Tonomura, K. (1972a). Metallic mercury-releasing enzyme in mercury-resistant *Pseudomonas*. *Agr. Biol. Chem.*, **36**, 217–226.

Furukawa, K., and Tonomura, K. (1972b). Induction of metallic mercury-releasing enzyme in mercury-resistant *Pseudomonas*. *Agr. Biol. Chem.*, **36**, 2441–2448.

Hayes, W. (1968). *The Genetics of Bacteria and Their Viruses*, 2nd ed. Blackwell Scientific Publications, Oxford.

Holloway, B. W., Krishnapillai, V., and Stanisich, V. (1971). *Pseudomonas* genetics. *Ann. Rev. Genet.*, **5**, 425–446.

Izaki, K., Tashiro, Y., and Funaba, T. (1974). Mechanism of mercuric chloride resistance in microorganisms. III. Purification and properties of a mercuric ion reducing enzyme from *Escherichia coli* bearing R factor. *J. Biochem.* (*Tokyo*), **75**, 591–599.

Jasper, P. L. P. (1975). Cation transport systems of *Rhodopseudomonas capsulata*. Ph.D. Thesis, Washington University, St. Louis, Missouri.

Jensen, S., and Jernelöv, A. (1969). Biological methylation of mercury in aquatic organisms. *Nature* (*London*), **223**, 753–754.

Jernelöv, A., and Martin, A. L. (1975). Ecological implications of metal metabolism by microorganisms. *Ann. Rev. Microbiol.*, **29**, in press.

Kondo, I., Ishikawa, T., and Nakahara, H. (1974). Mercury and cadmium resistances mediated by the penicillinase plasmid in *Staphylococcus aureus*. *J. Bacteriol.*, **117**, 1–7.

Laub-Kupersztejn, R., Thomas, J., and Pohl, P. (1974). Résistance des entérobactéries au mercure et aux antibiotiques. Études expérimentale chez le porc. *Ann. Microbiol.* (*Inst. Pasteur*), **125B**, 501–508.

Levy, S. B., and McMurry, L. (1974). Detection of an inducible membrane protein associated with R-factor-mediated tetracycline resistance. *Biochem. Biophys. Res. Commun.*, **56**, 1060–1068.

McCaull, J. (1971). Building a shorter life. *Environment*, **13** (7), 2–15.

Medveczky, N., and Rosenberg, H. (1971). Phosphate transport in *Escherichia coli*. *Biochim. Biophys. Acta*, **241**, 494–506.

Mitra, R. S., Gray, R. H., Chin, B., and Bernstein, I. A. (1975). Molecular mechanisms of accommodation in *Escherichia coli* to toxic levels of Cd^{2+}. *J. Bacteriol.*, **121**, 1180–1188.

Nelson, D. L., and Kennedy, E. P. (1972). Transport of magnesium by a repressible and nonrepressible system in *Escherichia coli*. *Proc. Natl. Acad. Sci. U.S.A.*, **69**, 1091–1093.

Nelson, J. D., Blair, W., Brinckman, F. E., Colwell, R. R., and Iverson, W. P. (1973). Biodegradation of phenylmercuric acetate by mercury-resistant bacteria. *Appl. Microbiol.*, **26**, 321–326.

Novick, R. P. (1969). Extrachromosomal inheritance in bacteria. *Bacteriol. Rev.*, **33**, 210–263.

Novick, R. P., and Bouanchaud, D. (1971). Extrachromosomal nature of drug resistance in *Staphylococcus aureus*. *Ann. N.Y. Acad. Sci.*, **182**, 279–294.

Novick, R. P., and Roth, C. (1968). Plasmid-linked resistance to inorganic salts in *Staphylococcus aureus*. *J. Bacteriol.*, **95**, 1335–1342.

Ohta, N., Galsworthy, P. R., and Pardee, A. B. (1971). Genetics of sulfate transport by *Salmonella typhimurium*. *J. Bacteriol.*, **105**, 1053–1062.

Schlessinger, D. (ed.) (1975). *Microbiology*-1974. American Society for Microbiology, Washington, D.C.

Schottel, J., Mandal, A., Clark, D., Silver, S., and Hedges, R. W. (1974a). Volatilisation of mercury and organomercurials determined by inducible R-factor systems in enteric bacteria. *Nature* (*London*), **251**, 335–337.

Schottel, J., Mandal, A., Toth, K., Clark, D., and Silver, S. (1974b). Mercury and mercurial resistance determined by plasmids in *Escherichia coli* and *Pseudomonas aeruginosa*. Proceedings of the International Conference on Transport of Persistent Chemicals in Aquatic Ecosystems, pp. II-65–II-71, National Research Council, Ottawa.

Silver, S., Johnseine, P., and King, K. (1970). Manganese active transport in *Escherichia coli*. *J. Bacteriol.*, **104**, 1229–1306.

Silver, S., Johnseine, P., Whitney, E., and Clark, D. (1972). Manganese-resistant mutants of *Escherichia coli*: physiological and genetic studies. *J. Bacteriol.*, **110**, 186–195.

Smith, K., and Novick, R. P. (1972). Genetic studies on plasmid-linked cadmium resistance in *Staphylococcus aureus*. *J. Bacteriol.*, **112**, 761–772.

Spangler, W. J., Spigarelli, J. L., Rose, J. M., Flippin, R. S., and Miller, H. H. (1973). Degradation of methylmercury by bacteria isolated from environmental samples. *Appl. Microbiol.*, **25**, 488–493.

Summers, A. O., and Silver, S. (1972). Mercury resistance in a plasmid-bearing strain of *Escherichia coli*. *J. Bacteriol.*, **112**, 1228–1236.

Summers, A. O., and Lewis, E. (1973). Volatilization of mercuric chloride by mercury-resistant plasmid-bearing strains of *Escherichia coli*, *Staphylococcus aureus* and *Pseudomonas aeruginosa*. *J. Bacteriol.*, **113**, 1070–1072.

Summers, A. O., and Sugarman, L. I. (1974). Cell-free mercury(II) reducing activity in a plasmid-bearing strain of *Escherichia coli*. *J. Bacteriol.*, **119**, 242–249.

Summers, A. O., Schottel, J., Clark, D., and Silver, S. (1975). Plasmid-borne Hg(II) and organomercurial resistance. In *Microbiology* 1974, pp. 219–226 (D. Schlessinger, ed.), American Society for Microbiology, Washington, D.C.

Tonomura, K., and Kanzaki, F. (1969). The reductive decomposition of organic mercurials by cell-free extract of a mercury-resistant pseudomonad. *Biochim. Biophys. Acta*, **184**, 227–229.

Tonomura, K., Furukawa, K., and Yamada, M. (1972). Microbial conversion of mercury compounds. In *Environmental Toxicology of Pesticides*, pp. 115–133 (F. Matsumura, G. M. Boush, and T. Misato, eds.), Academic Press, New York.

Tynecka, Z., Zajac, J., and Gos, Z. (1975). Plasmid dependent impermeability barrier to cadmium ions in *Staphylococcus aureus*. *Acta Microbiol. Pol.*, **7**, 11–20.

Vonk, J. W., and Sijpesteijn, A. K. (1973). Studies on the methylation of mercuric chloride by pure cultures of bacteria and fungi. *Ant. van Leeuwenhoek*, **39**, 505–513.

Walker, J. D., and Colwell, R. R. (1974). Mercury-resistant bacteria and petroleum degradation. *Appl. Microbiol.*, **27**, 285–287.

Wood, J. M. (1974). Biological cycles for toxic elements in the environment. *Science*, **183**, 1049–1052.

Wood, J. M., Kennedy, F. S., and Rosen, C. G. (1968). Synthesis of methyl-mercury compounds by extracts of a methanogenic bacterium. *Nature* (*London*), **220**, 173–174.

Yamada, M., and Tonomura, K. (1972). Formation of methylmercury compounds from inorganic mercury by *Clostridium cochlearium*. *J. Ferment. Technol.* (*Osaka*), **50**, 159–166.

APPROACHES TO THE STUDY OF MICROBIAL TRANSFORMATIONS OF METALS

F. E. BRINCKMAN, W. P. IVERSON and W. BLAIR

National Bureau of Standards, Washington, D.C. 20234, U.S.A.

Summary

An approach to the study of microbial transformations of metals existing in trace amounts in the environment is based primarily on the detection and speciation of volatile forms of these metals. Simple qualitative means for surveying the potential of microorganisms for volatilizing metals are described. Identification of transport agents has been accomplished primarily through the use of flameless atomic spectrophotometry coupled with gas chromatography. This technique permits characterization of volatile metals and organometallic compounds in the nanogram range so that microbial transformations of metals at environmental levels may be monitored. The abiotic transfer of methyl groups from biologically methylated metals to Hg(II) provides a useful tool for the speciation of volatile metal forms suspected of being methylated.

INTRODUCTION

The exact geological time at which biological activities became important in mineral transformation is uncertain although some evidence for extensive geomicrobial activity in proterozoic, palaeozoic, and mesozoic times is suggested by isotope fractionation studies and observations on fossil microorganisms from geological deposits of these eras (Silverman and Ehrlich, 1964; Swain, 1969).

Nevertheless, we know that mineral transformation is common to all cellular life. Microorganisms which are involved in transformation processes include bacteria, fungi, algae, and protozoa. Certain groups of microorganisms, however, contribute more significantly than others to the geological distribution of mineral matter (Silverman and Ehrlich, 1964; Zajic, 1969).

With the advent of the industrial revolution, gradually increasing amounts of heavy metals from anthropogenic sources have been discharged to the aquatic environment where locally heavy concentrations have accumulated. As a result of food chain cycling in the vicinity of these accumulations, one of these metals has been incorporated into man's foodstuffs as the highly toxic methylmercury, with disastrous effects at Minimata and Niigata (Dinman and Hartung, 1972). The discovery that microorganisms were able to methylate mercury stimulated investigations into the microbial transformations of this element. The significance of other similar heavy metal transformations with respect to the public health area is as yet undetermined. Moreover, the attack by microbes on structural metals and alloys is another important example of the public safety and economic significance of these transformations (Iverson, 1972a). The processing (leaching) of ores by microbiological transformation is yet another area which will become of increasing economic importance (Zajic, 1969).

1. Production of reactive metabolic products (*i.e.* acids, NH_3, H_2S, mercaptans, oxidants, reductants, etc.).
2. Adsorption to cell surfaces.
3. Production of oxygen and chemical concentration cells on metal surfaces.
4. Enzymatic hydrolysis of metal complexing agents.
5. Oxidation-reduction reactions.
6. Methylation reactions.

Fig. 1. Microbe-metal reactions.

Possible mechanisms of microbe-metal interactions are shown in Fig. 1. Essentially they are of two types: those processes that take place outside the cell (reactions 1, 2, 3, and 4) and those that occur within the cell as a result of intracellular enzymatic reactions (reactions 5 and 6). The majority of the metal transformations studied have been either of the oxidative-reductive type or of metabolic product reactions with the metal in the elemental or ionic form. The toxicity of the metal systems to the organisms investigated in the majority of these reactions studied was relatively low. Consequently, the reactants usually could be determined at milligram levels, using paper chromatographic techniques, weight loss measurements, colorimetric, and 'wet' quantitative analytical procedures. Highly metal-tolerant organisms or cell extracts are employed in studying relatively toxic systems employing compounds of arsenic, selenium, and tellurium.

With the demonstration that bacteria could produce the highly toxic methylmercury species from inorganic mercury at environmental concentrations (*e.g.* ppm and ppb) or levels well below those previously used in bacterial transformation studies (Jensen and Jernelöv, 1969), it became necessary to devise new analytical methodology. It was desired that such methodology should not only be capable of measuring the concentration of the metal element at nanogram levels but also of speciating it, *i.e.* characterizing the molecular properties of the active metal-containing species involved in biological transformations. This has been the primary goal of our investigations.

GENERAL APPROACHES TO METAL TRANSFORMATION

It is desirous and necessary to investigate pollutant toxic metal transformation at low levels, first, because they usually exist at low concentrations in the environment, and second, because the organisms characteristically only tolerate these low levels.

Organisms selected for investigation on metal transformations should, if possible, be isolated directly from the environment. Development of metal-tolerant strains of the organisms may be attempted in order to permit use of chemical methodology at high metal concentrations, but the reactions and products may vary importantly from those in a natural environment. A more desirous alternative would be to isolate metal-tolerant organisms from environmental locales where the highest concentration of the metal already exists (Colwell and Nelson, 1974). Thus, a direct relationship exists between the numbers of mercury-tolerant bacteria in the Chesapeake Bay and the mercury concentration in the sediment. Moreover, the medium to be selected should approach as closely as possible that of the natural environment, particularly with respect to the electrolyte balance. It may be desirable to use sterile or nonsterile field samples of sediment or water in the growth matrix.

The bacterial transformation reactions can be followed by the disappearance of the initial metal or its compound and the appearance of the reactive metal products. In the case of a metal in its elemental form, the rate of dissolution (or corrosion) may be followed by electrochemical techniques (Iverson, 1972b; Phelps, 1962). The end-products may be insoluble and nonvolatile, soluble with a low vapour pressure, or volatile. With compounds having a low vapour pressure, the formation of metal reaction products may be detected and measured using specific ion electrodes, atomic absorption techniques, neutron activation analysis, or radioisotopes, if available. If the initial starting metal is in the form of a soluble compound, other procedures employing some form of separation must be employed. Fortunately, with mercury, the toxic products of interest (elemental metal, methyl-, and dimethylmercury) are relatively volatile, thus eliminating a chemical isolation and/or concentration step, where the possibility of altering the chemical configuration always exists. The formation of volatile products has been considered to be a form of detoxification by the microorganisms (Vallee and Ulmer, 1972).

A technique for physical isolation and concentration of volatile substances, originally devised for determination of aromatic hydrocarbons in seawater, has been devised by Wasik and Boyd (1974). It may also prove to be of considerable future value in the isolation of volatile metal compounds provided the volatile material has a partition coefficient for hydrogen comparable to that for water. In this technique hydrocarbons are stripped from seawater under equilibrium conditions by very small bubbles of electrolytically generated hydrogen gas. The hydrocarbon concentration in the seawater is determined from the volume of the seawater and the hydrocarbon concentration in the head-space after a (coulometrically) measured volume of hydrogen has bubbled through the stripping cell. Since it was shown that some heavy metals concentrate preferentially in marine petroleum pollutants, this procedure may offer useful means of speciating the metals so incorporated (Walker and Colwell, 1974; Brinckman and Iverson, 1975).

A technique commonly employed at the National Bureau of Standards (NBS) for trapping and concentrating volatile substances consists of placing an absorbate solution and inoculated media in separate compartments of a closed vessel. The basic approach is conceptualized in the schematic diagram of Fig. 2. A sampling access is provided for analysis of the atmosphere, absorbate solution, or both, for kinetic studies. The absorbate solution may be distilled water, an indicator dye solution, or even another reactive metal ion solution. Transport of the volatile species to the absorbate solution will occur providing it is not adsorbed or modified by the vessel walls, and the partition coefficients between the surface of the growth medium, and the atmosphere and the absorbate solution are high enough to facilitate transport to the absorbate solution. It is clear that chemical modification across these interfaces can also occur just as it probably does in nature.

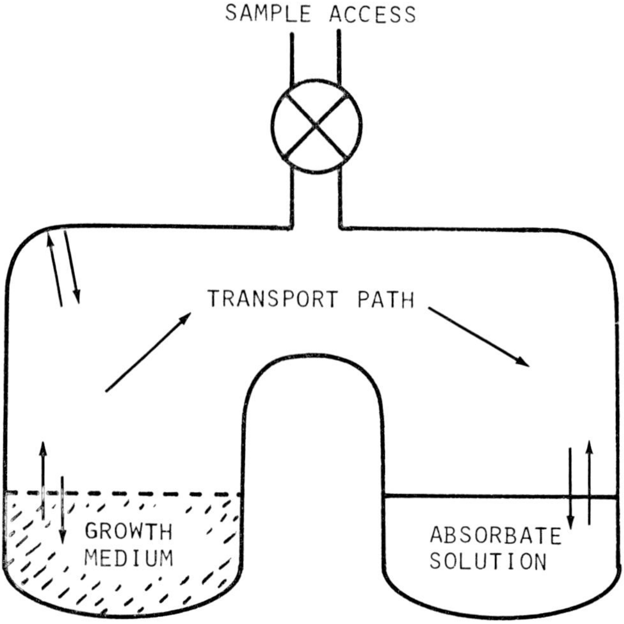

Fig. 2. Conceptual scheme for analysis of transport species in two-compartment bioreactor.

Finally, it is necessary to consider the available base of chemical information which can both direct attention to reasonable organometallic models aiding in interpretation of biotransformations as well as provide suitable calibration materials for speciation methods. The significance of the discovery that biomethylation of mercury or arsenic occurs in environmental settings was cast on a background of well-established descriptive chemistry, albeit not extended to aqueous solvent systems. For more recent work demonstrating biological transport of tin or cadmium (Huey *et al.*, 1974, 1975), considerably less information is available from the literature which might guide elucidation of aquatic reaction pathways. This matter is particularly prominent in

Table I

TRANSMETHYLATION REACTIONS ATTEMPTED IN AQUEOUS SOLUTIONS[a]

Methylator	$HgCl_2$	$NaAuCl_4$	Na_2PdCl_4	K_2PtCl_4	$PtCl_4$[d]	Me_2TlCl	Me_2SnCl_2	$CdCl_2$
Me_3Sn^+	+[b]	+	+	+	0			
Me_3Pb^+	+[b]	+	+	+		0	0	0
$MeHg^+$		+	+	+			0	
Me_2Tl^+	0		+					
Me_2Sn^{++}	0		0					
Me_3S^+	0[c]							

[a] Reactions were carried out in 5 mm O.D. NMR tubes at $\sim 10^{-2}$ M for each reactant. Plus signs indicate cases where reaction occurred and zeros indicate cases in which no reaction was observed over a 100-hr period.
[b] Rate of transmethylation retarded by presence of CN^- or SCN^-.
[c] No reaction up to 100°C.
[d] Conducted in degassed water.

assessing partition coefficients between gaseous organometal intermediates and their usual saline solutions for purposes of analysis by vapour detection schemes. Another facet of the programme at NBS has therefore dealt with extending basic descriptive information concerning, for example, the dynamics of transmethylation reactions between aqueous heavy metal ions and the important factors influencing these reactions (Jewett and Brinckman, 1974). Table I provides an overview of some typical reactions important to this field which have been examined by proton magnetic resonance spectrometry.

The procedure and techniques which will be described have been developed for detecting, speciating, and measuring volatile metals and/or compounds in the trace amounts. These can also be applied to liquid samples.

VOLATILE TRACE METAL DETECTION

The first approach in investigating the microbial transformations of toxic heavy metals has been to detect the formation of a volatile metal or metal-containing species. Radioactive ^{203}Hg has been used by some investigators (Nelson *et al.*, 1973; Magos *et al.*, 1964) to detect the presence of volatile mercury, but simple general survey methods applicable to a number of metals are desirable. Moreover, careful selection of a method might yield qualitative information in metal transport while offering means for speciating

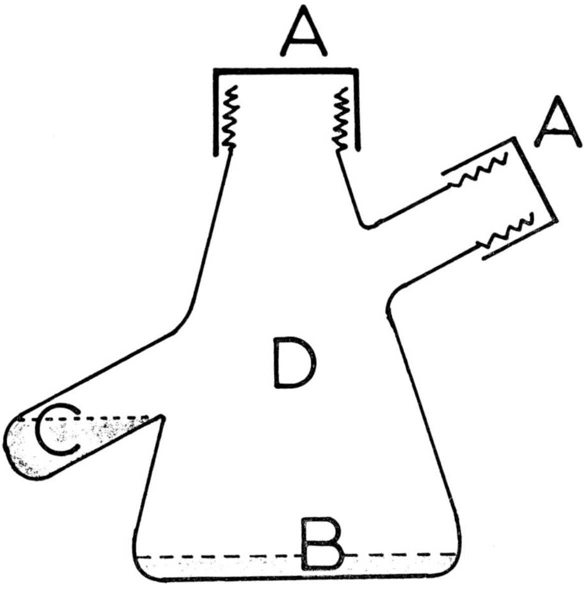

Fig. 3. Diagram of sidearm flask containing selective dye indicator for volatile organometals. A, septum screw caps; B, inoculated medium; C, selective dye indicator; D, captive atmosphere. (1) Dye indicator responsive to specific organometal, viz., *Me_2Sn^{+2}* vs. *Me_3Sn^{+}. (2) Indicator essentially unaffected by concurrently generated metabolites. (3) Effective indication under either aerobic or anaerobic atmospheres. (4) Dye indicator solution affords minimum or no perturbation of culture. (5) Dye indicator sensitive to low concentrations* (e.g. *100 ppb*) *of metals. (6) Indicator response sufficiently stable for rate studies over days.*

transport intermediates as well. A technique which may be used in screening for volatile metal-containing metabolites involves the use of a metal- or organometal-specific dye indicator which is in contact with the respirant atmosphere of the organism growing in the presence of a metal ion or compound. One such arrangement, indicated in Fig. 3, involves the use of a sealed sidearm flask. The dye system must have certain desirable properties which are enumerated. In some favourable cases, the dye indicator can function as an *in situ* monitor, where the buildup of transported metal agents can be determined in the sidearm spectrophotometrically.

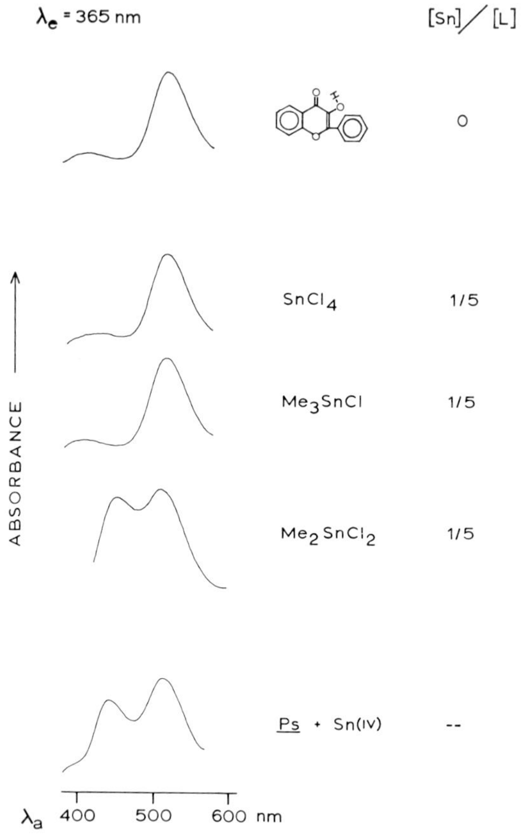

Fig. 4. Characteristic absorption spectra obtained for dilute (ppm) tin solutions by spectrofluorimetry employing excitation at 364 nm. Comparison of authentic tin(IV) or methyltin compounds plus flavinol at optimum ratios with metabolic products (bottom) suggests a dimethyltin product predominates (Huey et al., *1974).*

An excellent example of the colorimetric technique is illustrated in Fig. 4, where a dye, 8-hydroxyflavanol, selective to certain organotin species at appropriate wavelengths of incident irradiation, is employed to aid characterization of the methyltin metabolites volatilized from an estuarine *Pseudomonas* sp.

In another technique the head gas above inoculated plates of organisms growing on agar-containing inorganic mercury was circulated through an

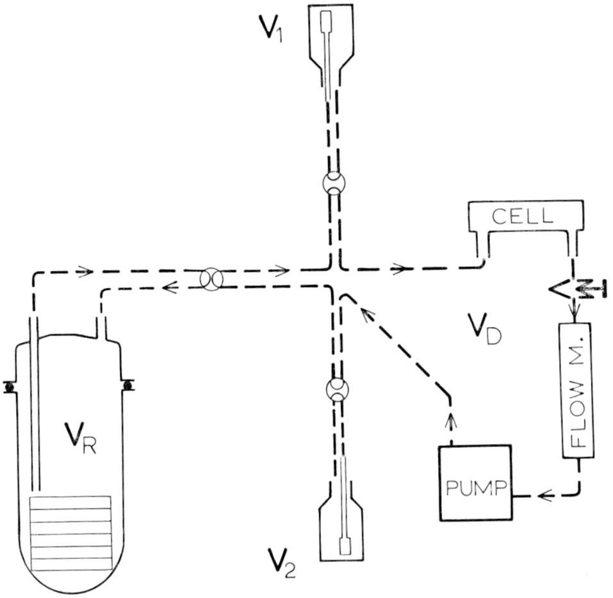

Fig. 5. *Flow diagram of atmosphere from bioreactor for AA analysis of mercury* (*Nelson* et al., *1973*).

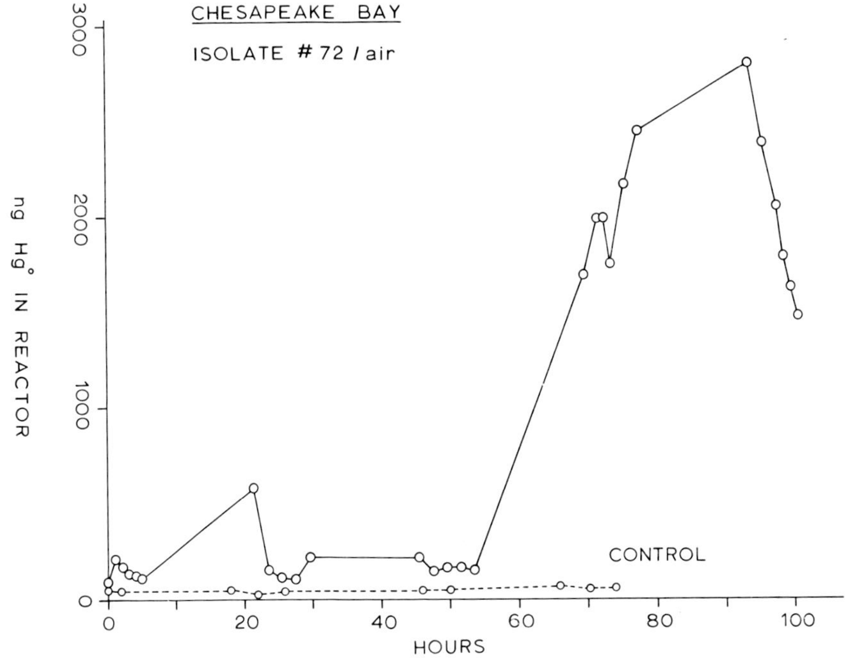

Fig. 6. *Production of elemental mercury from a Chesapeake Bay bacterial isolate grown on Nelson's agar containing phenylmercuric acetate (0·3 ppm) is determined as a function of time by the AA analyses.*

absorption cell of a mercury analyser, a flameless atomic absorption spectrophotometer (Fig. 5). In connection with the absorbing system shown, the total mercury as well as the metallic mercury product could be measured (Nelson *et al.*, 1973). The formation rate of Hg° from a bacterial isolate in this system is shown in Fig. 6.

For the atomic absorption (AA) detection of metals other than mercury, a flameless AA system cannot be employed as relatively high temperatures are needed to dissociate the element from its chemical bonds. The development of a new system for detection of metal species in the head gas will be described below.

An alternative procedure is to collect or trap the head gas volatiles in water over an extended period of time. Any number of two-compartment

Fig. 7. Flask with two sampling ports above the sidearm.

glass vessel configurations may be used. A simple inexpensive flask with a sidearm containing water and inoculated medium in the bottom of the flask is shown in Fig. 7. Two glass outlet tubes fitted with septa permit withdrawal of the water, while allowing the reaction to be followed for varying periods of time with minimum perturbation of the growth system.

Water samples so acquired, up to 100 μlitre, are then inserted into a graphite furnace of the AA unit. A graphite furnace permits the determination of most metallic elements with detection limits up to 1000 times better than a conventional flame, thus rivalling the detection limits of the neutron activation technique. Standard curves prepared for a number of tin compounds are indicated in Fig. 8, showing linearity at nanogram levels. Of particular importance is the fact that necessary drying and charring cycles prior to the atomization step in the graphite furnace do not lead to deviations resulting

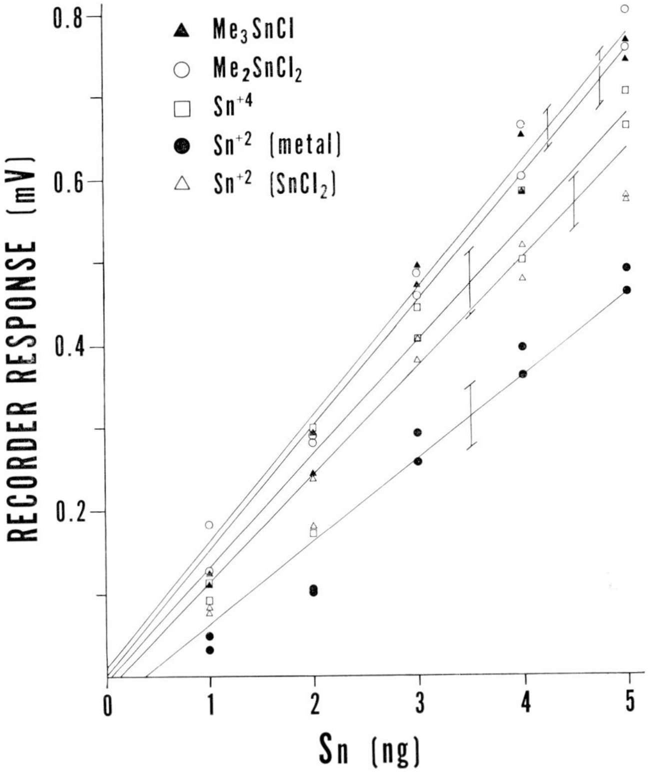

Fig. 8. *AA calibration curves of five tin compounds. Greater atomization efficiency (response) is indicated for Sn(IV) species as opposed to Sn(IV) ions.*

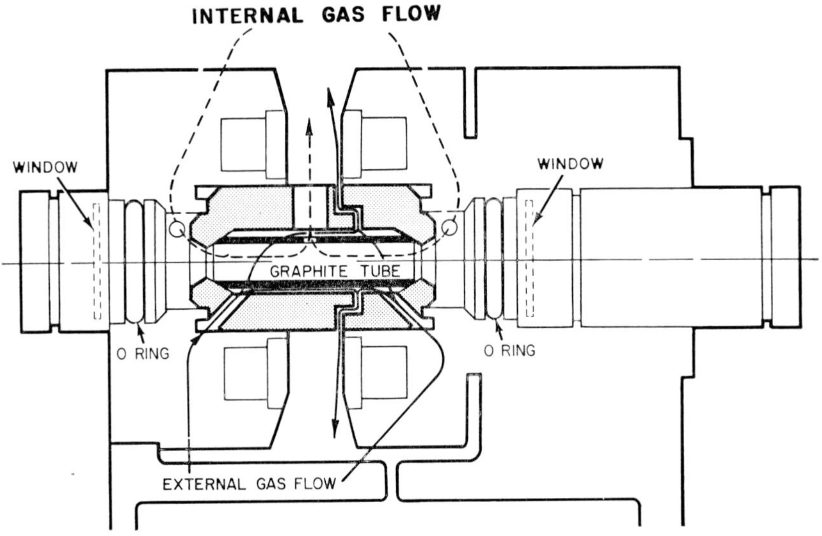

Fig. 9. *Cross-sectional drawing of graphite furnace (courtesy Perkin-Elmer Corp.).*

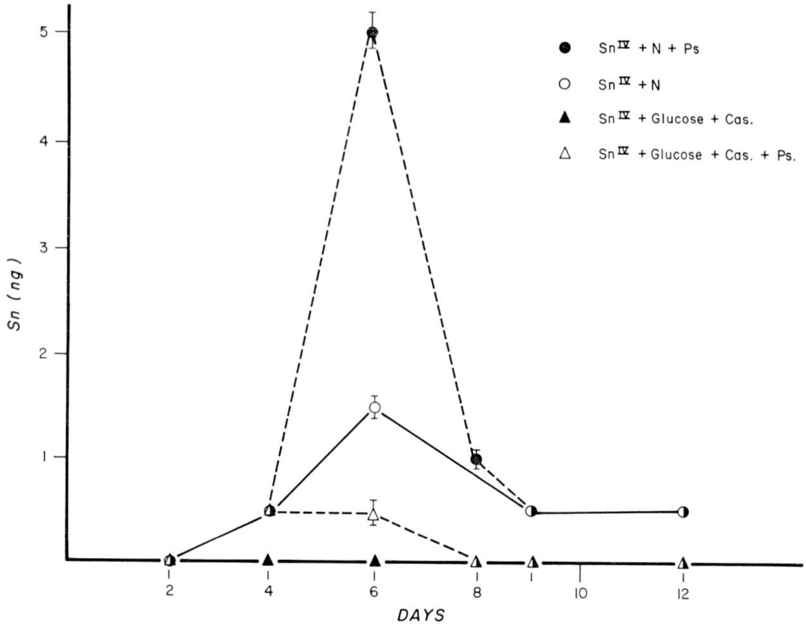

Fig. 10. Formation of a volatile tin compound by a Pseudomonas *strain in the presence of Sn(IV) is shown (●) by its accumulation in the water-absorbing compartment as a function of time.*

from volatilization of the organotins. Here, increased volumes of water solution (*e.g.* 1 ng/10 μlitre) of the tin compound show no significant fall-off in the calibration curve as the water volume is increased fivefold. A sectional drawing of the graphite furnace is shown in Fig. 9. It consists of a hollow graphite cylinder, about 28 mm long and 8 mm in diameter, positioned so that the lamp beam passes through it. The solution sample is pipetted through a 2-mm diameter hole in the top middle of the cylinder. Solid samples may be introduced through the end of the cylinder. By passing an electric current through its walls, the cylinder may be heated in three stages: a drying stage, a charring stage, and an atomizing stage in which the carbon cylinder is heated to incandescence.

Figure 10 depicts a kinetic survey plot utilizing the AA batch withdrawal method. Here, compared against an abiotic or nutrient control, the transport of tin from a *Pseudomonas* sp. growth medium to the water-absorbing compartment is monitored as a function of time. It should be noted that these experiments complement those described above employing spectrofluorimetric detection of volatilized methyltin products with this microorganism.

VOLATILE TRACE METAL SPECIATION

If, as a result of using these survey techniques, evidence of the volatilization of the metal is obtained, the problem of determining the exact nature of the metal-containing species and its rate of formation then remains.

Among the methods available, gas chromatography (GC) has by far the

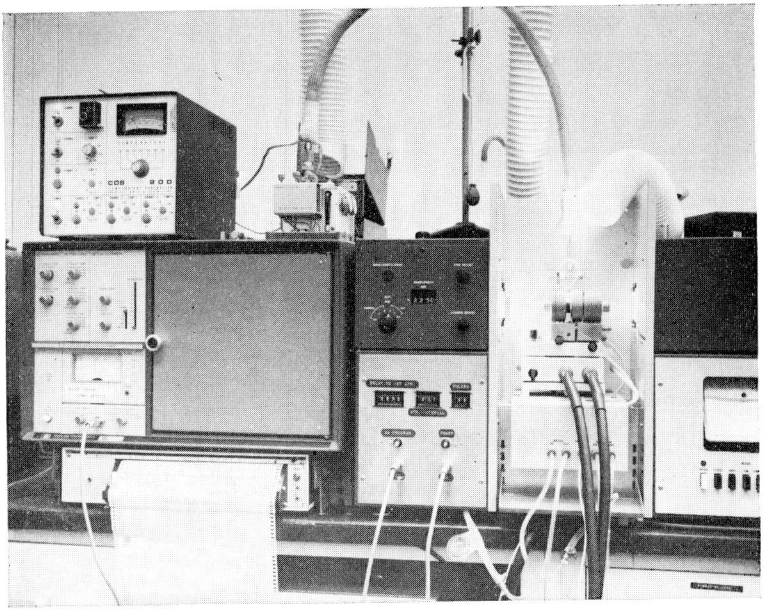

Fig. 13. Gas-chromatograph (left) coupled to an atomic absorption unit (right) via *thermostatically controlled tube (above both units) and electronic pulse programmer (lower centre).*

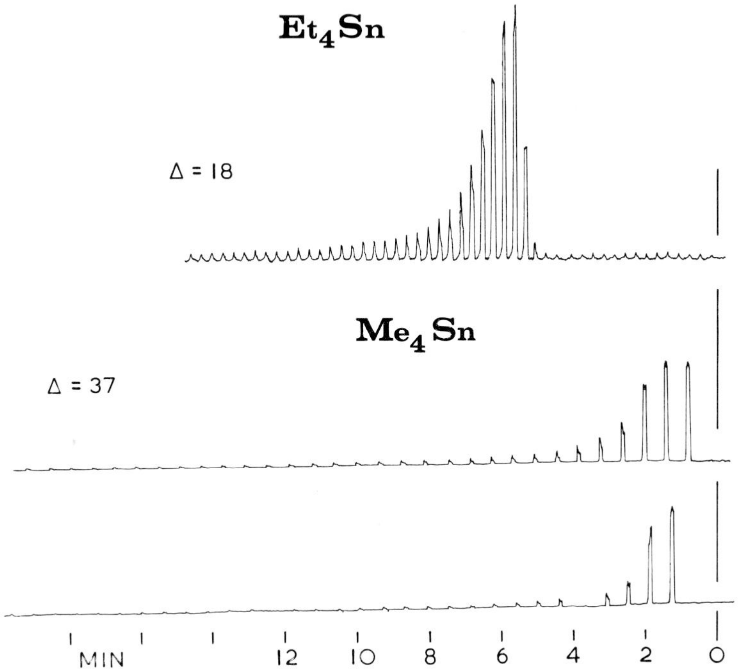

Fig. 14. Pulsed AA-chromatograms of gaseous samples of tetramethyltin and trimethyltin are compared, showing Sn-specific detection at λ = 224·6 nm.

under investigation is a procedure in which the effluent from a programmed pyrolyser is introduced into the pulsed GC-AA unit. It is expected through use of this technique to reduce partially and even completely eliminate the preconcentration stages.

In field studies, the GC-AA unit does not lend itself to easy portability although the GC-mercury analyser has been transported to a large research vessel, assembled, and placed in operation in a few hours.

CONCLUSIONS

It is clear that microbial transformations are responsible for substantial distribution of metals in the biosphere. At least in one case (*i.e.* mercury) anthropogenic inputs of metals can result in severe public health hazards as a result of such transformations. Other critical economic factors must be viewed in the light of biotransformations of metals, including biocorrosion of marine and terrestrial structures and bioprocessing of ore deposits. Means for understanding and controlling degradation processes affecting our industrial society will result from reliable measurement methodology.

Our present approach to surveying and speciating metal transport by microorganisms has relied until now principally on detecting volatile transport species incorporating the metal. Because we recognize that significant partitioning of the biotransformed species may occur between the vapour phase and the aqueous growth matrix, attention must be directed toward detecting any variability of the species between these two phases.

At present, the GC employing the AA spectrophotometer as a detector provides a useful system for detecting and speciating volatile metal-organic compounds at the nanogram level. Although this system is highly effective in studying microbial transformations in cultures where volatile metal species may be produced, it is expected to become of considerable utility, as well, in studying the degradation (physical, chemical, and biological) of condensed phases, such as paints and polymer coatings, incorporating organometal compounds. Such systems are employed extensively today in marine antifungal and antifouling coatings and will receive increased attention as general controlled-release biocides.

Recently, for example, organotin polymers which incorporate the metallic toxicant directly into the polymer chain have been employed for antifouling coatings (Dyckman *et al.*, 1974). Both from an environmental and materials design standpoint, it is of importance to know the metal species leached from the coating, in addition to its possible transformed species which may be generated as a result of microbial activity. Since the leach rate of the organometal polymers is very low, the present GC-AA system, because of its extreme sensitivity, will provide a means of detecting and speciating any compounds that may be leached or transformed at these low levels.

Acknowledgements

The authors wish to express their appreciation for the valuable assistance of the following NBS associates in various phases of the foregoing work: Mr C. W. Huey, Mr K. L. Jewett, Mr S. Wagner, Mr R. A. Thompson, and Dr S. I. Parks.

We gratefully acknowledge the continuing encouragement and partial financial support of the NBS Office of Air and Water Measurement, the Environmental Protection Agency, and the Naval Ship Research and Development Laboratory (Annapolis).

REFERENCES

Bache, C. A., and Liske, D. J. (1971). Gas chromatographic determination of organic mercury compounds by emission spectrometry in a helium plasma. Application to the analysis of methylmercuric salts in fish. *Anal. Chem.*, **43** (7), 950–952.

Baughman, G. C., Carter, M. N., Wolf, N. L., and Zepp, R. G. (1973). Gas–liquid chromatography–mass spectrometry of organomercury compounds. *J. Chromatogr.*, **76** (2), 471–476.

Blair, W., Iverson, W. P., and Brinckman, F. E. (1974). Application of a gas chromatograph-atomic absorption detection system to a survey of mercury transformations by Chesapeake Bay microorganisms. *Chemosphere*, No. 4, 167–174.

Brinckman, F. E., and Iverson, W. P. (1975). Chemical and bacterial cycling of heavy metals in the estuarine system. Am. Chem. Soc. Symp.: Marine Chemistry in the Coastal Environment, Philadelphia, Pa., in press.

Colwell, R. R., and Nelson, J. D., Jr. (1974). Bacterial Mobilization of Mercury in Chesapeake Bay. Proc. Int. Conf. on Transport of Persistent Chemicals in Aquatic Ecosystems III, pp. 1–10, Natl. Res. Council, Ottawa, Canada.

Dinman, B. D., and Hartung, R. (1972). *Environmental Mercury Contamination.* Ann Arbor Science Publishers, Inc., Ann Arbor, Michigan.

Dressman, R. C. (1972). New method for the gas chromatographic separation and detection of dialkylmercury compounds. *J. Chromatogr. Sci.*, **10**, 472–475.

Dyckman, E. J., Montemarano, J. A., and Fischer, E. C. (1974). Antifouling organometallic structural plastics. *Naval Eng. J.*, **86**, 59–64.

Fishbein, L. (1971). Chromatographic and biological aspects of inorganic mercury. *Chromatogr. Rev.*, **15**, 195–238.

Gonzalez, J. G., and Ross, R. T. (1972). Interfacing of an atomic absorption spectrophotometer with a gas–liquid chromatograph for the determination of trace quantities of alkylmercury compounds in fish tissue. *Anal. Lett.*, **5** (10), 683–694.

Huey, C. W., Brinckman, F. E., Grim, S. O., and Iverson, W. P. (1974). The role of tin in bacterial methylation of mercury. Proc. Int. Conf. on Transport of Persistent Chemicals in Aquatic Ecosystems II, pp. 73–78, Natl. Res. Council, Ottawa, Canada.

Huey, C. W., Brickman, F. E., Iverson, W. P., and Grim, S. O. (1975). Bacterial volatilization of cadmium. Proc. Int. Conf. on Heavy Metals in the Environment, Toronto, Canada.

Iverson, W. P. (1972a). The corrosion of mild steel by a marine strain of *Desulfovibrio*. Proc. 3rd Int. Congr. Marine Corros. and Fouling, pp. 61–82, Northwestern University Press, Evanston, Illinois.

Iverson, W. P. (1972b). Biological corrosion. *Adv. Corros. Sci.*, **2**, 1–42.

Jensen, S., and Jernelöv, A. (1969). Biological methylation of mercury in aquatic organisms. *Nature*, **223**, 753–754.

Jewett, K. L., and Brinckman, F. E. (1974). *Transmethylation of heavy metal ions in water*. Abst. Environ. Chem. Div., Am. Chem. Soc., Los Angeles, Calif., March 31–April 5.

Karasek, F. W. (1975). Detection limits in instrumental analysis. *Res. & Dev.*, **26** (7), 20–24.

Longbottom, J. E. (1972). Inexpensive mercury-specific gas chromatographic detector. *Anal. Chem.*, **44** (6), 1111.

Magos, L., Tuffery, A. A., and Clarkson, T. W. (1964). Volatilization of Mercury by Bacteria. *Brit. J. Ind. Med.*, **21**, 294–298.

Nelson, J. D., Blair, W., Brinckman, F. E., Colwell, R. R., and Iverson, W. P. (1973). Biodegradation of phenylmercuric acetate by mercury-resistant bacteria. *Appl. Microbiol.*, **26** (3), 321–326.

Phelps, E. H. (1962). Electrochemical techniques for measurement and interpretation of corrosion. *Corrosion*, **18**, 239t–246t.

Silverman, M. P., and Ehrlich, H. L. (1964). Microbial formation and degradation of minerals. *Adv. Appl. Microbiol.*, **6**, 153–206.

Swain, F. M. (1969). Paleomicrobiology. *Ann. Rev. Microbiol.*, **23**, 455–472.

Vallee, B. L., and Ulmer, D. D. (1972). Biochemical effects of mercury, cadmium and lead. *Ann. Rev. Biochem.*, **41**, 91–128.

Walker, J. D., and Colwell, R. R. (1974). Mercury-resistant bacteria and petroleum degradation. *Appl. Microbiol.*, **27** (1), 285–287.

Wasik, S. P., and Boyd, R. N. (1974). Determination of aromatic hydrocarbons in sea water using an electrical stripping cell. Marine Pollut. Monitoring (Petroleum) Symp. and Workshop, pp. 117–118, Natl. Bur. Standards Special Publ. 140.

Zajic, J. E. (1969). *Microbial Biogeochemistry*. Academic Press, New York.

BIODEGRADATION OF CHALCOPYRITE

A. E. TORMA

Mineral Research Centre (R & D), Q.D.N.R.,
Scientific Complex of Quebec, Ste-Foy, Quebec, Canada

Summary

The applicability of microbiological oxidation for recovery of copper from chalcopyrite-bearing materials has been investigated using a pure strain of *Thiobacillus ferrooxidans*. It has been shown that the commercially available surface-active agents (Tween 20, Tween 40, Tween 60, and Tween 80) reduce the chalcopyrite oxidation ability of these microorganisms. The percolator experiments, imitating the naturally occurring leaching process, resulted in 65% copper extraction from the chalcopyrite ore in 14 months. During the leaching, massive precipitations of jarosite and, in a limited extent, of antherite were produced. These precipitates were accompanied by sulphuric acid formation. Therefore, to avoid pollution problems on mine sites care has to be taken regarding the disposal of reject materials and of drainage water. The optimum substrate concentration has been determined from the shake-flask experiments to be 20% pulp density using the chalcopyrite concentrate. The larger scale experiments resulted after three stages of leaching copper extractions as high as 97·6%. The possibility of copper recovery from the leach liquor has also been indicated.

The role of *T. ferrooxidans* in the biodegradation of copper sulphides was recently reviewed by Trudinger (1971), Karavaiko *et al.* (1972), Dutrizac and MacDonald (1974), and Tuovinen and Kelly (1972, 1974). This chemolithotrophic bacterium derives the necessary energy for its life processes from oxidation of ferrous iron and of reduced valency inorganic sulphur compounds and utilizes carbon dioxide for growth when cultured on chalcopyrite. Malouf (1971), discussing the importance of this microorganism, estimated that about 5% of the yearly world copper production is achieved through biogenic activity.

The present study investigates the effect of surface-active agents on the degradability of chalcopyrite using a pure strain of *T. ferrooxidans*.

MATERIALS AND METHODS

Microorganism. A pure strain of *T. ferrooxidans* used in this study was originally isolated by Torma (1971) from acid mine waters in northern Quebec. It has been routinely maintained in our laboratories on a modified Silverman and Lundgren (1959) nutrient medium, containing a chalcopyrite concentrate as the source of energy.

Substrates. The chalcopyrite ore was broken to less than ½-in pieces; it contained 1·9% copper, 31·8% iron, 30·2% sulphur, and the gangue. The chalcopyrite concentrate was ball-milled to pass a 400-mesh sieve (Tayler). It assayed 15·7% copper, 37·8% iron, 30·1% sulphur, and contained the minerals chalcopyrite, pyrite, pyrrhotine, quartz, and blende.

Surfactants. The surface-active agents Tween 20, Tween 40, Tween 60, and Tween 80 were procured from the J. T. Baker Chemical Co., Phillipsburg, N.J., U.S.A.

Manometric technique. The conventional manometric technique described by Umbreit *et al.* (1972) was employed using 16-ml Warburg flasks and an apparatus, model 15-AD-8, from the Precision Scientific Co., Chicago, U.S.A. Each reaction flask contained a total volume of 2·3 ml: 200 mg of the chalcopyrite concentrate, 2 ml of basal salts nutrient medium containing different concentrations of surfactants, and 0·3 ml of a 10% (w/v) bacterial suspension containing 5·8 mg of protein. The centre well contained 0·2 ml of 20% KOH. These experiments were carried out at pH 1·9, temperature 35°C, and a speed of 130 strokes/min, during 3 hr. The reaction was started by tipping, after 15 min equilibration, the cell suspension into the main compartment of the flasks from one of the side arms.

Harvesting. Cells were cultured in 30 litres of leach suspensions containing chalcopyrite in iron-free nutrient medium as outlined in the larger scale leaching. *T. ferrooxidans* free of substrate was collected from the decanted leach solutions by centrifugation as described by Torma and Gabra (1975), using an apparatus, model RC2-B, of I. Sorvall Inc., Newton, Conn., U.S.A. The solutions were centrifuged at 5000 rpm and the supernatant fluid containing the bacteria was carefully removed so as not to disturb colloidal and solid particles. Then this fluid was subjected to centrifugation at 18 000 rpm. The packed cells were resuspended in iron-free nutrient medium (Silverman and Lundgren, 1959) in proportion to produce a 10% wet weight per volume (w/v) suspension and used for experimentation within six days.

Protein determination. Cellular protein was liberated by alkaline digestion (0·1 N NaOH) of the bacteria and estimated by the method of Lowry *et al.* (1951), using folin phenol reagent, crystalline bovine serum albumin as the standard and a spectrophotometer, model 70, from Bausch and Lomb, Rochester, N.Y., U.S.A.

The percolator experiments were carried out in an air-lift percolator utilizing 1·0 kg of ore and 3 litres of iron-free nutrient medium (Silverman and Lundgren, 1959) containing 10 ml of bacterial suspension as inoculum.

Shake flask tests were carried out in 250-ml Erlenmeyer flasks on a 'Gyrotory Incubator Shaker' apparatus, model G26, of the New Brunswick Sci. Co. Inc., N.J., U.S.A. The desired quantity of chalcopyrite concentrate and 70-ml iron-free nutrient medium (Silverman and Lundgren, 1959) was placed in the flasks and inoculated with 5 ml of an active culture of *T. ferrooxidans.* Instead of inoculum, 5 ml of a 2% alcoholic solution of thymol were added to the sterile control flasks as shown by Torma and Subramanian (1974). All experiments were carried out at pH 2·3, temperature 35°C, and with air containing 0·2% CO_2 used for aeration (Torma *et al.*, 1972). Periodically, the water lost through evaporation was replaced with distilled water and the pH adjusted with 1 N H_2SO_4 or 1 N NaOH when necessary.

Larger scale experiments were performed with 30 litres of leach suspensions containing 20% pulp density (mass of chalcopyrite concentrate × 100/volume of liquid medium) in a thermostatted, stirred tank leaching apparatus as shown in Fig. 1. These experiments were carried out at pH 2·3 and 35°C, and with air enriched to 0·2% in CO_2 for aeration.

The metal extractions were determined by periodically removing a 1-ml sample from the leach suspensions and analysing it for its metal content with an 'Atomic Absorption Spectrophotometer', model AA-4, of the Varian Techtron Pty, Ltd, Melbourne, Australia.

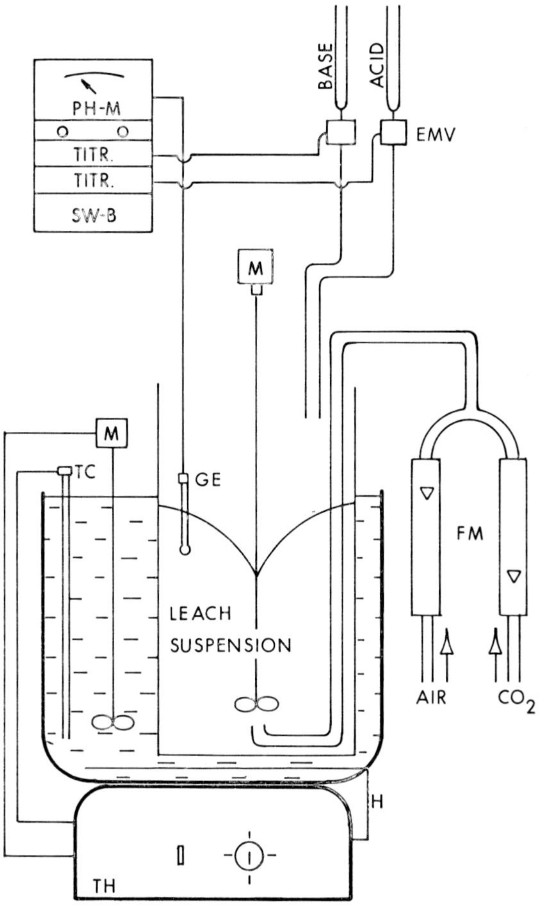

Fig. 1. Schematic representation of the thermostatted, stirred tank leaching apparatus. EMV, electromagnetic valve; FM, flowmeter; GE, combined glass-electrode; H, heater; M, motor; pH-M, pH meter; SW-B, switch-box; T, leach tank; TH, thermostat; Titr, titrator.

The rate of metal extraction was determined by linear regression analysis technique (Draper and Smith, 1966), using the experimental data situated on the straight line part of the leach curves showing the dissolved copper concentration as a function of time, and a computer, model PH9820A, from Hewlett-Packard, Loveland, Colorado, U.S.A.

RESULTS AND DISCUSSION

The effect of the surface-active agents Tween 20 [polyoxyethylene (20) sorbitan monolaurate], Tween 40 [polyoxyethylene (20) monopalmitate], Tween 60 [polyoxyethylene (20) monostearate], and Tween 80 [polyoxyethylene (20) monooleate] on chalcopyrite oxidation by *T. ferrooxidans* are shown in Table I, with respect to the specific rate of oxygen uptake (μl O_2/hr/mg protein).

Table I
SPECIFIC RATE OF OXYGEN UPTAKE BY *T. FERROOXIDANS* DURING OXIDATION OF THE CHALCOPYRITE CONCENTRATE AT pH 1·9 AND TEMPERATURE 35°C

Wetting agent concentration (ppm)	*Specific rate of oxygen uptake (μl $O_2/hr/mg$ protein)*[a]			
	Tween 20	*Tween* 40	*Tween* 60	*Tween* 80
0	71·7 ± 0·62	71·7 ± 0·62	71·7 ± 0·62	71·7 ± 0·62
10	35·6 ± 0·27	47·0 ± 0·23	39·5 ± 0·23	45·1 ± 0·23
100	30·1 ± 0·17	35·0 ± 0·23	36·9 ± 0·23	36·9 ± 0·17
1 000	18·2 ± 0·12	13·5 ± 0·06	15·3 ± 0·06	18·4 ± 0·12
10 000	18·1 ± 0·12	14·4 ± 0·10	14·9 ± 0·06	17·4 ± 0·15
sterile controls (no Tween, no bacteria)	5·6 ± 0·03	5·6 ± 0·03	5·6 ± 0·03	5·6 ± 0·03

[a] Each figure is the mean value of three individual experiments ± SEM (standard error of the mean).

In all four cases the bacterial activity was decreased as the concentration of the surfactant increased. Similar observation has been made by Torma and Gabra (1975). These data are in disagreement with those reported by Duncan and Trussell (1964), Duncan and Teather (1966, 1968), and Marjanovic and Salatic (1967), who claimed that the surface-active agents enhance the bacterial activity when grown on chalcopyrite. The decay in the chalcopyrite oxidation ability of *T. ferrooxidans* in the presence of surface-active substances is probably the consequence of the decrease in the surface tension of the leach solution, similarly as suggested by the experiments from Gibbs *et al.* (1926), Pizarro (1927), Alexander and Soltys (1946), Glassman (1948), and Von Riesen (1955) and in the value of the oxygen mass transfer (K_{1a}) mentioned by Aiba *et al.* (1965). Marshall (1924) has reported that in normal culture media, the surface tension is of the order of 45–54 dyn/cm as compared with 72 for water. When the surface tension is progressively reduced to 40 dyn/cm, the bacterial growth is increasingly inhibited and below about 28–30 dyn/cm no growth can be achieved as indicated by Alexander and Soltys (1946). The surface-active agents which are absorbed at the solid–liquid interface (Somasundaran *et al.*, 1964; Bull and Kempe, 1971) might affect the bacterial growth by increasing the wetting of the substrate and the bacterium and thus altering its environment or by absorption upon the bacterial envelope, thus influencing certain nutritive growth processes. Further, the decrease in the K_{1a}-value will result in a diminution in the oxygen concentration available for bacterial action. This unavoidably leads to growth limitation. Similar to the findings of this investigation, Imai *et al.* (1972), who studied the ferrous iron oxidation in the presence of sodium dodecylsulphate (SDS) surfactant, found that the ferrous iron oxidation ability of *T. ferrooxidans* was markedly depressed as the concentration of SDS increased. Also, it is known from the studies reported by Schnaitman and Lundgren (1965) that *T. ferrooxidans* secretes certain chelating and wetting agents into the leach solution, therefore, probably the microorganisms can cover their need in this respect.

Percolator leaching. The yield of copper extractions varied between 54 and 65% during 8–14 months of leaching. The X-ray analysis of the residues indicated the presence of insoluble basic ferric sulphate (jarosite) and basic copper sulphate (antherite). The findings are in agreement with previous observations (Torma and Legault, 1973; Ivarson, 1973; Silver and Torma, 1974) which can be described by the following equations:

$$2CuFeS_2 + 8\tfrac{1}{2}O_2 + H_2SO_4 \xrightarrow{\text{T. ferrooxidans}} 2CuSO_4 + Fe_2(SO_4)_3 + H_2O$$

Partial hydrolysis of ferric sulphate:

$$3Fe_2(SO_4)_3 + 12H_2O \rightleftharpoons \underset{\text{Jarosite}}{2\{H[Fe(SO_4)_2 \cdot 2Fe(OH)_3]\}} + 5H_2SO_4$$

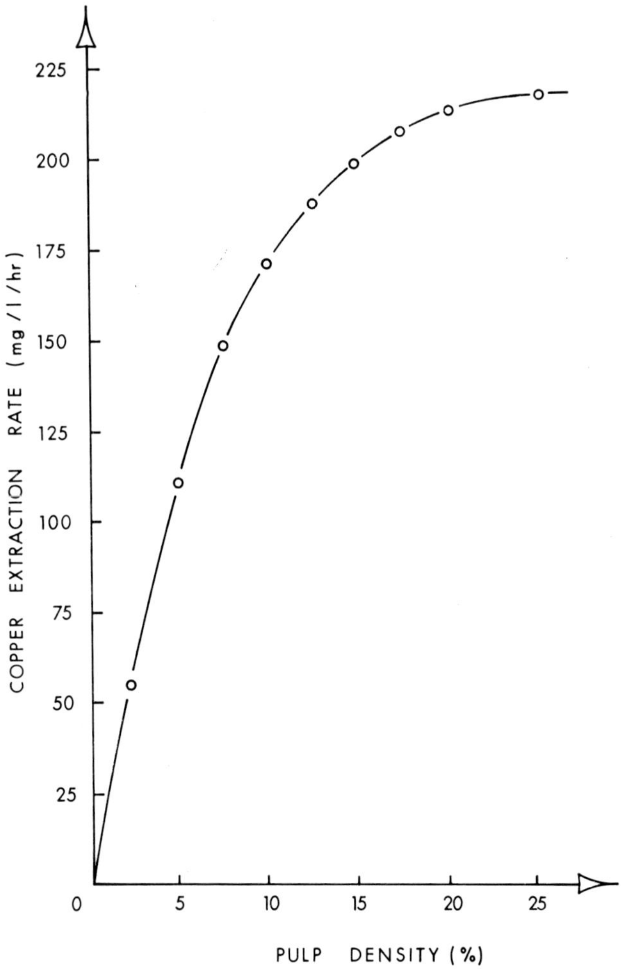

Fig. 2. Effect of pulp density on the rate of copper extraction from a chalcopyrite concentrate, at pH 2·3 and temperature 35°C by Thiobacillus ferrooxidans.

Partial hydrolysis of cupric sulphate:

$$3CuSO_4 + 4H_2O \rightleftharpoons \{CuSO_4 \cdot 2Cu(OH)_2\} + 2H_2SO_4$$

Antherite

The antherite in the residue represents a loss from the point of view of copper extraction. However, no attempt had been made to estimate how much microbiologically leached copper was associated with the leach residue or what techniques were required for its recovery. These experiments closely reproduce the situation which may occur in nature on mining sites. As has been shown, considerable copper and sulphuric acid concentration can be produced from such materials as low grade ore or as flotation residues. Therefore, much care has to be taken in handling the disposing of such materials. The mine drainage water has to be treated (neutralized) before releasing it in the streams and lakes in order to avoid environmental pollution.

Shake-flask leaching. The results derived in these experiments are shown in Fig. 2. The highest copper extraction was about 16 g/litre obtained with leach suspension of 20–25 % pulp density. The final extractions varied between 55 and 81 % in the inoculated samples and between 5 and 9 % in the sterile controls. The highest extractions were obtained with the lowest pulp density solutions. It can be seen that the optimum pulp density lies at about 20 %. The X-ray analysis detected in the residue the jarosite and the unreacted sulphides. During the leaching, the mineral surface has been covered by the jarosite precipitate which considerably inhibited the progress of the degradation reaction. This situation is shown in Fig. 3.

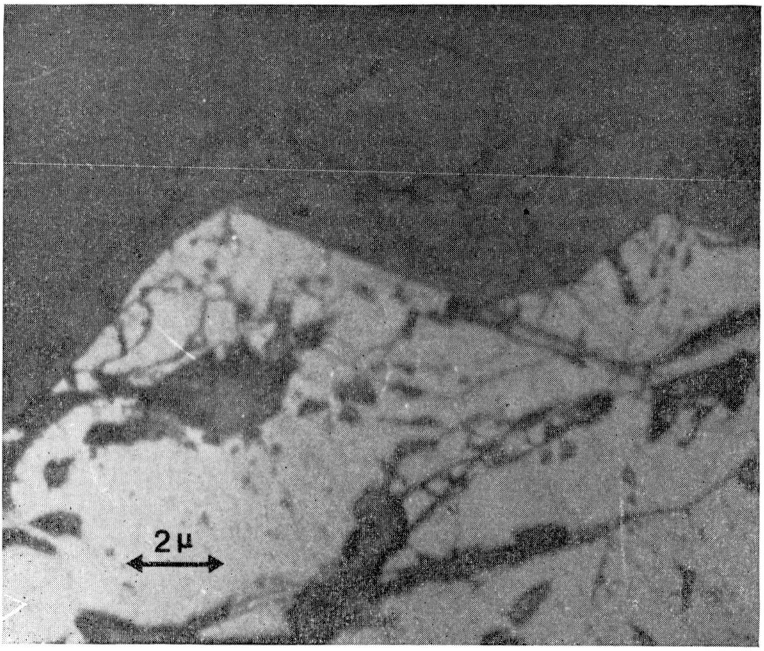

Fig. 3. Jarosite formation on the surface of chalcopyrite during leaching.

Tank leaching. The larger scale experiments were carried out in three consecutive leachings, including a regrinding of the leach residue after the first and the second leaching. The results are represented in Fig. 4. The final

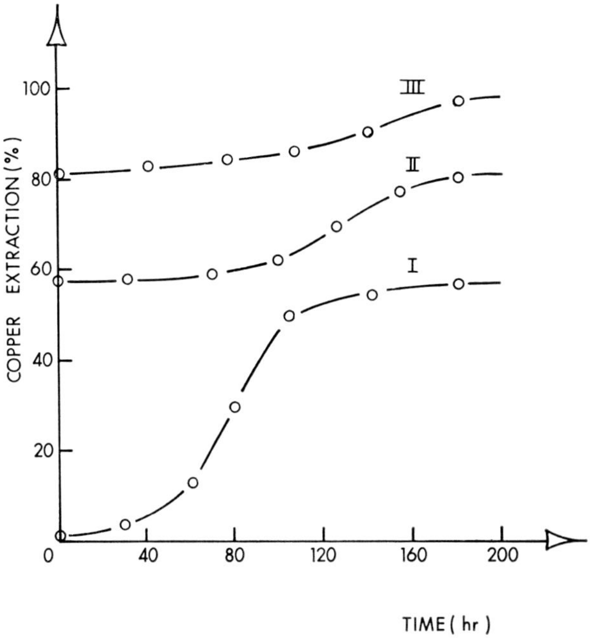

Fig. 4. Microbiological copper extraction by Thiobacillus ferrooxidans. *Experimental conditions: 30 litres of leach suspension containing 20% of chalcopyrite, pH 2·3, temperature 35°C, and 0·2% CO_2 in air. I, treatment of the concentrate; II, treatment of the residue of I; III, treatment of the residue of II.*

yield of copper extraction after the third leaching was 97·6%. The pH was controlled only during the lag time (the period of time preceding the rapid metal dissolution), then it was left to drop. At the end of the first treatment, the pH was stabilized at 1·9. Then the bacteria were harvested and utilized in the Warburg experiments. The grinding of the first and the second leach residue was necessary to liberate the solid substrate surface from the deposited jarosite precipitate. Similar treatment has been suggested by Torma (1975) for the biohydrometallurgical extraction of nickel and cobalt from sulphide-bearing materials.

On the basis of this experimental study, a flow-sheet has been suggested for microbiological leaching of chalcopyrite concentrates as shown in Fig. 5. This scheme includes a three-step leaching of the chalcopyrite concentrate and the recovery of copper from the leach solution by cementation (Kougioumoutzakis *et al.*, 1974) or by electrolysis which is coupled with the solvent extraction technique (Ritcey, 1973; Kougioumoutzakis *et al.*, 1975). After the copper recovery, the leach liquor is recycled, readjusted for its nutrient concentrations and pH, and used again.

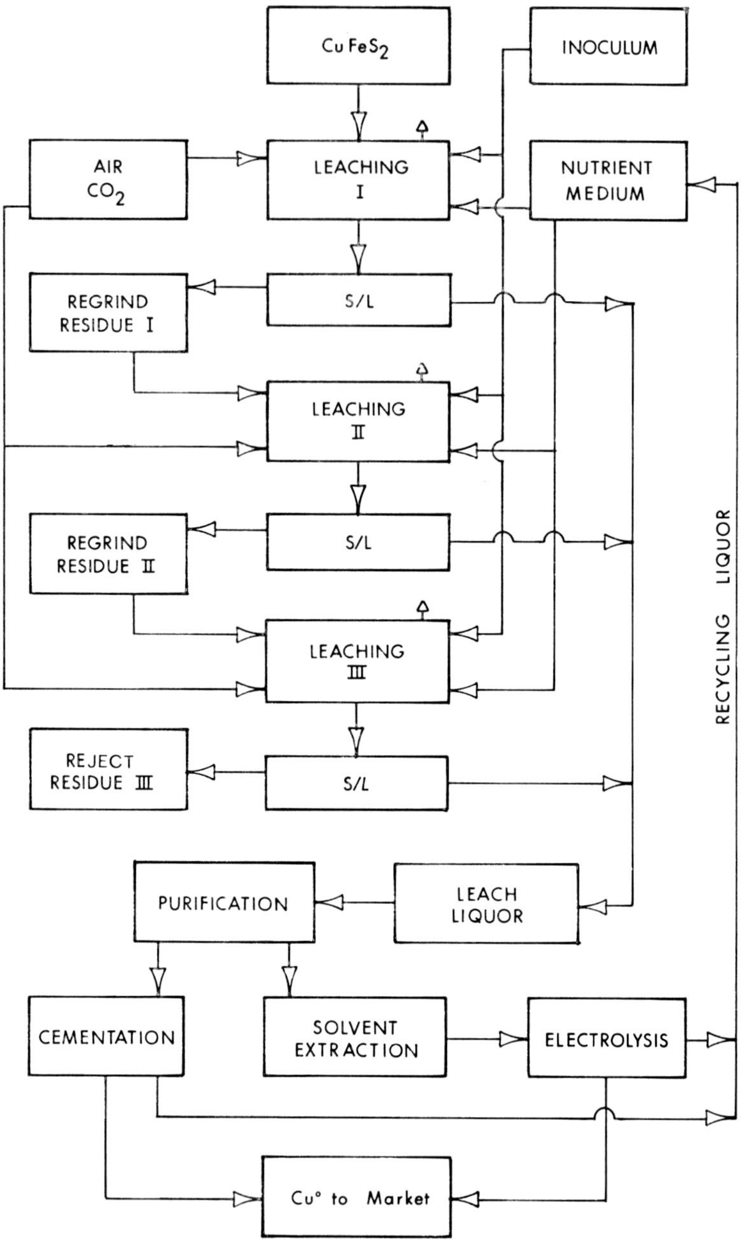

Fig. 5. Schematic representation of a cyclic microbiological metal sulphide batch leaching process.

Acknowledgements

The author wishes to thank the Minister of the Quebec Department of Natural Resources for his permission to publish this study. He also wishes to express his appreciation to Mr Yves St-Louis, who carried out the leaching tests, and to the personnel of the Analysis and Control Service of the Mineral Research Centre, for execution of the chemical and X-ray analyses associated with this investigation.

REFERENCES

Aiba, S., Humphrey, A. E., and Millis, N. F. (1965). *Biochemical Engineering*, pp. 153–156. Academic Press, New York and London.

Alexander, A. E., and Soltys, M. A. (1946). The influence of surface-active substances on the growth of acid-fast bacteria. *J. Pathol. Bacteriol.*, **58**, 37–42.

Bull, D. N., and Kempe, L. L. (1971). Influence of surface active agents on oxygen absorption to the free interface in a stirred fermentor. *Biotechnol. Bioeng.*, **13**, 529–547.

Draper, N. R., and Smith, H. (1966). *Applied Regression Analysis*, pp. 1–407. John Wiley, New York.

Duncan, D. W., and Teather, C. J. (1966). Method of extracting metals from sulfide ores using bacteria and an accelerating agent. U.S. Patent, No. 3,266,889, August 16.

Duncan, D. W., and Teather, C. J. (1968). Method of extracting metals from sulfide ore. Can. Patent, No. 780,405, March 12.

Duncan, D. W., and Trussell, P. C. (1964). Advances in the microbiological leaching of sulphide ore. *Can. Met. Quart.*, **3**, 43–55.

Dutrizac, J. E., and MacDonald, R. J. C. (1974). Ferric iron as a leaching medium. *Miner. Sci. Eng.*, **6**, 59–100.

Gibbs, W. M., Batchelor, H. W., and Sickels, T. N. (1926). Surface tension and bacterial growth. *J. Bacteriol.*, **11**, 393–406.

Glassman, H. N. (1948). Surface active agents and their application in bacteriology. *Bacteriol. Rev.*, **12**, 105–148.

Imai, K., Sugio, T., Yasuhara, T., and Tano, T. (1972). Iron oxidation by bacteria. Paper presented at the Joint Meeting of the Min. Met. Inst. Japan and the Am. Inst. Min. Eng. in Tokyo, Japan, May 24–27, pp. 1–8.

Ivarson, K. C. (1973). *Microbiological formation of basic ferric sulfates. Can. J. Soil Sci.*, **53**, 315–323.

Karavaiko, G. I., Kuznetsov, S. I., and Golomzik, A. E. (1972). Role of Microorganisms in Leaching of Metals from Ores. Nauka, Moskow, pp. 1–248.

Kougioumoutzakis, D., Torma, A. E., Ouellet, R., and Le Houillier, R. (1974). Récupération des métaux par cémentation à partir des solutions de lixiviation. Paper presented at the 52nd Annual Meeting of the 'Association Canadienne-Française Pour Avancement Des Sciences', in Quebec City, Canada, May 8–10, pp. 1–20.

Kougioumoutzakis, D., Torma, A. E., and Bolduc, J. P. (1975). Etude d'extraction du cuivre et du nickel par solvents organiques. Paper presented at the 43rd Annual Meeting of the 'Association Canadienne-Française Pour Avancement Des Sciences', in Moncton, N.B., Canada, May 7–9, pp. 1–32.

Lowry, O. H., Rosebrough, N. J., Farr, A. L., and Randall, R. J. (1951). Protein measurement with the folin phenol reagent. *J. Biol. Chem.*, **193**, 265–275.

Malouf, E. E. (1971). The role of microorganisms in chemical mining. *Min. Eng.*, **23**, 43–46.

Marjanovic, D., and Salatic, D. (1967). Microbiological leaching of copper by the application of the surface active agents. *Rudarski Glasmik*, **67**, 63–68.

Marshall, M. S. (1924). Surface tension of culture mediums. *J. Infect. Dis.*, **35**, 526–536.

Pizarro, D. R. (1927). The relation of surface tension to bacterial development. *J. Bacteriol.*, **13**, 387–408.

Ritcey, G. M. (1973). Recovery of copper from concentrated solution by solvent extraction using Kelex 100. *Can. Min. Met. Bull.*, April 75–83.

Schnaitman, C., and Lundgren, D. G. (1965). Organic compounds in the spent medium of *Ferrobacillus ferrooxidans*. *Can. J. Microbiol.*, **11**, 23–27.

Silver, M., and Torma, A. E. (1974). Oxidation of metal sulfides by *Thiobacillus ferrooxidans* grown on different substrates. *Can. J. Microbiol.*, **20**, 141–147.

Silverman, M. P., and Lundgren, D. G. (1959). Studies on the chemoautotrophic iron bacterium *Ferrobacillus ferrooxidans*. I. An improved medium and a harvesting procedure for securing high cell yields. *J. Bacteriol.*, **70**, 642–647.

Somasundaran, P., Healy, T. W., and Fuerstenau, D. W. (1964). Surfactant absorption on the solid–liquid interface: dependence of mechanism on chain length. *J. Phys. Chem.*, **68**, 3562–3566.

Torma, A. E. (1971). Microbiological oxidation of synthetic cobalt, nickel and zinc sulfides by *Thiobacillus ferrooxidans*. *Rev. Can. Biol.*, **30**, 209–216.

Torma, A. E. (1975). Microbiological extraction of cobalt and nickel from sulfide ores and concentrates. Can. Patent No. 960,463, January 7.

Torma, A. E., and Gabra, G. G. (1975). Effect of wetting agents on chalcopyrite oxidation ability of *Thiobacillus ferrooxidans*. *Int. Res. Commun. Syst. Med. Sci. Biochem. Microbiol. Parasitol. Infect. Dis.*, **3**, 228.

Torma, A. E., and Legault, G. (1973). Rôle de la surface des minerais sulfurés lors de leur biodégradation par *Thiobacillus ferrooxidans*. *Ann. Microbiol.* (*Inst. Pasteur*), **124A**, 111–121.

Torma, A. E., and Subramanian, K. N. (1974). Selective bacterial leaching of a lead sulphide concentrate. *Int. J. Miner. Process.*, **1**, 125–134.

Torma, A. E., Walden, C. C., Duncan, D. W., and Branion, R. M. R. (1972). Effect of carbon dioxide and particle surface area on the microbiological leaching of a zinc sulfide concentrate. *Biotech. Bioeng.*, **14**, 777–786.

Trudinger, P. A. (1971). Microbes, metals and minerals. *Miner. Sci. Eng.*, **3**, 13–25.

Tuovinen, O. H., and Kelly, D. P. (1972). Biology of *Thiobacillus ferrooxidans* in relation to the microbiological leaching of sulphide ores. *Z. Allg. Mikrobiol.*, **12**, 311–346.

Tuovinen, O. H., and Kelly, D. P. (1974). Use of micro-organisms for recovery of metals. *Int. Met. Rev.*, **19**, 21–31.

Umbreit, W. W., Burris, R. H., and Staufer, J. F. (1972). *Manometric and Biochemical Techniques*, pp. 1–116, 5th edn, Burgess Publishing Co., Minneapolis, Minn.

Von Riesen, V. L. (1955). Studies on bacteria-surface active agent relationships. I. Decomposition of anionic SAA by bacteria. *Trans. Kansas Acad. Sci.*, **58**, 337–344.

DESTRUCTIVE MARINE WOOD AND STONE BORERS IN MONTEREY BAY

E. C. HADERLIE

Naval Postgraduate School, Monterey, California, U.S.A.

Summary

Experimental wood, stone, and concrete panels designed to collect marine borers have been exposed at various depths from shallow harbour water down to 110 m in the open water of Monterey Bay for varying periods of time over the past 8 years. In addition, wooden and concrete piles and harbour timbers that have been in place for many years have been examined for borers. Natural shale and calcareous sandstone of varying hardness found on the bottom of Monterey Bay have been examined by divers and representative samples have been quarried for laboratory analysis of borers. Using radiographic techniques on experimental wooden panels that have been infested with borers, growth rates in the natural environment have been measured. Experimental panels of stone and concrete have been exposed for long time periods to try to determine seasons of settlement and the relatively slow growth rates of stone borers.

AREA OF STUDY

Over the past 8 years, a number of sites have been used for making observations and collections and for exposing experimental materials to marine borers. Figure 1 shows the location and configuration of Monterey Bay and the locations of the various sites used in these studies. Observations in the Monterey harbour area and in the shallow water off Del Monte Beach have been essentially continuous throughout the entire period whereas at other sites, particularly in deeper water below conventional diving depths, experimental materials on recoverable arrays have been kept in place from 1 to 3 years.

The water in the harbour area is flushed out regularly by tidal currents and, except for some oil spills, is relatively unpolluted. Water depth at the wharfs where many of these studies were made is 10 m maximum. Surface temperatures vary from highs of 17–18°C in September and October to lows of 10–11°C in January and February. Salinity of harbour water normally ranges between 32·0 and 33·8 $^{0}/_{00}$. In the open water of the bay, the surface temperatures normally range from 11·0–16·0°C and the salinity from 32·8–33·8 $^{0}/_{00}$. At depths of 66 m, the temperature varies from 8·0–11·0°C.

METHODS AND MATERIALS

In Monterey harbour, where wooden and concrete pilings and wooden sea walls have been in place for many years, surveys of boring organisms and

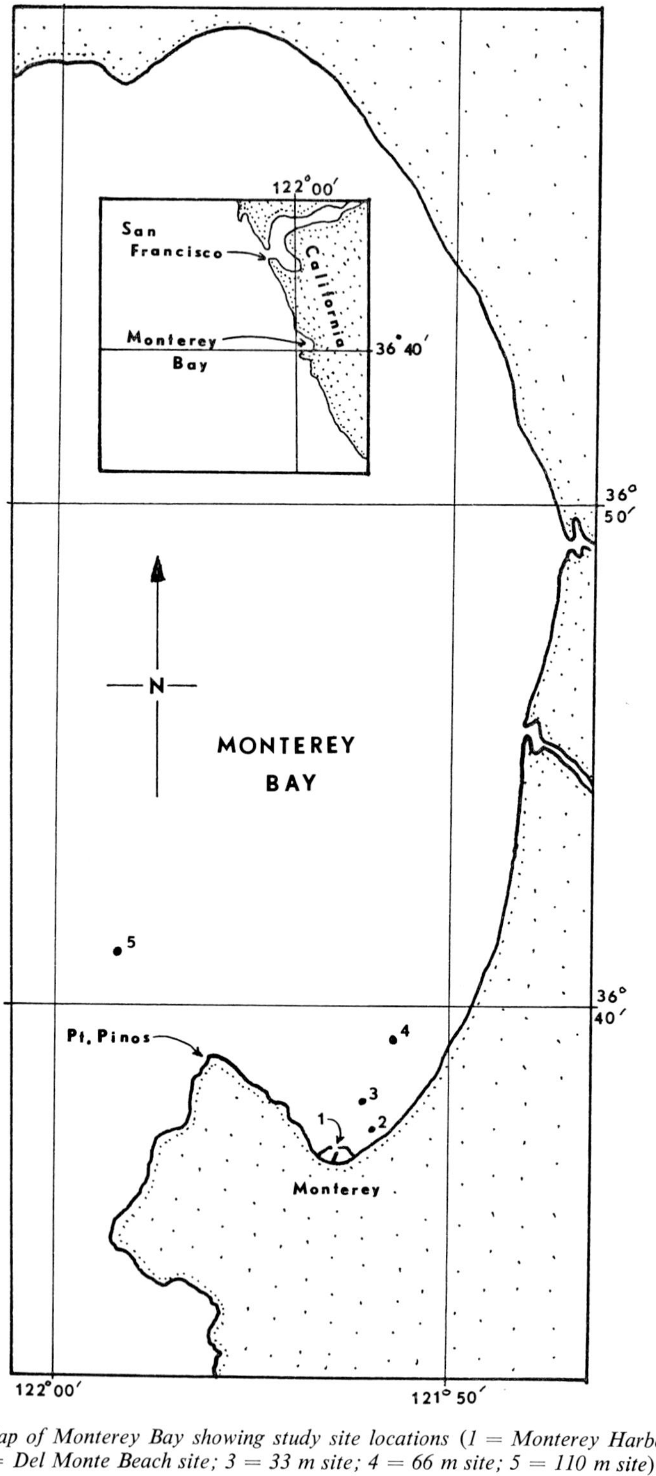

Fig. 1. *Map of Monterey Bay showing study site locations (1 = Monterey Harbour site; 2 = Del Monte Beach site; 3 = 33 m site; 4 = 66 m site; 5 = 110 m site).*

their activities have been made by carefully examining the piles and sea walls using standard SCUBA gear. On occasions, when wooden piles have been removed for replacement, the piles have been sectioned to determine the extent of destruction of the wood and the level of the attack.

In the shallow water off Del Monte Beach, where there is an extensive outcrop of Monterey shale on which a large kelp bed grows, the rocky bottom has been surveyed by divers out to depths of 25 m. Many of the stone-boring forms, such as some of the pholads, can be identified *in situ* from the appearance of their siphons. For a truly qualitative and quantitative analysis, however, it has been necessary to quarry representative stone samples and raise these samples with the winch on a ship. On shore the stone has been carefully broken apart and the borers identified and counted. In water too deep for diving, pipe dredges have been used to recover stone samples.

To gain knowledge of season of settlement of various borers, substrate preference, water depth preference, growth rates, and longevity of individual species, it has been necessary to use experimental techniques. In most cases experimental panels of several kinds of wood, stone, and concrete have been exposed to the marine environment at various places and depths and for differing time periods. These panels have been deployed in a number of different kinds of arrays. Racks (normally made of stainless steel or monel) holding a variety of panels have been designed for resting on the bottom or suspended from wharfs. Taut line arrays with an anchor on the bottom and a float on the surface have carried panels spaced along the vertical line. Wooden panels have generally been $30 \times 15 \times 1{\cdot}5$ cm in size, but in early studies plywood panels $25 \times 20 \times 0{\cdot}5$ cm were used. Stone not easily cut into standard panels has been placed as blocks in porous polypropylene cases and exposed on the bottom. In deep water arrays where no surface buoy for recovery could be left in place, acoustic release mechanisms and long ground lines for grappling have been employed to aid in the recovery of the experimental materials. In the shallow water of the harbour, experimental wooden piles extending from above the surface of the water to the bottom have been driven into the bottom and recovered later for analysis. The details of these various techniques have been reported on and will not be repeated here (*see* Haderlie, 1968a, 1968b, 1969, 1971, 1974; Haderlie and Mellor, 1973; Haderlie *et al.*, 1974).

Specifically in relation to the shipworm *Bankia setacea*, radiographic techniques have been used to monitor the growth rates of these animals in wooden panels. As will be noted later, these techniques have produced valuable data. Attempts to use radiographic techniques on the much slower growing stone borers in stone panels have not proved successful so far.

WOOD BORERS

Wooden structures and experimental wooden panels are extensively attacked by wood borers in Monterey harbour and in deeper waters of the bay and offshore. Yet to date only three species of wood borers have been found despite extensive and long-term sampling.

Limnoria quadripunctata is the only species of crustacean borer so far encountered in these studies. It does extensive damage in the harbour, particularly in older wooden pilings but also in sea wall timbers improperly impregnated with creosote. The attack is primarily in the mid-intertidal zone, but some gribbles in the superficial wood can be found at all water depths in the harbour.

In experimental panels placed at various depths in the harbour, and in experimental piles, *L. quadripunctata* was found throughout the year and settled on new wood whenever it was placed in the water. Newly released young were found only during May and June. Plankton nets pulled between pilings in the harbour periodically collected adult gribbles, indicating that they can swim from pile to pile on occasion. Thin plywood panels exposed in the intertidal zone were totally destroyed by the action of the gribbles after one year of exposure.

Limnoria quadripunctata were also found in panels exposed at the surface and at depths of 16 m off Del Monte Beach 1 km east of the harbour and near no other sources of wood, but they were present in small numbers and had slight destructive effect on the wood. None were found in any panels placed in water further offshore than the 16-m depth contour.

Old pilings in Monterey harbour that are still black in colour due to the original treatment with creosote were often found to be infested heavily with *Limnoria quadripunctata*, yet never were members of this species present on recently treated pilings.

Over the years, a special attempt has been made to determine if *Limnoria tripunctata* occurs in Monterey harbour, but without success. Menzies (1957) reported this gribble as being commonly found in creosote-treated piles along the Pacific coast from San Francisco to Mazatlan, Mexico; yet to date, the species has not been reported from Monterey. These studies have also found that the gammerid amphipod *Chelura terebrans* which is commonly associated with and living in the burrows of *Limnoria* in other localities is not common in Monterey harbour.

Bankia setacea was the only species of shipworm found boring into wood anywhere in Monterey Bay. The older wooden pilings of the harbour become damaged severely, particularly at the mudline, and must be replaced periodically, although many that were thoroughly creosoted have remained functional for up to 30 years. The hulls of wooden fishing vessels and small boats, if not properly cared for, become severely infested with *Bankia*. Over the past 8 years, a special effort has been made to study the biology of *Bankia* in our local waters (Haderlie and Mellor, 1973). In addition to making surveys of natural populations of *Bankia* in the wood structures of the harbour area, wooden panels have been exposed at all sites shown in Fig. 1 down to a water depth of 110 m off Pt Pinos. In addition, experimental piles were employed in Monterey harbour.

For a period of 5 years, and at sites out to 66 m depth, the initial settlement of *Bankia setacea* larvae was monitored. Although there were some differences with depth of water, in general, the season of settlement was found to be from late autumn to early summer, with a maximum settlement in the winter months when water temperatures were at their lowest level. The intensity of settlement increased with depth out to the 66-m site where, in a typical season, there was a settlement of 240 larvae per 225-cm^2 wooden surface. At the 110-m site, no *Bankia setacea* has been found so far, but wooden panels at this depth have been monitored for only one year. Studies presently underway are attempting to find the maximum depth in Monterey Bay where *Bankia setacea* can settle and survive.

In experimental piles of untreated Douglas fir exposed vertically in the harbour in water 8 m deep, shipworm larvae settled and bored into the wood at nearly all levels from the low intertidal zone to the mudline. However, the greatest settlement and ultimate destruction of the wood occurred within 2 m of the bottom. After an exposure of 6 months, a 10 × 10 cm timber could have up to 47 % of the wood destroyed at any one level; after 12 months, up to 82 % destroyed. In the open water of the bay at the deepest study site where *Bankia* occurred (66 m), and where settlement was most intense, Douglas fir panels and blocks were completely destroyed in less than 6 months if initially exposed during the season of *Bankia* settlement. At 33 m depth, panels were destroyed after 7 to 8 months.

In panels exposed in the harbour area, radiographic techniques have been used to monitor growth rates in infested wood. Control panels not subject to periodic X-ray were exposed in racks along with panels X-rayed monthly. Terminal comparison of the radiated and the control panels indicated that the level of radiation used had no detectable influence on either the rate of growth or the ultimate size of the radiated shipworms.

Crowding was found to significantly influence ultimate size of the individual shipworms. When all surfaces of an experimental panel were exposed to settling shipworm larvae, the number that settled and bored into the wood was often so great that crowding soon occurred and individuals could not achieve optimal growth potential. The maximum length attained by an individual shipworm under crowded conditions was 43 mm. Yet panels severely infested with these stunted forms could have up to 34 % of the wood destroyed in 3 months.

In experiments designed to restrict the number of shipworms entering any one panel, thin plastic sheets were wrapped around the wood, leaving only the ends exposed to water and borer attack. Under these conditions, fewer *Bankia* gained entrance to the wood and they were therefore less crowded and could achieve maximum growth. Under these conditions, shipworm burrows attained a maximum length of 976 mm and a diameter of 15 mm in approximately 9 months. On average, the monthly increment in length peaked at 74 mm between the third and fourth months. After 9 months of growth, 34 % of one panel had been destroyed by only three shipworms.

The third species of wood-boring animal encountered in Monterey Bay during the course of these studies was the pholad *Xyophaga washingtona*, which to date has been found boring into panels placed on the bottom in 110 m of water off Pt Pinos. Although much smaller than shipworms, these borers settled in numbers of 50 or more per cm^2 of wood and rapidly eroded the surface layers of Douglas fir.

STONE AND SHELL BORERS

In the studies being reported on here, an attempt has been made to survey the waters of Monterey harbour and the deeper waters of the bay for stone and shell borers, and experimental panels of stone and a variety of types of cement have been deployed in water of various depths to determine when stone borers settle and rates at which they grow. To date, these latter studies have been in progress for only a little over a year and many of the experimental panels examined so far have the initial stages of several borers in them. We do not have data as yet on season of settlement or growth rates.

The commonest rock outcropping around the south shore of Monterey Bay is Cretaceous Santa Lucia granodiorite and this rock is too hard to be penetrated by stone borers. In restricted areas, however, such as off Del Monte Beach and elsewhere, there are outcrops on the bottom of what is called 'Monterey shale'. This is really an outcrop within the bay of the Miocene marine unit known as the Monterey Formation and is a resistant, buff or grey siliceous mudstone composed mainly of diatomite and diatomaceous shale, interbedded with chert (Green, 1970). This shale varies from very soft mudstone to hard chert (7 on the Mohs hardness scale). From the intertidal zone out to depths of 24 m where we have examined this shale, it has been bored by a variety of pelecypods and several other species of invertebrates.

To date, 16 species of boring and nestling clams have been found in the natural shale outcrops in shallow water off Del Monte Beach where most of this study was concentrated. The species encountered were: *Parapholas californica*, *Chaceia ovoidea*, *Zirfaea pilsbryi*, *Barnea subtruncata*, *Netastoma rostrata*, *Penitella penita*, *P. gabbii*, *P. conradi*, *Adula falcata*, *A. californiensis*, *Lithophaga plumula*, *Kellia laperousii*, *Sphenia pholadidea*, *Hiatella arctica*, *Petricola carditoides*, and *Irus lamellifer*. In addition, the gastropod slipper limpet *Crepidula perforans* was common in vacated pholad holes. The distribution and density of each of these varied significantly from place to place and reasons for these differences are not yet clear. Nonetheless, nearly all of the submerged shale in the areas studied carried populations of these animals.

One of the most surprising discoveries was that most of the boring pelecypods found in softer shales were capable of boring into hard chert. In order to determine the relative numbers of each species in a fair-sized piece of chert (30 × 35 × 60 cm) dredged from a depth of 24 m off Del Monte Beach, the flint-hard rock was carefully broken apart and was found to contain the following normal-sized living borers: 60 *Lithophaga plumula*, 46 *Penitella gabbii*, 43 *Penitella conradi*, 5 *Adula falcata*, 4 *Penitella penita*, 2 *Chaceia ovoidea*, 2 *Parapholas californica*, and 1 *Netastoma rostrata*. In addition, many species of nestlers were found. Thus, many forms previously thought to bore only into soft shale or mudstone can penetrate much harder sedimentary rock. Most unexpected were the large numbers of the mytilid *Lithophaga plumula* that were found alive and snugly fitting in burrows of their own making in the hard siliceous chert. *Lithophaga* has been reported to be a chemical borer (Turner and Boss, 1962; Yonge, 1955) but the chert from Monterey contains no carbonates and does not give off bubbles when treated with acid.

The pholads in chert were all found to be of normal size, so they are not stunted by boring in hard materials. The two *Chaceia ovoidea* noted above were 8 cm in shell diameter, and including extended siphons were 45 cm long and occupied large bore holes in the hard rock.

In addition to the siliceous shale outcrops in Monterey Bay, large ellipsoid or spherical concretions of sandstone cemented by calcite can be found. These are probably Pliocene Purisma Formation and often lie free on the bottom as large boulders. In the shallow water off Del Monte Beach, these concretions are extensively bored by a number of animals. Cirratulid worms (*Caulleriella* sp.) penetrate and form a honeycombed outer 2 cm on many rocks. *Penitella gabbii*, *P. conradi*, *Lithophaga plumula*, and *Adula falcata* create large bore holes and living animals of each of these species have been recovered. Also, the nestlers *Hiatella arctica* and *Petricola carditoides* occupy vacated holes.

In the harbour area there are no exposed shale or sandstone outcrops. The concrete piles on the wharfs were made by using a good grade of cement and granite aggregate, therefore the piles show no evidence of damage by borers. Animals with calcareous shells moving about or permanently attached to the piles are, however, often damaged by borers of various kinds.

The boring sponges of the genus *Cliona* are very common in the intertidal and subtidal areas not only in the harbour but elsewhere. The sponges apparently bore by acid secretion. When they settle on the shell of an abolone or a barnacle, they can severely weaken and destroy the shell. In Monterey harbour *Cliona* spp. are often found penetrating the calcareous shells of such attached animals as the rock oyster *Pododesmus cepio* and the large acorn barnacle *Balanus nubilys*.

Other borers often found penetrating calcareous shells of other animals include a variety of spionid worms in the genus *Boccardia* and *Polydora* and the cirratulid *Dodecaceria fewkesi*.

REFERENCES

Green, H. G. (1970). Geology of southern Monterey Bay and its relationship to the ground water basin and salt water intrusion. U.S. Dept. Interior, Geological Survey, Open-file Report.

Haderlie, E. C. (1968a). Marine fouling and boring organisms in Monterey Harbor. *The Veliger*, **10** (4), 327–341.

Haderlie, E. C. (1968b). Marine boring and fouling organisms in open water of Monterey Bay, California, In *Biodeterioration of materials; microbiological and allied aspects*, pp. 658–679 (A. H. Walters and J. S. Elphick, eds.), Applied Sci. Publ. Ltd, London.

Haderlie, E. C. (1969). Marine fouling and boring organisms in Monterey Harbor. II. Second year of investigation. *The Veliger*, **12** (2), 182–192.

Haderlie, E. C. (1971). Marine fouling and boring organisms at 200 feet depth in open water of Monterey Bay, California, In *Biodeterioration of materials, Vol.* 2, pp. 432–442 (A. H. Walters and E. H. Heuck-van der Plas, eds.), Applied Sci. Publ. Ltd, London.

Haderlie, E. C. (1974). Growth rates, depth preference and ecological succession of some sessile marine invertebrates in Monterey Harbor. *The Veliger*, **17** (Suppl.), 1–35.

Haderlie, E. C., and Mellor, J. C. (1973). Settlement, growth rates and depth preference of the shipworm *Bankia setacea* (Tryon) in Monterey Bay. *The Veliger*, **15** (4), 265–286.

Haderlie, E. C., Mellor, J. C., Minter, C. S. III, and Booth, G. C. (1974). The sublittoral benthic fauna and flora off Del Monte Beach, Monterey, California. *The Veliger*, **17** (2), 185–264.

Menzies, R. J. (1957). The marine borer family Limnoriidae (Crustacea, Isopoda). Part I. Northern and Central America: Systematics, distribution, and ecology. *Bull. Marine Sci. Gulf and Caribbean*, **7** (2), 101–200.

Turner, R. D., and Boss, K. J. (1962). The genus *Lithophaga* in the Western Atlantic. *Johnsonia*, **4** (41), 81–116.

Yonge, C. M. (1955). Adaptation to rock boring in *Botula* and *Lithophaga* (Lamellibranchia, Mytilidae) with a discussion on the evolution of this habit. *Quart. J. Microsc. Sci.*, **96**, 383–410.

CELLULOSE AND WOOD DIGESTION IN THE MARINE MOLLUSK *BANKIA GOULDI* BARTSCH

ROBERT C. DEAN*

Department of Zoology, Duke University, Durham, North Carolina, U.S.A.
and
Duke University Marine Laboratory, Beaufort, North Carolina, U.S.A.

Summary

A model for the structure and chemistry of the wood cell wall and the mechanism of its degradation by the shipworm *Bankia gouldi* is presented. Fecal pellet analysis shows significant loss of cellulose and hemicelluloses from ingested wood. Degradation of hemicelluloses in wood fragments can be shown *in vitro* with cell-free digestive tract extracts, but degradation of native crystalline cellulose at a nutritionally significant rate cannot be shown. A method for demonstrating *in vivo* cellulose digestion is proposed, together with a hypothetical new mechanism for wood digestion by *Bankia gouldi*.

INTRODUCTION

An organism which eats wood may be viewed as a mechanism for reducing a tree to a pile of sawdust. Even if we assume the alteration to be purely mechanical, the role of wood-ingesting organisms in accelerating the rate of decomposition must be significant.

In the marine environment, wood is rapidly invaded by a variety of organisms that bore into it and convert it to fine particles (Jones and Eltringham, 1971). That such organisms have evolved in an environment nearly free of living wood argues that the input of wood from the terrestrial environment must be large; such organisms are present even in deep sea environments (Turner, 1966). The marine wood borers are the primary agents in reintegrating this wood into the carbon cycle. While the economic depradations of wood borers are well documented (Clapp and Kenk, 1963), their quantitative contribution to natural turnover processes is seldom discussed, but should be an important consideration when developing control measures.

One of the most important groups of marine wood borers are the bivalve mollusks in the family Teredinidae (Nair and Saraswathy, 1971; Turner, 1966; Turner and Johnson, 1971), the teredoes or shipworms. In North Carolina's coastal waters, *Bankia gouldi* Bartsch is the most abundant species, settling as larvae on wood surfaces. Metamorphosis to the adult form enables them to tunnel through the wood, scraping small fragments from the end of the tunnel; these wood fragments enter the digestive tract. This tissue has two unique structures involved in processing the wood fragments: a large caecum normally packed with wood and a specialized digestive gland that phagocytoses the fragments (Morton and McQuiston, 1974; Potts, 1923).

* Present address: Box 3206, Duke Medical Center, Durham, N.C. 27710, U.S.A.

Several lines of evidence suggest that wood is of primary nutritional significance to *Bankia gouldi* and other shipworms (Carpenter and Culliney, 1975; Lazier, 1924; Turner and Johnson, 1971). Fecal pellet analysis with *Teredo navalis* has provided more direct evidence for wood digestion (Dore and Miller, 1923). While wood-digesting enzyme activity in shipworms has been demonstrated *in vitro* (Boynton and Miller, 1927; Kadota, 1959), only traces of native crystalline cellulose-digesting activity have been shown (Bolker and Brenner, 1970; Cutter and Rosenberg, 1972; Kadota, 1959; Rosenberg and Cutter, 1973).

The central question I am interested in is: if *Bankia gouldi* can be shown to digest wood, what biochemical mechanisms are involved? Wood degradation in general is a poorly understood process (Dean, 1975; Hajny and Reese, 1969; Norkrans, 1967), largely due to the complexity of the substrate (Mark, 1967; Northcote, 1972; Shafizadeh and McGinnis, 1971). To understand the problems faced in this study, it is necessary to understand something of the structure and chemistry of the wood cell wall. Most of the mass of gymnosperm wood such as pine is in the secondary wall (S) of nonliving tracheids (Fig. 1). I have developed a simple model, derived from the literature (Dean, 1975), for the molecular architecture of the secondary wall, presented in cross section in Fig. 2. The wall is considered as a single cross-linked macromolecule, resolvable into three phases, corresponding to the three principal classes of wood components (Table I) and with structures observable with the electron microscope (Freundlich and Robards, 1975; Heyn, 1969).

Table I

COMPOSITION OF NORMAL WOOD

Component	*Percent dry weight* *Gymnosperms* (Pinus silvestris)	*Angiosperms* (Betula verrucosa)
Cellulose	40	40
Hemicellulose		
Glucomannan	18	3
Xylan	12	35
Pectins and others	1	2
Lignin	29	20

Data from Meier, 1964.

(1) *Crystalline phase*, consisting of imperfectly but highly crystalline fibrils of anhydrous, water-insoluble cellulose, wound helically through the cell wall (Muhlethaler, 1969; Ranby, 1969; Stockmann, 1972). This material has a unique crystal structure and is termed native crystalline cellulose.

(2) *Hydrophilic phase*, a cortex of 'sticky' hydrated hemicelluloses adsorbed to the fibril surface (Albersheim *et al.*, 1973; Mark, 1967; Northcote, 1972).

(3) *Hydrophobic phase*, in which the other phases are embedded, composed of lignin, a polyphenolic three-dimensional gel not unlike an epoxy resin; this phase is covalently bound to some of the hemicelluloses (Bolker and Brenner, 1970; Higuchi, 1971; Pew and Weyna, 1962).

There are several crucial implications of this model for understanding processes of wood degradation. The carbohydrate fraction of the wall is protected from interacting with water-soluble macromolecules by the hydro-

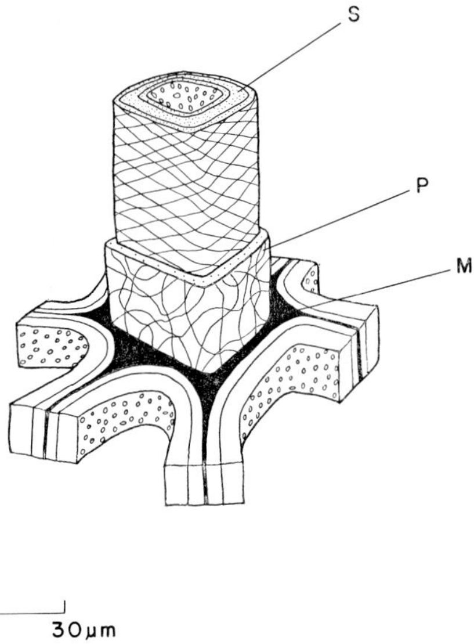

Fig. 1. Schematic drawing of a gymnosperm tracheid seen from the side. Three cell wall layers are shown: P. primary wall; S, secondary wall; M, middle lamella. Portions of adjacent tracheids are shown; microfibrils are suggested. After B. A. Meylan and B. G. Butterfield, Three-Dimensional Structure of Wood, *Chapman and Hall, London, 1972.*

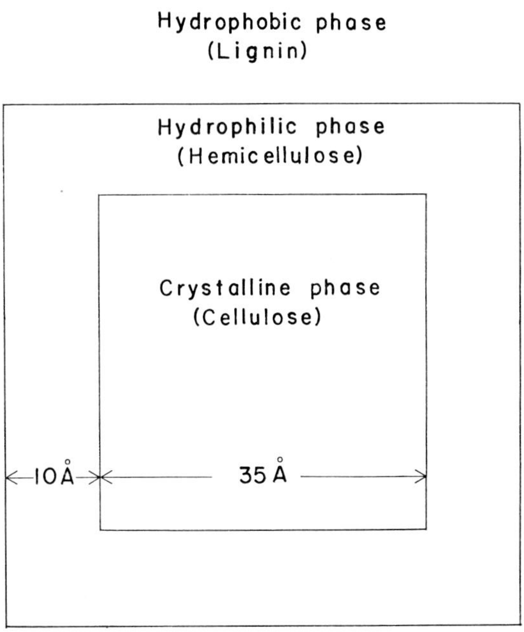

Fig. 2. Model for the molecular architecture of the secondary wall of a gymnosperm tracheid. A portion of the wall cross section, viewed from above (see Fig. 1), at high magnification, is represented. The crystalline phase is shown as a cross section of a single cellulose fibril.

phobic phase. Water can diffuse into the wall only through the hydrophilic phase, whose cross-sectional dimensions do not exceed 10–20 Å (Fig. 2) (Ranby, 1969; Stone *et al.*, 1969). Furthermore, much of the carbohydrate is in a highly crystalline or ordered state which imposes severe biochemical problems for digestion (Cowling and Brown, 1969; Norkrans, 1967; Whitaker, 1971).

An organism digesting wood cell walls therefore faces two principal problems: first, making the cell wall components accessible to degradative processes, and second, evolving the requisite degradative mechanisms for these components. My experimental work with *Bankia gouldi*, which has been reported in detail elsewhere (Dean, unpubl. data), was an effort to test the following simple model as a strategy for solving these problems.

(1) Accessibility of cell wall carbohydrates is increased primarily through mechanical disruption of the hydrophobic phase (Pew and Weyna, 1962; Wayman and Obiaga, 1974).

(2) Hemicellulose digestion proceeds *via* exo- and endo-hydrolases active on hydrated polysaccharides.

(3) Cellulose is digested with a fungal-type cellulolytic enzyme system (Almin *et al.*, 1975; Dean, 1975; Eriksson and Pettersson, 1975; Hajny and Reese, 1969).

SUMMARY OF RESULTS

The fecal pellets of *Bankia gouldi* consist almost entirely of wood fragments. The lignin content (hydrophobic phase) of fecal pellet material (FPM) was analysed according to standard techniques (Browning, 1967; Dean, 1975); by assuming the lignin content to be unaltered by digestion, minimum values for the loss of weight and carbohydrate content after passage through the digestive tract of *Bankia gouldi* were calculated (Table II), suggesting that

Table II

COMPOSITION OF FECAL PELLET MATERIAL (FPM) OF TEREDINIDAE COMPARED TO THE ORIGINAL WOOD (MW) (as percent dry weight)

Material	*No. of samples*	*Mean lignin content*	*Mean estimated weight loss from ingested wood*	*Mean estimated carbohydrate loss from ingested wood*
MW[a]	3	27·44 ± 0·35	—	—
FPM, *Bankia*	3	46·01 ± 6·22	43·06	57·92
FPM, *Teredo*	2	54·61	53·50	66·65

[a] Sawdust ground in Wiley mill.
Data from Dore and Miller, 1923.

more than half of the carbohydrate fraction of ingested wood fragments was removed by the digestive tract. *In vitro* incubation of FPM and milled wood (MW: sawdust ground in a Wiley mill) with enzymes from *Bankia gouldi* and the cellulolytic fungus *Trichoderma viride* ('cellulase'; Worthington) (Stone *et al.*, 1969) showed all enzyme-accessible carbohydrate labile to these enzymes had been removed during passage of wood through the digestive

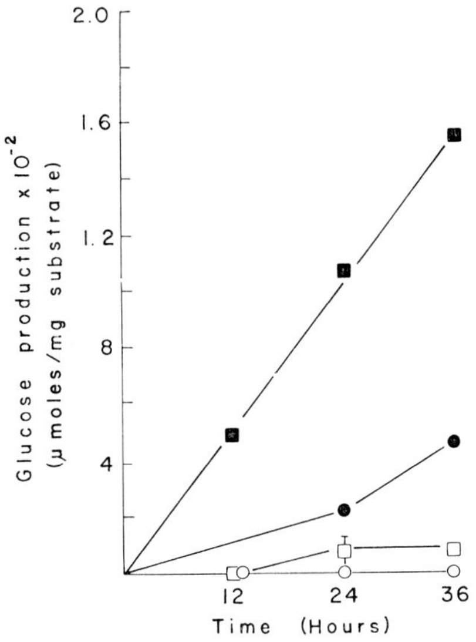

Fig. 3. Production of sugars through time from milled wood (MW) and fecal pellet material (FPM) by dialysed Bankia gouldi *homogenate. MW: 20 mg/mg protein; FPM: 9 mg/mg protein.* (■) *MW, reducing sugars (glucose equivalents);* (●) *MW, glucose;* (□) *FPM, reducing sugars;* (○) *FPM, glucose.*

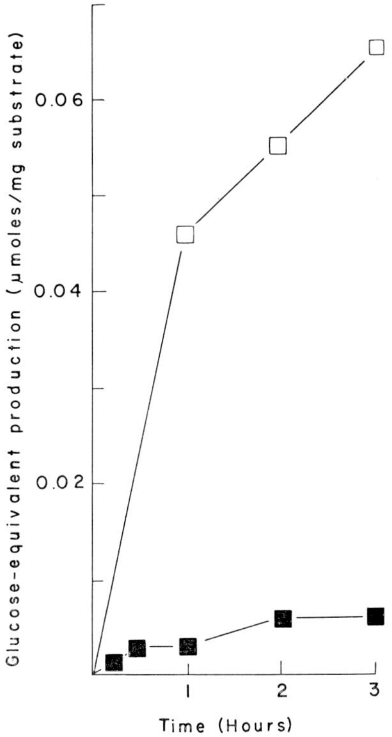

Fig. 4. Production of sugars through time from milled wood (MW) by dialysed Bankia gouldi *homogenate. 18·5 mg MW/mg protein.* (■) *glucose;* (□) *reducing sugars.*

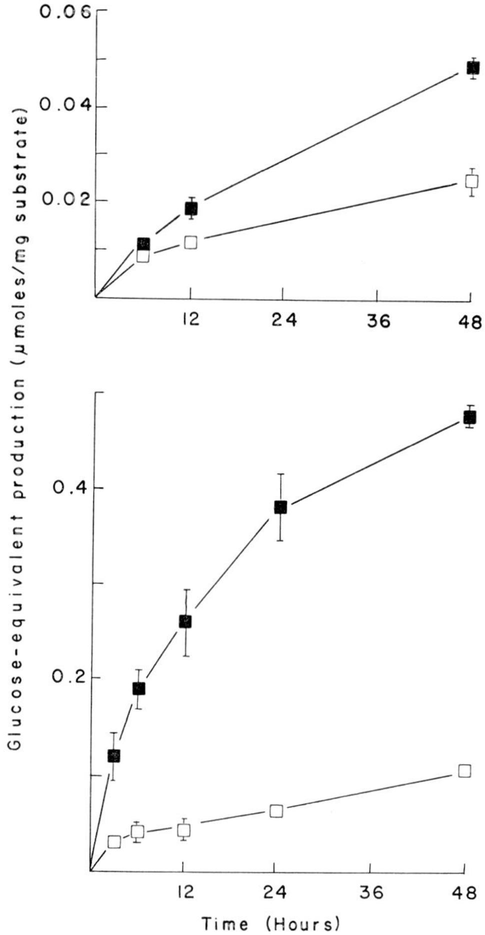

Fig. 5. Production of sugars through time from milled wood (MW) by dialysed Bankia gouldi *homogenate.* Upper: *MW, 37 mg/mg protein;* Lower: *delignified MW, 37 mg/mg protein.* (■) *Reducing sugars;* (□) *glucose.*

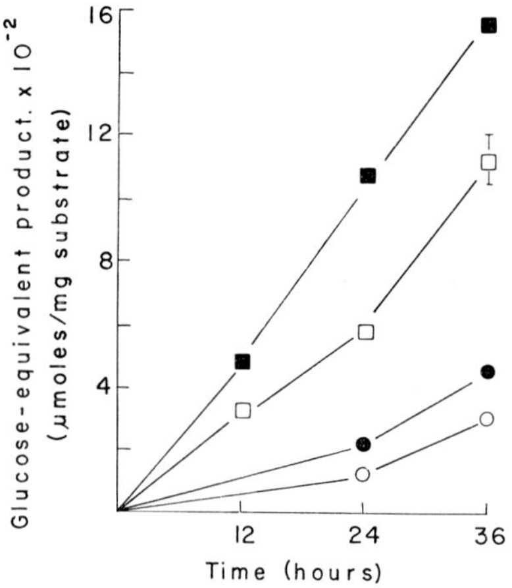

Fig. 6. Production of sugars through time from milled wood (MW) and α-cellulose (AC) by dialysed Bankia gouldi *homogenate. MW: 9·1 mg carbohydrate/mg protein; AC: 20 mg carbohydrate/mg protein.* (■) *MW, reducing sugars;* (●) *MW, glucose;* (□) *AC, reducing sugars;* (○) *AC, glucose.*

tract (Fig. 3). The residual carbohydrate was presumably still protected by the hydrophobic phase.

As a first step in revealing the mechanism of wood digestion in *Bankia gouldi*, the *in vitro* digestion of wood and cellulose, defined as the production of soluble reducing sugars from the carbohydrate fraction, by digestive tract enzymes was studied (Dean, 1975). The results indicated the following.

(1) Reducing sugars are produced from MW in significantly larger amounts than glucose, suggesting hemicellulose digestion is the major process observed (Fig. 4).

(2) Delignification by chemical means (Browning, 1967) of MW greatly enhances its digestibility (Fig. 5).

(3) Extraction of hemicelluloses with alkali from delignified wood (Browning, 1967) yields α-cellulose, which is significantly less digestible than untreated MW (Fig. 6).

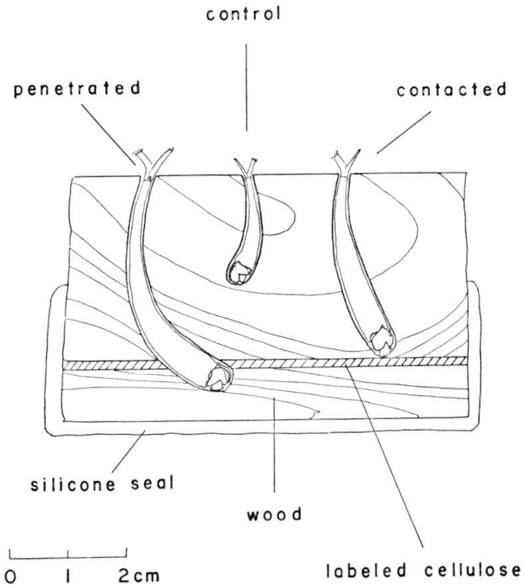

Fig. 7. Schematic diagram of means of introducing labelled cellulose into the digestive tract of Bankia gouldi. *The grain of the wood runs vertically; the larvae settle on the exposed wood and tunnel downward. The unit shown is mounted on another board before immersion.*

(4) Incubation of purified cellulose substrates with *Bankia gouldi* digestive tract extracts showed high levels of β-1,4-glucanase activity on water-soluble and other non-native celluloses. Only very low levels of activity on native crystalline cellulose were found. Comparison with commercial enzyme preparations from organisms able to survive on native crystalline cellulose and an organism limited to non-native cellulose suggests that my results with *Bankia gouldi* provide no evidence for native crystalline cellulose digestion at a nutritionally significant rate (Dean, unpubl. data). This is consistent with earlier results but not with earlier interpretations (Bolker and Brenner, 1970; Cutter and Rosenberg, 1972; Kadota, 1959; Rosenberg and Cutter, 1973).

I have devised a means for demonstrating cellulose digestion *in vivo* without making assumptions as to degradation products, by exposing *Bankia gouldi* to ^{14}C-cellulose and looking for incorporation of label into tissue components, especially isolated glycogen (Van Handel, 1965) (Figs. 7 and 8). The initial results were invalidated by the impurity of the purchased substrate (ICN Pharmaceuticals) (Dean, 1975), but additional experiments with a more purified substrate are planned.

The experimental results summarized here support the following conclusions. *Bankia gouldi* significantly removes carbohydrates, up to 60 %, from ingested wood, limited by the accessibility of this fraction to digestive enzymes. Significant cellulose digestion takes place. Accessibility is a function of disruption of the hydrophobic phase. *Bankia gouldi* contains hydrolytic enzymes active in digesting hemicelluloses, but a fungal-type cellulolytic enzyme system is probably not present. A direct demonstration of native crystalline cellulose digestion has not been achieved, but is within reach.

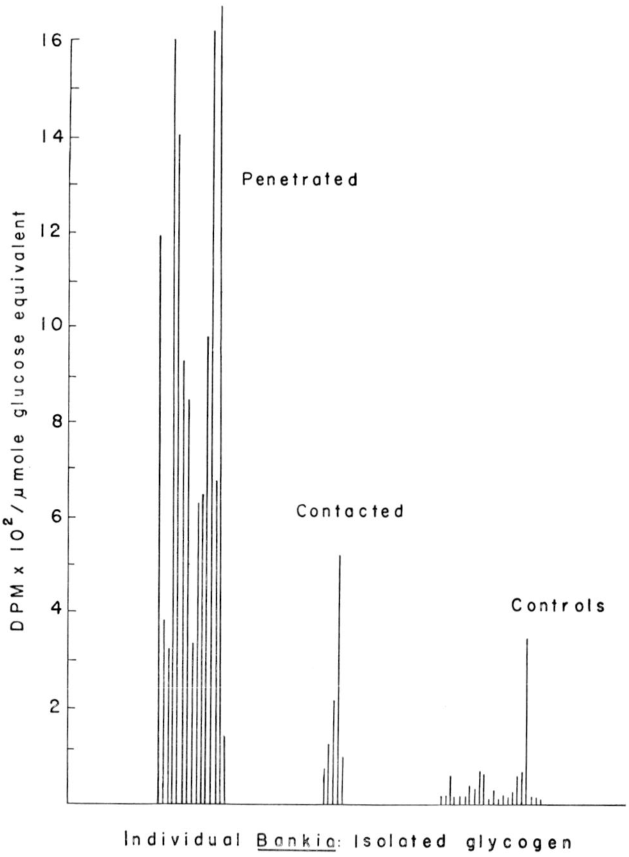

Fig. 8. Incorporation of ^{14}C *into mantle glycogen by* Bankia gouldi, *as disintegrations per minute (dpm) over background. See Fig. 7 for identification of the three groups.*

DISCUSSION

These results raise a major enigma. If the fecal pellet analysis clearly supports cellulose digestion while a fungal-type cellulolytic enzyme system cannot be demonstrated, how does *Bankia gouldi* digest cellulose? While the enzyme system might have been lost as an artifact of the experimental procedure, a more exciting solution to the enigma is the possibility of a previously unsuspected mechanism of cellulose and wood digestion. Oxidative degradation of native crystalline cellulose, possibly coupled with lignin degradation, by wood-rotting fungi has recently been demonstrated (Eriksson *et al.*, 1974; Westermark and Eriksson, 1974a; Westermark and Eriksson, 1974b). Depolymerization and solubilization of native crystalline cellulose by free radical oxidizing agents such as superoxide and hydroxyl radicals as well as hydrogen peroxide have been shown *in vitro* under conditions compatible with biological systems (Arthur and Hinajosa, 1971; Dore and Miller, 1923). In mammalian cells, phagocytosis is associated with the production of such free radical oxidizing agents at the cell surface and within the phagocytic vacuole (Babior *et al.*, 1973; McCord and Salin, 1975). In the teredinid shipworms, including *Bankia gouldi*, a specialized gland phagocytoses and apparently digests wood fragments (Morton and McQuiston, 1974; Potts, 1923). An oxidation of cellulose, with the added possibility of lignin degradation, is potentially the biochemical mechanism of wood digestion in these mollusks. That such processes would be so much a function of cellular structure and metabolism, together with the uncertainty of the degradation products formed, can explain the failure to show cellulose digestion *in vitro* with cell-free enzyme extracts.

Other questions closely related to the problem of wood digestion in *Bankia gouldi* deserve mention. It has been suggested that suspension feeding is an important process in shipworm nutrition, particularly as a source of nitrogen (Morton and McQuiston, 1974; Nair and Saraswathy, 1971; Turner, 1966). However, direct evidence of ctenidial suspension feeding is lacking, and some contradictory evidence has been published (Lane *et al.*, 1952; Turner and Johnson, 1971). I predict that future studies will show that suspension feeding is not important at least for the genera *Bankia*, *Teredo*, and *Lyrodus*.

Furthermore, recent evidence provides an alternate mechanism for meeting nitrogen requirements: nitrogen fixation has been demonstrated in four species of shipworms (Carpenter and Culliney, 1975). I have suggested (Dean, 1975) that such activity will be found to be associated with the 'gland of Deshayes', which recently has been shown to be aggregates of bacteria within the ctenidial filaments (Popham, 1975; Popham and Dickson, 1973).

I have carefully avoided consideration of the very difficult question of the role symbiotic microorganisms might play in wood and cellulose digestion in *Bankia gouldi*. In the absence of any convincing evidence for the presence of such organisms within the digestive tract, I feel pursuit of this question is premature. The failure to show the type of native crystalline cellulose-digesting mechanism characteristic of microorganisms certainly does not argue strongly for their presence.

Acknowledgement

Portions of this study were supported by a contract, NR 104-194, between Duke University and the Office of Naval Research.

REFERENCES

Albersheim, P. *et al.* (1973). In *Biogenesis of Plant Cell Wall Polysaccharides*, pp. 117 (F. Loewus, ed.), Academic Press, New York.

Almin, K. E. *et al.* (1975). *Eur. J. Biochem.*, **51**, 207.

Arthur, J. C., and Hinajosa, O. (1971). *J. Polymer Sci.*, **C36**, 53.

Babior, B. M. *et al.* (1973). *J. Clin. Invest.*, **52**, 741.

Bolker, H. L., and Brenner, H. S. (1970). *Science*, **170**, 173.

Boynton, L. C., and Miller, R. C. (1927). *J. Biol. Chem.*, **75**, 613.

Browning, B. L. (1967). *Methods of Wood Chemistry*. Interscience, New York.

Carpenter, E. J., and Culliney, J. L. (1975). *Science*, **187**, 551.

Clapp, W. F., and Kenk, R. (1963). Marine Borers, an Annotated Bibliography. Office of Naval Research, Washington, D.C.

Cowling, E. B., and Brown, W. (1969). *Cellulases and Their Applications*, pp. 152 (G. J. Hajny and E. T. Reese, eds.), American Chemical Society, Washington.

Cutter, J., and Rosenberg, F. A. (1972). *Material und Organismen*, **7**, 225.

Dean, R. C. (1975). *Digestion of Cellulose and Wood in the Shipworm* Bankia gouldi *Bartsch*, Ph.D. Dissertation, Duke Univ., Durham, N.C.

Dore, W. H., and Miller, R. C. (1923). *Univ. Calif. Publ. Zool.*, **22**, 383.

Eriksson, K.-E., and Pettersson, B. (1975). *Eur. J. Biochem.*, **51**, 193.

Eriksson, K.-E. *et al.* (1974). *FEBS Letters*, **49**, 282.

Freundlich, A., and Robards, A. W. (1975). *Cytobiologie*, **8**, 355.

Hajny, G. J., and Reese, E. T. (Eds.) (1969). *Cellulases and Their Applications*, Adv. Chem. Ser. No. 95, American Chemical Society, Washington, D.C.

Halliwell, G. (1965). *Biochem. J.*, **95**, 35.

Heyn, A. N. J. (1969). *J. Ultrastruct. Res.*, **26**, 52.

Higuchi, T. (1971). *Adv. Enzymes*, **34**, 207.

Jones, E. G. B., and Eltringham, S. K. (1971). *Marine Borers, Fungi and Fouling Organisms of Wood*. OECD, Paris.

Kadota, H. (1959). In *Marine Boring and Fouling Organisms*, pp. 332 (D. L. Ray, ed.), Univ. Washington, Seattle.

Lane, C. E. *et al.* (1952). *Bull. Mar. Sci. Gulf Carib.*, **2**, 385.

Lazier, E. L. (1924). *Univ. Calif. Publ. Zool.*, **22**, 455.

McCord, J. M., and Salin, M. L. (1975). In *Erythrocyte Structure and Function*, pp. 731 (G. J. Brewer, ed.), Alan R. Liss, New York.

Mark, R. E. (1967). *Cell Wall Mechanics of Tracheids*. Yale Univ., New Haven, Conn.

Meier, H. (1964). In *The Formation of Wood in Forest Trees*, pp. 137 (M. H. Zimmerman, ed.), Academic Press, New York.

Morton, B., and McQuiston, R. W. (1974). *Forma et Functio*, **7**, 58.

Muhlethaler, K. (1969). *J. Polymer Sci.*, **C28**, 305.

Nair, N. B., and Saraswathy, M. (1971). *Adv. Marine Biol.*, **9**, 335.

Norkrans, B. (1967). *Adv. Appl. Microbiol.*, **9**, 91.

Northcote, D. H. (1972). *Ann. Rev. Plant Physiol.*, **23**, 113.

Pew, J. C., and Weyna, P. (1962). *Tappi*, **45**, 247.

Popham, J. D. (1975). *Veliger*, **18**, 55.

Popham, J. D., and Dickson, M. R. (1973). *Mar. Biol.*, **19**, 338.

Potts, F. A. (1923). Proc. Cambridge Philos. Soc., *Biol. Sci.*, **1**, 1.

Ranby, B. (1969). In *Cellulases and Their Applications*, pp. 139 (G. J. Hajny and E. T. Reese, eds.), American Chemical Society, Washington.

Rosenberg, F. A., and Cutter, J. (1973). In *Proceedings of the Third International Congress on Marine Corrosion and Fouling*, pp. 778 (R. F. Acker *et al.*, eds.), Northwestern Univ., Evanston, Ill.

Shafizadeh, F., and McGinnis, G. D. (1971). *Adv. Carbohyd. Chem.*, **26**, 297.

Stockmann, V. E. (1972). *Biopolymers*, **11**, 251.

Stone, J. E. *et al.* (1969). In *Cellulases and Their Applications*, pp. 219 (G. J. Hajny and E. T. Reese, eds.), American Chemical Society, Washington.
Turner, R. D. (1966). *A Survey and Illustrated Catalogue of the Teredinidae.* Museum of Comparative Zoology, Cambridge, Mass.
Turner, R. D. (1973). *Science*, **180**, 1377.
Turner, R. D., and Johnson, A. C. (1971). In *Marine Borers, Fungi and Fouling Organisms of Wood*, pp. 259, OECD, Paris.
Van Handel, E. (1965). *Anal. Biochem.*, **11**, 266.
Wayman, M., and Obiaga, T. I. (1974). *Tappi*, **57**, 123.
Westermark, U., and Eriksson, K.-E. (1974a). *Acta Chem. Scand.*, **B28**, 204.
Westermark, U., and Eriksson, K.-E. (1974b). *Acta Chem. Scand.*, **B28**, 209.
Whitaker, D. R. (1971). In *The Enzymes*. 3rd Edn. Vol. 5, pp. 273 (P. D. Boyer, ed.), Academic Press, New York.

ANATOMICAL AND CHEMICAL ANALYSES OF UNTREATED WHITE PINE EXPOSED TO MARINE BORER DAMAGE

E. A. McGINNES, Jr., J. E. PHELPS and P. J-Y. LIEU

School of Forestry, Fisheries and Wildlife,
University of Missouri, Columbia, Missouri 65201, U.S.A.

Summary

A sample of white pine exposed to marine borer attack for 11 months, at a depth of 11 m off Hokkaido, Japan, was used for anatomical studies. The sample received had been damaged by *Limnoria*, *Bankia setacea*, and *Teredo navalis*. In addition to damage due to the above, degradation of parenchyma cells indicates possible bacterial action. With the exception of loss of parenchyma, anatomical integrity of the sample had been maintained; there was no evidence of fungal attack. Analyses of wood extractives, holocellulose content, lignin content, and selected inorganics were also undertaken. Both lignin and holocellulose content were below normal, indicating that the lower molecular weight fractions of these components were increased during the exposure period. Interaction of seawater, borer damage, and possible presence of microorganisms (bacteria) on wood structure are discussed along with the use of microscopy as a supporting tool in studies of marine borer damage to wood.

MATERIALS AND METHODS

A white pine board (approximately 15 × 31 × 2 cm) exposed to a variety of types of marine borer attack for an 11-month period in the sea just off the sea bed at a depth of 11 m off Hokkaido, Japan, was obtained for microscopic analyses through the courtesy of the U.S. Navy (DePalma, pers. comm.). The organisms which had attacked this sample were *Limnoria*, *Bankia setacea*, and *Teredo navalis*. Both face and end views of the sample were photographed prior to division for microscopic and chemical analyses; these macroscopic views are shown in Fig. 1. For light microscopic analyses, conventional paraffin embedding techniques proved to be futile. A nitrocellulose method (Humason, 1962) was found acceptable and cross, radial, and tangential sections approximately 30 μ thick were cut and stained in Safranin and Fast Green and mounted in Permount. Blocks approximately 1 × 2 × 0·5 cm were cut from the sample board for scanning electron microscopy (SEM) studies. Samples were taken from both the exterior and interior surfaces of the board. These samples were dried over a desiccant, coated with gold, and examined with a JEOL instrument operating at 10 kv. Extractives were determined in a Soxhlet apparatus, using water and then ethanol as extracting solvents. Extraction time was four hours for each solvent. Holocellulose was determined using a chlorite procedure (Wise *et al.*, 1946) and lignin content was determined by a standard 72% sulphuric acid method (Browning, 1967). Selected inorganics were obtained by neutron activation analyses utilizing the university reactor facility which operates on an approximate flux of 8×10^{13} thermal neutrons per cm^2/second.

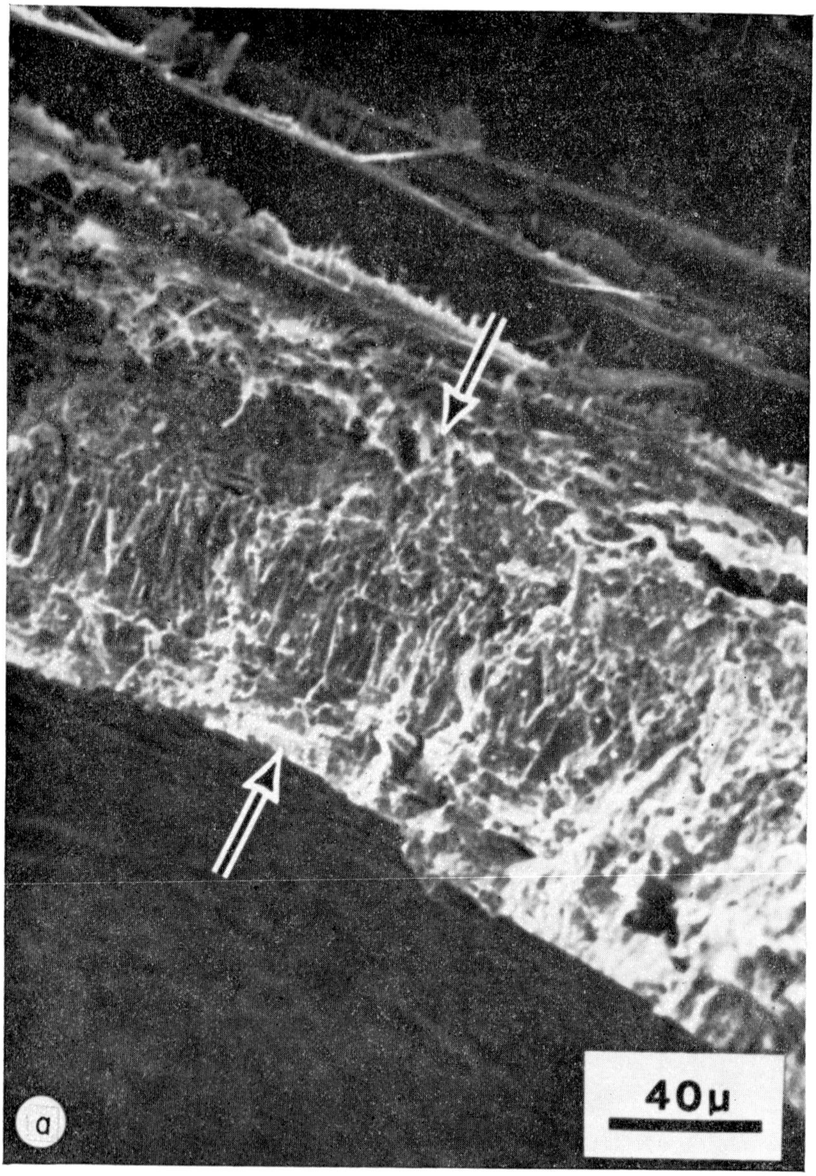

Fig. 1. Macroscopic views (a) and (c) of white pine sample showing damage on surface (a) due to Limnoria *plus attack (c) of* Bankia setacea *(smaller cross-sectional holes) and* Teredo navalis. *'Limey burrows' are evident (arrows). Light (b) and SEM (d) views of gribble damage on tangential face of sample.*

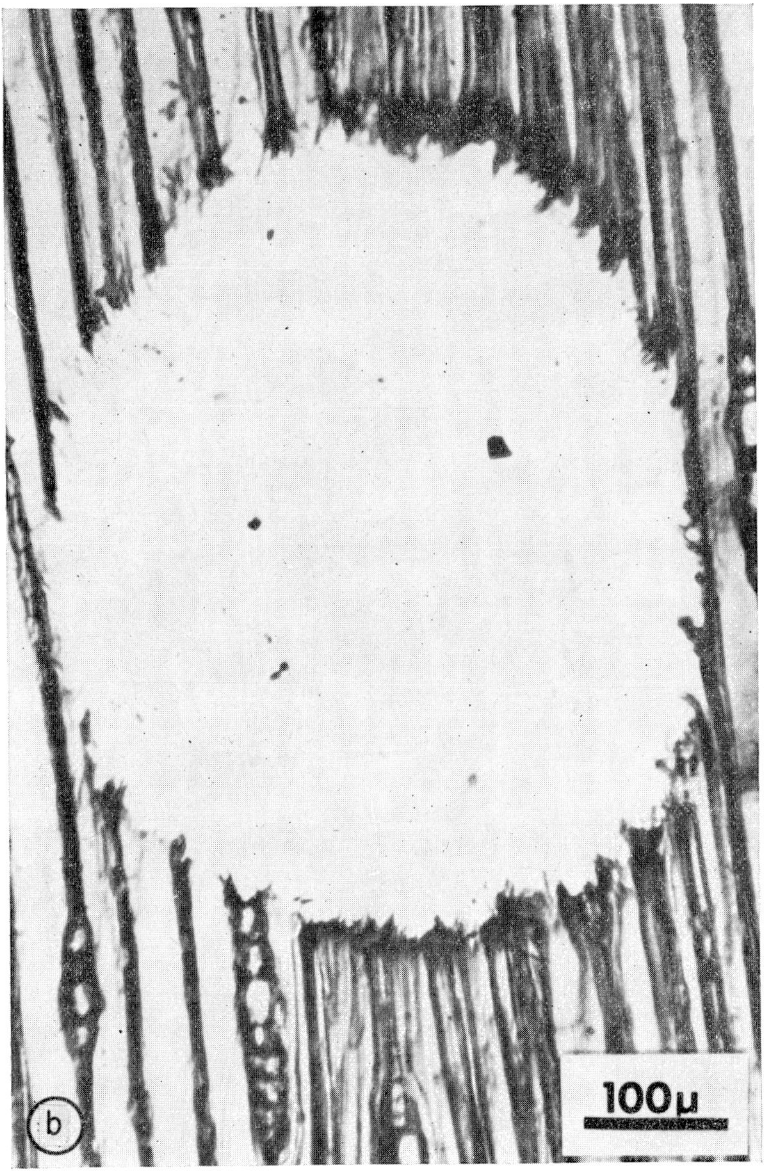

Fig. 2—contd.

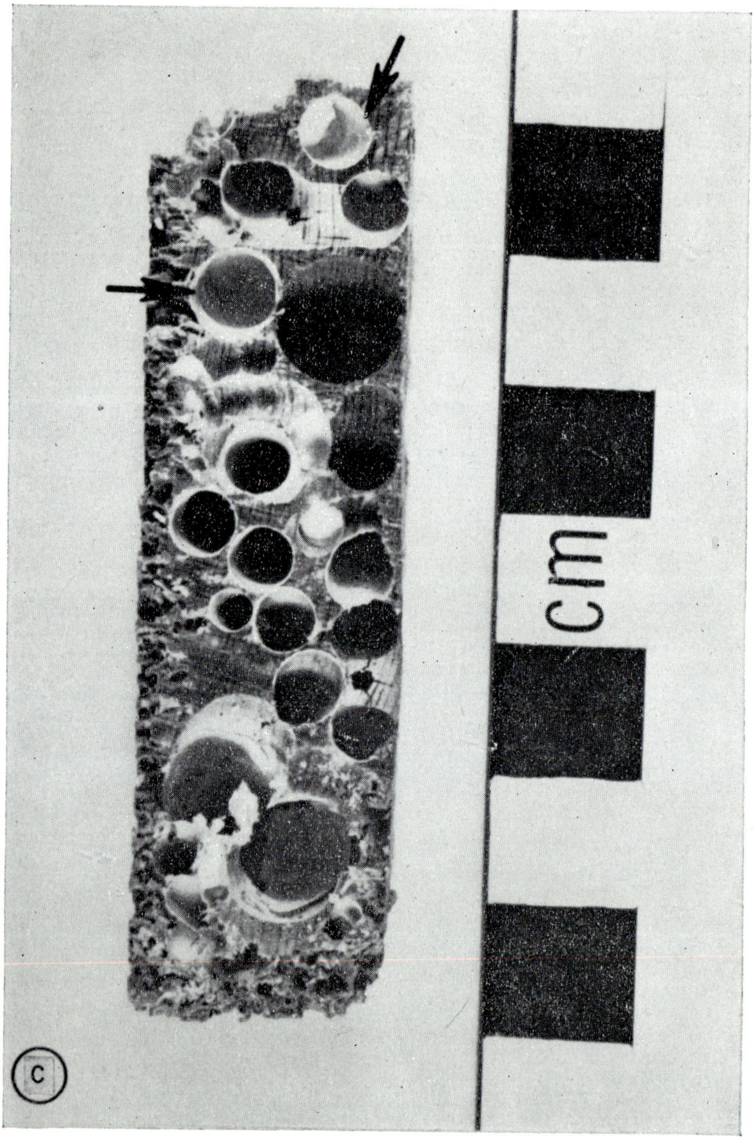

Fig. 2—contd.

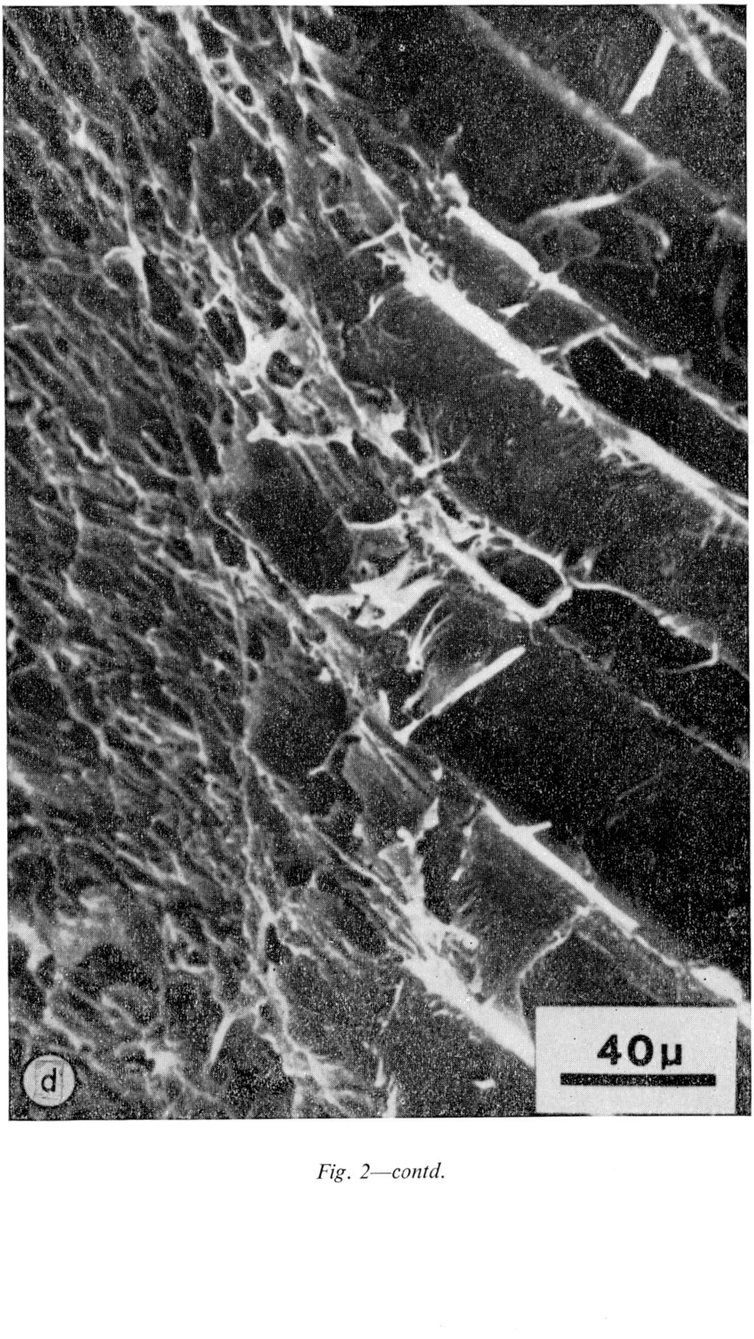

Fig. 2—contd.

RESULTS

Ray tissue and particularly parenchyma cells appear destroyed or in a badly degraded state throughout the sample as shown in Fig. 2. Although bordered pit structure retains anatomical integrity (Fig. 2), the SEM views of cell walls shown indicate some type of degradation since excessive, torn lamella exposure are atypical for normal wood of white pine species similarly prepared for microscopic observations. The chemical tabulations (Table I) on holo-

Table I

SELECTED CHEMICAL ANALYSES OF UNTREATED WHITE PINE EXPOSED TO MARINE BORER DAMAGE

Sample form	*Chemical analyses*						
	EtOH Extractives %		*HOH Extractives* %		*Lignin*[a] %	*Holocellulose*[a] %	
40 mesh, entire sample	1·44		5·57		21·78	59·18	
	Inorganics (PPM)[b]						
	Mg	*Cl*	*Na*	*Ca*	*Mn*	*I*	*Br*
Whole sample unextracted	1 940	3 370	2 817	7 579	30	59	50
Outer wood (primarily gribble zone) unextracted	8 903	26 564	16 549	3 341	239	475	428
Inner wood (primarily shipworm zone) unextracted	—	3 349	2 972	960	—	—	26
'Limey Burrows' unextracted	21 459	705	2 891	246 904	—	—	14

[a] Determined on extractive-free samples.
[b] Mg, Cl, Na, Ca, Mn, and I determined after 30 sec irradiation time, 10 min decay, 500 counts. Br determined by 20 min irradiation, 115 hr decay, 2000 counts. Elements such as Fe and Pb visible on the spectrum; however, no quantitative analyses were determined.

cellulose and lignin determinations of the sample support this view, since lignin content is only 2/3 that of typical softwoods and the holocellulose fraction is also considerably less than normal of extractive-free wood, comprising only 59% of the sample. This is approximately 85% of the normal value (70%) of holocellulose content for coniferous woods. The inner wood zone (shipworm area) contains less of the inorganics analysed than the outer zone (gribble area), with the obvious exception of the 'limey burrows'. There was no evidence of fungal attack on the sample.

DISCUSSION

Sources of information on wood damage due to marine borers generally lack any information on anatomical changes of the wood at the microscopic level (Findlay, 1953; Hunt and Garratt, 1967; Hochman, 1973; Panshin *et al.*, 1964; St. George, 1950). Actually, the use of light microscopy and SEM studies may provide useful information to supplement chemical and other analyses of the complex, interrelated happenings during exposure of wood to marine conditions. This study was initiated to explore results of marine borer damage at the microscopic level and, hopefully, to indicate that such analyses would provide an additional dimension to marine borer damage to wood in future research studies. The SEM is a useful tool for study of fragile and refractory materials and thus may provide additional insight to studies of wood degraded in service where light microscopy is limited. (Problems in sample preparation for observation and/or limited depth of focus and magnification ranges.) Damage due to gribbles and shipworms are shown macroscopically in Fig. 1. Viewed under both light and electron microscopy, the gribble burrows are cylindrical in cross section and lack incrustation buildup typically associated with shipworm damage (Figs. 1 and 2). Apparently, ray tissue and particularly parenchyma cells appear to be destroyed or else in a partly degraded state throughout the whole sample as exemplified in Fig. 2. This destruction of ray tissue is similar to that reported for bacterial action on wood (Wilcox, 1970), and chemical analyses alone would not demonstrate this point. Bordered pit structure retained anatomical integrity throughout the exposure period as shown in Fig. 2. However, SEM views of cell walls shown—coupled with lignin and holocellulose data in Table I—suggest some type of degradation during the exposure period because of excessive tearing of all secondary wall lamella and their resultant exposure (Fig. 2), such conditions being atypical for normal wood samples prepared for SEM observations. This degradation apparently was influenced by the solubilizing effect of seawater itself, the influence of marine borers, and an additional factor, presumably microorganisms (bacteria), cannot be ruled out. The chemical tabulations (Table I) on holocellulose and lignin determinations of this sample are indicative of some type of degradation. Presumably, the lower molecular weight component of the lignin fraction of the wood and the hemicellulose fraction of the holocellulose component of the wood were increased during the exposure period and then 'lost' in the resultant lignin and holocellulose analyses. It is well known that lower molecular weight 'lignins' are lost in the sulphuric acid analysis. The same condition holds for lower molecular weight components of the holocellulose fraction during conventional procedures for estimating the total carbohydrate content of wood. These observations do not clarify whether such degradation is caused predominantly by borer attack, seawater, exposure, microorganisms, or a combination thereof. Unfortunately, there is a lack of information in the literature to clarify this subject. Studies on feeding habits of marine organisms indicate that, although wood may be the nutrient source, there is evidence (Becker, 1959) that symbiosis exists between *Limnoria* and fungi while the studies of Ray (1959) indicate *Limnoria* feces are lignin-rich. Bacteria also attack the lignin-rich middle lamella zones of wood. The concentration (with the exception of 'limey burrows') of inorganics in the outer wood (gribble zone) may be attributed to greater seawater accessibility (wood permeability) on the outer wood surface in this zone; however, this is an assumption at present. The use of light and SEM techniques in this study fills a void in published English texts dealing with marine borer damage to wood at the cellular level and points out that microscopy is a valuable support tool in such studies by indicating preferential loss of ray parenchyma cells in the sample studies—such action possibly due to microorganism (bacteria) presence and thus resultant increased water permeability of the sample (through ray destruction) contributing to an increase of marine borer population and/or penetration.

REFERENCES

Becker, G. (1959). Biological investigations on marine borers in Berlin-Dahlem, In *Marine Boring and Fouling Organisms*, pp. 62–83 (D. L. Ray, ed.), Univ. of Washington Press, Seattle.

Browning, B. L. (1967). Determination of lignin, In *Methods of Wood Chemistry. Vol. II.* Chapt. 34, pp. 785–791. John Wiley & Sons, Inc., New York.

Findlay, W. P. K. (1953). *Dry rot and other timber troubles.* Hutchinson's Scientific and Technical Publications, London. 267 pp.

Hochman, H. (1973). Degradation and protection of wood from marine organisms, In *Wood Deterioration and Its Prevention by Preservative Treatments.* Vol. I, pp. 247–275 (D. D. Nicholas, ed.), Syracuse Univ. Press, Syracuse, N.Y.

Humason, G. L. (1962). *Animal Tissue Techniques*, Chapt. 7, Nitrocellulose method, pp. 70–79. W. H. Freeman & Co., San Francisco, Cal.

Hunt, G. M., and Garratt, G. A. (1967). *Wood Preservation*, 3rd Edn., pp. 62–68. McGraw-Hill, New York.

Panshin, A., DeZeeuw, C., and Brown, H. (1964). *Textbook of Wood Technology*, 2nd Edn., Chapt. 10, Wood deterioration and stains; natural durability of wood, pp. 378–382. McGraw-Hill, New York.

Ray, D. L. (1959). Nutritional physiology of limnoria, In *Marine Boring and Fouling Organisms*, pp. 46–61 (D. L. Ray, ed.), Univ. of Washington Press, Seattle.

St. George, R. A. (1950). Marine borers, In *Insect Enemies of Eastern Forests*, pp. 66–69. Misc. Pub. 657, USDA.

Wilcox, W. W. (1970). Anatomical changes in wood cell walls attacked by fungi and bacteria. *Bot. Rev.*, **36** (1), 1–28.

Wise, L. E., Murphy, M., and D'Addicco, A. A. (1946). Chlorite holocellulose—its fractionation and bearing on summative wood analyses and on studies on the hemicelluloses. *Paper Trade J.*, **122** (2), 11–19.

WOOD PRESERVATIVE DEGRADATION BY MARINE BACTERIA

WALLACE E. ESLYN

Forest Service, USDA, Forest Products Laboratory,
Madison, Wisconsin 53705, U.S.A.

Summary

Various bacteria, including species of *Bacillus*, *Pseudomonas*, and *Micrococcus*, were isolated from wood treated with creosote, copper chrome aersenate (CCA), ammoniacal copper arsenate (ACA), or combinations; all the treated wood had been exposed in the sea for varying times. To determine the significance of the bacteria in preservative detoxification, small wood blocks impregnated with these preservatives were incubated for one year in liquid cultures containing one or more species of the isolated bacteria. Blocks were then exposed to the action of wood decay fungi. Amount of decay wrought by the fungi indicated the effect of bacterial action upon the preservative involved. All tested bacteria substantially reduced toxicity in ACA-treated blocks and, to a much lesser extent, in blocks treated with creosote and creosote + CCA. However, CCA-treated blocks exposed to bacterial action decayed less than the controls.

To determine effectiveness and permanence of several wood preservatives in the marine environment, the Forest Products Laboratory initiated a study in 1964 in cooperation with the U.S. Navy and the Wm. F. Clapp Laboratories, Inc., Duxbury, Mass. Wood samples treated with various preservatives were exposed in the ocean for periods ranging from 6 months to 5 years. The wood samples were then examined for presence of marine borers and chemically analysed for preservative content (Baechler *et al.*, 1970b).

Additionally, isolations were attempted from the fresh samples to determine the microorganisms associated with the different chemical treatments. The present work describes the results of these isolations and considers the possible effects of selected bacterial isolates upon durability of preservative-treated wood.

MATERIALS AND METHODS

Samples of wood from southern pine (*Pinus* sp.) and Douglas fir [*Pseudotsuga menziesii* (Mirb.) Franco], 102 by 76 by 19 mm (4 by 3 by $\frac{3}{4}$ inches), were treated at several concentrations with creosote, creosote-coal tar solution, ammoniacal copper arsenate (ACA), chromated copper arsenate (CCA, Type 1), or with a mixture of creosote plus ACA or CCA. In all, 27 treatments were applied—13 to southern pine and 14 to Douglas fir wood samples. After treatment, half of the samples receiving treatments (1-18) were immersed about 0·9 m (3 ft) below the water surface in the harbour at Wrightsville Beach, N.C., and half about 0·6 m (2 ft) above bottom in the harbour at Boston, Mass., for periods up to 5 years. Treatments 19–27 received the highest preservative retentions, and these were exposed only in the Wrightsville Beach harbour. Preservative applications, immersion in seawater, and the schedule for sample removal for examination and testing are described by Baechler *et al.* (1970b).

When samples were removed from the sea, a cross section about 3·2 mm ($\frac{1}{8}$ in) thick was cut from one sample in each treatment group. The surface of each sample was flamed, split apart, and small pieces asceptically removed from their interiors. These pieces were placed on media of casamino acid (3·0 g casamino acids, 3·0 g yeast extract, 1·0 g dextrose, and 2·0 g agar per litre of distilled water) and artificial seawater-yeast-peptone (24·7 g NaCl, 0·7 g KCl, 6·3 g $MgSO_4 \cdot 7H_2O$, 4·6 g $MgCl_2 \cdot 6H_2O$, 1·0 g anhy. $CaCl_2$, 0·2 $NaHCO_3$, 5·0 g yeast extract, and 5·0 g peptone per litre of distilled water), and on malt agar. Isolations were made by Dr C. G. Duncan on the samples removed every 6 months during the first three and a half years' immersion in seawater. After her death, the author continued the isolations with material immersed in the sea for periods of 4 to 5 years.

Many isolates obtained by Duncan were lost. Those that survived and those isolated by the author were screened against concentrations of preservatives to determine the more preservative-tolerant and preservative-degrading isolates. Thus, isolates from wood treated with creosote, creosote + CCA, or creosote + ACA were inoculated into flasks of casamino acid medium (without agar) plus 10, 20, 40, 80, or 95% of the water-soluble components of creosote; from CCA or creosote + CCA treatments into similar media containing 1, 3, 5, or 10% CCA; and from ACA and creosote + ACA treatments into media containing similar concentrations of ACA instead of CCA.

From these inoculations, isolates most tolerant to the preservative treatments were selected for experiments to determine effects of marine bacteria upon preservative toxicity. Cubes (19 mm, or $\frac{3}{4}$ in) of southern pine wood were treated with marine grade creosote (128 kg^3 or 8 pcf, and 256 kg^3 or 16 pcf retentions), CCA (56 kg^3 or 3·5 pcf), ACA (32 kg^3 or 2·0 pcf), creosote + CCA (208 kg^3 + 32 kg^3 or 13 pcf + 2 pcf), or creosote + ACA (208 kg^3 + 16 kg^3 or 13 pcf + 1 pcf) according to the standard soil-block test method (ASTM, 1961). Each treated block was then placed into a bottle containing peptone-yeast-seawater medium which had been previously inoculated (except for controls) with one or two bacteria tolerant to the type of preservative with which the block was treated. Hence, creosoted blocks were inserted into cultures of bacteria tolerant to creosote, creosote + CCA, and creosote + ACA. The remainder were inserted into cultures in this fashion:

Wood block treatment	*Bacteria tolerant to*
Creosote + CCA	Creosote + CCA
Creosote + ACA	Creosote + ACA
CCA	CCA and Creosote + CCA
ACA	ACA and Creosote + ACA

Treated controls were placed in bottles containing sterile medium. Culture media were changed and, where necessary, reinoculated bi-weekly during the first 6 months of incubation. Thereafter, media were changed about once a month. At the end of 1 year's incubation the blocks were removed, conditioned at 80°F and 70% relative humidity, and weighed.

To determine whether bacteria had or had not degraded preservatives, the blocks were then autoclaved and subjected to degradation by wood decay fungi according to the soil-block procedure (ASTM, 1961). Exceptions were that the incubation periods were increased and *Lentinus lepideus* Fr. (Mad 534), a creosote-tolerant fungus, was used to test creosote-treated blocks, while *Poria vaillantii* (DC. ex Fr.) Cke. (MD 497), a fungus tolerant of CCA (DaCosta, 1959), was used to test remaining toxicity of ACA- and CCA-treated blocks. Test blocks treated with a combination of creosote + salt were inoculated with each of the above fungi.

After this exposure, weight losses were calculated and comparisons made between losses occurring in blocks exposed to bacterial action and losses occurring in control blocks. Although fungi toxicity was used to determine extent of detoxification of marine preservatives, the results are thought to apply to wood-destroying marine animals as well.

RESULTS

Isolations from wood samples exposed in the sea

The percentage of samples in each treatment from which bacteria were isolated from both locations after 48, 54, or 60 months' immersion in the sea are presented in Table I. The values are based upon wood samples treated

Table I

PERCENTAGE OF PRESERVATIVE-TREATED WOOD SAMPLES FROM WHICH BACTERIA WERE ISOLATED AFTER STORAGE IN SEAWATER AT TWO HARBOURS FOR 48–60 MONTHS

Wood treatment	*Percentage of samples bacterially infected*[a]	
	Wrightsville Beach	*Boston*
Creosote	67 (60)	21
CCA	80 (87)	58
ACA	62 (45)	33
Creosote + CCA	83 (50)	0
Creosote + ACA	33 (17)	0

[a] Values are for low retention only; numbers in parentheses represent combined high and low retentions.

with low retentions of preservatives (treatments 1–18), permitting a direct comparison of results obtained at both harbours which similarly treated wood samples. Bacteria were isolated much more frequently from wood immersed in Wrightsville Beach harbour than from that in Boston harbour, regardless of type of preservative treatment.

Bacteria were isolated most frequently from CCA-treated wood and least frequently from creosote + ACA-treated wood (Table I). Average number of isolations from creosote- and ACA-treated woods were similar and second in frequency to CCA-treated wood.

No fungi were isolated from the wood samples. In addition, no signs of fungal soft-rot attack were evident in slides made from each of the samples from which isolations were made.

Bacterial identifications

Twenty-three bacterial cultures containing strains of one or more species shown to be tolerant to one or more test preservatives were identified (Table II) and included in further tests. Among the eight species in the genus

Table II
IDENTITY AND PRESERVATIVE-TOLERANCE OF TEST BACTERIA

Bacterial species	*Isolate no.*	*Preservative tolerance*
Bacillus brevis Migula	1241	ACA
B. cereus Frankland & Frankland	1178	Creosote
B. circulans Jordan	866-1	Creosote + CCA
B. lentus Gibson	226	Creosote
	485, 491-1	Creosote + CCA
	287-2	ACA
B. megaterium de Bary	425	CCA
B. pumilus Meyer & Gottheil	1148	Creosote
	392, 1328	CCA
	488-2, 491-2	Creosote + CCA
B. sphaericus Meyer & Neide	326	ACA
	548	Creosote + ACA
B. subtilis (Ehrenberg) Cohn	359	CCA
	488-1	Creosote + CCA
Micrococcus colpogens Campbell & Williams or *M. varians* Migula	293	ACA
	544, 557	Creosote + ACA
	866-3	Creosote + CCA
Pseudomonas creosotensis O'Neill, Drisko & Hochman	ATCC-14582	Creosote
P. membranoformis? (ZoBell & Allen) ZoBell	386, 1331	CCA
P. sp. A	644	Creosote
P. sp. B	287-1	ACA
P. sp. C	1265	ACA

Bacillus, *B. lentus* and *B. pumilus* occurred most frequently. Collectively, the eight *Bacillus* species exhibited tolerance to all preservatives included in present tests. A *Micrococcus* species, either *colpogens* or *varians*, was associated with ACA, Creosote + ACA-, and creosote + CCA-treated woods. Four *Pseudomonas* species, one of which may be *P. membranoformis*, were tolerant to CCA, creosote, or ACA (Table II). *Pseudomonas creosotensis* was not isolated, but it has been associated with creosoted marine piling (O'Neill *et al.*, 1961). Therefore, a culture (ATCC-14582) was obtained from the American Type Culture Collections for comparison with tolerant or detoxifying bacteria isolated in these studies.

Effect of wood decay fungi upon preservative-treated blocks

The amounts of decay caused by *Lentinus lepideus* in wood blocks treated with 128 and 256 kg^3 of creosote and by *L. lepideus* + *Poria vaillantii* in blocks treated with 208 kg^3 creosote + 32 kg^3 CCA are given in Table III. Blocks exposed first to bacterial action underwent greater decay, except for blocks treated with 256 kg^3 creosote and exposed to *B. lentus* isolate 226. The largest average weight loss due to prior bacterial action occurred in blocks treated with 256 kg^3 creosote and exposed to a bacterial mixture of *B. lentus* + *B. pumilus*. Here, weight losses averaged 13·9% compared to 6·7% for the control.

Decay brought about by *P. vaillantii* in ACA-treated control blocks was

Table III
DECAY LOSSES IN CREOSOTE- AND CREOSOTE + CCA-TREATED WOOD PREVIOUSLY EXPOSED TO PRESERVATIVE-TOLERANT BACTERIA

Bacterial isolate	*Average weight loss*[a] (%)		
	8 *pcf Creosote*[b]	16 *pcf Creosote*[b]	*Creosote+CCA*[c]
Control	7·6	6·7	7·1
B. cereus (1178)	11·6	11·7	
B. circulans + *M. colpogens* (866-1, -2, -3)	9·8	12·7	9·6
B. lentus (226)	9·7	6·1	
B. lentus (485)	10·5	9·5	11·5
B. lentus + *B. pumilus* (491-1, -2)	11·2	13·9	10·9
B. pumilus (1148)	11·2	13·1	
B. pumilus + *B. subtilis* (488-1, -2)	8·9	11·1	11·0
B. sphaericus (548)	11·5	10·3	
M. colpogens (554)	8·8	10·7	
M. colpogens (557)	11·8	11·0	
P. creosotensis (ATCC 14582)	10·7	11·8	
Pseudomonas sp. A (644)	10·9	10·4	

[a] Average of four replications.
[b] Weight losses as caused by *L. lepideus*.
[c] Weight losses as caused by *L. lepideus* + *P. vaillantii*.

Table IV
DECAY LOSSES IN ACA-TREATED WOOD PREVIOUSLY EXPOSED TO PRESERVATIVE-TOLERANT BACTERIA

Bacterial isolate	*Average weight loss*[a] (%)
Control	21·7
B. brevis (1241)	33·9
B. lentus + *Pseudomonas* sp. B. (287-1, -2)	41·4
B. sphaericus (326)	41·8
B. sphaericus (548)	37·0
M. colpogens (293)	37·7
M. colpogens (554)	35·8
M. colpogens (557)	39·1
Pseudomonas sp. C (1265)	44·6

[a] Average of eight replications.

Table V
DECAY LOSSES IN CCA-TREATED WOOD PREVIOUSLY EXPOSED TO PRESERVATIVE-TOLERANT BACTERIA

Bacterial isolate	*Average weight loss*[a] (%)
Control	36·5
B. lentus (485)	24·3
B. megaterium (425)	16·7
B. pumilus (392)	11·1
B. pumilus (1328)	14·9
B. subtilis (359)	19·8
P. membranoformis? (386)	16·1
P. membranoformis? (1331)	19·1

[a] Average of eight replications.

high (21·7%), but similarly treated blocks that had been previously exposed to bacteria were decayed to a considerably greater degree (Table IV). The greatest weight loss was 44·6% and occurred in blocks first exposed to the action of *Pseudomonas* sp. C (isolate 1265).

Results with CCA-treated blocks were opposite to those obtained with creosote- and ACA-treated blocks. *P. vaillantii* reduced weight 36·5% in the control blocks and considerably less in blocks exposed first to bacteria, the greatest loss being 24·3% and least, 11·1% (Table V).

DISCUSSION

Bacillus, *Pseudomonas*, and *Micrococcus* species tolerant to creosote, CCA, and ACA were isolated from preservative-treated wood exposed in the sea. *Pseudomonas creosotensis* was not found among these isolates although it had previously predominated in isolations made from creosoted pilings located on both east and west coasts of the United States (O'Neill *et al.*, 1961). However, the species may have been among those isolated earlier by Duncan, and subsequently lost, as she had reported (unpublished) the frequent isolation of *Pseudomonas* species from Wrightsville Beach samples.

On the basis of isolations made on wood samples immersed in the sea for 48–60 months (Table I), greater populations and more species of bacteria were present in Wrightsville Beach harbour than in Boston harbour. For example, bacteria tolerant to dual treatments (creosote + salt) were apparently present in Wrightsville Beach but not in Boston waters (Table I). This may reflect differences in temperature or pollution between the two harbours, or differences in the depths to which the samples were immersed in the sea at the two locales.

Creosoted piling has been noted, at times, to lose resistance to marine borer attack although considerable quantities of preservative still remained in the wood (Drisko and O'Neill, 1966; Hochman, 1967). Mixed populations of unidentified marine bacteria were found to metabolize and chemically alter creosote; however, a culture of *P. creosotensis* alone was unable to metabolize whole creosote although it was capable of utilizing a number of neutral constituents of the creosote (Drisko and O'Neill, 1966).

In the present study, pure cultures of *P. creosotensis*, *Bacillus* spp., and *Micrococcus colpogens* (*varians*?) brought about consistent small decreases in toxicity to decay fungi of creosote- and creosote + CCA-treated wood. Considerably more bacteria may be capable of degrading creosote, as ZoBell (1946) reported wide distribution of hydrocarbon-oxidizing microorganisms, including bacteria, in soil and water.

To deter biodegradation of creosoted piling, Hochman (1967) suggested three alternatives: (a) incorporation of a bactericidal and *Limnora*-toxic additive to creosote; (b) double treatment of piling when chemicals not dissolvable in creosote are to be added; or (c) use of a barrier wrap about piling. Double- or dual-treatment of piling using inorganic water-borne salts in combination with creosote was suggested many years ago and has received renewed interest during recent years (Baechler *et al.*, 1970a). Recent field tests of wood dually treated with high retentions of creosote and copper-arsenic compounds indicate such treatments protect against marine organisms more effectively than does creosote alone (Baechler *et al.*, 1970b; Baechler *et al.*, 1970a; Johnson *et al.*, 1973).

In present tests, however, wood blocks treated to retentions of 208 kg^3 of creosote plus 32 kg^3 of CCA lost fungitoxicity at a rate similar to that occurring in blocks treated solely with creosote. Apparently, the marine environment contains bacteria capable of metabolizing or altering creosote and a combination of creosote and CCA at similar rates. Perhaps incorporation of a bactericide would prevent biological detoxification and even permit use of lower retentions of these preservatives. Because tests of wood treated with creosote + ACA have not yet been completed, it is not known whether bacterial detoxification of this preservative combination occurs.

In these trials, blocks treated with 32 kg^3 of ACA or with 56 kg^3 of CCA were subject to considerable decay whether exposed to bacteria and leaching or solely to leaching in sterile media. CCA controls, in fact, experienced substantially more decay than did bacterially exposed test blocks. The isolate of *P. vaillantii* used in these tests was subsequently found to be abnormally tolerant to CCA-treated wood, but some isolates of this species have been reported capable of degrading wood loaded to a retention of 77 kg^3 (Wallace, 1968) and sawdust at a retention of about 102 kg^3 of CCA (DaCosta, 1959). (Assuming that these loadings are not based on the current oxide method of calculating CCA retentions, conversion would still result in calculated retentions of at least 58% of the values shown by Wallace and DaCosta—retentions still sufficient for protection of wood.) Performance of CCA in field and service has, however, been much better than has been indicated by laboratory tests (Wallace, 1968). Scheffer (1964) has recommended that fungus strains showing median, rather than high or low, preservative tolerance be used in laboratory assays of wood preservatives. Presumably, this would help guard against occurrence of large differences in performance of preservatives in laboratory and field.

In view of the high CCA tolerance of the *P. vaillantii* strain used, and the resultant poor performance of the copper-arsenates undergoing test herein, new tests have been set up to determine the capability of marine bacteria to detoxify these preservatives. A different test fungus, less tolerant to copper-arsenate, will be used for this work with the results reported later.

Acknowledgements

The author is indebted to Mr William G. Moore, University of Wisconsin, Madison, for identification of the majority of the bacteria included herein, and to Dr Ruth E. Gordon, Rutgers University, New Brunswick, N.J., for identification of *Bacillus circulans*, *B. cereus*, and *B. sphaericus* isolates.

REFERENCES

American Society for Testing and Materials (1961). Standard method of testing wood preservatives by laboratory soil-block cultures. ASTM D 1413-61. Philadelphia, Pa.

Baechler, R. H., Gjovik, L. R., and Roth, H. G. (1970a). Marine tests on combination-treated round and sawed specimens. *Am. Wood-Preserver's Assoc. Proc.*, **66**, 249–257.

Baechler, R. H., Richards, B. R., Richards, A. P., and Roth, H. G. (1970b). Effectiveness and permanence of several preservatives in wood coupons exposed to sea water. *Am. Wood-Preserver's Assoc. Proc.*, **66**, 47–64.

DaCosta, E. W. B. (1959). Abnormal resistance of *Poria vaillantii* (DC. ex Fr.) Cke. strains to copper-chrome-arsenate wood preservatives. *Nature*, **183**, 910–911.

DaCosta, E. W. B. (1967). Laboratory evaluations of wood preservatives. Part 1. Effectiveness of waterborne preservatives against decay fungi after severe leaching. *Holzforschung*, **21** (2), 50–57.

Drisko, R. W., and O'Neill, T. B. (1966). Microbiological metabolism of creosote. *Forest Prod. J.*, **16** (7), 31–34.

Hochman, H. (1967). Creosoted wood in a marine environment—A summary report. *Am. Wood-Preserver's Assoc. Proc.*, **63**, 138–150.

Johnson, B. R., Gjovik, L. R., and Roth, H. G. (1973). Single- and dual-treated panels in a semi-tropical harbor: Preservative and retention variables and performance. *Am. Wood-Preserver's Assoc. Proc.*, **69**, 207–215.

O'Neill, T. B., Drisko, R. W., and Hochman, H. (1961). *Pseudomonas creosotensis* sp. n., a creosote-tolerant marine bacterium. *Appl. Microbiol.*, **9** (6), 472–474.

Scheffer, T. C. (1964). Biological observations of significance for improved preservative treatment. *Holzforschung*, **18** (3), 88–94.

Wallace, E. M. (1968). The copper-chrome-arsenate preservatives and their use in modern wood preservation. *Am. Wood-Preserver's Assoc. Proc.*, **64**, 50–59.

ZoBell, C. E. (1946). Action of microorganisms on hydrocarbons. *Bacteriol. Rev.*, **10**, 1–49.

A NUTRITIONAL BASIS FOR MICROFUNGAL SUCCESSION AND DECAY IN WOOD

B. KING and T. A. OXLEY

Biodeterioration Information Centre,
University of Aston in Birmingham, Birmingham, U.K.

Summary

The classification of microfungi colonising wood, on the basis of the effects they cause, has been reconsidered in the light of soluble nutrient distribution in green wood and redistribution of these nutrients to the surfaces through which wood loses water by evaporation when it dries. It is postulated that much of the apparently indeterminate behaviour of microfungi in wood in soil contact is determined by (a) nutrient deficiencies in wood, especially of nitrogen, and (b) the occurrence of raised nutrient levels at the evaporating surfaces, to an extent depending on the amount of water loss which has occurred and the amount of soluble nutrients which were available in the wood before drying.

INTRODUCTION

Wood, particularly when of large dimensions, is generally ultimately decayed only by basidiomycetes. Wood, however, may also be colonized and decayed by microfungi of the classes Fungi Imperfecti and Ascomycetes. At this laboratory nutritional aspects of microfungal colonization and succession in wood have been examined. In this paper the subject is discussed in relation to recent findings on nutrient (especially nitrogen) availability and distribution in wood.

NITROGEN IN WOOD AND CARBON TO NITROGEN (C:N) RATIOS

In contrast with other plant remains, the proportion of nitrogen in wood is small (Cowling and Merrill, 1966). The bulk of this nitrogen is in the form of cell wall protein (Merrill and Cowling, 1966) deposited in the primary wall middle lamella complex and is thus not immediately available to colonizing organisms. It has usually been considered that only small amounts of the nitrogen in wood are in a soluble form (Bletchly and Farmer, 1959; Baker *et al.*, 1970) and most of this is in the form of reserve food materials in parenchyma (Merrill and Cowling, 1966). King *et al.* (1974) described the distribution of nitrogen in wood in freshly felled, *i.e.* 'green' pine and spruce. They confirmed previously published data (Cowling and Merrill, 1966) showing that the highest total nitrogen content is in the outer sapwood (closest to cambium), with lesser values in the inner sapwood/heartwood regions. By determination of nitrogen before and after extraction with

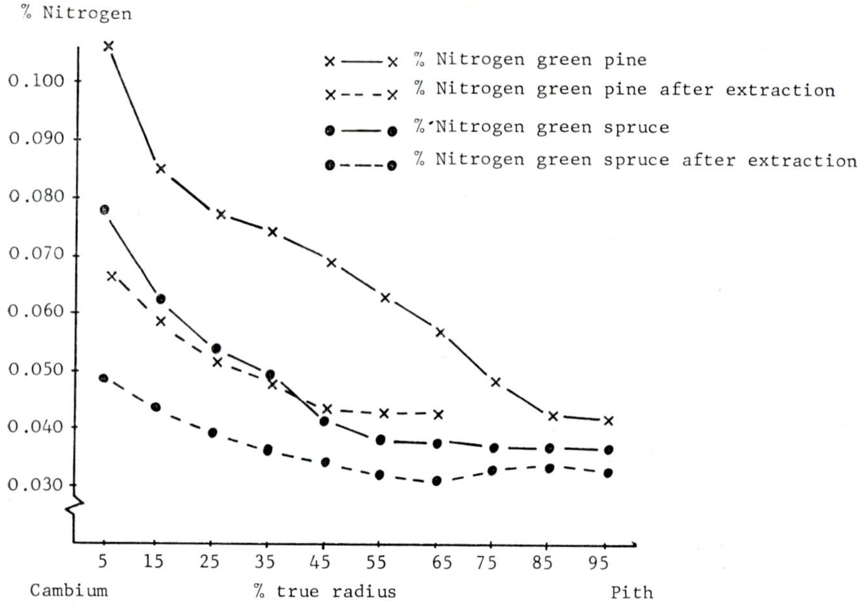

Fig. 1. *Distribution of total nitrogen in pine and spruce in the green condition and after extraction with hot alcohol/benzene, hot alcohol/water and cold water.*

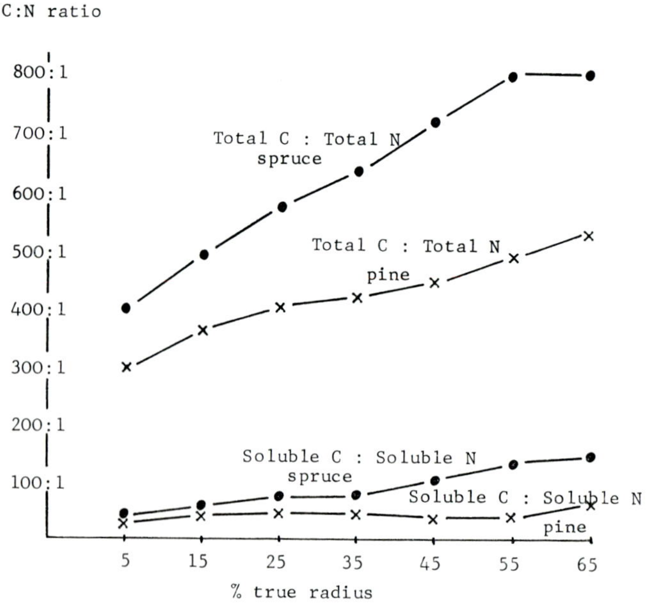

Fig. 2. *Carbon to nitrogen ratios in green spruce and pine.*

benzene, alcohol, and water, it was shown that substantial amounts (approximately 30%) of the total nitrogen content in the outer sapwood was in a soluble form (Fig. 1).

Further work (King, 1975) involving extraction of all soluble carbohydrates showed that these approximate to 2% of the dry weight of green spruce wood. The probable soluble carbon to soluble nitrogen ratios (C:N) in pine and spruce were calculated using a molecular formula of $C_6H_{12}O_6$ (for glucose) and presuming similar soluble carbohydrate levels in pine. These are shown in Fig. 2. This figure also shows the total (*i.e.* total C to total N) C:N ratios for pine and spruce in the green condition, calculating the weight of carbon on a molecular formula of $C_6H_{10}O_5$ (for cellulose) and presuming a 70% carbohydrate content in wood (*i.e.* omitting lignin because microfungi are generally not lignolytic). It can be seen that the total C:N ratios are too great

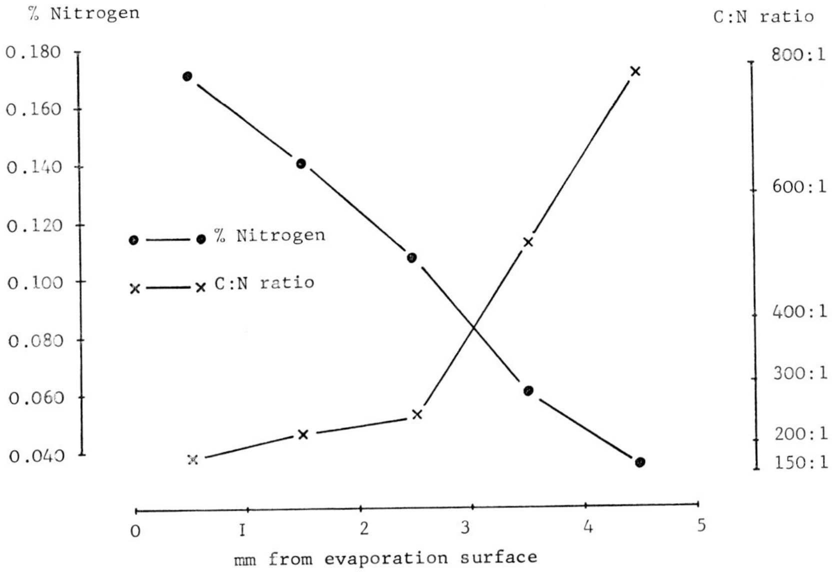

Fig. 3. Nitrogen contents and C:N ratios of 1 mm thick layers successively removed from a pine plank evaporation surface.

for cellulolytic activity by microfungi (Levi and Cowling, 1969) and even further removed from those shown by Butcher and Drysdale (1974) for optimal cellulose production by soft rot fungi. The soluble C:N ratios are much smaller than the total C:N ratios and are in fact similar to those in many laboratory media (Levi and Cowling, 1969). Such media provide optimal conditions for growth of microfungi, but there is no tendency for production of cellulases if carbon is present in soluble form (Bravery, 1968).

King *et al.* (1974) showed that during drying of wood, soluble nitrogenous materials and carbohydrates migrate with water to wood surfaces where they are deposited after evaporation. Figure 3 shows the nitrogen gradient from the radial evaporation surface of the sapwood (1–25 rings measured from the cambium) of a 55 mm thick quartersawn pine plank after drying at 50°C. Nitrogen was measured by a microkjeldahl technique (Humphries, 1956) on successive 1 mm layers removed by planing from the plank evaporation surfaces. It can be seen that the nitrogen content of the surface millimetre is over four times that of the fifth successive millimetre and that even the nitrogen content of the fourth millimetre is one and a half times that of the fifth. Previous studies (King, 1975) have shown that from the fifth millimetre through to the plank centre, the wood is depleted of soluble nitrogen.

Figure 3 also shows the total C:N ratio gradient from the plank surface (the total nitrogen near the surface is predominantly soluble). The C:N ratios in the outer 2 millimetres are similar to those considered by Levi and Cowling (1969) to be suitable for cellulase production by microfungi, but at 5 mm and deeper the C:N ratios are considerably in excess of those required for this purpose. Furthermore, the nitrogen is almost entirely nonsoluble. King *et al.* (1974) showed that after an extraction of evaporation surface layers of dried pine to a depth of 2 mm with hot alcohol/benzene, further extraction with hot alcohol/water and cold water solutions produced further weight losses in the wood of approximately 4%. Since pine does not contain large quantities of tannins (TAPPI, T13m54), these latter extracts were presumed to consist largely of soluble carbohydrate materials. By calculation of soluble C on the basis of the molecular formula $C_6H_{12}O_6$ for glucose and soluble N, deduced from the data in Fig. 3, it can be shown that the soluble C:N ratios at the evaporation surface of the wood are in the region of 14:1, *i.e.* lower than that for soluble nutrients found in the green material.

DERIVATION OF THE NOMENCLATURE OF MICROFUNGI WHICH COLONIZE WOOD

The nomenclature of microfungi which colonize wood has been based on the effects produced by them. Thus it has been usual to speak of 'staining' fungi, 'mould' fungi, and 'soft rot' fungi but these behavioural descriptions are not related to any taxonomic grouping.

The review by Scheffer (1973) has discussed, among others the 'mould' fungi and the 'stain' fungi. These generally colonize the sapwood of freshly felled material which has been stored under poor drying conditions. 'Mould' fungi usually produce a superficial discoloration by sporulation which may be easily planed or brushed off, but they may also penetrate deeply into the wood. The 'staining' fungi cause a deeply penetrating dark stain produced by dark pigmentation of the mycelium or by chemical reactions with the wood (Gadd, 1965). The intensity of mould and staining may vary with wood species and the conditions under which it develops. It may also occur to a lesser extent after the wood has been dried and rewetted (Findlay, 1966).

In contrast with mould and staining, 'soft-rot' is seldom produced in green wood but is produced in wood in service usually when basidiomycete colonization is precluded by preservation or high moisture levels, and especially in conditions particularly conducive to microbial colonization such as soil contact or cooling tower fill. Savory (1954) proposed the term 'soft-rot' for decay caused by cellulose-destroying microfungi to distinguish it from the brown and white rots caused by the wood-destroying basidiomycetes. He described cavities of distinctive shape but many different forms of cavity formation are now recognized, *e.g.* Courtois (1963). In the present context all are included in the term 'soft-rot'.

The differences between the microfungal groupings are becoming less obvious (Savory, 1968) as it has been shown (Krapivina, 1960; Merrill, 1966) that a number of staining and mould fungi can act as soft rot fungi. Liese (1970a,b) considered that several fungal species long allocated to the blue-staining and mould group may have to be transferred to the soft rot group because of their ability to degrade wood cell walls.

Work on the successions of fungi which colonize dried wood (Corbett, 1963; Merrill and French, 1966; Butcher, 1968; Levy, 1969; Banerjee and Levy, 1971) has shown that many mould-staining and soft rot fungi are intimately associated with the decay process. The work of Butcher (1968) and Banerjee and Levy (1971) has shown that soft rot is largely confined to the peripheral regions of fence-posts (up to 5 mm) but that basidiomycetes were responsible for decay in depth. Whereas maximal numbers of mould and staining species were isolated from the outer regions of the wood, only few fungal species succeeded in penetrating further than 5 mm into the wood.

Gorshin and Krapivina (1969) in experiments with microfungi in the absence of basidiomycetes showed that wooden stakes were eventually degraded. They suggested that microfungi which colonize wood are very adaptable to varying conditions and hence should not be classified on a behavioural basis.

RELATIONSHIP OF MICROFUNGAL NOMENCLATURE WITH SOLUBLE NUTRIENT DISTRIBUTION

Microfungi are capable of only slow decay production in the absence of additional nitrogenous materials (Savory, 1954). It has also been observed that soft rot is most active in timber which is in contact with soil or with water containing nutrients in solution (Findlay, 1966); and Garrett (1963) has suggested that woody stems absorb nitrogen from surrounding soil during decomposition. Levy (1968) considered that nutrient salts moved with water into wooden fence posts in ground contact. The consensus of opinion, therefore, is that microfungal decay requires nitrogen movement into wood presumably in a soluble form.

Most of the previous work on the role of wood nitrogen content in decay has been related to the total nitrogen content of the wood. However, since much of the nitrogen is in the form of cell wall protein, it is not available to microfungi until breakdown has occurred. Microfungi are therefore dependent on soluble nitrogen before they can decay wood, hence the form and distribution of nitrogen in wood probably exercises a controlling influence on their development. Thus the redistribution to evaporation surfaces may provide an explanation for the indeterminate behaviour of microfungi which colonize wood. On this basis, the behavioural differences in microfungal attack on wood may be interpreted as follows.

Wood in the green condition is rich in nutrients with a well-balanced soluble C:soluble N ratio, apparently fully adequate for rapid growth of microfungi. So long as soluble carbohydrates remain, there may be no stimulus for cellulase production (Bravery, 1968). By the time they are consumed, it is presumed that much, if not all, of the soluble nitrogen has also been consumed and the C:N ratio, considering the carbon of cellulose, is then far too high for production of cellulases. Thus the rapid colonization typical of microfungi in green wood (which if they are dermatiaceous results in staining) and the normal lack of soft rot is explained.

Several workers have shown that soluble nutrients in wood are apparently depleted on drying (Levy and Olofinboba, 1967). This has been attributed, in part, no doubt correctly, to respiration of living cells (Scheffer and Cowling, 1966). This depletion was considered by these authors and by Findlay (1966) to explain decreased susceptibility of wood to mould and staining fungi after drying. Johnson (1964), however, showed that starch accumulates at the ends of felled logs and, as stated earlier, King *et al.* (1974) have demonstrated migration of carbohydrates as well as nitrogen to evaporating surfaces of wood on drying. Thus dried wood is depleted of nutrients below the surface, which is perhaps part of the reason for the apparent loss on drying reported by earlier workers. The corresponding concentrations of nutrients at the evaporation surfaces is a general phenomenon which has hitherto escaped notice.

King *et al.* (1975) have correlated the depths to which soft rot cavities are formed in wood in soil contact with the presence of concentrated soluble nitrogen in those surfaces from which evaporation had occurred. They also showed that soluble nitrogen did not move into wood from the (nonfertilized) garden soil which they used. It would seem that the process of drying wood (whether artificial or natural) produces in outer layers a nutritional situation similar in terms of soluble C:N ratio to that in green wood, but with a much greater concentration of soluble nitrogen.

The result of this is to provide, when the wood is rewetted, an ecological situation similar to green wood, but to a very limited depth. This depth is easily colonizable by large numbers of microfungi, traditionally considered to colonize only green wood, but because of the highly concentrated nitrogenous materials, their nutritional requirements for soft rot production are also satisfied. Thus the frequent development of soft rot to a limited depth in wood which has been rewetted by soil contact is partially explained. If the exposed surfaces are those through which water evaporated when the wood was initially dried, soft rot production will be intense. If these surfaces have been removed, soft rot may be much delayed.

Much of the indeterminate behaviour of microfungi in wood, which results in their classification as ‘mould’, ‘staining’, ‘sapstain’ fungi which also cause ‘soft rot’, etc., may therefore be due less to differences between the organisms themselves than to the treatment of the wood after felling because of the changes, especially in soluble nutrient distribution and concentration, which have occurred during the drying process.

Acknowledgements

The authors are grateful to the Department of the Environment for a grant in support of this work and to the Forestry Commission for supplying the raw material used.

REFERENCES

Baker, J. M., Laidlaw, R. A., and Smith, G. A. (1970). Wood breakdown and nitrogen utilisation by *Anobium punctatum* Deg. feeding on Scots pine sapwood. *Holzforschung*, **24** (2), 46–54.

Banerjee, A. K., and Levy, J. F. (1971). Fungal succession in wooden fence posts. *Mat. u. Org.*, **6**, 1–25.

Bletchly, J. D., and Farmer, R. H. (1959). Some investigations into the susceptibility of Corsican and Scots pines and of European Oak to attack by the common furniture beetle, *Anobium punctatum* Deg. (*Col. Anobiidae*). *J. Inst. Wood Sci.*, **1** (3), 2–20.

Bravery, A. F. (1968). Microbiological breakdown of cellulose in the presence of alternative carbon sources. *J. Sci. Fd. Agr.*, **19**, 133–135.

Butcher, J. A. (1968). The ecology of fungi infecting untreated sapwood of *Pinus radiata*. *Can. J. Bot.*, **46**, 1577–1588.

Butcher, J. A., and Drysdale, J. (1974). Effect of nitrogen source and carbon: nitrogen ratio on cellulose activity and decay capability of certain soft rot fungi. *Mat. u. Org.*, **9**, 255–268.

Corbett, N. H. (1963). Anatomical, ecological and physiological studies on microfungi associated with decaying wood. Ph.D. Thesis. Univ. of London.

Courtois, H. (1963). Micromorphological symptoms of wood deterioration by soft rot fungi. D.S.I.R. Translation from *Holz Forschung u. Holz-Verwertung*, **15** (5), 88–101.

Cowling, E. B., and Merrill, W. (1966). Nitrogen in wood and its role in wood deterioration. *Can. J. Bot.*, **44**, 1539–1554.

Findlay, W. P. K. (1966). Ecology of wood-decaying and wood inhabiting fungi. *Mat. u. Org. Suppl.*, **1** (1), 199–211.

Gadd, O. (1965). Contribution to the knowledge of wood blueing. *Paperi Puu*, **47**, 667–669.

Garrett, S. D. (1963). *Soil Fungi and Soil Fertility*. Pergamon Press, London. 100 pp.

Gorschin, S. N., and Krapivina, I. G. (1969). The role of Ascomycetes and Imperfect fungi in effecting the degradation of wood. Trans. FSTC-HT-23-982-70. Defence Documentation Center. Virginia, ATTN. TSR 1. Trans. from *Mikologiya I Fitopatologiya*, **3** (5), 477–480.

Humphries, E. C. (1956). Mineral components and ash analysis. In *Modern Methods of Plant Analysis*, pp. 468–502 (K. Paech and M. V. Tracey, eds.), Springer Verlag, Berlin.

Johnson, N. E. (1964). Effects of different drying rates and two insecticides on beetle attacks in felled Douglas fir and Western hemlock. *Weyerhaeuser Company Forestry Research Note* (58) 16 pp.

King, B. (1975). Some ecological aspects of early microbial colonisation of wood. Ph.D. Thesis. Univ. of Aston in Birmingham. 250 pp.

King, B., Oxley, T. A., and Long, K. D. (1974). Soluble nitrogen in wood and its redistribution on drying. *Mat. u. Org.*, **9**, 241–254.

King, B., Oxley, T. A., and Long, K. D. (1975). Some biological effects of redistribution of soluble nutrients during drying of wood. *Mat. u. Org.* (in press).

Krapivina, I. G. (1960). Destruction of the secondary layer of the cell wall by blue staining fungi. Commonwealth Scientific and Industrial Research Organisation, Australia. Trans. No. 5329, Trans. from *Lesn. Zh.*, **3** (1), 130–133.

Levi, M. P., and Cowling, E. B. (1969). Role of nitrogen in wood deterioration. vii. Physiological adaptation of wood-destroying and other fungi to substrates deficient in nitrogen. *Phytopathology*, **59**, 460–468.

Levy, J. F. (1968). Studies of the ecology of fungi in wooden fence posts. In *Biodeterioration of Materials*, pp. 424–428 (A. H. Walters and J. J. Elphick, eds.), Applied Science Publishers, London.

Levy, J. F. (1969). The spectrum of inter-action between fungi and wood. *Rec. Ann. Conv. Brit. Wood Pres. Assoc.*, 1969, 71–82.

Levy, J. F., and Olofinboba, M. (1967). Blue stain of *Antiaris africana* and *Pycnanthus angolensis*. *Brit. Wood Pres. Ass. News Sheet*, No. 80.

Liese, W. (1970a). Ultrastructural aspects of woody tissue disintegration. *Ann. Rev. Phytopathol.*, **8**, 231–258.

Liese, W. (1970b). The action of fungi and bacteria during wood deterioration. *Rec. Ann. Conv. Brit. Wood Pres. Ass.*, 1970.

Merrill, W. (1966). Decay of wood and wood fibre boards by fungi imperfecti. *Mat. u. Org., Suppl.*, **1** (1), 69–76.

Merrill, W., and Cowling, E. B. (1966). Role of nitrogen in wood deterioration: amounts and distribution of nitrogen in tree stems. *Can. J. Bot.*, **44**, 1555–1580.

Merrill, W., and French, D. W. (1966). Colonisation of wood by soil fungi. *Phytopathology*, **56** (3), 301–303.

Savory, J. G. (1954). Breakdown of timber by ascomycetes and fungi imperfecti. *Ann. Appl. Biol.*, **41**, 336–347.

Savory, J. G. (1968). Microbial attack of timber and allied constructional materials. In *Biodeterioration of Materials*, pp. 403–407 (A. H. Walters and J. J. Elphick, eds.), Applied Science Publishers, London.

Scheffer, T. C. (1973). Microbiological degradation and the causal organisms. In *Wood deterioration and its prevention by preservation treatments.* 1. *Degradation and protection of wood*, pp. 31–106 (D. H. Nicholas, ed.), Syracuse Univ. Press, New York.

Scheffer, T. C., and Cowling, E. B. (1966). Natural resistance of wood to microbial deterioration. *Ann. Rev. Phytopathol.*, **4**, 147–170.

Tappi (1954). Official Standard T13 m-54. Lignin in wood. 1945. Corrected 1954.

FUTURE PROSPECTS FOR IMMOBILIZED ENZYMES IN BIODEGRADATION

RICHARD D. FALB

Battelle, Columbus Laboratories, 505 King Avenue,
Columbus, Ohio 43201, U.S.A.

Immobilization of enzymes continues to be a rapidly expanding area of scientific research. The number of scientific papers published on immobilized enzymes continues to grow year by year, and judging from the large number of symposia occurring each year, the interest in immobilized enzymes continues to be very active. In the last few years, a number of industrial processes based on immobilized enzymes has been developed and is now on stream. These include hydrolysis of starch, conversion of glucose to fructose, hydrolysis of lactose, and resolution of optical isomers of amino acids. However, the development of commercial processes is slower than many early workers in the field originally anticipated and this possibly has occurred because of some of the inherent imitations which are attendant to immobilized enzymes. One of the most important controlling factors in utilization of immobilized enzymes is economics. This factor, as well as several others, will be considered in this paper in an attempt to project the future use of enzymes in the field of biodegradation.

In the context of enzyme immobilization, the term 'biodegradation' is somewhat difficult to define. When an enzyme is placed on a solid carrier such as glass and used in a column, it bears more resemblance to a classical heterogeneous catalyst than to a 'biological process'. In order to avoid a review of the complete field of enzyme immobilization, we will consider here the present and future uses of enzymes which break down a constituent of a process stream or a constituent of wastewater or air.

BACKGROUND

The common configurations in which enzymes are immobilized are summarized as follows: (1) covalent bonding to a solid phase; (2) covalent bonding to soluble polymers; (3) physical absorption to a solid phase; (4) cross-linking at solid surfaces; (5) cross-linking with difunctional reagents; (6) inclusion in a gel phase; and (7) encapsulation.

It should be noted at the onset that the above procedures can be utilized with purified enzymes, crude enzymes, mixtures of several enzymes, subcellular particles, and even whole cells themselves. The largest amount of work on enzyme immobilization has been done utilizing covalent methods in which the enzyme is attached by a covalent bond to the matrix. A very large number of synthetic, natural polymers, and inorganic materials have been used for the immobilizing phase (Zaborsky, 1973). In general, the enzyme is attached by utilizing a reactive group on the immobilizing phase in the reaction with a functional group of the enzyme, such as carboxyl, amino, hydroxyl, sulphhydryl, or hydroxyphenyl. Attachment of the enzyme must be done in such a way so as to preserve the activity of the enzyme and generally obtain a high loading of active enzyme which results in a high activity per unit volume. As stated previously, economics is a prime consideration in determining the value of any immobilization procedure. In addition to retaining activity of the enzyme during immobilization, one also endeavours to attach most of the enzyme in the solution during the attachment process.

Another important economic consideration is the cost of the support material. At present, a number of relatively inexpensive materials such as silica, metal oxides, phenolic resins, cellulose, and many low-cost synthetic polymers can conveniently be used as enzyme supports. The attachment procedure should ideally be straightforward and simple and not involve multiple steps.

Immobilization of enzymes offers several advantages over the use of soluble enzymes, and one of the chief advantages is the variety of physical forms which immobilized enzymes can take. Incorporation of an enzyme on a solid enables the facile separation of the enzyme from the reaction mixture. Enzymes attached to particles can be used in continuous column or stirred tank reactors and also in batch processes. In addition, enzymes may be attached to the wall of reactor vessels on tubing or plates and interact with the solution as it flows through the reactor.

Another configuration of immobilized enzymes which has been used is the attachment of the enzyme to the walls of hollow-fibre reactors. In cases where contact of the enzyme with substrates causes a potential problem, enzymes can be attached to soluble polymers and separated from the reaction mixture by ultrafiltration devices (Wykes *et al.*, 1971). Enzymes, enzyme mixtures, and unicellular organisms can be incorporated in extruded fibres to produce adducts which have very high specific activity (Dinelli and Morisi, 1974). These authors report on a series of 39 different enzymes which have been immobilized by fibre entrapment and indicate that most of these have very high stability.

A novel configuration which permits easy separation of the immobilized enzyme from the reaction mixture has been developed by Robinson *et al.* (1973). In this process, the enzymes were attached to the outside of iron oxide powder and could be removed from the reaction mixture by a magnet. These particles could also be readily coated on the interior surface of a vessel or a conduit. The utilization of immobilized enzymes is not limited to the aqueous phase, and several examples of immobilized enzymes used in nonaqueous solvents have been demonstrated.

The incorporation of enzymes into microcapsules is another convenient form of immobilized enzymes. The use of microcapsules has particular value when one wishes to protect the enzyme from the surrounding milieu. The use of this configuration is limited to cases where the substrate can diffuse into the capsule and the product can diffuse out. Capsule walls can consist of any number of polymeric materials as well as liquid membranes (May and Li, 1974). One of the most novel techniques of enzyme immobilization is illustrated by the work of Emery (1972), who attached enyzmes to the surface of various microorganisms such as brewer's yeast, *B. subtilis*, and *E. coli*. In this way, enzymes not normally produced by these organisms could be supplied to them for various degradative processes. In some cases, it may be more practical and economically feasible to stabilize intact microorganisms or organelles. Several examples of this exist, and among these are the immobilization of *E. coli* for the production of L-aspartic acid (Chibata *et al.*, 1974). The immobilization of multienzyme complexes, membrane fragments, and cellular organelles such as chloroplasts is also possible and might be used to some advantage for biodegradation processes.

An application which illustrates the versatility of immobilized enzymes for biodegradation is provided by the work of Kirwan *et al.* (1974). Trypsin and ribonuclease have been attached to silica or glass wool to produce a product which can be used to remove airborne virus particles. Presumably, the virus particles are entrapped by the glass wool and are inactivated through lysis by the surface-attached enzymes.

LIMITATIONS OF IMMOBILIZED ENZYMES

To assess realistically the future potential for immobilized enzymes in biodegradation, one must consider several potential limitations. Since immobilized enzymes can be separated readily from the product stream, they can be repeatedly reused provided they are stable. The continued reuse of the enzymes can thus compensate for the cost involved in the immobilization process if they have sufficient stability. Quite often immobilization does result in enhanced stability, but this is certainly not always the case. The stability of immobilized enzyme processes for potential use in biodegradation processes of waste effluents may present a serious limitation. If immobilized enzymes were placed in direct contact with the waste effluent, bacterial contamination could result which would inactivate the enzyme. In addition, trace amounts of heavy metal salts could be bound by the enzymes and over a period of time accumulate such sufficient quantities as to render the enzyme inactive.

Another characteristic of immobilized enzymes which may limit the potential usefulness in biodegradation processes is the relative inefficiency with which they interact with high molecular weight or insoluble substrates. In the simplest sense, the situation can be reviewed as a logistics problem in which the active site of the enzyme has difficulty in binding to the appropriate bond in the substrate particle. For example, in the degradation of cellulose, the enzyme must bind intimately with the cellulose chain in order to effect cleavage. If the enzyme is attached to an insoluble particle, this intimate binding is very difficult to effect. This difficulty can be overcome partially by attaching the enzyme to a water-soluble polymer and retaining the adduct by means of an ultrafiltration membrane. An interesting example of 'immobilization' of cellulose is provided by the work of Ghose and Kostick (1970). These authors consider the formation of the enzyme adsorbed in soluble cellulose as an immobilized enzyme. As the cellulose is hydrolysed, the enzyme is released from the products and subsequently binds to another cellulose particle.

An additional possible limitation on the use of immobilized enzymes for biodegradation may be imposed by the fact that many degradation processes involve multiple steps. For example, if one wished to break down hydrocarbons or degrade alkyl-based detergents present in wastewater, a multistep process would be involved. To provide the enzymes necessary for such a sequence would be prohibitively expensive, certainly not practical in terms of present-day technology. In addition, one must mention here a further limitation of immobilization enzyme technology, which is the availability of coenzymes. A very large number of enzymes requires cofactors in their reactions, and a practical way of providing these has not yet been found. Until this problem is solved, a number of potentially useful biodegradative enzymes cannot be considered commercially feasible. With respect to the problem of multiple reaction and cofactors, the most practical approach could well involve the immobilization of intact organisms. In this case, the enzyme sequences as well as the cofactors could be contained within the microorganisms.

FUTURE PROSPECTS FOR BIODEGRADATION

A great many hydrolytic enzymes have been immobilized by a host of different techniques and many of these could have future potential in biodegradation processes (Zaborsky, 1973). For example, proteolytic enzymes which have been immobilized are aminopeptidase, bromelain, carboxypeptidase, cathepsin-c, chymotrypsin, pancreatin, pronase, papain, pepsin, plasmin, protease, rennin, subtilisin, thrombin, thermolysin, and trypsin. Hydrolytic enzymes acting on carbohydrates immobilized in various ways are amylase, amyloglucosidase, cellulase, dextranase, chitinase, polygalacturonase, fructofuranosidase, galactosidase, saccharase, and xylanase. In addition to this, other hydrolytic enzymes such as ribonuclease, lysozyme, and tanase have been immobilized by a variety of procedures. With the present state of knowledge in enzyme immobilization techniques, it is safe to say that almost any enzyme can be successfully immobilized. Thus if one identifies an enzyme which may have a particular use in a biodegradation process, the future technical problem will not be the immobilization as much as the application of the immobilized enzyme in an economical fashion. The constraints on immobilized enzymes listed above such as stability, coenzyme requirements, and difficulty in multistep operation remain yet to be solved. It is the opinion of this author that one of the most promising areas for immobilized enzymes for biodegradation will be in the immobilization of intact organisms. The next few years should yield considerable advances in this area and perhaps some of these advances will be of real advantage in biodegradation applications.

REFERENCES

Chibata, I., Tosa, T., Sato, T., Mori, T., and Yamamoto, K. (1974). *Enzyme Engineering*, 2nd ed. (E. K. Pye and L. B. Wingard, eds.), Plenum Press, New York.

Dinelli, D., and Morisi, F. (1974). *Enzyme Engineering*, 2nd ed., p. 293 (E. K. Pye and L. B. Wingard, eds.), Plenum Press, New York.
Emery, A. N. (1972). *Chem. Engr*, **71**.
Ghose, T. K., and Kostick, J. A. (1970). *Biotech. Bioeng.*, **12**, 921.
Kirwan, D. J., Enright, J. T., and Gainer, J. L. (1974). *Biotech. Bioeng.*, **16**, 551.
May, S. W., and Li, N. N. (1974). *Enzyme Engineering*, Vol. 2, p. 77 (E. K. Pye and L. B. Wingard, eds.), Plenum Press, New York.
Robinson, P. J., Dunhill, P., and Lilly, M. D. (1973). *Biotech. Bioeng.*, **15**, 603.
Wykes, J. R., Dunhill, P., and Lilly, M. D. (1971). *Biochim. Biophys. Acta*, **250**, 522.
Zaborsky, O. (1973). *Immobilized Enzymes*. CRC Press, Cleveland, Ohio.

PAPER MILL EFFLUENT CLARIFIED WITH IMMOBILIZED ALPHA-AMYLASE

K. L. SMILEY, B. T. HOFREITER, J. A. BOUNDY and S. P. ROGOVIN

Northern Regional Research Laboratory, Agricultural Research Service, U.S. Department of Agriculture,* Peoria, Illinois 61604, U.S.A.

INTRODUCTION

National pollution abatement efforts, particularly those reducing suspended solids in industrial effluents discharged to receiving streams, are of concern to manufacturers of fine-grade papers. Wastewater or 'white water' from machines making pigment-filled papers contains inorganic fillers and small fibre particles in colloidal suspension that are highly resistant to flocculation and sedimentation in primary clarifiers by such usual methods as treatment with either alum or polyelectrolytes. It is recognized by the industry that these intractable suspensions are stabilized to varying degrees by starch and its derivatives (Davis, 1964; Herrick, 1966; Rebhun *et al.*, 1967), which are employed ubiquitously in papermaking.

Wilhelm (1968) showed that hypochlorite-oxidized starches are 10 times more effective as filler dispersants than are other starches commonly used in papermaking, such as hydroxyethylated, acetylated, and enzyme-hydrolysed starch. He attributed enhanced activity of oxidized starch to the presence of carboxyl groups. He also suggested that mill effluents could be treated with amylolytic enzymes to alleviate flocculation difficulties since his experiments had shown that the dispersive powers of starch were directly related to its molecular weight.

In a discussion of possible solutions to the problem of the dispersive power of papermaking starches, Gillespie *et al.* (1970) considered the substitution of less dispersive cationic starches; process modifications, such as broke repulping separate from the principal furnish preparation, and careful selection of coating and sizing adhesives to minimize presence of highly active dispersant types. However, they concluded that universal application of these alternative procedures probably cannot be achieved. They pointed out that secondary treatments provide starch turbidity removal but that these are costly when used for only aesthetic improvement in areas where such degrees of clarification are not required by standards.

Subsequently, two reports (Schwonke and Davis, 1973; Wu, 1974) appeared in the literature describing the use of amylolytic enzymes to treat paper mill effluents containing various types of starch. One of these (Schwonke and Davis, 1973) described the treatment of approximately 3 million gal of effluent/day from a fine paper mill with 150 lb (6-ppm level) of a soluble amylolytic enzyme used in conjunction with alum flocculant. The other (Wu,

* The mention of firm names or trade products does not imply that they are endorsed or recommended by the Department of Agriculture over other firms or similar products not mentioned.

1974) was a laboratory study on effluents from book paper manufacture that showed the cost advantages of enzyme treatments, which reduced the amount of alum flocculant required. However, disadvantages of the soluble enzyme treatment are, first, loss of expensive active enzyme and, second, loss in efficiency of BOD_5-removal in primary clarification because of solubilization of starch that otherwise might be coagulated and removed along with the fibre fines and filler. Furthermore, soluble enzymes in recycled water could reduce the effectiveness of starches added to the pulp furnish to enhance inter-fibre bonding.

Developments in the technology of bound or immobilized enzymes in the early 1970s invited consideration of these systems for treatment of starch-stabilized colloids (Smiley *et al.*, 1974). Laboratory experiments by Smiley *et al.* (1974) showed that α-amylase, immobilized on several carriers, disrupted colloidal suspensions of a simulated paper machine effluent. The practicality of immobilized enzyme systems for treating paper mill wastewater by larger-scale testing was undertaken using a laboratory-prepared effluent. In this work, the enzyme carrier is a protein-adsorbant resin.

Initially, white water was treated batchwise with the enzyme-resin complex to determine the general relationships of flocculation with alum to effluent volume, contact time, and quantities of alum. Later, white water was treated with the bound enzyme in a continuous reactor to provide sufficient quantities of treated effluents on which settling path curves could be determined. From such curves, conclusions could be drawn relative to the efficacy of the enzyme treatment in terms of percentage of solids removed, overflow rates, and retention times that can be expected in primary clarifiers. Finally, stability of the bound enzyme was established during continuous reaction with effluents for 4 days.

EXPERIMENTAL

Materials

Enzyme

α-Amylase (1,4-α-D-glucan 4-glucanohydrolase; EC 3.2.1.1) 4x crystallised from *Bacillus subtilis* was purchased from Sigma Chemical Co., St Louis, Mo.

Thermamyl 120, a thermotolerant α-amylase, was furnished by Nova Enzyme Corp., Mamaroneck, N.Y.

Carrier resin

Duolite DS-73141 came from Diamond Shamrock Co., Redwood City, Calif.

Oxidized starch

Hypochloride-oxidized corn starch, Amaize-540, was provided by the American Maize Products Co., Roby, Ind. Intrinsic viscosity values were 0·55 and 0·48 dl/g in 90% aqueous dimethyl sulphoxide and N KOH solvents, respectively. Carboxyl content was 17·0 mmol/100 g, determined by the method of Mattisson and Legendre (1952).

Rosin size

Accobrite rosin is a product of the American Cyanamid Co., New York, N.Y.

Clay filler

A kaolin-type Georgia clay filler, Hi-White, is manufactured by the Huber Corp., Huber, Ga. Particle size, diameter in microns, is distributed 65·4, 18·8, 9·2, and 6·6% for the ranges 0–2, 2–5, 5–10, and above 10 μ, respectively.

Alum

A 1% alum solution was prepared from Fisher certified reagent [$Al_2(SO_4)_3 \cdot 18H_2O$].

Pulp

Pulp was obtained by mechanical disintegration in water of a filled bond paper that had been prepared at the Northern Laboratory on a pilot paper machine. This paper stock, containing 15% kaolin clay by weight, had been made from a furnish composed of equal parts of bleached hardwood and softwood kraft pulps and having the following percentages of additives (based on dry pulp weight): 20% kaolin clay, 0·4% rosin size, 2% alum, and 0·05% Reten 210 (a cationic retention aid manufactured by Hercules Inc., Wilmington, Del.).

Equipment

Settling column

The column used for evaluation of flocculant settling was identical to that described by Eckenfelder (1966). It is an 8-ft × 8-in lucite column with a 5·5-in diameter. The column capacity is about 12 gal.

Enzyme-resin reactors

A water-jacketed, double-sidearm spinner flask of 1-litre capacity (Bellco Glass Inc., Vineland, N.J.) served as a small-scale continuous reactor. The substrate (white water) was fed in one sidearm, and the liquid was maintained at a predetermined level by a vacuum filter tube that continuously removed through the other sidearm all fluid above the height selected. Temperature (40°C) was controlled by circulation of tempered water through the outer jacket.

A larger-scale continuous reactor was fabricated from 24 × 6-in plexiglass pipe sealed at one end and having a $\frac{1}{2}$-in inlet near the bottom and a $\frac{1}{2}$-in outlet 2 in from the top. Its working capacity was 8 litres. A 1 × 4-in stainless steel paddle provided agitation. The temperature was maintained at 40°C by a thermostatically controlled 1kW immersion heater.

Spectrophotometer

Turbidities of enzyme-treated and untreated white water and optical densities of the starch-iodine complexes were determined in a Beckman Model B spectrophotometer. Optically matched tubes of 1·5-cm inside diameter were used.

Procedures

Preparation of α-amylase-resin complex

The enzyme-binding procedure was a modification of that described by Olson and Stanley (1973). In 1 litre of 0·1 M acetate buffer, pH 5·0, 200 g Duolite DS 73141 (65% moisture) was suspended and allowed to soak overnight. The supernatant liquid was drained; 400 ml of 0·1 M acetate buffer, pH 5·0, containing 910 mg of crystalline α-amylase was added; and the suspension was stirred gently on a rotary shaker for 3 hr. The resin was allowed to settle and drain. A negative test for protein (Lowry *et al.*, 1951) indicated that all the amylase had been absorbed from the solution. In the same buffer 400 ml of 1·25% glutaraldehyde was added to the resin and reacted for 90 min at room temperature on a rotary shaker. The cross-linked enzyme-resin complex was separated by decanting excess glutaraldehyde and was washed by stirring with 1 litre of pH 5·0 acetate buffer, followed by three 1-litre washes of 0·1 M acetate buffer, pH 6·0, and then by 8–10 washes of 0·1% soluble starch in the pH 6·0 buffer. The enzyme-resin complex was stored under 0·1 M acetate buffer, pH 6·0.

Measurement of activity of α-amylase-resin complex

A suitable portion (7 g, d.b.) of the α-amylase-resin complex was added to 1 litre of 0·01% Amaize starch at pH 5·0 in a 2800-ml Fernbach flask. The mixture was agitated at 40°C in a rotary shaker water bath, and assayed periodically for apparent maltose by method C of Robyt *et al.* (1972). The reaction rate at this low starch concentration was 1·07 μm/min/g resin. Significantly higher activity at higher starch concentrations indicated substrate limitation at the 0·01% level. Even so, the activity in dilute starch more nearly represents the capacity of the bound enzyme under conditions prevailing in white water.

Preparation of simulated white water

Large batches of simulated white water were prepared: to 100 gal of 0·13% pulp slurry in tap water was added sequentially with stirring 0·4% Accobrite rosin size, 4·85% alum, 10% kaolin clay, sulphuric acid to pH 5·5, and 3·33% Amaize-50 oxystarch (percentages based on dry pulp weight). The slurry remained at room temperature for 72 hr. The supernatant liquid (*ca.* 65 gal), which was withdrawn with a siphon, had a suspended solids (SS) content of 70 mg/litre or 0·58 lb/1000 gal.

Batch treatments with immobilized enzymes

A portion of drained amylase-resin complex (7·0 g, d.b.) was added to 1 litre of simulated white water (pH 6·0, 40°C) in a 2800-ml Fernbach flask. The flask was gently shaken on a rotary water bath. At 10 min, 10 ml was removed, and at 20 min all the resin was removed by filtration through a coarse sintered glass filter. The sample and the filtrate were assayed for apparent maltose. After the resin complex was washed with water, it was returned to the Fernbach flask, and the procedure was repeated. Under these conditions, the production of apparent maltose was about 0·3 μm/min/g of dry resin. Filtrates from each of four consecutive runs in which contact time differed were examined for flocculation characteristics.

Continuous treatments with immobilized enzymes

In our 8-litre reactor, 56 g (d.b.) of the enzyme-resin complex was stirred in 8 litres of simulated white water at 40°C, pH 6·5, for 30 min. The stirring rate was sufficient to cause the resin complex to rise to three-quarters of the plexiglass cylinder height. White water was then fed into the bottom of the cylinder at a flow rate of 8 litres/hr. The effluent was directed into a 2-litre catch basin to trap escaping enzyme resin before being collected in a 50-gal stainless steel container. Settling characteristics of the effluent were studied in the 12-gal settling column.

Stability of the α-amylase-resin complex in the presence of white water was tested in a 1-litre stirred tank reactor. A single white water preparation, pH 6·0, was fed into the reactor containing 7·0 g (d.b.) of the resin complex. Flow rate was 1 litre/hr, and temperature was maintained at 40°C. A filter on the effluent outlet prevented the escape of resin particles. Samples of the effluent were collected periodically for evaluation of settling characteristics.

Total suspended solids

An accurate weight (*ca.* 1 g) of diatomaceous earth filter aid was added to 1 litre of simulated white water. The white water was then filtered through a Buchner funnel containing a Whatman No. 50 filter paper preweighed on an oven balance at 95°C. The filter paper and collected solids were dried at 105°C for $\frac{1}{2}$ hr and then weighed in the oven balance. Correction was made for the filter aid.

Flocculation and settling tests

Both test tube-scale and 12-gal settling columns were employed to test the effects of enzyme treatment on the clarification of simulated white water.

The extent of flocculation, as a function of contact time with bound α-amylase and level of alum, was evaluated in test tubes as follows: to 80 ml of white water in 150-ml beakers was added the desired amount of alum (1% solution) and the mixture was adjusted to pH 6·5 with NaOH. These mixtures were then poured into 25 × 200 mm test tubes. After 1 hr, 10-ml samples were carefully pipetted from the tops of the test tubes, and their turbidity was measured as optical density at 600 nm. The amount of SS in white water was directly proportional to optical density in the SS range studied.

A larger-scale evaluation of enzyme-treated white water was made in the 12-gal column at ambient temperature (approximately 23°C). White water (12 gal) at pH 6·5 was pumped into the column. Samples of approximately 15 ml each were withdrawn from the 2-, 4-, and 6-ft depths at various times, and optical densities were measured at 600 nm. Plots were made of the percentage SS removal $[(\mathrm{O.D.}_{\mathrm{initial}} - \mathrm{O.D.}_{t})/(\mathrm{O.D.}_{\mathrm{initial}}) \times 100]$ versus settling time (t). Data from these plots were used to construct settling path curves, which show extent of solids removal as a function of settling time and depth.

RESULTS AND DISCUSSION

Our primary purpose was to set up a model system for treating paper machine white water with immobilized α-amylase and to collect sufficient data for technical evaluation for its application to industrial paper mill effluents. In the following sections, we discuss white water characteristics, α-amylase immobilization on a protein-adsorbent resin, design of a fluid-bed-type reactor, evaluation of enzyme treatments for clarification of simulated white water, and extrapolation of results to mill-scale operations.

Simulated white water

Characteristics of white water differ considerably for a given grade of paper because furnish components vary throughout the industry, as well as efficiency of solids recovery before discharge. For example, a survey of 13 fine paper mills disclosed a range of SS from 53 to 1326 ppm (Gehm, 1973, p. 157). While some investigators of the dispersive effects of starches have used clay suspensions to simulate white water, others have used actual effluents from paper mills. Our preparations of simulated white water have total solids content of 70 ppm and are representative of difficult-to-flocculate components. In addition, it contains fibre fines normally present in paper-mill waste streams.

Marton (1974) estimated the average starch concentration in effluents from fine paper mills to be 0·015%. We selected a level of 0·01%, according to the following generalizations: the average water usage in a fine paper mill is about 32 000 gal/ton of product (Gehm, 1973, p. 157). Assuming that starch is added to the paper furnish at a level of 2% (dry pulp basis) and is only 50% retained, there would be 20 lb starch/32 000 gal or concentration of 0·007% in the effluent.

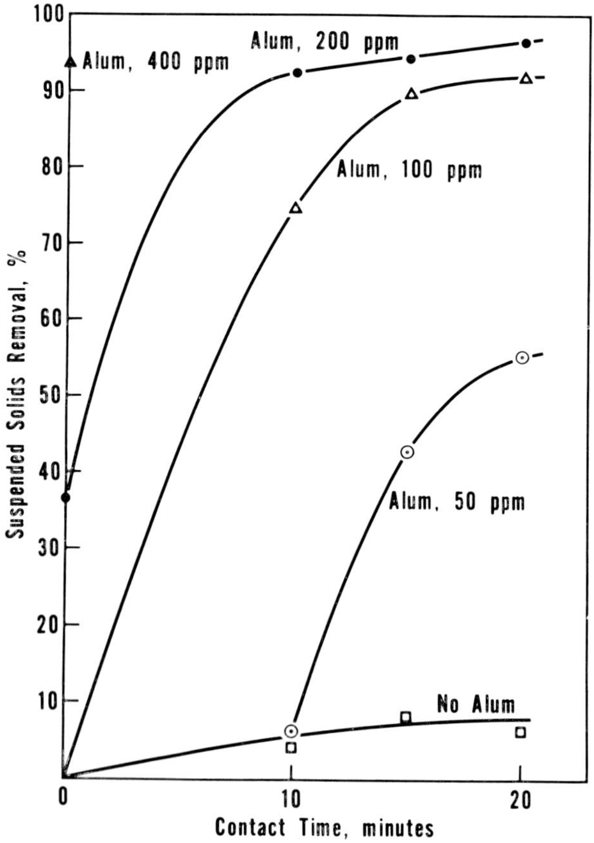

Fig. 1. Percentage of solids removed from 1000-ml white water by alum flocculation after batchwise treatment at various times with 7 g (dry) alpha-amylase-phenolformaldehyde resin complex.

Preliminary batch treatments

To determine the enzyme-white water contact times and levels of alum required for effective flocculation of SS, 1-litre portions of white water were treated with suitable quantities of the enzyme-resin complex. Flocculation was considered to be effective if 88% of the solids were removed in 1 hr following the addition of alum. This percentage was chosen because it is an average degree of clarification in fine paper mill effluents (Gehm, 1973, p. 39). For these effluents, however, the retention time in primary clarifiers is 3–4 hr.

There is a trade off between amounts of alum required to achieve 88% solids removal and the contact time with immobilized α-amylase (Fig. 1). It is desirable to keep the amount of alum low not only because of cost but also to minimize buildup of dissolved solids in recycled water. Such buildup has adverse effects on the papermaking process. Note particularly that 400 ppm of alum is required to flocculate effectively white water having no enzyme treatment.

Evidently, a flow rate of 1·13 gal white water/hr/7 g enzyme-resin complex will remove 88% of the SS after settling for 1 hr with 100 ppm alum. The initial flow rate used in the 8-litre continuous enzyme reactor was based on this computation.

Continuous reactor

The flow rate of white water through the 8-litre reactor, which contained 56 g dry enzyme-resin complex, was set at 2·11 gal/hr or 17·1 gal/hr/lb of dry resin. Settling path curves were determined for 50 and 75 ppm of alum in the effluent (Figs. 2 and 3).

The settling paths show the decrease in SS in time throughout 6 ft of the settling column, at alum levels of 50 ppm (Fig. 2) and 75 ppm (Fig. 3). There

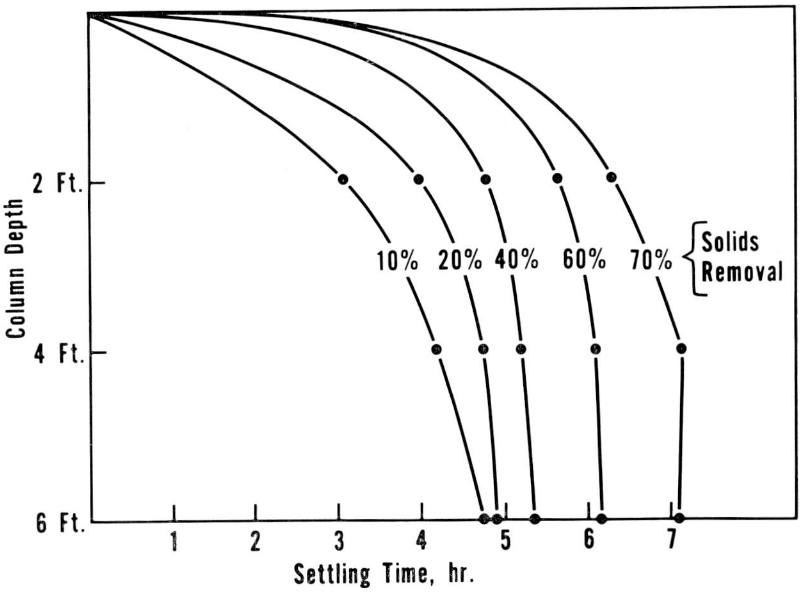

Fig. 2. Settling path curves of white water flocculated with 50 ppm alum at 23°C. White water treated in 8-litre continuous reactor with flow rate of 8-litre/hr at 40°C and pH 6·5.

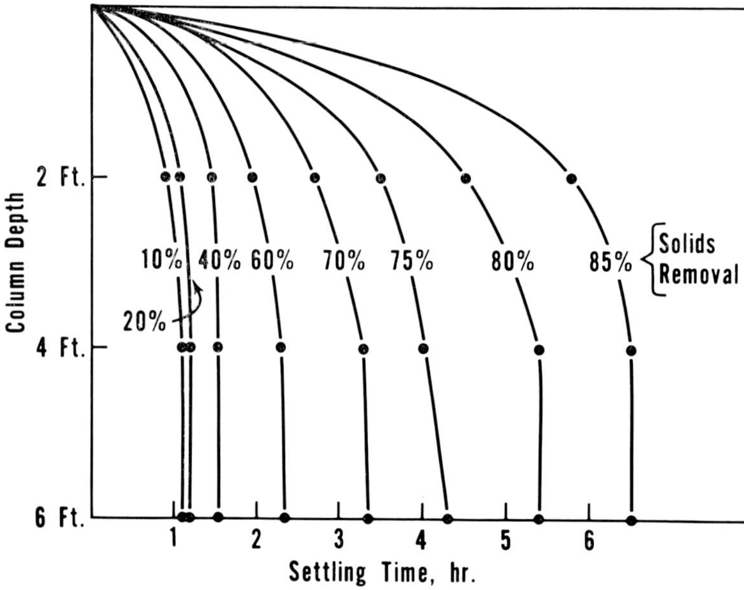

Fig. 3. *Settling path curves of white water flocculated with 75 ppm alum at 23°C. Conditions same as for Fig. 2.*

is a significant decrease in settling time when the alum level is increased from 50 to 75 ppm.

On the basis of the data in these figures, the percentage SS removal in the 6-ft column was calculated (Eckenfelder, 1966) for several retention times (Table I). These calculations involve summation of the percentage SS eliminated at the 6-ft depth and in each additional 10% SS removal-range where significant amounts of solids are flocculated. Overflow rates (gal/day/ft^2 of clarifier surface) are also given in Table I. The results show that increasing alum dosage from 50 to 75 ppm increases overflow rate by a factor of 2·6 for 64–65% SS removal. Although higher levels of clarification were achieved with 75 ppm alum, overflow rates were reduced. SS removal can also be

Table I

SETTLING DATA FOR SIMULATED PAPER-MILL EFFLUENT FROM AN 8-LITRE IMMOBILIZED ENZYME REACTOR[a]

Alum (ppm)	*Total removal of suspended solids*[b] *(%)*	*Retention time (hr)*	*Overflow rate (gpd/ft^2)*[c]
50	64	6·15	175
75	65	2·35	459
75	73	3·35	322
75	78	4·30	250

[a] Flow rate 8-litre/hr at 40°C and pH 6·5.
[b] Values calculated according to Eckenfelder (1966) represent the total percentage solids removed throughout a 6-ft depth of a settling column.
[c] Gallons/day/ft^2 of clarifier surface.

enhanced by increased reaction time with the bound enzyme (Fig. 1). The untreated white water was exceedingly stable and did not settle in 18 hr.

The stability of the enzyme-resin complex in continuous operation with white water was evaluated over a period of 98 hr. Previously, Smiley *et al.* (1974) used an enzyme-resin complex for 950 hr without loss of activity against a Lintner starch solution. With white water, however, it was anticipated that problems could arise from either abrasive action of clay particles or fouling of the resin surface with fibre fines. Two measurements were made of enzyme efficacy during the continuous operation: evaluation of SS removal over a range of alum levels on a test tube scale, as described under Flocculation and Settling Tests, and the decrease in optical density of the starch-iodine complex in the white water.

After 4 days' operation, SS removal ranged from 77 to 83% with 100 ppm of alum. Optical density of the iodine complex (at 600 nm) of the treated samples was about 0·10 compared with 0·68 for untreated white water. Data for stability evaluation are shown in Table II. After 168 hr, the level of alum

Table II

CONTINUOUS[a] TREATMENT OF SIMULATED PAPER-MILL EFFLUENT BY IMMOBILIZED ALPHA-AMYLASE

Running time (hr)	*Starch-iodine colour,[b] (O.D.)*		*Suspended solids,[c] (%)*	
	Untreated control	*Enzyme treated*	*Untreated control (150 ppm alum)*	*Enzyme treated (100 ppm alum)*
0			100	100
2			97·3	18·9
26	0·65	0·125	91·9	21·6
44	—	0·115	—	32·4
68	0·70	0·120	67·6	—
98	0·68	0·100	59·5	16·2
168	0·24		10·8	13·5

[a] An 8-litre reactor with a white water flow-through rate of 8 litre/hr at 40°C and pH 6·5.
[b] Samples clarified by centrifugation and 1·0 ml of 0·005 N I_2 added to 9-ml supernatant; colour read immediately at 600 nm.
[c] Determined 1 hr after addition of alum.

required to reduce SS in untreated white water to 80% after 1 hr settling shifted from 200 to 112 ppm, a decrease that indicates deterioration of the white water. Further evidence of starch degradation in untreated white water was the change in optical density of the starch-iodine complex from 0·68 to 0·24. The loss of colloidal starch may be due either to microbial attack or to its irreversible adsorption onto fibre fines. Such deterioration in the dispersive power of starches has been observed with an unspecified type of starch in book paper effluent (Wu, 1974) and with a pregelatinized starch (Rebhun *et al.*, 1967).

SUMMARY AND CONCLUSIONS

Immobilized α-amylase has been effective in laboratory trials for the continuous treatment of a white water typical of that from a paper mill making fine grades of paper. Approximately 78% of the SS was removed from enzyme-treated white water after settling 4·3 hr following treatment with 75 ppm alum. Because the dispersive power of starch in simulated white water deteriorated after 4 days' running time in a continuous enzyme reactor, higher levels of alum were not tried. However, additional alum in combination with increased reaction time with the bound enzyme would allow for clarification to 88%.

Technically, the system described is feasible for treating industrial effluents. We estimate that about 17 gal of white water, containing 0·01% oxystarch, could be treated in 1 hr by a pound of dry enzyme-resin complex to render the effluent susceptible to flocculation with nominal levels of alum. Assuming proportional enzyme activity when operating on a larger scale, approximately 2000 lb of enzyme-resin would be needed to treat 1 million gallons of paper mill effluent/day. Although a continuous fluid-bed-type reactor was used in this work, batch treatments, requiring little in the way of special equipment, should also be feasible for full-scale operation.

Industrial use has been reported (Schwonke and Davis, 1973) of soluble α-amylase to treat paper mill effluents at a level of 6 ppm or 50 lb of enzyme (dry) for a million gallons of effluent. The economic feasibility of superseding such a treatment with bound enzymes depends upon (1) amount of capital investment in a treatment facility; (2) costs of enzyme-resin preparation; (3) stability of the enzyme-resin complex; and (4) reuse of the resin carrier after enzyme activity has decreased. Research in progress at the Northern Laboratory is addressed to answering these questions.

Acknowledgement

We thank R. G. Fecht for his valuable assistance in preparing simulated white water and in performing flocculation and settling tests.

REFERENCES

Anon. (1971–72). NCASI-WMU research on enzymes to clarify water. *Paper Technol. News*, **16** (1). Paper Technology Foundation, Western Michigan University, Kalamazoo.

Davis, W. S., Jr. (1964). A laboratory method for comparing the effects of starches in white water. *Tappi*, **47** (8), 129A–130A.

Eckenfelder, W. W., Jr. (1966). *Industrial Water Pollution Control*, pp. 32–34, McGraw-Hill, New York.

Gehm, H. (1973). State-of-the-art review of pulp and paper waste treatment. U.S. Dept. Commer., NTIS, PB-221-434, Washington, D.C.

Gillespie, W. J., Mazzola, C. A., and Marshall, D. W. (1970). Challenge remains: Starch-related pollution problems. *Pulp Pap.*, **44** (3), 120–122.

Herrick, R. (1966). Starch effects on turbidity upheld. *Pulp Pap.*, **40** (17), 38.

Lowry, O. H., Rosebrough, N. J., Farr, A. L., and Randall, R. J. (1951). Protein measurement with the folin phenol reagent. *J. Biol. Chem.*, **193**, 265–275.

Marton, J. (1974). Fines and wet end chemistry. *Tappi*, **57** (12), 90–93.

Mattisson, M. F., and Legendre, K. A. (1952). Determination of the carboxyl content of oxidized starch. *Anal. Chem.*, **24**, 1942–1944.

Olson, A. C., and Stanley, W. L. (1973). Lactase and other enzymes bound to a phenol-formaldehyde resin with glutaraldehyde. *J. Agr. Food Chem.*, **21**, 440–445.

Rebhun, M., Sperber, H., and Saliternick, Ch. (1967). Purification of paper mill effluents by flocculation. *Tappi*, **50** (12), 62A–64A.

Robyt, J. F., Ackerman, R. J., and Keng, J. G. (1972). Reducing value methods for maltodextrins: II Automated methods and chain-length independence of alkaline ferricyanide. *Anal. Biochem.*, **45**, 517–524.

Schwonke, P. A., and Davis, W. S. (1973). Enzyme enhanced turbidity removal through primary treatment. *Tappi*, **56** (1), 93–95.

Smiley, K. L., Boundy, J. A., Hofreiter, B. T., and Rogovin, S. P. (1974). Immobilized *alpha*-amylase for clarification of colloidal starch-clay suspensions. In *Immobilized Enzymes In Food and Microbial Processes*, p. 133 (A. C. Olson and C. L. Conney, eds.), Plenum Publishing Corp., New York.

Wilhelm, L. K. (1968). The effects of oxidized starch dispersing power on turbidity removal from aqueous pigment suspensions by sedimentation and coagulation. *Tech. Bull.*, 216, Nat. Council Paper Ind. Air and Stream Impr., New York, N.Y.; *Paper Trade J.*, **152** (35), 85 (Summary).

Wu, Y. C. (1974). Clarification of book-mill effluent by enzyme-alum. *J. Environ. Eng. Div.*, 100 N EE3:541–548.

CHARACTERISTICS OF IMMOBILIZED LIPASE

PAUL MELIUS, W. J. BROWN and B. C. WANG

Department of Chemistry, Auburn University,
Auburn, Alabama 36830, U.S.A.

INTRODUCTION

Before discussing the immobilization of swine pancreatic lipase and the characteristics of the immobilized system, the nature of the free enzyme will be briefly presented. Lipase is of particular interest as its substrate in most instances is water insoluble and emulsification of substrate is required for optimal enzymic activity. Brockerhoff (1972) has summarized investigations of various laboratories indicating the nature of the catalytic site of the enzymes. The evidence indicates a Ser, Leu, His, and possibly an Asp at the active site of the enzyme. An analogy is made to chymotrypsin, with a difference being a greater steric specificity for the lipase. The enzyme contains a carbohydrate moiety, discovered by Verger (1970), which may be involved in the orientation of the enzyme at an interphase. Bile salts have been suggested by Fritz and Melius (1963) to activate the lipase by causing the diglyceride enzyme complex to dissociate so that enzyme can more rapidly interact with triglyceride substrate. Bile salts are also known to protect lipase from surface denaturation effects.

Recently, a co-lipase has been discovered by Maylie *et al.* (1971) which is a small protein (MW $\sim$10 000). It is heat-resistant and has five disulphide bonds and also has been completely sequenced. The inhibition of pancreatic lipase by high concentrations of bile salt is overcome by the action of co-lipase (Borgstrom and Erlanson, 1971). There is a pH shift of optimum activity of about three pH units from 9 to 6 in the presence of high concentrations of bile salt.

The triglyceride substrates for the pancreatic lipase are usually emulsified in a gum arabic solution in order to obtain optimal activities of the enzyme. Recently, we have found that the gum arabic solutions contain calcium and magnesium. Calcium ions have been reported by Benzonana and Desnuelle (1968) to prevent to some extent the inhibitory effect of long-chain soaps, thus increasing the overall rate of reaction but not appreciably affecting the initial rate of the hydrolysis. Calcium and magnesium ions have been removed from the gum arabic solutions by dialysis experiments.

Brandenberger (1956) attached lipase to the isocyanato derivative of poly-*p*-aminostyrene and Kitajima *et al.* (1969) entrapped the enzyme in microcapsule form. Melius and Wang (1974) coupled pure pancreatic lipase to polysaccharides using cyanogen bromide activation. Lipase immobilized on polyacrylamide by using a diazonium intermediate was reported by Datta and Ollis (1974).

Two purified lipase enzyme preparations have been immobilized in our studies. We have used enzyme prepared according to Melius and Simmons (1965) and also by the method of Verger (1970). The enzymes were then coupled to cyanogen bromide (CNBr) activated polysaccharides.

MATERIAL AND METHODS

Porcine pancreatic lipase (EC 3.1.1.3) was prepared by the methods of Melius and Simmons (1965) and Verger (1970) and freeze-dried. The coupling of the enzyme to cellulose, Sephadex, and Sepharose was carried out as described by Melius and Wang (1974). We have also modified that procedure in the following way:

> Cellulose (400 mg) was suspended in 16 ml of CNBr (25 ml/ml water) and the pH was maintained at 10·0 by addition of NaOH for 12 min. The activated polymer was washed with 300 ml cold 0·1 M $NaHCO_3$ to remove unreacted CNBr. Approximately 300 mg of the activated polymer were then suspended in 3·5 ml of $NaHCO_3$ buffer, pH 9·8, which contained 0·63 mg of purified lipase. The slurry was gently stirred for 24 hr, then washed with 300 ml of 0·1 M $NaHCO_3$, 30 ml of 1 mM HCl, 300 ml of 0·5 M NaCl, and 200 ml of distilled water. No enzyme activity was found in the original supernatant, indicating all the enzyme was coupled to the activated cellulose.

Assays of the immobilized lipase were carried out in a Gooch crucible which contained a coarse fritted disc and an outlet tube. The crucible was mounted in a constant temperature bath and the electrodes and NaOH delivery tube from a Radiometer pH Stat were fitted into the crucible. The assay system contained 5·0 ml of 10% tributyrin emulsified with Ca^{2+} free 10% gum arabic, 10 ml of water, and varying amounts of sodium taurocholate and calcium chloride. After the completion of the assay, the reaction contents were removed by aspiration leaving the immobilized enzyme on the fritted disc. The enzyme was washed with cold water and was ready for another assay procedure.

An estimate of the amount of enzyme activity bound to the carrier indicated that approximately 10% of the free enzyme activity was retained. The amount of lipase bound to the polysaccharides was also determined by ninhydrin analysis of a sample of immobilized enzyme after acid hydrolysis.

A carboxyl-Sepharose preparation was made by first treating 10 ml of Sepharose 6B with 1 g of CNBr in 10 ml of solution. The pH was maintained at 11 with 1 M NaOH and the temperature at 20°C. The gel was then washed with 300 ml of cold 0·1 M $NaHCO_3$ on a Buchner funnel and then 1 g of ε-aminocaproic acid in 10 ml of 0·1 M $NaHCO_3$ was added to the activated Sepharose on the funnel and mixed well. The mixture was gently stirred for 24 hr at 4°C and then washed with 0·1 M $NaHCO_3$ (carboxyl-Sepharose preparation).

Five millilitres of dimethylformamide (DMF) was added to the carboxyl-Sepharose preparation followed by 1 g of dicyclohexylcarbodiimide (DCC) in 5 ml of DMF. The pH was adjusted to 4·7 and the mixture was gently stirred for 48 hr at 4°C. The preparation was then washed with DMF, 300 ml water, and 300 ml of 0·1 M $NaHCO_3$. The enzyme was now added to the activated Sepharose and allowed to react for 24 hr at 4°C, after which it was washed with 400 ml of 0·1 M $NaHCO_3$ and 400 ml of cold water.

RESULTS

The degree of activation and coupling of the polysaccharide to the enzyme are pH dependent. The coupling yield was reduced considerably when the amount of enzyme used was doubled and the pH was kept constant. The cyanogen bromide activated polysaccharides react readily with proteins at alkaline conditions. For Sepharose 2B and Sephadex, the coupling yield was as high as 95% or more of the added enzyme. Compared with the Sepharose and Sephadex, the cellulose required a more alkaline environment for a high coupling yield. As shown in Table I, the coupling yield decreased very rapidly when the amount of enzyme used was increased. The same effect occurred for the total enzymic activity retained on the carrier.

Table I
CHEMICAL COUPLING OF LIPASE TO CYANOGEN BROMIDE-ACTIVATED POLYSACCHARIDES

Polymer (activated pH value)	*Amount of enzyme used per 25 mg dry polymer (mg) (coupling pH)*	*Enzyme bound per g dry conjugate (mg/g)*	*Coupling yield based on amount of enzyme added (%)*
Cellulose (10)	2·4(9·7)	76·4	84
Cellulose (10)	4·6(9·7)	86·1	50
Cellulose (10)	5·4(9·7)	59·5	30
Cellulose (10)	5·6(9·0)	79·5	38
Cellulose (10)	14·9(7·0)	82·1	15
Sephadex (10)	2·1(9·0)	78·0	100
Sephadex (10)	6·0(7·0)	166·0	83
Sepharose (2%) (9·1)	0·7(9·7)	25·9	95
Sepharose (2%) (9·1)	1·8(9·7)	54·5	80
Sepharose (2%) (9·1)	2·4(10·3)	65·4	73
Sepharose (2%) (9·5)	1·2(9·7)	40·9	89
Sepharose (2%) (9·5)	2·5(9·7)	72·5	77
Sepharose (6%) (11·0)	0·4(7·0)	1·1	7

Table II
ACTIVITY OF BOUND LIPASE TOWARD TRIBUTYRIN SUBSTRATE (TB) OR TRIOLEIN SUBSTRATE (TO)

Polymer (activation pH) value	*Activity toward TB or TO*		*Total activity retained (%)*
	pH optimum	*Activity ratio (%)*	
Cellulose (10)	10·4	26·2(TO)	22·0
Cellulose (10)	10·4	10·4(TO)	5·2
Cellulose (10)	10·4	13·3(TO)	4·0
Cellulose (10)	10·4	1·8(TO)	0·7
Sepharose (2%) (9·1)	9·0	9·9(TB)	9·4
Sepharose (2%) (9·1)	9·0	6·1(TB)	5·5
Sepharose (2%) (9·5)	9·0	4·8(TB)	4·3
Sepharose (2%) (9·5)	9·0	3·2(TB)	2·6
Sepharose (6%) (11·0)	9·0	29·2(TB)	2·2

From Table II it is apparent that the cellulose-enzyme conjugates had pH optimum displacements of 1·4 units to the alkaline side (9·0–10·4). There was no significant shift of optimum pH for the Sepharose-enzyme conjugates.

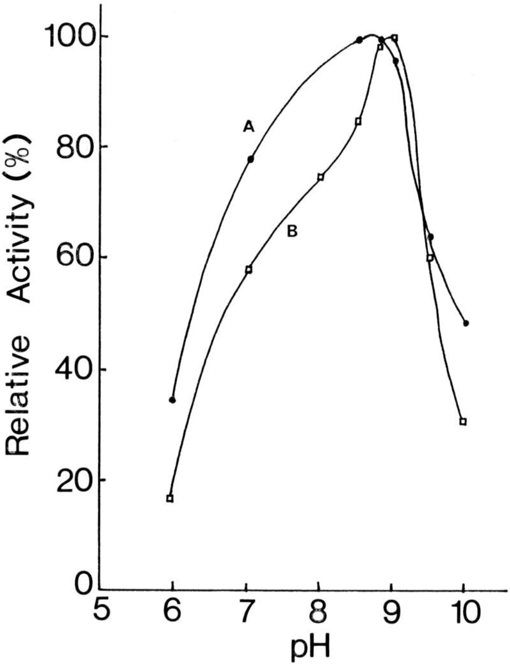

Fig. 1. pH-activity curves for free pancreatic lipase (curve A) and the Sepharose 6B-lipase conjugate (curve B). The immobilized lipase used was prepared by coupling the lipase to carboxyl-Sepharose 6B using dicyclohexylcarbodiimide as the coupling reagent in DMF.

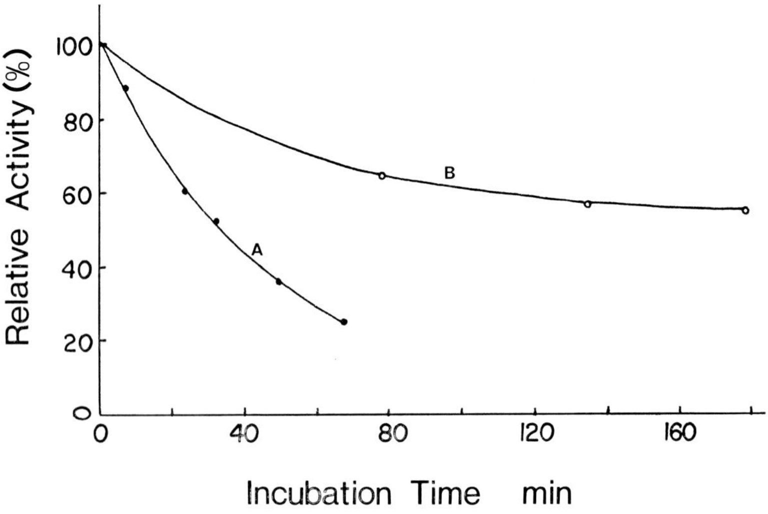

Fig. 2. Relative activity of free lipase (curve A) and immobilized lipase (curve B) with an incubation time of 34 min at 37°C and pH 9·0. The titrimetric assay method with tributyrin as substrate was used.

For the Sepharose immobilized enzyme, there were only small changes in pH optima and they were on the acid side of the pH optimum (Fig. 1). It is interesting to note that the profile of the immobilized lipase was sharper than that of the corresponding free enzyme. The pH optimum was displaced only a little to the alkaline side and was consistent with the pH-activity behaviour of most Sepharose immobilized enzymes presented in the literature, since Sepharose is a neutral or slightly anionic polymer.

In Fig. 2, the thermal stability of the free enzyme is compared to the immobilized enzyme. After 34 min at 37°C, 50% of the native free enzyme activity was lost, while it required 3 hr to lose 50% of the activity for the immobilized enzyme at 37°C. Also, the rate of decrease of enzyme activity after 3 hr for the immobilized enzyme was very small.

The Michaelis constant, K_m, for the free enzyme was found to be three times greater than the apparent K_m for the immobilized enzyme. Thus the substrate binding to the immobilized enzyme is apparently greater than for the free enzyme (Fig. 3). The apparent Michaelis constant for the free enzyme was $4{\cdot}17 \times 10^{-2}$ M and for the immobilized enzyme it was $1{\cdot}39 \times 10^{-2}$ M.

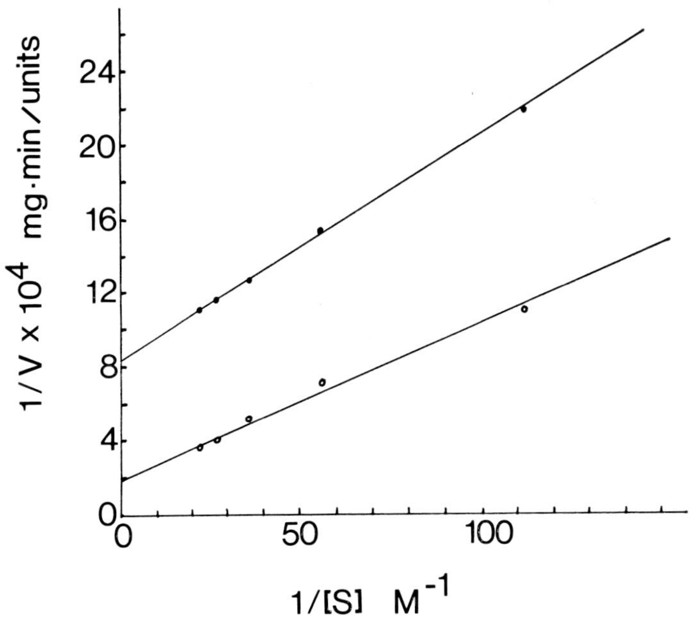

Fig. 3. Determination of K_m of the immobilized lipase with tributyrin as the substrate.

When immobilized enzyme was allowed to react with emulsified tributyrin at pH 9·0 without bile salt or Ca^{2+}, the results presented in Table III were obtained. The initial rate of the reaction was 2·15 units of activity. The reaction, contrary to what occurs in free enzyme reactions, was linear (constant rate of reaction) for 3 hr.

When a consecutive sequence of assays was carried out with the same quantity of enzyme, there was some loss of enzyme when Ca^{2+} was added to the reaction mixture. However, when Ca^{2+} was omitted in assays 6 and 7 (Table IV), the immobilized enzyme did not lose any more activity. Increas-

Table III
ACTIVITY OF IMMOBILIZED LIPASE VERSUS LENGTH OF ASSAY

Time (min)	*Units of activity*
0	2·15
60	1·81
120	1·85
180	1·68
240	1·85
300	1·85

Table IV
CALCIUM ION AND SODIUM TAUROCHOLATE EFFECTS ON IMMOBILIZED LIPASE

Assay no.	*Conditions*	*Units of activity*
1	6·6 mM Ca^{2+}	1·85
2	6·6 mM Ca^{2+}	1·42
3	6·6 mM Ca^{2+}	1·31
4	6·6 mM Ca^{2+}	1·24
5	6·6 mM Ca^{2+}	1·09
6	0 Ca^{2+}	1·05
7	0 Ca^{2+}	0·98
8	0 Ca^{2+}; 5·2 mM sodium taurocholate	0·93
9	0 Ca^{2+}; 18·2 mM NaTC	1·05
10	0 Ca^{2+}; 18·2 mM NaTC	1·00
11	0 Ca^{2+}; 78 mM NaTC	0·85

ing concentrations of sodium taurocholate did not affect the immobilized lipase.

DISCUSSION

As far as the pH-activity behaviour is concerned, an explanation based on the electrostatic field produced by a highly charged carrier was first offered by Goldstein *et al.* (1964). Based on the chemistry of the reaction of the enzyme with the imidocarbonate containing dextran, it is expected that the total number of positive groups in the enzyme will be decreased and therefore lead to an alkaline displacement of the pH-activity curve. Charged groups also might well be introduced in the cyanogen bromide activation step. Thus, for the neutral carrier Sepharose, the previous explanation could be applied to explain the slight change in the pH optimum. There was no pH optimum shift due to an enzymatically generated pH gradient between the domain of the enzyme-polymer conjugate and the external solution, since the pH optimum shift observed was very small. On the contrary, Axen and Ernback (1971) observed this kind of shift toward higher pH values with trypsin and chymotrypsin immobilized to Sepharose beads.

In most cases a very high yield of enzyme bound to the carrier was obtained, but the greatest retention of enzyme activity occurred when the enzyme to carrier ratio was very large. The use of a spacer arm such as 4,4′-methylenedianaline with Sepharose 2B did give a greater retention of enzyme activity, whereas ε-aminocaproic acid did not improve the yield of retained enzyme activity on the immobilized preparation.

When cellulose was used as a carrier, there was a 1·4 unit shift in the pH optimum for the immobilized enzyme. No pH optimum shift was observed in the Sepharose or Sephadex preparations. This pH shift with the cellulose preparation may be due to an increase in negative charges around the active site of the enzyme (Goldstein *et al.*, 1964).

The K_m of the immobilized enzyme preparations indicates a stronger binding with the substrate than the free enzyme. However, the K_m may have a much more complex significance in this case, and this may be too simplified a conclusion.

It is of interest that we have found significant quantities of Ca^2 and Mg^2 in gum arabic solutions from atomic absorption spectroscopy measurements. The substrate emulsions used in these studies have been prepared with calcium- and magnesium-free gum arabic solutions. Thus the finding that calcium is without any effect on the immobilized lipase seems significant as compared to most findings for the free enzyme reaction. Also, the lack of any inhibition or activation effects of sodium taurocholate indicates a different mechanism of action for the immobilized enzyme. Of even greater interest is the fact that the immobilized lipase produces a constant rate of hydrolysis over a period of 5 hr in the absence of bile salt or calcium. The free enzyme produces a nonconstant rate of hydrolysis within a few minutes unless bile salts and calcium are present. This behaviour fits the hypothesis presented by Fritz and Melius (1963) for the effect of bile salts on the lipase catalysed reaction. In the case of the immobilized enzyme, the enzyme will not bind to the emulsified product molecules as it is covalently bound to the water-insoluble matrix. Therefore, there is no requirement for bile salt to split and dissociate the enzyme-diglyceride complex. The reaction with the immobilized enzyme is first order with respect to triglyceride in the absence of bile salt and calcium ions.

REFERENCES

Axen, R., and Ernback, S. (1971). Chemical fixation of enzymes to cyanogen halide activated polysaccharide carriers. *Eur. J. Biochem.*, **18**, 351–356.

Benzonana, G., and Desnuelle, P. (1968). Action of some effestors on the hydrolysis of long-chain triglycerides by lipase. *Biochim. Biophys. Acta*, **164**, 47–58.

Borgstrom, B., and Erlanson, C. (1971). Pancreatic juice co-lipase: Physiological importance. *Biochim. Biophys. Acta*, **242**, 509–511.

Brandenberger, H. (1956). Methods for linking enzymes to insoluble carriers. *Rev. Ferment. Ind. Aliment.*, **11**, 237–241.

Brockerhoff, H. (1972). A model of pancreatic lipase and the orientation of enzymes at interfaces. *Chem. Phys. Lipids*, **10**, 215–222.

Datta, R., and Ollis, D. F. (1974). Optimization of activity of immobilized lipase. In *Immobilized Biochemicals and Affinity Chromatography*, pp. 293–315 (R. Bruce Dunlap, ed.), Plenum, New York.

Fritz, P. J., and Melius, P. (1963). Mechanism of activation of hog pancreatic lipase by sodium taurocholate. *Can. J. Biochem.*, **41**, 719–730.

Goldstein, L., Leivin, Y., and Katchalski, E. (1964). A water-insoluble polyanionic derivative of trypsin. *Biochemistry*, **3**, 1913–1919.

Kitajima, M., Miyano, S., and Kondo, A. (1969). Enzyme-containing microcapsules. *Kogyo Kagaku Zasshi*, **72**, 493–497.

Maylie, M. F., Charles, M., Gache, C., and Desnuelle, P. (1971). Isolation and partial identification of a pancreatic colipase. *Biochim. Biophys. Acta*, **229**, 286–292.

Melius, P., and Simmons, W. S. (1965). Utilization of 1-butanol for the preparation of pancreatic lipase. *Biochim. Biophys. Acta*, **105**, 600–602.

Melius, P., and Wang, B. C. (1974). Immobilization of lipase to cyanogen bromide activated polysaccharide carriers. In *Immobilized Biochemicals and Affinity Chromatography*, pp. 339–343 (R. Bruce Dunlap, ed.), Plenum, New York.

Verger, R. (1970). Ph.D. Dissertation, Faculte des Sciences de Marseilles.

CONTINUOUS PRODUCTION OF GLUCOSE FROM STARCH HYDROLYSATES BY IMMOBILIZED GLUCOAMYLASE

Y. Y. LEE,* D. D. LEE and G. T. TSAO†

Department of Chemical Engineering,
Iowa State University, Ames, Iowa, U.S.A.

Summary

Glucoamylase covalently attached to porous silica particles was investigated for its applicability as an industrial catalyst in glucose production from dextrin. The fundamental aspects regarding reactivity, enzyme loading, kinetics, diffusional effects, and thermal stability were studied and followed by a pilot scale test performed in a 1 ft^3 packed column reactor. The test results showed that the one-CF enzyme column can continuously produce approximately 700 lb of glucose per day at 38°C. The maximum glucose yield varied from 87 to 93·5% depending mainly on the feed dextrin. In 80 days continuous operation, no appreciable deactivation of enzyme was observed. Continuous heat sterilization of the feed stream virtually ensured sterile operation of the immobilized enzyme reactor.

MATERIALS AND METHODS

Preparation of immobilized glucoamylase

The enzyme support used in this work was Corning porous silica (SiO_2), 30–45 mesh with a mean pore diameter of 400 Å. The glucoamylase was obtained in a purified form from NOVO Enzyme Corp., Mammaroneck, N.Y. The enzyme was immobilized onto silanized enzyme carrier by Schiff base linkage using glutaraldehyde as a bifunctional reagent. The details of this method are given elsewhere (Messing and Weetall, 1970). The silica support used in the scale-up test was obtained as silanized form from Corning Glass Co. The alkylamine silica was packed in the column reactor and enzyme attachment thereafter was carried out by recirculating glutaraldehyde solution (2·5%) and enzyme solution through the column. For initial sterilization, saturated chloroform solution was filled in the column and left several hours, then thoroughly washed with distilled water.

Immobilized enzyme characteristics and batch reaction conversion

The pH and temperature characteristics of immobilized glucoamylase were studied using a recirculated batch reactor system which contained a ½ inch ID glass column reactor, peristaltic pump, heat equilibrating coil (stainless steel), constant temperature bath, and reactant reservoir. The detailed description of this setup is given elsewhere (Wun, 1974). The activity of the glucoamylase

* Current address: Department of Chemical Engineering, Auburn University, Auburn, Alabama 36830, U.S.A.
† Current address: Department of Chemical Engineering, Purdue University, West Lafayette, Indiana, U.S.A.

was taken by the rate of glucose formation from dextrin at low conversion state (less than 20%). The same reactor system was used in batch reaction conversion studies. In studies of enzyme deactivation, a jacketed glass column reactor (2·5 cm ID × 30 cm H) was used. Pre-warmed dextrin solution was continuously fed to the reactor at constant flow rate over an extended period until appreciable deactivation had occurred.

The output was intermittently analysed for conversion into glucose from which the activity of the enzyme was estimated. The flow rate and amount of the immobilized enzyme were adjusted so that the initial conversion into glucose was near 80%.

Analytical method

The feed dextrin and the product were analysed for glucose and its polymers by liquid chromatography (Waters Associates, Model ALC-201 with RI detector). The column ($\frac{3}{8}$ × 22 inches) packed with Bio-Rad cation exchange resin (AG 50W-X4, chloride form) was maintained at 80°C. A digital integrator (Infortronics) was incorporated with the LC for quantitative analysis. Separation of glucose polymers was nearly complete from glucose to G_8, and G_9 and above appeared in a single peak. When it was necessary, the glucose content was analysed separately by Beckman Glucose Analyser.

Methods in scale-up experiment

The flow sheet of the experimental setup for scale-up study is shown in Fig. 1. The packed column reactor was constructed with a 6 inch schedule 40 pipe (stainless steel) and the column height was 5 ft, giving an effective reactor volume of 0·98 ft^3. The estimated reactor void volume when packed with the immobilized enzyme was 0·3 ft^3 (pore volume of enzyme support excluded). Thus the reactor residence time at normal operating condition (0·25 gpm) was 9 min. At both ends of the reactor, small mixing chambers were allowed to avoid channelling in the reactor. Feed dextrin was dissolved into deionized water preheated to 60°C by a panel heat exchanger installed inside the make-up tank (200 gal). The pH was pre-adjusted to the optimum of 4·5 with HCl and rechecked before the reactor inlet by an on-line pH meter (Leeds and Northrup). The warm feed solution (30 wt %) was transferred to an insulated storage tank (200 gal) by a centrifugal pump. The feed stream from the storage tank was continuously pumped to the sterilizer by a diaphragm pump (Milton Roy, adjustable flow rate with 1 gpm max.). The temperature of the feed stream was raised to 117°C at the outlet of the steam heat exchanger and maintained at that temperature in the soaking coil for the residence time of 4 min. The stream from the sterilizer was then continuously cooled to the reaction temperature by a co-current water-cooled heat exchanger in which the temperature was controlled by a position-controller (Leeds and Northrup). After sterilization and temperature control, the feed stream was filtered through a fibre cartridge filter (Fram, 10 μ core with polypropylene casing) before it was pumped into the reactor by a Milton Roy diaphragm pump (flow rate adjustable with $\frac{1}{2}$ gpm max.). The direction of flow in the reactor was from top to bottom. From the inlet point of the reactor pump, a recycle line was connected to the feed storage tank in order to resolve the problems that arise when two positive displacement pumps are used in series. Thus the flow rate in the feed pump was maintained slightly higher than that of the reactor pump recycling the difference. For protection

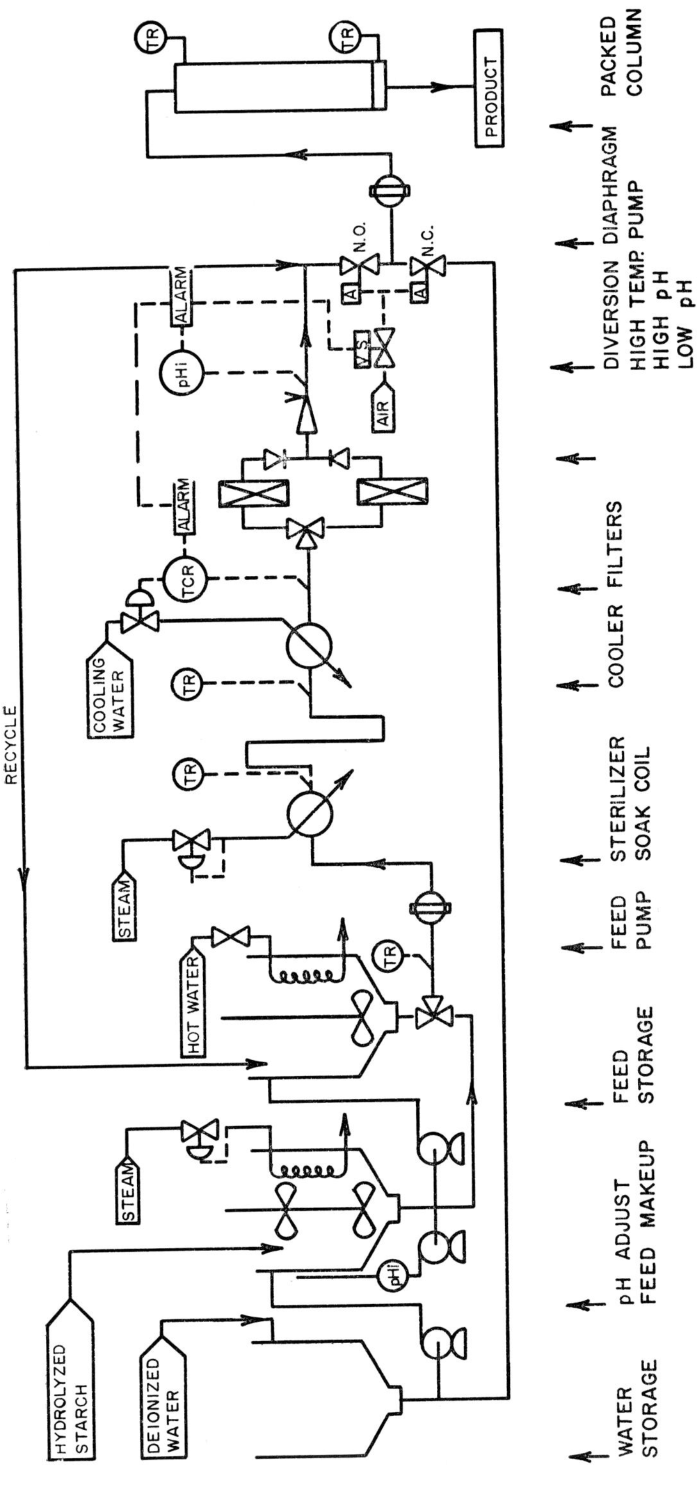

Fig. 1. Process flow diagram for immobilized enzyme pilot plant.

of the immobilized enzyme, pH, and temperature alarm, a flow diversion device was attached on the process line. This device was designed to switch the total process flow to the recycle line when the temperature and/or pH goes off the pre-set range while the reactor is fed with deionized water. The material used in this setup was stainless steel (316) throughout except the filter and on-line pH probe chamber where it was plastic. The process lines were mostly $\frac{3}{8}$ inch tubing except the soaking coil (1 inch tubing) and steam heat exchanger coil ($\frac{1}{4}$ inch tubing). Temperatures at five different locations including reactor inlet and outlet were measured by thermocouples and recorded.

The pressures at four locations across the process line and the pressure drop in the reactor were also measured.

RESULTS AND DISCUSSION

Characteristics of immobilized glucoamylase

A soluble glucoamylase showed initial activity of 60 000 units/g (1 unit = 1 μmole glucose/min at 60°C, pH 4·7). Four pounds of this enzyme were offered to a 31·8 lb reactor load of carrier (dry basis). By comparison of the activities of bulk enzyme solutions before and after the contact to the support, the amount of enzyme bound to the support was calculated to be 3·56 lb enzyme/31·8 lb carrier or 112 mg enzyme/g carrier. The apparent reactivity of the immobilized enzyme was approximately 3000 units/g (dry basis). This indicated the enzyme retained about 50% of the free enzyme activity when it was immobilized. Studies of the pH and temperature effects on the activity of the immobilized enzyme have been previously carried out in this laboratory and reported for glucoamylase bound to Corning Controlled Pore Glass (CPG 96% silica, 40–80 mesh, 550 Å pore diameter) (Marsh *et al.*, 1973). Similar tests were conducted for porous silica support and it was found that pH and temperature effects were almost identical to those with CPG. The effects of substrate diffusion in this heterogeneous reaction have also been tested and analysed for CPG-glucoamylase under similar experimental conditions as reported previously (Marsh *et al.*, 1973; Lee and Tsao, 1974), and it is judged that there is no significant difference for this enzyme support. For CPG support the effect of external film diffusion in a packed column reactor was not noticeable in the range of superficial flow rate (flow rate/column cross-sectional area) from 0·008 cm/sec to 0·2 cm/sec. The flow rate applied in the scale-up test of the present study was in the vicinity of 0·05 cm/sec in superficial velocity. Therefore, the reaction can be safely regarded as unaffected by external diffusion. In the case of pore diffusion, the effectiveness factor estimated from the result of CPG-glucoamylase was 0·6–0·8 under experimental conditions in the scale-up test. The effect of pore diffusion is therefore considered one of the important factors of decreasing the overall enzyme activity upon immobilization.

Thermal deactivation of immobilized glucoamylase

Thermal stability is perhaps the most important factor in utilization of an immobilized enzyme as an industrial catalyst. Loss of activity of immobilized enzymes under continuous use is a rather complex phenomenon resulting from various causes. As preliminary work the deactivation of immobilized gluco-

Table I

MAXIMUM GLUCOSE YIELD AND D.E. OBTAINED IN A RECIRCULATED BATCH REACTOR OF IMMOBILIZED GLUCOAMYLASE AND IN A FREE GLUCOAMYLASE BATCH REACTOR[a]

Origin	*D.E. of substrate*	*Enzyme type*	*Max. glucose yield*[b] (*wt %, dry basis*)	*Max. D.E.*
GPC	10	Immob.	92·8	95·2
		Free	95·1	97·1
GPC	15	Immob.	92·0	94·9
		Free	93·0	94·8
P&F[c]	18	Immob.	93·5	96·0 60°C
		Free	94·5	96·6 60°C
GPC	20	Immob.	92·2	94·9
		Free	93·4	95·8
Staley	24	Immob.	90·5	93·9
		Free	90·7	93·8
GPC	25	Immob.	92·8	95·4
		Free	93·0	95·4
Hubinger	25	Immob.	89·5	92·6 55°C
Staley	35	Immob.	88·4	91·9
		Free	88·8	92·5
Staley	42	Immob.	87·2	91·3
GPC	42	Immob.	86·1	90·8

[a] All reactions at 45°C and 30% by wt starch unless otherwise indicated.
[b] Wt % of glucose in product.
GPC = Grain Processing Company, Muskatine, Iowa (Enzyme Thinned Starch).
P&F = Penick & Ford Ltd, Cedar Rapids, Iowa.
Staley = A. E. Staley Mfg Co., Decatur, Illinois (Acid and Enzyme Thinned Starch).
Hubinger = Hubinger Co., Keokuk, Iowa (Acid Thinned Starch).
[c] Pearl Starch Thinned with α-amylase in our laboratory.

to 94% glucose syrups from 28 DE dextrin by employing α-amylase and glucoamylase (immobilized onto acrylamide gel) simultaneously in a packed column. A similar experiment was conducted in this work whereby the low DE powder dextrin was hydrolysed by two separate steps: first increasing the DE to 45 by soluble α-amylase and treating with immobilized glucoamylase. However, no improvement in glucose yield was achieved by this method. The maximum glucose conversion, in general, was higher for low DE substrates within the range of DE from 10 to 42. The optimum DE value in terms of maximum glucose conversion was found to be in the range of 10–20. Also the conversion depended to a great extent on the methods by which the feed dextrin was processed. For instance, straight enzyme processed dextrin (GPC Maltrine-25) gave about 2% higher conversion than acid-enzyme treated dextrin (Hubinger DE 25). Dextrin freshly prepared (DE 18) from pearl starch (Penick and Ford) by α-amylase (soluble) gave the conversion as high as 94·5% glucose by free glucoamylase and 93·5% by immobilized glucoamylase, which is still 1% higher than that from enzyme-thinned powder dextrin. The main difference between fresh soluble dextrin and dried dextrin (both enzyme thinned) is the degree of retrogradation. Thus, the difference in glucose yield between these two is logically explained by the effect of retrogradation on the conversion. It is therefore expected that by starting with starch, liquefying with α-amylase, 93–94% glucose in the product can be obtained by immobilized glucoamylase with little difficulty.

Scale-up studies

Conversion test

Experimental studies on continuous conversion and other practical aspects of the process were conducted by the 1 ft^3 reactor system described earlier. The substrate in this unit was a 30 wt % solution of commercial starch hydrolysate having 24 DE (Staley Star-Dri 24). In order to reduce the feed consumption, the base operating temperature was set at 38°C, which is not necessarily the optimum temperature. Actually, at the reactor inlet, the temperature was set at 37°C which gave an outlet temperature of 39°C due to heat of reaction. The temperature jump of approximately 2°C was in good agreement with an earlier study made by Marsh (1974). In the first series of experiments, the continuous glucose conversion was measured at varying feed flow rates or reactor contact time. The conversion-flow rate relation is given in Table II. The maximum glucose concentration obtained in this experiment

Table II
ANALYSIS OF FEED AND PRODUCT IN CONTINUOUS CONVERSION

	Flow rate[a] *(lb/day)*	*Degree of polymerization* ≥7	6	5	4	3	2	1	*Calc. D.E.*
Feed analysis (Staley Star-Dri 24-R)	—	49·5%	12·8%	6·9%	7·1%	9·9%	8·6%	5·2%	23·5
Product analysis	1 100	3·8%	0·6%	0·5%	0·5%	1·0%	3·8%	89·7%	92·6
	900	3·3%	0·6%	0·5%	0·5%	0·7%	4·4%	90·0%	93·0
	700	2·8%	0·6%	0·5%	0·5%	0·7%	4·5%	90·4%	93·5
	500	2·6%	0·7%	0·5%	0·5%	0·9%	4·7%	90·1%	93·4
	300	1·7%	0·6%	0·5%	0·4%	1·0%	7·7%	88·1%	92·7
Feed analysis (Staley Star-Dri 35-R)	—	42·5%	6·6%	7·3%	8·2%	9·9%	11·2%	14·3%	33·4
Product analysis	500	3·1%	1·2%	1·1%	1·1%	1·2%	5·3%	87·0%	91·0
Feed analysis (Staley Star-Dri 42-R)	—	35·6%	7·0%	7·7%	8·6%	11·2%	12·7%	17·2%	35·7
Product analysis	900	3·3%	1·2%	1·3%	0·7%	1·8%	5·2%	86·5%	90·7
	500	2·9%	1·2%	1·2%	0·7%	1·7%	5·7%	86·6%	91·0

[a] Based on dry product. Temperature 38°C.

was 90·4%, which agreed well with the batch reaction data. The reversion reaction was also noticed at a flow rate lower than optimum or excessive residence time. The production capacity based on the flow rate of maximum glucose was approximately 700 lb of dry dextrin per day. As indicated in the batch conversion data, the glucose level stayed relatively constant over a wide range of reaction time or flow rate in the continuous system. This was confirmed in this experiment; the glucose level stayed within 0·4% when the flow rate was changed from 300 lb solid/day to 900 lb. The production capacity given here thus conceives such flexibility. In view of the reactivity *vs.* temperature relation of glucoamylase, the production rate would be 850 lb at 45°C and 1130 lb at 50°C. The conversion data of 35 DE dextrin and 42 DE dextrin are also given in Table II.

Stability of immobilized enzyme and reactor contamination

To test the enzyme stability under realistic processing conditions the scale-up system was continuously operated for 80 days at 38°C, using Staley Star-Dri 24-R most of the time except for two occasions where it was not available and Star-Dri 35 DE and 42 DE were substituted. The system was operated for 40 days at flow rates of 1000 lb/day and then the next 40 days at 500 lb/day to conserve substrate. When the activity tests were made, the flow rate was raised back to 1000 lb/day and then the product distribution was compared with that of the initial product. During 80 days continuous operation, there was no sign of enzyme deactivation.

The microbial contamination of the enzyme column was also of prime concern in the scale-up test. To verify the contamination level, the microbial densities were measured by taking plate counting for bacteria and moulds at three different locations of the process (feed tank, recycle line, and reactor outlet). After 30 days operation, the product stream showed a bacterial count of 36/ml, recycle stream 93/ml, and feed 220/ml. There was no sign of mould contamination. Until 2 months' operation, the bacterial count stayed constant in the product, whereas it fluctuated in the feed batch by batch and steadily increased at the recycle line. Up to this point the continuous heat sterilization (117°C and 4 min holding) was quite satisfactory to maintain the continuous reactor virtually sterile. However, a steady increase of contamination at the open end of the recycle line indicates that it may result in the eventual contamination of the reactor by reverse penetration of microorganisms through the recycle stream. From this standpoint, using a single pump for removing the recycle line seems preferable to using two pumps. Shortly before the 70th day, the feed pump malfunctioned and air was drawn from the recycle line and pushed through the reactor for several hours by the reactor pump. After this incident, the bacterial count quickly rose to 1600/ml in the product and 4400/ml in the recycle line. Five days later the count jumped to 41 000/ml in the product and the pressure drop in the reactor had significantly increased, making the system almost inoperable (Table III). The entire system was then

Table III
BACTERIAL CONTAMINATION

Operation time	*Bacterial count/ml fluid*		
	Feed	*Recycle*	*Product*
30 days	220	93	36
70 days[a]	20	4 400	1 600
75 days			41 000
After flush			300

[a] After column was contaminated with air.

flushed with saturated chloroform solution, allowed to stand for 2 hr for sterilization, and operation was then resumed. The activity of the enzyme was found to be unaffected by this high level contamination but the bacterial count in the product was enhanced to 300/ml and maintained at that level. The enzyme activity was rechecked after 3 weeks' storage in a cold room (4°C) and still no deactivation was noticed.

Other process aspects

There was no major difficulty in operating the scale-up test system except occasional minor equipment malfunctioning such as pump, or cooling water control valve. The total process residence time from outlet of feed tank to the reactor outlet was about 20 min. During this period, there was no visible increase in colour formation. This should be considered as an important factor of cost-saving in colour removal. However, gradual development of colour (yellowish) and odour was observed in the feed tank which was refilled every 10 hr. This necessitated periodical cleaning of the tanks. The maximum pressure drop across the reactor was about 20 psi (fluctuated from 10 to 20) at normal operating conditions and was not, by any respect, a limiting factor for design of the reactor. A brief test on the degree dispersion in the reactor conducted by temperature disturbances indicated the flow in the reactor was very close to plug flow and far less than the level that can give measurable effect in conversion.

Acknowledgement

This research was supported by the National Science Foundation through Grant GI-38101 and by the Engineering Research Institute of Iowa State University.

REFERENCES

Baker, S. A., Emery, A. N., and Novais, T. M. (1971). *Process Biochem.*, 11 October.

Fratzke, A., Lee, Y. Y., and Tsao, G. T. (1974). Kinetics and mass transfer characteristics of glucose isomerase immobilized on porous glass. AIChE Symp. Proc., AIChe-VTG Joint Meeting, Munich, West Germany.

Gruesbeck, C., and Rase, H. F. (1972). *I & EC Res. Dev.*, **11**, 74.

Lee, Y. Y., and Tsao, G. T. (1974). *J. Food Sci.*, **38**, 667.

Marsh, D. R. (1974). Ph.D. Thesis. Iowa State University, Ames, Iowa.

Marsh, D. R., Lee, Y. Y., and Tsao, G. T. (1973). *Biotech. Bioeng.*, **15**, 483.

Messing, R. A., and Weetall, H. H. (1970). U.S. Patent 3,515,538.

Smiley, K. L. (1971). *Biotech. Bioeng.*, **13**, 309.

Walton, H. M., and Eastman, J. E. (1973). *Biotech. Bioeng.*, **15**, 951.

Weetall, H. H., and Havewala, N. B. (1972). *Enzyme Engineering* (L. B. Wingard, ed.), Interscience, New York.

Wun, K. (1974). M.S. Thesis. Iowa State University, Ames, Iowa.

protein/g support were bound, despite excess enzyme in the reaction mixture. The pH maximum was shifted from 7·0 to 6·5, but other parameters, including K_m and V, were unchanged.

Weetall *et al.* (1974a,b) investigated two fungal and one yeast lactase bound to ZrO_2-coated CPG (the ZrO_2 treatment increases the stability of CPG). The activity of the fungal lactase-M derivative ranged from 50 to 100 units/g glass; the yeast lactase ranged from 400 to 1100 units/g glass, and the fungal lactase-W, from 250 to 800 units/g glass. This latter represents 25–80% of the activity offered. In each case, there was a pH shift toward the acid side. The fungal lactases demonstrated pH shifts of 0·5 and 1·5 units, to a minimum at pH 3·0. The yeast lactase shifted from pH 6·25 to 3·0. The apparent K_m of the immobilized yeast lactase was lower than that for the soluble enzyme; there was no change in K_m upon immobilization for either fungal lactase. Galactose inhibition was demonstrated for yeast and fungal lactases.

Other support materials have been investigated for use in the glutaraldehyde immobilization of enzymes. Stanley and Palter (1973) tested a number of phenolic resins for lactase immobilization and achieved 60% initial efficiency using a phenol-formaldehyde resin, Duolite S-30. The same group, again using glutaraldehyde, bound an *A. niger* lactase to chitin, a cellulose-like polymer isolated from the exoskeleton of marine crustaceans (Stanley *et al.*, 1975). A pH shift from 4·0 to 3·0 or 3·5 was explained as due to the amino groups present in the support. Sixty percent of the original activity was immobilized; the highest loading efficiency was determined to be 30 mg lactase/g chitin.

Alumina and stainless steel have also been used as supports. The lactase was adsorbed to the carrier, and then cross-linked with glutaraldehyde. The binding efficiency for the alumina was 60–70%, for the titania-activated stainless steel, 10–20%. The pH and temperature profiles for these immobilized forms of acid lactase were similar to those for other reported immobilized lactases (Charles *et al.*, 1974a).

The excellent mechanical strength and hydrophilicity of collagen led to an investigation of its potential as an enzyme support in particulate form (Woychik *et al.*, 1974). Fibrous collagen was prepared from comminuted hide collagen by grinding, screening, and lyophilizing. These fibres swell rapidly and take up approximately 100% of their weight of buffered enzyme solution. Protein-protein interactions can lead to noncovalent binding of enzyme, but at very low levels; therefore, glutaraldehyde was used to form cross-links between the enzyme and collagen. A maximum of 8 mg lactase/g collagen was bound. The pH maximum was shifted slightly toward the acid side, similar to results obtained with the same lactase bound to CPG. Higher K_m values for CPG- and collagen-immobilized lactase can be attributed to steric factors resulting from covalent binding to the supports which may cause decreased binding of lactose to the active site. Maximum velocities were comparable for soluble and bound enzymes.

APPLICATIONS

Of all the lactases that have been investigated and immobilized, two seem to offer more promise for use in the dairy industry than any of the others. One of these is the acid lactase isolated from *A. niger*, which has the appropriate pH maximum (4·5) for use in wheys produced in the manufacture of cottage, cream, and Italian cheeses. At 55°C, 70–90% of the lactose in whey concentrates (21% lactose) can be hydrolysed in 4–5 hr, despite the product inhibition by galactose which slows the rate after 2 hr (Wierzbicki and Kosikowski, 1973b). Deproteinized and demineralized cottage cheese acid whey could be treated and concentrated to give clear, very sweet syrups, which could be used in yoghurt, fruit juices, puddings, and imitation maple syrup (Wierzbicki and Kosikowski, 1973a).

Immobilized on CPG, this lactase offers the possibility of continuous operations free of microbial growth because its optimal temperature is 55°C, a temperature fatal to most microorganisms. Because of the galactose inhibition, column operation was more efficient than batchwise procedures. A stirred enzyme suspension hydrolysed 56% of the lactose in 35 ml substrate in 1 hr, while the same amount of immobilized enzyme hydrolysed 73% of the lactose at 35 ml/hr. No loss of activity was found after storage at 4°C for 2 months. The addition of nonlactose whey solids to the substrate feed reduced the rate of lactose hydrolysis (Woychik and Wondolowski, 1973). Similar reductions in rate have been reported for other lactases (Wendorff *et al.*, 1971). An *A. niger* lactase immobilized on chitin had a calculated operational half-life of 27 days when used in a reverse-flow column at 60 ml/hr and 45°C (Stanley *et al.*, 1975). The same lactase on ZrO_2-CPG had a half-life of 44 days at 30°C (Weetall *et al.*, 1974b). When immobilized on phenol-formaldehyde resin, the *A. niger* lactase hydrolysed 99% of a 3% lactose solution at 45°C for 4 weeks with no loss of activity at 30 or at 60 ml/hr (Olson and Stanley, 1973).

The second lactase that shows promise for use in the dairy industry is that isolated from *S. lactis*. This lactase has a pH optimum of 6·5–7·0, which makes it useful in sweet wheys produced in the manufacture of cheddar, swiss, and blue cheeses. This enzyme has been immobilized on CPG and on comminuted collagen, and evaluated for activity in sweet whey and in skim milk (Woychik et al., 1974). The nonlactose solids content again was found to influence activity. At comparable pH's and lactose concentrations, the activity was decreased approximately 10% in sweet whey and 13% in skim milk when compared with the activity in buffered lactose solutions. Because the enzyme itself is labile above 30°C, room temperature or below must be used in processing. This increased the possibility of microbial growth, and treatment of the immobilized lactase with working concentrations of some standard sanitizing agents caused inactivation of the enzyme. No loss of activity was observed during 5 days of continuous operation using 5% lactose. Seventy grams of the collagen-lactase in a column operation hydrolysed 93% of the lactose at 2·3 litre/hr. Increasing the flow rate to 5·2 litre/hr decreased hydrolysis to 76%. A large-scale test of the collagen-lactase was made using 375 g at room temperature in a column treatment of sweet whey. At flow rates up to 18 litre/hr, 85% of the lactose was hydrolysed.

One problem encountered in a large-scale operation was packing of the collagen at flow rates above 1 litre/hr. This could be prevented by operating the column in reverse flow. The second problem is bacterial contamination.

The *S. lactis* lactase has been successfully used in batchwise procedures to treat milk before it is used in manufacture. A commercial lactase, added prior to the starter in cheddar cheese manufacture, hydrolysed 70% of the lactose in 2·5 hr. The presence of the free glucose speeded ripening by 15–20 min, and there was a decrease in cheddaring time required. The flavour, body, and texture of the experimental cheese was determined to be superior to controls. Similar prior treatment of milks used in making cottage cheese, buttermilk, and yoghurt resulted in reduced times for acid development and coagulation of buttermilk. The increased sweetness was found objectionable in buttermilk, but was covered by the cottage cheese dressing (Thompson and Brower, 1975a,b).

Many of the support materials are available commercially, activated and ready to use. Immobilized enzymes are also commercially available. The cost of using immobilized or soluble lactases must be balanced by the possibility of savings on sweeteners, production of new products such as food grade syrups, utilization of all the whey produced by cheese manufacture instead of costly dumping, and improvement in the efficiency and quality of current processes. Investigations into the problems of continuous operation, sanitation, engineering, and costs are continuing, and the future for the use of lactases, soluble or immobilized, is bright.

REFERENCES AND BIBLIOGRAPHY

Charles, M., Coughlin, R. W., Allen, B. R., Paruchuri, E. K., and Hasselberger, F. X. (1974a). Lactase immobilized on stainless steel and other dense metal and metal oxide supports. *Adv. Exp. Med. Biol.*, **42**, 213–234.

Charles, M., Coughlin, R. W., Tedman, R., and Beard, K. W. (1974b). Polymer-metal composite supports for immobilized enzyme catalysts. *Biotechnol. Bioeng.*, **16**, 1549–1551.

Charles, M., Coughlin, R. W., Paruchuri, E. K., Allen, B. R., and Hasselberger, F. X. (1975). Enzymes immobilized on alumina and stainless steel supports. *Biotechnol. Bioeng.*, **17**, 203–210.

Dahlqvist, A., Mattiasson, B., and Mosbach, K. (1973). Hydrolysis of β-galactosides using polymer-entrapped lactase. A study towards producing lactose-free milk. *Biotechnol. Bioeng.*, **15**, 395–402.

Giacin, J. R., Jakubowski, J., Leeder, J. G., Gilbert, S. G., and Kleyn, D. H. (1974). Characterization of lactase immobilized on collagen. *J. Food Sci.*, **39**, 751–754.

Hasselberger, F. X., Allen, B., Paruchuri, E. K., Charles, M., and Coughlin, R. W. (1974). Immobilized enzymes. Lactase bonded to stainless steel and other dense carriers for use in fluidized bed reactors. *Biochem. Biophys. Res. Commun.*, **57**, 1054–1062.

Hinberg, I., Korus, R., and O'Driscoll, K. F. (1974). Gel entrapped enzymes: Kinetic studies of immobilized β-galactosidase. *Biotechnol. Bioeng.*, **16**, 943–963.

Husted, G. O., Richardson, T., and Olson, N. F. (1973). Immobilization of β-galactosidase on an insoluble carrier with a polyisocyanate polymer. I. Preparation and properties. *J. Dairy Sci.*, **56**, 1111–1117.

Kay, G., and Crook, E. M. (1967). Coupling of enzymes to cellulose using chloro-*s*-triazines. *Nature*, **216**, 514–515.

Kay, G., Lilly, M. D., Sharp, A. K., and Wilson, R. J. H. (1968). Preparation and use of porous sheets with enzyme action. *Nature*, **217**, 641–642.

Line, W. F., Kwong, A., and Weetall, H. H. (1971). Pepsin insolubilized by covalent attachment to glass: Preparation and characterization. *Biochim. Biophys. Acta*, **242**, 194–202.

Maeda, H., Suzuki, H., Yamaguchi, A., and Sakimae, A. (1975). Preparation of immobilized enzymes from acrylic monomers under x-ray irradiation. *Biotechnol. Bioeng.*, **17**, 119–128.

Mattiasson, B., and Mosbach, K. (1971). Studies on a matrix-bound three-enzyme system. *Biochim. Biophys. Acta*, **235**, 253–257.

Morisi, F., Pastore, M., and Viglia, A. (1973). Reduction of lactose content of milk by entrapped β-galactosidase. I. Characteristics of β-galactosidase from yeast and *Escherichia coli*. *J. Dairy Sci.*, **56**, 1123–1127.

Nilsson, H., Mosbach, R., and Mosbach, K. (1972). The use of bead polymerization of acrylic monomers for immobilization of enzymes. *Biochim. Biophys. Acta*, **268**, 253–256.

Okos, E. S., and Harper, W. J. (1974). Activity and stability of β-galactosidase immobilized on porous glass. *J. Food Sci.*, **39**, 88–93.

Olson, A. C., and Stanley, W. L. (1973). Lactase and other enzymes bound to a phenol-formaldehyde resin with glutaraldehyde. *J. Agr. Food Chem.*, **21**, 440–445.

Østergaard, J. C. W., and Martiny, S. C. (1973). The immobilization of β-galactosidase through encapsulation in water-insoluble microcapsules. *Biotechnol. Bioeng.*, **15**, 561–563.

Paine, M. A., and Carbonell, R. G. (1975). Immobilization of β-galactosidase in collodion microcapsules. *Biotechnol. Bioeng.*, **17**, 617–619.

Robinson, P. J., Dunnill, P., and Lilly, M. D. (1971). Porous glass as a solid support for immobilization or affinity chromatography of enzymes. *Biochim. Biophys. Acta*, **242**, 659–661.

Rosensweig, N. S. (1969). Adult human milk intolerance and intestinal lactase deficiency. A review. *J. Dairy Sci.*, **52**, 585–587.

Silman, I. H., and Katchalski, E. (1966). Water insoluble derivatives of enzymes, antigens, and antibodies. *Ann. Rev. Biochem.*, **35**, 872–908.

Stanley, W. L., and Palter, R. (1973). Lactase immobilization on phenolic resins. *Biotechnol. Bioeng.*, **15**, 597–602.

Stanley, W. L., Watters, G. G., Chan, B., and Mercer, J. M. (1975). Lactase and other enzymes bound to chitin with glutaraldehyde. *Biotechnol. Bioeng.*, **17**, 315–326.

Thompson, M. P., and Brower, D. P. (1975a). Hydrolyzed lactose cultured dairy products. I. Cheddar cheese. *J. Dairy Sci.*, in press.

Thompson, M. P., and Brower, D. P. (1975b). Hydrolyzed lactose cultured dairy products. II. Yoghurt, buttermilk and cottage cheese. *J. Dairy Sci.*, in press.

Wang, S. S., and Vieth, W. R. (1973). Collagen-enzyme complex membranes and their performance in biocatalytic modules. *Biotechnol. Bioeng.*, **15**, 93–115.

Weetall, H. H. (1969). Trypsin and papain covalently coupled to porous glass: Preparation and characterization. *Science*, **166**, 615–617.

Weetall, H. H., Havewala, N. B., Pitcher, W. H., Jr., Detar, C. C., Vann, W. P., and Yaverbaum, S. (1974a). The preparation of immobilized lactase and its use in the enzymatic hydrolysis of acid whey. *Biotechnol. Bioeng.*, **16**, 295–313.

Weetall, H. H., Havewala, N. B., Pitcher, W. H., Jr., Detar, C. C., Vann, W. P., and Yaverbaum, S. (1974b). Preparation of immobilized lactase. Continued studies on the enzymatic hydrolysis of lactose. *Biotechnol. Bioeng.*, **16**, 689–696.

Wendorff, W. L., Amundson, C. H., Olson, N. F., and Garver, J. C. (1971). Use of yeast beta-galactosidase in milk and milk products. *J. Milk Food Technol.*, **34**, 294–299.

Wierzbicki, L. E., and Kosikowski, F. V. (1973a). Food syrups from acid whey treated with β-galactosidase of *Aspergillus niger*. *J. Dairy Sci.*, **56**, 1182–1184.

Wierzbicki, L. E., and Kosikowski, F. V. (1973b). Kinetics of lactose hydrolysis in acid whey by β-galactosidase from *Aspergillus niger*. *J. Dairy Sci.*, **56**, 1396–1399.

Wierzbicki, L. E., Edwards, V. H., and Kosikowski, F. V. (1973). Immobilization of microbial lactases by covalent attachment to porous glass. *J. Food Sci.*, **38**, 1070–1073.

Wondolowski, M. V., and Woychik, J. H. (1974). Characterization and immobilization of *E. coli* (ATCC-26) β-galactosidase. *Biotechnol. Bioeng.*, **16**, 1633–1644.

Worthington, C. (1974). Use of membrane-bound enzymes in an immobilized enzyme reactor. *Adv. Exp. Med. Biol.*, **42**, 235–240.

Woychik, J. H., and Wondolowski, M. V. (1972). Covalent bonding of fungal β-galactosidase to glass. *Biochim. Biophys. Acta*, **289**, 347–351.

Woychik, J. H., and Wondolowski, M. V. (1973). Lactose hydrolysis in milk and milk products by bound fungal beta-galactosidase. *J. Milk Food Technol.*, **36**, 31–33.

Woychik, J. H., Wondolowski, M. V., and Dahl, K. J. (1974). Preparation and application of immobilized β-galactosidase of *Saccharomyces lactis*, In *Immobilized Enzymes in Food and Microbial Processes*, pp. 41–49 (A. C. Olson and C. L. Conney, eds.), Plenum Publishing Corp., New York.

Zaborsky, O. R. (1973). *Immobilized Enzymes*. CRC Press, Cleveland, Ohio.

MECHANISM OF ACTION OF FUNGICIDES AND THEIR REACTIVITIES WITH CELLULAR AND ENVIRONMENTAL SUBSTRATES

H. P. BURCHFIELD and ELEANOR E. STORRS

Gulf South Research Institute, P.O. Box 1177,
New Iberia, Louisiana 70560, U.S.A.

Summary

Consideration of steric and electronic interactions is important in the design of fungicides which have the capacity for combining with cellular and environmental substrates by S_N2 reactions. These interactions can be both intramolecular and intermolecular. Examples are given of the effects of structure on chemical and biological activity of benzene and *s*-triazine derivatives. In general, compounds of this class react with environmental substrates containing amino and thiol groups by the same mechanisms through which they react with intercellular metabolites.

INTRODUCTION

All biologically active compounds including fungicides must function through the formation of chemical bonds with biological substrates. These bond types include Van der Waals, hydrogen, charge-transfer, ionic, dative, and covalent. Often, several bond types may be involved in potentiating the reactions of toxicants with enzymes and other macromolecules.

The formation of covalent bonds can be most easily studied directly, and has been best documented. However, these bonds can be formed by a variety of mechanisms including, among others, addition reactions, redox reactions, first order nucleophilic substitution reactions (S_N1 mechanism), and second order nucleophilic substitution reactions (S_N2 mechanism) (Burchfield and Storrs, 1969).

This paper is concerned with electronic and steric effects which govern the formation of covalent bonds between fungicides and various cellular and environmental substrates which react by the S_N2 mechanism. Two types of effects are considered, intramolecular and intermolecular. Intramolecular effects are concerned primarily with electronic and steric interactions within the fungicide molecule which influence its innate chemical reactivity. Intermolecular effects are concerned primarily with interactions between fungicide molecules and biological substrates. Of course, the two cannot be completely separated since transition states formed between reacting pairs of molecules are unique. Nevertheless, the following section on benzene derivatives is concerned primarily with intramolecular effects, and the section on *s*-triazines with intermolecular effects.

It is believed that a better understanding of these interactions for all classes of fungicides will lead to a systematic approach to the development of new biologically active compounds which are readily degradable in the environment.

BENZENE DERIVATIVES

The simplest but at the same time most successful fungicides which function by undergoing S_N2 reactions with biologic substrates are benzene derivatives. Fungicidal compounds in this series always contain a leaving group which is replaced during the reaction of the fungicide with various substrates, and activating groups which decrease electron density at the carbon atom to which the leaving group is attached through resonance and inductive effects. The activating groups are usually positioned *ortho* or *para* to the leaving group although *meta* substitution can modify reactivity through inductive effects. Activating *and* leaving groups are always highly electronegative, and their roles are determined by the overall electronic and steric configuration of the molecule. Examples of such groups are F, Cl, Br, NO_2, CN, SCN, RSO_2, $^+N(CH_3)_3$, and others.

Kinetic studies have shown that the most effective activating groups are those that are capable of stabilizing a negative charge by an electromeric process. This fact is readily explained by considering the reaction of 1-fluoro-2,4-dinitrobenzene (FDNB), a compound with fungicidal properties, with an ionized thiol such as cysteine. The rate controlling transition state is represented by IIa ↔ IIb. Structures Ib, IIb, and IIIb can make their contribution to these hybrids only if the nitro group can become coplanar to the benzene ring. Valence bond structures can also be written involving the *ortho* nitro group.

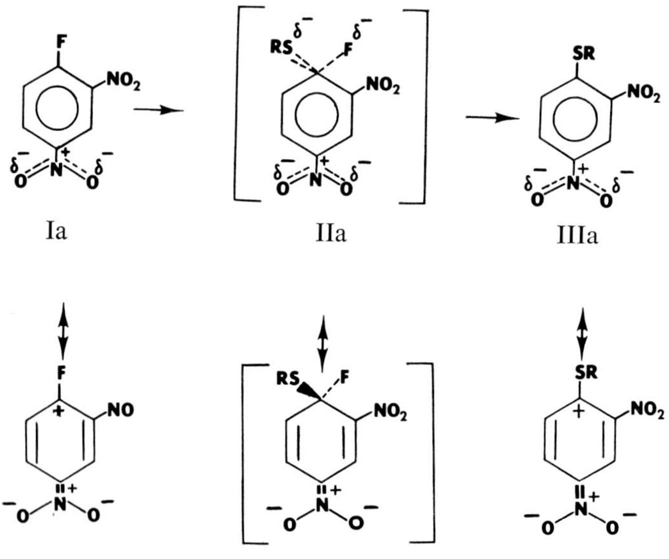

Ia IIa IIIa

Ib Reactants · IIb Transition state · IIIb Product

In the example given here, FDNB reacts rapidly with the ionized thiol group of cysteine to yield S-2,4-dinitrophenylcysteine (IV). However, this compound subsequently undergoes a Smiles' rearrangement (Burchfield, 1958), which results in the conversion to N-2,4-dinitrophenylcysteine (V).

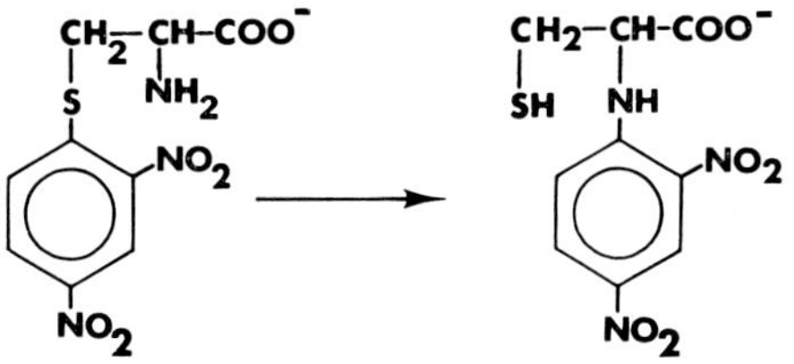

IV V

In the first step of this reaction, the steric requirement for coplanarity is much more important to the stabilization of the transition state than to the reactants or products. Thus, the more nearly planar the activating group can become with the ring in the transition state, the lower the energy of that transition state and the faster the reaction proceeds. For example, compound VI reacts only 4% as rapidly as compound VII with nucleophilic substrates since the methyl groups adjacent to the nitro group prevent it from becoming coplanar with the benzene ring in the reactant and transition states (Spitzer and Wheland, 1940).

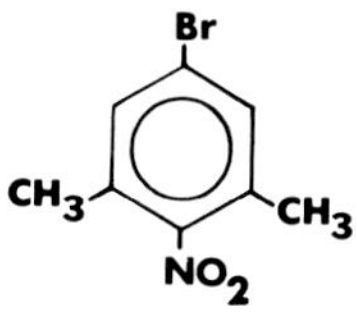

VI

VII

The interactions of steric and electronic configurations are critical in the design of molecules which have practical applications as fungicides since to achieve the desired level of biological activity at least 5 of the 6 positions in the benzene ring must contain substituent groups other than hydrogen. While FDND and its analogue 1-chloro-2,4-dinitrobenzene (CDNB) are toxic to fungus spores, they have not been successful commercial fungicides, presumably because their molecular weights are too low to obtain the required persistence and/or uptake by fungus spores. Benzene derivatives which can undergo S_N2 reactions with biological substrates and which have also proven to be practical fungicides are compounds VIII–XI.

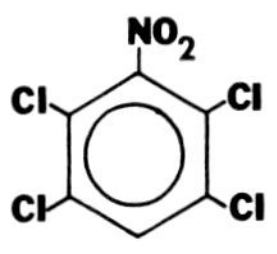

TCNB
VIII

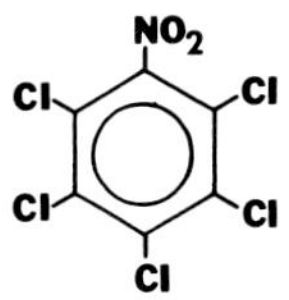

PCNB
IX

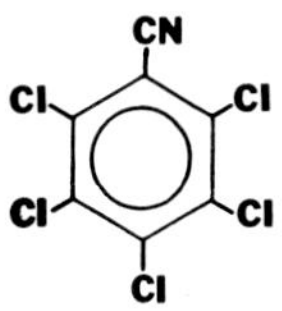

daconil
X

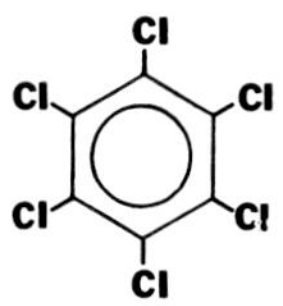

HCB
XI

In the case of PCNB (IX), the activating groups are the chlorine atoms and the leaving group is the nitro group as shown by the fact that alkaline hydrolysis of this compound yields nitrite ion and pentachlorophenol

(Wheland, 1955). Also, a major metabolite of this compound in rabbits is compound XII (Burchfield and Storrs, 1969).

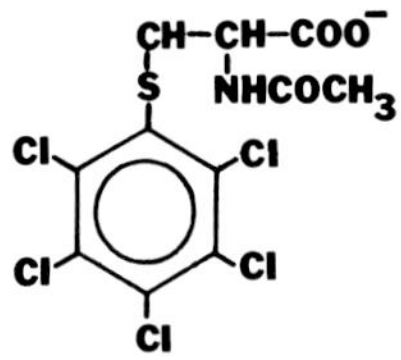

XII

As with compound VI, the substituents *ortho* to the nitro group prevent it from becoming coplanar with the aromatic ring in the transition state and thus reduce its activating power. However, rather than reducing the reactivity of the chlorine atoms, the nitro group becomes the leaving group. This is also the case for compound VIII.

Daconil (X) is the most successful fungicide of this class so far developed and is also one of the best examples for illustrating how the interaction of steric and electronic effects can be utilized in the design of new compounds with biological activity.

In this molecule the nitrile groups combined with chlorine atoms substitute for nitro groups as the activating moieties. Nitrile groups are almost as highly activating as nitro groups as shown by the fact that they increase the acid strengths of phenols nearly as much (Wheland, 1955). Moreover, unlike the nitro group, the nitrile group is linear; therefore, *ortho* substitution does not force it out of coplanarity with the benzene ring so that its activating power is maintained even in highly substituted molecules. Also, the atomic dimensions are small even compared with halogens. Thus the C-triple bond radius is 0·6 Å, the N-triple bond radius is 0·55 Å while the Cl-single bond radius is 1·0 Å. This combination of high electron withdrawing power, linearity with the benzene ring, and short covalent bond radii make the nitrile group nonpareil as an activating group.

The daconil molecule has four positions at which substitution could occur: (a) the two chlorine atoms *ortho* to nitrile groups and a chlorine atom; (b) the chlorine atom sandwiched between the two nitrile groups; (c) the chlorine atom *meta* to the nitrile groups; and (d) the nitrile groups themselves.

Actually, reaction could take place at all four sites. Vincent and Sisler (1968) have shown that four derivatives are obtained on causing daconil to react with 2-mercaptoethanol.

They proposed that the chlorine atoms are substituted sequentially, and that the final reaction product is

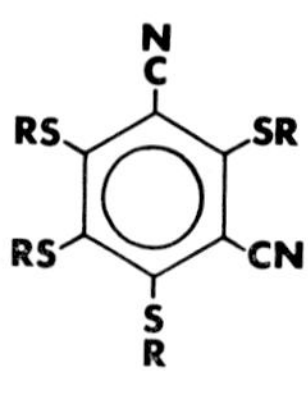

XIII

However, this does not seem likely on theoretical grounds, since each substitution of an electronegative halogen atom by an electron-releasing thiol group should greatly reduce the reactivities of the remaining halogens (*see* Ib for comparison).

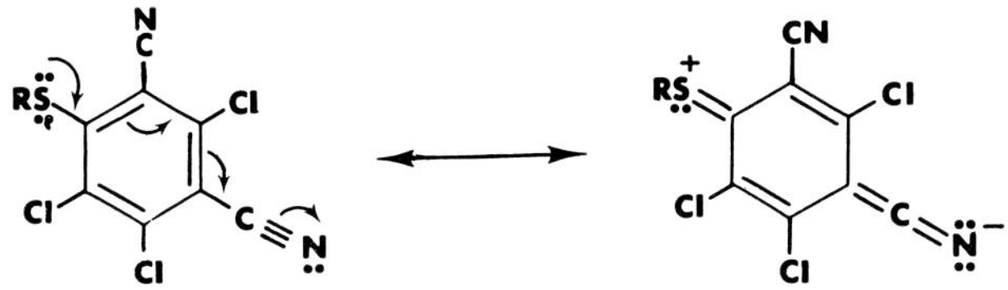

XIVa XIVb

The concept of multiple primary reaction sites rather than a series of consecutive reactions at different sites on the same molecule is confirmed by the studies of Turner and Battershell (1969, 1970), who investigated the reactivities of various halogen derivatives of isophthalonitrile with 4-nitrothiophenol.

The least reactive and also the least fungitoxic compounds investigated were:

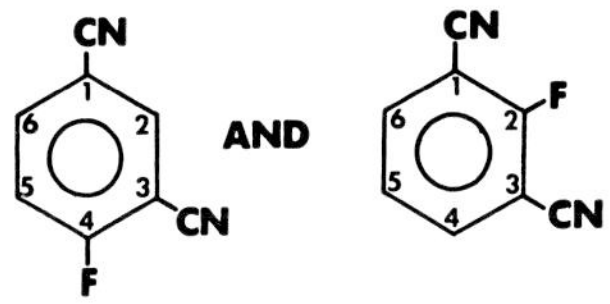

XV XVI

These compounds reacted with the substrate at approximately equal rates. Therefore, two reaction products (one from each compound) must have been produced in approximately equal amounts: one with a sulphide group *ortho* to both nitrile groups, and one with a sulphide group *ortho-para* to the nitrile groups. Therefore, if fluorine atoms were substituted in both the 2 and 4 positions of the same benzene ring, the formation of two reaction products would be expected. It should be noted that compound XV is the nitrile analogue of FDNB (Compound I).

The next two compounds with respect to increased chemical reactivity and fungitoxicity contained two chlorine atoms substituted in different positions on the benzene ring.

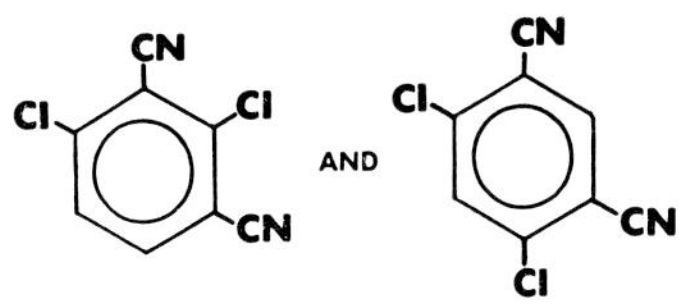

XVII XVIII

Again, the two compounds were almost identical with respect to fungitoxicity and chemical reactivity with the substrate. However, XVII should yield two isomers, and XVIII only one isomer following reaction with nucleophilic substrates.

Substitution of five of the six positions in the benzene ring with nitrile and halogen groups further increased chemical reactivity with the substrate as well as biological activity. These compounds are:

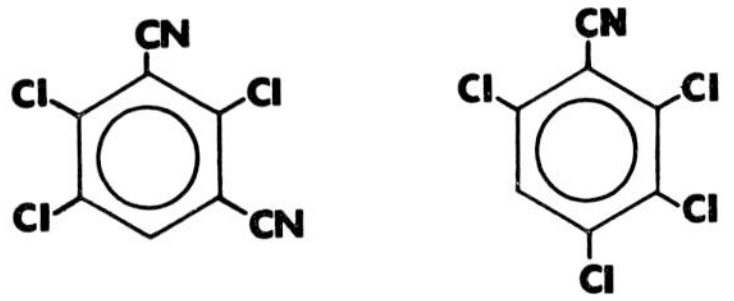

XIX XX

Compound XIX could produce three or four isomers as the result of primary S_N2 reactions, and compound XX only two or three, depending on whether or not the CN group is replaced.

Maximum chemical reactivity and fungitoxicity were achieved when all six positions of the benzene ring were substituted with nitrile and halogen groups. The most practical fungicide in this series for crop protection was daconil. However, this was not the most highly active compound with respect to both chemical reactivity with thiol groups and fungitoxicity. More reactive was a daconil analogue in which the halogen atom in the 2-position sandwiched between the nitrile groups was fluorine (XXI). This reflects relative freedom from steric effects of the nitrile groups as well as the high reactivity of fluorine atoms in aromatic S_N2 reactions.

Chemical reactivity with the substrate was augmented when the fluorine atom was located in the 6-position of the benzene ring (XXII), and when all of the chlorine atoms were replaced by fluorine (XXIII).

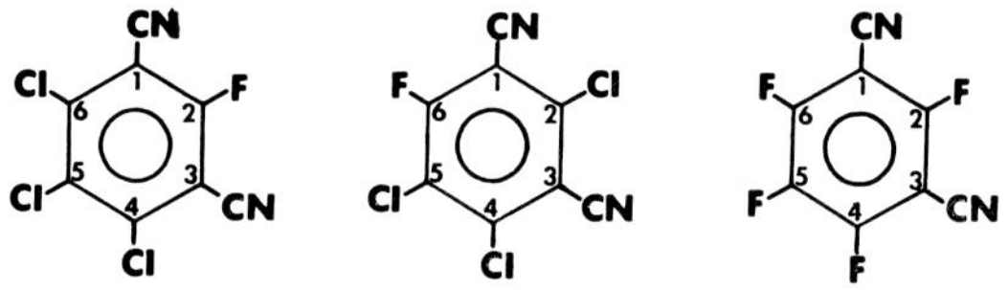

XXI XXII XXIII

Fungitoxicity was enhanced when two chlorine and two fluorine atoms were substituted in the benzene ring, although no further increase in chemical reactivity with the substrate was observed.

It is evident that the nitrile group is a more satisfactory activating group than the nitro group, and that many opportunities still exist for developing new fungicides having single, multiple, and condensed aromatic ring systems.

Chloraneb (XXIV) is a benzenoid fungicide which would *not* be expected to react with cellular and environmental substrates through S_N2 mechanisms. The oxygen atoms of the ether groups are electron releasing; hence the chlorine atoms should be even less reactive than those of chlorobenzene. Obviously, this compound must be fungitoxic through an entirely different mechanism *or* it must be metabolized by plants and/or fungus spores to a more reactive compound. It has been shown that chloraneb is converted to

XXVII by cotton and bean plants (Rhodes *et al.*, 1971). The amount found is small but this may result from the fact that chloroquinones react rapidly with substrates containing amino or thiol groups because of the electron withdrawing power of the conjugated carbonyl groups. It will be recalled that tetrachloro-*p*-benzoquinone (chloranil) and 2,3-dichloro-1,4-naphthoquinone (dichlone) were once important agricultural fungicides, and it is known that these compounds can undergo S_N2 reactions with cellular and environmental substrates. However, chloranil and to a lesser extent dichlone can undergo photochemical decomposition in the environment, resulting in poor persistence. Also, dichlone is somewhat phytotoxic, particularly when formulated with oils. Chloraneb should be photochemically stable until intracellular conversion to the reactive quinone. Thus in the design of new fungicides, the reactivities of both the parent compounds and potential metabolites should be considered. Many insecticides and carcinogens are known to be activated *in vivo*. In the parlance of cancer research, chloraneb would be designated the proximate fungicide, and the quinone the ultimate fungicide. This is a relatively unexplored principle in the design of new fungicides.

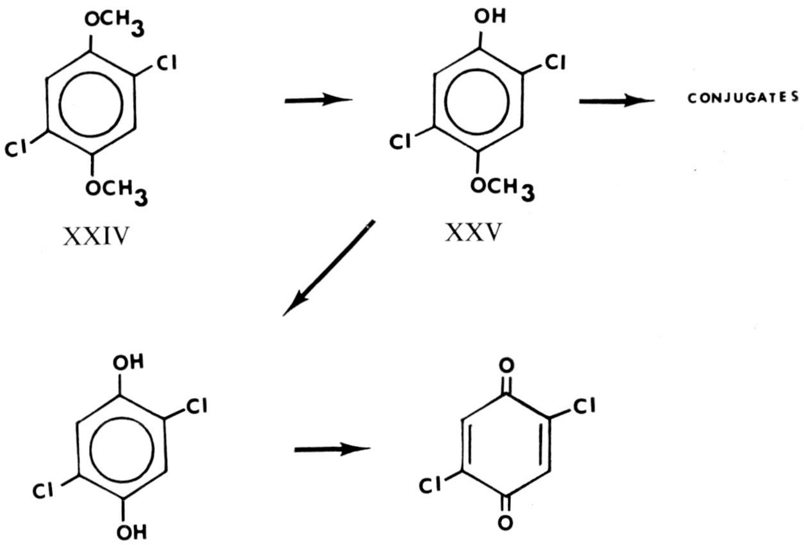

XXIV XXV XXVI XXVII

TRIAZINE DERIVATIVES

The interaction of steric and electronic effects is important in determining the biological activity of compounds containing a single aromatic ring. However, the situation is much more complicated in the *s*-triazine series of biocides in which an aromatic and a heterocyclic ring can interact and substituents on both rings can be changed. Also, steric effects which govern the binding of these compounds to protein surfaces and their uptake by fungus spores must be taken into account.

Compounds in the triazine series differ greatly with respect to their stabilities in the environment, reactivities with biological substrates, and biological specificity (Burchfield and Storrs, 1969). Typical members of this group are illustrated below.

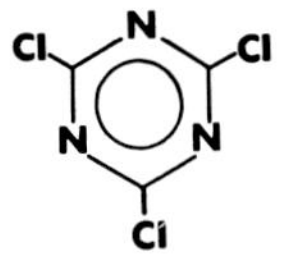

XXVIII

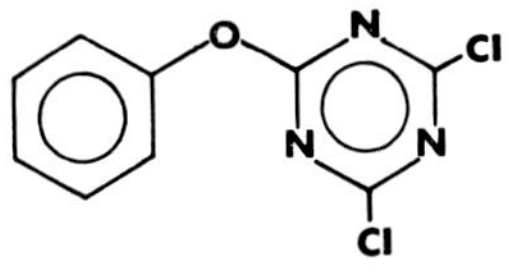

XXIX

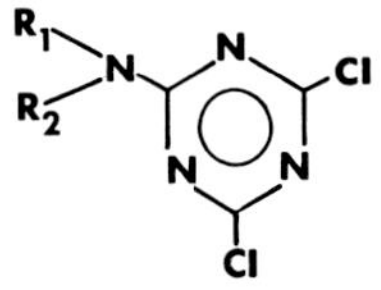

XXX

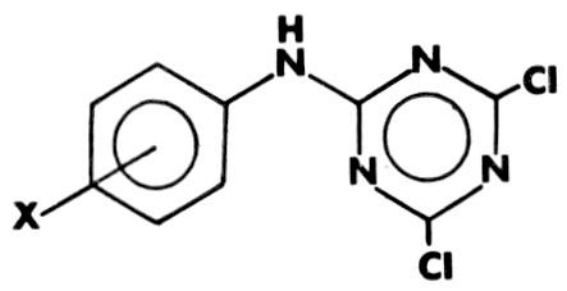

XXXI

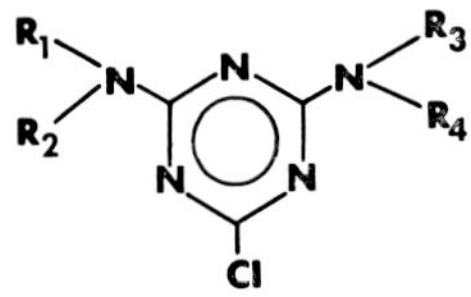

XXXII

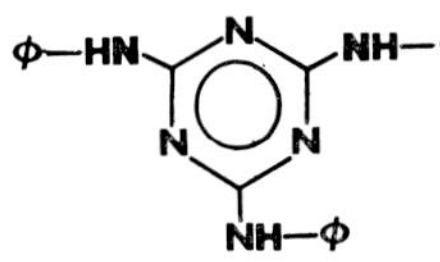

XXXIII

The parent compound, cyanuric chloride (XXVIII), contains three chlorine atoms substituted in the heterocyclic ring and hydrolyses almost immediately on exposure to water. Thus the compound does not survive in the environment for a long enough time to be effective as a fungicide. However, when a phenoxy group is substituted for a chlorine atom on the triazine ring, chemical stability is somewhat improved. Thus the compound illustrated above (XXIX) has a half-life of about one-half to one hour in buffered aqueous media. These compounds are fungitoxic, but only weakly so. Evidently, the enhanced chemical stability conferred by the phenoxy group is only moderately successful in promoting biological activity. The dichloroalkylamino-*s*-triazines (XXX) are somewhat more stable than the phenoxy compounds, but tend to be phytotoxic as well as fungicidal. The most useful fungicides of this group are represented by the dichloroanilino-*s*-triazines (XXXI). These compounds are stable in aqueous buffer for 20–30 days. However, their half-lives in the presence of dilute solutions of thiols (10^{-4} M) are of the order of one-half hour. Hence these compounds are sufficiently stable to survive in aqueous media, but on the other hand are highly reactive with compounds containing thiol, amino, and other nucleophilic groups. When two of the chlorine atoms in the heterocyclic ring are replaced by substituted amino groups (XXXII), a series of compounds is produced having an entirely different spectrum of biological activity. Whereas the dichloroanilino-*s*-triazines are fungicides, the monochloro-bis(alkylamino)-*s*-triazines are herbicides. These compounds survive hydrolysis almost indefinitely in water. Nevertheless, they have been observed to react with thiol groups, although at rates much smaller than for the dichloro-*s*-triazines. The compound XXXIII in which all three halogens are replaced by anilino groups is biologically inert, and does not react at observable rates with amino and thiol groups. Thus the reaction velocities of the *s*-triazines with the hydroxyl ion and with various metabolic substrates can be varied enormously by changes in the nature of the substituent groups. This series includes compounds which are essentially nontoxic, compounds which are good foliage fungicides, and compounds which are active as herbicides. Changes in reaction rates would appear to contribute at least in part to this broad spectrum of biological activity.

Smaller changes can be produced by varying substituents within a narrower range (Burchfield and Storrs, 1962). Thus in the dichloro-anilino-*s*-triazine series (XXVII), it is possible to introduce various substituents such as chlorine, bromine, methoxy, methyl, and ethoxy into the *meta* and *para* positions of the benzene ring. Some of these groups are activating compared to hydrogen due to inductive effects, while others are deactivating by resonance-polar effects. A Hammett plot of the second-order velocity coefficients for the reactions of these compounds with glycine and glutathione is shown in Fig. 1. It is evident that the reaction velocities are proportional

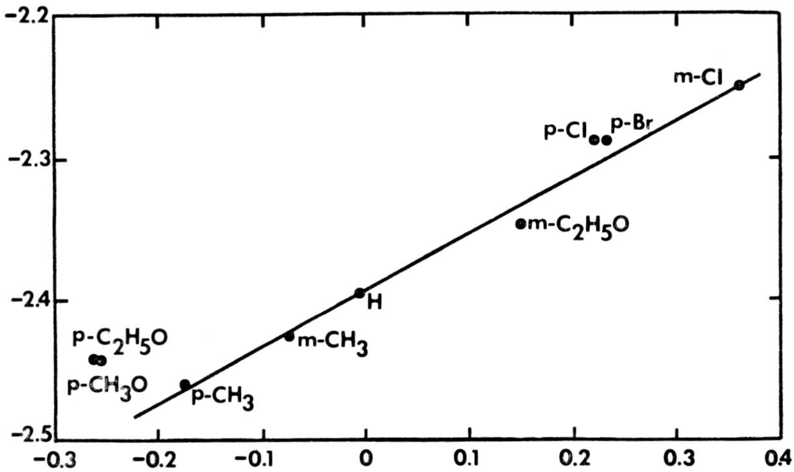

Fig. 1. Relation between Hammett substitution constants and second-order velocity coefficients for the s*-triazines. Ordinate: log* k *(mole litre*$^{-1}$ *sec*$^{-1}$).

to the $\sigma\rho$ substitution constants. Thus the reactivities of the chlorine atoms in the triazine ring in these compounds can be regulated by changing the substituent group in the benzene ring.

Substitution of a single chlorine in the benzene ring of 6-anilino-2,4-dichloro-*s*-triazine increases reactivity in the order

$$o\text{-chloro} > m\text{-chloro-} > p\text{-chloro}$$

However, chemical reactivity with simple metabolic substrates such as cysteine is believed to account only in part for the biological activities of members of this group of compounds, since secondary valence bonds may be involved in reactions with more complex molecules.

Some evidence for this has been obtained by studies on the reaction rates of mercaptalbumin with various dichloro-6-anilino-*s*-triazine derivatives (Storrs, 1958). Mercaptalbumin does not possess enzymatic activity; nevertheless, it contains a number of centres (3–13) which are highly reactive with compounds such as the *s*-triazines and FDNB. Thus the fungicide dyrene reacts with mercaptalbumin at a rate about a hundred times faster than would be predicted from the number and nature of the terminal and side chain functional groups of the protein. This indicates that intermediate complex formation is probably involved. Comparative results obtained on the kinetics of the reactions which take place between derivatives of 2,4-dichloro-6-anilino-*s*-triazine with glycine and mercaptalbumin serve to illustrate this. The reactivities of several of these compounds with glycine and the protein are compared in Table I. Reactivity with the parent triazine (hydrogen in the *para* position in the benzene ring) is assigned an arbitrary value of unity. It should be remembered that the reactivities of the triazines with glycine and mercaptalbumin differ by several orders of magnitude. The relative reactivities of the *s*-triazines with glycine decrease slightly when a methyl group is introduced into the *ortho*, *meta*, or *para* positions of the benzene ring. This is caused by the electron-releasing capacity of the methyl group. The reduction in reactivity is quite small and is not influenced greatly by the position of the methyl group on the benzene ring. However, the situation with respect to the

Table I
RELATIVE REACTIVITIES OF GLYCINE AND MERCAPTALBUMIN WITH DICHLORO-ANILINO-*s*-TRIAZINES CONTAINING METHYL GROUPS IN VARIOUS POSITIONS

Substituent	*Relative reactivity*	
	Glycine	*Mercaptalbumin*
H	1·0	1·0
m-CH_3	0·92	2·4
p-CH_3	0·86	3·0
o-CH_3	0·92	0·26

influence of methyl groups on reaction rates with mercaptalbumin is markedly different. The introduction of a methyl group in the *meta* or *para* positions of the benzene ring results in a two- to three-fold increase in chemical reactivity. This can be ascribed to increased binding of the triazines to the protein surface by hydrogen bonds and van der Waals forces. Thus substitution of a methyl group in the *meta* or *para* positions of the benzene ring tends to potentiate chemical reactivities with the protein, in contrast to the small reduction in reaction rates observed with the amino acid glycine.

However, it is equally important to note that substitution of a methyl group in the *ortho* position of the benzene ring (next to the bridge atom) tends to decrease reactivity substantially below that of the parent compound. Also, the *ortho* derivative is only about 10% as reactive with the protein as the *meta* derivative. Evidently, the presence of an *ortho* group in a position adjacent to the imino bridge connecting the benzene and heterocyclic ring inhibits the formation of an intermediate complex since the nucleophilicities of the three compounds are approximately the same. This suggests that the imino bridge of the *m*-methyl- and *p*-methyl-*s*-triazines potentiates reactivity by forming hydrogen bonds with functional groups on the protein surface such as with the OH group of tyrosine. Evidence supporting this theory has been obtained from studies on the reaction rates of *s*-triazine derivatives with free amino acids. It has been shown that *o*-, *m*-, and *p*-methyl derivatives react at substantially the same rates with all free amino acids with the exception of tyrosine. In the case of tyrosine, the *ortho* derivative is about 20% less reactive than the other isomers. This suggests that the reactivities of the *meta* and *para* compounds are potentiated by hydrogen bonding between the imino group of the triazine and the OH group of tyrosine. Hydrogen bonding would be reduced in the case of the *ortho* derivative due to shielding of the imino bridge by the methyl group.

In the case of mercaptalbumin, it is possible that a shift from a semi-planar to a nonplanar molecular conformation also contributes to the lower reactivity of the *ortho* derivative. Thus a methyl group adjacent to the imino bridge connecting the triazine ring with the benzene ring could reduce coplanarity by van der Waals repulsion forces. Molecular models show that steric hindrance is not sufficient to interfere with free rotation as is the case with certain *ortho*-substituted biphenyls. Nevertheless, the conformation with the lowest potential energy is probably nonplanar.

To test the validity of this theory, u.v. absorption spectra of these compounds were measured in acetonitrile solution (Fig. 2). It was found that the

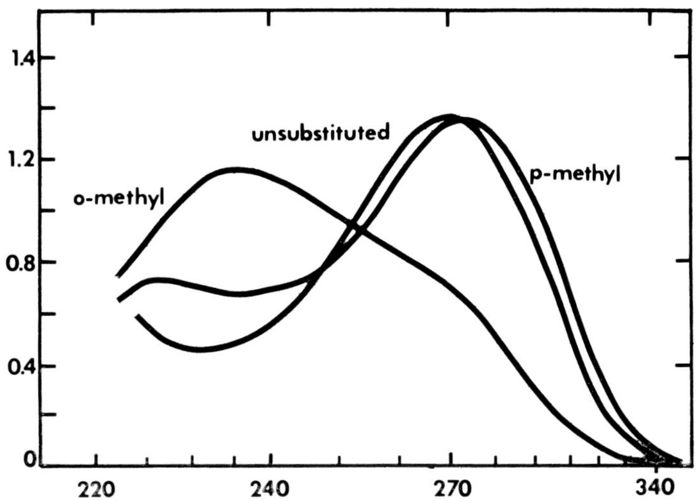

Fig. 2. Influence of substitution and isomerism on u.v. absorption spectra of dichloro-6-anilino-s-triazine derivatives in acetonitrile solution. Ordinate: absorbence.

spectra indicated a change in conformation (Burchfield and Storrs, 1960). Substitution of a methyl group in the *para* position of the benzene ring caused a slight red shift from 269 to 272 nm. By contrast, introduction of this same group into the *ortho* position suppressed the molecular absorbance in the 270 nm region and resulted in a new maximum at 237 nm. This could only have resulted from an uncoupling of the resonance interaction between the two rings caused by a shift from a semiplanar to a nonplanar molecular conformation (XXXIV).

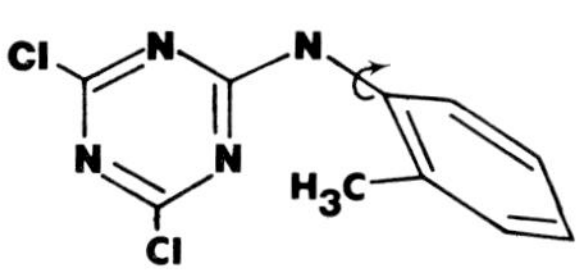

XXXIV

Therefore, it is evident that molecular conformation is an important factor in governing the affinity of these materials for mercaptalbumin. Evidently, binding of the *s*-triazine to the protein surface takes place more readily if the molecule tends to be planar.

The *ortho* effect has also been found to reduce the amount of fungicide which can be taken up by fungus spores, thus changing the effective dose available within the cells to inactivate metabolites.

To study this in more detail, uptake of several derivatives by conidia of *Neurospora sitophila* was measured at a series of doses, and the amount of fungicide taken up plotted against the dose applied (Fig. 3). A semi-logarithmic relationship between dose and uptake was obtained in all cases. However, the *para* methyl derivative was taken up very readily, the *ortho*

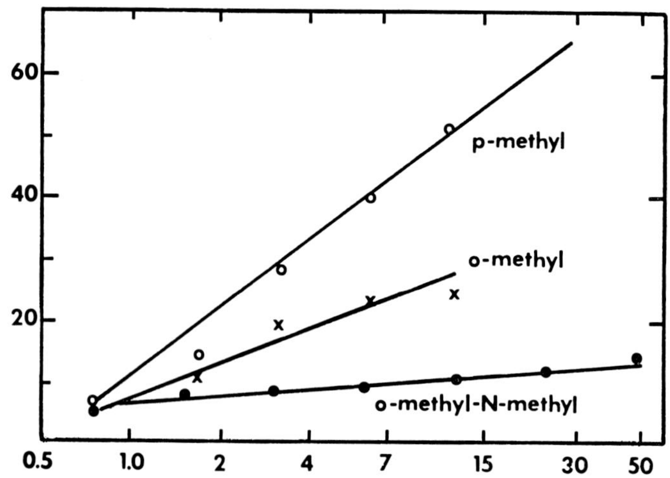

Fig. 3. Dose of dichloro-6-triazine derivatives versus *uptake for* Neurospora sitophila. *Ordinate: amount take up (μg/mg spores).*

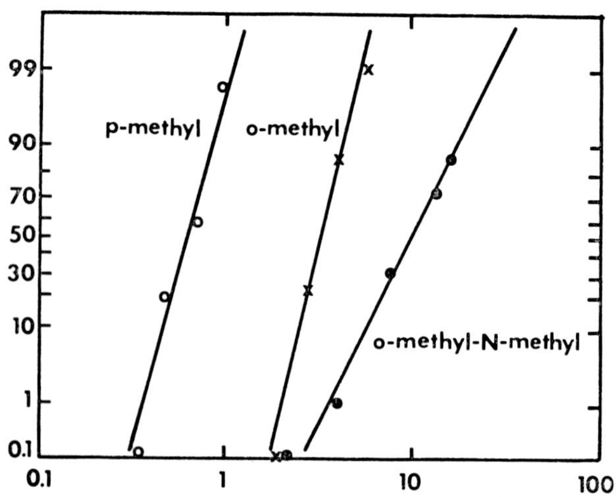

*Fig. 4. Influence of substitution and isomerism of dichloro-6-anilino-*s*-triazine derivatives on spore germination. Ordinate: Percent of spores ungerminated.*

derivative much less readily, and the *o*-methyl-N-methyl derivative in small amounts only. To determine if these uptake values were reflected in relative toxicities as measured by ED_{50} values, a series of germination tests on glass slides were made using conidia of *N. sitophila* as the test organism. The results were in very good agreement with those found in uptake studies (Fig. 4). The *p*-methyl derivative had an ED_{50} of about 0·5 ppm, while the *o*-methyl derivative had an ED_{50} of 10 ppm. The *o*-methyl derivative was intermediate in toxicity. Evidently, these three compounds differ in toxicity mainly because of the relative amounts taken up by fungus spores. The ED_{50} on a spore-weight basis is of the same order of magnitude for all three compounds. It is interesting to note that the effects involved here parallel very closely the affinities of the compounds for some protein surfaces as measured by their reaction rates with mercaptalbumin. Thus it is seen once more that substituent groups in the molecule can have a pronounced effect in altering biological specificity even though the mode of action of the compound remains essentially the same.

REFERENCES

Betts, J. J., James, S. P., and Thorpe, W. V. (1955). Metabolism of pentachloronitrobenzene and 2,3,4,6-tetrachloronitrobenzene and the formation of mercapturic acids in the rabbit. *Biochem. J.*, **61**, 611–617.

Burchfield, H. P. (1958). Molecular rearrangement in the reaction of cysteine with 1-fluoro-2,4-dinitrobenzene. *Nature*, **181**, 49.

Burchfield, H. P., and Storrs, E. E. (1960). An *ortho* effect influencing fungitoxicity, electronic absorption spectra, and chemical reactivity with mercaptalbumin. *Nature*, **187**, 593.

Burchfield, H. P., and Storrs, E. E. (1962). Applications of the Hammett equation to interannular group interactions in the *s*-triazine series. Southwest Regional Meeting, American Chemical Society, Dallas, Texas.

Burchfield, H. P., and Storrs, E. E. (1969). Electronic and steric aspects of fungicidal action. In *Fungicides*, Vol. II, Chapter 2 (D. C. Torgeson, ed.), Academic Press, New York.

Kolbezen, M. F. (1964). PCNB. In *Analytical Methods for Pesticides, Plant Growth Regulators and Food Additives*, Vol. III, Chapter 12 (G. Zweig, ed.), Academic Press, New York.

Rhodes, R. C., Pease, H. L., and Brantley, R. K. (1971). Fate of C^{14}-labelled chloroneb in plants and soils. *J. Agr. Food Chem.*, **19**, 745.

Spitzer, W. C., and Wheland, G. W. (1940). Steric inhibition of resonance (II) rates of reaction of nitro- and cyano-aryl halides. *J. Am. Chem. Soc.*, **62**, 2995.

Storrs, E. E. (1958). The interactions of mercaptalbumin fraction of bovine serum albumin with 2,4-dichloro-6-anilino-*s*-triazine derivatives and 1-fluoro-2,4-dinitrobenzene. M.S. Thesis, New York University.

Turner, N. J., and Battershell, R. D. (1969). The relative influence of chemical and physical properties on the fungitoxicity of tetrachloroisophthalonitrile and some of its analogues. *Contrib. Boyce Thompson Inst.*, **24**, 139–147.

Turner, N. J., and Battershell, R. D. (1970). The relationship between chemical reactivity, oil: water partitioning and temperature on the rate of fungicidal action of tetrachloroisophthalonitrile and some of its analogues. *Contrib. Boyce Thompson Inst.*, **24**, 203–212.

Vincent, P. G., and Sisler, H. D. (1968). Mechanisms of antifungal action of 2,4,5,6-tetrachloroisophthalonitrile. *Physiol. Plantarum*, **21**, 1249–1264.

Wheland, G. W. (1955). *Resonance in Organic Chemistry*, p. 368, Wiley, New York and London.

MUTATIONAL RESISTANCE TO SITE-SPECIFIC FUNGICIDES

S. G. GEORGOPOULOS

N.R.C. Demokritus, Aghia Paraskevi, Athens, Greece

Summary

Knowledge on the mechanisms involved in mutational resistance to fungicides acting by site-specific inhibition has increased in recent years. Mutational alteration of the sensitive cellular site seems to be responsible for resistance to benzimidazole fungicides in *Aspergillus nidulans*, to carboxamides in *Ustilago maydis* and *A. nidulans*, to cycloheximide in *Saccharomyces cerevisiae* and *Neurospora crassa* and, probably, to kasugamycin in *Piricularia oryzae*, and to dimethirimol in *Sphaerotheca fuliginea*. By contrast, resistance to blasticidin S in *P. oryzae* and to polyoxin D in *Alternaria kikuchiana* are reported to result from reduced membrane permeability.

Most of the agriculturally important fungicides which were used until the latter part of the 1960s were compounds of indiscriminate reactivity, therefore acting as general toxicants in fungal cells (Lukens, 1968). With the use of such compounds, resistant strains were very rare and very few studies on biochemical mechanisms of fungicide resistance were carried out. This situation was changed with the recent introduction of some site-specific antifungal antibiotics and some synthetic systemic fungicides which have acquired major importance. Mutants resistant to such compounds are usually easy to obtain in the laboratory and often have caused large-scale failures of disease control in the field. In some cases attempts to elucidate the biochemical mechanism of resistance to site-specific fungicides have been made and our knowledge on this subject today will be reviewed in this paper.

Benzimidazoles

This group, comprising the important systemics benomyl, thiabendazole (TBZ), fuberidazole (FBZ), and thiophanate-methyl, active against a great variety of fungi, is most seriously threatened by the development of resistant strains (Harding, 1972; Georgopoulos and Dovas, 1973; Berger, 1973; Wicks, 1974). A considerable amount of work has been done on the mode of action, particularly of the methyl ester of benzimidazole-2-ylcarbamic acid (MBC) which is the important breakdown product of benomyl and of thiophanate-methyl. In haploid cells of *Ustilago maydis*, *Saccharomyces cerevisiae*, and *Aspergillus nidulans* MBC prevents mitosis and its effects on biosynthetic processes and on cytocynesis are secondary (Hammerschlag and Sisler, 1973; Davidse, 1973). When heterozygous diploid colonies of *A. nidulans* are exposed to benzimidazole fungicides, at concentrations which do not completely prevent growth, somatic segregation is induced which apparently results from nondisjunction of occasional chromosome pairs (Hastie, 1970; Kappas *et al.*, 1974). The evidence available, therefore, indicates a colchicine-type effect, by interference with subunit assembly for the formation of spindle microtubuli.

Davidse (1974) studied the mechanism of MBC resistance in *A. nidulans*. In this fungus two loci conditioning sensitivity to benzimidazole derivatives are known, mutation at each resulting in a different tolerance level (Hastie and Georgopoulos, 1971). Van Tuyl *et al.* (1974) who studied a large number of mutants at the high resistance locus *ben*-1 made the very interesting observation that positively correlated cross-resistance between MBC and thiabendazole (TBZ) is not a rule without exception. Rare mutations within this locus may give resistance to MBC but not to TBZ or increased resistance to TBZ with increased sensitivity to MBC. This provided Davidse (1974) with two allelic mutants, one of high resistance and one of high sensitivity to compare to the wild type. It was found that gene-controlled sensitivity differences in *A. nidulans* correlate very well with the amount of MBC which binds to a macromolecule. The magnitude of these differences was approximately the same, independently of whether mycelium was incubated with MBC *in vivo* and then extracted or cell-free extracts of mycelium were incubated with MBC *in vitro*. Although the MBC-binding macromolecule has not been purified and fully characterized, there is good evidence that it may be identical with the protein subunit (tubulin) of spindle fibres. If this is true then mutations in the coding gene probably change this protein to cause lower or, in rare cases, higher affinity for MBC with the result of higher resistance or higher sensitivity (Davidse, 1974).

Greenaway and Ward (pers. comm.) recently studied TBZ resistance in *Penicillium digitatum*. It was found that sensitive and resistant mycelia take up approximately equal amounts of toxicant from the surrounding medium. Subsequent rinsing removes more TBZ from the resistant than from the sensitive strain, which can easily be accounted for by a differential binding to an internal site. What was rather unexpected is that removal of TBZ from *P. digitatum* mycelia appeared to involve an active secretion, while uptake seemed to be due to passive diffusion. This seems to indicate that resistance may result from increased active secretion (Greenaway and Ward, pers. comm.), but differential binding to an internal site cannot be excluded.

Carboxamides

Carboxin, oxycarboxin, and pyracarbolid are some of the commercial fungicides of this group. Most of the work on mode of action has been done with carboxin which blocks respiratory electron transport in the complex II region, *i.e.* between succinate and coenzyme Q (White, 1971; Ulrich and Mathre, 1972). In a recent paper White and Thorn (1975) report similar activity for a great variety of carboxamides. It is not yet known what component of complex II interacts with the toxicant to inhibit the transport of electrons. Solubilized succinic dehydrogenase is not inhibited by carboxin (Ulrich and Mathre, 1972) and if intact complex II activity is assayed with dichlorophenol indophenol (DCIP) in the presence of phenazine methosulphate (PMS), maximal inhibition does not exceed 50% (Georgopoulos *et al.*, 1975). It therefore seems that carboxin interacts with a component, located after the first PMS reaction site in complex II.

Fungal mutants resistant to carboxamides are easily obtained in the laboratory but do not seem to have been reported from nature. In *U. maydis* carboxin selects for mutants of low resistance (*ants*) which are alternate oxidase-negative and mutants of high resistance (*oxr*) which do possess the alternate oxidase system (Georgopoulos and Sisler, 1970; Georgopoulos *et al.*, 1972). A mutant (*oxr*-1) of the second type has been studied *in vitro*.

Mitochondrial preparations from this mutant are not only resistant to carboxin inhibition but also easily distinguishable from wild-type preparations in the absence of fungicide because they lose their succinic dehydrogenase activity at a much faster rate (Georgopoulos *et al.*, 1972, 1975). This instability is restricted to the carboxin-sensitive segment of the respiratory chain. Ageing does not differentially affect other enzyme activities of mutant and wild-type preparations. The same is true for the carboxin-insensitive portion (approximately 50 %) of complex II activity assayed in the presence of PMS. Probably the *oxr*-1 locus codes for a component of complex II, which is responsible for carboxin sensitivity and at the same time important for stability. It is worth noting that in a heterozygous diploid cell of *U. maydis* both the wild-type and the mutant component are produced and they appear to have an equal opportunity to participate in enzyme arrays of complex II (Georgopoulos *et al.*, 1975).

The mechanism of carboxin resistance has also been studied in *A. nidulans* (Gunatilleke *et al.*, pers. comm.). In this fungus also, mutation to resistance does not affect the permeability of the cytoplasmic or the outer mitochondrial membrane to the fungicidal molecule. All of three unlinked loci for resistance which have been recognized control carboxin sensitivity of succinic dehydrogenase *in vitro*. Mutational alteration of a component of complex II or of the mitochondrial inner membrane is considered responsible for carboxin resistance in *A. nidulans* (Gunatilleke *et al.*, pers. comm.).

Cycloheximide, blasticidin S, and kasugamycin

The three antibiotics of this group act as inhibitors of protein synthesis, but while cycloheximide is active only in eucaryotic cells (80 S ribosomes), blasticidin S and kasugamycin inhibit the synthesis of protein in both eucaryotic and procaryotic (70 S ribosomes) cells.

The mechanism of cycloheximide resistance has been studied in *Saccharomyces cerevisiae* and *Neurospora crassa*. In *S. cerevisiae*, out of eight genes which may mutate to resistance at least one, of a low resistance level, is expressed *in vitro* and is responsible for a mutational alteration of the ribosome (Cooper *et al.*, 1967). In *Neurospora crassa* all of three resistant genes studied give high resistance *in vitro* and all control ribosomal properties (Vomvoyanni, 1974). In wild-type *Saccharomyces fragilis* resistance is associated with the 60 S ribosomal subunit (Rao and Grollman, 1967). If the same is true in resistant mutants of *S. cerevisiae* and *N. crassa*, then components of this subunit must be controlled by the genes for cycloheximide resistance.

Only one fungal strain resistant to blasticidin S seems to have been studied. The protein-synthesizing system of this strain of *Piricularia oryzae* was not resistant to the antibiotic *in vitro*. Resistance was assumed to result from decreased permeability of the cytoplasmic membrane to the toxic molecule (Huang *et al.*, 1964). By contrast, a kasugamycin-resistant strain of the same fungus was found resistant *in vitro* (Misato and Ko, 1974) but details on the mechanism involved are not available. In the bacterium *Escherichia coli* kasugamycin resistance is determined by the action of an RNA methylase which dimethylates adjacent adenines of the 16 S ribosomal RNA (Helser *et al.*, 1972).

Polyoxins

The polyoxins act as competitive inhibitors for uridine diphosphate-N-acetylglucosamine in the chitin synthetase reaction, the last step in the biosynthesis of chitin (Endo *et al.*, 1970; Hori *et al.*, 1974a,b). Since chitin is absent from the walls of higher plants, this type of activity would be ideal for an agricultural fungicide. A mixture of polyoxins is widely used in Japan, but field development of resistant strains of *Alternaria kikuchiana* has become a problem (Nishimura *et al.*, 1973). The genetics of this resistance is not known. A recent study (Hori *et al.*, 1974a,b) indicates that with intact cells there is good correlation between the polyoxin sensitivity of a strain and the amount of inhibition of chitin synthesis or uridine diphosphate-N-acetylglucosamine accumulation. However, particulate chitin synthetase preparations from sensitive and resistant strains of *A. kikuchiana* were all strongly inhibited by the antibiotic. Polyoxin resistance, therefore, did not seem to result from modification of the sensitive enzyme but rather from decreased permeability (Hori *et al.*, 1974a,b).

Dimethirimol

Dimethirimol and ethirimol are pyrimidine derivatives active against powdery mildew fungi. They are supposed to act as noncompetitive enzyme inhibitors, interfering with tetrahydrofolate directed C-1 metabolism, which plays a role in the synthesis of purines, thymidine, and amino acids (Calderbank, 1971). Resistance to both compounds has developed in the field but only in the case of dimethirimol resistance of *Sphaerotheca fuliginea* has an attempt to elucidate the mechanism involved been made. Bent *et al.* (1971) observed that sensitive and resistant strains of this fungus take up approximately equal amounts of dimethirimol from impregnated membranes or treated leaf-discs. This seems to indicate no difference in permeability but further work is needed before alteration of internal site(s) can be suggested.

REFERENCES

Bent, K. J., Cole, A. M., Turner, J. A. W., and Woolner, M. (1971). Resistance of cucumber powdery mildew to dimethirimol. Proc. 6th Brit. Insectic. Fungic. Conf., Vol. 1, **1**, 274–282.

Berger, R. D. (1973). Disease progress of *Cercospora* apii resistant to benomyl. *Plant Dis. Reptr.*, **57**, 837–840.

Calderbank, A. (1971). Metabolism and mode of action of dimethirimol and ethirimol. *Acta Phytopathol. Acad. Sci. Hung.*, **6**, 355–363.

Cooper, D., Banthorpe, D. V., and Wilkie, D. (1967). Modified ribosomes conferring resistance to cycloheximide in mutants of *Saccharomyces cerevisiae. J. Mol. Biol.*, **26**, 347–350.

Davidse, L. C. (1973). Antimitotic activity of methyl benzimidazol-2-yl-carbamate (MBC) in *Aspergillus nidulans. Pestic. Biochem. Physiol.*, **3**, 317–325.

Davidse, L. C. (1974). Mode of action of methyl-benzimidazol-2-yl carbamate (MBC) and some biochemical aspects of acquired resistance against this fungicide in *Aspergillus nidulans*. Proc. Symp. Systemic Fungic. Reinhardsbrunn (East Germany) May 6–11, 1974 (in press).

Endo, A., Kakiki, K., and Misato, T. (1970). Mechanism of action of the antifungal agent polyoxin D. *J. Bacteriol.*, **104**, 189–196.

Georgopoulos, S. G., and Dovas, C. (1973). A serious outbreak of strains of *Cercospora beticola* resistant to benzimidazole fungicides in Northern Greece. *Plant Dis. Reptr.*, **57**, 321–324.

Georgopoulos, S. G., and Sisler, H. D. (1970). Gene mutation eliminating antimycin A-tolerant electron transport in *Ustilago maydis*. *J. Bacteriol.*, **103**, 745–750.
Georgopoulos, S. G., Alexandri, E., and Chrysayi, M. (1972). Genetic evidence for the action of oxathiin and thiazole derivatives on the succinic dehydrogenase system of *Ustilago maydis* mitochondria. *J. Bacteriol.*, **110**, 809–817.
Georgopoulos, S. G., Chrysayi, M., and White, G. A. (1975). Carboxin resistance in the haploid, the heterozygous diploid, and the plant-parasitic dicaryotic phase of *Ustilago maydis*. *Pestic. Biochem. Physiol.* (in press).
Hammerschlag, R. S., and Sisler, H. D. (1973). Benomyl and methyl-2-benzimidazole carbamate (MBC): biochemical, cytological and chemical aspects of toxicity to *Ustilago maydis* and *Saccharomyces cerevisiae*. *Pestic. Biochem. Physiol.*, **3**, 42–54.
Harding, P. R., Jr. (1972). Differential sensitivity to thiabendazole by strains of *Penicillium italicum* and *P. digitatum*. *Plant Dis. Reptr.*, **56**, 256–260.
Hastie, A. C. (1970). Benlate induced instability of *Aspergillus* diploids. *Nature*, **226**, 771.
Hastie, A. C., and Georgopoulos, S. G. (1971). Mutational resistance to fungitoxic benzimidazole derivatives in *Aspergillus nidulans*. *J. Gen. Microbiol.*, **67**, 371–373.
Helser, T. L., Davies, J. E., and Dahlberg, J. E. (1972). Mechanism of kasugamycin resistance in *Escherichia coli*. *Nature, New Biol.*, **235**, 6–9.
Hori, M., Kakiki, K., and Misato, T. (1974a). Further study on the relation of polyoxin structure to chitin synthetase inhibition. *Agr. Biol. Chem.*, **38**, 691–698.
Hori, M., Egughi, J., Kakiki, K., and Misato, T. (1974b). Studies on the mode of action of polyoxins. VI. Effect of polyoxin B on chitin synthesis in polyoxin-sensitive and resistant strains of *Alternaria kikuchiana*. *J. Antibiot.*, **27**, 260–266.
Huang, K. T., Misato, T., and Asuyama, H. (1964). Selective toxicity of blasticidin S to *Piricularia orysae* and *Pellicularia sasakii*. *J. Antibiot.*, **17**, 71–74.
Kappas, A., Georgopoulos, S. G., and Hastie, A. C. (1974). On the genetic activity of benzimidazole and thiophanate fungicides on diploid *Aspergillus nidulans*. *Mutation Res.*, **26**, 17–27.
Lukens, R. J. (1968). Fungitoxic action of non-metallic organic fungicides. Proc. 1st Int. Biodeterioration Symp., pp. 486–497, Elsevier Publ. Co., Barking, Essex.
Misato, T., and Ko, K. (1974). The development of resistance to agricultural antibiotics. Abstract Book, 3rd Int. Congr. Pestic. Chem., Helsinki, 3–9 July, 1974, abstr. 170.
Nishimura, S., Kohmoto, K., and Udagawa, H. (1973). Field emergence of fungicide-tolerant strains of *Alternaria kikuchiana* TANAKA. *Rept. Tottori Mycol. Inst. (Japan)*, **10**, 677–686.
Rao, S. S., and Grollman, A. P. (1967). Cycloheximide resistance in yeast: a property of the 60 S ribosomal subunit. *Biochem. Biophys. Res. Commun.*, **29**, 696–704.
Ulrich, J. T., and Mathre, D. E. (1972). Mode of action of oxathiin systemic fungicides. V. Effect on electron transport system of *Ustilago maydis* and *Saccharomyces cerevisiae*. *J. Bacteriol.*, **110**, 628–632.
Van Tuyl, J. M., Davidse, L. C., and Dekker, J. (1974). Lack of cross resistance to benomyl and thiabendazole in some strains of *Aspergillus nidulans*. *Neth. J. Plant Pathol.*, **80**, 165–168.
Vomvoyanni, V. (1974). Multigenic control of ribosomal properties associated with cycloheximide sensitivity in *Neurospora crassa*. *Nature*, **248**, 508–510.
White, G. A. (1971). A potent effect of 1,4-oxathiin systemic fungicides on succinate oxidation by a particulate preparation from *Ustilago maydis*. *Biochem. Biophys. Res. Commun.*, **44**, 1212–1219.
White, G. A., and Thorn, G. D. (1975). Structure-activity relationships of carboxamide fungicides and the succinic dehydrogenase complex of *Cryptococcus laurentii* and *Ustilago maydis*. *Pestic. Biochem. Physiol.* (in press).
Wicks, T. (1974). Tolerance of the apple scab fungus to benzimidazole fungicides. *Plant Dis. Reptr.*, **58**, 886–889.

POTENTIAL ARYL-VINYL-KETONES—A NEW CLASS OF MICROBIOCIDES WITH GOOD ENVIRONMENTAL PROPERTIES

W. PAULUS

Bayer AG, 4150 Krefeld-Uerdingen, German Federal Republic

Summary

Aryl-vinyl-ketones, Ar—CO—CR=CH_2 (I) offer a wide range of microbiocidal activities. However, being unstable and irritating to the skin and mucosa, the substances in this form are unsuitable for practical application. Aroyl-ethyl-compounds, Ar—CO—CHR—CH_2—LG (II), bearing a so-called leaving group (LG), *e.g.* Hal, $N(R)_2$, S—R, $P^+(Ph)_3$, are transport forms for I. Stability and antimicrobial effectiveness of II depend on the electronegativity of the hetero-atom of LG, but also on the nature of R; an alkyl group with more than one C-atom in the place of aryl (Ar) diminishes the microbiocidal activity of II significantly. In aqueous solutions, even at normal temperature, potential aryl-vinyl-ketones (II) allow aryl-vinyl-ketone (I) to be split off and display the microbiocidal effect of the latter. As a result of the reactivity of I, the microbiocidal effect of II-containing aqueous solutions will gradually decrease to such an extent that the use of such substances as slimicides involves no wastewater problems. Also the following substance classes represent potential aryl-vinyl-ketones:

$$\mathrm{Ar{-}CO{-}CHR_1{-}CH_2{-}S{-}\underset{\underset{\displaystyle NH_2^+X^-}{\|}}{C}{-}NHR_2}$$

and 2-amino-4-aryl-1,3-thiazines, resulting from the latter by cyclisation.

INTRODUCTION

On screening reactive substances with regard to microbiocidal action in order to find active ingredients capable of counteracting the biodegradation of materials without causing ecological problems, we came across aryl-vinyl-ketones. They are distinguished by a wide range of activity covering fungi and bacteria which are of considerable importance as far as the biodegradation of materials is concerned. In some cases, the effective concentrations are in the region of a few parts per million. Geiger and Conn (1945) already had found that in the antibiotics clavacin and penicillic acid, it is the vinyl-ketone-grouping and the α,β-unsaturated carbonyl function, respectively, that are responsible for the antibiotic effect, because the vinyl-ketone-grouping tends to addition reactions with nucleophiles of great biological importance such as amines and thiols, thus inactivating, for instance, succinic dehydrogenase (Fig. 1).

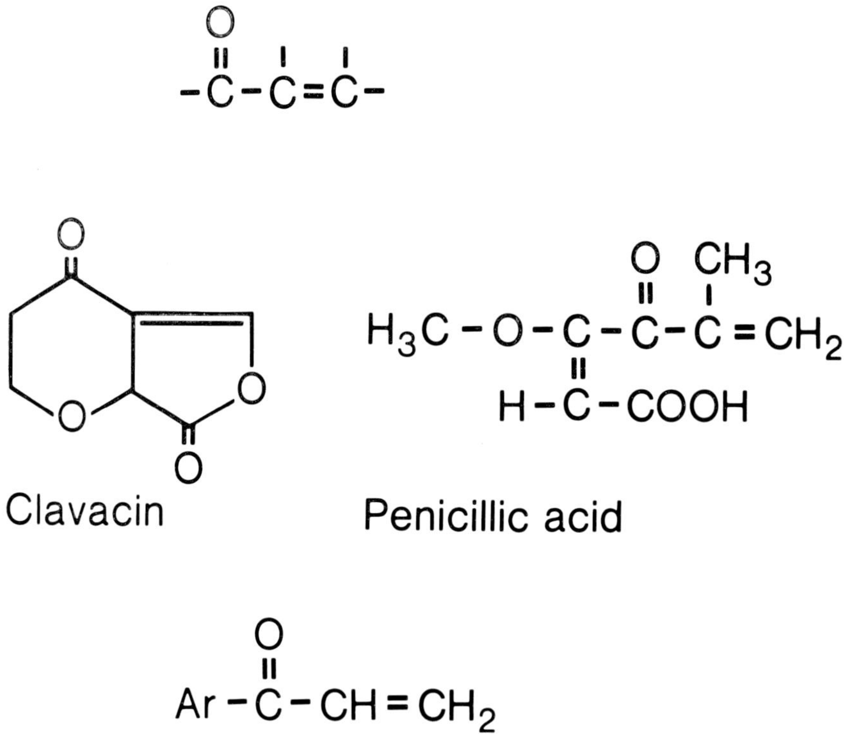

Fig. 1.

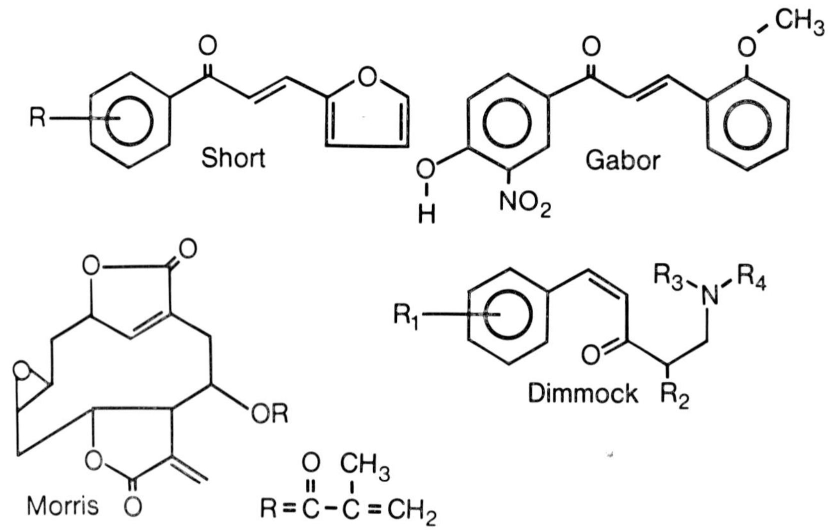

Fig. 2.

Geiger (1948) also detected that vinyl-ketones show a stronger microbiocidal effect if there is a grouping in a system stabilized by mesomerism as is the case with aryl-vinyl-ketone.

Short and Rockwood (1969) reported on the anti-inflammatory effect of aryl-(furyl-) vinyl-ketones, and Gabor *et al.* (1970) reported on antibacterial chalkones that can be regarded as aryl-(phenyl-) vinyl-ketones. It has been observed that the cytotoxic activity of certain sesquiterpenelactones is based on the presence of α,β-unsaturated carbonyl functions; this also applies to the aryl-vinyl-ketones with anti-tumour properties as described by Dimmock and Taylor (1974) (Fig. 2).

Vinyl-ketones—especially aryl-vinyl-ketones with strong microbiocidal effect, which are of particular interest to us—are in general unstable, easily polymerizing substances that irritate the skin and mucosa; hence they are in this form unsuitable for practical applications. Considering the paths of synthesis for aryl-vinyl-ketones, it is obvious that some syntheses are based on the splitting of groups that are known to be leaving groups (Fig. 3).

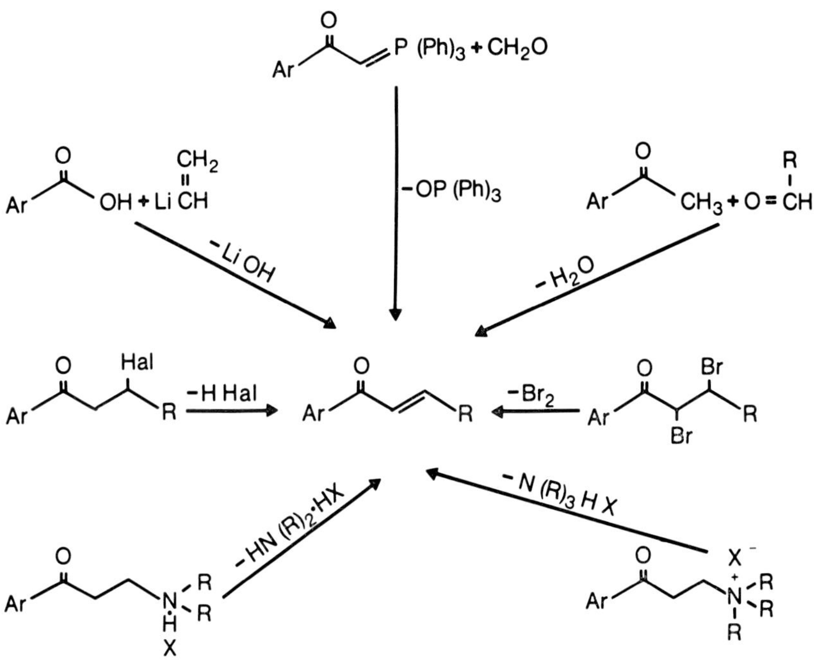

Fig. 3.

This raises two questions:

(1) Do the basic products containing leaving groups possess microbiocidal activity, such as the Hoffmann products or aryl-vinyl-ketones originating from them?

(2) Do the basic products containing leaving groups possess sufficient stability for practical microbiocidal application, and is their skin-irritating effect so low that they can be handled without hazard?

MATERIALS AND METHODS

The test substances were produced partly by known methods and identified by elementary analysis. As far as new test substances and original methods of preparation are concerned, they will be referred to on another occasion.

The slime bacteria described by Kato and Fukumura (1962) were used as test organisms. They can grow by utilizing ε-caprolactam or its cyclic dimer or trimer as the sole carbon source. They are easy to cultivate in Allen's (1952) nutritive solutions if the solutions contain 1% of ε-caprolactam. A dose which, after 14 days' incubation at room temperature, prevents any growth of test organisms, as far as is perceptible with the naked eye, represents the microbiostatic minimum concentration (clear nutritive solution).

The influence of active ingredients on the biocoenosis and on the activity of microorganisms in the activated sludge was determined by the Sapromat method (Pauli and Franke, 1972).

RESULTS

Table I shows leaving groups as already mentioned, corresponding to potential aryl-vinyl-ketones, and the microbiostatic minimum concentration of the substances as a measure of antibacterial effect. The potential aryl-vinyl-ketones are practically odourless, pulverulent substances; in the form

Table I

MICROBIOSTATIC MINIMUM CONCENTRATIONS (MMC) OF POTENTIAL ARYL-VINYL-KETONES (PAVK) WITH DIFFERENT LEAVING GROUPS (LG). TEST ORGANISMS: BACTERIA-DEGRADING ε-CAPROLACTAM

PAVK	*m.p.* (°*C*)	*LG*	*MMC* (*ppm*)
4-Cl—Ph—CO—CH_2—CH_2—Cl	51	—Cl	2
4-Cl—Ph—CO—CH_2—CH_2—$N(CH_3)_2 \cdot HCl$	142	—$N(CH_3)_2$	5
4-Cl—Ph—CO—CH_2—CH_2—S—CH_3	34	—S—CH_3	20
[4-Cl—Ph—CO—CH_2—CH_2—$P(Ph)_3]^+J^-$	170	—$P(Ph)_3$	50
—	—	—SO_2—CH_3	—
—	—	—O—SO_2—tol	—
4-Cl—Ph—CO—CH=CH_2	—	—	1–2

of their salts with acids or in quaternized form, they are soluble in water. If kept dry, they remain stable and in the form of powder cause no skin irritations. As can be noted from Table I, they exhibit a microbiocidal effect comparable to that of the aryl-vinyl-ketone they are based on. This effect is due to the splitting of the leaving group, the aryl-vinyl-ketone being set to produce its antimicrobial effect. The splitting already occurs under physiological conditions (room temperature, pH-values around the neutral point), as can be easily determined by common analytical methods (*e.g.* u.v. spectro-

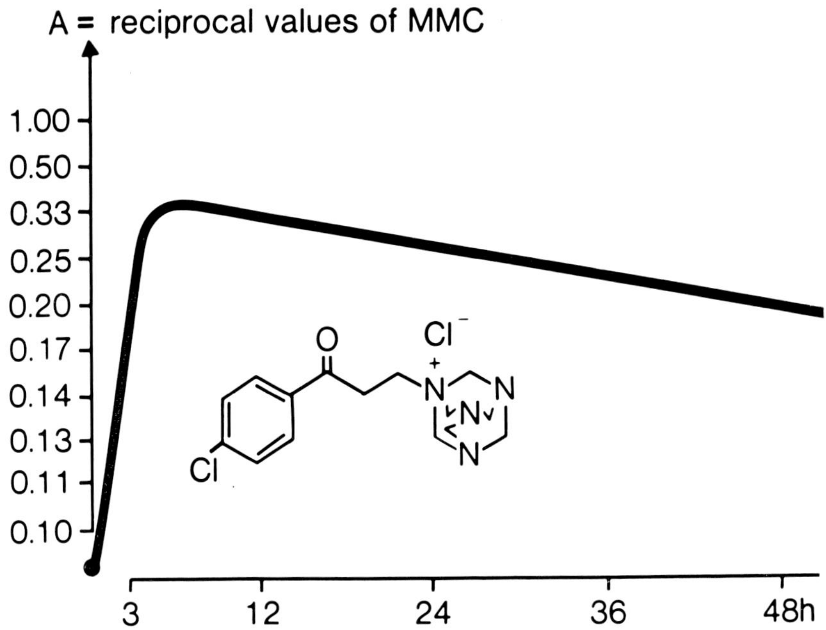

Fig. 4. *Antimicrobial activity (A) of an aqueous solution of aroylethylhexaminiumchloride in dependence on time (h) at pH 6·5.*

scopy, thin-layer chromatography). In general, the rate of splitting is many times higher than the rate of the reactions leading to the inactivation of the aryl-vinyl-ketone product of the splitting process. This has a favourable effect on the practical application of potential aryl-vinyl-ketones as microbiocides. A biokinetic test has led to a curve which illustrates the above statements (Fig. 4).

Depending on the leaving group, there are differences in the antimicrobial action of potential aryl-vinyl-ketones (*see* Table I). On recalling some basic facts of chemistry, however, this becomes quite understandable. Hetero-atoms attached to the C-atom, having a higher electronegativity than the C-atom itself, stabilize a positive charge at the neighbouring C-atom, thus making it liable to nucleophilic attack, whereas at the following C-atom a

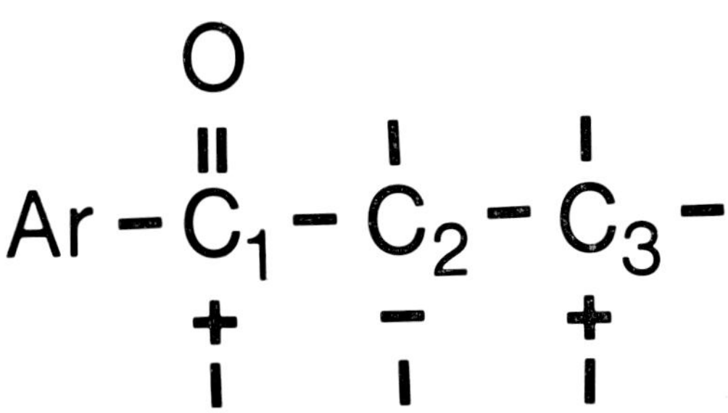

Fig. 5.

negative charge is stabilized, this atom being made liable to electrophilic attack. On considering the system of potential aryl-vinyl-ketones it is obvious that through the influence of two hetero-atoms, the stabilization of a negative charge at C-2 is particularly marked (Fig. 5).

This results in a higher tendency to carbanion formation, which precedes the splitting of the leaving group. The stability of the carbanion, hence the behaviour of the leaving group and the antimicrobial action, can be influenced by the choice of the hetero-atom at, *e.g.* C-3. The higher the electronegativity of the hetero-atom, the higher of course the stability of the carbanion. Substituents increasing the resonance, such as the aryl group, also produce a stabilizing effect on the carbanion.

The synthesis of various appropriate substances allowed these predictions to be confirmed, as can be seen from Table II.

Table II

MICROBIOSTATIC MINIMUM CONCENTRATIONS (MMC) OF $A{-}C({=}O){-}CH(R){-}CH_2{-}LG$ IN DEPENDENCE ON THE SUBSTITUENTS A, R, AND LG (LEAVING GROUP) AND ON THE ELECTRONEGATIVITY (E) OF THE HETERO-ATOM IN LG. TEST ORGANISMS: BACTERIA-DEGRADING ε-CAPROLACTAM

A	*R*	*LG*	*E*[a] *of hetero-atom in LG*	*m.p. (°C)*	*MMC (ppm)*	*Subst. no.*
4-Cl—Ph	H	Cl	3·16	51	2	1
4-Cl—Ph	H	$N(CH_3)_2$	3·04	142[b]	5	2
4-Cl—Ph	H	$S{-}CH_3$	2·44	34	20	3[e]
4-Cl—Ph	H	$P^{+}(Ph)_3$	2·06	170	50	4
Ph	H	$N(CH_3)_2$	3·04	156[b]	5	5
Ph	CH_3	$N(CH_3)_2$	3·04	152[b]	50	6
CH_3	Ph	$N(CH_3)_2$	3·04	157[b]	40	7
C_2H_5	CH_3	$N(CH_3)_2$	3·04	95[b]	>100	8
CH_3	H	Cl	3·16	Oil[c]	4	9
CH_3	H	$N(CH_3)_2$	3·04	97[b]	7	10
CH_3	H	$S^{+}(CH_3)(C_2H_5)$	2·44	Oil	15	11
CH_3	CH_3	$N(CH_3)_2$	3·04	Oil[d]	35	12

[a] Electronegativity by Pauling.
[b] m.p. of hydrochloride.
[c] b.p.$_{15}$ 53°C.
[d] b.p.$_{60}$ 84°C.
[e] Paulus *et al.*, 1974.

The lower the electronegativity of the hetero-atom of the leaving group, the higher the MMC values, hence the smaller the antimicrobial effect. The substituent R is also important to the antimicrobial action of the potential vinyl-ketones. The influence of the methyl group, which is known to reduce the carbanion stability, can be noticed in substance no. 6. If substituent A represents a nonresonance-increasing group—as in the case of an aryl rest—

an appreciable antimicrobial effect can no longer be observed (*see* substance no. 8). The aryl group, however, may be replaced by a methyl group without the antimicrobial effect being considerably impaired (*see* substances 9–12).

As shown by Figs. 6 and 7, it is possible that in a class of substances classified by a certain leaving group (Paulus *et al.*, 1975), both the substituents at the hetero-atom of the leaving group and the resonance-increasing group A may be varied at will without the antimicrobial action being impaired to a considerable extent.

Ar	R	X	MMC in ppm	mp in °C
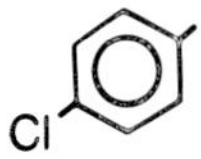	N $(CH_3)_3$	Cl	2	136
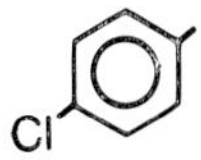	Hexamethylenetetramine	Cl	5	180
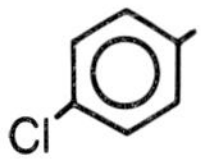	Tetraazatricyclododecane	Cl	2	76
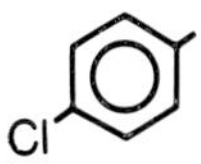		Cl	3	142
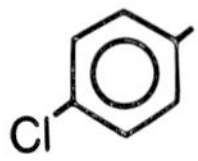		Cl	3	140
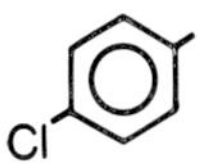	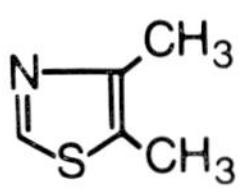	Cl	2	161
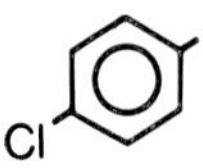	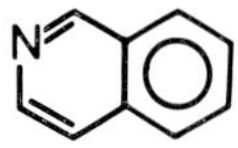	Cl	5	201
	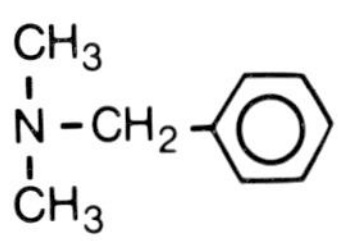	Cl	5	147

Fig. 6. *Microbiostatic minimum concentration* (*MMC*) *of* $[ArCO—(CH_2)_2—R]^+X^-$.

If the possible variables are taken into account—leaving groups, substituents at the leaving groups, group A increasing the resonance, substituents at this group—it will be noticed that among the potential vinyl-ketones with antimicrobial effect there is a very large number that still offers a wide scope for synthesis. Here is a possibility of synthesizing antimicrobial substances with widely differing properties intended for use in a variety of applications.

Potential aryl-vinyl-ketones containing a halogen atom as leaving group strongly irritate the skin, much like the aryl-vinyl-ketones they are based on. Potential aryl-vinyl-ketones containing an amino group as leaving group behave far better; their antimicrobial action is particularly rapid if the amino group is a quaternary one.

The substance shown in Fig. 8 was chosen to control slime formation in paper mills. The following criteria were considered: wide range of antimicrobial activity; stability in dry form; no skin irritation; simple and economical manufacturing process.

Ar	R	X	MMC in ppm	mp in °C
	Hexamethylenetetramine	Cl	5	162
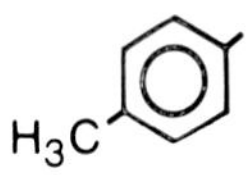	Hexamethylenetetramine	Cl	5	170
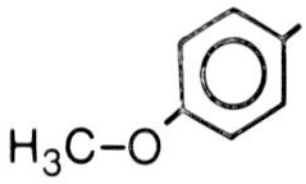	Hexamethylenetetramine	Cl	5	127
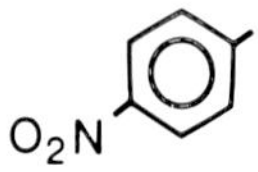	Hexamethylenetetramine	Cl	3	175
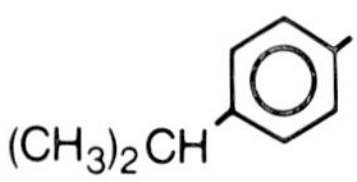	Hexamethylenetetramine	Cl	5	189
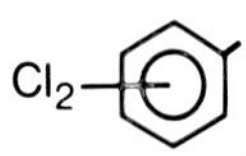	Hexamethylenetetramine	Cl	5	184
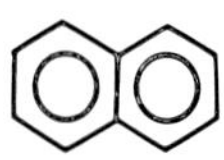	Hexamethylenetetramine	Cl	1	175
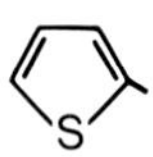	Hexamethylenetetramine	Cl	2	160

Fig. 7. *Microbiostatic minimum concentration (MMC) of* $[Ar{-}CO{-}(CH_2)_2{-}R]^+X^-$.

As a quaternary hexaminium salt, the substance splits off formaldehyde in aqueous solution, independent of the pH-value involved; this, however, is of no significant importance as far as the effect of the substance on slime-forming microorganisms is concerned. Meanwhile, the substance has proved its worth as an effective slimicide, especially for use in paper mills. At intact concentrations of above 10 ppm, the substance is toxic to fish; hence its fish toxicity is many times lower than that of organo mercurials, organo tin compounds, or pentachlorophenol. In addition, it has to be taken into account

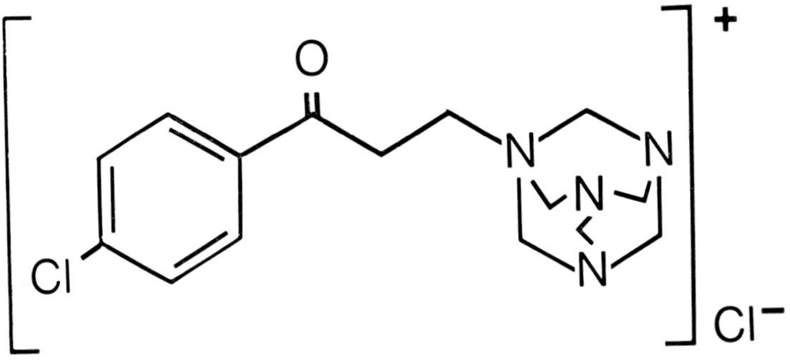

mp: 180°C (degrad.).
solubility i.H_2O: 20 g/l (degrad.)
LD_{50} p. o.: 1008 mg/kg rat.
fish-toxic in concentrations >5 ppm

Fig. 8.

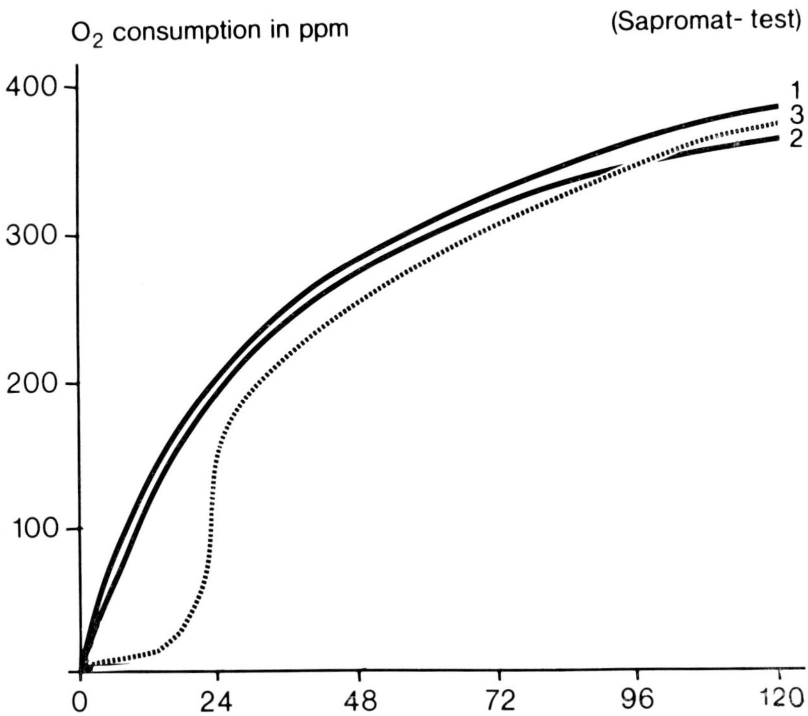

Fig. 9. Influence of (4-chloro-)benzoylethylhexaminiumchloride on the O_2-consumption of sewage bacteria. 1: Sewage water without microbiocide; 2/3: sewage water containing 10/25 ppm aroylethylhexaminiumchloride. Abscissa: hours.

that degradation occurs; so material amounts of toxic concentrations will hardly ever get into the wastewater. The wastewaters of paper mills using aroyl-ethyl-hexaminium salt as a slimicide cause no problems in sewage purification plants either. As can be seen from Fig. 9, only the first degradation step of the substance, 25 ppm in aqueous solutions, makes up for a negative effect first produced on the activity of microorganisms in the activated sludge. No effect is observed in the presence of 10 ppm of the substance.

A potential aryl-vinyl-ketone with interesting activities is also obtained by making aryl-(2-chloro)-ethyl-ketone react with thioureas (Paulus *et al.*, 1975) (Fig. 10).

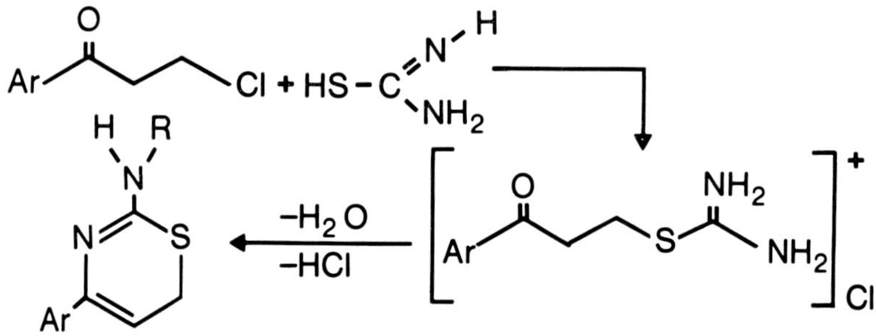

Fig. 10.

Under certain reaction conditions the isothiuronium salt, which is formed first, is cyclized to a 2-amino-4-aryl-1,3-thiazine. It is a most interesting fact that the antimicrobial effect of 2-amino-4-aryl-1,3-thiazines is not restricted to slime-forming bacteria and fungi, but also covers phytopathogen microorganisms and especially phytopathogen bacteria. Being stable in the form of hydrohalogenides as well as in the form of free bases, the 2-amino-4-aryl-1,3-thiazines are distinguished by properties that make them still more suitable, especially for crop protection applications.

REFERENCES

Allen, M. B. (1952). The cultivation of myxophyceae. *Arch. Microbiol.*, **17**, 34–53.

Dimmock, J. R., and Taylor, W. G. (1974). Synthesis and physical properties of 4-dimethylaminomethyl-1-phenyl-1-nonen-3-ones possessing antitumor properties. *J. Pharmceut. Sci.*, **63**, 69–74.

Gabor, M., Sallay, J., and Széll, T. (1970). Die antibakterielle Wirkung einiger neuer Benzyliden-Flavanone und -Acetophenone. *Arch. Pharmaz.*, **303** (7), 593–595.

Geiger, W. B. (1948). Antibacterial unsaturated ketones and their mode of action. *Arch. Biochem.*, **16**, 423–435.

Geiger, W. B., and Conn, J. E. (1945). The mechanism of the antibiotic action of clavacin and penicillic acid. *J. Am. Chem. Soc.*, **67**, 112–116.

Kato, K., and Fukumura, T. (1962). Degradation of ε-caprolactam by bacteria. *Chem. Ind.*, **25**, 1146.

Pauli, O., and Franke, G. (1972). Behaviour and degradation of technical preservatives in the biological purification of sewage. *Biodeterior. Materials*, **2**, 53–60, Halsted Press, New York.

Paulus, W., Oeckl, S., and Pauli, O. (1974). Benzoyläthylsulfoniumsalze als Mikrobizide. DOS 2.326.111 vom 12.12.1974.

Paulus, W., Hamburger, B., and Pauli, O. (1975). Mikrobizide Mittel. DAS 2.034.540 vom 10.4.1975.

Short, F. W., and Rockwood, G. M. (1969). Synthesis and chemical reactions on 6-aroyl-4-oxohexanoic acids and 5-aryl-2-furyl-propionic acids. Inflammation inhibitors. *J. Heterocyclic Chem.*, **6** (5), 713–722.

PROBLEMS ENCOUNTERED WITH FORMALDEHYDE-RELEASING COMPOUNDS USED AS PRESERVATIVES IN AQUEOUS SYSTEMS, ESPECIALLY LUBRICOOLANTS—POSSIBLE SOLUTIONS TO THE PROBLEMS

W. PAULUS

Bayer AG, 4150 Krefeld-Uerdingen, German Federal Republic

Summary

Formaldehyde-releasing compounds, which are widely used as preservatives for aqueous systems, show considerable differences in antimicrobial effectiveness and the effective spectrum. Benzylalcoholmono-(poly)hemiformal (BHF) is a preparation with a synergistic effect extending to bacteria, fungi, and yeasts. A desirable improvement of the antimicrobial effect of BHF, especially against fungi and formaldehyde-resistant bacteria, is obtained by combination with small amounts of phenolic microbiocides. unless the components are mixed together at pH levels exceeding 9–9·5, The mixtures have a marked synergistic effect against fungi and yeasts, yet not against formaldehyde-resistant bacteria which, although tolerating up to 4200 ppm formalin (29%), react pronouncedly to N-hydroxy-methyl-chloroacetamide, Na-pentachlorophenolate, 1,5-dimethyl-2-thiono-tetra-hydro-thiadiazine, 3-methyl-4-chlorophenol—the reaction increasing in this order.

INTRODUCTION

Ready-to-use dilutions of lubricoolants are aqueous systems whose pH allows the survival and proliferation of bacteria, yeasts, and fungi (including the pathogenic and mucigenous organisms) (Bennett, 1956).

The resistance of lubricoolants to microorganisms is important because it has a decisive influence on whether or not the cooling and lubricating properties, corrosion control (pH level), skin compatibility, and long service life will be retained without adverse effects on the production process. It is unavoidable that lubricoolants are contaminated by microorganisms during operation (Hill, 1967).

The contamination of lubricoolants not containing preservatives will result in irreversible microbiological decomposition; its consequences in lubricoolants include: unpleasant odour, clogged filters, generation of acid metabolism products, decreasing pH, corrosion, creaming of oil, and finally unserviceability of the lubricant, unscheduled cleaning operations and shutdowns, lost production time. Moreover, it should be noted that pathogenic microorganisms can proliferate in nonpreserved lubricoolants so that personnel handling these liquids may suffer from severe skin infections (Pivnik *et al.*, 1954; Weidle, 1973).

Requirements for a preservative to ensure the resistance of lubricoolants to microorganisms include the absence of any adverse effect on the properties of the lubricoolant (s.a.). In addition, the microbiocide should have a broad spectrum of effectiveness and favourable solubility properties; it should have low toxicity, be biodegradable and, in economical concentrations, have sustained effectiveness against both aerobic and anaerobic organisms. None of the numerous preservatives for lubricoolants meets all the requirements mentioned. Therefore, it is necessary to make a selection and to be ready to accept compromise solutions. Today, widespread opinion (Bennett, 1973) holds that, although formaldehyde-releasing compounds are not perfect, they are the most valuable preservatives available for lubricoolants: they have adequate effectiveness, are economical in use, and do not present waste-water problems since the substances generally have satisfactory biodegradability. Also the appropriate use of formaldehyde-releasing compounds will not cause any dermatologic problems. However, it should not be concealed that formaldehyde-releasing compounds have disadvantages, too. The instability of the active ingredients and the high reactivity of the formaldehyde can lead to a fast reduction of effectiveness; the irritative effect of formaldehyde-releasing compounds or formaldehyde on the mucous membranes can have unpleasant effects where lubricants are applied by spraying. Complaints have also been raised that the effect of formaldehyde-releasing compounds against fungi is less pronounced than against bacteria, and that problems result from the emergence of formaldehyde-resistant bacteria (Druskeit and Eggensperger, 1971).

The following report deals with approaches to overcome the inadequate effect of formaldehyde-releasing compounds and formaldehyde against fungi and formaldehyde-resistant bacteria.

MATERIALS AND METHODS

A measure of the antimicrobial effect of active ingredients and active ingredient combinations are the microbiostatic minimum concentrations (MMC's), which are determined as follows. From the test substance or test mixture, prepare a concentration series in a nutrient solution or nutrient substrate. At the end of 1 or 2 weeks after infection of the nutrient media with test organisms, record, as the microbiostatic minimum concentration, the lowest concentration of active ingredient that suppresses any growth of the test organisms (clear nutrient solution or no growth on the nutrient substrates that is recognizable with the naked eye). The nutrient media for bacteria (pH 7·5) contain meat extract powder, neutral meat peptone, and D-glucose, whereas the nutrient media for fungi and yeasts (pH 5·6) contain malt liquor and peptone.

Synergism is determined using the method described by Kull *et al.* (1961), with the following relationship applying:

$$\frac{Q_A}{Q_a} + \frac{Q_B}{Q_b} = X$$

$X = 1$: additivity; $X > 1$: antagonism; $X < 1$: synergism. Q_a = MMC of substance A; Q_b = MMC of substance B; Q_A = amount of substance A in the MMC of mixture A/B; Q_B = amount of substance B in the MMC of mixture A/B.

RESULTS

Figure 1 is a summary of the formaldehyde-releasing compounds that are used as preservatives for lubricoolants. This summary does not claim to be complete. Table I shows the effective spectrum of a benzylalcoholhemiformal preparation (formaldehyde content: approximately 29%). It shows clearly that the antimicrobial effect of the preparation against bacteria is far more

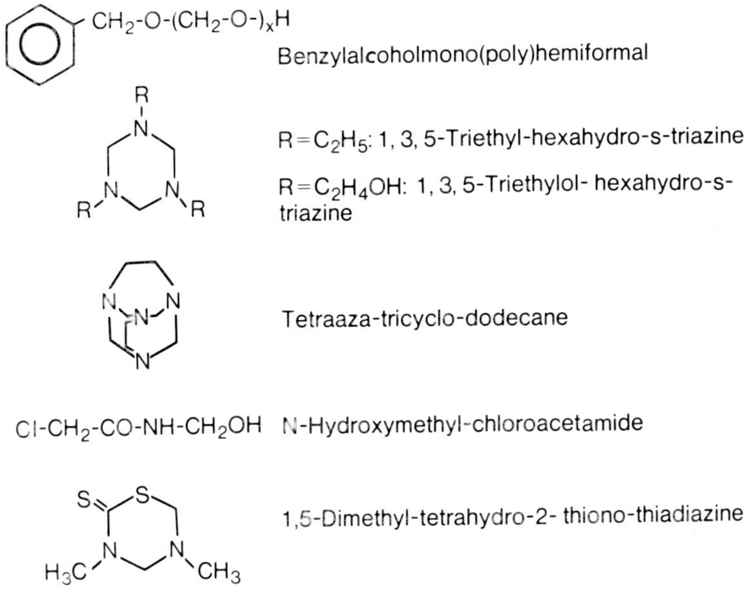

Fig. 1. Formaldehyde-releasing compounds as preservatives.

Table I
MICROBIOSTATIC MINIMUM CONCENTRATION (MMC) OF BENZYLALCOHOLMONO(POLY)HEMIFORMAL IN AGAR NUTRIENT SUBSTRATES

Microorganisms	*MMC (ppm)*
Staphlococcus aureus	70
Bacterium subtilis	50
Bacterium coli	100
Bacterium fluorescens	130
Bacterium punctatum	50
Bacterium proteus vulgare	50
Bacterium pyocyaneum	150
Bacterium mycoides	100
Desulfovibrio desulfuricans	50
Aspergillus niger	900
Aspergillus flavus Link	240
Rhizopus nigricans	600
Chaetomium globosum Kunze	240
Pullularia pullulans	500
Penicillium glaucum	1 500
Torula utilis	400
Candida crusei	600

pronounced than against fungi and yeasts. A more thorough investigation of the MMC values showed that they are based on a synergism (*see* Table II). The synergism is clearly found to extend to bacteria, fungi, and yeasts

Table II
SYNERGISM ($X < 1$) OF A BENZYLALCOHOLHEMIFORMAL PREPARATION (BHF); CH_2O CONTENT: 28%

Test organisms	*Microbiostatic minimum concentration (ppm)*			
	Benzylalcohol	*CH_2O*	*BHF*	*X*
Aerobacter aerogenes	6 000	150	150	0·30
Bacterium coli	6 000	50	100	0·57
Bacterium fluorescens	6 000	70	130	0·52
Pseudomonas aeruginosa	3 000	60	150	0·73
CH_2O-resistant bacteria	5 000	1 400	2 000	0·67
Candida crusei	6 000	200	500	0·76
Chaetomium globosum K.	2 500	250	300	0·43
Pullularia pullulans	2 000	400	700	0·74
Trichoderma viride	5 000	4 500	2 000	0·42

Thus it is possible to broaden the effective spectrum of formaldehyde by reacting formaldehyde with benzylalcohols and generally to improve the antimicrobial effectiveness. This result is not obtained by condensing formaldehyde with an amine not having appreciable antimicrobial effectiveness. Table III represents the most significant properties of the benzylalcohol-

Table III
PROPERTIES OF BENZYLALCOHOLMONO(POLY)HEMIFORMAL (CH_2O CONTENT: 29%)

Appearance:	water-clear liquid
Density:	1·11 at 20°C
Solubility:	oil- and water-soluble; solubility in H_2O: 25 g/litre at 20°C
Reaction in H_2O:	neutral
Corrosion behaviour:	noncorrosive
Flash point:	95°C
Vapour pressure:	34 mm Hg at 50°C
LD_{50} per os:	1 700 mg/kg rat
LD_{50} intravenously:	153 mg/kg rat
LD_{50} percutaneously:	$>$1 000 µl/kg rat after 7-day skin contact
Concentrated product:	skin irritation
0·2% solution:	no skin irritation, no sensibilization, no allergic effectiveness

hemiformal preparation. It should be kept in mind that the product shows neutral reaction since it contains no amines. Thus its use as a preservative in lubricoolants will not increase the alkalinity and its toxically degenerative influences on the skin.

When looking for possibilities of closing the fungi gap in the effective spectrum of formaldehyde-releasing compounds, it is self-suggesting to use the phenol derivatives as combination partners, which are known to have an excellent fungicidal effect. When combining benzylalcoholmono(poly)hemiformal with phenol derivatives, a remarkable synergism is also noticed, which especially extends to fungi, but also to yeasts. This synergism is so striking that even additions of 15–20% phenol derivatives to the mixture are sufficient to broaden the effective spectrum to the desired extent. Therefore, the risk of such a mixture presenting problems of ecology and skin compatibility is negligibly small. Table IV shows the synergism of a 5/1 combination based

Table IV
SYNERGISM (X DETERMINED BY THE METHOD OF KULL *ET AL.*, 1961) OF BENZYLALCOHOLMONO(POLY)HEMIFORMAL (BHF): PENTACHLOROPHENOL (P) = 5:1

Test organisms	*Microbiostatic minimum concentration (ppm)*			
	BHF	*P*	*BHF/P* = 5:1	*X*
Penicillium glaucum	1 500	110	320	0·65
Rhizopus nigricans	600	110	225	0·64
Aspergillus niger	900	800	225	0·67
Chaetomium globosum K.	240	35	120	0·97
Torula utilis	400	32	120	0·87
Torula rubra	3 500	100	320	0·62
Candida crusei	600	100	240	0·73
Candida albicans	2 500	100	320	0·64

on benzylalcoholmono(poly)hemiformal (formaldehyde content: 29%) and pentachlorophenol, as determined by the method of Kull *et al.* (1961).

Table V shows that phenol derivatives other than pentachlorophenol also produce a synergism when combined with benzylalcoholmono(poly)hemiformal.

Table V
SYNERGISM (X DETERMINED BY THE METHOD OF KULL *ET AL.*, 1961) OF 4/1 MIXTURES BASED ON BENZYLALCOHOLMONO(POLY)HEMIFORMAL (BHF) AND PHENOL DERIVATIVES

Phenol derivative	*Synergism: X* < 1	
	Rhizopus nigricans	*Aspergillus niger*
2-Phenyl-phenol	0·69	0·39
2-Benzyl-4-chlorophenol	0·47	0·68
(2,2′-Dihydroxy-5-5′-dichloro)diphenyl-methane	0·58	0·64
4-Chloro-3,5-dimethyl-phenol	0·69	0·72
Pentachlorophenol	0·57	0·50

However, phenols are known to be able to react with formaldehyde, and nucleus methylolations may occur under specific conditions. In this way the formaldehyde-releasing compounds are deprived of microbiocidal formaldehyde. In addition, the phenol derivatives methylolated in the phenyl ring are less effective than the corresponding starting products. However, these reactions, which reduce the antimicrobial effect of the mixing partners, can be suppressed as long as pH levels between 6 and 9, and max. 9·5, are available; above a pH of 9·5, phenol derivatives, if having corresponding, free reactive sites, will already react with formaldehyde, as indicated, at room temperature. This applies to all phenol derivatives that are used as microbiocides in practice; exceptions naturally are penta-substituted phenols. Thus, if advantage is to be taken of the above-mentioned synergism offered by the combination of a benzylalcoholhemiformal preparation with a phenol derivative, it is essential to keep the pH below 9·5; the pH level will be of no influence when using pentachlorophenol as the mixing partner.

Addition or condensation products on the basis of amine-formaldehyde also react with phenols; benzylamines are possible reaction products (Rosenbusch, 1955). It should be mentioned in this connection that Eggensperger *et al.* (1975) found out recently that a synergistic increase in effectiveness and a desirable broadening of the effective spectrum of benzylalcoholhemiformals as well as N-alkylolhexahydro-*s*-triazines are obtained when these substances are used in combination with 2-mercapto-pyridine.

As mentioned earlier, problems in the use of formaldehyde-releasing compounds as preservatives for aqueous formulations can also arise as a result of formaldehyde-resistant bacteria being selected. For example, we noticed that a number of germs in a lubricoolant increased from 1 million to 10 million/ml although benzylalcoholmono(poly)hemiformal was present in a concentration of 1000 ppm. It was found that the concentration of 1500 ppm BHF, which is generally used, would not have been sufficient to cover these bacteria. The bacteria still proliferated in a nutrient solution even in the presence of 3500 ppm formalin, 35%. It had previously been ensured analytically by the method described by Tanenbaum and Bricker (1951) that the concentration of free formaldehyde in the nutrient solution did not change significantly. On the other hand, as shown in Fig. 2, the formaldehyde

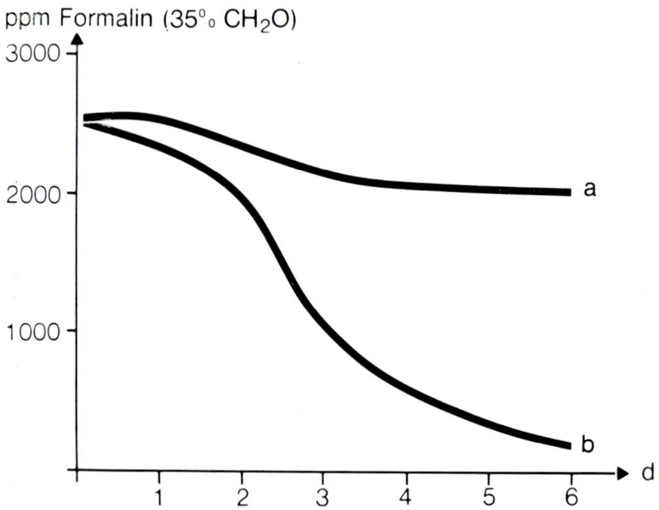

Fig. 2. Decrease of CH_2O-concentration in a nutrient solution: (a) infected; (b) not infected, with CH_2O-resistant bacteria. Initial concentration: 2500 ppm formalin (35% CH_2O).

concentration in the nutrient solution decreases in the presence of formaldehyde-resistant bacteria as the incubation period or number of germs increases. This suggests the conclusion that the bacteria isolated from the lubricoolant are capable of metabolizing formaldehyde even in relatively high concentration.

The data in Table VI show that, while having the same calculated content of formaldehyde, formulations of formaldehyde-releasing compounds yield different values of inhibition of formaldehyde-resistant bacteria.

Table VI
MICROBIOSTATIC MINIMUM CONCENTRATIONS (MMC's) OF FORMALDEHYDE-RELEASING SUBSTANCES WITH RESPECT TO CH_2O-RESISTANT BACTERIA. CALCULATED CONTENT OF FORMALDEHYDE OF THE FORMULATIONS: 29%

Formulation	*MMC (ppm)*
Formalin	4 200
n-Butanol(mono)polyhemiformal	4 000
Benzylalcohol(mono)polyhemiformal	2 500
1,3,5-Triethyl-hexahydro-*s*-triazine	3 600
1,3,5-Triethylol-hexahydro-*s*-triazine	3 000
N-Hydroxymethyl-chloroacetamide	1 600
1,5-Dimethyl-2-thiono-tetrahydro-thiadiazine	260

Although it does not matter very much whether formaldehyde is added to water or *n*-butanol, there will be a clear difference when benzylalcohol is substituted for water or *n*-butanol. In this case the synergism mentioned earlier makes itself felt. When adding or condensing formaldehyde to an amine, reaction products will be obtained that are only slightly more effective than formalin or *n*-butanolhemiformal. Much more suitable for the control of formaldehyde-resistant bacteria are substances obtained by reacting active ingredients such as chloroacetamide, especially N-methyl-dithiocarbamate, with formaldehyde. The 1,5-dimethyl-2-thiono-tetrahydro-thiadiazine is considerably more effective than a dithiocarbamate on which it is based (*see* Table VII).

Table VII
MICROBIOSTATIC MINIMUM CONCENTRATIONS (MMC) OF VARIOUS PRESERVATIVES WITH RESPECT TO CH_2O-RESISTANT BACTERIA

Substance	*MMC (ppm)*
Na-dimethyl-dithiocarbamate (40%)	1 000
Na-pentachlorophenolate	400
2-Phenyl-phenol	500
3-Methyl-4-chlorophenol	200

As the MMC values in Table VII prove, phenol derivatives, too, are suitable for the control of formaldehyde-resistant bacteria, 3-methyl-4-chlorophenol being characterized by particular effectiveness. In the case of the formaldehyde-resistant bacteria, a synergistic increase in effectiveness is not possible by combination of benzylalcoholhemiformal with phenol derivatives.

It can be maintained in summarizing that there are considerable differences between the formaldehyde-releasing compounds being used as preservatives with regard to the antimicrobial effectiveness and the effective spectrum. Reaction of benzylalcohols with formaldehyde leads to preparations with a synergistic effect, which extends to bacteria, fungi, and yeasts. In spite of synergism, the effectiveness of benzylalcoholhemiformal against fungi and formaldehyde-resistant bacteria still requires improvement. A very effective combination is that with small amounts of phenol derivatives when the mixing partners will not combine at pH levels exceeding 9–9·5. These mixtures have a synergistic effect against fungi and yeasts yet not against formaldehyde-resistant bacteria; nevertheless, it is especially 3-methyl-4-chlorophenol which suggests itself for intensifying the effect of benzylalcoholhemiformal against formaldehyde-resistant bacteria, where it is particularly effective; the same is true for 1,5-dimethyl-2-thiono-tetrahydro-thiadiazine.

REFERENCES

Bennett, E. O. (1956). Bacterial spoilage of emulsion oils. *Soap Chem. Specialties*, **32**, 47–48, 155.

Bennett, E. O. (1973). Formaldehyde preservatives for cutting fluids. *Int. Biodeterior. Bull.*, **9** (4), 95–100.

Druskeit, W., and Eggensperger, H. (1971). Beitrag zum Problem der Resistenzentwicklung von Bakterien gegenüber Konservierungszusätzen für wässrige Mineraloeleumulsionen. *Erdoel-Kohle-Erdgas-Petrochemie*, **24** (2), 103–105.

Eggensperger, H., Diehl, K. H., and Ramlau, P. (1975). Konservierungs- und Desinfektionsmittel- DOS 2.337.755.

Hill, E. C. (1967). The significance and control of microorganisms in rolling mill oils and emulsions. *Metals and Materials*, **1**, 294–297.

Kull, F. C., Eisman, P. C., Sylvestrowicz, H. D., and Mayer, R. L. (1961). Mixtures of quaternary ammonium compounds and long-chain fatty acids as antifungal agents. *Appl. Microbiol.*, **9**, 538–541.

Pivnick, H., Engelhard, W. E., and Thompson, T. L. (1954). The growth of pathogenic bacteria in soluble oil emulsions. *Appl. Microbiol.*, **2**, 140–143.

Rosenbusch, K. (1955). Studie über die Synthese von amphoteren Gerbstoffen. *Das Leder*, **6**, 80–86.

Tanenbaum, M., and Bricker, C. E. (1951). Microdetermination of free formaldehyde. *Anal. Chem.*, **23**, 354–357.

Weidle, H. (1973). Die technische Konservierung von Kühlmitteln in der metallverarbeitenden Industrie. *Seifen-Oele-Fette-Wachse*, **99**, 725–730, 741.

SUBLETHAL POLLUTANT EFFECTS UPON THE NATIVE MICROFLORA OF GEORGIA STRAIT

ALLAN S. DIETZ, TAINA TUOMINEN and LAWRENCE J. ALBRIGHT

Department of Biological Sciences, Simon Fraser University,
Burnaby, British Columbia, Canada

Summary

Glucose heterotrophic potential, bacterial productivity, and primary productivity of the native microflora of water from Georgia Strait were determined before and after treatment with crude oil, diesel fuel oil, primary sewage effluent, metallic salts, caustic pulp mill effluent, PCB (Arochlor 1260), and Fraser River silt. These pollutants were supplied at concentrations at which they might occur in Georgia Strait. Crude oil enhanced bacterial productivity and inhibited primary productivity and decreased heterotrophic potential. Diesel fuel oil had no effect upon these three biological parameters. Primary sewage effluent enhanced bacterial productivity and heterotrophic potential and had no effect upon primary productivity. Metallic salts had no effect upon these parameters. PCB Arochlor 1260 enhanced bacterial productivity but inhibited primary productivity and heterotrophic potential. Pump mill effluent decreased both bacterial and primary productivity with no effect upon heterotrophic potential. Fraser River silt had no effect upon these parameters.

INTRODUCTION

Various inorganic and organic components are known to enter the water of Georgia Strait from a variety of sources, including land drainage, sewage outfalls, ocean dumping, oil spills, bilge waters, and pulp mills (Hall *et al.*, 1974; Hoos and Packman, 1974). These substances, which are of both natural (*e.g.* Fraser River silt) and domestic, industrial, or agricultural origin (*e.g.* petroleum and its products, sewage, metallic salts, chlorinated hydrocarbons, pulp mill effluent) are however generally present in sublethal concentrations (Hall *et al.*, 1974; Hoos and Packman, 1974).

Much work has been done on the sublethal and lethal effects of these perturbants upon a variety of marine flora and fauna, including vertebrates, invertebrates, algae, and bacteria (Albright *et al.*, 1972; Davis, 1973; George, 1970; Reimer, 1975; Roberts, 1975; Sprague, 1969; Wurster, 1968). In this investigation, the influence of the above mentioned pollutants upon the *in situ* heterotrophic and autotrophic activities and biomasses of the native microflora of Georgia Strait water was observed. Attempts were made to treat this water with these perturbants at concentrations at which they might be found in this aquatic system.

MATERIALS AND METHODS

Water samples (*ca.* 60 litres) were removed from a depth of 1 m in Georgia Strait with the use of a Van Doren sampler at the approximate location of

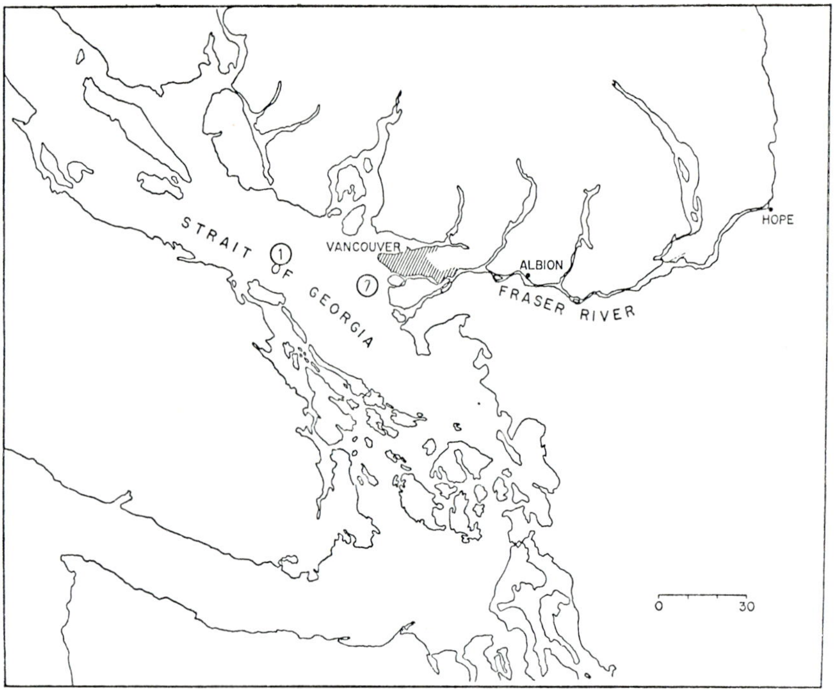

Fig. 1. Sampling stations in Georgia Strait.

Station 1 (Fig. 1). Various physical, chemical, and biological parameters of each sample were immediately determined (see below). The remaining water sample (*ca.* 50 litres) was then split into two 25-litre portions and placed in each of two 30-litre polyethylene carboys. In each instance one carboy was treated with a perturbant (*see* below) and the other acted as a control. Immediately following treatment, the carboys were attached to a metal frame, suspended from a hydrographic wire at a depth of 1 m below the water surface, and incubated for 24 hr. Following this incubation period, the waters contained within the carboys were assayed as described below.

The following perturbants were used: (a) Alberta crude or (b) diesel fuel oil layered to give a 1 cm layer on the contained water (325 ml); (c) primary sewage effluent, autoclaved to destroy biological activity and added to yield a final concentration of 0·45 and 0·60 mg/litre particulate carbon and nitrogen, respectively, and 1·0 mg/litre of soluble carbon; (d) metallic cation mixture (as chloride salts), mg/litre of Georgia Strait water: Cd, 0·07; Cr, 1·00; Cu, 0·20; Ni, 0·30; Pb, 2·00; Zn, 0·60; and Hg, 0·50 μg/litre. These are approximate values of major metallic cation concentrations in Iona Island primary treated sewage (Hall *et al.*, 1974); (e) PCB was added to a final concentration of 0·10 ppm using Arochlor 1260; (f) caustic pulp mill effluent was added to a final concentration of 2%; and (g) silt from the Fraser River obtained by centrifuging Fraser River water at 12 000 × *g* in a Sorvall KSB 'Szent-Gyorgi and Blum' continuous flow system (Ivan Sorvall Inc., Norwalk, Conn.). Silt harvested in this manner was autoclaved, dried at 65°C for 24 hr and then added to a final concentration of 26·3 mg/litre.

Soluble organic carbon was determined by the method of Menzel and Vaccaro (1964), whereas particulate carbon and nitrogen were assayed using a Beckman CNH analyser.

Heterotrophic marine bacterial numbers were determined using the spread plate technique, an incubation temperature of 15°C, and 2216E Marine agar (Difco). Heterotrophic terrestrial bacterial numbers were determined using the same technique and incubation temperature on nutrient agar (Difco).

Direct phytoplankton counts were determined by filtering 100 ml volumes of seawater through 0·22 μ membrane filters (Millipore), fixing the cells retained by the filters in formaldehyde fumes, followed by direct counts at 1000× magnification.

Adenosine-triphosphate (ATP) and chlorophyll a concentrations as well as $^{14}CO_2$ mediated phytoplankton productivity assays were determined using standard techniques as described by Strickland and Parsons (1972).

Heterotrophic potential (V_{max}), turnover time (T_t), and $K_t + S_n$ values for utilization of labelled glucose (D-glucose-C14 (U), 250 mCi/mM, Amersham/Searle) were determined using the heterotrophic activity technique as described by Wright and Hobbie (1966) and as modified by Albright *et al.* (1972).

The heterotrophic productivity assay has been described by Romanenko *et al.* (1972) and is based upon the dark uptake of $^{14}CO_2$ (as bicarbonate) (New England Nuclear, 8·4 mCi/mM).

RESULTS

In the majority of cases both the biomasses and activities of both the perturbant treated and untreated samples increased after incubation. This phenomenon has been previously noted by other investigators and is described as a 'bottle effect' (Albright and Wilson, 1974; Hobbie, 1969). Accordingly, all data noted for the perturbant-treated samples are compared with the untreated control samples and presented in Tables I and II. An effort was made to treat samples with sublethal exposures of the perturbant. From assays of bacterial and algal populations, it is apparent that all additions were sublethal.

Addition of crude oil to Georgia Strait water did not have a major impact upon bacterioplankton biomass nor chlorophyll a content. Phytoplankton numbers and total plankton biomass (ATP) were marginally reduced. Heterotrophic productivity was slightly increased but heterotrophic potential was slightly decreased. Primary productivity was significantly decreased (15×).

Addition of diesel fuel oil did not alter biomasses nor the metabolic parameters measured.

The effect of primary sewage upon bacterio- and phytoplankton biomasses was negligible. However, it did result in a slight decrease in total biomass. Heterotrophic potential was slightly enhanced, whereas heterotrophic productivity was significantly increased (80×). Primary productivity remained unchanged.

Metallic salt addition did not alter bacterio- nor phytoplankton numbers but slightly increased chlorophyll a content. Total plankton biomass decreased slightly. None of the metabolic parameters was affected.

Presence of PCB (Arochlor 1260) had no effect on bacterio- and phyto-

Table I
SUBLETHAL POLLUTANT EFFECTS UPON THE NATIVE MICROFLORA OF GEORGIA STRAIT

Sample	Date (d/m/y)	Temp. °C	Sal. ‰	pH	Sol. C. (mg/litre)	Part. C. (mg/litre)	Part. N. (mg/litre)	Plate Counts NA	Plate Counts 2216E	ATP (pop/ml)	Bact. prod. (μgC/litre/d)	Heterotrophic potential K_t+S_n (mg/litre)	Heterotrophic potential V_{max} (mg/litre/hr)	Heterotrophic potential T_t (hr)	1° prod. (μg/litre/hr)	Chlor. A (mg/m³)	Algal # (per ml)
Crude oil	3/12/74	8·0	24·5	7·9	3·09	0·137	0·016 5	$2{\cdot}7\times10^2$	$1{\cdot}6\times10^4$	$1{\cdot}6\times10^4$	63·3	0·002 7	$3{\cdot}4\times10^{-6}$	794	0·27	0·486	358
Control	3/12/74	8·0	24·5	7·9	3·80	0·210	0·029 4	$4{\cdot}0\times10^2$	$1{\cdot}5\times10^4$	$6{\cdot}8\times10^4$	27·9	0·003 2	$8{\cdot}8\times10^{-6}$	357	4·13	0·812	1316
Geor. Str. pre-incub.	2/12/74	8·0	24·5	7·9	5·58	0·263	0·040 3	$2{\cdot}5\times10^2$	$4{\cdot}0\times10^3$	$5{\cdot}0\times10^4$	7·2	0·000 3	$5{\cdot}6\times10^{-7}$	448	1·91	0·653	808
Diesel fuel oil	16/1/75	5·6	24·2	7·8	5·49	0·152	0·016 5	$5{\cdot}9\times10^2$	$5{\cdot}4\times10^3$	$2{\cdot}3\times10^4$	19·2	0·001 5	$1{\cdot}5\times10^{-5}$	98	1·157	0·605	295
Control	16/1/75	5·6	24·2	7·9	3·46	0·319	0·030 3	$1{\cdot}0\times10^3$	$6{\cdot}9\times10^3$	$4{\cdot}4\times10^4$	24·1	0·003 5	$1{\cdot}4\times10^{-5}$	259	0·788	0·540	326
Geor. Str. pre-incub.	15/1/75	5·6	24·2	8·0	3·95	0·287	0·033 5	$1{\cdot}8\times10^3$	$8{\cdot}8\times10^3$	$4{\cdot}4\times10^4$	16·7	0·002 5	$3{\cdot}5\times10^{-5}$	72	0·723	0·360	681
Primary sewage	20/2/75	6·8	27·2	7·9	12·37	0·331	0·045 0	$2{\cdot}5\times10^3$	$2{\cdot}7\times10^3$	$2{\cdot}1\times10^4$	203·5	0·007 2	$1{\cdot}2\times10^{-6}$	5 900	0·936	0·768	197
Control	20/2/75	6·8	27·2	7·9	10·28	0·132	0·019 0	$8{\cdot}5\times10^1$	$3{\cdot}9\times10^3$	$4{\cdot}7\times10^4$	2·6	0·000 8	$6{\cdot}7\times10^{-7}$	1 190	0·665	0·609	213
Geor. Str. pre-incub.	19/2/75	6·8	27·2	7·9	11·96	0·149	0·018 0	$4{\cdot}0\times10^1$	$8{\cdot}5\times10^2$	$3{\cdot}6\times10^4$	1·4	0·004 3	$6{\cdot}3\times10^{-7}$	6 860	0·612	0·731	237
Metallic salts	11/3/75	7·2	27·1	8·0	4·39	0·240	0·026 0	$2{\cdot}7\times10^2$	$6{\cdot}6\times10^3$	$1{\cdot}1\times10^4$	26·0	0·005 7	$1{\cdot}1\times10^{-5}$	533	1·131	0·757	282
Control	11/3/75	7·2	27·1	8·0	5·45	0·153	0·021 0	$2{\cdot}2\times10^2$	$1{\cdot}2\times10^4$	$3{\cdot}3\times10^4$	22·0	0·004 8	$1{\cdot}4\times10^{-5}$	343	1·234	0·197	320
Geor. Str. pre-incub.	10/3/75	7·2	27·1	7·9	10·47	0·278	0·031 0	$2{\cdot}9\times10^2$	$6{\cdot}0\times10^3$	$4{\cdot}2\times10^4$	8·8	0·005 5	$2{\cdot}6\times10^{-6}$	2 160	0·837	0·897	409
Arochlor 1260 (PCB)	8/4/75	8·8	23·3	8·0	4·08	0·234	0·038	$6{\cdot}5\times10^3$	$2{\cdot}1\times10^4$	$5{\cdot}6\times10^3$	58·4	0·010 3	$1{\cdot}1\times10^{-5}$	980	0·519	0·801	322
Control	8/4/75	8·8	23·3	7·9	6·69	0·245	0·043	$4{\cdot}9\times10^2$	$1{\cdot}6\times10^4$	$1{\cdot}9\times10^3$	23·0	0·005 1	$2{\cdot}2\times10^{-5}$	228	1·339	1·415	311
Geor. Str. pre-incub.	7/4/75	8·8	23·3	7·9	4·08	0·177	0·017	$7{\cdot}1\times10^2$	$8{\cdot}4\times10^3$	$1{\cdot}1\times10^3$	9·3	0·005 1	$3{\cdot}0\times10^{-6}$	1 680	0·723	1·290	281
Pulp mill effluent	7/5/75	9·5	28·0	8·2	2·97	0·435	0·029	$9{\cdot}5\times10^1$	$7{\cdot}5\times10^2$	$6{\cdot}0\times10^4$	7·5	0·003 8	$3{\cdot}2\times10^{-6}$	1 180	0·988	0·910	99
Control	7/5/75	9·5	28·0	8·0	1·97	0·147	0·022	$1{\cdot}2\times10^2$	$1{\cdot}1\times10^3$	$1{\cdot}5\times10^5$	21·4	0·004 0	$2{\cdot}9\times10^{-6}$	1 372	2·020	0·990	168
Geor. Str. pre-incub.	6/5/75	9·5	28·0	7·9	2·55	0·238	0·030	$1{\cdot}0\times10^1$	$7{\cdot}3\times10^2$	$2{\cdot}3\times10^5$	75·3	0·007 1	$2{\cdot}3\times10^{-6}$	3 080	2·480	0·780	141
Fraser R. silt	30/7/75	18·5	18·0	8·3	12·19	0·281	0·039	$4{\cdot}0\times10^2$	$1{\cdot}4\times10^4$	$1{\cdot}6\times10^5$	47·3	0·005 5	$1{\cdot}5\times10^{-5}$	366	5·420	3·270	394
Control	30/7/75	18·5	18·0	8·3	6·51	0·411	0·042	$2{\cdot}1\times10^3$	$1{\cdot}7\times10^4$	$2{\cdot}1\times10^5$	54·2	0·003 7	$1{\cdot}8\times10^{-5}$	203	4·490	2·530	375
Geor. Str. pre-incub.	29/7/75	18·5	18·0	8·2	5·87	1·005	0·131	$1{\cdot}8\times10^2$	$5{\cdot}4\times10^2$	$9{\cdot}6\times10^4$	13·3	0·004 6	$2{\cdot}6\times10^{-6}$	1 730	1·840	1·140	322

Table II
SUBLETHAL POLLUTANT EFFECTS UPON THE NATIVE MICROFLORA OF GEORGIA STRAIT—SUMMARY

	Crude oil	*Diesel oil*	*Primary sewage*	*Metal mixture*	*PCB (Arochlor* 1260)	*Pulp mill effluent*	*Fraser River silt*
Heterotrophic activity	(−)	0	0	0	(−)	0	0
Dark productivity	(+)	0	+	0	(+)	(−)	0
Light productivity	—	0	0	0	(−)	(−)	0
Plate counts	0	0	(+)	0	0	0	0
ATP	(−)	0	(−)	(−)	(+)	(−)	0
Algal ♯	(−)	0	0	0	0	0	0

0 = no difference from control (less than a factor of 2).
(+)/(−) = 2–10 × greater/less than control.
+/− = 10 or more × greater/less than control.

plankton biomasses but did cause a slight increase in total biomass. Heterotrophic potential and primary productivity decreased marginally.

Bacterial, algal, and total biomasses remained unchanged after incubation with 2% pulp mill effluent. Primary productivity and heterotrophic productivity decreased slightly, while heterotrophic potential remained unchanged.

Addition of Fraser River silt had no effect upon the biomasses or activities assayed.

DISCUSSION

Previous investigations of bacterial degradation of petroleum in coastal waters (Atlas and Bartha, 1972a,b; Mulkins-Phillips and Stewart, 1974; Walker and Colwell, 1974) have generally concluded that a significant portion of the bacteria present are able to metabolize most components of crude petroleum and that few are adversely affected by the presence of various crude oils. Thus the increase in heterotrophic potential noted in the presence of crude oil is probably a reflection of the utilization of crude oil components by the bacteria present in this water.

Lacaze (1974) noted that within 24 hr of addition of crude oil to an *in situ* experimental marine ecosystem, a 50% decrease in primary productivity occurred and a 60% or greater decrease occurred after incubation periods greater than 24 hr. He attributed this decrease to the presence of toxic light crude oil fractions, which were subsequently eliminated from the seawater surface by evaporation.

Since the addition of diesel fuel oil did not affect the microbial biomasses or activities, it would appear that this crude oil fraction contains neither readily assimilable nutrients nor appreciably toxic components.

Bacterial and algal biomasses were not affected by addition of primary sewage, whereas heterotrophic potential increased marginally (5×) and bacterial productivity increased by almost two orders of magnitude (80×). These responses are a reflection of the soluble and particulate organic nutrients present in the sewage. Previous work has shown that the major chemico-physical parameters influencing activities of the heterotrophic microbes within Georgia Strait are temperature and soluble carbon concentrations (Albright, unpubl. data). The result of these increased activities may be an increase in the standing crop of bacteria and/or increases in nutrient and energy flow to higher trophic levels.

Previous work by Albright *et al.* (1972) noted that addition of metallic salts to freshwater systems caused an immediate and marked inhibition of heterotrophic potential, whereas this marine system showed little or no response in biomass or activities. This would indicate that each aquatic system has its own unique response to metallic salts addition. Further experimental information is needed.

Previous work by Oloffs *et al.* (1972) with PCB Arochlor 1254 at a concentration of 0·10 ppm demonstrated no apparent adverse effects upon bacterioplankton. At the present time, no explanation can be offered for the approximate doubling of bacterial numbers. Previous investigators showed that chlorinated hydrocarbons inhibit phytoplankton productivity in both fresh- and seawater (Glooschenko, 1971; Menzel *et al.*, 1970; Wurster, 1968). However, chlorinated hydrocarbon analysis of Fraser River and Georgia Strait waters indicates that these compounds are not present at concentrations above the detection level of 1 ppb (Hall *et al.*, 1974; Oloffs *et al.*, 1972). Thus, these compounds are not presently of ecological significance to the plankton trophic levels in this system.

Since the final pH of the water sample with 2% caustic pulp mill effluent was 8·17, inhibition of bacterio- and phytoplankton productivity probably did not result from unfavourable pH. Bagge and Lehmusluoto (1971) noted a similar effect of pulp mill effluent on primary productivity. They attribute this decrease to presence of toxic effect(s) of the pulp mill waste liquor. However, they also noted relatively high bacterial productivity.

As early as 1943, ZoBell (1943) reported increased activity of bacteria when supplied with a solid surface. Both Fraser River silt treated and untreated samples were exposed to the solid surface of the carboy walls. This solid surface area may have been sufficient to allow maximum 'bottle effect' activity and addition of silt, and even more surface area did not further enhance microbial numbers or activities.

These experiments describe relatively short-term effects on bacterio- and phytoplankton activities and biomasses. Longer-term effects, which we did not study, may cause altered growth rates and activities resulting in a shift in species composition, activities, and biomasses.

The influences of combinations of two or more perturbants were not studied. Sprague and Ramsay (1965) and Albright and Wilson (1974) showed that the combination of Cu and Zn cations causes an additive effect. Thus, phytoplankton productivity may be additively inhibited by crude oil and pulp mill effluent, whereas bacterial productivity would be less stimulated by the above combination than by crude oil alone.

Precise knowledge of combination effects requires further investigation. Also, the presence of other trophic levels in the experimental system would have impact upon the results.

REFERENCES

Albright, L. J., and Wilson, E. M. (1974). Sublethal effects of several metallic salts-organic compounds combinations upon the heterotrophic microflora of a natural water. *Water Res.*, **8**, 101–105.

Albright, L. J., Wentworth, J. W., and Wilson, E. M. (1972). Technique for measuring metallic salt effects upon the indigenous heterotrophic microflora of a natural water. *Water Res.*, **6**, 1589–1596.

Atlas, R. M., and Bartha, R. (1972a). Biodegradation of petroleum in seawater at low temperatures. *Can. J. Microbiol.*, **18**, 1851–1855.

Atlas, R. M., and Bartha, R. (1972b). Degradation and mineralization of petroleum by two bacteria isolated from coastal waters. *Biotechnol. Bioeng.*, **14**, 297–308.

Bagge, P., and Lehmusluoto, P. O. (1971). Phytoplankton primary production in some Finnish coastal areas in relation to pollution. *Merentutkimuslait. Julk./Havsforskningsinst Skr.*, **235**, 3–18.

Davis, J. C. (1973). Sublethal effects of bleached Kraft pulp mill effluent on respiration and circulation in sockeye salmon (*Oncorhynchus nerka*). *J. Fish. Res. Bd. Can.*, **30**, 369–377.

George, J. D. (1970). Sublethal effects on living organisms. *Mar. Pollut. Bull.*, **1**, 107–109.

Glooschenko, W. A. (1971). The effect of DDT and dieldrin upon ^{14}C uptake by *in situ* phytoplankton in Lakes Erie and Ontario. Proc. 14th Conf. Great Lakes Res., pp. 219–223.

Hall, K. J., Koch, J. A., and Yesaki, I. (1974). Further investigations into water quality conditions in the lower Fraser River system. Tech. Rept no. 4. Westwater Res. Centre, University of British Columbia.

Hobbie, J. E. (1969). A method for studying heterotrophic bacteria, In A Manual on Methods for Measuring Primary Production in Aquatic Environments, pp. 146–151 (R. A. Vollenweider, ed.), Blackwell Scientific Publications, Oxford.

Hoos, L. M., and Packman, G. A. (1974). The Fraser River estuary: status of environmental knowledge to 1974. Special Estuary Ser. no. 1, Environment Canada, Ottawa.

Lacaze, J.-C. (1974). Ecotoxicity of crude oils and the use of experimental marine ecosystems. *Mar. Pollut. Bull.*, **5**, 153–156.

Menzel, D. W., and Vacarro, R. F. (1964). The measurement of dissolved organic and particulate carbon in seawater. *Limnol. Oceanogr.*, **9**, 138–142.

Menzel, D. W., Anderson, J., and Randtke, A. (1970). Marine phytoplankton vary in their response to chlorinated hydrocarbons. *Science*, **167**, 1724–1726.

Mulkins-Phillips, G. J., and Stewart, J. E. (1974). Distribution of hydrocarbon-utilizing bacteria in Northwestern Atlantic waters and coastal sediments. *Can. J. Microbiol.*, **20**, 955–962.

Oloffs, P. C., Albright, L. J., and Szeto, S. Y. (1972). Fate and behavior of five chlorinated hydrocarbons in three natural waters. *Can. J. Microbiol.*, **18**, 1393–1398.

Reimer, A. A. (1975). Effects of crude oil on corals. *Mar. Pollut. Bull.*, **6**, 39–43.

Roberts, D. (1975). Sublethal effects of chlorinated hydrocarbons on bivalves. *Mar. Pollut. Bull.*, **6**, 20–24.

Romanenko, V. I., Overbeck, J., and Sorokin, Y. I. (1972). Estimation of production and decomposition in fresh waters, pp. 82–85 (Y. I. Sorokin and J. Kadota, eds.), Blackwell Scientific Publications, Oxford.

Sprague, J. B. (1969). Measurement of pollutant toxicity to fish—III. Sub-lethal concentrations and 'safe' concentrations. *Water Res.*, **5**, 245–266.

Sprague, J. B., and Ramsay, B. A. (1965). Lethal levels of mixed copper-zinc solutions for juvenile salmon. *J. Fish. Res. Bd. Can.*, **22**, 425–432.

Strickland, J. D. R., and Parsons, T. R. (1972). A practical handbook of seawater analysis. *Bull.* 167 (2nd ed.). *Fish. Res. Bd. Can.*, Ottawa.

Walker, J. D., and Colwell, R. R. (1974). Microbial degradation of model petroleum at low temperatures. *J. Microbial. Ecol.*, **1**, 59–91.

Wright, R. T., and Hobbie, J. E. (1966). Use of glucose and acetate by bacteria and algae in aquatic ecosystems. *Ecology*, **47**, 447–464.

Wurster, G. F. (1968). DDT reduces photosynthesis by marine phytoplankton. *Science*, **159**, 1474–1475.

ZoBell, C. E. (1943). The effect of solid surfaces upon bacterial activity. *J. Bacteriol.*, **46**, 39–56.

FUNGAL SPORE WALL AS A POSSIBLE BARRIER AGAINST POTENTIAL ANTIFUNGAL AGENTS

HERMAN GERSHON

Boyce Thompson Institute for Plant Research,
Yonkers, New York 10701, U.S.A.

Summary

A hypothesis is presented that the fungal spore wall may act as a barrier against certain of the copper(II) bischelates of the substituted 8-quinolinols. Steric and electrostatic factors may be involved in the passage of the complexes through the pores of the spore wall. If the geometry and charge distribution of a molecule are not compatible with the geometry and distribution of charge around the periphery of the pores in the fungal spore wall, the compound cannot penetrate the wall and cause toxic reactions in the spore. The effect on antifungal activity of simultaneously varying the substituents on the 5,5′ and 7,7′ axes of the copper(II) bischelate of 8-quinolinol was studied. Calculations of approximate long and short dimensions of the pores of the spore walls of four fungi were made based on the geometry of the molecules in conjunction with their fungitoxicities. They were found to be: *Aspergillus niger*, 15·0, 10·8 Å; *Trichoderma viride*, 16·6, 10·7 Å; *Aspergillus oryzae*, 15·0, 10·8 Å; *Trichophyton mentagrophytes*, 16·6, 10·7 Å.

INTRODUCTION

8-Quinolinol has been known as a potent antifungal agent for many years. Its bischelate with copper(II) was found to be more fungitoxic than the ligand (Powell, 1946; Mason, 1948). A mechanism of antibacterial action that was also extended to antifungal action was proposed by Albert *et al.* (1953). The bischelate of 8-quinolinol with copper(II), which is able to penetrate the cell, dissociates into the 1:1 chelate and free 8-quinolinol. The 1:1 chelate could then become the toxic entity by combining with and blocking metal binding sites in enzymes.

The preparation and antimicrobial activity of 1:1:1 chelates of copper(II) with 8-quinolinol and arylhydroxycarboxylic acids were reported by Gershon *et al.* (1962), and it was found that the mixed ligand chelates containing 8-quinolinol, copper(II), and the arylhydroxy acids, salicylic and 3-hydroxy-2-naphthoic acids, possessed the same antifungal activity as bis(8-quinolinolato)copper(II) (Gershon and Parmegiani, 1963). Since both types of chelates contained the 1:1 complex of copper(II) with 8-quinolinol, experimental support was lent to the hypothesis of Albert *et al.* (1953). The dissociation of (8-quinolinolato)(salicylato)copper(II) to [(8-quinolinolato)copper(II)]$^+$ and [salicylate]$^-$ was observed by Schulman *et al.* (1968).

To study the effect of substituents on the fungitoxicity of these copper(II) chelates, three series of compounds were prepared. These were the 1:2 chelates of copper(II) with 5- and 5,7-substituted 8-quinolinols and 1:1:1 chelates of copper(II) with 5-substituted 8-quinolinols and 3,5-diiodosalicylic of 4-bromo-3-hydroxynaphthoic acids. The substituents in the 5-position of 8-quinolinol were H, F, Cl, Br and I, and the 5,7-substituted compounds also contained the same substituents (Fig. 1). The activity against five fungi (*Aspergillus niger*, *Aspergillus oryzae*, *Trichoderma viride*, *Myrothecium verrucaria* and *Trichophyton mentagrophytes*) indicated that all the chelates containing 8-quinolinol (Ia–IIIa) and the 5-fluoro derivative (Ib–IIIb) were toxic to the five fungi. Bis(5-chloro-8-quinolinolato)copper(II) (Ic) inhibited

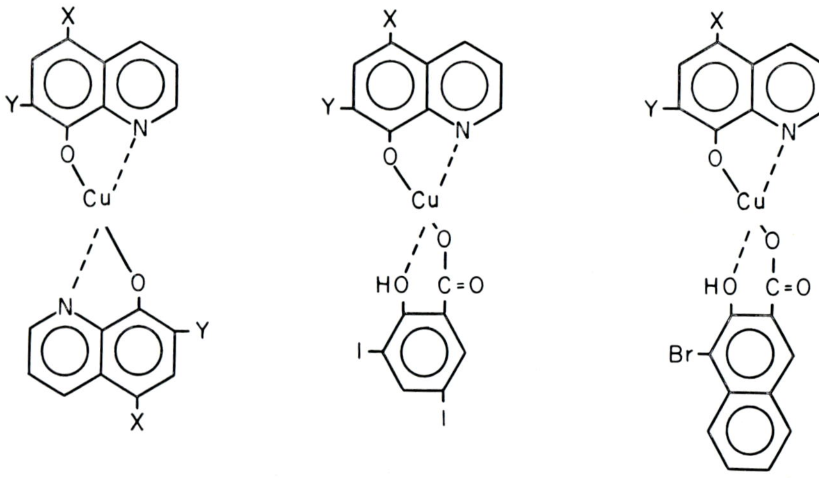

I II III

Ia–IIIa, X = H; Y = H	Ij, X = F; Y = NO_2
Ib–IIIb, X = F; Y = H	Ik, X = Cl; Y = NO_2
Ic–IIIc, X = Cl; Y = H	Il, X = Br; Y = NO_2
Id–IIId, X = Br; Y = H	Im, X = I; Y = NO_2
Ie–IIIe, X = I; Y = H	In, X = NO_2; H = H
If X = Cl; Y = Cl	Io, X = NO_2; Y = F
Ig X = Br; Y = Br	Ip, X = NO_2; Y = Cl
Ih X = I; Y = I	Iq, X = NO_2; Y = Br
Ii X = H; Y = NO_2	Ir, X = NO_2; Y = I

Fig. 1. Bischelates of 8-quinolinols with copper(II) and mixed ligand chelates of 5-halogeno-8-quinolinols with copper(II) and aromatic hydroxy acids

only three fungi, and the 5-bromo and 5-iodo (Id and Ie) analogs did not inhibit any of the fungi. The 5,7-dihalo-8-quinolinol copper(II) chelates (If–Ih) did not inhibit any of the fungi except for bis(5,7-dichloro-8-quinolinolato)copper(II) (If) which inhibited only one fungus. Of the mixed ligand chelates, those containing 8-quinolinol (IIa and IIIa), 5-fluoro (IIb and IIIb), 5-chloro (IIc and IIIc) and 5-bromo-8-quinolinol (IId and IIId) were toxic to five fungi, and the chelates containing 5-iodo-8-quinolinol (IIe and IIIe) were nontoxic to all of the fungi (Gershon *et al.*, 1966).

It was found that the compounds were toxic at low concentrations or inactive. Many of the chelates were tested at 1000 μg/ml, and good growth was observed. It thus appeared that this was an 'all or none' phenomenon which could be explained by the following hypothesis (Gershon *et al.*, 1966, and Gershon, 1972). Bis(8-quinolinolato)copper(II) was reported to be square-planar (Kanamaru *et al.*, 1963), and the 1:2 chelates, although they possess small negative charges at the opposite ends of the long axes, are nonpolar. It is reasonable to consider that the fungal spore wall is porous in nature, in order for materials to pass between the environment and the interior of the spore. Although the net charge of the pores is zero, it is composed of alternate + and − charges around the periphery of each pore. The chelate molecules float around in the liquid medium at random until they approach the spore wall. They are then oriented parallel to the pores in the wall by electrostatic attraction. If the long axes of the chelates are shorter than the diameters of the pores, the chelates can penetrate the spore and manifest their toxic reactions. The copper(II) bischelates containing 8-quinolinol (Ia), 5-fluoro-8-quinolinol (Ib), and in some cases 5-chloro-8-quinolinol (Ic) are fungitoxic presumably, because their long axes are shorter than the diameters of the pores in the fungal spore wall and penetration can take place. The inactivity of the remaining bischelates of the 5-substituted 8-quinolinols (Id and Ie) can thus be explained on the basis that the increase in the length of the long axes, due to the increase in size of the 5,5′-substituents, causes the molecule to be too large to pass through the pores of the spore wall.

The mixed ligand chelates containing 8-quinolinol (IIa and IIIa), 5-fluoro-8-quinolinol (IIb and IIIb), 5-chloro-8-quinolinol (IIc and IIIc) and 5-bromo-8-quinolinol (IId and IIId) were toxic to all five fungi. Although the long axes of these chelates are, in most cases, longer than those of the corresponding bischelates (Table I), there is no symmetrical charge distribution at

Table I

LENGTHS OF LONG AXES OF 1:2 CHELATES OF COPPER(II) WITH 8-QUINOLINOLS AND 1:1:1 CHELATES OF COPPER(II) WITH 8-QUINOLINOLS AND AROMATIC HYDROXY ACIDS (Å)[a]

Substituent in 5 position of 8-quinolinol	*1:2 Complex (I)*	*1:1:1 Complex with 4-BrBON(II)*[b]	*1:1:1 Complex with 3,5-I_2Sal(III)*[c]
H	14·2	16·9	16·1
F	15·0	17·4	16·6
Cl	16·6	18·2	17·4
Br	17·4	18·6	17·8
I	18·0	18·9	18·1

[a] Computations were based on Kanamaru *et al.* (1963) and Pauling (1960).
[b] 4-BrBON = 4-bromo-3-hydroxy-2-naphthoic acid.
[c] 3,5-I_2Sal = 3,5-diiodosalicylic acid.

the opposite ends of the long axes, and the bischelates possess a dipole. Consequently, on approaching the pores of the spore wall, these molecules do not approach parallel, but at an angle, which should be related to the net charge at one end of the long axis of the unsymmetrical chelate. The greater the charge, the greater will be the angle the chelate makes with the wall, and the shorter will be the projected long axis of the molecule. This could be expressed by the equation: $P = L \cos \phi$ (P = projected axis of the chelate, L = long axis of the chelate, ϕ = angle between L and spore wall). When P does not exceed the diameter of the pores, penetration can occur.

It is of interest that this 'all or none' observation was reported by Byrde *et al.* (1958) in a study of the fungitoxicity of a series of bischelates of copper(II) with 5-alkyl-8-quinolinols in which the alkyl group varied in length between zero and nine carbon atoms. The cutoff was at the three carbon chain. No interpretation was offered.

That the spore wall is assumed to be the barrier rather than the membrane is based on the following. When effects such as partition coefficient and enzyme–substrate complex formation are considered, a series with graded alterations in structure such as stepwise increase in size of an alkyl side chain or gradation in halogen substituent, the expected change in biological effect would be gradual and not sudden. Penetration of membrane is influenced by partition coefficient and an 'all or none' effect would not be expected. No claim is made in this hypothesis that fungal spore wall replaces membrane in all its functions. The membrane carries out the usual functions attributed to it; however, fungal spore wall may also act as a barrier.

This hypothesis which claims that the nontoxicity of certain compounds is due to exclusion from the spore because the long axis of the test compound is too long to allow for passage through the pores of the fungal spore wall was tested as follows. Since bis(5-nitro-8-quinolinolato)copper(II) (In) was observed to be nontoxic to the five fungi in this test system (Gershon *et al.*, 1966), it would also be expected that substituents placed in other positions of this chelate would not yield fungitoxic derivatives. To verify this, the copper(II) bischelates of 5-nitro (In), 7-fluoro-5-nitro (Io), 7-chloro-5-nitro (Ip), 7-bromo-5-nitro (Iq), and 7-iodo-5-nitro-8-quinolinol (Ir) were prepared and tested. As anticipated, these chelates were inactive (Gershon *et al.*, 1969a and 1969b; Gershon and McNeil, 1971b; Gershon, 1972 and 1974).

A second axis of the chelate was studied to determine whether the pores of the fungal spore wall could possess a geometrical configuration other than a circle. The series of bischelates of copper(II) with 7-nitro-8-quinolinol (Ii) and the 5-fluoro (Ij), 5-chloro (Ik), 5-bromo (Il), and 5-iodo (Im) analogs were prepared and tested against the five fungi (Gershon, 1968; Gershon *et al.*, 1969a; Gershon and McNeil, 1971b; Gershon, 1972 and 1974). All of the chelates were nontoxic.

To explain the inactivity of the copper(II) bischelates of the 5-substituted 7-nitro-8-quinolinols (Ii–Im) and of the corresponding chelates of the 5,7-dihalo-8-quinolinols, it was necessary to modify the earlier hypothesis. Table II shows the computed lengths of the axes formed by the substituted 5,5′ and 7,7′ positions and the angles between them, assuming square planarity of the chelates. If the pores of the fungal spore wall were circular, Ii–Ik would be expected to inhibit all five fungi. This was not observed. If the pores are not circular and could possibly be elliptical or hexagonal, they would have a second dimension which would be smaller than the diameter of the circle. The modified hypothesis would then be able to account for the inactivity of the bis(7-nitro-8-quinolinolato)copper(II) chelates as being due to exclusion from the spore.

To further develop information on the sizes and shapes of the pores of the fungal spore walls, a study was made on the 5,5′ and 7,7′ axes of the copper(II) bischelates of substituted 8-quinolinols by simultaneously varying the substituents from H to F to Cl to Br to I. The following derivatives of compound I were studied: a, X = H; Y = H, ab, X = H; Y = F, ac, X = I; Y = Cl, ad, X = H; Y = Br, ae, X = H; Y = I, b, X = F; Y = H, bb, X = F; Y = F, bc, X = F; Y = Cl, bd, X = F; Y = Br, be, X = F; Y = I, c, X = Cl; Y = H, cb, X = Cl; Y = F, f, X = Cl; Y = Cl, cd, X = Cl; Y = Br, ce, X = Cl; Y = I, d, X = Br; Y = H, db, X = Br; Y = F, dc, X = Br; Y = Cl, g, X = Br; Y = Br, de, X = Br; Y = I, e, X = I; Y = H, eb, X = I; Y = F, ec, X = I; Y = Cl, ed, X = I; Y = Br, h, X = I; Y = I.

Table II

LENGTHS AND ANGLES FORMED BY 5,5′ AND 7,7′ AXES OF COPPER(II) CHELATES OF 5- AND 7-SUBSTITUTED 8-QUINOLINOLS

5,5′ *axis*		7,7′ *axis*		*Angle formed between axes (degrees)*
Substituent	*Length* (Å)	*Substituent*	*Length* (Å)	
H	14·2	H	11·5	59
H	14·2	NO_2	14·9	70
F	15·0	H	11·5	60
F	15·0	NO_2	14·9	71
Cl	16·6	H	11·5	62
Cl	16·6	NO_2	14·9	73
Br	17·4	H	11·5	63
Br	17·4	NO_2	14·9	74
I	18·0	H	11·5	64
I	18·0	NO_2	14·9	75
NO_2	18·1	H	11·5	64
NO_2	18·1	NO_2	14·9	75

MATERIALS AND METHODS

Compounds

The ligands of the chelates listed in Table III were obtained as follows: Ia, Ic, Id, If, Ig, and Ih were purchased from Aldrich Chemical Co., Milwaukee, Wisconsin, and the remaining ligands were prepared according to published methods: Iab and Iec (Gershon and McNeil, 1971b), Iac and Iad (Gershon *et al.*, 1969a), Iae (Gershon and McNeil, 1971a), Ib and Ibe (Helin and Vanderwerf, 1952), Ibb, Ibc, Ibd, Icb, Idb, Ide, Ieb, and Ied Gershon *et al.*, 1972a), Icd (Gesellschaft, 1930), Ice (Bose, 1945), Idc (Boedecker and Cassel, 1951), and Ie (Coll and Coll, 1951a and 1951b). The copper(II) bischelates of the ligands were prepared as reported by Gershon (1974).

Microbiological methods

Antifungal tests were carried out in Sabouraud dextrose broth (Difco) on *Aspergillus niger* van Tiegham, *Aspergillus oryzae* (Ahlburg) Cohn, *Trichoderma viride* Persoon ex Fries, *Myrothecium verrucaria* Ditmar ex Fries, and *Trichophyton mentagrophytes* Robin. The tests were conducted in shake culture at 28°C for 6 days, and the presence or absence of growth was recorded (Gershon and Parmegiani, 1963; Gershon *et al.*, 1966).

RESULTS

The antifungal data obtained on the 25 chelates are listed in Table III. Bis(8-quinolinolato)copper(II) (Ia) and the 5-fluoro analog (Ib) are toxic to five fungi. The remaining 7-substituted chelates (Iab, Iac, Iad, and Iae) are toxic to *T. viride*, *M. verrucaria*, and *T. mentagrophytes*. Of the 5-fluoro series, the 7-chloro (Ibc), 7-bromo (Ibd), and 7-iodo (Ibe) are inhibitory to the same three fungi, whereas bis(5,7-difluoro-8-quinolinolato)copper(II) is toxic to *M. verrucaria* and *T. mentagrophytes*. The copper(II) bischelate of

Table III

MINIMAL FUNGISTATIC ACTIVITY (mmoles/litre) OF COPPER(II) COMPLEXES OF 5-, 7- AND 5,7-SUBSTITUTED 8-QUINOLINOLS IN SABOURAUD DEXTROSE BROTH AT 28°C IN SHAKE FLASKS AFTER 6 DAYS[a]

Compound	*X*	*Y*	A. niger	T. viride	A. oryzae	M. verrucaria	T. mentagrophytes
Ia	H	H	0·008 5	0·008 5	0·008 5	0·008 5	0·005 7
Iab	H	F	NA[b]	0·036	NA	0·036	0·018
Iac	H	Cl	NA	<0·002 4[c]	NA	<0·002 4	<0·002 4
Iad	H	Br	NA	<0·002 0	NA	<0·002 0	<0·002 0
Iae	H	I	NA	<0·001 7	NA	<0·001 7	<0·001 7
Ib	F	H	0·010	0·007 8	0·013	0·007 8	0·005 7
Ibb	F	F	NA	NA	NA	0·019	0·019
Ibc	F	Cl	NA	0·035	NA	<0·002 2	<0·002 2
Ibd	F	Br	NA	0·018	NA	<0·018	<0·001 8
Ibe	F	I	NA	0·031	NA	<0·001 6	<0·001 6
Ic	Cl	H	NA	0·007 2	NA	0·007 2	0·012
Icb	Cl	F	NA	NA	NA	0·006 5	NA
If	Cl	Cl	NA	NA	NA	0·012	NA
Icd	Cl	Br	NA	NA	NA	NA	NA
Ice	Cl	I	NA	NA	NA	NA	NA
Id	Br	H	NA	NA	NA	NA	NA
Idb	Br	F	NA	NA	NA	NA	NA
Idc	Br	Cl	NA	NA	NA	NA	NA
Ig	Br	Br	NA	NA	NA	NA	NA
Ide	Br	I	NA	NA	NA	NA	NA
Ie	I	H	NA	NA	NA	NA	NA
Ieb	I	F	NA	NA	NA	NA	NA
Iec	I	Cl	NA	NA	NA	NA	NA
Ied	I	Br	NA	NA	NA	NA	NA
Ih	I	I	NA	NA	NA	NA	NA

[a] Data taken from Gershon *et al.* 1966 and Gershon, 1974.
[b] NA = not active below 100 ppm.
[c] < = inhibitory at 1 ppm, lowest level tested.

5-chloro-8-quinolinol (Ib) is toxic to *T. viride*, *M. verrucaria*, and *T. mentagrophytes*, and the corresponding chelates of 5-chloro-7-fluoro (Icb) and 5,7-dichloro-8-quinolinol (If) are toxic to *M. verrucaria* only.

Table IV shows the lengths of the 5,5′ and 7,7′ axes and the angles between them, assuming that all the bischelates are square planar. The calculations were based on the X-ray crystallographic studies on bis(8-quinolinolato) copper(II) (Ia) (Kanamaru *et al.*, 1963) together with the van der Waals radii of the substituents (Pauling, 1960). It should be noted that the chelates

Table IV

LENGTHS AND ANGLES BETWEEN THE 5,5′ AND 7,7′ AXES OF THE BIS-CHELATES OF COPPER(II) WITH 8-QUINOLINOLS DUE TO SUBSTITUENTS

5,5′ axis		*7,7′ axis*		*Angle*
Substituent	*Length* (Å)	*Substituent*	*Length* (Å)	*(degrees)*
H	14·2	H	11·5	59
F	15·0	H	11·5	60
Cl	16·6	H	11·5	62
Br	17·4	H	11·5	63
I	18·0	H	11·5	64
H	14·2	F	12·2	62
F	15·0	F	12·2	63
Cl	16·6	F	12·2	65
Br	17·4	F	12·2	66
I	18·0	F	12·2	67
H	14·2	Cl	13·5	67
F	15·0	Cl	13·5	68
Cl	16·6	Cl	13·5	70
Br	17·4	Cl	13·5	71
I	18·0	Cl	13·5	72
H	14·2	Br	14·1	69
F	15·0	Br	14·1	70
Cl	16·6	Br	14·1	72
Br	17·4	Br	14·1	73
I	18·0	Br	14·1	74
H	14·2	I	14·7	70
F	15·0	I	14·7	71
Cl	16·6	I	14·7	73
Br	17·4	I	14·7	74
I	18·0	I	14·7	75

containing 7-bromo and 7-iodo substituents are not square planar, as determined by stability constant measurements. The chloro derivatives might also be nonplanar (Gershon *et al.*, 1972b).

DISCUSSION

The antifungal data (Table III) show that for the Aspergilli when H is in the 7 and 7′ positions, the pores will allow passage of compounds in which F is in the 5 and 5′ positions, and when the 7 and 7′ positions contain F, only those chelates containing H in the 5 and 5′ positions can pass through the pores of the fungal spore wall. As for *T. viride*, when the 7 and 7′ positions contain H, the chelate containing Cl in the 5 and 5′ positions can pass through the pores, but when the 7 and 7′ positions contain F, chelates containing H in the 5 and 5′ positions can be accommodated by the pores. With respect to *T. mentagrophytes*, when the 7 and 7′ positions are occupied by H, the 5 and 5′ position holding Cl can pass through the pores, and when the 7 and 7′ positions are filled with F, the 5 and 5′ positions containing F can pass into the fungal spore. An analysis of the comparable data for

M. verrucaria leads to indefinite conclusions, due to the uncertainty of the square planarity of the 7 and 7′ chloro derivatives.

Thus by shortening the long axis of the chelate, longer short axes can be accommodated by the pores of the spore wall. This can be visualized by considering the copper atom of the chelate as a pivotal point in the centre of the pore, and as the long axis becomes shorter, the molecule can rotate about the point to accommodate longer short axes.

Based on the geometry of the square planar chelates which possess the largest dimensions and are fungitoxic, it is possible to calculate approximate dimensions that are perpendicular to the long dimensions of the pores in the fungal walls of the four fungi,. *A. niger*, *T. viride*, *A. oryzae* and *T. mentagrophytes*. Knowing the lengths of the two axes of the toxicant and the angle between them (Table IV), they can be circumscribed by a symmetrical geometric figure. An ellipse would be suitable as a first approximation in which the major axis would be related to the long axis of the toxicant and the shorter axis of the toxicant would be a diameter. The minor axis of the ellipse would be related to the shortest diameter of the pore in the spore wall, and it would also be perpendicular to the long axis of the pore. The semi-minor axis of the ellipse can be calculated by rearranging the equation of the ellipse to

$$b = [a^2y^2/(a^2 - x^2)]^{\frac{1}{2}}$$

where b = semiminor axis of the ellipse;
a = semimajor axis = half of the length of the 5,5′ axis of the bischelate;
x and y are coordinates of the point on the circumference of the ellipse which intersects with the diameter $2r$, the 7,7′ axis of the chelate;
$x = r \cos \theta$;
$y = r \sin \theta$;
θ = angle between 5,5′ and 7,7′ axes.

By appropriate substitutions for x and y in terms of the 7,7′ axis ($2r$) and the angle (θ) between the 5,5′ and 7,7′ axes, the formula

$$b = [a^2r^2 \sin^2 \theta/(a^2 - r^2 \cos^2 \theta)]^{\frac{1}{2}}$$

can be obtained from which the shortest diameters ($2b$) of the pores in the fungal spore wall can be computed.

Table V shows the approximate long and perpendicular short dimensions of the pores in the spore walls of the four fungi.

Table V
APPROXIMATE LONG AND SHORT DIMENSIONS OF PORES IN FUNGAL SPORE WALL

Organism	*Length* (Å)	*Width* (Å)
A. niger	15·0	10·8
T. viride	16·6	10·7
A. oryzae	15·0	10·8
T. mentagrophytes	16·6	10·7

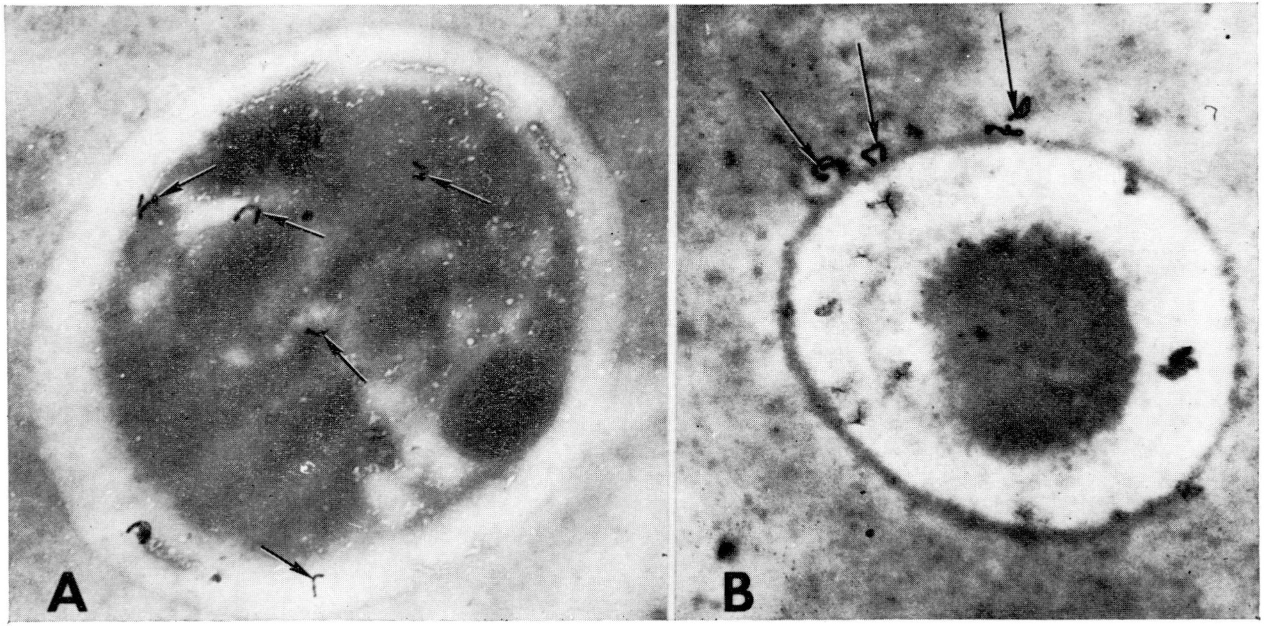

Fig. 2. Electron microscopic autoradiograms of A. niger *spores exposed to tritiated copper(II) chelates. A: bis(8-quinolinolato)copper(II); B: bis(5,7-diiodo-8-quinolinolato) copper(II)*

Although the short diameters of the pores in the fungal spore walls were obtained by means of calculations using ellipses, it does not imply that the pores are elliptical. It is anticipated that further study with the 4,4′ together with the 5,5′ axes would allow the assignment of shape to the pores. It should be noted that all these proposed chelates are expected to be square planar.

Although conclusive proof establishing the presence of the toxic bischelate within the cell for toxic compounds and the exclusion of the nontoxic bischelate from the cell as the basis for nontoxicity is lacking, preliminary results indicate that this consideration is correct. Upon exposing spores of *A. niger* to tritiated bis(8-quinolinolato)copper(II) and bis(5,7-diiodo-8-quinolinolato)copper(II), it can be seen as a result of the methods of electron microscopic autoradiography that the toxic copper(II) chelate of 8-quinolinol had penetrated the spore and the nontoxic copper(II) chelate of 5,7-diiodo-8-quinolinol was excluded (Hirumi and Gershon, 1972) (Fig. 2).

The significance of this work is that it is not unreasonable to believe that fungal spore wall is perforated and may behave as a barrier against certain potential antifungal agents. If sound data concerning the dimensions and electrical properties of the pores in the fungal spore wall were developed, such information could be used in the design of antifungal agents. If the geometry and distribution of charge around a molecule are incompatible with the geometry and charge distribution around the peripheries of the pores in the fungal spore wall, penetration of the wall by the agent cannot take place, and toxic reactions within the spore will not occur. Ultimately, this leads to the concept that with respect to the fungal spore wall of each species only those compounds possessing certain limitations in geometry and dipole moments need be prepared and studied. Reasonable estimates of such data can be expected from calculations using bond lengths, angles, and moments, prior to synthesis of compounds.

Acknowledgement

This work was supported in part by U.S. Public Health Service Grant No. AI-05808.

REFERENCES

Albert, A., Gibson, M. I., and Rubbo, S. D. (1953). The influence of chemical constitution on antibacterial activity; Part IV: The bactericidal action of 8-hydroxyquinoline (oxine). *Brit. J. Exp. Pathol.*, **34**, 119–130.

Boedecker, F., and Cassel, L. (1951). Stepwise chlorination or bromination of 8-hydroxyquinoline. German Patent 767,098.

Bose, M. K. (1945). Preparation of iodochlorohydroxyquinoline. *J. Indian Chem. Soc.*, **22**, 169–170.

Byrde, R. J. W., Clifford, D. R., and Woodcock, D. (1958). Fungicidal activity and chemical constitution. VI. The activity of some substituted 8-hydroxyquinolines (oxines). *Ann. Appl. Biol.*, **46**, 167–177.

Coll, A.-L., and Coll, G. P. (1951a). Contribucion al estudio de los derivados halogenados de la 8-oxiquinolina 7 acidos 7.halogeno-8.oxiquinolin-5.sulfonicos. *Afinidad*, **28**, 101–109.

Coll, A.-L., and Coll, G. P. (1951b). Contribucion al estudio de los derivados halogenados de la 8-oxiquinolina y acidos 7.halogeno-8.oxiquinolin-5.sulfonicos. II—Analisis. *Afinidad*, **28**, 163–170.

Gershon, H. (1968). 7-Nitro-8-quinolinols and their copper(II) complexes. Implications of the fungal spore wall as a possible barrier against potential antifungal agents. *J. Med. Chem.*, **11**, 1094–1096.

Gershon, H. (1972). Implications of fungal spore wall as a possible barrier against potential antifungal agents. Proc. 2nd Int. Congr. Pestic. Chem., pp. 363–377.

Gershon, H. (1974). Antifungal activity of bischelates of 5-, 7- and 5,7-halogenated 8-quinolinols with copper(II). Determination of approximate dimensions of the long and short axes of the pores in fungal spore wall. *J. Med. Chem.*, **17**, 824–827.

Gershon, H., and McNeil, M. W. (1971a). 7-Iodo-8-quinolinol from 5-amino-8-quinolinol via a Gattermann reaction. *J. Heterocycl. Chem.*, **8**, 129–131.

Gershon, H., and McNeil, M. W. (1971b). Nitration of 5-halogeno-8-quinolinols and accompanying Reverdin rearrangement. *J. Heterocycl. Chem.*, **8**, 821–824.

Gershon, H., and Parmegiani, R. (1963). Antimicrobial activity of 8-quinolinol, its salts with salicylic acid and 3-hydroxy-2-naphthoic acid, and the respective copper(II) chelates in liquid culture. *Appl. Microbiol.*, **11**, 62–65.

Gershon, H., Parmegiani, R., and Nickerson, W. J. (1962). Antimicrobial activity of metal chelates of salts of 8-quinolinols with aromatic hydroxycarboxylic acids. *Appl. Microbiol.*, **10**, 556–560.

Gershon, H., Parmegiani, R., Weiner, A., and D'Ascoli, R. (1966). Fungal spore wall as a possible barrier against potential antifungal agents of the group, copper(II) complexes of 5-halogeno- and 5-nitro-8-quinolinols. *Contrib. Boyce Thompson Inst.*, **23**, 219–228.

Gershon, H., McNeil, M. W., and Grefig, A. T. (1969a). Reinvestigation of the action of N-halosuccinimides on bis(8-quinolinolato)copper(II). *J. Org. Chem.*, **34**, 3268–3270.

Gershon, H., McNeil, M. W., and Hinds, Y. J. (1969b). 5-Nitro-8-quinolinols and their copper(II) complexes. Implications of the fungal spore wall as a possible barrier against potential antifungal agents. *J. Med. Chem.*, **12**, 1115–1117.

Gershon, H., McNeil, M. W., Parmegiani, R., and Godfrey, P. K. (1972a). Antifungal activity of 7- and 5,7-substituted 8-quinolinols. *J. Med. Chem.*, **15**, 987–989.

Gershon, H., McNeil, M. W., Schulman, S. G., and Parkes, J. W. (1972b). Stability constants of some sterically hindered bis(8-quinolinolato)copper(II) complexes. *Anal. Chim. Acta*, **62**, 43–47.

Gesellschaft für Chemische Industrie in Basel (1930). Verfahren zur Darstellung halogenhaltiger basischer Äther. German Patent 556,324.

Helin, A. F., and Vanderwerf, C. A. (1952). Synthesis of medicinals derived from 5-fluoro-8-hydroxyquinoline. *J. Org. Chem.*, **17**, 229–232.

Hirumi, H., and Gershon, H. (1972). Unpublished results.

Kanamaru, F., Ogawa, K., and Nitta, I. (1963). The crystal structures of metal 8-hydroxyquinolinate. I. Copper 8-hydroxyquinolinate. *Bull. Chem. Soc. Japan*, **36**, 422–427.

Mason, C. L. (1948). A study of the fungicidal action of 8-quinolinol and some of its derivatives. *Phytopathology*, **38**, 740–751.

Pauling, L. (1960). *The Nature of the Chemical Bond*, 3rd edn. Cornell University Press, Ithaca, N.Y., pp. 221–264.

Powell, D. (1946). Copper 8-quinolinolate, a promising fungicide. *Phytopathology*, **36**, 572–573.

Schulman, S. G., Kilroy, W. P., and Gershon, H. (1968). Mixed ligand chelates of copper(II) with 8-quinolinol and arylhydroxycarboxylic acids. I. Electronic absorption. spectra. *J. Phys. Chem.*, **72**, 3372–3375.

A RADIORESPIROMETRIC TECHNIQUE FOR MEASURING THE BIODEGRADATION OF SPECIFIC COMPONENTS IN A COMPLEX EFFLUENT

R. T. BELLY and C. T. GOODHUE

Eastman Kodak Company, Research Laboratories,
Rochester, New York 14650, U.S.A.

Summary

A radiorespirometric technique was used to study the biodegradation of specific chemicals and their end products in a complex photographic processing effluent. Degradation was followed by adding ^{14}C-labelled chemicals to a small volume of a simulated effluent solution, and monitoring $^{14}CO_2$ evolution in the presence of acclimated sludge microorganisms.

INTRODUCTION

Present techniques for evaluating the extent of biodegradation include the standard BOD_5 test and the conventional respirometric procedures using Warburg procedures. BOD_5 determinations have inherent limitations since adapted sludges containing a large number of diverse microorganisms cannot easily be used in this procedure. Although degradation by acclimated sludges can be conveniently studied in a Warburg respirometer, such systems have a limited capacity and require considerable time and effort during the course of a long-term experiment. Furthermore, both the BOD_5 and the Warburg respirometric methods cannot be easily applied to study the degradation of a particular chemical in an effluent where other compounds may also be metabolized.

Radiorespirometric techniques have been applied to the study of microbial activities in soils and natural waters (Brock, 1971; Hubbard, 1973). Biodegradability is based upon the measurement of $^{14}CO_2$ formed when ^{14}C-labelled materials are metabolized by microorganisms. Radiorespirometric methods have several distinct advantages, including relative simplicity, speed, and a high degree of sensitivity. In addition, such techniques are not subject to many of the chemical interferences that one encounters with other analytical procedures. Furthermore, radiorespirometric techniques can be applied to the evaluation of biodegradation by cometabolism, by mixed cultures of microorganisms, by microorganisms which cannot be isolated in pure culture, and the effects of other compounds on biodegradation can easily be determined.

The present paper describes a radiorespirometric technique for measuring $^{14}CO_2$ release during the biodegradation of ^{14}C-labelled industrial chemicals. This technique has permitted us to determine the biodegradability of a number of industrial chemicals by acclimated sludges, and to study the biodegradation of specific chemicals and their end products in a simulated complex industrial effluent.

MATERIALS AND METHODS

Radiochemicals. Hydroquinone-2,3,5,6-^{14}C (0·030 mC/mg) was obtained from Schwartz/Mann (Orangeburg, N.Y.). Benzoic acid-1-^{14}C (0·026 mC/mg), benzyl alcohol-7-^{14}C (carbinol-^{14}C) (0·037 mC/mg), formaldehyde-^{14}C (0·330 mC/mg), glycerol-U-^{14}C (0·111 mC/mg), phenol-1-^{14}C (0·114 mC/mg), sodium acetate-U-^{14}C (0·658 mC/mg), sodium cyanide (0·139 mC/mg), and sodium ferrocyanide-^{14}C (0·026 mC/mg) were obtained from New England Nuclear Corporation (Boston, Mass.).

Sludge samples. Fresh sludge was obtained daily from a 20 000-gallon-per-day-capacity activated sludge unit described by Dagon (1973). This unit has been operating continuously with photographic processing effluent feed since 1969.

Sludge was centrifuged at 5000 × *g* for 10 min; approximately two parts of distilled water were mixed with one part of sludge pellet, and the sludge mixture was homogenized with a Teflon-tissue homogenizer.

Simulated industrial effluent. The composition of the simulated industrial effluent used in the present study has been described by Dagon (1973). Sodium sulphite, sodium thiosulphite, and formalin were added from freshly prepared solutions immediately prior to the addition of radioactive substrate.

Radiorespirometric incubations. One or two ml of sludge was placed in a 30-ml-capacity serum bottle (Wheaton Glass Co., Cat. No. 223743) containing either 3 ml of salt solution (Allen, 1959) or simulated effluent and 0·2 ml of 1 M potassium phosphate buffer, pH 7·3. The serum bottles were sealed with rubber stoppers, and either 0·1 or 0·2 ml of a solution of radioactive substrate (27 000–400 000 dpm) was injected through the stopper. To prevent leakage of gas from the vial, the vials were coated with 731RTV adhesive (Dow Corning Co.). Samples were incubated routinely in the dark at room temperature. The reactions were stopped by the addition of 2 ml of 1 M perchloric acid to each serum vial.

$^{14}CO_2$ was trapped with phenethylamine in scintillation fluid according to procedures described previously (Belly *et al.*, 1975). In experiments in which $Na_4Fe(CN)_6$-^{14}C or NaCN-^{14}C was used as a substrate, reactions were stopped by the addition of 2 ml of 4% (w/v) perchloric acid and $^{14}CO_2$ was trapped with 10 ml of 1 M NaOH. After bubbling with air for 7 min, the contents of both vials were mixed, and 0·3 ml of $BaCl_2$ (10% w/v) as well as 25 μl of carrier $NaHCO_3$ (10% w/v) were added. The resultant $Ba^{14}CO_3$ precipitate was recovered by filtering the suspension through a 0·22-μ Millipore filter. The filters were then washed with 40 ml of 0·001 M NaOH, glued to a planchet, and counted on a gas flow end-window counter with a Tracerlab Compu/matic Scaler. The efficiency of counting with this system was 12·8%.

Radioactive liquid samples were counted using a Packard Tricarb 3375 liquid scintillation spectrometer. In early studies, AQUASOL Universal LSC Cocktail (New England Nuclear Corp., Boston, Mass.) was used for counting aqueous samples. However, in later studies EASTMAN Ready-to-Use II (Eastman Kodak Co., Rochester, N.Y.) scintillation fluid was used.

RESULTS

Respirometric studies using an industrial sludge. Initially, the radiorespirometric technique was used to evaluate the treatability of several pure chemicals by activate sludge receiving industrial wastes (Table I).

Table I
RADIORESPIROMETRIC STUDIES USING ^{14}C-CHEMICALS IN THE PRESENCE AND ABSENCE OF AN INDUSTRIAL EFFLUENT

Chemical	*% $^{14}CO_2$ recovery without effluent 5 days*	*% $^{14}CO_2$ recovery in presence of effluent 5 days*	*% Degradation based on BOD_5 values*[a]
Benzoic acid-1-^{14}C	68·2	86·9	80·2–83·8
Benzyl alcohol-7-^{14}C	85·7	88·9	60·8
Formaldehyde-U-^{14}C	18·2	10·1	3·4–66·1
Glycerol-U-^{14}C	75·2	N.D.[b]	43·5–52·9
Hydroquinone-2,3,5,6-^{14}C	65·2	46·3	25·4
Phenol-1-^{14}C	18·1	46·5	22·7–88·2
Sodium acetate-U-^{14}C	93·2	90·8	34·6–85·9
Sodium cyanide-^{14}C	25·7	N.D.	N.D.
Sodium ferrocyanide-^{14}C	1·7	N.D.	0·3

[a] See Heukelekian and Rand, 1965.
[b] Not determined by these workers.

Several chemicals, including benzoic acid-1-^{14}C, benzyl alcohol-7-^{14}C, glycerol-U-^{14}C, and sodium acetate-U-^{14}C, demonstrated $^{14}CO_2$ recoveries greater than 65%. In contrast, phenol-1-^{14}C and sodium cyanide-^{14}C resulted in $^{14}CO_2$ recoveries of less than 26% and ferrocyanide-^{14}C produced a $^{14}CO_2$ recovery of only 1·7%. Control vials measured at 0 hr routinely demonstrated recoveries of less than 1·5%.

Biodegradation studies with simulated industrial effluents. Studies were also performed to determine the effect of the addition of a simulated photographic processing effluent on the biodegradability of certain ^{14}C-chemicals. In these studies, ^{14}C-chemical was added to 3 ml of simulated effluent solution, and aged for 2 hr prior to the addition of sludge.

Certain ^{14}C-chemicals, including benzyl alcohol-7-^{14}C and sodium acetate-U-^{14}C, demonstrated no apparent change in $^{14}CO_2$ formation in the presence or absence of simulated effluent (Table I). Hydroquinone-2,3,5,6-^{14}C and formaldehyde-^{14}C demonstrated detectable decreases in $^{14}CO_2$ formation in the presence of the effluent solution; and phenol-1-^{14}C and benzoic acid-1-^{14}C showed a marked stimulation in $^{14}CO_2$ formation under similar incubation conditions.

DISCUSSION

The present technique for measuring $^{14}CO_2$ generated during the decomposition of ^{14}C-labelled chemicals has several advantages: (1) the extent of degradation of a large number of samples can be determined in a relatively short period of time; (2) this technique is simple to perform and has the sensitivity inherent in radioisotopic methods and a precision of $\pm 1{\cdot}4\%$ with significant conversions; (3) this technique makes it possible to study the effects on biodegradation of various chemical reactions and environmental factors occurring in industrial effluents; (4) this procedure readily permits the determination of biodegradation with acclimated and nonacclimated soil, aquatic or sludge microorganisms; and (5) the effects of competing substrates and cometabolism processes can be ascertained with this method.

One possible disadvantage of this technique is that measurements of $^{14}CO_2$ recovery usually will represent an underestimation of total biodegradation. This underestimation is due to incorporation of the initially added radioactivity into cellular materials or end-products. Additional studies measuring radioactivity remaining in solution after biodegradation or incorporated into cellular materials can be performed should material balances be required. For example, one study performed with ^{14}C-glucose and sludge resulted in an 82% conversion to $^{14}CO_2$ with an 11% incorporation into cellular materials, and 7% of the radioactivity remaining in the medium after 5 days.

In view of the extensive differences in methodologies and microbial sources used, a surprisingly good correlation was observed in the biodegradability of all chemicals except hydroquinone and formaldehyde with the present radiorespirometric technique and previously reported BOD_5 determinations (*see* Table I) (Heukelekian and Rand, 1965). Higher levels of $^{14}CO_2$ recovery than would be expected based on BOD_5 values for hydroquinone and formalin may be explained by possible acclimatization of the sludge used in our study to these photographic processing chemicals.

Because of the chemical reactivity of many industrial effluents as well as the many physical and chemical interactions occurring in the sludge environment, the determination of the biodegradability of a pure chemical may not be representative of its actual fate during biological treatment in mixed waste. For this reason, experiments were also performed in which ^{14}C-chemicals were added to a simulated photographic processing effluent, and aged for 2 hr prior to the addition of sludge. Studies performed with either acetate-^{14}C or benzyl alcohol-7-^{14}C showed no apparent change in $^{14}CO_2$ formation in the presence or absence of effluent. These results suggest that acetate and benzyl alcohol do not undergo a significant conversion to a recalcitrant product in the presence of photographic effluent, and that photographic processing effluents are not inhibitory to the metabolism of these chemicals. In contrast to acetate or benzyl alcohol, $^{14}CO_2$ formation from phenol-1-^{14}C and benzoic acid-1-^{14}C showed a marked stimulation in the presence of this industrial effluent. Neither of these chemicals are used in photographic processing. Therefore, this stimulation may be due in part to the utilization of carbon sources present in the effluent as an energy source for microbial growth or metabolism of these chemicals. Finally, in the case of formaldehyde-^{14}C and hydroquinone-2,3,5,6-^{14}C, a marked decrease in biodegradation was observed in the presence of the synthetic effluent. Presumably, formaldehyde can undergo reaction with a number of compounds such as ammonia, amines, and organics present in photographic processing effluents (Hickinbotton, 1962). The oxidation of hydroquinone in the presence of sulphite to form hydroquinone mono- or disulphonates has been well studied (Lee, 1966).

Our studies indicate that the complex chemical reactions and the additional carbon sources present in some effluents may have an important influence on biodegradability.

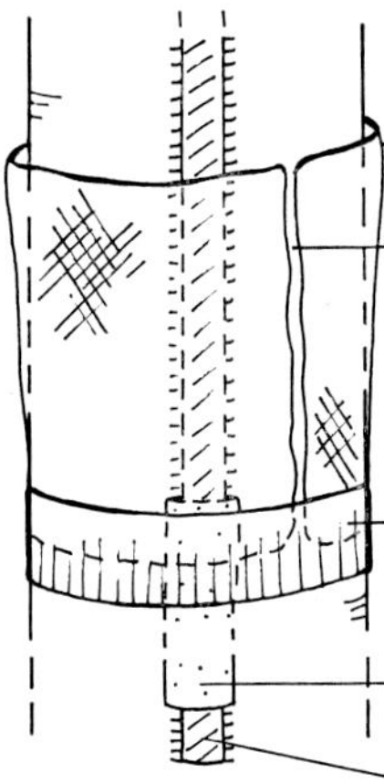

Fig. 3.

The textile strips are arranged round the tube in such a manner as to leave one larger gap between strips in order to facilitate the fixing of the glass-fibre screen (*see* Figs. 2 and 3).

Aluminium foil is then wrapped over the upper portion of the tube, overlapping the glass-fibre screen slightly, and folded over the tube top. The junction of the foil is then sealed with adhesive tape, as is the vertical join in the glass-fibre screen. A cap is then made for the tube by folding a piece of foil and edging it with adhesive tape (*see* Fig. 4).

After partially filling the tube with the desired liquid nutrients, the wicks are folded up flat against the sides of the tube. The tubes are then autoclaved in metal test-tube racks. After autoclaving and cooling, the contents are then topped-up from a sterile stock solution. Some solution is lost during autoclaving as the textile strips become wet, and if the tubes are overfilled initially, spillage may occur. Before burial in the soil, the tail wicks (foil-covered portion of the textile strips) are folded partially down, as shown in Figs. 5 and 6.

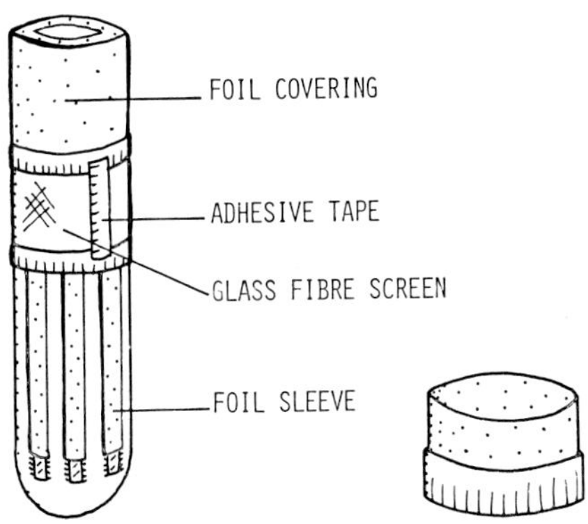

Fig. 4.

Two features of this assembly should be noted.

(1) Where the substrate used is a textile or similar material capable of conducting liquids by capillarity, the specimen acts as its own wick and no additional glass-fibre wicks are needed. This speeds and simplifies construction greatly and helps even flow with no risk of breaks in the system.

(2) The use of aluminium foil as opposed to some form of autoclavable rubber or plastic tubing as a sleeving material speeds construction and is much cheaper. The foil is stiff when folded and thus the final bends made in the wicks are maintained sufficiently to ease burial of the tubes.

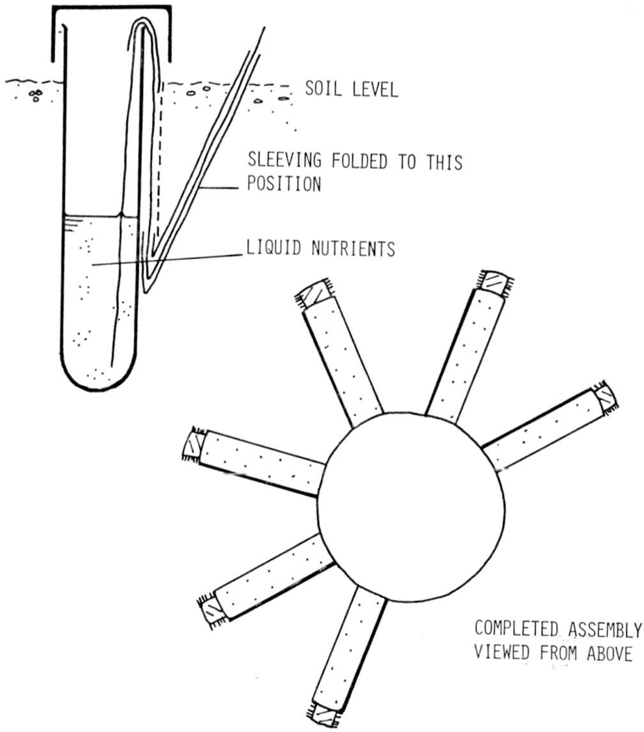

Fig. 5.

MAINTENANCE AND EXTRACTION OF SAMPLES

After burial in soil, either in the laboratory or in the field, the tubes require little attention except for the occasional replenishment of nutrients. The rate of flow depends upon external conditions of temperature, air flow, humidity, and soil moisture, and experience will give an indication of the frequency of attention required. Under ordinary laboratory soil burial conditions using six 1-cm wide strips per tube, replenishment of nutrients may be required every 10–14 days. It is advisable not to fill tubes more than two-thirds full to avoid an over-rapid siphon-like flow of nutrients.

Although it is possible to remove single strips from a tube, it is difficult to do this without introducing soil, and it is recommended that all strips on a tube are removed at the same time. Loose soil should be brushed from the tube prior to cutting away the glass-fibre screen. The central portion of the textile strips from beneath the screen are then cut away with a sterile scalpel blade and transferred to a sterile petri dish prior to plating out on agar media. If strips are required for strength testing, they should be removed intact by unfastening the adhesive tape and dismantling the tube components.

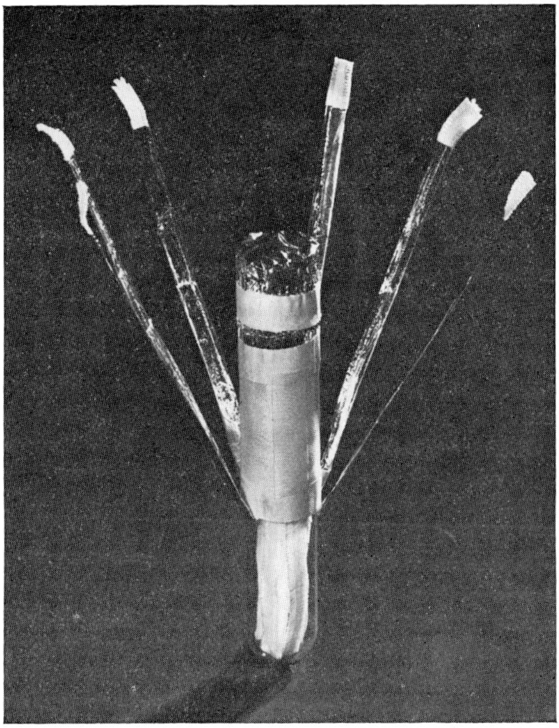

Fig. 6. Completed perfusion isolation assembly before burial in soil.

DISCUSSION

This technique has been used in studies of soil fungi, where it has proved a most useful tool. Each tube yields several textile strips and thus a number of subsequent agar platings or textile strength measurements can be made. The construction of the tubes and the capacity of the reservoir, giving up to several weeks' service without attention, dependent upon conditions of use, lend this technique to field studies as well as laboratory investigations.

The technique could also be employed in environments other than soil provided that the tube can be buried, for example, in composting town waste or in agricultural wastes such as straw.

The technique is put forward as a routine tool for the isolation of fungi onto solid substrates, where it is especially useful under slow colonization conditions, as in the case of biocidally protected materials. It is also proposed as a test technique where conditions more stable than in many soil burial situations are required, or where a change in nutrient or other perfusate components need to be changed during incubation, for example, the effects of a soluble material on fungal decay may be investigated by this method.

It is hoped that the simplicity of this technique will lend itself not only to established research laboratories but also to students in schools and colleges for basic studies in biodeterioration and biodegradation.

Acknowledgements

This work was carried out as part of a postgraduate research programme sponsored by Catomance Limited of Welwyn Garden City, U.K., at the University of Aston in Birmingham, U.K. The authors wish to thank that company for their generous help and support.

REFERENCES

Allsopp, D., and Eggins, H. O. W. (1972). A demonstration of the effects of substrate screening on the colonisation and decay of cotton textiles in soil burial tests. *Mycopathol. Mycol. Appl.*, **47** (4), 331–336.

Allsopp, D., Eggins, H. O. W., and David, J. (1970). A new perfusion technique for rapid initial screening of biocidally protected materials. *Int. Biodeterior. Bull.*, **6** (3), 115–117.

Allsopp, D., Hueck, H. J., Eggins, H. O. W., and Lloyd, A. O. (1970b). Evidence of the hydrolysis of pentachlorophenyl laurate by soil micro-organisms. *J. Text. Inst.*, **61** (10), 509–511.

Eggins, H. O. W., and Lloyd, A. O. (1968). Cellulolytic fungi isolated by the screened substrate method. *Experientia*, **24**, 749.

Eggins, H. O. W., Malik, K. A., and Sharp, R. F. (1968). Some techniques to investigate the colonisation of cellulosic and wood substrates. In *Biodeterioration of Materials*, pp. 120–130 (A. H. Walters and J. J. Elphick, eds.), Applied Science Publishers, London.

Eggins, H. O. W., Szilvinyi, A. von, and Allsopp, D. (1972). The isolation of actively growing thermophilic fungi from insolated soils. *Int. Biodeterior. Bull.*, **8** (2), 53–58.

A STUDY ON THE EFFECTS OF pH ON THE DECAY OF COTTON TEXTILES IN VARIOUS SOILS, USING A PERFUSION TECHNIQUE

D. ALLSOPP

Biodeterioration Information Centre,
University of Aston, Birmingham, U.K.

and

A. R. M. BARR

Catomance Limited, Welwyn Garden City, Hertfordshire, U.K.

Summary

In order to clarify further the acknowledged but little documented major variable of pH in the soil-burial testing of textiles, various soil samples of differing pH were investigated using a perfusion technique. Aggressiveness of the soils toward unprotected and chemically protected cotton textile was measured by subsequent strength testing, the textile samples being incubated in the presence of soil at the normal soil pH and higher and lower pH. The substrate pH was regulated by means of the liquid perfusate. The effect of pH and changes in pH as an environmental factor in the fungal decay process is demonstrated, together with variations in the spectra of fungi isolated.

INTRODUCTION

It is well recognized that different soils may possess different degrees of aggressiveness in regard to the decay of cotton textiles and cellulosic materials in general. This is usually thought to be due to variations in soils with regard to their abilities to support a varied and active flora of deteriogenic micro-organisms, especially fungi. The main variations in soils are usually considered to be those of nutrition and aeration (Lloyd, 1968; Stolzy and Van Gundy, 1968; Turner, 1972). In order to maximize and regulate these factors, test soils are often amended from their natural state. This amendment may take the form of addition of compost or manure to increase the available nutrients and often to improve drainage, to make water holding more even, and to improve aeration. In cases of heavy soils, sand may be added to improve drainage and aeration. Thus, it is often the policy of a test laboratory to prepare an aggressive but probably unique soil stock to suit its own particular requirements, as constant as possible regarding aeration and nutrients.

The pH of this soil may often be noted, but attempts to investigate the effects of changes in one soil or variations between different soils are not usually made; changing and maintaining the pH of a soil in bulk are very difficult to accomplish. The variation of pH within any one soil or between different soils is a recognized important variable in testing work and in field conditions (Wessels and Adema, 1968). For this reason, it was decided to carry out the following work.

METHOD

A perfusion isolation technique (Allsopp, 1973) was devised to isolate fungi from soils at differing pH, not by altering the soil itself, but by maintaining the substrates in the soil at various pH levels. This technique is described in detail by Allsopp and Barr elsewhere in these Proceedings. This technique minimizes the initial shock to the mycoflora of the soil as a whole because the soil pH remains unaltered, but species which were able to colonize the cotton textile at a changed pH were encouraged and selectively isolated.

The experiment was designed to show the variations in fungal species and the decay resulting from their colonization of substrates from different soils, the normal pH of each soil being known. Also shown was the effect of changes in pH of substrates in each soil within a range of 1·5 units below or above the normal soil pH. Soils were kept in each case near their normal maximum water-holding capacity, but at a constant temperature of 26°C, and so far as is possible at a uniform packing density under test conditions.

For burial in each soil, perfusion isolation tubes containing untreated cotton textile and also 1·5% pentachlorophenyl laurate (PCPL)-treated textile were prepared. The tubes were filled with nutrient salts solution (Eggins and Pugh, 1962), adjusted to each pH required and autoclaved at 15 psi for 20 min—tubes were then buried in the soils listed below.

1. *From Catomance Ltd.* Clay topsoil from waste ground close to a lauric acid plant. pH 6·0.

2. *Garden soil.* Sandy topsoil from a fallow plot. pH 7·4.

3. *From Oldbury, Worcestershire.* A dark, gritty topsoil from waste ground in an industrial area. pH 6·4.

4. *From Clent, Worcestershire.* Sandy topsoil from an uncultivated field. pH 6·5.

5. *Woodland soil.* Topsoil from a mixed woodland containing a large amount of decaying leaf litter. pH 5·2.

After 10 days' incubation, tubes were removed from soils and the textile samples used for strength testing and plating-out on cellulose and glucose-stored media (at the same pH as that maintained in the perfused substrate when buried).

RESULTS

These experiments yielded a large amount of data which is summarized in the following conclusions.

Fungal isolations

In general. Fungi were isolated from both untreated and PCPL-treated textile samples from each soil, the greatest number of species occurring on untreated textiles.

For individual soils

Woodland soil (perfusate pH: normal 5·4, high 6·7, low 4·1). From untreated textile, more species were isolated at the normal pH than at the higher and lower levels. The predominant species were *Dicoccum* sp., *Trichoderma viride*, and *Rhizopus* sp. From treated textile, most species were isolated at the lowest pH. Fungal species were fewer than from untreated textile. The predominant species was *Trichoderma viride*.

Catomance soil (perfusate pH: normal 6·1, high 7·3, low 4·9). From untreated textile, more species were isolated at the normal pH than at higher and lower levels. The predominant species were *Humicola grisea*, *Rhizopus* sp., and *Trichoderma viride*. From treated textile, most species were isolated at the lowest pH. Fungal species were fewer than from untreated textile. The predominant species were *Fusarium* sp. and *Trichoderma viride*.

Clent soil (perfusate pH: normal 6·6, high 7·8, low 5·3). From untreated textile, most species were isolated at the lowest pH (one more than at the normal pH). The predominant species were *Arthrobotrys* sp., especially where nematodes were present, and *Trichoderma viride*. From treated textile, most species were isolated at the lowest pH. Fungal species isolated were fewer than from untreated textile. The predominant species was *Fusarium* sp., the only species isolated at all three pH levels.

Garden soil (perfusate pH: normal 7·2, high 9·1, low 6·1). From untreated textile, the same number of fungal species were isolated at each pH. The most predominant species was *Humicola grisea*. From treated textile, most species were isolated at the lowest pH. The most predominant species was *Fusarium* sp.

Oldbury soil (perfusate pH: normal 6·3, high 7·6, low 5·1). From untreated textile, most species were isolated at the normal pH. The predominant species was *Rhizopus* sp.; several cellulolytic fungi were also isolated but somewhat less frequently. From treated textile, most species were isolated at the normal pH. Fewer fungal species were isolated than from untreated textile. The most predominant species was *Fusarium* sp.

Regarding pH

In general, for untreated textile, most species were isolated at the normal pH, but in the case of PCPL-treated textile, most species were isolated at the lowest pH.

Regarding cellulolytic ability

The predominant species mentioned above, with the exception of *Rhizopus* species, were all cellulolytic, shown by their ability to grow on, and clear, cellulose agar.

Regarding nematodes and bacteria

These were noted on isolation plates and found to occur commonly. It was also noted that the fungus *Arthrobotrys* sp. and nematodes often occurred together.

Strength losses (*see* Fig. 1)

Regarding treatment

Although both untreated and PCPL-treated textile samples lost strength in

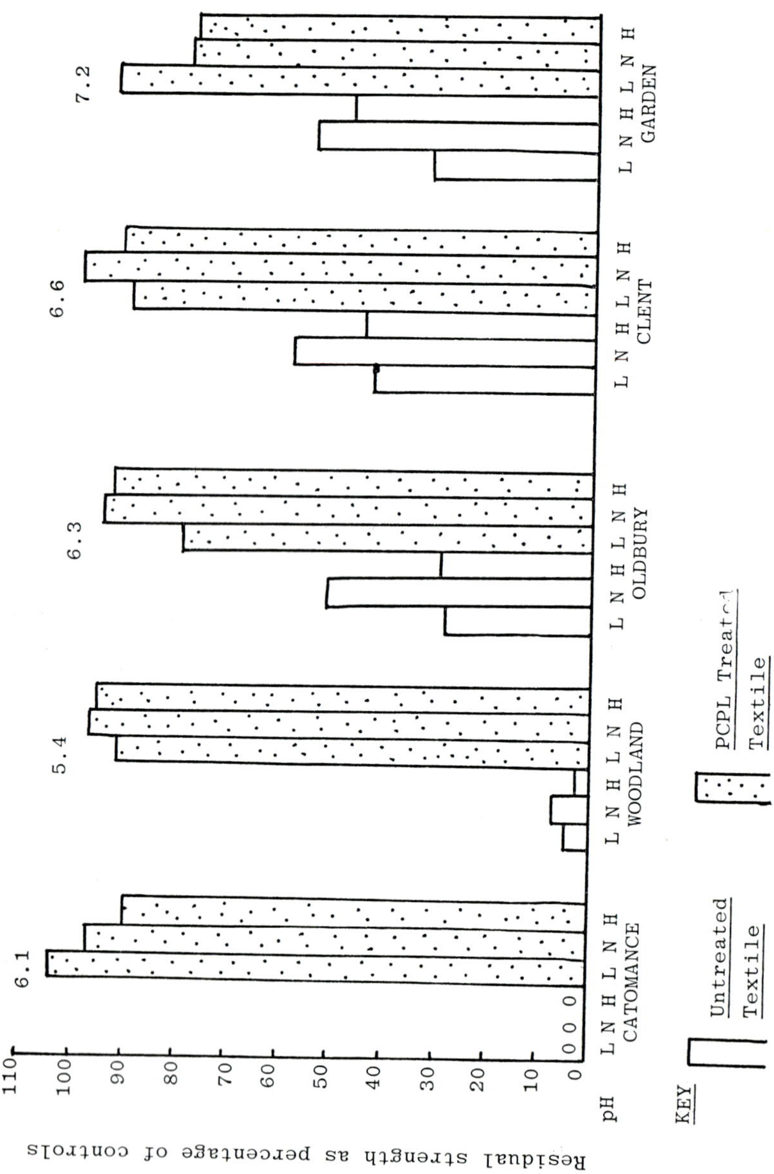

Fig. 1. *Residual strength at various pH after 10 days in soil at 26°C. Normal (soil) perfusate pH is noted above each group of results.*

soil, there was a very marked variation in decay, the treated samples showing only small losses in strength.

Regarding soils and pH

The soils showing the greatest aggressiveness by far to untreated textile were the Catomance soil and the woodland soil. These had the lowest pH of the soils investigated, 6·0 and 5·2, respectively.

Regarding modified pH

Untreated textile showed the least losses in strength at the normal pH for each soil. In addition to losing much less strength than untreated samples, the treated samples showed no clear pattern regarding the effect on strength losses as related to modification of pH. In three cases least loss occurred at the normal pH of the range for the soil, and in the other two, at the lowest pH of the range.

It was to be expected that more fungal species and most wood decay would occur on untreated textile than on that treated with the PCPL fungicide. Considering the differing natures of the soils used, it was also to be expected that different species in different numbers and producing different amounts of decay would be found.

This experiment was designed to investigate the effects of pH and fungal species of soils in the aggressiveness in textile decay. From the results it would appear that there is some correlation between these factors. It is generally accepted that fungi tend to prefer acid conditions, and with regard to their decay abilities, this does in fact appear to be the case here. The two soils which had the lowest pH were markedly more aggressive than the others at higher pH. Although different in nature and water-holding, these two soils had a similar mycoflora, with similar predominances of vigorous cellulolytic species. *Trichoderma viride* was common in both soils, and from previous work (Allsopp, 1973) it can be seen that this fungus is very active at low pH. It is possible that as pH is depressed two factors concerning fungal species become more important. First, those fungi which prefer a lower pH become more widespread and, as a result of this, become more vigorous due to lack of competition.

One may hypothesize, therefore, that a lowering of pH would tend to increase fungal decay activity. When pH was lowered for each soil, this proved to be the case; however, decay activity also increased as pH was raised. It is possible that a different mechanism is at work here. In a soil in its natural state, changes are usually slow and gradual, and fluctuations in the decay ability of the mycoflora will gradually even out as a mature steady state is reached. In this experiment, however, a sharp change in substrate pH occurred at a single point in time, and the experiment itself continued only for a short time. There was therefore some degree of shock imposed on the mycoflora. It is probable that less vigorous species growing within the soil were less able to adapt to this change in the short time allowed. Thus the more vigorous primary colonizers, cellulolytic in nature, encountered less competition and therefore were able to colonize the substrate more effectively than at the normal pH of the soil, being also less affected themselves by the change in pH due to their normal vigour and ability to grow over wider ranges of pH.

This effect is less marked when considering the decay of the treated textiles, but as fungal growth on these samples was small and is often erratic on protected samples, this is perhaps to be expected.

Several trends may be seen concerning individual fungal species. *Trichoderma viride* showed its common nature and vigour, especially at the three lower pH levels on untreated textile, extending its range when the textile was treated with PCPL. This shows that this species is not an anomalously aggressive fungus, restricted to the Clent soil used in other experiments. *Rhizopus* sp. occurred together with *Trichoderma viride* on untreated textile, and it may be filling the niche of a secondary colonizer here. Another species which was active at lower pH on untreated textile was *Dicoccum* sp., but this does not appear to have the fungicide resistance of *Trichoderma viride*. *Humicola grisea*, found over the pH range on untreated textile, also appears to be less tolerant to the fungicide than *Trichoderma viride*. *Fusarium* sp. appeared to fill the niche left by the less fungicide-tolerant species on treated textile, especially at the higher pH levels. This is of interest because this fungus is a common and troublesome deteriogen, not only being cellulolytic but also being responsible for much pink-staining of cotton.

A final point of interest concerning fungal species is the coincidence of *Arthrobotrys* sp. with nematode worms at the three higher pH levels on untreated textile. This fungal species is known to live in association with nematodes on which it may prey. Its occurrence here serves as a reminder of the complex ecological situation in soil, and the need for wider considerations to be taken into account in future work.

As regards the technique employed in this experiment, it proved simple in use. Since it combines several principles which encourage the growth of vigorous fungi under steady and adaptable conditions, and is nearer to natural conditions than some techniques such as agar plate techniques, it is suggested that this technique could be employed usefully in further work on the variable parameters of soil-burial testing.

Acknowledgements

This work was carried out as part of a postgraduate research programme sponsored by Catomance Limited of Welwyn Garden City, U.K., at the University of Aston in Birmingham, U.K. The authors wish to thank that company for their generous help and support.

REFERENCES

Allsopp, D. (1973). Some aspects of the colonisation and decay of fungicidally-protected cotton textiles by soil fungi. Ph.D. Thesis. University of Aston in Birmingham. 306 pp.

Eggins, H. O. W., and Pugh, G. J. F. (1962). Isolation of cellulose decomposing fungi from the soil. *Nature*, **193**, 94–95.

Lloyd, A. O. (1968). The evaluation of rot resistance of cellulosic textiles. In *Biodeterioration of Materials*, pp. 170–177 (A. H. Walters and J. J. Elphick, eds.), Applied Science Publishers, London.

Stolzy, L. H., and Van Gundy, S. D. (1968). The soil as an environment for microflora and microfauna. *Phytopathology*, **58** (7), 889–899.

Turner, R. L. (1972). Important factors in the soil burial test applied to rot-proof textiles. In *Biodeterioration of Materials*, Vol. 2, pp. 218–226 (E. H. Hueck-van der Plas and A. H. Walters, eds.), Applied Science Publishers, London.

Wessels, J. M. C., and Adema, D. M. M. (1968). Some data on the relationship between fungicidal protection and pH. In *Biodeterioration of Materials*, pp. 517–523 (A. H. Walters and J. J. Elphick, eds.), Applied Science Publishers, London.

INDEX